Lecture Notes in Computer Science 13680

More information about this series at https://link.springer.com/bookseries/558

Shai Avidan · Gabriel Brostow ·
Moustapha Cissé · Giovanni Maria Farinella ·
Tal Hassner (Eds.)

Computer Vision – ECCV 2022

17th European Conference
Tel Aviv, Israel, October 23–27, 2022
Proceedings, Part XX

 Springer

Editors
Shai Avidan
Tel Aviv University
Tel Aviv, Israel

Gabriel Brostow
University College London
London, UK

Moustapha Cissé
Google AI
Accra, Ghana

Giovanni Maria Farinella
University of Catania
Catania, Italy

Tal Hassner
Facebook (United States)
Menlo Park, CA, USA

ISSN 0302-9743   ISSN 1611-3349 (electronic)
Lecture Notes in Computer Science
ISBN 978-3-031-20043-4   ISBN 978-3-031-20044-1 (eBook)
https://doi.org/10.1007/978-3-031-20044-1

This Springer imprint is published by the registered company Springer Nature Switzerland AG
The registered company address is: Gewerbestrasse 11, 6330 Cham, Switzerland

Foreword

Organizing the European Conference on Computer Vision (ECCV 2022) in Tel-Aviv during a global pandemic was no easy feat. The uncertainty level was extremely high, and decisions had to be postponed to the last minute. Still, we managed to plan things just in time for ECCV 2022 to be held in person. Participation in physical events is crucial to stimulating collaborations and nurturing the culture of the Computer Vision community.

There were many people who worked hard to ensure attendees enjoyed the best science at the 16th edition of ECCV. We are grateful to the Program Chairs Gabriel Brostow and Tal Hassner, who went above and beyond to ensure the ECCV reviewing process ran smoothly. The scientific program includes dozens of workshops and tutorials in addition to the main conference and we would like to thank Leonid Karlinsky and Tomer Michaeli for their hard work. Finally, special thanks to the web chairs Lorenzo Baraldi and Kosta Derpanis, who put in extra hours to transfer information fast and efficiently to the ECCV community.

We would like to express gratitude to our generous sponsors and the Industry Chairs, Dimosthenis Karatzas and Chen Sagiv, who oversaw industry relations and proposed new ways for academia-industry collaboration and technology transfer. It's great to see so much industrial interest in what we're doing!

Authors' draft versions of the papers appeared online with open access on both the Computer Vision Foundation (CVF) and the European Computer Vision Association (ECVA) websites as with previous ECCVs. Springer, the publisher of the proceedings, has arranged for archival publication. The final version of the papers is hosted by SpringerLink, with active references and supplementary materials. It benefits all potential readers that we offer both a free and citeable version for all researchers, as well as an authoritative, citeable version for SpringerLink readers. Our thanks go to Ronan Nugent from Springer, who helped us negotiate this agreement. Last but not least, we wish to thank Eric Mortensen, our publication chair, whose expertise made the process smooth.

October 2022

Rita Cucchiara
Jiří Matas
Amnon Shashua
Lihi Zelnik-Manor

Preface

Welcome to the proceedings of the European Conference on Computer Vision (ECCV 2022). This was a hybrid edition of ECCV as we made our way out of the COVID-19 pandemic. The conference received 5804 valid paper submissions, compared to 5150 submissions to ECCV 2020 (a 12.7% increase) and 2439 in ECCV 2018. 1645 submissions were accepted for publication (28%) and, of those, 157 (2.7% overall) as orals.

846 of the submissions were desk-rejected for various reasons. Many of them because they revealed author identity, thus violating the double-blind policy. This violation came in many forms: some had author names with the title, others added acknowledgments to specific grants, yet others had links to their github account where their name was visible. Tampering with the LaTeX template was another reason for automatic desk rejection.

ECCV 2022 used the traditional CMT system to manage the entire double-blind reviewing process. Authors did not know the names of the reviewers and vice versa. Each paper received at least 3 reviews (except 6 papers that received only 2 reviews), totalling more than 15,000 reviews.

Handling the review process at this scale was a significant challenge. To ensure that each submission received as fair and high-quality reviews as possible, we recruited more than 4719 reviewers (in the end, 4719 reviewers did at least one review). Similarly we recruited more than 276 area chairs (eventually, only 276 area chairs handled a batch of papers). The area chairs were selected based on their technical expertise and reputation, largely among people who served as area chairs in previous top computer vision and machine learning conferences (ECCV, ICCV, CVPR, NeurIPS, etc.).

Reviewers were similarly invited from previous conferences, and also from the pool of authors. We also encouraged experienced area chairs to suggest additional chairs and reviewers in the initial phase of recruiting. The median reviewer load was five papers per reviewer, while the average load was about four papers, because of the emergency reviewers. The area chair load was 35 papers, on average.

Conflicts of interest between authors, area chairs, and reviewers were handled largely automatically by the CMT platform, with some manual help from the Program Chairs. Reviewers were allowed to describe themselves as senior reviewer (load of 8 papers to review) or junior reviewers (load of 4 papers). Papers were matched to area chairs based on a subject-area affinity score computed in CMT and an affinity score computed by the Toronto Paper Matching System (TPMS). TPMS is based on the paper's full text. An area chair handling each submission would bid for preferred expert reviewers, and we balanced load and prevented conflicts.

The assignment of submissions to area chairs was relatively smooth, as was the assignment of submissions to reviewers. A small percentage of reviewers were not happy with their assignments in terms of subjects and self-reported expertise. This is an area for improvement, although it's interesting that many of these cases were reviewers handpicked by AC's. We made a later round of reviewer recruiting, targeted at the list of authors of papers submitted to the conference, and had an excellent response which helped provide enough emergency reviewers. In the end, all but six papers received at least 3 reviews.

The challenges of the reviewing process are in line with past experiences at ECCV 2020. As the community grows, and the number of submissions increases, it becomes ever more challenging to recruit enough reviewers and ensure a high enough quality of reviews. Enlisting authors by default as reviewers might be one step to address this challenge.

Authors were given a week to rebut the initial reviews, and address reviewers' concerns. Each rebuttal was limited to a single pdf page with a fixed template.

The Area Chairs then led discussions with the reviewers on the merits of each submission. The goal was to reach consensus, but, ultimately, it was up to the Area Chair to make a decision. The decision was then discussed with a buddy Area Chair to make sure decisions were fair and informative. The entire process was conducted virtually with no in-person meetings taking place.

The Program Chairs were informed in cases where the Area Chairs overturned a decisive consensus reached by the reviewers, and pushed for the meta-reviews to contain details that explained the reasoning for such decisions. Obviously these were the most contentious cases, where reviewer inexperience was the most common reported factor.

Once the list of accepted papers was finalized and released, we went through the laborious process of plagiarism (including self-plagiarism) detection. A total of 4 accepted papers were rejected because of that.

Finally, we would like to thank our Technical Program Chair, Pavel Lifshits, who did tremendous work behind the scenes, and we thank the tireless CMT team.

October 2022

Gabriel Brostow
Giovanni Maria Farinella
Moustapha Cissé
Shai Avidan
Tal Hassner

Organization

General Chairs

Rita Cucchiara — University of Modena and Reggio Emilia, Italy
Jiří Matas — Czech Technical University in Prague, Czech Republic
Amnon Shashua — Hebrew University of Jerusalem, Israel
Lihi Zelnik-Manor — Technion – Israel Institute of Technology, Israel

Program Chairs

Shai Avidan — Tel-Aviv University, Israel
Gabriel Brostow — University College London, UK
Moustapha Cissé — Google AI, Ghana
Giovanni Maria Farinella — University of Catania, Italy
Tal Hassner — Facebook AI, USA

Program Technical Chair

Pavel Lifshits — Technion – Israel Institute of Technology, Israel

Workshops Chairs

Leonid Karlinsky — IBM Research, Israel
Tomer Michaeli — Technion – Israel Institute of Technology, Israel
Ko Nishino — Kyoto University, Japan

Tutorial Chairs

Thomas Pock — Graz University of Technology, Austria
Natalia Neverova — Facebook AI Research, UK

Demo Chair

Bohyung Han — Seoul National University, Korea

Social and Student Activities Chairs

Tatiana Tommasi   Italian Institute of Technology, Italy
Sagie Benaim   University of Copenhagen, Denmark

Diversity and Inclusion Chairs

Xi Yin   Facebook AI Research, USA
Bryan Russell   Adobe, USA

Communications Chairs

Lorenzo Baraldi   University of Modena and Reggio Emilia, Italy
Kosta Derpanis   York University & Samsung AI Centre Toronto,
                                    Canada

Industrial Liaison Chairs

Dimosthenis Karatzas   Universitat Autònoma de Barcelona, Spain
Chen Sagiv   SagivTech, Israel

Finance Chair

Gerard Medioni   University of Southern California & Amazon,
                                    USA

Publication Chair

Eric Mortensen   MiCROTEC, USA

Area Chairs

Lourdes Agapito   University College London, UK
Zeynep Akata   University of Tübingen, Germany
Naveed Akhtar   University of Western Australia, Australia
Karteek Alahari   Inria Grenoble Rhône-Alpes, France
Alexandre Alahi   École polytechnique fédérale de Lausanne,
                                    Switzerland
Pablo Arbelaez   Universidad de Los Andes, Columbia
Antonis A. Argyros   University of Crete & Foundation for Research
                                  and Technology-Hellas, Crete
Yuki M. Asano   University of Amsterdam, The Netherlands
Kalle Åström   Lund University, Sweden
Hadar Averbuch-Elor   Cornell University, USA

Hossein Azizpour	KTH Royal Institute of Technology, Sweden
Vineeth N. Balasubramanian	Indian Institute of Technology, Hyderabad, India
Lamberto Ballan	University of Padova, Italy
Adrien Bartoli	Université Clermont Auvergne, France
Horst Bischof	Graz University of Technology, Austria
Matthew B. Blaschko	KU Leuven, Belgium
Federica Bogo	Meta Reality Labs Research, Switzerland
Katherine Bouman	California Institute of Technology, USA
Edmond Boyer	Inria Grenoble Rhône-Alpes, France
Michael S. Brown	York University, Canada
Vittorio Caggiano	Meta AI Research, USA
Neill Campbell	University of Bath, UK
Octavia Camps	Northeastern University, USA
Duygu Ceylan	Adobe Research, USA
Ayan Chakrabarti	Google Research, USA
Tat-Jen Cham	Nanyang Technological University, Singapore
Antoni Chan	City University of Hong Kong, Hong Kong, China
Manmohan Chandraker	NEC Labs America, USA
Xinlei Chen	Facebook AI Research, USA
Xilin Chen	Institute of Computing Technology, Chinese Academy of Sciences, China
Dongdong Chen	Microsoft Cloud AI, USA
Chen Chen	University of Central Florida, USA
Ondrej Chum	Vision Recognition Group, Czech Technical University in Prague, Czech Republic
John Collomosse	Adobe Research & University of Surrey, UK
Camille Couprie	Facebook, France
David Crandall	Indiana University, USA
Daniel Cremers	Technical University of Munich, Germany
Marco Cristani	University of Verona, Italy
Canton Cristian	Facebook AI Research, USA
Dengxin Dai	ETH Zurich, Switzerland
Dima Damen	University of Bristol, UK
Kostas Daniilidis	University of Pennsylvania, USA
Trevor Darrell	University of California, Berkeley, USA
Andrew Davison	Imperial College London, UK
Tali Dekel	Weizmann Institute of Science, Israel
Alessio Del Bue	Istituto Italiano di Tecnologia, Italy
Weihong Deng	Beijing University of Posts and Telecommunications, China
Konstantinos Derpanis	Ryerson University, Canada
Carl Doersch	DeepMind, UK

Matthijs Douze	Facebook AI Research, USA
Mohamed Elhoseiny	King Abdullah University of Science and Technology, Saudi Arabia
Sergio Escalera	University of Barcelona, Spain
Yi Fang	New York University, USA
Ryan Farrell	Brigham Young University, USA
Alireza Fathi	Google, USA
Christoph Feichtenhofer	Facebook AI Research, USA
Basura Fernando	Agency for Science, Technology and Research (A*STAR), Singapore
Vittorio Ferrari	Google Research, Switzerland
Andrew W. Fitzgibbon	Graphcore, UK
David J. Fleet	University of Toronto, Canada
David Forsyth	University of Illinois at Urbana-Champaign, USA
David Fouhey	University of Michigan, USA
Katerina Fragkiadaki	Carnegie Mellon University, USA
Friedrich Fraundorfer	Graz University of Technology, Austria
Oren Freifeld	Ben-Gurion University, Israel
Thomas Funkhouser	Google Research & Princeton University, USA
Yasutaka Furukawa	Simon Fraser University, Canada
Fabio Galasso	Sapienza University of Rome, Italy
Jürgen Gall	University of Bonn, Germany
Chuang Gan	Massachusetts Institute of Technology, USA
Zhe Gan	Microsoft, USA
Animesh Garg	University of Toronto, Vector Institute, Nvidia, Canada
Efstratios Gavves	University of Amsterdam, The Netherlands
Peter Gehler	Amazon, Germany
Theo Gevers	University of Amsterdam, The Netherlands
Bernard Ghanem	King Abdullah University of Science and Technology, Saudi Arabia
Ross B. Girshick	Facebook AI Research, USA
Georgia Gkioxari	Facebook AI Research, USA
Albert Gordo	Facebook, USA
Stephen Gould	Australian National University, Australia
Venu Madhav Govindu	Indian Institute of Science, India
Kristen Grauman	Facebook AI Research & UT Austin, USA
Abhinav Gupta	Carnegie Mellon University & Facebook AI Research, USA
Mohit Gupta	University of Wisconsin-Madison, USA
Hu Han	Institute of Computing Technology, Chinese Academy of Sciences, China

Bohyung Han	Seoul National University, Korea
Tian Han	Stevens Institute of Technology, USA
Emily Hand	University of Nevada, Reno, USA
Bharath Hariharan	Cornell University, USA
Ran He	Institute of Automation, Chinese Academy of Sciences, China
Otmar Hilliges	ETH Zurich, Switzerland
Adrian Hilton	University of Surrey, UK
Minh Hoai	Stony Brook University, USA
Yedid Hoshen	Hebrew University of Jerusalem, Israel
Timothy Hospedales	University of Edinburgh, UK
Gang Hua	Wormpex AI Research, USA
Di Huang	Beihang University, China
Jing Huang	Facebook, USA
Jia-Bin Huang	Facebook, USA
Nathan Jacobs	Washington University in St. Louis, USA
C.V. Jawahar	International Institute of Information Technology, Hyderabad, India
Herve Jegou	Facebook AI Research, France
Neel Joshi	Microsoft Research, USA
Armand Joulin	Facebook AI Research, France
Frederic Jurie	University of Caen Normandie, France
Fredrik Kahl	Chalmers University of Technology, Sweden
Yannis Kalantidis	NAVER LABS Europe, France
Evangelos Kalogerakis	University of Massachusetts, Amherst, USA
Sing Bing Kang	Zillow Group, USA
Yosi Keller	Bar Ilan University, Israel
Margret Keuper	University of Mannheim, Germany
Tae-Kyun Kim	Imperial College London, UK
Benjamin Kimia	Brown University, USA
Alexander Kirillov	Facebook AI Research, USA
Kris Kitani	Carnegie Mellon University, USA
Iasonas Kokkinos	Snap Inc. & University College London, UK
Vladlen Koltun	Apple, USA
Nikos Komodakis	University of Crete, Crete
Piotr Koniusz	Australian National University, Australia
Philipp Kraehenbuehl	University of Texas at Austin, USA
Dilip Krishnan	Google, USA
Ajay Kumar	Hong Kong Polytechnic University, Hong Kong, China
Junseok Kwon	Chung-Ang University, Korea
Jean-Francois Lalonde	Université Laval, Canada

Ivan Laptev	Inria Paris, France
Laura Leal-Taixé	Technical University of Munich, Germany
Erik Learned-Miller	University of Massachusetts, Amherst, USA
Gim Hee Lee	National University of Singapore, Singapore
Seungyong Lee	Pohang University of Science and Technology, Korea
Zhen Lei	Institute of Automation, Chinese Academy of Sciences, China
Bastian Leibe	RWTH Aachen University, Germany
Hongdong Li	Australian National University, Australia
Fuxin Li	Oregon State University, USA
Bo Li	University of Illinois at Urbana-Champaign, USA
Yin Li	University of Wisconsin-Madison, USA
Ser-Nam Lim	Meta AI Research, USA
Joseph Lim	University of Southern California, USA
Stephen Lin	Microsoft Research Asia, China
Dahua Lin	The Chinese University of Hong Kong, Hong Kong, China
Si Liu	Beihang University, China
Xiaoming Liu	Michigan State University, USA
Ce Liu	Microsoft, USA
Zicheng Liu	Microsoft, USA
Yanxi Liu	Pennsylvania State University, USA
Feng Liu	Portland State University, USA
Yebin Liu	Tsinghua University, China
Chen Change Loy	Nanyang Technological University, Singapore
Huchuan Lu	Dalian University of Technology, China
Cewu Lu	Shanghai Jiao Tong University, China
Oisin Mac Aodha	University of Edinburgh, UK
Dhruv Mahajan	Facebook, USA
Subhransu Maji	University of Massachusetts, Amherst, USA
Atsuto Maki	KTH Royal Institute of Technology, Sweden
Arun Mallya	NVIDIA, USA
R. Manmatha	Amazon, USA
Iacopo Masi	Sapienza University of Rome, Italy
Dimitris N. Metaxas	Rutgers University, USA
Ajmal Mian	University of Western Australia, Australia
Christian Micheloni	University of Udine, Italy
Krystian Mikolajczyk	Imperial College London, UK
Anurag Mittal	Indian Institute of Technology, Madras, India
Philippos Mordohai	Stevens Institute of Technology, USA
Greg Mori	Simon Fraser University & Borealis AI, Canada

Vittorio Murino	Istituto Italiano di Tecnologia, Italy
P. J. Narayanan	International Institute of Information Technology, Hyderabad, India
Ram Nevatia	University of Southern California, USA
Natalia Neverova	Facebook AI Research, UK
Richard Newcombe	Facebook, USA
Cuong V. Nguyen	Florida International University, USA
Bingbing Ni	Shanghai Jiao Tong University, China
Juan Carlos Niebles	Salesforce & Stanford University, USA
Ko Nishino	Kyoto University, Japan
Jean-Marc Odobez	Idiap Research Institute, École polytechnique fédérale de Lausanne, Switzerland
Francesca Odone	University of Genova, Italy
Takayuki Okatani	Tohoku University & RIKEN Center for Advanced Intelligence Project, Japan
Manohar Paluri	Facebook, USA
Guan Pang	Facebook, USA
Maja Pantic	Imperial College London, UK
Sylvain Paris	Adobe Research, USA
Jaesik Park	Pohang University of Science and Technology, Korea
Hyun Soo Park	The University of Minnesota, USA
Omkar M. Parkhi	Facebook, USA
Deepak Pathak	Carnegie Mellon University, USA
Georgios Pavlakos	University of California, Berkeley, USA
Marcello Pelillo	University of Venice, Italy
Marc Pollefeys	ETH Zurich & Microsoft, Switzerland
Jean Ponce	Inria, France
Gerard Pons-Moll	University of Tübingen, Germany
Fatih Porikli	Qualcomm, USA
Victor Adrian Prisacariu	University of Oxford, UK
Petia Radeva	University of Barcelona, Spain
Ravi Ramamoorthi	University of California, San Diego, USA
Deva Ramanan	Carnegie Mellon University, USA
Vignesh Ramanathan	Facebook, USA
Nalini Ratha	State University of New York at Buffalo, USA
Tammy Riklin Raviv	Ben-Gurion University, Israel
Tobias Ritschel	University College London, UK
Emanuele Rodola	Sapienza University of Rome, Italy
Amit K. Roy-Chowdhury	University of California, Riverside, USA
Michael Rubinstein	Google, USA
Olga Russakovsky	Princeton University, USA

Mathieu Salzmann	École polytechnique fédérale de Lausanne, Switzerland
Dimitris Samaras	Stony Brook University, USA
Aswin Sankaranarayanan	Carnegie Mellon University, USA
Imari Sato	National Institute of Informatics, Japan
Yoichi Sato	University of Tokyo, Japan
Shin'ichi Satoh	National Institute of Informatics, Japan
Walter Scheirer	University of Notre Dame, USA
Bernt Schiele	Max Planck Institute for Informatics, Germany
Konrad Schindler	ETH Zurich, Switzerland
Cordelia Schmid	Inria & Google, France
Alexander Schwing	University of Illinois at Urbana-Champaign, USA
Nicu Sebe	University of Trento, Italy
Greg Shakhnarovich	Toyota Technological Institute at Chicago, USA
Eli Shechtman	Adobe Research, USA
Humphrey Shi	University of Oregon & University of Illinois at Urbana-Champaign & Picsart AI Research, USA
Jianbo Shi	University of Pennsylvania, USA
Roy Shilkrot	Massachusetts Institute of Technology, USA
Mike Zheng Shou	National University of Singapore, Singapore
Kaleem Siddiqi	McGill University, Canada
Richa Singh	Indian Institute of Technology Jodhpur, India
Greg Slabaugh	Queen Mary University of London, UK
Cees Snoek	University of Amsterdam, The Netherlands
Yale Song	Facebook AI Research, USA
Yi-Zhe Song	University of Surrey, UK
Bjorn Stenger	Rakuten Institute of Technology
Abby Stylianou	Saint Louis University, USA
Akihiro Sugimoto	National Institute of Informatics, Japan
Chen Sun	Brown University, USA
Deqing Sun	Google, USA
Kalyan Sunkavalli	Adobe Research, USA
Ying Tai	Tencent YouTu Lab, China
Ayellet Tal	Technion – Israel Institute of Technology, Israel
Ping Tan	Simon Fraser University, Canada
Siyu Tang	ETH Zurich, Switzerland
Chi-Keung Tang	Hong Kong University of Science and Technology, Hong Kong, China
Radu Timofte	University of Würzburg, Germany & ETH Zurich, Switzerland
Federico Tombari	Google, Switzerland & Technical University of Munich, Germany

James Tompkin   Brown University, USA
Lorenzo Torresani   Dartmouth College, USA
Alexander Toshev   Apple, USA
Du Tran   Facebook AI Research, USA
Anh T. Tran   VinAI, Vietnam
Zhuowen Tu   University of California, San Diego, USA
Georgios Tzimiropoulos   Queen Mary University of London, UK
Jasper Uijlings   Google Research, Switzerland
Jan C. van Gemert   Delft University of Technology, The Netherlands
Gul Varol   Ecole des Ponts ParisTech, France
Nuno Vasconcelos   University of California, San Diego, USA
Mayank Vatsa   Indian Institute of Technology Jodhpur, India
Ashok Veeraraghavan   Rice University, USA
Jakob Verbeek   Facebook AI Research, France
Carl Vondrick   Columbia University, USA
Ruiping Wang   Institute of Computing Technology, Chinese
                                 Academy of Sciences, China
Xinchao Wang   National University of Singapore, Singapore
Liwei Wang   The Chinese University of Hong Kong,
                                 Hong Kong, China
Chaohui Wang   Université Paris-Est, France
Xiaolong Wang   University of California, San Diego, USA
Christian Wolf   NAVER LABS Europe, France
Tao Xiang   University of Surrey, UK
Saining Xie   Facebook AI Research, USA
Cihang Xie   University of California, Santa Cruz, USA
Zeki Yalniz   Facebook, USA
Ming-Hsuan Yang   University of California, Merced, USA
Angela Yao   National University of Singapore, Singapore
Shaodi You   University of Amsterdam, The Netherlands
Stella X. Yu   University of California, Berkeley, USA
Junsong Yuan   State University of New York at Buffalo, USA
Stefanos Zafeiriou   Imperial College London, UK
Amir Zamir   École polytechnique fédérale de Lausanne,
                                 Switzerland
Lei Zhang   Alibaba & Hong Kong Polytechnic University,
                                 Hong Kong, China
Lei Zhang   International Digital Economy Academy (IDEA),
                                 China
Pengchuan Zhang   Meta AI, USA
Bolei Zhou   University of California, Los Angeles, USA
Yuke Zhu   University of Texas at Austin, USA

Todd Zickler   Harvard University, USA
Wangmeng Zuo   Harbin Institute of Technology, China

Technical Program Committee

Davide Abati
Soroush Abbasi
    Koohpayegani
Amos L. Abbott
Rameen Abdal
Rabab Abdelfattah
Sahar Abdelnabi
Hassan Abu Alhaija
Abulikemu Abuduweili
Ron Abutbul
Hanno Ackermann
Aikaterini Adam
Kamil Adamczewski
Ehsan Adeli
Vida Adeli
Donald Adjeroh
Arman Afrasiyabi
Akshay Agarwal
Sameer Agarwal
Abhinav Agarwalla
Vaibhav Aggarwal
Sara Aghajanzadeh
Susmit Agrawal
Antonio Agudo
Touqeer Ahmad
Sk Miraj Ahmed
Chaitanya Ahuja
Nilesh A. Ahuja
Abhishek Aich
Shubhra Aich
Noam Aigerman
Arash Akbarinia
Peri Akiva
Derya Akkaynak
Emre Aksan
Arjun R. Akula
Yuval Alaluf
Stephan Alaniz
Paul Albert
Cenek Albl

Filippo Aleotti
Konstantinos P.
    Alexandridis
Motasem Alfarra
Mohsen Ali
Thiemo Alldieck
Hadi Alzayer
Liang An
Shan An
Yi An
Zhulin An
Dongsheng An
Jie An
Xiang An
Saket Anand
Cosmin Ancuti
Juan Andrade-Cetto
Alexander Andreopoulos
Bjoern Andres
Jerone T. A. Andrews
Shivangi Aneja
Anelia Angelova
Dragomir Anguelov
Rushil Anirudh
Oron Anschel
Rao Muhammad Anwer
Djamila Aouada
Evlampios Apostolidis
Srikar Appalaraju
Nikita Araslanov
Andre Araujo
Eric Arazo
Dawit Mureja Argaw
Anurag Arnab
Aditya Arora
Chetan Arora
Sunpreet S. Arora
Alexey Artemov
Muhammad Asad
Kumar Ashutosh

Sinem Aslan
Vishal Asnani
Mahmoud Assran
Amir Atapour-Abarghouei
Nikos Athanasiou
Ali Athar
ShahRukh Athar
Sara Atito
Souhaib Attaiki
Matan Atzmon
Mathieu Aubry
Nicolas Audebert
Tristan T.
    Aumentado-Armstrong
Melinos Averkiou
Yannis Avrithis
Stephane Ayache
Mehmet Aygün
Seyed Mehdi
    Ayyoubzadeh
Hossein Azizpour
George Azzopardi
Mallikarjun B. R.
Yunhao Ba
Abhishek Badki
Seung-Hwan Bae
Seung-Hwan Baek
Seungryul Baek
Piyush Nitin Bagad
Shai Bagon
Gaetan Bahl
Shikhar Bahl
Sherwin Bahmani
Haoran Bai
Lei Bai
Jiawang Bai
Haoyue Bai
Jinbin Bai
Xiang Bai
Xuyang Bai

Yang Bai
Yuanchao Bai
Ziqian Bai
Sungyong Baik
Kevin Bailly
Max Bain
Federico Baldassarre
Wele Gedara Chaminda
    Bandara
Biplab Banerjee
Pratyay Banerjee
Sandipan Banerjee
Jihwan Bang
Antyanta Bangunharcana
Aayush Bansal
Ankan Bansal
Siddhant Bansal
Wentao Bao
Zhipeng Bao
Amir Bar
Manel Baradad Jurjo
Lorenzo Baraldi
Danny Barash
Daniel Barath
Connelly Barnes
Ioan Andrei Bârsan
Steven Basart
Dina Bashkirova
Chaim Baskin
Peyman Bateni
Anil Batra
Sebastiano Battiato
Ardhendu Behera
Harkirat Behl
Jens Behley
Vasileios Belagiannis
Boulbaba Ben Amor
Emanuel Ben Baruch
Abdessamad Ben Hamza
Gil Ben-Artzi
Assia Benbihi
Fabian Benitez-Quiroz
Guy Ben-Yosef
Philipp Benz
Alexander W. Bergman

Urs Bergmann
Jesus Bermudez-Cameo
Stefano Berretti
Gedas Bertasius
Zachary Bessinger
Petra Bevandić
Matthew Beveridge
Lucas Beyer
Yash Bhalgat
Suvaansh Bhambri
Samarth Bharadwaj
Gaurav Bharaj
Aparna Bharati
Bharat Lal Bhatnagar
Uttaran Bhattacharya
Apratim Bhattacharyya
Brojeshwar Bhowmick
Ankan Kumar Bhunia
Ayan Kumar Bhunia
Qi Bi
Sai Bi
Michael Bi Mi
Gui-Bin Bian
Jia-Wang Bian
Shaojun Bian
Pia Bideau
Mario Bijelic
Hakan Bilen
Guillaume-Alexandre
    Bilodeau
Alexander Binder
Tolga Birdal
Vighnesh N. Birodkar
Sandika Biswas
Andreas Blattmann
Janusz Bobulski
Giuseppe Boccignone
Vishnu Boddeti
Navaneeth Bodla
Moritz Böhle
Aleksei Bokhovkin
Sam Bond-Taylor
Vivek Boominathan
Shubhankar Borse
Mark Boss

Andrea Bottino
Adnane Boukhayma
Fadi Boutros
Nicolas C. Boutry
Richard S. Bowen
Ivaylo Boyadzhiev
Aidan Boyd
Yuri Boykov
Aljaz Bozic
Behzad Bozorgtabar
Eric Brachmann
Samarth Brahmbhatt
Gustav Bredell
Francois Bremond
Joel Brogan
Andrew Brown
Thomas Brox
Marcus A. Brubaker
Robert-Jan Bruintjes
Yuqi Bu
Anders G. Buch
Himanshu Buckchash
Mateusz Buda
Ignas Budvytis
José M. Buenaposada
Marcel C. Bühler
Tu Bui
Adrian Bulat
Hannah Bull
Evgeny Burnaev
Andrei Bursuc
Benjamin Busam
Sergey N. Buzykanov
Wonmin Byeon
Fabian Caba
Martin Cadik
Guanyu Cai
Minjie Cai
Qing Cai
Zhongang Cai
Qi Cai
Yancheng Cai
Shen Cai
Han Cai
Jiarui Cai

Bowen Cai

Mu Cai

Qin Cai

Ruojin Cai

Weidong Cai

Weiwei Cai

Yi Cai

Yujun Cai

Zhiping Cai

Akin Caliskan

Lilian Calvet

Baris Can Cam

Necati Cihan Camgoz

Tommaso Campari

Dylan Campbell

Ziang Cao

Ang Cao

Xu Cao

Zhiwen Cao

Shengcao Cao

Song Cao

Weipeng Cao

Xiangyong Cao

Xiaochun Cao

Yue Cao

Yunhao Cao

Zhangjie Cao

Jiale Cao

Yang Cao

Jiajiong Cao

Jie Cao

Jinkun Cao

Lele Cao

Yulong Cao

Zhiguo Cao

Chen Cao

Razvan Caramalau

Marlène Careil

Gustavo Carneiro

Joao Carreira

Dan Casas

Paola Cascante-Bonilla

Angela Castillo

Francisco M. Castro

Pedro Castro

Luca Cavalli

George J. Cazenavette

Oya Celiktutan

Hakan Cevikalp

Sri Harsha C. H.

Sungmin Cha

Geonho Cha

Menglei Chai

Lucy Chai

Yuning Chai

Zenghao Chai

Anirban Chakraborty

Deep Chakraborty

Rudrasis Chakraborty

Souradeep Chakraborty

Kelvin C. K. Chan

Chee Seng Chan

Paramanand Chandramouli

Arjun Chandrasekaran

Kenneth Chaney

Dongliang Chang

Huiwen Chang

Peng Chang

Xiaojun Chang

Jia-Ren Chang

Hyung Jin Chang

Hyun Sung Chang

Ju Yong Chang

Li-Jen Chang

Qi Chang

Wei-Yi Chang

Yi Chang

Nadine Chang

Hanqing Chao

Pradyumna Chari

Dibyadip Chatterjee

Chiranjoy Chattopadhyay

Siddhartha Chaudhuri

Zhengping Che

Gal Chechik

Lianggangxu Chen

Qi Alfred Chen

Brian Chen

Bor-Chun Chen

Bo-Hao Chen

Bohong Chen

Bin Chen

Ziliang Chen

Cheng Chen

Chen Chen

Chaofeng Chen

Xi Chen

Haoyu Chen

Xuanhong Chen

Wei Chen

Qiang Chen

Shi Chen

Xianyu Chen

Chang Chen

Changhuai Chen

Hao Chen

Jie Chen

Jianbo Chen

Jingjing Chen

Jun Chen

Kejiang Chen

Mingcai Chen

Nenglun Chen

Qifeng Chen

Ruoyu Chen

Shu-Yu Chen

Weidong Chen

Weijie Chen

Weikai Chen

Xiang Chen

Xiuyi Chen

Xingyu Chen

Yaofo Chen

Yueting Chen

Yu Chen

Yunjin Chen

Yuntao Chen

Yun Chen

Zhenfang Chen

Zhuangzhuang Chen

Chu-Song Chen

Xiangyu Chen

Zhuo Chen

Chaoqi Chen

Shizhe Chen

Xiaotong Chen

Xiaozhi Chen

Dian Chen

Defang Chen

Dingfan Chen

Ding-Jie Chen

Ee Heng Chen

Tao Chen

Yixin Chen

Wei-Ting Chen

Lin Chen

Guang Chen

Guangyi Chen

Guanying Chen

Guangyao Chen

Hwann-Tzong Chen

Junwen Chen

Jiacheng Chen

Jianxu Chen

Hui Chen

Kai Chen

Kan Chen

Kevin Chen

Kuan-Wen Chen

Weihua Chen

Zhang Chen

Liang-Chieh Chen

Lele Chen

Liang Chen

Fanglin Chen

Zehui Chen

Minghui Chen

Minghao Chen

Xiaokang Chen

Qian Chen

Jun-Cheng Chen

Qi Chen

Qingcai Chen

Richard J. Chen

Runnan Chen

Rui Chen

Shuo Chen

Sentao Chen

Shaoyu Chen

Shixing Chen

Shuai Chen

Shuya Chen

Sizhe Chen

Simin Chen

Shaoxiang Chen

Zitian Chen

Tianlong Chen

Tianshui Chen

Min-Hung Chen

Xiangning Chen

Xin Chen

Xinghao Chen

Xuejin Chen

Xu Chen

Xuxi Chen

Yunlu Chen

Yanbei Chen

Yuxiao Chen

Yun-Chun Chen

Yi-Ting Chen

Yi-Wen Chen

Yinbo Chen

Yiran Chen

Yuanhong Chen

Yubei Chen

Yuefeng Chen

Yuhua Chen

Yukang Chen

Zerui Chen

Zhaoyu Chen

Zhen Chen

Zhenyu Chen

Zhi Chen

Zhiwei Chen

Zhixiang Chen

Long Chen

Bowen Cheng

Jun Cheng

Yi Cheng

Jingchun Cheng

Lechao Cheng

Xi Cheng

Yuan Cheng

Ho Kei Cheng

Kevin Ho Man Cheng

Jiacheng Cheng

Kelvin B. Cheng

Li Cheng

Mengjun Cheng

Zhen Cheng

Qingrong Cheng

Tianheng Cheng

Harry Cheng

Yihua Cheng

Yu Cheng

Ziheng Cheng

Soon Yau Cheong

Anoop Cherian

Manuela Chessa

Zhixiang Chi

Naoki Chiba

Julian Chibane

Kashyap Chitta

Tai-Yin Chiu

Hsu-kuang Chiu

Wei-Chen Chiu

Sungmin Cho

Donghyeon Cho

Hyeon Cho

Yooshin Cho

Gyusang Cho

Jang Hyun Cho

Seungju Cho

Nam Ik Cho

Sunghyun Cho

Hanbyel Cho

Jaesung Choe

Jooyoung Choi

Chiho Choi

Changwoon Choi

Jongwon Choi

Myungsub Choi

Dooseop Choi

Jonghyun Choi

Jinwoo Choi

Jun Won Choi

Min-Kook Choi

Hongsuk Choi

Janghoon Choi

Yoon-Ho Choi

Haisong Ding
Hui Ding
Jiahao Ding
Jian Ding
Jian-Jiun Ding
Shuxiao Ding
Tianyu Ding
Wenhao Ding
Yuqi Ding
Yi Ding
Yuzhen Ding
Zhengming Ding
Tan Minh Dinh
Vu Dinh
Christos Diou
Mandar Dixit
Bao Gia Doan
Khoa D. Doan
Dzung Anh Doan
Debi Prosad Dogra
Nehal Doiphode
Chengdong Dong
Bowen Dong
Zhenxing Dong
Hang Dong
Xiaoyi Dong
Haoye Dong
Jiangxin Dong
Shichao Dong
Xuan Dong
Zhen Dong
Shuting Dong
Jing Dong
Li Dong
Ming Dong
Nanqing Dong
Qiulei Dong
Runpei Dong
Siyan Dong
Tian Dong
Wei Dong
Xiaomeng Dong
Xin Dong
Xingbo Dong
Yuan Dong

Samuel Dooley
Gianfranco Doretto
Michael Dorkenwald
Keval Doshi
Zhaopeng Dou
Xiaotian Dou
Hazel Doughty
Ahmad Droby
Iddo Drori
Jie Du
Yong Du
Dawei Du
Dong Du
Ruoyi Du
Yuntao Du
Xuefeng Du
Yilun Du
Yuming Du
Radhika Dua
Haodong Duan
Jiafei Duan
Kaiwen Duan
Peiqi Duan
Ye Duan
Haoran Duan
Jiali Duan
Amanda Duarte
Abhimanyu Dubey
Shiv Ram Dubey
Florian Dubost
Lukasz Dudziak
Shivam Duggal
Justin M. Dulay
Matteo Dunnhofer
Chi Nhan Duong
Thibaut Durand
Mihai Dusmanu
Ujjal Kr Dutta
Debidatta Dwibedi
Isht Dwivedi
Sai Kumar Dwivedi
Takeharu Eda
Mark Edmonds
Alexei A. Efros
Thibaud Ehret

Max Ehrlich
Mahsa Ehsanpour
Iván Eichhardt
Farshad Einabadi
Marvin Eisenberger
Hazim Kemal Ekenel
Mohamed El Banani
Ismail Elezi
Moshe Eliasof
Alaa El-Nouby
Ian Endres
Francis Engelmann
Deniz Engin
Chanho Eom
Dave Epstein
Maria C. Escobar
Victor A. Escorcia
Carlos Esteves
Sungmin Eum
Bernard J. E. Evans
Ivan Evtimov
Fevziye Irem Eyiokur
    Yaman
Matteo Fabbri
Sébastien Fabbro
Gabriele Facciolo
Masud Fahim
Bin Fan
Hehe Fan
Deng-Ping Fan
Aoxiang Fan
Chen-Chen Fan
Qi Fan
Zhaoxin Fan
Haoqi Fan
Heng Fan
Hongyi Fan
Linxi Fan
Baojie Fan
Jiayuan Fan
Lei Fan
Quanfu Fan
Yonghui Fan
Yingruo Fan
Zhiwen Fan

Zicong Fan

Sean Fanello

Jiansheng Fang

Chaowei Fang

Yuming Fang

Jianwu Fang

Jin Fang

Qi Fang

Shancheng Fang

Tian Fang

Xianyong Fang

Gongfan Fang

Zhen Fang

Hui Fang

Jiemin Fang

Le Fang

Pengfei Fang

Xiaolin Fang

Yuxin Fang

Zhaoyuan Fang

Ammarah Farooq

Azade Farshad

Zhengcong Fei

Michael Felsberg

Wei Feng

Chen Feng

Fan Feng

Andrew Feng

Xin Feng

Zheyun Feng

Ruicheng Feng

Mingtao Feng

Qianyu Feng

Shangbin Feng

Chun-Mei Feng

Zunlei Feng

Zhiyong Feng

Martin Fergie

Mustansar Fiaz

Marco Fiorucci

Michael Firman

Hamed Firooz

Volker Fischer

Corneliu O. Florea

Georgios Floros

Wolfgang Foerstner

Gianni Franchi

Jean-Sebastien Franco

Simone Frintrop

Anna Fruehstueck

Changhong Fu

Chaoyou Fu

Cheng-Yang Fu

Chi-Wing Fu

Deqing Fu

Huan Fu

Jun Fu

Kexue Fu

Ying Fu

Jianlong Fu

Jingjing Fu

Qichen Fu

Tsu-Jui Fu

Xueyang Fu

Yang Fu

Yanwei Fu

Yonggan Fu

Wolfgang Fuhl

Yasuhisa Fujii

Kent Fujiwara

Marco Fumero

Takuya Funatomi

Isabel Funke

Dario Fuoli

Antonino Furnari

Matheus A. Gadelha

Akshay Gadi Patil

Adrian Galdran

Guillermo Gallego

Silvano Galliani

Orazio Gallo

Leonardo Galteri

Matteo Gamba

Yiming Gan

Sujoy Ganguly

Harald Ganster

Boyan Gao

Changxin Gao

Daiheng Gao

Difei Gao

Chen Gao

Fei Gao

Lin Gao

Wei Gao

Yiming Gao

Junyu Gao

Guangyu Ryan Gao

Haichang Gao

Hongchang Gao

Jialin Gao

Jin Gao

Jun Gao

Katelyn Gao

Mingchen Gao

Mingfei Gao

Pan Gao

Shangqian Gao

Shanghua Gao

Xitong Gao

Yunhe Gao

Zhanning Gao

Elena Garces

Nuno Cruz Garcia

Noa Garcia

Guillermo
    Garcia-Hernando

Isha Garg

Rahul Garg

Sourav Garg

Quentin Garrido

Stefano Gasperini

Kent Gauen

Chandan Gautam

Shivam Gautam

Paul Gay

Chunjiang Ge

Shiming Ge

Wenhang Ge

Yanhao Ge

Zheng Ge

Songwei Ge

Weifeng Ge

Yixiao Ge

Yuying Ge

Shijie Geng

Zhengyang Geng
Kyle A. Genova
Georgios Georgakis
Markos Georgopoulos
Marcel Geppert
Shabnam Ghadar
Mina Ghadimi Atigh
Deepti Ghadiyaram
Maani Ghaffari Jadidi
Sedigh Ghamari
Zahra Gharaee
Michaël Gharbi
Golnaz Ghiasi
Reza Ghoddoosian
Soumya Suvra Ghosal
Adhiraj Ghosh
Arthita Ghosh
Pallabi Ghosh
Soumyadeep Ghosh
Andrew Gilbert
Igor Gilitschenski
Jhony H. Giraldo
Andreu Girbau Xalabarder
Rohit Girdhar
Sharath Girish
Xavier Giro-i-Nieto
Raja Giryes
Thomas Gittings
Nikolaos Gkanatsios
Ioannis Gkioulekas
Abhiram
   Gnanasambandam
Aurele T. Gnanha
Clement L. J. C. Godard
Arushi Goel
Vidit Goel
Shubham Goel
Zan Gojcic
Aaron K. Gokaslan
Tejas Gokhale
S. Alireza Golestaneh
Thiago L. Gomes
Nuno Goncalves
Boqing Gong
Chen Gong

Yuanhao Gong
Guoqiang Gong
Jingyu Gong
Rui Gong
Yu Gong
Mingming Gong
Neil Zhenqiang Gong
Xun Gong
Yunye Gong
Yihong Gong
Cristina I. González
Nithin Gopalakrishnan
   Nair
Gaurav Goswami
Jianping Gou
Shreyank N. Gowda
Ankit Goyal
Helmut Grabner
Patrick L. Grady
Ben Graham
Eric Granger
Douglas R. Gray
Matej Grcić
David Griffiths
Jinjin Gu
Yun Gu
Shuyang Gu
Jianyang Gu
Fuqiang Gu
Jiatao Gu
Jindong Gu
Jiaqi Gu
Jinwei Gu
Jiaxin Gu
Geonmo Gu
Xiao Gu
Xinqian Gu
Xiuye Gu
Yuming Gu
Zhangxuan Gu
Dayan Guan
Junfeng Guan
Qingji Guan
Tianrui Guan
Shanyan Guan

Denis A. Gudovskiy
Ricardo Guerrero
Pierre-Louis Guhur
Jie Gui
Liangyan Gui
Liangke Gui
Benoit Guillard
Erhan Gundogdu
Manuel Günther
Jingcai Guo
Yuanfang Guo
Junfeng Guo
Chenqi Guo
Dan Guo
Hongji Guo
Jia Guo
Jie Guo
Minghao Guo
Shi Guo
Yanhui Guo
Yangyang Guo
Yuan-Chen Guo
Yilu Guo
Yiluan Guo
Yong Guo
Guangyu Guo
Haiyun Guo
Jinyang Guo
Jianyuan Guo
Pengsheng Guo
Pengfei Guo
Shuxuan Guo
Song Guo
Tianyu Guo
Qing Guo
Qiushan Guo
Wen Guo
Xiefan Guo
Xiaohu Guo
Xiaoqing Guo
Yufei Guo
Yuhui Guo
Yuliang Guo
Yunhui Guo
Yanwen Guo

Akshita Gupta

Ankush Gupta

Kamal Gupta

Kartik Gupta

Ritwik Gupta

Rohit Gupta

Siddharth Gururani

Fredrik K. Gustafsson

Abner Guzman Rivera

Vladimir Guzov

Matthew A. Gwilliam

Jung-Woo Ha

Marc Habermann

Isma Hadji

Christian Haene

Martin Hahner

Levente Hajder

Alexandros Haliassos

Emanuela Haller

Bumsub Ham

Abdullah J. Hamdi

Shreyas Hampali

Dongyoon Han

Chunrui Han

Dong-Jun Han

Dong-Sig Han

Guangxing Han

Zhizhong Han

Ruize Han

Jiaming Han

Jin Han

Ligong Han

Xian-Hua Han

Xiaoguang Han

Yizeng Han

Zhi Han

Zhenjun Han

Zhongyi Han

Jungong Han

Junlin Han

Kai Han

Kun Han

Sungwon Han

Songfang Han

Wei Han

Xiao Han

Xintong Han

Xinzhe Han

Yahong Han

Yan Han

Zongbo Han

Nicolai Hani

Rana Hanocka

Niklas Hanselmann

Nicklas A. Hansen

Hong Hanyu

Fusheng Hao

Yanbin Hao

Shijie Hao

Udith Haputhanthri

Mehrtash Harandi

Josh Harguess

Adam Harley

David M. Hart

Atsushi Hashimoto

Ali Hassani

Mohammed Hassanin

Yana Hasson

Joakim Bruslund Haurum

Bo He

Kun He

Chen He

Xin He

Fazhi He

Gaoqi He

Hao He

Haoyu He

Jiangpeng He

Hongliang He

Qian He

Xiangteng He

Xuming He

Yannan He

Yuhang He

Yang He

Xiangyu He

Nanjun He

Pan He

Sen He

Shengfeng He

Songtao He

Tao He

Tong He

Wei He

Xuehai He

Xiaoxiao He

Ying He

Yisheng He

Ziwen He

Peter Hedman

Felix Heide

Yacov Hel-Or

Paul Henderson

Philipp Henzler

Byeongho Heo

Jae-Pil Heo

Miran Heo

Sachini A. Herath

Stephane Herbin

Pedro Hermosilla Casajus

Monica Hernandez

Charles Herrmann

Roei Herzig

Mauricio Hess-Flores

Carlos Hinojosa

Tobias Hinz

Tsubasa Hirakawa

Chih-Hui Ho

Lam Si Tung Ho

Jennifer Hobbs

Derek Hoiem

Yannick Hold-Geoffroy

Aleksander Holynski

Cheeun Hong

Fa-Ting Hong

Hanbin Hong

Guan Zhe Hong

Danfeng Hong

Lanqing Hong

Xiaopeng Hong

Xin Hong

Jie Hong

Seungbum Hong

Cheng-Yao Hong

Seunghoon Hong

Yi Hong
Yuan Hong
Yuchen Hong
Anthony Hoogs
Maxwell C. Horton
Kazuhiro Hotta
Qibin Hou
Tingbo Hou
Junhui Hou
Ji Hou
Qiqi Hou
Rui Hou
Ruibing Hou
Zhi Hou
Henry Howard-Jenkins
Lukas Hoyer
Wei-Lin Hsiao
Chiou-Ting Hsu
Anthony Hu
Brian Hu
Yusong Hu
Hexiang Hu
Haoji Hu
Di Hu
Hengtong Hu
Haigen Hu
Lianyu Hu
Hanzhe Hu
Jie Hu
Junlin Hu
Shizhe Hu
Jian Hu
Zhiming Hu
Juhua Hu
Peng Hu
Ping Hu
Ronghang Hu
MengShun Hu
Tao Hu
Vincent Tao Hu
Xiaoling Hu
Xinting Hu
Xiaolin Hu
Xuefeng Hu
Xiaowei Hu

Yang Hu
Yueyu Hu
Zeyu Hu
Zhongyun Hu
Binh-Son Hua
Guoliang Hua
Yi Hua
Linzhi Huang
Qiusheng Huang
Bo Huang
Chen Huang
Hsin-Ping Huang
Ye Huang
Shuangping Huang
Zeng Huang
Buzhen Huang
Cong Huang
Heng Huang
Hao Huang
Qidong Huang
Huaibo Huang
Chaoqin Huang
Feihu Huang
Jiahui Huang
Jingjia Huang
Kun Huang
Lei Huang
Sheng Huang
Shuaiyi Huang
Siyu Huang
Xiaoshui Huang
Xiaoyang Huang
Yan Huang
Yihao Huang
Ying Huang
Ziling Huang
Xiaoke Huang
Yifei Huang
Haiyang Huang
Zhewei Huang
Jin Huang
Haibin Huang
Jiaxing Huang
Junjie Huang
Keli Huang

Lang Huang
Lin Huang
Luojie Huang
Mingzhen Huang
Shijia Huang
Shengyu Huang
Siyuan Huang
He Huang
Xiuyu Huang
Lianghua Huang
Yue Huang
Yaping Huang
Yuge Huang
Zehao Huang
Zeyi Huang
Zhiqi Huang
Zhongzhan Huang
Zilong Huang
Ziyuan Huang
Tianrui Hui
Zhuo Hui
Le Hui
Jing Huo
Junhwa Hur
Shehzeen S. Hussain
Chuong Minh Huynh
Seunghyun Hwang
Jaehui Hwang
Jyh-Jing Hwang
Sukjun Hwang
Soonmin Hwang
Wonjun Hwang
Rakib Hyder
Sangeek Hyun
Sarah Ibrahimi
Tomoki Ichikawa
Yerlan Idelbayev
A. S. M. Iftekhar
Masaaki Iiyama
Satoshi Ikehata
Sunghoon Im
Atul N. Ingle
Eldar Insafutdinov
Yani A. Ioannou
Radu Tudor Ionescu

Umar Iqbal
Go Irie
Muhammad Zubair Irshad
Ahmet Iscen
Berivan Isik
Ashraful Islam
Md Amirul Islam
Syed Islam
Mariko Isogawa
Vamsi Krishna K. Ithapu
Boris Ivanovic
Darshan Iyer
Sarah Jabbour
Ayush Jain
Nishant Jain
Samyak Jain
Vidit Jain
Vineet Jain
Priyank Jaini
Tomas Jakab
Mohammad A. A. K.
  Jalwana
Muhammad Abdullah
  Jamal
Hadi Jamali-Rad
Stuart James
Varun Jampani
Young Kyun Jang
YeongJun Jang
Yunseok Jang
Ronnachai Jaroensri
Bhavan Jasani
Krishna Murthy
  Jatavallabhula
Mojan Javaheripi
Syed A. Javed
Guillaume Jeanneret
Pranav Jeevan
Herve Jegou
Rohit Jena
Tomas Jenicek
Porter Jenkins
Simon Jenni
Hae-Gon Jeon
Sangryul Jeon

Boseung Jeong
Yoonwoo Jeong
Seong-Gyun Jeong
Jisoo Jeong
Allan D. Jepson
Ankit Jha
Sumit K. Jha
I-Hong Jhuo
Ge-Peng Ji
Chaonan Ji
Deyi Ji
Jingwei Ji
Wei Ji
Zhong Ji
Jiayi Ji
Pengliang Ji
Hui Ji
Mingi Ji
Xiaopeng Ji
Yuzhu Ji
Baoxiong Jia
Songhao Jia
Dan Jia
Shan Jia
Xiaojun Jia
Xiuyi Jia
Xu Jia
Menglin Jia
Wenqi Jia
Boyuan Jiang
Wenhao Jiang
Huaizu Jiang
Hanwen Jiang
Haiyong Jiang
Hao Jiang
Huajie Jiang
Huiqin Jiang
Haojun Jiang
Haobo Jiang
Junjun Jiang
Xingyu Jiang
Yangbangyan Jiang
Yu Jiang
Jianmin Jiang
Jiaxi Jiang

Jing Jiang
Kui Jiang
Li Jiang
Liming Jiang
Chiyu Jiang
Meirui Jiang
Chen Jiang
Peng Jiang
Tai-Xiang Jiang
Wen Jiang
Xinyang Jiang
Yifan Jiang
Yuming Jiang
Yingying Jiang
Zeren Jiang
ZhengKai Jiang
Zhenyu Jiang
Shuming Jiao
Jianbo Jiao
Licheng Jiao
Dongkwon Jin
Yeying Jin
Cheng Jin
Linyi Jin
Qing Jin
Taisong Jin
Xiao Jin
Xin Jin
Sheng Jin
Kyong Hwan Jin
Ruibing Jin
SouYoung Jin
Yueming Jin
Chenchen Jing
Longlong Jing
Taotao Jing
Yongcheng Jing
Younghyun Jo
Joakim Johnander
Jeff Johnson
Michael J. Jones
R. Kenny Jones
Rico Jonschkowski
Ameya Joshi
Sunghun Joung

Felix Juefei-Xu
Claudio R. Jung
Steffen Jung
Hari Chandana K.
Rahul Vigneswaran K.
Prajwal K. R.
Abhishek Kadian
Jhony Kaesemodel Pontes
Kumara Kahatapitiya
Anmol Kalia
Sinan Kalkan
Tarun Kalluri
Jaewon Kam
Sandesh Kamath
Meina Kan
Menelaos Kanakis
Takuhiro Kaneko
Di Kang
Guoliang Kang
Hao Kang
Jaeyeon Kang
Kyoungkook Kang
Li-Wei Kang
MinGuk Kang
Suk-Ju Kang
Zhao Kang
Yash Mukund Kant
Yueying Kao
Aupendu Kar
Konstantinos Karantzalos
Sezer Karaoglu
Navid Kardan
Sanjay Kariyappa
Leonid Karlinsky
Animesh Karnewar
Shyamgopal Karthik
Hirak J. Kashyap
Marc A. Kastner
Hirokatsu Kataoka
Angelos Katharopoulos
Hiroharu Kato
Kai Katsumata
Manuel Kaufmann
Chaitanya Kaul
Prakhar Kaushik

Yuki Kawana
Lei Ke
Lipeng Ke
Tsung-Wei Ke
Wei Ke
Petr Kellnhofer
Aniruddha Kembhavi
John Kender
Corentin Kervadec
Leonid Keselman
Daniel Keysers
Nima Khademi Kalantari
Taras Khakhulin
Samir Khaki
Muhammad Haris Khan
Qadeer Khan
Salman Khan
Subash Khanal
Vaishnavi M. Khindkar
Rawal Khirodkar
Saeed Khorram
Pirazh Khorramshahi
Kourosh Khoshelham
Ansh Khurana
Benjamin Kiefer
Jae Myung Kim
Junho Kim
Boah Kim
Hyeonseong Kim
Dong-Jin Kim
Dongwan Kim
Donghyun Kim
Doyeon Kim
Yonghyun Kim
Hyung-Il Kim
Hyunwoo Kim
Hyeongwoo Kim
Hyo Jin Kim
Hyunwoo J. Kim
Taehoon Kim
Jaeha Kim
Jiwon Kim
Jung Uk Kim
Kangyeol Kim
Eunji Kim

Daeha Kim
Dongwon Kim
Kunhee Kim
Kyungmin Kim
Junsik Kim
Min H. Kim
Namil Kim
Kookhoi Kim
Sanghyun Kim
Seongyeop Kim
Seungryong Kim
Saehoon Kim
Euyoung Kim
Guisik Kim
Sungyeon Kim
Sunnie S. Y. Kim
Taehun Kim
Tae Oh Kim
Won Hwa Kim
Seungwook Kim
YoungBin Kim
Youngeun Kim
Akisato Kimura
Furkan Osman Kınlı
Zsolt Kira
Hedvig Kjellström
Florian Kleber
Jan P. Klopp
Florian Kluger
Laurent Kneip
Byungsoo Ko
Muhammed Kocabas
A. Sophia Koepke
Kevin Koeser
Nick Kolkin
Nikos Kolotouros
Wai-Kin Adams Kong
Deying Kong
Caihua Kong
Youyong Kong
Shuyu Kong
Shu Kong
Tao Kong
Yajing Kong
Yu Kong

Zishang Kong
Theodora Kontogianni
Anton S. Konushin
Julian F. P. Kooij
Bruno Korbar
Giorgos Kordopatis-Zilos
Jari Korhonen
Adam Kortylewski
Denis Korzhenkov
Divya Kothandaraman
Suraj Kothawade
Iuliia Kotseruba
Satwik Kottur
Shashank Kotyan
Alexandros Kouris
Petros Koutras
Anna Kreshuk
Ranjay Krishna
Dilip Krishnan
Andrey Kuehlkamp
Hilde Kuehne
Jason Kuen
David Kügler
Arjan Kuijper
Anna Kukleva
Sumith Kulal
Viveka Kulharia
Akshay R. Kulkarni
Nilesh Kulkarni
Dominik Kulon
Abhinav Kumar
Akash Kumar
Suryansh Kumar
B. V. K. Vijaya Kumar
Pulkit Kumar
Ratnesh Kumar
Sateesh Kumar
Satish Kumar
Vijay Kumar B. G.
Nupur Kumari
Sudhakar Kumawat
Jogendra Nath Kundu
Hsien-Kai Kuo
Meng-Yu Jennifer Kuo
Vinod Kumar Kurmi

Yusuke Kurose
Keerthy Kusumam
Alina Kuznetsova
Henry Kvinge
Ho Man Kwan
Hyeokjun Kweon
Heeseung Kwon
Gihyun Kwon
Myung-Joon Kwon
Taesung Kwon
YoungJoong Kwon
Christos Kyrkou
Jorma Laaksonen
Yann Labbe
Zorah Laehner
Florent Lafarge
Hamid Laga
Manuel Lagunas
Shenqi Lai
Jian-Huang Lai
Zihang Lai
Mohamed I. Lakhal
Mohit Lamba
Meng Lan
Loic Landrieu
Zhiqiang Lang
Natalie Lang
Dong Lao
Yizhen Lao
Yingjie Lao
Issam Hadj Laradji
Gustav Larsson
Viktor Larsson
Zakaria Laskar
Stéphane Lathuilière
Chun Pong Lau
Rynson W. H. Lau
Hei Law
Justin Lazarow
Verica Lazova
Eric-Tuan Le
Hieu Le
Trung-Nghia Le
Mathias Lechner
Byeong-Uk Lee

Chen-Yu Lee
Che-Rung Lee
Chul Lee
Hong Joo Lee
Dongsoo Lee
Jiyoung Lee
Eugene Eu Tzuan Lee
Daeun Lee
Saehyung Lee
Jewook Lee
Hyungtae Lee
Hyunmin Lee
Jungbeom Lee
Joon-Young Lee
Jong-Seok Lee
Joonseok Lee
Junha Lee
Kibok Lee
Byung-Kwan Lee
Jangwon Lee
Jinho Lee
Jongmin Lee
Seunghyun Lee
Sohyun Lee
Minsik Lee
Dogyoon Lee
Seungmin Lee
Min Jun Lee
Sangho Lee
Sangmin Lee
Seungeun Lee
Seon-Ho Lee
Sungmin Lee
Sungho Lee
Sangyoun Lee
Vincent C. S. S. Lee
Jaeseong Lee
Yong Jae Lee
Chenyang Lei
Chenyi Lei
Jiahui Lei
Xinyu Lei
Yinjie Lei
Jiaxu Leng
Luziwei Leng

Jan E. Lenssen
Vincent Lepetit
Thomas Leung
María Leyva-Vallina
Xin Li
Yikang Li
Baoxin Li
Bin Li
Bing Li
Bowen Li
Changlin Li
Chao Li
Chongyi Li
Guanyue Li
Shuai Li
Jin Li
Dingquan Li
Dongxu Li
Yiting Li
Gang Li
Dian Li
Guohao Li
Haoang Li
Haoliang Li
Haoran Li
Hengduo Li
Huafeng Li
Xiaoming Li
Hanao Li
Hongwei Li
Ziqiang Li
Jisheng Li
Jiacheng Li
Jia Li
Jiachen Li
Jiahao Li
Jianwei Li
Jiazhi Li
Jie Li
Jing Li
Jingjing Li
Jingtao Li
Jun Li
Junxuan Li
Kai Li

Kailin Li
Kenneth Li
Kun Li
Kunpeng Li
Aoxue Li
Chenglong Li
Chenglin Li
Changsheng Li
Zhichao Li
Qiang Li
Yanyu Li
Zuoyue Li
Xiang Li
Xuelong Li
Fangda Li
Ailin Li
Liang Li
Chun-Guang Li
Daiqing Li
Dong Li
Guanbin Li
Guorong Li
Haifeng Li
Jianan Li
Jianing Li
Jiaxin Li
Ke Li
Lei Li
Lincheng Li
Liulei Li
Lujun Li
Linjie Li
Lin Li
Pengyu Li
Ping Li
Qiufu Li
Qingyong Li
Rui Li
Siyuan Li
Wei Li
Wenbin Li
Xiangyang Li
Xinyu Li
Xiujun Li
Xiu Li

Xu Li
Ya-Li Li
Yao Li
Yongjie Li
Yijun Li
Yiming Li
Yuezun Li
Yu Li
Yunheng Li
Yuqi Li
Zhe Li
Zeming Li
Zhen Li
Zhengqin Li
Zhimin Li
Jiefeng Li
Jinpeng Li
Chengze Li
Jianwu Li
Lerenhan Li
Shan Li
Suichan Li
Xiangtai Li
Yanjie Li
Yandong Li
Zhuoling Li
Zhenqiang Li
Manyi Li
Maosen Li
Ji Li
Minjun Li
Mingrui Li
Mengtian Li
Junyi Li
Nianyi Li
Bo Li
Xiao Li
Peihua Li
Peike Li
Peizhao Li
Peiliang Li
Qi Li
Ren Li
Runze Li
Shile Li

Sheng Li
Shigang Li
Shiyu Li
Shuang Li
Shasha Li
Shichao Li
Tianye Li
Yuexiang Li
Wei-Hong Li
Wanhua Li
Weihao Li
Weiming Li
Weixin Li
Wenbo Li
Wenshuo Li
Weijian Li
Yunan Li
Xirong Li
Xianhang Li
Xiaoyu Li
Xueqian Li
Xuanlin Li
Xianzhi Li
Yunqiang Li
Yanjing Li
Yansheng Li
Yawei Li
Yi Li
Yong Li
Yong-Lu Li
Yuhang Li
Yu-Jhe Li
Yuxi Li
Yunsheng Li
Yanwei Li
Zechao Li
Zejian Li
Zeju Li
Zekun Li
Zhaowen Li
Zheng Li
Zhenyu Li
Zhiheng Li
Zhi Li
Zhong Li

Zhuowei Li
Zhuowan Li
Zhuohang Li
Zizhang Li
Chen Li
Yuan-Fang Li
Dongze Lian
Xiaochen Lian
Zhouhui Lian
Long Lian
Qing Lian
Jin Lianbao
Jinxiu S. Liang
Dingkang Liang
Jiahao Liang
Jianming Liang
Jingyun Liang
Kevin J. Liang
Kaizhao Liang
Chen Liang
Jie Liang
Senwei Liang
Ding Liang
Jiajun Liang
Jian Liang
Kongming Liang
Siyuan Liang
Yuanzhi Liang
Zhengfa Liang
Mingfu Liang
Xiaodan Liang
Xuefeng Liang
Yuxuan Liang
Kang Liao
Liang Liao
Hong-Yuan Mark Liao
Wentong Liao
Haofu Liao
Yue Liao
Minghui Liao
Shengcai Liao
Ting-Hsuan Liao
Xin Liao
Yinghong Liao
Teck Yian Lim

Che-Tsung Lin
Chung-Ching Lin
Chen-Hsuan Lin
Cheng Lin
Chuming Lin
Chunyu Lin
Dahua Lin
Wei Lin
Zheng Lin
Huaijia Lin
Jason Lin
Jierui Lin
Jiaying Lin
Jie Lin
Kai-En Lin
Kevin Lin
Guangfeng Lin
Jiehong Lin
Feng Lin
Hang Lin
Kwan-Yee Lin
Ke Lin
Luojun Lin
Qinghong Lin
Xiangbo Lin
Yi Lin
Zudi Lin
Shijie Lin
Yiqun Lin
Tzu-Heng Lin
Ming Lin
Shaohui Lin
SongNan Lin
Ji Lin
Tsung-Yu Lin
Xudong Lin
Yancong Lin
Yen-Chen Lin
Yiming Lin
Yuewei Lin
Zhiqiu Lin
Zinan Lin
Zhe Lin
David B. Lindell
Zhixin Ling

Zhan Ling
Alexander Liniger
Venice Erin B. Liong
Joey Litalien
Or Litany
Roee Litman
Ron Litman
Jim Little
Dor Litvak
Shaoteng Liu
Shuaicheng Liu
Andrew Liu
Xian Liu
Shaohui Liu
Bei Liu
Bo Liu
Yong Liu
Ming Liu
Yanbin Liu
Chenxi Liu
Daqi Liu
Di Liu
Difan Liu
Dong Liu
Dongfang Liu
Daizong Liu
Xiao Liu
Fangyi Liu
Fengbei Liu
Fenglin Liu
Bin Liu
Yuang Liu
Ao Liu
Hong Liu
Hongfu Liu
Huidong Liu
Ziyi Liu
Feng Liu
Hao Liu
Jie Liu
Jialun Liu
Jiang Liu
Jing Liu
Jingya Liu
Jiaming Liu

Jun Liu
Juncheng Liu
Jiawei Liu
Hongyu Liu
Chuanbin Liu
Haotian Liu
Lingqiao Liu
Chang Liu
Han Liu
Liu Liu
Min Liu
Yingqi Liu
Aishan Liu
Bingyu Liu
Benlin Liu
Boxiao Liu
Chenchen Liu
Chuanjian Liu
Daqing Liu
Huan Liu
Haozhe Liu
Jiaheng Liu
Wei Liu
Jingzhou Liu
Jiyuan Liu
Lingbo Liu
Nian Liu
Peiye Liu
Qiankun Liu
Shenglan Liu
Shilong Liu
Wen Liu
Wenyu Liu
Weifeng Liu
Wu Liu
Xiaolong Liu
Yang Liu
Yanwei Liu
Yingcheng Liu
Yongfei Liu
Yihao Liu
Yu Liu
Yunze Liu
Ze Liu
Zhenhua Liu

Zhenguang Liu
Lin Liu
Lihao Liu
Pengju Liu
Xinhai Liu
Yunfei Liu
Meng Liu
Minghua Liu
Mingyuan Liu
Miao Liu
Peirong Liu
Ping Liu
Qingjie Liu
Ruoshi Liu
Risheng Liu
Songtao Liu
Xing Liu
Shikun Liu
Shuming Liu
Sheng Liu
Songhua Liu
Tongliang Liu
Weibo Liu
Weide Liu
Weizhe Liu
Wenxi Liu
Weiyang Liu
Xin Liu
Xiaobin Liu
Xudong Liu
Xiaoyi Liu
Xihui Liu
Xinchen Liu
Xingtong Liu
Xinpeng Liu
Xinyu Liu
Xianpeng Liu
Xu Liu
Xingyu Liu
Yongtuo Liu
Yahui Liu
Yangxin Liu
Yaoyao Liu
Yaojie Liu
Yuliang Liu

Yongcheng Liu
Yuan Liu
Yufan Liu
Yu-Lun Liu
Yun Liu
Yunfan Liu
Yuanzhong Liu
Zhuoran Liu
Zhen Liu
Zheng Liu
Zhijian Liu
Zhisong Liu
Ziquan Liu
Ziyu Liu
Zhihua Liu
Zechun Liu
Zhaoyang Liu
Zhengzhe Liu
Stephan Liwicki
Shao-Yuan Lo
Sylvain Lobry
Suhas Lohit
Vishnu Suresh Lokhande
Vincenzo Lomonaco
Chengjiang Long
Guodong Long
Fuchen Long
Shangbang Long
Yang Long
Zijun Long
Vasco Lopes
Antonio M. Lopez
Roberto Javier
  Lopez-Sastre
Tobias Lorenz
Javier Lorenzo-Navarro
Yujing Lou
Qian Lou
Xiankai Lu
Changsheng Lu
Huimin Lu
Yongxi Lu
Hao Lu
Hong Lu
Jiasen Lu

Juwei Lu
Fan Lu
Guangming Lu
Jiwen Lu
Shun Lu
Tao Lu
Xiaonan Lu
Yang Lu
Yao Lu
Yongchun Lu
Zhiwu Lu
Cheng Lu
Liying Lu
Guo Lu
Xuequan Lu
Yanye Lu
Yantao Lu
Yuhang Lu
Fujun Luan
Jonathon Luiten
Jovita Lukasik
Alan Lukezic
Jonathan Samuel Lumentut
Mayank Lunayach
Ao Luo
Canjie Luo
Chong Luo
Xu Luo
Grace Luo
Jun Luo
Katie Z. Luo
Tao Luo
Cheng Luo
Fangzhou Luo
Gen Luo
Lei Luo
Sihui Luo
Weixin Luo
Yan Luo
Xiaoyan Luo
Yong Luo
Yadan Luo
Hao Luo
Ruotian Luo
Mi Luo

Tiange Luo
Wenjie Luo
Wenhan Luo
Xiao Luo
Zhiming Luo
Zhipeng Luo
Zhengyi Luo
Diogo C. Luvizon
Zhaoyang Lv
Gengyu Lyu
Lingjuan Lyu
Jun Lyu
Yuanyuan Lyu
Youwei Lyu
Yueming Lyu
Bingpeng Ma
Chao Ma
Chongyang Ma
Congbo Ma
Chih-Yao Ma
Fan Ma
Lin Ma
Haoyu Ma
Hengbo Ma
Jianqi Ma
Jiawei Ma
Jiayi Ma
Kede Ma
Kai Ma
Lingni Ma
Lei Ma
Xu Ma
Ning Ma
Benteng Ma
Cheng Ma
Andy J. Ma
Long Ma
Zhanyu Ma
Zhiheng Ma
Qianli Ma
Shiqiang Ma
Sizhuo Ma
Shiqing Ma
Xiaolong Ma
Xinzhu Ma

Gautam B. Machiraju
Spandan Madan
Mathew Magimai-Doss
Luca Magri
Behrooz Mahasseni
Upal Mahbub
Siddharth Mahendran
Paridhi Maheshwari
Rishabh Maheshwary
Mohammed Mahmoud
Shishira R. R. Maiya
Sylwia Majchrowska
Arjun Majumdar
Puspita Majumdar
Orchid Majumder
Sagnik Majumder
Ilya Makarov
Farkhod F.
   Makhmudkhujaev
Yasushi Makihara
Ankur Mali
Mateusz Malinowski
Utkarsh Mall
Srikanth Malla
Clement Mallet
Dimitrios Mallis
Yunze Man
Dipu Manandhar
Massimiliano Mancini
Murari Mandal
Raunak Manekar
Karttikeya Mangalam
Puneet Mangla
Fabian Manhardt
Sivabalan Manivasagam
Fahim Mannan
Chengzhi Mao
Hanzi Mao
Jiayuan Mao
Junhua Mao
Zhiyuan Mao
Jiageng Mao
Yunyao Mao
Zhendong Mao
Alberto Marchisio

Diego Marcos
Riccardo Marin
Aram Markosyan
Renaud Marlet
Ricardo Marques
Miquel Martí i Rabadán
Diego Martin Arroyo
Niki Martinel
Brais Martinez
Julieta Martinez
Marc Masana
Tomohiro Mashita
Timothée Masquelier
Minesh Mathew
Tetsu Matsukawa
Marwan Mattar
Bruce A. Maxwell
Christoph Mayer
Mantas Mazeika
Pratik Mazumder
Scott McCloskey
Steven McDonagh
Ishit Mehta
Jie Mei
Kangfu Mei
Jieru Mei
Xiaoguang Mei
Givi Meishvili
Luke Melas-Kyriazi
Iaroslav Melekhov
Andres Mendez-Vazquez
Heydi Mendez-Vazquez
Matias Mendieta
Ricardo A. Mendoza-León
Chenlin Meng
Depu Meng
Rang Meng
Zibo Meng
Qingjie Meng
Qier Meng
Yanda Meng
Zihang Meng
Thomas Mensink
Fabian Mentzer
Christopher Metzler

Gregory P. Meyer
Vasileios Mezaris
Liang Mi
Lu Mi
Bo Miao
Changtao Miao
Zichen Miao
Qiguang Miao
Xin Miao
Zhongqi Miao
Frank Michel
Simone Milani
Ben Mildenhall
Roy V. Miles
Juhong Min
Kyle Min
Hyun-Seok Min
Weiqing Min
Yuecong Min
Zhixiang Min
Qi Ming
David Minnen
Aymen Mir
Deepak Mishra
Anand Mishra
Shlok K. Mishra
Niluthpol Mithun
Gaurav Mittal
Trisha Mittal
Daisuke Miyazaki
Kaichun Mo
Hong Mo
Zhipeng Mo
Davide Modolo
Abduallah A. Mohamed
Mohamed Afham
   Mohamed Aflal
Ron Mokady
Pavlo Molchanov
Davide Moltisanti
Liliane Momeni
Gianluca Monaci
Pascal Monasse
Ajoy Mondal
Tom Monnier

Aron Monszpart
Gyeongsik Moon
Suhong Moon
Taesup Moon
Sean Moran
Daniel Moreira
Pietro Morerio
Alexandre Morgand
Lia Morra
Ali Mosleh
Inbar Mosseri
Sayed Mohammad
    Mostafavi Isfahani
Saman Motamed
Ramy A. Mounir
Fangzhou Mu
Jiteng Mu
Norman Mu
Yasuhiro Mukaigawa
Ryan Mukherjee
Tanmoy Mukherjee
Yusuke Mukuta
Ravi Teja Mullapudi
Lea Müller
Matthias Müller
Martin Mundt
Nils Murrugarra-Llerena
Damien Muselet
Armin Mustafa
Muhammad Ferjad Naeem
Sauradip Nag
Hajime Nagahara
Pravin Nagar
Rajendra Nagar
Naveen Shankar Nagaraja
Varun Nagaraja
Tushar Nagarajan
Seungjun Nah
Gaku Nakano
Yuta Nakashima
Giljoo Nam
Seonghyeon Nam
Liangliang Nan
Yuesong Nan
Yeshwanth Napolean

Dinesh Reddy
    Narapureddy
Medhini Narasimhan
Supreeth
    Narasimhaswamy
Sriram Narayanan
Erickson R. Nascimento
Varun Nasery
K. L. Navaneet
Pablo Navarrete Michelini
Shant Navasardyan
Shah Nawaz
Nihal Nayak
Farhood Negin
Lukáš Neumann
Alejandro Newell
Evonne Ng
Kam Woh Ng
Tony Ng
Anh Nguyen
Tuan Anh Nguyen
Cuong Cao Nguyen
Ngoc Cuong Nguyen
Thanh Nguyen
Khoi Nguyen
Phi Le Nguyen
Phong Ha Nguyen
Tam Nguyen
Truong Nguyen
Anh Tuan Nguyen
Rang Nguyen
Thao Thi Phuong Nguyen
Van Nguyen Nguyen
Zhen-Liang Ni
Yao Ni
Shijie Nie
Xuecheng Nie
Yongwei Nie
Weizhi Nie
Ying Nie
Yinyu Nie
Kshitij N. Nikhal
Simon Niklaus
Xuefei Ning
Jifeng Ning

Yotam Nitzan
Di Niu
Shuaicheng Niu
Li Niu
Wei Niu
Yulei Niu
Zhenxing Niu
Albert No
Shohei Nobuhara
Nicoletta Noceti
Junhyug Noh
Sotiris Nousias
Slawomir Nowaczyk
Ewa M. Nowara
Valsamis Ntouskos
Gilberto Ochoa-Ruiz
Ferda Ofli
Jihyong Oh
Sangyun Oh
Youngtaek Oh
Hiroki Ohashi
Takahiro Okabe
Kemal Oksuz
Fumio Okura
Daniel Olmeda Reino
Matthew Olson
Carl Olsson
Roy Or-El
Alessandro Ortis
Guillermo Ortiz-Jimenez
Magnus Oskarsson
Ahmed A. A. Osman
Martin R. Oswald
Mayu Otani
Naima Otberdout
Cheng Ouyang
Jiahong Ouyang
Wanli Ouyang
Andrew Owens
Poojan B. Oza
Mete Ozay
A. Cengiz Oztireli
Gautam Pai
Tomas Pajdla
Umapada Pal

Simone Palazzo
Luca Palmieri
Bowen Pan
Hao Pan
Lili Pan
Tai-Yu Pan
Liang Pan
Chengwei Pan
Yingwei Pan
Xuran Pan
Jinshan Pan
Xinyu Pan
Liyuan Pan
Xingang Pan
Xingjia Pan
Zhihong Pan
Zizheng Pan
Priyadarshini Panda
Rameswar Panda
Rohit Pandey
Kaiyue Pang
Bo Pang
Guansong Pang
Jiangmiao Pang
Meng Pang
Tianyu Pang
Ziqi Pang
Omiros Pantazis
Andreas Panteli
Maja Pantic
Marina Paolanti
Joao P. Papa
Samuele Papa
Mike Papadakis
Dim P. Papadopoulos
George Papandreou
Constantin Pape
Toufiq Parag
Chethan Parameshwara
Shaifali Parashar
Alejandro Pardo
Rishubh Parihar
Sarah Parisot
JaeYoo Park
Gyeong-Moon Park

Hyojin Park
Hyoungseob Park
Jongchan Park
Jae Sung Park
Kiru Park
Chunghyun Park
Kwanyong Park
Sunghyun Park
Sungrae Park
Seongsik Park
Sanghyun Park
Sungjune Park
Taesung Park
Gaurav Parmar
Paritosh Parmar
Alvaro Parra
Despoina Paschalidou
Or Patashnik
Shivansh Patel
Pushpak Pati
Prashant W. Patil
Vaishakh Patil
Suvam Patra
Jay Patravali
Badri Narayana Patro
Angshuman Paul
Sudipta Paul
Rémi Pautrat
Nick E. Pears
Adithya Pediredla
Wenjie Pei
Shmuel Peleg
Latha Pemula
Bo Peng
Houwen Peng
Yue Peng
Liangzu Peng
Baoyun Peng
Jun Peng
Pai Peng
Sida Peng
Xi Peng
Yuxin Peng
Songyou Peng
Wei Peng

Weiqi Peng
Wen-Hsiao Peng
Pramuditha Perera
Juan C. Perez
Eduardo Pérez Pellitero
Juan-Manuel Perez-Rua
Federico Pernici
Marco Pesavento
Stavros Petridis
Ilya A. Petrov
Vladan Petrovic
Mathis Petrovich
Suzanne Petryk
Hieu Pham
Quang Pham
Khoi Pham
Tung Pham
Huy Phan
Stephen Phillips
Cheng Perng Phoo
David Picard
Marco Piccirilli
Georg Pichler
A. J. Piergiovanni
Vipin Pillai
Silvia L. Pintea
Giovanni Pintore
Robinson Piramuthu
Fiora Pirri
Theodoros Pissas
Fabio Pizzati
Benjamin Planche
Bryan Plummer
Matteo Poggi
Ashwini Pokle
Georgy E. Ponimatkin
Adrian Popescu
Stefan Popov
Nikola Popović
Ronald Poppe
Angelo Porrello
Michael Potter
Charalambos Poullis
Hadi Pouransari
Omid Poursaeed

Shraman Pramanick
Mantini Pranav
Dilip K. Prasad
Meghshyam Prasad
B. H. Pawan Prasad
Shitala Prasad
Prateek Prasanna
Ekta Prashnani
Derek S. Prijatelj
Luke Y. Prince
Véronique Prinet
Victor Adrian Prisacariu
James Pritts
Thomas Probst
Sergey Prokudin
Rita Pucci
Chi-Man Pun
Matthew Purri
Haozhi Qi
Lu Qi
Lei Qi
Xianbiao Qi
Yonggang Qi
Yuankai Qi
Siyuan Qi
Guocheng Qian
Hangwei Qian
Qi Qian
Deheng Qian
Shengsheng Qian
Wen Qian
Rui Qian
Yiming Qian
Shengju Qian
Shengyi Qian
Xuelin Qian
Zhenxing Qian
Nan Qiao
Xiaotian Qiao
Jing Qin
Can Qin
Siyang Qin
Hongwei Qin
Jie Qin
Minghai Qin

Yipeng Qin
Yongqiang Qin
Wenda Qin
Xuebin Qin
Yuzhe Qin
Yao Qin
Zhenyue Qin
Zhiwu Qing
Heqian Qiu
Jiayan Qiu
Jielin Qiu
Yue Qiu
Jiaxiong Qiu
Zhongxi Qiu
Shi Qiu
Zhaofan Qiu
Zhongnan Qu
Yanyun Qu
Kha Gia Quach
Yuhui Quan
Ruijie Quan
Mike Rabbat
Rahul Shekhar Rade
Filip Radenovic
Gorjan Radevski
Bogdan Raducanu
Francesco Ragusa
Shafin Rahman
Md Mahfuzur Rahman
    Siddiquee
Hossein Rahmani
Kiran Raja
Sivaramakrishnan
    Rajaraman
Jathushan Rajasegaran
Adnan Siraj Rakin
Michaël Ramamonjisoa
Chirag A. Raman
Shanmuganathan Raman
Vignesh Ramanathan
Vasili Ramanishka
Vikram V. Ramaswamy
Merey Ramazanova
Jason Rambach
Sai Saketh Rambhatla

Clément Rambour
Ashwin Ramesh Babu
Adín Ramírez Rivera
Arianna Rampini
Haoxi Ran
Aakanksha Rana
Aayush Jung Bahadur
    Rana
Kanchana N. Ranasinghe
Aneesh Rangnekar
Samrudhdhi B. Rangrej
Harsh Rangwani
Viresh Ranjan
Anyi Rao
Yongming Rao
Carolina Raposo
Michalis Raptis
Amir Rasouli
Vivek Rathod
Adepu Ravi Sankar
Avinash Ravichandran
Bharadwaj Ravichandran
Dripta S. Raychaudhuri
Adria Recasens
Simon Reiß
Davis Rempe
Daxuan Ren
Jiawei Ren
Jimmy Ren
Sucheng Ren
Dayong Ren
Zhile Ren
Dongwei Ren
Qibing Ren
Pengfei Ren
Zhenwen Ren
Xuqian Ren
Yixuan Ren
Zhongzheng Ren
Ambareesh Revanur
Hamed Rezazadegan
    Tavakoli
Rafael S. Rezende
Wonjong Rhee
Alexander Richard

Christian Richardt
Stephan R. Richter
Benjamin Riggan
Dominik Rivoir
Mamshad Nayeem Rizve
Joshua D. Robinson
Joseph Robinson
Chris Rockwell
Ranga Rodrigo
Andres C. Rodriguez
Carlos Rodriguez-Pardo
Marcus Rohrbach
Gemma Roig
Yu Rong
David A. Ross
Mohammad Rostami
Edward Rosten
Karsten Roth
Anirban Roy
Debaditya Roy
Shuvendu Roy
Ahana Roy Choudhury
Aruni Roy Chowdhury
Denys Rozumnyi
Shulan Ruan
Wenjie Ruan
Patrick Ruhkamp
Danila Rukhovich
Anian Ruoss
Chris Russell
Dan Ruta
Dawid Damian Rymarczyk
DongHun Ryu
Hyeonggon Ryu
Kwonyoung Ryu
Balasubramanian S.
Alexandre Sablayrolles
Mohammad Sabokrou
Arka Sadhu
Aniruddha Saha
Oindrila Saha
Pritish Sahu
Aneeshan Sain
Nirat Saini
Saurabh Saini

Takeshi Saitoh
Christos Sakaridis
Fumihiko Sakaue
Dimitrios Sakkos
Ken Sakurada
Parikshit V. Sakurikar
Rohit Saluja
Nermin Samet
Leo Sampaio Ferraz
    Ribeiro
Jorge Sanchez
Enrique Sanchez
Shengtian Sang
Anush Sankaran
Soubhik Sanyal
Nikolaos Sarafianos
Vishwanath Saragadam
István Sárándi
Saquib Sarfraz
Mert Bulent Sariyildiz
Anindya Sarkar
Pritam Sarkar
Paul-Edouard Sarlin
Hiroshi Sasaki
Takami Sato
Torsten Sattler
Ravi Kumar Satzoda
Axel Sauer
Stefano Savian
Artem Savkin
Manolis Savva
Gerald Schaefer
Simone Schaub-Meyer
Yoni Schirris
Samuel Schulter
Katja Schwarz
Jesse Scott
Sinisa Segvic
Constantin Marc Seibold
Lorenzo Seidenari
Matan Sela
Fadime Sener
Paul Hongsuck Seo
Kwanggyoon Seo
Hongje Seong

Dario Serez
Francesco Setti
Bryan Seybold
Mohamad Shahbazi
Shima Shahfar
Xinxin Shan
Caifeng Shan
Dandan Shan
Shawn Shan
Wei Shang
Jinghuan Shang
Jiaxiang Shang
Lei Shang
Sukrit Shankar
Ken Shao
Rui Shao
Jie Shao
Mingwen Shao
Aashish Sharma
Gaurav Sharma
Vivek Sharma
Abhishek Sharma
Yoli Shavit
Shashank Shekhar
Sumit Shekhar
Zhijie Shen
Fengyi Shen
Furao Shen
Jialie Shen
Jingjing Shen
Ziyi Shen
Linlin Shen
Guangyu Shen
Biluo Shen
Falong Shen
Jiajun Shen
Qiu Shen
Qiuhong Shen
Shuai Shen
Wang Shen
Yiqing Shen
Yunhang Shen
Siqi Shen
Bin Shen
Tianwei Shen

Xi Shen
Yilin Shen
Yuming Shen
Yucong Shen
Zhiqiang Shen
Lu Sheng
Yichen Sheng
Shivanand Venkanna
   Sheshappanavar
Shelly Sheynin
Baifeng Shi
Ruoxi Shi
Botian Shi
Hailin Shi
Jia Shi
Jing Shi
Shaoshuai Shi
Baoguang Shi
Boxin Shi
Hengcan Shi
Tianyang Shi
Xiaodan Shi
Yongjie Shi
Zhensheng Shi
Yinghuan Shi
Weiqi Shi
Wu Shi
Xuepeng Shi
Xiaoshuang Shi
Yujiao Shi
Zenglin Shi
Zhenmei Shi
Takashi Shibata
Meng-Li Shih
Yichang Shih
Hyunjung Shim
Dongseok Shim
Soshi Shimada
Inkyu Shin
Jinwoo Shin
Seungjoo Shin
Seungjae Shin
Koichi Shinoda
Suprosanna Shit

Palaiahnakote
   Shivakumara
Eli Shlizerman
Gaurav Shrivastava
Xiao Shu
Xiangbo Shu
Xiujun Shu
Yang Shu
Tianmin Shu
Jun Shu
Zhixin Shu
Bing Shuai
Maria Shugrina
Ivan Shugurov
Satya Narayan Shukla
Pranjay Shyam
Jianlou Si
Yawar Siddiqui
Alberto Signoroni
Pedro Silva
Jae-Young Sim
Oriane Siméoni
Martin Simon
Andrea Simonelli
Abhishek Singh
Ashish Singh
Dinesh Singh
Gurkirt Singh
Krishna Kumar Singh
Mannat Singh
Pravendra Singh
Rajat Vikram Singh
Utkarsh Singhal
Dipika Singhania
Vasu Singla
Harsh Sinha
Sudipta Sinha
Josef Sivic
Elena Sizikova
Geri Skenderi
Ivan Skorokhodov
Dmitriy Smirnov
Cameron Y. Smith
James S. Smith
Patrick Snape

Mattia Soldan
Hyeongseok Son
Sanghyun Son
Chuanbiao Song
Chen Song
Chunfeng Song
Dan Song
Dongjin Song
Hwanjun Song
Guoxian Song
Jiaming Song
Jie Song
Liangchen Song
Ran Song
Luchuan Song
Xibin Song
Li Song
Fenglong Song
Guoli Song
Guanglu Song
Zhenbo Song
Lin Song
Xinhang Song
Yang Song
Yibing Song
Rajiv Soundararajan
Hossein Souri
Cristovao Sousa
Riccardo Spezialetti
Leonidas Spinoulas
Michael W. Spratling
Deepak Sridhar
Srinath Sridhar
Gaurang Sriramanan
Vinkle Kumar Srivastav
Themos Stafylakis
Serban Stan
Anastasis Stathopoulos
Markus Steinberger
Jan Steinbrener
Sinisa Stekovic
Alexandros Stergiou
Gleb Sterkin
Rainer Stiefelhagen
Pierre Stock

Ombretta Strafforello
Julian Straub
Yannick Strümpler
Joerg Stueckler
Hang Su
Weijie Su
Jong-Chyi Su
Bing Su
Haisheng Su
Jinming Su
Yiyang Su
Yukun Su
Yuxin Su
Zhuo Su
Zhaoqi Su
Xiu Su
Yu-Chuan Su
Zhixun Su
Arulkumar Subramaniam
Akshayvarun Subramanya
A. Subramanyam
Swathikiran Sudhakaran
Yusuke Sugano
Masanori Suganuma
Yumin Suh
Yang Sui
Baochen Sun
Cheng Sun
Long Sun
Guolei Sun
Haoliang Sun
Haomiao Sun
He Sun
Hanqing Sun
Hao Sun
Lichao Sun
Jiachen Sun
Jiaming Sun
Jian Sun
Jin Sun
Jennifer J. Sun
Tiancheng Sun
Libo Sun
Peize Sun
Qianru Sun

Shanlin Sun
Yu Sun
Zhun Sun
Che Sun
Lin Sun
Tao Sun
Yiyou Sun
Chunyi Sun
Chong Sun
Weiwei Sun
Weixuan Sun
Xiuyu Sun
Yanan Sun
Zeren Sun
Zhaodong Sun
Zhiqing Sun
Minhyuk Sung
Jinli Suo
Simon Suo
Abhijit Suprem
Anshuman Suri
Saksham Suri
Joshua M. Susskind
Roman Suvorov
Gurumurthy Swaminathan
Robin Swanson
Paul Swoboda
Tabish A. Syed
Richard Szeliski
Fariborz Taherkhani
Yu-Wing Tai
Keita Takahashi
Walter Talbott
Gary Tam
Masato Tamura
Feitong Tan
Fuwen Tan
Shuhan Tan
Andong Tan
Bin Tan
Cheng Tan
Jianchao Tan
Lei Tan
Mingxing Tan
Xin Tan

Zichang Tan
Zhentao Tan
Kenichiro Tanaka
Masayuki Tanaka
Yushun Tang
Hao Tang
Jingqun Tang
Jinhui Tang
Kaihua Tang
Luming Tang
Lv Tang
Sheyang Tang
Shitao Tang
Siliang Tang
Shixiang Tang
Yansong Tang
Keke Tang
Chang Tang
Chenwei Tang
Jie Tang
Junshu Tang
Ming Tang
Peng Tang
Xu Tang
Yao Tang
Chen Tang
Fan Tang
Haoran Tang
Shengeng Tang
Yehui Tang
Zhipeng Tang
Ugo Tanielian
Chaofan Tao
Jiale Tao
Junli Tao
Renshuai Tao
An Tao
Guanhong Tao
Zhiqiang Tao
Makarand Tapaswi
Jean-Philippe G. Tarel
Juan J. Tarrio
Enzo Tartaglione
Keisuke Tateno
Zachary Teed

Ajinkya B. Tejankar
Bugra Tekin
Purva Tendulkar
Damien Teney
Minggui Teng
Chris Tensmeyer
Andrew Beng Jin Teoh
Philipp Terhörst
Kartik Thakral
Nupur Thakur
Kevin Thandiackal
Spyridon Thermos
Diego Thomas
William Thong
Yuesong Tian
Guanzhong Tian
Lin Tian
Shiqi Tian
Kai Tian
Meng Tian
Tai-Peng Tian
Zhuotao Tian
Shangxuan Tian
Tian Tian
Yapeng Tian
Yu Tian
Yuxin Tian
Leslie Ching Ow Tiong
Praveen Tirupattur
Garvita Tiwari
George Toderici
Antoine Toisoul
Aysim Toker
Tatiana Tommasi
Zhan Tong
Alessio Tonioni
Alessandro Torcinovich
Fabio Tosi
Matteo Toso
Hugo Touvron
Quan Hung Tran
Son Tran
Hung Tran
Ngoc-Trung Tran
Vinh Tran

Phong Tran
Giovanni Trappolini
Edith Tretschk
Subarna Tripathi
Shubhendu Trivedi
Eduard Trulls
Prune Truong
Thanh-Dat Truong
Tomasz Trzcinski
Sam Tsai
Yi-Hsuan Tsai
Ethan Tseng
Yu-Chee Tseng
Shahar Tsiper
Stavros Tsogkas
Shikui Tu
Zhigang Tu
Zhengzhong Tu
Richard Tucker
Sergey Tulyakov
Cigdem Turan
Daniyar Turmukhambetov
Victor G. Turrisi da Costa
Bartlomiej Twardowski
Christopher D. Twigg
Radim Tylecek
Mostofa Rafid Uddin
Md. Zasim Uddin
Kohei Uehara
Nicolas Ugrinovic
Youngjung Uh
Norimichi Ukita
Anwaar Ulhaq
Devesh Upadhyay
Paul Upchurch
Yoshitaka Ushiku
Yuzuko Utsumi
Mikaela Angelina Uy
Mohit Vaishnav
Pratik Vaishnavi
Jeya Maria Jose Valanarasu
Matias A. Valdenegro Toro
Diego Valsesia
Wouter Van Gansbeke
Nanne van Noord

Simon Vandenhende
Farshid Varno
Cristina Vasconcelos
Francisco Vasconcelos
Alex Vasilescu
Subeesh Vasu
Arun Balajee Vasudevan
Kanav Vats
Vaibhav S. Vavilala
Sagar Vaze
Javier Vazquez-Corral
Andrea Vedaldi
Olga Veksler
Andreas Velten
Sai H. Vemprala
Raviteja Vemulapalli
Shashanka
    Venkataramanan
Dor Verbin
Luisa Verdoliva
Manisha Verma
Yashaswi Verma
Constantin Vertan
Eli Verwimp
Deepak Vijaykeerthy
Pablo Villanueva
Ruben Villegas
Markus Vincze
Vibhav Vineet
Minh P. Vo
Huy V. Vo
Duc Minh Vo
Tomas Vojir
Igor Vozniak
Nicholas Vretos
Vibashan VS
Tuan-Anh Vu
Thang Vu
Mårten Wadenbäck
Neal Wadhwa
Aaron T. Walsman
Steven Walton
Jin Wan
Alvin Wan
Jia Wan

Jun Wan
Xiaoyue Wan
Fang Wan
Guowei Wan
Renjie Wan
Zhiqiang Wan
Ziyu Wan
Bastian Wandt
Dongdong Wang
Limin Wang
Haiyang Wang
Xiaobing Wang
Angtian Wang
Angelina Wang
Bing Wang
Bo Wang
Boyu Wang
Binghui Wang
Chen Wang
Chien-Yi Wang
Congli Wang
Qi Wang
Chengrui Wang
Rui Wang
Yiqun Wang
Cong Wang
Wenjing Wang
Dongkai Wang
Di Wang
Xiaogang Wang
Kai Wang
Zhizhong Wang
Fangjinhua Wang
Feng Wang
Hang Wang
Gaoang Wang
Guoqing Wang
Guangcong Wang
Guangzhi Wang
Hanqing Wang
Hao Wang
Haohan Wang
Haoran Wang
Hong Wang
Haotao Wang

Hu Wang
Huan Wang
Hua Wang
Hui-Po Wang
Hengli Wang
Hanyu Wang
Hongxing Wang
Jingwen Wang
Jialiang Wang
Jian Wang
Jianyi Wang
Jiashun Wang
Jiahao Wang
Tsun-Hsuan Wang
Xiaoqian Wang
Jinqiao Wang
Jun Wang
Jianzong Wang
Kaihong Wang
Ke Wang
Lei Wang
Lingjing Wang
Linnan Wang
Lin Wang
Liansheng Wang
Mengjiao Wang
Manning Wang
Nannan Wang
Peihao Wang
Jiayun Wang
Pu Wang
Qiang Wang
Qiufeng Wang
Qilong Wang
Qiangchang Wang
Qin Wang
Qing Wang
Ruocheng Wang
Ruibin Wang
Ruisheng Wang
Ruizhe Wang
Runqi Wang
Runzhong Wang
Wenxuan Wang
Sen Wang

Shangfei Wang
Shaofei Wang
Shijie Wang
Shiqi Wang
Zhibo Wang
Song Wang
Xinjiang Wang
Tai Wang
Tao Wang
Teng Wang
Xiang Wang
Tianren Wang
Tiantian Wang
Tianyi Wang
Fengjiao Wang
Wei Wang
Miaohui Wang
Suchen Wang
Siyue Wang
Yaoming Wang
Xiao Wang
Ze Wang
Biao Wang
Chaofei Wang
Dong Wang
Gu Wang
Guangrun Wang
Guangming Wang
Guo-Hua Wang
Haoqing Wang
Hesheng Wang
Huafeng Wang
Jinghua Wang
Jingdong Wang
Jingjing Wang
Jingya Wang
Jingkang Wang
Jiakai Wang
Junke Wang
Kuo Wang
Lichen Wang
Lizhi Wang
Longguang Wang
Mang Wang
Mei Wang

Min Wang
Peng-Shuai Wang
Run Wang
Shaoru Wang
Shuhui Wang
Tan Wang
Tiancai Wang
Tianqi Wang
Wenhai Wang
Wenzhe Wang
Xiaobo Wang
Xiudong Wang
Xu Wang
Yajie Wang
Yan Wang
Yuan-Gen Wang
Yingqian Wang
Yizhi Wang
Yulin Wang
Yu Wang
Yujie Wang
Yunhe Wang
Yuxi Wang
Yaowei Wang
Yiwei Wang
Zezheng Wang
Hongzhi Wang
Zhiqiang Wang
Ziteng Wang
Ziwei Wang
Zheng Wang
Zhenyu Wang
Binglu Wang
Zhongdao Wang
Ce Wang
Weining Wang
Weiyao Wang
Wenbin Wang
Wenguan Wang
Guangting Wang
Haolin Wang
Haiyan Wang
Huiyu Wang
Naiyan Wang
Jingbo Wang

Jinpeng Wang
Jiaqi Wang
Liyuan Wang
Lizhen Wang
Ning Wang
Wenqian Wang
Sheng-Yu Wang
Weimin Wang
Xiaohan Wang
Yifan Wang
Yi Wang
Yongtao Wang
Yizhou Wang
Zhuo Wang
Zhe Wang
Xudong Wang
Xiaofang Wang
Xinggang Wang
Xiaosen Wang
Xiaosong Wang
Xiaoyang Wang
Lijun Wang
Xinlong Wang
Xuan Wang
Xue Wang
Yangang Wang
Yaohui Wang
Yu-Chiang Frank Wang
Yida Wang
Yilin Wang
Yi Ru Wang
Yali Wang
Yinglong Wang
Yufu Wang
Yujiang Wang
Yuwang Wang
Yuting Wang
Yang Wang
Yu-Xiong Wang
Yixu Wang
Ziqi Wang
Zhicheng Wang
Zeyu Wang
Zhaowen Wang
Zhenyi Wang

Zhenzhi Wang
Zhijie Wang
Zhiyong Wang
Zhongling Wang
Zhuowei Wang
Zian Wang
Zifu Wang
Zihao Wang
Zirui Wang
Ziyan Wang
Wenxiao Wang
Zhen Wang
Zhepeng Wang
Zi Wang
Zihao W. Wang
Steven L. Waslander
Olivia Watkins
Daniel Watson
Silvan Weder
Dongyoon Wee
Dongming Wei
Tianyi Wei
Jia Wei
Dong Wei
Fangyun Wei
Longhui Wei
Mingqiang Wei
Xinyue Wei
Chen Wei
Donglai Wei
Pengxu Wei
Xing Wei
Xiu-Shen Wei
Wenqi Wei
Guoqiang Wei
Wei Wei
XingKui Wei
Xian Wei
Xingxing Wei
Yake Wei
Yuxiang Wei
Yi Wei
Luca Weihs
Michael Weinmann
Martin Weinmann

Congcong Wen
Chuan Wen
Jie Wen
Sijia Wen
Song Wen
Chao Wen
Xiang Wen
Zeyi Wen
Xin Wen
Yilin Wen
Yijia Weng
Shuchen Weng
Junwu Weng
Wenming Weng
Renliang Weng
Zhenyu Weng
Xinshuo Weng
Nicholas J. Westlake
Gordon Wetzstein
Lena M. Widin Klasén
Rick Wildes
Bryan M. Williams
Williem Williem
Ole Winther
Scott Wisdom
Alex Wong
Chau-Wai Wong
Kwan-Yee K. Wong
Yongkang Wong
Scott Workman
Marcel Worring
Michael Wray
Safwan Wshah
Xiang Wu
Aming Wu
Chongruo Wu
Cho-Ying Wu
Chunpeng Wu
Chenyan Wu
Ziyi Wu
Fuxiang Wu
Gang Wu
Haiping Wu
Huisi Wu
Jane Wu

Jialian Wu
Jing Wu
Jinjian Wu
Jianlong Wu
Xian Wu
Lifang Wu
Lifan Wu
Minye Wu
Qianyi Wu
Rongliang Wu
Rui Wu
Shiqian Wu
Shuzhe Wu
Shangzhe Wu
Tsung-Han Wu
Tz-Ying Wu
Ting-Wei Wu
Jiannan Wu
Zhiliang Wu
Yu Wu
Chenyun Wu
Dayan Wu
Dongxian Wu
Fei Wu
Hefeng Wu
Jianxin Wu
Weibin Wu
Wenxuan Wu
Wenhao Wu
Xiao Wu
Yicheng Wu
Yuanwei Wu
Yu-Huan Wu
Zhenxin Wu
Zhenyu Wu
Wei Wu
Peng Wu
Xiaohe Wu
Xindi Wu
Xinxing Wu
Xinyi Wu
Xingjiao Wu
Xiongwei Wu
Yangzheng Wu
Yanzhao Wu

Yawen Wu
Yong Wu
Yi Wu
Ying Nian Wu
Zhenyao Wu
Zhonghua Wu
Zongze Wu
Zuxuan Wu
Stefanie Wuhrer
Teng Xi
Jianing Xi
Fei Xia
Haifeng Xia
Menghan Xia
Yuanqing Xia
Zhihua Xia
Xiaobo Xia
Weihao Xia
Shihong Xia
Yan Xia
Yong Xia
Zhaoyang Xia
Zhihao Xia
Chuhua Xian
Yongqin Xian
Wangmeng Xiang
Fanbo Xiang
Tiange Xiang
Tao Xiang
Liuyu Xiang
Xiaoyu Xiang
Zhiyu Xiang
Aoran Xiao
Chunxia Xiao
Fanyi Xiao
Jimin Xiao
Jun Xiao
Taihong Xiao
Anqi Xiao
Junfei Xiao
Jing Xiao
Liang Xiao
Yang Xiao
Yuting Xiao
Yijun Xiao

Yao Xiao

Zeyu Xiao

Zhisheng Xiao

Zihao Xiao

Binhui Xie

Christopher Xie

Haozhe Xie

Jin Xie

Guo-Sen Xie

Hongtao Xie

Ming-Kun Xie

Tingting Xie

Chaohao Xie

Weicheng Xie

Xudong Xie

Jiyang Xie

Xiaohua Xie

Yuan Xie

Zhenyu Xie

Ning Xie

Xianghui Xie

Xiufeng Xie

You Xie

Yutong Xie

Fuyong Xing

Yifan Xing

Zhen Xing

Yuanjun Xiong

Jinhui Xiong

Weihua Xiong

Hongkai Xiong

Zhitong Xiong

Yuanhao Xiong

Yunyang Xiong

Yuwen Xiong

Zhiwei Xiong

Yuliang Xiu

An Xu

Chang Xu

Chenliang Xu

Chengming Xu

Chenshu Xu

Xiang Xu

Huijuan Xu

Zhe Xu

Jie Xu

Jingyi Xu

Jiarui Xu

Yinghao Xu

Kele Xu

Ke Xu

Li Xu

Linchuan Xu

Linning Xu

Mengde Xu

Mengmeng Frost Xu

Min Xu

Mingye Xu

Jun Xu

Ning Xu

Peng Xu

Runsheng Xu

Sheng Xu

Wenqiang Xu

Xiaogang Xu

Renzhe Xu

Kaidi Xu

Yi Xu

Chi Xu

Qiuling Xu

Baobei Xu

Feng Xu

Haohang Xu

Haofei Xu

Lan Xu

Mingze Xu

Songcen Xu

Weipeng Xu

Wenjia Xu

Wenju Xu

Xiangyu Xu

Xin Xu

Yinshuang Xu

Yixing Xu

Yuting Xu

Yanyu Xu

Zhenbo Xu

Zhiliang Xu

Zhiyuan Xu

Xiaohao Xu

Yanwu Xu

Yan Xu

Yiran Xu

Yifan Xu

Yufei Xu

Yong Xu

Zichuan Xu

Zenglin Xu

Zexiang Xu

Zhan Xu

Zheng Xu

Zhiwei Xu

Ziyue Xu

Shiyu Xuan

Hanyu Xuan

Fei Xue

Jianru Xue

Mingfu Xue

Qinghan Xue

Tianfan Xue

Chao Xue

Chuhui Xue

Nan Xue

Zhou Xue

Xiangyang Xue

Yuan Xue

Abhay Yadav

Ravindra Yadav

Kota Yamaguchi

Toshihiko Yamasaki

Kohei Yamashita

Chaochao Yan

Feng Yan

Kun Yan

Qingsen Yan

Qixin Yan

Rui Yan

Siming Yan

Xinchen Yan

Yaping Yan

Bin Yan

Qingan Yan

Shen Yan

Shipeng Yan

Xu Yan

Yan Yan
Yichao Yan
Zhaoyi Yan
Zike Yan
Zhiqiang Yan
Hongliang Yan
Zizheng Yan
Jiewen Yang
Anqi Joyce Yang
Shan Yang
Anqi Yang
Antoine Yang
Bo Yang
Baoyao Yang
Chenhongyi Yang
Dingkang Yang
De-Nian Yang
Dong Yang
David Yang
Fan Yang
Fengyu Yang
Fengting Yang
Fei Yang
Gengshan Yang
Heng Yang
Han Yang
Huan Yang
Yibo Yang
Jiancheng Yang
Jihan Yang
Jiawei Yang
Jiayu Yang
Jie Yang
Jinfa Yang
Jingkang Yang
Jinyu Yang
Cheng-Fu Yang
Ji Yang
Jianyu Yang
Kailun Yang
Tian Yang
Luyu Yang
Liang Yang
Li Yang
Michael Ying Yang

Yang Yang
Muli Yang
Le Yang
Qiushi Yang
Ren Yang
Ruihan Yang
Shuang Yang
Siyuan Yang
Su Yang
Shiqi Yang
Taojiannan Yang
Tianyu Yang
Lei Yang
Wanzhao Yang
Shuai Yang
William Yang
Wei Yang
Xiaofeng Yang
Xiaoshan Yang
Xin Yang
Xuan Yang
Xu Yang
Xingyi Yang
Xitong Yang
Jing Yang
Yanchao Yang
Wenming Yang
Yujiu Yang
Herb Yang
Jianfei Yang
Jinhui Yang
Chuanguang Yang
Guanglei Yang
Haitao Yang
Kewei Yang
Linlin Yang
Lijin Yang
Longrong Yang
Meng Yang
MingKun Yang
Sibei Yang
Shicai Yang
Tong Yang
Wen Yang
Xi Yang

Xiaolong Yang
Xue Yang
Yubin Yang
Ze Yang
Ziyi Yang
Yi Yang
Linjie Yang
Yuzhe Yang
Yiding Yang
Zhenpei Yang
Zhaohui Yang
Zhengyuan Yang
Zhibo Yang
Zongxin Yang
Hantao Yao
Mingde Yao
Rui Yao
Taiping Yao
Ting Yao
Cong Yao
Qingsong Yao
Quanming Yao
Xu Yao
Yuan Yao
Yao Yao
Yazhou Yao
Jiawen Yao
Shunyu Yao
Pew-Thian Yap
Sudhir Yarram
Rajeev Yasarla
Peng Ye
Botao Ye
Mao Ye
Fei Ye
Hanrong Ye
Jingwen Ye
Jinwei Ye
Jiarong Ye
Mang Ye
Meng Ye
Qi Ye
Qian Ye
Qixiang Ye
Junjie Ye

Sheng Ye
Nanyang Ye
Yufei Ye
Xiaoqing Ye
Ruolin Ye
Yousef Yeganeh
Chun-Hsiao Yeh
Raymond A. Yeh
Yu-Ying Yeh
Kai Yi
Chang Yi
Renjiao Yi
Xinping Yi
Peng Yi
Alper Yilmaz
Junho Yim
Hui Yin
Bangjie Yin
Jia-Li Yin
Miao Yin
Wenzhe Yin
Xuwang Yin
Ming Yin
Yu Yin
Aoxiong Yin
Kangxue Yin
Tianwei Yin
Wei Yin
Xianghua Ying
Rio Yokota
Tatsuya Yokota
Naoto Yokoya
Ryo Yonetani
Ki Yoon Yoo
Jinsu Yoo
Sunjae Yoon
Jae Shin Yoon
Jihun Yoon
Sung-Hoon Yoon
Ryota Yoshihashi
Yusuke Yoshiyasu
Chenyu You
Haoran You
Haoxuan You
Yang You

Quanzeng You
Tackgeun You
Kaichao You
Shan You
Xinge You
Yurong You
Baosheng Yu
Bei Yu
Haichao Yu
Hao Yu
Chaohui Yu
Fisher Yu
Jin-Gang Yu
Jiyang Yu
Jason J. Yu
Jiashuo Yu
Hong-Xing Yu
Lei Yu
Mulin Yu
Ning Yu
Peilin Yu
Qi Yu
Qian Yu
Rui Yu
Shuzhi Yu
Gang Yu
Tan Yu
Weijiang Yu
Xin Yu
Bingyao Yu
Ye Yu
Hanchao Yu
Yingchen Yu
Tao Yu
Xiaotian Yu
Qing Yu
Houjian Yu
Changqian Yu
Jing Yu
Jun Yu
Shujian Yu
Xiang Yu
Zhaofei Yu
Zhenbo Yu
Yinfeng Yu

Zhuoran Yu
Zitong Yu
Bo Yuan
Jiangbo Yuan
Liangzhe Yuan
Weihao Yuan
Jianbo Yuan
Xiaoyun Yuan
Ye Yuan
Li Yuan
Geng Yuan
Jialin Yuan
Maoxun Yuan
Peng Yuan
Xin Yuan
Yuan Yuan
Yuhui Yuan
Yixuan Yuan
Zheng Yuan
Mehmet Kerim Yücel
Kaiyu Yue
Haixiao Yue
Heeseung Yun
Sangdoo Yun
Tian Yun
Mahmut Yurt
Ekim Yurtsever
Ahmet Yüzügüler
Edouard Yvinec
Eloi Zablocki
Christopher Zach
Muhammad Zaigham
    Zaheer
Pierluigi Zama Ramirez
Yuhang Zang
Pietro Zanuttigh
Alexey Zaytsev
Bernhard Zeisl
Haitian Zeng
Pengpeng Zeng
Jiabei Zeng
Runhao Zeng
Wei Zeng
Yawen Zeng
Yi Zeng

Yiming Zeng
Tieyong Zeng
Huanqiang Zeng
Dan Zeng
Yu Zeng
Wei Zhai
Yuanhao Zhai
Fangneng Zhan
Kun Zhan
Xiong Zhang
Jingdong Zhang
Jiangning Zhang
Zhilu Zhang
Gengwei Zhang
Dongsu Zhang
Hui Zhang
Binjie Zhang
Bo Zhang
Tianhao Zhang
Cecilia Zhang
Jing Zhang
Chaoning Zhang
Chenxu Zhang
Chi Zhang
Chris Zhang
Yabin Zhang
Zhao Zhang
Rufeng Zhang
Chaoyi Zhang
Zheng Zhang
Da Zhang
Yi Zhang
Edward Zhang
Xin Zhang
Feifei Zhang
Feilong Zhang
Yuqi Zhang
GuiXuan Zhang
Hanlin Zhang
Hanwang Zhang
Hanzhen Zhang
Haotian Zhang
He Zhang
Haokui Zhang
Hongyuan Zhang

Hengrui Zhang
Hongming Zhang
Mingfang Zhang
Jianpeng Zhang
Jiaming Zhang
Jichao Zhang
Jie Zhang
Jingfeng Zhang
Jingyi Zhang
Jinnian Zhang
David Junhao Zhang
Junjie Zhang
Junzhe Zhang
Jiawan Zhang
Jingyang Zhang
Kai Zhang
Lei Zhang
Lihua Zhang
Lu Zhang
Miao Zhang
Minjia Zhang
Mingjin Zhang
Qi Zhang
Qian Zhang
Qilong Zhang
Qiming Zhang
Qiang Zhang
Richard Zhang
Ruimao Zhang
Ruisi Zhang
Ruixin Zhang
Runze Zhang
Qilin Zhang
Shan Zhang
Shanshan Zhang
Xi Sheryl Zhang
Song-Hai Zhang
Chongyang Zhang
Kaihao Zhang
Songyang Zhang
Shu Zhang
Siwei Zhang
Shujian Zhang
Tianyun Zhang
Tong Zhang

Tao Zhang
Wenwei Zhang
Wenqiang Zhang
Wen Zhang
Xiaolin Zhang
Xingchen Zhang
Xingxuan Zhang
Xiuming Zhang
Xiaoshuai Zhang
Xuanmeng Zhang
Xuanyang Zhang
Xucong Zhang
Xingxing Zhang
Xikun Zhang
Xiaohan Zhang
Yahui Zhang
Yunhua Zhang
Yan Zhang
Yanghao Zhang
Yifei Zhang
Yifan Zhang
Yi-Fan Zhang
Yihao Zhang
Yingliang Zhang
Youshan Zhang
Yulun Zhang
Yushu Zhang
Yixiao Zhang
Yide Zhang
Zhongwen Zhang
Bowen Zhang
Chen-Lin Zhang
Zehua Zhang
Zekun Zhang
Zeyu Zhang
Xiaowei Zhang
Yifeng Zhang
Cheng Zhang
Hongguang Zhang
Yuexi Zhang
Fa Zhang
Guofeng Zhang
Hao Zhang
Haofeng Zhang
Hongwen Zhang

Hua Zhang	Zhizhong Zhang	Bowen Zhao
Jiaxin Zhang	Qilong Zhangli	Pu Zhao
Zhenyu Zhang	Bingyin Zhao	Bingchen Zhao
Jian Zhang	Bin Zhao	Borui Zhao
Jianfeng Zhang	Chenglong Zhao	Fuqiang Zhao
Jiao Zhang	Lei Zhao	Hanbin Zhao
Jiakai Zhang	Feng Zhao	Jian Zhao
Lefei Zhang	Gangming Zhao	Mingyang Zhao
Le Zhang	Haiyan Zhao	Na Zhao
Mi Zhang	Hao Zhao	Rongchang Zhao
Min Zhang	Handong Zhao	Ruiqi Zhao
Ning Zhang	Hengshuang Zhao	Shuai Zhao
Pan Zhang	Yinan Zhao	Wenda Zhao
Pu Zhang	Jiaojiao Zhao	Wenliang Zhao
Qing Zhang	Jiaqi Zhao	Xiangyun Zhao
Renrui Zhang	Jing Zhao	Yifan Zhao
Shifeng Zhang	Kaili Zhao	Yaping Zhao
Shuo Zhang	Haojie Zhao	Zhou Zhao
Shaoxiong Zhang	Yucheng Zhao	He Zhao
Weizhong Zhang	Longjiao Zhao	Jie Zhao
Xi Zhang	Long Zhao	Xibin Zhao
Xiaomei Zhang	Qingsong Zhao	Xiaoqi Zhao
Xinyu Zhang	Qingyu Zhao	Zhengyu Zhao
Yin Zhang	Rui Zhao	Jin Zhe
Zicheng Zhang	Rui-Wei Zhao	Chuanxia Zheng
Zihao Zhang	Sicheng Zhao	Huan Zheng
Ziqi Zhang	Shuang Zhao	Hao Zheng
Zhaoxiang Zhang	Siyan Zhao	Jia Zheng
Zhen Zhang	Zelin Zhao	Jian-Qing Zheng
Zhipeng Zhang	Shiyu Zhao	Shuai Zheng
Zhixing Zhang	Wang Zhao	Meng Zheng
Zhizheng Zhang	Tiesong Zhao	Mingkai Zheng
Jiawei Zhang	Qian Zhao	Qian Zheng
Zhong Zhang	Wangbo Zhao	Qi Zheng
Pingping Zhang	Xi-Le Zhao	Wu Zheng
Yixin Zhang	Xu Zhao	Yinqiang Zheng
Kui Zhang	Yajie Zhao	Yufeng Zheng
Lingzhi Zhang	Yang Zhao	Yutong Zheng
Huaiwen Zhang	Ying Zhao	Yalin Zheng
Quanshi Zhang	Yin Zhao	Yu Zheng
Zhoutong Zhang	Yizhou Zhao	Feng Zheng
Yuhang Zhang	Yunhan Zhao	Zhaoheng Zheng
Yuting Zhang	Yuyang Zhao	Haitian Zheng
Zhang Zhang	Yue Zhao	Kang Zheng
Ziming Zhang	Yuzhi Zhao	Bolun Zheng

Haiyong Zheng
Mingwu Zheng
Sipeng Zheng
Tu Zheng
Wenzhao Zheng
Xiawu Zheng
Yinglin Zheng
Zhuo Zheng
Zilong Zheng
Kecheng Zheng
Zerong Zheng
Shuaifeng Zhi
Tiancheng Zhi
Jia-Xing Zhong
Yiwu Zhong
Fangwei Zhong
Zhihang Zhong
Yaoyao Zhong
Yiran Zhong
Zhun Zhong
Zichun Zhong
Bo Zhou
Boyao Zhou
Brady Zhou
Mo Zhou
Chunluan Zhou
Dingfu Zhou
Fan Zhou
Jingkai Zhou
Honglu Zhou
Jiaming Zhou
Jiahuan Zhou
Jun Zhou
Kaiyang Zhou
Keyang Zhou
Kuangqi Zhou
Lei Zhou
Lihua Zhou
Man Zhou
Mingyi Zhou
Mingyuan Zhou
Ning Zhou
Peng Zhou
Penghao Zhou
Qianyi Zhou

Shuigeng Zhou
Shangchen Zhou
Huayi Zhou
Zhize Zhou
Sanping Zhou
Qin Zhou
Tao Zhou
Wenbo Zhou
Xiangdong Zhou
Xiao-Yun Zhou
Xiao Zhou
Yang Zhou
Yipin Zhou
Zhenyu Zhou
Hao Zhou
Chu Zhou
Daquan Zhou
Da-Wei Zhou
Hang Zhou
Kang Zhou
Qianyu Zhou
Sheng Zhou
Wenhui Zhou
Xingyi Zhou
Yan-Jie Zhou
Yiyi Zhou
Yu Zhou
Yuan Zhou
Yuqian Zhou
Yuxuan Zhou
Zixiang Zhou
Wengang Zhou
Shuchang Zhou
Tianfei Zhou
Yichao Zhou
Alex Zhu
Chenchen Zhu
Deyao Zhu
Xiatian Zhu
Guibo Zhu
Haidong Zhu
Hao Zhu
Hongzi Zhu
Rui Zhu
Jing Zhu

Jianke Zhu
Junchen Zhu
Lei Zhu
Lingyu Zhu
Luyang Zhu
Menglong Zhu
Peihao Zhu
Hui Zhu
Xiaofeng Zhu
Tyler (Lixuan) Zhu
Wentao Zhu
Xiangyu Zhu
Xinqi Zhu
Xinxin Zhu
Xinliang Zhu
Yangguang Zhu
Yichen Zhu
Yixin Zhu
Yanjun Zhu
Yousong Zhu
Yuhao Zhu
Ye Zhu
Feng Zhu
Zhen Zhu
Fangrui Zhu
Jinjing Zhu
Linchao Zhu
Pengfei Zhu
Sijie Zhu
Xiaobin Zhu
Xiaoguang Zhu
Zezhou Zhu
Zhenyao Zhu
Kai Zhu
Pengkai Zhu
Bingbing Zhuang
Chengyuan Zhuang
Liansheng Zhuang
Peiye Zhuang
Yixin Zhuang
Yihong Zhuang
Junbao Zhuo
Andrea Ziani
Bartosz Zieliński
Primo Zingaretti

Nikolaos Zioulis
Andrew Zisserman
Yael Ziv
Liu Ziyin
Xingxing Zou
Danping Zou
Qi Zou

Shihao Zou
Xueyan Zou
Yang Zou
Yuliang Zou
Zihang Zou
Chuhang Zou
Dongqing Zou

Xu Zou
Zhiming Zou
Maria A. Zuluaga
Xinxin Zuo
Zhiwen Zuo
Reyer Zwiggelaar

Contents – Part XX

tSF: Transformer-Based Semantic Filter for Few-Shot Learning

Jinxiang Lai, Siqian Yang, Wenlong Liu, Yi Zeng, Zhongyi Huang, Wenlong Wu, Jun Liu, Bin-Bin Gao, and Chengjie Wang$^{(\boxtimes)}$

Youtu Lab, Tencent, Shenzhen, China
{jinxianglai,seasonsyang,sylviazeng,ezrealwu,
jasoncjwang}@tencent.com, csgaobb@gmail.com

Abstract. Few-Shot Learning (FSL) alleviates the data shortage challenge via embedding discriminative target-aware features among plenty seen (base) and few unseen (novel) labeled samples. Most feature embedding modules in recent FSL methods are specially designed for corresponding learning tasks (e.g., classification, segmentation, and object detection), which limits the *utility* of embedding features. To this end, we propose a light and universal module named transformer-based Semantic Filter (tSF), which can be applied for different FSL tasks. The proposed tSF redesigns the inputs of a transformer-based structure by a semantic filter, which not only embeds the knowledge from whole base set to novel set but also filters semantic features for target category. Furthermore, the parameters of tSF is equal to half of a standard transformer block (less than $1M$). In the experiments, our tSF is able to boost the performances in different classic few-shot learning tasks (about 2% improvement), especially outperforms the state-of-the-arts on multiple benchmark datasets in few-shot classification task.

1 Introduction

Few-Shot Learning (FSL) aims to recognize unseen objects with plenty known data (base) and few labeled unknown samples (novel). Due to the shortage of novel data, FSL tasks suffer from weak representation problem. Hence, researchers [17,24,40,50,59, 65,76,77] manage to design a embedding network to make extracted features robust and fine-grained enough in unseen instances recognition. To deal with different FSL tasks (e.g., classification, segmentation, and object detection), researchers propose different feature embedding modules, e.g., FEAT [69], CTX [6] for classification, HSNet [30] for segmentation, and FSCE [47] for detection, respectively.

Nevertheless, these methods are limited by the purposes of different tasks. In the classification task, methods put emphasis on locating the representative prototype for each class. Methods in the detection task aim to distinguish the similar objects and correct the bounding box, while in the segmentation task, methods manage to generate a precise mask. In consequence, classification task requires more robust features, while detection task and segmentation task need more fine-grained features. To satisfy the demands of different tasks concurrently, *a target-aware and image-aware feature embedding method is inevitable.*

© The Author(s), under exclusive license to Springer Nature Switzerland AG 2022
S. Avidan et al. (Eds.): ECCV 2022, LNCS 13680, pp. 1–19, 2022.
https://doi.org/10.1007/978-3-031-20044-1_1

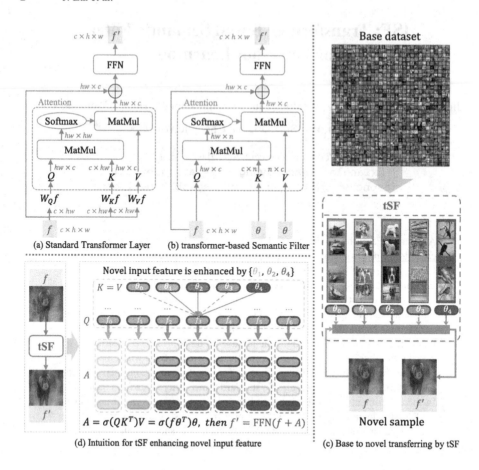

Fig. 1. (a) Standard Transformer layer. (b) transformer-based Semantic Filter (tSF), where θ is learnable semantic filter. (c) Base to novel transferring by tSF. After training, the semantic info of base dataset are embedded into θ, e.g. there are $n = 5$ semantic groups and θ_1 represents dog-like group. Then, given a novel input sample, tSF enhances its regions which are semantic similar to θ. (d) Intuition for tSF enhancing novel input feature.

Throughout the recent investigations, transformer-base structure [53] brings an important significance at computer vision field, which works on almost all common tasks due to its sensibility on both big dataset (macro) and a single image (micro). Specifically, transformer is able to store the information of whole dataset modeling spatio-temporal correlations among instances. The property exactly satisfies the purpose of the feature embedding operation. Besides, observing from Transformer [53], Feat [69], SuperGlue [42], CTX [6] and DETR [4], we notice that the transformer layer could perform different learning behaviors with different input forms of $\{Q, K, V\}$. In common, a traditional transformer structure needs big training dataset to achieve high performances. However, in few-shot learning field, it may fall into overfitting problem due to the shortage of data without a carefully designed framework.

To this end, we propose a light and general feature embedding module, named transformer based Semantic Filter (tSF), as illustrated in Fig. 1(b). We redesign the inputs Q, K, V of traditional transformer as f, θ, θ, where f is the extracted feature and θ is a learnable weight, named semantic filter. The tSF uses the correlation matrix between (f, θ) to re-weight θ. The average of the re-weighted θ is involved the dataset-attention response, which enhances the feature f from the views of both global dataset and local image. In this way, tSF can be trained without big data, while keeping the macro and micro information at the same time. Intuitively, Fig. 1(c) and Fig. 1(d) show that the proposed tSF is able to enhance the foreground regions of input novel feature via the embedded semantic info of base dataset. Besides, to further show the efficient of tSF, we insert the tSF into a strong few-shot classification framework, named PatchProto. The experimental results show that tSF helps PatchProto achieve SOTA performance with about 2% improvement. In addition, the parameter size of tSF is less than $1M$, half of that in traditional transformer. Moreover, to prove that tSF can suit different FSL frameworks, we conduct massive experiments on different few-shot learning tasks, and the results show that tSF can make $2\% - 3\%$ improvements on classification, detection and segmentation tasks.

In summary, our contributions are listed as follows:

- An effective and light module named transformer-based Semantic Filter (tSF) is proposed, which is helpful to learn a generalized-well embedding for novel targets (unseen in model training phase). The tSF leverages dataset-attention mechanism to realize information interaction between single input instance and whole dataset.
- A strong baseline framework called PatchProto is introduced for few-shot classification. Experimental results show our PatchProto+tSF approach outperforms the state-of-the-arts on multiple benchmark datasets such as miniImageNet and tiered-ImageNet.
- As a universal module, tSF is applied in different few-shot learning tasks, including classification, semantic segmentation and object detection. The massive experiments demonstrate that tSF is able to boost their performances.

2 Related Work

Few-Shot Classification. FSL algorithms first pre-train a base classifier with abundant samples (seen images), then learn to recognize novel classes (unseen images) with a few labeled samples. According to recent investigations, there are four representative directions: optimization-based, parameter-generating based, metric-learning based, and embedding-based methods. *Optimization-based methods* are able to perform rapid adaption with a few training samples for new classes by learning a good optimizer [2,38] or learning a well-initialized model [12,34,41]. *Parameter-generating methods* [3,14,31,32,35] focus on learning a parameter generating network. *Metric-learning based methods* learn to compare to tackle the few-shot classification problem. The main idea is classifying a new input image by computing the similarity compared with labeled instances [15,17,45,49,54,62,65,70]. These methods design carefully to embedding network to match their corresponding distance metrics. *Embedding-based* *methods* [24,40,50,59,76,77] firstly focus on learning a generalize-well embedding with supervised or self-supervised learning tasks, and then freeze this embedding and further train a linear classifier or design a metric classifier on novel classes.

Auxiliary Task for Few-Shot Classification. Some recent works gain a performance improvement by training few-shot models with supervised and self-supervised auxiliary tasks. The supervised task for FSL simply performs global classification on the base dataset as in CAN [17]. The effectiveness of self-supervised learning for FSL has been demonstrated, such as contrastive learning in either unsupervised pre-training [29] or episodic training [6,24], and auxiliary rotation prediction task [13,40,46].

Few-Shot Semantic Segmentation. Early few-shot semantic segmentation methods apply a dual-branch architecture [7,37,43], one segmenting query-images with the prototypes learned by the other branch. In recently, the dual-branch architecture is unified into a single-branch, using the same embedding for support and query images [26,44,56,67,74]. These methods aim to leverage better guidance for the segmentation of query-images [33,55,71,74], via learning better class-specific representations [25,26,44,56,67] or iteratively refining [72].

Few-Shot Object Detection. Existing few-shot object detection approaches can be divided into two paradigms: meta-learning based [10,18,20,63] and transfer learning based [11,36,47,58,61]. The majority of meta-learning approaches adopt *feature reweighting* or its variants to aggregate query and support features, which predict detections conditioned on support sets. Differently, the transfer learning based approaches firstly train the detectors on base set, then fine-tune the task-head layer on novel set, which achieve competitive results comparing to meta-learning approaches.

Transformer. Transformer is an attention-based network architecture that is widely applied in natural language processing area [5,53]. Due to its power in learning representation, it has been introduced in many computer vision tasks, such as image classification [8,52,57], detection [4,73,79], segmentation [22,64,75], image matching [42,48] and few-shot learning [6,27,68].

To best of our knowledge, there is no general feature embedding method, which can be applied on multiple few-shot learning tasks.

3   Transformer-Based Semantic Filter (tSF)

3.1   Related Transformer

Transformer-Based Self-Attention The architecture of a standard Transformer [53] layer is presented in Fig. 1(a), which consists of Attention and Feed-Forward Network (FFN). Given input feature $f \in \mathbb{R}^{c \times h \times w}$, the Transformer layer outputs $f' \in \mathbb{R}^{c \times h \times w}$. The key operation *Attention* is expressed as:

$$Attention(Q, K, V) = \sigma(QK^T)V, \tag{1}$$

where, $\{Q, K, V\}$ are known as query, key and value respectively, σ is softmax function. The forms of $\{Q, K, V\}$ in *transformer-based self-attention* are:

$$Q = W_Q f, \quad K = W_K f, \quad V = W_V f, \tag{2}$$

where, $\{Q, K, V\} \in \mathbb{R}^{hw \times c}$ and some feature-reshaping operations are omitted for simplicity, W_Q, W_K, W_V are learnable weights (e.g. convolution layers). For few-shot classification, based on self-attention mechanism, Feat [69] used the standard transformer layer as a set-to-set function to perform embedding adaptation, formally:

$$Q = W_Q f_{set}, \quad K = W_K f_{set}, \quad V = W_V f_{set}, \tag{3}$$

where f_{set} is a set of features of all the instances in the support set. Both the standard transformer [53] and Feat [69] are based on self-attention mechanism which learns to model the relationship between the feature-nodes insight the input features. Differently, as illustrated in Fig. 1(b) and Eq. 6, our Transformer-based Semantic Filter (tSF) applies *Transformer-based Dataset-Attention*, which makes information interaction between single input sample and whole dataset.

Transformer-Based Cross-Attention. The standard Transformer performs self-attention behavior, and SuperGlue [42] found that Transformer can be used to make cross-attention between pair-features. Given input pair-features (f_{ref}, f), it can obtain a cross-correlation matrix between (f_{ref}, f) which is used to re-weight f and achieves the cross-attention implementation. Specifically, $\{Q, K, V\}$ forms in *transformer-based cross-attention* are:

$$Q = W_Q f_{ref}, \quad K = W_K f, \quad V = W_V f. \tag{4}$$

For few-shot classification, based on cross-attention mechanism, CTX [6] used the transformer layer to generate query-aligned prototype for support class.

Transformer-Based Decoder. The forms of $\{Q, K, V\}$ in transformer-based decoder of DETR [4] are:

$$Q = \varphi, \quad K = V = f, \tag{5}$$

where $\varphi \in \mathbb{R}^{u \times c}$ is learnable weights which is called as object queries. Given input feature $f \in \mathbb{R}^{c \times h \times w}$, the DETR [4] decoder outputs $\varphi' \in \mathbb{R}^{u \times c}$ which is used to locate the objects. The dimension u of object queries φ represents the maximum number of objects insight a image.

3.2 tSF Methodology

In few-shot learning task, obtaining a generalized-well embedding for novel categories is one of the key problem. To this end, we plan to model the whole dataset information and then transfer the knowledge from base set to novel set. Benefiting from the property of modeling spatio-temporal correlations among instances, transformer is able to learn the whole dataset information. Besides, observing from Transformer [53], Feat [69], SuperGlue [42], CTX [6] and DETR [4], we notice that the transformer layer performs different learning behavior with different input forms of $\{Q, K, V\}$. Therefore, in order to transfer the knowledge from whole base set to novel set, we propose a transformer-based Semantic Filter (tSF) as illustrated in Fig. 1(b), where the forms of $\{Q, K, V\}$ are designed as:

$$Q = f, \quad K = V = \theta, \tag{6}$$

where, $Q \in \mathbb{R}^{hw \times c}$ after feature-reshaping, and $\theta \in \mathbb{R}^{n \times c}$ is learnable weights which is called as semantic filter. Formally, tSF is expressed as:

$$f^{'} = tSF(f) = FFN\left(f + \sigma\left(f\theta^{T}\right)\theta\right). \tag{7}$$

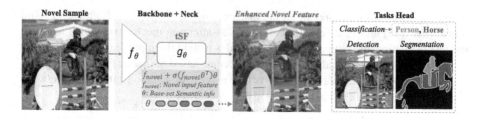

Fig. 2. The tSF for few-shot learning tasks such as classification, semantic segmentation and object detection.

The input and output features of tSF are $\{f, f^{'}\} \in \mathbb{R}^{c \times h \times w}$ respectively, which is consistent with the standard Transformer. The tSF can be utilized as a feature-to-feature function, e.g. stacking tSF as a neck after the backbone architecture for few-shot learning as illustrated in Fig. 2 and Fig. 3(a).

Formally, let's define C_{base} and C_{novel} as categories of base and novel respectively. Although $C_{base} \cap C_{novel} = \emptyset$, we assume C_{base} consists of sub-sets C_{base}^{sim} and C_{base}^{diff} which are semantically similar and different from C_{novel} respectively. Then, tSF transfers knowledge from base to novel:

$$Base\,Training: f^{'}_{base} = FFN\left(f_{base} + \sigma\left(f_{base}\theta^{T}\right)\theta\right),$$
$$Novel\,Testing: f^{'}_{novel} = FFN\left(f_{novel} + \sigma\left(f_{novel}\theta^{T}\right)\theta\right). \tag{8}$$

After training on base, semantic info of C_{base} are embedded into θ of which dimension n are interpreted as projected semantic groups, i.e. θ also consists of sub-sets θ^{sim} and θ^{diff} which are similar and different from C_{novel} respectively. Given a novel image, tSF enhances its regions which are similar to θ^{sim} while θ^{diff} doesn't. For example, if 'dog' in base, tSF enhances novel 'wolf' due to their high similarity relation. In Fig. 1(d), θ^{sim} is $\{\theta_1, \theta_2, \theta_4\}$, θ^{diff} is $\{\theta_0, \theta_3\}$.

Intuitively, as illustrated in Fig. 1(c), with the model training on base set, the semantic filter θ learns the whole dataset information. In the model testing on novel set, according to Eq. 7 and Fig. 1(d), the tSF uses the correlation matrix between (f, θ) to re-weight θ. The average of the re-weighted θ is the dataset-attention response A, which is semantically similar to f and can be used to enhance the target object as $f^{'} = FFN(f + A)$. f and θ contains the info of one sample and whole dataset respectively, and the tSF can collect the target information (i.e. A which is semantically similar to f) from θ to enhance f. Therefore the foreground region of the input novel sample feature can be enhanced by the dataset-attention response.

3.3   Discussions

Visualizations Under the few-shot classification framework of PatchProto+tSF introduced in Sect. 4.2, we give the visualizations of feature response map and t-SNE for tSF as shown in Fig. 3(c) and Fig. 3(b) respectively. In detail, $f' = g_\theta(f)$ and g_θ is the proposed tSF, the correlation matrix $R = \sigma(f_{novel}\theta^T) \in \mathbb{R}^{hw \times n}$ and $R\_\theta_i \in \mathbb{R}^{hw}$ represents the correlation vector between f_{novel} and $\theta_i \in \mathbb{R}^c$ (i^{th} position of $\theta \in \mathbb{R}^{n \times c}$). In Fig. 3(c), comparing to f, f' obtains more accurate and complete response map focusing on the foreground region, which indicates that tSF is able to transfer semantic knowledge from base set to novel set. The visualizations of the correlation vector R_θ_i show that θ_i learns semantic information from base set, specifically, these targets foreground are enhanced mainly contributed from $\{\theta_1, \theta_2, \theta_4\}$. In addition, the t-SNE visualizations in Fig. 3(b) show that f' obtains more clear category boundaries than f, which demonstrates that tSF is able to produce more discriminative embedding features for novel categories.

Properties The properties of tSF are as follows: (i) Generalization ability: Based on dataset-attention mechanism, the tSF models the whole dataset information and then transfer the knowledge from base set to novel set. The tSF makes information interaction between input sample and whole dataset, while self-attention based transformer interacts info insight input sample itself which leads to overfitting problem due to insufficient information interaction. (ii) Representation ability: The low dimension semantic filter $\theta \in \mathbb{R}^{n \times c}$ learns high-level semantic information from whole dataset. (iii) Efficiency: The computational complexity of tSF is less than self-attention based transformer. The complexity of transformer in calculating *Attention* by Eq. 1 is $O(transformer) = (h \times w)^3 \times c$, while our tSF is only $O(tSF) = (h \times w)^2 \times n \times c$, i.e. $\frac{O(transformer)}{O(tSF)} = \frac{h \times w}{n} \gg 1$.

Comparisons Comparing our tSF with Transformer and DETR decoder, in model testing on novel set, the input novel feature f_{novel} is enhanced by different information. Our tSF enhances f_{novel} by base dataset info (i.e. the semantic filter θ learned on base set) as defined in Eq. 8, which fulfils base to novel transferring. Differently, the standard Transformer enhances f_{novel} by itself instance info:

$$Transformer: \quad f'_{novel} = FFN\left(f_{novel} + \sigma\left(f_{novel}f_{novel}^T\right)f_{novel}\right). \quad (9)$$

Besides, the DETR decoder also enhances f_{novel} by itself instance info:

$$DETR: \quad f'_{novel} = FFN\left(f_{novel} + \sigma\left(\theta f_{novel}^T\right)f_{novel}\right). \quad (10)$$

As illustrated in Table 2, the experimental comparisons show that our tSF obtains obvious performance gains, while Transformer and DETR show performance degradation due to overfitting problem.

4   tSF for Few-Shot Classification

4.1   Problem Definition

N-way M-shot task to learn a classifier for N unseen classes with M labeled samples. Formally, we have three mutually disjoint datasets: a base set X_{base} for training, a validation set X_{val}, and a novel set X_{novel} for testing. Following [17,49,54,65], the episodic training strategy is adopted to mimic the few-shot learning setting, which has shown that it can effectively train a meta-learner (i.e., a few-shot classification model). Each episode contains N classes with M samples per class as the support set $\mathcal{S} = \{(x_i^s, y_i^s)\}_{i=1}^{m_s}$ ($m_s = N \times M$), and a fraction of the rest samples as the query set $\mathcal{Q} = \{(x_i^q, y_i^q)\}_{i=1}^{m_q}$. And the support subset of the kth class is denoted as $\mathcal{S}^k$.

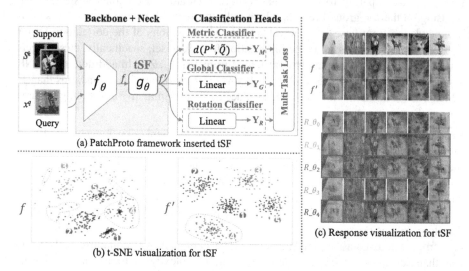

Fig. 3. (a) The PatchProto framework inserted tSF for few-shot classification. (b) The t-SNE visualization comparison for PatchProto+tSF, where $f' = g_\theta(f)$ and g_θ is the proposed tSF. (c) The visualizations of response map with the input of novel sample, where R_θ_i is the correlation vector between (f, θ_i), and the dimension n of $\theta \in \mathbb{R}^{n \times c}$ is set to 5.

4.2   PatchProto Framework with tSF

As illustrated in Fig. 3, the proposed PatchProto network consists of five components: feature extractor backbone f_θ, transformer-based Semantic Filter (tSF) g_θ, Metric classifier f_M for few-shot classification, and Global f_G and Rotation f_R classifiers for auxiliary tasks which are only used to assist model training.

The input image x^q in the query set $\mathcal{Q} = \{(x_i^q, y_i^q)\}_{i=1}^{m_q}$ is rotated with [0°, 90°, 180°, 270°] and outputs a rotated $\tilde{\mathcal{Q}} = \{(\tilde{x}_i^q, \tilde{y}_i^q)\}_{i=1}^{m_q \times 4}$. Each support subset $\mathcal{S}^k$ and a rotated query image $\tilde{x}^q$ are fed through the feature extractor backbone f_θ and tSF g_θ, and produces the class feature $P^k = \frac{1}{|\mathcal{S}^k|} \sum_{x_i^s \in \mathcal{S}^k} g_\theta(f_\theta(x_i^s))$ and query feature $\tilde{Q} = g_\theta(f_\theta(\tilde{x}^q)) \in \mathbb{R}^{c \times h \times w}$, respectively. Then the Metric classifier f_M makes classification via measuring the similarity between each pair-features $(P^k, \tilde{Q})$. Finally, PatchProto network is trained by optimizing the multi-task loss contributing from meta loss and auxiliary loss. In inductive inference phase, with the embedding learned in train set, the Metric classifier predicts the query x^q as Y_M based on cosine similarity measurement.

Objective Functions. *Meta loss:* As a metric-based meta learner, the Metric classifier predicts the query into N support categories by measuring cosine similarity. Following [17], we adopt the patch-wise classification mechanism to generate precise embeddings. Specifically, each local feature $\tilde{Q}_m$ at m^{th} spatial position of $\tilde{Q}$, is predicted into N categories. Formally, the probability of recognizing $\tilde{Q}_m$ as k^{th} category is:

$$\hat{Y}(y = k|\tilde{Q}_m) = \sigma\left(d\left(\tilde{Q}_m, GAP\left(P^k\right)\right)\right), \tag{11}$$

where GAP denotes global average pooling, d is cosine distance. Then the metric classification loss with the few-shot label $\tilde{y}^q$ is:

$$\mathcal{L}_M = -\sum_{i=1}^{m_q}\sum_{m=1}^{h \times w} \log \hat{Y}(y = \tilde{y}_i^q|(\tilde{Q}_m)_i). \tag{12}$$

Auxiliary loss: The Global classifier predicts the query into all C categories of train set, thus its loss is:

$$\mathcal{L}_G = PCE(\tilde{Q}, C^q) = -\sum_{i=1}^{m_q}\sum_{n=1}^{h \times w} C_i^q \log\left(\sigma(W_l(\tilde{Q}_m)_i)\right). \tag{13}$$

where, W_l is a linear layer, C_i^q is the global category of $\tilde{x}_i^q$ with all C categories, and PCE denotes the patch-wise cross-entropy function. Similarly, the loss of Rotation classifier is derived by $\mathcal{L}_R = PCE(\tilde{Q}, B^q)$, where B_i^q is the rotation category of $\tilde{x}_i^q$ with four categories.

Multi-Task Loss: Therefore, inspired by [19], the overall classification loss is:

$$\mathcal{L} = \frac{1}{2}\mathcal{L}_M + \sum_{j=G,R}\left((\lambda + w_j)\mathcal{L}_j + \log\frac{1}{(\lambda + w_j)}\right), \tag{14}$$

where, $w = \frac{1}{2\alpha^2}$, α is learnable variable, λ is a hyper-parameter to balance the effects of the losses of few-shot task and auxiliary tasks. The influence of λ is studied in Table 3.

5   tSF for Few-Shot Segmentation and Detection

As shown in Fig. 2, the proposed tSF is stacked after the backbone architecture (i.e. Backbone + tSF) for few-shot learning tasks, such as classification, segmentation and detection. To verify the effectiveness and the university of our tSF module, we insert the tSF into the current state-of-the-art few-shot segmentation and detection methods. And the details are introduced in the next.

RePRI+tSF for Segmentation. RePRI (Region Proportion Regularized Inference) [28] approach leverages the statistics of unlabeled pixels for the input image. It optimizes three complementary loss terms, including cross-entropy on labeled support pixels, Shannon entropy on unlabeled query pixels and a global KL-divergence regularizer on

Table 1. Comparison with existing methods on 5-way classification task on benchmark miniImageNet and tieredImageNet datasets.

Model	Backbone	miniImageNet		tieredImageNet	
		1-shot	5-shot	1-shot	5-shot
MatchingNet [54]	Conv4	43.44 ± 0.77	60.60 ± 0.71	–	–
ProtoNet [45]	Conv4	49.42 ± 0.78	68.20 ± 0.66	53.31 ± 0.89	72.69 ± 0.74
RelationNet [49]	Conv4	50.44 ± 0.82	65.32 ± 0.70	54.48 ± 0.93	71.32 ± 0.78
PatchProto	Conv4	54.71 ± 0.46	70.67 ± 0.38	56.90 ± 0.51	71.47 ± 0.42
PatchProto+tSF	Conv4	$\mathbf{57.39 \pm 0.47}$	$\mathbf{73.34 \pm 0.37}$	$\mathbf{59.79 \pm 0.51}$	$\mathbf{74.56 \pm 0.41}$
CAN [17]	ResNet-12	63.85 ± 0.48	79.44 ± 0.34	69.89 ± 0.51	84.23 ± 0.37
MetaOpt+ArL [16]	ResNet-12	65.21 ± 0.58	80.41 ± 0.49	-	-
DeepEMD [70]	ResNet-12	65.91 ± 0.82	82.41 ± 0.56	71.16 ± 0.87	86.03 ± 0.58
IENet [40]	ResNet-12	66.82 ± 0.80	$\mathbf{84.35 \pm 0.51}$	71.87 ± 0.89	$\mathbf{86.82 \pm 0.58}$
DANet [65]	ResNet-12	67.76 ± 0.46	82.71 ± 0.31	71.89 ± 0.52	85.96 ± 0.35
PatchProto	ResNet-12	68.46 ± 0.47	82.65 ± 0.31	70.50 ± 0.50	83.60 ± 0.37
PatchProto+tSF	ResNet-12	$\mathbf{69.74 \pm 0.47}$	83.91 ± 0.30	$\mathbf{71.98 \pm 0.50}$	85.49 ± 0.35
wDAE-GNN [14]	WRN-28	61.07 ± 0.15	76.75 ± 0.11	68.18 ± 0.16	83.09 ± 0.12
LEO [41]	WRN-28	61.76 ± 0.08	77.59 ± 0.12	66.33 ± 0.05	81.44 ± 0.09
PSST [76]	WRN-28	64.16 ± 0.44	80.64 ± 0.32	–	–
FEAT [69]	WRN-28	65.10 ± 0.20	81.11 ± 0.14	70.41 ± 0.23	84.38 ± 0.16
CA [1]	WRN-28	65.92 ± 0.60	82.85 ± 0.55	74.40 ± 0.68	86.61 ± 0.59
PatchProto	WRN-28	69.34 ± 0.46	83.46 ± 0.30	73.40 ± 0.50	86.85 ± 0.35
PatchProto+tSF	WRN-28	$\mathbf{70.23 \pm 0.46}$	$\mathbf{84.55 \pm 0.29}$	$\mathbf{74.87 \pm 0.49}$	$\mathbf{88.05 \pm 0.32}$

predicted foreground. RePRI achieves state-of-the-art on few-shot segmentation benchmark PASCAL-5^i built from PASCAL VOC [9]. Based on the RePRI framwork, we simply stack our tSF behind its backbone which obtains the RePRI+tSF approach.

DeFRCN+tSF for Detection. DeFRCN (Decoupled Faster R-CNN) [36] is a simple yet effective fine-tuned approach for few-shot object detection, which proposes Gradient Decoupled Layer and Prototypical Calibration Block to alleviate the contradictions of Faster R-CNN in FSL scenario. Due to its simplicity and effective, DeFRCN achieves state-of-the-art on PASCAL VOC [9] and COCO [23]. To verify the effectiveness of tSF module in few shot object detection task, we use DeFRCN as baseline, and insert the tSF into the ResNet-101 backbone to obtain the DeFRCN+tSF approach, specifically, tSF module is followed in the 5th residual block. For the hyperparameter settings, we use the default parameter as same with DeFRCN.

6 Experiments

To prove the effectiveness and universality of the proposed tSF, massive experiments are conducted on differnet few-shot learning tasks, including classification, semantic segmentation and object detection. Overall, the results show that tSF can make $2\% - 3\%$ improvements on these tasks.

6.1 Few-Shot Classification

Datasets *mini*ImageNet dataset is a subset of ImageNet [21], which consists of 100 classes. We split the 100 classes following the setting in [17,49,65], i.e. 64, 16 and 20 classes for training, validation and test respectively. *tiered*ImageNet dataset [39] is also a subcollection of ImageNet [21]. It contains 608 classes, which are separated into 351 classes for training, 97 for validation and 160 for testing.

Evaluation and Implementation Details. We conduct experiments under 5-way 1-shot and 5-shot settings. We report the *average accuracy* and 95% *confidence interval* over 2000 episodes sampled from the test set. Horizontal flipping, random cropping, random erasing [78] and color jittering are employed for data augmentation in training. According to the ablation study results in Table 3, the hyperparameter λ in Eq. 14 is set to 0.5 and 1.5 for ResNet-12 and WRN-28 respectively. The detailed info of optimizer, learning-rate and training-epochs are referred to our public source code.

Comparison with State-of-the-Arts. Table 1 compares our methods with existing few-shot classification algorithms on miniImageNet and tieredImageNet, which shows that the proposed PatchProto and PatchProto+tSF mthods outperform the existing SOTAs under different backbones. And PatchProto+tSF shows obviously accuracy improvements under different backbones on 1-shot and 5-shot tasks than the strong baseline PatchProto, which demonstrates the effectiveness of the proposed tSF. The proposed PatchProto+tSF performs better than the optimization-based MetaOpt+ArL [16] and parameter-generating method wDAE-GCNN [14], with an improvement up to 4.53% and 9.16% respectively. Comparing to the competitive metric-based DeepEMD [70], PatchProto+tSF achieves 3.83% higher accuracy. Some metric-based methods [17,65]

Table 2. The results on 5-way miniImageNet classification about the structure (refers to Fig. 1 (a) and Fig. 1 (b)) influence of tSF, which utilize PatchProto+tSF framework under ResNet-12 backbone. The dimension n of θ in tSF is set to 5. The Metric and Global loss weights are set to 0.5 and 1.0 respectively, and the Rotation classifier is not applied.

Neck	Q, K, V	Attention heads	Param	miniImageNet	
				1-shot	5-shot
None	–	–	7.75M	67.47 ± 0.47	81.85 ± 0.32
Transformer	$Q=K=V=f$	1	8.75M	63.25 ± 0.45	79.44 ± 0.33
Transformer	$Q=W_Q f, K=W_K f, V=W_V f$	1	9.75M	62.96 ± 0.47	78.92 ± 0.33
Transformer	$Q=K=V=f$	4	8.75M	62.68 ± 0.47	78.98 ± 0.34
Transformer	$Q=W_Q f, K=W_K f, V=W_V f$	4	9.75M	62.70 ± 0.47	78.33 ± 0.34
DETR	$Q=\theta, K=V=f$	1	8.75M	63.55 ± 0.45	79.65 ± 0.33
tSF-V	$Q=K=f, V=\theta$	1	9.00M	61.84 ± 0.48	76.11 ± 0.36
tSF-K	$Q=V=f, K=\theta$	1	9.00M	64.73 ± 0.45	80.43 ± 0.33
tSF	$Q=f, K=V=\theta$	1	8.75M	68.37 ± 0.46	83.08 ± 0.31
tSF	$Q=W_Q f, K=W_K \theta, V=W_V \theta$	1	9.75M	68.27 ± 0.47	83.01 ± 0.31
tSF	$Q=f, K=V=\theta$	4	8.75M	$\mathbf{68.60 \pm 0.47}$	$\mathbf{83.26 \pm 0.31}$
tSF	$Q=W_Q f, K=W_K \theta, V=W_V \theta$	4	9.75M	68.49 ± 0.47	82.95 ± 0.31
tSF	$Q=f, K=V=\theta$	8	8.75M	68.46 ± 0.46	83.14 ± 0.31
tSF	$Q=W_Q f, K=W_K \theta, V=W_V \theta$	8	9.75M	68.42 ± 0.47	83.12 ± 0.31

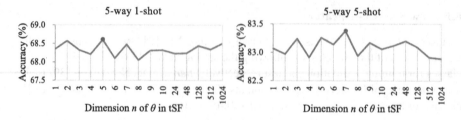

Fig. 4. The results on miniImageNet classification about the influence of dimension n of $\theta \in \mathbb{R}^{n \times c}$ in tSF, which utilize PatchProto+tSF framework under ResNet-12 backbone without Rotation classifier.

employing cross attention mechanism, and our PatchProto+tSF still surpasses the best DANet [65] with an performance improvement up to 1.98% on 1-shot. Overall our PatchProto+tSF obtains a new SOTA performance on both 1-shot and 5-shot classification tasks on miniImageNet and tieredImageNet, which demonstrates the strength of our framework and the effectiveness of the proposed tSF.

Ablation Study. *Structure Influence of tSF:* As shown in Table 2, by comparing our tSF to the baseline without neck in first row, it shows consistent improvements on 1-shot and 5-shot tasks, because tSF can transfer the knowledge from base set to novel set and generates more discriminative representations via focusing on the foreground regions. Comparing to the baseline without neck, the self-attention based transformer shows performance degradation due to overfitting on base dataset. Instead of behaving self-attention as standard transformer do, the proposed dataset-attention based tSF is able to prevent overfitting and generalize well on novel task, which is illustrated by the large accuracy improvement of tSF.

Table 3. The results on 5-way miniImageNet classification about the influence of multi-task loss employed in PatchProto+tSF under ResNet-12 and WRN-28 backbones. As introduced in Eq. 14, λ is the hyper-parameter, and w_G, w_R are learnable weights.

λ	Loss weights			ResNet-12		WRN-28	
	Metric	Global	Rotation	1-shot	5-shot	1-shot	5-shot
–	0.5	–	–	62.76 ± 0.49	80.07 ± 0.34	61.71 ± 0.50	77.53 ± 0.36
–	0.5	–	1.0	65.57 ± 0.49	80.81 ± 0.34	63.97 ± 0.50	79.25 ± 0.37
–	0.5	1.0	–	68.60 ± 0.47	82.98 ± 0.31	67.73 ± 0.47	82.59 ± 0.31
–	0.5	1.0	1.0	$\mathbf{69.41 \pm 0.46}$	$\mathbf{83.82 \pm 0.30}$	$\mathbf{69.64 \pm 0.47}$	$\mathbf{84.01 \pm 0.30}$
0.0	0.5	$\lambda + w_G$	$\lambda + w_R$	68.19 ± 0.47	83.00 ± 0.31	68.55 ± 0.48	83.30 ± 0.31
0.5	0.5	$\lambda + w_G$	$\lambda + w_R$	$\mathbf{69.74 \pm 0.47}$	$\mathbf{83.91 \pm 0.30}$	69.20 ± 0.46	84.03 ± 0.30
1.0	0.5	$\lambda + w_G$	$\lambda + w_R$	69.44 ± 0.46	83.90 ± 0.31	70.05 ± 0.46	84.17 ± 0.29
1.5	0.5	$\lambda + w_G$	$\lambda + w_R$	69.30 ± 0.45	83.82 ± 0.30	$\mathbf{70.23 \pm 0.46}$	$\mathbf{84.55 \pm 0.29}$
2.0	0.5	$\lambda + w_G$	$\lambda + w_R$	69.50 ± 0.45	83.86 ± 0.30	70.02 ± 0.45	83.61 ± 0.30

Table 4. The results on 1-way PASCAL-5^i segmentation using mean-IoU. The dimension n of θ in tSF is set to 5.

Method	Backbone	1 shot					5 shot				
		Fold-0	Fold-1	Fold-2	Fold-3	Mean	Fold-0	Fold-1	Fold-2	Fold-3	Mean
PANet [56]	VGG-16	42.3	58.0	51.1	41.2	48.1	51.8	64.6	59.8	46.5	55.7
RPMM [26]		47.1	**65.8**	50.6	**48.5**	53.0	50.0	66.5	51.9	47.6	54.0
RePRI [28]	VGG-16	49.7	63.4	58.2	42.8	53.5	54.5	67.2	63.7	48.8	58.6
RePRI+tSF		**53.0**	65.3	**58.3**	44.2	**55.2**	**57.0**	**67.9**	**63.9**	**50.8**	**59.9**
CANet [72]	ResNet-50	52.5	65.9	51.3	**51.9**	55.4	55.5	67.8	51.9	53.2	57.1
PGNet [71]		56.0	66.9	50.6	50.4	56.0	57.7	68.7	52.9	54.6	58.5
RPMM [67]		55.2	66.9	52.6	50.7	56.3	56.3	67.3	54.5	51.0	57.3
PPNet [26]		47.8	58.8	53.8	45.6	51.5	58.4	67.8	64.9	56.7	62.0
RePRI [28]	ResNet-50	60.8	67.8	60.9	47.5	59.3	66.0	70.9	65.9	56.4	64.8
RePRI+tSF		**62.4**	**68.6**	**61.4**	49.4	**60.5**	**66.4**	**71.1**	**66.4**	**58.3**	**65.6**

Dimension Influence of θ in tSF: As shown in Fig. 4, with a wide range $[1, 1024]$ dimension n of $\theta \in \mathbb{R}^{n \times c}$ in tSF, the accuracy differences are within 0.5% on 1-shot and 5-shot tasks, i.e. our PatchProto+tSF framework is not sensitive to the dimension of θ. Considering the accuracy performance and computational complexity, we recommend to set the dimension $n = 5$. The n is interpreted as number of semantic groups. As n going larger, semantic groups become more fine-grained. The setting of $n = 1$ represents one foreground group and is still able to obtain impressive performance.

Influence of Multi-task Loss: As illustrated in Table 3, the proposed PatchProto+tSF obtains its best results as setting λ to 0.5 and 1.5 under ResNet-12 and WRN-28 backbones respectively. Comparing to the baseline (i.e. with Metric classification task only) in first row, our multi-task framework achieves large improvements on 1-shot and 5-shot tasks under different backbones. These results indicate that the auxiliary tasks (i.e. Global classification and Rotation classification) are useful for training a more robust embedding leading to an accuracy improvement, and the weights of the auxiliary tasks have a great influence on the overall few-shot classification performance.

Table 5. The results on 1-way COCO-20^i segmentation using mean-IoU. The dimension n of θ in tSF is set to 5.

Method	Backbone	1 shot					5 shot				
		Fold-0	Fold-1	Fold-2	Fold-3	Mean	Fold-0	Fold-1	Fold-2	Fold-3	Mean
PPNet [26]	ResNet-50	34.5	25.4	24.3	18.6	25.7	**48.3**	30.9	35.7	30.2	36.2
RPMM [67]		29.5	36.8	29.0	27.0	30.6	33.8	42.0	33.0	33.3	35.5
PFENet [51]		36.5	38.6	34.5	33.8	35.8	36.5	43.3	37.8	38.4	39.0
RePRI [28]	ResNet-50	36.1	40.0	34.0	36.1	36.6	43.3	48.7	44.0	44.9	45.2
RePRI+tSF		**38.4**	**41.3**	**35.2**	**37.7**	**38.2**	45.0	**49.9**	**45.5**	**45.6**	**46.5**

Table 6. The results on VOC dataset. we evaluate the performance(AP_{50}) of DeFRCN under ResNet-101 with tSF module on three novel splits over multiple runs. The term w/G denotes whether using G-$FSOD$ setting [58]. The dimension n of θ in tSF is set to 5.

Method / Shots	w/G	Novel Set 1					Novel Set 2					Novel Set 3				
		1	2	3	5	10	1	2	3	5	10	1	2	3	5	10
MetaDet [60]	✗	18.9	20.6	30.2	36.8	49.6	21.8	23.1	27.8	31.7	43.0	20.6	23.9	29.4	43.9	44.1
TFA [58]	✗	39.8	36.1	44.7	55.7	56.0	23.5	26.9	34.1	35.1	39.1	30.8	34.8	42.8	49.5	49.8
DeFRCN [36]	✗	**53.6**	57.5	**61.5**	**64.1**	**60.8**	30.1	38.1	**47.0**	**53.3**	**47.9**	48.4	**50.9**	52.3	**54.9**	57.4
DeFRCN+tSF	✗	**53.6**	**58.1**	**61.5**	63.8	**60.8**	**31.5**	**39.3**	**47.0**	52.1	47.3	**48.5**	50.5	**52.8**	54.5	**58.0**
TFA [58]	✔	25.3	36.4	42.1	47.9	52.8	18.3	27.	30.9	34.1	39.5	17.9	27.2	34.3	40.8	45.6
FSDetView [63]	✔	24.2	35.3	42.2	49.1	57.4	21.6	24.6	31.9	37.0	45.7	21.2	30.0	37.2	43.8	49.6
DeFRCN [36]	✔	40.2	53.6	58.2	63.6	**66.5**	29.5	39.7	43.4	48.1	**52.8**	35.0	38.3	52.9	57.7	60.8
DeFRCN+tSF	✔	**43.6**	**57.4**	**61.2**	**65.1**	65.9	**31.0**	**40.3**	**45.3**	**49.6**	52.5	**39.3**	**51.4**	**54.8**	**59.8**	**62.1**

6.2   Few-Shot Semantic Segmentation

Datasets and Setting *PASCAL-5^i and COCO-20^i Datasets:* PASCAL-5^i is built from PASCAL VOC [9]. The 20 object categories are split into 4 folds. For each fold, 15 categories are utilized for training and the remaining 5 classes for testing. COCO-20^i is built from MS-COCO [23]. COCO-20^i dataset is divided into 4 folds with 60 base classes and 20 test classes in each fold.

Evaluation Setting: Following [26], the mean Intersection over Union (mIoU) is adopted for evaluation, and we report the average mIoU over 5 runs of 1000 tasks.

Comparison with State-of-the-Arts Table 4 and Table 5 present our evaluation results on PASCAL-5^i and COCO-20^i. Comparing with existing few-shot semantic segmentation methods, the RePRI+tSF approach achieves new state-of-the-art results. With the help of our tSF module, the RePRI+tSF obtains consistent performance improvement than RePRI, on 1-way 1-shot and 5-shot tasks under VGG-16 and ResNet-50 backbones.

6.3   Few-Shot Object Detection

Datasets and Setting. *PASCAL VOC and COCO Datasets:* PASCAL VOC [9] are randomly sampled into 3 splits, and each contains 20 categories. For each split, there are 15 base and 5 novel categories. Each novel class has $K = 1, 2, 3, 5, 10$ objects sampled from the train/val set of VOC2007 and VOC2012 for training, and the test set of VOC2007 for testing. COCO [23] use 60 categories disjoint with VOC as base set, and the remaining 20 categories are novel set with $K = 1, 2, 3, 5, 10, 30$ shots. The total 5k images randomly sampled from the validation set are utilized for testing, while the rest for training.

Evaluation Setting: Following [20,36,58, 63,66], we conduct experiments on two evaluation protocols: few-shot object detection *(FSOD)* and generalized few-shot object detection *(G-FSOD)*. The setting of FSOD only observes the performance of novel set. More comprehensively, that of G-FSOD considers both novel set and base set.

Comparison with State-of-the-Arts Table 6 and Table 7 present our evaluation results on VOC and COCO. Comparing with existing few-shot object detection methods, the DeFRCN+tSF approach achieves new state-of-the-art on VOC and COCO. On VOC

Table 7. The results on COCO dataset. we report the performance (mAP) of DeFRCN under ResNet-101 with tSF module over multiple runs. The dimension n of θ in tSF is set to 5.

Method / Shots	w/G	Shot Number					
		1	2	3	5	10	30
TFA [58]	✗	4.4	5.4	6.0	7.7	10.0	13.7
FSDetView [63]	✗	4.5	6.6	7.2	10.7	12.5	14.7
DeFRCN [36]	✗	**9.3**	**12.9**	**14.8**	**16.1**	**18.5**	**22.6**
DeFRCN+tSF	✗	**9.9**	**13.5**	**14.8**	**16.3**	18.3	22.5
TFA [58]	✔	1.9	3.9	5.1	7.0	9.1	12.1
FSDetView [63]	✔	3.2	4.9	6.7	8.1	10.7	15.9
DeFRCN [36]	✔	4.8	8.5	10.7	**13.6**	**16.8**	**21.2**
DeFRCN+tSF	✔	**5.0**	**8.7**	**10.9**	**13.6**	16.6	20.9

three different data splits, under the FSOD and G-FSOD setting, the indicators of all shots have been improved to a certain extent afer adding tSF module. We achieve around $2.6AP$ improvement over the DeFRCN in all shots under G-FSOD setting. On COCO dataset, DeFRCN+tSF consistently outperforms DeFRCN in 1, 2, 3 and 5 shots.

7 Conclusions

In this paper, we propose a transformer-based semantic filter (tSF) for few-shot learning problem. tSF leverages a well-designed transformer-based structure to encode the knowledge from base dataset and novel samples. In this way, a target-aware and image-aware feature can be generated via tSF. Moreover, tSF is a universal module, which can be applied into multiple few-shot learning tasks, e.g., classification, segmentation and detection. In addition, the parameter size of tSF is half of that of a standard transformer (less than $1M$). The experimental results show that tSF is able to improve the performances about 2% in multiple classic few-shot learning tasks.

References

1. Afrasiyabi, A., Lalonde, J.-F., Gagné, C.: Associative alignment for few-shot image classification. In: Vedaldi, A., Bischof, H., Brox, T., Frahm, J.-M. (eds.) ECCV 2020. LNCS, vol. 12350, pp. 18–35. Springer, Cham (2020). https://doi.org/10.1007/978-3-030-58558-7_2

2. Andrychowicz, M., et al.: Learning to learn by gradient descent by gradient descent. In: NeurIPS (2016)

3. Bertinetto, L., Henriques, J.F., Valmadre, J., Torr, P., Vedaldi, A.: Learning feed-forward one-shot learners. In: NeurIPS (2016)

4. Carion, N., Massa, F., Synnaeve, G., Usunier, N., Kirillov, A., Zagoruyko, S.: End-to-end object detection with transformers. In: Vedaldi, A., Bischof, H., Brox, T., Frahm, J.-M. (eds.) ECCV 2020. LNCS, vol. 12346, pp. 213–229. Springer, Cham (2020). https://doi.org/10.1007/978-3-030-58452-8_13

5. Devlin, J., Chang, M.W., Lee, K., Toutanova, K.: Bert: Pre-training of deep bidirectional transformers for language understanding. In: NAACL-HLT (2020)

6. Doersch, C., Gupta, A., Zisserman, A.: Crosstransformers: spatially-aware few-shot transfer. In: NeurIPS (2020)

7. Dong, N., Xing, E.: Few-shot semantic segmentation with prototype learning. In: British Machine Vision Conference (BMVC) (2018)

8. Dosovitskiy, A., et al.: An image is worth 16×16 words: Transformers for image recognition at scale. In: ICLR (2021)

9. Everingham, M., Van Gool, L., Williams, C.K., Winn, J., Zisserman, A.: The pascal visual object classes (voc) challenge. Int. J. Comput. Vision **88**(2), 303–338 (2010)

10. Fan, Q., Zhuo, W., Tang, C.K., Tai, Y.W.: Few-shot object detection with attention-rpn and multi-relation detector. In: CVPR (2020)

11. Fan, Z., Ma, Y., Li, Z., Sun, J.: Generalized few-shot object detection without forgetting. In: CVPR (2021)

12. Finn, C., Abbeel, P., Levine, S.: Model-agnostic meta-learning for fast adaptation of deep networks. In: ICML (2017)

13. Gidaris, S., Bursuc, A., Komodakis, N., Pérez, P., Cord, M.: Boosting few-shot visual learning with self-supervision. In: ICCV (2019)

14. Gidaris, S., Komodakis, N.: Generating classification weights with gnn denoising autoencoders for few-shot learning. In: CVPR (2019)

15. Gregory, K., Richard, Z., Ruslan, S.: Siamese neural networks for one-shot image recognition. In: ICML Workshops (2015)

16. Hongguang, Z., Piotr, K., Songlei, J., Hongdong, L., Philip, H. S., T.: Rethinking class relations: Absolute-relative supervised and unsupervised few-shot learning. In: CVPR (2021)

17. Hou, R., Chang, H., Bingpeng, M., Shan, S., Chen, X.: Cross attention network for few-shot classification. In: NeurIPS (2019)

18. Hu, H., Bai, S., Li, A., Cui, J., Wang, L.: Dense relation distillation with context-aware aggregation for few-shot object detection. In: CVPR (2021)

19. Jinxiang, L., et al.: Rethinking the metric in few-shot learning: from an adaptive multi-distance perspective. In: ACMMM (2022)

20. Kang, B., Liu, Z., Wang, X., Yu, F., Feng, J., Darrell, T.: Few-shot object detection via feature reweighting. In: ICCV (2019)

21. Krizhevsky, A., Sutskever, I., Hinton, G.E.: Imagenet classification with deep convolutional neural networks. In: NeurIPS (2012)

22. Liang, J., Homayounfar, N., Ma, W.C., Xiong, Y., Hu, R., Urtasun, R.: Polytransform: deep polygon transformer for instance segmentation. In: CVPR (2020)

23. Lin, T.-Y., et al.: Microsoft COCO: Common objects in context. In: Fleet, D., Pajdla, T., Schiele, B., Tuytelaars, T. (eds.) ECCV 2014. LNCS, vol. 8693, pp. 740–755. Springer, Cham (2014). https://doi.org/10.1007/978-3-319-10602-1_48

24. Liu, C., Fu, Y., Xu, C., Yang, S., Li, J., Wang, C., Zhang, L.: Learning a few-shot embedding model with contrastive learning. In: AAAI (2021)

25. Liu, W., Zhang, C., Lin, G., Liu, F.: CRNet: Cross-reference networks for few-shot segmentation. In: CVPR (2020)

26. Liu, Y., Zhang, X., Zhang, S., He, X.: Part-aware prototype network for few-shot semantic segmentation. In: Vedaldi, A., Bischof, H., Brox, T., Frahm, J.-M. (eds.) ECCV 2020. LNCS, vol. 12354, pp. 142–158. Springer, Cham (2020). https://doi.org/10.1007/978-3-030-58545-7_9

27. Lu, Z., He, S., Zhu, X., Zhang, L., Song, Y.Z., Xiang, T.: Simpler is better: Few-shot semantic segmentation with classifier weight transformer. In: ICCV (2021)

28. Malik, B., Hoel, K., Ziko, I.M., Pablo, P., Ismail, B.A., Jose, D.: Few-shot segmentation without meta-learning: A good transductive inference is all you need? In: CVPR (2021)

29. Medina, C., Devos, A., Grossglauser, M.: Self-supervised prototypical transfer learning for few-shot classification. arXiv:2006.11325 (2020)

30. Min, J., Kang, D., Cho, M.: Hypercorrelation squeeze for few-shot segmentation. In: ICCV (2021)
31. Munkhdalai, T., Yu, H.: Meta networks. In: ICML (2017)
32. Munkhdalai, T., Yuan, X., Mehri, S., Trischler, A.: Rapid adaptation with conditionally shifted neurons. In: ICML (2018)
33. Nguyen, K., Todorovic, S.: Feature weighting and boosting for few-shot segmentation. In: ICCV (2019)
34. Nichol, A., Achiam, J., Schulman, J.: On first-order meta-learning algorithms. arXiv:1803.02999 (2018)
35. Qi, C., Yingwei, P., Ting, Y., Chenggang, Y., Tao, M.: Memory matching networks for one-shot image recognition. In: CVPR (2018)
36. Qiao, L., Zhao, Y., Li, Z., Qiu, X., Wu, J., Zhang, C.: Defrcn: Decoupled faster r-cnn for few-shot object detection. In: ICCV (2021)
37. Rakelly, K., Shelhamer, E., Darrell, T., Efros, A., Levine, S.: Conditional networks for few-shot semantic segmentation. In: ICLR Workshop (2018)
38. Ravi, S., Larochelle, H.: Optimization as a model for few-shot learning. In: ICLR (2017)
39. Ren, M., et al.: Meta-learning for semi-supervised few-shot classification. In: ICLR (2018)
40. Rizve, M.N., Khan, S., Khan, F.S., Shah, M.: Exploring complementary strengths of invariant and equivariant representations for few-shot learning. In: CVPR (2021)
41. Rusu, A.A., et al.: Meta-learning with latent embedding optimization. In: ICLR (2019)
42. Sarlin, P.E., DeTone, D., Malisiewicz, T., Rabinovich, A.: Superglue: Learning feature matching with graph neural networks. In: CVPR (2020)
43. Shaban, A., Bansal, S., Liu, Z., Essa, I., Boots, B.: One-shot learning for semantic segmentation. In: BMVC (2018)
44. Siam, M., Oreshkin, B.N., Jagersand, M.: AMP: Adaptive masked proxies for few-shot segmentation. In: ICCV (2019)
45. Snell, J., Swersky, K., Zemel, R.: Prototypical networks for few-shot learning. In: NeurIPS (2017)
46. Su, J.-C., Maji, S., Hariharan, B.: When does self-supervision improve few-shot learning? In: Vedaldi, A., Bischof, H., Brox, T., Frahm, J.-M. (eds.) ECCV 2020. LNCS, vol. 12352, pp. 645–666. Springer, Cham (2020). https://doi.org/10.1007/978-3-030-58571-6_38
47. Sun, B., Li, B., Cai, S., Yuan, Y., Zhang, C.: Fsce: Few-shot object detection via contrastive proposal encoding. In: CVPR (2021)
48. Sun, J., Shen, Z., Wang, Y., Bao, H., Zhou, X.: Loftr: Detector-free local feature matching with transformers. In: CVPR (2021)
49. Sung, F., Yang, Y., Zhang, L., Xiang, T., Torr, P.H., Hospedales, T.M.: Learning to compare: Relation network for few-shot learning. In: CVPR (2018)
50. Tian, Y., Wang, Y., Krishnan, D., Tenenbaum, J.B., Isola, P.: Rethinking few-shot image classification: a good embedding is all you need? In: Vedaldi, A., Bischof, H., Brox, T., Frahm, J.-M. (eds.) ECCV 2020. LNCS, vol. 12359, pp. 266–282. Springer, Cham (2020). https://doi.org/10.1007/978-3-030-58568-6_16
51. Tian, Z., Zhao, H., Shu, M., Yang, Z., Li, R., Jia, J.: Prior guided feature enrichment network for few-shot segmentation. In: TPAMI (2020)
52. Touvron, H., Cord, M., Douze, M., Massa, F., Sablayrolles, A., Jégou, H.: Training data-efficient image transformers & distillation through attention. In: ICML (2021)
53. Vaswani, A., et al.: Attention is all you need. In: NeurIPS (2017)
54. Vinyals, O., Blundell, C., Lillicrap, T., Wierstra, D., et al.: Matching networks for one shot learning. In: NeurIPS (2016)

55. Wang, H., Zhang, X., Hu, Y., Yang, Y., Cao, X., Zhen, X.: Few-shot semantic segmentation with democratic attention networks. In: Vedaldi, A., Bischof, H., Brox, T., Frahm, J.-M. (eds.) ECCV 2020. LNCS, vol. 12358, pp. 730–746. Springer, Cham (2020). https://doi.org/10.1007/978-3-030-58601-0_43

56. Wang, K., Liew, J.H., Zou, Y., Zhou, D., Feng, J.: Panet: Few-shot image semantic segmentation with prototype alignment. In: ICCV (2019)

57. Wang, W., et al.: Pyramid vision transformer: A versatile backbone for dense prediction without convolutions. In: ICCV (2021)

58. Wang, X., Huang, T.E., Darrell, T., Gonzalez, J.E., Yu, F.: Frustratingly simple few-shot object detection. arXiv:2003.06957 (2020)

59. Wang, Y., Chao, W.L., Weinberger, K.Q., van der Maaten, L.: Simpleshot: Revisiting nearest-neighbor classification for few-shot learning. arXiv:1911.04623 (2019)

60. Wang, Y.X., Ramanan, D., Hebert, M.: Meta-learning to detect rare objects. In: ICCV (2019)

61. Wu, J., Liu, S., Huang, D., Wang, Y.: Multi-scale positive sample refinement for few-shot object detection. In: Vedaldi, A., Bischof, H., Brox, T., Frahm, J.-M. (eds.) ECCV 2020. LNCS, vol. 12361, pp. 456–472. Springer, Cham (2020). https://doi.org/10.1007/978-3-030-58517-4_27

62. Wu, Z., Li, Y., Guo, L., Jia, K.: Parn: Position-aware relation networks for few-shot learning. In: ICCV (2019)

63. Xiao, Y., Marlet, R.: Few-shot object detection and viewpoint estimation for objects in the wild. In: Vedaldi, A., Bischof, H., Brox, T., Frahm, J.-M. (eds.) ECCV 2020. LNCS, vol. 12362, pp. 192–210. Springer, Cham (2020). https://doi.org/10.1007/978-3-030-58520-4_12

64. Xie, E., Wang, W., Yu, Z., Anandkumar, A., Alvarez, J.M., Luo, P.: Segformer: Simple and efficient design for semantic segmentation with transformers. In: NeurIPS (2021)

65. Xu, C., et al.: Learning dynamic alignment via meta-filter for few-shot learning. In: CVPR (2021)

66. Yan, X., Chen, Z., Xu, A., Wang, X., Liang, X., Lin, L.: Meta r-cnn: Towards general solver for instance-level low-shot learning. In: ICCV (2019)

67. Yang, B., Liu, C., Li, B., Jiao, J., Ye, Q.: Prototype mixture models for few-shot semantic segmentation. In: Vedaldi, A., Bischof, H., Brox, T., Frahm, J.-M. (eds.) ECCV 2020. LNCS, vol. 12353, pp. 763–778. Springer, Cham (2020). https://doi.org/10.1007/978-3-030-58598-3_45

68. Yang, Z., Wang, Y., Chen, X., Liu, J., Qiao, Y.: Context-transformer: tackling object confusion for few-shot detection. In: AAAI (2020)

69. Ye, H.J., Hu, H., Zhan, D.C., Sha, F.: Few-shot learning via embedding adaptation with set-to-set functions. In: CVPR (2020)

70. Zhang, C., Cai, Y., Lin, G., Shen, C.: Deepemd: Few-shot image classification with differentiable earth mover's distance and structured classifiers. In: CVPR (2020)

71. Zhang, C., Lin, G., Liu, F., Guo, J., Wu, Q., Yao, R.: Pyramid graph networks with connection attentions for region-based one-shot semantic segmentation. In: ICCV (2019)

72. Zhang, C., Lin, G., Liu, F., Yao, R., Shen, C.: CANet: Class-agnostic segmentation networks with iterative refinement and attentive few-shot learning. In: CVPR (2019)

73. Zhang, D., Zhang, H., Tang, J., Wang, M., Hua, X., Sun, Q.: Feature pyramid transformer. In: Vedaldi, A., Bischof, H., Brox, T., Frahm, J.-M. (eds.) ECCV 2020. LNCS, vol. 12373, pp. 323–339. Springer, Cham (2020). https://doi.org/10.1007/978-3-030-58604-1_20

74. Zhang, X., Wei, Y., Yang, Y., Huang, T.S.: SG-one: Similarity guidance network for one-shot semantic segmentation. IEEE Trans. Cybern. **50**, 3855–3865 (2020)

75. Zheng, S., et al.: Rethinking semantic segmentation from a sequence-to-sequence perspective with transformers. In: CVPR (2021)

76. Zhengyu, C., Jixie, G., Heshen, Z., Siteng, H., Donglin, W.: Pareto self-supervised training for few-shot learning. In: CVPR (2021)

77. Zhiqiang, S., Zechun, L., Jie, Q., Marios, S., Kwang-Ting, C.: Partial is better than all:revisiting fine-tuning strategy for few-shot learning. In: AAAI (2021)
78. Zhong, Z., Zheng, L., Kang, G., Li, S., Yang, Y.: Random erasing data augmentation. In: AAAI (2020)
79. Zhu, X., Su, W., Lu, L., Li, B., Wang, X., Dai, J.: Deformable detr: Deformable transformers for end-to-end object detection. In: ICLR (2020)

Adversarial Feature Augmentation for Cross-domain Few-Shot Classification

Yanxu Hu[1] and Andy J. Ma[1,2,3](✉) (iD)

[1] School of Computer Science and Engineering, Sun Yat-sen University,
Guangzhou, China
huyx69@mail2.sysu.edu.cn, majh8@mail.sysu.edu.cn
[2] Guangdong Province Key Laboratory of Information Security Technology,
Guangzhou, China
[3] Key Laboratory of Machine Intelligence and Advanced Computing,
Guangzhou, China

Abstract. Few-shot classification is a promising approach to solving the problem of classifying novel classes with only limited annotated data for training. Existing methods based on meta-learning predict novel-class labels for (target domain) testing tasks via meta knowledge learned from (source domain) training tasks of base classes. However, most existing works may fail to generalize to novel classes due to the probably large domain discrepancy across domains. To address this issue, we propose a novel adversarial feature augmentation (AFA) method to bridge the domain gap in few-shot learning. The feature augmentation is designed to simulate distribution variations by maximizing the domain discrepancy. During adversarial training, the domain discriminator is learned by distinguishing the augmented features (unseen domain) from the original ones (seen domain), while the domain discrepancy is minimized to obtain the optimal feature encoder. The proposed method is a plug-and-play module that can be easily integrated into existing few-shot learning methods based on meta-learning. Extensive experiments on nine datasets demonstrate the superiority of our method for cross-domain few-shot classification compared with the state of the art. Code is available at https://github.com/youthhoo/AFA_For_Few_shot_learning.

Keywords: Few-shot classification · Domain adaptation · Adversarial learning · Meta-learning

1 Introduction

The development of deep convolutional neural networks (DCNNs) has achieved great success in image/video classification [16,22,39,47]. The impressive performance improvement relies on the continuously upgrading computing devices

Supplementary Information The online version contains supplementary material available at https://doi.org/10.1007/978-3-031-20044-1_2.

and manual annotations of large-scale datasets. To ease the heavy annotation burdens for training DCNNs, few-shot classification [21] has been proposed to recognize instances from novel classes with only limited labeled samples. Among various recent methods to address the few-shot learning problem, the meta-learning approach [8,10,24,34,36,38,42,46] have received a lot of attention due to its effectiveness. In general, meta-learning divides the training data into a series of tasks and learns an inductive distribution bias of these tasks to alleviate the negative impact of the imbalance between base and novel classes.

Meta-learning is good at generalizing the base-class model to novel classes under the condition that the training distribution of base classes is almost equal to the testing one of novel classes. Nevertheless, when the distributions of the training (source domain) and the testing (target domain) data differ from each other, the performance of the meta-learning model will degrade as justified by existing works [5,15]. Figure 1 illustrates the domain shift problem in which the target dataset (e.g. CUB) is different from the source domain (e.g. mini-ImageNet). In this scenario, the distribution of the target domain features extracted by the encoder E may greatly deviate from the source domain distribution.

With the distribution misalignment, the class discriminator D_c cannot make a correct decision for classifying novel-class data. Domain adaptation (DA) [43] can learn domain-invariant features by adversarial training [12] to bridge the domain gap. While DA assumes a lot of unlabelled samples are available in the target domain for training, the domain generalization (DG) approach [23] can generalize from source domains to target domain without accessing the target data. Differently, in few-shot learning, novel classes in the target domain do not overlap with base classes in the source domain and only very limited number of training samples are available for each class. As a result, existing DA methods are not applicable for cross-domain few-shot classification.

To mitigate the domain shift under the few-shot setting, the adversarial task augmentation (ATA) method [44] is proposed to search for the worst-case problem around the source task distribution. While the task augmentation lacks of the capacity of simulating various feature distributions across domains, the feature-wise transformation (FT) [40] is designed for feature augmentation using affine transforms. With multiple source domains for training, the hyper-parameters in the FT are optimized to capture variations of the feature distributions. When there is only single source domain, these hyper-parameters are empirically determined for training. Though the FT achieves convincing performance improvement for both the base and novel classes, the empirical setting of hyper-parameters in the FT is sub-optimal. Consequently, it cannot fully imitate the distribution mismatch under single source domain adaptation.

To overcome the limitations in existing works, we propose a novel adversarial feature augmentation (AFA) method for domain-invariant feature learning in cross-domain few-shot classification. Different from the ATA, our method performs data augmentation in features (instead of tasks) to *simulate the feature distribution mismatch across domains*. Unlike the FT using multiple source domains

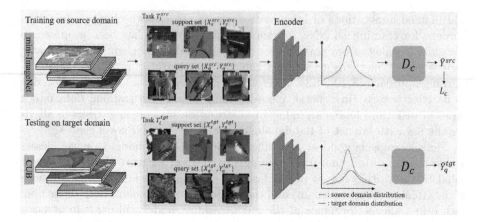

Fig. 1. Feature distribution misalignment problem in cross-domain few-shot classification. In meta-learning methods, it consists of an feature encoder E and a prediction head D_c. There may be domain shift between the training (source domain) data of base classes and testing (target domain) data of novel classes. In this case, the distribution of the features extracted by the source domain encoder (blue) differs from the target domain features (red). Due to the distribution misalignment, *the meta-learned prediction head may not be able to correctly classify samples of novel classes from the target domain*. Moreover, *the feature distribution in the target domain can hardly be estimated due to the limited number of novel-class sample*. In this paper, we propose a novel **adversarial feature augmentation (AFA)** method to learn domain-invariant features for cross-domain few-shot classification. (Color figure online)

to determine the optimal solution, the proposed AFA aligns the cross-domain feature distributions by *adversarial learning based on single source domain*.

In our method, we design a feature augmentation module to transform the features extracted by the encoder E according to sufficient statistics of normal distribution. Considering the original and the augmented features as two different domains (seen and unseen respectively), the feature augmentation module is trained by maximizing the domain discrepancy across domains. Moreover, the feature augmentation module is inserted into multiple layers of the encoder, such that the difference between the distributions of the seen and unseen domains are enlarged. The distance between the gram matrices of multi-layer features from seen and unseen domains is used to measure the domain discrepancy. During domain adversarial training, both the feature augmentation module and the domain discriminator is trained to distinguish the seen domain from the unseen one, while the encoder is learned by confusing the two different domains.

In summary, the contributions of this work are in three folds:

1. We propose a model-agnostic feature augmentation module based on sufficient statistics of normal distribution. The feature augmentation module can generate various feature distributions to better simulate the domain gap by maximizing the domain discrepancy under the cross-domain few-shot setting.

2. We develop a novel adversarial feature augmentation (AFA) method for distribution alignment without accessing to the target domain data. During adversarial training, the domain discriminator is learned by recognizing the augmented features (unseen domain) from the original ones (seen domain). At the same time, the domain discrepancy is maximized to train the feature augmentation module, while it is minimized to obtain the optimal feature encoder. In this way, the domain gap is reduced under the few-shot setting.

3. The proposed AFA is a plug-and-play module which can be easily integrated into existing few-shot learning methods based on meta-learning including matching net (MN) [42], graph neural network (GNN) [31], transductive propagation network (TPN) [25], and so on. We experimentally evaluate the performance on the proposed method combined with the MN, GNN and TPN under the cross-domain few-shot setting. Experimental results demonstrate that our method can improve the classification performance over the few-shot learning baselines and outperform the state-of-the-art cross-domain few-shot classification methods in most cases.

2 Related Work

Few-Shot Classification. Few-shot classification [7,10,15,24,25,46] aims to recognize novel classes objects with few labeled training samples. MatchingNet [42] augments neural networks with external memories via LSTM module and maps a few labelled support samples and an unlabelled query samples to its label, while GNN [31] assimilates generic message-passing inference algorithms with their neural-network counterparts to interact the information between the labelled data and unlabelled data by graph. TPN [25] learns a graph construction module that exploits the manifold structure in the data to propagate labels from labeled support images to unlabelled query instances, which can well alleviate the few-shot classification problem. However, these meta-learning methods fail to generalize to target domains since the distribution of image features may vary largely due to the domain shift. Our work improves the generalization ability of the meta-learning model with the proposed adversarial feature augmentation (ATA) to better recognize target domain samples.

Domain Adaptation. Existing domain adaptation (DA) methods can be divided into three categories, i.e., discrepancy-based [26,50], reconstruction-based approaches [6,13] and adversarial-based [11,12,19,41]. For the discrepancy-based methods, DAN [26] measures the distance between the distribution of source and target domain and the domain discrepancy is further reduced using an optimal multi-kernel selection method for mean embedding matching. The reconstruction-based method DRCN [13] proposes a constructor to reconstruct target domain data, the more similar between the original data and constructed data, the more effective the feature learned by encoder are. While the adversarial-based method DANN [12] learns domain-variance features by adversarial progress between encoder and domain discriminators. Nevertheless, these DA methods take the unlabelled data in the target domain as inputs for training, which is unavailable in the training stage under cross-domain few-shot classification setting.

Adversarial Training. Adversarial training [14,27,32] is a powerful training module to improve the robustness of deep neural networks. To the end, Madry et al. [27] develop projected gradient descent as a universal "first-order adversary" and use it to train model in adversarial way. Sinha et al. [33] provide a training procedure that updates model parameters with worst-case perturbations of training data to perturb the data distribution, which has been referred by ATA [44] to generate virtual 'challenging' tasks to improve the robustness of models. In this work, we generate the "bad-case perturbations" in feature level via adversarial feature augmentation, which can simulate various feature distributions, to improve the generalization ability of various meta-learning methods.

Cross-Domain Few-Shot Classification. Different from the few-shot domain adaptation works [30,49], the unlabelled data from target domain isn't used for training and the categories vary from training set to the testing set in cross-domain few-shot classification (CDFSC) problems. Compared to the few-shot classification, in the CDFSC, base classes are not share the same domain with novel classes. To improve the generalization of meta-learning methods, LRP [37] develops a explanation-guided training strategy that emphasizes the features which are important for the predictions. While Wang et al. [44] focus on elevating the robustness of various inductive bias of training data via enlarging the task distribution space. And Feature Transformation [40] try to improve generalization to the target domain of metric-base meta-learning methods through modelling the various different distribution with feature-wise transformation layer. Compared to the above methods, The CNAPs-based approaches [1–4,29] developed from different perspectives, which is proposed based on Feature-wise Linear Modulation (FiLM) for efficient adaptation to new task at test time. Different from their approaches, we aim to simulate the various distributions in the feature-level with adversarial training and take it as feature augmentation to learn an encoder for extracting domain-invariant features.

3   Proposed Method

In this section, the preliminaries and the overall network architecture of our method are first introduced. Then, the feature augmentation module and the adversarial training process are presented.

3.1   Preliminaries

Following the few-shot classification setting [28], the novel categories in the testing stage C_{test} are different from base classes C_{train} used in the training stage, i.e., $C_{train} \cap C_{test} = \emptyset$, then data for training and testing is divided into a series of tasks T. Each task contains a support set T_s and a query set T_q. In a n-way

k-shot support set T_s, the number of categories and labelled samples of each category is n and k, respectively. The query set T_q consists of the samples sharing the same classes as in T_s. During meta-learning, the training process on T_s and the testing process on T_q are called meta-training and meta-testing. For each task, the goal is to correctly classify samples from T_q by learning from T_s.

With the domain shift problem in cross-domain few-shot classification, the training dataset (e.g. mini-ImageNet) is different from the testing data (e.g. CUB [45] or Cars [20]). In this work, we focus on adapting the meta model from single source domain to various target domains. In other words, only one source domain dataset is used for training while testing can be performed on different datasets. Notice that T_s and T_q of each task is from the same domain. Since the labelled data from the target domain is very limited and the target novel classes are not overlapped with the source base classes, we propose to augment the features of each task by adversarial training to bridge the domain gap.

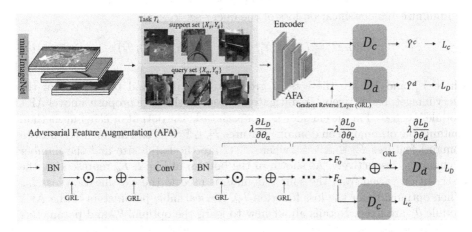

Fig. 2. Top: Network Architecture. The network architecture of our method consists of a feature encoder E, a class discriminator and a domain discriminator D_d. In the feature encoder E, a novel adversarial feature augmentation (AFA) module is embedded after each batch normalization layer. **Bottom: Adversarial Feature Augmentation.** The AFA module generates augmented features F_a (unseen domain) to simulate distribution variations. By using adversarial training through inserting the gradient reverse layer (GRL) into the AFA module, the discrepancy between the distributions of F_a and original features F_o (seen domain) is maximized. At the same time, the domain discriminator D_d is learned by distinguishing the seen domain from the unseen one, while the discrepancy E is minimized to obtain the optimal feature encoder. Parameters of the AFA module, D_d and E are denoted as θ_a, θ_e and θ_d, respectively.

3.2   Network Architecture

As shown in Fig. 2, the network architecture of the proposed method contains a feature encoder E and a class discriminator D_c similar to meta-learning models. Different from the traditional feature encoder, a novel adversarial feature augmentation (AFA) module is embedded after each batch normalization layer to simulate various feature distributions with details introduced in Sect. 3.3 & 3.4. With the augmented features (unseen domain), a domain discriminator D_d is trained to distinguish the unseen domain from the seen one (original features).

The training procedures of our method follows the meta-learning approach to learn the inductive bias over the feature distribution from a series of tasks. By doing this, a class discriminator D_c is learned and transferred to target tasks in the testing stage. For meta-training in each task, the base learner $\mathcal{B}$ outputs the optimal class discriminator D_c based on the support set T_s and the feature encoder E, i.e., $D_c = \mathcal{B}(E(T_s; \theta_e); \theta_c)$, where θ_c, θ_e denote the learnable parameters of D_c and E, respectively. During meta-testing, the objective function is to minimize the classification loss of the query set T_q, i.e.,

$$\min_{\theta_c, \theta_e} L_c = L_c(Y_q^c, \hat{Y}_q^c), \hat{Y}_q^c = D_c(E(T_q; \theta_e); \theta_c) \tag{1}$$

where Y_q^c and $\hat{Y}_q^c$ are the sets of ground-truth labels and predictions of the query images, respectively. To mitigate the domain shift, we propose a novel AFA module integrated in the encoder E. For each task, the output of E in our method contains the original (seen domain) features $F_o \in \mathbb{R}^{N \times C}$ and augmented (unseen domain) features $F_a \in \mathbb{R}^{N \times C}$, where N, C are the batch size and the number of channels, respectively. As shown in the bottom of Fig. 2, F_a representing the distribution varied from the source domain is used in the classification loss L_c. When optimizing for the loss function L_c, the learnable parameters of the AFA module θ_a are fixed. Details about how to learn the optimal θ_a and parameters θ_d of the domain discriminator D_d are given in the following two subsections.

3.3   Feature Augmentation

To simulate various feature distributions, we design the feature augmentation function via disturbing the sufficient statistics of the original (seen domain) feature distribution. Given a specified mean and variance, normal distribution best represents the current state of knowledge with maximum entropy. As a results, we assume the feature maps in a training batch follows multivariate normal distribution and is independent and identically distributed. Denote f as any element in the feature map and $f_1, \ldots, f_N$ as the corresponding observations in a batch. Since the marginal distribution of multivariate normal distribution is still normal, the probability density of a batch of f in can be estimated by,

$$p(f) = \prod_{i=1}^{N} \frac{1}{\sqrt{2\pi}\sigma} \exp\left(-\frac{(f_i - \mu)^2}{2\sigma^2}\right) \tag{2}$$

where μ, σ are the mean and variance of f. Then the probability density function can be decomposed into the part that is relevant to the overall distribution and is independent of the overall distribution, By simplifying the product in right-hand side of Eq. (2), we have

$$p(f) = (2\pi\sigma)^{-\frac{N}{2}} \exp\left(-\frac{1}{2\sigma^2}\sum_i (f_i^2 - 2\mu f_i + \mu^2)\right) \tag{3}$$

By Eq. (3), the probability density $p(f)$ can be decomposed into the form of a factor which does not depend on the distribution parameters μ, σ multiplied by the other factor depending on μ, σ and statistics $\sum_i f_i^2, \sum_i f_i$. According to the Fisher-Neyman factorization theorem [9,35], $\sum_i f_i^2$ and $\sum_i f_i$ are sufficient statistics of normal distribution. Moreover, the mean and variance are also sufficient statistics of normal distribution, because the statistics $\sum_i f_i^2$ and $\sum_i f_i$ can be calculated by them. The sufficient statistics of feature distribution include all the information of the distribution. Thus, we propose to simulate various feature distributions by disturbing the mean and variance of the original features.

For this purpose, we insert a linear perturbation function with learnable parameters after each batch normalization layer. Denote the original intermediate features from a certain batch normalization layer as $m^o \in \mathbb{R}^{C \times H \times W}$, where H, W are the spatial resolutions of the feature map. We initialize the scaling parameter $\gamma \in \mathbb{R}^C$ (for variance perturbation) and bias term $\beta \in \mathbb{R}^C$ (for mean disturbance) by normal distribution similar to [40]. Then, the augmented feature $m^a \in \mathbb{R}^{C \times H \times W}$ is computed by,

$$m^a_{c,h,w} = \gamma_c \times m^o_{c,h,w} + \beta_c \tag{4}$$

The learnable parameters γ, β are optimized by adversarial training which will be elaborated in the next subsection.

3.4 Adversarial Feature Augmentation

The augmentation module would not explore the distribution space that would enable the encoder to better handle tasks from a completely different domain, if γ and β is directly learned by solving the optimization problem in Eq. (1). In this case, the ATA module fails to simulate the domain variations for domain-invariant feature learning. In our method, we optimize the parameters in the AFA module, the domain discriminator D_d and the feature encoder E by adversarial training. Let us consider the original features F_o and the augmented features F_a as the seen and unseen domain, respectively. Denote the domain label of the input features as y_i^d. If the input features are from the seen domain, then $y_i^d = 0$. Otherwise, $y_i^d = 1$. As in the DANN [12], the domain discriminator $D_d(\cdot) : \mathbb{R}^C \to [0,1]$ is defined as a logistic regressor, i.e.,

$$\hat{y}^d = D_d(F; \boldsymbol{\mu}, b) = h(\boldsymbol{\mu}^T F + b), \boldsymbol{\mu} \in \mathbb{R}^C, F = F_o \text{ or } F_a \tag{5}$$

where $h(\cdot)$ is the sigmoid function, μ, b are learnable parameters in D_d, and $\hat{Y}^d$ is the predicted labeled of the original features or the augmented features. Then, the loss function given by cross-entropy is,

$$L_d(\hat{Y}^d, Y^d) = \frac{1}{2N} \sum_i [-y_i^d \log(\hat{y}_i^d) - (1 - y_i^d) \log(1 - \hat{y}_i^d)], 1 \le i \le 2N \quad (6)$$

where $Y^d, \hat{Y}^d$ are the sets of all the ground-truth and predicted domain labels. The domain discriminator can be trained by minimizing the loss function Eq. (6) to distinguish between the seen and unseen domains.

Besides the domain similarity measured by the final output features from the encoder in Eq. (6), we also measure the domain discrepancy of the AFA module inserted after each batch normalization layer. The gram matrices representing domain information of m_o and m_a are calculated as follows,

$$\hat{m} = Flatten(m), \hat{m} \in \mathbb{R}^{C \times S}, S = HW \quad (7)$$

$$G(m) = \hat{m} \times \hat{m}^T, G(m) \in \mathbb{R}^{C \times C}, m = m_o \text{ or } m_a \quad (8)$$

Then, the domain discrepancy between the intermediate features m_o and m_a is determined by the distance between $G(m_o)$ and $G(m_a)$, i.e.,

$$L_g = \frac{1}{4S^2 C^2} \sum_{i,j} (G_{i,j}(m_a) - G_{i,j}(m_o))^2 \quad (9)$$

By maximizing the gram-matrix loss L_g, the AFA module is trained to ensure that the augmented intermediate features in each layer are different from the original ones to better mimic the target feature distribution.

For adversarial training, the domain similarity loss L_d is maximized and the domain discrepancy loss L_g is minimized to learn the feature encoder E for distribution alignment. In summary, the optimization problem for adversarial training is given as follows,

$$\max_{\theta_e} \min_{\theta_d, \theta_a} L_D = L_d - L_g \quad (10)$$

The min-max optimization problem in Eq. (10) can be solved as gradient reverse layers (GRLs) introudced in the DANN [12], which reverse the gradients during back propagation. As shown in the bottom of Fig. 2, the gradients of the domain discriminator D_d, the encoder E and the AFA module are updated by $\lambda \partial L_D / \partial \theta_d$, $-\lambda \partial L_D / \partial \theta_e$ and $\lambda \partial L_D / \partial \theta_a$, respectively, where λ is a hyperparameter set empirically as in the DANN.

Comparing to FT [40]. Both the feature-wise transformation (FT) [40] and our method aim at transforming image features to simulate various feature distributions. Our method takes full advantages of the original and augmented features to explicitly bridge the domain gap by adversarial training. Thus, the distribution variations can be imitated by using only single source domain for training. Nevertheless, FT relies on multiple source domains to learn the optimal feature transformation parameters. Under the single source domain setting, the transformation parameters are set as constants, such that FT may suffer from the problem of performance drop as shown in our experiments.

Comparing to ATA [44]. The adversarial task augmentation (ATA) method employs adversarial training to search for the worst-case tasks around the source task distribution. In this way, the space of the source task distribution could be enlarged, so that it may be closer to the task distribution in the target domain. Nevertheless, the perturbation on source tasks would degrade the performance on the unseen classes of source domain compared to other competitive models. Different from task augmentation, we propose feature augmentation with adversarial training via the gradient reverse layers to learn domain-invariant features without the problem of performance degradation. Moreover, the ATA may not be able to fully utilize the available information in which only one of the generated tasks or the original tasks is used for training. In our method, both the original and augmented feature are used to train the domain discriminator. At the same time, the proposed gram-matrix loss helps to generate unseen augmented features through maximizing the difference compared to the original features. In addition, ATA is more computational expensive to find the worst-case tasks via gradient ascents as shown in the complexity comparison in the supplementary.

4   Experiment

4.1   Implementation

In this section, we evaluate the proposed adversarial feature augmentation (AFA) module inserted into the Matching Network (MN) [42], Graph Neural Network (GNN) [31] and Transductive Propagation Network (TPN) [25]. We compare our method with the feature-wise transformation (FT) [40], explanation-guide training (LRP) [37] and Adversarial Task augmentation (ATA) [44].

4.2 Experimental Setting

Datasets. In this work, nine publicly available benchmarks are used for experiments, i.e., mini-ImageNet [42], CUB [45], Cars [20], Places [51], Plantae [18], CropDiseases, EuroSAT, ISIC and ChestX. Following the experimental setting of previous works [40,44], we split these datasets into train/val/test sets, which is further divided into k-shot-n-class support sets and the same n-class query sets. We use the mini-ImageNet dataset as the source domain, and select the models with best accuracy on the validation set of mini-ImageNet for testing.

Implementation Details. Our model can be integrated into existing meta-learning methods, e.g., MN [42], GNN [31], TPN [25]. In these methods, we use the ResNet-10 [17] with the proposed AFA module as the feature encoder. The scaling term $\gamma \sim N(1, softplus(0.5))$ and bias term $\beta \sim N(0, softplus(0.3))$ are sampled from normal distribution for initialization. To ensure fair comparison with the FT [40], LRP [37], ATA [44] and the baseline methods. We follow the training protocol from [5]. Empirically, the proposed model is trained with the learning rate 0.001 and 40,000 iterations. The performance measure is the average of the 2000 trials with randomly sampled batches. There are 16 query samples and 5-way 5-shot/1-shot support samples for each trial.

Pre-trained Feature Encoder. Before the few-shot training stage, we apply an additional pre-training strategy as in FT [40], LRP [37] and ATA [44] for fair comparison. The pre-trained feature encoder is minimized by the standard cross-entropy classification loss on the 64 training categories (the same as the training categories in few-shot training) in the mini-ImageNet dataset.

4.3 Results on Benchmarks

We train each model using the mini-ImageNet as the source domain and evaluate the model on the other eight target domains, i.e., CUB, Cars, Places, Plantae, CropDiseases, EuroSAT, ISIC and ChestX. In our method, the AFA module is inserted after each batch normalization layer of the feature encoder during the training stage. All the results are shown in Table 1. We have following observations from these results: **i.** Our method outperforms the state of the art for almost all the datasets and different-shot settings in different meta-learning methods. For 1-shot classification, our method improves the baselines by 3.45% averagely over the eight datasets in different models. In 5-shot setting, the average improvement is 4.25% compared to the baselines. **ii.** Compared to the competitive ATA [44], our method integrated with the proposed AFA achieves an average improvement of about 1%.

Table 1. Few-shot classification accuracy (%) of 5-way 5-shot/1-shot setting trained on the mini-ImageNet dataset, and tested on various datasets from target domains. The best results in different settings are in **Bold**.

Method/shot	CUB		Cars		Places		Planae	
	1-shot	5-shot	1-shot	5-shot	1-shot	5-shot	1-shot	5-shot
MN [42]	$35.89_{\pm0.5}$	$51.37_{\pm0.8}$	$30.77_{\pm0.5}$	$38.99_{\pm0.6}$	$49.86_{\pm0.8}$	$63.16_{\pm0.8}$	$32.70_{\pm0.6}$	$46.53_{\pm0.7}$
w/ FT [40]	$36.61_{\pm0.5}$	$55.23_{\pm0.8}$	$29.82_{\pm0.4}$	$41.24_{\pm0.7}$	$51.07_{\pm0.7}$	$64.55_{\pm0.8}$	$34.48_{\pm0.5}$	$41.69_{\pm0.6}$
w/ ATA [44]	$39.65_{\pm0.4}$	$57.53_{\pm0.4}$	$32.22_{\pm0.4}$	$45.73_{\pm0.4}$	$53.63_{\pm0.5}$	$67.87_{\pm0.4}$	$36.42_{\pm0.4}$	$51.05_{\pm0.4}$
Ours	$\mathbf{41.02}_{\pm0.4}$	$\mathbf{59.46}_{\pm0.4}$	$\mathbf{33.52}_{\pm0.4}$	$\mathbf{46.13}_{\pm0.4}$	$\mathbf{54.66}_{\pm0.5}$	$\mathbf{68.87}_{\pm0.4}$	$\mathbf{37.60}_{\pm0.4}$	$\mathbf{52.43}_{\pm0.4}$
GNN [31]	$44.40_{\pm0.5}$	$62.87_{\pm0.5}$	$31.72_{\pm0.4}$	$43.70_{\pm0.4}$	$52.42_{\pm0.5}$	$70.91_{\pm0.5}$	$33.60_{\pm0.4}$	$48.51_{\pm0.4}$
w/ FT [40]	$45.50_{\pm0.5}$	$64.97_{\pm0.5}$	$32.25_{\pm0.4}$	$46.19_{\pm0.4}$	$53.44_{\pm0.5}$	$70.70_{\pm0.5}$	$32.56_{\pm0.4}$	$49.66_{\pm0.4}$
w/ LRP [37]	$43.89_{\pm0.5}$	$62.86_{\pm0.5}$	$31.46_{\pm0.4}$	$46.07_{\pm0.4}$	$52.28_{\pm0.5}$	$71.38_{\pm0.5}$	$33.20_{\pm0.4}$	$50.31_{\pm0.4}$
w/ ATA [44]	$45.00_{\pm0.5}$	$66.22_{\pm0.5}$	$33.61_{\pm0.4}$	$49.14_{\pm0.4}$	$53.57_{\pm0.5}$	$75.48_{\pm0.4}$	$34.42_{\pm0.4}$	$52.69_{\pm0.4}$
Ours	$\mathbf{46.86}_{\pm0.5}$	$\mathbf{68.25}_{\pm0.5}$	$\mathbf{34.25}_{\pm0.4}$	$\mathbf{49.28}_{\pm0.5}$	$\mathbf{54.04}_{\pm0.6}$	$\mathbf{76.21}_{\pm0.5}$	$\mathbf{36.76}_{\pm0.4}$	$\mathbf{54.26}_{\pm0.4}$
TPN [25]	$48.30_{\pm0.4}$	$63.52_{\pm0.4}$	$32.42_{\pm0.4}$	$44.54_{\pm0.4}$	$56.17_{\pm0.5}$	$71.39_{\pm0.4}$	$37.40_{\pm0.4}$	$50.96_{\pm0.4}$
w/ FT [40]	$44.24_{\pm0.5}$	$58.18_{\pm0.5}$	$26.50_{\pm0.3}$	$34.03_{\pm0.4}$	$52.45_{\pm0.5}$	$66.75_{\pm0.5}$	$32.46_{\pm0.4}$	$43.20_{\pm0.5}$
w/ ATA [44]	$50.26_{\pm0.5}$	$65.31_{\pm0.4}$	$34.18_{\pm0.4}$	$46.95_{\pm0.4}$	$57.03_{\pm0.5}$	$72.12_{\pm0.4}$	$39.83_{\pm0.4}$	$55.08_{\pm0.4}$
Ours	$\mathbf{50.85}_{\pm0.4}$	$\mathbf{65.86}_{\pm0.4}$	$\mathbf{38.43}_{\pm0.4}$	$\mathbf{47.89}_{\pm0.4}$	$\mathbf{60.29}_{\pm0.5}$	$\mathbf{72.81}_{\pm0.4}$	$\mathbf{40.27}_{\pm0.4}$	$\mathbf{55.67}_{\pm0.4}$

	CropDiseases		EuroSAT		ISIC		ChestX	
	1-shot	5-shot	1-shot	5-shot	1-shot	5-shot	1-shot	5-shot
MN [42]	$57.57_{\pm0.5}$	$73.26_{\pm0.5}$	$54.19_{\pm0.5}$	$67.50_{\pm0.5}$	$29.62_{\pm0.3}$	$32.98_{\pm0.3}$	$22.30_{\pm0.2}$	$22.85_{\pm0.2}$
w/ FT [40]	$54.21_{\pm0.5}$	$70.56_{\pm0.5}$	$55.62_{\pm0.4}$	$63.33_{\pm0.5}$	$30.64_{\pm0.3}$	$35.73_{\pm0.3}$	$21.50_{\pm0.2}$	$22.88_{\pm0.2}$
w/ ATA [44]	$55.57_{\pm0.5}$	$79.28_{\pm0.4}$	$56.44_{\pm0.5}$	$68.83_{\pm0.4}$	$31.48_{\pm0.3}$	$40.53_{\pm0.3}$	$21.52_{\pm0.2}$	$\mathbf{23.19}_{\pm0.2}$
Ours	$\mathbf{60.71}_{\pm0.4}$	$\mathbf{80.07}_{\pm0.4}$	$\mathbf{61.28}_{\pm0.5}$	$\mathbf{69.63}_{\pm0.5}$	$\mathbf{32.32}_{\pm0.3}$	$39.88_{\pm0.3}$	$\mathbf{22.11}_{\pm0.2}$	$23.18_{\pm0.2}$
GNN [31]	$59.19_{\pm0.5}$	$83.12_{\pm0.4}$	$54.61_{\pm0.5}$	$78.69_{\pm0.4}$	$30.14_{\pm0.3}$	$42.54_{\pm0.4}$	$21.94_{\pm0.2}$	$23.87_{\pm0.2}$
w/ FT [40]	$60.74_{\pm0.5}$	$87.07_{\pm0.4}$	$55.53_{\pm0.5}$	$78.02_{\pm0.4}$	$30.22_{\pm0.3}$	$40.87_{\pm0.4}$	$22.00_{\pm0.2}$	$24.28_{\pm0.2}$
w/ LRP [37]	$59.23_{\pm0.5}$	$86.15_{\pm0.4}$	$54.99_{\pm0.5}$	$77.14_{\pm0.4}$	$30.94_{\pm0.3}$	$44.14_{\pm0.4}$	$22.11_{\pm0.2}$	$24.53_{\pm0.3}$
w/ ATA [44]	$67.45_{\pm0.5}$	$\mathbf{90.59}_{\pm0.3}$	$61.35_{\pm0.5}$	$83.75_{\pm0.4}$	$\mathbf{33.21}_{\pm0.4}$	$44.91_{\pm0.4}$	$22.10_{\pm0.2}$	$24.32_{\pm0.4}$
Ours	$\mathbf{67.61}_{\pm0.5}$	$88.06_{\pm0.3}$	$\mathbf{63.12}_{\pm0.5}$	$\mathbf{85.58}_{\pm0.4}$	$\mathbf{33.21}_{\pm0.3}$	$\mathbf{46.01}_{\pm0.4}$	$\mathbf{22.92}_{\pm0.2}$	$\mathbf{25.02}_{\pm0.2}$
TPN [25]	$68.39_{\pm0.6}$	$81.91_{\pm0.5}$	$63.90_{\pm0.5}$	$77.22_{\pm0.4}$	$35.08_{\pm0.4}$	$45.66_{\pm0.3}$	$21.05_{\pm0.2}$	$22.17_{\pm0.2}$
w/ FT [40]	$56.06_{\pm0.7}$	$70.06_{\pm0.7}$	$52.68_{\pm0.6}$	$65.69_{\pm0.5}$	$29.62_{\pm0.3}$	$36.96_{\pm0.4}$	$20.46_{\pm0.1}$	$21.22_{\pm0.1}$
w/ ATA [44]	$\mathbf{77.82}_{\pm0.5}$	$\mathbf{88.15}_{\pm0.5}$	$65.94_{\pm0.5}$	$79.47_{\pm0.3}$	$\mathbf{34.70}_{\pm0.4}$	$45.83_{\pm0.3}$	$21.67_{\pm0.2}$	$\mathbf{23.60}_{\pm0.2}$
Ours	$72.44_{\pm0.6}$	$85.69_{\pm0.4}$	$\mathbf{66.17}_{\pm0.4}$	$\mathbf{80.12}_{\pm0.4}$	$34.25_{\pm0.4}$	$\mathbf{46.29}_{\pm0.3}$	$\mathbf{21.69}_{\pm0.1}$	$23.47_{\pm0.2}$

4.4 Ablation Experiments

Effect of the Domain Discriminator. As mentioned in Sect. 3.1, we apply the domain discriminator to maximize the discrepancy between the augmented features and original features. In this experiment, we perform ablation experiments of the domain discriminator through training the AFA via the classification loss function L_c but without using the domain discriminator. The classification accuracy on various datasets are reported in the second line of Table 2. Based on the results, we have the following observations: *i.* When using the AFA without the domain discriminator, the performance degrades. This indicates that it can improve the performance on various datasets under the settings with different number of shots by training with the domain discriminator, (e.g. an average 2.48% improvement on the 1-shot setting). *ii.* Compared to the baseline (MN), the AFA without the domain discriminator also help to generalize to various domains in most cases. These results demonstrate that the adversarial training can alleviate the antagonistic action between the class discriminator and the feature augmentation module. **iii.** Although training with the AFA module with the classification loss can improve the performance in most datasets, it leads to a decline in the CropDiseases dataset comparing with the baseline.

Table 2. Accuracy (%) of ablation experiments under 1-shot/5-shot 5-way few-shot classification on the target domains datasets. **w/o** D_d is the experiment without the domain discriminator. **w/o** L_g is the ablation experiment of the gram-matric loss calculated by Eq. (7). **Non-linear** indicate that the linear transformation of adversarial feature augmentation is replaced by convolution layer as the non-linear transformation. Here we use the matching network (MN) as the baseline for experiments.

Method/shot	CUB		Cars		Places		Planae	
	1-shot	5-shot	1-shot	5-shot	1-shot	5-shot	1-shot	5-shot
MN [42]	$35.89_{\pm0.5}$	$51.37_{\pm0.8}$	$30.77_{\pm0.5}$	$38.99_{\pm0.6}$	$49.86_{\pm0.8}$	$63.16_{\pm0.8}$	$32.70_{\pm0.6}$	$46.53_{\pm0.7}$
w/o D_d	$38.83_{\pm0.4}$	$58.06_{\pm0.4}$	$32.35_{\pm0.4}$	$45.92_{\pm0.4}$	$51.46_{\pm0.5}$	$65.45_{\pm0.4}$	$36.80_{\pm0.4}$	$49.00_{\pm0.4}$
w/o L_g	$40.49_{\pm0.4}$	$56.44_{\pm0.4}$	$31.08_{\pm0.3}$	$44.78_{\pm0.4}$	$51.98_{\pm0.5}$	$66.60_{\pm0.4}$	$35.03_{\pm0.4}$	$50.56_{\pm0.4}$
Non-linear	$34.42_{\pm0.4}$	$50.17_{\pm0.4}$	$28.77_{\pm0.3}$	$42.04_{\pm0.4}$	$49.92_{\pm0.4}$	$59.00_{\pm0.4}$	$34.27_{\pm0.4}$	$50.90_{\pm0.4}$
Ours	$\mathbf{41.02}_{\pm0.4}$	$\mathbf{59.46}_{\pm0.4}$	$\mathbf{33.52}_{\pm0.4}$	$\mathbf{46.13}_{\pm0.4}$	$\mathbf{54.66}_{\pm0.5}$	$\mathbf{68.87}_{\pm0.4}$	$\mathbf{37.60}_{\pm0.4}$	$\mathbf{52.43}_{\pm0.4}$
	CropDiseases		EuroSAT		ISIC		ChestX	
	1-shot	5-shot	1-shot	5-shot	1-shot	5-shot	1-shot	5-shot
MN [42]	$57.57_{\pm0.5}$	$73.26_{\pm0.5}$	$54.19_{\pm0.5}$	$67.50_{\pm0.5}$	$29.62_{\pm0.3}$	$32.98_{\pm0.3}$	$22.30_{\pm0.2}$	$22.85_{\pm0.2}$
w/o D_d	$56.87_{\pm0.5}$	$71.02_{\pm0.5}$	$54.78_{\pm0.5}$	$67.66_{\pm0.4}$	$30.19_{\pm0.3}$	$38.83_{\pm0.3}$	$21.70_{\pm0.2}$	$22.86_{\pm0.2}$
w/o L_g	$55.43_{\pm0.5}$	$74.62_{\pm0.5}$	$57.41_{\pm0.5}$	$65.50_{\pm0.4}$	$30.78_{\pm0.3}$	$36.87_{\pm0.3}$	$21.23_{\pm0.2}$	$22.96_{\pm0.2}$
Non-linear	$56.01_{\pm0.5}$	$79.95_{\pm0.4}$	$56.90_{\pm0.5}$	$\mathbf{72.15}_{\pm0.4}$	$29.19_{\pm0.3}$	$\mathbf{40.20}_{\pm0.3}$	$20.95_{\pm0.2}$	$\mathbf{23.58}_{\pm0.2}$
Ours	$\mathbf{60.71}_{\pm0.5}$	$\mathbf{80.07}_{\pm0.4}$	$\mathbf{61.28}_{\pm0.5}$	$69.63_{\pm0.5}$	$\mathbf{32.32}_{\pm0.3}$	$39.88_{\pm0.3}$	$\mathbf{22.11}_{\pm0.2}$	$23.18_{\pm0.2}$

Effect of the Gram-Matrix Loss. The gram-matrix loss function is to measure the difference between augmented and original features in each AFA module. The accuracy on various datasets are reported on the third line of Table 2. Compared with the last line results, we can find that the gram-matrix loss brings about 2.57% improvement. It leads to the best results for novel classes by combining the domain discriminator and the gram-matrix loss. The main reason behind is that these two modules contribute to a complementary improvement on global and local discrepancy between the augmented and original features.

How About Non-Linear Transformation? In Sect. 3.2, we introduce linear perturbation in the AFA module to mimic various feature distributions via disturbing the sufficient statistics of original feature distribution. Here, we replace the linear transformation with the non-linear transformation (convolution) layers to generate unseen feature distribution. The classification accuracy are report in the forth line of Table 2. As we can see, the non-linear transformation cannot bring obvious improvement or even performs worse. It verify the theoretical justification of our method based on sufficient statistic such that the disturbance to the mean and variance is better for generalizing to target domain.

4.5 Results of Base Classes and Novel Classes

Since meta-learning methods may show inconsistent results for *base* and *novel* classes, we report results of both novel and base classes accuracy (%) for comparison in Fig. 3. Here the performance of the novel classes is the average accuracy of the eight datasets, i.e., CUB, Cars, Places, Plantae, CropDiseases, EuroSAT, ISIC and ChestX. The base classes are the rest categories of the mini-ImageNet dataset different from the categories used for training. Our proposed modules with GNN perform better than the baseline (GNN) on both the base and novel classes, which indicates that the AFA module does not sacrifice the base classes performance to make do with cross-domain few-shot learning. The red dashed line of the base classes on 1-shot setting show that the Graph Convolution Network (GNN) with feature-wise transformation [40] has the slight improvement over our model on the base classes, but the performance on the novel classes degrades significantly. Moreover, compared to the competitive methods ATA [44], our method remarkably improve the performance on the base classes. Our method suppresses all the related works by the performance on the novel classes and also achieves competitive results on base classes. All these results demonstrate that our method can give the best balance between base and novel classes and classify the samples of novel classes well.

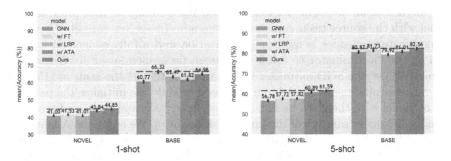

Fig. 3. Accuracy (%) of baseline (GNN), FT, LRP, ATA and our model for 1/5-shot cross-domain classification on both novel classes and base classes.

Table 3. Accuracy (%) of fine-tuning with the augmented support dataset from the target domain and our model for 1/5-shot 5-way classification on the target domains. ∗ means the method fine-tuned with target tasks generated through data augmentation. **Bold** indicates the best results.

Method/shot	CUB		Cars		Places		Planae	
	1-shot	5-shot	1-shot	5-shot	1-shot	5-shot	1-shot	5-shot
Fine-tuning	$43.53_{\pm0.4}$	$63.76_{\pm0.4}$	$35.12_{\pm0.4}$	$51.21_{\pm0.4}$	$50.57_{\pm0.4}$	$70.68_{\pm0.4}$	$38.77_{\pm0.4}$	$56.45_{\pm0.4}$
MN+*Ours*∗	$43.62_{\pm0.4}$	$68.73_{\pm0.4}$	$36.83_{\pm0.4}$	$52.53_{\pm0.4}$	$52.82_{\pm0.5}$	$71.56_{\pm0.4}$	$38.56_{\pm0.4}$	$56.50_{\pm0.4}$
GNN+*Ours*∗	$47.40_{\pm0.5}$	$\mathbf{70.33_{\pm0.5}}$	$36.50_{\pm0.4}$	$\mathbf{55.75_{\pm0.5}}$	$55.34_{\pm0.6}$	$\mathbf{76.92_{\pm0.4}}$	$39.97_{\pm0.4}$	$\mathbf{59.58_{\pm0.5}}$
TPN+*Ours*∗	$\mathbf{48.05_{\pm0.5}}$	$67.78_{\pm0.4}$	$\mathbf{38.45_{\pm0.4}}$	$54.89_{\pm0.4}$	$\mathbf{57.27_{\pm0.5}}$	$73.06_{\pm0.4}$	$\mathbf{40.85_{\pm0.4}}$	$59.04_{\pm0.4}$
	CropDiseases		EuroSAT		ISIC		ChestX	
	1-shot	5-shot	1-shot	5-shot	1-shot	5-shot	1-shot	5-shot
Fine-tuning	$73.43_{\pm0.5}$	$89.84_{\pm0.3}$	$66.17_{\pm0.5}$	$81.59_{\pm0.3}$	$34.60_{\pm0.3}$	$49.51_{\pm0.3}$	$22.13_{\pm0.2}$	$25.37_{\pm0.2}$
MN+*Ours*∗	$74.67_{\pm0.4}$	$90.53_{\pm0.3}$	$66.48_{\pm0.5}$	$82.00_{\pm0.3}$	$34.58_{\pm0.3}$	$48.46_{\pm0.3}$	$22.29_{\pm0.2}$	$\mathbf{25.80_{\pm0.3}}$
GNN+*Ours*∗	$74.80_{\pm0.5}$	$\mathbf{95.66_{\pm0.2}}$	$69.64_{\pm0.6}$	$\mathbf{89.56_{\pm0.4}}$	$\mathbf{35.33_{\pm0.4}}$	$\mathbf{50.44_{\pm0.4}}$	$22.25_{\pm0.2}$	$24.96_{\pm0.2}$
TPN+*Ours*∗	$\mathbf{81.89_{\pm0.5}}$	$93.67_{\pm0.2}$	$\mathbf{70.37_{\pm0.5}}$	$86.68_{\pm0.2}$	$34.88_{\pm0.4}$	$50.17_{\pm0.3}$	$\mathbf{22.65_{\pm0.2}}$	$24.79_{\pm0.2}$

4.6 Comparison with Fine-Tuning

As mentioned by Guo et al. [15], when coming across the domain shift, traditional pre-training and fine-tuning methods perform better than meta-learning methods in few-shot setting. This experiment is to verify that the superiority of the meta-learning methods with our module over the traditional pre-training and fine-tuning under the cross-domain few-shot setting. For a fair comparison, we follow the way of Wang et al. [44], i.e., using data augmentation for fine-tuning in target tasks. Given an target task T formed by the k-shot n-way samples as support set and $n \times 15$ pseudo samples as query set. The pseudo samples of query set are generated by the support samples using the data augmentation method from [48]. For pre-training and fine-tuning, we first pre-train the model with the source tasks composed of the mini-ImageNet dataset. Then, the trained feature encoder is used for initialization and a fully connected layer is used as the discriminator to fulfill the unseen tasks mentioned above for fine-tuning. We use the SGD optimizer with learning rate 0.01 the same as [15]. For the meta-learning methods with our proposed module, we initialize the parameters of model with the meta-learning on the source tasks and then used the same support and query samples of the target task as above. We apply the Adam optimizer with the learning rate 0.001. Both fine-tuning and meta-learning methods are fine-tuned for 50 epoch under the 5-shot/1-shot 5-way setting. Since the data used for training are consistent in all models, it is a fair comparison. As shown in the Table 3, our method consistently outperforms the traditional pre-training and fine-tuning.

5   Conclusions

In this paper, we present a novel method namely Adversarial Feature Augmentation (AFA) which can generate augmented features to simulate domain variations and improve the generalization ability of meta-learning models. Based on sufficient statistics of normal distribution, the feature augmentation module is designed by perturbation on feature mean and variance. By adversarial training, the ATA module is learned by maximizing the domain discrepancy with the domain discriminator, while the feature encoder is optimized by confusing the seen and unseen domains. Experimental results on nine datasets show that the proposed AFA improves the performance of meta-learning baselines and outperforms existing works for cross-domain few-shot classification in most cases.

Acknowledgments. This work was supported partially by NSFC (No. 61906218), Guangdong Basic and Applied Basic Research Foundation (No. 2020A1515011497), and Science and Technology Program of Guangzhou (No. 202002030371).

References

1. Bateni, P., Barber, J., van de Meent, J., Wood, F.: Enhancing few-shot image classification with unlabelled examples. In: WACV (2022)
2. Bateni, P., Goyal, R., Masrani, V., Wood, F., Sigal, L.: Improved few-shot visual classification. In: CVPR (2020)
3. Bronskill, J., Gordon, J., Requeima, J., Nowozin, S., Turner, R.E.: Tasknorm: Rethinking batch normalization for meta-learning. In: ICML (2020)
4. Bronskill, J., Massiceti, D., Patacchiola, M., Hofmann, K., Nowozin, S., Turner, R.: Memory efficient meta-learning with large images. In: NeurIPS (2021)
5. Chen, W., Liu, Y., Kira, Z., Wang, Y.F., Huang, J.: A closer look at few-shot classification. In: ICLR (2019)
6. Deng, W., et al.: Deep ladder reconstruction-classification network for unsupervised domain adaptation. Pattern Recognit. Lett. **152**, 398–405 (2021)
7. Finn, C., Abbeel, P., Levine, S.: Model-agnostic meta-learning for fast adaptation of deep networks. In: ICML, pp. 1126–1135 (2017)
8. Finn, C., Abbeel, P., et al.: Model-agnostic meta-learning for fast adaptation of deep networks. In: ICML, vol. 70, pp. 1126–1135 (2017)
9. Fisher, R.A.: On the mathematical foundations of theoretical statistics. Philos. Trans. Royal Soc. London. Ser. A Cont. Papers Mathem. Phys. Charact. **222**(594–604), 309–368 (1922)
10. Frikha, A., Krompaß, D., Köpken, H., Tresp, V.: Few-shot one-class classification via meta-learning. In: AAAI, pp. 7448–7456 (2021)
11. Ganin, Y., Lempitsky, V.S.: Unsupervised domain adaptation by backpropagation. In: Bach, F.R., Blei, D.M. (eds.) ICML, vol. 37, pp. 1180–1189 (2015)
12. Ganin, Y., et al.: Domain-adversarial training of neural networks. In: CVPR, pp. 189–209 (2017)
13. Ghifary, M., Kleijn, W.B., Zhang, M., Balduzzi, D., Li, W.: Deep reconstruction-classification networks for unsupervised domain adaptation. In: Leibe, B., Matas, J., Sebe, N., Welling, M. (eds.) ECCV 2016. LNCS, vol. 9908, pp. 597–613. Springer, Cham (2016). https://doi.org/10.1007/978-3-319-46493-0_36
14. Goodfellow, I.J., Shlens, J., Szegedy, C.: Explaining and harnessing adversarial examples. In: ICLR (2015)
15. Guo, Y., et al.: A broader study of cross-domain few-shot learning. In: Vedaldi, A., Bischof, H., Brox, T., Frahm, J.-M. (eds.) ECCV 2020. LNCS, vol. 12372, pp. 124–141. Springer, Cham (2020). https://doi.org/10.1007/978-3-030-58583-9_8

16. He, D., et al.: Stnet: Local and global spatial-temporal modeling for action recognition. In: AAAI, pp. 8401–8408 (2019)
17. He, K., Zhang, X., Ren, S., Sun, J.: Deep residual learning for image recognition. In: CVPR, pp. 770–778 (2016)
18. Horn, G.V., et al.: The inaturalist species classification and detection dataset. In: CVPR, pp. 8769–8778 (2018)
19. Hsu, H., et al.: Progressive domain adaptation for object detection. In: WACV, pp. 738–746. IEEE (2020)
20. Krause, J., Stark, M., Deng, J., Fei-Fei, L.: 3d object representations for fine-grained categorization. In: ICCV, pp. 554–561. IEEE Computer Society (2013)
21. Lake, B.M., Salakhutdinov, R., Tenenbaum, J.B.: Human-level concept learning through probabilistic program induction. Science **350**(6266), 1332–1338 (2015)
22. Li, X., Wang, W., Hu, X., Yang, J.: Selective kernel networks. In: CVPR, pp. 510–519 (2019)
23. Li, Y., Yang, Y., Zhou, W., Hospedales, T.M.: Feature-critic networks for heterogeneous domain generalization. In: ICML, vol. 97, pp. 3915–3924 (2019)
24. Liu, B., et al.: Negative margin matters: understanding margin in few-shot classification. In: Vedaldi, A., Bischof, H., Brox, T., Frahm, J.-M. (eds.) ECCV 2020. LNCS, vol. 12349, pp. 438–455. Springer, Cham (2020). https://doi.org/10.1007/978-3-030-58548-8_26
25. Liu, Y., et al.: Learning to propagate labels: Transductive propagation network for few-shot learning. In: ICLR (2019)
26. Long, M., Cao, Y., Wang, J., Jordan, M.I.: Learning transferable features with deep adaptation networks. In: ICML, pp. 97–105 (2015)
27. Madry, A., Makelov, A., Schmidt, L., Tsipras, D., Vladu, A.: Towards deep learning models resistant to adversarial attacks. In: ICLR (2018)
28. Ravi, S., Larochelle, H.: Optimization as a model for few-shot learning. In: ICLR (2017)
29. Requeima, J., Gordon, J., Bronskill, J., Nowozin, S., Turner, R.E.: Fast and flexible multi-task classification using conditional neural adaptive processes. In: NeurIPS (2019)
30. Saito, K., Kim, D., Sclaroff, S., Darrell, T., Saenko, K.: Semi-supervised domain adaptation via minimax entropy. In: ICCV, pp. 8049–8057. IEEE (2019)
31. Satorras, V.G., Estrach, J.B.: Few-shot learning with graph neural networks. In: ICLR (2018)
32. Shafahi, A., et al.: Adversarial training for free! In: NeurIPS, pp. 3353–3364 (2019)
33. Sinha, A., Namkoong, H., Duchi, J.C.: Certifying some distributional robustness with principled adversarial training. In: ICLR (2018)
34. Snell, J., Swersky, K., Zemel, R.S.: Prototypical networks for few-shot learning. In: NeurIPS, pp. 4077–4087 (2017)
35. Splawa-Neyman, J., Dabrowska, D.M., Speed, T.: On the application of probability theory to agricultural experiments. essay on principles. section 9. Statistical Science, pp. 465–472 (1990)
36. Sui, D., Chen, Y., Mao, B., Qiu, D., Liu, K., Zhao, J.: Knowledge guided metric learning for few-shot text classification. In: NAACL-HLT, pp. 3266–3271. Association for Computational Linguistics (2021)
37. Sun, J., Lapuschkin, S., Samek, W., Zhao, Y., Cheung, N., Binder, A.: Explanation-guided training for cross-domain few-shot classification. In: ICPR, pp. 7609–7616 (2020)

38. Sung, F., Yang, Y., Zhang, L., Xiang, T., Torr, P.H., Hospedales, T.M.: Learning to compare: Relation network for few-shot learning. In: CVPR, pp. 1199–1208 (June 2018)
39. Tan, M., Le, Q.V.: Efficientnet: Rethinking model scaling for convolutional neural networks. In: ICML, vol. 97, pp. 6105–6114 (2019)
40. Tseng, H., Lee, H., Huang, J., Yang, M.: Cross-domain few-shot classification via learned feature-wise transformation. In: ICLR (2020)
41. Tzeng, E., Hoffman, J., Saenko, K., Darrell, T.: Adversarial discriminative domain adaptation. In: CVPR, pp. 2962–2971 (2017)
42. Vinyals, O., Blundell, C., Lillicrap, T., kavukcuoglu, k., Wierstra, D.: Matching networks for one shot learning. In: NeurIPS, pp. 3630–3638 (2016)
43. Volpi, R., Namkoong, H., Sener, O., Duchi, J.C., Murino, V., Savarese, S.: Generalizing to unseen domains via adversarial data augmentation. In: NeurIPS, pp. 5339–5349 (2018)
44. Wang, H., Deng, Z.: Cross-domain few-shot classification via adversarial task augmentation. In: Zhou, Z. (ed.) IJCAI, pp. 1075–1081 (2021)
45. Welinder, P., et al.: Caltech-ucsd birds 200 (2010)
46. Wu, F., Smith, J.S., Lu, W., Pang, C., Zhang, B.: Attentive prototype few-shot learning with capsule network-based embedding. In: Vedaldi, A., Bischof, H., Brox, T., Frahm, J.-M. (eds.) ECCV 2020. LNCS, vol. 12373, pp. 237–253. Springer, Cham (2020). https://doi.org/10.1007/978-3-030-58604-1_15
47. Wu, W., He, D., Lin, T., Li, F., Gan, C., Ding, E.: Mvfnet: Multi-view fusion network for efficient video recognition. In: AAAI, pp. 2943–2951 (2021)
48. Yeh, J., Lee, H., Tsai, B., Chen, Y., Huang, P., Hsu, W.H.: Large margin mechanism and pseudo query set on cross-domain few-shot learning. CoRR abs/2005.09218 (2020)
49. Yue, X., et al.: Prototypical cross-domain self-supervised learning for few-shot unsupervised domain adaptation. In: CVPR, pp. 13834–13844. IEEE (2021)
50. Zellinger, W., Grubinger, T., Lughofer, E., Natschläger, T., Saminger-Platz, S.: Central moment discrepancy (CMD) for domain-invariant representation learning. In: ICLR (2017)
51. Zhou, B., Lapedriza, A., Khosla, A., Oliva, A., Torralba, A.: Places: A 10 million image database for scene recognition. IEEE Trans. Pattern Anal. Mach. Intell. **40**(6), 1452–1464 (2017)

Constructing Balance from Imbalance for Long-Tailed Image Recognition

Yue Xu[1], Yong-Lu Li[1,2], Jiefeng Li[1], and Cewu Lu[1(✉)]

[1] Shanghai Jiao Tong University, Shanghai, China
{silicxyue,yonglu_li,ljf_likit,lucewu}@sjtu.edu.cn
[2] Hong Kong University of Science and Technology, Hong Kong, China

Abstract. Long-tailed image recognition presents massive challenges to deep learning systems since the imbalance between majority (head) classes and minority (tail) classes severely skews the data-driven deep neural networks. Previous methods tackle with data imbalance from the viewpoints of data distribution, feature space, and model design, etc. In this work, instead of directly learning a recognition model, we suggest confronting the bottleneck of head-to-tail bias before classifier learning, from the previously omitted perspective of **balancing label space**. To alleviate the head-to-tail bias, we propose a concise paradigm by progressively adjusting label space and dividing the head classes and tail classes, *dynamically constructing balance from imbalance* to facilitate the classification. With flexible data filtering and label space mapping, we can easily **embed** our approach to most classification models, especially the decoupled training methods. Besides, we find the separability of head-tail classes varies among different features with different inductive biases. Hence, our proposed model also provides a *feature evaluation* method and paves the way for long-tailed **feature** learning. Extensive experiments show that our method can boost the performance of state-of-the-arts of different types on widely-used benchmarks. Code is available at https://github.com/silicx/DLSA.

Keywords: Image classification · Long-tailed recognition · Normalizing flows

1 Introduction

Deep learning shows its superiority in various computer vision tasks [17,32, 53], especially in balanced data scenarios. Though, real-world data is usually

---

Y. Xu and Y.-L. Li—Contributed equally.
C. Lu—Member of Qing Yuan Research Institute and Shanghai Qi Zhi Institute.

Supplementary Information The online version contains supplementary material available at https://doi.org/10.1007/978-3-031-20044-1_3.

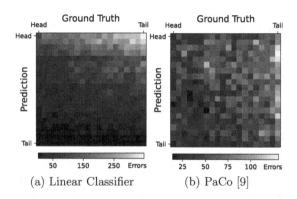

(a) Linear Classifier   (b) PaCo [9]

Fig. 1. Confusion matrices of models on ImageNet-LT [38] test set, indicating the severe head-tail bias and that tail classes are particularly prone to confusion with head classes (high density at right-top). **The correct samples are omitted for clarity.** The classes are ordered by their frequency and merged into 20 bins.

severely imbalanced, following a long-tailed distribution [34,35,55,71], *i.e.*, very few frequent classes take up the majority of data (head) while most classes are infrequent (tail). The highly biased data skews classifier learning and leads to performance drop on tail classes. As shown in Fig. 1, most errors stem from the *head-to-tail bias* and a large number of tail samples are misclassified as head classes, even with the very recent long-tail learning technique [9].

Many approaches have been proposed to re-balance long-tail learning by balancing the data distribution [1,2,6,15,24,33], balancing the output logits [41,67], balancing the training losses [3,19,20,46,50,54,60], balancing the feature space [9,25,26,58,63] or with balanced training strategy [22,27,70]. However, as it is the definition of class labels to blame for the long-tailed data distribution, there are very few works tackling long-tailed recognition from the perspective of balancing the *label space*. Samuel *et al.* [48] decompose the class labels into semantic class descriptors and exploit its familiarity effect, which is commonly used in zero-shot learning. Wu *et al.* [61] reorganize the class labels into a tree hierarchy by realistic taxonomy. Both methods partly alleviate the label imbalance but are restricted by the *class setting* (semantics, hierarchy). Here, we want to dynamically adjust the label space according to the imbalance of *realistic data distribution* to fit long-tail learning.

We can speculate from Fig. 1 that if the head and tail classes are ideally separated, the tail-to-head errors and the complexity of imbalanced learning can be significantly reduced. Thus, we conduct a simple probing to analyze the effect of head-tail separation in Fig. 2 (detailed in supplementary materials: Supp. Sect. 1). We separate the test samples into two groups (head: classes with >50 samples; tail: the rest classes) before individually classifying the groups. Figure 2 shows that accurate head-tail separation can help long-tail learning. In light of this, we incorporate a **D**ynamic **L**abel **S**pace **A**djustment (**DLSA**) method as shown in Fig. 3. We propose to first confront the *bottleneck* of head-to-tail bias and deal with long-tailed recognition in a **two-stage** fashion: first

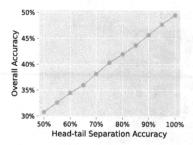

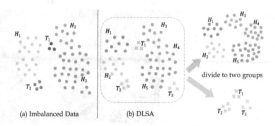

(a) Imbalanced Data | (b) DLSA | divide to two groups

Fig. 2. The performance curve of linear classifier on ImageNet-LT [38] with different head-tail separation accuracy (50%: **random** separation; 100%: **ideal** separation). The separation model divides the classes into head and tail, and then we classify within the two individual groups. **Better** separation leads to a **higher** overall accuracy.

Fig. 3. Demonstration of our dynamic label space adjustment (DLSA). H_i stands for a head class and T_j is a tail class. We re-define label space for imbalanced data and divide head and tail classes into two groups progressively to reduce head-to-tail bias, seeking balance from the new data sub-space. DLSA not only separates head and tail but also **redefines** the label space for the convenience of classifiers. So some of the head samples may be tail classes in new label space due to the $\mathcal{L}_{bal}$ constraint.

adjust the label space and separate the head and tail samples, and then apply classification for the divided groups respectively. In virtue of the *inductive bias* of deep learning feature models (*e.g.*, pretrained backbone), we could define the new label space as the initial clusters of features and approach *head-tail separation*. Moreover, we hope the clusters in the new label space are balanced and contain only a few classes. We formulate these assumptions on the latent space as: 1) **Head-Tail Separability**: the head and tail are separated during the adjustment; 2) **Cluster Balancedness**: the clusters have balanced sizes; 3) **Purity**: samples in each cluster are pure, *i.e.*, belong to as few classes as possible.

Specifically, we propose a *plug-and-play* module which can be embedded in most two-stage learning methods [27]. We use Gaussian mixture to model the pretrained features and produce clusters. Then the samples are divided into two groups and classified independently, where head and tail classes are desired to be separated. In practice, multiple modules can be linked to a cascade model to progressively separate head and tail classes and reduce the imbalance bias. Experiments show that the dataset partition and label re-assignment can effectively alleviate label bias. On top of that, we find backbone models with various inductive biases have different head-tail class separability. With DLSA, we can qualitatively evaluate the feature learning of the backbones from the perspective of head-to-tail bias, which can facilitate long-tailed recognition in practice.

Our main contributions are: 1) Proposing a dynamic label space adjustment paradigm for long-tail learning. 2) Our method is plug-and-play and can boost the performances of the state-of-the-arts on widely-adopted benchmarks. 3) Our method can also act as an evaluation of feature model selection to guide long-tailed feature learning and recognition.

2   Related Work

2.1   Long-Tailed Recognition

Real-world data are long-tailed distributed, which skews machine learning models. Numerous tasks face the challenge of long-tailed data, including object detection [14], attribute recognition [36] and action understanding [4,37]. There is increasing literature trying to alleviate the bias brought by imbalanced data.

Some small adjustments to the model components can alleviate imbalanced learning. The most intuitive approach is re-balance the **data distribution**, either by over-sampling the minority [1,2], or under-sampling of majority [1,15,24]. Data **augmentation** and **generation** [6,7,33] also flattens the long-tailed distribution and helps tail learning. Re-balancing by **loss function** adjust the significance of samples on each class [3,19,20,46,50,54,60]. Instead of weighting the losses, some methods balance the **output logits** [29,41,67] after training.

There are also methods to modify the whole training process. One of the paths is **knowledge transfer** from head classes to tail classes [22,27,30,38,64,70]. Liu et al. [38] enhance the feature of minority classes with a memory module. Zhang et al. [64] distill the tail-centric teacher models into the general student network to facilitate tail learning while keeping the head performance. Some works involve specific balanced **training strategy** for imbalanced learning. Kang et al. [27] exploit the two-stage approach by first training a feature extractor and then re-train a balanced classifier. Zhou et al. [70] combine uniform sampling and reversed sampling in a curriculum learning fashion. Jamal et al. [22] estimate class weights with meta-learning to modulate classification loss.

Recently, self-supervised learning is applied in long-tailed recognition for **feature space** re-balancing, among which contrastive learning is the most trendy technique [5,16,42]. Yang and Xu [63] first use self-supervised pretraining to overcome long-tailed label bias. Kang et al. [26] systematically compare contrastive learning with traditional supervised learning and find that the contrastive model learns a more balanced feature space which alleviates the bias brought by data imbalance. Here, our DLSA provides another strong proof for this conclusion (Sect. 4.4). Wang et al. [58] combine contrastive feature learning and cross-entropy-based classifier training with a curriculum learning strategy. Jiang et al. [25] exploit the contrast of the network and its pruned competitor to overcome the forgetting of the minority class. Cui et al. [9] propose parametric contrastive learning to rebalance the sampling process in contrastive learning.

However, very few works deal with long-tail bias from the perspective of balancing **label space**. Samuel et al. [48] incorporate semantic class descriptors and enhance minority learning with the familiarity effect of the descriptors.

Wu *et al.* [61] reorganize the label space into tree hierarchy with semantics. In this paper, we propose DLSA to filter the balance subset from imbalanced data.

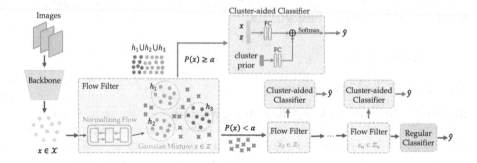

Fig. 4. Method overview. The features x are extracted by pretrained backbones and sent to **Gaussian Mixture Flow Filter**. The Flow Filter maps the samples to a Gaussian mixture latent space $\mathcal{Z}$ with three constraints: **cluster balancedness**, **purity**, and **head-tail separability** to balance the label space and separate head and tail classes. Then we re-group the data according to the likelihood of each sample. The samples that are more conforming to the Gaussian mixture (dots) are classified by **Cluster-aided Classifier** with cluster prior information. The samples with low likelihoods (red crosses) are *progressively* solved by Flow Filters or classified by a regular classifier at the last layer. (Color figure online)

2.2 Normalizing Flows

Normalizing flows [12,31,43,47] are a family of invertible networks and are widely adopted in density estimation and generative models. Specifically, a normalizing flow g is a cascaded transforms data $x \in \mathcal{X}$ to representation u in latent space $\mathcal{U}$: $u = g^{-1}(x)$. By assuming a tractable latent distribution, we can recover the data distribution with the *change of variable formula*:

$$P(x) = P(u) \cdot \left| \mathbf{J} \left(g^{-1}(x) \right) \right|, \tag{1}$$

where $\mathbf{J}(\cdot)$ is the Jacobian determinant of a transformation and is a crucial concern in the design of normalizing flows due to its high complexity. Rezende and Mohamed [47] propose some cascaded normalizing flows and their basic blocks have linear-time Jacobian computation. Real NVP [12] updates only part of the input vector in the normalizing flow block with a simple bijection, which is more capable of modeling high-dimension data distribution. IAF [31] and MAF [43] are more general than Real NVP, incorporating autoregressive transformations. In practical applications of density estimation, the latent distribution $\mathcal{U}$ is usually unit Gaussian for simplicity. Some works [21] exploit Gaussian mixture to explicitly capture the cluster structure of data. In this work, we also use a unit Gaussian mixture to simultaneously transform data and clusters.

3   Approach

3.1   Preliminaries

Long-tailed recognition is to learn a model that predict labels $y = 1, 2, \cdots, \mathcal{C}$ of the data sample $x \in \mathcal{X}$, while the training sample numbers of each class $n_i = |\{x|y = i, \ x \in X_{train}\}|$ is highly imbalanced. The imbalanceness of data are measured by imbalance factor $\beta = \max(n_i)/\min(n_i)$ and the β of typical long-tailed datasets ranges from 10 to 1000.

Since the **head-tail separation** is a well-known key for long-tail learning, our idea is to separate head and tail samples recursively and classify in a divide-and-conquer manner. So the longtailedness is reduced in each partition of data and, in this sense, we construct *balance from imbalance*. We first divide head and tail classes, by re-defining a more balanced label space $h = 1, 2, \cdots, K$. As discussed, if we can ideally map the data and labels to a balanced space and separate head and tail classes, it will be easier to learn classifiers. However, it is impractical to achieve an *absolutely* balanced space. So we propose to *progressively* map and cluster samples, and filter out the samples that are well subject to the balanced distributions. During the clustering, we force the model to learn head-tail class separability. In this paradigm, the complexity of long-tail learning is partially *transferred* to the label space learning and latent space mapping process, while these can be more easily settled with the help of the inductive biases of pretrained features and the transformation ability of normalizing flows [47].

Backbone Models. Different inductive biases exist in the various deep learning models and training schemes. Cross-entropy-based supervised learning imposes class separability on deep feature learning. As previously studied [27], for deep supervised learning, the data imbalance is mainly detrimental to the classifiers while the feature learning suffers less. Self-supervised contrastive learning methods [5,42] introduce instance-level transformation invariance to deep models, leading to a more balanced feature since there are no labels involved. Further supervised contrastive learning [26,63] combines the two aspects and incorporates both instance-level multi-view invariance and class-wise contrast. We compare the main-stream pretrained models with different inductive biases in Sect. 4.4 and supervised contrastive learning shows its superiority over the rest.

Normalizing Flow. As introduced in Sect. 2.2, normalizing flows are invertible transformations and are suitable for density estimation and distribution mapping. We utilize normalizing flows to simultaneously map the data x to representations z in a more balanced latent space and estimate the new class label h. The mathematical beauty of normalizing flow enables the maximum likelihood estimation and the learning of data mapping. We also design multiple constraints for the normalizing flows to re-balance the latent space.

3.2   Overview

Fig. 4 depicts the pipeline of our method. We apply our DLSA in feature space and the initial features are extracted by pretrained self-supervised models. There are mainly two types of modules arranged in a cascaded manner. **Gaussian Mixture Flow Filter** (Sect. 3.3) clusters the input data in latent space and assigns new cluster labels. Then, the samples are divided into two groups. The well-clustered samples are filtered out to mitigate the head-to-tail bias and learning complexity, and are sent to a dedicated **Cluster-aided Classifier** (Sect. 3.4) which exploits the cluster information in classification. The rest outliers are either forwarded to the next normalizing flow or regularly classified.

Specifically, the input features are transformed by normalizing flows to a Gaussian mixture distribution to obtain cluster labels. For better separability in the long-tailed scenario, besides the maximum likelihood objective of Gaussian mixture, the flow model is constrained by three additional **objectives**: 1) Head-tail Separability: head and tail classes should be separated; 2) Cluster Balancedness: the sizes of latent clusters should be as balanced as possible; 3) Purity: the samples in each cluster should belong to as few classes as possible.

3.3   Gaussian Mixture Flow Filter

Gaussian Mixture Flow Filter modules transform data to a desired latent distribution, perform clustering and filter the samples. Without loss of generality, we take the first Flow Filter (magnified in Fig. 4) as an example.

The filter module incorporates a normalizing flow model $g_\theta^{-1} : \mathcal{X} \to \mathcal{Z}$ with trainable parameter θ, mapping data samples $x \in \mathcal{X}$ to latent distribution $z \in \mathcal{Z}$. To cluster the input samples into K clusters, we assume the latent distribution $P(z)$ is a Gaussian mixture and each component corresponds to a cluster:

$$P(z) = \sum_{k=1}^{K} P(h = k)P(z|h = k), \tag{2}$$

where h is the random variable of cluster index. $P(z|h = k) = \mathcal{N}(z|\mu_k, I)$ is a Gaussian probability density with mean vectors μ_k and identity covariance matrix. The μ_k are randomly sampled from a normal distribution and fixed during training [21]. A proper assumption on *prior probability* $P(h)$ is required for a tractable optimization methods of NF with GMM. Since $P(h)$ in GMM determines the "size" of a Gaussian component, with the **balancedness** assumptions, the prior probability $P(h)$ can be set to uniform distribution, thus:

$$P(z) = \frac{1}{K} \sum_{k=1}^{K} P(z|h = k), \tag{3}$$

and the prediction of sample x is given by Bayes' theorem:

$$P(k|x) = \frac{P(z|h = k)}{\sum_{k'=1}^{K} P(z|h = k')}, \quad \hat{h} = \underset{k}{\mathrm{argmax}}\, P(k|x). \tag{4}$$

We train the normalizing flow model with end-to-end gradient descent optimization with the following objectives.

Maximum Likelihood Loss and Head-Tail Separability. We propose a weighted maximum likelihood loss to simultaneously train the cluster model and learn head-tail class separation. With the elegant invertible property of normalizing flow and change of variable formula Eq. 1, we can painlessly obtain the Gaussian mixture likelihood of data sample x:

$$L(\theta; x) = \sum_{k=1}^{K} \frac{1}{K} \mathcal{N}(g_\theta^{-1}(x)|\mu_k, I) \left| \mathbf{J}\left(g_\theta^{-1}(x)\right)\right|. \tag{5}$$

The maximum likelihood estimation (MLE) of parameter θ will produce a clustering model. To enhance the separability of head and tail classes of the Flow Filter, we incorporate *sample weighting* and impose higher weights on tail samples than head samples. In DLSA, samples in class i are weighted by:

$$\omega(i) = \frac{n_i^{-q}}{\sum_j n_j^{-q}}, \tag{6}$$

where n_i is the training sample size of class i. q is a positive number usually ranging from 1 to 2, hence the tail samples receive larger weights. The weighted $L(\theta; x)$ will impose higher likelihoods on tail samples than head samples, therefore tail classes are more likely to be filtered out by Flow Filter and separated with head classes. The weighted maximum likelihood loss is constructed with the negative log-likelihood of a data batch $\mathcal{B}$:

$$\mathcal{L}_{MLE}(\mathcal{B}) = -\sum_{x \in \mathcal{B}} \omega(y) \log L(\theta; x). \tag{7}$$

Training the flow model with a single MLE constraint may result in trivial solutions or a latent distribution that is not suitable for long-tailed recognition, e.g., g_θ^{-1} maps all samples to one cluster (Figs. 5 and 6). To avoid *model collapse* and *degeneration*, we introduce two more objectives for the flow filter.

Cluster Balancedness Loss. Though the prior distribution $P(h)$ has been set to uniform, the posterior after observing data samples X:

$$P(h = k|X) = \sum_{x \in X} P(x|X)P(h = k|x) = \sum_{x \in X} \frac{1}{|X|} P(h = k|x), \tag{8}$$

can still be imbalanced; if so, the actual cluster sizes would be biased or even severely long-tailed distributed. To promote cluster balance, we introduce

$$\mathcal{L}_{bal}(X) = \mathbb{E}_h\left[\log P(h|X)\right], \tag{9}$$

a cluster balancedness loss which is the negative entropy of the posterior. It is also equivalent to the KL divergence between $P(k|X)$ and a discrete uniform distribution, whose simple proof is shown in Supp. Sect. 2.1. So minimizing the loss will force the cluster sizes to be more even. As shown in Fig. 5, balancedness loss can significantly reduce the unbalancedness of the cluster sizes.

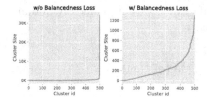

Fig. 5. Cluster sizes of DLSA w/ or w/o $\mathcal{L}_{bal}$ on ImageNet-LT [27].

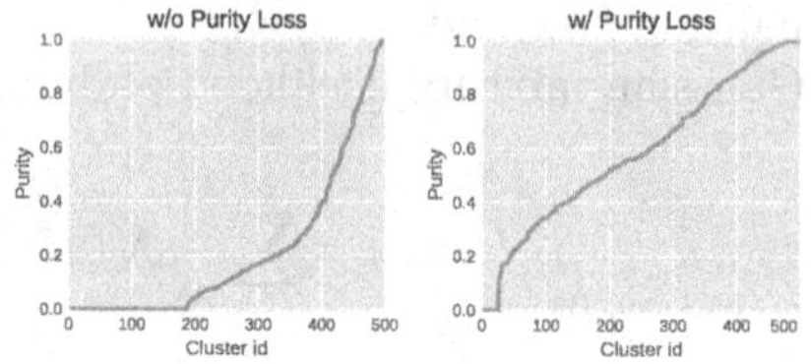

Fig. 6. Cluster purity of DLSA w/ or w/o $\mathcal{L}_{pure}$ on ImageNet-LT [27].

In practice, the optimization of $\mathcal{L}_{bal}$ can be unstable, especially with mini-batch based optimizer [28]. For each mini-batch $\mathcal{B}_t$ at t^{th} iteration, $\hat{p} = P(h = k|\mathcal{B}_t)$ is an unbiased estimator of the posterior $P(h = k|X)$. But small batch size can lead to large variance of $\hat{p}$. Therefore, we propose to exploit the *history information* and incorporate a momentum-based estimator: $\tilde{p}_t = \eta\tilde{p}_{t-1} + (1 - \eta)P(h = k|\mathcal{B}_t)$, where η is decay factor. $\tilde{p}_{t-1}$ is the estimation of last batch and we set $\tilde{p}_0$ to 0. Thus, the bias-corrected estimator $\frac{\tilde{p}_t}{1-\eta^t}$ is an unbiased and more efficient estimator than $\hat{p}$, and is more stable during mini-batch training. The proof of its unbiasedness and efficiency is shown in Supp. Sect. 2.2.

Purity Loss. We define a scalar "purity" $Purity(k)$ as the sample proportion of the largest class in the cluster k. To enhance the class purity in each clusters, we repeatedly randomly sample a pair of samples x_i, x_j from two different classes i, j and suppressing their similarity in cluster predictions:

$$\mathcal{L}_{pure}(x_i, x_j) = \sum_{k=1}^{K} P(h = k|x_i) \log P(h = k|x_j). \tag{10}$$

By minimizing the purity loss, we increase the cross-entropy between $P(h|x_i)$ and $P(h|x_j)$, thus pushing the two samples farther. Figure 6 indicates purity loss effectively increases the purity of the clusters. Different sampling strategies can be applied to the purity loss. Our experiments show that the best practice is sampling with equal probability for each class. Overall, the total loss of Flow Filter module is:

$$\mathcal{L}_{total} = \mathcal{L}_{MLE} + \lambda_{bal}\mathcal{L}_{bal} + \lambda_{pure}\mathcal{L}_{pure}. \tag{11}$$

Among the constraints, $\mathcal{L}_{MLE}$ is for learning the data distribution and its sample-wise weights make the model lean to head samples, enabling head-tail separation with sample likelihood $P(x)$. $\mathcal{L}_{pure}$ ensures that each cluster in GMM belongs to only one class, so we can use cluster information to enhance classification. $\mathcal{L}_{bal}$ reduces the longtailedness in clusters and prevents the model from trivial solutions, *e.g.*, most samples go to one cluster. $\mathcal{L}_{pure}$ and $\mathcal{L}_{bal}$ ensure the training stability and enhance the separatability learned by $\mathcal{L}_{MLE}$.

Filtering. The Flow Filter implements the divide step in the DLSA pipeline. After training, some samples do not belong to any cluster and become outliers of the Gaussian mixture. We use a likelihood threshold α to separate the well-clustered samples and outliers. As shown in Fig. 4, the samples with higher confidence ($P(x) \leq \alpha$, green and blue) are closer to the cluster centers and the rest ($P(x) > \alpha$, red) are outliers of any clusters so they are sent to next adjustment module to further decrease the training difficulty. Accordingly, the samples are divided into two groups to separate head and tail classes.

In practice, we set the threshold α to a quantile of $P(x)$ on training data, so that a certain proportion of data will be filtered out.

3.4   Cluster-Aided Classifier

The high-confidence samples filtered out by the Flow Filter are sent to a Cluster-aided Classifier. Under ideal conditions, each cluster h in the Flow Filter contains only one class y, so the prediction can be directly obtained by a simple label-class mapping. However, the learned Gaussian mixture is rather noisy and besides a majority class y in one cluster, there are some samples from other classes. To compensate for these noise samples, we introduce a softer label-class mapping method. As demonstrated in Fig. 4, for each sample x_i, we first compute cluster prior $P(y|h = \hat{k}_i)$, which is the class frequency of training samples belonging to the sample's predicted cluster $\hat{k}_i$. The cluster prior vector and the concatenation of feature x_i and latent representation z_i are forwarded to two independent fully-connected layers (FC). Their outputs are added and we take its Softmax as output probability: $P(y|x, z) = Softmax\left(FC[x, z] + FC[P(y|h = \hat{k}_i)]\right)$.

4   Experiment

4.1   Datasets

We evaluate our approach on three main-stream benchmarks: ImageNet-LT [38], Places-LT [38], and iNaturalist18 [56]. ImageNet-LT [38] and Places-LT [38] are long-tailed subsets of ImageNet [11] and Places-365 [69], with 1,000 and 365 categories and about 186 K and 106 K images respectively. Both datasets have Pareto distributed train sets and *balanced test sets* and the imbalanced factor β is 256 and 996. iNaturalist18 [56] is a fine-grained image classification dataset which is naturally highly imbalanced ($\beta = 500$), with 8,142 categories and over 437 K images. Following [38], we report overall accuracy on all datasets, and Many-shot (classes with over 100 images), Medium-shot (classes with 20–100 images), Few-shot (classes with fewer than 20 images) accuracy. We also use Matthews correlation coefficient (MCC) [40] and normalized mutual information (NMI) [10] to measure the performance in long-tailed scenario.

4.2   Baselines

We embed and evaluate our models in multiple two-stage [27] long-tail learning methods, with different trending feature learning and classifier learning. The involved methods cover various long-tail learning strategies, including data rebalancing, loss balancing, feature space balancing, and decision boundary adjusting.

Feature Learning. (1) **Supervised Cross-Entropy (CE)**: a baseline backbone trained with vanilla cross-entropy loss. (2) **Class-balance Sampler (CBS)** [52]: a balanced data sampler where each class has equal sampling probability. (3) **PaCo** [9]: supervised contrastive learning model based on MoCo [16] It manages to alleviate the sampling bias in contrastive learning. PaCo is currently the state-of-the-art approach on the two mentioned benchmarks. Following the original implementation, it is integrated with Balanced Softmax [46]. We reproduce the two-stage training version of PaCo and RandAug [8] is not applied in classifier learning for a fair comparison with other features.

Classifier Learning. (1) **BalSoftmax** [46]: an adjusted Softmax-cross-entropy loss for imbalanced learning, by adding $\log(n_j)$ to logit of class j during training, where n_j is the training sample size of class j. (2) **Class-balance Sampler (CBS)** [52]: same as CBS in Feature Learning. (3) **Cosine classifier (Cosine)** [13,44]: predicting according to the cosine distance of features x and class embeddings w_i. Or namely, normalize the features x and weight vector w_i of each class i of a linear classifier: $\hat{y} = \operatorname{argmin}_i\{\cos\langle w_i,\ x\rangle\}$ (4) **M2M** [30] transfers head information to tail by resampling strategies. (5) **RIDE** [59] trains multiple diversified expert classifier and dynamically select the best expert for each sample. We use RIDE with two experts.

4.3   Implementation Details

We use 2-layered MAF [43] as the Flow Filter model. The Gaussian mixture centers are randomly sampled from $\mathcal{N}(0, 0.05^2)$ and the details of variance selection please refer to Supp. Sect. 3. The model is trained with a SGD optimizer and the decay rate for balancedness loss momentum $\eta = 0.7$. By default, the weight q in the $\mathcal{L}_{MLE}$ is 2.0, and 30% samples are filtered to Cluster-aided Classifiers in each division step. **We reproduce the baselines with decoupled strategy** [27]. The detailed hyper-parameters of each dataset are listed in Supp. Sect. 5.

4.4   Results

We evaluate the methods on some baselines in Tables 1, 2 and 3. Supp. Sect. 6 shows the detailed results on iNaturalist18. Our method promotes the accuracy of all the baselines. On state-of-art method PaCo [9]+BalSoftmax [46], we achieve accuracy gain of 1.4%, 1.2%, 1.0% on the datasets respectively. Especially, on few-shot classes, the method brings over 1.5% accuracy bonus. Though iNaturalist18 is a large-scale, realistic, and severely imbalanced dataset, DLSA still brings 1% improvement with PaCo representations. DLSA also brings about 1% and 0.5% gain on MCC and NMI score.

Table 1. Results on ImageNet-LT [38] with ResNet-50 [17] backbone.

Feature	Classifier	Overall	Many	Medium	Few	MCC	MNI
CE	BalSoftmax [46]	42.8	54.1	39.4	23.2	42.7	70.1
	BalSoftmax [46] + DLSA	**43.9 (+1.1)**	**54.5**	**41.0**	**24.0**	**43.8**	**70.3**
CBS [52]	BalSoftmax [46]	42.2	**55.8**	38.3	17.6	42.2	69.9
	BalSoftmax [46] + DLSA	**43.1 (+0.9)**	55.3	**40.2**	**18.9**	**43.1**	**70.2**
PaCo [9]	CBS [52]	54.4	61.7	52.0	42.5	54.4	74.5
	CBS [52] + DLSA	**55.6 (+1.2)**	**62.9**	**52.7**	**45.1**	**55.6**	**74.9**
	BalSoftmax [46]	54.9	67.0	50.1	38.0	54.9	74.9
	BalSoftmax [46]+ DLSA	**56.3 (+1.4)**	**67.2**	**52.1**	**40.2**	**56.1**	**75.4**
	Cosine [13,44]	55.7	**64.9**	53.0	39.5	55.7	75.2
	Cosine [13,44] + DLSA	**56.9 (+1.2)**	64.6	**54.9**	**41.8**	**56.8**	**75.7**
	M2M [30]	55.8	67.3	52.1	36.5	55.8	75.3
	M2M [30] + DLSA	**56.7 (+0.9)**	**68.0**	**52.8**	**38.2**	**56.6**	**75.7**
	RIDE [59]	56.5	67.3	53.3	37.3	56.5	75.6
	RIDE [59] + DLSA	**57.5 (+1.0)**	**67.8**	**54.5**	**38.8**	**57.5**	**75.9**

Comparison of Classifiers. In Tables 1 and 2, we compare different classifier learning models on same pretrained features (last 10 rows of Tables 1 and rows beginning with "CE+" in Tables 2). On each dataset, all the classifier methods use a same trained Flow Filter model since we incorporate two-stage training in DLSA and the learning of Flow Filters only depends on the feature models. DLSA brings comparable improvement on different classifiers (1.2%, 1.4%, 1.2%, 0.9%, 1.0% on ImageNet-LT; 1.2%, 1.2%, 1.0% on Places-LT). The performance improvement of our model mainly depends on the initial structure of the features.

Feature Model Comparison and Evaluation. With DLSA, models with different feature models have comparable overall accuracy improvement (1.1%, 0.9%, 1.4% with BalSoftmax classifier on ImageNet-LT), but the improvement on few-shot classes differs (0.8%, 1.3%, 2.2% with BalSoftmax classifier). This is due to the different *head-tail separability* of pretrained features with different inductive biases. Thus, we compare the

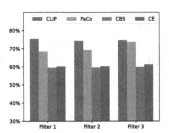

Fig. 7. The accuracy of head-tail separation of DLSA on different feature models. The threshold of head/tail is 50.

head-tail separation accuracy of DLSA on the three feature models. We analyze the samples filtered out by Flow Filters and compute the accuracy of these samples being tail classes. As discussed in Sect. 1, an ideal separation with 100% accuracy will significantly boost the long-tailed recognition. On PaCo+BalSoftmax, 100% accurate head-tail separation brings ~10% improvement. Figure 7 shows the accuracy of head-tail classes separation of models with different features. At all three Flow Filter layers, features of PaCo show superior separability than feature models pretrained with CE or CBS, thus PaCo has more potential of enhancing tail recognition. We also evaluate the DLSA on CLIP [45] features of ImageNet-LT. It achieves **65.2%** overall accuracy and over **75%** separation accuracy, surpassing all baselines, showing the **potential** of DLSA in the latest feature spaces. These also strongly support the conclusion of [26], that contrastive learning generates a more balanced feature space and enhances generalization capability. The positive correlation between head-tail separability and performance gain indicates that our proposed DLSA is a practical tool for feature analysis and evaluation with long-tailed data.

Table 2. Results on Places-LT [38] with ImageNet [11]-pretrained ResNet-152 [17].

Feature	Overall	Many	Medium	Few	MCC	NMI
Joint (baseline) [27]	30.2	45.7	27.3	8.2	–	–
cRT [27]	36.7	42.0	37.6	24.9	–	–
τ-norm [27]	37.9	37.8	40.7	31.8	–	–
LWS [27]	37.6	40.6	39.1	28.6	–	–
MetaDA [23]	37.1	–	–	–	–	–
LFME [62]	36.2	39.3	39.6	24.2	–	–
FeatAug [7]	36.4	42.8	37.5	22.7	–	–
BALMS [46]	38.7	41.2	39.8	31.6	–	–
RSG [57]	39.3	41.9	41.4	32.0	–	–
DisAlign [65]	39.3	40.4	42.4	30.1	–	–
CE+BalSoftmax [46]	37.8	40.0	**40.2**	28.5	37.7	58.1
CE+BalSoftmax [46]+ DLSA	**39.0** (+1.2)	**42.0**	39.8	**31.3**	**38.8**	**58.6**
CE+Cosine [13,44]	37.0	39.2	37.7	31.4	36.8	57.8
CE+Cosine [13,44]+ DLSA	**38.2** (+1.2)	**39.7**	**39.2**	**33.3**	**38.1**	**58.3**
CE+RIDE [59]	41.2	44.4	**43.2**	31.1	41.1	59.7
CE+RIDE [59]+ DLSA	**42.2** (+1.0)	**45.8**	43.0	**33.7**	**42.0**	**60.2**
PaCo+BalSoftmax [9]	40.9	**44.5**	42.7	30.4	40.8	59.6
PaCo+BalSoftmax [9]+ DLSA	**42.1** (+1.2)	44.4	**44.6**	**32.3**	**42.0**	**59.7**

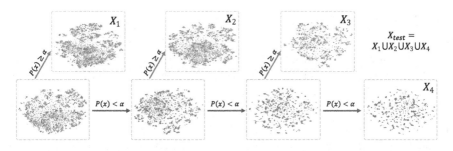

Fig. 8. Sample separation process of PaCo [9]+BalSoftmax [46] on ImageNet-LT [38] test set. The red and **blue** points are head and **tail** samples respectively. The grey boxes are Flow Filters and blue boxes are Cluster-aided Classifiers. (Color figure online)

4.5   Visualization

We visualize our cascade division process in Fig. 8 via t-SNE [39]. We take the PaCo+BalSoftmax as the example. Our method can effectively divide the samples of head and tail classes gradually. The samples filtered out to Cluster-aided Classifiers (in the blue rectangles) are mainly tail samples, while the samples passed to the next Flow Filters (in the grey rectangles) are mixed.

Table 3. Results on iNaturalist-18 [56] with ResNet-50 [17].

Feature	Overall
LWS [27]	69.5
MetaDA [23]	67.6
Deep-RTC [61]	64.0
Remix [6]	70.5
FeatAug [7]	65.9
smDRAGON [49]	69.1
KCL [26]	68.6
MiSLAS [68]	71.6
DisAlign [65]	70.6
Hybrid-PSC [58]	70.4
FSR [66]	65.5
DRO-LT [51]	69.7
DiVE [18]	71.7
PaCo [9]	71.8
PaCo [9] + Ours	**72.8 (+1.0)**

Table 4. Ablation study on ImageNet-LT [38] with PaCo [9]+BalSoftmax [46] model and ResNet-50 [17] backbone.

Method	Overall	Many	Medium	Few
Full model	**56.3**	**67.2**	**52.1**	**40.2**
w/o $\mathcal{L}_{MLE}$	54.7	66.1	50.8	36.3
w/o $\mathcal{L}_{bal}$	55.2	65.8	51.2	38.9
w/o $\mathcal{L}_{pure}$	55.4	66.1	51.5	39.2
300 clusters	55.5	66.1	51.3	39.3
1000 clusters	55.1	66.5	50.3	39.6
2 blocks	55.4	66.3	51.2	39.3
4 blocks	55.8	66.5	51.7	39.2

4.6   Ablation Study

We conduct ablation studies and look into the model components on the test set of ImageNet-LT [38] with PaCo [9]+BalSoftmax [46] model, from the following aspects. The results are reported in Tables 4. The extended ablation studies on Places-LT are shown in Supp. Sect. 7.

Objectives. We evaluate the three objectives $\mathcal{L}_{MLE}$, $\mathcal{L}_{bal}$, $\mathcal{L}_{pure}$ by removing one of them. Removing any loss leads to a significant performance drop, which indicates the necessity of all objectives. Among these losses, w/o $\mathcal{L}_{pure}$ shows the least degradation since the pretrained backbones have moderately learned intra-class similarity and $\mathcal{L}_{pure}$ aim to explicitly enhance the compactness.

Cluster Number. Models with less/more clusters perform worse than default 500 clusters. Larger cluster number results in slow training and inference too.

Filter Block Number. Models with more Flow Filter blocks can separate head and tail classes more finely. But excessive division operations lead to few training samples for the clustering model and classifiers. Due to this trade-off, the default model with 3 blocks outperforms that with 2 or 4 blocks.

5   Conclusions

In this paper, we propose to confront the head-to-tail bias by re-balancing the label space and separating head and tail classes. We present a plug-and-play module DLSA, which automatically adjusts the data distribution and constructs new label space to facilitate the recognition. We embed DLSA in various types of long-tailed recognition state-of-the-arts and boost their performances. We observe that DLSA is also capable of evaluating the different feature learning models. Our future work may extend to combining our paradigm with end-to-end models and more tasks, *e.g.*, object detection and segmentation.

Acknowledgement. This work was supported by the National Key R&D Program of China (No. 2021ZD0110700), Shanghai Municipal Science and Technology Major Project (2021SHZDZX0102), Shanghai Qi Zhi Institute, and SHEITC (2018-RGZN-02046).

References

1. Buda, M., Maki, A., Mazurowski, M.A.: A systematic study of the class imbalance problem in convolutional neural networks. Neural Netw. **106**, 249–259 (2018)
2. Byrd, J., Lipton, Z.: What is the effect of importance weighting in deep learning? In: International Conference on Machine Learning, pp. 872–881. PMLR (2019)
3. Cao, K., Wei, C., Gaidon, A., Arechiga, N., Ma, T.: Learning imbalanced datasets with label-distribution-aware margin loss. arXiv preprint arXiv:1906.07413 (2019)

4. Chao, Y.W., Wang, Z., He, Y., Wang, J., Deng, J.: HICO: a benchmark for recognizing human-object interactions in images. In: Proceedings of the IEEE International Conference on Computer Vision, pp. 1017–1025 (2015)
5. Chen, T., Kornblith, S., Norouzi, M., Hinton, G.: A simple framework for contrastive learning of visual representations. In: International Conference on Machine Learning, pp. 1597–1607. PMLR (2020)
6. Chou, Hsin-Ping., Chang, Shih-Chieh., Pan, Jia-Yu., Wei, Wei, Juan, Da-Cheng.: Remix: rebalanced mixup. In: Bartoli, Adrien, Fusiello, Andrea (eds.) ECCV 2020. LNCS, vol. 12540, pp. 95–110. Springer, Cham (2020). https://doi.org/10.1007/978-3-030-65414-6_9
7. Chu, P., Bian, X., Liu, S., Ling, H.: Feature space augmentation for long-tailed data. In: Vedaldi, A., Bischof, H., Brox, T., Frahm, J.-M. (eds.) ECCV 2020. LNCS, vol. 12374, pp. 694–710. Springer, Cham (2020). https://doi.org/10.1007/978-3-030-58526-6_41
8. Cubuk, E.D., Zoph, B., Shlens, J., Le, Q.V.: RandAugment: practical automated data augmentation with a reduced search space. In: Proceedings of the IEEE/CVF Conference on Computer Vision and Pattern Recognition Workshops, pp. 702–703 (2020)
9. Cui, J., Zhong, Z., Liu, S., Yu, B., Jia, J.: Parametric contrastive learning. arXiv preprint arXiv:2107.12028 (2021)
10. Danon, L., Diaz-Guilera, A., Duch, J., Arenas, A.: Comparing community structure identification. J. Stat. Mech Theory Exp. **2005**(09), P09008 (2005)
11. Deng, J., Dong, W., Socher, R., Li, L.J., Li, K., Fei-Fei, L.: ImageNet: a large-scale hierarchical image database. In: 2009 IEEE Conference on Computer Vision and Pattern Recognition, pp. 248–255. IEEE (2009)
12. Dinh, L., Sohl-Dickstein, J., Bengio, S.: Density estimation using real NVP. arXiv preprint arXiv:1605.08803 (2016)
13. Gidaris, S., Komodakis, N.: Dynamic few-shot visual learning without forgetting. In: Proceedings of the IEEE Conference on Computer Vision and Pattern Recognition, pp. 4367–4375 (2018)
14. Gupta, A., Dollar, P., Girshick, R.: LVIS: a dataset for large vocabulary instance segmentation. In: Proceedings of the IEEE/CVF conference on computer vision and pattern recognition, pp. 5356–5364 (2019)
15. He, H., Garcia, E.A.: Learning from imbalanced data. IEEE Trans. Knowl. Data Eng. **21**(9), 1263–1284 (2009)
16. He, K., Fan, H., Wu, Y., Xie, S., Girshick, R.: Momentum contrast for unsupervised visual representation learning. arXiv preprint arXiv:1911.05722 (2019)
17. He, K., Zhang, X., Ren, S., Sun, J.: Deep residual learning for image recognition. In: Proceedings of the IEEE Conference on Computer Vision and Pattern Recognition, pp. 770–778 (2016)
18. He, Y.Y., Wu, J., Wei, X.S.: Distilling virtual examples for long-tailed recognition. arXiv preprint arXiv:2103.15042 (2021)
19. Huang, C., Li, Y., Loy, C.C., Tang, X.: Learning deep representation for imbalanced classification. In: Proceedings of the IEEE Conference on Computer Vision and Pattern Recognition, pp. 5375–5384 (2016)
20. Huang, C., Li, Y., Loy, C.C., Tang, X.: Deep imbalanced learning for face recognition and attribute prediction. IEEE Trans. Pattern Anal. Mach. Intell. **42**(11), 2781–2794 (2019)
21. Izmailov, P., Kirichenko, P., Finzi, M., Wilson, A.G.: Semi-supervised learning with normalizing flows. In: International Conference on Machine Learning, pp. 4615–4630. PMLR (2020)

22. Jamal, M.A., Brown, M., Yang, M.H., Wang, L., Gong, B.: Rethinking class-balanced methods for long-tailed visual recognition from a domain adaptation perspective. In: Proceedings of the IEEE/CVF Conference on Computer Vision and Pattern Recognition, pp. 7610–7619 (2020)
23. Jamal, M.A., Brown, M., Yang, M.H., Wang, L., Gong, B.: Rethinking class-balanced methods for long-tailed visual recognition from a domain adaptation perspective. In: Proceedings of the IEEE/CVF Conference on Computer Vision and Pattern Recognition, pp. 7610–7619 (2020)
24. Japkowicz, N., Stephen, S.: The class imbalance problem: a systematic study. Intell. Data Anal. **6**(5), 429–449 (2002)
25. Jiang, Z., Chen, T., Mortazavi, B., Wang, Z.: Self-damaging contrastive learning. arXiv preprint arXiv:2106.02990 (2021)
26. Kang, B., Li, Y., Xie, S., Yuan, Z., Feng, J.: Exploring balanced feature spaces for representation learning. In: International Conference on Learning Representations (2020)
27. Kang, B., Xie, S., Rohrbach, M., Yan, Z., Gordo, A., Feng, J., Kalantidis, Y.: Decoupling representation and classifier for long-tailed recognition. arXiv preprint arXiv:1910.09217 (2019)
28. Kiefer, J., Wolfowitz, J.: Stochastic estimation of the maximum of a regression function. Ann. Math. Statist. **23**(3), 462–466 (1952)
29. Kim, B., Kim, J.: Adjusting decision boundary for class imbalanced learning. IEEE Access **8**, 81674–81685 (2020)
30. Kim, J., Jeong, J., Shin, J.: M2m: imbalanced classification via major-to-minor translation. In: Proceedings of the IEEE/CVF Conference on Computer Vision and Pattern Recognition, pp. 13896–13905 (2020)
31. Kingma, D.P., Salimans, T., Jozefowicz, R., Chen, X., Sutskever, I., Welling, M.: Improved variational inference with inverse autoregressive flow. Adv. Neural. Inf. Process. Syst. **29**, 4743–4751 (2016)
32. Krizhevsky, A., Sutskever, I., Hinton, G.E.: ImageNet classification with deep convolutional neural networks. Adv. Neural. Inf. Process. Syst. **25**, 1097–1105 (2012)
33. Li, S., Gong, K., Liu, C.H., Wang, Y., Qiao, F., Cheng, X.: MetaSAug: meta semantic augmentation for long-tailed visual recognition. In: Proceedings of the IEEE/CVF Conference on Computer Vision and Pattern Recognition, pp. 5212–5221 (2021)
34. Li, Y.L., et al.: Hake: a knowledge engine foundation for human activity understanding. arXiv preprint arXiv:2202.06851 (2022)
35. Li, Y.L., et al.: PaStaNet: toward human activity knowledge engine. In: CVPR (2020)
36. Li, Y.L., Xu, Y., Mao, X., Lu, C.: Symmetry and group in attribute-object compositions. In: CVPR (2020)
37. Li, Y.L., et al.: Transferable interactiveness knowledge for human-object interaction detection. In: CVPR (2019)
38. Liu, Z., Miao, Z., Zhan, X., Wang, J., Gong, B., Yu, S.X.: Large-scale long-tailed recognition in an open world. In: Proceedings of the IEEE/CVF Conference on Computer Vision and Pattern Recognition, pp. 2537–2546 (2019)
39. Van der Maaten, L., Hinton, G.: Visualizing data using t-SNE. J. Mach. Learn. Res. **9**(11) (2008)
40. Matthews, B.W.: Comparison of the predicted and observed secondary structure of t4 phage lysozyme. Biochimica et Biophysica Acta (BBA)-Protein Structure **405**(2), 442–451 (1975)

41. Menon, A.K., Jayasumana, S., Rawat, A.S., Jain, H., Veit, A., Kumar, S.: Long-tail learning via logit adjustment. arXiv preprint arXiv:2007.07314 (2020)
42. Oord, A.V.D., Li, Y., Vinyals, O.: Representation learning with contrastive predictive coding. arXiv preprint arXiv:1807.03748 (2018)
43. Papamakarios, G., Pavlakou, T., Murray, I.: Masked autoregressive flow for density estimation. arXiv preprint arXiv:1705.07057 (2017)
44. Qi, H., Brown, M., Lowe, D.G.: Low-shot learning with imprinted weights. In: Proceedings of the IEEE conference on computer vision and pattern recognition, pp. 5822–5830 (2018)
45. Radford, A., et al.: Learning transferable visual models from natural language supervision. In: International Conference on Machine Learning, pp. 8748–8763. PMLR (2021)
46. Ren, J., et al.: Balanced meta-softmax for long-tailed visual recognition. arXiv preprint arXiv:2007.10740 (2020)
47. Rezende, D., Mohamed, S.: Variational inference with normalizing flows. In: International conference on machine learning, pp. 1530–1538. PMLR (2015)
48. Samuel, D., Atzmon, Y., Chechik, G.: From generalized zero-shot learning to long-tail with class descriptors. In: Proceedings of the IEEE/CVF Winter Conference on Applications of Computer Vision, pp. 286–295 (2021)
49. Samuel, D., Atzmon, Y., Chechik, G.: From generalized zero-shot learning to long-tail with class descriptors. In: Proceedings of the IEEE/CVF Winter Conference on Applications of Computer Vision, pp. 286–295 (2021)
50. Samuel, D., Chechik, G.: Distributional robustness loss for long-tail learning. arXiv preprint arXiv:2104.03066 (2021)
51. Samuel, D., Chechik, G.: Distributional robustness loss for long-tail learning. arXiv preprint arXiv:2104.03066 (2021)
52. Shen, Li., Lin, Zhouchen, Huang, Qingming: Relay backpropagation for effective learning of deep convolutional neural networks. In: Leibe, Bastian, Matas, Jiri, Sebe, Nicu, Welling, Max (eds.) ECCV 2016. LNCS, vol. 9911, pp. 467–482. Springer, Cham (2016). https://doi.org/10.1007/978-3-319-46478-7_29
53. Simonyan, K., Zisserman, A.: Very deep convolutional networks for large-scale image recognition. arXiv preprint arXiv:1409.1556 (2014)
54. Sinha, S., Ohashi, H., Nakamura, K.: Class-wise difficulty-balanced loss for solving class-imbalance. In: Proceedings of the Asian Conference on Computer Vision (2020)
55. Spain, M., Perona, P.: Measuring and predicting importance of objects in our visual world (2007)
56. Van Horn, G., et al.: The inaturalist species classification and detection dataset. In: Proceedings of the IEEE conference on computer vision and pattern recognition, pp. 8769–8778 (2018)
57. Wang, J., Lukasiewicz, T., Hu, X., Cai, J., Xu, Z.: Rsg: A simple but effective module for learning imbalanced datasets. In: Proceedings of the IEEE/CVF Conference on Computer Vision and Pattern Recognition, pp. 3784–3793 (2021)
58. Wang, P., Han, K., Wei, X.S., Zhang, L., Wang, L.: Contrastive learning based hybrid networks for long-tailed image classification. In: Proceedings of the IEEE/CVF Conference on Computer Vision and Pattern Recognition, pp. 943–952 (2021)
59. Wang, X., Lian, L., Miao, Z., Liu, Z., Yu, S.X.: Long-tailed recognition by routing diverse distribution-aware experts. arXiv preprint arXiv:2010.01809 (2020)

60. Wang, Y.X., Ramanan, D., Hebert, M.: Learning to model the tail. In: Proceedings of the 31st International Conference on Neural Information Processing Systems, pp. 7032–7042 (2017)
61. Wu, Tz-Ying., Morgado, Pedro, Wang, Pei, Ho, Chih-Hui., Vasconcelos, Nuno: Solving long-tailed recognition with deep realistic taxonomic classifier. In: Vedaldi, Andrea, Bischof, Horst, Brox, Thomas, Frahm, Jan-Michael. (eds.) ECCV 2020. LNCS, vol. 12353, pp. 171–189. Springer, Cham (2020). https://doi.org/10.1007/978-3-030-58598-3_11
62. Xiang, Liuyu, Ding, Guiguang, Han, Jungong: Learning from multiple experts: self-paced knowledge distillation for long-tailed classification. In: Vedaldi, Andrea, Bischof, Horst, Brox, Thomas, Frahm, Jan-Michael. (eds.) ECCV 2020. LNCS, vol. 12350, pp. 247–263. Springer, Cham (2020). https://doi.org/10.1007/978-3-030-58558-7_15
63. Yang, Y., Xu, Z.: Rethinking the value of labels for improving class-imbalanced learning. arXiv preprint arXiv:2006.07529 (2020)
64. Zhang, S., Chen, C., Hu, X., Peng, S.: Balanced knowledge distillation for long-tailed learning. arXiv preprint arXiv:2104.10510 (2021)
65. Zhang, S., Li, Z., Yan, S., He, X., Sun, J.: Distribution alignment: a unified framework for long-tail visual recognition. In: Proceedings of the IEEE/CVF Conference on Computer Vision and Pattern Recognition, pp. 2361–2370 (2021)
66. Zhang, Z., Pfister, T.: Learning fast sample re-weighting without reward data. In: Proceedings of the IEEE/CVF International Conference on Computer Vision, pp. 725–734 (2021)
67. Zhong, Z., Cui, J., Liu, S., Jia, J.: Improving calibration for long-tailed recognition. In: Proceedings of the IEEE/CVF Conference on Computer Vision and Pattern Recognition, pp. 16489–16498 (2021)
68. Zhong, Z., Cui, J., Liu, S., Jia, J.: Improving calibration for long-tailed recognition. In: Proceedings of the IEEE/CVF Conference on Computer Vision and Pattern Recognition, pp. 16489–16498 (2021)
69. Zhou, B., Lapedriza, A., Khosla, A., Oliva, A., Torralba, A.: Places: a 10 million image database for scene recognition. IEEE Trans. Pattern Anal. Mach. Intell. **40**(6), 1452–1464 (2017)
70. Zhou, B., Cui, Q., Wei, X.S., Chen, Z.M.: BBN: bilateral-branch network with cumulative learning for long-tailed visual recognition. In: Proceedings of the IEEE/CVF Conference on Computer Vision and Pattern Recognition, pp. 9719–9728 (2020)
71. Zipf, G.K.: The psycho-biology of language: an introduction to dynamic philology. Routledge (2013)

On Multi-Domain Long-Tailed Recognition, Imbalanced Domain Generalization and Beyond

Yuzhe Yang[1]([✉]), Hao Wang[2], and Dina Katabi[1]

[1] MIT CSAIL, Cambridge, USA
yuzhe@mit.edu
[2] Rutgers University, Newark, USA

Abstract. Real-world data often exhibit imbalanced label distributions. Existing studies on data imbalance focus on single-domain settings, i.e., samples are from the same data distribution. However, natural data can originate from distinct domains, where a minority class in one domain could have abundant instances from other domains. We formalize the task of Multi-Domain Long-Tailed Recognition (MDLT), which learns from multi-domain imbalanced data, addresses *label imbalance*, *domain shift*, and *divergent label distributions across domains*, and generalizes to all domain-class pairs. We first develop the *domain-class transferability graph*, and show that such transferability governs the success of learning in MDLT. We then propose BoDA, a theoretically grounded learning strategy that tracks the upper bound of transferability statistics, and ensures *balanced* alignment and calibration across imbalanced domain-class distributions. We curate five MDLT benchmarks based on widely-used multi-domain datasets, and compare BoDA to twenty algorithms that span different learning strategies. Extensive and rigorous experiments verify the superior performance of BoDA. Further, as a byproduct, BoDA establishes new state-of-the-art on Domain Generalization benchmarks, highlighting the importance of addressing data imbalance across domains, which can be crucial for improving generalization to unseen domains. Code and data are available at: https://github.com/YyzHarry/multi-domain-imbalance.

1 Introduction

Real-world data often exhibit label imbalance – i.e., instead of a uniform label distribution over classes, in reality, data are by their nature imbalanced: a few classes contain a large number of instances, whereas many others have only a few instances [5,6,52]. This phenomenon poses a challenge for deep recognition models, and has motivated several prior solutions [6,10,33,39,52,53]. Such prior solutions focus on *single domain* scenarios, i.e., samples are from the same data

Supplementary Information The online version contains supplementary material available at https://doi.org/10.1007/978-3-031-20044-1_4.

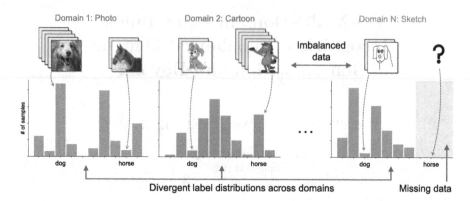

Fig. 1. Multi-Domain Long-Tailed Recognition (MDLT) aims to learn from imbalanced data from multiple distinct domains, tackle label imbalance, domain shift, and divergent label distributions across domains, and generalize to all domain-class pairs.

distribution; they propose techniques for learning from imbalanced training data and generalizing to a balanced test set.

In contrast, this paper formulates the problem of *Multi-Domain Long-Tailed Recognition* (MDLT) as learning from multi-domain imbalanced data, with each domain having its own imbalanced label distribution, and generalizing to a test set that is balanced over all domain-class pairs. MDLT is a natural extension of the single domain case. It arises in real-world scenarios, where data targeted for one task can originate from different domains. For example, in visual recognition problems, minority classes from "photo" images could be complemented with potentially abundant samples from "sketch" images. Similarly, in autonomous driving, the minority accident class in "real" life could be enriched with accidents generated in "simulation". Also, in medical diagnosis, data from distinct populations could enhance each other, where minority samples from one institution could be enriched with instances from others. In the above examples, different data types act as distinct *domains*, and such multi-domain data could be leveraged to tackle the inherent data imbalance within each domain.

We note that MDLT has key differences from its single-domain counterpart:

– First, the label distribution for each domain is likely different from other domains. For example, in Fig. 1, both "Photo" and "Cartoon" domains exhibit imbalanced label distributions; Yet, the "horse" class in "Cartoon" has many more samples than in "Photo". This creates challenges with *divergent label distributions across domains*, in addition to in-domain data imbalance.
– Second, multi-domain data inherently involves *domain shift*. Simply treating different domains as a whole and applying traditional data-imbalance methods is unlikely to yield the best results, as the domain gap can be arbitrarily large.
– Third, MDLT naturally motivates *zero-shot generalization within and across domains* – i.e., to generalize to both in-domain missing classes (Fig. 1 right part), as well as new domains with no training data, where the latter case is typically denoted as Domain Generalization (DG).

To deal with the above issues, we first develop the *domain-class transferability graph*, which quantifies the transferability between different domain-class pairs under data imbalance. In this graph, each node refers to a domain-class pair, and each edge refers to the distance between two domain-class pairs in the embedding space. We show that the transferability graph dictates the performance of imbalanced learning across domains. Inspired by this, we design BoDA (Balanced Domain-Class Distribution Alignment), a new loss function that encourages similarity between features of the same class in different domains, and penalizes similarity between features of different classes within and across domains. BoDA does so while accounting for that different classes have very different number of samples, and hence the statistics of their features are intrinsically imbalanced. Analytically, we prove that minimizing the BoDA loss optimizes an upper bound of the *balanced* transferability statistics, corroborating the effectiveness of BoDA for learning multi-domain imbalanced data.

For MDLT evaluation, we curate five MDLT benchmarks based on datasets widely used for domain generalization (DG). These datasets naturally exhibit heavy class imbalance within each domain and data shift across domains, highlighting that the MDLT problem is widely present in current benchmarks. We compare BoDA against twenty algorithms that span different learning strategies. Extensive experiments across benchmarks and algorithms verify that BoDA consistently outperforms all these baselines on all datasets.

Additionally, we examine how BoDA performs in the DG setting. We show that combining BoDA with the DG state-of-the-art (SOTA) consistently brings further gains, yielding a new SOTA for DG. These results shed light on how label imbalance can affect out-of-distribution generalization and highlight the importance of integrating label imbalance into practical DG algorithm design.

Our contributions are as follows:

- We formulate the MDLT problem as learning from multi-domain imbalanced data and generalizing across all domain-class pairs.
- We introduce the domain-class transferability graph, a unified model for investigating MDLT. We further show that the transferability statistics induced from such graph are crucial and govern the success of MDLT algorithms.
- We design BoDA, a simple, effective, and interpretable loss function for MDLT. We prove theoretically that minimizing the BoDA loss is equivalent to optimizing an upper bound of balanced transferability statistics.
- Extensive experiments on benchmark datasets verify the superior and consistent performance of BoDA. Further, combined with DG algorithms, BoDA establishes a new SOTA on DG benchmarks, highlighting the importance of tackling cross-domain data imbalance for domain generalization.

2 Related Work

Long-Tailed Recognition. The literature is rich with research on long-tailed recognition [33,57]. Proposed solutions include re-balancing the data by over-sampling/under-sampling [9,20], re-weighting or adjusting the loss functions [6,

10,12,22], as well as leveraging relevant learning paradigms such as transfer learning [33], metric learning [55], meta-learning [43], two-stage training [23], ensemble learning [48,56], and self-supervised learning [30,52]. Recent studies have also explored imbalanced regression [53]. In contrast to these past works, we extend long-tailed recognition to the multi-domain setting, and introduce new techniques suitable for learning from multi-domain imbalanced data.

Multi-Domain Learning. Multi-domain learning (MDL) aims to learn a model of minimal risk from datasets drawn from different underlying distributions [13], and is a specific case of transfer learning [37]. In contrast to domain adaptation (DA) [3,37], which aims to minimize the risk over a single "target" domain, MDL minimizes the risk over all "source" domains, and considers both average and worst risks over all distributions [41]. Past solutions for MDL include designing shared and domain-specific models [13,49], leveraging multi-task learning [51], and learning domain-invariant features [15,31,41,45]. Our work falls under the MDL framework, but considers the practical and realistic setting where the label distribution is imbalanced within each domain and across domains.

Domain Generalization. Unlike MDL which focuses on in-domain generalization, domain generalization (DG) aims to learn from multiple training domains and generalize to unseen domains [59]. Previous approaches include learning domain-invariant features [15,31,34], learning transferable model parameters using meta-learning [27,54], data augmentation [7,60], and capturing causal relationships [1,25]. Past work on DG has not investigated label imbalance within a domain and across domains. This paper shows that label imbalance plays a crucial role in DG, and that by combating data imbalance, we substantially boost DG performance on standard benchmarks.

3   Domain-Class Transferability Graph

When learning from MDLT, a natural question arises: How do we model MDLT in the presence of both *domain shift* and *class imbalance within and across domains*? We argue that in contrast to single-domain imbalanced learning where the basic unit one cares about is a *class* (i.e., minority *vs.* majority classes), in MDLT, the basic unit naturally translates to a **domain-class pair**.

Problem Setup. Given a multi-domain classification task with a discrete label space $\mathcal{C} = \{1, \ldots, C\}$ and a domain space $\mathcal{D} = \{1, \ldots, D\}$, let $\mathcal{S} = \{(\mathbf{x}_i, c_i, d_i)\}_{i=1}^{N}$ be the training set, where $\mathbf{x}_i \in \mathbb{R}^l$ denotes the input, $c_i \in \mathcal{C}$ is the class label, and $d_i \in \mathcal{D}$ is the domain label. We denote as $\mathbf{z} = f(\mathbf{x}; \theta)$ the representation of $\mathbf{x}$, where $f : \mathcal{X} \to \mathcal{Z}$ maps the input into a representation space $\mathcal{Z} \subseteq \mathbb{R}^h$. The final prediction $\widehat{c} = g(\mathbf{z})$ is given by a classification function $g : \mathcal{Z} \to \mathcal{C}$. We denote the set of samples belonging to domain d and class c (i.e., the domain-class pair (d, c)) as $\mathcal{S}_{d,c} \subseteq \mathcal{S}$, with $N_{d,c} \triangleq |\mathcal{S}_{d,c}|$ as the number of samples. Similarly, $\mathcal{Z}_{d,c} \subseteq \mathcal{Z}$ denotes the representation set for (d, c). We use $\mathcal{M} = \mathcal{D} \times \mathcal{C} := \{(d, c) : d \in \mathcal{D}, c \in \mathcal{C}\}$ to denote the set of all domain-class pairs.

Definition 1 (Transferability). *Given a learned model and a distance function* $d : \mathbb{R}^h \times \mathbb{R}^h \to \mathbb{R}$ *in the feature space, the transferability from domain-class pair* (d, c) *to* (d', c') *is:*

$$\text{trans}\big((d, c), (d', c')\big) \triangleq \mathbb{E}_{\mathbf{z} \in \mathcal{Z}_{d,c}} \big[d\left(\mathbf{z}, \boldsymbol{\mu}_{d',c'}\right) \big],$$

where $\boldsymbol{\mu}_{d',c'} \triangleq \mathbb{E}_{\mathbf{z'} \in \mathcal{Z}_{d',c'}} [\mathbf{z'}]$ *is the first order statistics (i.e., mean) of* (d', c').

Intuitively, the transferability between two domain-class pairs is the average distance between their learned representations, characterizing how close they are in the feature space. By default, d is chosen as the Euclidean distance, but it can also represent the higher order statistics of (d, c). For example, the Mahalanobis distance [11] uses the covariance $\boldsymbol{\Sigma}_{d,c} \triangleq \mathbb{E}_{\mathbf{z} \in \mathcal{Z}_{d,c}} \big[(\mathbf{z} - \boldsymbol{\mu}_{d,c})(\mathbf{z} - \boldsymbol{\mu}_{d,c})^{\top} \big]$. In the remainder of the paper, with a slight abuse of the notation, we allow $\boldsymbol{\mu}_{d,c}$ to represent both the first and higher order statistics for (d, c).

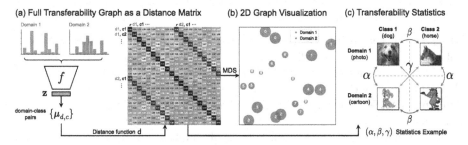

Fig. 2. Overall framework of transferability graph. **(a)** Distribution statistics $\{\boldsymbol{\mu}_{d,c}\}$ is computed for all domain-class pairs, by which we generate a full transferability matrix. **(b)** MDS is used to project the graph into a 2D space for visualization. **(c)** We define (α, β, γ) transferability statistics to further describe the whole transferability graph.

Definition 2 (Transferability Graph). *The transferability graph for a learned model is defined as* $\mathcal{G} = (\mathcal{V}, \mathcal{E})$, *where the vertices,* $\mathcal{V} \subseteq \{\boldsymbol{\mu}_{d,c}\}$, *represents the domain-class pairs, and the edges,* $\mathcal{E} \subseteq \mathcal{V} \times \mathcal{V}$, *are assigned weights equal to* $\text{trans}\left((d, c), (d', c')\right)$.

Transferability Graph Visualization. It is convenient to visualize the transferability graph of a learned model in a 2D Cartesian space. To do so, we use the average of $\text{trans}\left((d, c), (d', c')\right)$ and $\text{trans}\left((d', c'), (d, c)\right)$ as a similarity measure between them. We can then visualize this similarity and the underlying transferability graph using multidimensional scaling (MDS) [8]. Figures 2a and 2b show this process, where for each (d, c) pair, we estimate its distribution statistics $\{\boldsymbol{\mu}_{d,c}\}$ from the learned model, then compute the model transferability graph as a distance matrix. We then use MDS to project it into a 2D space, where each dot refers to one (d, c), and the distance represents transferability.

Definition 3 $((\alpha, \beta, \gamma)$ **Transferability Statistics**)**. *The transferability graph can be summarized by the following transferability statistics:*

$$\text{Different domains, same class:} \quad \alpha = \mathbb{E}_c \mathbb{E}_d \mathbb{E}_{d' \neq d} \left[\text{trans}((d, c), (d', c))\right].$$

$$\text{Same domain, different classes:} \quad \beta = \mathbb{E}_d \mathbb{E}_c \mathbb{E}_{c' \neq c} \left[\text{trans}((d, c), (d, c'))\right].$$

$$\text{Different domains, different classes:} \quad \gamma = \mathbb{E}_d \mathbb{E}_{d' \neq d} \mathbb{E}_c \mathbb{E}_{c' \neq c} \left[\text{trans}((d, c), (d', c'))\right].$$

As illustrated in Fig. 2c, (α, β, γ) captures the similarity between features of the same class across domains and different classes within and across domains.

4   What Makes for Good Representations in MDLT?

4.1   Divergent Label Distributions Hamper Transferable Features

MDLT has to deal with differences between the label distributions across domains. To understand the implications of this issue we start with an example.

Motivating Example. We construct `Digits-MLT`, a two-domain toy MDLT dataset that combines two digit datasets: MNIST-M [15] and SVHN [36]. The task is 10-class digit classification. Details of the datasets are in Appendix D. We manually vary the number of samples for each domain-class pair to simulate different label distributions, and train a plain ResNet-18 [21] using empirical risk minimization (ERM) for each case. We keep all test sets balanced and identical.

The results in Fig. 3 reveal interesting observations. When the per-domain label distributions are balanced and *identical* across domains, although a domain gap exists, it does not prohibit the model from learning discriminative features of high accuracy (90.5%), as shown in Fig. 3a. If the label distributions are imbalanced but *identical*, as in Fig. 3b, ERM is still able to align similar classes in the two domains, where majority classes (e.g., class 9) are closer in terms of

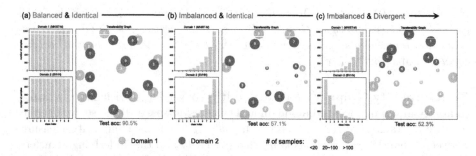

Fig. 3. The evolving pattern of transferability graph when varying label proportions of `Digits-MLT`. **(a)** Label distribution for two domains are balanced and identical. **(b)** Label distribution for two domains are imbalanced but identical. **(c)** Label distribution for two domains are imbalanced and *divergent*.

transferability than minority classes (e.g., class 0). In contrast, when the labels are both imbalanced and *mismatched* across domains, as in Fig. 3c, the learned features are no longer transferable, resulting in a clear gap across domains and the worst accuracy. This is because *divergent label distributions* across domains produce an undesirable shortcut; the model can minimize the classification loss simply by separating the two domains.

Transferable Features are Desirable. As the results indicate, *transferable* features across (d, c) pairs are needed, especially when imbalance occurs. In particular, the transferability link between the same class across domains should be greater than that between different classes within or across domains. This can be captured via the (α, β, γ) transferability statistics, as we show next.

4.2   Transferability Statistics Characterize Generalization

Motivating Example. Again, we use `Digits-MLT` with varying label distributions. We consider three imbalance types to compose different label configurations: (1) **Uniform** (i.e., balanced labels), (2) **Forward-LT**, where the labels exhibit a long tail over class ids, and (3) **Backward-LT**, where labels are inversely long-tailed with respect to the class ids. For each configuration, we train 20 ERM models with varying hyperparameters. We then calculate the (α, β, γ) statistics for each model, and plot its classification accuracy against $(\beta + \gamma) - \alpha$.

Figure 4 reveals the following findings: (1) *The (α, β, γ) statistics characterize a model's performance in MDLT*. In particular, the $(\beta + \gamma) - \alpha$ quantity displays a very strong correlation with test performance across the entire range and every label configuration. (2) *Data imbalance increases the risk of learning less transferable features*. When the label distributions are similar across domains (Fig. 4a), the models are robust to varying parameters, clustering in the upper-right region. However, as the labels become imbalanced (Figs. 4b, 4c) and further

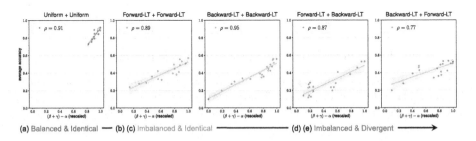

Fig. 4. Correspondence between $(\beta + \gamma) - \alpha$ quantity and test accuracy across different label configurations of `Digits-MLT`. Each plot refers to specific label distributions for two domains (e.g., (a) employs "Uniform" for domain 1 and "Uniform" for domain 2). Each point corresponds to a model trained with ERM using different hyperparameters.

divergent (Figs. 4d, 4e), chances that the model learns non-transferable features (i.e., lower $(\beta + \gamma) - \alpha$) increase, leading to a large drop in performance. We provide further evidence in Appendix H.4 showing that these observations hold regardless of datasets and training regimes.

4.3   A Loss that Bounds the Transferability Statistics

We use the above findings to design a new loss function particularly suitable for MDLT. We will first introduce the loss function then prove that it minimizes an upper bound of the (α, β, γ) statistics. We start from a simple loss inspired by the metric learning objective [17,44]. We call this loss $\mathcal{L}_{DA}$ since it aims for Domain Alignment, i.e., aligning the features of the same class across domains. Let $(\mathbf{x}_i, c_i, d_i)$ denote a sample with feature $\mathbf{z}_i$. Given a set of training samples with feature set $\mathcal{Z}$, we have

$$\mathcal{L}_{DA}(\mathcal{Z}, \{\boldsymbol{\mu}\}) = \sum_{\mathbf{z}_i \in \mathcal{Z}} \frac{-1}{|\mathcal{D}|-1} \sum_{d \in \mathcal{D} \setminus \{d_i\}} \log \frac{\exp(-\mathsf{d}(\mathbf{z}_i, \boldsymbol{\mu}_{d,c_i}))}{\sum_{(d',c') \in \mathcal{M} \setminus \{(d_i,c_i)\}} \exp(-\mathsf{d}(\mathbf{z}_i, \boldsymbol{\mu}_{d',c'}))}. \quad (1)$$

Intuitively, $\mathcal{L}_{DA}$ tackles label *divergence*, as (d, c) pairs that share same class would be pulled closer, and vice versa. It is also related to (α, β, γ) because the numerator represents *positive* cross-domain pairs (α), and the denominator represents *negative* cross-class pairs (β, γ). A detailed probabilistic interpretation of $\mathcal{L}_{DA}$ is provided in Appendix B.2.

But, $\mathcal{L}_{DA}$ does not address label *imbalance*. Note that (α, β, γ) is defined in a *balanced* way, independent of the number of samples of each (d, c). However, given an imbalanced dataset, most samples will come from majority domain-class pairs, which would dominate $\mathcal{L}_{DA}$ and cause minority pairs to be overlooked.

To tackle data imbalance across (d, c) pairs, we modify the loss in Eqn. (1) to the BoDA loss:

$$\mathcal{L}_{BoDA}(\mathcal{Z}, \{\boldsymbol{\mu}\}) = \sum_{\mathbf{z}_i \in \mathcal{Z}} \frac{-1}{|\mathcal{D}|-1} \sum_{d \in \mathcal{D} \setminus \{d_i\}} \log \frac{\exp(-\widetilde{\mathsf{d}}(\mathbf{z}_i, \boldsymbol{\mu}_{d,c_i}))}{\sum_{(d',c') \in \mathcal{M} \setminus \{(d_i,c_i)\}} \exp(-\widetilde{\mathsf{d}}(\mathbf{z}_i, \boldsymbol{\mu}_{d',c'}))}, \quad \widetilde{\mathsf{d}}(\mathbf{z}_i, \boldsymbol{\mu}_{d,c}) = \frac{\mathsf{d}(\mathbf{z}_i, \boldsymbol{\mu}_{d,c})}{N_{d_i,c_i}}. \quad (2)$$

BoDA scales the original d by a factor of $1/N_{d_i,c_i}$, i.e., it counters the effect of imbalanced domain-class pairs by introducing a *balanced* distance measure $\widetilde{\mathsf{d}}$.

Theorem 1 ($\mathcal{L}_{BoDA}$ as an Upper Bound). *Given a multi-domain long-tailed dataset $\mathcal{S}$ with domain label space $\mathcal{D}$ and class label space $\mathcal{C}$ satisfying $|\mathcal{D}| > 1$ and $|\mathcal{C}| > 1$, let $\mathcal{Z}$ be the representation set of all training samples, and (α, β, γ) be the transferability statistics for $\mathcal{S}$ defined in Definition 3. It holds that*

$$\mathcal{L}_{BoDA}(\mathcal{Z}, \{\boldsymbol{\mu}\}) \geq N \log \left(|\mathcal{D}| - 1 + |\mathcal{D}|(|\mathcal{C}| - 1) \exp \left(\frac{|\mathcal{C}||\mathcal{D}|}{N} \cdot \alpha - \frac{|\mathcal{C}|}{N} \cdot \beta - \frac{|\mathcal{C}|(|\mathcal{D}|-1)}{N} \cdot \gamma \right) \right). \quad (3)$$

The proof of Theorem 1 is in Appendix A.2. Theorem 1 has the following interesting implications: (1) $\mathcal{L}_{BoDA}$ *upper-bounds* (α, β, γ) *statistics in a desired form that naturally translates to better performance.* By minimizing $\mathcal{L}_{BoDA}$, we ensure a low α (attract same classes) and high β, γ (separate different classes), which are essential conditions for generalization in MDLT. (2) *The constant factors correspond to how much each component contributes to the transferability graph.* Zooming on the arguments of $\exp(\cdot)$, we observe that the objective is proportional to $\alpha - (\frac{1}{|\mathcal{D}|}\beta + \frac{|\mathcal{D}|-1}{|\mathcal{D}|}\gamma)$. According to Definition 3, we note that α summarizes data similarity for the same class, while $(\frac{1}{|\mathcal{D}|}\beta + \frac{|\mathcal{D}|-1}{|\mathcal{D}|}\gamma)$ summarizes data similarity across different classes, using the weighted average of β and γ, where their weights are proportional to the number of associated domains (i.e., 1 for β, $(|\mathcal{D}| - 1)$ for γ).

4.4 Calibration for Data Imbalance Leads to Better Transfer

BoDA works by encouraging feature transfer for similar classes across domains, i.e., if (d, c) and (d', c) refer to the same class in different domains, then we want to transfer their features to each other. But, minority domain-class pairs naturally have worse $\boldsymbol{\mu}_{d,c}$ estimates due to data scarcity, and forcing other pairs to transfer to them hurts learning. Thus, when bringing two domain-class pairs closer in the embedding space, we want the minority (d, c) to transfer to majority ones, not the inverse. The following example further clarifies this point.

Motivating Example. We use `Digits-MLT` with divergent labels (Fig. 5). We focus on *feature discrepancy*, i.e., the distance between training and test features for the same class. For each class in domain 1, we compute the distance in the feature space between the means of the training set and test set (solid line). We also compute the distance between the training data of domain 2 and test data of domain 1 (dashed line), for the same class.

As shown by the solid orange line in Fig. 5b, for minority domain-class pairs such as class "8" and "9" in domain 1, the distance in the feature space between training and testing is large. In fact, the test set of these minority domain-class pairs is closer to the training data for "8" and "9" in domain 2 than in their own domain, as shown by the dashed purple line. This example indicates that a better training would try to transfer the features of minority domain-class pairs to majority pairs with which they share the same class, as shown by the grey arrow in Fig. 5b. Such transfer will improve generalization to the test set.

BoDAwith Calibrated Distance. The above discussion motivates a modification to BoDA to favor transfer to majority domain-class pairs:

$$\widetilde{\mathcal{L}}_{\text{BoDA}}(\mathcal{Z}, \{\boldsymbol{\mu}\}) = \sum_{\mathbf{z}_i \in \mathcal{Z}} \frac{-1}{|\mathcal{D}| - 1} \sum_{d \in \mathcal{D} \setminus \{d_i\}} \log \frac{\exp\left(-\lambda_{d_i,c_i}^{d,c_i} \widetilde{d}(\mathbf{z}_i, \boldsymbol{\mu}_{d,c_i})\right)}{\sum_{(d',c') \in \mathcal{M} \setminus \{(d_i,c_i)\}} \exp\left(-\lambda_{d_i,c_i}^{d',c'} \widetilde{d}(\mathbf{z}_i, \boldsymbol{\mu}_{d',c'})\right)},$$

$$\lambda_{d,c}^{d',c'} = \left(\frac{N_{d',c'}}{N_{d,c}}\right)^{\nu}, \tag{4}$$

where ν is a constant that allows for a sublinear relation (default $\nu = 1$). $\lambda_{d,c}^{d',c'}$ indicates how much we would like to transfer (d, c) to (d', c'), based on their

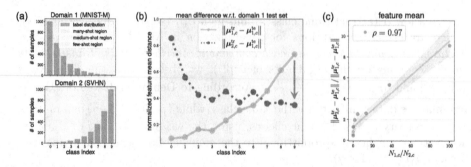

Fig. 5. The need for *calibration*. (**a**) Per-domain label distribution. (**b**) Distance between training and test data. Solid plots the distance between training and test data from the same domain-class pairs. Dashed plots the distance between test data from a particular domain-class pair and the training data with which it shares the same class but differs in the domain. The blue and red background colors refer to majority and minority domain-class pairs, respectively. (**c**) Correspondence between the feature distance ratio and the sample size ratio for two domain-class pairs. (Color figure online)

relative sample size. Figure 5c verifies that the ratio of the sample size is highly correlated with the ratio of the distance between testing and training. Further, Theorem 2 in Appendix A shows that $\widetilde{\mathcal{L}}_{\text{BoDA}}$ is an upper bound of the calibrated transferability statistics.

Variants of BoDA: Matching Higher Order Statistics. The distance d can be set to the Euclidean distance $d(\mathbf{z}, \boldsymbol{\mu}_{d,c}) = \sqrt{(\mathbf{z} - \boldsymbol{\mu}_{d,c})^\top (\mathbf{z} - \boldsymbol{\mu}_{d,c})}$, which captures the first order statistics. To match higher order statistics such as covariance, we set $d(\mathbf{z}, \{\boldsymbol{\mu}_{d,c}, \boldsymbol{\Sigma}_{d,c}\}) = \sqrt{(\mathbf{z} - \boldsymbol{\mu}_{d,c})^\top \boldsymbol{\Sigma}_{d,c}^{-1} (\mathbf{z} - \boldsymbol{\mu}_{d,c})}$, resembling the Mahalanobis distance [11]. We refer to these variants as $\widetilde{\mathcal{L}}_{\text{BoDA}}$ and $\widetilde{\mathcal{L}}_{\text{BoDA-M}}$.

Joint Loss. BoDA serves as a representation learning scheme for MDLT, which operates over $\mathcal{Z}$. For classification, we train deep networks by combining $\widetilde{\mathcal{L}}_{\text{BoDA}}$ and the standard cross-entropy (CE) loss in an end-to-end fashion, where CE is applied to the output layer, and BoDA is applied to the latent features. We combine the losses as $\mathcal{L}_{\text{CE}} + \omega \widetilde{\mathcal{L}}_{\text{BoDA}}$, with ω as a trade-off hyperparameter.

5   What Makes for Good Classifiers in MDLT?

In the long-tailed recognition literature, an important finding is that decoupling *representation learning* and *classifier learning* leads to better results [23,58]. In particular, instance-balanced sampling is used during the first stage of learning, while class-balanced sampling is used for re-training the classifier (with the representation fixed) in the second stage [23]. Motivated by this, we explore whether a similar decoupling benefits MDLT. We use three learning algorithms, ERM [46],

DANN [31], and CORAL [45]. We train each algorithm with and without the second stage classifier learning, and report the average accuracy over all MDLT datasets (presented later).

As Table 1 shows, similar to what has been observed in the single domain case [23,58], regardless of algorithm, decoupling the classifier learning consistently improves performance. Since BoDA can support both coupled and decoupled classifier learning, we use BoDA$_r$ to refer to models that couple representation and classifier learning, and BoDA$_{r,c}$ for models that decouple representation from classifier learning. In the classifier learning stage, we simply use class-balanced sampling.

Table 1. The benefits of decoupling the classifier.

Algorithm	w/o decouple	w/ decouple
ERM [46]	77.6 ±0.2	**79.2** ±0.3
DANN [15]	77.7 ±0.6	**79.0** ±0.1
CORAL [45]	78.0 ±0.1	**79.6** ±0.2

6   Benchmarking MDLT

Datasets. We curate five multi-domain datasets typically used in DG and adapt them for MDLT evaluation. To do so, for each dataset, we create two balanced datasets one for validation and the other for testing, and leave the rest for training. The size of the validation and test data sets is 5% and 10% of original data, respectively. Table 10 in Appendix D provides the statistics of each MDLT dataset. Figure 6 shows the label distributions across domains in the five datasets.

1. VLCS-MLT. We construct VLCS-MLT using the VLCS dataset [14], which is an object recognition dataset with 10,729 images from 4 domains and 5 classes.
2. PACS-MLT. PACS-MLT is constructed from the PACS dataset [28], an object recognition dataset with 9,991 images from 4 domains and 7 classes.
3. OfficeHome-MLT. We set up OfficeHome-MLT using the OfficeHome dataset [47] which contains 15,588 images from 4 domains and 65 classes.
4. TerraInc-MLT. TerraInc-MLT is created from TerraIncognita [2], a species classification dataset including 24,788 images from 4 domains and 10 classes.
5. DomainNet-MLT. We construct DomainNet-MLT using DomainNet [38], a large-scale multi-domain dataset for object recognition. It contains 586,575 images from 345 classes and 6 domains.

Network Architectures. For experiments on the synthetic Digits-MLT dataset, we use a simple CNN architecture as in [19]. For the MDLT datasets, we follow [19], and use ResNet-50 [21] for all algorithms.

Competing Algorithms. We compare BoDA to a large number of algorithms that span different learning strategies and categories, including (1) *vanilla:*

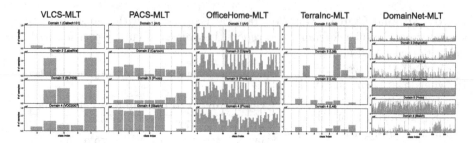

Fig. 6. Overview of training set label distribution for five MDLT datasets. We set up MDLT benchmarks from datasets traditionally used for DG, and make validation/test sets balanced across all domain-class pairs. More details are provided in Appendix D.

Table 2. Results on VLCS-MLT.

Algorithm	Accuracy (by domain)		Accuracy (by shot)			
	Average	Worst	Many	Medium	Few	Zero
ERM [46]	76.3 ±0.4	53.6 ±1.1	84.6 ±0.5	76.6 ±0.4	–	32.9 ±0.4
IRM [1]	76.5 ±0.2	52.3 ±0.7	85.3 ±0.6	75.5 ±1.0	–	33.5 ±1.0
GroupDRO [40]	76.7 ±0.4	54.1 ±1.3	85.3 ±0.9	76.2 ±1.0	–	34.5 ±2.0
Mixup [50]	75.9 ±0.1	52.7 ±1.3	84.4 ±0.2	77.1 ±0.6	–	29.2 ±1.4
MLDG [27]	76.9 ±0.2	53.6 ±0.5	84.9 ±0.3	77.5 ±1.0	–	34.4 ±0.9
CORAL [45]	75.9 ±0.5	51.6 ±0.7	84.3 ±0.6	75.5 ±0.5	–	34.5 ±0.8
MMD [29]	76.8 ±0.6	53.4 ±0.3	84.5 ±0.8	77.1 ±0.5	–	32.7 ±0.3
DANN [15]	77.5 ±0.1	54.1 ±0.3	85.9 ±0.5	76.0 ±0.4	–	38.0 ±0.3
CDANN [31]	76.6 ±0.4	53.6 ±0.4	84.4 ±0.7	77.3 ±0.8	–	35.0 ±0.8
MTL [4]	76.3 ±0.3	52.9 ±0.5	84.8 ±0.9	76.2 ±0.6	–	33.3 ±1.4
SagNet [35]	76.3 ±0.2	52.3 ±0.2	85.3 ±0.3	75.1 ±0.2	–	32.9 ±0.3
Fish [42]	77.5 ±0.3	54.3 ±0.4	86.2 ±0.5	76.0 ±0.4	–	35.6 ±2.2
Focal [32]	75.6 ±0.4	52.3 ±0.2	84.0 ±0.2	75.5 ±0.6	–	32.7 ±0.9
CBLoss [10]	76.8 ±0.3	52.5 ±0.5	84.8 ±0.7	77.5 ±1.4	–	33.2 ±1.6
LDAM [6]	77.5 ±0.1	52.9 ±0.2	86.5 ±0.4	75.5 ±0.5	–	35.2 ±0.6
BSoftmax [39]	76.7 ±0.5	52.9 ±0.9	84.4 ±0.9	78.2 ±0.6	–	34.3 ±0.9
SSP [52]	76.1 ±0.3	52.3 ±1.0	83.8 ±0.3	76.0 ±1.2	–	37.1 ±0.7
CRT [23]	76.3 ±0.2	51.4 ±0.3	84.5 ±0.1	77.3 ±0.0	–	31.7 ±1.0
BoDA$_r$	76.9 ±0.5	51.4 ±0.3	85.3 ±0.3	77.3 ±0.2	–	33.3 ±0.5
BoDA-M$_r$	76.3 ±0.3	53.4 ±0.3	85.8 ±0.2	77.3 ±0.2	–	35.7 ±0.7
BoDA$_{r,c}$	77.3 ±0.3	53.4 ±0.3	85.3 ±0.8	78.0 ±0.2	–	38.6 ±0.7
BoDA-M$_{r,c}$	**78.2** ±0.4	**55.4** ±0.5	85.3 ±0.3	**79.3** ±0.6	–	**43.3** ±1.1
BoDA vs. ERM	+1.9	+1.8	+0.7	+2.7	–	+10.4

Table 3. Results on PACS-MLT.

Algorithm	Accuracy (by domain)		Accuracy (by shot)			
	Average	Worst	Many	Medium	Few	Zero
ERM [46]	97.1 ±0.1	95.8 ±0.2	97.1 ±0.0	97.0 ±0.0	98.0 ±0.9	–
IRM [1]	96.7 ±0.2	95.2 ±0.4	96.8 ±0.2	96.7 ±0.7	94.7 ±1.4	–
GroupDRO [40]	97.0 ±0.1	95.3 ±0.4	97.3 ±0.1	95.3 ±1.2	94.7 ±3.6	–
Mixup [50]	96.7 ±0.2	95.1 ±0.2	97.0 ±0.1	96.7 ±0.3	91.3 ±2.7	–
MLDG [27]	96.6 ±0.1	94.1 ±0.3	96.8 ±0.1	96.3 ±0.7	92.7 ±0.5	–
CORAL [45]	96.6 ±0.5	94.3 ±0.7	96.6 ±0.5	97.0 ±0.8	94.7 ±0.5	–
MMD [29]	96.9 ±0.1	96.2 ±0.2	96.9 ±0.2	97.0 ±0.0	96.7 ±0.5	–
DANN [15]	96.5 ±0.0	94.3 ±0.1	96.5 ±0.1	98.0 ±0.0	94.7 ±2.4	–
CDANN [31]	96.1 ±0.1	94.5 ±0.2	96.1 ±0.1	96.3 ±0.5	94.0 ±0.9	–
MTL [4]	96.7 ±0.2	94.5 ±0.6	96.8 ±0.1	95.3 ±1.7	97.3 ±1.1	–
SagNet [35]	**97.2** ±0.1	95.2 ±0.3	**97.4** ±0.1	96.7 ±0.5	95.3 ±0.5	–
Fish [42]	96.9 ±0.2	95.2 ±0.2	97.0 ±0.1	97.0 ±0.5	94.7 ±1.1	–
Focal [32]	96.5 ±0.2	94.6 ±0.7	96.6 ±0.1	95.0 ±1.7	96.7 ±0.5	–
CBLoss [10]	96.9 ±0.1	95.1 ±0.4	96.8 ±0.2	97.0 ±1.2	**100.0** ±0.0	–
LDAM [6]	96.5 ±0.0	94.7 ±0.2	96.6 ±0.1	95.7 ±1.4	96.0 ±0.0	–
BSoftmax [39]	96.9 ±0.3	95.6 ±0.3	96.6 ±0.4	**98.7** ±0.7	99.3 ±0.5	–
SSP [52]	96.9 ±0.2	95.4 ±0.4	96.7 ±0.2	98.3 ±0.5	98.0 ±0.0	–
CRT [23]	96.3 ±0.1	94.9 ±0.1	96.3 ±0.1	97.3 ±0.3	94.0 ±0.9	–
BoDA$_r$	97.0 ±0.1	95.1 ±0.4	97.0 ±0.1	96.3 ±0.5	98.0 ±0.9	–
BoDA-M$_r$	97.1 ±0.1	94.9 ±0.1	97.3 ±0.1	96.3 ±0.5	96.0 ±0.0	–
BoDA$_{r,c}$	**97.2** ±0.1	95.7 ±0.3	**97.4** ±0.1	97.0 ±0.0	94.7 ±1.1	–
BoDA-M$_{r,c}$	97.1 ±0.2	**96.3** ±0.1	97.1 ±0.0	97.0 ±0.8	96.0 ±0.0	–
BoDA vs. ERM	+0.1	+0.5	+0.3	+0.0	-2.0	–

ERM [46], (2) *distributionally robust optimization:* **GroupDRO** [40], (3) *data augmentation:* **Mixup** [50], **SagNet** [35], (4) *meta-learning:* **MLDG** [27], (5) *domain-invariant feature learning:* **IRM** [1], **DANN** [15], **CDANN** [31], **CORAL** [45], **MMD** [29], (6) *transfer learning:* **MTL** [4], (7) *multi-task learning:* **Fish** [42], and (8) *imbalanced learning:* **Focal** [32], **CBLoss** [10], **LDAM** [6], **BSoftmax** [39], **SSP** [52], **CRT** [23]. We provide detailed descriptions in Appendix E.

Implementation and Evaluation Metrics. For a fair evaluation, following [19], for each algorithm we conduct a random search of 20 trials over a joint distribution of all hyperparameters (see Appendix E.2 for details). We then use the validation set to select the best hyperparameters for each algorithm, fix them and rerun the experiments under three different random seeds to report the final average accuracy with standard deviation. Such process ensures the comparison is best-versus-best, and the hyperparameters are optimized for all algorithms.

In addition to the average accuracy across domains, we also report the worst accuracy over domains, and further divide all domain-class pairs into *many-shot* (pairs with over 100 training samples), *medium-shot* (pairs with 20~100 training samples), *few-shot* (pairs with under 20 training samples), and *zero-shot* (pairs with no training data), and report the results for these subsets.

6.1 Main Results

We report the main results in this section for all MDLT datasets. The complete results and all additional experiments are provided in Appendix F and H.

Benchmark Results on MDLT Datasets. The performance of all methods on VLCS-MLT, PACS-MLT, OfficeHome-MLT, TerraInc-MLT and DomainNet-MLT are in Table 2, 3, 4, 5 and 6, respectively. We highlight rows in gray for BoDA and its variants, and bolden the best result in each column. First, as all tables indicate, BoDA consistently achieves the best average accuracy across all datasets. It also achieves the best worst-case accuracy most of the time. Moreover, on certain datasets (e.g., OfficeHome-MLT), MDL methods perform better (e.g., CORAL), while on others (e.g., TerraInc-MLT), imbalanced methods achieve higher gains (e.g., CRT); Nevertheless, regardless of dataset, BoDA outperforms all methods, highlighting its effectiveness for the MDLT task. Finally, compared to ERM, BoDA slightly improves the average and many-shot performance, while substantially boosting the performance for the medium-shot, few-shot, and zero-shot pairs. Table 7 summarizes the averaged accuracy across all datasets, where BoDA brings large overall improvements of $\sim 3\%$.

A Closer Look at Accuracy Gains. We further explore how BoDA performs across *all* domain-class pairs. Figure 7 shows the absolute accuracy gains of BoDA

Table 4. Results on OfficeHome-MLT.

Algorithm	Average	Worst	Many	Medium	Few	Zero
	Accuracy (by domain)		**Accuracy (by shot)**			
ERM [46]	80.7 ±0.0	71.3 ±0.1	87.8 ±0.2	81.0 ±0.2	63.1 ±0.1	63.3 ±7.2
IRM [1]	80.6 ±0.4	70.7 ±0.2	87.6 ±0.4	81.5 ±0.4	61.1 ±0.9	56.7 ±1.4
GroupDRO [40]	80.1 ±0.3	68.7 ±0.9	88.1 ±0.2	80.8 ±0.4	59.8 ±1.2	51.7 ±3.6
Mixup [50]	81.2 ±0.2	72.3 ±0.6	87.9 ±0.4	81.8 ±0.1	64.1 ±0.4	60.0 ±4.1
MLDG [27]	80.4 ±0.2	70.2 ±0.6	87.1 ±0.1	81.3 ±0.3	61.3 ±1.0	61.7 ±1.4
CORAL [45]	81.9 ±0.1	**72.7 ±0.6**	87.9 ±0.1	83.0 ±0.1	63.5 ±0.7	65.0 ±2.4
MMD [29]	78.4 ±0.4	67.7 ±0.8	85.2 ±0.2	79.4 ±0.7	58.8 ±0.4	56.7 ±3.6
DANN [15]	79.2 ±0.2	70.2 ±0.9	86.2 ±0.1	80.0 ±0.1	60.3 ±1.1	61.7 ±5.9
CDANN [31]	79.0 ±0.2	69.4 ±0.3	86.4 ±0.6	79.8 ±0.1	59.3 ±0.8	50.0 ±4.7
MTL [4]	79.5 ±0.2	69.8 ±0.6	87.3 ±0.3	79.8 ±0.2	61.1 ±0.2	51.7 ±2.7
SagNet [35]	80.9 ±0.1	70.5 ±0.5	87.8 ±0.4	81.9 ±0.1	61.2 ±0.9	56.7 ±3.6
Fish [42]	81.3 ±0.3	71.3 ±0.7	**88.2 ±0.2**	81.9 ±0.3	63.2 ±0.8	61.7 ±1.4
Focal [32]	77.9 ±0.0	67.6 ±0.4	86.5 ±0.3	78.3 ±0.1	57.4 ±0.3	46.7 ±3.6
CBLoss [10]	79.8 ±0.0	69.5 ±0.7	86.6 ±0.4	80.6 ±0.2	61.1 ±1.4	65.0 ±2.4
LDAM [6]	80.3 ±0.2	69.9 ±0.5	87.1 ±0.2	81.3 ±0.3	61.1 ±0.2	51.7 ±2.7
BSoftmax [39]	80.1 ±0.2	70.9 ±0.5	86.7 ±0.5	81.3 ±0.3	62.4 ±1.0	60.0 ±4.1
SSP [52]	81.1 ±0.3	71.1 ±0.3	87.3 ±0.6	82.3 ±0.5	61.6 ±0.7	63.3 ±1.4
CRT [23]	81.2 ±0.0	72.5 ±0.2	87.7 ±0.1	81.8 ±0.1	64.0 ±0.1	65.0 ±2.4
BoDA$_r$	81.5 ±0.1	71.8 ±0.1	87.7 ±0.2	82.3 ±0.1	**64.2 ±0.3**	63.8 ±1.4
BoDA-M$_r$	81.9 ±0.3	71.6 ±0.2	87.3 ±0.3	83.4 ±0.2	62.8 ±0.3	65.0 ±2.4
BoDA$_{r,e}$	82.3 ±0.1	72.3 ±0.3	87.1 ±0.2	**83.9 ±0.3**	63.2 ±0.2	65.0 ±2.4
BoDA-M$_{r,e}$	**82.4 ±0.2**	72.3 ±0.3	87.7 ±0.1	**83.9 ±0.6**	**64.2 ±0.2**	**66.7 ±2.7**
BoDA *vs.* ERM	+1.7	+1.0	-0.1	+2.9	+1.1	+3.4

Table 5. Results on TerraInc-MLT.

Algorithm	Average	Worst	Many	Medium	Few	Zero
	Accuracy (by domain)		**Accuracy (by shot)**			
ERM [46]	75.3 ±0.3	67.4 ±0.3	85.6 ±0.8	69.6 ±3.2	66.1 ±2.4	14.4 ±2.8
IRM [1]	73.3 ±0.7	64.3 ±1.3	83.5 ±0.6	70.0 ±1.8	58.3 ±3.4	20.1 ±1.4
GroupDRO [40]	72.0 ±0.4	66.6 ±0.2	84.7 ±1.1	64.6 ±4.7	38.9 ±1.2	13.5 ±1.1
Mixup [50]	71.1 ±0.7	60.4 ±1.1	82.2 ±0.7	60.0 ±0.6	56.1 ±3.0	12.2 ±2.1
MLDG [27]	76.6 ±0.2	66.9 ±0.5	86.1 ±0.6	73.8 ±3.9	70.6 ±3.7	18.8 ±2.4
CORAL [45]	76.4 ±0.5	67.8 ±0.9	86.3 ±0.3	77.5 ±3.1	66.1 ±2.0	11.0 ±1.4
MMD [29]	73.3 ±0.4	63.7 ±1.1	84.0 ±0.4	67.9 ±2.7	60.6 ±1.6	13.6 ±2.6
DANN [15]	68.7 ±0.9	61.1 ±1.0	79.6 ±1.2	62.5 ±8.1	48.9 ±2.8	13.3 ±1.1
CDANN [31]	70.3 ±0.5	63.9 ±1.0	83.5 ±0.8	50.0 ±4.2	43.9 ±4.7	20.4 ±3.1
MTL [4]	75.0 ±0.7	67.7 ±1.4	85.2 ±0.7	73.8 ±1.6	61.1 ±2.8	12.4 ±4.0
SagNet [35]	75.1 ±1.6	66.5 ±2.1	85.5 ±0.9	77.1 ±5.0	57.8 ±4.3	13.0 ±3.4
Fish [42]	75.3 ±0.5	66.3 ±0.5	85.8 ±0.2	73.3 ±3.9	61.1 ±3.0	13.7 ±3.3
Focal [32]	75.7 ±0.4	65.3 ±1.1	85.7 ±0.3	76.2 ±3.9	68.9 ±3.2	12.6 ±1.9
CBLoss [10]	78.0 ±0.4	68.3 ±2.0	85.0 ±0.1	89.2 ±1.2	83.9 ±2.5	9.3 ±3.9
LDAM [6]	74.7 ±0.9	64.1 ±1.4	85.1 ±0.6	70.8 ±3.5	67.8 ±2.1	13.1 ±0.4
BSoftmax [39]	76.7 ±1.0	65.6 ±1.3	83.4 ±0.8	90.8 ±0.7	78.3 ±3.9	12.6 ±2.4
SSP [52]	78.5 ±0.7	67.3 ±0.4	85.5 ±1.0	87.8 ±0.9	82.6 ±1.2	13.2 ±2.8
CRT [23]	81.6 ±0.1	70.0 ±0.4	**89.7 ±0.2**	90.4 ±0.3	83.9 ±0.6	12.9 ±0.0
BoDA$_r$	78.6 ±0.4	68.5 ±0.3	86.4 ±0.1	85.0 ±1.0	80.0 ±0.9	13.7 ±2.1
BoDA-M$_r$	79.4 ±0.6	71.3 ±0.4	88.4 ±0.3	76.2 ±2.7	88.3 ±1.6	14.4 ±1.4
BoDA$_{r,e}$	82.3 ±0.3	68.5 ±0.6	89.2 ±0.2	**92.5 ±0.9**	88.3 ±1.2	21.3 ±0.7
BoDA-M$_{r,e}$	**83.0 ±0.4**	**74.6 ±0.7**	89.2 ±0.2	91.2 ±0.6	**91.7 ±2.0**	**21.7 ±1.4**
BoDA *vs.* ERM	+7.7	+7.2	+3.6	+22.9	+25.6	+7.3

Table 6. Results on DomainNet-MLT.

Algorithm	Accuracy (by domain)		Accuracy (by shot)			
	Average	Worst	Many	Medium	Few	Zero
ERM [46]	58.6 ±0.2	29.4 ±0.3	66.0 ±0.1	56.1 ±0.1	35.9 ±0.3	27.6 ±0.3
IRM [1]	57.1 ±0.1	27.6 ±0.1	64.7 ±0.1	54.3 ±0.3	33.5 ±0.3	25.8 ±0.3
GroupDRO [40]	53.6 ±0.1	25.9 ±0.2	61.8 ±0.1	49.1 ±0.3	30.7 ±0.7	22.0 ±0.1
Mixup [50]	57.6 ±0.1	28.7 ±0.0	64.9 ±0.2	54.5 ±0.1	35.6 ±0.2	27.3 ±0.3
MLDG [27]	58.5 ±0.0	28.7 ±0.2	66.0 ±0.1	55.7 ±0.1	35.3 ±0.2	26.9 ±0.3
CORAL [45]	59.4 ±0.1	30.1 ±0.4	66.4 ±0.1	57.1 ±0.0	37.7 ±0.6	29.9 ±0.2
MMD [29]	56.7 ±0.0	27.2 ±0.2	64.2 ±0.1	54.0 ±0.0	33.9 ±0.2	25.4 ±0.2
DANN [15]	55.8 ±0.1	26.9 ±0.4	63.0 ±0.1	52.7 ±0.1	34.2 ±0.4	26.8 ±0.4
CDANN [31]	56.0 ±0.1	27.7 ±0.1	63.2 ±0.0	52.7 ±0.2	34.3 ±0.6	27.6 ±0.1
MTL [4]	58.6 ±0.1	29.3 ±0.2	65.9 ±0.1	56.0 ±0.4	35.4 ±0.1	28.2 ±0.3
SagNet [35]	58.9 ±0.0	29.4 ±0.2	66.3 ±0.1	56.4 ±0.0	36.2 ±0.3	27.2 ±0.4
Fish [42]	59.6 ±0.1	29.1 ±0.1	67.1 ±0.1	57.2 ±0.1	36.8 ±0.4	27.8 ±0.3
Focal [32]	57.8 ±0.2	27.5 ±0.1	65.2 ±0.2	55.1 ±0.2	35.8 ±0.1	26.3 ±0.1
CBLoss [10]	58.9 ±0.1	30.1 ±0.1	64.3 ±0.0	61.0 ±0.3	42.5 ±0.4	28.1 ±0.2
LDAM [6]	59.2 ±0.0	29.2 ±0.2	66.6 ±0.0	57.0 ±0.0	37.1 ±0.2	27.8 ±0.3
BSoftmax [39]	58.9 ±0.1	29.9 ±0.1	64.3 ±0.1	60.9 ±0.3	42.4 ±0.6	28.2 ±0.1
SSP [52]	59.7 ±0.0	31.6 ±0.2	64.3 ±0.1	62.6 ±0.1	45.0 ±0.3	30.5 ±0.0
CRT [23]	60.4 ±0.2	31.6 ±0.1	66.8 ±0.0	61.6 ±0.1	45.7 ±0.1	29.7 ±0.1
BoDA$_r$	60.1 ±0.2	32.6 ±0.1	65.7 ±0.2	60.6 ±0.1	42.6 ±0.3	30.5 ±0.2
BoDA-M$_r$	60.1 ±0.2	32.2 ±0.2	65.9 ±0.2	60.7 ±0.1	42.9 ±0.3	30.0 ±0.1
BoDA$_{r,c}$	**61.7** ±0.1	**33.4** ±0.1	**67.0** ±0.2	62.7 ±0.1	46.0 ±0.2	**32.2** ±0.3
BoDA-M$_{r,c}$	**61.7** ±0.2	33.3 ±0.3	**67.0** ±0.1	**63.0** ±0.3	**46.6** ±0.4	31.8 ±0.2
BoDA vs. ERM	+3.1	+4.0	+1.0	+6.9	+10.7	+4.6

Table 7. Results over all MDLT benchmarks.

Algorithm	VLCS-MLT	PACS-MLT	OfficeHome-MLT	TerraInc-MLT	DomainNet-MLT	Avg
ERM [46]	76.3 ±0.4	97.1 ±0.1	80.7 ±0.0	75.3 ±0.3	58.6 ±0.2	77.6
IRM [1]	76.6 ±0.3	97.0 ±0.1	80.6 ±0.4	73.3 ±0.7	57.1 ±0.1	76.8
GroupDRO [40]	76.7 ±0.4	97.0 ±0.1	80.1 ±0.3	72.0 ±0.4	53.6 ±0.1	75.9
Mixup [50]	75.9 ±0.1	96.7 ±0.2	81.2 ±0.2	71.1 ±0.7	57.6 ±0.1	76.5
MLDG [27]	76.9 ±0.2	96.6 ±0.1	80.4 ±0.2	76.6 ±0.2	58.5 ±0.0	77.8
CORAL [45]	75.9 ±0.5	96.6 ±0.5	81.9 ±0.1	76.4 ±0.5	59.4 ±0.1	78.0
MMD [29]	76.3 ±0.6	96.9 ±0.1	78.4 ±0.4	73.3 ±0.4	56.7 ±0.0	76.3
DANN [15]	77.5 ±0.1	96.5 ±0.0	79.2 ±0.2	68.7 ±0.9	55.8 ±0.1	75.5
CDANN [31]	76.6 ±0.4	96.1 ±0.1	79.0 ±0.2	70.3 ±0.5	56.0 ±0.1	75.6
MTL [4]	76.3 ±0.2	96.7 ±0.2	79.5 ±0.2	75.0 ±0.7	58.6 ±0.1	77.2
SagNet [35]	76.3 ±0.2	**97.2** ±0.1	80.9 ±0.1	75.1 ±1.6	58.9 ±0.0	77.7
Fish [42]	77.5 ±0.3	96.9 ±0.2	81.3 ±0.3	75.3 ±0.5	59.6 ±0.1	78.1
Focal [32]	75.6 ±0.2	96.5 ±0.2	77.9 ±0.0	75.7 ±0.4	57.8 ±0.2	76.7
CBLoss [10]	76.8 ±0.4	96.9 ±0.1	79.8 ±0.2	78.0 ±0.4	58.9 ±0.1	78.1
LDAM [6]	77.5 ±0.1	96.5 ±0.0	80.3 ±0.2	74.7 ±0.9	59.2 ±0.0	77.7
BSoftmax [39]	76.7 ±0.5	96.9 ±0.3	80.4 ±0.2	76.7 ±1.0	58.9 ±0.1	77.9
SSP [52]	76.1 ±0.3	96.9 ±0.2	81.1 ±0.3	78.5 ±0.7	59.7 ±0.0	78.5
CRT [23]	76.3 ±0.2	96.3 ±0.1	81.2 ±0.0	81.6 ±0.1	60.4 ±0.2	79.2
BoDA$_r$	76.9 ±0.5	97.0 ±0.1	81.5 ±0.1	78.6 ±0.4	60.1 ±0.2	78.8
BoDA-M$_r$	77.5 ±0.3	97.1 ±0.1	81.9 ±0.2	79.4 ±0.6	60.1 ±0.2	79.2
BoDA$_{r,c}$	77.3 ±0.2	**97.2** ±0.1	82.3 ±0.1	82.3 ±0.3	**61.7** ±0.1	80.2
BoDA-M$_{r,c}$	**78.2** ±0.4	97.1 ±0.2	**82.4** ±0.2	**83.0** ±0.4	**61.7** ±0.2	**80.5**
BoDA vs. ERM	+1.9	+0.1	+1.7	+7.7	+3.1	+2.9

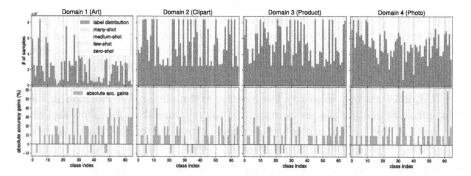

Fig. 7. The absolute accuracy gains of BoDA *vs.* ERM over all domain-class pairs on OfficeHome-MLT. BoDA establishes large improvements w.r.t. all regions, especially for the few-shot and zero-shot ones. Results for other datasets are in Appendix H.2.

over ERM on OfficeHome-MLT, where BoDA consistently improves the performance over all domains. The improvements are especially large for domain "Art", where most of the classes lie in the *few-shot* region. For certain classes, BoDA can improve up to 50% accuracy, indicating its effectiveness on tackling MDLT.

Ablation Studies on BoDA Components (Appendix H.1). We study the effects of (1) adding balanced distance (i.e., BoDA *vs.* vanilla DA), and (2) different choices of distance calibration coefficient $\lambda_{d,c}^{d',c'}$ in BoDA. We observe that BoDA improves over DA by a large margin (2.3% on average over all MDLT datasets), highlighting the importance of using *balanced* distance. Interestingly, as for $\lambda_{d,c}^{d',c'}$, we find that BoDA is pretty robust to different choices within a given range, and obtain similar gains (1.9% to 2.9% over ERM).

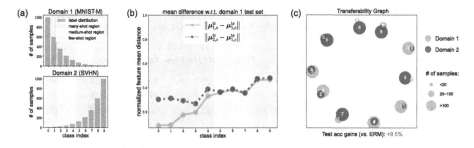

Fig. 8. BoDA analysis. **(a)** Label distribution setup. **(b)** Distance of feature mean between train and test data. BoDA enables better learned tail (d, c) with smaller feature discrepancy. **(c)** BoDA learns features that are more aligned across domains even in the presence of divergent labels, and significantly improves upon ERM by 9.5%.

6.2 Understanding the Behavior of BoDA on MDLT

To better understand how the design of BoDA contributes to its superior performance, we revisit the `Digits-MLT` dataset and run BoDA as opposed to ERM.

Better Learned Representations for Minority Data. Similar to Fig. 5, we plot in Fig. 8b the feature mean distance between training and test data for BoDA on `Digits-MLT`. The plot shows that, in BoDA, the distance between training and test data for minority classes (class "8" and "9") becomes smaller.

Improved Transferability against Severe Imbalance. Fig. 8c plots the transferability graph induced by BoDA for the label distributions in Fig. 8a. It shows that even in the presence of severe and divergent label imbalance, BoDA learns transferable features. Further, BoDA learns a *balanced* feature space that separates different classes away. The better learned features translates to better accuracy (9.5% gains vs. ERM in Fig. 3c). More results are in Appendix H.3.

Tightness of the Bound. We study whether the BoDA bound derived in Theorem 1 is tight. We train a ResNet-18 on `Digits-MLT` for 5,000 steps to ensure convergence. We compute the loss over all samples, and combine the results over 3 random seeds. Table 8 confirms the bound is empirically tight.

Table 8. BoDA bound.

	$\mathcal{L}_{\text{BoDA}}$
Empirical	2.92947 ±7.3e-3
Theoretical	2.92513 ±7.8e-3

Table 9. BoDA strengthens performance on Domain Generalization (DG) benchmarks. The full tables including detailed results for each dataset are in Appendix G.

Algorithm	VLCS	PACS	OfficeHome	TerraInc	DomainNet	Avg
ERM	77.5 ±0.4	85.5 ±0.2	66.5 ±0.3	46.1 ±1.8	40.9 ±0.1	63.3
Current SOTA [45]	**78.8** ±0.6	86.2 ±0.3	68.7 ±0.3	47.6 ±1.0	41.5 ±0.1	64.5
BoDA$_{r,c}$	78.5 ±0.3	**86.9** ±0.4	**69.3** ±0.1	**50.2** ±0.4	**42.7** ±0.1	**65.5**
BoDA$_{r,c}$ + Current SOTA [45]	79.1 ±0.1	87.9 ±0.5	69.9 ±0.2	50.7 ±0.6	43.5 ±0.3	66.2
BoDA *vs.* ERM	+1.6	+2.4	+3.4	+4.6	+2.6	+2.9

7 Beyond MDLT: (Imbalanced) Domain Generalization

Domain Generalization (DG) refers to learning from multiple domains and generalizing to unseen domains. Since naturally the learning domains differ in their label distributions and may even have class imbalance within each domain, we study whether BoDA can improve performance for DG. Note that all datasets we adapted for MDLT are standard benchmarks for DG, which confirms that data imbalance is an intrinsic problem in DG, but has been overlooked by past works.

To test BoDA, we follow the DG evaluation protocol in [19], and compare to the current SOTA [45]. Table 9 reveals the following findings: First, BoDA alone can improve upon the current SOTA on four out of the five datasets, and achieves notable average performance gains. Moreover, combined with the current SOTA, BoDA further boosts the result by a notable margin across all datasets, suggesting that label imbalance is orthogonal to existing DG-specific algorithms. Finally, similar to MDLT, the gains depend on how severe the imbalance is within a dataset – e.g., TerraInc exhibits the most severe label imbalance across domains, on which BoDA achieves the highest gains. The intriguing results shed light on the importance of integrating label imbalance for practical DG algorithm design.

8 Conclusion

We formalize MDLT as learning from multi-domain imbalanced data, and generalizing to all domain-class pairs. We introduce the domain-class transferability graph, and propose BoDA, a theoretically grounded loss that tackles MDLT. Extensive results on real-world MDLT benchmarks verify its superiority. Furthermore, BoDA establishes a new SOTA on DG benchmarks. Our work opens up new avenues for realistic multi-domain learning in the presence of data imbalance.

Acknowledgments. This work was supported by the GIST-MIT Research Collaboration grant funded by GIST.

References

1. Arjovsky, M., Bottou, L., Gulrajani, I., Lopez-Paz, D.: Invariant risk minimization. arXiv preprint arXiv:1907.02893 (2019)
2. Beery, S., Van Horn, G., Perona, P.: Recognition in terra incognita. In: Ferrari, V., Hebert, M., Sminchisescu, C., Weiss, Y. (eds.) ECCV 2018. LNCS, vol. 11220, pp. 472–489. Springer, Cham (2018). https://doi.org/10.1007/978-3-030-01270-0_28
3. Ben-David, S., Blitzer, J., Crammer, K., Kulesza, A., Pereira, F., Vaughan, J.W.: A theory of learning from different domains. Mach. Learn. **79**(1), 151–175 (2010)
4. Blanchard, G., Deshmukh, A.A., Dogan, U., Lee, G., Scott, C.: Domain generalization by marginal transfer learning. J. Mach. Learn. Res. **22**(2), 1–55 (2021)
5. Buda, M., Maki, A., Mazurowski, M.A.: A systematic study of the class imbalance problem in convolutional neural networks. Neural Netw. **106**, 249–259 (2018)
6. Cao, K., Wei, C., Gaidon, A., Arechiga, N., Ma, T.: Learning imbalanced datasets with label-distribution-aware margin loss. In: NeurIPS (2019)
7. Carlucci, F.M., D'Innocente, A., Bucci, S., Caputo, B., Tommasi, T.: Domain generalization by solving Jigsaw puzzles. In: CVPR (2019)
8. Carroll, J.D., Arabie, P.: Multidimensional scaling. In: Measurement, Judgment and Decision Making, pp. 179–250 (1998)
9. Chawla, N.V., Bowyer, K.W., Hall, L.O., Kegelmeyer, W.P.: Smote: synthetic minority over-sampling technique. J. Artif. Intell. Res. **16**, 321–357 (2002)
10. Cui, Y., Jia, M., Lin, T.Y., Song, Y., Belongie, S.: Class-balanced loss based on effective number of samples. In: CVPR (2019)
11. De Maesschalck, R., Jouan-Rimbaud, D., Massart, D.L.: The Mahalanobis distance. Chemom. Intell. Lab. Syst. **50**(1), 1–18 (2000)
12. Dong, Q., Gong, S., Zhu, X.: Imbalanced deep learning by minority class incremental rectification. IEEE Trans. Pattern Anal. Mach. Intell. **41**(6), 1367–1381 (2019)
13. Dredze, M., Kulesza, A., Crammer, K.: Multi-domain learning by confidence-weighted parameter combination. Mach. Learn. **79**(1), 123–149 (2010)
14. Fang, C., Xu, Y., Rockmore, D.N.: Unbiased metric learning: on the utilization of multiple datasets and web images for softening bias. In: ICCV (2013)
15. Ganin, Y., et al.: Domain-adversarial training of neural networks. J. Mach. Learn. Res. **17**(1), 2030–2096 (2016)
16. Globerson, A., Chechik, G., Pereira, F., Tishby, N.: Euclidean embedding of co-occurrence data. In: NeurIPS (2004)
17. Goldberger, J., Hinton, G.E., Roweis, S., Salakhutdinov, R.R.: Neighbourhood components analysis. In: NeurIPS (2004)
18. Gretton, A., Borgwardt, K.M., Rasch, M.J., Schölkopf, B., Smola, A.: A kernel two-sample test. J. Mach. Learn. Res. **13**(1), 723–773 (2012)
19. Gulrajani, I., Lopez-Paz, D.: In search of lost domain generalization. In: ICLR (2021)
20. He, H., Bai, Y., Garcia, E.A., Li, S.: AdaSyn: adaptive synthetic sampling approach for imbalanced learning. In: IEEE International Joint Conference on Neural Networks, pp. 1322–1328 (2008)
21. He, K., Zhang, X., Ren, S., Sun, J.: Deep residual learning for image recognition. In: CVPR (2016)
22. Huang, C., Li, Y., Chen, C.L., Tang, X.: Deep imbalanced learning for face recognition and attribute prediction. IEEE Trans. Pattern Anal. Mach. Intell. (2019)

23. Kang, B., et al.: Decoupling representation and classifier for long-tailed recognition. In: ICLR (2020)
24. Kingma, D.P., Ba, J.: Adam: a method for stochastic optimization. In: ICLR (2015)
25. Krueger, D., et al.: Out-of-distribution generalization via risk extrapolation (REX). arXiv preprint arXiv:2003.00688 (2020)
26. LeCun, Y., Bottou, L., Bengio, Y., Haffner, P.: Gradient-based learning applied to document recognition. Proc. IEEE **86**(11), 2278–2324 (1998)
27. Li, D., Yang, Y., Song, Y.Z., Hospedales, T.: Learning to generalize: meta-learning for domain generalization. In: AAAI (2018)
28. Li, D., Yang, Y., Song, Y.Z., Hospedales, T.M.: Deeper, broader and artier domain generalization. In: ICCV (2017)
29. Li, H., Pan, S.J., Wang, S., Kot, A.C.: Domain generalization with adversarial feature learning. In: CVPR (2018)
30. Li, T., et al.: Targeted supervised contrastive learning for long-tailed recognition. arXiv preprint arXiv:2111.13998 (2021)
31. Li, Y., Gong, M., Tian, X., Liu, T., Tao, D.: Domain generalization via conditional invariant representations. In: AAAI (2018)
32. Lin, T.Y., Goyal, P., Girshick, R., He, K., Dollár, P.: Focal loss for dense object detection. In: ICCV, pp. 2980–2988 (2017)
33. Liu, Z., Miao, Z., Zhan, X., Wang, J., Gong, B., Yu, S.X.: Large-scale long-tailed recognition in an open world. In: CVPR (2019)
34. Muandet, K., Balduzzi, D., Schölkopf, B.: Domain generalization via invariant feature representation. In: ICML (2013)
35. Nam, H., Lee, H., Park, J., Yoon, W., Yoo, D.: Reducing domain gap by reducing style bias. In: CVPR (2021)
36. Netzer, Y., Wang, T., Coates, A., Bissacco, A., Wu, B., Ng, A.Y.: Reading digits in natural images with unsupervised feature learning. In: NIPS Workshop on Deep Learning and Unsupervised Feature Learning (2011)
37. Pan, S.J., Yang, Q.: A survey on transfer learning. IEEE Trans. Knowl. Data Eng. **22**(10), 1345–1359 (2009)
38. Peng, X., Bai, Q., Xia, X., Huang, Z., Saenko, K., Wang, B.: Moment matching for multi-source domain adaptation. In: ICCV (2019)
39. Ren, J., Yu, C., Ma, X., Zhao, H., Yi, S., et al.: Balanced meta-softmax for long-tailed visual recognition. In: NeurIPS (2020)
40. Sagawa, S., Koh, P.W., Hashimoto, T.B., Liang, P.: Distributionally robust neural networks for group shifts: on the importance of regularization for worst-case generalization. In: ICLR (2020)
41. Schoenauer-Sebag, A., Heinrich, L., Schoenauer, M., Sebag, M., Wu, L.F., Altschuler, S.J.: Multi-domain adversarial learning. In: ICLR (2019)
42. Shi, Y., et al.: Gradient matching for domain generalization. arXiv preprint arXiv:2104.09937 (2021)
43. Shu, J., et al.: Meta-weight-net: learning an explicit mapping for sample weighting. arXiv preprint arXiv:1902.07379 (2019)
44. Sohn, K.: Improved deep metric learning with multi-class n-pair loss objective. In: NeurIPS (2016)
45. Sun, B., Saenko, K.: Deep CORAL: correlation alignment for deep domain adaptation. In: Hua, G., Jégou, H. (eds.) ECCV 2016. LNCS, vol. 9915, pp. 443–450. Springer, Cham (2016). https://doi.org/10.1007/978-3-319-49409-8_35
46. Vapnik, V.N.: An overview of statistical learning theory. IEEE Trans. Neural Netw. **10**(5), 988–999 (1999)

47. Venkateswara, H., Eusebio, J., Chakraborty, S., Panchanathan, S.: Deep hashing network for unsupervised domain adaptation. In: CVPR (2017)
48. Wang, X., Lian, L., Miao, Z., Liu, Z., Yu, S.: Long-tailed recognition by routing diverse distribution-aware experts. In: ICLR (2021)
49. Xiao, T., Li, H., Ouyang, W., Wang, X.: Learning deep feature representations with domain guided dropout for person re-identification. In: CVPR (2016)
50. Xu, M., et al.: Adversarial domain adaptation with domain mixup. In: AAAI (2020)
51. Yang, Y., Hospedales, T.M.: A unified perspective on multi-domain and multi-task learning. In: ICLR (2015)
52. Yang, Y., Xu, Z.: Rethinking the value of labels for improving class-imbalanced learning. In: NeurIPS (2020)
53. Yang, Y., Zha, K., Chen, Y.C., Wang, H., Katabi, D.: Delving into deep imbalanced regression. In: ICML (2021)
54. Zhang, M., Marklund, H., Gupta, A., Levine, S., Finn, C.: Adaptive risk minimization: a meta-learning approach for tackling group shift. arXiv preprint arXiv:2007.02931 (2020)
55. Zhang, X., Fang, Z., Wen, Y., Li, Z., Qiao, Y.: Range loss for deep face recognition with long-tailed training data. In: ICCV (2017)
56. Zhang, Y., Hooi, B., Hong, L., Feng, J.: Test-agnostic long-tailed recognition by test-time aggregating diverse experts with self-supervision. arXiv preprint arXiv:2107.09249 (2021)
57. Zhang, Y., Kang, B., Hooi, B., Yan, S., Feng, J.: Deep long-tailed learning: a survey. arXiv preprint arXiv:2110.04596 (2021)
58. Zhou, B., Cui, Q., Wei, X.S., Chen, Z.M.: BBN: bilateral-branch network with cumulative learning for long-tailed visual recognition. In: CVPR (2020)
59. Zhou, K., Liu, Z., Qiao, Y., Xiang, T., Loy, C.C.: Domain generalization in vision: a survey. arXiv preprint arXiv:2103.02503 (2021)
60. Zhou, K., Yang, Y., Qiao, Y., Xiang, T.: Domain generalization with mixstyle. In: ICLR (2021)

Few-Shot Video Object Detection

Qi Fan[1], Chi-Keung Tang[1(✉)], and Yu-Wing Tai[1,2]

[1] The Hong Kong University of Science and Technology, Hong Kong, China
cktang@cs.ust.hk
[2] Kuaishou Technology, Beijing, China

Abstract. We introduce Few-Shot Video Object Detection (FSVOD) with three contributions to real-world visual learning challenge in our highly diverse and dynamic world: 1) a large-scale video dataset FSVOD-500 comprising of 500 classes with class-balanced videos in each category for few-shot learning; 2) a novel Tube Proposal Network (TPN) to generate high-quality video tube proposals for aggregating feature representation for the target video object which can be highly dynamic; 3) a strategically improved Temporal Matching Network (TMN+) for matching representative query tube features with better discriminative ability thus achieving higher diversity. Our TPN and TMN+ are jointly and end-to-end trained. Extensive experiments demonstrate that our method produces significantly better detection results on two few-shot video object detection datasets compared to image-based methods and other naive video-based extensions. Codes and datasets are released at https://github.com/fanq15/FewX.

Keywords: Few-shot video object detection · Object indexing/retrieval · Tube proposal network · Temporal matching network

1 Introduction

We ask the following question: *Given a bunch of videos, how can we index and localize all novel objects of interest as video clips?* See Fig. 1.

This problem is becoming increasingly essential with massive video collections in this media era: movies, YouTube videos, TikTok streaming videos, surveillance videos, just to name a few. The video objects of interests can be highly novel, often personalized, and thus are not covered by any existing datasets. Marvel fans may want to collect all Iron Man or Hulk clips from all

This research was supported in part by Kuaishou Technology, and the Research Grant Council of the Hong Kong SAR under grant No. 16201420.

Supplementary Information The online version contains supplementary material available at https://doi.org/10.1007/978-3-031-20044-1_5.

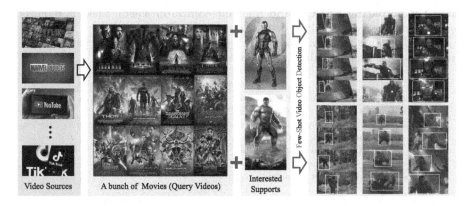

Fig. 1. Given only a few support objects of interest, our FSVOD detects all objects of the same category in query videos. Note that FSVOD enables object indexing/retrieval in a bunch of query videos to extract video clips containing the target objects.

Marvel movies, while warfare collectors want to create a TikTok video consisting of tank clip collections from war movies. We may not even know which videos contain the interested objects.

No existing tasks or solutions can solve this real-world challenge. Notably, multiple object tracking [13,90], image/video object detection [3,36,88,96] are all restricted in fixed and limited training classes. Single object tracking [5,67] can track new classes, but it requires user-provided template for every video and can only track the target template object. Few-shot learning seems a good candidate solution. But existing few-shot object detection [31,134] and few-shot classification [58,114] are specifically designed for still images and they will produce numerous false positive results in videos. Few-shot video classification [10,55,146] does not target at instance recognition.

This real-world challenge motivates few-shot video object detection (FSVOD): given only a few *support images* of the target object in an unseen class, FSVOD detects all the objects belonging to the same class in a given *query video*. The given support images can be arbitrary objects of interest, and FSVOD works on arbitrary videos for indexing and localization. The key to successful FSVOD is simultaneously modeling both high dynamics and high diversity of our dynamic and diverse world, while other existing tasks can only contribute either high dynamic or high diversity, as summarized in Table 1, and thus falling short of the real-world challenge.

The technical contributions of FSVOD, namely, Temporal Proposal Network (TPN) for high object dynamics and Temporal Matching Network (TMN+) for high object diversity, will be detailed. The core idea is to perform temporal matching between the tube-aggregated query features and supports, which enables high-quality detection based on the representative tube features and eliminates ghost objects (false positive predictions) which heavily suffers the few-shot image object detection methods.

Table 1. Comparing **FSVOD** and relevant computer vision tasks in terms of *dyn*amic and *div*ersity capabilities: detecting *box* for *novel* object classes and/or *multiple* objects, and whether *temporal* information is considered. Number of '+' indicates how diversity each task can contribute. 'S.A.' means scene adaptation.

Task	Dyn.	Div.	Box	Nov.	Mul.	Temp.	S.A.
Image Object Detection (IOD)	no	+	✓	✗	✓	✗	✓
Video Object Detection (VOD)	yes	+	✓	✗	✓	✓	✓
Multiple Object Tracking (MOT)	yes	+	✓	✗	✓	✓	✗
Single Object Tracking (SOT)	yes	+	✓	✓	✗	✓	✗
Few-Shot Classification (FSC)	no	++	✗	✓	✓	✗	✓
Few-Shot Object Detection (FSOD)	no	+++	✓	✓	✓	✗	-✓
Few-Shot Video Object Detection	yes	+++	✓	✓	✓	✓	✓

The other contribution of this paper consists of a large-scale dataset that enables new research on few-shot video object detection. Our dataset contains 500 classes with a small and balanced number of high-quality videos in each class. The numerous classes with class-balanced videos enable the trained model to learn a general relation metric for novel classes. Note that this dataset contributes not only as the first benchmark for FSVOD, but also as a useful benchmark for other important vision tasks, such as multi-object tracking and video object detection which are still in lack of a well-constructed, class-balanced video benchmark on par in the number of classes as FSVOD-500.

2   Related Work

The FSVOD task is related to few-shot learning, object detection and video understanding. Table 1 summarizes its relationship with closely related tasks.

Few-Shot Classification (FSC). Optimization-based works learn task-agnostic knowledge on model parameters [4,35,65] for fast adaptation to new tasks on limited training data, using only a few gradient update steps. Some works [42,117] hallucinate new images for novel classes from limited labeled data. Metric-based methods exploit a weight-shared network [58] to extract features of the support and query images before feeding them to a transferable distance metric. Such matching strategy [104,114,130,137] captures inherent variety between supports and queries irrespective of classes and thus can be directly applied for classifying novel classes.

Few-Shot Object Detection (FSOD). With encouraging progress made in the few-shot classification, few-shot learning has continued to contribute to important computer vision tasks [26,30,39,48,70,79,84] at a fast pace especially for object detection [92,118,134]. In LSTD [11] the gap between the source and target domain is minimized. RepMet [53] learns the multi-modal distribution of the training classes in the embedding space. FR [50] exploits a meta feature learner to quickly adapt to novel classes. Some works exploit semantic relation reasoning [144], restore negative information [132], feature hallucination [140] or other techniques [32,47,66,69,72,105,116,122,126,129,138] to facilitate few-shot object detection. All of the above methods however require fine-tuning on novel classes. In FSOD [31] the authors proposed to learn a matching metric with attention RPN and multi-relation detector to detect novel classes.

Our FSVOD extends FSOD task to the temporal domain, with the technical approach motivated by the matching network [114] and FSOD network [31] to detect novel classes without fine-tuning.

Image Object Detection (IOD). Existing object detection methods can be mainly categorized to the two-stage approach [36,73,96] and one-stage approach [74,76,77,94,95,139], based on whether a region-of-interest proposal step is used. The two-stage approach was pioneered by R-CNN [37]. In recent years, this approach has been improved by various excellent works and achieved remarkable performance [1,8,9,14,43,71,103]. The one-stage approach on the other hand discards the proposal generation procedure in lieu of higher computational efficiency and faster inference speed with anchor-based [60,101,141,147] or anchor-free detectors [27,59,64,78,80,111,133,143].

Video Object Detection (VOD). Video object detection aims at detecting objects of pre-defined classes in a given video. Some enhance the quality of per-frame features by integrating temporal information locally [3,24,115,124], globally [22,102,121] or both [12,88,119,120,127], while others follow the "sequential detection tracking" paradigm [34,51,107,148–150] to associate and rescore detected boxes on individual frames. The above work in intensive supervision and cannot be applied readily to detect novel classes. VOD variants include e.g., video object segmentation (VOS) [91,128], video instance segmentation (VIS) [131] and video panoptic segmentation (VPS) [57].

Both IOD and VOD are restricted to pre-defined classes making it hard for them to detect novel classes. FSVOD eliminates this restriction with its detection generality on novel classes in videos.

Single Object Tracking (SOT). Given an arbitrary target with its location in the first frame, single object tracking aims to infer its location in subsequent frames of the given video. Thanks to the construction of new benchmark datasets [29,123] and annually held tracking challenges [61–63], we have witnessed rapid performance boost in the last decade. The correlation filter based trackers [17,19,20,46] achieve superb performance with efficient inference speed. The recent emerging siamese network based trackers [5,40,45,67,68,109,112] have drawn much attention due to their well-balanced performance and efficiency.

Although SOT models can track unseen objects, they heavily rely on the provided template and can only track one target object. The online tracking trackers [6,7,16–20,38,83] can be finetuned/updated on the first frame, but they focus on tracking single object with the video-specific annotated first frame. On other hand, our FSVOD focuses on detecting arbitrary novel objects in videos based on given video-agnostic support images even from other images/videos and can be reused for all input videos.

Multiple Object Tracking (MOT). This task [82,113] requires simultaneous prediction of spatio-temporal location and classification of video objects into pre-defined classes. Current mainstream trackers [2,13,33,56,81,97,108,135,136, 142] adopt tracking-by-detection (TBD) by first performing per-frame detection and then associating the detected boxes in the temporal dimension. Some works leverage trajectories or tubes to capture motion trails of targets [52,89,90,100, 145].

While MOT models can simultaneously track multiple objects, they cannot generalize to novel classes. FSVOD can detect novel classes in videos. Our technical approach is inspired by these previous methods, especially tube-based MOT and VOD methods, e.g., CPN [107] and CTracker [90], which are restricted in limited training classes.

3   Proposed Method

Few-shot video object detection aims at detecting **novel** classes unseen in the training set. Given a **support image** containing one object of the support class c and a **query video** sequence with T frames, the task is to detect all the objects belonging to the support class c in every frame. Suppose the support set contains N classes with K samples for each class, the problem is defined as N-way K-shot detection. Specifically, during inference, if all the support classes are exploited for detection, it is dubbed full-way evaluation.

3.1   Overview

Technically, it is non-trivial to transfer few-shot learning [35,58,98] to the video object detection domain for simultaneously modeling the dynamic and diverse world. Few-shot learning requires a large-scale, class-balanced dataset with numerous base classes to train a class-agnostic metric capable of generalizing to novel classes [31,70,99]. Besides, videos present additional data challenges caused by e.g., motion blur, occlusion and deformation of objects, making infeasible straightforward extension of few-shot image to few-shot video object detection without adequate temporal consideration.

This paper extends the traditional video object detection to detect novel classes in a few-shot learning setting which is not a straightforward problem. We propose a novel tube-based few-shot video object detection model for detecting novel classes in a given video, without any fine-tuning or retraining. We make the following contributions:

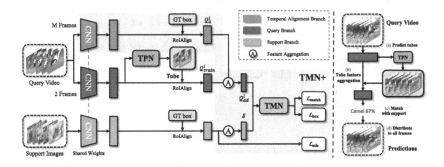

Fig. 2. Network architecture at training (left) and testing (right) stages. The query video and support images are processed by the weight-shared backbone. The query branch only processes two query images. The temporal alignment branch (TAB) is used for query feature alignment, and a classification module is introduced to produce representative support features. For clarity we show the detection on a single object, while our model can perform multi-object detection with corresponding tubes.

We first model *dynamic objects* by generating temporal tubes using our novel Tube Proposal Network (TPN) exploiting spatial adjacency and appearance similarity in the neighboring frames. Specifically, by introducing novel inter-frame proposals to detect objects in consecutive frames, TPN can capture potential objects in the query video while filtering out background and ghost objects (the false positive objects detected in isolated frames). We argue that the aggregated features across frames can better represent the target objects which leads to significant improvement on the detection performance.

Then we model *diversity of objects* using subsequent Temporal Matching Network (TMN+), which is specially designed and strategically improved to match support features and the aggregated query features from temporal tube proposals generated by TPN. Our proposed TMN+ effectively leverages the representative tube features by bridging the gap between training and inference via our novel temporal alignment branch. Furthermore, a new support classification loss is used to learn a highly discriminative feature, and a label-smoothing regularization is used for better generalization on novel unseen classes. Consequently, our TMN+ boosts matching performance on novel classes without extra computation overhead at inference.

The TPN and TMN+ are integrated into one unified network and jointly optimized in an end-to-end manner to simultaneously handle high dynamics and diversity in visual object detection.

3.2 Few-Shot Video Object Detection Network

Figure 2 shows the network architecture. We propose a novel temporal detection network that exploits tubes to locate and represent objects in the temporal domain, which are then matched with support features.

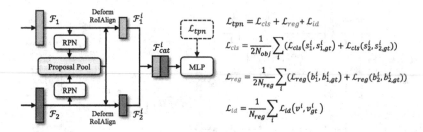

$$\mathcal{L}_{tpn} = \mathcal{L}_{cls} + \mathcal{L}_{reg} + \mathcal{L}_{id}$$

$$\mathcal{L}_{cls} = \frac{1}{2N_{obj}} \sum_i (\mathcal{L}_{cls}(s_1^i, s_{1,gt}^i) + \mathcal{L}_{cls}(s_2^i, s_{2,gt}^i))$$

$$\mathcal{L}_{reg} = \frac{1}{2N_{reg}} \sum_i (\mathcal{L}_{reg}(b_1^i, b_{1,gt}^i) + \mathcal{L}_{reg}(b_2^i, b_{2,gt}^i))$$

$$\mathcal{L}_{id} = \frac{1}{N_{reg}} \sum_i \mathcal{L}_{id}(v^i, v_{gt}^i)$$

Fig. 3. Tube Proposal Network (TPN) and the loss function. The $*_{gt}$ is the ground-truth label for the corresponding prediction, $\mathcal{L}_{cls}$ and $\mathcal{L}_{id}$ are both cross-entropy loss and $\mathcal{L}_{reg}$ is the smooth $\mathcal{L}_1$ loss. N_{obj} and N_{reg} are respectively the number of proposals and foreground proposals.

Tube Proposal Network. In image object detection, region proposal network RPN [96] has become a classical module to produce proposals for potential objects while filtering out the background. These proposals are fed to the R-CNN head for finer classification and localization.

We extend RPN to the temporal domain to generate *tube proposals* to locate and represent objects across frames. The resulting network is our novel tube proposal network (Fig. 3) which exploits the high likelihood that the same object in neighboring frames tend to have *similar location and appearance*.

To utilize the location cue in adjacent frames, we propose the novel *inter-frame proposals* by feeding the same proposals to two adjacent frames. Note that proposals usually serve as a coarse prediction prior for later finer regression. The predicted boxes regressed from the same proposals indicate the same objects and therefore inter-frame proposals can associate objects across frames. However, it is also possible that objects with large motion may locate far away in adjacent frames, or the locations are occupied by other objects in the next frame. To address this problem, we adopt the deformable RoIAlign [15] operator to enlarge the search region for the target objects by adapting the sampling bins conditioned on the input feature. To exploit the appearance cue in neighboring frames to address the second problem, we verify the same object by predicting the identification score of the predicted boxes regressed from the same proposal.

Specifically, given two adjacent frames $\{I_1, I_2\}$, we first use RPN to generate proposals for each frame and collect both frame proposals to construct the proposal pool. Each proposal p_i in the proposal pool is simultaneously fed to the two frames to extract proposal features $\{\mathcal{F}_1^i, \mathcal{F}_2^i\}$ with the deformable RoIAlign operator. These proposal features from individual frames are concatenated as $\mathcal{F}_{cat}^i = \text{concat}(\mathcal{F}_1^i, \mathcal{F}_2^i)$, which is then fed to the following multilayer perceptron (MLP) layer to perform objectness classification $\{s_1^i, s_2^i\}$, box regression $\{b_1^i, b_2^i\}$ for each frame, and identify verification v^i. The 2-frame tube prediction is trained with the TPN loss $\mathcal{L}_{tpn}$, as shown in Fig. 3.

During inference, the TPN needs to connect all frames in the given video by repeating the 2-frame tube prediction. Consider the 3-frame case where the

T-frame ($T > 3$) can be generalized[1]: given $\{I_1, I_2, I_3\}$, we first send $\{I_1, I_2\}$ to the model to generate a 2-frame tube $\{b_1, b_2\}$. Then we feed the pre-computed tube box b_2 to $\{I_2, I_3\}$ as the inter-frame proposal to generate the tube box b_3 for frame I_3 to construct another 2-frame tube $\{b_2, b_3\}$. We can construct a 3-frame tube $\{b_1, b_2, b_3\}$ by linking $\{b_1, b_2\}$ and $\{b_2, b_3\}$ through the inter-frame proposal b_2. The overlapped frame I_2 is used to verify the same objects between two frame pairs and its feature are reused in the process to avoid repeating computation as in CTracker [90]. Thus, we can sequentially detect tube boxes for all the frames and generate tube proposals.

Temporal Matching Network. After obtaining tube proposals, we extract and aggregate tube features and compare them with support features using a matching network, where the matching results are then distributed to the tube proposals in all frames. We re-design the matching network (MN) in the temporal domain to take advantage of tube features. Consequently, our discriminative temporal matching network TMN+ and TPN which share backbone features are jointly trained for better optimization. Below we detail the design rationale on a single instance/track i, starting from MN, TMN and finally TMN+, and it is easy to apply them on multiple objects of different classes.

MN. From $\{I_1, I_2\}$, the **query branch** of backbone extracts query features $\{\mathcal{Q}_1^i, \mathcal{Q}_2^i\}$ for each proposal p_i of instance i with RoIAlign operator. The **support branch** extracts the support features $\mathcal{S}$ in the ground-truth boxes of the support images. The MN then computes the distance between $\mathcal{Q} = \frac{1}{2}(\mathcal{Q}_1^i + \mathcal{Q}_2^i)$ and $\mathcal{S}$ and classifies $\mathcal{Q}$ to the nearest support neighbor. We adopt the multi-relation head with contrastive training strategy from FSOD [31] as our matching network (MN) for its high discriminative power. Refer to the supplementary material for more details about its architecture.

TMN. The above MN is however designed for image object detection and is unsuitable to be applied in the temporal domain. The main problem is the misalignment between training and inference for the query features $\mathcal{Q}$: In the training stage, $\mathcal{Q}_{train} = \frac{1}{2}(\mathcal{Q}_1^i + \mathcal{Q}_2^i)$ only involves the proposal feature in two frames, limited by the GPU memory and the joint training with TPN. While in the inference stage, $\mathcal{Q}_{test} = \frac{1}{T}(\mathcal{Q}_1^i + \mathcal{Q}_2^i + ... + \mathcal{Q}_T^i)$ is derived from all the frames in the tube proposal. This misalignment can produce bad matching result and overall performance degradation.

To bridge this training and inference gap, we propose a novel temporal matching network (TMN) by introducing a **temporal alignment branch (TAB)** for query feature alignment. Specifically, for proposal p_i of the target object i, $\mathcal{Q}_{train} = \frac{1}{2}(\mathcal{Q}_1^i + \mathcal{Q}_2^i)$ involves two frames $\{I_1, I_2\}$, and the TAB randomly selects

---

[1] The operations are parallel conducted for each instance/track i and we omit the instance notion for simplicity.

images[2] from remaining frames $\{I_3, I_4, ..., I_T\}$ and extracts the aligning features $\mathcal{Q}_a = \frac{1}{M}(\mathcal{Q}_3^i + \mathcal{Q}_4^i + ... + \mathcal{Q}_M^i)$ for the target object i, where M is the number of selected aligning query images. Then we generate the aligned query feature $\mathcal{Q}_{ad}^i = \alpha \mathcal{Q}_{train}^i + (1 - \alpha)\mathcal{Q}_a^i$ as the feature aggregation to represent the target object and perform matching with supports in the training stage. Our TMN thus bridges this gap without disrupting the design of TPN and without introducing additional computational overhead by removing TAB at inference time.

The loss function is $\mathcal{L}_{\text{tmn}} = \mathcal{L}_{\text{match}} + \mathcal{L}_{\text{box}}$, where $\mathcal{L}_{\text{match}}$ is the cross-entropy loss for binary matching and $\mathcal{L}_{\text{box}}$ is the smooth $\mathcal{L}_1$ loss for box regression.

TMN+. To enhance discriminative ability, TMN+ incorporates label-smoothing regularization [106] into TMN for better generalization and a jointly optimized support classification module for more representative feature.

We first introduce label smoothing to the matching loss $\mathcal{L}_{\text{match}}$ of TMN, which is widely used to prevent overfitting in the classification task [87,110] by changing the ground-truth label y_i to $y_i^* = (1 - \varepsilon)y_i + \frac{\varepsilon}{\beta}$, where ε is the constant smoothing parameter and β is the number of classes. This prevents the model from being overconfident to the training classes and is therefore inherently suitable for the few-shot learning models focusing on the generalization on novel classes. Then, we add a support classification module (classifier) to the support branch to enhance the intra-class compactness and inter-class separability in the Euclidean space and thus generate more representative features for matching in TMN. We adopt cross-entropy loss as its loss function $\mathcal{L}_{\text{scls}}$.

During training, the TPN and TMN+ are jointly and end-to-end optimized with the weight-shared backbone network by integrating all the aforementioned loss functions:

$$\mathcal{L} = \lambda_1 \mathcal{L}_{\text{tpn}} + \lambda_2 \mathcal{L}_{\text{tmn}} + \lambda_3 \mathcal{L}_{\text{scls}} \tag{1}$$

where λ_1, λ_2, and λ_3 are hyper-parameter weights to balance the loss functions and are set to 1 in our experiments.

4   FSVOD-500 Dataset

There exist a number of public datasets with box-level annotations for different video tasks: ImageNet-VID [23] for video object detection; LaSOT [29], GOT-10k [49], Youtube-BB [93], and TrackingNet [86] for single object tracking; MOT [85], TAO [21], Youtube-VOS [128] and Youtube-VIS [131] for multi-object tracking. However, none of these datasets meet the requirement of our proposed few-shot video object detection task. Some datasets (Youtube-BB [93], TrackingNet [86], ImageNet-VID [23], Youtube-VOS [128] and Youtube-VIS [131]) have many videos but limited classes, whereas a sufficiently large number of base classes is essential to few-shot learning. On the other hand, although other datasets (GOT-10k [49] and TAO [29]) contain diverse classes, not all instances

---

[2]   The random selection can be regarded as data augmentation to imitate the imperfect tube features during inference.

of the same target class are annotated in a video, and therefore are not suitable for the few-shot task. Last but not least, all of these datasets are not specifically designed for few-shot learning whose train/test/val sets are class-overlapping and cannot be used to evaluate the generality on unseen classes.

Thus, we design and construct a new dataset for the development and evaluation of few-shot video object detection task. The design criteria are:

- The dataset should consist of *highly-diversified classes* for learning a general relation metric for novel classes.
- The dataset should be *class-balanced* where each class has similar number of samples to avoid overfitting to any classes, given the long-tailed distribution of many novel classes in the real world [41].
- The train/test/val sets should contain *disjoint* classes to evaluate the generality of models on novel classes.

To save human annotation effort as much as possible, rather than building our dataset from scratch, we exploit existing large-scale video datasets for supervised learning, *i.e.*, LaSOT [29], GOT-10k [49], and TAO [21] to construct our dataset subject to the above three criteria. The dataset construction pipeline is consist of dataset filtering, balancing and splitting.

Dataset Filtering. Note that the above datasets cannot be directly used since they are only partially annotated for tracking task: although multiple objects of a given class are present in the video, only some or as few as one of them is annotated while the others are not annotated. Thus, we filter out videos with non-exhaustive labels while keeping those with high-quality labels covering all objects in the same class (target class). We also remove videos containing extremely small objects which are usually in bad visual quality and thus unsuitable for few-shot learning. Note that exhaustive annotation for all possible classes in such a large dataset is expensive and infeasible [21,41]. Therefore, only the target classes are exhaustively annotated for each video while non-target classes are categorically ignored.

Dataset Balancing. It is essential to maintain good data balancing in the few-shot learning dataset, so that sufficient generality to novel classes can be achieved without overfitting to any dominating training classes. Thus, we remove 'person' and 'human face' from the dataset which are in massive quantities (and they have already been extensively studied in many works and tasks [25,28,54,125]). Then, we manually remove easy samples for those classes with more than 30 samples. Finally, each class in our dataset has at least 3 videos and no more than 30.

Table 2. Dataset statistics of FSVOD-500 and FSYTV-40. "Class Overlap" denotes the class overlap with MS COCO [75] dataset.

	FSVOD-500			FSYTV-40	
	Train	Val	Test	Train	Test
label FPS	1	1	1	6	6
# Class	320	80	100	30	10
# Video	2553	770	949	1627	608
# Track	2848	793	1022	2777	902
# Frame	60432	14422	21755	41986	19843
# Box	65462	15031	24002	66601	27924
Class Overlap	Yes	No	No	Yes	No
Exhaustive	Only target classes			All classes	

Dataset Splitting. We summarize a four-level label system (shown in the supplementary material) to merge these datasets by grouping their leaf labels with the same semantics (*e.g.*, truck and lorry) into one class. Then, we select third-level node classes similar to the COCO [75] classes and exploit their leaf node as the training classes. The remaining classes are very distinct from COCO classes, and thus used to construct the test/val sets by randomly splitting the node classes. In this way, we can take advantage of the pre-training model on COCO dataset, while the test/val classes are rare novel classes and thus complying to the few-shot setting. We follow three guidelines for dataset split:

G1: The split should be in line with the few-shot learning setting, *i.e.*, train set should contain common classes in the real world, while test/val sets should contain rare classes. **G2:** To take advantage of pre-training on other datasets, the train set should have a large overlap with existing datasets while the test/val sets should have largely no overlap. **G3:** The train and test/val sets should have different node classes to evaluate the generality on novel classes in a challenging setting to avoid the influence of similar classes across sets, *e.g.*, if the train set has 'Golden Retriever', it is much easier to detect 'Labrador Retriever' in the test set, which is undesirable.

Consequently, *FSVOD-500* is the first benchmark specially designed for few-shot video object detection in evaluating the performance of a given model on novel classes.

5 Experiments

We conduct extensive experiments to validate the effectiveness of our proposed approach. Since this is the first paper on FSVOD, we compare with state-of-the-art (SOTA) methods of related tasks by adapting them to the FSVOD task.

Training. Our model is trained on four GeForce GTX 1080Ti GPUs using the SGD optimizer with 45,000 iterations. The initial learning rate is set to 0.002 which decays by a factor of 10 respectively in 30,000 and 40,000 iterations. Each GPU contains five cropped support images, two query images and M cropped aligning query images in the same video, where M is randomly sampled from $[1, 10]$. We use ResNet50 [44] as our backbone which is pre-trained on ImageNet [23] and MS COCO [75][3] for stable low-level features extraction and better convergence. The model is trained with 2-way 5-shot contrastive training strategy proposed in FSOD [31]. Other hyper-parameters are set as $\alpha = 0.5, \varepsilon = 0.2, \beta = 2$ in our experiments.

Evaluation. We adopt the full-way 5-shot evaluation (exploit all classes in the test/val set with 5 images per class as supports for evaluation) in our experiments with standard object detection evaluation metrics, *i.e.*, AP, AP_{50}, and AP_{75}. The evaluations are conducted 5 times on randomly sampled support sets and the mean and standard deviation are reported. Refer to the supplemental material for more training and evaluation details.

FSYTV-40. To validate model generalization on datasets with different characteristics, we construct another dataset built on Youtube-VIS dataset [131] for the FSVOD task. FSYTV-40 is vastly different from FSVOD-500 with only 40 classes (30/10 train/test class split following the same dataset split guidelines above, with instances of all classes are exhaustively annotated in each video), more videos in each class and more objects in each video. Table 2 tabulates the detailed statistics of both datasets.

5.1   Comparison with Other Methods

With no recognized previous work on FSVOD, we adapt representative models from related tasks to perform FSVOD, such as image object detection (Faster R-CNN [96], and FSOD [31]), video object detection (MEGA [12] and RDN [24]) and multiple object tracking (CTracker [90], and FairMOT [136], and Center-Track [142]). Only FSOD model can be directly applied frame-by-frame to perform FSVOD. For others, we exploit their models to generate class-agnostic boxes and adopt the multi-relation head trained in the FSOD [31] model to evaluate the distance between the query boxes and supports. We first perform comparison on FSVOD-500, and then generalize to FSYTV-40 (Table 3).

Comparison with IOD-Based Methods. FSOD serves as a strong baseline with its high recall of attention-RPN and powerful generalization of multi-relation head. With the same matching network, Faster R-CNN produces inferior

---

[3] There is no overlap between MS COCO and the val/test sets of both FSVOD-500 and FSYTV-40 datasets.

Table 3. Experimental results on FSVOD-500 and FSYTV-40 test set for novel classes with the full-way 5-shot evaluation.

Method	Tube	FSVOD-500			FSYTV-40		
		AP	AP_{50}	AP_{75}	AP	AP_{50}	AP_{75}
FR-CNN [96]	✗	$18.2_{\pm0.4}$	$26.4_{\pm0.4}$	$19.6_{\pm0.5}$	$9.3_{\pm1.4}$	$15.4_{\pm1.7}$	$9.6_{\pm1.7}$
FSOD [31]	✗	$21.1_{\pm0.6}$	$31.3_{\pm0.5}$	$22.6_{\pm0.7}$	$12.5_{\pm1.4}$	$20.9_{\pm1.8}$	$13.0_{\pm1.5}$
MEGA [12]	✗	$16.8_{\pm0.3}$	$26.4_{\pm0.5}$	$17.7_{\pm0.3}$	$7.8_{\pm1.1}$	$13.0_{\pm1.9}$	$8.3_{\pm1.1}$
RDN [24]	✗	$18.2_{\pm0.4}$	$27.9_{\pm0.4}$	$19.7_{\pm0.5}$	$8.1_{\pm1.1}$	$13.4_{\pm2.0}$	$8.6_{\pm1.1}$
CTracker [90]	✓	$20.1_{\pm0.4}$	$30.6_{\pm0.7}$	$21.0_{\pm0.8}$	$8.9_{\pm1.4}$	$14.4_{\pm2.5}$	$9.1_{\pm1.3}$
FairMOT [136]	✓	$20.3_{\pm0.6}$	$31.0_{\pm1.0}$	$21.2_{\pm0.8}$	$9.6_{\pm1.6}$	$16.0_{\pm2.2}$	$9.5_{\pm1.4}$
CenterTrack [142]	✓	$20.6_{\pm0.4}$	$30.5_{\pm0.9}$	$21.9_{\pm0.4}$	$9.5_{\pm1.6}$	$15.6_{\pm2.0}$	$9.7_{\pm1.3}$
Ours	✓	$\mathbf{25.1}_{\pm0.4}$	$\mathbf{36.8}_{\pm0.5}$	$\mathbf{26.2}_{\pm0.7}$	$\mathbf{14.6}_{\pm1.6}$	$\mathbf{21.9}_{\pm2.0}$	$\mathbf{16.1}_{\pm2.1}$

performance due to the lower recall of its generated boxes. With the representative aggregated query feature from TPN and discriminative TMN+ in the temporal domain, our FSVOD model outperforms FSOD by a large margin.

Comparison with VOD-Based Methods. VOD-based methods operate similarly to IOD-based methods in its per-frame object detection followed by matching with supports and thus both suffer from noisy proposals and less powerful features. Interestingly, we find that VOD-based methods have a worse performance because they produce excessive proposals which heavily burden the subsequent matching procedure despite their higher recalls.

Comparison with MOT-Based Methods. MOT-based methods have a similar detection mechanism to our approach, by first generating tubes for query objects and representing them with the aggregated tube features, followed by matching between query tube features and support features. Thus, even with much lower recalls ($\sim$70.0% *v.s.* $\sim$80.0%), they still have better performance than VOD-based methods by taking advantage of temporal matching. However, our approach still outperforms MOT-based methods by a significant margin leveraging our jointly optimized TPN and TMN+ with more representative features and powerful matching network.

Generalization on FSYTV-40 Dataset. This dataset is very different from FSVOD-500 with the former having significantly less classes but more videos in each class, more tracks in each video, and higher annotation FPS. Although our method still outperforms other methods on this dataset, a substantial performance degradation in comparison with FSVOD-500 is resulted, which is caused by the much reduced class diversity for the matching network to learn a general relation metric for novel classes. To verify this, we train our model on the FSVOD-500 train set and evaluate it on the FSYTV-40 test set[4]. It can promote the performance from 14.6 to 17.8 AP. The resulting large performance boost

---

[4] There is no overlapping or similar classes between them.

again validates the importance of high diversity of training classes, one of the desirable properties of our FSVOD-500 for few-shot video object learning.

5.2 Ablation Studies

Table 4 tabulates the ablation studies on the proposal box generation network and matching (classification) network. Compared to RPN, our proposed TPN improves the performance by 3.7 AP with the same matching network. Although RPN and TPN have similar recall performance (76.2% vs 76.8%), TPN has a better classification performance due to its discriminative and aggregated temporal features, and therefore producing better detection and matching performance.

Table 4. Ablation experimental results on FSVOD-500 val set for 80 novel classes with the full-way 5-shot evaluation. "LSR" denotes label-smoothing regularization and "SCM" denotes support classification module.

Box	Matching	AP	AP_{50}	AP_{75}
RPN	RN	$10.1_{\pm0.5}$	$14.0_{\pm0.6}$	$11.1_{\pm0.7}$
	MN	$19.5_{\pm0.9}$	$27.4_{\pm1.2}$	$21.8_{\pm1.1}$
TPN	MN	$23.2_{\pm1.2}$	$32.7_{\pm1.5}$	$25.6_{\pm1.5}$
	TMN	$26.4_{\pm1.5}$	$37.2_{\pm1.4}$	$29.5_{\pm1.6}$
	TMN w/ LSR	$27.9_{\pm1.3}$	$39.6_{\pm1.2}$	$30.8_{\pm1.5}$
	TMN w/ SCM	$29.4_{\pm0.8}$	$41.8_{\pm1.1}$	$31.9_{\pm1.2}$
	TMN+	$\mathbf{30.0_{\pm0.8}}$	$\mathbf{43.6_{\pm1.2}}$	$\mathbf{32.9_{\pm1.1}}$

For the matching network, the RN (Relation Network [130]) based baseline performs worst which is limited by its weak matching ability. Replacing RN by the more powerful multi-relation MN [31] can significantly improve the performance. When cooperating with TPN, our proposed TMN outperforms MN by 3.2 AP in the temporal domain using aligned query features. The improved TMN+ reaches 30.0 AP performance by capitalizing on better generalization and representative feature, which is optimized with the label-smoothing regularization and support classification module, bringing about respectively 1.5 and 3.0 performance increase. Note that our support classification module is fundamentally different from the meta-loss in Meta R-CNN [129] which requires training on novel classes to avoid prediction ambiguity in object attentive vectors, while our method targets at generating more representative features in the Euclidean space to generalize better on novel classes without any fine-tuning.

5.3 Advantages of Temporal Matching

Temporal matching has two substantial advantages over image-based matching:

Fig. 4. Qualitative 5-shot detection results on novel classes of FSVOD dataset. Our tube-based approach successfully detects objects in novel classes, while other methods miss or misclassify target objects or detect ghost objects.

Ghost Proposal Removal. Image-based matching suffers heavily from "ghost proposals" which are hard background proposals with similar appearance to foreground proposals. It is difficult to filter them out by the RPN in the spatial domain due to appearance ambiguity, while much easier to distinguish in the temporal domain due to their intermittent "ghost" or discontinuous appearances across frames. Our TPN takes this advantage to get rid of ghost proposals and thus obtains better detection performance.

Representative Feature. From the feature perspective, image-based matching exploits proposal features from each query frame to match with supports individually. Such independent query feature is inadequate in representing a target video object, especially those in bad visual quality due to *e.g.*, large deformation, motion blur or heavy occlusion, thus is liable to bad comparison results in the subsequent matching procedure and leading to bad predictions. In contrast, our temporal matching aggregates object features across frames in the tube proposal into a robust representative feature for the target video object, which helps the subsequent matching procedure to produce better result.

Validation. We show quantitatively and qualitatively the above advantages of our temporal matching. Specifically, we transform our tube-based matching to the image-based matching by performing per-frame detection and matching during inference. With the same trained model, the performance drastically drops from 30.0 to 25.8 after replacing tube-based feature by image-based feature. The large performance gap indicates the effectiveness of tube-based matching in the FSVOD task. In Fig. 4, the image-based methods produce ghost proposals and fails the target object matching, while our approach produces much better performance without suffering from ghost proposals.

5.4   Object Indexing in Massive Videos

Our FSVOD task enables models properly solving the object indexing/retrieval problem in massive videos, which is infeasible or extreme hard for other computer vision tasks. Specifically, we retrieve video clips for the target support class if there exists a detected box with the class score larger than 0.05. Thanks to the full-way evaluation, our FSVOD actually performs indexing for every class in the entire video set. We use the widely-used F_1 score to evaluate the retrieval performance. Our FSVOD model achieves 0.414 F_1 score on FSVOD-500 test set, while the classic few-shot object detection model [31] only obtains 0.339 F_1 score because of its numerous false positive predictions in videos. More details are in the supplementary material.

6   Conclusion

This paper proposes FSVOD for detecting objects in novel classes in a query *video* given *only a few support images*. FSVOD can be applied in high diversity/dynamic scenarios for solving relevant real-world problem that is infeasible or hard for other computer vision tasks. We contribute a new large-scale, class-balanced FSVOD dataset, which contains 500 classes of objects in high diversity with high-quality annotations. Our tube proposal network and aligned matching network effectively employ the temporal information in proposal generation and matching. Extensive comparison have been performed to compare related methods on two datasets to validate that our FSVOD method produces the best performance. We hope this paper will kindle future FSVOD research.

References

1. Bell, S., Lawrence Zitnick, C., Bala, K., Girshick, R.: Inside-outside net: detecting objects in context with skip pooling and recurrent neural networks. In: CVPR (2016)
2. Bergmann, P., Meinhardt, T., Leal-Taixe, L.: Tracking without bells and whistles. In: ICCV (2019)
3. Bertasius, G., Torresani, L., Shi, J.: Object detection in video with spatiotemporal sampling networks. In: Ferrari, V., Hebert, M., Sminchisescu, C., Weiss, Y. (eds.) ECCV 2018. LNCS, vol. 11216, pp. 342–357. Springer, Cham (2018). https://doi.org/10.1007/978-3-030-01258-8_21
4. Bertinetto, L., Henriques, J.F., Torr, P.H., Vedaldi, A.: Meta-learning with differentiable closed-form solvers. In: ICLR (2019)
5. Bertinetto, L., Valmadre, J., Henriques, J.F., Vedaldi, A., Torr, P.H.S.: Fully-convolutional Siamese networks for object tracking. In: Hua, G., Jégou, H. (eds.) ECCV 2016. LNCS, vol. 9914, pp. 850–865. Springer, Cham (2016). https://doi.org/10.1007/978-3-319-48881-3_56
6. Bhat, G., Danelljan, M., Gool, L.V., Timofte, R.: Learning discriminative model prediction for tracking. In: ICCV (2019)

7. Bhat, G., Danelljan, M., Van Gool, L., Timofte, R.: Know your surroundings: exploiting scene information for object tracking. In: Vedaldi, A., Bischof, H., Brox, T., Frahm, J.-M. (eds.) ECCV 2020. LNCS, vol. 12368, pp. 205–221. Springer, Cham (2020). https://doi.org/10.1007/978-3-030-58592-1_13

8. Cai, Z., Fan, Q., Feris, R.S., Vasconcelos, N.: A unified multi-scale deep convolutional neural network for fast object detection. In: Leibe, B., Matas, J., Sebe, N., Welling, M. (eds.) ECCV 2016. LNCS, vol. 9908, pp. 354–370. Springer, Cham (2016). https://doi.org/10.1007/978-3-319-46493-0_22

9. Cai, Z., Vasconcelos, N.: Cascade R-CNN: delving into high quality object detection. In: CVPR (2018)

10. Cao, K., Ji, J., Cao, Z., Chang, C.Y., Niebles, J.C.: Few-shot video classification via temporal alignment. In: CVPR (2020)

11. Chen, H., Wang, Y., Wang, G., Qiao, Y.: LSTD: a low-shot transfer detector for object detection. In: AAAI (2018)

12. Chen, Y., Cao, Y., Hu, H., Wang, L.: Memory enhanced global-local aggregation for video object detection. In: CVPR (2020)

13. Chu, P., Ling, H.: FAMNet: joint learning of feature, affinity and multi-dimensional assignment for online multiple object tracking. In: ICCV (2019)

14. Dai, J., Li, Y., He, K., Sun, J.: R-FCN: object detection via region-based fully convolutional networks. In: NeurIPS (2016)

15. Dai, J., Qi, H., Xiong, Y., Li, Y., Zhang, G., Hu, H., Wei, Y.: Deformable convolutional networks. In: ICCV (2017)

16. Danelljan, M., Bhat, G., Khan, F.S., Felsberg, M.: ATOM: accurate tracking by overlap maximization. In: CVPR (2019)

17. Danelljan, M., Bhat, G., Shahbaz Khan, F., Felsberg, M.: Eco: Efficient convolution operators for tracking. In: CVPR (2017)

18. Danelljan, M., Gool, L.V., Timofte, R.: Probabilistic regression for visual tracking. In: CVPR (2020)

19. Danelljan, M., Hager, G., Shahbaz Khan, F., Felsberg, M.: Learning spatially regularized correlation filters for visual tracking. In: ICCV (2015)

20. Danelljan, M., Shahbaz Khan, F., Felsberg, M., Van de Weijer, J.: Adaptive color attributes for real-time visual tracking. In: CVPR (2014)

21. Dave, A., Khurana, T., Tokmakov, P., Schmid, C., Ramanan, D.: TAO: a large-scale benchmark for tracking any object. In: Vedaldi, A., Bischof, H., Brox, T., Frahm, J.-M. (eds.) ECCV 2020. LNCS, vol. 12350, pp. 436–454. Springer, Cham (2020). https://doi.org/10.1007/978-3-030-58558-7_26

22. Deng, H., Hua, Y., Song, T., Zhang, Z., Xue, Z., Ma, R., Robertson, N., Guan, H.: Object guided external memory network for video object detection. In: ICCV (2019)

23. Deng, J., Dong, W., Socher, R., Li, L.J., Li, K., Fei-Fei, L.: ImageNet: a large-scale hierarchical image database. In: CVPR (2009)

24. Deng, J., Pan, Y., Yao, T., Zhou, W., Li, H., Mei, T.: Relation distillation networks for video object detection. In: ICCV (2019)

25. Dollár, P., Wojek, C., Schiele, B., Perona, P.: Pedestrian detection: a benchmark. In: CVPR (2009)

26. Dong, N., Xing, E.P.: Few-shot semantic segmentation with prototype learning. In: BMVC (2018)

27. Duan, K., Bai, S., Xie, L., Qi, H., Huang, Q., Tian, Q.: Centernet: keypoint triplets for object detection. In: ICCV (2019)

28. Ess, A., Leibe, B., Schindler, K., Van Gool, L.: A mobile vision system for robust multi-person tracking. In: CVPR (2008)

29. Fan, H., et al.: LaSOT: a high-quality benchmark for large-scale single object tracking. In: CVPR (2019)

30. Fan, Q., Ke, L., Pei, W., Tang, C.-K., Tai, Y.-W.: Commonality-parsing network across shape and appearance for partially supervised instance segmentation. In: Vedaldi, A., Bischof, H., Brox, T., Frahm, J.-M. (eds.) ECCV 2020. LNCS, vol. 12353, pp. 379–396. Springer, Cham (2020). https://doi.org/10.1007/978-3-030-58598-3_23

31. Fan, Q., Zhuo, W., Tang, C.K., Tai, Y.W.: Few-shot object detection with attention-RPN and multi-relation detector. In: CVPR (2020)

32. Fan, Z., Ma, Y., Li, Z., Sun, J.: Generalized few-shot object detection without forgetting. In: CVPR (2021)

33. Fang, K., Xiang, Y., Li, X., Savarese, S.: Recurrent autoregressive networks for online multi-object tracking. In: WACV (2018)

34. Feichtenhofer, C., Pinz, A., Zisserman, A.: Detect to track and track to detect. In: ICCV (2017)

35. Finn, C., Abbeel, P., Levine, S.: Model-agnostic meta-learning for fast adaptation of deep networks. In: ICML (2017)

36. Girshick, R.: Fast R-CNN. In: ICCV (2015)

37. Girshick, R., Donahue, J., Darrell, T., Malik, J.: Rich feature hierarchies for accurate object detection and semantic segmentation. In: CVPR (2014)

38. Bhat, G., et al.: Learning what to learn for video object segmentation. In: Vedaldi, A., Bischof, H., Brox, T., Frahm, J.-M. (eds.) ECCV 2020. LNCS, vol. 12347, pp. 777–794. Springer, Cham (2020). https://doi.org/10.1007/978-3-030-58536-5_46

39. Gui, L.-Y., Wang, Y.-X., Ramanan, D., Moura, J.M.F.: Few-shot human motion prediction via meta-learning. In: Ferrari, V., Hebert, M., Sminchisescu, C., Weiss, Y. (eds.) ECCV 2018. LNCS, vol. 11212, pp. 441–459. Springer, Cham (2018). https://doi.org/10.1007/978-3-030-01237-3_27

40. Guo, Q., Feng, W., Zhou, C., Huang, R., Wan, L., Wang, S.: Learning dynamic Siamese network for visual object tracking. In: ICCV (2017)

41. Gupta, A., Dollar, P., Girshick, R.: LVIS: a dataset for large vocabulary instance segmentation. In: CVPR (2019)

42. Hariharan, B., Girshick, R.: Low-shot visual recognition by shrinking and hallucinating features. In: ICCV (2017)

43. He, K., Gkioxari, G., Dollár, P., Girshick, R.: Mask R-CNN. In: ICCV (2017)

44. He, K., Zhang, X., Ren, S., Sun, J.: Deep residual learning for image recognition. In: CVPR (2016)

45. Held, D., Thrun, S., Savarese, S.: Learning to track at 100 fps with deep regression networks. In: Leibe, B., Matas, J., Sebe, N., Welling, M. (eds.) ECCV 2016. LNCS, vol. 9905, pp. 749–765. Springer, Cham (2016). https://doi.org/10.1007/978-3-319-46448-0_45

46. Henriques, J.F., Caseiro, R., Martins, P., Batista, J.: High-speed tracking with kernelized correlation filters. IEEE TPAMI (2014)

47. Hu, H., Bai, S., Li, A., Cui, J., Wang, L.: Dense relation distillation with context-aware aggregation for few-shot object detection. In: CVPR (2021)

48. Hu, T., Pengwan, Zhang, C., Yu, G., Mu, Y., Snoek, C.G.M.: Attention-based multi-context guiding for few-shot semantic segmentation. In: AAAI (2019)

49. Huang, L., Zhao, X., Huang, K.: Got-10k: a large high-diversity benchmark for generic object tracking in the wild. IEEE TPAMI (2019)

50. Kang, B., Liu, Z., Wang, X., Yu, F., Feng, J., Darrell, T.: Few-shot object detection via feature reweighting. In: ICCV (2019)

51. Kang, K., Li, H., Xiao, T., Ouyang, W., Yan, J., Liu, X., Wang, X.: Object detection in videos with tubelet proposal networks. In: CVPR (2017)
52. Kang, K., Ouyang, W., Li, H., Wang, X.: Object detection from video tubelets with convolutional neural networks. In: CVPR (2016)
53. Karlinsky, L., et al.: RepMet: representative-based metric learning for classification and few-shot object detection. In: CVPR (2019)
54. Kemelmacher-Shlizerman, I., Seitz, S.M., Miller, D., Brossard, E.: The megaface benchmark: 1 million faces for recognition at scale. In: CVPR (2016)
55. Khodadadeh, S., Boloni, L., Shah, M.: Unsupervised meta-learning for few-shot image classification. NeurIPS (2019)
56. Kim, C., Li, F., Ciptadi, A., Rehg, J.M.: Multiple hypothesis tracking revisited. In: ICCV (2015)
57. Kim, D., Woo, S., Lee, J.Y., Kweon, I.S.: Video panoptic segmentation. In: CVPR (2020)
58. Koch, G., Zemel, R., Salakhutdinov, R.: Siamese neural networks for one-shot image recognition. In: ICMLW (2015)
59. Kong, T., Sun, F., Liu, H., Jiang, Y., Li, L., Shi, J.: FoveaBox: beyound anchor-based object detection. IEEE TIP (2020)
60. Kong, T., Sun, F., Yao, A., Liu, H., Lu, M., Chen, Y.: Ron: Reverse connection with objectness prior networks for object detection. In: CVPR (2017)
61. Kristan, M., et al.: The visual object tracking vot2017 challenge results. In: ICCVW (2017)
62. Kristan, M., et al.: The sixth visual object tracking VOT2018 challenge results. In: Leal-Taixé, L., Roth, S. (eds.) ECCV 2018. LNCS, vol. 11129, pp. 3–53. Springer, Cham (2019). https://doi.org/10.1007/978-3-030-11009-3_1
63. Kristan, M., et al.: The visual object tracking vot2015 challenge results. In: ICCVW (2015)
64. Law, H., Deng, J.: CornerNet: detecting objects as paired keypoints. In: Ferrari, V., Hebert, M., Sminchisescu, C., Weiss, Y. (eds.) Computer Vision – ECCV 2018. LNCS, vol. 11218, pp. 765–781. Springer, Cham (2018). https://doi.org/10.1007/978-3-030-01264-9_45
65. Lee, Y., Choi, S.: Gradient-based meta-learning with learned layerwise metric and subspace. In: ICML (2018)
66. Li, A., Li, Z.: Transformation invariant few-shot object detection. In: CVPR (2021)
67. Li, B., Wu, W., Wang, Q., Zhang, F., Xing, J., Yan, J.: SiamRPN++: evolution of Siamese visual tracking with very deep networks. In: CVPR (2019)
68. Li, B., Yan, J., Wu, W., Zhu, Z., Hu, X.: High performance visual tracking with Siamese region proposal network. In: CVPR (2018)
69. Li, B., Yang, B., Liu, C., Liu, F., Ji, R., Ye, Q.: Beyond max-margin: class margin equilibrium for few-shot object detection. In: CVPR (2021)
70. Li, X., Wei, T., Chen, Y.P., Tai, Y.W., Tang, C.K.: FSS-1000: a 1000-class dataset for few-shot segmentation. In: CVPR (2020)
71. Li, Y., Chen, Y., Wang, N., Zhang, Z.: Scale-aware trident networks for object detection. In: ICCV (2019)
72. Li, Y., et al.: Few-shot object detection via classification refinement and distractor retreatment. In: CVPR (2021)
73. Lin, T.Y., Dollár, P., Girshick, R., He, K., Hariharan, B., Belongie, S.: Feature pyramid networks for object detection. In: CVPR (2017)
74. Lin, T.Y., Goyal, P., Girshick, R., He, K., Dollár, P.: Focal loss for dense object detection. In: ICCV (2017)

75. Lin, T.-Y., et al.: Microsoft COCO: common objects in context. In: Fleet, D., Pajdla, T., Schiele, B., Tuytelaars, T. (eds.) ECCV 2014. LNCS, vol. 8693, pp. 740–755. Springer, Cham (2014). https://doi.org/10.1007/978-3-319-10602-1_48

76. Liu, S., Huang, D., Wang, Y.: Receptive field block net for accurate and fast object detection. In: Ferrari, V., Hebert, M., Sminchisescu, C., Weiss, Y. (eds.) ECCV 2018. LNCS, vol. 11215, pp. 404–419. Springer, Cham (2018). https://doi.org/10.1007/978-3-030-01252-6_24

77. Liu, W., et al.: SSD: single shot multibox detector. In: Leibe, B., Matas, J., Sebe, N., Welling, M. (eds.) ECCV 2016. LNCS, vol. 9905, pp. 21–37. Springer, Cham (2016). https://doi.org/10.1007/978-3-319-46448-0_2

78. Liu, W., Liao, S., Ren, W., Hu, W., Yu, Y.: High-level semantic feature detection: a new perspective for pedestrian detection. In: CVPR (2019)

79. Liu, Y., Zhang, X., Zhang, S., He, X.: Part-aware prototype network for few-shot semantic segmentation. In: Vedaldi, A., Bischof, H., Brox, T., Frahm, J.-M. (eds.) ECCV 2020. LNCS, vol. 12354, pp. 142–158. Springer, Cham (2020). https://doi.org/10.1007/978-3-030-58545-7_9

80. Lu, X., Li, B., Yue, Y., Li, Q., Yan, J.: Grid R-CNN. In: CVPR (2019)

81. Lu, Z., Rathod, V., Votel, R., Huang, J.: RetinaTrack: online single stage joint detection and tracking. In: CVPR (2020)

82. Luiten, J., et al.: HOTA: a higher order metric for evaluating multi-object tracking. IJCV (2021)

83. Mayer, C., Danelljan, M., Paudel, D.P., Gool, L.V.: Learning target candidate association to keep track of what not to track. In: ICCV (2021)

84. Michaelis, C., Bethge, M., Ecker, A.S.: One-shot segmentation in clutter. In: ICML (2018)

85. Milan, A., Leal-Taixé, L., Reid, I., Roth, S., Schindler, K.: MOT16: a benchmark for multi-object tracking. arXiv preprint arXiv:1603.00831 (2016)

86. Müller, M., Bibi, A., Giancola, S., Alsubaihi, S., Ghanem, B.: TrackingNet: a large-scale dataset and benchmark for object tracking in the wild. In: Ferrari, V., Hebert, M., Sminchisescu, C., Weiss, Y. (eds.) ECCV 2018. LNCS, vol. 11205, pp. 310–327. Springer, Cham (2018). https://doi.org/10.1007/978-3-030-01246-5_19

87. Müller, R., Kornblith, S., Hinton, G.E.: When does label smoothing help? In: NeurIPS (2019)

88. Oh, S.W., Lee, J.Y., Xu, N., Kim, S.J.: Video object segmentation using space-time memory networks. In: ICCV (2019)

89. Pang, B., Li, Y., Zhang, Y., Li, M., Lu, C.: TubeTK: adopting tubes to track multi-object in a one-step training model. In: CVPR (2020)

90. Peng, J., et al.: Chained-Tracker: chaining paired attentive regression results for end-to-end joint multiple-object detection and tracking. In: Vedaldi, A., Bischof, H., Brox, T., Frahm, J.-M. (eds.) ECCV 2020. LNCS, vol. 12349, pp. 145–161. Springer, Cham (2020). https://doi.org/10.1007/978-3-030-58548-8_9

91. Perazzi, F., Pont-Tuset, J., McWilliams, B., Van Gool, L., Gross, M., Sorkine-Hornung, A.: A benchmark dataset and evaluation methodology for video object segmentation. In: CVPR (2016)

92. Perez-Rua, J.M., Zhu, X., Hospedales, T.M., Xiang, T.: Incremental few-shot object detection. In: CVPR (2020)

93. Real, E., Shlens, J., Mazzocchi, S., Pan, X., Vanhoucke, V.: Youtube-boundingboxes: a large high-precision human-annotated data set for object detection in video. In: CVPR (2017)

94. Redmon, J., Divvala, S., Girshick, R., Farhadi, A.: You only look once: unified, real-time object detection. In: CVPR (2016)
95. Redmon, J., Farhadi, A.: YOLO9000: better, faster, stronger. In: CVPR (2017)
96. Ren, S., He, K., Girshick, R., Sun, J.: Faster R-CNN: towards real-time object detection with region proposal networks. In: NeurIPS (2015)
97. Sadeghian, A., Alahi, A., Savarese, S.: Tracking the untrackable: learning to track multiple cues with long-term dependencies. In: ICCV (2017)
98. Santoro, A., Bartunov, S., Botvinick, M., Wierstra, D., Lillicrap, T.: Meta-learning with memory-augmented neural networks. In: ICML (2016)
99. Sbai, O., Couprie, C., Aubry, M.: Impact of base dataset design on few-shot image classification. In: Vedaldi, A., Bischof, H., Brox, T., Frahm, J.-M. (eds.) ECCV 2020. LNCS, vol. 12361, pp. 597–613. Springer, Cham (2020). https://doi.org/10.1007/978-3-030-58517-4_35
100. Shao, D., Xiong, Yu., Zhao, Y., Huang, Q., Qiao, Yu., Lin, D.: Find and focus: retrieve and localize video events with natural language queries. In: Ferrari, V., Hebert, M., Sminchisescu, C., Weiss, Y. (eds.) ECCV 2018. LNCS, vol. 11213, pp. 202–218. Springer, Cham (2018). https://doi.org/10.1007/978-3-030-01240-3_13
101. Shen, Z., Liu, Z., Li, J., Jiang, Y.G., Chen, Y., Xue, X.: DSOD: learning deeply supervised object detectors from scratch. In: ICCV (2017)
102. Shvets, M., Liu, W., Berg, A.C.: Leveraging long-range temporal relationships between proposals for video object detection. In: ICCV (2019)
103. Singh, B., Najibi, M., Davis, L.S.: Sniper: Efficient multi-scale training. In: NeurIPS (2018)
104. Snell, J., Swersky, K., Zemel, R.: Prototypical networks for few-shot learning. In: NeurIPS (2017)
105. Sun, B., Li, B., Cai, S., Yuan, Y., Zhang, C.: FSCE: few-shot object detection via contrastive proposal encoding. In: CVPR (2021)
106. Szegedy, C., Vanhoucke, V., Ioffe, S., Shlens, J., Wojna, Z.: Rethinking the inception architecture for computer vision. In: CVPR (2016)
107. Tang, P., Wang, C., Wang, X., Liu, W., Zeng, W., Wang, J.: Object detection in videos by high quality object linking. IEEE TPAMI (2019)
108. Tang, P., Wang, X., Bai, X., Liu, W.: Multiple instance detection network with online instance classifier refinement. In: CVPR (2017)
109. Tao, R., Gavves, E., Smeulders, A.W.: Siamese instance search for tracking. In: CVPR (2016)
110. Thulasidasan, S., Chennupati, G., Bilmes, J.A., Bhattacharya, T., Michalak, S.: On mixup training: improved calibration and predictive uncertainty for deep neural networks. In: NeurIPS (2019)
111. Tian, Z., Shen, C., Chen, H., He, T.: FCOS: fully convolutional one-stage object detection. In: ICCV (2019)
112. Valmadre, J., Bertinetto, L., Henriques, J., Vedaldi, A., Torr, P.H.: End-to-end representation learning for correlation filter based tracking. In: CVPR (2017)
113. Valmadre, J., Bewley, A., Huang, J., Sun, C., Sminchisescu, C., Schmid, C.: Local metrics for multi-object tracking. arXiv preprint arXiv:2104.02631 (2021)
114. Vinyals, O., et al.: Matching networks for one shot learning. In: NeurIPS (2016)
115. Wang, S., Zhou, Y., Yan, J., Deng, Z.: Fully motion-aware network for video object detection. In: Ferrari, V., Hebert, M., Sminchisescu, C., Weiss, Y. (eds.) ECCV 2018. LNCS, vol. 11217, pp. 557–573. Springer, Cham (2018). https://doi.org/10.1007/978-3-030-01261-8_33

116. Wang, X., Huang, T.E., Darrell, T., Gonzalez, J.E., Yu, F.: Frustratingly simple few-shot object detection. In: ICML (2020)

117. Wang, Y.X., Girshick, R., Hebert, M., Hariharan, B.: Low-shot learning from imaginary data. In: CVPR (2018)

118. Wang, Y.X., Ramanan, D., Hebert, M.: Meta-learning to detect rare objects. In: CVPR (2019)

119. Woo, S., Kim, D., Cho, D., Kweon, I.S.: LinkNet: relational embedding for scene graph. In: NeurIPS (2018)

120. Wu, C.Y., Feichtenhofer, C., Fan, H., He, K., Krahenbuhl, P., Girshick, R.: Long-term feature banks for detailed video understanding. In: CVPR (2019)

121. Wu, H., Chen, Y., Wang, N., Zhang, Z.: Sequence level semantics aggregation for video object detection. In: ICCV (2019)

122. Wu, J., Liu, S., Huang, D., Wang, Y.: Multi-scale positive sample refinement for few-shot object detection. In: Vedaldi, A., Bischof, H., Brox, T., Frahm, J.-M. (eds.) ECCV 2020. LNCS, vol. 12361, pp. 456–472. Springer, Cham (2020). https://doi.org/10.1007/978-3-030-58517-4_27

123. Wu, Y., Lim, J., Yang, M.H.: Online object tracking: a benchmark. In: CVPR (2013)

124. Xiao, F., Lee, Y.J.: Video object detection with an aligned spatial-temporal memory. In: Ferrari, V., Hebert, M., Sminchisescu, C., Weiss, Y. (eds.) ECCV 2018. LNCS, vol. 11212, pp. 494–510. Springer, Cham (2018). https://doi.org/10.1007/978-3-030-01237-3_30

125. Xiao, T., Li, S., Wang, B., Lin, L., Wang, X.: Joint detection and identification feature learning for person search. In: CVPR (2017)

126. Xiao, Y., Marlet, R.: Few-shot object detection and viewpoint estimation for objects in the wild. In: Vedaldi, A., Bischof, H., Brox, T., Frahm, J.-M. (eds.) ECCV 2020. LNCS, vol. 12362, pp. 192–210. Springer, Cham (2020). https://doi.org/10.1007/978-3-030-58520-4_12

127. Xu, J., Cao, Y., Zhang, Z., Hu, H.: Spatial-temporal relation networks for multi-object tracking. In: ICCV (2019)

128. Xu, N., et al.: YouTube-VOS: sequence-to-sequence video object segmentation. In: Ferrari, V., Hebert, M., Sminchisescu, C., Weiss, Y. (eds.) ECCV 2018. LNCS, vol. 11209, pp. 603–619. Springer, Cham (2018). https://doi.org/10.1007/978-3-030-01228-1_36

129. Yan, X., Chen, Z., Xu, A., Wang, X., Liang, X., Lin, L.: Meta R-CNN: towards general solver for instance-level low-shot learning. In: ICCV (2019)

130. Yang, F.S.Y., Zhang, L., Xiang, T., Torr, P.H., Hospedales, T.M.: Learning to compare: relation network for few-shot learning. In: CVPR (2018)

131. Yang, L., Fan, Y., Xu, N.: Video instance segmentation. In: ICCV (2019)

132. Yang, Y., Wei, F., Shi, M., Li, G.: Restoring negative information in few-shot object detection. In: NeurIPS (2020)

133. Yang, Z., Liu, S., Hu, H., Wang, L., Lin, S.: RepPoints: point set representation for object detection. In: ICCV (2019)

134. Yang, Z., Wang, Y., Chen, X., Liu, J., Qiao, Y.: Context-transformer: tackling object confusion for few-shot detection. In: AAAI (2020)

135. Yu, F., Li, W., Li, Q., Liu, Yu., Shi, X., Yan, J.: POI: multiple object tracking with high performance detection and appearance feature. In: Hua, G., Jégou, H. (eds.) ECCV 2016. LNCS, vol. 9914, pp. 36–42. Springer, Cham (2016). https://doi.org/10.1007/978-3-319-48881-3_3

136. Zhan, Y., Wang, C., Wang, X., Zeng, W., Liu, W.: A simple baseline for multi-object tracking. IJCV (2021)

137. Zhang, C., Cai, Y., Lin, G., Shen, C.: DeepEMD: few-shot image classification with differentiable earth mover's distance and structured classifiers. In: CVPR (2020)

138. Zhang, L., Zhou, S., Guan, J., Zhang, J.: Accurate few-shot object detection with support-query mutual guidance and hybrid loss. In: CVPR (2021)

139. Zhang, S., Chi, C., Yao, Y., Lei, Z., Li, S.Z.: Bridging the gap between anchor-based and anchor-free detection via adaptive training sample selection. In: CVPR (2020)

140. Zhang, W., Wang, Y.X.: Hallucination improves few-shot object detection. In: CVPR (2021)

141. Zhang, Z., Qiao, S., Xie, C., Shen, W., Wang, B., Yuille, A.L.: Single-shot object detection with enriched semantics. In: CVPR (2018)

142. Zhou, X., Koltun, V., Krähenbühl, P.: Tracking objects as points. In: Vedaldi, A., Bischof, H., Brox, T., Frahm, J.-M. (eds.) ECCV 2020. LNCS, vol. 12349, pp. 474–490. Springer, Cham (2020). https://doi.org/10.1007/978-3-030-58548-8_28

143. Zhou, X., Zhuo, J., Krahenbuhl, P.: Bottom-up object detection by grouping extreme and center points. In: CVPR (2019)

144. Zhu, C., Chen, F., Ahmed, U., Savvides, M.: Semantic relation reasoning for shot-stable few-shot object detection. In: CVPR (2021)

145. Zhu, J., Yang, H., Liu, N., Kim, M., Zhang, W., Yang, M.-H.: Online multi-object tracking with dual matching attention networks. In: Ferrari, V., Hebert, M., Sminchisescu, C., Weiss, Y. (eds.) ECCV 2018. LNCS, vol. 11209, pp. 379–396. Springer, Cham (2018). https://doi.org/10.1007/978-3-030-01228-1_23

146. Zhu, L., Yang, Y.: Compound memory networks for few-shot video classification. In: Ferrari, V., Hebert, M., Sminchisescu, C., Weiss, Y. (eds.) ECCV 2018. LNCS, vol. 11211, pp. 782–797. Springer, Cham (2018). https://doi.org/10.1007/978-3-030-01234-2_46

147. Zhu, R., Zhang, S., Wang, X., Wen, L., Shi, H., Bo, L., Mei, T.: ScratchDet: training single-shot object detectors from scratch. In: CVPR (2019)

148. Zhu, X., Dai, J., Yuan, L., Wei, Y.: Towards high performance video object detection. In: CVPR (2018)

149. Zhu, X., Wang, Y., Dai, J., Yuan, L., Wei, Y.: Flow-guided feature aggregation for video object detection. In: ICCV (2017)

150. Zhu, X., Xiong, Y., Dai, J., Yuan, L., Wei, Y.: Deep feature flow for video recognition. In: CVPR (2017)

Worst Case Matters for Few-Shot Recognition

Minghao Fu⑩, Yun-Hao Cao⑩, and Jianxin Wu$^{(\boxtimes)}$⑩

State Key Laboratory for Novel Software Technology, Nanjing University,
Nanjing, China
{fumh,caoyh}@lamda.nju.edu.cn, wujx2001@gmail.com

Abstract. Few-shot recognition learns a recognition model with very few (e.g., 1 or 5) images per category, and current few-shot learning methods focus on improving the average accuracy over many episodes. We argue that in real-world applications we may often only try one episode instead of many, and hence maximizing the worst-case accuracy is more important than maximizing the average accuracy. We empirically show that a high average accuracy not necessarily means a high worst-case accuracy. Since this objective is not accessible, we propose to reduce the standard deviation and increase the average accuracy simultaneously. In turn, we devise two strategies from the bias-variance tradeoff perspective to implicitly reach this goal: a simple yet effective stability regularization (SR) loss together with model ensemble to reduce variance during fine-tuning, and an adaptability calibration mechanism to reduce the bias. Extensive experiments on benchmark datasets demonstrate the effectiveness of the proposed strategies, which outperforms current state-of-the-art methods with a significant margin in terms of not only average, but also worst-case accuracy.

1 Introduction

Most people have the ability to learn to recognize new patterns via one or a few samples (e.g., images), thanks to the accumulated knowledge. Naturally, *few-shot learning* [35] aims at learning from scarce data, which is already studied long before the deep learning era. In this paper, we focus on the image recognition task, also known as *few-shot image classification*, a widely studied few-shot task [1,2,5,7,9,21,23,29,30,33,36]. Deep learning techniques have further pushed few-shot learning's *average* accuracy over multiple runs (i.e., *episodes*) towards a high level that appears to be already applicable to real-world applications.

In this task, a pretrained model is first derived from the *base set*, a large set of labeled images. Then, given some unseen categories and very few training images per category (the *novel set*), the model must learn to adapt to classifying new examples from these novel categories. Differently sampled novel sets lead to different episodes, different trained models and test accuracies. The common evaluation criterion is to run a large number of (usually 500 to 10000) episodes for the same task, and report the average accuracy and its 95% confidence interval.

© The Author(s), under exclusive license to Springer Nature Switzerland AG 2022
S. Avidan et al. (Eds.): ECCV 2022, LNCS 13680, pp. 99–115, 2022.
https://doi.org/10.1007/978-3-031-20044-1_6

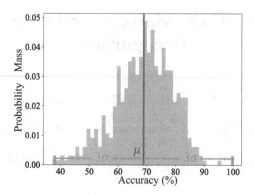

Fig. 1. Distribution of accuracy rates of 500 episodes in our experiments, with 5-way 1-shot experiments of the LR-DC [36] method on *mini*-ImageNet. Best viewed in color. (Color figure online)

We aim at making few-shot recognition more practical, too. But, we argue that both metrics (mean accuracy and 95% confidence interval) are *not helping us towards reaching this goal*, given the recent progress in this task. Figure 1 shows the distribution of accuracy of 500 episodes for the same task, whose average accuracy estimate is 68.96% and the 95% confidence interval is [68.07, 69.85]—this interval has 95% chance of the true average accuracy landing on it, not the chance of a single episode's accuracy dropping inside! In other words, both metrics are used to describe *the mean accuracy of 500 episodes, not the accuracy of a single experiment*.

In fact, the worst episode among these 500 in Fig. 1 has only 37.33% accuracy while that of the best is close to 100%. That is, few-shot learning is very *unstable*, and the accuracy varies dramatically. The *worst-case lags far behind the average*.

Furthermore, now that we aim at making few-shot learning practical, we have to accept that in most real-world applications, we can run the experiment *only once*, but do not have the luxury of running 500 (or more) experiments and pick the best or average episode among them. In other words, *the worst-case scenario* (if one is unlucky) is naturally more important than the average case—in commercial applications, we welcome a product with an acceptable worst-case performance more than one that performs well on average but occasionally goes calamitous. We argue that we need to pay attention to one episode instead of the average of 500 episodes, and to *maximize the worst case's accuracy* instead of the mean.

But, it is in general very challenging to explicitly optimize the worst-case scenario even when there are many training examples, let alone in the few-shot setting. We propose an alternative, indirect line of attack. As shown in Fig. 1, the accuracy distribution often fits well to a Gaussian. Then, the worst-case accuracy is naturally estimated by the 3σ rule as $\mu - 3\sigma$, where μ is the average accuracy and σ is the accuracy's standard deviation—in a normal distribution, the chance to fall on the left of $\mu - 3\sigma$ is only 0.135%. Figure 1 clearly supports this approximation. In other words, we do need to increase the mean accuracy μ (as in current methods), but also need to *simultaneously minimize the standard deviation* σ.

Unfortunately, directly optimizing σ is not plausible, because it involves different episodes and different models, which is beyond the capability of end-to-end deep learning. To tackle this issue, we propose to resort to the bias-variance decomposition theory [14], which states that the expected error $(1 - \mu)$ is the sum of squared bias plus variance (σ^2). From this classic theory, we know that *stable* learning machines lead to small σ, and hence propose a novel *stability regularization* (SR) to utilize the base set to achieve stability. We also incorporate model ensemble, because it is well-known for its effect of reducing both bias and variance. Similarly, this theory also states that a model with larger capacity often lead to smaller bias, hence we propose an *adaptability calibration* (AC) strategy to properly increase the model capacity without overfitting, so as to indirectly reduce the bias and to increase the average accuracy μ in turn.

Our contributions can be summarized as follows:

1. To the best of our knowledge, we are the first to emphasize the importance and to advocate the adoption of worst case accuracy in few-shot learning.
2. Motivated by Fig. 1, we argue that in addition to maximizing the average accuracy μ, we must also *simultaneously* reduce the standard deviation σ.
3. We propose to achieve this goal from the bias-variance tradeoff perspective. We propose a simple yet effective stability regularization (SR) loss together with model ensemble to reduce variance during fine-tuning. The SR loss is computed in an unsupervised fashion, and can be easily generalized to data beyond the base set. We also propose adaptability calibration (AC) to vary the number of learnable parameters to reduce bias.

As a result, our method not only achieves higher average accuracy than current state-of-the-art methods, but more importantly enjoys significantly higher worst case accuracy on benchmark few-shot recognition datasets.

2   Related Work

Generalizing from base categories, few-shot learning aims at designing effective recognition systems with limited training data from novel categories. Current few-shot learning research can be roughly divided into two branches: those based on meta learning and those based on transfer learning. We now give a brief introduction of representative methods within both branches. Besides, since we propose using the ensemble technique for better regularizing the model stability, model ensemble methods in few-shot problems will be discussed as well.

Meta Learning. This line of methods model the training process of few-shot learning by pretending the episodes (which are usually large in quantity) are to be learned in a way of "learning to learn". One line of work [9,10,20,22,28] learns a set of good initial weights to fast adapt to unseen episodes with a limited number of gradient descent steps. Another line of work [4,17,29,30,33] leverages the characteristics of different distance metrics to classify unknown samples by comparing with embeddings or their variants (e.g., prototypes) derived from the training examples.

Transfer Learning. These methods [1–3,5,7,8,12,13,16,21,23,36] take advantage of the standard transfer learning pipeline, which first pretrain a model on base classes, then revise the feature embeddings output by the pretrained model with limited novel samples. [5] used cosine classifier to normalize the magnitude of both embeddings and classification weights for compacting the intra-class intensity. [12] proposed to pretrain with self-supervised auxiliary tasks for boosting few-shot learning. [21] introduced a negative margin loss during pretraining to increase the discriminability of features. [1] used the associative alignment strategy to learn novel features by aligning them with closely related samples from the base set. [23] explored the way to pretrain with manifold mixup [32] for good generalization. [7] proposed a baseline for transductive fine-tuning with novel samples. [16] introduced a new pipeline, which first preprocessed features, then classified them with an optimal-transport inspired algorithm. [36] adjusted novel distributions using base categories to generate imaginary training samples.

Model Ensemble. It is a well-known strategy [14] to incorporate a number of weak models to build a strong model, which will effectively increase the stability and robustness during inference. Very recently, for the few-shot classification problem, [8] harmonized the cooperation and diversity between deep models to pursue better ensemble performance. E^3BM [22] tackled few-shot problems with the ensemble of epoch-wise base-learners whose hyper-parameters were generated by task-specific information. Different from these methods, our ensemble stability regularization directly increases the diversity of deep models by dealing with different parts of data from the base set, keeping simplicity while achieving superior performance.

3   The Worst-case Accuracy and Its Surrogate

We first give some background information about how a few-shot problem is defined, then describe the relationship between the commonly used 95% confidence interval and the standard deviation (σ). Lastly, our solution that advocates indirectly optimizing for the worst-case will be presented.

In the few-shot recognition setting, there exists a dataset with abundant labeled images called the base set, denoted as $D_b = \{x_i^b, y_i^b\}_{i=1}^{N_b}$, where $x_i^b \in R^D$ is the i-th training image, $y_i^b \in \mathcal{Y}_b$ is its corresponding category label, and N_b is the number of examples. In addition, there also exists another dataset with scarce labeled images from new categories (usually 1 or 5 per category) called the novel set, denoted as $D_n = \{x_j^n, y_j^n\}_{j=1}^{N_n}$, where $y_j^n \in \mathcal{Y}_n$ and $\mathcal{Y}_b \cap \mathcal{Y}_n = \emptyset$, with N_n being the number of examples it contains. In every episode, N categories will be sampled to constitute the novel set, and K examples will be sampled from each of those categories (called N-way K-shot). Learned with this tiny set

Table 1. 5-way 1-shot recognition results of existing methods on *mini*-ImageNet (500 episodes). $\flat$ means that σ is calculated by our implementation due to undisclosed n in published papers. ACC_m, ACC_1 and ACC_{10}: higher is better; $Z_{95\%}$ and σ: lower is better. The best results are shown in boldface.

Method	ACC_m	$Z_{95\%}$	σ	ACC_1	ACC_{10}
Negative-Cosine [21]	61.72	0.81	10.12	24.27	36.13
MixtFSL [2]	64.31	0.79	9.87	30.67	35.07
S2M2$_R$ [23]	64.93	**0.18**	**9.18**	37.58	42.87
PT+NCM [16]	65.35	0.20	10.20	32.00	38.13
CGCS [11]	67.02	0.20	10.20	**38.70**	**44.00**
LR-DC [36]	**68.57**	0.55	$10.28^{\flat}$	37.33	42.72

of images (and optionally D_b), a few-shot recognizer needs to recognize unseen examples from the N categories in this episode.

3.1 Existing and Proposed Metrics

Existing methods evaluate their performance by averaging the accuracy of n episodes, including the average accuracy μ and a 95% confidence interval $[\mu - Z_{95\%}, \mu + Z_{95\%}]$. As aforementioned, both metrics are estimates for the average accuracy random variable, not estimating the accuracy of one episode. Hence, the interval radius $Z_{95\%}$ is often surprisingly small, as shown in Table 1.

As established in Sect. 1, we need to focus more on the worst-case accuracy among all n episodes, which is denoted as ACC_1. In addition, we also report the average accuracy of the 10 worst cases as ACC_{10}. The empirical average accuracy is denoted as ACC_m. Although it is a general trend that ACC_m and ACC_1 are positively correlated, the 6 methods rank significantly differently using ACC_m and ACC_1. For example, S2M2$_R$ is 3.64% lower than LR-DC in terms of μ (ACC_m), but 0.25% higher in ACC_1 (worst-case accuracy). That is, although maximizing μ is useful, it is far from being enough. We argue that for few-shot recognition to be practically usable, we need to maximize ACC_1 instead.

The $Z_{95\%}$ metric is also misleading, as it measures uncertainty in estimating μ. In fact, based on its definition, we have

$$\sigma = Z_{95\%} \cdot \frac{\sqrt{n}}{1.96}, \tag{1}$$

where σ is the standard deviation of accuracy across episodes. Hence, a small $Z_{95\%}$ may well be because n is large (e.g., $n = 10000$), instead of due to a small σ, which is illustrated clearly in Table 1. For example, MixtFSL has almost 3x larger $Z_{95\%}$ than that of CGCS, but has a smaller σ within the pair. And different papers often use different n values, which renders $Z_{95\%}$ difficult to interpret.

Furthermore, the proposed worst-case accuracy ACC_1 is not only semantically more meaningful than ACC_m (μ), but also more stable. We use ACC_k to denote the average accuracy of the k worst episodes. Results in Table 1 exhibits that ACC_1 and ACC_{10} ranks the methods consistently.

3.2   Implicitly Optimizing the Worst-Case Accuracy

As briefly introduced in Sect. 1, we can only implicitly maximize ACC_1, and we propose to use $\mu - 3\sigma$ as a surrogate. As empirically shown in Fig. 1, when the accuracy distributes as a Gaussian, $\mu - 3\sigma$ is a perfect surrogate, as it pins to the 0.135-th percentile of ACC_1, because $\Phi(-3) = 0.00135$ in which Φ is the cumulative distribution function of the standard normal distribution $N(0, 1)$. Even if the accuracy distribution is highly non-normal (which empirical data suggests otherwise), the one-sided Chebyshev inequality also guarantees that $\mu - 3\sigma$ is no larger than the 10th percentile (i.e., no better than the 10% worst cases), because the inequality states that $\Pr(X \leq \mu - k\sigma) \leq \frac{1}{1+k^2}$ for any distribution X and any $k > 0$, while $k = 3$ in our case.

But, this surrogate loss is still a qualitative one instead of a variable that can be directly maximized, because σ non-linearly involves all n episodes, including the n models and the n training sets. Hence, we transform our objective to *simultaneously maximize μ and minimize σ*, and propose to use the bias-variance tradeoff for achieving both objectives indirectly.

3.3   The Bias-Variance Tradeoff in the Few-Shot Scenario

The bias-variance decomposition states that the expectation of error rate (i.e., $1 - \mu$) equals the sum of the squared bias and the variance [14].[1] Although the definitions for bias and variance of a classifier are not unique, the following qualitative properties are commonly agreed upon [14]:

1. A classifier has large variance if it is unstable (small changes in the input cause large changes in the prediction), and vice versa; Hence, we expect *a smaller σ (equivalently, variance) if we make the recognizer more stable*.
2. A classifier with larger capacity (e.g., more learnable weights) in general has a smaller bias, and vice versa. Hence, we prefer *a model with larger capacity* to reduce the bias.
3. Although minimizing both bias and variance simultaneously amounts to maximizing the expected accuracy, the two terms are often contradictory to each other. Larger models are more prone to overfitting and in turn larger variance, too. Hence, *a properly calibrated capacity increase* is necessary to strike a balance.

3.4   Reducing Variance: Stability Regularization

To fulfill these goals, our framework (cf. Fig. 2) follows the common practice to train a backbone network $f(\mathbf{x})$ (its classification head discarded) using the base set D_b. Then, in one episode, a N-way K-shot training set is sampled as the novel set D_n to fine-tune the backbone (with a randomly initialized classification head W) into $\hat{f}(\mathbf{x})$ using a usual cross entropy classification loss $\mathcal{L}_C$.

---

[1] Here μ is the population mean, but we also use the same notation for sample mean. They can be easily distinguished by the context.

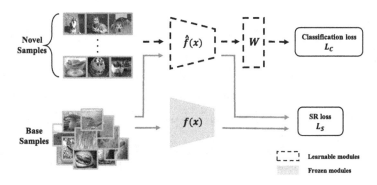

Fig. 2. The proposed few-shot recognition framework. A backbone network $f(\mathbf{x})$ is trained using the base set samples and its classification head is discarded. Using images from the novel set, the backbone is fine-tuned into $\hat{f}(\mathbf{x})$, plus a new classification head W. Stability of $\hat{f}(\mathbf{x})$ is maintained by the stability regularization (SR) loss $\mathcal{L}_S$, while the capacity increase (adaptability) is calibrated by fine-tuning a selected subset of layers in the backbone $\hat{f}(\mathbf{x})$. The classification loss $\mathcal{L}_C$ minimizes error (i.e., to increase the average accuracy μ). Best viewed in color. (Color figure online)

The danger is: because K is very small (mostly 1 or 5), the fine-tuning process easily gets overfit and is extremely *unstable*—different sampled training sets lead to dramatically different prediction accuracy. To increase the stability (and thus to reduce the variance σ^2), we must not allow the weights in $\hat{f}(\mathbf{x})$ to be dominated by the small novel training set and completely forget the representation learned from the base set, i.e., the knowledge in $f(\mathbf{x})$.

Because the base classification head has been discarded, the stability is regularized by requiring the original learned representation ($f(\mathbf{x})$ for an input $\mathbf{x}$) and the fine-tuned representation ($\hat{f}(\mathbf{x})$ for the same input) to remain similar to each other, or at least not to deviate dramatically. A simple negative cosine similarity loss function realizes this stability requirement:

$$\mathcal{L}_S(\mathbf{x}) = -\frac{f(\mathbf{x}) \cdot \hat{f}(\mathbf{x})}{\|f(\mathbf{x})\|\|\hat{f}(\mathbf{x})\|} . \tag{2}$$

In order to avoid overfitting, the $\mathbf{x}$ in our stability regularization (SR) loss is not sampled from the novel set. In each mini-batch, we randomly sample 256 images from the base set with replacement to calculate $\mathcal{L}_S$ and back-propagate to $\hat{f}(\mathbf{x})$ (but $f(\mathbf{x})$ is frozen). The proposed SR loss is minimized if $\hat{f}(\mathbf{x})$ and $f(\mathbf{x})$ are the same (modulo a scale factor), which makes $\hat{f}(\mathbf{x})$ produce similar representations to $f(\mathbf{x})$ for *all* base set images, hence we can expect that it will be stable given different training sets sampled from the novel split.

It is worth noting that the proposed SR loss is very flexible since only unlabeled images are required to compute the loss. Hence, it can be easily extended to use other unlabeled images. Results on using images from other than the base or novel set are presented in Sect. 4.4.

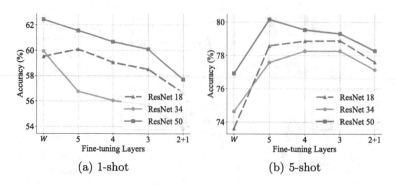

(a) 1-shot   (b) 5-shot

Fig. 3. Results of 5-way accuracy on *mini*-ImageNet w.r.t. the number of learnable blocks using the ResNet backbone [15]. The left-most column means only W is learnable. The column '5' means W plus the 'res5' group in ResNet are learnable, etc. The whole backbone together with W are updated at the right-most column '2+1'.

The other loss function $\mathcal{L}_C$ in Fig. 2 is a regular cross entropy loss, which aims at maximizing the average accuracy μ. Hence, the overall objective is $\mathcal{L} = \mathcal{L}_C + \alpha \cdot \mathcal{L}_S$, where α is always 0.1 in all experiments.

3.5   Reducing Bias: Adaptability Calibration

As aforementioned, to reduce the bias, we adjust the capacity of our model. In many few-shot learning methods [2,16,21,29,36], either the features extracted by the backbone $f(\mathbf{x})$ are directly used, or the backbone is freezed during fine-tuning. In other words, $\hat{f}(\mathbf{x}) \equiv f(\mathbf{x})$ and the capacity of the model is only determined by the linear classification head W (or its alike), which has very low capacity, and in turn leads to high bias according to the bias-variance tradeoff. Other methods [7] fine-tune the entire backbone network, i.e., all parameters in the backbone $f(\mathbf{x})$ are learnable and $\hat{f}(\mathbf{x})$ is completely different from $f(\mathbf{x})$. Such a high capacity model inevitably leads to overfitting and ultra high variance, despite having a small bias. As aforementioned, we need to strike a good balance.

We propose a simple remedy called adaptability calibration (AC), which freezes *part of the layers* in the backbone $f(\mathbf{x})$ and fine-tune the rest layers to form $\hat{f}(\mathbf{x})$. The less layers frozen, the higher is the model's adaptability. The calibration is experimentally determined.

Backbone models in the ResNet [15] family have 5 groups of residual blocks, denoted as 'res1' (close to the input) till 'res5' (close to the classification head W). In Fig. 3, from left to right we gradually make 'res5' to 'res1' learnable. In the 5-shot scenario, allowing both W and 'res5' updatable in fact consistently strikes the best tradeoff between bias and variance. However, in the difficult 1-shot case, only learning W seems to be the best while $W+$'res5' is the runner-up. Hence, considering both cases, our adaptability calibration chooses to update both W and 'res5' (the column '5' in Fig. 3).

3.6   Reducing Both Variance and Bias: Model Ensemble

It is well-known that model ensemble is an effective way in reducing both the variance and the bias of a classifier system [38], and hence it is potentially useful for our task. Diversity among models in the ensemble (i.e., models need to predict differently for the same input) is crucial to the success of ensemble learning, while diversity is often obtained by sampling different training sets (or example weights) for different models on the *same* task (e.g., in one episode of few-shot learning).

However, the sampled novel sets for fine-tuning few-shot recognizers are so small that it leaves almost no space for sampling training sets once more for ensemble learning (e.g., consider a 1-shot task). With the proposed stability regularization and its inclusion of the base set D_b, ensemble learning becomes possible and natural.

We randomly divide the base set D_b into $M = 4$ disjoint subsets, D_b^m ($m = 1, 2, 3, 4$). Then we train M few-shot recognizers for *one* episode, each trained with the sampled novel set and stability regularized by one D_b^m. During inference, the classification probabilities output by these M models are averaged to generate recognition results of the ensemble model.

4   Experiments

4.1   Implementation Details

Datasets. We evaluate the proposed method on three benchmark datasets, which are *mini*-ImageNet [33], CUB-200-2011 (CUB) [34] and CIFAR-FS [4]. *mini*-ImageNet consists of 100 categories randomly selected from the ImageNet dataset [27] with each category containing 600 images sized 84×84. Following [26], it is further split to 64 base, 16 validation and 20 novel categories. CUB is composed of 200 classes (11,788 images) with size of 84×84, which is split into 100 base, 50 validation and 50 novel categories. CIFAR-FS is produced by arbitrarily dividing CIFAR-100 [19] into 64 base, 16 validation and 20 novel classes. There are 600 images with size of 32×32 pixels within each class.

Training Details. To validate the effectiveness of the proposed method, we take WRN-28-10 [37] or models in the ResNet series [15] as the backbone $f(\mathbf{x})$, and the cosine classifier used in [5] as the classification head W. We follow [23] to conduct pretraining. At the fine-tuning stage, we train $\hat{f}(\mathbf{x})$ and W by adapting the pretrained model $f(\mathbf{x})$ to novel categories. For each few-shot episode, the sampled novel set is used to fine-tune the model for 100 epochs with the label smoothed cross-entropy loss [31] (ϵ=0.1), by using SGD with learning rate 0.1, weight decay 1e-4 and momentum 0.9.

Evaluation Protocols. We conduct evaluation in the form of N-way K-shot Q-query, where firstly N categories from the novel split of different datasets are selected arbitrarily, then mutually exclusive K and Q samples of each category are further chosen in the same manner, respectively. The $N \times K$ samples are regarded as D_n to adapt the pretrained model to these new categories, afterwards the model is evaluated on the $N \times Q$ test samples. The above process

Table 2. 5-way accuracy (%) on *mini*-ImageNet, CUB and CIFAR-FS with WRN-28-10 as the backbone. ACC_m is the average accuracy copied from published papers if not otherwise noted.

Dataset	Method	1-shot				5-shot			
		ACC_m	ACC_1	ACC_{10}	ACC_{100}	ACC_m	ACC_1	ACC_{10}	ACC_{100}
mini-ImageNet	ProtoNet[†] [29]	54.16	19.76	26.08	37.62	73.68	43.74	49.78	59.46
	Negative-Cosine [21]	61.72	24.27	36.13	46.92	81.79	53.30	58.12	68.86
	MixtFSL [2]	64.31	30.67	35.07	46.68	81.66	46.67	60.13	71.23
	S2M2$_R$ [23]	64.93	37.58	42.87	53.40	83.18	58.66	66.21	74.73
	PT+NCM [16]	65.35	32.00	38.13	48.41	83.87	56.00	64.00	73.89
	Transductive-FT [7]	65.73	24.00	33.73	47.60	78.40	50.67	53.73	63.09
	CGCS [11]	67.02	38.70	44.00	53.50	82.32	49.30	56.30	67.30
	LR-DC [36]	68.57	37.33	42.72	53.54	82.88	60.52	64.98	74.24
	AC+SR (ours)	69.38	40.52	44.51	54.97	85.87	63.20	66.51	76.28
	AC+EnSR (ours)	**69.59**	**40.52**	**44.94**	**55.32**	**85.97**	**63.48**	**66.74**	**76.40**
CUB	Negative-Cosine [21]	72.66	36.00	42.20	55.22	89.40	70.70	72.69	79.51
	ProtoNet[†] [29]	72.99	28.00	35.14	47.90	86.64	53.33	60.65	70.20
	MixtFSL [2]	73.94	40.00	44.93	55.76	86.01	57.33	66.40	76.65
	CGCS [11]	74.66	50.67	56.00	68.59	88.37	57.33	63.33	72.76
	LR-DC [36]	79.56	44.00	54.80	66.52	90.67	68.80	76.16	84.30
	PT+NCM [16]	80.57	40.00	52.67	65.83	91.15	69.33	76.13	84.67
	S2M2$_R$ [23]	80.68	52.00	56.56	69.70	90.85	73.86	77.72	85.47
	AC+SR (ours)	85.14	52.78	57.46	71.31	94.42	76.00	80.83	87.76
	AC+EnSR (ours)	**85.42**	**53.04**	**58.10**	**71.58**	**94.53**	**77.58**	**81.12**	**87.99**
CIFAR-FS	ProtoNet[◊] [29]	61.60	25.34	30.62	43.08	79.08	51.20	57.40	66.62
	Inductive-FT [7]	68.72	33.33	38.40	50.32	86.11	56.00	61.47	71.73
	Negative-Cosine[◊] [21]	68.90	26.66	39.76	52.58	83.82	52.00	61.54	71.88
	MixtFSL[◊] [2]	69.42	29.33	38.53	51.51	81.05	57.33	62.40	69.19
	LR-DC[◊] [36]	72.52	32.26	42.08	54.72	83.92	57.60	63.30	72.48
	CGCS [11]	73.00	33.33	41.33	55.53	85.80	56.00	62.13	69.16
	PT+NCM [16]	74.64	34.67	42.53	56.60	**87.64**	54.67	62.13	71.57
	S2M2$_R$ [23]	74.81	35.74	42.06	56.49	87.47	56.26	63.38	72.17
	AC+SR (ours)	74.00	**36.54**	42.73	56.83	86.65	56.26	65.97	75.23
	AC+EnSR (ours)	**74.85**	34.96	**42.91**	**57.70**	87.27	**59.48**	**66.61**	**75.84**

$◊$ ACC_m from our implementation $†$ ACC_m from [5]

is collectively called one episode. We keep N=5, Q=15 and vary K=1 or 5 to conduct experiments. We report the average of all (Acc_m), worst 1 (Acc_1), average of worst 10 (Acc_{10}), average of worst 100 (Acc_{100}) accuracy and standard deviation (σ) over 500 pre-sampled episodes by 5 runs then average. Specifically for previous methods with only Acc_m reported, we get other results following their official implementations. All experiments we conduct on one dataset are based on the same 500 episodes, which makes the comparison fair.

4.2   Comparing with State-of-the-art Methods

We compare 5-way accuracy of our method with state-of-the-art methods in Table 2. 'AC' and 'SR' stand for adaptability calibration and stability regularization, respectively, and 'EnSR' means ensemble of SR.

The proposed method (denoted as AC+SR) outperforms existing methods using the traditional ACC_m metric by 1–5% in both 1-shot and 5-shot scenarios on the first two datasets, and is slightly ($< 1\%$) worse on CIFAR-FS. When model ensemble is used, our AC+EnSR almost consistently outperforms existing methods.

More importantly, in the worst-case (ACC_1) and near-worst case average metrics (ACC_{10}, ACC_{100}), our methods are almost consistently better, and mostly higher by $> 2\%$ margins. In the worst-case metrics, model ensemble (AC+EnSR) is again better than single model. And, the ranking from ACC_m is significantly different from that based on ACC_1, but the ranking orders are almost consistent among the 3 worst-case metrics. Hence, the worst-case metric is more *stable* than the average accuracy metric.

4.3   Relationships: μ, σ, $\mu - 3\sigma$, ACC_1 and ACC_m

Table 3 presents a few important evaluation metrics, including the commonly used average accuracy ACC_m (μ), the proposed worst-case accuracy ACC_1 that we advocate, the standard deviation σ, and the surrogate $\mu - 3\sigma$ that motivates the proposed method.

As aforementioned, because the worst-case accuracy ACC_1 is unable to be directly modeled or optimized using few-shot samples, we alternatively resort to the surrogate $\mu - 3\sigma$ inspired by the 3σ rule. This objective is again difficult because σ is unable to be directly accessed, and we resort to the bias-variance decomposition: reducing σ and increasing μ simultaneously.

As Table 3 shows, both μ and σ are correlated with ACC_1, but not tightly correlated. For example, S2M2$_R$ [23] often has the smallest σ, but because its μ (ACC_m) is not the most competitive, its worst-case ACC_1 is significantly lower than ours. We do need to optimize both μ and σ simultaneously.

$\mu - 3\sigma$, which motivates our method, correlates better with ACC_1 than σ or μ, as Table 3 shows. However, this quantity is still far from being a perfect surrogate. It is still a great challenge to model the worst-case ACC_1 end-to-end.

4.4   Ablation Analyses

Switching Backbone. Table 4 shows the results of 5-way classification on *mini*-ImageNet by switching to different backbones. Our adaptability calibration (AC) alone is sometimes harmful in the 1-shot scenario, but always beneficial for 5-shot, which coincides well with the observations in Fig. 3.

On the other hand, the proposed stability regularization (SR) is consistently helpful in all cases and all metrics, especially in the 1-shot case. The proposed model ensemble method (EnSR) is almost consistently improving all evaluation metrics on all backbone models, in particular for the difficult 1-shot scenario.

Switching Pretraining Methods. We further conduct experiments to explore the influence of pretraining methods on the proposed strategies. Specifically, we additionally adopt two pretraining methods from [5], denoted as Baseline and Baseline++, respectively, for simplicity. As shown in Table 5, when pretrained

Table 3. Key evaluation metrics, including the commonly used average accuracy ACC_m (μ), the proposed worst-case accuracy ACC_1, the standard deviation σ, and the surrogate $\mu - 3\sigma$ that motivates the proposed method.

Dataset	Method	1-shot				5-shot			
		ACC_1	$\mu - 3\sigma$	σ	ACC_m	ACC_1	$\mu - 3\sigma$	σ	ACC_m
mini-ImageNet	ProtoNet[†] [29]	19.76	23.41	10.25	54.16	43.74	49.32	8.12	73.68
	Transductive-FT [7]	24.00	32.82	10.97	65.73	50.67	53.23	8.39	78.40
	Negative-Cosine [21]	24.27	31.36	10.12	61.72	53.30	61.18	6.87	81.79
	MixtFSL [2]	30.67	34.70	9.87	64.31	46.67	59.16	7.50	81.66
	PT+NCM [16]	32.00	34.75	10.20	65.35	56.00	63.98	6.63	83.87
	LR-DC [36]	37.33	37.73	10.28	68.57	60.52	63.02	6.62	82.88
	S2M2$_R$ [23]	37.58	37.39	**9.18**	64.93	58.66	66.35	**5.61**	83.18
	CGCS [11]	38.70	36.42	10.20	67.02	49.30	60.90	7.14	82.32
	AC+SR (ours)	40.52	40.25	9.71	69.38	63.20	66.46	6.47	85.87
	AC+EnSR (ours)	**40.52**	**40.67**	9.64	**69.59**	**63.48**	**66.71**	6.42	**85.97**
CUB	ProtoNet[†] [29]	28.00	39.99	11.00	72.99	53.33	67.53	6.37	86.64
	Negative-Cosine [21]	36.00	40.80	10.62	72.66	70.70	73.29	5.37	89.40
	PT+NCM [16]	40.00	49.97	10.20	80.57	69.33	75.85	5.10	91.15
	MixtFSL [2]	40.00	32.69	13.75	73.94	57.33	67.26	6.25	86.01
	LR-DC [36]	44.00	49.74	9.94	79.56	68.80	75.49	5.06	90.67
	CGCS [11]	50.67	42.53	10.71	74.66	57.33	70.01	6.12	88.37
	S2M2$_R$ [23]	52.00	50.32	10.12	80.68	73.86	74.35	5.50	90.85
	AC+SR (ours)	52.78	58.44	8.90	85.14	76.00	81.85	4.19	94.42
	AC+EnSR (ours)	**53.04**	**58.84**	**8.86**	**85.42**	**77.58**	**82.14**	**4.13**	**94.53**
CIFAR-FS	ProtoNet[◊] [29]	25.34	23.38	12.74	61.60	51.20	53.58	8.50	79.08
	Negative-Cosine[◊] [21]	26.66	34.49	11.47	68.90	52.00	60.21	7.87	83.82
	MixtFSL[◊] [2]	29.33	33.09	12.11	69.42	57.33	56.48	8.19	81.05
	LR-DC[◊] [36]	32.26	36.94	11.86	72.52	57.60	60.94	7.66	83.92
	Inductive-FT [7]	33.33	36.29	10.81	68.72	56.00	63.37	7.58	86.11
	CGCS [11]	33.33	39.13	11.29	73.00	56.00	61.59	8.07	85.80
	PT+NCM [16]	34.67	42.51	10.71	74.64	54.67	64.69	7.65	**87.64**
	S2M2$_R$ [23]	35.74	**45.74**	**9.69**	74.81	56.26	**67.58**	**6.63**	87.47
	AC+SR (ours)	**36.54**	39.35	11.55	74.00	56.26	64.45	7.40	86.65
	AC+EnSR (ours)	34.96	40.14	11.57	**74.85**	**59.48**	65.16	7.37	87.27

◊ ACC_m from our implementation   † ACC_m from [5]

with Baseline, after applying AC and SR, ACC_1 and ACC_{10} are increased (on average by 2% in 1-shot and 1% in 5-shot), although ACC_m is slightly decreased (about 0.2%). This observation not only shows our AC+SR's effectiveness, but also indicates that average and worst-case accuracy are not aligned in general. With AC+EnSR, the accuracy for all metrics gains significant margins than the naive Baseline method, in particular, 4-5% for 1-shot.

When pretraining with Baseline++, the accuracy in all cases are generally improved by gradually deploying AC+SR or AC+EnSR, too. The proposed techniques are compatible with various pretraining methods.

Calibrating Adaptability Differently. Table 6 shows the results by varying the number of learnable convolutional groups in ResNet 18 to experiment with different calibration of adaptability. First, there is a clear distinction between the first row (W only) and other rows, indicating that it is harmful to freeze the entire backbone and only fine-tune the classification head W.

Table 4. Ablation results of 5-way classification on *mini*-ImageNet with different backbones. The first row is the baseline and our AC or SR are gradually incorporated. Note that EnSR always includes SR.

Backbone	AC	SR	EnSR	1-shot					5-shot				
				ACC_m	σ	ACC_1	ACC_{10}	ACC_{100}	ACC_m	σ	ACC_1	ACC_{10}	ACC_{100}
ResNet 18				59.53	**9.99**	30.42	35.09	45.05	73.61	8.16	44.54	50.16	61.36
	✓			60.09	10.13	25.60	34.55	45.24	78.58	7.62	45.86	56.34	67.32
	✓	✓		62.33	10.27	32.00	37.03	47.36	79.02	7.68	46.14	56.48	67.62
	✓	✓	✓	**62.76**	10.16	**32.78**	**37.55**	**47.84**	**79.23**	7.62	46.14	**56.96**	**67.91**
ResNet 34				59.94	**9.64**	30.94	35.73	46.08	74.63	7.68	45.60	52.25	63.17
	✓			56.76	9.81	24.56	32.09	42.51	77.58	7.54	47.48	55.54	66.38
	✓	✓		61.42	9.87	32.54	37.17	47.00	78.59	7.43	51.46	57.76	67.43
	✓	✓	✓	**61.80**	9.90	**33.86**	**37.76**	**47.26**	**78.74**	**7.41**	**51.74**	**58.03**	**67.58**
ResNet 50				62.44	10.04	29.32	36.64	47.74	76.92	7.93	50.66	54.67	65.04
	✓			61.56	9.85	29.06	36.69	47.25	80.15	**7.21**	52.26	**59.07**	**69.29**
	✓	✓		63.60	9.90	30.94	37.36	49.00	80.33	7.46	53.06	57.89	68.93
	✓	✓	✓	**63.79**	9.84	**31.67**	**38.05**	**49.36**	**80.43**	7.36	**53.30**	58.89	69.24
WRN-28-10				67.86	9.89	37.58	42.87	53.40	84.57	6.50	58.66	66.21	74.73
	✓			67.36	9.79	36.80	41.44	52.80	85.53	6.50	60.54	65.63	75.95
	✓	✓		69.38	9.71	40.52	44.51	54.97	85.87	6.47	63.20	66.51	76.28
	✓	✓	✓	**69.59**	**9.64**	40.52	**44.94**	**55.32**	**85.97**	**6.42**	**63.48**	**66.74**	**76.40**

Table 5. Ablation results of 5-way classification on *mini*-ImageNet with the ResNet 18 backbone pretrained by other methods.

Pretraining Method	AC	SR	EnSR	1-shot					5-shot				
				ACC_m	σ	ACC_1	ACC_{10}	ACC_{100}	ACC_m	σ	ACC_1	ACC_{10}	ACC_{100}
Baseline [5]	✓	✓		50.37	10.59	18.42	24.91	35.43	74.13	**8.14**	44.82	51.20	62.15
				50.27	10.37	21.86	26.45	35.43	73.92	8.22	46.94	51.49	61.85
	✓	✓	✓	**54.42**	10.26	**23.20**	**29.39**	**39.61**	**74.88**	8.16	**47.46**	**52.89**	**62.78**
Baseline++ [5]	✓	✓		57.59	**9.50**	29.06	33.58	43.99	75.35	7.66	49.08	55.66	64.17
				59.30	9.91	29.58	34.08	44.77	77.57	7.63	54.68	58.36	66.19
	✓	✓	✓	**59.58**	9.82	**30.66**	**34.83**	**45.20**	**77.73**	7.60	**56.80**	**58.94**	**66.51**

Second, when more residual groups are made learnable, there is not a clear winner, especially when we consider different evaluation metrics. For example, every row in Table 6's last 3 rows has been the champion for at least one evaluation metric in 5-shot. But, if we consider 1-shot, it is obviously observed that the second row (fine-tuning $W+$'res5') is significantly better. This is the adaptability calibration we choose in our experiments.

Generalize SR to Other Data. Since only raw images are required to compute the SR loss, it enjoys the flexibility to use others images (i.e., those not from D_b) for its computation (i.e., being generalizable). We use CUB [34], Cars [18], Describable Textures (DTD) [6], Pets [25], VGG Flower (Flower) [24] and CIFAR-100 [19] as the images to compute our stability regularization loss when the base set is *mini*-ImageNet. All these images are resized to 84×84.

Results in Table 7 reveals an interesting observation. No matter what images are used to compute the SR loss, the accuracy is consistently higher than the

Table 6. Ablation results of 5-way classification on *mini*-ImageNet with ResNet 18 as the backbone by varying the number of learnable blocks (✓ means learnable) to control the degree of AC. The number beneath the column AC stands for the index of convolutional groups in ResNet 18. AC+EnSR is used in all experiments.

W	AC				1-shot					5-shot				
	5	4	3	2+1	ACC_m	σ	ACC_1	ACC_{10}	ACC_{100}	ACC_m	σ	ACC_1	ACC_{10}	ACC_{100}
✓					59.53	**9.99**	30.42	35.09	45.05	73.61	8.16	44.54	50.16	61.36
✓	✓				62.76	10.16	**32.78**	**37.55**	**47.84**	79.23	7.62	46.14	56.96	67.91
✓	✓	✓			62.77	10.38	30.12	35.44	47.46	79.91	**7.54**	45.60	**57.84**	68.77
✓	✓	✓	✓		**62.80**	10.50	22.96	34.93	47.27	**80.04**	7.57	45.58	57.74	**68.85**
✓	✓	✓	✓	✓	62.19	10.50	25.60	35.29	46.68	79.87	7.58	**48.00**	57.41	68.67

Table 7. 5-way 1-shot results on *mini*-ImageNet by deploying AC+SR with ResNet 18 as the backbone, while the SR loss is computed with images from *different* datasets, denoted as D_{sr}. The first row ('−') is the baseline with only AC, and the second row is AC+SR, where SR uses the base set (*mini*-ImageNet).

D_{sr}	ACC_m	σ	ACC_1	ACC_{10}	ACC_{100}
−	60.09	10.13	25.60	34.55	45.24
mini-ImageNet	**62.33**	10.27	32.00	**37.03**	**47.36**
CUB [34]	60.66	9.99	30.14	36.43	46.29
Cars [18]	61.05	10.29	30.94	36.64	46.20
DTD [6]	61.99	10.23	31.20	36.75	47.23
Pets [25]	61.57	**9.95**	32.00	36.62	47.27
Flower [24]	61.03	10.07	**32.26**	36.51	46.37
CIFAR-100 [19]	60.47	10.04	30.68	36.69	46.02

baseline (AC without SR), regardless of which metric is considered. That is, the stability regularization is indeed generalizable.

Moreover, although D_b (the base set) performs the best in general, using other images for SR computation leads to results that are on par with it. That is, we *do not* require images used in this regularization to be visually similar or semantically correlated. Our stability regularization is *consistently useful*.

5   Conclusions

This paper advocated to use the worst-case accuracy in optimizing and evaluating few-shot learning methods, which is a better fit to real-world applications than the commonly used average accuracy. Worst-case accuracy, however, is much more difficult to work on. We designed a surrogate loss inspired by the 3σ rule, and in turn proposed two strategies to implicitly optimize this surrogate: a stability regularization (SR) loss together with model ensemble to reduce the variance, and an adaptability calibration (AC) to vary the number of learnable parameters to reduce the bias.

The proposed strategies have achieved significantly higher worst-case (and also average) accuracy than existing methods. In the future, we will design more direct attacks to reduce the worst-case error, because the current surrogate is not a highly accurate approximation of the worst-case performance yet.

Acknowledgments. This research was partly supported by the National Natural Science Foundation of China under Grant 61921006 and Grant 61772256.

References

1. Afrasiyabi, A., Lalonde, J.F., Gagné, C.: Associative alignment for few-shot image classification. In: European Conference on Computer Vision (2020)
2. Afrasiyabi, A., Lalonde, J.F., Gagné, C.: Mixture-based feature space learning for few-shot image classification. In: IEEE/CVF International Conference on Computer Vision (2021)
3. Bateni, P., Goyal, R., Masrani, V., Wood, F., Sigal, L.: Improved few-shot visual classification. In: IEEE/CVF Conference on Computer Vision and Pattern Recognition (2020)
4. Bertinetto, L., Henriques, J.F., Torr, P.H., Vedaldi, A.: Meta-learning with differentiable closed-form solvers. In: International Conference on Learning Representations (2019)
5. Chen, W.Y., Liu, Y.C., Kira, Z., Wang, Y.C.F., Huang, J.B.: A closer look at few-shot classification. In: International Conference on Learning Representations (2019)
6. Cimpoi, M., Maji, S., Kokkinos, I., Mohamed, S., Vedaldi, A.: Describing textures in the wild. In: IEEE Conference on Computer Vision and Pattern Recognition (2014)
7. Dhillon, G.S., Chaudhari, P., Ravichandran, A., Soatto, S.: A baseline for few-shot image classification. In: International Conference on Learning Representations (2020)
8. Dvornik, N., Schmid, C., Mairal, J.: Diversity with cooperation: Ensemble methods for few-shot classification. In: IEEE/CVF International Conference on Computer Vision (2019)
9. Finn, C., Abbeel, P., Levine, S.: Model-agnostic meta-learning for fast adaptation of deep networks. In: International Conference on Machine Learning (2017)
10. Finn, C., Xu, K., Levine, S.: Probabilistic model-agnostic meta-learning. In: Advances in Neural Information Processing Systems (2018)
11. Gao, Z., Wu, Y., Jia, Y., Harandi, M.: Curvature generation in curved spaces for few-shot learning. In: IEEE/CVF International Conference on Computer Vision (2021)
12. Gidaris, S., Bursuc, A., Komodakis, N., Pérez, P., Cord, M.: Boosting few-shot visual learning with self-supervision. In: IEEE/CVF International Conference on Computer Vision (2019)
13. Gidaris, S., Komodakis, N.: Generating classification weights with gnn denoising autoencoders for few-shot learning. In: IEEE/CVF Conference on Computer Vision and Pattern Recognition (2019)
14. Hastie, T., Tibshirani, R., Friedman, J.H., Friedman, J.H.: The elements of statistical learning: data mining, inference, and prediction, vol. 2. Springer (2009). https://doi.org/10.1007/978-0-387-84858-7

15. He, K., Zhang, X., Ren, S., Sun, J.: Deep residual learning for image recognition. In: IEEE Conference on Computer Vision and Pattern Recognition (2016)
16. Hu, Y., Gripon, V., Pateux, S.: Leveraging the feature distribution in transfer-based few-shot learning. In: International Conference on Artificial Neural Networks (2021)
17. Kim, J., Oh, T.H., Lee, S., Pan, F., Kweon, I.S.: Variational prototyping-encoder: One-shot learning with prototypical images. In: IEEE/CVF Conference on Computer Vision and Pattern Recognition (2019)
18. Krause, J., Stark, M., Deng, J., Fei-Fei, L.: 3d object representations for fine-grained categorization. In: IEEE International Conference on Computer Vision Workshops (2013)
19. Krizhevsky, A., Hinton, G.: Learning multiple layers of features from tiny images. Tech. rep., University of Toronto (2009)
20. Lee, K., Maji, S., Ravichandran, A., Soatto, S.: Meta-learning with differentiable convex optimization. In: IEEE/CVF Conference on Computer Vision and Pattern Recognition (2019)
21. Liu, B., et al.: Negative margin matters: Understanding margin in few-shot classification. In: Vedaldi, A., Bischof, H., Brox, T., Frahm, J.-M. (eds.) ECCV 2020. LNCS, vol. 12349, pp. 438–455. Springer, Cham (2020). https://doi.org/10.1007/978-3-030-58548-8_26
22. Liu, Y., Schiele, B., Sun, Q.: An ensemble of epoch-wise empirical bayes for few-shot learning. In: Vedaldi, A., Bischof, H., Brox, T., Frahm, J.-M. (eds.) ECCV 2020. LNCS, vol. 12361, pp. 404–421. Springer, Cham (2020). https://doi.org/10.1007/978-3-030-58517-4_24
23. Mangla, P., Kumari, N., Sinha, A., Singh, M., Krishnamurthy, B., Balasubramanian, V.N.: Charting the right manifold: Manifold mixup for few-shot learning. In: IEEE/CVF Winter Conference on Applications of Computer Vision (2020)
24. Nilsback, M.E., Zisserman, A.: Automated flower classification over a large number of classes. In: 2008 Sixth Indian Conference on Computer Vision, Graphics & Image Processing (2008)
25. Parkhi, O.M., Vedaldi, A., Zisserman, A., Jawahar, C.: Cats and dogs. In: IEEE Conference on Computer Vision and Pattern Recognition (2012)
26. Ravi, S., Larochelle, H.: Optimization as a model for few-shot learning. In: International Conference on Learning Representations (2017)
27. Russakovsky, O., et al.: Imagenet large scale visual recognition challenge. Int. J. Comput. Vis. **115**(3), 211–252 (2015)
28. Rusu, A.A., et al.: Meta-learning with latent embedding optimization. In: International Conference on Learning Representations (2019)
29. Snell, J., Swersky, K., Zemel, R.: Prototypical networks for few-shot learning. In: Advances in Neural Information Processing Systems (2017)
30. Sung, F., Yang, Y., Zhang, L., Xiang, T., Torr, P.H., Hospedales, T.M.: Learning to compare: Relation network for few-shot learning. In: IEEE Conference on Computer Vision and Pattern Recognition (2018)
31. Szegedy, C., Vanhoucke, V., Ioffe, S., Shlens, J., Wojna, Z.: Rethinking the inception architecture for computer vision. In: IEEE Conference on Computer Vision and Pattern Recognition (2016)
32. Verma, V., et al.: Manifold mixup: Better representations by interpolating hidden states. In: International Conference on Machine Learning (2019)
33. Vinyals, O., Blundell, C., Lillicrap, T., Wierstra, D., et al.: Matching networks for one shot learning. In: Advances in Neural Information Processing Systems (2016)

34. Wah, C., Branson, S., Welinder, P., Perona, P., Belongie, S.: The Caltech-UCSD birds-200-2011 dataset. Tech. Rep. CNS-TR-2011-001, California Institute of Technology (2011)
35. Wang, Y., Yao, Q., Kwok, J.T., Ni, L.M.: Generalizing from a few examples: A survey on few-shot learning. ACM Comput. Surv. **53**(3), 1–34 (2020)
36. Yang, S., Liu, L., Xu, M.: Free lunch for few-shot learning: Distribution calibration. In: International Conference on Learning Representations (2021)
37. Zagoruyko, S., Komodakis, N.: Wide residual networks. In: British Machine Vision Conference (2016)
38. Zhou, Z.H.: Ensemble Methods: Foundations and Algorithms. Chapman and Hall/CRC (2012)

Exploring Hierarchical Graph Representation for Large-Scale Zero-Shot Image Classification

Kai Yi[1]([✉]) [iD], Xiaoqian Shen[1] [iD], Yunhao Gou[1,2] [iD], and Mohamed Elhoseiny[1] [iD]

[1] King Abdullah University of Science and Technology (KAUST),
Thuwal, Saudi Arabia
{kai.yi,xiaoqian.shen,yunhao.gou,mohamed.elhoseiny}@kaust.edu.sa
[2] University of Electronic Science and Technology of China (UESTC), Sichuan, China

Abstract. The main question we address in this paper is how to scale up visual recognition of unseen classes, also known as zero-shot learning, to tens of thousands of categories as in the ImageNet-21K benchmark. At this scale, especially with many fine-grained categories included in ImageNet-21K, it is critical to learn quality visual semantic representations that are discriminative enough to recognize unseen classes and distinguish them from seen ones. We propose a *H*ierarchical *G*raphical knowledge *R*epresentation framework for the confidence-based classification method, dubbed as HGR-Net. Our experimental results demonstrate that HGR-Net can grasp class inheritance relations by utilizing hierarchical conceptual knowledge. Our method significantly outperformed all existing techniques, boosting the performance by 7% compared to the runner-up approach on the ImageNet-21K benchmark. We show that HGR-Net is learning-efficient in few-shot scenarios. We also analyzed our method on smaller datasets like ImageNet-21K-P, 2-hops and 3-hops, demonstrating its generalization ability. Our benchmark and code are available at https://kaiyi.me/p/hgrnet.html.

Keywords: Zero-shot learning · Semantic hierarchical graph · Large-scale knowledge transfer · Vision and language

1 Introduction

Zero-Shot Learning (ZSL) is the task of recognizing images from unseen categories with the model trained only on seen classes. Nowadays, ZSL relies on semantic information to classify images of unseen categories and can be formulated as a visual semantic understanding problem. In other words, given candidate text descriptions of a class that has not been seen during training, the goal

Supplementary Information The online version contains supplementary material available at https://doi.org/10.1007/978-3-031-20044-1_7.

is to identify images of that unseen class and distinguish them from seen ones and other unseen classes based on their text descriptions.

In general, current datasets contain two commonly used semantic information including attribute descriptions (e.g., AWA2 [35], SUN [22], and CUB [34]), and more challenging unstructured text descriptions (e.g., CUB-wiki[5], NAB-wiki [6]). However, these datasets are all small or medium-size with up to a few hundred classes, leaving a significant gap to study generalization at a realistic scale. In this paper, we focus on large-scale zero-shot image classification. More specifically, we explore the learning limits of a model trained from 1K seen classes and transfer it to recognize more than 10 million images from 21K unseen candidate categories from ImageNet-21K [4], which is the largest available image classification dataset to the best of our knowledge.

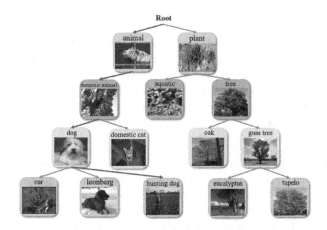

Fig. 1. Intuitive illustration of our proposed HGR-Net. Suppose the ground truth is Hunting Dog, then we can find the real-label path Root → Animal → Domestic Animal → Dog → Hunting Dog. Our goal is to efficiently leverage semantic hierarchical information to help better understand the visual-language pairs.

A few works of literature explored zero-shot image classification on ImageNet-21K. However, the performance has plateaued to a few percent Hit@1 performances on ImageNet-21K zero-shot classification benchmark ([7,13,20,32]). We believe the key challenge is distinguishing among 21K highly fine-grained classes. These methods represents class information by GloVe [23] or Skip-Gram [17] to align the vision-language relationships. However, these lower-dimensional features from GloVe or Skip-Gram are not representative enough to distinguish among 21K classes, especially since they may collapse for fine-grained classes. Besides, most existing works train a held-out classifier to categorize images of unseen classes with different initialization schemes. One used strategy is to initialize the classifier weights with semantic attributes [12,27,36,38], while another is to conduct fully-supervised training with generated unseen images. However, the trained MLP-like classifier is not representative enough to capture fine-grained differences to classify the image into a class with high confidence.

To resolve the challenge of large-scale zero-shot image classification, we proposed a novel *H*ierarchical *G*raph knowledge *R*epresentation network (denoted as HGR-Net). We explore the conceptual knowledge among classes to prompt the distinguishability. In Fig. 1, we state the intuition of our proposed method. Suppose the annotated image class label is `Hunting Dog`. The most straightforward way is to extract the semantic feature and train the classifier with cross-entropy. However, our experiments find that better leveraging hierarchical conceptual knowledge is important to learn discriminative text representation. We know the label as `Hunting Dog`, but all the labels from the root can also be regarded as the real label. We incorporate conceptual semantic knowledge to enhance the network representation.

Moreover, inspired by the recent success of pre-trained models from large vision-language pairs such as CLIP [24] and ALIGN [9], we adopt a dynamic confidence-based classification scheme, which means we multiply a particular image feature with candidate text features and then select the most confident one as the predicted label. Unlike traditional softmax-based classifier, this setting is dynamic, and no need to train a particular classifier for each task. Besides, the confidence-based scheme can help truly evaluate the vision-language relationship understanding ability. For better semantic representation, we adopt Transformer [29] as the feature extractor, and follow-up experiments show Transformer-based text encoder can significantly boost the classification performance.

Contributions. We consider the most challenging large-scale zero-shot image classification task on ImageNet-21K and proposed a novel hierarchical graph representation network, HGR-Net, to model the visual-semantic relationship between seen and unseen classes. Incorporated with a confidence-based learning scheme and a Transformers to represent class semantic information, we show that HGR-Net achieved new state-of-the-art performance with significantly better results than baselines. We also conducted few-shot evaluations of HGR, and we found our method can learn very efficiently by accessing only one example per class. We also conducted extensive experiments on the variants of ImageNet-21K, and the results demonstrate the effectiveness of our HGR-Net. To better align with our problem, we also proposed novel matrices to reflect the conceptual learning ability of different models.

2 Related Work

2.1 Zero-/Few-Shot Learning

Zero-Shot Learning (ZSL) is recognizing images of unseen categories. Our work is more related to semantic-based methods, which learn an alignment between different modalities (i.e., visual and semantic modalities) to facilitate classification [12,27,36,38]. CNZSL [27] proposed to map attributes into the visual space by normalization over classes. In contrast to [27], we map both the semantic text and the images into a common space and calculate the confidence. Experimental studies are conducted to show that mapping to a common space achieves higher accuracy. We also explore the Few-Shot Learning (FSL) task, which focuses on classification with only accessing a few testing examples during training [28,37,39]. Unlike [33] which defines the FSL task as extracting few training data from all classes, we took all images from seen classes and selected only a few samples from unseen classes during training. Our main goal here is to analyze how the performance differs from zero to one-shot.

2.2 Large-Scale Graphical Zero-Shot Learning

Graphical Neural Networks [11] are widely applied to formulate zero-shot learning, where each class is associated with a graph node, and a graph edge represents each inter-class relationship. For example, [32] trains a GNN based on the WordNet knowledge to generate classifiers for unseen classes. Similarly, [10] uses fewer convolutional layers but one additional dense connection layer to propagate features towards distant nodes for the same graph. More recently, [19] adopts a transformer graph convolutional network (TrGCN) for generating class representations. [31] leverages additional neighbor information in the graph with a contrastive objective. Unlike these methods, our method utilizes fruitful information of a hierarchical structure based on class confidence and thus grasps hierarchical relationships among classes to distinguish hard negatives. Besides, some works exploit graphical knowledge without explicitly training a GNN. For example, [15] employs semantic vectors of the class names using multidimensional scaling (MDS) [3] on the WordNet to learn a joint visual-semantic embedding for classification; [12] learns similarity between the image representation and the class representations in the hyperbolic space.

2.3 Visual Representation Learning from Semantic Supervision

Visual representation learning is a challenging task and has been widely studied with supervised or self-supervised methods. Considering semantic supervision from large-scale unlabeled data, learning visual representation from text representation [24] is a promising research topic with the benefit of large-scale visual and linguistic pairs collected from the Internet. These methods train a separate encoder for each modality (i.e., visual and language), allowing for extended to unseen classes for zero-shot learning. Upon these methods, [2] improves the data efficiency during training, [9] enables learning from larger-scale noisy image-text pairs, [40] optimizes the language prompts for better classifier generation. Our work adopts the pre-trained encoders of [24] but tackles the problem of large-scale zero-shot classification from a candidate set of 22K classes instead of at most 1K as in [24].

3 Method

3.1 Problem Definition

Zero-Shot Learning. Let $\mathcal{C}$ denote the set of all classes. $\mathcal{C}_s$ and $\mathcal{C}_u$ to be the unseen and seen classes, respectively, where $\mathcal{C}_s \cap \mathcal{C}_u = \emptyset$, and $\mathcal{C} = \mathcal{C}_s \cup \mathcal{C}_u$. For each class $c_i \in \mathcal{C}$, a d-dimensional semantic representation vector $t(c_i) \in \mathbb{R}^d$ is provided. We denote the training set $\mathcal{D}_{tr} = \{(\mathbf{x}_i, c_i, t(c_i))\}_{i=1}^N$, where $\mathbf{x}_i$ is the i-th training image. In ZSL setting, given testing images $\mathbf{x}_{te}$, we aim at learning a mapping function $\mathbf{x}_{te} \rightarrow \mathcal{C}_u$. In a more challenging setting, dubbed as generalized ZSL, we not only aim at classifying images from unseen categories but also seen categories, where we learn $\mathbf{x}_{te} \rightarrow \mathcal{C}_u \cup \mathcal{C}_s$ covering the entire prediction space.

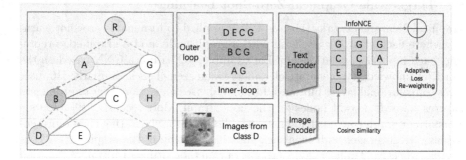

Fig. 2. HGR-Net: Suppose the annotated single label is D and we can find the tracked label path $R \cdots \rightarrow A \rightarrow B \rightarrow D$ from the semantic graph extended from WordNet. We first set D as the positive anchor and contrast with negatives which are sampled siblings of its ancestors (i.e., $\{E, C, G\}$) layer by layer. Then we iterate to set the positive anchor to be controlled depth as B, A, which has layer-by-layer negatives $\{C, G\}$ and G, respectively. Finally, we use a memory-efficient adaptive re-weighting strategy to fuse knowledge from different conceptual level.

Semantic Hierarchical Structure. We assume access to a semantic Directed Acyclic Graph (DAG), $\mathcal{G} = (\mathcal{V}, \mathcal{E})$, where $\mathcal{V} = \mathcal{C} \cup \{R\}$ and $\mathcal{E} \subseteq \{(x, y) \mid (x, y) \in \mathcal{C}^2, \ x \neq y\}$. Here the two-tuple (x, y) represents an parenting relationship between x and y, which means y is a more abstract concept than x. Here we manually add a root node R with a in-degree of 0 into $\mathcal{G}$. For simplicity, given any node c_i, we denote the ordered set of all its ancestors obtained by shortest path search from R to c_i as $\mathcal{A}^{c_i} = \{a(c_i)_j\}_{j=1}^{N_i^a} \subseteq \mathcal{C}$. Similarly, we denote the set of all siblings of c_i as $\mathcal{S}^{c_i} = \{s(c_i)_j\}_{j=1}^{N_i^s} \subseteq \mathcal{C}$. Finally, $d(c_i) \triangleq |\mathcal{A}^{c_i}|$ is defined as depth of node c_i.

3.2 HGR-Net: Large-Scale ZSL with Hierarchical Graph Representation Learning

We mainly focus on zero-shot learning on the variants of ImageNet-21K, the current largest image classification dataset to our knowledge. Previous strategies [7,13,20,32] adopt a N-way classification as the training task on all the N seen classes. However, we argue that this is problematic, especially in using a Transformer as the text encoder to obtain class semantic representations.

First, when conducting N-way classification, all the classes except the single ground-truth are regarded as negative ones. Even though this helps build a well-performing classifier in a fully-supervised learning scenario, we argue that this is harmful to knowledge transfer from seen to unseen classes in ZSL. Second, a batch of N samples is fed to the text Transformer [29] to obtain their corresponding text representations and to compute the class logits afterward. This strategy can be acceptable for datasets with a small number of classes. However, when the number of classes scales to tens of thousands, as in our case, it becomes formidable to implement the operations mentioned above. Therefore, we propose a memory-efficient hierarchical contrastive objective to learn transferable and discriminative representations for ZSL.

Intuitively, as illustrated in Fig. 2, suppose we have an image sample with annotated ground-truth label D according to ImageNet-21K. Then, we could find a shortest path $R - \cdots \rightarrow A \rightarrow B \rightarrow D$ to be the tracked true-label path $\mathcal{T}_E$. With our definition of the hierarchical structure, the true labels for the image sample are defined by all the nodes along this path through different levels of conceptions, from abstract to concrete in our case. Therefore, to better leverage this hierarchical semantic knowledge, we propose a hierarchical contrastive loss that conducts two levels of pre-defined degrees of bottom-up contrasting.

Specifically, for node D with a depth of $d(D)$. In the outer-level loop, we iterate ground-truth labels of different levels, along the ancestor path $\mathcal{A}^D$, we traverse from itself bottom-up $D - \cdots \rightarrow B \rightarrow A$ until reaching one of its ancestors of a depth of $Kd(D)$, where K is the outer ratio. In the inner-level loop, fixing the ground-truth label, we conduct InfoNCE loss [21] layer by layer in a similar bottom-up strategy with an inner ratio M (e.g., when fixing current ground truth node as B in Fig. 2, for inner loop we consider $\langle B, C \rangle, \langle B, G \rangle$). We provide more details in Algorthim 1.

Formally, given an image of x from class c_i, we define its loss as:

$$\mathcal{L}_{\text{cont}} = \sum_{j=k_s}^{k_e} g(j, \mathcal{L}_j), \quad \mathcal{L}_j = \frac{1}{m_e - m_s + 1} \sum_{l=m_s}^{m_e} \mathcal{L}_{j,l}, \tag{1}$$

where $g(\cdot, \cdot)$ is an adaptive attention layer to dynamically re-weight the importance of labels given different levels j, $j \in [k_s, k_e]$ and $l \in [m_s, m_e]$ are the outer-level and inner-level loop respectively. k_s, k_e represents the start layer and the end layer for outer loop while m_s, m_e are the start layer and the end layer for the inner loop.

$$\mathcal{L}_{j,l} = -\log \frac{\text{pos}^j}{\text{pos}^j + \text{neg}^{j,l}}, \tag{2}$$

where

$$\text{pos}^j = \exp\left(\text{sim}\left(T(c_j^+), V(x)\right)/\tau\right) \tag{3}$$

$$\text{neg}^{j,l} = \sum_{q=1}^{n_l} \exp\left(\text{sim}\left(T(c_{j,l,q}^-), V(x)\right)/\tau\right) \tag{4}$$

where, $\text{sim}(\cdot)$ is the measure of similarity, τ is the temperature value. $V(\cdot)$ and $T(\cdot)$ are the visual and image encoders, $c_j^+ = a(c_i)_j$ is the selected positive label on the tracked lable path at layer l. $c_{j,l,q}^-$ is the q-th sibling of the j-th ground-truth at level l; see Agorithm 1.

Algorithm 1. Hierarchical Graph Representation Net (HGR-Net)

Require: Training set $\mathcal{D}_{\text{tr}}$, text encoder T, visual encoder V, inner ratio M, outer ratio K, per layer sampling number threshold ϵ, training label set $\mathcal{C}$, hierarchical graph $\mathcal{G}$

Sample a batch of data $\mathbf{X}$ from class c_i

Obtain its ancestor path $\mathcal{A}^{c_i}$

Set the outer loop range $k_s = Kd(c_i), k_e = d(c_i)$

for $j = k_s, k_s + 1, \ldots, k_e$ **do**

    Set the current ground-truth label $c_j^+ = a(c_i)_j$

    Prepare pos^j according to Eq. 3

    Set the inner-loop ranges $m_s = Md(c^+), m_e = d(c^+)$

    for $l = m_s, m_s + 1, \ldots, m_e$ **do**

        Prepare sibling set $\mathcal{S}^{c_j}$

        $n_l = \max(\epsilon, |\mathcal{S}^{c_j}|)$

        Sample n_l negative sibling set $\left\{ c_{j,l,q}^- \right\}_{q=1}^{n_l}$

        Prepare neg^{j,l} according to Eq. 4

        Compute $\mathcal{L}_{j,l}$ according to Eq. 2

    end for

    Compute $\mathcal{L}_j$ according to Eq. 1 right part

end for

Compute $\mathcal{L}_{\text{cont}}$ according to Eq. 1 left part

---

4   Experiments

4.1   Datasets and the Hierarchical Structure

ImageNet [4] is a widely used large-scale benchmark for ZSL organized according to the WordNet hierarchy [18], which can lead our model to learn the hierarchical relationship among classes. However, the original hierarchical structure is not a DAG (Directed Acyclic Graph), thus not suitable when implementing our method. Therefore, to make all of the classes fit into an appropriate location in the hierarchical DAG, we reconstruct the hierarchical structure by removing some classes from the original dataset, which contains seen classes from the ImageNet-1K and unseen classes from the ImageNet-21K (winter-2021 release), resulting a modified dataset ImageNet-21K-D (D for Directed Acyclic Graph).

It is worth noticing that although there are 12 layers in the reconstructed hierarchical tree in total, most nodes reside between 2nd and 6th layers. Our class-wise dataset split is based on GBU [35], which provides new dataset splits for ImageNet-21K with 1K seen classes for training and the remaining 20, 841 classes as test split. Moreover, GBU [35] splits the unseen classes into three different levels, including "2-hop", "3-hops" and "All" based on WordNet hierarchy [18]. More specifically, the "2-hops" unseen concepts are within 2-hops from the known concepts. After the modification above, the training is then conducted on the processed ImageNet-1K with seen 983 classes, while 17,295 unseen classes from the processed ImageNet-21K are for ZSL testing, and 1533 and 6898 classes for "2-hops" and "3-hops" respectively. Please note that there is no overlap between the seen and unseen classes. The remaining 983 seen classes make our training setting more difficult because our model gets exposed to fewer images than the original 1k seen classes. Please refer to the supplementary materials for more detailed descriptions of the dataset split and reconstruction procedure.

4.2   Implementation Details

We use a modified ResNet-50 [8] from [24] as the image encoder, which replaces the global average pooling layer with an attention mechanism, to obtain visual representation with feature dimensions of 1024. Text descriptions are encoded into tokens and bracketed with start tokens and end tokens based on byte pair encoding (BPE) [26] with the max length of 77. For text embedding, we use CLIP [24] Transformer to extract semantic vectors with the same dimensions as feature representation. We obtain the logits with L2-normalized image and text features and calculate InfoNCE loss [21] layer by layer with an adaptive reweighting strategy. More specifically, a learnable parameter with a size equivalent to the depth of the hierarchical tree is used to adjust the weights adaptively.

Training Details. We use the AdamW optimizer [14] applied to all weights except the adaptive attention layer with a learning rate 3e-7. We use the SGD optimizer for the adaptive layer with a learning rate of 1e-4. When computing the matmul product of visual and text features, a learnable temperature parameter τ is initialized as 0.07 from [30] to scale the logits and clips gradient norm of the parameters to prevent training instability. To accelerate training and avoid additional memory, mixed-precision [16] is used, and the weights of the model are only transformed into float32 for optimization. Our proposed HGR model is implemented in PyTorch, and training and testing are conducted on a Tesla V100 GPU with a batch size of 256 and 512, respectively.

4.3   Large-Scale ZSL Performance

Comparison Approaches. We compare with the following approaches:
– **DeViSE** [7] linearly maps visual information to the semantic word-embedding space. The transformation is learned using a hinge ranking loss.
– **HZSL** [12] learns similarity between the image representation and the class representations in the hyperbolic space.
– **SGCN** [10] uses an asymmetrical normalized graph Laplacian to learn the class representations.

- **DGP** [10] separates adjacency matrix into ancestors and descendants and propagates knowledge in two phases with one additional dense connection layer based on the same graph as in GCNZ [32].
- **CNZSL** [27] utilizes a simple but effective class normalization strategy to preserve variance during a forward pass.
- **FREE** [1] incorporates semantic-visual mapping into a unified generative model to address cross-dataset bias.

Evaluation Protocols. We use the typical Top@K criterion, but we also introduce additional metrics. Since it could be more desirable to have a relatively general but correct prediction rather than a more specific but wrong prediction, the following three metrics evaluate a given model's ability to learn the hierarchical relationship between the ground truth and its general classes.

- **Top-Overlap Ratio (TOR).** In this metric, we take a further step to also cover all the ancestor nodes of the ground truth class. More concretely, for an image x_j from class c_i of depth q_{c_i}, TOR is defined as:

$$TOR(x_j) = \frac{|p_{x_j} \cap \{A_{c_i}, c_i\}|}{q_{c_i}} \qquad (5)$$

where c_i is the corresponding class to image x_j. A_{c_i} is the union of all the ancestors of class c_i and p_{x_j} is the predicted class of x_j. In other words, this metric consider the predicted class correct if it is an ancestor of the ground truth.

- **Point-Overlap Ratio (POR).** In this setting, we let the model predict labels layer by layer. POR is defined as:

$$POR(x_j) = \frac{|P_{x_j} \cap P_{c_i}|}{q_{c_i}}, \qquad (6)$$

where $P_{c_i} = \{c_{i_1}, c_{i_2}, c_{i_3}, \cdots, c_{i_{q_{c_i}-1}}, c_i\}$ is the union of classes from the root to the ground truth through all the ancestors, and P_{x_j} is the union of classes predicted by our model layer by layer. q_{c_i} is count of all the ancestors including the ground truth label, which is tantamount to the depth of node c_i. The intersection calculates the overlap between correct and predicted points for image x_j.

Results Analysis. Table 1 demonstrates the performance of different models on ImageNet-21K ZSL setting on Top@K and above-mentioned three hierarchical evaluation. Our proposed model outperforms SoTA methods in all metrics, including hierarchical measures, proving the ability to learn the hierarchical relationship between the ground truth and its ancestor classes. We also attach the performance on 2-hops and 3-hops in the supplementary.

4.4 Ablation Studies

Different Attributes. Conventional attribute-based ZSL methods use GloVe [23] or Skip-Gram [17] as text models, while CLIP [24] utilizes prompts (i.e., text description) template: "a photo of a [CLASS]", and take advantage of Transformer to extract text feature. Blindly adding Transformer to some attribute-based methods like HZSL [12] which utilizes unique techniques to improve their performance in the attribute setting result in unreliable results. Therefore, we conducted experiments comparing three selected methods with different attributes. The result in Table 2 shows that methods based on text embedding extracted by CLIP transformer outperform traditional attribute-based ones since the low dimension representations (500-D) from w2v [17] is not discriminative enough to distinguish unseen classes, while higher dimension (1024-D) text representations significantly boost classification performance. Our HGR-Net gained significant improvement by utilizing Transformer compared to the low dimension representation from w2v [17].

Table 1. Top@k accuracy, Top-Overlap Ratio (TOR), and Point-Overlap Ratio (POR) for different models on the ImageNet-21K-D only testing on unseen classes. Tr means text encoder is CLIP Transformer.

Method	Hit@ k(%)					TOR	POR
	1	2	5	10	20		
Devise [7]	1.0	1.8	3.0	15	23.8	–	–
HZSL [12]	3.7	5.9	10.3	13.0	16.4	–	–
SGCN(w2v) [10]	2.79	4.49	8.26	13.05	19.49	4.97	10.01
SGCN(Tr) [10]	4.83	8.17	14.61	21.23	29.42	8.33	14.69
DGP(w2v) [10]	3.00	5.12	9.49	14.28	20.55	7.07	11.71
DGP(Tr) [10]	5.78	9.57	16.89	24.09	32.62	12.39	15.50
CNZSL(w2v) [27]	1.94	3.17	5.88	9.12	13.73	3.93	4.03
CNZSL(Tr) [27]	5.77	9.48	16.49	23.25	31.00	8.32	7.22
FREE(w2v) [1]	2.87	4.91	9.54	13.28	20.36	4.89	5.37
FREE(Tr) [1]	5.76	9.54	16.71	23.65	31.81	8.59	9.68
CLIP [24]	15.22	22.54	33.43	42.13	50.93	18.55	14.68
HGR-Net(Ours)	**16.39**	**24.19**	**35.66**	**44.68**	**53.71**	**18.90**	**16.19**

Different Outer and Inner Ratio. Fig. 3 demonstrate the Top1, Top-Overlap Ratio (TOR) and Point-Overlap Ratio (POR) metrics of different K and M, where K and M $\in [0, 1]$. K and M are outer and inner ratio that determine how many samples is considered in the inner and outer loop respectively as earlier illustrated.

We explore different K and M in this setting and observe how performance differs under three evaluations. Please note that when K or M is 0.0, it means only the current node is involved in a loop. As K increases, the model is prone to

Table 2. Different attributes. DGP(w/o) means without separating adjacency matrix into ancestors and descendants, `CN` and `INIT` in CNZSL means class normalization and proper initialization respectively.

Attributes	Methods	Hit@ k(%)					TOR	POR
		1	2	5	10	20		
w2v	SGCN [10]	2.79	4.49	8.26	13.05	19.49	4.97	10.01
	DGP(w/o) [10]	2.90	4.86	8.91	13.67	20.18	3.96	11.49
	DGP [10]	3.00	5.12	9.49	14.28	20.55	7.07	11.71
	CNZSL(w/o CN) [27]	0.83	1.47	3.03	5.08	8.27	1.98	2.05
	CNZSL(w/o INIT) [27]	1.84	3.13	6.08	9.47	14.13	3.04	4.05
	CNZSL [27]	1.94	3.17	5.88	9.12	13.73	3.93	4.03
	FREE [1]	2.87	4.91	9.54	13.28	20.36	4.89	5.37
	HGR-Net(Ours)	2.35	3.69	7.03	11.46	18.27	4.38	5.76
Transformer(CLIP)	SGCN [10]	4.83	8.17	14.61	21.23	29.42	8.33	14.69
	DGP(w/o) [10]	5.42	9.16	16.01	22.92	31.20	7.80	15.29
	DGP [10]	5.78	9.57	16.89	24.09	32.62	12.39	15.50
	CNZSL(w/o CN) [27]	1.91	3.45	6.74	10.55	15.51	3.19	3.43
	CNZSL(w/o INIT) [27]	5.65	9.33	16.24	22.88	30.63	8.32	7.03
	CNZSL [27]	5.77	9.48	16.49	23.25	31.00	7.97	7.22
	FREE [1]	5.76	9.54	16.71	23.65	31.81	8.59	9.68
	HGR-Net(Ours)	**16.39**	**24.19**	**35.66**	**44.68**	**53.71**	**18.95**	**16.19**

obtain higher performance on hierarchical evaluation. An intuitive explanation is that more conceptual knowledge about ancestor nodes facilitates hierarchical learning relationships among classes.

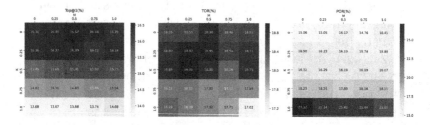

Fig. 3. Different outer ratio (K) and inner ratio (M)

Different Negative Sampling Strategies. We explore various sampling strategies for choosing negative samples and observe how they differ in performance. *Random* randomly samples classes from all the classes. *TopM* samples neighbour nodes from $(q_{c_i} - M)$ to q_{c_i} layers, where q_{c_i} is the depth of inner anchor c_i, and we set M as 1. *Similarity* calculates the similarity of text features and chooses the top similar samples with the positive sample as hard negatives. *Sibling* samples sibling nodes of the target class. Table 3 indicates that *TopM*

Table 3. Analysis of sampling strategies

Strategy	Hit@ k(%)					TOR	POR
	1	2	5	10	20		
Random	15.72	23.33	34.69	43.68	52.73	16.12	13.04
Sibling	16.25	23.95	35.29	44.16	53.09	17.91	13.46
Similarity	16.35	24.04	35.33	44.17	53.07	18.60	14.78
TopM(default)	**16.39**	**24.19**	**35.66**	**44.68**	**53.71**	**18.90**	**16.19**

outperforms other sampling strategies. Therefore, we adopt the *TopM* sampling strategy in the subsequent ablation studies.

Different Weighting Strategies. Orthogonal to negative sampling methods, we explore in this ablation the influence of different weighting strategies across the levels of the semantic hierarchy. The depth of the nodes in the hierarchical structure is not well-balanced, and the layers are not accessible for all objects. Therefore, it is necessary to focus on the importance of different layers. In this case, we experimented with 6 different weighting strategies and observed how they differ in multiple evaluations. As Table 4 shows, *Increasing* gives more weights to deeper layers in a linear way and ↑ *non-linear* is exponentially increasing weights to deeper layers. To balance the Top@K and hierarchical evaluations, the adaptive weighting method is proposed to obtain a comprehensive result. More specifically, *Adaptive* uses a learnable parameter with a size equivalent to the depth of the hierarchical tree to adjust the weights adaptively. We attached the exact formulation of different weighting strategies in the supplementary.

Table 4. Analysis of the weighting strategies when re-weighting in both inner and outer loop with K=0.25 and M=0.5.

Weighting	Hit@ k(%)					TOR	POR
	1	2	5	10	20		
Adaptive(default)	**16.39**	**24.19**	**35.66**	**44.68**	**53.71**	**18.90**	**16.19**
Equal	15.97	23.65	35.02	43.97	52.97	17.82	13.71
Increasing ↑	15.85	23.50	34.85	43.83	52.83	17.80	13.81
Decreasing ↓	16.08	23.77	35.16	44.10	53.09	17.84	13.59
↑ (non-linear)	15.58	23.13	34.43	43.44	52.46	17.79	14.12
↓ (non-linear)	16.19	23.89	35.26	44.18	53.13	17.87	13.47

Experiment on ImageNet-21K-P [25]. ImageNet-21K-P [25] is a pre-processed dataset from ImageNet21K by removing infrequent classes, reducing the number of total numbers by half but only removing only 13% of the original images, which contains 12,358,688 images from 11,221 classes. We select the intersection of this dataset with our modified ImageNet21K dataset to ensure DAG structure consistency. The spit details (class and sample wise) are demonstrated in the supplementary.

We show experimental results on ImageNet-21K-P comparing our method to different SoTA variants. Our model performs better in this smaller dataset compared to the original larger one in Table 1 and outstrips all the previous ZSL methods. We presented important results in Table 5 and we attached more results in the supplementary.

4.5 Qualitative Results

Figure 4 shows several retrieved images by implementing our model in the ZSL setting on ImageNet-21K-D. The task is to retrieve images from an unseen class with its semantic representation. Each row demonstrates three correct retrieved images and one incorrect image with its true label. Although our algorithm retrieves images from the wrong class, they are still visually similar to ground truth. For instance, the true label hurling and the wrong class American football belong to sports games, and images from both contain several athletes wearing helmets against a grass background. We also show some prediction examples in Fig. 5 to present Point-Overlap results.

Table 5. Result of ImageNet21K-P [25]. DGP(w/o) [10] means without separating adjacency matrix into ancestors and descendants, CN and INIT in CNZSL [27] means class normalization and proper initialization respectively, and Tr is Transformer of CLIP for short.

Models	Hit@ k(%)					TOR	POR
	1	2	5	10	20		
CNZSL(Tr w/o CN) [27]	3.27	5.59	10.69	16.17	23.33	5.32	7.68
CNZSL(Tr w/o INIT) [27]	7.90	12.77	21.40	29.50	38.63	11.23	12.56
CNZSL(Tr) [27]	7.97	12.81	21.75	29.92	38.97	11.50	12.62
FREE(Tr) [1]	8.15	12.90	21.37	30.29	40.62	11.82	13.34
CLIP [24]	19.33	28.07	41.66	53.77	61.23	20.08	20.27
HGR-Net (Ours)	**20.08**	**29.35**	**42.49**	**52.47**	**62.00**	**23.43**	**23.22**

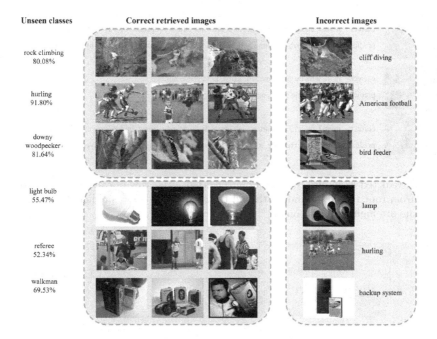

Fig. 4. Zero-shot retrieved images. The first column represents unseen class names and corresponding confidence, the middle shows correct retrieval, and the last demonstrates incorrect images and their true labels.

4.6   Low-shot Classification on Large-Scale Dataset

Apart from zero-shot experiments being our primary goal in this paper, we also explore the effectiveness of our method in the low-shot setting compared to several baselines. Unlike pure few-shot learning, our support set comprises two parts. To be consistent with ZSL experiments, all the training samples of 983 seen classes are for low-shot training. For the 17, 295 unseen classes used in the ZSL setting, k-shots (1,2,3,5,10) images are randomly sampled for training in the low-shot setting, and the remaining images are used for testing. The main goal of this experiment is to show how much models could improve from zero to one shot and whether our proposed hierarchical-based method could generalize well in the low-shot scenario. Figure 6 illustrated the few-shots results comparing our model to various SoTA methods. Although our approach gains trivial Top@k improvements from 1 to 10 shots, the jump from 0 to 1 shot is two times that from 1 to 10, proving that our model is an efficient learner.

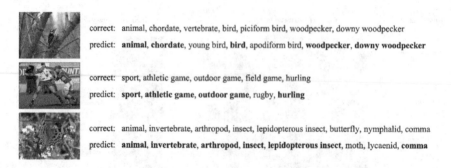

correct: animal, chordate, vertebrate, bird, piciform bird, woodpecker, downy woodpecker
predict: **animal, chordate**, young bird, **bird**, apodiform bird, **woodpecker, downy woodpecker**

correct: sport, athletic game, outdoor game, field game, hurling
predict: **sport, athletic game, outdoor game**, rugby, **hurling**

correct: animal, invertebrate, arthropod, insect, lepidopterous insect, butterfly, nymphalid, comma
predict: **animal, invertebrate, arthropod, insect, lepidopterous insect**, moth, lycaenid, **comma**

Fig. 5. Predicted examples to show Point-Overlap. First row of each image is correct points from root to the ground truth and the second row show predicted points. The hit points are highlighted in bold.

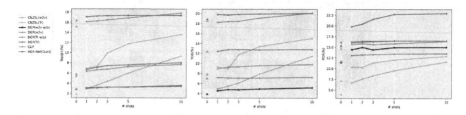

Fig. 6. Few shots comparison. DGP(w/o) [10] means without separating adjacency matrix into ancestors and descendants, CN and INIT in CNZSL [27] means class normalization and proper initialization respectively, and Tr is Transformer of CLIP [24] for short.

5   Conclusions

This paper focuses on scaling-up visual recognition of unseen classes to tens of thousands of categories. We proposed a novel hierarchical graphic knowledge representation framework for confidence-based classification and demonstrated significantly better performance than baselines over Image-Net-21K-D and Image-Net-21K-P benchmarks, achieving new SOTA. We hope our work help ease future research of zero-shot learning and pave a steady way to understand large-scale visual-language relationships with limited data.

Acknowledgments. Research reported in this paper was supported by King Abdullah University of Science and Technology (KAUST), BAS/1/1685-01-01.

References

1. Chen, S., et al.: Free: Feature refinement for generalized zero-shot learning. In: Proceedings of the IEEE/CVF International Conference on Computer Vision (ICCV), pp. 122–131 (2021)

2. Cheng, R.: Data efficient language-supervised zero-shot recognition with optimal transport distillation (2021)
3. Cox, M.A., Cox, T.F.: Multidimensional scaling. In: Handbook of data visualization, pp. 315–347. Springer (2008). https://doi.org/10.1007/978-3-642-28753-4_101322
4. Deng, J., Dong, W., Socher, R., Li, L.J., Li, K., Fei-Fei, L.: Imagenet: a large-scale hierarchical image database. In: 2009 IEEE Conference on Computer Vision and Pattern Recognition, pp. 248–255. IEEE (2009)
5. Elhoseiny, M., Saleh, B., Elgammal, A.: Write a classifier: Zero-shot learning using purely textual descriptions. In: Proceedings of the IEEE International Conference on Computer Vision, pp. 2584–2591 (2013)
6. Elhoseiny, M., Zhu, Y., Zhang, H., Elgammal, A.: Link the head to the" beak": Zero shot learning from noisy text description at part precision. In: 2017 IEEE Conference on Computer Vision and Pattern Recognition (CVPR), pp. 6288–6297. IEEE (2017)
7. Frome, A., et al.: Devise: A deep visual-semantic embedding model (2013)
8. He, K., Zhang, X., Ren, S., Sun, J.: Deep residual learning for image recognition. In: Proceedings of the IEEE Conference on Computer Vision and Pattern Recognition, pp. 770–778 (2016)
9. Jia, C., et al.: Scaling up visual and vision-language representation learning with noisy text supervision. arXiv preprint arXiv:2102.05918 (2021)
10. Kampffmeyer, M., Chen, Y., Liang, X., Wang, H., Zhang, Y., Xing, E.P.: Rethinking knowledge graph propagation for zero-shot learning. In: Proceedings of the IEEE/CVF Conference on Computer Vision and Pattern Recognition, pp. 11487–11496 (2019)
11. Kipf, T.N., Welling, M.: Semi-supervised classification with graph convolutional networks. arXiv preprint arXiv:1609.02907 (2016)
12. Liu, S., Chen, J., Pan, L., Ngo, C.W., Chua, T.S., Jiang, Y.G.: Hyperbolic visual embedding learning for zero-shot recognition. In: Proceedings of the IEEE/CVF Conference on Computer Vision and Pattern Recognition, pp. 9273–9281 (2020)
13. Long, Y., Shao, L.: Describing unseen classes by exemplars: Zero-shot learning using grouped simile ensemble. In: 2017 IEEE Winter Conference on Applications of Computer Vision (WACV), pp. 907–915. IEEE (2017)
14. Loshchilov, I., Hutter, F.: Decoupled weight decay regularization. In: International Conference on Learning Representations (2019)
15. Lu, Y.: Unsupervised learning on neural network outputs: with application in zero-shot learning. arXiv preprint arXiv:1506.00990 (2015)
16. Micikevicius., et al.: Mixed precision training. arXiv preprint arXiv:1710.03740 (2017)
17. Mikolov, T., Sutskever, I., Chen, K., Corrado, G.S., Dean, J.: Distributed representations of words and phrases and their compositionality. In: Advances in Neural Information Processing Systems, pp. 3111–3119 (2013)
18. Miller, G.A.: Wordnet: a lexical database for english. Commun. ACM 38(11), 39–41 (1995)
19. Nayak, N.V., Bach, S.H.: Zero-shot learning with common sense knowledge graphs. arXiv preprint arXiv:2006.10713 (2020)
20. Norouzi, M., et al.: Zero-shot learning by convex combination of semantic embeddings. arXiv preprint arXiv:1312.5650 (2013)
21. Van den Oord, A., Li, Y., Vinyals, O.: Representation learning with contrastive predictive coding. arXiv e-prints pp. arXiv-1807 (2018)

22. Patterson, G., Hays, J.: Sun attribute database: Discovering, annotating, and recognizing scene attributes. In: 2012 IEEE Conference on Computer Vision and Pattern Recognition (CVPR), pp. 2751–2758. IEEE (2012)
23. Pennington, J., Socher, R., Manning, C.D.: Glove: Global vectors for word representation. In: Proceedings of the 2014 Conference on Empirical Methods in Natural Language Processing (EMNLP), pp. 1532–1543 (2014)
24. Radford, A., et al.: Learning transferable visual models from natural language supervision. arXiv preprint arXiv:2103.00020 (2021)
25. Ridnik, T., Ben-Baruch, E., Noy, A., Zelnik-Manor, L.: Imagenet-21k pretraining for the masses. arXiv preprint arXiv:2104.10972 (2021)
26. Sennrich, R., Haddow, B., Birch, A.: Neural machine translation of rare words with subword units. In: Proceedings of the 54th Annual Meeting of the Association for Computational Linguistics (2016)
27. Skorokhodov, I., Elhoseiny, M.: Class normalization for zero-shot learning. In: International Conference on Learning Representations (2021). https://openreview.net/forum?id=7pgFL2Dkyyy
28. Sun, Q., Liu, Y., Chen, Z., Chua, T.S., Schiele, B.: Meta-transfer learning through hard tasks. IEEE Trans. Pattern Anal. Mach. Intell. (2020)
29. Vaswani, A., et al.: Attention is all you need. In: Advances in Neural Information Processing Systems, pp. 5998–6008 (2017)
30. Veeling, B.S., Linmans, J., Winkens, J., Cohen, T., Welling, M.: Rotation equivariant cnns for digital pathology. CoRR (2018)
31. Wang, J., Jiang, B.: Zero-shot learning via contrastive learning on dual knowledge graphs. In: Proceedings of the IEEE/CVF International Conference on Computer Vision, pp. 885–892 (2021)
32. Wang, X., Ye, Y., Gupta, A.: Zero-shot recognition via semantic embeddings and knowledge graphs. In: Proceedings of the IEEE Conference on Computer Vision and Pattern Recognition, pp. 6857–6866 (2018)
33. Wang, Y., Yao, Q., Kwok, J.T., Ni, L.M.: Generalizing from a few examples: a survey on few-shot learning. ACM Comput. Surv. (CSUR) 53(3), 1–34 (2020)
34. Welinder, P., et al.: Caltech-ucsd birds 200 (2010)
35. Xian, Y., Lampert, C.H., Schiele, B., Akata, Z.: Zero-shot learning-a comprehensive evaluation of the good, the bad and the ugly. In: PAMI (2018)
36. Xie, G.S., et al.: Attentive region embedding network for zero-shot learning. In: Proceedings of the IEEE/CVF Conference on Computer Vision and Pattern Recognition, pp. 9384–9393 (2019)
37. Ye, H.J., Hu, H., Zhan, D.C.: Learning adaptive classifiers synthesis for generalized few-shot learning. Int. J. Comput. Vision 129(6), 1930–1953 (2021)
38. Yu, Y., Ji, Z., Fu, Y., Guo, J., Pang, Y., Zhang, Z.M.: Stacked semantics-guided attention model for fine-grained zero-shot learning. In: NeurIPS (2018)
39. Zhang, C., Cai, Y., Lin, G., Shen, C.: Deepemd: Few-shot image classification with differentiable earth mover's distance and structured classifiers. In 2020 IEEE CVF Conference on Computer Vision and Pattern Recognition, pp. 12200–12210 (2020)
40. Zhou, K., Yang, J., Loy, C.C., Liu, Z.: Learning to prompt for vision-language models. arXiv preprint arXiv:2109.01134 (2021)

Doubly Deformable Aggregation of Covariance Matrices for Few-Shot Segmentation

Zhitong Xiong[1](✉) ⓘ, Haopeng Li[2] ⓘ, and Xiao Xiang Zhu[1,3] ⓘ

[1] Data Science in Earth Observation, Technical University of Munich (TUM), Munich, Germany
zhitong.xiong@tum.de
[2] School of Computing and Information Systems, University of Melbourne, Melbourne, Australia
[3] Remote Sensing Technology Institute (IMF), German Aerospace Center (DLR), Weßling, Germany

Abstract. Training semantic segmentation models with few annotated samples has great potential in various real-world applications. For the few-shot segmentation task, the main challenge is how to accurately measure the semantic correspondence between the support and query samples with limited training data. To address this problem, we propose to aggregate the learnable covariance matrices with a deformable 4D Transformer to effectively predict the segmentation map. Specifically, in this work, we first devise a novel hard example mining mechanism to learn covariance kernels for the Gaussian process. The learned covariance kernel functions have great advantages over existing cosine similarity-based methods in correspondence measurement. Based on the learned covariance kernels, an efficient doubly deformable 4D Transformer module is designed to adaptively aggregate feature similarity maps into segmentation results. By combining these two designs, the proposed method can not only set new state-of-the-art performance on public benchmarks, but also converge extremely faster than existing methods. Experiments on three public datasets have demonstrated the effectiveness of our method. (Code: https://github.com/ShadowXZT/DACM-Few-shot.pytorch)

Keywords: Deep kernel learning · Few-shot segmentation · Gaussian process · Similarity measurement · Transformer

1 Introduction

Semantic segmentation at the pixel level [6,19,41,46] is one of the fundamental tasks in computer vision and has been extensively studied for decades. In

Supplementary Information The online version contains supplementary material available at https://doi.org/10.1007/978-3-031-20044-1_8.

recent years, the performance of semantic segmentation tasks has been significantly improved due to the substantial progress in deep learning techniques. Large-scale convolutional neural networks (CNNs) [8,28], vision Transformers [3,18], and MLP-based deep networks [15] have greatly improved the ability of visual representation learning, and significantly enhanced the performance of downstream tasks such as semantic segmentation [46,47].

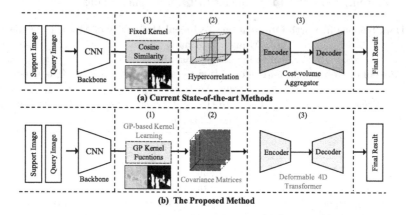

Fig. 1. Comparison of the proposed DACM (b) with the framework of existing state-of-the-art methods (a). There are three main differences. (1) We propose to utilize GP-based kernel functions for similarity measurement. (2) We use covariance matrices instead of cosine similarity as the cost volume. (3) A DDT module is designed for effective cost volume aggregation.

However, these advances currently rely heavily on large-scale annotated datasets that require extensive manual annotation efforts. This is not conducive to real-world computer vision applications, as obtaining pixel-level annotations is expensive and time-consuming. To remedy this issue, the few-shot segmentation task has emerged and attracted more and more research attention [10,17,37,40,42,43]. For few-shot segmentation, only a handful of annotated samples (support samples) for each class are provided for the training of the model, which largely mitigates the reliance on large-scale manual annotations. Although an attempt has been made [1] to handle few-shot segmentation in a transductive way, most existing works model the few-shot segmentation in a meta-learning setting [26,34] to avoid overfitting brought by insufficient training data. As support samples are the most important guidance for the prediction of query samples, the key to few-shot segmentation is to effectively exploit information from the support set.

Prototype-based few-shot segmentation methods attempt to extract a prototype vector to represent the support sample [31,35,40]. Despite its effectiveness, compressing the features of support samples into a single vector is prone to noise and neglects the spatial structure of objects in the support image. To maintain the geometric structure of objects, visual correspondence-based methods

[10,20,43] opt to utilize dense, pixel-wise semantic correlations between support and query images. Sophisticated pixel-wise attention-based methods [44] have also been devised to leverage dense correlations between support and query samples. However, directly using pixel-level dense information is a double-edged sword, which brings two new challenges: feature ambiguities caused by repetitive patterns, and background noise. To handle these issues, recent work [20] shows that aggregating dense correlation scores (also called *cost volume*) with 4D convolution can attain outstanding performance for few-shot segmentation. To better model the interaction among correlation scores, [9] proposes to combine 4D convolutions and 4D swin Transformer [18] in a coarse-to-fine manner. Although recent cost aggregation-based methods [9,20] attain state-of-the-art performance, they still suffer from two limitations: 1) *lack of flexibility in computing correlation scores*, which results in an extremely slow convergence speed; 2) *lack of flexibility in aggregating high-dimensional correlation scores*, which leads to limited performance with high computational cost.

To tackle the first limitation, we propose a Gaussian process (GP) based kernel learning method to learn flexible kernel functions for a more accurate similarity measurement. We find that existing state-of-the-art methods [9,20] mainly focus on the cost volume aggregation module and directly rely on the hyper-correlation tensors constructed by the cosine similarity. However, we argue that directly using the cosine similarity lacks flexibility and cannot faithfully reflect the semantic similarities between support and query pixels. As shown in Fig. 1 (a), the match between the cosine similarity map and the ground truth label is poor. To handle this problem, we design a hard example mining mechanism to dynamically sample hard samples to train the covariance kernel functions. As shown in Fig. 1 (b), using the learned covariance kernels can generate a more reasonable similarity map and greatly improve the convergence speed.

Another limitation is mainly caused by the cost volume aggregation module. Prior works have shown that a better cost volume aggregation method can attain superior few-shot segmentation performance. However, the existing 4D convolution-based method [20] lacks the ability to model longer distance relations between elements in the cost volume. VAT [9] proposes to combine 4D convolutions with 4D swin Transformers [18] for cost volume aggregation. Although outstanding performance can be achieved, it requires a notable GPU memory consumption and suffers from a slow convergence speed. Considering this, we propose a doubly deformable 4D Transformer based aggregation method, with deformable attention mechanisms on both the support and query dimensions of the 4D cost volume input. Compared with 4D convolutions, it can not only model longer-distance interactions between pixels owing to the Transformer network design, but also learn to selectively attend to more informative regions of the support and query information.

To sum up, we propose a novel few-shot segmentation framework using doubly deformable aggregation of covariance matrices (DACM), which aggregates learnable covariance matrices with a doubly deformable 4D Transformer to predict segmentation results effectively. The proposed GP-based kernel learning can learn more accurate similarity measurements for cost volume generation. In what follows, the designed doubly deformable 4D Transformer can effectively and efficiently aggregate the multi-scale cost volume into the final segmentation result. Specifically, our contributions can be summarized as follows:

- Towards a more accurate similarity measurement, a GP-based kernel learning method is proposed to learn flexible covariance kernel functions, with a novel hard example mining mechanism. To our knowledge, we are the first to use learnable covariance functions instead of cosine similarity for the few-shot segmentation task.
- Towards a more flexible cost volume aggregation, we propose a doubly deformable 4D Transformer (DDT) module, which utilizes deformable attention mechanisms on both the support and query dimensions of the 4D cost volume input. DDT can enhance representation learning by selectively attending to more informative regions.
- By combining these two modules, the proposed DACM method can attain new state-of-the-art performance on three public datasets with an extremely fast convergence speed. We also provide extensive visualization to better understand the proposed method.

2   Related Work

2.1   Few-Shot Segmentation

Mainstream methods for few-shot segmentation can be roughly categorized into prototype-based methods [2,17,37,40] and correlation-based methods [10,36,42, 43]. Prototype-based methods aim to generate a prototype representation [29] for each class based on the support sample, and then predict segmentation maps by measuring the distance between the prototype and the representations of the query image densely [2,37,45]. How to generate the prototype representation is the core of such methods. For example, a feature weighting and boosting model was proposed for prototype learning in [21]. Considering the limitations of using a single holistic representation for each class, a prototype learner was proposed in [40] to learn multiple prototypes based on the support images and the corresponding object masks. Then, the query images and the prototypes were used to generate the final object masks. To capture diverse and fine-grained object characteristics, Part-aware Prototype Network (PPNet) was developed in [17] to learn a set of part-aware prototypes for each semantic class. A prototype alignment regularization between support and query samples was proposed in Prototype Alignment Net-work (PANet) [37] to fully exploit semantic information from the support images and achieve better generalization on unseen classes. To handle the intra-class variations and inherent uncertainty, probabilistic latent variable-based prototype modeling was designed in [31,35], which leveraged probabilistic modeling to enhance the segmentation performance.

As prototype-based methods suffer from the loss of spatial structure due to average pooling [45], correlation-based methods were proposed to model the pairwise semantic correspondences densely between the support and query images.

For instance, Attention-based Multi-Context Guiding (A-MCG) network [10] was proposed to fuse the multi-scale context features extracted from the support and query branch. Specifically, a residual attention module was introduced into A-MCG to enrich the context information. To retain the structure representations of segmentation data, attentive graph reasoning [43] was exploited to transfer the class information from the support set to the query set, where element-to-element correspondences were captured at multiple semantic levels. Democratic Attention Networks (DAN) was introduced in [36] for few-shot semantic segmentation, where the democratized graph attention mechanism was applied to establish a robust correspondence between support and query images. As proven in existing state-of-the-art methods, cost volume aggregation plays an important role in the few-shot segmentation task. To this end, 4D convolutions and Transformers were designed in [9,20] and achieved highly competitive few-shot segmentation results.

2.2 Gaussian Process

Recently, the combination of the Gaussian process and deep neural networks has drawn more and more research attention due to the success of deep learning [23, 30,33,38]. For instance, scalable deep kernels were introduced in [38] to combine the non-parametric flexibility of GP and the structural properties of deep models. Furthermore, adaptive deep kernel learning [33] was proposed to learn a family of kernels for numerous few-shot regression tasks, which enabled the determination of the appropriate kernel for specific tasks. Deep Kernel Transfer (DKT) [23] was presented to learn a kernel that can be transferred to new tasks. Such a design can be implemented as a single optimizer and can provide reliable uncertainty measurement. GP has been introduced to few-shot segmentation in [11], where they proposed to learn the output space of the GP via a neural network that encoded the label mask. Although there is an attempt to incorporate GP, it is still in its infancy and far from fully exploiting the potential of GP for few-shot segmentation.

3   Methodology

The whole architecture of the proposed DACM framework is illustrated in Fig. 2. Given the input support and query images, multi-level deep features are first extracted by the fixed backbone network pre-trained on ImageNet [25]. Three levels of deep features with different spatial resolutions form a pyramidal design. For each level, average pooling is used to aggregate multiple features into a single one. Then, three different GP models are trained for computing the covariance matrices (4D cost volume) for each level. Next, the proposed Deformable 4D Transformer Aggregator (DTA) is combined with the weight-sparsified 4D convolution for cost volume aggregation. Finally, a decoder is used to predict the final segmentation result for the input query image.

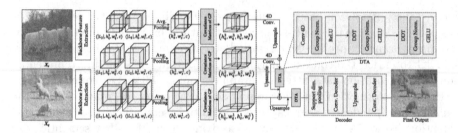

Fig. 2. The whole architecture of our proposed DACM framework. Under a pyramidal design, DACM consists of three parts: 1) feature extraction by the backbone network; 2) cost volume generation based on covariance kernels; 3) cost volume aggregation using DTA modules and 4D convolutions [20].

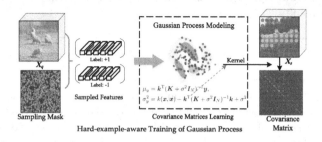

Fig. 3. Illustration of the GP-based kernel learning process. Owing to Gaussian modeling, covariance functions can measure the similarity in a continuous feature space.

3.1 Preliminaries: Gaussian Process

The covariance kernels of GP specify the statistical relationship between two points at the input space (*e.g.*, support and query features). Namely, the learned covariance matrices can be naturally used to measure the correlations between support and query samples. Formally, the data set consists of N samples of dimension D, *i.e.*, $\{(x_i, y_i)\}_{i=1}^N$, where $x_i \in \mathbb{R}^D$ is a data point and y_i is the corresponding label. The GP regression model assumes that the outputs y_i can be regarded as certain deterministic latent function $f(x_i)$ with zero-mean Gaussian noise ε, *i.e.*, $y_i = f(x_i) + \varepsilon$, where $\varepsilon \sim \mathcal{N}(0, \sigma^2)$. GP sets a zero-mean prior on f, with covariance $k(x_i, x_j)$. The covariance function k reflects the smoothness of f. The most widely-used covariance function is the Automatic Relevance Determination Squared Exponential (ARD SE), *i.e.*,

$$k(x_i, x_j) = \sigma_0^2 \exp\left\{ -\frac{1}{2} \sum_{d=1}^D \frac{((x_i)_d - (x_j)_d)^2}{l_d^2} \right\}, \tag{1}$$

where $\sigma^2, \sigma_0^2, \{l_d\}_{d=1}^D$ are hyper-parameters. As shown in Fig. 3, after the optimization of hyper-parameters, we can compute the covariance matrices using the learned kernel k.

3.2 Hard Example Mining-based GP Kernel Learning

For few-shot segmentation, the whole dataset is split into the meta-training, meta-validation, and meta-testing subsets. Each *episode* [26,34] in the subsets consists of a support set $\mathcal{S} = \left\{ (\boldsymbol{I}_s^i, \boldsymbol{M}_s^i) \right\}_{i=1}^{K}$ and a query image $\boldsymbol{I}_q$ of the same class, where $\boldsymbol{I}_s^i \in \mathbb{R}^{3 \times H \times W}$ is the support image and $\boldsymbol{M}_s^i \in \{0, 1\}^{H \times W}$ is its binary object mask for a certain class. K is the number of the support images. We discuss the one-shot segmentation ($K = 1$) case in the following sections.

Given the support image-mask pair $(\boldsymbol{I}_s, \boldsymbol{M}_s)$ and the query image-mask pair $(\boldsymbol{I}_q, \boldsymbol{M}_q)$ where $\boldsymbol{I}_s, \boldsymbol{I}_q \in \mathbb{R}^{3 \times H \times W}$ and $\boldsymbol{M}_s, \boldsymbol{M}_q \in \{0, 1\}^{H \times W}$, a pretrained backbone is used to extract features from $\boldsymbol{I}_s$ and $\boldsymbol{I}_q$, and the feature maps at layer l are denoted as $\boldsymbol{F}_s^l, \boldsymbol{F}_q^l \in \mathbb{R}^{c_l \times h_l \times w_l}$, where c_l, h_l, w_l denote that the features are of channel c_l, height h_l and width w_l. Note that we view the feature vectors at all spatial positions on feature maps of the query images as the training data. As shown in Fig. 3, our motivation is to dynamically select training samples from $\boldsymbol{F}_q^l$ for training the GP at layer l with a hard example mining mechanism. To this end, we propose a hard example aware sampling strategy based on the similarity (cost volume) between support and query samples $\boldsymbol{F}_s^l$ and $\boldsymbol{F}_q^l$. The 4D covariance matrices $\mathcal{C}^l \in \mathbb{R}^{h_l \times w_l \times h_l \times w_l}$, *i.e.*, the cost volume, can be computed with the GP using an initialized covariance kernel by

$$\mathcal{C}^l(i, j) = \mathrm{ReLU} \left(k \left(\frac{\boldsymbol{F}_q^l(i)}{\left\| \boldsymbol{F}_q^l(i) \right\|}, \frac{\boldsymbol{F}_s^l(j)}{\left\| \boldsymbol{F}_s^l(j) \right\|} \right) \right), \tag{2}$$

where i and j denote spatial positions of 2D feature maps. In what follows, we reshape the 4D cost volume into $\mathcal{C}^l \in \mathbb{R}^{h_l \times w_l \times h_l w_l}$ and sum up it along the third dimension. Then we can get a 2D similarity map $\mathcal{S}^l \in \mathbb{R}^{h_l \times w_l}$ of the query image. An example 2D similarity map can be found in Fig. 1. Ideally, we expect the high values in $\mathcal{S}^l$ to totally lie in the ground truth mask. The negative positions with high similarity values in the 2D similarity map $\mathcal{S}^l$ can be viewed as hard examples. Thus, we utilize $\mathcal{S}^l$ to generate a probability map for sampling the training samples from $\boldsymbol{F}_q^l$. Specifically, we can obtain a 2D probability map for sampling training data by

$$\hat{\mathcal{S}}^l = \mathcal{S}^l + \lambda \boldsymbol{M}_q^l, \tag{3}$$

$$p^l = \frac{\hat{\mathcal{S}}^l - \min(\hat{\mathcal{S}}^l)}{\max(\hat{\mathcal{S}}^l) - \min(\hat{\mathcal{S}}^l)} \tag{4}$$

$$t^l \sim \mathrm{Bernoulli}(p^l). \tag{5}$$

In Eq. 3, we add $\mathcal{S}^l$ and the scaled query mask $\boldsymbol{M}_q^l$ with λ to ensure that adequate number of positive samples can be selected for training at the early stage. Then, in Eq. 4, we normalize $\mathcal{S}^l$ to represent the probability that the sample will be selected or not. After the softmax operation, the 2D probability map $p^l \in [0, 1]^{h_l \times w_l}$ can be obtained with each position representing the probability of being selected. Finally, a binary mask $t^l \in \{0, 1\}^{h_l \times w_l}$ can be sampled from a Bernoulli distribution, where 0 means unselected and 1 means selected.

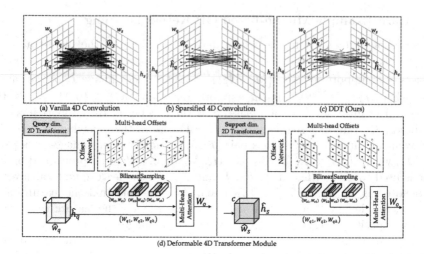

(a) Vanilla 4D Convolution   (b) Sparsified 4D Convolution   (c) DDT (Ours)

(d) Deformable 4D Transformer Module

Fig. 4. Illustration of the proposed DDT module. DDT utilizes doubly deformable attention mechanisms on both the support and query dimensions of the 4D cost volume. As shown in (b), the red dashed line indicates that the path is dropped by weight sparsification. While the dropped path can be recovered with the proposed DDT module if it is important, as presented in (c). (Color figure online)

Such a sampling manner aims to sample query features which are likely to be incorrectly classified during the training stage. Then, we aim to use the sampled hard example-aware features on the query image to optimize the kernel. Suppose the sampled N training data at feature layer l is $\{(x_j^q, y_j^q)\}_{j=1}^N$ ($x_j^q \in \mathbb{R}^{c_l}$). Then these samples are used to learn the hyper-parameters $\left\{\sigma^2, \sigma_0^2, \{l_d\}_{d=1}^D\right\}$ of GP for computing the covariance matrices by maximizing the marginal likelihood. Please refer to § 4 of the supplementary material for a more detailed definition.

3.3   Doubly Deformable 4D Transformer for Cost Volume Aggregation

To clearly describe the proposed DDT module, we will first introduce the sparsified 4D convolution proposed in [20]. As shown in Fig. 4, weight-sparsified 4D convolution drops majority paths and only considers the activations at positions of either one of 2-dimensional centers. Formally, given 4D position $(\mathbf{u}, \mathbf{u}') \in \mathbb{R}^4$, by only considering its neighbors adjacent to either $\mathbf{u}$ or $\mathbf{u}'$, the weight-sparsified 4D convolution can be approximated by two 2-dimensional convolutions performed on 2D slices of hypercorrelation tensor. Different from prior works, the proposed DDT module aims to 1) utilize deformable attention mechanisms [39,48] on both the support and query dimensions of the 4D input to learn more flexible aggregation; 2) provide the ability to model longer distance relations between elements of the cost volume. Specifically, we can formulate the

DDT module for aggregating 4D tensor input $\mathcal{C}$ (we omit layer l for simplicity) as follows:

$$\mathrm{DDT}(\mathcal{C}, \mathbf{u}, \mathbf{u}') = \mathrm{SDT}\left(\mathcal{C}, (\mathbf{u}, \mathbf{p}')\right) + \mathrm{QDT}(\mathcal{C}, (\mathbf{p}, \mathbf{u}')),$$
$$\mathbf{p} := \mathcal{P}\left(\mathbf{u}\right), \, \mathbf{p}' := \mathcal{P}\left(\mathbf{u}'\right), \tag{6}$$

where $\mathbf{p}$ and $\mathbf{p}'$ denote the neighbour pixels centered at position $\mathbf{u}$ and $\mathbf{u}'$. SDT denotes the deformable Transformer on the support dimension (2D slice) of the cost volume. Similarly, QDT represents the deformable Transformer on the query dimension (2D slice) of $\mathcal{C}$.

As QDT shares similar computation process with SDT, we will only introduce the computation of SDT for the sake of simplicity. Suppose the 2D slice $\mathcal{C}(\mathbf{u}, \mathbf{p}') \in \mathbb{R}^{c \times h \times w}$ of the cost volume is with channel dimension c, height h and width w. Note that $c = 1$ in this case. We first normalize the coordinates to the range $[-1, +1]$. $(-1, -1)$ indicates the top-left corner and $(+1, +1)$ represents the bottom-right corner. Then, an offset network is utilized to generate n offset maps $\{\Delta \mathbf{p}_i'\}_{i=1}^n$ using two convolution layers. Each offset map has the spatial dimension (h, w) and is responsible for a head in the multi-head attention module. By adding the original coordinates $\mathbf{p}'$ and $\Delta \mathbf{p}'$, we can obtain the shifted positions for the 2D slice input. In what follows, the differentiable bilinear sampling is used to obtain vectors $\tilde{x} \in \mathbb{R}^{c \times h \times w}$ at the shifted positions by:

$$\tilde{x} = \phi(\mathcal{C}(\mathbf{u}, :), \mathbf{p}' + \Delta \mathbf{p}'). \tag{7}$$

Then, we project the sample feature vectors into keys $k \in \mathbb{R}^{c' \times hw}$ and values $v \in \mathbb{R}^{c' \times hw}$ using learnable W_k and W_v as follows: $k = \tilde{x} W_k$, $v = \tilde{x} W_v$. Where c' is the projected channel dimension. Next, the query tokens $\mathcal{C}(\mathbf{u}, \mathbf{p}')$ in the query features are projected to queries $q \in \mathbb{R}^{c' \times hw}$. For each head of the multi-head attention, we can compute the output $z \in \mathbb{R}^{c' \times h \times w}$ using the self-attention mechanism as follows:

$$q = \mathcal{C}\left(\mathbf{u}, \mathbf{p}'\right) W_q, \, z = \mathrm{softmax}\left(qk^{\mathrm{T}}/\sqrt{d}\right) v, \tag{8}$$

where d denotes the dimension of each head, W_q is a learnable matrix. The whole computation process of DDT is illustrated in Fig. 4. To sum up, the proposed light-weight, plug-and-play DDT module utilizes flexible deformable modeling to compensate the dropped informative activations in weight-sparsified 4D convolution. It can also enjoy the long-distance interactions modeling brought by Transformers.

4 Experiments

4.1 Experiment Settings

Datasets. Three few-shot segmentation datasets are exploited to evaluate the proposed method: PASCAL-5^i [26], COCO-20^i [14], and FSS-1000 [13].

PASCAL-5^i is re-created from the data in PASCAL VOC 2012 [4] and the augmented mask annotations from [7]. PASCAL-5^i contains 20 object classes and is split into 4 folds. COCO-20^i consists of 80 object classes and is also split into 4 folds. Following prior experimental settings [17,21,32,36] on PASCAL-5^i and COCO-20^i datasets, for each fold i, the data in the rest folds are used for training, and 1,000 episodes randomly sampled from the fold i are used for evaluation. FSS-1000 contains 1,000 classes and is split into 520, 240, and 240 classes for training, validation, and testing, respectively.

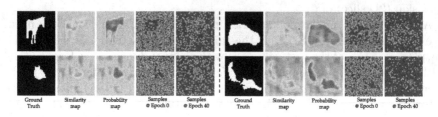

Fig. 5. Some qualitative examples of the proposed hard example-aware sampling strategy for training the GP models.

Evaluation Metrics. Mean intersection over union (mIoU) and foreground-background IoU (FB-IoU) are reported for comparisons. mIoU is the average of IoU over all classes in a fold, *i.e.*, mIoU $= \frac{1}{|\mathcal{C}_{test}|} \sum_{c \in \mathcal{C}} \text{IoU}_c$. As for FB-IoU, the object classes are not considered in this metric. The average of foreground IoU and background IoU is computed, *i.e.*, FB-IoU $= \frac{1}{2}(\text{IoU}_F + \text{IoU}_B)$. Compared with FB-IoU, mIoU can better reflect the generalization ability of segmentation methods to unseen classes.

Implementation Details. The proposed method is implemented using PyTorch [22] and GPyTorch [5]. VGG16 [28], ResNet50 and ResNet101 [8] pre-trained on ImageNet are adopted as the backbone networks. To construct pyramidal features, three levels of features with resolutions of 50×50, 25×25, 13×13 (12×12 for VGG backbone) are used. The GP models and cost volume aggregation models are trained in an end-to-end manner. Adam optimizer [12] is used for the optimization of the covariance matrices aggregation part as well as the GP models. The initial learning rate for the covariance matrices aggregation part is set to $1e - 3$, while the learning rate for three GP models is set to $1e - 2$. Only **50 epochs** are used for training the proposed DACM on all three datasets. It is worth noting that 50 epochs are significantly less than existing methods that usually require more than 200 epochs for model training.

4.2   Results and Analysis

PASCAL-5^i. We compare our method with existing state-of-the-art methods. Table 1 presents the 1-shot and 5-shot segmentation results. "DACM (Ours)"

Table 1. Performance comparisons on PASCAL-5^i dataset in mIoU and FB-IoU.

Backbone	Methods	1-shot Segmentation						5-shot Segmentation						#Learnable Params
		5^0	5^1	5^2	5^3	Mean	FB-IoU	5^0	5^1	5^2	5^3	Mean	FB-IoU	
VGG16 [28]	co-FCN [24]	36.7	50.6	44.9	32.4	41.1	60.1	37.5	50.0	44.1	33.9	41.4	60.2	34.2M
	AMP-2 [27]	41.9	50.2	46.7	34.7	43.4	61.9	40.3	55.3	49.9	40.1	46.4	62.1	15.8M
	PANet [37]	42.3	58.0	51.1	41.2	48.1	66.5	51.8	64.6	59.8	46.5	55.7	70.7	14.7M
	PFENet [32]	56.9	**68.2**	54.4	52.4	58.0	72.0	59.0	69.1	54.8	52.9	59.0	72.3	10.4M
	HSNet [20]	59.6	65.7	59.6	54.0	59.7	73.4	64.9	69.0	64.1	58.6	64.1	76.6	**2.6M**
	DACM (Ours)	**61.8**	67.8	**61.4**	**56.3**	**61.8**	**75.5**	**66.1**	**70.6**	**65.8**	**60.2**	**65.7**	**77.8**	3.0M
ResNet50 [8]	PPNet [17]	48.6	60.6	55.7	46.5	52.8	69.2	58.9	68.3	66.8	58.0	63.0	75.8	31.5M
	PFENet [32]	61.7	69.5	55.4	56.3	60.8	73.3	63.1	70.7	55.8	57.9	61.9	73.9	10.8M
	RePRI [1]	59.8	68.3	62.1	48.5	59.7	—	64.6	71.4	**71.1**	59.3	66.6	—	—
	HSNet [20]	64.3	70.7	60.3	60.5	64.0	76.7	70.3	73.2	67.4	67.1	69.5	80.6	**2.6M**
	CyCTR [44]	67.8	72.8	58.0	58.0	64.2	—	71.1	73.2	60.5	57.5	65.6	—	—
	VAT [9]	67.6	71.2	62.3	60.1	65.3	77.4	72.4	73.6	68.6	65.7	70.0	80.9	3.2M
	DACM (Ours)	66.5	72.6	62.2	61.3	65.7	77.8	72.4	73.7	69.1	68.4	70.9	81.3	3.0M
	DACM (VAT)	**68.4**	**73.1**	**63.5**	**62.2**	**66.8**	**78.6**	**73.8**	**74.7**	70.3	68.1	**71.7**	**81.7**	3.3M
ResNet101 [8]	FWB [21]	51.3	64.5	56.7	52.2	56.2	—	54.8	67.4	62.2	55.3	59.9	—	43.0M
	PPNet [17]	52.7	62.8	57.4	47.7	55.2	70.9	60.3	70.0	69.4	60.7	65.1	77.5	50.5M
	DAN [36]	54.7	68.6	57.8	51.6	58.2	71.9	57.9	69.0	60.1	54.9	60.5	72.3	—
	PFENet [32]	60.5	69.4	54.4	55.9	60.1	72.9	62.8	70.4	54.9	57.6	61.4	73.5	10.8M
	RePRI [1]	59.6	68.6	62.2	47.2	59.4	—	66.2	71.4	67.0	57.7	65.6	—	—
	HSNet [20]	67.3	72.3	62.0	63.1	66.2	77.6	71.8	74.4	67.0	68.3	70.4	80.6	**2.6M**
	CyCTR [44]	69.3	72.7	56.5	58.6	64.3	72.9	73.5	74.0	58.6	60.2	66.6	75.0	—
	VAT [9]	68.4	72.5	64.8	64.2	67.5	78.8	73.3	75.2	68.4	69.5	71.6	82.0	3.3M
	DACM (Ours)	68.7	73.5	63.4	64.2	67.5	78.9	72.7	75.3	68.3	69.2	71.4	81.5	3.1M
	DACM (VAT)	**69.9**	**74.1**	**66.2**	**66.0**	**69.1**	**79.4**	**74.2**	**76.4**	**71.1**	**71.6**	**73.3**	**83.1**	3.4M

demotes the model presented in Fig. 2. Although DACM is trained with only **50 epochs**, it can still significantly outperform existing state-of-the-art models [9, 20] with three different backbones. The comparison results demonstrate that the proposed GP-based kernel learning and DDT module are beneficial for few-shot segmentation task. In addition, we also visualize the sampled locations (described in Eq. 5) during the GP kernel training in Fig. 5 for a better understanding of the designed hard example-aware sampling strategy. It can be seen that hard examples are selected at the later stage of the training process. Some comparative visualization results of "DACM (Ours)" model on this dataset are presented in Fig. 6.

The proposed GP-based kernel learning and DDT module can be plugged in any cost volume aggregation-based model. "DACM+VAT" denotes the combination of DACM with VAT by 1) replacing the cosine similarity-based cost volume with our covariance matrices; 2) replacing the last 4D convolution layer with our DDT. We can see that the combined method "DACM (VAT)" can achieve better segmentation performance than others. This demonstrates the superiority of the proposed framework.

Table 2. Performance comparisons on COCO-20^i dataset in mIoU and FB-IoU.

Backbone	Methods	1-shot Segmentation						5-shot Segmentation					
		5^0	5^1	5^2	5^3	Mean	FB-IoU	5^0	5^1	5^2	5^3	Mean	FB-IoU
ResNet50 [8]	PMM [40]	29.3	34.8	27.1	27.3	29.6	—	33.0	40.6	30.3	33.3	34.3	—
	RPMM [40]	29.5	36.8	28.9	27.0	30.6	—	33.8	42.0	33.0	33.3	35.5	—
	PFENet [32]	36.5	38.6	35.0	33.8	35.8	—	36.5	43.3	38.0	38.4	39.0	—
	RePRI [1]	32.0	38.7	32.7	33.1	34.1	—	39.3	45.4	39.7	41.8	41.6	—
	HSNet [20]	36.3	43.1	38.7	38.7	39.2	68.2	43.3	51.3	48.2	45.0	46.9	70.7
	CyCTR [44]	38.9	43.0	39.6	39.8	40.3	—	41.1	48.9	45.2	_47.0_	45.6	—
	VAT [9]	_39.0_	43.8	_42.6_	39.7	_41.3_	68.8	44.1	51.1	_50.2_	46.1	47.9	_72.4_
	DACM (Ours)	37.5	_44.3_	40.6	_40.1_	40.6	_68.9_	_44.6_	_52.0_	49.2	46.4	_48.1_	71.6
	DACM (VAT)	**41.2**	**45.2**	**44.1**	**41.3**	**43.0**	**69.4**	**45.2**	**52.2**	**51.5**	**47.7**	**49.2**	**72.9**

COCO-20^i. The comparison results on the COCO-20^i dataset are reported in Table 2. The results reveal that "DACM (Ours)" clearly outperform the HSNet baseline under both 1-shot and 5-shot settings. "DACM (VAT)" model achieves the best results compared with existing methods. COCO-20^i is a more difficult few-shot segmentation dataset. Basically, taking HSNet [20] as the baseline, DACM can achieve an clear improvement. It is worth noting that DACM requires less epochs for the model training, while it can still outperform VAT [9] and set new state-of-the-art results on the COCO-20^i dataset.

Table 3. Performance Comparisons on the FSS-1000 dataset.

Backbone	Methods	mIoU	
		1-shot	5-shot
ResNet50 [8]	FSOT [16]	82.5	83.8
	HSNet [20]	85.5	87.8
	VAT [9]	_89.5_	_90.3_
	DACM (Ours)	**90.7**	**91.6**
ResNet101 [8]	DAN [36]	85.2	88.1
	HSNet [20]	86.5	88.5
	VAT [9]	_90.0_	_90.6_
	DACM (Ours)	**90.8**	**91.7**

Table 4. Ablation study on PASCAL-5^i dataset.

Backbone	Methods	mIoU	
		1-shot	5-shot
VGG16 [28]	Baseline [20]	59.6	64.9
	+ GP-KL	60.8	65.2
	+ DDT-1	61.2	65.7
	+ DDT-2	**61.8**	**66.1**
	+ DDT-3	_61.6_	_65.8_

FSS-1000. FSS-1000 is simpler than the other two datasets. The comparison results are shown in Table 3. Under both the 1-shot setting and 5-shot setting, DACM can obtain a clear new state-of-the-art performance in terms of mIoU.

This indicates the effectiveness of our method. Comparison results on these three public datasets demonstrate the effectiveness of the proposed DACM method.

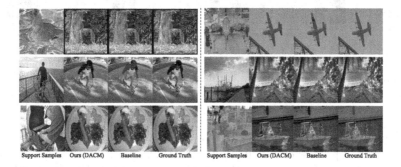

Fig. 6. Some visualization examples of the proposed method for few-shot segmentation. We compare DACM with the baseline (HSNet) method.

4.3 Ablation Study and Analysis

Ablation Study. To validate the effects of the proposed two modules, *i.e.*, GP-based kernel learning (GP-KL) module and DDT module, we conduct ablation studies. In addition, we compare the performance of using 1, 2 and 3 DDT layers (DDT-1, DDT-2 and DDT-3) to further study the effects of the number of DDT layers. The mIoU results on the PASCAL-5^i dataset are presented in Table 4. We can see that using 2 DDT layers can obtain the best result. The reason is that more DDT layers may bring the overfitting problem on several folds of the dataset. We also find that combining the GP-based kernel learning and DDT modules together results in the best few-shot segmentation performance.

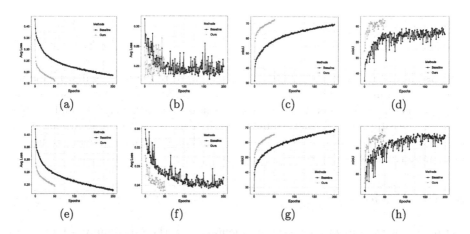

Fig. 7. (a), (b), (e), and (f) show the training/validation loss curves on the fold 0 and fold 1 of PASCAL-5i dataset. (c), (d), (g), and (h) present the training/validation mIoU curves.

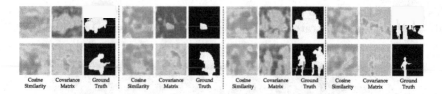

Fig. 8. Visualization of some examples of the covariance matrices. Compared with the commonly used cosine similarity, our method can learn more reasonable similarity measurements for few-shot segmentation.

Convergence Speed. In Fig. 7, we plot the loss curves and mIoU values during the training process on two folds of the PASCAL-5^i dataset. Obviously, our DACM can converge much faster than the baseline method (HSNet). Although trained with only 50 epochs, DACM can obtain much better training and validation mIoU performance than the baseline method. This clearly demonstrate the effectiveness of the proposed modules.

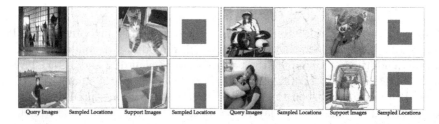

Fig. 9. Visualization of some examples of the learned deformable offsets for both the query and support samples.

Covariance Matrices Visualization. To better understand the proposed DACM method, we further visualize and compare the learned covariance matrices. As shown in Fig. 8, our method can get more reasonable similarity maps than cosine similarity. This also explains why DACM can converge faster than existing methods. Furthermore, we also visualize the learned deformable offsets for the support and query dimensions in Fig. 9. It can be seen that the deformable attention mechanism tends to attend on informative regions to learn more powerful representations.

5   Conclusions

In this paper, we present a few-shot segmentation method by doubly deformable aggregating covariance matrices. The proposed method aggregates learnable covariance matrices with a doubly deformable 4D Transformer to predict segmentation results effectively. Specifically, we make the following contributions. 1) We devise a novel hard example mining mechanism for learning covariance kernels of Gaussian process to enable a more accurate correspondence measurement. 2) We design a doubly deformable 4D Transformer to effectively and efficiently aggregate the multi-scale cost volume into the final segmentation results. 3) By combining these two modules, the proposed method can achieve state-of-the-art few-shot segmentation performance with a fast convergence speed.

Acknowledgement. This work is jointly supported by the European Research Council (ERC) under the European Union's Horizon 2020 research and innovation programme (grant agreement No. [ERC-2016-StG-714087], Acronym: *So2Sat*), by the Helmholtz Association through the Framework of Helmholtz AI (grant number: ZT-I-PF-5-01) - Local Unit "Munich Unit @Aeronautics, Space and Transport (MASTr)" and Helmholtz Excellent Professorship "Data Science in Earth Observation - Big Data Fusion for Urban Research"(grant number: W2-W3-100), by the German Federal Ministry of Education and Research (BMBF) in the framework of the international future AI lab "AI4EO – Artificial Intelligence for Earth Observation: Reasoning, Uncertainties, Ethics and Beyond" (grant number: 01DD20001) and by German Federal Ministry of Economics and Technology in the framework of the "national center of excellence ML4Earth" (grant number: 50EE2201C).

References

1. Boudiaf, M., Kervadec, H., Masud, Z.I., Piantanida, P., Ben Ayed, I., Dolz, J.: Few-shot segmentation without meta-learning: A good transductive inference is all you need? In: Proceedings of the IEEE/CVF Conference on Computer Vision and Pattern Recognition, pp. 13979–13988 (2021)
2. Dong, N., Xing, E.P.: Few-shot semantic segmentation with prototype learning. In: BMVC, vol. 3 (2018)
3. Dosovitskiy, A., et al.: An image is worth 16x16 words: Transformers for image recognition at scale. arXiv preprint arXiv:2010.11929 (2020)
4. Everingham, M., et al.: The pascal visual object classes challenge: A retrospective. Int. J. Comput. Vision **111**(1), 98–136 (2015)
5. Gardner, J.R., Pleiss, G., Bindel, D., Weinberger, K.Q., Wilson, A.G.: Gpytorch: Blackbox matrix-matrix gaussian process inference with gpu acceleration. arXiv preprint arXiv:1809.11165 (2018)
6. Hao, S., Zhou, Y., Guo, Y.: A brief survey on semantic segmentation with deep learning. Neurocomputing **406**, 302–321 (2020)
7. Hariharan, B., Arbeláez, P., Girshick, R., Malik, J.: imultaneous detection and segmentation. In: Fleet, D., Pajdla, T., Schiele, B., Tuytelaars, T. (eds.) ECCV 2014. LNCS, vol. 8695, pp. 297–312. Springer, Cham (2014). https://doi.org/10.1007/978-3-319-10584-0_20
8. He, K., Zhang, X., Ren, S., Sun, J.: Deep residual learning for image recognition. In: Proceedings of the IEEE Conference on Computer Vision and Pattern Recognition, pp. 770–778 (2016)
9. Hong, S., Cho, S., Nam, J., Kim, S.: Cost aggregation is all you need for few-shot segmentation. arXiv preprint arXiv:2112.11685 (2021)

10. Hu, T., Yang, P., Zhang, C., Yu, G., Mu, Y., Snoek, C.G.: Attention-based multi-context guiding for few-shot semantic segmentation. In: Proceedings of the AAAI Conference on Artificial Intelligence, vol. 33, pp. 8441–8448 (2019)

11. Johnander, J., Edstedt, J., Felsberg, M., Khan, F.S., Danelljan, M.: Dense gaussian processes for few-shot segmentation. arXiv preprint arXiv:2110.03674 (2021)

12. Kingma, D.P., Ba, J.: Adam: A method for stochastic optimization. arXiv preprint arXiv:1412.6980 (2014)

13. Li, X., Wei, T., Chen, Y.P., Tai, Y.W., Tang, C.K.: Fss-1000: A 1000-class dataset for few-shot segmentation. In: Proceedings of the IEEE/CVF Conference on Computer Vision and Pattern Recognition, pp. 2869–2878 (2020)

14. Lin, T.-Y., et al.: Microsoft COCO: Common objects in context. In: Fleet, D., Pajdla, T., Schiele, B., Tuytelaars, T. (eds.) ECCV 2014. LNCS, vol. 8693, pp. 740–755. Springer, Cham (2014). https://doi.org/10.1007/978-3-319-10602-1_48

15. Liu, H., Dai, Z., So, D., Le, Q.: Pay attention to mlps. In: Advances in Neural Information Processing Systems, vol. 34 (2021)

16. Liu, W., Zhang, C., Ding, H., Hung, T.Y., Lin, G.: Few-shot segmentation with optimal transport matching and message flow. arXiv preprint arXiv:2108.08518 (2021)

17. Liu, Y., Zhang, X., Zhang, S., He, X.: Part-aware prototype network for few-shot semantic segmentation. In: Vedaldi, A., Bischof, H., Brox, T., Frahm, J.-M. (eds.) ECCV 2020. LNCS, vol. 12354, pp. 142–158. Springer, Cham (2020). https://doi.org/10.1007/978-3-030-58545-7_9

18. Liu, Z., et al.: Swin transformer: Hierarchical vision transformer using shifted windows. In: Proceedings of the IEEE/CVF International Conference on Computer Vision, pp. 10012–10022 (2021)

19. Long, J., Shelhamer, E., Darrell, T.: Fully convolutional networks for semantic segmentation. In: Proceedings of the IEEE Conference on Computer Vision and Pattern Recognition, pp. 3431–3440 (2015)

20. Min, J., Kang, D., Cho, M.: Hypercorrelation squeeze for few-shot segmentation. arXiv preprint arXiv:2104.01538 (2021)

21. Nguyen, K., Todorovic, S.: Feature weighting and boosting for few-shot segmentation. In: Proceedings of the IEEE/CVF International Conference on Computer Vision, pp. 622–631 (2019)

22. Paszke, A., et al.: Pytorch: An imperative style, high-performance deep learning library. Adv. Neural. Inf. Process. Syst. **32**, 8026–8037 (2019)

23. Patacchiola, M., Turner, J., Crowley, E.J., O'Boyle, M., Storkey, A.J.: Bayesian meta-learning for the few-shot setting via deep kernels. In: Advances in Neural Information Processing Systems, vol. 33 (2020)

24. Rakelly, K., Shelhamer, E., Darrell, T., Efros, A., Levine, S.: Conditional networks for few-shot semantic segmentation (2018)

25. Russakovsky, O., et al.: Imagenet large scale visual recognition challenge. Int. J. Comput. Vision **115**(3), 211–252 (2015)

26. Shaban, A., Bansal, S., Liu, Z., Essa, I., Boots, B.: One-shot learning for semantic segmentation. arXiv preprint arXiv:1709.03410 (2017)

27. Siam, M., Oreshkin, B., Jagersand, M.: Adaptive masked proxies for few-shot segmentation. arXiv preprint arXiv:1902.11123 (2019)

28. Simonyan, K., Zisserman, A.: Very deep convolutional networks for large-scale image recognition. arXiv preprint arXiv:1409.1556 (2014)

29. Snell, J., Swersky, K., Zemel, R.S.: Prototypical networks for few-shot learning. arXiv preprint arXiv:1703.05175 (2017)

30. Snell, J., Zemel, R.: Bayesian few-shot classification with one-vs-each pólya-gamma augmented gaussian processes. arXiv preprint arXiv:2007.10417 (2020)

31. Sun, H., et al.: Attentional prototype inference for few-shot semantic segmentation. arXiv preprint arXiv:2105.06668 (2021)

32. Tian, Z., Zhao, H., Shu, M., Yang, Z., Li, R., Jia, J.: Prior guided feature enrichment network for few-shot segmentation. IEEE Trans. Pattern Anal. Mach. Intel. **01**, 1–1 (2020)

33. Tossou, P., Dura, B., Laviolette, F., Marchand, M., Lacoste, A.: Adaptive deep kernel learning. arXiv preprint arXiv:1905.12131 (2019)

34. Vinyals, O., Blundell, C., Lillicrap, T., Wierstra, D., et al.: Matching networks for one shot learning. Adv. Neural. Inf. Process. Syst. **29**, 3630–3638 (2016)

35. Wang, H., Yang, Y., Cao, X., Zhen, X., Snoek, C., Shao, L.: Variational prototype inference for few-shot semantic segmentation. In: Proceedings of the IEEE/CVF Winter Conference on Applications of Computer Vision, pp. 525–534 (2021)

36. Wang, H., Zhang, X., Hu, Y., Yang, Y., Cao, X., Zhen, X.: Few-shot semantic segmentation with democratic attention networks. In: Vedaldi, A., Bischof, H., Brox, T., Frahm, J.-M. (eds.) ECCV 2020. LNCS, vol. 12358, pp. 730–746. Springer, Cham (2020). https://doi.org/10.1007/978-3-030-58601-0_43

37. Wang, K., Liew, J.H., Zou, Y., Zhou, D., Feng, J.: PANet: Few-shot image semantic segmentation with prototype alignment. In: Proceedings of the IEEE/CVF International Conference on Computer Vision, pp. 9197–9206 (2019)

38. Wilson, A.G., Hu, Z., Salakhutdinov, R., Xing, E.P.: Deep kernel learning. In: Artificial Intelligence and Statistics, pp. 370–378. PMLR (2016)

39. Xia, Z., Pan, X., Song, S., Li, L.E., Huang, G.: Vision transformer with deformable attention. arXiv preprint arXiv:2201.00520 (2022)

40. Yang, B., Liu, C., Li, B., Jiao, J., Ye, Q.: Prototype mixture models for few-shot semantic segmentation. In: Vedaldi, A., Bischof, H., Brox, T., Frahm, J.-M. (eds.) ECCV 2020. LNCS, vol. 12353, pp. 763–778. Springer, Cham (2020). https://doi.org/10.1007/978-3-030-58598-3_45

41. Yang, Y., Soatto, S.: Fda: Fourier domain adaptation for semantic segmentation. In: Proceedings of the IEEE/CVF Conference on Computer Vision and Pattern Recognition, pp. 4085–4095 (2020)

42. Yang, Y., Meng, F., Li, H., Wu, Q., Xu, X., Chen, S.: A new local transformation module for few-shot segmentation. In: Ro, Y.M., et al. (eds.) MMM 2020. LNCS, vol. 11962, pp. 76–87. Springer, Cham (2020). https://doi.org/10.1007/978-3-030-37734-2_7

43. Zhang, C., Lin, G., Liu, F., Guo, J., Wu, Q., Yao, R.: Pyramid graph networks with connection attentions for region-based one-shot semantic segmentation. In: Proceedings of the IEEE/CVF International Conference on Computer Vision, pp. 9587–9595 (2019)

44. Zhang, G., Kang, G., Yang, Y., Wei, Y.: Few-shot segmentation via cycle-consistent transformer. In: Advances in Neural Information Processing Systems, vol. 34 (2021)

45. Zhang, X., Wei, Y., Yang, Y., Huang, T.S.: Sg-one: Similarity guidance network for one-shot semantic segmentation. IEEE Trans. Cybern. **50**(9), 3855–3865 (2020)

46. Zheng, S., et al.: Rethinking semantic segmentation from a sequence-to-sequence perspective with transformers. In: Proceedings of the IEEE/CVF Conference on Computer Vision and Pattern Recognition, pp. 6881–6890 (2021)

47. Zhu, F., Zhu, Y., Zhang, L., Wu, C., Fu, Y., Li, M.: A unified efficient pyramid transformer for semantic segmentation. In: Proceedings of the IEEE/CVF International Conference on Computer Vision, pp. 2667–2677 (2021)
48. Zhu, X., Su, W., Lu, L., Li, B., Wang, X., Dai, J.: Deformable detr: Deformable transformers for end-to-end object detection. arXiv preprint arXiv:2010.04159 (2020)

Dense Cross-Query-and-Support Attention Weighted Mask Aggregation for Few-Shot Segmentation

Xinyu Shi[1], Dong Wei[2], Yu Zhang[1(✉)], Donghuan Lu[2(✉)], Munan Ning[2], Jiashun Chen[1], Kai Ma[2], and Yefeng Zheng[2]

[1] School of Computer Science and Engineering, Key Lab of Computer Network and Information Integration (Ministry of Education), Southeast University, Nanjing, China
{shixinyu,zhang_yu,jiashunchen}@seu.edu.cn
[2] Tencent Jarvis Lab, Shenzhen, China
{donwei,caleblu,masonning,kylekma,yefengzheng}@tencent.com

Abstract. Research into Few-shot Semantic Segmentation (FSS) has attracted great attention, with the goal to segment target objects in a query image given only a few annotated support images of the target class. A key to this challenging task is to fully utilize the information in the support images by exploiting fine-grained correlations between the query and support images. However, most existing approaches either compressed the support information into a few class-wise prototypes, or used partial support information (e.g., only foreground) at the pixel level, causing non-negligible information loss. In this paper, we propose Dense pixel-wise Cross-query-and-support Attention weighted Mask Aggregation (DCAMA), where both foreground and background support information are fully exploited via multi-level pixel-wise correlations between paired query and support features. Implemented with the scaled dot-product attention in the Transformer architecture, DCAMA treats every query pixel as a token, computes its similarities with all support pixels, and predicts its segmentation label as an additive aggregation of all the support pixels' labels—weighted by the similarities. Based on the unique formulation of DCAMA, we further propose efficient and effective one-pass inference for n-shot segmentation, where pixels of all support images are collected for the mask aggregation at once. Experiments show that our DCAMA significantly advances the state of the art on standard FSS benchmarks of PASCAL-5^i, COCO-20^i, and FSS-1000, e.g., with 3.1%, 9.7%, and 3.6% absolute improvements in 1-shot mIoU over previous best records. Ablative studies also verify the design DCAMA.

Keywords: Few-shot segmentation · Dense cross-query-and-support attention · Attention weighted mask aggregation

---

X. Shi and D. Wei—Equal contributions and the work was done at Tencent Jarvis Lab.

Supplementary Information The online version contains supplementary material available at https://doi.org/10.1007/978-3-031-20044-1_9.

S. Avidan et al. (Eds.): ECCV 2022, LNCS 13680, pp. 151–168, 2022.
https://doi.org/10.1007/978-3-031-20044-1_9

1   Introduction

Recent years Deep Neural Networks (DNNs) have achieved remarkable progress in semantic segmentation [25,37], one of the fundamental tasks in computer vision. However, the success of DNNs relies heavily on large-scale datasets, where abundant training images are available for every target class to segment. In the extreme low-data regime, DNNs' performance may degrade quickly on previously unseen classes with only few examples due to poor generalization. Humans, in contrast, are capable of learning new tasks rapidly in the low-data scenario, utilizing prior knowledge accumulated from life experience [16]. Few-Shot Learning (FSL) [11,12] is a machine learning paradigm that aims at imitating such generalizing capability of human learners, where a model can quickly adapt for novel tasks given only a few examples. Specifically, a *support* set containing novel classes with limited samples is given for the model adaption, which is subsequently evaluated on a *query* set containing samples of the same classes. FSL has been actively explored in the field of computer vision, such as image classification [42], image retrieval [39], image captioning and visual question answering [8], and semantic segmentation [7,20,23,24,26,33,35,38,43–46,48–50]. In this paper, we tackle the problem of Few-shot Semantic Segmentation (FSS).

The key challenge of FSS is to fully exploit the information contained in the small support set. Most previous works followed the notion of prototyping [33], where the information contained in the support images was abstracted into class-wise prototypes via class-wise average pooling [44,48,50] or clustering [20,45], against which the query features were matched for segmentation label prediction. Being originally proposed for image classification tasks, however, the prototyping may result in great loss of the precious information contained in the already scarce samples when applied to FSS. Given the dense nature of segmentation tasks, [24,43,47] recently proposed to explore pixel-wise correlations between the query features and the foreground support features, avoiding the information compression in prototyping. However, these methods totally ignored the abundant information contained in the background regions of the support images. Zhang et al. [49] took into account background support features as well when computing the pixel-level correlations; yet they only considered sparse correlations with uniformly sampled support pixels, causing potential information loss of another kind. Thus, no previous work has fully investigated dense pixel-wise correlations of the query features with both of the foreground and background support features for FSS.

In this work, we propose Dense pixel-wise Cross-query-and-support Attention weighted Mask Aggregation (DCAMA) for FSS, which fully exploits all available foreground and background features in the support images. As shown in Fig. 1, we make a critical observation that the mask value of a query pixel can be predicted by an additive aggregation of the support mask values, proportionally weighted by its similarities to the corresponding support image pixels—including both foreground and background. This is intuitive: if the query pixel is semantically close to a foreground support pixel, the latter will vote for foreground as the mask value for the former, and vice versa—an embodiment of metric learn-

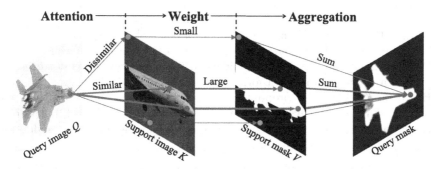

Fig. 1. Conceptual overview of our approach. The query mask is directly predicted by a pixel-wise additive aggregation of the support mask values, weighted by the dense cross-query-and-support attention.

ing [15]. In addition, we notice that the DCAMA computation pipeline for all pixels of a query image can be readily implemented with the dot-product attention mechanism in the Transformer architecture [40], where each pixel is treated as a token, flattened features of all query pixels compose the query matrix Q,[1] those of all support image pixels compose the key matrix K, and flattened label values of the support mask pixels compose the value matrix V. Then, the query mask can be readily and efficiently computed by softmax$(QK^T)V$. For practical implementation, we follow the common practices of multi-head attention [40], and multi-scale [18] and multi-layer [24] feature correlation; besides, the aggregated masks are mingled with skip-connected support and query features for refined query label prediction. As we will show, the proposed approach not only yields superior performance to that of the previous best performing method [24], but also demonstrates higher efficiency in training.

Besides, previous works adopting the pipeline of pixel-level correlation paid little attention to extension from 1-shot to few-shot segmentation: they either performed separate 1-shot inferences followed by ensemble [24], or used a subset of uniformly sampled support pixels for inference [49]. Both solutions led to a loss of pixel-level information, due to the independent inferences prior to ensemble and drop of potentially useful pixels, respectively. In contrast, we make full use of the support set, by using all pixels of all the support images and masks to compose the K and V matrices, respectively. Meanwhile, we use the same 1-shot trained model for testing in different few-shot settings. This is not only computationally economic but also reasonable, because what the model actually learns from training is a metric space for cross-query-and-support attention. As long as the metric space is well learned, extending from 1-shot to few-shot is simply aggregating the query mask from more support pixels.

In summary, we make the following contributions:

- We innovatively model the FSS problem as a novel paradigm of Dense pixel-wise Cross-query-and-support Attention weighted Mask Aggregation

---

[1] Caution: not to confuse the query in FSL with that in Transformer.

(DCAMA), which fully exploits the foreground and background support information. Being essentially an embodiment of metric learning, the paradigm is nonparametric in mask aggregation and expected to generalize well.

- We implement the new FSS paradigm with the well-developed dot-product attention mechanism in Transformer, for simplicity and efficiency.
- Based on DCAMA, we propose an approach to n-shot inference that not only fully utilizes the available support images at the pixel level, but also is computationally economic without the need for training n-specific models for different few-shot settings.

The comparative experimental results on three standard FSS benchmarks of PASCAL-5^i [31], COCO-20^i [26], and FSS-1000 [17] show that our DCAMA sets new state of the art on all the three benchmarks and in both 1- and 5-shot settings. In addition, we conduct thorough ablative experiments to validate the design of DCAMA.

2   Related Work

Semantic Segmentation. Semantic segmentation is a fundamental task in computer vision, whose goal is to classify each pixel of an image into one of the predefined object categories. In the past decade, impressive progress has been made with the advances in DNNs [25,37]. The cornerstone Fully Convolutional Network (FCN) [22] proposed to replace the fully connected output layers in classification networks with convolutional layers to efficiently output pixel-level dense prediction for semantic segmentation. Since then, prevailing segmentation DNN models [3,30,52] have evolved to be dominated by FCNs with the generic encoder-decoder architecture, where techniques like skip connections [30] and multi-scale processing [4,52] are commonly employed for better performance. Recently, inspired by the success of the Vision Transformer (ViT) [9], we have witnessed a surge of active attempts to apply the Transformer architecture to semantic segmentation [34,53]. Notably, the Swin Transformer, a general-purpose computer-vision backbone featuring a hierarchical architecture and shifted windows, achieved new state-of-the-art (SOTA) performance on the ADE20K semantic segmentation benchmark [21]. Although these methods demonstrated their competency given abundant training data and enlightened our work, they all suffered from the inability to generalize for the low-data regime.

Few-Shot Learning. Few-shot learning [11] is a paradigm that aims to improve the generalization ability of machine learning models in the low-data regime. Motivated by the pioneering work ProtoNet [33], most previous works [7,20,44,45,48,50] on FSS followed a metric learning [7] pipeline, where the information contained in support images was compressed into abstract prototypes, and the query images were classified based on certain distances from the prototypes in the metric space. Dong and Xing [7] extended the notion of prototype for FSS by computing class-wise prototypes via global average pooling on features of masked support images. Instead of masking the input images,

PANet [44] performed average pooling on masked support features for proto-typing and introduced prototype alignment as regularization. CANet [48] also relied on masked feature pooling but followed the Relation Network [36] to learn a deep metric using DNNs. PFENet [38] further proposed a training-free prior mask as well as a multi-scale feature enrichment module. Realizing the limited representation power of a single prototype, [5,20,45,46] all proposed to repre-sent a class with multiple prototypes. These prototype-based methods jointly advanced the research into FSS; however, compressing all available information in a support image into merely one or few concentrated prototypes unavoidably resulted in great information loss.

More recently, researchers started to exploit pixel-level information for FSS, to better utilize the support information and align with the dense nature of the task. PGNet [47] and DAN [43] modeled pixel-to-pixel dense connections between the query and support images with the graph attention [41], whereas HSNet [24] constructed 4D correlation tensors to represent dense correspondences between the query and support images. Notably, HSNet proposed center-pivot 4D con-volutions for efficient high-dimensional convolutions, and achieved SOTA per-formance on three public FSS benchmarks by large margins. However, these methods all masked out background regions in the support images, ignoring the rich information thereby. In contrast, our DCAMA makes equal use of both fore-ground and background information. In addition, implementing straightforward metric learning with the multi-head attention [40], our DCAMA is easier to train than HSNet, converging to higher performance in much fewer epochs and much less time. Lastly, instead of the ensemble of separate 1-shot inferences [24] or training n-specific models [48] for n-shot inference, DCAMA constructs the key and value matrices with pixels of all the support images and masks, and infers only once reusing the 1-shot trained model.

Vision Transformers for FSS. Inspired by the recent success of the Trans-former architecture in computer vision [9,21], researchers also started to explore their applications to FSS lately. Sun et al. [35] proposed to employ standard multi-head self-attention Transformer blocks for global enhancement. Lu et al. [23] designed the Classifier Weight Transformer (CWT) to dynamically adapt the classifier's weights for each query image. However, both of them still followed the prototyping pipeline thus did not fully utilize fine-grained support informa-tion. The Cycle-Consistent Transformer (CyCTR) [49] may be the most related work to ours in terms of: (i) pixel-level cross-query-and-support similarity com-putation with the dot-product attention mechanism, and (ii) use of both fore-ground and background support information. The main difference lies in that CyCTR used the similarity to guide the reconstruction of query features from support features, which were then classified into query labels by a conventional FCN. In contrast, our DCAMA can directly predict the query labels by aggre-gating the support labels weighted by this similarity, which is metric learning and expected to generalize well for its nonparametric form. Another distinction is that CyCTR subsampled the support pixels thus was subject to potential information loss dependent on the sampling rate, whereas our DCAMA makes full use of all available support pixels.

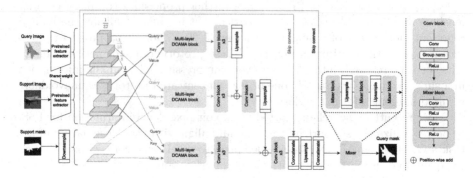

Fig. 2. Pipeline of the proposed framework, shown in 1-shot setting. DCAMA: Dense Cross-query-and-support Attention weighted Mask Aggregation.

3   Methodology

In this section, we first introduce the problem setup of Few-shot Semantic Segmentation (FSS). Then, we describe our Dense Cross-query-and-support Attention weighted Mask Aggregation (DCAMA) framework in 1-shot setting. Lastly, we extend the framework for n-shot inference.

3.1   Problem Setup

In a formal definition, a 1-way n-shot FSS task $\mathcal{T}$ comprises a support set $\mathcal{S} = \{(I^s, M^s)\}$, where I^s and M^s are a support image and its ground truth mask, respectively, and $|S| = n$; and similarly a query set $\mathcal{Q} = \{(I^q, M^q)\}$, where $\mathcal{S}$ and $\mathcal{Q}$ are sampled from the same class. The goal is to learn a model to predict M^q for each I^q given the support set $\mathcal{S}$, subject to n being small for few shots. For method development, suppose we have two image sets $\mathcal{D}_{\text{train}}$ and $\mathcal{D}_{\text{test}}$ for model training and evaluation, respectively, where $\mathcal{D}_{\text{train}}$ and $\mathcal{D}_{\text{test}}$ do not overlap in classes. We adopt the widely used meta-learning paradigm called episodic training [42], where each episode is designed to mimic the target task by subsampling the classes and images in the training set. Specifically, we repeatedly sample new episodic tasks $\mathcal{T}$ from $\mathcal{D}_{\text{train}}$ for model training. The use of episodes is expected to make the training process more faithful to the test environment and thereby improve generalization [29]. For testing, the trained model is also evaluated with episodic tasks but sampled from $\mathcal{D}_{\text{test}}$.

3.2   DCAMA Framework for 1-Shot Learning

Overview. The overview of our DCAMA framework is shown in Fig. 2. For simplicity, we first describe our framework for 1-shot learning. The input to the framework is the query image, and the support image and mask. First, both the query and support images are processed by a pretrained feature extractor, yielding multi-scale query and support features. Meanwhile, the support mask is downsampled to multiple scales matching the image features. Second, the query features, support features, and support mask at each scale are input to the multi-layer DCAMA block of the same scale as Q, K, and V for multi-head attention [40] and aggregation of the query mask. The query masks aggregated at multiple scales are processed and combined with convolutions, upsampling (if needed), and element-wise additions. Third, the output of the previous stage (multi-scale DCAMA) is concatenated with multi-scale image features via skip connections, and subsequently mingled by a mixer to produce the final query mask. In the following, we describe each of these three stages in turn, with an emphasis on the second—which is our main contribution.

Feature Extraction and Mask Preparation. First of all, both the query and support images are input to a pretrained feature extractor to obtain the collections of their multi-scale multi-layer feature maps $\{F_{i,l}^q\}$ and $\{F_{i,l}^s\}$, where i is the scale of the feature maps with respect to the input images and $i \in \{\frac{1}{4}, \frac{1}{8}, \frac{1}{16}, \frac{1}{32}\}$ for the feature extractor we use, and $l \in \{1, \ldots, L_i\}$ is the index of all layers of a specific scale i. Unlike most previous works that only used the last-layer feature map of each scale, i.e., F_{i,L_i}, we follow Min et al. [24] to make full use of all intermediate-layer features, too. Meanwhile, support masks of different scales $\{M_i^s\}$ are generated via bilinear interpolation from the original support mask. The query features, support features, and support masks of scales $i \in \{\frac{1}{8}, \frac{1}{16}, \frac{1}{32}\}$ are input to the multi-layer DCAMA block,[2] as described next.

Multi-scale Multi-Layer Cross Attention Weighted Mask Aggregation. The scaled dot-product attention is the core of the Transformer [40] architecture, and is formulated as:

$$\text{Attn}(Q, K, V) = \text{softmax}\left(QK^T / \sqrt{d}\right)V, \tag{1}$$

where Q, K, V are sets of query, key, and value vectors packed into matrices, and d is the dimension of the query and key vectors. In this work, we adopt Eq. (1) to compute dense pixel-wise attention across the query and support features, and subsequently weigh the query mask aggregation process from the support mask with the attention values. Without loss of generality, we describe the mechanism with a pair of generic query and support feature maps $F^q, F^s \in \mathbb{R}^{h \times w \times c}$, where h, w, and c are the height, width, and channel number, respectively, and a generic support mask $M^s \in \mathbb{R}^{h \times w \times 1}$ of the same size. As shown in Fig. 3, we first flatten the two-dimensional (2D) inputs to treat each pixel as a token, and then generate the Q and K matrices from the flattened F^q and F^s after adding positional

---

[2] The $\frac{1}{4}$ scale features are not cross-attended due to hardware constraint.

encoding and linear projections. We follow the original Transformer [40] to use sine and cosine functions of different frequencies for positional encoding, and employ the multi-head attention. As to the support mask, it only needs flattening to construct V. After that, the standard scaled dot-product attention in Eq. (1) can be readily computed for each head. Lastly, we average the outputs of the multiple heads for each token, and reshape the tensor to 2D to obtain $\hat{M}^q \in \mathbb{R}^{h \times w \times 1}$, which is the aggregated query mask.

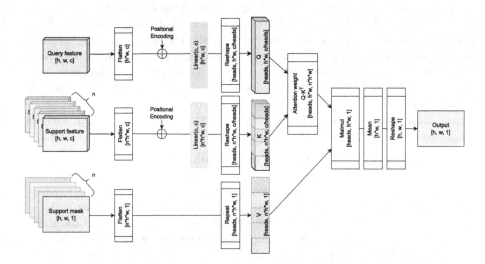

Fig. 3. Dense Cross-query-and-support Attention weighted Mask Aggregation (DCAMA) for generic n-shot settings ($n \geq 1$).

Remarks. It is worth explaining the physical meaning of the DCAMA process. For a specific query pixel, QK^T measures its similarities to all the support pixels, and the subsequent multiplication with V aggregates its mask value from the support mask, weighted by the similarities. Intuitively, if it is more similar (closer) to the foreground than background pixels, the weighted aggregation process will vote for foreground for the pixel, and vice versa. In this way, our DCAMA utilizes all support pixels—both foreground and background—for effective metric learning.

In practical implementation, we conduct DCAMA separately on query-support feature pairs $(F_{i,l}^s, F_{i,l}^q)$ for all the intermediate and last layers of a specific scale i, and concatenate the set of independently aggregated query masks to get $\hat{M}_i^q = \text{concat}\{\hat{M}_{i,l}^q | l = 1, \ldots, L_i\}$. The DCAMA operations on all-layer features of a specific scale followed by the concatenation compose a multi-layer DCAMA block (see Fig. 2), and we have three such blocks for scales $i \in \{\frac{1}{8}, \frac{1}{16}, \frac{1}{32}\}$, respectively. Then, $\hat{M}_i^q$ is processed by three Conv blocks which gradually increase its channel number from L_i to 128, upsampled by bilinear interpolation, combined with the counterpart of the one time larger scale by element-wise addition, and again processed by another three Conv blocks of a constant channel number. The first three Conv blocks prepare $\hat{M}_i^q$ for effective inter-scale integration with that of the larger scale by the second three Conv blocks. This process repeats from $i = \frac{1}{32}$ up to $\frac{1}{8}$, yielding a collection of intermediate query masks to be fused with skip-connected image features for final prediction.

Mask-Feature Mixer. Motivated by the success of the skip connection design in generic semantic segmentation [30,52], we also propose to skip connect the image features to the output (upsampled when needed) of the previous stage via concatenation (Fig. 2). Specifically, we skip connect the last-layer features at the $\frac{1}{4}$ and $\frac{1}{8}$ scales based on our empirical experiments (included in the supplementary material). Then, the concatenated intermediate query masks and image features are fused by three mask-feature mixer blocks, each containing two series of convolution and ReLU operations. The mixer blocks gradually decrease the number of output channels to 2 (for foreground and background, respectively) for 1-way segmentation, with two interleaved upsampling operations to restore the output size to that of the input images.

3.3   Extension to n-Shot Inference

So far we have introduced our DCAMA framework for 1-shot segmentation, next we extend it for n-shot setting, too. Although it is possible to develop and train a specific model for each different value of n (e.g., [48]), it is computationally prohibitive to do so. In contrast, many previous works extended the 1-shot trained model for n-shot inference without retraining [24,38,46]. The most common method is to perform n 1-shot inferences separately with each of the support images, followed by certain ensembles of the individual inferences [24,46]. However, this method unavoidably lost the pixel-level subtle cues across support images, as it treated each support image independently for inference. In this work, we also reuse the 1-shot trained model for n-shot inference for computational efficiency, but meanwhile utilize all the pixel-wise information of all support images simultaneously during inference.

Thanks to the problem formulation of DCAMA, the extension is straightforward. First, we obtain the multi-scale image features and masks for all the support images. Next, we simply treat the additional pixels in the extra support features and masks as more tokens in K and V, with a proper reshaping of the tensors (Fig. 3). Then, the entire DCAMA process (cross attention, mask aggregation, etc.) stays the same as the 1-shot setting. This is feasible as the core of DCAMA is the scaled dot-product attention in Eq. (1), which is parameter-free. Thus, the DCAMA process is actually n-agnostic and can be applied for inference with arbitrary n.[3] Intuitively, the label of a query pixel is jointly determined by all available support pixels at once, irrespective of the exact number of support images. This one-pass inference is distinct from the ensemble of individual inferences, where image-level predictions are first obtained with each support

---

[3] Despite that, it is still computationally prohibitive to *train* n-specific models for $n > 1$, given both time and GPU memory considerations.

image independently and then merged to produce a set-level final prediction. It is also different from some prototype-based n-shot approaches [7,45], where features of all support images were processed simultaneously but compressed into one or few prototypes, losing pixel-level granularity. Lastly, a minor adaption is to max-pool the skip connected support features across the support images, such that the entire DCAMA framework shown in Fig. 2 becomes applicable to generic n-shot settings, where $n \geq 1$.

4   Experiments and Results

Datasets and Evaluation. We evaluate the proposed method on three standard FSS benchmarks: PASCAL-5^i [31], COCO-20^i [26] and FSS-1000 [17]. PASCAL-5^i is created from PASCAL VOC 2012 [10] and SDS [13] datasets. It contains 20 classes, which are evenly split into four folds, i.e., five classes per fold. COCO-20^i is a larger and more challenging benchmark created from the COCO [19] dataset. It includes 80 classes, again evenly divided into four folds. For both PASCAL-5^i and COCO-20^i, the evaluation is done by cross-validation, where each fold is selected in turn as $\mathcal{D}_{\text{test}}$, with the other three folds as $\mathcal{D}_{\text{train}}$; 1,000 testing episodes are randomly sampled from $\mathcal{D}_{\text{test}}$ for evaluation [24]. FSS-1000 [17] comprises 1,000 classes that are divided into training, validation, and testing splits of 520, 240, and 240 classes, respectively, with 2,400 testing episodes sampled from the testing split for evaluation [24]. For the metrics, we adopt mean intersection over union (mIoU) and foreground-background IoU (FB-IoU) [24]. For PASCAL-5^i and COCO-20^i, the mIoUs on individual folds, and the averaged mIoU and FB-IoU across the folds are reported; for FSS-1000, the mIoU and FB-IoU on the testing split are reported. Note that we attempt to follow the common practices adopted by previous works [24,26,38,49] for fair comparisons.

Implementation Details. All experiments are conducted with the PyTorch [28] framework (1.5.0). For the backbone feature extractor, we employ ResNet-50 and ResNet-101 [14] pretrained on ImageNet [6] for their prevalent adoption in previous works.[4] In addition, we also experiment with the base Swin Transformer model (Swin-B) pretrained on ImageNet-1K [21], to evaluate the generalization of our method on the non-convolutional backbone. We use three multi-layer DCAMA blocks for $\frac{1}{8}$, $\frac{1}{16}$, and $\frac{1}{32}$ scales, respectively, resulting in three pyramidal levels of cross attention weighted mask aggregation. Unless otherwise specified, the last-layer features of the $\frac{1}{4}$ and $\frac{1}{8}$ scales are skip connected. The input size of both the support and query images is 384×384 pixels. The mean binary cross-entropy loss is employed: $\mathcal{L}_{\text{BCE}} = -\frac{1}{N} \sum [y \log p + (1-y) \log(1-p)]$, where N is the total number of pixels, $y \in \{0, 1\}$ is the pixel label (0 for background and 1 for foreground), and p is the predicted probability. We impose $\mathcal{L}_{\text{BCE}}$ only on the final output to train our model, with the backbone parameters frozen. The SGD

---

[4] While several works (e.g., [38]) reported superior performance of ResNet-50 and ResNet-101, VGG-16 [32] mostly produced inferior performance to the ResNet family in previous works on FSS. Therefore, we do not include VGG-16 in our experiment.

optimizer is employed, with the learning rate, momentum, and weight decay set to 10^{-3}, 0.9, and 10^{-4}, respectively. The batch size is set to 48, 48, and 40 for PASCAL-5^i, COCO-20^i, and FSS-1000, respectively. We follow HSNet [24] to train our models without data augmentation and until convergence, for a fair comparison with the previous best performing method. The training is done on four NVIDIA Tesla V100 GPUs and the inference is on an NVIDIA Tesla T4 GPU. Our code is available at https://github.com/pawn-sxy/DCAMA.git.

4.1 Comparison with State of the Art

In Table 1 and Table 2, we compare the performance of our proposed DCAMA framework with that of SOTA approaches to FSS published since 2020 on PASCAL-5^i, COCO-20^i, and FSS-1000, respectively. Unless otherwise specified, reported numbers of other methods are from the original papers; when results with different backbones are available, we report the higher ones only to save space.

Above all, our method is very competitive: it achieves the best performance in terms of both mIoU and FB-IoU for almost all combinations of the backbone networks (ResNet-50, ResNet-101, and Swin-B) and few-shot settings (1- and 5-shot) on all the three benchmark datasets. The only exception is on PASCAL-5^i with ResNet-101, where our DCAMA and HSNet [24] share the top two spots for mIoU and FB-IoU in 1- and 5-shot settings with comparable performance. When employing Swin-B as the backbone feature extractor, DCAMA significantly advances the system-level SOTA on all the three benchmarks upon the previous three-benchmark SOTA HSNet (ResNet-101), e.g., by 3.1% (1-shot) and 4.5% (5-shot) on PASCAL-5^i, 9.7% and 8.8% on COCO-20^i, and 3.6% and 1.9% FSS-1000, respectively, in terms of mIoU. Second, although the performance of HSNet improves with the Swin-B backbone, yet still suffers considerable disadvantages of 2–3.6% (1-shot) and 1.5–3.3% (5-shot) in mIoU from that of DCAMA. In addition, our DCAMA is also more stable than HSNet across the folds, observing the lower standard deviations. These results demonstrate DCAMA's applicability to both convolution- and attention-based backbones. Third, the performance of the methods based on pixel-wise correlations (except for DAN [43]) is generally better than that of those based on prototyping, confirming the intuition of making use of the fine-grained pixel-level information for the task of FSS. Last but not least, it is worth noting that our DCAMA demonstrates consistent advantages over the other three methods (CWT [23], TRFS [35], and CyCTR [49]) that also implemented the dot-product attention of the Transformer [40]. Fig. 4(a) visualizes some segmentation results by DCAMA in challenging cases. More results and visualizations, including region-wise over- and under-segmentation measures [51], are given in the supplementary material.

Remarks. Although the three best performing methods in Table 1 (HSNet [24], CyCTR [49], and ours) all rely on pixel-level cross-query-and-support similarities, their underlying concepts of query label inference are quite different and

Table 1. Performance on PASCAL-5^i (top) and COCO-20^i (bottom). HSNet†: our reimplementation based on official codes; *: methods implementing the dot-product attention [40]. Bold and underlined numbers highlight the best and second best performance (if necessary) for each backbone, respectively.

PASCAL-5^i [31]

Backbone	Methods	Type	1-shot						5-shot					
			Fold-0	Fold-1	Fold-2	Fold-3	mIoU	FB-IoU	Fold-0	Fold-1	Fold-2	Fold-3	mIoU	FB-IoU
ResNet-50	PPNet [20]	Prototype	52.7	62.8	57.4	47.7	55.2 (5.6)	–	60.3	70.0	69.4	60.7	65.1 (4.6)	–
	PMM [45]		52.0	67.5	51.5	49.8	55.2 (7.2)	–	55.0	68.2	52.9	51.1	56.8 (6.7)	–
	RPMM [45]		55.2	66.9	52.6	50.7	56.3 (6.3)	–	56.3	67.3	54.5	51.0	57.3 (6.1)	–
	RePRI [2]		59.8	68.3	62.1	48.5	59.7 (7.2)	–	64.6	71.4	71.1	59.3	66.6 (5.0)	–
	PEFNet [38]		61.7	69.5	55.4	56.3	60.8 (5.6)	73.3	63.1	70.7	55.8	57.9	61.9 (5.7)	73.9
	SCL [46]		63.0	70.0	56.5	57.7	61.8 (5.3)	71.9	64.5	70.9	57.3	58.7	62.9 (5.4)	72.8
	TRFS* [35]		62.9	70.7	56.5	57.5	61.9 (5.6)	–	65.0	71.2	55.5	60.9	63.2 (5.7)	–
	DCAMA (Ours)*	Pixel-wise	**67.5**	**72.3**	59.6	**59.0**	**64.6 (5.6)**	**75.7**	**70.5**	**73.9**	63.7	**65.8**	**68.5 (4.0)**	**79.5**
ResNet-101	CWT* [23]	Prototype	56.9	65.2	61.2	48.8	58.0 (6.1)	–	62.6	70.2	68.8	57.2	64.7 (5.2)	–
	DoG-LSTM [1]		57.0	67.2	56.1	54.3	58.7 (5.0)	–	57.3	68.5	61.5	56.3	60.9 (4.8)	–
	DAN [43]	Pixel-wise	54.7	68.6	57.8	51.6	58.2 (6.4)	71.9	57.9	69.0	60.1	54.9	60.5 (5.3)	72.3
	CyCTR* [49]		**69.3**	**72.7**	56.5	58.6	64.3 (6.9)	–	**73.5**	74.0	58.6	60.2	66.6 (7.2)	–
	HSNet [24]		67.3	72.3	62.0	63.1	**66.2 (4.1)**	77.6	71.8	**74.4**	67.0	68.3	**70.4 (2.9)**	80.6
	DCAMA (Ours)*		65.4	71.4	**63.2**	58.3	64.6 (4.7)	77.6	70.7	73.7	66.8	61.9	68.3 (4.4)	**80.8**
Swin-B	HSNet† [24]	Pixel-wise	67.9	**74.0**	60.3	67.0	67.3 (4.9)	77.9	72.2	**77.5**	64.0	72.6	71.6 (4.8)	81.2
	DCAMA (Ours)*		**72.2**	73.8	**64.3**	**67.1**	**69.3 (3.8)**	**78.5**	**75.7**	77.1	**72.0**	**74.8**	**74.9 (1.8)**	**82.9**

COCO-20^i [26]

Backbone	Methods	Type	1-shot						5-shot					
			Fold-0	Fold-1	Fold-2	Fold-3	mIoU	FB-IoU	Fold-0	Fold-1	Fold-2	Fold-3	mIoU	FB-IoU
ResNet-50	PPNet [20]	Prototype	36.5	26.5	26.0	19.7	27.2 (6.0)	–	**48.9**	31.4	36.0	30.6	36.7 (7.3)	–
	PMM [45]		29.3	34.8	27.1	27.3	29.6 (3.1)	–	33.0	40.6	30.3	33.3	34.3 (3.8)	–
	RPMM [45]		29.5	36.8	28.9	27.0	30.6 (3.7)	–	33.8	42.0	33.0	33.3	35.5 (3.7)	–
	TRFS* [35]		31.8	34.9	36.4	31.4	33.6 (2.1)	–	35.4	41.7	42.3	36.1	38.9 (3.1)	–
	RePRI [2]		31.2	38.1	33.3	33.0	34.0 (2.6)	-	38.5	46.2	40.0	43.6	42.1 (3.0)	–
	CyCTR* [49]	Pixel-wise	38.9	43.0	39.6	39.8	40.3 (1.6)	-	41.1	48.9	45.2	47.0	45.6 (2.9)	–
	DCAMA (Ours)*		**41.9**	**45.1**	**44.4**	**41.7**	**43.3 (1.5)**	69.5	45.9	**50.5**	**50.7**	46.0	**48.3 (2.3)**	71.7
ResNet-101	CWT* [23]	Prototype	30.3	36.6	30.5	32.2	32.4 (2.5)	–	38.5	46.7	39.4	43.2	42.0 (3.3)	–
	SCL [46]		36.4	38.6	37.5	35.4	37.0 (1.2)	–	38.9	40.5	41.5	38.7	39.9 (1.2)	–
	PEFNet [38]		36.8	41.8	38.7	36.7	38.5 (2.1)	63.0	40.4	46.8	43.2	40.5	42.7 (2.6)	65.8
	DAN [43]	Pixel-wise	–	–	–	–	24.4 (–)	62.3	–	–	–	–	29.6 (–)	63.9
	HSNet [24]		37.2	44.1	42.4	41.3	41.2 (2.5)	69.1	45.9	53.0	51.8	47.1	49.5 (3.0)	72.4
	DCAMA (Ours)*		**41.5**	**46.2**	**45.2**	41.3	**43.5 (2.2)**	69.9	**48.0**	**58.0**	**54.3**	47.1	**51.9 (4.5)**	73.3
Swin-B	HSNet† [24]	Pixel-wise	43.6	49.9	49.4	46.4	47.3 (2.5)	72.5	50.1	58.6	56.7	55.1	55.1 (3.2)	76.1
	DCAMA (Ours)*		**49.5**	**52.7**	**52.8**	**48.7**	**50.9 (1.8)**	73.2	**55.4**	**60.3**	**59.9**	**57.5**	**58.3 (2.0)**	76.9

Fig. 4. (a) Qualitative results on PASCAL-5^i in 1-shot setting, in the presence of intra-class variations, size differences, complex background, and occlusions. (b) Multi-scale intermediate query masks aggregated by the multi-layer DCAMA blocks for a 1-shot task sampled from PASCAL-5^i.

Table 2. Performance on FSS-1000 [17]. HSNet[†]: our reimplementation based on the official codes. Bold and underlined numbers highlight the best and second best performance (if necessary) for each backbone, respectively.

Backbone	Methods	Type	1-shot		5-shot	
			mIoU	FB-IoU	mIoU	FB-IoU
ResNet-50	DCAMA (Ours)	Pixel-wise	88.2	92.5	88.8	92.9
ResNet-101	DoG-LSTM [1]	Prototype	80.8	–	83.4	–
	DAN [43]	Pixel-wise	85.2	–	88.1	–
	HSNet [24]		<u>86.5</u>	–	<u>88.5</u>	–
	DCAMA (Ours)		**88.3**	92.4	**89.1**	93.1
Swin-B	HSNet[†] [24]	Pixel-wise	86.7	91.8	88.9	93.2
	DCAMA (Ours)		**90.1**	**93.8**	**90.4**	**94.1**

worth clarification. HSNet predicted the query label based on the similarities of a query pixel to all the foreground support pixels (while ignoring the background); intuitively, the more similar a query pixel is to the foreground support pixels, the more likely it is foreground. CyCTR first reconstructed the query features from the support features based on the similarities to subsets of both foreground and background support pixels, then trained a classifier on the reconstructed query features. Our DCAMA directly aggregates the query label from the support mask weighted by the query pixel's similarities to all support pixels, representing a totally different concept.

Training and Inference Efficiency. We compare the training efficiency of our method to that of HSNet [24]. As both methods incur the computational complexity of $O(N^2)$ for pixel-wise correlation, they also take comparable amounts of time for training per epoch (e.g., around four minutes with our hardware and training setting on COCO-20^i). However, as shown in Fig. 5, our method takes much fewer training epochs to converge. Therefore, our DCAMA is also more efficient in terms of training time, in addition to achieving higher performance than the previous SOTA method HSNet. As to inference, DCAMA runs at comparable speed with HSNet using the same backbone, e.g., about 8 and 20 frames per second (FPS) with Swin-B and ResNet-101, respectively, for 1-shot segmentation on an entry-level NVIDIA Tesla T4 GPU. In contrast, CyCTR [49] runs at about three FPS with the ResNet-50 backbone.

4.2 Ablation Study

We conduct thorough ablation studies on the PASCAL-5^i [31] dataset, to gain deeper understanding of our proposed DCAMA framework and verify its design. Swin-B [21] is used as the backbone feature extractor for the ablation studies.

Intermediate Query Masks Aggregated by the Multi-layer DCAMA Blocks. We first validate the physical meaning of the proposed mask aggrega-

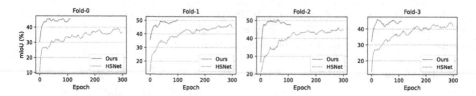

Fig. 5. The mIoU curves on the validation set during training on COCO-20^i. The curves of HSNet [24] are produced with the official codes released by the authors.

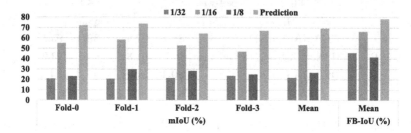

Fig. 6. Performance of the multi-scale intermediate query masks aggregated by the multi-layer DCAMA blocks and the final prediction (1-shot on PASCAL-5^i).

tion paradigm, by verifying that the outputs of the multi-layer DCAMA blocks (Fig. 3) are indeed meaningful segmentations. For this purpose, we sum $\hat{M}_i^q$ along the layer dimension for scales $i \in \{\frac{1}{8}, \frac{1}{16}, \frac{1}{32}\}$, binarize the sum with the Otsu's method [27], and resize the resulting masks to evaluate against the ground truth for 1-shot segmentation. As shown in Fig. 6, the mIoU and FB-IoU of the $\frac{1}{16}$ scale masks are fairly high, close to those of some comparison methods in Table 1. Meanwhile, those of the $\frac{1}{8}$ and $\frac{1}{32}$ scale masks are much lower, which may be because: the $\frac{1}{8}$ scale features have not learned enough high-level semantics, and the $\frac{1}{32}$ scale features are too abstract/class-specific and coarse. The final prediction effectively integrates the multi-scale intermediate masks and achieves optimal performance. For intuitive perception, we also visualize the intermediate masks for a specific 1-shot task in Fig. 4(b), where the overlays clearly demonstrate that they are valid segmentations, albeit less accurate than the final prediction. These results verify that the proposed DCAMA indeed functions as designed, and that the multi-scale strategy is effective.

Effect of Background Support Information. A notable difference between our DCAMA and the previous SOTA HSNet [24] is that HSNet ignored the background support features whereas DCAMA uses them for full utilization of the available information. To evaluate the actual effect of the difference, we conduct an ablative experiment where the background pixels are zeroed out from the support feature maps before fed to the multi-layer DCAMA blocks for training and inference—similar to HSNet, for 1-shot segmentation. The results in Table 3 show that ignoring background support information leads to decreases

Table 3. Ablation study on the effect of the background support information (1-shot on PASCAL-5^i).

Background	Fold-0	Fold-1	Fold-2	Fold-3	mIoU	FB-IoU
✓	72.2	**73.8**	**64.3**	67.1	**69.3**	**78.5**
✗	**73.3**	72.7	53.6	**69.8**	67.3	76.2

Table 4. Ablation study on strategies for n-shot segmentation (5-shot on PASCAL-5^i).

Strategy	Voting	Averaging	HSNet [24]	SCL [46]	Ours
mIoU	74.0	74.0	73.9	74.1	**74.9**
FB-IoU	82.0	82.0	81.8	82.0	**82.9**

of 2.0% and 2.3% in mIoU and FB-IoU, respectively, suggesting the value of fully utilizing all the information available in the support set.

Strategy for n-Shot Inference. To verify the effectiveness of our proposed one-pass n-shot inference, we compare its 5-shot performance to the following ensembles of five 1-shot predictions: naive voting and averaging, the normalized voting in HSNet [24], and cross-guided averaging in SCL [46]. As shown in Table 4, our strategy outperforms all the ensembles, suggesting the advantage in collectively utilizing all available support features together for FSS.

5 Conclusion

In this work, we proposed a new paradigm based on metric learning for Few-shot Semantic Segmentation (FSS): Dense Cross-query-and-support Attention weighted Mask Aggregation (DCAMA). In addition, we implemented the DCAMA framework with the scaled dot-production attention in the Transformer structure, for simplicity and efficiency. The DCAMA framework was distinct from previous works in three aspects: (i) it directly predicted the mask value of a query pixel as an additive aggregation of the mask values of all support pixels, weighted by pixel-wise similarities between the query and support features; (ii) it fully utilized all support pixels, including both foreground and background; and (iii) it proposed efficient and effective one-pass n-shot inference which considered pixels from all support images simultaneously. Experiments showed that our DCAMA framework set the new state of the art on all three commonly used FSS benchmarks. For future research, it is tempting to adapt the paradigm for other few-shot learning tasks that involve dense predictions such as detection.

Acknowledgement. This work was supported by the National Key R&D Program of China (2018AAA0100104, 2018AAA0100100), Natural Science Foundation of Jiangsu Province (BK20211164).

References

1. Azad, R., Fayjie, A.R., Kauffmann, C., Ben Ayed, I., Pedersoli, M., Dolz, J.: On the texture bias for few-shot CNN segmentation. In: Proceedings of the IEEE/CVF Winter Conference on Applications of Computer Vision, pp. 2674–2683 (2021)

2. Boudiaf, M., Kervadec, H., Masud, Z.I., Piantanida, P., Ben Ayed, I., Dolz, J.: Few-shot segmentation without meta-learning: A good transductive inference is all you need? In: Proceedings of the IEEE/CVF Conference on Computer Vision and Pattern Recognition, pp. 13979–13988 (2021)

3. Chen, L.C., Papandreou, G., Kokkinos, I., Murphy, K., Yuille, A.L.: DeepLab: semantic image segmentation with deep convolutional nets, atrous convolution, and fully connected CRFs. IEEE Trans. Pattern Anal. Mach. Intell. **40**(4), 834–848 (2017)

4. Chen, L.-C., Zhu, Y., Papandreou, G., Schroff, F., Adam, H.: Encoder-decoder with atrous separable convolution for semantic image segmentation. In: Ferrari, V., Hebert, M., Sminchisescu, C., Weiss, Y. (eds.) ECCV 2018. LNCS, vol. 11211, pp. 833–851. Springer, Cham (2018). https://doi.org/10.1007/978-3-030-01234-2_49

5. Cui, H., Wei, D., Ma, K., Gu, S., Zheng, Y.: A unified framework for generalized low-shot medical image segmentation with scarce data. IEEE Trans. Med. Imaging **40**(10), 2656–2671 (2021)

6. Deng, J., Dong, W., Socher, R., Li, L.J., Li, K., Fei-Fei, L.: ImageNet: A large-scale hierarchical image database. In: 2009 IEEE Conference on Computer Vision and Pattern Recognition, pp. 248–255. IEEE (2009)

7. Dong, N., Xing, E.P.: Few-shot semantic segmentation with prototype learning. In: British Machine Vision Conference, vol. 3 (2018)

8. Dong, X., Zhu, L., Zhang, D., Yang, Y., Wu, F.: Fast parameter adaptation for few-shot image captioning and visual question answering. In: Proceedings of the ACM International Conference on Multimedia, pp. 54–62 (2018)

9. Dosovitskiy, A., et al.: An image is worth 16×16 words: Transformers for image recognition at scale. In: International Conference on Learning Representations (2021)

10. Everingham, M., Van Gool, L., Williams, C.K., Winn, J., Zisserman, A.: The PASCAL visual object classes (VOC) challenge. Int. J. Comput. Vision **88**(2), 303–338 (2010)

11. Fei-Fei, L., Fergus, R., Perona, P.: One-shot learning of object categories. IEEE Trans. Pattern Anal. Mach. Intell. **28**(4), 594–611 (2006)

12. Fink, M.: Object classification from a single example utilizing class relevance metrics. Adv. Neural. Inf. Process. Syst. **17**, 449–456 (2005)

13. Hariharan, B., Arbeláez, P., Girshick, R., Malik, J.: Simultaneous detection and segmentation. In: Fleet, D., Pajdla, T., Schiele, B., Tuytelaars, T. (eds.) ECCV 2014. LNCS, vol. 8695, pp. 297–312. Springer, Cham (2014). https://doi.org/10.1007/978-3-319-10584-0_20

14. He, K., Zhang, X., Ren, S., Sun, J.: Deep residual learning for image recognition. In: Proceedings of the IEEE Conference on Computer Vision and Pattern Recognition, pp. 770–778 (2016)

15. Kulis, B., et al.: Metric learning: A survey. Found. Trends® Mach, Learn. **5**(4), 287–364 (2013)

16. Lake, B., Salakhutdinov, R., Gross, J., Tenenbaum, J.: One shot learning of simple visual concepts. In: Proceedings of the Annual Meeting of the Cognitive Science Society, vol. 33, pp. 2568–2573 (2011)

17. Li, X., Wei, T., Chen, Y.P., Tai, Y.W., Tang, C.K.: FSS-1000: A 1000-class dataset for few-shot segmentation. In: Proceedings of the IEEE/CVF Conference on Computer Vision and Pattern Recognition, pp. 2869–2878 (2020)

18. Lin, T.Y., Dollár, P., Girshick, R., He, K., Hariharan, B., Belongie, S.: Feature pyramid networks for object detection. In: Proceedings of the IEEE/CVF Conference on Computer Vision and Pattern Recognition, pp. 2117–2125 (2017)

19. Lin, T.-Y., et al.: Microsoft COCO: common objects in context. In: Fleet, D., Pajdla, T., Schiele, B., Tuytelaars, T. (eds.) ECCV 2014. LNCS, vol. 8693, pp. 740–755. Springer, Cham (2014). https://doi.org/10.1007/978-3-319-10602-1_48
20. Liu, Y., Zhang, X., Zhang, S., He, X.: Part-aware prototype network for few-shot semantic segmentation. In: Vedaldi, A., Bischof, H., Brox, T., Frahm, J.-M. (eds.) ECCV 2020. LNCS, vol. 12354, pp. 142–158. Springer, Cham (2020). https://doi.org/10.1007/978-3-030-58545-7_9
21. Liu, Z., et al.: Swin Transformer: Hierarchical vision Transformer using shifted windows. In: Proceedings of the IEEE/CVF International Conference on Computer Vision, pp. 10012–10022 (2021)
22. Long, J., Shelhamer, E., Darrell, T.: Fully convolutional networks for semantic segmentation. In: Proceedings of the IEEE/CVF Conference on Computer Vision and Pattern Recognition, pp. 3431–3440 (2015)
23. Lu, Z., He, S., Zhu, X., Zhang, L., Song, Y.Z., Xiang, T.: Simpler is better: Few-shot semantic segmentation with classifier weight Transformer. In: Proceedings of the IEEE/CVF International Conference on Computer Vision, pp. 8741–8750 (2021)
24. Min, J., Kang, D., Cho, M.: Hypercorrelation squeeze for few-shot segmentation. In: Proceedings of the IEEE/CVF International Conference on Computer Vision (2021)
25. Minaee, S., Boykov, Y.Y., Porikli, F., Plaza, A.J., Kehtarnavaz, N., Terzopoulos, D.: Image segmentation using deep learning: A survey. IEEE Trans. Pattern Anal. Mach. Intell. **14**, 3523–3542 (2021). https://doi.org/10.1109/TPAMI.2021.3059968
26. Nguyen, K., Todorovic, S.: Feature weighting and boosting for few-shot segmentation. In: Proceedings of the IEEE/CVF International Conference on Computer Vision, pp. 622–631 (2019)
27. Otsu, N.: A threshold selection method from gray-level histograms. IEEE Trans. Syst. Man Cybern. **9**(1), 62–66 (1979)
28. Paszke, A., et al.: PyTorch: An imperative style, high-performance deep learning library. Adv. Neural. Inf. Process. Syst. **32**, 8026–8037 (2019)
29. Ravi, S., Larochelle, H.: Optimization as a model for few-shot learning. In: International Conference on Learning Representations (2017)
30. Ronneberger, O., Fischer, P., Brox, T.: U-Net: convolutional networks for biomedical image segmentation. In: Navab, N., Hornegger, J., Wells, W.M., Frangi, A.F. (eds.) MICCAI 2015. LNCS, vol. 9351, pp. 234–241. Springer, Cham (2015). https://doi.org/10.1007/978-3-319-24574-4_28
31. Shaban, A., Bansal, S., Liu, Z., Essa, I., Boots, B.: One-shot learning for semantic segmentation. In: Proceedings of the British Machine Vision Conference, pp. 167.1–167.13 (2017)
32. Simonyan, K., Zisserman, A.: Very deep convolutional networks for large-scale image recognition. arXiv preprint arXiv:1409.1556 (2014)
33. Snell, J., Swersky, K., Zemel, R.: Prototypical networks for few-shot learning. In: Advances in Neural Information Processing Systems, pp. 4080–4090 (2017)
34. Strudel, R., Garcia, R., Laptev, I., Schmid, C.: Segmenter: Transformer for semantic segmentation. In: Proceedings of the IEEE/CVF International Conference on Computer Vision, pp. 7262–7272 (2021)
35. Sun, G., Liu, Y., Liang, J., Van Gool, L.: Boosting few-shot semantic segmentation with Transformers. arXiv preprint arXiv:2108.02266 (2021)
36. Sung, F., Yang, Y., Zhang, L., Xiang, T., Torr, P.H., Hospedales, T.M.: Learning to compare: Relation network for few-shot learning. In: Proceedings of the IEEE/CVF Conference on Computer Vision and Pattern Recognition, pp. 1199–1208 (2018)

37. Taghanaki, S.A., Abhishek, K., Cohen, J.P., Cohen-Adad, J., Hamarneh, G.: Deep semantic segmentation of natural and medical images: A review. Artif. Intell. Rev. **54**(1), 137–178 (2021)

38. Tian, Z., Zhao, H., Shu, M., Yang, Z., Li, R., Jia, J.: Prior guided feature enrichment network for few-shot segmentation. IEEE Trans. Pattern Anal. Mach. Intell. **44**, 1050–1065 (2020)

39. Triantafillou, E., Zemel, R., Urtasun, R.: Few-shot learning through an information retrieval lens. In: Advances in Neural Information Processing Systems, pp. 2252–2262 (2017)

40. Vaswani, A., et al.: Attention is all you need. In: Advances in Neural Information Processing Systems, pp. 5998–6008 (2017)

41. Veličković, P., Cucurull, G., Casanova, A., Romero, A., Liò, P., Bengio, Y.: Graph attention networks. In: International Conference on Learning Representations (2018). https://openreview.net/forum?id=rJXMpikCZ

42. Vinyals, O., Blundell, C., Lillicrap, T., Wierstra, D., et al.: Matching networks for one shot learning. Adv. Neural. Inf. Process. Syst. **29**, 3630–3638 (2016)

43. Wang, H., Zhang, X., Hu, Y., Yang, Y., Cao, X., Zhen, X.: Few-shot semantic segmentation with democratic attention networks. In: Vedaldi, A., Bischof, H., Brox, T., Frahm, J.-M. (eds.) ECCV 2020. LNCS, vol. 12358, pp. 730–746. Springer, Cham (2020). https://doi.org/10.1007/978-3-030-58601-0_43

44. Wang, K., Liew, J.H., Zou, Y., Zhou, D., Feng, J.: PANet: Few-shot image semantic segmentation with prototype alignment. In: Proceedings of the IEEE/CVF International Conference on Computer Vision, pp. 9197–9206 (2019)

45. Yang, B., Liu, C., Li, B., Jiao, J., Ye, Q.: Prototype mixture models for few-shot semantic segmentation. In: Vedaldi, A., Bischof, H., Brox, T., Frahm, J.-M. (eds.) ECCV 2020. LNCS, vol. 12353, pp. 763–778. Springer, Cham (2020). https://doi. org/10.1007/978-3-030-58598-3_45

46. Zhang, B., Xiao, J., Qin, T.: Self-guided and cross-guided learning for few-shot segmentation. In: Proceedings of the IEEE/CVF Conference on Computer Vision and Pattern Recognition, pp. 8312–8321 (2021)

47. Zhang, C., Lin, G., Liu, F., Guo, J., Wu, Q., Yao, R.: Pyramid graph networks with connection attentions for region-based one-shot semantic segmentation. In: Proceedings of the IEEE/CVF International Conference on Computer Vision, pp. 9587–9595 (2019)

48. Zhang, C., Lin, G., Liu, F., Yao, R., Shen, C.: CANet: Class-agnostic segmentation networks with iterative refinement and attentive few-shot learning. In: Proceedings of the IEEE/CVF Conference on Computer Vision and Pattern Recognition, pp. 5217–5226 (2019)

49. Zhang, G., Kang, G., Wei, Y., Yang, Y.: Few-shot segmentation via cycle-consistent Transformer. arXiv preprint arXiv:2106.02320 (2021)

50. Zhang, X., Wei, Y., Yang, Y., Huang, T.S.: SG-One: similarity guidance network for one-shot semantic segmentation. IEEE Trans. Cybern. **50**(9), 3855–3865 (2020)

51. Zhang, Y., Mehta, S., Caspi, A.: Rethinking semantic segmentation evaluation for explainability and model selection. arXiv preprint arXiv:2101.08418 (2021)

52. Zhao, H., Shi, J., Qi, X., Wang, X., Jia, J.: Pyramid scene parsing network. In: Proceedings of the IEEE/CVF Conference on Computer Vision and Pattern Recognition, pp. 2881–2890 (2017)

53. Zhu, F., Zhu, Y., Zhang, L., Wu, C., Fu, Y., Li, M.: A unified efficient pyramid Transformer for semantic segmentation. In: Proceedings of the IEEE/CVF International Conference on Computer Vision, pp. 2667–2677 (2021)

Rethinking Clustering-Based Pseudo-Labeling for Unsupervised Meta-Learning

Xingping Dong[1], Jianbing Shen[2]([⊠]), and Ling Shao[3]

[1] Inception Institute of Artificial Intelligence, Abu Dhabi, United Arab Emirates
[2] SKL-IOTSC, Computer and Information Science, University of Macau, Zhuhai, China
shenjianbingcg@gmail.com
[3] Terminus Group, Beijing, China
ling.shao@ieee.org

Abstract. The pioneering method for unsupervised meta-learning, CA-CTUs, is a clustering-based approach with pseudo-labeling. This approach is model-agnostic and can be combined with supervised algorithms to learn from unlabeled data. However, it often suffers from label inconsistency or limited diversity, which leads to poor performance. In this work, we prove that the core reason for this is lack of a clustering-friendly property in the embedding space. We address this by minimizing the inter- to intra-class similarity ratio to provide clustering-friendly embedding features, and validate our approach through comprehensive experiments. Note that, despite only utilizing a simple clustering algorithm (k-means) in our embedding space to obtain the pseudo-labels, we achieve significant improvement. Moreover, we adopt a progressive evaluation mechanism to obtain more diverse samples in order to further alleviate the limited diversity problem. Finally, our approach is also model-agnostic and can easily be integrated into existing supervised methods. To demonstrate its generalization ability, we integrate it into two representative algorithms: MAML and EP. The results on three main few-shot benchmarks clearly show that the proposed method achieves significant improvement compared to state-of-the-art models. Notably, our approach also outperforms the corresponding supervised method in two tasks. The code and models are available at https://github.com/xingpingdong/PL-CFE.

Keywords: Meta-learning · Unsupervised learning · Clustering-friendly

Supplementary Information The online version contains supplementary material available at https://doi.org/10.1007/978-3-031-20044-1_10.

1   Introduction

Recently, few-shot learning has attracted increasing attention in the *machine learning* and *computer vision* communities [14,32,34,40,42]. It is also commonly used to evaluate meta-learning approaches [16,22,44]. However, most of the existing literature focuses on the supervised few-shot classification task, which is built upon datasets with human-specified labels. Thus, most previous works cannot make use of the rapidly increasing amount of unlabeled data from, for example, the internet.

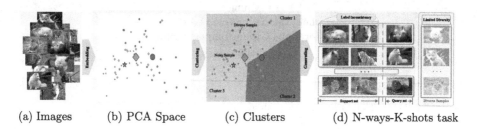

(a) Images   (b) PCA Space   (c) Clusters   (d) N-ways-K-shots task

Fig. 1. Illustration of the *label inconsistency* and *limited diversity* issues in the clustering-based CACTUs [25]. (a) The unlabeled images. (b) The 2D mapping space of the embedding features, generated via principal component analysis (PCA). Each mark (or color) represents one class, and the larger marks are the class centers (*i.e.* the average features in a class). (c) The noisy clustering labels generated by *k-means*, with many inconsistent or noisy samples in each cluster. For example, we use the green class label for cluster 3, since cluster 3 has the most green-samples. Thus, the samples with inconsistent labels, like the orange samples in cluster 3, are regarded as noisy samples. Further, in the unsupervised setting, the green samples in other regions (cluster 1 and cluster 2) cannot be used to build few-shot tasks. This leads to the unsupervised few-shot tasks being less diverse than the supervised tasks. We term this issue *limited diversity* and refer to these green samples as diverse samples. (d) The *label inconsistency* and *limited diversity* problems in the few-shot task, which are caused by the noisy clustering labels, as shown in (c). The red box highlights an incorrect sample in the few-shot task, and the green box illustrates the limited diversity problem, *i.e.*, the diverse samples cannot be used to build few-shot tasks.

To solve this issue, the pioneering work CACTUs [25], was introduced to automatically construct the few-shot learning tasks by assigning pseudo-labels to the samples from unlabeled datasets. This approach partitions the samples into several clusters using a clustering algorithm (*k-means*) on their embedding features, which can be extracted by various unsupervised methods [4,15,21,38, 49]. Unlabeled samples in the same clusters are assigned the same pseudo-labels. Then, the supervised meta-learning methods, MAML [16] and ProtoNets [46], are applied to the few-shot tasks generated by this pseudo-labeled dataset. It is worth mentioning that CACTUs is model-agnostic and any other supervised methods can be used in this framework.

However, this approach based on clustering and pseudo-label generation suffers from *label inconsistency*, *i.e.* samples with the same pseudo-label may have different human-specified labels in the few-shot tasks (*e.g.* Fig. 1(d)). This is caused by noisy clustering labels. Specifically, as shown in Fig. 1(c), several samples with different human-specified labels are partitioned into the same cluster, which leads to many noisy samples (with different labels) in the few-shot tasks. This is one reason for performance degeneration. Besides, noisy clustering labels will also partition samples with the same label into different clusters, which results in a lack of diversity for the few-shot tasks based on pseudo-labels compared with supervised methods. As shown in Fig. 1 (c)(d), CACTUs ignores the diversity among partitioned samples with the same human label. This is termed the *limited diversity* problem. How to utilize the diversity is thus one critical issue for performance improvement.

To overcome the above problems, we first analyze the underlying reasons for the noisy clustering labels in depth, via a qualitative and quantitative comparison bewteen the embedding features of CACTUs. As shown in Fig. 1(b), we find that the embedding features extracted by unsupervised methods are not clustering-friendly. In other words, most samples are far away from their class centers, and different class centers are close to each other in the embedding space. Thus, a standard clustering algorithms cannot effectively partition samples from the same class into one cluster, leading to noisy clustering labels. For example, cluster 3 in Fig. 1(c) contains many noisy samples from different classes. Furthermore, we propose an inter- to intra-class similarity ratio to measure the clustering-friendly property. In the quantitative comparison, we observe that the accuracy of the few-shot task is inversely proportional to the similarity ratio. This indicates that reducing the similarity ratio is critical for performance improvement.

According to these observations, a novel pseudo-labeling framework based on a clustering-friendly feature embedding (PL-CFE) is proposed to construct few-shot tasks on unlabeled datasets, in order to alleviate the *label inconsistency* and *limited diversity* problems. Firstly, we introduce a new unsupervised training method to extract clustering-friendly embedding features. Since the similarity ratio can only be applied to labeled datasets, we simulate a labeled set via data augmentation and try to minimize the similarity ratio on this to provide a clustering-friendly embedding function. Given our embedding features, we can run *k-means* to generate several clusters for pseudo-labeling and build clean few-shot tasks, reducing both the *label inconsistency* and *limited diversity* problems. Secondly, we present a progressive evaluation mechanism to obtain more divisive samples and further alleviate the *limited diversity*, by utilizing additional clusters for the task construction. Specifically, for each cluster, which we call a base cluster, in the task construction, we choose its *k-nearest* clusters as candidates. We use an evaluation model based on previous meta-learning models to measure the entropy of the candidates, and select the one with the highest entropy as the additional cluster for building a hard task, as it contains newer information for the current meta-learning model.

To evaluate the effectiveness of the proposed PL-CFE, we incorporate it into two representative supervised meta-learning methods: MAML [25] and EP [42], termed as PL-CFE-MAML and PL-CFE-EP, respectively. We conduct extensive experiments on Omniglot [30], *mini*ImageNet [39], *tiered*ImageNet [41]. The results demonstrate that our approach achieves significant improvement compared with state-of-the-art model-agnostic unsupervised meta-learning methods. In particular, our PL-CFE-MAML outperforms the corresponding supervised MAML method on *mini*ImageNet in both the 5-ways-20-shots and 5-ways-50-shots tasks. Notably, we achieve a gain of 1.75% in the latter task.

2   Related Work

2.1   Meta-Learning for Few-Shot Classification

Meta-learning, whose inception goes as far back as the 1980s [22,44], is usually interpolated as *fast weights* [3,22], or *learning-to-learn* [1,23,44,48]. Recent meta-learning methods can be roughly split into three main categories. The first kind of approaches is metric-based [7,29,42,46,50], which attempt to learn discriminative similarity metrics to distinguish samples from the same class. The second kind of approaches are memory-based [39,43], investigating the storing of key training examples with effective memory architectures or encoding fast adaptation methods. The last kind of approaches are optimization-based methods [16,17], searching for adaptive initialization parameters, which can be quickly adjusted for new tasks. Most meta-learning methods are evaluated under supervised few-shot classification, which requires a large number of manual labels. In this paper, we explore a new approach of few-shot task construction for unsupervised meta-learning to reduce this requirement.

2.2   Unsupervised Meta-Learning

Unsupervised learning aims to learn previously unknown patterns in a dataset without manual labels. These patterns learned during unsupervised pre-training can be used to more efficiently learn various downstream tasks, which is one of the more practical applications [4,15,21,38,49]. Unsupervised pre-training has achieved significant success in several fields, such as image classification [19,55], speech recognition [54], machine translation [37], and text classification [10,24,36].

Hsu *et al.* [25] proposed an *unsupervised meta-learning method* based on clustering, named CACTUs, to explicitly extract effective patterns from small amounts of data for a variety of tasks. However, as mentioned, this method suffers from the *label inconsistency* and *limited diversity* problems, which our approach can significantly reduce. Subsequently, Khodadadeh *et al.* [27] proposed the UMTRA method to build synthetic training tasks in the meta-learning phase, by using random sampling and augmentation. Stemming from this, several works have been introduced to synthesize more generative and divisive meta-training tasks via various techniques, such as introducing a distribution shift between the support and query set [35], using latent space interpolation in generative models to generate divisive samples [28], and utilizing a variational autoencoder with a mixture of Gaussian for producing meta tasks and training [31]. Compared to these synthetic methods, our approach can obtain harder few-shot tasks. Specifically, these previous methods only utilize the differences between augmented samples to increase the diversity of tasks, while our method can use the differences inside the class. Besides, some researchers explore to apply clustering methods for meta-training, by prototypical transfer learning [33] or progressive clustering [26]. These methods are not model-agnostic, while our approach can be incorporated into any supervised meta-learning method. This will bridge the unsupervised and supervised methods in meta-learning.

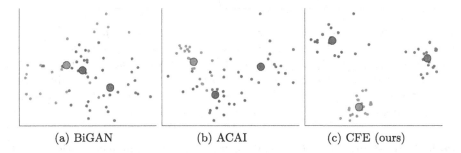

(a) BiGAN   (b) ACAI   (c) CFE (ours)

Fig. 2. Illustrations of 2D mapping of different embeddings, including BiGAN [11], ACAI [5], and our clustering-friendly embedding (CFE) features. We randomly select three classes from Omniglot [30] and map the embeddings of their samples in 2D space via principal component analysis (PCA). Each color represents one class, and large circles are the class centers.

3   In-Depth Analysis of Clustering-Based Unsupervised Methods

CACTUs [25] is a clustering-based and model-agnostic algorithm, which is easily incorporated into supervised methods for the unsupervised meta-learning task. However, there is still a large gap in performance between CACTUs and the supervised methods. We believe that the main reason is that the unsupervised embedding algorithms, such as InfoGAN [8], BiGAN [11], ACAI [5] and DC [6], in CACTUs are not suitable for the clustering task, which is a core step of CACTUs. This is because these unsupervised methods were originally designed for the pre-training stage, where the extracted features can be further fine-tuned in the downstream tasks. Thus, they do not need to construct a clustering-friendly feature space, where the samples in the same class are close to their class center and each class center is far away from the samples in other classes. However, the clustering-friendly property is very important for CACTUs, since it directly clusters the embedding features without fine-tuning.

We first provide an intuitive analysis by visualizing the embedding features. We collect samples from three randomly selected classes of a dataset and map them into 2D space with principal component analysis (PCA). For example, we observe the BiGAN and ACAI embedding features on Omniglot [30]. As shown in Fig. 2(a)(b), many samples are far away from their class centers, and the class centers are close to each other. Thus, we can see that many samples in different classes are close to each other. These samples are difficult for a simple clustering method to partition.

In order to observe the embedding features in the whole dataset, we define two metrics, *intra-similarity* and *inter-similarity*, to measure the clustering performance. We define the *intra-similarity* for each class as follows:

$$s_i^{\text{intra}} = exp(\sum_{j=1}^{N_s} \boldsymbol{\mu}_i \cdot \mathbf{z}_{ij}/(\tau N)), \tag{1}$$

where $\mathbf{z}_{ij}$ is the embedding feature of the j-th sample in class i, $\boldsymbol{\mu}_i = \frac{1}{N_s}\sum_{j=1}^{N_s}\mathbf{z}_{ij}$ is the class center, $\cdot$ is the dot product, N_s is the number of samples in a class, and τ is a temperature hyperparameter [53]. The *intra-similarity* is the average similarity between the samples and their class center. It is used to evaluate the compactness of a class in the embedding space. A large value indicates that the class is compact and most samples are close to the class center.

Table 1. Illustrations of the relationship between similarity ratio R and classification accuracy (Acc). We present the average *intra-similarity* s^{intra}, *inter-similarity* s^{inter}, and similarity ratio R of different embedding methods on Omniglot [30](Omni) and *mini*ImageNet [39](Mini). We also report the accuracy of MAML [16] based on these embeddings for the 5-ways-1-shot task. All Acc values except ours are sourced from CACTUs [25].

Embedding	Dataset	s^{inter} ↓	s^{intra} ↑	R ↓	Acc (%) ↑
BIGAN [11]	Omni	**1.032**	2.690	0.449	58.18
ACAI [5]	Omni	5.989	16.33	0.413	68.84
CFE (ours)	Omni	1.316	**20.10**	**0.081**	**91.32**
DC [6]	Mini	**0.999**	1.164	0.862	39.90
CFE (ours)	Mini	1.048	**1.680**	**0.641**	**41.99**

The other metric, *inter-similarity*, is defined as follows:

$$s_{ij}^{\text{inter}} = exp(\boldsymbol{\mu}_i \cdot \boldsymbol{\mu}_j/\tau), \; j \neq i. \tag{2}$$

A low value of s_{ij}^{inter} indicates that two classes are far away from each other.

To combine the above two similarities, we use the inter- to intra-class similarity ratio $r_{ij} = s_{ij}^{\text{inter}}/s_i^{\text{intra}}$ to represent the clustering performance. Finally, the average similarity ratio R over the whole dataset is denoted as:

$$R = \frac{1}{C}\sum_{i=1}^{C}\frac{\sum_{j\neq i}^{C} s_{ij}^{\text{inter}}}{(C-1)s_i^{\text{intra}}}, \tag{3}$$

where C is the number of classes. The lower the value of R, the better the clustering performance. In addition to R, we also calculate the average *intra-similarity* $s^{\text{intra}} = \sum s_i^{\text{intra}}/C$ and average *inter-similarity* $s^{\text{inter}} = \sum s_{ij}^{\text{inter}}/(N_s C)$ for complete analysis, where N_s is the samples number of a class.

As shown in Table 1, we apply the above three criteria, R, s^{intra} and s^{inter}, to the different embedding features (BiGAN [11] and ACAI [5]) in the Omniglot [30] dataset, and report the accuracy on the 5-ways-1-shot meta-task with the CACTUs-MAML [25] method. We find that the accuracy is inversely proportional to the similarity ratio R, but has no explicit relationship with the individual similarities s^{intra} or s^{inter}. This indicates that minimizing R is critical for better accuracy.

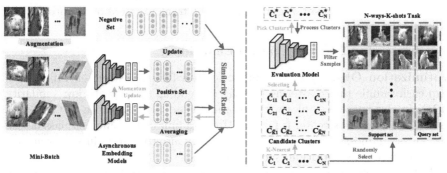

(a) Clustering-Friendly Feature Embedding   (b) Progressive Evaluation Mechanism

Fig. 3. The frameworks of our clustering-friendly feature embedding and progressive evaluation mechanism. (a) The embedding feature extraction. We first augment the samples in one mini-batch and feed them into the asynchronous embedding models to build a positive set. We also use the negative set storing the historical embedding features to obtain more varied features. Then we update our main embedding model (green) with backpropagation (green arrows), by minimizing the similarity ratio between the average features (purple ellipses), and the features in the positive and negative sets. The historical encoder is momentum updated with the main model. **(b) The progressive evaluation mechanism.** We randomly select N clusters as base clusters ($\bar{C}_i$), and choose the $\bar{K}$-nearest neighbors as the candidates. Then, we use an evaluation model to select the clusters ($\bar{C}_i^*$) with the highest entropy and filter out noisy samples.

Therefore, we propose a novel clustering-friendly embedding method (CFE) to reduce the similarity ratio in the following subsection §4.1. In the visualization comparison of Fig. 2, the proposed method provides more compact classes than the previous BiGAN and ACAI. As shown in Table 1, our CFE approach also has significantly reduced R on both the Omniglot and *mini*ImageNet. These results indicate that our method is more clustering-friendly. To investigate the relationship between the clustering-friendly property and the accuracy on the meta-task, we firstly use *k-means* to split the samples into different clusters, and assign the same pseudo-label to samples in one cluster. Then we run the supervised MAML method on the 5-ways-1-shot task constructed by pseudo-labeled samples. Compared with CACTUs, which is based on previous embedding methods, our approach achieves significant improvement on both Omniglot and *mini*ImageNet. Notably, the gain on Omniglot is more than 20%. The results clearly support our claim: a clustering-friendly embedding space can benefit clustering-based unsupervised meta-learning methods.

4 Our Approach

We propose a novel clustering-based pseudo-labeling for unsupervised meta-learning, which includes a clustering-friendly feature embedding and a progressive evaluation mechanism. Figure 3 shows the frameworks of our approach.

4.1 Clustering-Friendly Feature Embedding

Optimization Objective. We aim to learn a feature extraction function to map each sample to a clustering-friendly embedding space (with low similarity ratio). In this space, samples with the same class label are close to their class center, and each class center is far away from samples with different class labels. We can assign most samples from the same class into the same cluster, even with a simple clustering algorithm, such as *k-means*.

In the unsupervised setting, we do not have access to the class labels. Thus, we need to simulate a labeled dataset. To do so, we first randomly select $N_p \ll N_{all}$ samples from the unlabeled dataset $\mathcal{D}$ to build a positive set $\mathcal{D}_p$, where N_{all} is the size of $\mathcal{D}$. For each sample $\mathbf{x}_i \in \mathcal{D}_p$, we produce N_a augmented samples $\{\mathbf{x}_{ij}^+, j \in [1, N_a]\}$ via data augmentation methods, such as random color jittering, random horizontal flipping, and random grayscale conversion. Samples augmented from the same original sample are regarded as belonging to the same class. To involve more samples for training, we also construct a negative set by randomly selecting N_n samples from $\mathcal{D}$, without including the N_p positive samples. We augment each sample once to obtain the final negative set $\{\mathbf{x}_k^-, k \in [1, N_n]\}$. Given the positive ('labeled') set and negative set, we can reformulate the *intra-similarity* in Eq. (1) and *inter-similarity* in Eq. (2), for the final optimization objective. Maximizing the former will force the samples to be close to their class center, while minimizing the latter will pull the class centers far away from the other class center and negative samples.

We rewrite the *intra-similarity* for each class:

$$s_i^{\text{intra}} = exp(\sum_{j=1}^{N_a} \boldsymbol{\mu}_i \cdot \mathbf{z}_{ij}^+/(\tau N_a)), \tag{4}$$

where $\mathbf{z}_{ij}^+ = \phi(\mathbf{x}_{ij}^+; \boldsymbol{\omega})$ is the embedding feature and $\boldsymbol{\mu}_i = \frac{1}{N_a}\sum_j^{N_a} \mathbf{z}_{ij}^+$ is the class center. ϕ represents the embedding function and $\boldsymbol{\omega}$ is its parameter. This embedding function can be any learnable module, such as a convolutional network. The rewritten *inter-similarity* includes a negative-similarity measuring the similarity between the class center and negative sample, and a center-similarity calculating the similarity bewteen the class centers. The formulations are as follows:

$$s_{ik}^{\text{neg}} = exp(\boldsymbol{\mu}_i \cdot \mathbf{z}_k^- / \tau), \ s_{ij}^{\text{cen}} = exp(\boldsymbol{\mu}_i \cdot \boldsymbol{\mu}_j / \tau), \tag{5}$$

where $\mathbf{z}_k^- = \phi(\mathbf{x}_k^-; \boldsymbol{\omega})$ is the negative embedding feature.

To utilize the common losses from standard deep learning tools, such as PyTorch [47], we incorporate the logarithm function with the similarity ratio R in Eq. (3) to obtain the minimization objective, as follows:

$$L = \frac{1}{N_p} \sum_{i=1}^{N_p} \log \frac{1}{N_o} (1 + \frac{\sum_{j \neq i}^{N_p} s_{ij}^{\text{cen}} + \sum_{k=1}^{N_n} s_k^{\text{neg}}}{s_i^{\text{intra}}}), \tag{6}$$

where $N_o = N_p + N_n - 1$. Similar to R in Eq. (3), a low value of L in Eq. (6) means that the current embedding space is clustering-friendly. Thus, we can partition most samples from the same class into the same cluster, using a simple clustering algorithm. A low value of L can also reduce the label inconsistency problem in the clustering-based unsupervised meta-learning.

Asynchronous Embedding. If we only use a single encoder function ϕ for embedding learning, it is difficult to produce various embedding features for the samples augmented from the same original sample. Thus, we propose to utilize two asynchronous encoders, $\phi(\cdot; \boldsymbol{\omega})$ and $\bar{\phi}(\cdot; \bar{\boldsymbol{\omega}})$ for training to avoid fast convergence caused by similar positive samples. The first function ϕ is the main encoder, which is updated with the current samples (in a mini-batch). In other words, we use gradient backpropagation to update its parameter $\boldsymbol{\omega}$. The latter $\bar{\phi}$ is the history encoder, which collects information from the main encoders in previous mini-batches. To ensure that the history encoders on-the-fly are smooth, i.e., the difference among these encoders can be made small [19], we use a momentum coefficient $m \in [0, 1)$ to update the encoder parameter $\bar{\boldsymbol{\omega}} = m\bar{\boldsymbol{\omega}} + \boldsymbol{\omega}$.

To reduce the computational load, only one sample from each class is encoded by the main encoder ϕ, and the others are encoded by the history encoder $\bar{\phi}$. Without loss of generality, we encode the first augmented sample in each class. Then, the embedding features of the positive dataset $\{\mathbf{x}_{ij}^+, i \in [1, N_p], j \in [1, N_a]\}$ are reformulated as follows:

$$\mathbf{z}_{ij}^+ = \begin{cases} \phi(\mathbf{x}_{ij}^+; \boldsymbol{\omega}), j = 1, \\ \bar{\phi}(\mathbf{x}_{ij}^+; \bar{\boldsymbol{\omega}}), j \neq 1. \end{cases} \tag{7}$$

For the negative set, a naive approach would be to randomly select samples from the unlabeled dataset and then encode them with the history encoder. However, this solution is time-consuming and does not make full use of the different history encoders from previous mini-batches. Inspired by [19], we use a queue to construct the negative set by inserting the embedding features encoded by the history encoders. Specifically, we only insert one historical embedding of each class into the queue to maintain the diversity of the negative set. We remove the features of the oldest mini-batch to maintain the size of the negative set, since the oldest features are most outdated and least inconsistent with the new ones. Using the queue mechanism can reduce the computational cost and also remove the limit on the mini-batch size.

Clustering and Meta-Learning. With the above training strategy, we can provide embedding features for the samples in the unlabeled dataset. Similar to CACTUs [25], we can then apply a simple clustering algorithm, *e.g. k-means*, to these features to split the samples into different clusters, denoted as $\{C_1, C_2, \cdots, C_{N_s}\}$. We assign the same pseudo-label to all samples in a cluster. Thus, we obtain the pseudo-labeled dataset $\{(\mathbf{x}_i, \bar{y}_i), \mathbf{x}_i \in \mathcal{D}\}$, where $\bar{y}_i$ is the pseudo-label obtained by the clusters, *i.e.* $\bar{y}_i = k$ if $\mathbf{x}_i \in C_k$. Then, we can use any supervised meta-learning algorithm on this dataset to learn an efficient model for the meta-tasks.

4.2 Progressive Evaluation Mechanism

Although our embedding approach yields promising improvements, it still suffers from the *limited diversity* problem. In other words, we cannot generate meta-tasks using the divisive samples, which are far away from their class center, and easily assigned the wrong label by the simple clustering algorithm. However, the supervised methods can utilize these samples to generate hard meta-tasks to enhance the model's discrimination ability. Thus, we propose a progressive evaluation mechanism to produce similar hard meta-tasks with supervised methods.

As mentioned before, our embedding approach will push samples close to their class centers. For each cluster, denoted as the base cluster, we assume that the divisive samples are in its $\bar{K}$-nearest-neighbors, which are denoted as the candidate clusters. Thus, we can try to use the samples in the candidate clusters to generate hard meta-tasks. To do so, we build an evaluation model using the meta-learning models in previous iterations to evaluate the candidates. To obtain a stable evaluation model, only the meta-training models trained at the end of an epoch are utilized. We denote the evaluation model as $f(\mathbf{x}; \bar{\boldsymbol{\theta}})$, where $\bar{\boldsymbol{\theta}}$ is the corresponding model parameter. Then, we select those with high information entropy as the final clusters, and choose the samples from the base and final clusters to generate hard meta-tasks, by filtering out noise.

Selecting Clusters with High Information Entropy. To construct an N-ways-K-shots meta-task, we need to randomly choose N_c clusters denoted as base clusters $\bar{C}_i, i \in [1, N_c]$. We use the cluster center similarity $\bar{s}_{ik} = \bar{\boldsymbol{\mu}}_i \cdot \bar{\boldsymbol{\mu}}_k$ to obtain the $\bar{K}$ most similar neighbors as the candidate clusters $\bar{C}_{i\bar{k}}$, $k \in [1, \bar{K}]$, where $\bar{\boldsymbol{\mu}}_i$ is the average embedding feature of cluster $\bar{C}_i$, and $\bar{K} = 5$ in this paper. We randomly select K samples from each base cluster to build the support set, which we then use to finetune the evaluation model $f(\cdot; \bar{\boldsymbol{\theta}})$ with the same meta-training method.

For simplification, we first define the entropy of a cluster C based on the fine-tuned evaluation model. The N-ways classification label of each sample $\mathbf{x}_i \in C$ is computed as $l_i = \arg\max_{j=1}^N \mathbf{l}_i[j]$, where $\mathbf{l}_i = f(\mathbf{x}_i; \bar{\boldsymbol{\theta}})$. Then we can provide the probability p_j of selecting a sample with label j from the cluster C by computing the frequency of the label j, *i.e.*, $p_j = N_j/\dot{N}_l$, where N_j is the occurrence number of label j and $\dot{N}_l$ is the length of cluster $\mathcal{C}$. Then the entropy of $\mathcal{C}$ is formulated as:

$$H(\mathcal{C}) = -\sum_{j=1}^{N} p_j \log p_j. \tag{8}$$

According to Eq. 8, we choose the cluster $\bar{\mathcal{C}}_i^*$ with the highest entropy as the final cluster to construct the query set. The formulation is as follows:

$$\bar{\mathcal{C}}_i^* = \bar{\mathcal{C}}_{ik^*}, k^* = \arg\max_{\bar{k}=1}^{\bar{K}} H(\bar{\mathcal{C}}_{i\bar{k}}) \tag{9}$$

In our case, a low entropy for a cluster indicates that it is certain or information-less for the evaluation model, since the label outcome of a cluster can be regarded as a variable based on our evaluation model. In other words, the information in a cluster with low entropy has already been learned in the previous training epochs and is thus not new for the current evaluation model. In contrast, a cluster with high entropy can provide unseen information for further improvement.

Filtering Out Noisy Samples. Notice that the cluster $\bar{\mathcal{C}}_i^*$ contains many noisy samples (with inconsistent labels compared with the support samples). Thus, we use the proposed evaluation model $f(\cdot; \bar{\theta})$ to filter out several noisy samples and build the query set. First, we run $f(\cdot; \bar{\theta})$ on each sample $\mathbf{x}_{ij}$ in $\bar{\mathcal{C}}_i^*$ to provide the *probabilities* $\mathbf{l}_{ij}$ of N-ways classification, *i.e.*, $\mathbf{l}_{ij} = f(\mathbf{x}_{ij}; \bar{\theta})$. We take the *probability* of the i-th classification $\mathbf{l}_{ij}[i]$ as the evaluation score of the sample $\mathbf{x}_{ij}$. According to these scores, we re-sort the cluster $\bar{\mathcal{C}}_i^*$ in descending order to obtain a new cluster $\dot{\mathcal{C}}_i$. The noisy samples are placed at the end of the cluster. Thus, we can filter out the noisy samples by removing the ones at the end of the new cluster $\dot{\mathcal{C}}_i^*$. The *keep rate* $\beta \in (0,1)$ is used to control the number of samples removed. Specifically, we define the removing operation as $\dot{\mathcal{C}}_i = \dot{\mathcal{C}}_i[1 : \lfloor \beta \bar{N}_l \rfloor]$, where $\bar{N}_l$ is the length of $\dot{\mathcal{C}}_i$, and $\lfloor \cdot \rfloor$ is the *floor* operation. Finally, we randomly select Q samples from the cluster $\dot{\mathcal{C}}_i$ as the query set.

In particular, during training, we employ a random value $\eta \in (0,1)$ for each mini-batch. Only when $\eta > 0.9$, we use the progressive evaluation mechanism to build the meta-tasks. Since our pseudo-labels contain the most useful information for training, we need to utilize this information as much as possible.

5   Experiments

5.1   Datasets and Implementation Details

Datasets. We evaluate the proposed approach on three popular datasets: Omniglot [30], *mini*ImageNet [39], and *tiered*ImageNet [41]. Following the setting in [25], for all three datasets, we only use the unlabeled data in the training subset to construct the unsupervised few-shot tasks.

Embedding Functions and Hyperparameters. For Omniglot, we adopt a 4-Conv model (the same as the one in MAML [16]), as the backbone, and add two fully connected (FC) layers with 64 output dimensions (64-FC). For miniImageNet and tieredImageNet, we use the same embedding function, which includes a ResNet-50 [20] backbone and two 128-FC layers, and train it on the whole ImageNet. The other training details are the same for the two models. The number of training epochs is set to 200. We use the same data augmentation and cosine learning rate schedule as [9]. The other hyperparameters for training are set as $N_p = 256$, $N_a = 2$, $N_n = 65536$, $\tau = 0.2$, $m = 0.999$. We use the same number of clusters as CACTUs for fair comparison, $i.e.$ $N_c = 500$. The $keep\ rate$ β in the progressive evaluation is set as 0.75 to filter out very noisy samples.

Table 2. Influence of key components in our model. We evaluate different variants in terms of the accuracy for N-ways-K-shots (N,K) tasks. The A/D labels represent ACAI [5] on Omniglot [30] and DC [6] on miniImageNet [39]. MC and M are short names of MoCo-v2 [9] and MAML [16]. Similarly, CFE is our embedding method and E represents the EP [42]. The results of A/D-M are taken from CACTUs [25].

	Pos	Neg	AE	SC	FN	Omniglot				miniImageNet			
						(5,1)	(5,5)	(20,1)	(20,5)	(5,1)	(5,5)	(5,20)	(5,50)
A/D-M						68.84	87.78	48.09	73.36	39.90	53.97	63.84	69.64
MC-M						–	–	–	–	39.83	55.62	67.27	73.63
CFE-M	✓					89.97	97.50	74.33	92.44	39.75	57.31	69.68	75.64
	✓	✓				69.88	89.68	40.50	72.50	25.35	34.91	47.80	55.92
	✓		✓			90.67	97.90	75.18	92.75	41.04	57.86	69.62	75.67
	✓	✓	✓			91.32	97.96	76.19	93.30	41.99	58.66	69.64	75.46
	✓	✓	✓	✓		92.01	98.27	78.29	94.27	43.39	59.70	70.06	75.38
	✓	✓	✓		✓	91.40	97.94	78.61	94.46	42.85	59.31	69.61	75.25
	✓	✓	✓	✓	✓	**92.81**	**98.46**	**80.86**	**95.05**	**43.60**	**59.84**	**71.19**	**75.75**
A/D-E						78.23	88.97	53.50	72.37	39.73	52.69	59.75	62.06
MC-E						–	–	–	–	43.26	56.67	63.19	65.07
CFE-E	✓					93.31	97.31	78.59	90.09	43.44	56.98	64.06	66.65
	✓	✓				60.70	73.37	37.93	49.33	26.68	32.36	37.17	38.86
	✓		✓			93.48	97.40	79.75	90.67	43.55	57.73	64.17	66.17
	✓	✓	✓			93.88	97.46	79.55	91.13	47.55	62.13	68.89	71.11
	✓	✓	✓	✓		95.31	98.01	83.17	92.50	48.25	62.54	69.76	71.58
	✓	✓	✓		✓	94.93	98.00	82.57	91.97	48.02	62.56	69.85	**72.31**
	✓	✓	✓	✓	✓	**95.48**	**98.04**	**83.67**	**92.54**	**49.13**	**62.91**	**70.47**	71.79

Supervised Meta-Learning Methods. We combine the proposed pseudo-labeling based on clustering-friendly embedding (PL-CFE) with two supervised methods including classical model-agnostic meta-learning (MAML) [16], and the recently proposed embedding propagation (EP) [42]. The corresponding methods are denoted as PL-CFE-MAML and PL-CFE-EP, respectively. In the following experiments, we only train the meta-learning models (MAML and EP) on the one-shot task and directly test them on the other few-shot tasks.

5.2   Ablation Study

To analyze the effectiveness of the key components in our framework, we conduct extensive experiments on Omniglot and miniImageNet with MAML and

EP. First, we choose the unsupervised method ACAI [5]/DC [6] as the baselines of the embedding function, which achieves the best accuracy for CACTUs. Specifically, we apply *k-means* to the ACAI/DC embedding features to provide the pseudo-labels, and then run MAML and EP over eight few-shot tasks. These two baselines are denoted as A/D-M and A/D-E, respectively. The results of A/D-M come from the original CACTUs paper.

The key components in our clustering-friendly embedding (CFE) are the positive set (Pos), negative set (Neg), and asynchronous embedding (AE). We also investigate the main parts of our progressive evaluation, including selecting the cluster (SC) and filtering out noise (FN). First, we minimize the similarity ratio on the positive set. As shown in Table 2, our approach achieves impressive improvement for both MAML and EP in terms of all four tasks on Omniglot, compared with the baselines. In particular, in the 5-ways-1-shot task, the highest accuracy gain reaches 21.15%. Then, we add the negative set for training. However, the performance drops dramatically. The reason may be that the rapidly changing embedding models reduces the consistency of negative embedding features. Thus, we add AE to alleviate this issue. The increased performance (Pos+Neg+AE vs. Pos) indicates that Neg+AE can obtain more divisive samples for better training. We also use AE on the positive set (Pos+AE) and achieve expected accuracy gains for the two supervised methods in terms of all tasks. Finally, we use the SC strategy to select diverse samples for constructing few-shot tasks, yielding performance gains in all tasks. Then, the FN strategy is added to filter out noisy samples (SC+FN), which further improves the accuracy scores of all tasks. We also directly use our FN strategy on the original clusters (Pos+Neg+AE+FN) and achieve promising improvement in all tasks. This indicates that the proposed FN strategy can reduce the *label inconsistency* issue for improved performance.

Besides, we also compare the recent contrast learning method MoCo-v2 [9], since our model has a similar training strategy but different training loss. We apply the pretrained MoCo-v2 model (200 epochs) to extract embeddings and provide the pseudo-labels, and then run MAML and EP over four few-shot tasks on *mini*ImageNet. These are denoted as MC-M and MC-E, respectively. As shown in Table 2, even our methods without progress evaluation (Pos+Neg+AE) can outperform the MoCo-based methods in all tasks. This clearly demonstrates the effectiveness of our clustering-friendly embedding. Notice that our progress evaluation (Pos+Neg+AE+SC+FN) can further improve the performance.

5.3   Comparison with Other Algorithms

Results on Omniglot and *mini*ImageNet. We compare our PL-CFE-MAML and PL-CFE-EP with model-agnostic methods: CACTUs [25], AAL [2], ULDA [35], and LASIUM [28].

As shown in Table 3, our PL-CFE-MAML outperforms all model-agnostic methods in eight few-shot tasks on two datasets. Specifically, for the 5-ways-5-shots task on the Omniglot dataset, our method achieves a very high score

Table 3. Accuracy (%) of N-ways-K-shots (N,K) tasks. A/D represents ACAI [5] on Omniglot [30] and DC [6] on *mini*ImageNet [39]. The best values are in bold.

Algorithm (N, K)	Clustering	Omniglot				*mini*ImageNet			
		(5,1)	(5,5)	(20,1)	(20,5)	(5,1)	(5,5)	(5,20)	(5,50)
Training from scratch [25]	–	52.50	74.78	24.91	47.62	27.59	38.48	51.53	59.63
CACTUs-MAML [25]	BiGAN	58.18	78.66	35.56	58.62	36.24	51.28	61.33	66.91
CACTUs-ProtoNets [25]	BiGAN	54.74	71.69	33.40	50.62	36.62	50.16	59.56	63.27
CACTUs-MAML [25]	A/D	68.84	87.78	48.09	73.36	39.90	53.97	63.84	69.64
CACTUs-ProtoNets [25]	A/D	68.12	83.58	47.75	66.27	39.18	53.36	61.54	63.55
AAL-ProtoNets [2]	-	84.66	89.14	68.79	74.28	37.67	40.29	–	–
AAL-MAML++ [2]	-	88.40	97.96	70.21	88.32	34.57	49.18	–	–
ULDA-ProtoNets [35]	-	91.00	98.14	78.05	94.08	40.63	56.18	64.31	66.43
ULDA-MetaOptNet [35]	-	90.51	97.60	76.32	92.48	40.71	54.49	63.58	67.65
LASIUM-MAML [28]	-	83.26	95.29	–	–	40.19	54.56	65.17	69.13
LASIUM-ProtoNets [28]	-	80.15	91.1	–	–	40.05	52.53	59.45	61.43
CACTUs-EP	A/D	78.23	88.97	53.50	72.37	39.73	52.69	59.75	62.06
PL-CFE-MAML (ours)	CFE	92.81	**98.46**	80.86	**95.05**	43.38	60.00	**70.64**	**75.52**
PL-CFE-EP (ours)	CFE	**95.48**	98.04	**83.67**	92.54	**49.13**	**62.91**	70.47	71.79
MAML (supervised)	–	94.64	98.90	87.90	97.50	48.03	61.78	70.20	73.77
EP (supervised)	–	98.24	99.23	92.38	97.18	57.15	71.27	77.46	78.77

of 98.46%, which is very close to the supervised result of 98.90%. Surprisingly, the proposed PL-CFE-MAML even outperforms the corresponding supervised method under the 5-ways-20-shots and 5-ways-50-shots settings on the *mini*ImageNet dataset. In addition, compared with the baseline CACTUs-MAML, we achieve significant accuracy gains, which reach 32.77% (for 20-ways-1-shot) on Omniglot and 6.8% (for 5-ways-20-shots) on *mini*ImageNet.

In the experiments for our PL-CFE-EP, we first create the CACTUs-EP baseline by combining CACTUs with EP. Our PL-CFE-EP provides impressive improvements in performance compared with CACTUs-EP. In particular, in the 20-ways-1-shot task on Omniglot, we achieve a huge gain of 30.17%. On *mini*ImageNet, the gains reach 10.72% (in 5-ways-20-shots). Compared with the other existing methods, our PL-CFE-EP obtains superior performance in six tasks and comparable performance in the other two tasks. In particular, we achieve a significant accuracy gain of 8.42% in 5-ways-1-shot on *mini*ImageNet.

Results on *tiered*ImageNet. We compare our PL-CFE-MAML and PL-CFE-EP with the recent ULDA [35] on *tiered*ImageNet. As shown in Table 4, our models outperform the compared methods in four few-shot tasks. Specifically, for the 5-ways-1-shots and 5-ways-5-shots tasks, our PL-CFE-EP achieves the highest scores. Compared with the best previous method, ULDA-MetaOptNet, our method obtains accuracy gains of 7.74% and 7.35% on these two tasks, respectively. Our PL-CFE-MAML outperforms all compared methods in the 5-ways-20-shots and 5-ways-50-shots tasks. It also achieves significant improvements, with gains of 3.98% and 4.36%, respectively, compared with ULDA-MetaOptNet.

In summary, the results demonstrate the effectiveness of our method.

Table 4. Accuracy (%) of N-ways-K-shots (N,K) tasks on *tiered*ImageNet [41]. We show the results of supervised methods MAML [25] and EP [42] for complete comparison. The best values are in bold.

	(5,1)	(5,5)	(5,20)	(5,50)
Training from scratch [35]	26.27	34.91	38.14	38.67
ULDA-ProtoNets [35]	41.60	56.28	64.07	66.00
ULDA-MetaOptNet [35]	41.77	56.78	67.21	71.39
PL-CFE-MAML (ours)	43.60	59.84	**71.19**	**75.75**
PL-CFE-EP (ours)	**49.51**	**64.31**	70.98	73.06
MAML (supervised) [25]	50.10	66.79	75.61	79.16
EP (supervised) [42]	58.21	71.73	77.40	78.78

6   Conclusion

In this paper, we introduce a new framework of pseudo-labeling based on clustering-friendly embedding (PL-CFE) to automatically construct few-shot tasks from unlabeled datasets for meta-learning. Specifically, we present an unsupervised embedding approach to provide clustering-friendly features for few-shot tasks, which significantly reduces the *label inconsistency* and *limited diversity* problems. Moreover, a progressive evaluation is designed to build hard tasks to further alleviate *limited diversity* issue. We successfully integrate the proposed method into two representative supervised models to demonstrate its generality. Finally, extensive empirical evaluations clearly demonstrate and the effectiveness of our PL-CFE, which outperforms the corresponding supervised meta-learning methods in two few-shot tasks. In the future, we will utilize our model to more computer vision tasks, such as object tracking [12,18,45] and segmentation [13,51,52], to explore label-free or label-less solutions.

References

1. Andrychowicz, M., et al.: Learning to learn by gradient descent by gradient descent. In: NeurIPS (2016)
2. Antoniou, A., Storkey, A.: Assume, augment and learn: Unsupervised few-shot meta-learning via random labels and data augmentation. arXiv preprint arXiv: 1902.09884 (2019)
3. Ba, J., Hinton, G.E., Mnih, V., Leibo, J.Z., Ionescu, C.: Using fast weights to attend to the recent past. In: NeurIPS (2016)
4. Bengio, Y., Lamblin, P., Popovici, D., Larochelle, H.: Greedy layer-wise training of deep networks. In: NeurIPS (2007)
5. Berthelot, D., Raffel, C., Roy, A., Goodfellow, I.: Understanding and improving interpolation in autoencoders via an adversarial regularizer. In: ICLR (2019)

6. Caron, M., Bojanowski, P., Joulin, A., Douze, M.: Deep clustering for unsupervised learning of visual features. In: Ferrari, V., Hebert, M., Sminchisescu, C., Weiss, Y. (eds.) Computer Vision – ECCV 2018. LNCS, vol. 11218, pp. 139–156. Springer, Cham (2018). https://doi.org/10.1007/978-3-030-01264-9_9

7. Chen, W.Y., Liu, Y.C., Kira, Z., Wang, Y.C.F., Huang, J.B.: A closer look at few-shot classification. In: ICLR (2019)

8. Chen, X., Duan, Y., Houthooft, R., Schulman, J., Sutskever, I., Abbeel, P.: Infogan: Interpretable representation learning by information maximizing generative adversarial nets. In: NeurIPS (2016)

9. Chen, X., Fan, H., Girshick, R., He, K.: Improved baselines with momentum contrastive learning. arXiv preprint arXiv:2003.04297 (2020)

10. Dai, A.M., Le, Q.V.: Semi-supervised sequence learning. In: NeurIPS (2015)

11. Donahue, J., Krähenbühl, P., Darrell, T.: Adversarial feature learning. In: ICLR (2017)

12. Dong, X., Shen, J., Shao, L., Porikli, F.: Clnet: A compact latent network for fast adjusting siamese trackers. In: Vedaldi, A., Bischof, H., Brox, T., Frahm, J.-M. (eds.) ECCV 2020. LNCS, vol. 12365, pp. 378–395. Springer, Cham (2020). https://doi.org/10.1007/978-3-030-58565-5_23

13. Dong, X., Shen, J., Shao, L., Van Gool, L.: Sub-markov random walk for image segmentation. In: IEEE T-IP (2015)

14. Dvornik, N., Schmid, C., Mairal, J.: Diversity with cooperation: Ensemble methods for few-shot classification. In: ICCV (2019)

15. Erhan, D., Bengio, Y., Courville, A., Manzagol, P.A., Vincent, P., Bengio, S.: Why does unsupervised pre-training help deep learning? In: JMLR, pp. 625–660 (2010)

16. Finn, C., Abbeel, P., Levine, S.: Model-agnostic meta-learning for fast adaptation of deep networks. In: ICML (2017)

17. Finn, C., Xu, K., Levine, S.: Probabilistic model-agnostic meta-learning. In: NeurIPS (2018)

18. Han, W., Dong, X., Khan, F.S., Shao, L., Shen, J.: Learning to fuse asymmetric feature maps in siamese trackers. In: CVPR (2021)

19. He, K., Fan, H., Wu, Y., Xie, S., Girshick, R.: Momentum contrast for unsupervised visual representation learning. In: CVPR (2020)

20. He, K., Zhang, X., Ren, S., Sun, J.: Delving deep into rectifiers: Surpassing human-level performance on imagenet classification. In: ICCV (2015)

21. Hinton, G.E., Osindero, S., Teh, Y.W.: A fast learning algorithm for deep belief nets. Neural Comput. 18(7), 1527–1554 (2006)

22. Hinton, G.E., Plaut, D.C.: Using fast weights to deblur old memories. In: CCSS (1987)

23. Hochreiter, S., Younger, A.S., Conwell, P.R.: Learning to learn using gradient descent. In: ICANN (2001)

24. Howard, J., Ruder, S.: Universal language model fine-tuning for text classification. In: ACL (2018)

25. Hsu, K., Levine, S., Finn, C.: Unsupervised learning via meta-learning. In: ICLR (2019)

26. Ji, Z., Zou, X., Huang, T., Wu, S.: Unsupervised few-shot feature learning via self-supervised training. Frontiers Comput. Neurosci. (2020)

27. Khodadadeh, S., Boloni, L., Shah, M.: Unsupervised meta-learning for few-shot image classification. In: NeurIPS (2019)

28. Khodadadeh, S., Zehtabian, S., Vahidian, S., Wang, W., Lin, B., Bölöni, L.: Unsupervised meta-learning through latent-space interpolation in generative models. In: ICLR (2021)

29. Koch, G., Zemel, R., Salakhutdinov, R.: Siamese neural networks for one-shot image recognition. In: ICML workshop (2015)
30. Lake, B., Salakhutdinov, R., Gross, J., Tenenbaum, J.: One shot learning of simple visual concepts. In: CogSci (2011)
31. Lee, D.B., Min, D., Lee, S., Hwang, S.J.: Meta-GMVAE: Mixture of Gaussian VAE for Unsupervised Meta-Learning. In: ICLR (2021)
32. Li, X., Sun, Q., Liu, Y., Zhou, Q., Zheng, S., Chua, T.S., Schiele, B.: Learning to self-train for semi-supervised few-shot classification. In: NeurIPS (2019)
33. Medina, C., Devos, A., Grossglauser, M.: Self-supervised prototypical transfer learning for few-shot classification. arXiv preprint arXiv:2006.11325 (2020)
34. Peng, Z., Li, Z., Zhang, J., Li, Y., Qi, G.J., Tang, J.: Few-shot image recognition with knowledge transfer. In: ICCV (2019)
35. Qin, T., Li, W., Shi, Y., Gao, Y.: Unsupervised few-shot learning via distribution shift-based augmentation. arXiv preprint arXiv:2004.05805 (2020)
36. Radford, A., Narasimhan, K., Salimans, T., Sutskever, I.: Improving language understanding by generative pre-training. Preprint (2018)
37. Ramachandran, P., Liu, P.J., Le, Q.V.: Unsupervised pretraining for sequence to sequence learning. In: EMNLP (2017)
38. Ranzato, M., Poultney, C., Chopra, S., Cun, Y.L.: Efficient learning of sparse representations with an energy-based model. In: NeurIPS (2007)
39. Ravi, S., Larochelle, H.: Optimization as a model for few-shot learning. In: ICLR (2017)
40. Ravichandran, A., Bhotika, R., Soatto, S.: Few-shot learning with embedded class models and shot-free meta training. In: ICCV (2019)
41. Ren, M., Triantafillou, E., et al.: Meta-learning for semi-supervised few-shot classification. In: ICLR (2018)
42. Rodríguez, P., Laradji, I., Drouin, A., Lacoste, A.: Embedding propagation: Smoother manifold for few-shot classification. In: Vedaldi, A., Bischof, H., Brox, T., Frahm, J.-M. (eds.) ECCV 2020. LNCS, vol. 12371, pp. 121–138. Springer, Cham (2020). https://doi.org/10.1007/978-3-030-58574-7_8
43. Santoro, A., Bartunov, S., Botvinick, M., Wierstra, D., Lillicrap, T.: Meta-learning with memory-augmented neural networks. In: ICML (2016)
44. Schmidhuber, J.: Evolutionary principles in self-referential learning, or on learning how to learn: the meta-meta-... hook. Ph.D. thesis, Technische Universität München (1987)
45. Shen, J., Liu, Y., Dong, X., Lu, X., Khan, F.S., Hoi, S.C.: Distilled siamese networks for visual tracking. In: IEEE T-PAMI (2021)
46. Snell, J., Swersky, K., Zemel, R.: Prototypical networks for few-shot learning. In: NeurIPS (2017)
47. Steiner, B., et al.: Pytorch: An imperative style, high-performance deep learning library. In: NeurIPS. pp. 8026–8037 (2019)
48. Thrun, S., Pratt, L.: Learning to learn: Introduction and overview. In: Learning to Learn, pp. 3–17. Springer (1998).https://doi.org/10.1007/978-1-4615-5529-2_1
49. Vincent, P., Larochelle, H., Bengio, Y., Manzagol, P.A.: Extracting and composing robust features with denoising autoencoders. In: ICML (2008)
50. Vinyals, O., Blundell, C., Lillicrap, T., Wierstra, D., et al.: Matching networks for one shot learning. In: NeurIPS (2016)
51. Wang, W., Shen, J., Dong, X., Borji, A., Yang, R.: Inferring salient objects from human fixations. In: IEEE T-PAMI (2019)
52. Wu, D., Dong, X., Shao, L., Shen, J.: Multi-level representation learning with semantic alignment for referring video object segmentation. In: CVPR (2022)

53. Wu, Z., Xiong, Y., Yu, S.X., Lin, D.: Unsupervised feature learning via non-parametric instance discrimination. In: CVPR (2018)
54. Yu, D., Deng, L., Dahl, G.: Roles of pre-training and fine-tuning in context-dependent dbn-hmms for real-world speech recognition. In: NeurIPS Workshop (2010)
55. Zhang, R., Isola, P., Efros, A.A.: Split-brain autoencoders: Unsupervised learning by cross-channel prediction. In: CVPR (2017)

CLASTER: Clustering with Reinforcement Learning for Zero-Shot Action Recognition

Shreyank N. Gowda[1(✉)], Laura Sevilla-Lara[1], Frank Keller[1], and Marcus Rohrbach[2]

[1] University of Edinburgh, Edinburgh, UK
S.Narayana-Gowda@sms.ed.ac.uk
[2] Meta AI, Menlo Park, USA

Abstract. Zero-Shot action recognition is the task of recognizing action classes without visual examples. The problem can be seen as learning a representation on seen classes which generalizes well to instances of unseen classes, without losing discriminability between classes. Neural networks are able to model highly complex boundaries between visual classes, which explains their success as supervised models. However, in Zero-Shot learning, these highly specialized class boundaries may overfit to the seen classes and not transfer well from seen to unseen classes. We propose a novel cluster-based representation, which regularizes the learning process, yielding a representation that generalizes well to instances from unseen classes. We optimize the clustering using reinforcement learning, which we observe is critical. We call the proposed method CLASTER and observe that it consistently outperforms the state-of-the-art in all standard Zero-Shot video datasets, including UCF101, HMDB51 and Olympic Sports; both in the standard Zero-Shot evaluation and the generalized Zero-Shot learning. We see improvements of up to 11.9% over SOTA.

Project Page: https://sites.google.com/view/claster-zsl/home

Keywords: Zero-shot · Clustering · Action recognition

1 Introduction

Research on action recognition in videos has made rapid progress in the last years, with models becoming more accurate and even some datasets becoming saturated. Much of this progress has depended on large scale training sets. However, it is often not practical to collect thousands of video samples for a new class. This idea has led to research in the Zero-Shot learning (ZSL) domain,

Supplementary Information The online version contains supplementary material available at https://doi.org/10.1007/978-3-031-20044-1_11.

where training occurs in a set of seen classes, and testing occurs in a set of unseen classes. In particular, in the case of video ZSL, each class label is typically enriched with semantic embeddings. These embeddings are sometimes manually annotated, by providing attributes of the class, and other times computed automatically using language models of the class name or class description. At test time the semantic embedding of the predicted seen class is used to search for a nearest neighbor in the space of semantic embeddings of unseen classes.

While ZSL is potentially a very useful technology, this standard pipeline poses a fundamental representation challenge. Neural networks have proven extraordinarily powerful at learning complex discriminative functions of classes with many modes. In other words, instances of the same class can be very different and still be projected by the neural network to the same category. While this works well in supervised training, it can be a problem in Zero-Shot recognition, where the highly specialized discriminative function might not transfer well to instances of unseen classes. In this work, we adress this representation problem using three main ideas.

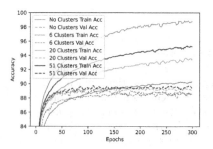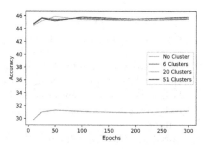

Fig. 1. Left: learning curve for the seen classes. Right: learning curve for the unseen classes. The clustering-based representation avoids overfitting, which in the case of seen classes means that the gap between validation and training accuracy is smaller than in the vanilla representation. This regularization effect improves the validation accuracy in unseen classes.

First, we turn to clustering, and use the centroids of the clusters to represent a video. We argue that centroids are more robust to outliers, and thus help regularize the representation, avoiding overfitting to the space of seen classes. Figure 1 shows that the gap between training and validation accuracy is smaller when using clustering in seen classes (left). As a result, the learned representation is more general, which significantly improves accuracy in unseen classes (right).

Second, our representation is a combination of a visual and a semantic representation. The standard practice at training time is to use a visual representation, and learn a mapping to the semantic representation. Instead, we use both cues, which we show yields a better representation. This is not surprising, since both visual and semantic information can complement each other.

Third, we use the signal from classification as direct supervision for clustering, by using Reinforcement Learning (RL). Specifically, we use the REINFORCE algorithm to directly update the cluster centroids. This optimization improves the clustering significantly and leads to less noisy and more compact representations for unseen classes.

These three pieces are essential to learn a robust, generalizable representation of videos for Zero-Shot action recognition, as we show in the ablation study. Crucially, none of them have been used in the context of Zero-Shot or action recognition. They are simple, yet fundamental and we hope they will be useful to anyone in the Zero-Shot and action recognition communities.

We call the proposed method CLASTER, for *CLustering with reinforcement learning for Action recognition in zero-ShoT lEaRning*, and show that it significantly outperforms all existing methods across all standard Zero-Shot action recognition datasets and tasks.

2   Related Work

Fully Supervised Action Recognition. This is the most widely studied setting in action recognition, where there is a large amount of samples at training time and the label spaces are the same at training and testing time. A thorough survey is beyond our scope, but as we make use of these in the backbone of our model, we mention some of the most widely used work. The seminal work of Simonyan and Zisserman [39] introduced the now standard two-stream deep learning framework, which combines spatial and temporal information. Spatiotemporal CNNs [5,35,41] are also widely used as backbones for many applications, including this work. More recently, research has incorporated attention [14,42] and leveraged the multi-modal nature of videos [2]. In this work, we use the widely used I3D [5].

Zero-Shot Learning. Early approaches followed the idea of learning semantic classifiers for seen classes and then classifying the visual patterns by predicting semantic descriptions and comparing them with descriptions of unseen classes. In this space, Lampert et al. [21] propose attribute prediction, using the posterior of each semantic description. The SJE model [1] uses multiple compatibility functions to construct a joint embedding space. ESZSL [37] uses a Frobenius norm regularizer to learn an embedding space. Repurposing these methods for action classification is not trivial. In videos, there are additional challenges: action labels need more complex representations than objects and hence give rise to more complex manual annotations.

ZSL for Action Recognition. Early work [36] was restricted to cooking activities, using script data to transfer to unseen classes. Gan et al. [11] consider each action class as a domain, and address semantic representation identification as a multi-source domain generalization problem. Manually specified semantic representations are simple and effective [49] but labor-intensive to annotate. To overcome this, the use of label embeddings has proven popular, as only category names are needed. Some approaches use common embedding space between class labels and video features [45,46], pairwise relationships between classes [9], error-correcting codes [34], inter-class relationships [10], out-of-distribution detectors [26], and Graph Neural networks [13]. In contrast, we are learning to optimize centroids of visual semantic representations that generalize better to unseen classes.

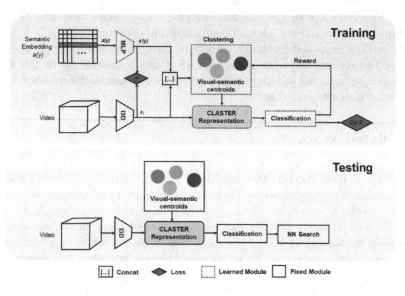

Fig. 2. Overview of CLASTER. We map the semantic embedding $a(y_i)$ to the space of visual features x_i, and concatenate both to obtain a visual-semantic representation. We cluster these visual-semantic representations with K-means to obtain initial cluster centroids. Each video is represented as the sum of the visual-semantic representation and the centroid clusters, weighted by their distance (see Sect. 3.3 and Fig. 3). This is used as input for classification (Sect. 3.4 and Eq. 3). Based on the classification result, we send a reward to optimize the cluster centroids using REINFORCE (Sect. 3.5). At test time, we first perform classification on the seen classes and then do a nearest neighbor (NN) search to predict the unseen class.

Reinforcement Learning for Zero-Shot Learning. RL for ZSL in images was introduced by Liu et al. [23] by using a combination of ontology and RL. In Zero-Shot text classification, Ye et al. [47] propose a self-training method to leverage unlabeled data. RL has also been used in the Zero-Shot setting for task generalization [32], active learning [7], and video object segmentation [16]. To the best of our knowledge, there is no previous work using RL for optimizing centroids in Zero-Shot recognition.

Deep Approaches to Centroid Learning for Classification. Since our approach learns cluster centroids using RL, it is related to the popular cluster learning strategy for classification called Vector of Locally Aggregated Descriptors (VLAD) [3]. The more recent NetVLAD [3] leverages neural networks which helps outperform the standard VLAD by a wide margin. ActionVLAD [15] aggregates NetVLAD over time to obtain descriptors for videos. ActionVLAD uses clusters that correspond to spatial locations in a video while we use joint visual semantic embeddings for the entire video. In general, VLAD uses residuals with respect to cluster centroids as representation while CLASTER uses a weighting of the centroids. The proposed CLASTER outperforms NetVLAD by a large margin on both HMDB51 and UCF101.

3 CLASTER

We now describe the proposed CLASTER, which leverages clustering of visual and semantic features for video action recognition and optimizes the clustering with RL. Figure 2 shows an overview of the method.

3.1 Problem Definition

Let S be the training set of seen classes. S is composed of tuples $(x, y, a(y))$, where x represents the spatio-temporal features of a video, y represents the class label in the set of Y_S seen class labels, and $a(y)$ denotes the category-specific semantic representation of class y. These semantic representations are either manually annotated or computed using a language-based embedding of the category name, such as word2vec [28] or sentence2vec [33].

Let U be the set of pairs $(u, a(u))$, where u is a class in the set of unseen classes Y_U and $a(u)$ are the corresponding semantic representations. The seen classes Y_S and the unseen classes Y_U do not overlap.

In the Zero-Shot Learning (ZSL) setting, given an input video the task is to predict a class label in the unseen classes, as $f_{ZSL} : X \rightarrow Y_U$. In the related generalized Zero-Shot learning (GZSL) setting, given an input video, the task is to predict a class label in the union of the seen and unseen classes, as $f_{GZSL} : X \rightarrow Y_S \cup Y_U$.

3.2 Visual-Semantic Representation

Given video i, we compute visual features x_i and a semantic embedding $a(y_i)$ of their class y_i (see Sect. 4 for details). The goal is to map both to the same space, so that they have the same dimensionality and magnitude, and therefore they will have a similar weight during clustering. We learn this mapping with a simple multi-layer perceptron (MLP) [48], trained with a least-square loss. This loss minimizes the distance between x_i and the output from the MLP, which we call $a'(y)$. Finally, we concatenate x_i and $a'(y)$ to obtain the visual-semantic representations that will be clustered. The result is a representation which is not only aware of the visual information but also the semantic.

3.3 CLASTER Representation

We now detail how we represent videos using the proposed *CLASTER* representation, which leverages clustering as a form of regularization. In other words, a representation w.r.t centroids is more robust to outliers, which is helpful since all instances of the unseen classes are effectively outliers w.r.t. the training distribution.

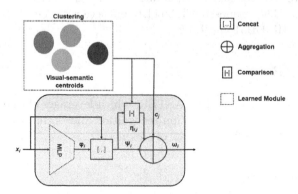

Fig. 3. The proposed CLASTER Representation in detail (see Fig. 2 for the overview of the full method). The visual feature is mapped to match the space of the visual-semantic cluster centroids with an MLP and concatenation. Based on the distances to the cluster centroids the final representation ω is a weighted representation of the centroids, more robust to the out-of-distribution instances of the unseen test classes. Details in Sect. 3.3 and Eq. 1.

We initialize the clustering of the training set S using K-means [8]. Each resulting cluster j has a centroid c_j, that is the average of all visual-semantic samples in that particular cluster. The CLASTER representation of a given video is the sum of the visual-semantic representation and the centroids, weighted by the inverse of the distance, such that closer clusters will have more weight. Figure 3 shows this process in detail.

Specifically, given video i, we compute the visual representation x_i. We estimate the semantic vector ϕ_i using an MLP. This is a necessary step, as during test time we do not have any semantic information. Concatenating the visual x_i and semantic ϕ_i we obtain the intermediate representation ψ_i, which is in the same space as the cluster centroids.

We compute the Euclidean distance $d_{i,j}$ between the visual-semantic point ψ_i and each cluster j, which we refer to as $d_{i,j}$. We take the inverse $1/d_{i,j}$ and normalize them using their maximum and minimum values, such that they are between 0 and 1. We refer to these normalized values as $\eta_{i,j}$, and they are used as the weights of each cluster centroid in the final CLASTER representation ω_i:

$$\omega_i = \psi_i + \sum_{j=1}^{k} \eta_{i,j} c_j. \tag{1}$$

3.4   Loss Function

Given the CLASTER representation ω_i we predict a seen class using a simple MLP, V. Instead of the vanilla softmax function, we use semantic softmax [18], which includes the semantic information $a(y_i)$ and thus can transfer better to Zero-Shot classes:

$$\hat{y}_i = \frac{e^{a(y_i)^T V(\omega_i)}}{\sum_{j=1}^{S} e^{a(y_j)^T V(\omega_i)}}. \tag{2}$$

The output $\hat{y}_i$ is a vector with a probability distribution over the S seen classes. We train the classifier, which minimizes the cross-entropy loss with a regularization term:

$$\min_W \sum_{i=1}^{N} \mathcal{L}(x_i) + \lambda \|W\|, \tag{3}$$

where W refers to all weights in the network.

3.5   Optimization with Reinforcement Learning

As there is no ground truth for clustering centroids, we use the classification accuracy as supervision. This makes the problem non-differentiable, and therefore we cannot use traditional gradient descent. Instead, we use RL to optimize the cluster centroids. For this, we compute two variables that will determine each centroid update: the reward, which measures whether the classification is correct and will determine the direction of the update, and the classification score, which measures how far the prediction is from the correct answer and will determine the magnitude of the update.

Given the probabilistic prediction $\hat{y}_i$ and the one-hot representation of the ground truth class y_i, we compute the classification score as the dot product of the two: $z_i = y_i \hat{y}_i$. To obtain the reward, we check if the maximum of $\hat{y}_i$ and y_i lie in the same index:

$$r = \begin{cases} 1 & if \ \arg\max \hat{y}_i = \arg\max y_i \\ -1 & otherwise \end{cases} \tag{4}$$

This essentially gives a positive reward if the model has predicted a correct classification and a negative reward if the classification was incorrect. This formulation is inspired by Likas [22], which was originally proposed for a different domain and the problem of competitive learning.

For each data point ψ_i we only update the closest cluster centroid c_j. We compute the update Δc_j using the REINFORCE [22] algorithm as:

$$\Delta c_j = \alpha \ r \ (z_i - p_j) \ (\psi_i - c_j). \tag{5}$$

For further details on this derivation, please see the Supplementary Material as well as Likas [22]. The main difference between our model and Likas' is that we do not consider cluster updates to be Bernoulli units. Instead, we modify the cluster centroid with the classification score z_i, which is continuous in the range between 0 and 1.

4 Implementation Details

Visual Features. We use RGB and flow features extracted from the *Mixed 5c* layer of an I3D network pre-trained on the Kinetics [5] dataset. The *Mixed 5c* output of the flow network is averaged across the temporal dimension and pooled by four in the spatial dimension and then flattened to a vector of size 4096. We then concatenate the two.

Network Architecture. The MLP that maps the semantic features to visual features consists of two fully-connected (FC) layers and a ReLU. The MLP in the CLASTER Representation module, which maps the visual feature to the semantic space is a two-layer FC network, whose output after concatenation with the video feature has the same dimensions as the cluster representatives. The size of the FC layers is 8192 each. The final classification MLP (represented as a classification block in Fig. 2) consists of two convolutional layers and two FC layers, where the last layer equals the number of seen classes in the dataset we are looking at. All the modules are trained with the Adam optimizer with a learning rate of 0.0001 and weight decay of 0.0005.

Number of Clusters. Since the number of clusters is a hyperparameter, we evaluate the effect of the number of clusters on the UCF101 dataset for videos and choose 6 after the average performance stablizes as can be seen in the supplementary material. We then use the same number for the HMDB51 and Olympics datasets.

RL Optimization. We use 10,000 iterations and the learning rate α is fixed to 0.1 for the first 1000 iterations, 0.01 for the next 1000 iterations and then drop it to 0.001 for the remaining iterations.

Semantic Embeddings. We experiment with three types of embeddings as semantic representations of the classes. We have human-annotated semantic representations for UCF101 and the Olympic sports dataset of sizes 40 and 115 respectively. HMDB51 does not have such annotations. Instead, we use a skip-gram model trained on the news corpus provided by Google to generate word2vec embeddings. Using action classes as input, we obtain a vector representation of 300 dimensions. Some class labels contain multiple words. In those cases, we use the average of the word2vec embeddings. We also use sentence2vec embeddings, trained on Wikipedia. These can be obtained for both single words and multi-word expressions. The elaborate descriptions are taken from [6] and only evaluated for fair comparison to them.

For the elaborative descriptions, we follow ER [6] and use the provided embeddings in their codebase.

Rectification of the Semantic Embedding. Sometimes, in ZSL, certain data points tend to appear as nearest-neighbor of many other points in the projection space. This is referred to as the hubness problem [38]. We avoid this problem using semantic rectification [24], where the class representation is modified by averaging the output generated by the projection network, which in our case is the penultimate layer of the classification MLP. Specifically, for the unseen classes, we perform rectification by first using the MLP trained on the seen classes to project the semantic embedding to the visual space. We add the average of projected semantic embeddings from the k-nearest neighbors of the seen classes, specifically as follows:

$$\hat{a}(y_i) = a'(y_i) + \frac{1}{k} \sum_{n \in N} cos\left(a'(y_i), n\right) \cdot n, \tag{6}$$

where $a'(y)$ refers to the embedding after projection to the visual space, $cos(a, n)$ refers to the cosine similarity between a and n, the operator $\cdot$ refers to the dot product and N refers to the k-nearest neighbors of $a'(y_{u_i})$.

Nearest Neighbor Search. At test time in ZSL, given a test video, we predict a seen class and compute or retrieve its semantic representation. After rectification, we find the nearest neighbor in the set of unseen classes. In the GZSL task, class predictions may be of seen or unseen classes. Thus, we first use a bias detector [12] which helps us detect if the video belongs to the seen or unseen class. If it belongs to a seen class, we predict the class directly from our model, else we proceed as in ZSL.

5   Experimental Analysis

In this section, we look at the qualitative and quantitative performance of the proposed model. We first describe the experimental settings, and then show an ablation study, that explores the contribution of each component. We then compare the proposed method to the state-of-the-art in the ZSL and GZSL tasks, and give analytical insights into the advantages of CLASTER.

5.1   Datasets

We choose the Olympic Sports [31], HMDB-51 [20] and UCF-101 [40], so that we can compare to recent state-of-the-art models [11,26,34]. We follow the commonly used 50/50 splits of Xu et al. [45], where 50 percent are seen classes and 50 are unseen classes. Similar to previous approaches [11,19,27,34,49], we report average accuracy and standard deviation over 10 independent runs. We report results on the split proposed by [44], in the standard inductive setting. We also report on the recently introduced TruZe [17]. This split accounts for the fact that some classes present on the dataset used for pre-training (Kinetics [5]) overlap with some of the unseen classes in the datasets used in the Zero-Shot setting, therefore breaking the premise that those classes have not been seen.

5.2 Ablation Study

Table 1 shows the impact of using the different components of CLASTER.

Impact of Clustering. We consider several baselines. First, omitting clustering, which is the equivalent of setting $\omega_i = \psi_i$, in Eq. 1. This is, ignoring the cluster centroids in the representation. This is referred to in Table 1 as "No clustering". Second, we use random clustering, which is assigning each instance to a random cluster. Finally, we use the standard K-means. We observe that using clusters is beneficial, but only if they are meaningful, as in the case of K-means.

Impact of Using a Visual-Semantic Representation. We compare to the standard representation, which only includes visual information, and keep everything the same. This is, clustering and classification are done using only the visual features. This is referred to in the table as "CLASTER w/o SE". We observe that there is a very wide gap between using and not using the semantic features at training time. This effect is present across all datasets, suggesting it is a general improvement in the feature learning. We also show a comparison of aggregation strategies and interaction between visual and semantic features in the supplementary material.

Impact of Different Optimization Choices. We make cluster centroids learnable parameters and use the standard SGD to optimize them ('CLASTER w/o RL") We also test the use of the related work of NetVLAD to optimize the cluster ("CLASTER w/ NetVLAD"). We see that the proposed model outperforms NetVLAD by an average of 4.7% and the CLASTER w/o RL by 7.3% on the UCF101 dataset. A possible reason for this difference is that the loss is back-propagated through multiple parts of the model before reaching the centroids. However, with RL the centroids are directly updated using the reward signal. Section 5.6 explores how the clusters change after the RL optimization. In a nutshell, the RL optimization essentially makes the clusters cleaner, moving most instances in a class to the same cluster.

5.3 Results on ZSL

Table 2 shows the comparison between CLASTER and several state-of-the-art methods: the out-of-distribution detector method (OD) [26], a generative approach to Zero-Shot action recognition (GGM) [30], the evaluation of output embeddings (SJE) [1], the feature generating networks (WGAN) [43], the end-to-end training for realistic applications approach (E2E) [4], the inverse autoregressive flow (IAF) based generative model, bi-directional adversarial GAN(Bi-dir GAN) [29] and prototype sampling graph neural network (PS-GNN) [13]. To make results directly comparable, we use the same backbone across all of them, which is the I3D [5] pre-trained on Kinetics.

Table 1. Results of the ablation study of different components of CLASTER ZSL. The study shows the effect of clustering, using visual-semantic representations, and optimizing with different methods. All three components show a wide improvement over the various baselines, suggesting that they are indeed complementary to improve the final representation.

Component	HMDB51	Olympics	UCF101
No clustering	25.6 ± 2.8	57.7 ± 3.1	31.6 ± 4.6
Random clustering (K=6)	20.2 ± 4.2	55.4 ± 3.1	24.1 ± 6.3
K-means (K=6)	27.9 ± 3.7	58.6 ± 3.5	35.3 ± 3.9
CLASTER w/o SE	27.5 ± 3.8	55.9 ± 2.9	39.4 ± 4.4
CLASTER w/o RL	30.1 ± 3.4	60.5 ± 1.9	39.1 ± 3.2
CLASTER w/ NetVLAD	33.2 ± 2.8	62.6 ± 4.1	41.7 ± 3.8
CLASTER	**36.8 ± 4.2**	**63.5 ± 4.4**	**46.4 ± 5.1**

We observe that the proposed CLASTER consistently outperforms all other state-of-the-art methods across all datasets. The improvements are significant: up to 3.5% on HMDB51 and 13.5% on UCF101 with manual semantic embedding. We also measure the impact of different semantic embeddings, including using sentence2vec instead of word2vec. We show that sentence2vec significantly improves over using word2vec, especially on UCF101 and HMDB51. Combination of embeddings resulted in average improvements of 0.3%, 0.8% and 0.9% over the individual best performing embedding of CLASTER.

5.4   Results on GZSL

We now compare to the same approaches in the GZSL task in Table 3, the reported results are the harmonic mean of the seen and unseen class accuracies. Here CLASTER outperforms all previous methods across different modalities. We obtain an improvement on average of 2.6% and 5% over the next best performing method on the Olympics dataset using manual representations and word2vec respectively. We obtain an average improvement of 6.3% over the next best performing model on the HMDB51 dataset using word2vec. We obtain an improvement on average performance by 1.5% and 4.8% over the next best performing model on the UCF101 dataset using manual representations and word2vec respectively. Similarly to ZSL, we show generalized performance improvements using sentence2vec. We also report results on the combination of embeddings. We see an improvement of 0.3%, 0.6% and 0.4% over the individual best embedding for CLASTER. The seen and unseen accuracies are shown in the Supplemental Material.

Table 2. Results on ZSL. SE: semantic embedding, M: manual representation, W: word2vec embedding, S: sentence2vec, C: Combination of embeddings. The proposed CLASTER outperforms previous state-of-the-art across tasks and datasets.

Method	SE	Olympics	HMDB51	UCF101
SJE [1]	M	47.5 ± 14.8	–	12.0 ± 1.2
Bi-Dir GAN [29]	M	53.2 ± 10.5	–	24.7 ± 3.7
IAF [29]	M	54.9 ± 11.7	–	26.1 ± 2.9
GGM [30]	M	57.9 ± 14.1	–	24.5 ± 2.9
OD [26]	M	65.9 ± 8.1	–	38.3 ± 3.0
WGAN [43]	M	64.7 ± 7.5	–	37.5 ± 3.1
CLASTER (ours)	M	**67.4 ± 7.8**	–	**51.8 ± 2.8**
SJE [1]	W	28.6 ± 4.9	13.3 ± 2.4	9.9 ± 1.4
IAF [29]	W	39.8 ± 11.6	19.2 ± 3.7	22.2 ± 2.7
Bi-Dir GAN [29]	W	40.2 ± 10.6	21.3 ± 3.2	21.8 ± 3.6
GGM [30]	W	41.3 ± 11.4	20.7 ± 3.1	20.3 ± 1.9
WGAN [43]	W	47.1 ± 6.4	29.1 ± 3.8	25.8 ± 3.2
OD [26]	W	50.5 ± 6.9	30.2 ± 2.7	26.9 ± 2.8
PS-GNN [13]	W	61.8 ± 6.8	32.6 ± 2.9	43.0 ± 4.9
E2E [4]*	W	61.4 ± 5.5	33.1 ± 3.4	46.2 ± 3.8
CLASTER (ours)	W	**63.8 ± 5.7**	**36.6 ± 4.6**	**46.7 ± 5.4**
CLASTER (ours)	S	**64.2 ± 3.3**	**41.8 ± 2.1**	**50.2 ± 3.8**
CLASTER (ours)	C	**67.7 ± 2.7**	**42.6 ± 2.6**	**52.7 ± 2.2**
ER [6]	ED	60.2 ± 8.9	35.3 ± 4.6	51.8 ± 2.9
CLASTER (ours)	ED	**68.4 ± 4.1**	**43.2 ± 1.9**	**53.9 ± 2.5**

5.5   Results on TruZe

We also evaluate on the more challenging TruZe split. The proposed UCF101 and HMDB51 splits have 70/31 and 29/22 classes (represented as training/testing). We compare to WGAN [43], OD [26] and E2E [4] on both ZSL and GZSL scenarios. Results are shown in Table 4.

5.6   Analysis of the RL Optimization

We analyze how optimizing with RL affects clustering on the UCF101 training set. Figure 4 shows the t-SNE [25] visualization. Each point is a video instance in the unseen classes, and each color is a class label. As it can be seen, the RL optimization makes videos of the same class appear closer together.

We also do a quantitative analysis of the clustering. For each class in the training set, we measure the distribution of clusters that they belong to, visualized in the Fig 5. We observe that after the RL optimization, the clustering

Table 3. Results on GZSL. SE: semantic embedding, M: manual representation, W: word2vec embedding, S: sentence2vec, C: combination of embeddings. The seen and unseen class accuracies are listed in the supplementary material.

Method	SE	Olympics	HMDB51	UCF101
Bi-Dir GAN [29]	M	44.2 ± 11.2	–	22.7 ± 2.5
IAF [29]	M	48.4 ± 7.0	–	25.9 ± 2.6
GGM [30]	M	52.4 ± 12.2	–	23.7 ± 1.2
WGAN [43]	M	59.9 ± 5.3	–	44.4 ± 3.0
OD [26]	M	66.2 ± 6.3	–	49.4 ± 2.4
CLASTER (ours)	M	**68.8 ± 6.6**	–	**50.9 ± 3.2**
IAF [29]	W	30.2 ± 11.1	15.6 ± 2.2	20.2 ± 2.6
Bi-Dir GAN [29]	W	32.2 ± 10.5	7.5 ± 2.4	17.2 ± 2.3
SJE [1]	W	32.5 ± 6.7	10.5 ± 2.4	8.9 ± 2.2
GGM [30]	W	42.2 ± 10.2	20.1 ± 2.1	17.5 ± 2.2
WGAN [43]	W	46.1 ± 3.7	32.7 ± 3.4	32.4 ± 3.3
PS-GNN [13]	W	52.9 ± 6.2	24.2 ± 3.3	35.1 ± 4.6
OD [26]	W	53.1 ± 3.6	36.1 ± 2.2	37.3 ± 2.1
CLASTER (ours)	W	**58.1 ± 2.4**	**42.4 ± 3.6**	**42.1 ± 2.6**
CLASTER (ours)	S	**58.7 ± 3.1**	**47.4 ± 2.8**	**48.3 ± 3.1**
CLASTER (ours)	C	**69.1 ± 5.4**	**48.0 ± 2.4**	**51.3 ± 3.5**

becomes "cleaner". This is, most instances in a class belong to a dominant cluster. This effect can be measured using the purity of the cluster:

$$Purity = \frac{1}{N} \sum_{i=1}^{k} max_j \left| c_i \cap t_j \right|, \tag{7}$$

where N is the number of data points (video instances), k is the number of clusters, c_i is a cluster in the set of clusters, and t_j is the class which has the maximum count for cluster c_i. Poor clustering results in purity values close to 0, and a perfect clustering will return a purity of 1. Using K-means, the purity is 0.77, while optimizing the clusters with RL results in a purity of 0.89.

Table 4. Results on TruZe. For ZSL, we report the mean class accuracy and for GZSL, we report the harmonic mean of seen and unseen class accuracies. All approaches use sen2vec annotations as the form of semantic embedding.

Method	UCF101		HMDB51	
	ZSL	GZSL	ZSL	GZSL
WGAN	22.5	36.3	21.1	31.8
OD	22.9	42.4	21.7	35.5
E2E	45.5	45.9	31.5	38.9
CLASTER	**45.8**	**47.3**	**33.2**	**44.5**

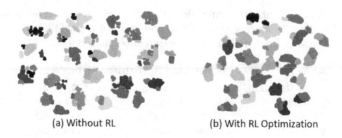

(a) Without RL   (b) With RL Optimization

Fig. 4. CLASTER improves the representation and clustering in unseen classes. The figure shows t-SNE [25] of video instances, where each color corresponds to a unique unseen class label. The RL optimization improves the representation by making it more compact: in (b) instances of the same class, i.e. same color, are together and there are less outliers for each class compared to (a).

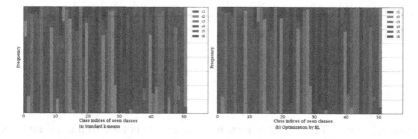

Fig. 5. Analysis of how RL optimization changes the cluster to which an instance belongs. The frequencies are represented as percentages of instances in each cluster. We can see that the clusters are a lot "cleaner" after the optimization by RL.

Finally, we observe another interesting side effect of clustering. Some of the most commonly confused classes before clustering (e.g. "Baby crawling" vs. "Mopping floor", "Breaststroke" vs. "front crawl", "Rowing vs. front crawl") are assigned to different clusters after RL, resolving confusion. This suggests that clusters are also used as a means to differentiate between similar classes.

6   Conclusion

Zero-Shot action recognition is the task of recognizing action classes without any visual examples. The challenge is to map the knowledge of seen classes at training time to that of novel unseen classes at test time. We propose a novel model that learns clustering-based representation of visual-semantic features, optimized with RL. We observe that all three of these components are essential. The clustering helps regularizing, and avoids overfitting to the seen classes. The visual-semantic representation helps improve the representation. And the RL yields better, cleaner clusters. The results is remarkable improvements across datasets and tasks over all previous state-of-the-art, up to 11.9% absolute improvement on HMDB51 for GZSL.

References

1. Akata, Z., Reed, S., Walter, D., Lee, H., Schiele, B.: Evaluation of output embeddings for fine-grained image classification. In: Proceedings of the IEEE Conference on Computer Vision and Pattern Recognition, pp. 2927–2936 (2015)
2. Alwassel, H., Mahajan, D., Torresani, L., Ghanem, B., Tran, D.: Self-supervised learning by cross-modal audio-video clustering. arXiv preprint arXiv:1911.12667 (2019)
3. Arandjelovic, R., Gronat, P., Torii, A., Pajdla, T., Sivic, J.: Netvlad: CNN architecture for weakly supervised place recognition. In: Proceedings of the IEEE Conference on Computer Vision and Pattern Recognition, pp. 5297–5307 (2016)
4. Brattoli, B., Tighe, J., Zhdanov, F., Perona, P., Chalupka, K.: Rethinking zero-shot video classification: end-to-end training for realistic applications. In: Proceedings of the IEEE/CVF Conference on Computer Vision and Pattern Recognition, pp. 4613–4623 (2020)
5. Carreira, J., Zisserman, A.: Quo vadis, action recognition? a new model and the kinetics dataset. In: IEEE Conference on Computer Vision and Pattern Recognition (2017)
6. Chen, S., Huang, D.: Elaborative rehearsal for zero-shot action recognition. In: Proceedings of the IEEE/CVF International Conference on Computer Vision, pp. 13638–13647 (2021)
7. Fan, Y., Tian, F., Qin, T., Bian, J., Liu, T.Y.: Learning what data to learn. arXiv preprint arXiv:1702.08635 (2017)
8. Forgy, E.W.: Cluster analysis of multivariate data: efficiency versus interpretability of classifications. Biometrics **21**, 768–769 (1965)
9. Gan, C., Lin, M., Yang, Y., De Melo, G., Hauptmann, A.G.: Concepts not alone: exploring pairwise relationships for zero-shot video activity recognition. In: Thirtieth AAAI Conference on Artificial Intelligence (2016)
10. Gan, C., Lin, M., Yang, Y., Zhuang, Y., Hauptmann, A.G.: Exploring semantic inter-class relationships (sir) for zero-shot action recognition. In: Proceedings of the National Conference on Artificial Intelligence (2015)
11. Gan, C., Yang, T., Gong, B.: Learning attributes equals multi-source domain generalization. In: Proceedings of the IEEE Conference on Computer Vision and Pattern Recognition, pp. 87–97 (2016)
12. Gao, J., Zhang, T., Xu, C.: I know the relationships: zero-shot action recognition via two-stream graph convolutional networks and knowledge graphs. In: Proceedings of the AAAI Conference on Artificial Intelligence, vol. 33, pp. 8303–8311 (2019)
13. Gao, J., Zhang, T., Xu, C.: Learning to model relationships for zero-shot video classification. IEEE Trans. Pattern Anal. Mach. Intell. **43**(10), 3476–3491 (2020)
14. Girdhar, R., Ramanan, D.: Attentional pooling for action recognition. In: Advances in Neural Information Processing Systems, pp. 34–45 (2017)
15. Girdhar, R., Ramanan, D., Gupta, A., Sivic, J., Russell, B.: Actionvlad: learning spatio-temporal aggregation for action classification. In: Proceedings of the IEEE Conference on Computer Vision and Pattern Recognition, pp. 971–980 (2017)
16. Gowda, S.N., Eustratiadis, P., Hospedales, T., Sevilla-Lara, L.: Alba: reinforcement learning for video object segmentation. arXiv preprint arXiv:2005.13039 (2020)
17. Gowda, S.N., Sevilla-Lara, L., Kim, K., Keller, F., Rohrbach, M.: A new split for evaluating true zero-shot action recognition. arXiv preprint arXiv:2107.13029 (2021)

18. Ji, Z., Sun, Y., Yu, Y., Guo, J., Pang, Y.: Semantic softmax loss for zero-shot learning. Neurocomputing **316**, 369–375 (2018)
19. Kodirov, E., Xiang, T., Fu, Z., Gong, S.: Unsupervised domain adaptation for zero-shot learning. In: Proceedings of the IEEE International Conference on Computer Vision, pp. 2452–2460 (2015)
20. Kuehne, H., Jhuang, H., Garrote, E., Poggio, T., Serre, T.: Hmdb: a large video database for human motion recognition. In: 2011 International Conference on Computer Vision, pp. 2556–2563. IEEE (2011)
21. Lampert, C.H., Nickisch, H., Harmeling, S.: Learning to detect unseen object classes by between-class attribute transfer. In: 2009 IEEE Conference on Computer Vision and Pattern Recognition, pp. 951–958. IEEE (2009)
22. Likas, A.: A reinforcement learning approach to online clustering. Neural Comput. **11**(8), 1915–1932 (1999)
23. Liu, B., Yao, L., Ding, Z., Xu, J., Wu, J.: Combining ontology and reinforcement learning for zero-shot classification. Knowl.-Based Syst. **144**, 42–50 (2018)
24. Luo, C., Li, Z., Huang, K., Feng, J., Wang, M.: Zero-shot learning via attribute regression and class prototype rectification. IEEE Trans. Image Process. **27**(2), 637–648 (2017)
25. Maaten, L.V.D., Hinton, G.: Visualizing data using t-sne. J. Mach. Learn. Res. **9**, 2579–2605 (2008)
26. Mandal, D., Narayan, S., Dwivedi, S.K., Gupta, V., Ahmed, S., Khan, F.S., Shao, L.: Out-of-distribution detection for generalized zero-shot action recognition. In: Proceedings of the IEEE Conference on Computer Vision and Pattern Recognition, pp. 9985–9993 (2019)
27. Mettes, P., Snoek, C.G.: Spatial-aware object embeddings for zero-shot localization and classification of actions. In: Proceedings of the IEEE International Conference on Computer Vision, pp. 4443–4452 (2017)
28. Mikolov, T., Sutskever, I., Chen, K., Corrado, G.S., Dean, J.: Distributed representations of words and phrases and their compositionality. In: Advances in Neural Information Processing Systems, pp. 3111–3119 (2013)
29. Mishra, A., Pandey, A., Murthy, H.A.: Zero-shot learning for action recognition using synthesized features. Neurocomputing **390**, 117–130 (2020)
30. Mishra, A., Verma, V.K., Reddy, M.S.K., Arulkumar, S., Rai, P., Mittal, A.: A generative approach to zero-shot and few-shot action recognition. In: 2018 IEEE Winter Conference on Applications of Computer Vision (WACV), pp. 372–380. IEEE (2018)
31. Niebles, J.C., Chen, C.-W., Fei-Fei, L.: Modeling temporal structure of decomposable motion segments for activity classification. In: Daniilidis, K., Maragos, P., Paragios, N. (eds.) ECCV 2010. LNCS, vol. 6312, pp. 392–405. Springer, Heidelberg (2010). https://doi.org/10.1007/978-3-642-15552-9_29
32. Oh, J., Singh, S., Lee, H., Kohli, P.: Zero-shot task generalization with multi-task deep reinforcement learning. In: International Conference on Machine Learning, pp. 2661–2670 (2017)
33. Pagliardini, M., Gupta, P., Jaggi, M.: Unsupervised learning of sentence embeddings using compositional n-gram features. In: Proceedings of NAACL-HLT, pp. 528–540 (2018)
34. Qin, J., Liu, L., Shao, L., Shen, F., Ni, B., Chen, J., Wang, Y.: Zero-shot action recognition with error-correcting output codes. In: Proceedings of the IEEE Conference on Computer Vision and Pattern Recognition, pp. 2833–2842 (2017)

35. Qiu, Z., Yao, T., Mei, T.: Learning spatio-temporal representation with pseudo-3d residual networks. In: proceedings of the IEEE International Conference on Computer Vision, pp. 5533–5541 (2017)

36. Rohrbach, M., Regneri, M., Andriluka, M., Amin, S., Pinkal, M., Schiele, B.: Script data for attribute-based recognition of composite activities. In: Fitzgibbon, A., Lazebnik, S., Perona, P., Sato, Y., Schmid, C. (eds.) ECCV 2012. LNCS, vol. 7572, pp. 144–157. Springer, Heidelberg (2012). https://doi.org/10.1007/978-3-642-33718-5_11

37. Romera-Paredes, B., Torr, P.: An embarrassingly simple approach to zero-shot learning. In: International Conference on Machine Learning, pp. 2152–2161 (2015)

38. Shigeto, Y., Suzuki, I., Hara, K., Shimbo, M., Matsumoto, Y.: Ridge regression, hubness, and zero-shot learning. In: Appice, A., Rodrigues, P.P., Santos Costa, V., Soares, C., Gama, J., Jorge, A. (eds.) ECML PKDD 2015. LNCS (LNAI), vol. 9284, pp. 135–151. Springer, Cham (2015). https://doi.org/10.1007/978-3-319-23528-8_9

39. Simonyan, K., Zisserman, A.: Two-stream convolutional networks for action recognition in videos. In: Advances in Neural Information Processing Systems, pp. 568–576 (2014)

40. Soomro, K., Zamir, A.R., Shah, M.: Ucf101: a dataset of 101 human actions classes from videos in the wild. arXiv preprint arXiv:1212.0402 (2012)

41. Tran, D., Bourdev, L., Fergus, R., Torresani, L., Paluri, M.: Learning spatiotemporal features with 3d convolutional networks. In: Proceedings of the IEEE International Conference on Computer Vision, pp. 4489–4497 (2015)

42. Wang, X., Girshick, R., Gupta, A., He, K.: Non-local neural networks. In: Proceedings of the IEEE Conference on Computer Vision and Pattern Recognition, pp. 7794–7803 (2018)

43. Xian, Y., Lorenz, T., Schiele, B., Akata, Z.: Feature generating networks for zero-shot learning. In: Proceedings of the IEEE Conference on Computer Vision and Pattern Recognition, pp. 5542–5551 (2018)

44. Xian, Y., Schiele, B., Akata, Z.: Zero-shot learning-the good, the bad and the ugly. In: Proceedings of the IEEE Conference on Computer Vision and Pattern Recognition, pp. 4582–4591 (2017)

45. Xu, X., Hospedales, T., Gong, S.: Transductive zero-shot action recognition by word-vector embedding. Int. J. Comput. Vis. **123**(3), 309–333 (2017)

46. Xu, X., Hospedales, T.M., Gong, S.: Multi-task zero-shot action recognition with prioritised data augmentation. In: Leibe, B., Matas, J., Sebe, N., Welling, M. (eds.) ECCV 2016. LNCS, vol. 9906, pp. 343–359. Springer, Cham (2016). https://doi.org/10.1007/978-3-319-46475-6_22

47. Ye, Z., et al.: Zero-shot text classification via reinforced self-training. In: Proceedings of the 58th Annual Meeting of the Association for Computational Linguistics, pp. 3014–3024 (2020)

48. Zhang, L., Xiang, T., Gong, S.: Learning a deep embedding model for zero-shot learning. In: Proceedings of the IEEE Conference on Computer Vision and Pattern Recognition, pp. 2021–2030 (2017)

49. Zhu, Y., Long, Y., Guan, Y., Newsam, S., Shao, L.: Towards universal representation for unseen action recognition. In: Proceedings of the IEEE Conference on Computer Vision and Pattern Recognition, pp. 9436–9445 (2018)

Few-Shot Class-Incremental Learning for 3D Point Cloud Objects

Townim Chowdhury[1] , Ali Cheraghian[2,3] , Sameera Ramasinghe[4] ,
Sahar Ahmadi[5] , Morteza Saberi[6] , and Shafin Rahman[1](✉)

[1] Department of Electrical and Computer Engineering, North South University,
Dhaka, Bangladesh
{townim.faisal,shafin.rahman}@northsouth.edu
[2] School of Engineering, Australian National University, Canberra, Australia
ali.cheraghian@anu.edu.au
[3] Data61, Commonwealth Scientific and Industrial Research Organisation,
Canberra, Australia
[4] Australian Institute for Machine Learning, University of Adelaide,
Adelaide, Australia
sameera.ramasinghe@adelaide.edu.au
[5] Business School, The University of New South Wales, Sydney, Australia
sahar.ahmadi@unsw.edu.au
[6] School of Computer Science and DSI, University of Technology Sydney,
Sydney, Australia
morteza.saberi@uts.edu.au

Abstract. Few-shot class-incremental learning (FSCIL) aims to incrementally fine-tune a model (trained on base classes) for a novel set of
classes using a few examples without forgetting the previous training.
Recent efforts address this problem primarily on 2D images. However,
due to the advancement of camera technology, 3D point cloud data has
become more available than ever, which warrants considering FSCIL on
3D data. This paper addresses FSCIL in the 3D domain. In addition
to well-known issues of catastrophic forgetting of past knowledge and
overfitting of few-shot data, 3D FSCIL can bring newer challenges. For
example, base classes may contain many synthetic instances in a realistic
scenario. In contrast, only a few real-scanned samples (from RGBD sensors) of novel classes are available in incremental steps. Due to the data
variation from synthetic to real, FSCIL endures additional challenges,
degrading performance in later incremental steps. We attempt to solve
this problem using Microshapes (orthogonal basis vectors) by describing
any 3D objects using a pre-defined set of rules. It supports incremental
training with few-shot examples minimizing synthetic to real data variation. We propose new test protocols for 3D FSCIL using popular synthetic datasets (ModelNet and ShapeNet) and 3D real-scanned datasets
(ScanObjectNN and CO3D). By comparing state-of-the-art methods, we
establish the effectiveness of our approach in the 3D domain. Code is
available at: https://github.com/townim-faisal/FSCIL-3D.

Keywords: 3D point cloud · Few-shot class-incremental learning

Supplementary Information The online version contains supplementary material
available at https://doi.org/10.1007/978-3-031-20044-1_12.

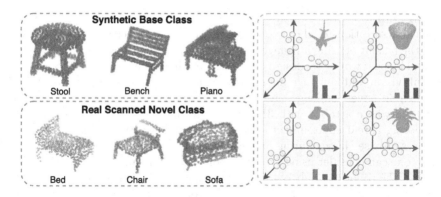

Fig. 1. *(Left)* A realistic setup of FSCIL may consider synthetic stool, bench, and piano as base classes and real-scanned bed, chair and sofa as novel (few-shot) classes. One can notice that shared 'leg' looks (size and shape) different in base and novel classes, which is a source of domain gap. This gap intensified when the synthetic 'legs' needed to generalize the real-scanned (noisy) 'legs'. Therefore, we propose to describe all 'legs' under a common description, named Microshape. Note that Microshapes are abstract property that may not always have semantically conceivable meaning like 'legs'. They are orthogonal vectors expressing shared semantics of any object. *(Right)* High dimensional representations of 3D points are projected onto Microshapes (red, green and blue axis) and calculated to what extent (colored bars) an object contains those Microshapes, which construct a semantic description of the given object. (Color figure online)

1   Introduction

Humans have a remarkable ability to gradually expand their knowledge while keeping past insights intact. Similarly, natural learning systems are incremental in nature, where new knowledge is continually learned over time while preserving existing knowledge [15]. Therefore, it is important to design systems that have the ability to learn incrementally when exposed to novel data. The above task becomes more challenging in realistic scenarios, where only a few samples of novel classes are available for training. This restriction makes incremental learning further difficult due to two reasons: *a*) overfitting of novel classes with few training samples and *b*) catastrophic forgetting of the previous knowledge. This problem setting is known as Few-shot class incremental learning (FSCIL) in the literature [1–3,17]. While incremental learning for image data has been studied to a certain extent, its extension to 3D data remains unexplored. To our knowledge, this is the first work that aims to tackle FSCIL on 3D point clouds.

FSCIL methods usually train a base model initially using abundantly available base class instances. Then, new data samples of few-shot incremental tasks are added over time to train the network. During incremental training, models typically tend to overfit few-shot data and forget previously trained class knowledge. Existing works on FSCIL have addressed the latter issue using a small memory module that contains few examples of the previous tasks [36] (during learning with few-shot examples of novel classes) and the former issue by prototypical descriptions of class representation [49]. More recent works [8,9] advocated the benefit of using language prototype instead of vision/feature prototype. Inspired by past works, we design our FSCIL method based on a memory module, and link vision and language information in an end-to-end manner. But, FSCIL on 3D entail novel challenges in addition to overfitting and forgetting. Assume a scenario where a robot with 3D sensors and incremental learning capability is exploring a real environment. The robot is already well-trained with many instances of base classes during its construction. Because of the abundant availability, one can consider using synthetic point cloud instances to train the robot. However, the robot can only obtain a few samples of real-world scanned data for incremental stages. It is a challenge for the robot trained on base classes (with synthetic data) to gather knowledge of incremental classes with newly scanned real-world data. Previous works [37] showed that synthetic data-based training and real-world data-based testing leads to a significant performance gap because of the variability of feature distribution (domain gap) of objects (see Fig. 1). Similarly, the transition from the synthetic data dependant base task to real-world data dependant incremental tasks makes feature-prototype alignments cumbersome during the incremental steps, eventually magnifying forgetting and overfitting issues. As a result, the models experience a drastic drop in performance in the subsequent incremental steps. This paper attempts to address this issue using a set of Microshape descriptions obtained from base classes instances.

Our goal is to represent synthetic and real objects in a way that helps to align 3D point cloud features with language prototype of classes. This representation should be robust to noisy real-scanned objects, so that object features describe a meaningful set of attributes. To this end, based on many base class instances, we calculate a set of Microshape descriptions (vectors orthogonal to each other) working as a building block for any 3D shape. Each Microshape describes a particular aspect of a 3D shape that might be shared across different objects. We project a high dimensional representation of 3D points onto each Microshape and find to what extent each Microshape is present in the given object. Aggregating (average pool) the strength of all Microshape presented in a given shape, we create a single feature vector describing that 3D shape. Representing all class objects under a common set of Microshapes, the overall object representation becomes relatively more noise-tolerant, minimizes the domain gap from synthetic to real data, and better aligns features to prototypes. Also, such representation can be equally useful for related problem like 3D recognition and dynamic few-shot learning. Moreover, we have proposed new experimental testbeds for testing FSCIL on 3D point cloud objects based on two 3D synthetic datasets, ModelNet [42] and ShapeNet [4], and two 3D real world-scanned datasets, ScanObjectNN [37] and Common Objects in 3D (CO3D) [29]. We benchmark popular FSCIL methods on this new setup and show the superiority of our proposed Microshape based method over the existing works.

In summary, the main contributions of our paper are as follow: *(a)* To the best of our knowledge, we are the first to report few-shot class incremental learning results for 3D point cloud objects. Moreover, we propose a novel and realistic problem setup for 3D objects where base classes are originated from synthetic data, and later few-shot (novel) classes obtained from real-world scanned objects are incrementally added over time (in future tasks). The motivation is that synthetic 3D point cloud objects may not be available for rare class instances, whereas few real-world scanned data may be readily available. *(b)* We propose a new backbone for feature extraction from 3D point cloud objects that can describe both synthetic and real objects based on a common set of descriptions, called Microshapes. It helps to minimize the domain gap between synthetic and real objects retaining old class knowledge. *(c)* We propose experimental testbeds for 3D class-incremental learning based on two 3D synthetic datasets, and two 3D real-scanned datasets. We also perform extensive experiments using existing and proposed methods by benchmarking performances on the proposed setups.

2   Related Work

3D Point Cloud Object Recognition: Thanks to recent advent of 3D sensors [14,31,44], many works have been proposed to classify 3D point cloud objects directly [16,19,24–27,40,43]. Qi *et al.* [25] proposed Pointnet, as the pioneer work, to process 3D point cloud with multi-layer perceptron (MLP) networks. This approach ignores local structures of the input data. Later methods are developed to handle this limitation [16,19,24,26,27,40,43]. [26] introduced PointNet++ to take advantages of local features by extracting features hierarchically. [16,19,24,27,40,43] proposed several convolution strategies to extract local information. [40] introduced PointConv to define convolution operation as locating a Monte Carlo estimation of the hidden continuous 3D convolution with respect to an important sampling strategy. [27] introduced graph convolution on spherical points, which are projected from input point clouds. Some other works [38,39,46] consider each point cloud as a graph vertex to learn features in spatial or spectral domains. [39] proposed DGCNN to construct a graph in feature space and update it dynamically using MLP for each edge. [46] suggested a method to construct the graph using k-nearest neighbors from a point cloud to capture the local structure and classify the point cloud. As the points in 3D point clouds represent positional characteristics, [47] proposed to employ self-attention into a 3D point cloud recognition network. [21] suggested replacing the explicit self-attention process with a mix of spatial transformer and multi-view convolutional networks. To provide permutation invariance, Guo *et al.* [12] presented offset-attention with an implicit Laplace operator and normalizing refinement. In this paper, we propose a novel permutation invariant feature extraction process.

Incremental Learning: Incremental learning methods are separated into three categories, task-incremental [5,23,30], domain-incremental [32,45], and class-incremental [3,13,28,41] learning. Here, we are interested in the class-incremental learning. Rebuffi *et al.* [28] used a memory bank called "episodic memory", of the old classes. For novel classes, the nearest-neighbor classifiers are incrementally accommodated. Castro *et al.* [3] employed a knowledge distillation cost function to store information of the previously seen classes and a classification cost function to learn the novel classes. Hou *et al.* [13] introduced a novel method for learning a unified classifier that reduces the imbalance between old and novel classes. Wu *et al.* [41] fine-tuned the bias in the model's output with the help of a linear model. They proposed a class-incremental learning method working on the low data regime. Simon *et al.* [33] presented a knowledge distillation loss using geodesic flow to consider gradual change among incremental tasks. Liu *et al.* [18] introduced Adaptive Aggregation Networks to aggregate two feature outputs from two residual blocks at each residual level achieving a better stability-plasticity trade-off. Zhu *et al.* [49] proposed a non-exemplar-based method consisting of prototype augmentation and self-supervision to avoid memory limitations and task-level overfitting.

Few-Shot Class-Incremental Learning: FSCIL problem setting was proposed by Tao *et al.* [36]. They proposed a neural gas network to minimize the forgetting issue by learning and preserving the topology of the feature from different classes. Chen *et al.* [6] suggested a non-parametric method to squeeze knowledge of the old tasks in a small quantized vector space. They also consider less forgetting regularization, intra-class variation, and calibration of quantized vectors to minimize the catastrophic forgetting. After that, [20] introduced a method that selects a few parameters of the model for learning novel classes to reduce the overfitting issue. Also, by freezing the important parameters in the model, they prevent catastrophic forgetting. [8] employed class semantic information from text space with a distillation technique to mitigate the impact of catastrophic forgetting. Additionally, they utilize an attention mechanism to reduce the overfitting issue on novel tasks, where only a tiny amount of training samples are available for each class. [9] proposed a method that creates multiple subspaces based on the training data distribution, where each subspace is created based on one specific part of the training distribution, which ends to unique subspaces. The preceding approaches were only explored on 2D image data, whereas our method investigates FSCIL in 3D point cloud data.

3   Few-Shot Class-Incremental Learning for 3D

Problem Formulation. FSCIL models learn novel classes with a small amount of training samples over time in a variety of (incremental) tasks. Each task contains novel classes which have not been observed by the model beforehand. Assume a sequence of T tasks, $\mathcal{H} = \{h^1, h^2, ..., h^T\}$, where $\mathcal{Y}^t$ is the set of classes in the task h^t, and $\mathcal{Y}^i \cap \mathcal{Y}^j = \emptyset$. In addition, a set of d-dimensional semantic prototype for each class of all tasks are assigned as $\mathcal{S}^t$. That is, each task can be represented with a tuple $h^t = \{\mathcal{X}_i^t, \mathbf{y}_i^t, \mathbf{s}_i^t\}_{i=1}^{n_t}$, where, $\mathcal{X}_i^t = \{\mathbf{x}_i^t\}_{i=1}^{l}$ denotes a 3D point cloud object with coordinates $\mathbf{x}_i^t \in \mathbb{R}^3$. Further, $\mathbf{y}_i^t \in \mathcal{Y}^t$ and $\mathbf{s}_i^t \in \mathcal{S}^t$ are the label of the point cloud and the corresponding semantic embedding feature, respectively. In the proposed FSCIL setting, h^1 is the base task ($t = 1$) where the model is trained on a large scale synthetic 3D dataset.

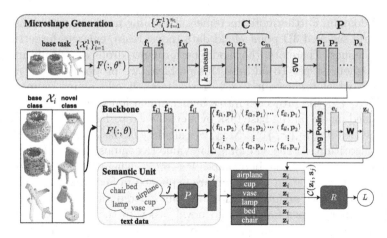

Fig. 2. Overall architecture. **Microshape Generation:** All training samples of the base task are used to calculate Microshapes. To this end, the base training set $\{\mathcal{X}_i^1\}_{i=1}^{n_1}$ are fed into $F(:, \theta^*)$ in order to extract features for all points $\{\mathcal{F}_i^1\}_{i=1}^{n_1}$. Next, we apply K-means to find m centers $\mathbf{C} \in \mathbb{R}^{q \times m}$ of the point cloud features. Ultimately, we employ SVD to form Microshape matrix $\mathbf{P} \in \mathbb{R}^{q \times u}$. **Backbone:** Our proposed backbone for generating features of point cloud input is presented. To be more specific, given an instance from 3D point cloud data $\mathcal{X}_i$, we map it into a higher dimensional space using $F(:, \theta)$. After that, we calculate the inner product between all points and Microshape basis $\langle \mathbf{f}_i, \mathbf{p}_k \rangle$, where $\mathbf{f}_i, \mathbf{p}_k \in \mathbb{R}^q$. At the end, we obtain the feature embedding $\mathbf{z}_i \in \mathbb{R}^d$ after Avg. pooling and a projection $\mathbf{W} \in \mathbb{R}^{d \times u}$ module. $F(:, \theta)$ and $\mathbf{W}$ are trainable at the first incremental step but remain fixed in later steps. **Semantic Unit:** Semantic prototypes, $\mathbf{s}_j$ of all novel and base classes are generated by a language model, P. At the end, a pair of point cloud features and semantic prototype of classes are provided into the relation module, which calculates a class label based on the similarity score.

For $t > 1$, the training data are sampled from real world 3D point clouds, and only consist of few instances, *i.e.*, $n^t \ll n^1$. The model is trained sequentially over the tasks $t = 1, \ldots, T$. However, during t-th task, the model sees $\mathcal{X}^t$, y^t and $\{\mathcal{S}^1, \mathcal{S}^2, ..., \mathcal{S}^t\}$. At inference, the trained model on the current task h^t must classify test samples of current and old tasks *i.e.*, $\{h^1, h^2, ..., h^t\}$.

3.1 Model Overview

To solve the FSCIL problem on the 2D domain, we need to design an algorithm that can address the catastrophic forgetting of old classes and the overfitting issue on novel classes [36]. Likewise, we require to follow the same procedure to develop an FSCIL model for 3D point cloud data. In this paper, we use the exemplar approach [1–3,13,28], which is an established method in the literature, to address the forgetting issue. To this end, for old tasks, we save a sample, chosen randomly for each class in a tiny memory $\mathcal{M}$. Additionally, we employ semantic embedding features in our proposed baseline model to tackle the overfitting issue. Such semantics have been used recently [8,9] for the FSCIL in the 2D domain. Furthermore, incremental learning from synthetic (base task) to real-world scanned 3D point cloud data (novel task), enlarges forgetting and overfitting issues. Therefore, the model performance drops drastically in subsequent increment steps. To address this problem, we propose a novel method, Microshape, which is shown in Fig. 2. In this method, all training samples of the base task $\{\mathcal{X}_i^1\}_{i=1}^{n_1}$ are projected into a higher dimension space by $F(:, \theta^*)$, which is a pretained model on the base task by a cross-entropy loss and consist of a few MLP layers, leading to $\{\mathcal{F}_i^1\}_{i=1}^{n_1}$. Next, we calculate m class centers $\mathbf{C} \in \mathbb{R}^{q \times m}$ by K-means algorithm. At the end, Singular Value Decomposition (SVD) is used to choose the most important Microshapes $\mathbf{P} \in \mathbb{R}^{q \times u}$, where $u < m$. The details of the Microshape generation algorithm are explained in Sect. 3.2. After this step, we train our proposed architecture for 3D point cloud data, shown in Fig. 2, with the help of Microshapes for base and novel tasks. In the point cloud pipeline, there are two kinds of classes: novel and old. It is important to mention that the old class samples are from the tiny memory $\mathcal{M}$. In the text pipeline, there is one representation per class for old and novel classes. While training the baseline model, we forward a 3D point cloud sample, $\mathcal{X}_i$, into the projection module $F(:, \theta)$ to extract point cloud features $\mathcal{F}_i = \{\mathbf{f}_i\}_{i=1}^n$, where $\mathbf{f}_i \in \mathbb{R}^q$. Then, we find the inner product $\langle , \rangle$ between all point cloud features $\mathbf{f}_i$ and Microshape $\mathbf{p}_k \in \mathbb{R}^q$. After that, we use average pooling for each Microshape to form a feature vector representation $\mathbf{e}_i \in \mathbb{R}^u$ of the point cloud input $\mathcal{X}_i$. Next, we map the feature vector $\mathbf{e}_i$ into semantic embedding space by $\mathbf{W} \in \mathbb{R}^{d \times u}$, which is a fully connected layer, to form $\mathbf{z}_i \in \mathbb{R}^d$. In the text domain, all current and previous class labels are fed into a semantic embedding module P, which can generate a feature embedding $\mathbf{s}_j \in \mathbb{R}^d$ for each class. Ultimately, we forward $\mathbf{z}_i$ and $\mathbf{s}_j$ in the relation network [34] R, consist of a few fully connected layers, which gives a prediction given input $\mathcal{X}_i$. In the end, we use a binary cross-entropy loss to train the model. It is important to mention that $F(:, \theta)$ and $\mathbf{W}$ are trained only on the base task and kept frozen for the incremental tasks. While the relation module R is fine-tuned for all incremental tasks.

3.2   Microshapes

Incremental learning faces newer challenges, from 2D images to 3D point cloud objects. *First,* it is more difficult to obtain noise-free real-world 3D data reconstructed from RGB-D scan than natural 2D images. 3D objects are susceptible to partial observations, occlusions, reconstruction errors, and cluttered backgrounds. An effect of the noise-free version of 3D (synthetic) and real-world scanned objects is shown in Fig. 1 (left). *Second,* it is realistic to assume that 3D (synthetic) objects are relatively abundant for base classes but scarce for novel classes. A practical incremental learning setup should consider synthetic objects for training base classes (1st incremental step) and real-scanned objects for the rest of the incremental steps containing few-shot classes. The 3D-feature and language-prototype alignment learned using base instances could not generalize well for real-scanned data of novel classes. It increases the domain gap between base and novel classes. Now, we formally present the following hypothesis.

Hypothesis 1. *Every 3D shape can be adequately represented as a particular combination of smaller entities. These entities are common across various 3D shapes, while the combinations might vary.*

Based on the above hypothesis, we strive to mitigate the domain gap between synthetic and real 3D point cloud data by proposing a novel feature extraction method, called Microshapes. In the proposed method, the common properties of samples are selected to form a feature representation which is robust to domain shift between synthetic and real data. The assumption is that there are many similar points among different point cloud objects that can form as a cluster and can be termed as Microshape. After that, we extract feature representation of all samples based on Microshapes.

Microshape Generation. Let $\{\mathcal{X}_i^1\}_{i=1}^{n_1}$ be the set of 3D samples used to train the model in the base task h^1. We forward $\{\mathcal{X}_i^1\}_{i=1}^{n_t}$ through the backbone F and extract M features $\{\mathbf{f}_i\}_{i=1}^M$, where $f_i \in \mathbb{R}^q$, $M = l \times n_1$, and l is the number of points in a point cloud. For the backbone, we use a pre-trained PointNet (trained using $\{\mathcal{X}_i^1\}_{i=1}^{n_t}$), after removing the pooling and classification layers. Then, we obtain m cluster centers by applying K-means on the set $\{\mathbf{f}_i\}_{i=1}^M$. Let us denote the cluster centers by $\mathbf{C} \in \mathbb{R}^{q \times m}$. Our intention is to define a vector space spanned by the column vectors of $\mathbf{C}$ which then can be used to represent novel 3D point clouds as a vector in this space. However, the column vectors of $\mathbf{C}$ is not necessarily a basis since they might not be linearly independent. On the other hand, the number of cluster centers m is a hand picked hyperparameter, which might not be optimal. Therefore, we use Singular Value Decomposition (SVD) to choose a set of basis vectors that approximately spans $\mathbf{C}$. We decompose the matrix consisting of samples within a cluster as $\mathbf{C} = \mathbf{U}\mathbf{D}\mathbf{V}^\top$. Then, the u leading left singular vectors $\mathbf{U}$ form an orthogonal basis which we define by $\mathbf{P}$, i.e., $\mathbb{R}^{q \times u} \ni \mathbf{P} = [\mathbf{p}_1, ..., \mathbf{p}_u]$; $\mathbf{P}^\top \mathbf{P} = \mathbf{I}_n$ (see Fig. 2).

Definition 1. *We define each column vector of P as a Microshape.*

Role of SVD. Since the Microshapes are mutually orthogonal, they contain minimal mutual information, which allows us to encode maximum amount of total information within them. In comparison, the raw cluster centers obtained using k-means might be linearly dependant, representing duplicate information. Choosing orthogonal basis vectors based on the 95% of energy of singular values allows to find the number of Microshapes required to describe a dataset in a principled manner. Further, since the number of cluster centers is a hand picked parameter, they can result in either redundant or insufficient information encoding. In contrast, using SVD allows us eliminate the effect of this hyperparameter.

Microshape Feature Extraction. To form a feature vector that describes a particular 3D shape, we use these Microshapes. Consider a point cloud $\mathcal{X}_i = \{\mathbf{x}_{i1}, ..., \mathbf{x}_{il}\}$. We project it into a higher dimensional space using F, which gives $\mathcal{F}_i = \{\mathbf{f}_{i1}, ..., \mathbf{f}_{il}\}$. Next, for each Microshape, we calculate the average similarity value to create the feature vector,

$$\mathbf{e}_i = \frac{1}{l} \sum_{b=1}^{l} [\mathbf{f}_{ib} \cdot \mathbf{p}_1, \mathbf{f}_{ib} \cdot \mathbf{p}_2, \ldots, \mathbf{f}_{ib} \cdot \mathbf{p}_u], \qquad \mathbf{z}_i = ReLU(\mathbf{W}.\mathbf{e}_i), \qquad (1)$$

where, $\mathbf{e}_i \in \mathbb{R}^u$. Note that $\mathbf{e}_i$ is a permutation invariant representation, since it is calculated using all the projected points $\mathbf{f}_i$'s of the original point cloud, and is independent of the point ordering. An FC layer ($\mathbf{W}$) converts $\mathbf{e}_i$ to $\mathbf{z}_i \in \mathbb{R}^d$. Next, we describe our training pipeline utilizing the obtained $\mathbf{z}_i$ of each objects.

Benefit of Mircoshape Features. FSCIL allows only a few examples of real-scanned objected during incremental stages. Inherent noise/occlusion/clutter background presented in 3D models results in overfitting to unnecessary information the incremental learning steps. In this context, Microshpae based feature representation has benefits over general features such as PointNet representations. Below, we list several key benefits of Mircoshape features: *(a) Reduced impact of noise:* Microshapes allows the model to describe any object based on common entities. As explained earlier, traditional backbone (such as Pointnet) features can represent redundant information. In contrast, Microshapes only try to estimate to what extent these pre-defined properties are present in a given input and helps to discard unnecessary information. Such an approach is beneficial to describe novel objects, especially in cross-domain (synthetic to real) situations. *(b) Aligning with the prototypes:* We use language prototypes that contain semantic regularities such that similar concepts like (table, chair, etc.) are located in nearby positions in a semantic space. By describing objects based on their similarity to pre-defined Microshapes, features can describe different semantics/attributes of a given object. As language prototypes and 3D features are based on object semantics rather than low-level (CNN) features, aligning them becomes an easier task for the relation network. *(c) Microshapes as feature prototypes:* For any few-shot learning problem, Microshapes can serve the role of creating feature prototypes (instead of language prototypes used in this paper). One can consider the average of Microshape-based features of all available instances belonging to any class, as a class prototype. This could be useful to describe fine-grained objects (e.g., China/Asian Dragon, Armadillo) for which language prototypes are less reliable. In the experiments, we deal with common objects where language prototypes work better than feature prototypes.

3.3   Training Pipeline

First, we train the backbone F on the base task h^1, with a large number of synthetic 3D samples. For the novel tasks, $h^{t>1}$, the backbone F is kept frozen. The semantic class backbone P, which is a pretrained model, *e.g.* BERT [10] and w2v [22], is kept frozen during the entire training stage. Additionally, for each old class, we randomly select a training sample which is stored on a tiny memory $\mathcal{M}$. Let us define the feature vector obtained using Eq. 1 for the ith point cloud of the tth task as $\mathbf{z}_i^t$. To train the proposed model for a task h^t, at

Table 1. Summary of our experimental setups.

Experiment setups	# Base classes	# Novel classes	# Tasks	#Train in base	#Test in base	#Test in novel
ModelNet40	20	20	5	7438	1958	510
ShapeNet	25	30	7	36791	9356	893
CO3D	25	25	6	12493	1325	407
ModelNet40 → ScanObjectNN	26	11	4	4999	1496	475
ShapeNet → ScanObjectNN	44	15	4	22797	5845	581
ShapeNet → CO3D	39	50	11	26287	6604	1732

first, we generate the features $\mathbf{z}_i^t$'s using Eq. 1 for the n_t training samples of the current task h^t. Then, the features $\mathbf{z}_i^t$ and semantic class embedding $\mathbf{s}_j$ of the tasks $h^1, \ldots, h^t$ are forwarded into a relation module R which provides a score between $[0, 1]$, representing the similarity between them. In other words, for each training sample, we generate a score against each of the classes in both the novel and previous tasks as, $r_{ij}^t = \gamma \circ R \circ (\mathbf{z}_i^t \oplus \mathbf{s}_j), j \in \mathcal{Y}_{tl}$ where, $\mathcal{Y}_{tl} = \bigcup_{i=1}^t \mathcal{Y}^i$, $\oplus$ is the concatenation operator, R is the relation module, and γ is the sigmoid function. Finally, for each feature r_{ij}, and the corresponding ground truth $\mathbf{y}_i$, we employ a binary cross entropy cost function to train the model as,

$$L = -\frac{1}{|\mathcal{Y}_{tl}||\mathcal{S}|} \sum_{j \in \mathcal{Y}_{tl}} \sum_{y_i \in \mathcal{S}} \left(1\!\!1_{(y_i=j)} \log(r_{ij}) + (1 - 1\!\!1_{(y_i=j)}) \log(1 - r_{ij}) \right) \qquad (2)$$

where $\mathcal{S}$ is the set of true labels in the current task and the memory $\mathcal{M}$.

Inference. During inference, given the trained model an unlabeled sample $\mathcal{X}^c$, $c \in \mathcal{Y}_{tl}$, the prediction of the label is calculated by $y^* = \arg\max_{j \in \mathcal{Y}_{tl}} \left(R \circ (\mathbf{z}_c \oplus \mathbf{s}_j) \right)$, where $\mathbf{z}_c$ is the feature calculated using the microshapes for the sample $\mathcal{X}^c$.

4   Experiments

Datasets. We experiment on two 3D synthetic datasets, (ModelNet [42], ShapeNet [4]) and two 3D real-scanned datasets (ScanObjectNN [37] and CO3D [29]).

Setups. We propose two categories of experiments for 3D incremental learning: within- and cross-dataset experiments. Within dataset experiments are done on both synthetic and real datasets individually where the base and incremental classes come from the same dataset. For cross-dataset experiments, base and incremental classes come from synthetic and real-world scanned datasets, respectively. In Table 1, we summarize three within and three cross-dataset experimental setups proposed in this paper. We utilize the class distribution of ModelNet, ShapeNet, and CO3D datasets to select base and incremental few-shot classes for within dataset experiments. First, we sort all classes in descending order based on instance frequency. Then top 50% of total classes (having many available

Table 2. Overall FSCIL results for within dataset experiments

Method	ModelNet						CO3D							ShapeNet							
	20	25	30	35	40	Δ↓	25	30	35	40	45	50	Δ↓	25	30	35	40	45	50	55	Δ↓
FT	89.8	9.7	4.3	3.3	3.0	96.7	76.7	11.2	3.6	3.2	1.8	0.8	99.0	87.0	25.7	6.8	1.3	0.9	0.6	0.4	99.5
Joint	89.8	88.2	87.0	83.5	80.5	10.4	76.7	69.4	64.8	62.7	60.7	59.8	22.0	87.0	85.2	84.3	83.0	82.5	82.2	81.3	6.6
LwF [17]	89.8	36.0	9.1	3.6	3.1	96.0	76.7	14.7	4.7	3.5	2.3	1.0	98.7	87.0	60.8	33.5	15.9	3.8	3.1	1.8	97.9
IL2M [1]	89.8	65.5	58.4	52.3	53.6	40.3	76.7	31.5	27.7	18.1	27.1	21.9	71.4	87.0	58.6	45.7	40.7	50.1	49.4	49.3	43.3
ScaIL [2]	89.8	66.8	64.5	58.7	56.5	37.1	76.7	39.5	34.1	24.1	30.1	27.5	64.1	87.0	56.6	51.8	44.3	50.3	46.3	45.4	47.8
EEIL [3]	89.8	75.4	67.2	60.1	55.6	38.1	76.7	61.4	52.4	42.8	39.5	32.8	57.2	87.0	77.7	73.2	69.3	66.4	65.9	65.8	22.4
FACT [48]	90.4	81.3	77.1	73.5	65.0	28.1	77.9	67.1	59.7	54.8	50.2	46.7	40.0	87.5	75.3	71.4	69.9	67.5	65.7	62.5	28.6
Sem-aware [8]	91.3	82.2	74.3	70.0	64.7	29.1	76.8	66.9	59.2	53.6	49.1	42.9	44.1	87.2	74.9	68.1	69.0	68.1	66.9	63.8	26.8
Ours	**93.6**	**83.1**	**78.2**	**75.8**	**67.1**	**28.3**	**78.5**	**67.3**	**60.1**	**56.1**	**51.4**	**47.2**	**39.9**	**87.6**	**83.2**	**81.5**	**79.0**	**76.8**	**73.5**	**72.6**	**17.1**

instances) are chosen as base classes, and the rest (having relatively fewer available examples) are added incrementally in our experiments. The motivation is that rare categories are the realistic candidate for novel (few-shot) classes. In the ModelNet40 experiment, there is one base task with 20 classes and four incremental tasks with another 20 classes. ShapeNet and CO3D experiments have 25 base classes, whereas incremental classes are 30 and 25, respectively, divided into 6 and 4 tasks. Among cross-dataset experimental setups, we choose base classes from a synthetic dataset and later add incremental classes from a real-world scanned dataset. For ModelNet40 → ScanObjectNN experiment, we follow the selection of base and incremental (novel) classes from [7], and it has total 4 tasks. ShapeNet → ScanObjectNN has four tasks where 44 non-overlapping ShapeNet and 15 ScanObjectNN classes are used as the base and incremental classes, respectively. ShapeNet → CO3D experiment has a sequence of 11 tasks with 44 non-overlapped classes from ShapeNet as base classes and 50 classes from CO3D as incremental classes. This setup is the most challenging and realistic among all experiments because of its vast number of tasks, classes, and object instances. Unless mentioned explicitly, we use randomly selected five 3D models as few-shot samples/class in all FSCIL setups and allow one exemplar 3D model/class (randomly chosen) as the memory of previous tasks. For synthetic and real-scanned object classes, we use synthetic and real-scanned data as few-shot or exemplar data, respectively. See the supplementary material for more details about the setups.

Semantic Embedding. We use 300-dim. word2vec [22] as semantic representation for each classes of datasets. The word2vec is produced from word vectors trained in an unsupervised manner on an unannotated text corpus, Wikipidia.

Validation Strategy. We make a validation strategy for within dataset experiments. We randomly divide the set of base classes into val-base and val-novel. We choose 60% classes from base classes as val-base classes and the rest as val-novel classes to find the number of centroids for generating Microshapes. We find the number of centroids, $m = 1024$ performing across experimental setup.

Implementation Details. In all experiments and compared approaches, we use PointNet [25] as a base feature extractor to build the centroids for microshapes. We use the farthest 1024 points from 3D point cloud objects as input for all samples. We train the base feature extractor for 100 epochs using Adam optimizer with a learning rate of 0.0001. For microshape formulation, we use k-means clustering algorithms to initialize centroids. We randomly shift and scale points in the input point cloud during base and incremental class training with randomly dropping out points. During the training of all base classes, we employ the Adam optimizer with a learning rate of 0.0001 and batch sizes of 64. For novel classes, we choose the batch size of 16 and the learning rate of 0.00005. The feature vector size from backbone is 300 dimensional, similar to the dimension of the semantic prototypes. In the relation network, we utilize three fully connected layers of (600,300,1) using LeakyReLU activations, except the output layer uses Sigmoid activation. We use the *PyTorch* framework to perform our experiments.

Evaluation Metrics. We calculate the accuracy after each incremental step by combining both base and novel classes. Finally, as suggested in [35], we calculate the relative accuracy dropping rate, $\Delta = \frac{|acc_T - acc_0|}{acc_0} \times 100$, where, acc_T and acc_0 represent the last and first incremental task's accuracy, respectively. Δ summarizes the overall evaluation of methods. Lower relative accuracy indicates a better performance. We report the mean accuracy after ten different runs with random initialization.

Table 3. Overall results for cross dataset experiments

Method	ShapeNet → CO3D												ModelNet → ScanObjectNN					ShapeNet → ScanObjectNN				
	39	44	49	54	59	64	69	74	79	84	89	Δ↓	26	30	34	37	Δ↓	44	49	54	59	Δ↓
FT	81.0	20.2	2.3	1.7	0.8	1.0	1.3	0.9	0.5	1.6	98.0		88.4	6.4	6.0	1.9	97.9	81.4	38.7	4.0	0.9	98.9
Joint	81.0	79.5	78.3	75.2	75.1	74.8	72.3	71.3	70.0	68.8	67.3	16.9	88.4	79.7	74.0	71.2	19.5	81.4	82.5	79.8	78.7	3.3
LwF [17]	81.0	57.4	19.3	2.3	1.0	0.9	0.8	1.3	1.1	0.8	1.9	97.7	88.4	35.8	5.8	2.5	97.2	81.4	47.9	14.0	5.9	92.8
IL2M [1]	81.0	45.6	36.8	35.1	31.8	33.3	34.0	31.5	30.6	32.3	30.0	63.0	88.4	58.2	52.9	52.0	41.2	81.4	53.2	43.9	45.8	43.7
ScaIL [2]	81.0	50.1	45.7	39.1	39.0	37.9	38.0	36.0	33.7	33.0	35.2	56.5	88.4	56.5	55.9	52.9	40.2	81.4	49.0	46.7	40.0	50.9
EEIL [3]	81.0	75.2	69.3	63.2	60.5	57.9	53.0	51.9	51.3	47.8	47.6	41.2	88.4	70.2	61.0	56.8	35.7	81.4	74.5	69.8	63.4	22.1
FACT [48]	81.4	76.0	70.3	68.1	65.8	63.5	63.0	60.1	58.2	57.5	55.9	31.3	89.1	72.5	68.3	63.5	28.7	82.3	74.6	69.9	66.8	18.8
Sem-aware [8]	80.6	69.5	66.5	62.9	63.2	63.0	61.2	58.3	58.1	57.2	55.2	31.6	88.5	**73.9**	67.7	64.2	27.5	81.3	70.6	65.2	62.9	22.6
Ours	**82.6**	**77.9**	**73.9**	**72.7**	**67.7**	**66.2**	**65.4**	**63.4**	**60.6**	**58.1**	**57.1**	**30.9**	**89.3**	73.2	**68.4**	**65.1**	**27.1**	**82.5**	**74.8**	**71.2**	**67.1**	**18.7**

Table 4. Ablation study on using Microshapes and semantic prototypes.

Microshape	Prototype	ShapeNet → CO3D												ShapeNet							
		39	44	49	54	59	64	69	74	79	84	89	Δ↓	25	30	35	40	45	50	55	Δ↓
No	Feature	79.4	70.5	68.1	65.5	62.9	60.7	59.1	57.6	56.2	54.3	51.8	34.8	87.4	79.1	76.9	72.9	59.9	67.1	57.0	34.8
Yes	Feature	80.4	71.9	70.2	66.1	64.6	62.5	60.6	58.4	57.7	55.0	53.4	33.6	87.5	75.6	76.3	66	66.6	63.5	62.0	29.1
No	Language	80.6	75.9	66.3	66.1	66.0	63.9	62.8	60.0	56.5	54.1	53.6	33.5	87.3	79.4	76.8	72.8	68.6	66.1	63.7	27.0
Yes	Language	82.6	77.9	73.9	72.7	67.7	66.2	65.4	63.4	60.6	58.1	57.1	30.9	87.6	83.2	81.5	79.0	76.8	73.5	72.6	17.1

4.1   Main Results

Compared Methods. We compare our method with the following approaches. (1) *Fine-tuning (FT):* It is vanilla fine-tuning wherein each incremental task, the model is initialized with the previous task's weight and only uses a few samples from incremental classes. Note no exemplar is used in this method. (2)

Joint: All incremental classes are jointly trained using all samples belonging to those classes. *FT* and *Joint* are considered as lower and upper bound results, respectively. (3) State-of-the-art methods, e.g., IL2M [1], ScaIL [2], EEIL [3], LwF [17], FACT [48], and Sem-aware [8]. These methods originally reported results on 2D datasets. We replace CNN features with PointNet features and use their official implementation to produce results for 3D.

Analysis. We present the comparative results for within and cross dataset experiments in Table 2 and 3, respectively. Our observations are as follows: *(1)* All methods perform poorly in general in cross dataset experiments in comparison to within dataset case. This is due to the noisy data presented in 3D real datasets. *(2)* FT gets the lowest performance among all experimental setups because of catastrophic forgetting that occurred during training due to not utilizing any memory sample. In contrast, *Joint* achieves the best results because it uses all samples and trains all few-shot classes jointly at a time. This upper bound setup is not FSCIL. *(3)* Other state-of-the-art approaches (IL2M, ScaIL, EEIL, LwF, FACT, and Sem-aware) could not perform well in most of the experimental setups. IL2M [1] and ScaIL [2] propose a special training mechanism for 2D image examples. LwF [17] and EEIL [3] both apply knowledge distillation in loss but EEIL uses exemplar on the top of [17]. We use exemplar without any knowledge distillation. Interestingly, FACT [48] and Sem-aware [8] address few-shot class incremental learning in particular. Among them, Sem-aware [8] successfully applies class-semantic embedding information during training. In general, past approaches are designed aiming at 2D image data where challenges of 3D data are not addressed. *(4)* Our approach beats state-of-the-art methods on within and across dataset setups. We achieve superior results in each incremental task and relative accuracies. This success originates from Microshape descriptions and their ability to align with semantic prototypes, minimizing domain gaps. *(5)* Moreover, one can notice for all experimental setups, the performance of existing methods on base classes (1st incremental stages) are similar because of using the same PointNet features. Our approach beats them consistently. This success comes from Microshape based feature extraction. In addition, unlike other methods, our method gets benefited from language prototype and relation network. *(6)* Shapenet → CO3D setup has the longest sequence of tasks, arguably the most complicated of all setups. Other competing methods perform poorly in this particular setup. Our method outperforms the second best method, FACT, by up to 1.2% accuracy in the final task, with a relative performance drop rate of 30.9%, which is 0.4% lower than FACT's result.

4.2   Ablation Studies

Effect of Microshape. In Table 4, we discuss the effect of using Microshapes and (language or Microshape feature-based) prototype vectors discussed in Sect. 3.2. For no Microshape case, we use PointNet-like backbone features instead of Microshape-based features. We report results for different combinations of features and prototypes. One can notice that using Microshape or language prototypes individually outperforms no Microshape case. The reason is the inclusion

Table 5. Impact of *(Left)* loss, SVD, freezing and *(Right)* number of centroids.

Criteria	39	44	49	54	59	64	69	74	79	84	89	Δ↓
No freezing	82.3	75.8	69.4	65.1	62.7	61.5	58.3	57.3	57.9	52.2	50.7	38.4
No SVD	82.0	76.7	72.5	69.7	67.3	64.5	61.7	63.0	61.7	55.2	54.6	33.4
L_{mse}	82.4	77.6	73.1	71.0	67.5	66.1	62.5	61.1	60.3	56.4	55.2	33.0
L_{cross}	82.6	77.9	73.9	72.7	67.7	66.2	65.4	63.4	60.6	58.1	57.1	30.9

#	39	44	49	54	59	64	69	74	79	84	89	Δ↓
256	81.6	77.1	71.2	60.2	57.8	55.9	53.1	52.7	50.8	46.6	44.8	45.1
512	82.0	77.4	70.4	60.5	58.9	56.5	53.5	54.1	52.7	47.6	47.8	41.7
1024	82.6	77.9	73.9	72.7	67.7	66.2	65.4	63.4	60.6	58.1	57.1	30.9
2048	82.7	78	73.7	72.3	67.8	65.4	64.9	63.9	60.8	58.4	57.3	30.7
4096	82.3	77.2	72.9	72.0	67.1	66.1	64.8	63.0	60.8	57.9	57.0	30.7

of semantic description, transferring knowledge from base to novel classes. Also, we work on common object classes where rich language semantics are available. It could be a reason why language prototypes work better than feature prototypes. Finally, utilizing both Microshape features and language prototypes yields the best results aligning with our final recommendation.

Impact of Freezing, SVD and Loss. In Table 5 *(Left)*, We ablate our proposed method in terms of freezing, SVD, and loss. The performance gets reduced while keeping the full network trainable (no freezing) during all incremental steps. Training the full network using few-shot data during few-shot incremental steps promotes overfitting, which reduces performance. Not using the SVD step also degrades the performance because without SVD microshapes contain redundant information and do not guarantee orthogonality of microshapes. SVD step assembles the microshapes mutually exclusive and orthogonal to each other. Unlike the suggestion of [34] regarding RelationNet architecture, we notice that the cross-entropy loss (L_{cross}) in Eq. 2 performs better than MSE loss (L_{mse}). A possible reason could be we are using RelationNet for classification, not directly to compare the relation among features and language prototypes.

Hyperparameter Sensitivity. In Table 5 *(Right)*, we perform experiments varying numbers of centroids (of k-means clustering) on ShapeNet → CO3D setting. We find that a low amount of centroids (256 and 512) performs poorly in our proposed approach. However, employing 1024 (and above) centroids, the last task's accuracy and relative performance become stable. Increasing centroids up to 2048 or 4096 could not significantly differ in performance. The reason might be that applying SVD to select important centroids removes duplicate centroids, resulting in similar results for the higher number of centroids.

4.3 Beyond FSCIL

In Table 6, we experiment on dynamic few-shot learning (DFSL), proposed by [11], for Model-Net → ScanObjectNN setup. DFSL is equivalent to FSCIL with only two sequences of tasks instead of

Table 6. Dynamic few-shot learning

Microshape	Prototype	1-shot	5-shot
No	Feature	39.90 ± 0.71	51.65 ± 0.35
Yes	Feature	44.41 ± 1.05	64.54 ± 0.71
Yes	Language	**72.23 ± 0.52**	**72.53 ± 0.68**

various sequences in FSCIL. Note that traditionally DFSL is investigated on 2D images, but here we show results on 3D DFSL. The base classes, consisting of 26 classes, are chosen from ModelNet, but disjoint classes of the ScanObjectNN dataset, 11 classes, are used as novel classes. In each episode, the support set is built by selecting 1 or 5 examples of the novel classes, representing a 1-shot or 5-shot scenario. The query set consists of instances of the base and novel tasks. After averaging over 2000 randomly produced episodes from the test set, we evaluate few-shot classification accuracy. It has been demonstrated that utilizing more few-shot samples increases accuracy. Nevertheless, our approach (using Microshape and language prototype) can sufficiently address the DFSL problem.

5   Conclusions

This paper investigates FSCIL in the 3D object domain. It proposes a practical experimental setup where the base and incremental classes include synthetic and real-scanned objects, respectively. The proposed setup exhibits domain gaps related to class semantics and data distribution. To minimize the domain gap, we propose a solution to describe any 3D object with a common set of Microshapes. Each Microshape describes different aspects of 3D objects shared across base and novel classes. It helps to extract relevant features from synthetic and real-scanned objects uniquely that better aligns with class semantics. We propose new experimental protocols based on both within and cross-dataset experiments on synthetic and real object datasets. Comparing state-of-the-art methods, we show the superiority of our approach in the proposed setting.

Acknowledgement. This work was supported by North South University Conference Travel and Research Grants 2020–2021 (Grant ID: CTRG-20/SEPS/04).

References

1. Belouadah, E., Popescu, A.: IL2M: class incremental learning with dual memory. In: CVPR (2019)
2. Belouadah, E., Popescu, A.: ScaIL: classifier weights scaling for class incremental learning. In: WACV (2020)
3. Castro, F.M., Marín-Jiménez, M.J., Guil, N., Schmid, C., Alahari, K.: End-to-end incremental learning. In: Ferrari, V., Hebert, M., Sminchisescu, C., Weiss, Y. (eds.) ECCV 2018. LNCS, vol. 11216, pp. 241–257. Springer, Cham (2018). https://doi.org/10.1007/978-3-030-01258-8_15
4. Chang, A.X., et al.: ShapeNet: an information-rich 3D model repository. arXiv preprint arXiv:1512.03012 (2015)
5. Chaudhry, A., Dokania, P.K., Ajanthan, T., Torr, P.H.S.: Riemannian walk for incremental learning: understanding forgetting and intransigence. In: Ferrari, V., Hebert, M., Sminchisescu, C., Weiss, Y. (eds.) ECCV 2018. LNCS, vol. 11215, pp. 556–572. Springer, Cham (2018). https://doi.org/10.1007/978-3-030-01252-6_33
6. Chen, K., Lee, C.G.: Incremental few-shot learning via vector quantization in deep embedded space. In: ICLR (2021)
7. Cheraghian, A., Rahman, S., Chowdhury, T.F., Campbell, D., Petersson, L.: Zero-shot learning on 3D point cloud objects and beyond. arXiv preprint arXiv:2104.04980 (2021)

8. Cheraghian, A., Rahman, S., Fang, P., Roy, S.K., Petersson, L., Harandi, M.: Semantic-aware knowledge distillation for few-shot class-incremental learning. In: CVPR (2021)

9. Cheraghian, A., et al.: Synthesized feature based few-shot class-incremental learning on a mixture of subspaces. In: ICCV (2021)

10. Devlin, J., Chang, M.W., Lee, K., Toutanova, K.: BERT: pre-training of deep bidirectional transformers for language understanding. In: Proceedings of the 2019 Conference of the North American Chapter of the Association for Computational Linguistics: Human Language Technologies, Volume 1 (Long and Short Papers). Association for Computational Linguistics (2019)

11. Gidaris, S., Komodakis, N.: Dynamic few-shot visual learning without forgetting. In: CVPR (2018)

12. Guo, M.-H., Cai, J.-X., Liu, Z.-N., Mu, T.-J., Martin, R.R., Hu, S.-M.: PCT: point cloud transformer. Comput. Vis. Media 7(2), 187–199 (2021). https://doi.org/10.1007/s41095-021-0229-5

13. Hou, S., Pan, X., Loy, C.C., Wang, Z., Lin, D.: Learning a unified classifier incrementally via rebalancing. In: CVPR (2019)

14. Korres, G., Eid, M.: Haptogram: Ultrasonic point-cloud tactile stimulation. IEEE Access 4, 7758–7769 (2016)

15. Legg, S., Hutter, M.: Universal intelligence: a definition of machine intelligence. Minds Mach. 17, 391–444 (2008). https://doi.org/10.1007/s11023-007-9079-x

16. Li, Y., Bu, R., Sun, M., Wu, W., Di, X., Chen, B.: PointCNN: convolution on X-transformed points. In: NeurIPS (2018)

17. Li, Z., Hoiem, D.: Learning without forgetting. IEEE Trans. Pattern Anal. Mach. Intell. 40, 2935–2947 (2018)

18. Liu, Y., Schiele, B., Sun, Q.: Adaptive aggregation networks for class-incremental learning. In: CVPR (2021)

19. Liu, Y., Fan, B., Xiang, S., Pan, C.: Relation-shape convolutional neural network for point cloud analysis. In: CVPR (2019)

20. Mazumder, P., Singh, P., Rai, P.: Few-shot lifelong learning. In: AAAI (2021)

21. Mazur, K., Lempitsky, V.: Cloud transformers: a universal approach to point cloud processing tasks. In: ICCV (2021)

22. Mikolov, T., Sutskever, I., Chen, K., Corrado, G.S., Dean, J.: Distributed representations of words and phrases and their compositionality. In: NIPS (2013)

23. Nguyen, C.V., Li, Y., Bui, T.D., Turner, R.E.: Variational continual learning. In: ICLR (2018)

24. Poulenard, A., Rakotosaona, M.J., Ponty, Y., Ovsjanikov, M.: Effective rotation-invariant point CNN with spherical harmonics kernels. In: 3DV (2019)

25. Qi, C.R., Su, H., Mo, K., Guibas, L.J.: PointNet: deep learning on point sets for 3D classification and segmentation. In: CVPR (2017)

26. Qi, C.R., Yi, L., Su, H., Guibas, L.J.: PointNet++: deep hierarchical feature learning on point sets in a metric space. In: NeurIPS (2017)

27. Rao, Y., Lu, J., Zhou, J.: Spherical fractal convolutional neural networks for point cloud recognition. In: CVPR (2019)

28. Rebuffi, S.A., Kolesnikov, A., Sperl, G., Lampert, C.H.: iCaRL: incremental classifier and representation learning. In: CVPR (2017)

29. Reizenstein, J., Shapovalov, R., Henzler, P., Sbordone, L., Labatut, P., Novotny, D.: Common objects in 3D: large-scale learning and evaluation of real-life 3D category reconstruction. In: ICCV (2021)

30. Riemer, M., et al.: Learning to learn without forgetting by maximizing transfer and minimizing interference. In: ICLR (2018)

31. Schumann, O., Hahn, M., Dickmann, J., Wöhler, C.: Semantic segmentation on radar point clouds. In: FUSION (2018)
32. Shin, H., Lee, J.K., Kim, J., Kim, J.: Continual learning with deep generative replay. In: NeurIPS (2017)
33. Simon, C., Koniusz, P., Harandi, M.: On learning the geodesic path for incremental learning. In: CVPR (2021)
34. Sung, F., Yang, Y., Zhang, L., Xiang, T., Torr, P.H., Hospedales, T.M.: Learning to compare: relation network for few-shot learning. In: CVPR (2018)
35. Tan, Z., Ding, K., Guo, R., Liu, H.: Graph few-shot class-incremental learning. In: Proceedings of the Fifteenth ACM International Conference on Web Search and Data Mining (2022)
36. Tao, X., Hong, X., Chang, X., Dong, S., Wei, X., Gong, Y.: Few-shot class-incremental learning. In: CVPR (2020)
37. Uy, M.A., Pham, Q.H., Hua, B.S., Nguyen, D.T., Yeung, S.K.: Revisiting point cloud classification: a new benchmark dataset and classification model on real-world data. In: ICCV (2019)
38. Wang, C., Samari, B., Siddiqi, K.: Local spectral graph convolution for point set feature learning. In: Ferrari, V., Hebert, M., Sminchisescu, C., Weiss, Y. (eds.) ECCV 2018. LNCS, vol. 11208, pp. 56–71. Springer, Cham (2018). https://doi.org/10.1007/978-3-030-01225-0_4
39. Wang, Y., Sun, Y., Liu, Z., Sarma, S.E., Bronstein, M.M., Solomon, J.M.: Dynamic graph CNN for learning on point clouds. ACM Trans. Graph. (TOG) **38**, 1–12 (2019)
40. Wu, W., Qi, Z., Fuxin, L.: PointCONV: deep convolutional networks on 3D point clouds. In: CVPR (2019)
41. Wu, Y., et al.: Large scale incremental learning. In: CVPR (2019)
42. Wu, Z., et al.: 3D ShapeNets: a deep representation for volumetric shapes. In: CVPR (2015)
43. Xu, Y., Fan, T., Xu, M., Zeng, L., Qiao, Yu.: SpiderCNN: deep learning on point sets with parameterized convolutional filters. In: Ferrari, V., Hebert, M., Sminchisescu, C., Weiss, Y. (eds.) ECCV 2018. LNCS, vol. 11212, pp. 90–105. Springer, Cham (2018). https://doi.org/10.1007/978-3-030-01237-3_6
44. Yue, X., Wu, B., Seshia, S.A., Keutzer, K., Sangiovanni-Vincentelli, A.L.: A LiDAR point cloud generator: from a virtual world to autonomous driving. In: Proceedings of the 2018 ACM on International Conference on Multimedia Retrieval (2018)
45. Zenke, F., Poole, B., Ganguli, S.: Continual learning through synaptic intelligence. In: International Conference on Machine Learning, pp. 3987–3995. PMLR (2017)
46. Zhang, Y., Rabbat, M.: A graph-CNN for 3D point cloud classification. In: ICASSP (2018)
47. Zhao, H., Jiang, L., Jia, J., Torr, P.H., Koltun, V.: Point transformer. In: ICCV (2021)
48. Zhou, D.W., Wang, F.Y., Ye, H.J., Ma, L., Pu, S., Zhan, D.C.: Forward compatible few-shot class-incremental learning. In: CVPR (2022)
49. Zhu, F., Zhang, X.Y., Wang, C., Yin, F., Liu, C.L.: Prototype augmentation and self-supervision for incremental learning. In: CVPR (2021)

Meta-Learning with Less Forgetting on Large-Scale Non-Stationary Task Distributions

Zhenyi Wang[1], Li Shen[2(✉)], Le Fang[1], Qiuling Suo[1], Donglin Zhan[3], Tiehang Duan[4], and Mingchen Gao[1(✉)]

[1] State University of New York at Buffalo, Buffalo, USA
`{zhenyiwa,lefang,qiulings,mgao8}@buffalo.edu`
[2] JD Explore Academy, Beijing, China
`mathshenli@gmail.com`
[3] Columbia University, New York, NY, USA
`dz2478@columbia.edu`
[4] Meta, Seattle, WA, USA
`tiehang.duan@gmail.com`

Abstract. The paradigm of machine intelligence moves from purely supervised learning to a more practical scenario when many loosely related unlabeled data are available and labeled data is scarce. Most existing algorithms assume that the underlying task distribution is stationary. Here we consider a more realistic and challenging setting in that task distributions evolve over time. We name this problem as **S**emi-supervised meta-learning with **E**volving **T**ask di**S**tributions, abbreviated as **SETS**. Two key challenges arise in this more realistic setting: (i) how to use unlabeled data in the presence of a large amount of unlabeled out-of-distribution (OOD) data; and (ii) how to prevent catastrophic forgetting on previously learned task distributions due to the task distribution shift. We propose an **O**OD **R**obust and knowle**D**ge pres**E**rved semi-supe**R**vised meta-learning approach (**ORDER**) (we use ORDER to denote the task distributions sequentially arrive with some ORDER), to tackle these two major challenges. Specifically, our ORDER introduces a novel mutual information regularization to robustify the model with unlabeled OOD data and adopts an optimal transport regularization to remember previously learned knowledge in feature space. In addition, we test our method on a very challenging dataset: SETS on large-scale non-stationary semi-supervised task distributions consisting of (at least) 72K tasks. With extensive experiments, we demonstrate the proposed ORDER alleviates forgetting on evolving task distributions and is more robust to OOD data than related strong baselines.

Supplementary Information The online version contains supplementary material available at https://doi.org/10.1007/978-3-031-20044-1_13.

S. Avidan et al. (Eds.): ECCV 2022, LNCS 13680, pp. 221–238, 2022.
https://doi.org/10.1007/978-3-031-20044-1_13

1   Introduction

Meta-learning [51] focuses on learning lots of related tasks to acquire common knowledge adaptable to new unseen tasks, and has achieved great success in many computer vision problems [26, 27, 61]. Recently, it has been shown that additional unlabeled data is beneficial for improving the meta-learning model performance [35, 36, 47] especially when labeled data is scarce. One common assumption of these works [35, 36, 47] is that the task distribution (i.e., a family of tasks) is stationary during meta training. However, real-world scenarios are more complex and often require learning across different environments.

One instance is in personalized self-driving systems [7]. Each task refers to learning a personal object recognition subsystem for an individual user. The labeled data (annotated objects) for each user is scarce and expensive, but unlabeled data (e.g., images captured by their in-car cameras) is abundant. Learning useful information from these unlabeled data would be very helpful for learning the system. Moreover, the task distributions are not stationary, as users' driving behaviors are different across regions. Assume the learning system is first deployed in Rochester and later extended to New York. The original dataset for Rochester may be no longer available when training on New York users due to storage or business privacy constraints. Similar scenarios occur when Amazon extends the market for the personalized conversation AI product, Alexa, to different countries [19, 29].

As can be seen from the above examples, the first significant challenge is that task distributions may evolve during the learning process; the learning system could easily forget the previously learned knowledge. Second, despite that the additional unlabeled data are potentially beneficial to improve model performance [35, 36, 47], blindly incorporating them may bring negative effects. This is due to the fact that unlabeled data could contain both useful (In-Distribution (ID)) and harmful (Out-Of-Distribution (OOD)) data. How to leverage such mixed (ID and OOD) unlabeled data in meta-learning is rarely explored in previous literature.

In general, these challenges can be summarized into a new realistic problem setting that the model learns on a sequence of distinct task distributions without prior information on task distribution shift. We term this problem setup as *semi-supervised meta-learning on evolving task distributions* (**SETS**). To simulate the task distribution shift in SETS, we construct a dataset by composing 6 datasets, including *CIFARFS* [12], *AIRCRAFT* [38], *CUB* [67], *Miniimagenet* [56], *Butterfly* [16] and *Plantae* [28]. A model will be sequentially trained on the datasets in a meta-learning fashion, simulating the real scenarios of evolving environment. However, at the end of training on the last domain, the meta learned model could easily *forget* the learned knowledge to solve *unseen* tasks sampled from the task distributions trained previously.

In this paper, we aim to tackle the SETS problem, which requires the meta-learning model to adapt to a new task distribution and retain the knowledge learned on previous task distributions. To better use unlabeled data of each task, we propose a regularization term driven by mutual information, which measures the mutual dependency between two variables. We first propose an OOD detection method to separate the unlabeled data for each task into ID and OOD data, respectively. Then, we minimize the mutual information between class prototypes and OOD data to reduce the effect of OOD on the class prototypes. At the same time, we maximize the mutual information between class prototypes and ID data to maximize the information transfer. We derive tractable sample-based variational bounds specialized for SETS for mutual information maximization and minimization for practical computation since the closed-form of probability distribution in mutual information is not available.

To mitigate catastrophic forgetting during the learning process, we propose a novel and simple idea to reduce forgetting in feature space. Specifically, in addition to storing the raw data as used in standard continual learning, we also store the data features of memory tasks. To achieve this goal, we use optimal transport (OT) to compute the distance between the feature distribution of memory tasks at the current iteration and the feature distribution of memory tasks stored in the memory. This ensures minimal feature distribution shift during the learning process. The proposed OT framework can naturally incorporate the unlabeled data features of previous tasks to further improve the memorization of previous knowledge. Besides, as the feature size is much smaller than the raw data, this makes it scalable to large-scale problems with a very long task sequence.

Our contributions are summarized as three-fold:

- To our best knowledge, this is the first work considering the practical and challenging *semi-supervised meta-learning on evolving task distributions* (SETS) setting.
- We propose a novel algorithm for SETS setting to exploit the unlabeled data and a mutual information inspired regularization to tackle the OOD problem and an OT regularization to mitigate forgetting.
- Extensive experiments on the constructed dataset with at least 72K tasks demonstrate the effectiveness of our method compared to strong baselines.

2 Related Work

Meta Learning [51] focuses on rapidly adapting to new unseen tasks by learning prior knowledge through training on a large number of similar tasks. Various approaches have been proposed [4,5,21,22,24,34,39,40,50,54,56,65,66,69]. All of these methods work on the simplified setting where the model is meta trained on a stationary task distribution in a fully supervised fashion, which is completely different from our proposed SETS. On the other hand, online meta-learning [23] aims to achieve better performance on future tasks and stores all the previous tasks to avoid forgetting with a small number of tasks. Furthermore, [62,63] has proposed a meta learning method to solve large-scale task stream with sequential domain shift. However, our method is different from the previous work in several aspects. Specifically (i) we additionally focus on leveraging unlabeled data and improving model robustness; (ii) we only store a small number of tasks to enable our method to work in our large-scale setting; (iii) we consider that the task distributions could be evolving and OOD-perturbed during meta training under the semi-supervised learning scenario. In summary, SETS is more general and realistic than online meta-learning and other settings.

Continual Learning (CL) aims to maintain previous knowledge when learning on a sequence of tasks with distribution shift. Many works mitigate catastrophic forgetting during the learning process [2,3,10,14,20,31,37,42,48,64,68]. Semi-supervised continual learning (SSCL) [33,53,60] is a recent extension of CL to the semi-supervised learning setting. SETS is completely different from CL and SSCL in several aspects: (i) our SETS focuses on a large-scale task sequence (more than 72K), while CL and SSCL are not scalable to this large-scale setting; (ii) our goal is to enable the model to generalize to the *unseen* tasks from all the previous task distributions, while SSCL works on the generalization to the testing data of previous tasks.

Semi-supervised Few-Shot Learning (SSFSL) [35,36,47] aims to generalize to unseen semi-supervised tasks when meta training on semi-supervised few-shot learning tasks in an *offline* learning manner. It assumes that the few-shot learning task distributions are stationary during meta training. Unlike these works, we work on a more challenging and realistic setting of evolving task distributions in an *online* learning manner. Meanwhile, SETS also aims at mitigating catastrophic forgetting for the previous learned tasks during online learning process as well, which makes it more challenging and practical to real-world scenarios.

We summarize and compare different benchmarks in Table 1, including continual learning (CL), semi-supervised learning (SSL), continual few-shot learning (CFSL), semi-supervised continual learning (SSCL), semi-supervised few-shot learning (SSFSL), and SETS (Ours) in terms of whether they consider catastrophic forgetting

Table 1. Comparisons among different benchmarks.

Settings	CF	Unlabeled	Few-Shot	Unseen Tasks
CL [31]	✓	✗	✗	✗
SSL [11]	✗	✓	✗	✗
CFSL [6]	✓	✗	✓	✗
SSCL [60]	✓	✓	✗	✗
SSFSL [47]	✗	✓	✓	✓
SETS (Ours)	✓	✓	✓	✓

(CF), use unlabeled data, few-shot learning, and generalize to unseen tasks. In summary, our setting is more comprehensive and practical and covers more learning aspects in real applications.

3   Problem Setting

General Setup. SETS (Fig. 1 for illustration), online meta-learns on a sequence of task distributions, $\mathcal{D}_1, \mathcal{D}_2, \ldots, \mathcal{D}_L$. For time $t = 1, \cdots, \tau_1$, we randomly sample mini-batch semi-supervised tasks $\mathcal{T}_t$ at each time t from task distribution $P(\mathcal{D}_1)$; for time $t = \tau_1 + 1, \cdots, \tau_2$, we randomly sample mini-batch semi-supervised tasks $\mathcal{T}_t$ at each time t from task distribution $P(\mathcal{D}_2)$; for time $t = \tau_i + 1, \cdots, \tau_{i+1}$, we randomly sample mini-batch semi-supervised tasks $\mathcal{T}_t$ at each time t from task distribution $P(\mathcal{D}_i)$, where $P(\mathcal{D}_i)$ is the task distribution (a collection of

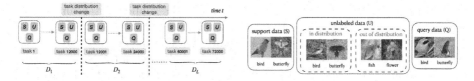

Fig. 1. Illustration of SETS on a large-scale non-stationary semi-supervised task distributions $\mathcal{D}_1, \mathcal{D}_2, \cdots, \mathcal{D}_L$ (top) with at least 72K tasks. Each task (bottom) contains labeled support data (S), unlabeled data (U) (in-distribution and out-of-distribution data) and query data (Q).

a large number of tasks) in $\mathcal{D}_i$. Thus, in SETS, each task from each domain sequentially arrives. Time $t = \tau_1, \tau_2, \cdots, \tau_i, \cdots$ is the time when task distribution shift happens, but we do not assume any prior knowledge about them. This is a more practical and general setup. Each time interval $|\tau_i - \tau_{i-1}|$ is large enough to learn a large number of tasks from each task distribution adequately. Each task $\mathcal{T}$ is divided into support, unlabeled and query data $\{\mathcal{S}, \mathcal{U}, \mathcal{Q}\}$. The support set $\mathcal{S}$ consists of a collection of labeled examples, $\{(\mathbf{x}^k, y^k)\}$, where $\mathbf{x}^k$ is the data and y^k is the label. The unlabeled set $\mathcal{U}$ contains only inputs: $\mathcal{U} = \{\tilde{\mathbf{x}}^1, \tilde{\mathbf{x}}^2, \cdots, \}$. The goal is to predict labels for the examples in the task's query set $\mathcal{Q}$. Our proposed learning system maintains a memory buffer $\mathcal{M}$ to store a small number of training tasks from previous task distributions. We use reservoir sampling [57] (RS) to maintain tasks in the memory. RS assigns equal probability for each incoming task without needing to know the total number of training tasks in advance. The total number of meta training tasks is much larger than the capacity of memory buffer, making it infeasible to store all the tasks. More details about memory buffer update and maintenance are provided in Appendix B. We evaluate model performance on the unseen tasks sampled from both current and all previous task distributions.

Evolving Semi-supervised Few-Shot Episode Distributions Construction. In most existing works, the meta-learning model is trained in a stationary setting. By contrast, in SETS, the model is meta trained on an evolving sequence of datasets. A large number of semi-supervised tasks sampled from each dataset form task distribution $\mathcal{D}_i$. We thus have *a sequence of semi-supervised task distributions* $\mathcal{D}_1, \mathcal{D}_2, \cdots, \mathcal{D}_L$. In episodic meta-learning, a task is usually denoted as an episode. When training on task distribution $\mathcal{D}_i$, to sample an N-way K-shot training episode, we first uniformly sample N classes from the set of training classes $\mathcal{C}_i^{train}$. For the labeled support set $\mathcal{S}$ (training data), we then sample K images from the labeled split of each of these classes. For the unlabeled set $\mathcal{U}$, we sample Z images from the unlabeled split of each of these classes as ID data. When including OOD in $\mathcal{U}$, we additionally sample R images from external datasets as OOD. In practical scenarios, the OOD data could dominate the unlabeled portion $\mathcal{U}$. To be more realistic, we construct the unlabeled set with different proportions of OOD to simulate real-world scenarios. The query data (testing data) $\mathcal{Q}$ of the episode is comprised of a fixed number of images from the labeled split of each of the N chosen classes. The support set $\mathcal{S}$ and query set $\mathcal{Q}$ contain different inputs but with the same class set.

At the end of meta training on task distribution $\mathcal{D}_L$, the meta trained model is expected to generalize to the *unseen* tasks of each dataset in the sequence. The evaluation on *unseen* classes of previous datasets is to measure the forgetting. We adopt the N-way K-shot classification on *unseen* classes of each dataset. The semi-supervised testing episodes are constructed similarly as above. The accuracy is averaged on the query set $\mathcal{Q}$ of many meta-testing episodes from all the trained datasets.

Constructed Dataset. To simulate realistic evolving semi-supervised task distributions, we construct a new large-scale dataset and collect 6 datasets, including **CIFARFS** [12], **AIRCRAFT** [38], **CUB** [67], **Miniimagenet** [56], **Butterfly** [16] and **Plantae** [28]. The OOD data for each task are sampled from another three datasets, including **Omniglot** [32], **Vggflower** [43] and **Fish** [70]. For each dataset, we randomly sample 12K tasks, we thus have 72K tasks in total. This task sequence is much larger than existing continual learning models considered. We perform standard 5-way-1-shot and 5-way-5-shot learning on this new dataset. All the datasets are publicly available with more details provided in Appendix A due to the space limitation.

4   Methodology

In the following, we focus on the N-way K-shot setting of SETS as described above. We first describe the standard SSFSL task in Sect. 4.1, then present our method for handling unlabeled data, especially OOD data in Sect. 4.2. Next, we present our method for mitigating catastrophic forgetting in Sect. 4.3. At last, we summarize our proposed methodology in Sect, 4.4.

4.1   Standard Semi-supervised Few-Shot Learning

Prototypical network [54] has been extended to semi-supervised few-shot learning by [47]. We denote the feature embedding function as $h_\theta(\mathbf{x})$ with parameter $\boldsymbol{\theta}$. The class-wise prototype for support set $\mathcal{S}$ is calculated as : $\mathbf{p}_c = \sum_{\mathbf{x} \in \mathcal{S}_c} h_\theta(\mathbf{x}) / \sum_{\mathbf{x} \in \mathcal{S}_c} 1$, where $\mathcal{S}_c$ is the set of data in $\mathcal{S}$ belonging to category c. Each class prototype is refined by unlabeled data as:

$$\mathbf{p}'_c = \frac{\sum_{\mathbf{x} \in \mathcal{S}_c} h_\theta(\mathbf{x}) + \sum_{\tilde{\mathbf{x}} \in \mathcal{U}} h_\theta(\tilde{\mathbf{x}}) \mu_c(\tilde{\mathbf{x}})}{\sum_{\mathbf{x} \in \mathcal{S}_c} 1 + \sum_{\tilde{\mathbf{x}} \in \mathcal{U}} \mu_c(\tilde{\mathbf{x}})}$$

by assigning each unlabeled data $\tilde{\mathbf{x}} \in \mathcal{U}$ with a soft class probability $\mu_c(\tilde{\mathbf{x}}) = \frac{exp(-||h_\theta(\tilde{\mathbf{x}}) - \mathbf{p}_c||_2^2)}{\sum_{c'} exp(-||h_\theta(\tilde{\mathbf{x}}) - \mathbf{p}'_c||_2^2)}$. To work with OOD data, they use an additional class with zero mean in addition to the normal c classes in the support set. The learning objective is to simply maximize the average log-probability of the correct class assignments over query examples $\mathcal{Q}$ and rigorously defined as following:

$$\mathcal{L}_{meta} = P(\mathcal{Q}|\boldsymbol{\theta}, \mathcal{S}; \mathcal{U}) = P(\mathcal{Q}|\boldsymbol{\theta}, \{\mathbf{p}'_c\}) = \sum_{(\mathbf{x}_i, y_i) \in \mathcal{Q}} P(y_i|\mathbf{x}_i, \boldsymbol{\theta}, \{\mathbf{p}'_c\}). \quad (1)$$

4.2 Mutual-Information for Unlabeled Data Handling

To be more realistic, the unlabeled data contains OOD data for each task, whose data categories are not present in the support set. In practice, the unlabeled OOD data could potentially negatively impact the performance. It has been shown that existing works [35,36,47] are particularly sensitive to OOD data. How to properly use unlabeled data is challenging when OOD data dominates the unlabeled data, which is common in practical scenarios. Instead of sidestepping this problem by simply discarding the detected OOD data, we propose to exploit discriminative information implied by unlabeled OOD samples. To explicitly address this problem, we propose to learn the representations that are robust to OOD data from an information-theoretical perspective. We first give the detailed definitions for in-distribution (ID) and out-of-distribution (OOD) below.

Definition 1. *Suppose a task $\mathcal{T}$ with data $\{\mathcal{S}, \mathcal{U}, \mathcal{Q}\}$, contains labeled categories $\mathcal{C}$ in $\mathcal{S}$. For unlabeled data $\mathcal{U}$, in-distribution unlabeled data is defined as the set of data $\mathbf{x}$ whose categories $c \in \mathcal{C}$, i.e., $\mathcal{U}_{id} = \{\mathbf{x} | \mathbf{x} \in \mathcal{U}, c \in \mathcal{C}\}$; similarly, the out-of-distribution data is defined as the set of data $\mathbf{x}$ whose categories $c \notin \mathcal{C}$, i.e., $\mathcal{U}_{ood} = \{\mathbf{x} | \mathbf{x} \in \mathcal{U}, c \notin \mathcal{C}\}$.*

OOD sample detection. We propose to automatically divide the unlabeled data for each task into ID and OOD. The existing confidence-based approach is highly unreliable for detecting OOD data in the few-shot learning setting since the network can produce incorrect high-confidence predictions for OOD samples [41]. We instead propose a metric-based OOD sample detection method in feature embedding space. First, we calculate the class-wise prototypes $\mathbf{p}_c$ for each class c. We define the distance of data $\mathbf{x}$ to the prototypes set as:

Algorithm 1. OOD sample detection.

Require: A task data $\{\mathcal{S}, \mathcal{U}, \mathcal{Q}\}$; unlabeled query $\mathcal{Q}$ serves as ID-calibration set;
1: calculate prototypes $\mathbf{p}_c$ for each class c with support data $\mathcal{S}$.
2: **for** $\mathbf{x}_i \in \mathcal{Q}$ **do**
3:     $\mathbf{w}_i = h_\theta(\mathbf{x}_i) / \|h_\theta(\mathbf{x}_i)\|_2$ (normalization)
4:     $d_i = d(\mathbf{w}_i, \{\mathbf{p}_c\}_{c=1}^N) \stackrel{def}{=} \min_c d(\mathbf{w}_i, \mathbf{p}_c)$
5: **end for**
6: thresh $= \mu(\{d_i\}) + \sigma(\{d_i\})$ [$\mu(\{d_i\})$ is the mean of $\{d_i\}$ and $\sigma(\{d_i\})$ is the standard deviation of $\{d_i\}$]
7: **for** $\mathbf{x}_i \in \mathcal{U}$ **do**
8:     $\mathbf{w}_i = h_\theta(\mathbf{x}_i) / \|h_\theta(\mathbf{x}_i)\|_2$ (normalization)
9:     $d_i = d(\mathbf{w}_i, \{\mathbf{p}_c\}_{c=1}^N) \stackrel{def}{=} \min_c d(\mathbf{w}_i, \mathbf{p}_c)$
10:     **if** $d_i >$ thresh **then**
11:         $\mathcal{U}_{ood} = \mathcal{U}_{ood} \bigcup \mathbf{x}_i$
12:     **else**
13:         $\mathcal{U}_{id} = \mathcal{U}_{id} \bigcup \mathbf{x}_i$
14:     **end if**
15: **end for**
16: **return** ID data $\mathcal{U}_{id}$ and OOD data $\mathcal{U}_{ood}$

$$d(h_\theta(\mathbf{x}), \{\mathbf{p}_c\}_{c=1}^N) \stackrel{def}{=} \min_c d(h_\theta(\mathbf{x}), \mathbf{p}_c),$$

i.e., the smallest distance to all the prototypes in the embedding space; where we use Euclidean distance as the distance metric d for simplicity. Intuitively, the data closer to the class prototypes are more likely to be ID data, the data farther to the class prototypes are more likely to be OOD. We adopt a method similar to the state-of-the-art few-shot OOD detection algorithm, named cluster-conditioned detection [52]. We use the unlabeled query data $\mathcal{Q}$ for each task as an ID calibration set to determine the threshold between OOD and ID data. The OOD sample detection is presented in Algorithm 1. The threshold is based on the Gaussian assumption of $d(h_\theta(\mathbf{x}), \{\mathbf{p}_c\}_{c=1}^N)$ (3-Sigma rule).

We adopt mutual information (MI) to separately handle ID data and OOD data, where MI describes the mutual dependence between the two variables and is defined as follows.

Definition 2. *The mutual information between variables* $\mathbf{a}$ *and* $\mathbf{b}$ *is defined as*

$$\mathcal{I}(\mathbf{a}, \mathbf{b}) = \mathbb{E}_{p(\mathbf{a},\mathbf{b})} \left[\log \frac{p(\mathbf{a}, \mathbf{b})}{p(\mathbf{a})p(\mathbf{b})} \right], \tag{2}$$

where $p(\mathbf{a})$ *and* $p(\mathbf{b})$ *are marginal distribution of* $\mathbf{a}$ *and* $\mathbf{b}$, $p(\mathbf{a}, \mathbf{b})$ *is the joint distribution.*

In information theory, $\mathcal{I}(\mathbf{a}, \mathbf{b})$ is a measure of the mutual dependence between the two variables $\mathbf{a}$ and $\mathbf{b}$. More specifically, it quantifies the "amount of information" about the variable $\mathbf{a}$ (or $\mathbf{b}$) captured via observing the other variable $\mathbf{b}$ (or $\mathbf{a}$). The higher dependency between two variables, the higher MI is. Intuitively, the higher dependency between embedded OOD data and class prototypes, the higher the negative impact on the model performance. On the contrary, more dependency between embedded ID data and class prototypes could improve the performance. To enhance the effect of ID data on the class prototypes, we maximize the MI between class prototypes and ID data. To reduce the negative effect of OOD data on the class prototypes, we further consider minimizing the MI between class prototypes and OOD data:

$$\mathcal{L}_{total} = \mathcal{L}_{meta} + \lambda \left(\mathcal{I}(\mathbf{e}_{ood}, \mathbf{p}_{c'}) - \mathcal{I}(\mathbf{e}_{id}, \mathbf{p}_{c'}) \right), \tag{3}$$

where $\mathcal{L}_{meta}$ is defined in Eq. (1); $\mathbf{e}_{ood} = h_\theta(\mathbf{x}), \mathbf{x} \in \mathcal{U}_{ood}$ and $\mathbf{e}_{id} = h_\theta(\mathbf{x}), \mathbf{x} \in \mathcal{U}_{id}$. $c' = \mathrm{argmin}_c \|\mathbf{e} - \mathbf{p}_c\|$; $\mathcal{I}(\mathbf{e}_{ood}, \mathbf{p}_{c'})$ is the mutual information between embedded OOD data and class prototypes; $\mathcal{I}(\mathbf{e}_{id}, \mathbf{p}_{c'})$ is the mutual information between embedded ID data and class prototypes; λ is a hyperparameter to control the relative importance of different terms. Since this objective is to minimize $-\mathcal{I}(\mathbf{e}_{id}, \mathbf{p}_{c'})$, it is equivalent to maximize $\mathcal{I}(\mathbf{e}_{id}, \mathbf{p}_{c'})$. However, the computation of MI values is intractable [18,44], since the form of joint probability distribution $p(\mathbf{a}, \mathbf{b})$ in Eq. (2) is not available in our case and only samples from the joint distribution are accessible. To solve this problem, we derive the variational lower bound of MI for MI maximization and variational upper bound of MI for MI minimization specialized for SETS, respectively. Due to the limited space, all proofs for Lemmas and Theorems are provided in Appendix D.

Theorem 1. *For a task* $\mathcal{T} = \{\mathcal{S}, \mathcal{U}, \mathcal{Q}\}$, *suppose the unlabeled OOD data embedding* $\mathbf{e}_{ood} = h_\theta(\mathbf{x}), \mathbf{x} \in \mathcal{U}_{ood}$. *The nearest prototype corresponds to* $\mathbf{e}_{ood}$ *is* $\mathbf{p}_{c'}$, *where* $c' = \mathrm{argmin}_c \|\mathbf{e}_{ood} - \mathbf{p}_c\|$. *Given a collection of samples* $\{(\mathbf{e}_{ood}^i, \mathbf{p}_{c'}^i)\}_{i=1}^{i=L} \sim P(\mathbf{e}_{ood}, \mathbf{p}_{c'})$, *the variational upper bound of mutual information* $\mathcal{I}(\mathbf{e}_{ood}, \mathbf{p}_{c'})$ *is:*

$$\mathcal{I}(\mathbf{e}_{ood}, \mathbf{p}_{c'}) \le \sum_{i=1}^{i=L} \log P(\mathbf{p}_{c'}^i | \mathbf{e}_{ood}^i) - \sum_{i=1}^{i=L}\sum_{j=1}^{j=L} \log P(\mathbf{p}_{c'}^i | \mathbf{e}_{ood}^j) = \mathcal{I}_{ood}. \tag{4}$$

The bound is tight (equality holds) when $\mathbf{p}_{c'}$ and $\mathbf{e}_{ood}$ are independent.

This upper bound requires the unknown conditional probability distribution $P(\mathbf{p}_{c'}|\mathbf{e}_{ood})$ and thus it is intractable. In this work, we approximate it with variational distribution $Z_\phi(\mathbf{p}_{c'}|\mathbf{e}_{ood})$ parameterized by a neural network.

Lemma 1. *For a task $\mathcal{T} = \{\mathcal{S}, \mathcal{U}, \mathcal{Q}\}$, suppose the unlabeled ID data embedding $\mathbf{e}_{id} = h_\theta(\mathbf{x}), \mathbf{x} \in \mathcal{U}_{id}$. The nearest prototype corresponds to $\mathbf{e}_{id}$ is $\mathbf{p}_{c'}$, where $c' = \arg\min_c \|\mathbf{e}_{id} - \mathbf{p}_c\|$. Given a collection of samples $\{(\mathbf{e}_{id}^i, \mathbf{p}_{c'}^i)\}_{i=1}^{i=L} \sim P(\mathbf{e}_{id}, \mathbf{p}_{c'})$, the variational lower bound of $\mathcal{I}(\mathbf{e}_{id}, \mathbf{p}_{c'})$ is:*

$$\mathcal{I}(\mathbf{e}_{id}, \mathbf{p}_{c'}) \geqslant \sum_{i=1}^{i=L} f(\mathbf{e}_{id}^i, \mathbf{p}_{c'}^i) - \sum_{j=1}^{j=L} \log \sum_{i=1}^{i=L} e^{f(\mathbf{e}_{id}^i, \mathbf{p}_{c'}^j)}. \tag{5}$$

This bound is tight if $f(\mathbf{e}_{id}, \mathbf{p}_{c'}) = \log P(\mathbf{p}_{c'}|\mathbf{e}_{id}) + c(\mathbf{p}_{c'})$.

Based on Lemma 1, the variational lower bound is derived as in Theorem 2.

Theorem 2. *For a task $\mathcal{T} = \{\mathcal{S}, \mathcal{U}, \mathcal{Q}\}$, suppose the unlabeled ID data embedding $\mathbf{e}_{id} = h_\theta(\mathbf{x}), \mathbf{x} \in \mathcal{U}_{id}$. The nearest prototype corresponds to $\mathbf{e}_{id}$ is $\mathbf{p}_{c'}$, where $c' = \arg\min_c \|\mathbf{e}_{id} - \mathbf{p}_c\|$. Given a collection of samples $\{(\mathbf{e}_{id}^i, \mathbf{p}_{c'}^i)\}_{i=1}^{i=L} \sim P(\mathbf{e}_{id}, \mathbf{p}_{c'})$, the variational lower bound of $\mathcal{I}(\mathbf{e}_{id}, \mathbf{p}_{c'})$ is:*

$$\mathcal{I}(\mathbf{e}_{id}, \mathbf{p}_{c'}) \geqslant \mathcal{I}_{id} := \sum_{i=1}^{i=L} f(\mathbf{e}_{id}^i, \mathbf{p}_{c'}^i) - \sum_{j=1}^{j=L} \left[\frac{\sum_{i=1}^{i=L} e^{f(\mathbf{e}_{id}^i, \mathbf{p}_{c'}^j)}}{a(\mathbf{p}_{c'}^j)} + \log(a(\mathbf{p}_{c'}^j)) - 1 \right]. \tag{6}$$

The bound is tight (equality holds) when $f(\mathbf{e}_{id}, \mathbf{p}_{c'}) = \log P(\mathbf{p}_{c'}|\mathbf{e}_{id}) + c(\mathbf{p}_{c'})$ and $a(\mathbf{p}_{c'}) = \mathbb{E}_{p(\mathbf{e}_{id})} e^{f(\mathbf{e}_{id}, \mathbf{p}_{c'})}$. $a(\mathbf{p}_{c'})$ is any function that $a(\mathbf{p}_{c'}) > 0$.

Hence, by combining Theorems 1 and 2, we obtain an surrogate for the loss function Eq. (3) as:

$$\mathcal{L}_{total} = \mathcal{L}_{meta} + \lambda(\mathcal{I}(\mathbf{e}_{ood}, \mathbf{p}_{c'}) - \mathcal{I}(\mathbf{e}_{id}, \mathbf{p}_{c'})) \leq \mathcal{L}_{meta} + \lambda(\mathcal{I}_{ood} - \mathcal{I}_{id}). \tag{7}$$

4.3   Mitigate CF by Optimal Transport

A straightforward idea of mitigating catastrophic forgetting on previous task distributions is to replay the memory tasks when training on current task $\mathcal{T}_t$, i.e., we randomly sample a small batch tasks $\mathcal{B}$ from the memory buffer $\mathcal{M}$. Thus the loss function becomes $\mathcal{H}(\theta) = \mathcal{L}(\mathcal{T}_t) + \sum_{\mathcal{T}_i \in \mathcal{B}} \mathcal{L}(\mathcal{T}_i)$. However, simple experience replay only forces the model to remember a limited number of labeled data, i.e., a much larger number of previous unlabeled data is totally neglected, which causes the model to sub-optimally remember the previous knowledge. To further avoid forgetting, we directly regularize the network outputs on both labeled and unlabeled data since what ultimately matters is the network output, in addition to remembering output labels on memory tasks. To achieve this goal, we propose a novel constrained loss via optimal transport to reduce the *output (feature) distribution shift* on both labeled and unlabeled data.

Optimal Transport. A discrete distribution $\mu = \sum_{i=1}^{n} u_i \delta_{e_i}(\cdot)$, where $u_i \geq 0$ and $\sum_i u_i = 1$, $\delta_e(\cdot)$ denotes a spike distribution located at $\mathbf{e}$. Given two discrete distributions, $\mu = \sum_{i=1}^{n} u_i \delta_{e_i}$ and $\nu = \sum_{j=1}^{m} v_j \delta_{g_j}$, respectively, the OT distance [8] between μ and ν is defined as the optimal value of the following problem:

$$\mathcal{L}_{OT} = \min_{W \in \Pi(\mu,\nu)} \sum_{i=1}^{n} \sum_{j=1}^{m} W_{ij} \cdot d(e_i, g_j) , \tag{8}$$

where W_{ij} is defined as the joint probability mass function on (e_i, g_j) and $d(\mathbf{e}, \mathbf{g})$ is the cost of moving $\mathbf{e}$ to $\mathbf{g}$ (matching $\mathbf{e}$ and $\mathbf{g}$). $\Pi(\mu, \nu)$ is the set of joint probability distributions with two marginal distributions equal to μ and ν, respectively. The exact calculation of the OT loss is generally difficult [25]. Therefore, for simplicity, we use CVXPY [1] for efficient approximation.

When storing a semi-supervised episode into the memory, we store the feature embedding for each data point in addition to the raw data. Memory features are the collection of $\{h_\theta(\mathbf{x}), \mathbf{x} \in \bigcup_{\{S,\mathcal{U}_{id}\} \in \mathcal{M}} \{S, \mathcal{U}_{id}\}\}$, the concatenation of labeled and unlabeled ID data embedding. Suppose the previously stored data features are $\{g_i\}_{i=1}^{G}$ with distribution ν, where G is the number of data examples stored in memory buffer. The data features generated with model parameters θ_t of current iteration are $\{e_i\}_{i=1}^{G}$ with distribution μ. Then, the loss function becomes minimizing on the optimal transport between $\{e_i\}_{i=1}^{G}$ and $\{g_i\}_{i=1}^{G}$.

Algorithm 2. ORDER Algorithm

Require: evolving episodes $\mathcal{T}_1, \mathcal{T}_2, \ldots, \mathcal{T}_N$, with $\{\tau_1, < \cdots, \tau_i, \cdots, < \tau_{L-1}\}$ the time steps when task distribution shift; memory buffer $\mathcal{M} = \{\}$; model parameters θ; memory task features $\mathcal{E} = \{\}$.

1: **for** $t = 1$ to N **do**
2: sample tasks $\mathcal{B}$ from $\mathcal{M}$ and $\mathcal{B} = \mathcal{B} \bigcup \mathcal{T}_t$.
3: **for** $\mathcal{T} \in \mathcal{B}$ **do**
4:   $\mathcal{T} = \{S, \mathcal{U}, \mathcal{Q}\}$, OOD detection to divide the unlabeled data $\mathcal{U}$ into $\mathcal{U}_{ood}$ and $\mathcal{U}_{id}$
5:   update parameters θ by minimizing $\mathcal{H}(\theta)$
6: **end for**
7: **if** reservoir sampling: **then**
8:   $\mathcal{M} = \mathcal{M} \bigcup \mathcal{T}_t$ // store raw task data
9:   $\mathcal{E} = \mathcal{E} \bigcup \mathcal{E}(\mathcal{T}_t)$ // store task data features
10: **end if**
11: **end for**

4.4 Overall Learning Objective

By unifying the above-derived novel mutual information driven unlabeled data handling technique and optimal transport driven forgetting mitigation, we propose an OOD robust and knowledge preserved semi-supervised meta-learning algorithm (ORDER) to tackle the challenges in the SETS setting. Combining the OT loss from Eq. (8) with Eq. (7) and assigning λ and β as two hyperparameters for controlling the relative importance of the corresponding terms, we obtain the following final loss function:

$$\mathcal{H}(\theta) = \mathcal{L}_{meta}(\mathcal{T}_t) + \lambda(\mathcal{I}_{ood} - \mathcal{I}_{id}) + \sum_{\mathcal{T}_i \in \mathcal{B}} \mathcal{L}(\mathcal{T}_i) + \beta \mathcal{L}_{OT}. \tag{9}$$

The benefits of the second term are two-fold: (i) promote the effect of ID data on class prototypes by maximizing the MI between class prototypes and ID data; (ii) reduce the negativity of OOD data on class prototypes by minimizing the MI between class prototypes and OOD data. The last term is to force the model to remember in feature space. The detailed procedure of ORDER is shown in Algorithm 2.

5   Experiments

To show the benefit of using unlabeled data, we first compare semi-supervised meta-learning using unlabeled data with meta-learning without unlabeled data in Sect. 5.1. To evaluate the effectiveness of the proposed method, ORDER, for learning useful knowledge in SETS, we compare to SOTA meta-learning models in Sect. 5.2. Next, we compare ORDER to SOTA

Table 2. 5-way, 1-shot and 5-shot accuracy compared to meta-learning baselines.

Algorithm	1-Shot	5-Shot
ProtoNet	31.96 ± 0.93	42.62 ± 0.73
ANIL	30.58 ± 0.81	42.24 ± 0.86
MSKM	33.79 ± 1.05	45.41 ± 0.58
LST	34.81 ± 0.81	46.09 ± 0.69
TPN	34.52 ± 0.83	46.23 ± 0.60
ORDER (Ours with only MI)	36.93 ± 0.90	48.85 ± 0.53
ORDER (Ours)	$\mathbf{41.25 \pm 0.64}$	$\mathbf{53.96 \pm 0.71}$

continual learning baselines to show the effectiveness of ORDER for mitigating forgetting in SETS in Sect. 5.3. For the evaluation, we calculate the average testing accuracy on previously unseen 600 testing tasks sampled from the unseen categories of each dataset in the dataset sequence to evaluate the effectiveness of the tested methods. Moreover, we conduct extensive ablation studies in Sect. 5.4. Due to the space limitation, we put detailed implementation details in Appendix B.

5.1   Benefit of Using Unlabeled Data in SETS

To verify the benefit of using unlabeled data in SETS, we compare with various meta-learning methods including supervised and semi-supervised ones. All of these methods are adapted to SETS setting directly. Supervised meta-learning methods *without unlabeled data*, including (1) gradient-based meta-learning **ANIL** [45], which is a simplified model of MAML [22]; (2) metric-based meta-learning **Prototypical Network (ProtoNet)** [54]. In addition, we compare to representative SSFSL models *using unlabeled data*: (3) Masked Soft k-Means (**MSKM**) [47]; (4) Transductive Propagation Network (**TPN**) [36]; (5) Learning to Self-Train (**LST**) [35]. LST and MSKM can be viewed as extension of MAML(ANIL) and ProtoNet to SETS by using additional unlabeled data respectively. Table 2 shows that with additional unlabeled data, semi-supervised meta-learning methods substantially outperform corresponding meta-learning methods without using unlabeled data, demonstrating that unlabeled data could further improve performance in SETS. We also include a reduced version of our method, which uses Eq. (7) as the objective function, and keep other components the same as baselines without considering mitigating CF. We can see that with only the guidance of MI, our method can outperform baselines by more than 2.6% for 5-shot learning and 2.1% for 1-shot learning, which verifies the effectiveness of the proposed MI for handling unlabeled data.

5.2 Comparison to Meta-Learning

To compare to SOTA meta-learning methods, similar to the baselines in Sect. 5.1, Table 2 also shows the advantage of ORDER. We observe that our method significantly outperforms baselines for 5-shot learning by 7.7% and for 1-shot learning by 6.4%. This improvement shows the effectiveness of our method in this challenging setting. LST [35] performs relatively worse in SETS, probably due to the challenging OOD data and noise of pseudo-labels.

Table 3. 5-way, 1-shot and 5-shot classification accuracy compared to continual learning baselines.

Algorithm	1-Shot	5-Shot
Semi-Seq	33.79 ± 1.05	45.41 ± 0.58
Semi-ER	37.62 ± 0.97	50.35 ± 0.76
Semi-AGEM	36.97 ± 1.16	50.29 ± 0.69
Semi-MER	37.90 ± 0.83	50.46 ± 0.74
Semi-GPM	36.53 ± 0.81	49.78 ± 0.65
Semi-DEGCL	36.78 ± 0.89	50.07 ± 0.61
ORDER (Ours OT only)	39.21 ± 0.61	51.72 ± 0.61
ORDER (Ours)	$\mathbf{41.25 \pm 0.64}$	$\mathbf{53.96 \pm 0.71}$
Joint-training	49.91 ± 0.79	61.78 ± 0.75

5.3 Comparison to Continual Learning

To evaluate the effectiveness of ORDER for mitigating forgetting in SETS, we compare to SOTA CL baselines, including Experience replay (ER) [15], A-GEM [14], MER [48], GPM [49] and DEGCL [13]. Note that all of these methods are originally designed for mitigating forgetting for *a small number of tasks*, we adapt these methods to SETS by combing them with meta-learning methods for mitigating forgetting. Here, we combine CL methods with MSKM for illustration since it is more efficient and performs comparably in SETS than the other two meta-learning methods. The combination baselines are named as Semi-, etc. We also compare to (1) **sequential training (Seq)**, which trains the latent task distributions sequentially without any external mechanism and helps us understand the model forgetting behavior; (2) **joint offline training**, which learns all the task distributions jointly in a stationary setting and provides the performance upper bound.

The memory buffer is updated by reservoir sampling mentioned in Sect. 3. More details about how to update the memory buffer is provided in Appendix B. All the baselines and our method maintain 200 tasks in memory and the number of OOD data $R = 50$ for each task. Table 3 shows the results. We also include a version of our method with memory replay and only OT regularization, without

MI regularization, and other components are kept as the same as baselines. Our method can outperform baselines by 1.3% for 1-shot learning and 1.2% for 5-shot learning. Our full method significantly outperforms baselines for 5-shot learning by 3.1% and for 1-shot learning by 3.3%. This improvement shows the effectiveness of ORDER for mitigating forgetting in SETS.

5.4 Ablation Study

Robustness to OOD. As for the OOD data of each episode, we randomly sample OOD from the mixture of the OOD dataset as described in Sect. 3. Furthermore, it is more natural for the unlabeled set to include more OOD samples than ID samples. To investigate the sensitivity of the baselines and our method to OOD data, we fix the number of ID data for each task and gradually increase the amount of OOD data in the unlabeled set. We set R (number of OOD data) $= 50, 100, 150, 200, 250$, respectively. The results are shown in Fig. 2 in Appendix C. Interestingly, we can observe that our method is much more robust to OOD data than baselines, even when the number of OOD data is large.

To study the individual effect of mutual information regularization and optimal transport, we conduct ablation studies to justify (i) whether adding mutual information regularization helps to be robust to OOD data; (ii) whether adding OT helps mitigate catastrophic forgetting. The results of both ablation studies are shown in Table 11 in Appendix C. The ✓ means the corresponding column component is used and ✗ means do not use the corresponding column component. Specifically, "MI" means using mutual information regularization, "OT" means using optimal transport constraint loss, "$\mathcal{U}_{id}$" means using additional ID unlabeled data features in addition to labeled data for OT loss. We observe that the improvement of robustness to OOD becomes more significant, especially with a larger amount of OOD data. When the number of OOD is 250 for each task, the improvement is nearly 4%. This demonstrates the effectiveness of mutual information regularization. The optimal transport (OT) component improves model performance by more than 2% with 250 OOD data, demonstrating the effectiveness of remembering in feature space with both labeled and unlabeled data. Additionally, we compare OT regularization with only labeled data features to baseline. We find the performance is moderately improved in this case. This shows the benefits of using unlabeled data for remembering.

To investigate whether using OOD data is helpful, we performed comparisons to cases where we only use ID-data for training and prediction with OOD data filtered out. The results are shown in Table 12 in Appendix C. The results show that filtering out OOD data is already helpful, but with additional OOD data and MI regularization, the performance is improved further. Results on other ablation settings are also available in Table 12.

Sensitivity to Dataset Ordering. To investigate the sensitivity of baselines and our method to dataset order, we also performed comparisons on two other dataset sequences, including (i) Butterfly, CUB, CIFARFS, Plantae, MiniImagenet, Aircraft and (ii) CUB, CIFARFS, Plantae, MiniImagenet, Butterfly, Aircraft. The results are shown in Appendix C (see details in Table 4 and Table 5 for order (i), Table 6 and Table 7 for order (ii)). In all cases, our method substantially outperforms the baselines and demonstrates its superiority.

Sensitivity of Hyperparameter. The sensitivity analysis of hyperparameters λ and β of ORDER are provided in Appendix C (see details in Table 8). β controls the magnitude of optimal transport regularization. Results indicate the model performance is positively correlated with β when it increases until optimal trade-off is reached after which performance deteriorates with over regularization when β reaches 0.01. A similar trend is observed in mutual information regularization λ.

More Results. To investigate the effect of memory buffer size, we provide more experimental results with a memory buffer size of 50, 100, 200 tasks in Appendix C (see details in Table 9). With increasing memory buffer size, the performance improves further. The current size of the memory buffer is negligible compared to the total amount of 72K tasks.

6   Conclusion

In this paper, we step towards a challenging and realistic problem scenario named SETS. The proposed method, i.e., ORDER, is designed to leverage unlabeled data to alleviate the forgetting of previously learned knowledge and improve the robustnesss to OOD. Our method first detects unlabeled OOD data from ID data. Then with the guidance of mutual information regularization, it further improves the accuracy. The method is also shown to be less sensitive to OOD. The OT constraint is adopted to mitigate the forgetting in feature space. We compare different methods on the constructed dataset. For the future work direction, we are working on adapting the method to the unsupervised scenario that no labeled data in new coming tasks, which requires the model with stronger robustness and capability of learning feature representations.

Acknowledgement. We thank all the anonymous reviewers for their thoughtful and insightful comments. This research was supported in part by NSF through grant IIS-1910492.

References

1. Agrawal, A., Amos, B., Barratt, S., Boyd, S., Diamond, S., Kolter, Z.: Differentiable convex optimization layers. In: Advances in Neural Information Processing Systems (2019)
2. Aljundi, R., Babiloni, F., Elhoseiny, M., Rohrbach, M., Tuytelaars, T.: Memory aware synapses: learning what (not) to forget. In: Ferrari, V., Hebert, M., Sminchisescu, C., Weiss, Y. (eds.) ECCV 2018. LNCS, vol. 11207, pp. 144–161. Springer, Cham (2018). https://doi.org/10.1007/978-3-030-01219-9_9

3. Aljundi, R., Kelchtermans, K., Tuytelaars, T.: Task-free continual learning. In: Proceedings of the IEEE Conference on Computer Vision and Pattern Recognition (CVPR) (2019)
4. Andrychowicz, M., et al.: Learning to learn by gradient descent by gradient descent. In: Advances in Neural Information Processing Systems (2016)
5. Antoniou, A., Edwards, H., Storkey, A.: How to train your maml. In: International Conference on Learning Representations (2019)
6. Antoniou, A., Patacchiola, M., Ochal, M., Storkey, A.: Defining benchmarks for continual few-shot learning (2020). https://arxiv.org/abs/2004.11967
7. Bae, I., et al.: Self-driving like a human driver instead of a robocar: Personalized comfortable driving experience for autonomous vehicles. In: NeurIPS Workshop (2019)
8. Balaji, Y., Chellappa, R., Feizi, S.: Robust optimal transport with applications in generative modeling and domain adaptation. In: Advances in Neural Information Processing Systems, pp. 12934–12944 (2020)
9. Barber, D., Agakov, F.: The im algorithm: A variational approach to information maximization (2003)
10. Belouadah, E., Popescu, A.: Il2m: Class incremental learning with dual memory. In: 2019 IEEE/CVF International Conference on Computer Vision (ICCV), pp. 583–592 (2019)
11. Berthelot, D., Carlini, N., Goodfellow, I., Papernot, N., Oliver, A., Raffel, C.: Mixmatch: A holistic approach to semi-supervised learning (2019)
12. Bertinetto, L., Henriques, J.F., Torr, P.H.S., Vedaldi, A.: Meta-learning with differentiable closed-form solvers. In: International Conference on Learning Representations (2019)
13. Buzzega, P., Boschini, M., Porrello, A., Abati, D., Calderara, S.: Dark experience for general continual learning: a strong, simple baseline. In: 34th Conference on Neural Information Processing Systems (2020)
14. Chaudhry, A., Ranzato, M., Rohrbach, M., Elhoseiny, M.: Efficient lifelong learning with a-gem. In: Proceedings of the International Conference on Learning Representations (2019)
15. Chaudhry, A., et al.: Continual learning with tiny episodic memories (2019). https://arxiv.org/abs/1902.10486
16. Chen, T., Wu, W., Gao, Y., Dong, L., Luo, X., Lin, L.: Fine-grained representation learning and recognition by exploiting hierarchical semantic embedding. In: ACM International Conference on Multimedia (2018)
17. Chen, W.Y., Liu, Y.C., Kira, Z., Wang, Y.C., Huang, J.B.: A closer look at few-shot classification. In: International Conference on Learning Representations (2019)
18. Cheng, P., Hao, W., Dai, S., Liu, J., Gan, Z., Carin, L.: Club: A contrastive log-ratio upper bound of mutual information. In: Proceedings of the 37th International Conference on Machine Learning (2020)
19. Diethe, T.: Practical considerations for continual learning (Amazon) (2020)
20. Ebrahimi, S., Meier, F., Calandra, R., Darrell, T., Rohrbach, M.: Adversarial continual learning. In: Vedaldi, A., Bischof, H., Brox, T., Frahm, J.-M. (eds.) ECCV 2020. LNCS, vol. 12356, pp. 386–402. Springer, Cham (2020). https://doi.org/10.1007/978-3-030-58621-8_23
21. Edwards, H., Storkey, A.: Towards a neural statistician. arXiv: 6060.2185 (2017)
22. Finn, C., Abbeel, P., Levine, S.: Model-agnostic meta-learning for fast adaptation of deep networks. In: International Conference on Machine Learning (2017)
23. Finn, C., Rajeswaran, A., Kakade, S., Levine, S.: Online meta-learning. In: Proceedings of International Conference on Machine Learning (2019)

24. Finn, C., Xu, K., Levine, S.: Probabilistic model-agnostic meta-learning. In: Advances in Neural Information Processing Systems (2018)
25. Genevay, A., Peyré, G., Cuturi, M.: Learning generative models with sinkhorn divergences (2018)
26. Guo, J., Zhu, X., Zhao, C., Cao, D., Lei, Z., Li, S.Z.: Learning meta face recognition in unseen domains. In: Conference on Computer Vision and Pattern Recognition (CVPR) (2020)
27. Guo, X., Yang, C., Li, B., Yuan, Y.: Metacorrection: Domain-aware meta loss correction for unsupervised domain adaptation in semantic segmentation. In: Conference on Computer Vision and Pattern Recognition (CVPR) (2021)
28. Horn, G.V., et al.: The inaturalist species classification and detection dataset. In: IEEE Conference on Computer Vision and Pattern Recognition (CVPR) (2018)
29. Huang, X., Qi, J., Sun, Y., Zhang, R.: Semi-supervised dialogue policy learning via stochastic reward estimation. In: Proceedings of the 58th Annual Meeting of the Association for Computational Linguistics (2020)
30. Kingma, D.P., Ba, J.: Adam: A method for stochastic optimization. In: International Conference on Learning Representations (2014)
31. Kirkpatrick, J., et al.: Overcoming catastrophic forgetting in neural networks. Proceedings of the National Academy of Sciences (2017)
32. Lake, B., Salakhutdinov, R., Gross, J., Tenenbaum, J.: One shot learning of simple visual concepts. In: Conference of the Cognitive Science Society (2011)
33. Lee, K., Lee, K., Shin, J., Lee, H.: Overcoming catastrophic forgetting with unlabeled data in the wild. In: 2019 IEEE/CVF International Conference on Computer Vision (ICCV) (2019)
34. Lee, K., Maji, S., Ravichandran, A., Soatto, S.: Meta-learning with differentiable convex optimization. In: Proceedings of the IEEE Conference on Computer Vision and Pattern Recognition (CVPR) (2019)
35. Li, X., et al.: Learning to self-train for semi-supervised few-shot classification. In: Proceedings of the Advances in Neural Information Processing Systems (2019)
36. Liu, Y., et al.: Learning to propagate labels: Transductive propagation network for few-shot learning. In: International Conference on Learning Representations (2019)
37. Lopez-Paz, D., Ranzato, M.: Gradient episodic memory for continual learning. In: Advances in Neural Information Processing Systems (2017)
38. Maji, S., Rahtu, E., Kannala, J., Blaschko, M., Vedaldi, A.: Fine-grained visual classification of aircraft (2013). https://arxiv.org/abs/1306.5151
39. Mishra, N., Rohaninejad, M., Chen, X., Abbeel, P.: A simple neural attentive meta-learner. In: International Conference on Learning Representations (2018)
40. Munkhdalai, T., Yu, H.: Meta networks. In: Proceedings of the 34th International Conference on Machine Learning (2017)
41. Nguyen, A., Yosinski, J., Clune, J.: Deep neural networks are easily fooled: High confidence predictions for unrecognizable images. In: IEEE Conference on Computer Vision and Pattern Recognition (CVPR) (2015)
42. Nguyen, C.V., Li, Y., Bui, T.D., Turner, R.E.: Variational continual learning. In: Proceedings of the International Conference on Learning Representations (2018)
43. Nilsback, M.E., Zisserman, A.: Automated flower classification over a large number of classes (2008)
44. Poole, B., Ozair, S., van den Oord, A., Alemi, A.A., Tucker, G.: On variational bounds of mutual information (2019)

45. Raghu, A., Raghu, M., Bengio, S., Vinyals, O.: Rapid learning or feature reuse? towards understanding the effectiveness of maml. In: International Conference on Learning Representations (2020)

46. Ravi, S., Larochelle, H.: Optimization as a model for few-shot learning. In: International Conference on Learning Representations (2017)

47. Ren, M., et al.: Meta-learning for semi-supervised few-shot classification. In: International Conference on Learning Representations (2018)

48. Riemer, M., et al.: Learning to learn without forgetting by maximizing transfer and minimizing interference. In: International Conference on Learning Representations (2019)

49. Saha, G., Garg, I., Roy, K.: Gradient projection memory for continual learning. In: International Conference on Learning Representations (2021)

50. Santoro, A., Bartunov, S., Botvinick, M., Wierstra, D., Lillicrap, T.: Meta-learning with memory-augmented neural networks. In: Proceedings of the 34th International Conference on Machine Learning (2016)

51. Schmidhuber, J.: A neural network that embeds its own meta-levels. In: IEEE International Conference on Neural Networks (1993)

52. Sehwag, V., Chiang, M., Mittal, P.: Ssd: A unified framework for self-supervised outlier detection. In: International Conference on Learning Representations (2021)

53. Smith, J., Balloch, J., Hsu, Y.C., Kira, Z.: Memory-efficient semi-supervised continual learning: The world is its own replay buffer (2021). https://arxiv.org/abs/2101.09536

54. Snell, J., Swersky, K., Zemel, R.S.: Prototypical networks for few-shot learning. In: Advances in Neural Information Processing Systems (2017)

55. Tseng, H.Y., Lee, H.Y., Huang, J.B., Yang, M.H.: Cross-domain few-shot classification via learned feature-wise transformation. In: Proceedings of the International Conference on Learning Representations (2020)

56. Vinyals, O., Blundell, C., Lillicrap, T., Kavukcuoglu, K., Wierstra, D.: Matching networks for one shot learning (2016). https://arxiv.org/pdf/1606.04080.pdf

57. Vitter, J.S.: Random sampling with a reservoir. Association for Computing Machinery (1985)

58. Vitter, J.S.: Random sampling with a reservoir. ACM Trans. Math. Software **11**, 37–57 (1985)

59. Vuorio, R., Sun, S.H., Hu, H., Lim, J.J.: Multimodal model-agnostic meta-learning via task-aware modulation. In: Proceedings of the Advances in Neural Information Processing Systems (2019)

60. Wang, L., Yang, K., Li, C., Hong, L., Li, Z., Zhu, J.: Ordisco: Effective and efficient usage of incremental unlabeled data for semi-supervised continual learning. In: IEEE Conference on Computer Vision and Pattern Recognition (CVPR) (2021)

61. Wang, Y.X., Ramanan, D., Hebert, M.: Meta-learning to detect rare objects. In: Proceedings of the IEEE/CVF International Conference on Computer Vision (ICCV) (October 2019)

62. Wang, Z., Duan, T., Fang, L., Suo, Q., Gao, M.: Meta learning on a sequence of imbalanced domains with difficulty awareness. In: Proceedings of the IEEE/CVF International Conference on Computer Vision, pp. 8947–8957 (2021)

63. Wang, Z., Shen, L., Duan, T., Zhan, D., Fang, L., Gao, M.: Learning to learn and remember super long multi-domain task sequence. In: Proceedings of the IEEE/CVF Conference on Computer Vision and Pattern Recognition, pp. 7982–7992 (2022)

64. Wang, Z., Shen, L., Fang, L., Suo, Q., Duan, T., Gao, M.: Improving task-free continual learning by distributionally robust memory evolution. In: International Conference on Machine Learning, pp. 22985–22998 (2022)
65. Wang, Z., et al.: Meta-learning without data via wasserstein distributionally-robust model fusion. In: The Conference on Uncertainty in Artificial Intelligence (2022)
66. Wang, Z., Zhao, Y., Yu, P., Zhang, R., Chen, C.: Bayesian meta sampling for fast uncertainty adaptation. In: International Conference on Learning Representations (2020)
67. Welinder, P., et al.: Caltech-UCSD Birds 200 (2010)
68. Yoon, J., Yang, E., Lee, J., Hwang, S.J.: Lifelong learning with dynamically expandable networks. In: International Conference on Learning Representations (2018)
69. Zhou, Y., Wang, Z., Xian, J., Chen, C., Xu, J.: Meta-learning with neural tangent kernels. In: International Conference on Learning Representations (2021)
70. Zhuang, P., Wang, Y., Qiao, Y.: Wildfish: A large benchmark for fish recognition in the wild (2018)

DnA: Improving Few-Shot Transfer Learning with Low-Rank Decomposition and Alignment

Ziyu Jiang[1,2], Tianlong Chen[3], Xuxi Chen[3], Yu Cheng[1], Luowei Zhou[1], Lu Yuan[1], Ahmed Awadallah[1], and Zhangyang Wang[3(✉)]

[1] Microsoft Corporation, Redmond, USA
[2] Texas A&M University, College Station, USA
[3] University of Texas at Austin, Austin, USA
atlaswang@utexas.edu

Abstract. Self-supervised (SS) learning has achieved remarkable success in learning strong representation for in-domain few-shot and semi-supervised tasks. However, when transferring such representations to downstream tasks with domain shifts, the performance degrades compared to its supervised counterpart, especially at the few-shot regime. In this paper, we proposed to boost the transferability of the self-supervised pre-trained models on cross-domain tasks via a novel self-supervised alignment step on the target domain using only unlabeled data before conducting the downstream supervised fine-tuning. A new reparameterization of the pre-trained weights is also presented to mitigate the potential catastrophic forgetting during the alignment step. It involves low-rank and sparse decomposition, that can elegantly balance between preserving the source domain knowledge without forgetting (via fixing the low-rank subspace), and the extra flexibility to absorb the new out-of-the-domain knowledge (via freeing the sparse residual). Our resultant framework, termed Decomposition-and-Alignment (**DnA**), significantly improves the few-shot transfer performance of the SS pre-trained model to downstream tasks with domain gaps. (The code is released at https://github.com/VITA-Group/DnA).

Keywords: Self-supervised learning · Transfer few-shot · Low-rank

1 Introduction

Employing Self-Supervised (SS) models pre-trained on large datasets for boosting downstream tasks performance has become de-facto for many applications [10], given it could save the expensive annotation cost and yield strong

---

Z. Jiang—Work done during an intership at Microsoft Corporation.

Supplementary Information The online version contains supplementary material available at https://doi.org/10.1007/978-3-031-20044-1_14.

S. Avidan et al. (Eds.): ECCV 2022, LNCS 13680, pp. 239–256, 2022.
https://doi.org/10.1007/978-3-031-20044-1_14

performance boosting for downstream tasks [6,8,17]. Recent advance in the SS pre-training method points out its potential on surpassing its supervised counterpart for few-shot and semi-supervised downstream tasks [7,39].

For the transfer learning of SS models, most previous works followed the many-shot setting [6,14,17]. However, recent discoveries state that when the target domain has a domain gap with the source data and it has only limited label samples, the transferability of SS models is still inferior to its supervised counter part [28]. The authors argued that comparing to SS pre-training, supervised pre-training encourages the learned representation to be more compactly distributed, and the label supervision also enforces stronger alignment across different images. As a result, the supervised representations display better clustering properties on the target data, facilitating the few-shot learning of classifier boundaries. The authors thus proposed to progressively sample and mix the unlabeled target data into the unsupervised pretraining stage, for several rounds. Such "target-aware" unsupervised pretraining (TUP) improves few-shot SS transfer performance. Yet it would be too expensive if we re-conduct pre-training (even in part) for every downstream purpose. Moreover, when pre-training is conducted on privileged data that is inaccessible to downstream users, the above solution will also become practically infeasible.

As previous findings reveal the few-shot transfer performance degradation due to the source-target domain discrepancy, in this paper, we are inspired to boost the transferability through a *self-supervised domain adaptation* perspective. Following the assumption of [28,32] that a small target dataset can be available, instead of "mixing" it with the pre-training data, we "fine-tune" the general pre-trained model with the small target data, under self-supervision. This extra step between pre-training and downstream fine-tuning, called **Alignment**, incurs a much smaller overhead compared to re-conducting pre-training on the mixed data [28], and avoids accessing the pre-training data.

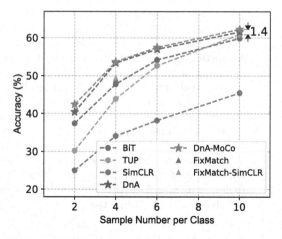

Fig. 1. Comparison with State-Of-The-Art (SOTA) methods on few-shot transfer tasks (source: ImageNet-1k; target: CIFAR100, with different labeled sample numbers per class). While the SimCLR performance struggles when directly transferring to cross-domain down-stream tasks, the proposed DnA method (implemented on top of SimCLR backbone) can significantly improve it by a large margin (>14%). DnA also remarkably surpasses the previous SOTAs (FixMatch [35], BiT [26] and TUP [28]) when combining with MoCo (DnA-MoCo) by at least 1.4%. FixMatch-SimCLR denotes FixMatch initilized from SimCLR pre-training.

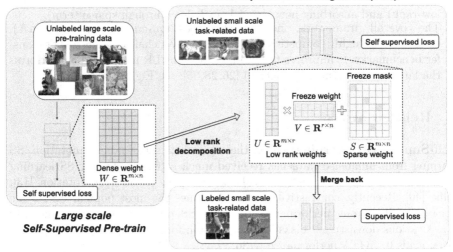

Fig. 2. The overview of the proposed DnA framework. It is applied on top of any self-supervised pre-trained model, to boost its few-shot transfer performance for the downstream tasks on the target data with a domain shift from the pre-training source data.

One specific challenge arising from the alignment step is the potential catastrophic forgetting of the pre-training knowledge [25]. To mitigate this risk, we introduce a **Decomposition** of the pre-trained weights before the alignment step, which involves no re-training. Specifically, we re-parameterize the pre-trained weight into the sum of the low-rank term (involving the produce of two matrix factors), and a sparse residual term. That is inspired by the findings that big pre-trained models have a low "intrinsic dimension" [1,20]. Then during the alignment, we freeze the low-rank subspace (but different subspace dimensions can be reweighted) in order to preserve the "in-domain" pre-training knowledge, while allowing the sparse residual to freely change for encapsulating the "out-of-domain" target knowledge. Although low-rank and sparse decomposition is a canonical idea [3,47,55], this is its first time to be connected the large model fine-tuning, to our best knowledge.

Our contributions can be summarized as following:

– We present a simple and effective self-supervised *alignment* method for mitigating the domain gaps between pre-training and downstream transfer, in order to enhance the few-shot transferability of self-supervised pre-trained models, without the (expensive and often infeasible) re-training with the source domain data.
– We further present a novel *decomposition* of the pre-trained weights, to mitigate the potential catastrophic forgetting during the alignment. It draws inspirations from the classical low rank and sparse decomposition algorithm,

and gracefully balances between preserving pre-training knowledge (through low-rank) and absorbing new target knowledge (through sparse term).
- The overall framework, named *Decomposition-and-Alignment* (**DnA**), demonstrates highly competitive performance over challenging few-shot transfer benchmarks. It improves the pre-trained SimCLR model and outperforms the latest state-of-the-arts (SOTA) [26, 28, 35]: see Fig. 1 for an example.

2   Related Works

Self-Supervised Learning: Given the expensive cost of label annotation, SS learning from unlabeled data has received much attention. Earlier SS learning methods employs proxy task like colorization [49], jigsaw [31], rotation [13], selfie [40]. Recently, contrastive learning becomes the most popular SS regime because of its strong performance [6, 14, 17]. SS pre-training can significantly boost various downstream tasks [5, 24, 39, 42, 45]; for semi-supervised learning, it often leads to SOTA performance [7, 36, 39, 48].

Transferability of Self-Supervised Models: For transferring SS models, most previous works studied the many-shot benchmarks [6, 14, 17]. The more practical yet challenging few-shot transferability of SS models is under-explored. Pioneer works investigated the effect of downstream few-shot transfer learning for different pre-training opinions [11, 22]. [22] reveals the improvement room of SS transferability via combining supervised learning, but it can only work on labeled data. TUP [28] explores to improve few-shot transferability via minimizing the domain discrepancy between pre-training and downstream tasks. It significantly boosts few-shot transferability by mixing the pre-training dataset with small-scale unlabeled samples progressively acquired from the target domain.

Low Rank and Sparsity in Deep Networks: Low-rank has been studied with a long history in deep networks [23, 33, 34, 37, 50, 53], for multiple contexts including model compression, multi-task learning and efficient training. Recent literature [1] reveals that the low-rank structure exists in the pre-trained model, which motivates the parameter efficient tuning [20]. The same prolific research can be found in the field of sparsity for deep networks, which is perhaps best known as a model compression means [12, 16]. Sparsity also effectively regularizes few-shot learning [4, 51, 56]; and naturally emerges during fine-tuning [15, 52]. We note that the current fine-tuning works relying on either low rank or sparsity [15, 20, 52] are all in the natural language processing (NLP) domain, and none of them operates in the few-shot setting with domain shifts.

The marriage of low rank and sparsity is well known as the robust principal component analysis (RPCA) algorithm [3, 55]. In deep networks, the most relevant work to this idea is perhaps [47], which reconstructed the weight matrices by using sparse plus low-rank approximation, for model compression - an orthogonal purpose to ours. Other applications include combining those two priors in deep compressive sensing [21]. To the best of our knowledge, no previous work has linked low rank and sparse decomposition to transfer learning.

3   Method

3.1   Overview

In this paper, we employ SimCLR [6] as a strong SS pre-training backbone. SimCLR [6] learns visual representation via enforcing the consistency between different augmented views while enlarging the difference from other samples. Formally, the loss of SimCLR is

$$\mathcal{L}_{\mathrm{CL},i} = - \log \frac{s^\tau \left(v_i^1, v_i^2\right)}{s^\tau \left(v_i^1, v_i^2\right) + \sum_{v^- \in V} s^\tau \left(v_i^1, v^-\right)} \tag{1}$$

where v_i^1 and v_i^2 are the normalized features of two augmented views for the same image, while V is the set of negative samples for ith image, which is composed by the features for other images in the same batch. All features are calculated sequentially with the feature encoder and projection head. s^τ is the feature similarity function with temperature τ that can be formalized as

$$s^\tau \left(v_i^1, v_i^2\right) = \exp(\frac{v_i^1 \cdot v_i^2}{\tau}) \tag{2}$$

As an overview, our holistic framework is demonstrated in Fig. 2: DnA first decomposes the SS pre-trained weight to the low-rank terms (U and V) and sparse term S. Afterwards, the model is aligned by self-supervised tuning over small target domain data, by fixing V while tuning U and S in this step. The aligned model then goes through the typical supervised fine-tuning for the downstream few-shot task. Below we present step-by-step method details.

3.2   Basic Alignment Step

In this work, we leverage unlabeled training data from the downstream dataset , yet avoiding the (expensive and often infeasible) re-pre-training for every downstream transfer, and assuming no access to the pre-training data. To reduce the discrepancy between source and target domains, we design the extra **Alignment** step between pre-training and fine-tuning: we continue to tune the pre-trained model on a small-scale unlabeled dataset from the target domain, with self-supervised loss (here we use the same loss of SimCLR). Note that we follow [28] to assume that small-scale unlabeled data from the domain of the target few-shot task is available. (e.g., the unlabeled training set of the downstream dataset.)

The proposed alignment step is rather simple and efficient. Notably, it only requires the pre-trained model but not the pre-training dataset, which we believe is a more practical setting. Perhaps surprisingly, we observe that this vanilla alignment already suffices to outperform TUP [28] in some experiments.

Challenge: However, the alignment might run into the risk of *catastrophic forgetting* of the pre-training knowledge, due to tuning with only the target domain data. That will also damage the transferred model's generalization. We started by trying off-the-shelf learning-without-forgetting strategies, such as enforcing the ℓ_2 norm similarity between the pre-trained model with the aligned model weights [44]. That indeed yields empirical performance improvements, but mostly only marginal. We are hence motivated to look into more effective mitigation for the fine-tuning scenarios.

3.3 Decomposition Before Alignment

Inspired by the findings that the weight of the pre-trained model resides with a low "intrinsic dimension" [1,20], we propose to leverage the low-rank subspace assumption to effectively "lock in" the pre-training knowledge, with certain flexibility to adjust the subspace coefficients. However, the low-rank structure alone might be too restricted to learn the target domain knowledge that might lie out of this low-rank subspace, and we extend another sparse residual term to absorb such. Adopting the decomposed weight form for alignment is hence assumed to well balance between memorizing the source-domain knowledge and flexibly accommodating the new out-of-the-domain knowledge.

This idea of combining low rank and sparsity is a canonical one [3,55], yet has not been introduced to fine-tuning before. In the following sections, we first introduce how to decompose the weight in Sect. 3.3. Then, we discuss the details of applying in Sect. 3.3.

Low Rank and Sparse Weight Decomposition. For a convolutional or a fully connected layer, the forward process can be formalized as

$$\mathbf{y} = W\mathbf{x} \tag{3}$$

where $\mathbf{y} \in \mathbf{R}^m$ is the output, $W \in \mathbf{R}^{m \times k}$ is the weight matrix, and $\mathbf{x} \in \mathbf{R}^k$ is the input. It is worth noting that, for the convolutional layer, we follow [38] to reshape the 4D convolutional kernel $W \in \mathbf{R}^{C_{in} \times H \times W \times C_{out}}$ to a 2D matrix $W \in \mathbf{R}^{(C_{in}H) \times (WC_{out})}$.

The resultant 2D weight W can be decomposed as

$$W = UV + S \tag{4}$$

where $U \in \mathbf{R}^{m \times r}$ and $V \in \mathbf{R}^{r \times k}$ are two low rank matrix factors with $r < \min(m, k)$, and together denotes the low-rank subspace component of the pre-trained weight. Meanwhile S denotes the sparse residual.

The low rank and sparse decomposition were found to well capture the long-tail structure of trained weights [47]. To solve this decomposition, we resort to the fast and data-free matrix decomposition method called GreBsmo [54]. GreBsmo formalizes the low rank decomposition as:

$$\min_{U,V,S} \|W - UV - S\|_F^2$$
$$\text{s.t. } \mathrm{rank}(U) = \mathrm{rank}(V) \leq r, P_\Omega S = 0 \tag{5}$$

where $1 - P_\Omega$ is the sparse mask of S, defined as

$$P_\Omega = \begin{cases} 1, (i,j) \in \Omega \\ 0, (i,j) \in \Omega^C \end{cases} \tag{6}$$

where Ω denotes the set of points that has value, (i, j) indicates the coordinate of a point in 2d weight matrix. GreBsmo then solves this optimization with an alternative updating algorithm (its derivation can be found at [54]) as:

$$\begin{cases} U_k = Q, D_{\mathrm{QR}}\left((W - S_{k-1})V_{k-1}^T\right) = QR \\ V_k = Q^T\left(W - S_{k-1}\right) \\ S_k = \mathcal{S}_\lambda\left(W - U_kV_k\right) \end{cases} \tag{7}$$

where D_{QR} is the fast QR decomposition algorithm to generate two inter-media matrix Q and R. The subscription k of U, V, S indicates the U, V, S in the kth iteration. $\mathcal{S}_\lambda$ is an element-wise soft threshold function formally as:

$$\mathcal{S}_\lambda W = \{\mathrm{sgn}\left(W_{ij}\right)\max\left(|W_{ij}| - \lambda, 0\right) : (i, j) \in [m] \times [k]\} \tag{8}$$

where sgn x would output the sign for x.

Practically, the decomposed weight would inevitably have a slight loss $E = W - UV - S$ compared to the origin pre-trained weight, due to the limited optimization precision. Such difference is usually very small at each layer, but might be amplified during the forward pass. To mitigate that, we treat the E term as a fixed bias for each layer, and never tune it during the alignment step.

How to Align over the Decomposed Weights. After decomposition, we freeze the low-rank subspace V as the "fixed support" from the pre-trained weight. For flexibility, we consider U as representation coefficients over this subspace, and allow that to be "alignable" (i.e., we allow for "re-composing" the existing knowledge). Meanwhile, S represents the out-of-the-subspace component and is always set to be "alignable" for new target knowledge.

Hence, the alignment step over the decomposed weights could be represented as:

$$\begin{aligned} W' &= (U + \Delta U)V + S + \Delta S \\ \text{s.t. } &\mathrm{rank}(U) = \mathrm{rank}(V) \leq r \\ &P_\Omega S = 0, P_\Omega \Delta S = 0 \end{aligned} \tag{9}$$

where ΔU and ΔS are the two variables that will be tuned in the alignment step; the original U, V, S are all fixed for alignment meanwhile. Note that ΔU has the same dimension as the low-rank factor U, and ΔS has the same sparse support (i.e., locations of non-zero elements) as S.

After the alignment step, we re-combine the decomposed form into one weight matrix W, and proceed to the fine-tuning step as normal.

4 Experiments

4.1 Settings

Datasets: To evaluate the proposed method, we exploit the following datasets: iNaturalist [41], CIFAR100 [27], EuroSAT [19], Food101 [2]. Particularly, we adopt a random sampled 1000 classes subset of iNaturalist (Denoted as iNaturalist-1k) to validate the proposed method's performance on an imbalanced dataset. As shown in Table 1, these datasets are different in terms of resolutions (ranging from 32–224) and the number of classes (ranging from 10–1000). In addition, we consider both general classification tasks (e.g., CIFAR100) and fine-grain classification tasks (e.g., iNaturalist-1k, Food101). The imbalanced dataset is also included (e.g., iNaturalist-1k). The datasets also have different levels of similarity to the ImageNet (ranging from natural images like iNaturalist to the satellite images like EuroSAT). These differences indicate the chosen datasets can represent practical cases. We report the Top1 accuracy for all datasets. For all datasets, we upsample/downsample it to 224×224 following the common practice of applying ImageNet pre-trained model [6,28]. We follow [28] and assume that the available unlabeled dataset is the full training set for each dataset.

Training Settings: For all the experiments, we employ the network architecture of Resnet-50 [18]. When conducting SS, the fully connected layer is replaced with a two-layer projection head as in Sim-CLR [6]. For SS pre-trained model, we employ

Table 1. Summary of the datasets employed in this work for resolution, number of train images (#train images), and number of categories (#Categories).

Dataset	Resolution	#Train images	#Categories
iNaturalist-1k	>224×224	51984	1000
CIFAR100	32×32	50000	100
EuroSAT	64×64	19000	10
Food101	>224×224	75750	101

the official SimCLR and Moco model pre-trained on ImageNet-1k for 800 epochs.

For all DnA experiments, we follow SimCLR to use LARS optimizer [46], augmentation settings and cosine learning rate decay. We employ a batch size of 256 instead of larger batch size for ensuring the proposed method's practicability. On iNaturalist-1K, we train 200 epochs for ablation study. On other datasets, We train the model for 800 epochs following [28]. The initial learning rate is set as 0.2. For few-shot fine-tuning, we follow the setting of SimCLR-v2 [7] with LARS optimizer, cosine learning rate decay, and tune from the first layer of the projection head for 200 epochs.

We also consider Linear Probing (LP) performance for SimCLR, where a linear classifier is trained on the frozen features, following the setting of [39].

Configuration for the GreBsmo Algorithm: To improve the practical speed, GreBsmo invokes a greedy rank r for both U and V. It starts from a very small rank of r_0 for V, Then $V \in \mathbf{R}^{r_0 \times k}$ iterates Eq. 7 for M times. Afterwards, the rank of V would increase to $r_1 = r_0 + \Delta r$ by adding Δr extra rows to V, where Δr is the rank step size. The added Δr rows are selected greedily as the top Δr row basis that can minimize the objective in Eq. 5, which are obtained with a small SVD [54].

Benefiting from the aforementioned, the weight decomposition for the entire ResNet50 can be done in less than one hour on a single 1080 Ti, which is a small time overhead compared to either pre-training or finetuning. Moreover, the decomposition can be reused among different downstream datasets, further amortizing the computation overhead.

Empirically, we initialize with $U_0 = 0$, $V_0 \sim \mathcal{N}(0, 0.02)$ $S_0 = 0$. r_0, Δr set as 1 and iterations M is set as 100. As Resnet-50 [18] has a large variety for the weight dimension for different layers, we thus set the rank adaptively with the size of weight matrix as $r = \min(\lceil \alpha_r \cdot m \rceil, k)$, α_r is the rank ratio by default set as 0.25. The soft sparsity threshold λ is by default 0.2. For the stability of training, we further fix sparsity to 99.7% after decomposing via assigning the top 0.3% large magnitude parameters in $W - UV$ as S.

4.2   DnA Improves Few-Shot Performance

DnA Improves the Transfer Few-shot Performance. To verify the effectiveness of the proposed methods, we study each component at iNaturalist-1k in terms of the 5-shot accuracy. As illustrated in Table 2a, the model without any pre-training yields a performance of 12.7%. By leveraging the SimCLR model pre-trained on ImageNet, the performance increases to 45.0%. Further, the proposed Align can significantly improve the performance by 1.4%. Finally, when combining the weight decomposition with the Align, the resultant DnA can further improve the accuracy over Align by an obvious margin of 0.9%.

DnA Surpasses the Semi-supervised Methods. For the proposed DnA framework, we assume the existence of a small scale unlabeled dataset following [28]. This setting is close to semi-supervised learning, which has been widely studied in previous works [7,35,43]. Therefore, we further conduct experiments to compare our methods against well-established semi-supervised methods. Here, we exploit the SOTA semi-supervised strategy introduced in SimCLR-v2 [7] for comparison. As illustrated in Table 2a, when employing the semi-supervised

Table 2. (a) The few-shot and semi-supervised performance on iNaturalist-1k under the 5-shot setting. We employ the *distillation with unlabeled examples* method introduced in [7] as the semi-supervised method. (b) Comparing with FixMatch [35] on CIFAR100 4-shot in terms of semi-supervised performance.

Pre-train	Semi-supervised	Accuracy
None		12.7
	✓	13.0
SimCLR		45.0
	✓	46.3
Aligned (Ours)		47.4
	✓	48.5
DnA (Ours)		48.3
	✓	49.6

(a)

Pre-train	Method	Accuracy
None	FixMatch	48.9
SimCLR	FixMatch	49.4
SimCLR	DnA (ours)	53.4
SimCLR	DnA+Semi (ours)	56.4

(b)

method, the performance consistently improves compared to the corresponding few-shot performance. However, the improvement is marginal when the few-shot performance is not optimized (e.g. when the pre-training is not applied). While semi-supervised learning can improve the SimCLR model by 1.3%, it is still inferior by 1.1% compared to the Aligned model, indicating that the proposed Align method can yield better performance than semi-supervised method when using the same amount of unlabeled and labeled data. Besides, when combined with semi-supervised learning, the performance of both Aligned and DnA could further improve by 1.1% and 1.3%, respectively. Further, we compare with the state-of-the-art semi-supervised method FixMatch [35] on the official CIFAR100 4-shot benchmark (corresponding to 400 labels setting in the original paper). As shown in Table 2b, while FixMatch achieves a promising performance of 48.9% when training from scratch, initializing FixMatch with self-supervised pre-training can hardly improve its performance even though we have tried smaller learning rate and warm-up. This may be because the forgetting problem is serious given it requires a long training schedule. In contrast, the proposed method can yield a significantly higher performance of 52.6 and 56.4 for w/o and w/semi-supervised methods, respectively.

Weight Decomposition Better Prevents Forgetting. As the effectiveness of the proposed weight decomposition method works by preventing forgetting source information, we also include a baseline for comparing with the previous learning without forgetting methods. We choose L2-SP [44], Delta [29] and BSS [9] as our baselines. The results are shown in Table 3: for 5-shot accuracy in iNaturalist-1k, while applying [L2-SP, Delta, Delta+BSS] for Align yield an improvement over the naive Aligned by

Table 3. Compare with learning without forgetting methods in terms of the 5-shot performance on iNaturalist-1k.

Method	Accuracy
Align+L2-SP	47.7
Align+Delta	48.1
Align+Delta+BSS	48.3
DnA (Ours)	48.3
DnA+Delta+BSS(Ours)	48.5

[0.3%, 0.7%, 1.0%], respectively. While the performance Align+Delta+BSS is on par with DnA, the proposed DnA can also be combined with Delta+BSS and further yield accuracy of 48.5%.

DnA Surpass the State-of-The-Art (SOTA). To further study the effectiveness of the proposed method, we compare it with the SOTA supervised method BiT and SOTA self-supervised method TUP [28] on three different datasets. As illustrated in the Table 4, while TUP [28] can surpass the state-of-the-art supervised method BiT in terms of the Mean accuracy by [1.0%, 1.9%, 3.5%, 4.2%] for [2-shot, 4-shot, 6-shot, 10-shot] performance, respectively. The proposed DnA can further yield an improvement with a significant margin of [8.3%, 8.7%, 5.0%, 3.5%] compared to TUP. For a fair comparison with TUP, we also employ DnA with pre-training of MoCo v2, termed as DnA-Moco. DnA-Moco also yields consistent improvement. It's worth noting that the proposed method is also more efficient than TUP as DnA i) has no sampling step ii) conducts the

Table 4. The few-shot fine-tuning performance comparison on different datasets for different pre-trained models. SimCLR here represents the SimCLR pre-trained on ImageNet-1k. The average accuracy and standard deviation (%) on five different labeled subsets are reported. #Few-shot means the number of few-shot samples for each class. The performance of BiT [26] and TUP [28] are from [28]. We consider few-shot performance on the random sampled few-shot subsets, which corresponds to the oracle few-label transfer setting of [28]. LP denotes linear probing. DnA-MoCo is the combination of ours and MoCo v2.

#Few-shot	Method	CIFAR100	EuroSAT	Food101	Mean
2	BiT [26]	37.4	68.3	24.5	43.4
	TUP [28]	30.2	68.8	34.3	44.4
	SimCLR	15.2±1.4	57.9±1.3	13.6±0.4	28.9
	SimCLR - LP	25.0±0.9	69.5±3.1	20.0±0.8	38.2
	DnA (Ours)	<u>40.4 ± 1.6</u>	77.5 ± 1.6	<u>40.2 ± 1.4</u>	<u>52.7</u>
	DnA-MoCo (Ours)	**42.4±1.3**	**80.7±2.2**	**42.7±1.5**	**55.3**
4	BiT [26]	47.8	79.1	35.3	54.1
	TUP [28]	43.9	76.2	48.0	56.0
	SimCLR	26.6±1.5	65.7±2.8	22.7±0.2	38.3
	SimCLR - LP	34.1±0.4	77.7±2.4	27.7±0.6	46.5
	DnA (Ours)	<u>53.4 ± 0.8</u>	**86.5±0.7**	**54.2±0.4**	**64.7**
	DnA-MoCo (Ours)	**53.7±0.6**	85.1 ± 2.1	<u>53.9 ± 0.9</u>	<u>64.2</u>
6	BiT [26]	54.2	82.6	41.3	59.4
	TUP [28]	52.7	80.6	55.4	62.9
	SimCLR	34.7±0.9	72.0±1.7	28.3±0.7	45.0
	SimCLR - LP	38.2±0.8	81.2±1.2	32.6±0.3	50.7
	DnA (Ours)	<u>57.0 ± 0.3</u>	**87.8±1.0**	**59.0±0.7**	**67.9**
	DnA-MoCo (Ours)	**57.5±0.4**	<u>87.2 ± 0.9</u>	<u>58.9 ± 0.6</u>	**67.9**
10	BiT [26]	59.9	86.3	48.8	65.0
	TUP [28]	60.9	84.1	62.6	69.2
	SimCLR	45.5±1.1	76.6±1.7	37.3±0.6	53.1
	SimCLR - LP	43.2±0.5	83.4±1.3	38.9±0.5	55.2
	DnA (Ours)	<u>61.6 ± 0.4</u>	**89.4±1.2**	<u>63.3 ± 0.5</u>	<u>71.4</u>
	DnA-MoCo (Ours)	**62.3±0.3**	<u>89.3 ± 0.3</u>	**65.0±0.4**	**72.2**

training only on the target dataset, which is 5 times smaller than the dataset employed in TUP.

Besides, applying DnA for the SS pre-trained model can make a large difference in these datasets as DnA can surpass its start point, the SimCLR model, by a large margin of [23.8%, 26.4%, 22.9%, 18.3%] and [14.5%, 18.2%, 17.2%, 16.2%] for fine-tuning and linear probing setting, respectively. In comparison, the improvement on iNaturalist-1k is milder, the intuition behind this is that the domain difference between these datasets and ImageNet is much larger.

4.3 Ablation Study

Does SS ImageNet Pre-training Help? As the proposed DnA improves a lot based on the pre-trained model, a natural question arises: would the in-domain data be enough for few-shot learning? How many benefits can ImageNet pre-training bring? This motivates us to compare the proposed DnA method with SS pre-training on the target dataset. As illustrated in Table 5, when training without ImageNet pre-trained model, in CIFAR100, the performance would degrade from [53.4%, 61.6%] to [32.8%, 49.2%] for [4-shot, 10-shot] performance,

Table 5. Comparing the few-shot performance on CIFAR100 and Food101 between DnA (with SS ImageNet pre-trained model) and in-domain SS pre-training (without SS ImageNet pre-trained model).

Dataset	Pretrain	4-shot	10-shot
CIFAR100		32.8	49.2
	✓	53.4	61.6
Food101		29.7	44.9
	✓	54.2	63.3

respectively. The observation on Food-101 is also consistent, demonstrating the importance of the SS ImageNet pre-trained model. Also, this observation further motivates us to employ the learning with forgetting method.

The Fix Components Choosing for DnA. In this part, we ablation study the effect of fixing different components of the weight decomposition. As illustrated in Table 6a, when we free the fixing weight and mask for every component of the three terms, the performance would degrade to 47.4%, which is equal to the performance of Align. This is because the tunable parameters are even more than the original pre-trained model, which means this architecture can not prevent forgetting. By adding the fixing mask on S, the performance could improve to 47.9%, showing that only applying the mask can prevent forgetting. By fixing both sparse mask and low-rank sub-spaces via freezing V, the performance could further improve to 48.3% with a small variance of 0.1%, showing that fixing low-rank subspace could further prevent information loss. However, when switching the fixing low-rank component from U to V, the performance could decrease by 0.5%. The intuition behind this is that, for resnet-50, the size of $U \in \mathbf{R}^{C_{in}H \times r}$ is usually smaller or equal to $V \in \mathbf{R}^{r \times WC_{out}}$, indicating choosing V as fixed bases could preserve more information. When fixing S and tuning U, the performance would decrease by 0.3% compared to tuning U and S, showing the S can capture the "out-of-the-domain" knowledge when subspace is fixed. When fixing every term, it would fail back to the origin SimCLR model. Last but not least, as shown in Table 6b, when employing the decomposition strategy with only the low-rank component as $W = UV$, the performance would be weaker to 'tuning U,S' of $W = UV + S$ for both 'tuning U, V' and 'tuning U'. Because $W = UV$ needs a higher rank to minimize the decomposition loss and thus fail to prevent forgetting.

Table 6. Ablation study for different decomposition and fixing strategies of weight on iNaturalist-1k in terms of 5-shot performance. The ✓ under U, V denotes the corresponding terms is adjustable. The ✓ under S denotes the value of S is adjustable while sparse mask is fixed. In contrast, ✓✓ for S means S is adjustable for both value and mask. (a) and (b) employ different decomposition strategies. (a) The employed $W = UV + S$ decomposition, (b) $W = UV$, the decomposition strategy without low rank term.

U	V	S	5-shot
✓	✓	✓✓	47.4
✓	✓	✓	47.9±0.2
✓		✓	**48.3±0.1**
	✓	✓	47.7
✓	✓		47.8±0.3
✓			48.0±0.3
			45.0

(a)

U	V	5-shot
✓	✓	47.6
✓		47.9

(b)

Hyper-Parameters for Weight Decomposition. We further report the ablation study on the selection of the sparsity threshold s and the value r (We remove the sparsity fixing step here). The performance with different sparsity levels is shown in Fig. 3a. We can see that either a too large sparsity or a too small sparsity would lead to inferior performance. Especially, when the sparsity decrease from 99.7% to 70%, the performance could decrease very fast from 48.3% to 47.5%, which is very close to the Align performance of 47.4%. This is because too large flexibility of the S would override the low-rank component and make DnA degrade to Align method. The sweet point is sparsity of 99.7%, showing the "out-of-domain" knowledge can be efficiently encapsulated with only 0.3% free parameters.

The study of the different number of ranks is shown in Fig. 3b. When choosing rank ratio α_r ranging from 0.05 to 0.35, the DnA could yield an improvement of at least 0.5%, demonstrating the choice of rank is not sensitive. For a too large rank of 0.45, the performance would decrease to the level of Align because of too much flexibility. It is also worth noting the proposed DnA can even achieve a performance of 48.1% with a very small α_r of 0.05. Combining a very sparse mask of 99.7%, the proposed DnA can adapt the SS pre-trained model with very less parameters compared to the original network, showing its parameter efficiency.

Representation Visualization. In this section, we utilize t-SNE [30] to analyze the feature distribution before and after applying the proposed DnA. As illustrated in Fig. 4, for SimCLR pre-trained on the ImageNet, the features of samples from different classes would overlap with each other. Only a small portion of samples from the same class form a cluster. In contrast, after applying

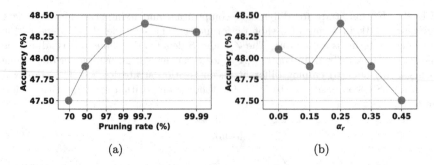

(a)   (b)

Fig. 3. Ablation study for hyper-parameter selection on iNaturalist-1k in terms of the 5-shot accuracy. (a) studies the influence different pruning rates, the sparsity of [68.10%, 88.83%, 97.39%, 99.74%, 99.99%] shown in the figure correspond to s of [0.02, 0.05, 0.1, 0.2, 0.5], respectively. (b) studies the different selection for rank ratio α_r.

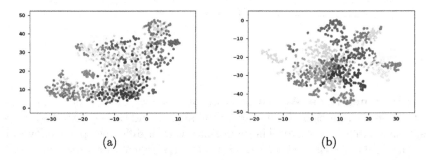

(a)   (b)

Fig. 4. t-SNE visualization for SimCLR representations before and after applying DnA on the test dataset of CIFAR100 (We random sample 20 classes for visualization). (a) t-SNE before applying DnA (SimCLR) (b) t-SNE after applying DnA. Different color stands for different classes.

DnA, the overlap can be significantly alleviated. Many samples in the same class are clustered together. Some clusters even have a gap with a large margin to other clusters. We believe the strong few-shot performance of DnA is closely related to the good learned representations.

5   Conclusion

When transferring SS pre-trained model to the downstream task with a domain discrepancy, the few-shot performance of the SS model could degraded more compared to its supervised counterpart. In this paper, we tackle this problem with DnA framework, which is designed from a domain adaptation perspective. The proposed DnA achieved new SOTA performance on three different datasets (CIFAR100, NeurSAT and Food101), demonstrating its effectiveness. We believe our technique is beneficial for realistic few-shot classification.

References

1. Aghajanyan, A., Zettlemoyer, L., Gupta, S.: Intrinsic dimensionality explains the effectiveness of language model fine-tuning. arXiv preprint arXiv:2012.13255 (2020)
2. Bossard, L., Guillaumin, M., Van Gool, L.: Food-101 – mining discriminative components with random forests. In: Fleet, D., Pajdla, T., Schiele, B., Tuytelaars, T. (eds.) ECCV 2014. LNCS, vol. 8694, pp. 446–461. Springer, Cham (2014). https://doi.org/10.1007/978-3-319-10599-4_29
3. Candès, E.J., Li, X., Ma, Y., Wright, J.: Robust principal component analysis? J. ACM (JACM) **58**(3), 1–37 (2011)
4. Chen, T., Cheng, Y., Gan, Z., Liu, J., Wang, Z.: Data-efficient GAN training beyond (just) augmentations: a lottery ticket perspective. arXiv preprint arXiv:2103.00397 (2021)
5. Chen, T., Liu, S., Chang, S., Cheng, Y., Amini, L., Wang, Z.: Adversarial robustness: from self-supervised pre-training to fine-tuning. In: Proceedings of the IEEE/CVF Conference on Computer Vision and Pattern Recognition, pp. 699–708 (2020)
6. Chen, T., Kornblith, S., Norouzi, M., Hinton, G.: A simple framework for contrastive learning of visual representations. In: International Conference on Machine Learning, pp. 1597–1607. PMLR (2020)
7. Chen, T., Kornblith, S., Swersky, K., Norouzi, M., Hinton, G.: Big self-supervised models are strong semi-supervised learners. arXiv preprint arXiv:2006.10029 (2020)
8. Chen, X., Fan, H., Girshick, R., He, K.: Improved baselines with momentum contrastive learning. arXiv preprint arXiv:2003.04297 (2020)
9. Chen, X., Wang, S., Fu, B., Long, M., Wang, J.: Catastrophic forgetting meets negative transfer: batch spectral shrinkage for safe transfer learning. Adv. Neural Inf. Process. Syst. **32** (2019)
10. Devlin, J., Chang, M.W., Lee, K., Toutanova, K.: Bert: pre-training of deep bidirectional transformers for language understanding. arXiv preprint arXiv:1810.04805 (2018)
11. Ericsson, L., Gouk, H., Hospedales, T.M.: How well do self-supervised models transfer? In: Proceedings of the IEEE/CVF Conference on Computer Vision and Pattern Recognition, pp. 5414–5423 (2021)
12. Frankle, J., Carbin, M.: The lottery ticket hypothesis: finding sparse, trainable neural networks. In: International Conference on Learning Representations (2019)
13. Gidaris, S., Singh, P., Komodakis, N.: Unsupervised representation learning by predicting image rotations. arXiv preprint arXiv:1803.07728 (2018)
14. Grill, J.B., et al.: Bootstrap your own latent: a new approach to self-supervised learning. arXiv preprint arXiv:2006.07733 (2020)
15. Guo, D., Rush, A.M., Kim, Y.: Parameter-efficient transfer learning with diff pruning. arXiv preprint arXiv:2012.07463 (2020)
16. Han, S., Pool, J., Tran, J., Dally, W.J.: Learning both weights and connections for efficient neural networks. arXiv preprint arXiv:1506.02626 (2015)
17. He, K., Fan, H., Wu, Y., Xie, S., Girshick, R.: Momentum contrast for unsupervised visual representation learning. In: Proceedings of the IEEE/CVF Conference on Computer Vision and Pattern Recognition, pp. 9729–9738 (2020)

18. He, K., Zhang, X., Ren, S., Sun, J.: Deep residual learning for image recognition. In: Proceedings of the IEEE Conference on Computer Vision and Pattern Recognition, pp. 770–778 (2016)
19. Helber, P., Bischke, B., Dengel, A., Borth, D.: Eurosat: a novel dataset and deep learning benchmark for land use and land cover classification. IEEE J. Sel. Top. Appl. Earth Obs. Remote Sens. **12**(7), 2217–2226 (2019)
20. Hu, E.J., et al.: Lora: low-rank adaptation of large language models. arXiv preprint arXiv:2106.09685 (2021)
21. Huang, W., et al.: Deep low-rank plus sparse network for dynamic MR imaging (2021)
22. Islam, A., Chen, C.F., Panda, R., Karlinsky, L., Radke, R., Feris, R.: A broad study on the transferability of visual representations with contrastive learning. arXiv preprint arXiv:2103.13517 (2021)
23. Jaderberg, M., Vedaldi, A., Zisserman, A.: Speeding up convolutional neural networks with low rank expansions. arXiv preprint arXiv:1405.3866 (2014)
24. Jiang, Z., Chen, T., Chen, T., Wang, Z.: Robust pre-training by adversarial contrastive learning. In: NeurIPS (2020)
25. Kirkpatrick, J., et al.: Overcoming catastrophic forgetting in neural networks. Proc. Nat. Acad. Sci. **114**(13), 3521–3526 (2017)
26. Kolesnikov, A., et al.: Big transfer (BiT): general visual representation learning. In: Vedaldi, A., Bischof, H., Brox, T., Frahm, J.-M. (eds.) ECCV 2020. LNCS, vol. 12350, pp. 491–507. Springer, Cham (2020). https://doi.org/10.1007/978-3-030-58558-7_29
27. Krizhevsky, A., Hinton, G., et al.: Learning multiple layers of features from tiny images (2009)
28. Li, S., et al.: Improve unsupervised pretraining for few-label transfer. In: Proceedings of the IEEE/CVF International Conference on Computer Vision, pp. 10201–10210 (2021)
29. Li, X., et al.: Delta: deep learning transfer using feature map with attention for convolutional networks. arXiv preprint arXiv:1901.09229 (2019)
30. Van der Maaten, L., Hinton, G.: Visualizing data using t-SNE. J. Mach. Learn. Res. **9**(11), 2579–2605 (2008)
31. Noroozi, M., Favaro, P.: Unsupervised learning of visual representations by solving jigsaw puzzles. In: Leibe, B., Matas, J., Sebe, N., Welling, M. (eds.) ECCV 2016. LNCS, vol. 9910, pp. 69–84. Springer, Cham (2016). https://doi.org/10.1007/978-3-319-46466-4_5
32. Phoo, C.P., Hariharan, B.: Self-training for few-shot transfer across extreme task differences. arXiv preprint arXiv:2010.07734 (2020)
33. Povey, D., et al.: Semi-orthogonal low-rank matrix factorization for deep neural networks. In: Interspeech, pp. 3743–3747 (2018)
34. Sainath, T.N., Kingsbury, B., Sindhwani, V., Arisoy, E., Ramabhadran, B.: Low-rank matrix factorization for deep neural network training with high-dimensional output targets. In: 2013 IEEE International Conference on Acoustics, Speech and Signal Processing, pp. 6655–6659. IEEE (2013)
35. Sohn, K., et al.: Fixmatch: simplifying semi-supervised learning with consistency and confidence. arXiv preprint arXiv:2001.07685 (2020)
36. Su, J.-C., Maji, S., Hariharan, B.: When does self-supervision improve few-shot learning? In: Vedaldi, A., Bischof, H., Brox, T., Frahm, J.-M. (eds.) ECCV 2020. LNCS, vol. 12352, pp. 645–666. Springer, Cham (2020). https://doi.org/10.1007/978-3-030-58571-6_38

37. Sun, M., Baytas, I.M., Zhan, L., Wang, Z., Zhou, J.: Subspace network: deep multi-task censored regression for modeling neurodegenerative diseases. In: Proceedings of the 24th ACM SIGKDD International Conference on Knowledge Discovery & Data Mining, pp. 2259–2268 (2018)

38. Tai, C., Xiao, T., Zhang, Y., Wang, X., et al.: Convolutional neural networks with low-rank regularization. arXiv preprint arXiv:1511.06067 (2015)

39. Tian, Y., Wang, Y., Krishnan, D., Tenenbaum, J.B., Isola, P.: Rethinking few-shot image classification: a good embedding is all you need? In: Vedaldi, A., Bischof, H., Brox, T., Frahm, J.-M. (eds.) ECCV 2020. LNCS, vol. 12359, pp. 266–282. Springer, Cham (2020). https://doi.org/10.1007/978-3-030-58568-6_16

40. Trinh, T.H., Luong, M.T., Le, Q.V.: Selfie: self-supervised pretraining for image embedding. arXiv preprint arXiv:1906.02940 (2019)

41. Van Horn, G., et al.: The inaturalist species classification and detection dataset. In: Proceedings of the IEEE Conference on Computer Vision and Pattern Recognition, pp. 8769–8778 (2018)

42. Xie, E., et al.: Detco: unsupervised contrastive learning for object detection. In: Proceedings of the IEEE/CVF International Conference on Computer Vision, pp. 8392–8401 (2021)

43. Xie, Q., Luong, M.T., Hovy, E., Le, Q.V.: Self-training with noisy student improves imagenet classification. In: Proceedings of the IEEE/CVF Conference on Computer Vision and Pattern Recognition, pp. 10687–10698 (2020)

44. Xuhong, L., Grandvalet, Y., Davoine, F.: Explicit inductive bias for transfer learning with convolutional networks. In: International Conference on Machine Learning, pp. 2825–2834. PMLR (2018)

45. Yang, Y., Xu, Z.: Rethinking the value of labels for improving class-imbalanced learning. arXiv preprint arXiv:2006.07529 (2020)

46. You, Y., Gitman, I., Ginsburg, B.: Large batch training of convolutional networks. arXiv preprint arXiv:1708.03888 (2017)

47. Yu, X., Liu, T., Wang, X., Tao, D.: On compressing deep models by low rank and sparse decomposition. In: Proceedings of the IEEE Conference on Computer Vision and Pattern Recognition, pp. 7370–7379 (2017)

48. Zhai, X., Oliver, A., Kolesnikov, A., Beyer, L.: S4l: self-supervised semi-supervised learning. In: Proceedings of the IEEE/CVF International Conference on Computer Vision, pp. 1476–1485 (2019)

49. Zhang, R., Isola, P., Efros, A.A.: Colorful image colorization. In: Leibe, B., Matas, J., Sebe, N., Welling, M. (eds.) ECCV 2016. LNCS, vol. 9907, pp. 649–666. Springer, Cham (2016). https://doi.org/10.1007/978-3-319-46487-9_40

50. Zhang, Y., Chuangsuwanich, E., Glass, J.: Extracting deep neural network bottleneck features using low-rank matrix factorization. In: 2014 IEEE International Conference on Acoustics, Speech and Signal Processing (ICASSP), pp. 185–189. IEEE (2014)

51. Zhang, Z., Chen, X., Chen, T., Wang, Z.: Efficient lottery ticket finding: Less data is more. In: International Conference on Machine Learning, pp. 12380–12390. PMLR (2021)

52. Zhao, M., Lin, T., Mi, F., Jaggi, M., Schütze, H.: Masking as an efficient alternative to finetuning for pretrained language models. arXiv preprint arXiv:2004.12406 (2020)

53. Zhao, Y., Li, J., Gong, Y.: Low-rank plus diagonal adaptation for deep neural networks. In: 2016 IEEE International Conference on Acoustics, Speech and Signal Processing (ICASSP), pp. 5005–5009. IEEE (2016)

54. Zhou, T., Tao, D.: Greedy bilateral sketch, completion & smoothing. In: Artificial Intelligence and Statistics, pp. 650–658. PMLR (2013)
55. Zhou, Z., Li, X., Wright, J., Candes, E., Ma, Y.: Stable principal component pursuit. In: 2010 IEEE International Symposium on Information Theory, pp. 1518–1522. IEEE (2010)
56. Zhu, H., Wang, Z., Zhang, H., Liu, M., Zhao, S., Qin, B.: Less is more: domain adaptation with lottery ticket for reading comprehension. In: Findings of the Association for Computational Linguistics: EMNLP 2021, pp. 1102–1113 (2021)

Learning Instance and Task-Aware Dynamic Kernels for Few-Shot Learning

Rongkai Ma[1], Pengfei Fang[1,2,3]($\boxtimes$), Gil Avraham[4], Yan Zuo[3],
Tianyu Zhu[1], Tom Drummond[5], and Mehrtash Harandi[1,3]

[1] Monash University, Melbourne, Australia
`pengfei.fang@monash.edu`
[2] Australian National University, Canberra, Australia
[3] CSIRO, Canberra, Australia
[4] Amazon Australia, Melbourne, Australia
[5] The University of Melbourne, Melbourne, Australia

Abstract. Learning and generalizing to novel concepts with few samples (Few-Shot Learning) is still an essential challenge to real-world applications. A principle way of achieving few-shot learning is to realize a model that can rapidly adapt to the context of a given task. Dynamic networks have been shown capable of learning content-adaptive parameters efficiently, making them suitable for few-shot learning. In this paper, we propose to learn the dynamic kernels of a convolution network as a function of the task at hand, enabling faster generalization. To this end, we obtain our dynamic kernels based on the entire task and each sample, and develop a mechanism further conditioning on each individual channel and position independently. This results in dynamic kernels that simultaneously attend to the global information whilst also considering minuscule details available. We empirically show that our model improves performance on few-shot classification and detection tasks, achieving a tangible improvement over several baseline models. This includes state-of-the-art results on four few-shot classification benchmarks: *mini*-ImageNet, *tiered*-ImageNet, CUB and FC100 and competitive results on a few-shot detection dataset: MS COCO-PASCAL-VOC.

1 Introduction

Despite the great success of the modern deep neural networks (DNNs), in many cases, the problem of adapting a DNN with only a handful of labeled data is still challenging. Few-Shot Learning (FSL) aims to address the inefficiencies of modern machine learning frameworks by adapting models trained on large databases for novel tasks with limited data [13,31,32,44,58].

Early approaches of FSL learned a fixed embedding function to encode samples into a latent space, where they could be categorized by their semantic relationships [43,44,58]. However, such fixed approaches do not account for category

Supplementary Information The online version contains supplementary material available at https://doi.org/10.1007/978-3-031-20044-1_15.

differences [36], which may exist between already learned tasks and novel tasks. Ignoring these discrepancies can severely limit the adaptability of a model as well as its ability to scale in the FSL setting. Although methods that adapt embeddings [30,48,57] attempt to address this issue, they still utilize fixed models which lack full adaptability and are constrained to previously learned tasks. Another group of approaches, such as [13,32], adapt models with a few optimization steps. Given the complexity of the loss landscape of a DNN, such methods come short compared to metric-based solutions [52].

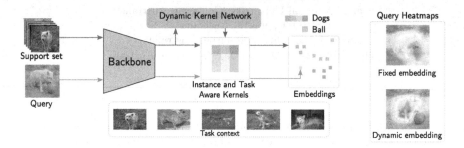

Fig. 1. Typical FSL models learn fixed embeddings, which are not flexible enough to rapidly adapt to novel tasks. Our method instead uses a dynamic kernel network to produce a set of dynamic parameters which are both instance and task-aware

Dynamic kernel approaches have been shown to be computationally efficient [60] relative to optimization-based solutions, resulting in the models capable of encoding the novel information at the parameter level [20]. A very recent study [54] showed that dynamic kernel methods are effective when applied to FSL via the learning of input-conditioned filters, enabling the realization of adaptive models; a key limitation of this dynamic kernel method is that only the per-class level information is utilized via class prototypes.

Arguably, when learning new tasks with sparse data, it is vital to fully utilize information on both an instance and task-level for efficiency. As an example, when given the task of differentiating between dog breeds (Fig. 1), both task and instance-level information is required; it is important that we are sensitive enough to distinguish the minuscule details between different dog breeds (instance-level), yet can still focus on the global knowledge required to filter out irrelevant, non-task related objects (task-level).

In this paper, we propose a novel, dynamic approach for FSL tasks. Our method is realized by a dynamic kernel that can encode both instance-level and task-level features via jointly learning a dynamic kernel generator and a context learning module. The dynamic kernel generator adaptively produces an instance kernel by exploiting the information along the spatial and channel dimensions of a feature map in a decoupled manner. It further incorporates information from the frequency domain resulting in a rich set of descriptive representations. The context learning module refines the support features into a task-specific representation which is used to produce the task-specific kernel. The resulting instance and task-specific kernels are fused to obtain a dynamic kernel which is INStance and Task-Aware (INSTA). Our method differentiates from approaches such as FEAT [57] and GLoFa [30] by learning a set of dynamic parameters adaptive to the novel tasks instead of employing fixed models during inference. Furthermore, in contrast to optimization-based methods [13], our approach can adapt the model parameters without the requirements of backpropagation during the inference stage. We offer the following contributions:

- We propose a novel FSL approach to extract both instance and task-specific information using dynamic kernels.
- We offer the first FSL framework capable of being evaluated on both classification *and* detection tasks.
- Empirically, we offer substantial improvements over several FSL baselines, including optimization-based and metric-based approaches.

2   Related Work

The family of few-shot learning literature is broad and diverse. However, those related to this work are mainly the family of optimization-based methods [3,13–15,23,32,37,42,61] and metric-based methods [4,10,41,43,44,48,57,58].

Optimization Based Methods. Optimization-based methods such as MAML [13] or Reptile [32] focus on learning a set of initial model parameters that can generalize to new tasks (or environments) with a small number of optimization steps and samples without severe over-fitting. Furthermore, in most cases, this group of methods present a framework trained with a bi-level optimization setting [15], which provides a feasible solution to adapt the model to the test set from the initialized model.

Our proposed framework is similar to the optimization-based methods [1,2, 6,24,29] in the sense that the model parameters are task-adaptive. However, our solution does not require backpropagation during inference to achieve so. Furthermore, our method can be incorporated with optimization-based methods, and empirically we observed that such construction yields performance improvement. This will be demonstrated in Sect. 4.

Metric-based Methods. In few-shot learning literature, the metric-based methods aim to define a metric to measure the dis/similarity between samples from a few observations [8,22,27,28,38,53,55,59]. ProtoNet [43] achieves this by learning a fixed latent space where the class representations (*i.e.*, prototype), obtained by averaging the features from the same class, are distinctive for different classes. DeepEMD [58] formulates the query-support matching as an optimal transport problem and adopts Earth Mover's distance as the metric. One commonality in the aforementioned methods is the fact that all employ a fixed embedding space when facing novel tasks, which essentially limits their adaptability. On the other hand, many previous methods suggested to adapt the embeddings to the novel tasks [12,17,30,33,40,46,57]. CTM [25] proposes to produce a mask to disregard the uninformative features from the support and query embeddings during inference. MatchingNet [48] uses a memory module to perform sample-wise embedding adaptation and determines the query label by a cosine distance. TADAM [33] proposes to learn a dynamic feature extractor by applying a linear transformation to the embeddings to learn a set of scale and shift parameters. FEAT [57] and GLoFa [30] provide an inspiring way to perform the embedding adaptation using a set function.

Our contribution is complementary to the aforementioned methods. We aim to learn a set of dynamic kernels via exploiting the instance and task-level information according to the task at hand. This results in a more distinctive and descriptive representation, which effectively boosts the performance of these methods directly relying on constructing a metric space.

Dynamic Kernels. The application of dynamic kernels solutions within the domain of few-shot learning is less explored in the current literature. However, it has been demonstrated useful when labels are abundant [5,7,16,19,45,47,50,60]. Zhou *et al.* [60] have proposed to use decoupled attention over the channel and spatial dimensions. This results in a content-adaptive and efficient method that provides a feasible way to achieve task-adaptive FSL.

To leverage the effectiveness of the dynamic kernel into few-shot learning tasks, [54] propose to learn dynamic filters for the spatial and channels separately via grouping strategy. The resulted kernels are then applied to the query to produce a support-aligned feature. Due to the usage of grouping, the performance might sacrifice for efficiency [60]. Inspired by these methods, our INSTA produces dynamic kernels that are both instance-aware and task (or episodic)-aware while also incorporating valuable frequency patterns; as such, our method produces more informative and representative features.

3   Method

In this section, we introduce our proposed INSTA. Note that while we present our method in terms of few-shot classification, the proposed approach is generic and can be seamlessly used to address other few-shot problems, including structured-prediction tasks such as few-shot object detection (see Subsect. 4.2 for details).

3.1   Problem Formulation

In what follows, we will give a brief description of the problem formulation for few-shot classification. The vectors and matrices (or high-dimensional tensors) are denoted by bold lower-case letters (*e.g.*, x) and bold upper-case letters (*e.g.*, X) throughout this paper. FSL aims to generalize the knowledge acquired by a model from a set of examples $\mathcal{D}_{train} = \{(X_i, y_i)|y_i \in \mathcal{C}_{train}\}$, to novel and unseen tasks $\mathcal{D}_{test} = \{(X_i, y_i)|y_i \in \mathcal{C}_{test}\}, \mathcal{C}_{train} \cap \mathcal{C}_{test} = \emptyset$, in a low data regime.

We follow the meta-learning protocol to formulate FSL with episodic training and testing. Specifically, an episode $\mathcal{E}$ consists of a support set $\mathcal{X}^s =$

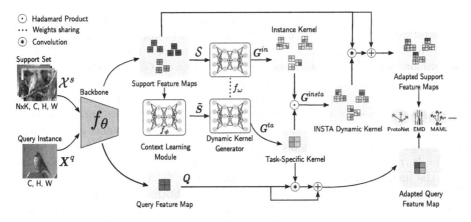

Fig. 2. The framework of our method. Given the support set $\mathcal{X}^s$ and the query sample $\boldsymbol{X}^q$, the backbone network f_θ first encodes them into a representation space as $\mathcal{S} = \{\boldsymbol{S}_{11}, \ldots, \boldsymbol{S}_{NK}\}$ and $\boldsymbol{Q}$. Then a dynamic generator f_ω is used to produce an instance kernel for each support sample. Meanwhile, a context learning module f_ϕ is used to refine and aggregate the context of the entire support feature maps to produce a task-specific representation, which is then used as the input of f_ω to obtain the task-specific kernel. Note the instance kernel and task-specific kernel share the parameters across f_ω. Finally, the INSTA dynamic kernel that is both instance-aware and task-aware is applied to the support feature maps, while only the task-specific kernel is applied to the query feature map to obtain the adapted representations

$\{(\boldsymbol{X}^s_{ij}, y^s_i)|i = 1, \ldots, N, j = 1, \ldots, K, y^s_i \in \mathcal{C}_{train}\}$, where $\boldsymbol{X}^s_{ij}$ denotes the j-th sample in the class y^s_i and the query set $\mathcal{X}^q = \{(\boldsymbol{X}^q_i, y^q_i)|i = 1, \ldots, N\}$, where $\boldsymbol{X}^q_i$ denotes a query example[1] sampled from class y^q_i (the test setup is the same as training but the episodes are sampled from $\mathcal{D}_{test}$). Such a formulation is known as N-way K-shot, where the goal is to utilize the support samples and their labels to obtain Θ^*, the optimal parameters of the model, such that each query is classified correctly into one of classes in the support set. The object of network training follows that:

$$\Theta^* = \arg\min_{\Theta} \sum_{\boldsymbol{X}^q_i \in \mathcal{X}^q} \mathcal{L}(f_\Theta(\boldsymbol{X}^q_i|\mathcal{X}^s), y^q_i), \tag{1}$$

where f_Θ represents the entire model parameterized by Θ.

3.2   Model Overview

We first provide an overview of our model (see the sketch of the pipeline in Fig. 2). Overall, our framework is composed of three modules, the backbone network f_θ, the dynamic kernel generator f_ω, and the context learning module f_ϕ. The

---

[1] Without losing generality, we use one sample per class as a query for presenting our method. In practice, each episode contains multiple samples per query class.

backbone network first extracts the feature maps from the input images, followed by a two-branch dynamic kernel network to obtain our proposed dynamic kernels. Specifically, the dynamic kernel generator independently refines the features in the support set to produce the instance kernel (*i.e.*, G^{in}) in one branch. In another branch, the context learning module f_ϕ first produces a task-specific representation by refining and summarizing the features from the entire support set, which is then used as the input of the dynamic kernel generator to produce the task-specific kernel (*i.e.*, G^{ta}) adaptively. The dynamic kernel generator is shared across these two branches to generate the instance kernel and the task-specific kernel. Such a two-branch design enables the network to be aware of both the instance-level feature and global context feature of the support set. The design choice is justified in Sect. 4. Finally, both the instance and task-specific kernels are fused and employed to boost the discrimination of instance features in the support set. The task-specific kernel is applied to the query feature maps for refining the representation without extracting a query instance kernel. Given the adapted support and query features, any post-matching algorithm (*e.g.*, metric-based or optimization-based FSL) can be employed seamlessly to achieve the few-shot classification task.

3.3   Dynamic Kernel Generator

In this part, we provide a detailed description of the dynamic kernel generator. The central component of our model is the dynamic kernel generator, which receives a feature map as input and produces a kernel adaptively. An essential problem one may face is that the feature map extracted by modern DNNs usually has a large size (*e.g.*, $640 \times 5 \times 5$ for ResNet-12). As such, we need to identify $c_{out} \times c \times k \times k$ amount of parameters for the dynamic kernel, where the c_{out} is the output channel size (we consider $c = c_{out}$ in our paper), c is the input channel size, and k is the kernel size. This poses a significant issue since the generated dynamic kernel is prone to overfitting given the limited data of FSL. To tackle this problem, we consider designing our dynamic kernel generator in a decomposed manner [60]. As such, we develop a channel kernel network and a spatial kernel network to produce a kernel for each channel and each spatial location independently. This design exploits the information per data-sample to a large extent while reducing the number of parameters of a dynamic kernel, which greatly fits the low data regime of FSL. Figure 3 illustrates the pipeline of the proposed dynamic kernel generator. In what follows, we detail out the operations of the channel and spatial kernel networks.

Channel Kernel Network. Given a feature map $S \in \mathbb{R}^{c \times h \times w}$, where c, h, and w denote the size of channel, height and width, a common practice to realize the channel kernel network follows the SE block [18], as done for example in [7]. In the SE block, global average pooling (GAP) is performed over the feature map to encode the global representation, which is fed to the following sub-network for channel-wise scaling. The drawback of this design is that the GAP operator mainly preserves the low frequency components of the feature map, as shown in [34] (averaging is equivalent to low-pass filtering in the frequency domain). Thus, GAP discards important signal patterns in the feature map to a great degree. Clearly, low-frequency components cannot fully characterize the information encoded in a feature map, especially in the low data regime. We will empirically show this in Subsect. 4.3. To mitigate this issue, we opt for multi-spectral attention (MSA) to make better use of high-frequency patterns in the feature map. Given a feature map S, we equally split the feature map into n smaller tensors along the channel dimension (in our experiments $n = 16$), as $\{S^0, S^1, \ldots, S^{n-1} | S^i \in \mathbb{R}^{\frac{c}{n} \times h \times w}\}$. Then, each channel of the tensor S^i is processed by a basis function of a 2D-Discrete Cosine Transform (DCT) following the work of Qin et $al.$ [34]. As a result, we obtain a real-valued feature vector in the frequency domain as $\tau^i = \text{DCT}(S^i), \tau^i \in \mathbb{R}^{c/n}$, which is the frequency-encoded vector corresponding to S^i. The frequency-encode vector for S is obtained by concatenating τ^is as:

$$\tau = \text{concat}(\tau^0, \tau^1, \ldots, \tau^{n-1}), \tag{2}$$

where $\tau \in \mathbb{R}^c$. Please refer to the supplementary material for the theoretical aspects of the MSA module. As compared to the global feature obtained by GAP, frequency-encoded feature τ contains more diverse information patterns, which brings extra discriminative power to our model. This will be empirically discussed in Subsect. 4.3.

Once we obtain τ, a light-weight network is used to produce the channel kernel values adaptively. This network is realized by a two-layer MLP (or 1×1 convolution), with architecture as $c \to \sigma \times c \to \text{ReLu} \to k^2 \times c$, where $0 < \sigma < 1$, and k controls the size of receptive field in the dynamic kernel (practically, we set $\sigma = 0.2$ and $k = 3$). Then we reshape the output vector into a $c \times k \times k$-sized tensor, denoted by G^{ch}. We can obtain the final channel dynamic kernel $\hat{G}^{ch} \in \mathbb{R}^{c \times h \times w \times k \times k}$ via applying batch normalization (BN) and spatial-wise broadcast to G^{ch}.

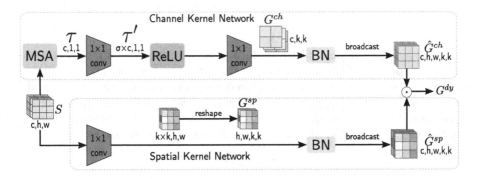

Fig. 3. The architecture of dynamic kernel generator. We adopt a decomposed architecture to produce dynamic kernels for the channels and spatial dimensions independently. This results in a light-weight kernel, which greatly fits the low-data regime of FSL

Spatial Kernel Network. Independent to the channel kernel network, our spatial kernel network adaptively produces a convolution kernel $G_{a,b}^{sp} \in \mathbb{R}^{k \times k}$ for each spatial location of a feature map (*i.e.*, $S_{:,a,b}$). This is achieved by using a 1×1 convolution layer, with the architecture of $(c, k^2, 1, 1)$ and the reshape operation. Therefore, the spatial kernel G^{sp} for all the spatial locations of a feature map S has a size of $h \times w \times k \times k$. We then obtain the final spatial dynamic kernel $\hat{G}^{sp} \in \mathbb{R}^{c \times h \times w \times k \times k}$ by applying the BN and channel-wise broadcast to G^{sp}.

Finally, to unify the channel and spatial dynamic kernels, we apply the Hadamard product between the $\hat{G}^{ch}$ and $\hat{G}^{sp}$ to obtain the dynamic kernel G^{dy} as:

$$G^{dy} = \hat{G}^{sp} \odot \hat{G}^{ch}, \tag{3}$$

where $G^{dy} \in \mathbb{R}^{c \times h \times w \times k \times k}$. Notably, the number of parameters of our dynamic kernel is much less than a normal convolution kernel ($c \times h \times w \times k^2 \ll c_{out} \times c \times k^2$ given $c_{out} = c = 640$ and $h = w = 5$).

3.4   Dynamic Kernel

In this part, we discuss the process of obtaining the proposed INSTA dynamic kernel, which consists of the instance kernel and the task-specific kernel.

Instance Kernel. We defined the instance kernel as the dynamic kernel extracted from each support feature map. Formally, given the support set images $\mathcal{X}^s = \{X_{11}^s, \ldots, X_{NK}^s | X_{ij}^s \in \mathbb{R}^{C \times H \times W}\}$, where C, H, and W indicate the channel, height, and width of an image, respectively, the backbone network first extracts a feature map from each image, as $\mathcal{S} = f_\theta(\mathcal{X}^s)$, with $\mathcal{S} = \{S_{11}, \ldots, S_{NK} | S_{ij} \in \mathbb{R}^{c \times h \times w}\}$. The feature map for the query sample can be obtained in a similar fashion, as $Q_i = f_\theta(X_i^q)$. We then use the dynamic kernel generator f_ω to independently produce a dynamic kernel $G_{ij}^{in} \in \mathbb{R}^{c \times h \times w \times k \times k}$ for each support feature map S_{ij}, following the process described in Subsect. 3.3.

Task-Specific Kernel. Adapting according to the whole context of support set is essential for FSL since it contains essential information of the task [33]. Following this intuition, we propose to learn the task-specific kernel to represent the knowledge of the task encoded in the support set. We achieve this by first using a context learning module f_ϕ to produce a fully context-aware representation for the support set by refining and aggregating the intermediate features. To be specific, we use four 1×1 convolution layers with a summation layer in the middle of the network to aggregate the $N \times K$ support features into one task-specific representation $\tilde{S}$ (Please refer to the supplementary material for the conceptual diagram of f_ϕ). Formally, this task representation can be obtained by:

$$\tilde{S} = f_\phi(\mathcal{S}), \tag{4}$$

where $\tilde{S} \in \mathbb{R}^{c \times h \times w}$ and the $\mathcal{S}$ denotes the entire support set. Then this task-specific representation $\tilde{S}$ is used as the input to the dynamic kernel generator, following the process described in Subsect. 3.3 to adaptively produce our task-specific kernel $G^{ta} \in \mathbb{R}^{c \times h \times w \times k \times k}$.

INSTA Dynamic Kernel. Once we have the instance and task-specific kernels, we fuse each instance kernel $\boldsymbol{G}_{ij}^{in}$ with task-specific kernel $\boldsymbol{G}^{ta}$ using the Hadamard product, such that each dynamic kernel considers both the instance-level and task-level features. Formally, it can be described as:

$$\boldsymbol{G}_{ij}^{insta} = \boldsymbol{G}_{ij}^{in} \odot \boldsymbol{G}^{ta}, \tag{5}$$

where $\boldsymbol{G}_{ij}^{insta}$ is the final fused kernel corresponding to the ij-th sample in the support set.

After $\boldsymbol{G}_{ij}^{insta}$ is obtained, we apply it on its corresponding support feature map $\boldsymbol{S}_{ij}$. This is equivalent to first apply the task-specific kernel to extract the features, which is relevant to the task and then apply the instance kernel to further increase the discriminatory power of the resulting feature maps. While we only apply the task-specific kernel to the query feature map. This design choice is justified in Subsect. 4.3, where we compare our current design with a variant where we also extract instance kernel from the query feature maps. The dynamic convolution can be implemented using the Hadamard product between the unfolded feature map and the dynamic kernel. In doing so, the unfold operation first samples a $k \times k$ spatial region of the input feature map at each time and then stores the sampled region into extended dimensions, thereby obtaining an unfolded support feature map $\boldsymbol{S}_{ij}^{u} \in \mathbb{R}^{c \times h \times w \times k \times k}$ and an unfolded query feature map $\boldsymbol{Q}_i^u \in \mathbb{R}^{c \times h \times w \times k \times k}$. Then the adapted support and query feature maps are obtained as:

$$\begin{aligned}
\dot{\boldsymbol{S}}_{ij} &= \text{AvgPool2d}(\boldsymbol{S}_{ij}^u \odot \boldsymbol{G}_{ij}^{insta}) + \boldsymbol{S}_{ij}, \\
\dot{\boldsymbol{Q}}_i &= \text{AvgPool2d}(\boldsymbol{Q}_i^u \odot \boldsymbol{G}^{ta}) + \boldsymbol{Q}_i,
\end{aligned} \tag{6}$$

where $\dot{\boldsymbol{S}}_{ij} \in \mathbb{R}^{c \times h \times w}$ and $\dot{\boldsymbol{Q}}_i \in \mathbb{R}^{c \times h \times w}$. Note that the convolution operation between the dynamic kernel and the feature map in Fig. 2 is equivalent to the average of the Hadamard product between the unfolded feature map and the dynamic kernel over the $k \times k$ dimension (see Fig. 4). In our design, we adopt a residual connection between the original and the updated feature maps to obtain the final adapted features. Having the adapted support and query feature maps, any post-matching algorithms can be adopted to achieve FSL tasks. In Sect. 4, we demonstrate that our algorithm can further boost the performance on various few-shot classification models (*e.g.*, MAML [13], ProtoNet [43] and EMD [58]) and the few-shot detection model [9].

Remark 1. The proposed dynamic kernels are convolutional filters and hence by nature differ from attention masks. As shown in Fig. 4, the attention mask merely re-weights each element of a feature map [11,18]. In contrast, our proposal in Eq. (6) realizes the dynamic convolution by first performing the element-wise multiplication between the unfolded feature map and the dynamic kernels, whose results are then averaged over the unfolded dimension. This is essentially equivalent to convolution operation (see Fig. 4).

4   Experiments

In this section, we first evaluate our method across four standard few-shot classification benchmarks: *mini*-ImageNet [35], *tiered*-ImageNet [36], CUB [49], and FC100 [33]. Full details of the implementation are provided in the supplementary material. Furthermore, we evaluate the effectiveness of our framework for the few-shot detection task on the MS COCO and PASCAL VOC datasets [9]. Finally, we provide an ablation study to discuss the effect of each module in our framework. Please refer to supplementary material for a detailed description of each dataset.

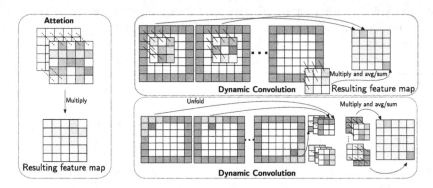

Fig. 4. Schematic comparison of the dynamic convolution and attention mechanism. The 2-D average of the element-wise multiplication between unfolded features and dynamic kernels is equivalent to using adaptive kernels sliding over the original feature map

4.1   Few-Shot Classification

We conduct few-shot classification experiments on three different state-of-the-art models, including MAML [13], ProtoNet [43] and DeepEMD [58] as baselines, and employ the proposed INSTA on top of them. We use the ResNet-10 backbone for the MAML and the ResNet-12 backbone for the other two baselines across all four benchmarks. For a fair comparison, we implement all the baseline models to report the results indicated by "*" across Table 1–Table 2. Notably, for DeepEMD experiments, we adopt the *open-cv* solver instead of the *qpth* solver originally used in the paper to train the network due to resource capacity, which is indicated by "♣". Moreover, following the same evaluation protocols in our baseline frameworks [6,58], we report the mean accuracy with 95% confidence interval (Please refer to supplementary material for more implementation details).

mini-**ImageNet.** As shown in Table 1, our method improves the performance of all the baseline models by a noticeable margin. It is worthwhile to mention

Table 1. Few-shot classification accuracy and 95% confidence interval on *mini*-ImageNet and *tiered*-ImageNet with ResNet backbones

Model	Backbone	*mini*-ImageNet		*tiered*-ImageNet	
		5-way 1-shot	5-way 5-shot	5-way 1-shot	5-way 5-shot
MetaOptNet [23]	ResNet-12	62.64 ± 0.61	78.63 ± 0.46	68.23 ± 0.23	84.03 ± 0.56
TADAM [33]	ResNet-12	58.50 ± 0.30	76.70 ± 0.30	–	–
FEAT [57]	ResNet-12	66.78 ± 0.20	82.05 ± 0.14	70.80 ± 0.23	84.79 ± 0.16
CAN [17]	ResNet-12	63.85 ± 0.48	79.44 ± 0.34	69.89 ± 0.51	84.23 ± 0.37
FRN [53]	ResNet-12	66.45 ± 0.19	82.83 ± 0.13	72.06 ± 0.22	86.89 ± 0.14
InfoPatch [27]	ResNet-12	67.67 ± 0.45	82.44 ± 0.31	71.51 ± 0.52	85.44 ± 0.35
GLoFA [30]	ResNet-12	66.12 ± 0.42	81.37 ± 0.33	69.75 ± 0.33	83.58 ± 0.42
DMF [54]	ResNet-12	67.76 ± 0.46	82.71 ± 0.31	71.89 ± 0.52	85.96 ± 0.35
MAML* [13]	ResNet-10	54.73 ± 0.87	66.72 ± 0.81	59.85 ± 0.97	73.20 ± 0.81
INSTA-MAML*	ResNet-10	**56.41 ± 0.87**	**71.56 ± 0.75**	**63.34 ± 0.92**	**78.01 ± 0.71**
ProtoNet* [43]	ResNet-12	62.29 ± 0.33	79.46 ± 0.48	68.25 ± 0.23	84.01 ± 0.56
INSTA-ProtoNet*	ResNet-12	**67.01 ± 0.30**	**83.13 ± 0.56**	**70.65 ± 0.33**	**85.76 ± 0.59**
DeepEMD*♣ [58]	ResNet-12	67.37 ± 0.45	83.17 ± 0.75	73.19 ± 0.32	86.79 ± 0.61
INSTA-DeepEMD*♣	ResNet-12	**68.46 ± 0.48**	**84.21 ± 0.82**	**73.87 ± 0.31**	**88.02 ± 0.61**

that our INSTA can boost the performance of the baseline model ProtoNet by 4.72% and 3.67%, and achieves 67.01% and 83.13% for 5-way 1-shot and 5-way 5-shot settings, which outperforms many recent published models. Furthermore, we can show an improvement over a strong baseline model, *i.e.* DeepEMD, and achieve state-of-the-art performance on this dataset. We provide more comparison between our approach and other state-of-the-art methods in our supplementary material for *mini*-ImageNet and *tiered*-ImageNet.

tiered-**ImageNet.** As shown in Table 1, our model consistently brings the performance gain over baseline models. Among baseline models, our model shows a greater performance improvement over MAML than ProtoNet and DeepEMD. Notably, a significant improvement on 5-way 5-shot can be seen over DeepEMD. We attribute this improvement to our model's context module leveraging the categorical nature of this dataset. The results also show that the DeepEMD exhibits a performance gain with our design and outperforms many other recent models, which achieves the state-of-the-art result on this dataset.

CUB. The result on the CUB dataset is shown in Table 2, where the performance of all the three baseline models is improved by integrating the proposed INSTA. For the ProtoNet baseline with ResNet-12 backbone, as an example, 4.29% and 3.72% improvements can be observed for 5-way 1-shot and 5-way 5-shot, respectively. Moreover, among the improved models, our INSTA-DeepEMD achieves the state-of-the-art results, which are 75.26% and 88.12% for 5-way 1-shot and 5-way 5-shot settings, respectively. Please refer to supplementary material for additional results of the INSTA-ProtoNet with ResNet-18 backbone.

FC100. Consistent with the observation in the other benchmarks, the results in Table 2 again show that a performance improvement can be achieved on FC100. Furthermore, both INSTA-ProtoNet and INSTA-DeepEMD achieve comparable performance with the recent state-of-the-art models on FC100, which vividly shows the effectiveness of our approach.

Table 2. Few-shot classification accuracy and 95% confidence interval on CUB and FC100 with ResNet backbones

Model	Backbone	CUB		FC100	
		5-way 1-shot	5-way 5-shot	5-way 1-shot	5-way 5-shot
RelationNet [44]	ResNet-18	67.59 ± 1.02	82.75 ± 0.58	–	–
Chen et al. [6]	ResNet-18	67.02	83.58	–	–
SimpleShot [51]	ResNet-18	70.28	86.37	–	–
Neg-Margin [26]	ResNet-18	72.66 ± 0.85	89.40 ± 0.43	–	–
P-transfer [39]	ResNet-12	73.88 ± 0.87	87.81 ± 0.48	–	–
TADAM [33]	ResNet-12	–	–	40.10 ± 0.40	56.10 ± 0.40
ConstellationNet [55]	ResNet-12	–	–	43.80 ± 0.20	59.70 ± 0.20
MAML* [13]	ResNet-10	70.46 ± 0.97	80.15 ± 0.73	34.50 ± 0.69	47.31 ± 0.68
INSTA-MAML*	ResNet-10	$\mathbf{73.08 \pm 0.97}$	$\mathbf{84.26 \pm 0.66}$	$\mathbf{38.02 \pm 0.70}$	$\mathbf{51.20 \pm 0.68}$
ProtoNet* [43]	ResNet-12	64.76 ± 0.33	77.99 ± 0.68	41.54 ± 0.76	57.08 ± 0.76
INSTA-ProtoNet*	ResNet-12	$\mathbf{69.05 \pm 0.32}$	$\mathbf{81.71 \pm 0.63}$	$\mathbf{44.12 \pm 0.26}$	$\mathbf{62.04 \pm 0.75}$
DeepEMD*♣ [58]	ResNet-12	74.55 ± 0.30	87.55 ± 0.54	45.12 ± 0.26	61.46 ± 0.70
INSTA-DeepEMD*♣	ResNet-12	$\mathbf{75.26 \pm 0.31}$	$\mathbf{88.12 \pm 0.54}$	$\mathbf{45.42 \pm 0.26}$	$\mathbf{62.37 \pm 0.68}$

4.2   Few-Shot Detection

Problem Definition. Given a query and several support images (each support image only contains one object), a few-shot detection task is to output labels and corresponding bounding boxes for all objects in the query image that belong to support categories. Specifically, few-shot detection follows the N-way K-shot setting, where the support set contains N categories, with each category having K samples. In the following, we will discuss the experiment setup of this task. The implementation details on incorporating our INSTA with the baseline model proposed by Fan et al. [9] are included in the supplementary material.

Experiment Setup. In this experiment, we follow the training and testing settings in [9], where a 2-way 9-shot contrastive training strategy is adopted, and a 20-way 10-shot setting is used for final testing.

Results.
We compare our model against the baseline model and other relevant state-of-the-art few-shot detection models. As shown in Table 3, by incorporating the proposed INSTA on the baseline model, improved performance can be seen across all of the metrics. Specifically, as compared to the baseline model, the AP_{50} is improved by 3.2% and achieves 23.6%, which empirically shows that our dynamic kernels indeed extract more informative and representative features. Moreover, the proposed INSTA improves the performance of detection for objects of all the scales (refer to AP_s, AP_m, and AP_l), which again illustrates the importance of our context module when adapting to different object scales in a given task.

Table 3. The few-shot detection results of 6 different average precision (AP) on the test set, including MS COCO and PASCAL VOC datasets with 20-way 10-shot setting

Model	AP	AP_{50}	AP_{75}	AP_s	AP_m	AP_l
FR [21]	5.6	12.3	4.6	–	–	–
Meta [56]	8.7	19.1	6.6	–	–	–
Fan *et al.* [9]	11.1	20.4	10.6	2.8	12.3	20.7
Fan *et al.* +INSTA	**12.5**	**23.6**	**12.1**	**3.3**	**13.2**	**21.4**

4.3   Ablation Study

In the following section, we conduct ablation studies to discuss and verify the effect of each component of our framework, including the instance-kernel, task-specific kernel, MSA *vs.* GAP, and we also evaluate some variants of the proposed INSTA to justify the selection of our current framework. In this study, we use the ProtoNet as the baseline, and the ResNet-12 is adopted as its backbone. Additionally, this study is conducted on the *mini*-ImageNet under the 5-way 5-shot setting. The results are summarized in Table 4.

Table 4. The ablation study of each component in our framework

ID	Model	Apply to S_{ij}	Apply to Q	5-way 5-shot
(i)	ProtoNet	–	–	79.46 ± 0.48
(ii)	ProtoNet + G^{ta}	G^{ta}	–	81.56 ± 0.57
(iii)	ProtoNet + G^{ta}	G^{ta}	G^{ta}	82.51 ± 0.58
(iv)	ProtoNet + G^{in}	G^{in}	–	80.81 ± 0.60
(v)	ProtoNet + G^{insta}	G^{ta}, G^{in}	–	81.74 ± 0.56
(vi)	INSTA-ProtoNet + GAP	G^{ta}, G^{in}	G^{ta}	82.07 ± 0.56
(vii)	INSTA-ProtoNet + G_Q^{in}	G^{ta}, G^{in}	G_Q^{in}, G^{ta}	82.24 ± 0.56
(viii)	INSTA-ProtoNet w/o sharing f_ω	G^{ta}, G^{in}	G^{ta}	81.36 ± 0.59
(ix)	INSTA-ProtoNet	G^{ta}, G^{in}	G^{ta}	$\mathbf{83.13 \pm 0.56}$

Effectiveness of Task-Specific Kernel. We study the effect of the task-specific kernel in this experiment (*i.e.*, (ii) in Table 4), where we only apply the task-specific kernel to support samples and leave the query unchanged. By comparing (i) and (ii), we can observe that our task-specific kernel improves the performance of ProtoNet by 2.10%. Moreover, we apply the task-specific kernel on the query in setting (iii), and we can observe a further improvement over the setting (ii), which verifies the effectiveness of our task-specific kernel.

Effectiveness of Instance Kernel. In setting (iv), we enable the instance kernel based on ProtoNet (*i.e.*, (i)) and the performance of ProtoNet is improved by 1.35%. Moreover, comparing setting (ii), where the only task-specific kernel is enabled, to setting (v), where both task-specific and instance kernels are enabled, a performance improvement can also be observed. Both cases illustrate the effectiveness of our instance kernel.

Effectiveness of Query Adaptation. The purpose of this experiment is to study the effect of the adaptation on the query feature. In setting (iii), we enable the task-specific kernel on both support and query features but disable the instance kernel. Compared to setting (ii), where the task-specific kernel is enabled only on the support features, setting (iii) yields a better performance. Furthermore, the comparison between setting (v) and (ix) highlights the importance of adapting the query feature map using the task information for FSL tasks.

Effectiveness of MSA. In this study, we show the performance gap between using GAP and MSA (Subsect. 3.3). In setting (vi) we replace the MSA in our final design (ix) with GAP. As the results in Table 4 showed, the MSA indeed helps with learning a more informative and representative dynamic kernel than GAP.

Instance Kernel for Query. In our framework, we only use task-specific kernels on the query feature map since the instance kernels obtained from support samples might not be instance level representative for the query sample but provide useful task information. Therefore, we perform extra experiments to verify whether the instance kernel obtained from the query itself can extract better features. As the result in (vii) indicates, the instance kernel extracted from the query feature map does not improve the performance, as compared to our final design.

Shared Dynamic Kernel Generator. We hypothesize that sharing the Dynamic Kernel Network for the task-specific kernel and instance kernel encourages learning a more representative instance kernel. To verify this, we further conduct an experiment where two independent dynamic kernel generators are used to produce the task-specific kernel and instance kernel. As the comparison between (ix) and (viii) shows, inferring the instance and task kernels using a shared weight dynamic kernel generator is an essential design choice.

5 Conclusion

In this paper, we propose to learn a dynamic embedding function realized by a novel dynamic kernel, which extracts features at both instance-level and task-level while encoding important frequency patterns. Our method improves the performance of several FSL models. This is demonstrated on 4 public few-shot classification datasets, including *mini*-ImageNet, *tiered*-ImageNet, CUB, and FC100 and a few-shot detection dataset, namely MS COO-PASCAL-VOC.

References

1. Andrychowicz, M., et al.: Learning to learn by gradient descent by gradient descent. In: Advances in Neural Information Processing Systems, pp. 3981–3989 (2016)
2. Antoniou, A., Edwards, H., Storkey, A.: How to train your MAML. In: International Conference on Learning Representations (2019)
3. Bertinetto, L., Henriques, J.F., Torr, P., Vedaldi, A.: Meta-learning with differentiable closed-form solvers. In: International Conference on Learning Representations (2018)
4. Bertinetto, L., Henriques, J.F., Valmadre, J., Torr, P.H., Vedaldi, A.: Learning feedforward one-shot learners. In: Proceedings of the 30th International Conference on Neural Information Processing Systems, pp. 523–531 (2016)
5. Bolukbasi, T., Wang, J., Dekel, O., Saligrama, V.: Adaptive neural networks for efficient inference. In: International Conference on Machine Learning, pp. 527–536. PMLR (2017)
6. Chen, W.Y., Liu, Y.C., Kira, Z., Wang, Y.C.F., Huang, J.B.: A closer look at few-shot classification. arXiv preprint arXiv:1904.04232 (2019)
7. Chen, Y., Dai, X., Liu, M., Chen, D., Yuan, L., Liu, Z.: Dynamic convolution: attention over convolution kernels. In: Proceedings of the IEEE/CVF Conference on Computer Vision and Pattern Recognition, pp. 11030–11039 (2020)
8. Choi, J., Krishnamurthy, J., Kembhavi, A., Farhadi, A.: Structured set matching networks for one-shot part labeling. In: Proceedings of the IEEE Conference on Computer Vision and Pattern Recognition, pp. 3627–3636 (2018)
9. Fan, Q., Zhuo, W., Tang, C.K., Tai, Y.W.: Few-shot object detection with attention-RPN and multi-relation detector. In: Proceedings of the IEEE/CVF Conference on Computer Vision and Pattern Recognition, pp. 4013–4022 (2020)
10. Fang, P., Harandi, M., Petersson, L.: Kernel methods in hyperbolic spaces. In: Proceedings of the IEEE/CVF International Conference on Computer Vision, pp. 10665–10674 (2021)
11. Fang, P., Zhou, J., Roy, S.K., Ji, P., Petersson, L., Harandi, M.: Attention in attention networks for person retrieval. IEEE Trans. Pattern Anal. Mach. Intell. **44**(9), 4626–4641 (2021)
12. Fei, N., Lu, Z., Xiang, T., Huang, S.: Melr: meta-learning via modeling episode-level relationships for few-shot learning. In: International Conference on Learning Representations (2020)
13. Finn, C., Abbeel, P., Levine, S.: Model-agnostic meta-learning for fast adaptation of deep networks. arXiv preprint arXiv:1703.03400 (2017)
14. Flennerhag, S., Rusu, A.A., Pascanu, R., Visin, F., Yin, H., Hadsell, R.: Meta-learning with warped gradient descent. arXiv preprint arXiv:1909.00025 (2019)
15. Franceschi, L., Frasconi, P., Salzo, S., Grazzi, R., Pontil, M.: Bilevel programming for hyperparameter optimization and meta-learning. arXiv preprint arXiv:1806.04910 (2018)
16. Ha, D., Dai, A., Le, Q.V.: Hypernetworks. arXiv preprint arXiv:1609.09106 (2016)
17. Hou, R., Chang, H., Ma, B., Shan, S., Chen, X.: Cross attention network for few-shot classification. arXiv preprint arXiv:1910.07677 (2019)
18. Hu, J., Shen, L., Sun, G.: Squeeze-and-excitation networks. In: Proceedings of the IEEE Conference on Computer Vision and Pattern Recognition, pp. 7132–7141 (2018)
19. Huang, G., Chen, D., Li, T., Wu, F., van der Maaten, L., Weinberger, K.: Multiscale dense networks for resource efficient image classification. In: International Conference on Learning Representations (2018)

20. Jia, X., De Brabandere, B., Tuytelaars, T., Gool, L.V.: Dynamic filter networks. Adv. Neural Inf. Process. Syst. **29**, 667–675 (2016)
21. Kang, B., Liu, Z., Wang, X., Yu, F., Feng, J., Darrell, T.: Few-shot object detection via feature reweighting. In: Proceedings of the IEEE/CVF International Conference on Computer Vision, pp. 8420–8429 (2019)
22. Koch, G., Zemel, R., Salakhutdinov, R., et al.: Siamese neural networks for one-shot image recognition. In: ICML Deep Learning Workshop, vol. 2. Lille (2015)
23. Lee, K., Maji, S., Ravichandran, A., Soatto, S.: Meta-learning with differentiable convex optimization. In: Proceedings of the IEEE Conference on Computer Vision and Pattern Recognition, pp. 10657–10665 (2019)
24. Lee, Y., Choi, S.: Gradient-based meta-learning with learned layerwise metric and subspace. In: International Conference on Machine Learning, pp. 2927–2936. PMLR (2018)
25. Li, H., Eigen, D., Dodge, S., Zeiler, M., Wang, X.: Finding task-relevant features for few-shot learning by category traversal. In: Proceedings of the IEEE Conference on Computer Vision and Pattern Recognition, pp. 1–10 (2019)
26. Liu, B., et al.: Negative margin matters: understanding margin in few-shot classification. arXiv preprint arXiv:2003.12060 (2020)
27. Liu, C., et al.: Learning a few-shot embedding model with contrastive learning. In: Proceedings of the AAAI Conference on Artificial Intelligence, vol. 35, pp. 8635–8643 (2021)
28. Liu, Y., et al.: Learning to propagate labels: transductive propagation network for few-shot learning. arXiv preprint arXiv:1805.10002 (2018)
29. Liu, Y., Schiele, B., Sun, Q.: An ensemble of epoch-wise empirical bayes for few-shot learning. In: Vedaldi, A., Bischof, H., Brox, T., Frahm, J.-M. (eds.) ECCV 2020. LNCS, vol. 12361, pp. 404–421. Springer, Cham (2020). https://doi.org/10.1007/978-3-030-58517-4_24
30. Lu, S., Ye, H.J., Zhan, D.C.: Tailoring embedding function to heterogeneous few-shot tasks by global and local feature adaptors. In: Proceedings of the AAAI Conference on Artificial Intelligence, vol. 35, pp. 8776–8783 (2021)
31. Ma, R., Fang, P., Drummond, T., Harandi, M.: Adaptive poincaré point to set distance for few-shot classification. In: Proceedings of the AAAI Conference on Artificial Intelligence, vol. 36, pp. 1926–1934 (2022)
32. Nichol, A., Achiam, J., Schulman, J.: On first-order meta-learning algorithms. arXiv preprint arXiv:1803.02999 (2018)
33. Oreshkin, B.N., Rodriguez, P., Lacoste, A.: Tadam: task dependent adaptive metric for improved few-shot learning. In: Proceedings of the 32nd International Conference on Neural Information Processing Systems, pp. 719–729 (2018)
34. Qin, Z., Zhang, P., Wu, F., Li, X.: Fcanet: frequency channel attention networks. arXiv preprint arXiv:2012.11879 (2020)
35. Ravi, S., Larochelle, H.: Optimization as a model for few-shot learning. In: ICLR (2017)
36. Ren, M., et al.: Meta-learning for semi-supervised few-shot classification. arXiv preprint arXiv:1803.00676 (2018)
37. Rusu, A.A., et al.: Meta-learning with latent embedding optimization. arXiv preprint arXiv:1807.05960 (2018)
38. Satorras, V.G., Estrach, J.B.: Few-shot learning with graph neural networks. In: International Conference on Learning Representations (2018)
39. Shen, Z., Liu, Z., Qin, J., Savvides, M., Cheng, K.T.: Partial is better than all: revisiting fine-tuning strategy for few-shot learning. In: Proceedings of the AAAI Conference on Artificial Intelligence, vol. 35, pp. 9594–9602 (2021)

40. Shyam, P., Gupta, S., Dukkipati, A.: Attentive recurrent comparators. In: International Conference on Machine Learning, pp. 3173–3181. PMLR (2017)

41. Simon, C., Koniusz, P., Nock, R., Harandi, M.: Adaptive subspaces for few-shot learning. In: Proceedings of the IEEE/CVF Conference on Computer Vision and Pattern Recognition, pp. 4136–4145 (2020)

42. Simon, C., Koniusz, P., Nock, R., Harandi, M.: On modulating the gradient for meta-learning. In: Vedaldi, A., Bischof, H., Brox, T., Frahm, J.-M. (eds.) ECCV 2020. LNCS, vol. 12353, pp. 556–572. Springer, Cham (2020). https://doi.org/10.1007/978-3-030-58598-3_33

43. Snell, J., Swersky, K., Zemel, R.: Prototypical networks for few-shot learning. In: Advances in Neural Information Processing Systems, pp. 4077–4087 (2017)

44. Sung, F., Yang, Y., Zhang, L., Xiang, T., Torr, P.H., Hospedales, T.M.: Learning to compare: relation network for few-shot learning. In: Proceedings of the IEEE Conference on Computer Vision and Pattern Recognition, pp. 1199–1208 (2018)

45. Teerapittayanon, S., McDanel, B., Kung, H.T.: Branchynet: fast inference via early exiting from deep neural networks. In: 2016 23rd International Conference on Pattern Recognition (ICPR), pp. 2464–2469. IEEE (2016)

46. Triantafillou, E., Zemel, R., Urtasun, R.: Few-shot learning through an information retrieval lens. In: Proceedings of the 31st International Conference on Neural Information Processing Systems, pp. 2252–2262 (2017)

47. Veit, A., Belongie, S.: Convolutional networks with adaptive inference graphs. In: Proceedings of the European Conference on Computer Vision (ECCV), pp. 3–18 (2018)

48. Vinyals, O., Blundell, C., Lillicrap, T., Wierstra, D., et al.: Matching networks for one shot learning. In: Advances in Neural Information Processing Systems, pp. 3630–3638 (2016)

49. Wah, C., Branson, S., Welinder, P., Perona, P., Belongie, S.: The caltech-ucsd birds-200-2011 dataset (2011)

50. Wang, X., Yu, F., Dou, Z.Y., Darrell, T., Gonzalez, J.E.: Skipnet: learning dynamic routing in convolutional networks. In: Proceedings of the European Conference on Computer Vision (ECCV), pp. 409–424 (2018)

51. Wang, Y., Chao, W.L., Weinberger, K.Q., van der Maaten, L.: Simpleshot: revisiting nearest-neighbor classification for few-shot learning. arXiv preprint arXiv:1911.04623 (2019)

52. Wang, Y., Yao, Q., Kwok, J.T., Ni, L.M.: Generalizing from a few examples: a survey on few-shot learning. ACM Comput. Surv. (CSUR) **53**(3), 1–34 (2020)

53. Wertheimer, D., Tang, L., Hariharan, B.: Few-shot classification with feature map reconstruction networks. In: Proceedings of the IEEE/CVF Conference on Computer Vision and Pattern Recognition, pp. 8012–8021 (2021)

54. Xu, C., et al.: Learning dynamic alignment via meta-filter for few-shot learning. In: Proceedings of the IEEE/CVF Conference on Computer Vision and Pattern Recognition, pp. 5182–5191 (2021)

55. Xu, W., Wang, H., Tu, Z., et al.: Attentional constellation nets for few-shot learning. In: International Conference on Learning Representations (2020)

56. Yan, X., Chen, Z., Xu, A., Wang, X., Liang, X., Lin, L.: Meta r-cnn: towards general solver for instance-level low-shot learning. In: Proceedings of the IEEE/CVF International Conference on Computer Vision, pp. 9577–9586 (2019)

57. Ye, H.J., Hu, H., Zhan, D.C., Sha, F.: Few-shot learning via embedding adaptation with set-to-set functions. In: Proceedings of the IEEE/CVF Conference on Computer Vision and Pattern Recognition, pp. 8808–8817 (2020)

58. Zhang, C., Cai, Y., Lin, G., Shen, C.: Deepemd: few-shot image classification with differentiable earth mover's distance and structured classifiers. In: Proceedings of the IEEE/CVF Conference on Computer Vision and Pattern Recognition, pp. 12203–12213 (2020)

59. Zhao, J., Yang, Y., Lin, X., Yang, J., He, L.: Looking wider for better adaptive representation in few-shot learning. In: Proceedings of the AAAI Conference on Artificial Intelligence, vol. 35, pp. 10981–10989 (2021)

60. Zhou, J., Jampani, V., Pi, Z., Liu, Q., Yang, M.H.: Decoupled dynamic filter networks. In: Proceedings of the IEEE/CVF Conference on Computer Vision and Pattern Recognition, pp. 6647–6656 (2021)

61. Zhu, T., Ma, R., Harandi, M., Drummond, T.: Learning online for unified segmentation and tracking models. In: 2021 International Joint Conference on Neural Networks (IJCNN), pp. 1–8 (2021)

Open-World Semantic Segmentation via Contrasting and Clustering Vision-Language Embedding

Quande Liu[1], Youpeng Wen[2], Jianhua Han[3], Chunjing Xu[3], Hang Xu[3(✉)], and Xiaodan Liang[2(✉)]

[1] The Chinese University of Hong Kong, Shenzhen, China
`qdliu@cse.cuhk.edu.hk`
[2] Shenzhen Campus of Sun Yat-sen University, Shenzhen, China
`xdliang328@gmail.com`
[3] Huawei Noah's Ark Lab, Montreal, Canada
`{hanjianhua4,xuchunjing,xu.hang}@huawei.com`

Abstract. To bridge the gap between supervised semantic segmentation and real-world applications that acquires one model to recognize arbitrary new concepts, recent zero-shot segmentation attracts a lot of attention by exploring the relationships between unseen and seen object categories, yet requiring large amounts of densely-annotated data with diverse base classes. In this paper, we propose a new open-world semantic segmentation pipeline that makes the first attempt to learn to segment semantic objects of various open-world categories without any efforts on dense annotations, by purely exploiting the image-caption data that naturally exist on the Internet. Our method, **Vi**sion-language-driven Semantic **Seg**mentation (ViL-Seg), employs an image and a text encoder to generate visual and text embeddings for the image-caption data, with two core components that endow its segmentation ability: First, the image encoder is jointly trained with a vision-based contrasting and a cross-modal contrasting, which encourage the visual embeddings to preserve both fine-grained semantics and high-level category information that are crucial for the segmentation task. Furthermore, an online clustering head is devised over the image encoder, which allows to dynamically segment the visual embeddings into distinct semantic groups such that they can be classified by comparing with various text embeddings to complete our segmentation pipeline. Experiments show that without using any data with dense annotations, our method can directly segment objects of arbitrary categories, outperforming zero-shot segmentation methods that require data labeling on three benchmark datasets.

1 Introduction

As a crucial problem in computer vision, semantic segmentation [30] aims to assign a class label to each pixel in the image. Most existing semantic segmentation methods [5,26,28,37,50] are only capable of segmenting base categories appearing in the training dataset. However, the number of object classes in

S. Avidan et al. (Eds.): ECCV 2022, LNCS 13680, pp. 275–292, 2022.
https://doi.org/10.1007/978-3-031-20044-1_16

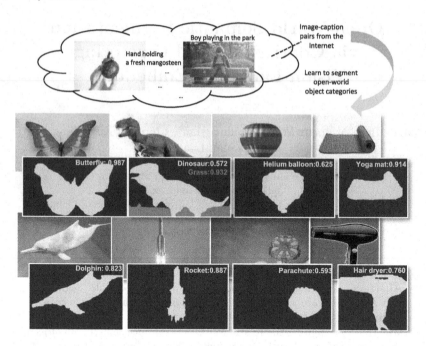

Fig. 1. By purely utilizing the image-caption pairs from the Internet (without using any data with dense annotations), ViL-Seg is able to segment various object categories in the open world even though they are never labeled in existing segmentation datasets.

existing semantic segmentation datasets [2,10,31] is limited due to the costly pixel-wise annotations, e.g., PASCAL VOC [10] with 20 categories and COCO Stuff [2] with 183 categories, which is far away from the number of object categories that exist in reality. The usual way to increase the category number is by annotating more images of the novel categories, which, however, not only requires tremendous human labeling efforts but also faces difficulty to collect enough samples given the extremely large class number in open-world [14].

Recently, zero-shot segmentation methods [1,9,13,45] have been proposed to generalize the semantic segmentation model to unseen classes by leveraging the word embeddings to discover the implicit relationship between base and novel classes. However, since all these methods rely on the *training on a specific dataset containing some base classes*, the developed segmentation model would be biased towards either seen classes or the training scenes [13], which will hurt the segmentation performance on novel classes and the transfer ability to other datasets in real-world applications.

Inspired by the recent advance of vision-language pre-training methods [27,36], we aim to learn a model that can segment various object categories in open-world by purely leveraging the vision-language data that exists naturally on the Internet (cf. Fig. 1). Compared with traditional manually-annotated datasets, image-caption data from the Internet [4,40] is much easier to collect and needs no more costly human labeling process. Besides, given the tremendous data resources on the Internet, these data can easily scale up to tens or hundreds of millions level and greatly increase the diversity of object categories [7], which paves the way for the model to handle object classes that are never labeled in existing datasets but exist in reality. Recently, there have been some studies [12, 49] to exploit the large-scale vision-language data to solve some downstream tasks, such as image classification [36] or captioning [39]. Zareian et al. [12] also proposed to leverage the cross-modal data to address unseen-class object detection problem by distilling the knowledge from a pre-trained zero-shot classification model into an object detector. However, how to leverage these web-based image-caption data to address the semantic segmentation problem for open-world object categories remains unsolved, which is also highly challenging given that the caption only contains a global semantic description for the image which is insufficient for the segmentation task that requires dense semantic understanding.

In this paper, we present Vision-language-driven Semantic Segmentation (ViL-Seg), a new open-world annotation-free semantic segmentation pipeline that makes the first attempt to learn to segment semantic objects of various open-world categories by purely exploiting the vision-language data from the Internet. In detail, ViL-Seg utilizes an image encoder and a text encoder to generate visual and text embeddings for two different modalities (i.e., image and caption). To preserve the fine-grained semantics and high-level category information which are two key properties for the visual embeddings in segmentation task, the image encoder has been trained under the supervision of two complementary objectives, i.e., a) a vision-based contrasting by comparing global and local image patches to learn local to global correspondence; b) a cross-modal contrasting to exploit the category information from natural language supervision. Furthermore, an online clustering head is further designed over the image encoder, which segments the fine-grained visual embeddings into distinct semantic groups such that they can be classified by comparing the alignment with text embeddings of various open-world object categories. This online clustering design also makes the training and inference of ViL-Seg end-to-end.

Our main contributions are summarized as follows:

- We present Vision-language-driven Semantic Segmentation (ViL-Seg), which to our knowledge is the first attempt to use the image-caption pairs from the Internet to learn to segment objects of various open-world categories without using any densely-annotated data.
- To explore the segmentation-related knowledge from image-caption data, ViL-Seg employs two complementary contrastive objectives to promote the quality of visual embeddings, with an online clustering head to dynamically divide the visual embeddings into different semantic regions. Both the training and inference of ViL-Seg are performed end-to-end.
- Experiments show that without using any data with dense annotations, our ViL-Seg can segment various open-world object categories, and outperform state-of-the-art zero-shot segmentation methods that require data labeling on three benchmark datasets, e.g., 5.56% mIoU increase on PASCAL VOC.

2   Related Work

2.1   Zero-Shot Semantic Segmentation

Zero-shot semantic segmentation [1] denotes segmenting unseen categories without training with any instances of them. For the past few years, some methods [21,23] have been proposed via learning word embeddings between seen and unseen categories. For instance, SPNet [45] utilizes a generator to generate synthetic features from word embedding to match the corresponding vision features, while ZS3Net [1] projects visual semantic embedding into class probability via a fixed word embedding matrix of different classes. To mitigate the seen categories' bias in SPNet and the model collapse problem in ZS3Net, CaGNet [13] proposes a contextual module to generate more diverse and context-aware word embeddings. Based on these methods, SIGN [9] further adopts and improves standard positional encoding to integrate spatial information of feature level and proposes annealed self-training to assign different importance to pseudo labels according to their confidence.

There are also several works [11,32,35] concentrating on the open-set recognition problem [38], which aim to distinguish whether the sample is from novel classes without providing a specific unseen category name. A variety of works on unsupervised semantic segmentation [18,42,48] also tend to learn dense semantic representations without using segmentation labels. However, these methods can only provide semantic groups by using clustering methods like K-Means [22] as post-processing on the network features, yet cannot provide the category name for each semantic group. Different from these methods, by exploiting the vision-language data from the internet [4], our method is capable of predicting the class name for each image pixel without using any data with dense annotations.

2.2   Vision-Language Pre-training

Vision-language pre-training [16,19,24,25,43] with massive image-text pairs from the Internet has attracted more and more attention in recent years. By using contrastive pre-training to predict the correct pairs of image and text samples, CLIP [36] achieves competitive results compared with the fully-supervised baseline on several downstream classification tasks. Some works [8,27] also introduce language-modeling-like objectives, including masked language/region modeling, image captioning and text-denoising to further improve the performance of vision-language models. Moreover, several methods [17,39] adopt a pre-trained object detector to obtain a sequence of object embeddings as the visual features.

Very recently, some studies [12,46,49] have proposed to leverage the pre-trained vision-language model to address the open-vocabulary object detection task, which aims at training a model to detect any object from a given vocabulary of classes. Zareian et al. [49] propose to learn a vision to language (V2L) layer during pre-training, and utilize it to initialize a Faster-RCNN model. ViLD [12] distills the knowledge from a pre-trained zero-shot classifier into a two-stage

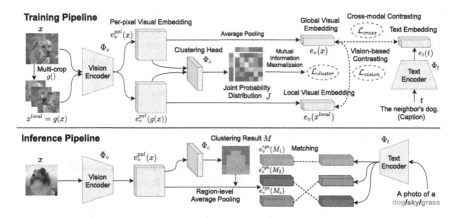

Fig. 2. Overall architecture of ViL-Seg. The image encoder is trained with two complementary objectives, i.e., the vision-based and cross-modal contrastive losses, aiming to promote the fine-grained semantics and high-level category information in the visual embeddings. Besides, an online clustering head is built over the image encoder to segment the pixel-wise visual embeddings into distinct semantic groups, which are trained with mutual information maximization. During inference, the segmentation is performed by comparing the feature pooled from each clustered region with different word embeddings. Both the training and inference are performed end-to-end.

detector. Based on ViLD, ZSD-YOLO [46] further expands the thought of distillation into YOLOv5 [20]. There are also several studies [33,47] that tend to leverage the vision-language models, e.g., CLIP, to reduce the annotation cost in semantic segmentation task. However, these studies either rely on the annotated data on seen classes for training [47], or can only support unsupervised segmentation that simply separates the image pixels into variant semantic clusters without providing the corresponding class labels [33]. In contrast, we aim to develop an complete semantic segmentation pipeline that can segment various open-world objects, by purely utilizing the image-caption data from the internet without using any densely-annotated data.

3 Method

Figure 2 overviews our proposed Vision-language-driven Semantic Segmentation (ViL-Seg) method. In this section, we first briefly introduce its framework and training objective in Sect. 3.1. Then, we describe the two complementary contrastive learning strategies which are used to enhance the visual embeddings in Sect. 3.2, and present how to segment per-pixel visual embeddings into different semantic groups with the online clustering head in Sect. 3.3.

3.1 ViL-Seg Framework

The base of ViL-Seg is a vision encoder Φ_v and a text encoder Φ_t to embed the image and its caption from the paired web data. We denote $e_v \in \mathbb{R}^D$ as the extracted global visual feature, $e_v^{pxl} \in \mathbb{R}^{HW \cdot D}$ as the per-pixel visual embeddings, e.g., embeddings before last pooling layer; and denote $e_t \in \mathbb{R}^D$ as the encoded text feature. To perform image segmentation over this framework, we also construct an online clustering head Φ_c over the image encoder, which is responsible for segmenting the per-pixel visual embeddings e_v^{pxl} into C semantic clusters.

The whole framework of ViL-Seg is trained in an end-to-end manner, using the objective function as follows:

$$\mathcal{L}(\Phi_{v,t,c}) = \mathcal{L}_{vision}(\Phi_v) + \mathcal{L}_{cross}(\Phi_{v,t}) + \mathcal{L}_{cluster}(\Phi_c) \qquad (1)$$

which is composed of the vision-based contrastive learning $\mathcal{L}_{vision}$ and the cross-modal contrastive alignment $\mathcal{L}_{cross}$ to enhance the fine-grained semantics and the high-level category information in the visual embeddings respectively; and an unsupervised clustering objective $\mathcal{L}_{cluster}$ optimized w.r.t. Φ_c to promote reasonable clustering results. Next, we will describe each part in detail.

3.2 Vision-Based and Cross-Modal Contrasting

As a dense classification task, semantic segmentation requires the learned visual embeddings to contain both *fine-grained semantics* and *high-level category information*. To this end, we have employed a vision-based contrasting and cross-modal contrasting to enhance the two properties of the visual representations respectively.

Vision-Based Contrasting of Global and Local Views: Self-supervision with contrastive learning has shown promising results in representation learning [6]. To meet the requirement of dense semantic understanding in segmentation, we devise a vision-based self-supervised learning strategy by contrasting local and global image patches to learn local to global semantic correspondence.

Specifically, given an input visual image, we first transform it into different distorted views or local patches, using the multi-crop strategy [3], denoted as function $g(\cdot)$. This generates an image set of different views, which in our case contains one global view x and k local views $x^{local} = g(x) = [x^{l1}, x^{l2}, ..., x^{lk}]$ of low resolution. All these images are then fed into the visual encoder, resulting in a global feature $e_v(x)$ of the global view x, and a local feature $e_v(x^{local})$ which is the concatenation of features of all local views $[e_v(x^{l1}), e_v(x^{l2}), ..., e_v(x^{lk})]$. Considering that imposing the regularization directly onto the image features might be too strict to impede the convergence, we pass the global and local features to a projection function Φ_a before computing the loss function, which is composed of a linear projection layer and a softmax activation layer inspired by knowledge distillation [15]. Our vision-based contrastive learning mechanism finally encourages the consistency of semantic information between the global and local features, aiming to encourage the model to capture the local to global correspondence and hence promote the fine-grained semantics of visual embeddings for the dense classification task. The objective function is expressed as:

$$\mathcal{L}_{vision} = H(\Phi_a(e_v(x)), \Phi_a(e_v(x^{local}))) \tag{2}$$

where $H(\cdot)$ denotes the cross-entropy loss.

Cross-Modal Contrasting of Natural Language Supervision: Learning from natural language supervision has been demonstrated with effectiveness in large-scale vision-language pre-training tasks [19,24,25]. Our ViL-Seg inherits the cross-modal contrastive learning strategy, aiming to learn the visual embeddings e_v and text embeddings e_t such that they can be close to each other if they are from the paired image and caption, and far away if not.

Specifically, given a minibatch containing b image-text pairs $\{x_j, t_j\}_{j=1}^b$, the image feature $e_v(x_m)$ and text feature $e_t(t_n)$ is a positive pair if $m = n$, and otherwise a negative pair. Then, the cross-modal contrastive alignment is performed over each positive pair in the minibatch as:

$$\ell(x_m, \{t_n\}_{n=1}^b) = -\log \frac{exp(e_v(x_m) \odot e_t(t_m)/\tau)}{\sum_{n=1}^b exp(e_v(x_m) \odot e_t(t_n)/\tau)}, \tag{3}$$

where $\odot$ denotes the cosine similarity: $a \odot b = \frac{\langle a,b \rangle}{||a||_2||b||_2}$; τ denotes the temperature parameter. The final objective function $\mathcal{L}_{cross}$ is the average of ℓ over all positive pairs:

$$\mathcal{L}_{cross} = \sum_{m=1}^b \frac{1}{b}\ell(x_m, \{t_n\}_{n=1}^b), \tag{4}$$

By aligning the visual and text embeddings as Eq. 4, the category information contained in the captions can be successfully transferred to the visual embeddings space, therefore allowing us to classify visual features by comparing their similarity with the word embeddings of different categories.

3.3   Online Clustering of Visual Embeddings

Semantic segmentation requires assigning a label to each image pixel. However, the cross-modal alignment above can only provide classification ability over the global visual feature e_v, instead of per-pixel embeddings e_v^{pxl}. To address this problem, we propose to cluster the per-pixel visual features into distinct groups according to their semantics. Then, the features of each semantic region can be respectively abstracted as a region-level feature for cross-modal alignment to fulfill the dense classification pipeline.

Specifically, we employ an online clustering strategy to efficiently separate the visual embeddings by maximizing the mutual information across cluster assignments. Given the per-pixel visual embeddings $e_v^{pxl} \in \mathbb{R}^{HW \cdot D}$, we aim to cluster these features into clustering space $Y = \{1, 2, \ldots, C\}$. To this end, we construct a clustering head Φ_c over the image encoder, which is composed of a convolution layer with C channel followed by a softmax function. Denote $q, q' \in \mathbb{R}^{1 \cdot D}$ as a pair of pixel embeddings from e_v^{pxl} which contain the same semantic, the goal of our clustering head is to preserve what is common between q and q' while removing their instance-specific information, which is equivalent to maximizing their mutual information as:

$$\max_{\Phi_c} \quad I(\Phi_c(q), \Phi_c(q')) \tag{5}$$

In our case, the paired embeddings (q, q') are unavailable since the category of each image pixel is unknown. Therefore, we adopt generated embedding pairs to compute the clustering objective, by extracting the embeddings for the input image x and its transformed image $g(x)$ respectively, obtaining $e_v^{pxl}(x)$ and $e_v^{pxl}(g(x))$. It is worthy to mention that $g(\cdot)$ here do not adopt the multi-crop strategy, but the random additive and multiplicative colour transformations with horizontal flipping, which are all affine transformations. Since $g(\cdot)$ contains geometric transformation, the embedding $e_v^{pxl}(x)_i$ at pixel i will correspond to $g^{-1}(e_v^{pxl}(g(x)))_i$. This is because translating the input image will also change the geometric order of the output feature. We need to undo the geometric function by applying $g^{-1}(\cdot)$ over the feature of transformed image such that it could be paired with $e_v^{pxl}(x)$ pixel-by-pixel. Please note that the reason we compute clustering loss between pixels of different views instead of pixels of the same class is that the class information of pixels are unknown in our case, since no dense annotations are provided. Besides, maximizing the common information between transformed views is an effective strategy to promote clustering samples of the same class, as demonstrated in unsupervised learning [6], which meets our goal to perform semantic segmentation task without dense annotations.

We now describe how to compute the mutual information of Eq. 5. For simplicity of description, we denote (q_i, q_i') as a pair of embeddings at pixel i of $e_v^{pxl}(x)$ and $g^{-1}(e_v^{pxl}(g(x)))$. Since our clustering head outputs soft label distributions using softmax activation function, the mutual information between q_i and q_i' (i.e., the probability of predicting q_i from q_i' and vice versa) is given by their joint probability distribution $J_i \in [0, 1]^{C \times C}$:

$$I(\Phi_c(q_i), \Phi_c(q_i')) = I(J_i), \quad J_i = \Phi_c(q_i) \cdot \Phi_c(q_i')^T \tag{6}$$

where $J_i^{cc'} = P(\Phi_c(q_i) = c, \Phi_c(q_i') = c')$. In each minibatch, the joint probability distributions J is computed as:

$$J = \frac{1}{BHWD} \sum_{i=1}^{BHWD} \Phi_c(q_i) \cdot \Phi_c(q_i')^T \tag{7}$$

Finally, the clustering objective is equivalent to maximizing the mutual information [41] of matrix J, and is extended to:

$$\mathcal{L}_{cluster} = \max \ I(J) = \max \ \sum_{c=1}^{C} \sum_{c'=1}^{C} J^{cc'} \cdot \ln \frac{J^{cc'}}{J^c \cdot J^{c'}} \tag{8}$$

where $J^c = P(\Phi_c(q_i) = c)$ and $J^{c'} = P(\Phi_c(q_i') = c')$ are computed by summing over the c-th row and c'-th column of the matrix J respectively.

We take the relation between mutual information and entropy [34] to explain why maximizing the mutual information can promote reasonable clustering results. Given $I(\Phi_c(q_i), \Phi_c(q_i')) = E(\Phi_c(q_i)) - E(\Phi_c(q_i)|\Phi_c(q_i'))$, maximizing the mutual information is equivalent to maximizing the individual clustering results entropy $E(\Phi_c(q_i))$ while minimizing the conditional clustering results entropy $E(\Phi_c(q_i)|\Phi_c(q_i'))$. The smallest value of the latter is attained when $E(\Phi_c(q_i)|\Phi_c(q_i')) = 0$, i.e., the cluster assignments for q_i and q_i' are predictable for each other. Therefore, it encourages embeddings with similar semantics to be assigned to the same cluster. Furthermore, the largest value of the $E(\Phi_c(q_i))$ is attained when all clusters are assigned in equal possibility among all embeddings in the whole dataset, hence it may avoid the degenerated solution that all features are assigned to the same cluster.

Inference Pipeline: During inference, the segmentation for an input image x can be produced by feeding it to the image encoder to extract per-pixel visual embeddings $e_v^{pxl}(x)$, which are then passed to the clustering head to obtain the clustering mask $M \in \{0,1\}^{H \times W \times C}$ with C clusters using argmax function. According to the semantic region indicated by each cluster $M_c \in \{0,1\}^{H \times W}$, we can extract its region-level feature $e_v^{rgn}(M_c)$ by filtering and averaging the per-pixel visual embeddings in pixel indexes where $M_c = 1$ (cf. the region-level averaging pooling in Fig. 2), i.e., $e_v^{rgn}(M_c) = \frac{\sum e_v^{pxl}(x) \cdot M_c}{\sum M_c}$. Finally, the category name of each region M_c is given by comparing its region-level feature $e_v^{rgn}(M_c)$ with the word embeddings of different classes, using prompt "a photo of a category" as CLIP [36].

4   Experiments

4.1   Experimental Setup

Dataset and Evaluation Protocol: Following the literature of zero-shot segmentation [9,13,45], we conduct experiments on three datasets, including PASCAL VOC [10], PASCAL Context [31], and COCO Stuff [2]. For PASCAL VOC and PASCAL Context datasets, we evaluate our method on their validation set containing 1449 images and 5105 images respectively. For COCO Stuff datasets, we adopt the setting in [9] to use 5000 images for testing.

Since there is not a standard evaluation protocol for our open-world semantic segmentation task without using any dense annotations, we follow the zero-shot segmentation settings defined in [13,45] to compare the segmentation performance on the unseen classes of the three datasets. Specifically, the unseen classes contain: 5 classes (potted plant, sheep, sofa, train, tv-monitor) out of the 20 object categories in PASCAL VOC; 4 classes (cow, motorbike, sofa, cat) out of the 59 object categories in PASCAL Context; and 15 classes (frisbee, skateboard, cardboard, carrot, scissors, suitcase, giraffe, cow, road, wall concrete, tree, grass, river, clouds, playingfield) out of the 183 object categories in COCO Stuff dataset. We adopt the standard metrics including mean intersection-over-union (mIoU) [28] and pixel accuracy (pix. acc.) to evaluate the segmentation results.

Implementation Detail: We adopt the transformer architecture (ViT-B/16) for the image encoder and text encoder, following the popular vision-language learning framework [36], with the embedding dimension of 512. The cluster number C in the online clustering head is set as 25, and we shall study this hyperparameter in detail in the ablation analysis. In vision-based contrasting, we crop 6 local patches with the multi-crop strategy, and the output dimension of the projection layer is 2048. We train the model with Adam [29] optimizer, using learning rate of 5e−4, weight decay coefficient of 0.04, and warm-up iterations of 4000. The ViL-Seg model is trained with no other data but CC12M dataset [4], which contains about 12 million image-caption pairs collected from the Internet. The whole framework is trained using 48 T V100 16 GB with batch size 768.

Table 1. Comparison of unseen-class segmentation results with zero-shot segmentation methods on Pascal VOC, Pascal Context and COCO Stuff datasets. "ST" stand for self-training.

Method	PASCAL VOC		PASCAL context		COCO stuff	
	mIoU [%]	pix. acc. [%]	mIoU [%]	pix. acc. [%]	mIoU [%]	pix. acc. [%]
SPNet [44]	15.63	-	4.00	-	8.73	-
ZS3 [1]	17.65	21.47	7.68	19.22	9.53	22.75
CaGNet (pi) [13]	26.59	42.97	14.42	39.76	12.23	25.45
CaGNet (pa) [13]	29.90	51.76	14.98	39.81	13.89	29.62
SIGN [9]	28.86	-	14.93	-	15.47	-
CLIP + Seg	27.40	48.35	14.52	37.48	13.20	28.75
ViL-Seg (Ours)	$\mathbf{34.42}_{(+5.56)}$	$\mathbf{76.03}_{(+24.27)}$	$\mathbf{16.32}_{(+1.39)}$	$\mathbf{45.64}_{(+5.83)}$	$\mathbf{16.43}_{(+0.96)}$	$\mathbf{32.58}_{(+2.96)}$
ZS3 + ST	21.15	-	9.53	-	10.55	-
CaGNet + ST	30.31	-	16.30	-	13.40	-
SIGN + ST	33.12	-	16.71	-	15.15	-
ViL-Seg + ST	$\mathbf{37.30}_{(+4.18)}$	85.62	$\mathbf{18.94}_{(+2.13)}$	50.14	$\mathbf{18.05}_{(+2.90)}$	35.23

4.2 Comparison with Other Methods

Experimental Setting: Due to the lack of previous study that purely utilizes web-based image-caption data to learn to segment novel object categories, we compare our method with several popular zero-shot segmentation (ZSS) methods, which also segment new object categories but via exploiting the relationships between the word embeddings of seen base classes and unseen class. Specifically, the comparison methods include (1) SPNet [44], a semantic projection network which maps each image pixel to a semantic word embedding space for ZSS; (2) ZS3 [1], which addresses unseen-class segmentation by combining a segmentation model with an approach to generate visual representations from semantic word embedding; (3) CaGNet [13], which devises a contextual module into the segmentation network to capture more diverse contextual information from semantic word embedding; and (4) SIGN [9], a very latest ZSS method which incorporates spatial information into semantic features using positional encodings to improve the segmentation of unseen classes. (5) CLIP [36] + Seg, we simply use the CLIP's image encoder(ViT-B/16) with its global attention pooling layer removed, to serve as a backbone for semantic segmentation. Classification for dense prediction can be directly obtained from the text embeddings of CLIP's text encoder. All these methods follow the same zero-shot segmentation setting described in Sect. 4.1, and for a fair comparison, we compare the performance of all these methods under both scenarios of using or without using self-training as followup. For each comparison method, the results are either referenced from their official paper or the number reproduced by other previous works.

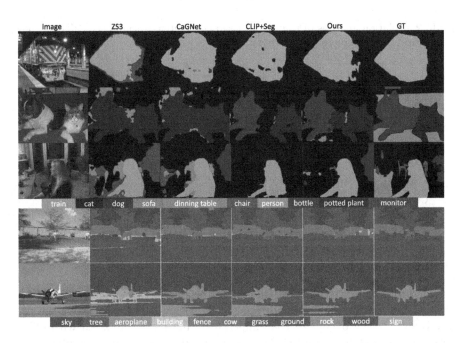

Fig. 3. Qualitative comparison with baseline and other methods. The top three samples are from PASCAL VOC and the bottom two samples are from PASCAL Context.

Comparison Results: Table 1 presents the comparison results of these methods on PASCAL VOC [10], PASCAL Context [31] and COCO stuff [2] dataset ("-" denote the result was not reported in their paper). From this table, we may draw the following observations: (1) Our ViL-Seg outperforms these zero-shot segmentation methods on all three datasets in terms of both mIoU and pixel accuracy. This confirms the feasibility to exploit the naturally-existing image-caption pairs from the Internet to learn the segmentation model that can segment various open-world object categories. It is notable that these ZSS methods need to be trained on the densely-annotated training sets containing diverse base categories, but our ViL-Seg does not use any data with dense annotations for training. (2) ViL-Seg shows a larger increase on PASCAL VOC over other methods compared with the other two datasets. A plausible reason is that PASCAL VOC only contains 15 seen bases classes for these ZSS methods to train the model, which is relatively less than the 55 and 168 seen classes in PASCAL Context and COCO Stuff. In such case, our larger improvements in PASCAL VOC may reflect the limitation of those ZSS methods that require a wide range of base categories with dense annotations to attain a good performance, and further confirms the advantage of ViL-Seg that requires no data labeling. Figure 3 shows a qualitative comparison between ViL-Seg and baselines (SIGN [9] does not release its code). We can see that ViL-Seg achieves high accuracy.

Table 2. Ablation analysis of the vision-based contrastive learning (i.e., $\mathcal{L}_{vision}$), and online clustering design on the three datasets.

		ViL-Seg w/o $\mathcal{L}_{vision}$	Offline (K-means)	ViL-Seg
Params		-	86.19M	86.27M
Speed (case/s)		-	8.5	9.8
PASCAL VOC	mIoU [%]	22.05	30.97	33.61
	pix. acc. [%]	50.76	69.88	75.97
PASCAL Context	mIoU [%]	13.14	14.82	15.89
	pix. acc. [%]	38.90	41.64	43.54
COCO Stuff	mIoU [%]	13.52	15.81	16.41
	pix. acc. [%]	28.07	30.45	31.20

4.3 Ablation Analysis of ViL-Seg

We conduct ablation studies on the three datasets to investigate several key questions of ViL-Seg: **1)** the importance of the vision-based contrastive learning in ViL-Seg; **2)** the benefit of the online clustering head compared with offline clustering method like K-means; **3)** the choice and effect of cluster number in the online clustering head; **4)** the performance of ViL-Seg on different unseen classes. In ablation analysis, all object categories in the three datasets are considered as unseen classes. The performance on each dataset is the average over all its contained classes.

Importance of Vision-Based Contrasting: Apart from the cross-modal contrasting to align the visual and text embedding space, the image encoder in our framework is further supervised with a self-supervision signal by contrasting local and global image patches. From the qualitative segmentation results in Fig. 4, we can clearly see that without using this vision-based contrasting (second column), the clustering results cannot accurately separate the semantic object from the background region. Besides, the quantitative results in Table 2 show that removing this supervision (ViL-Seg w/o $\mathcal{L}_{vision}$) will lead to large performance decreases on all three datasets. These results reflect that the cross-modal contrasting can only guarantee the semantics of global image feature which is insufficient for the dense classification problem, while our additional self-supervision signal with vision-based contrasting is crucial to promote the fine-grained semantics in the visual embeddings.

Online Clustering *v.s.* **Offline Clustering:** Traditionally, the usual way to segment a group of features into distinct clusters is the offline methods like K-means [22]. Table 2 compares our online clustering design with traditional offline method, by replacing our online clustering head with K-means to cluster the per-pixel visual embeddings. We may draw three observations: (1) Our online clustering design attains higher segmentation performance than the offline method on all three datasets. We consider that the online clustering head is tightly-coupled with the visual encoder and can learn to improve the quality of visual embeddings as the training goes on, which is what the offline methods cannot attain. The qualitative results in Fig. 4 can also reflect that our online method (the fourth column) can better refine the learned visual embeddings and produce more smooth segmentation masks than the offline method (the third column). (2) The framework with our online clustering design also achieves a higher inference speed than the offline K-means method (8.5 *v.s.* 9.8 cases/s). This is because K-means needs to be performed offline as post-processing on the network features, which would limit the inference efficiency. In contrast, our online clustering design makes the training and inference of our method end-to-end and allows us to adaptively cluster the visual embeddings for each sample.

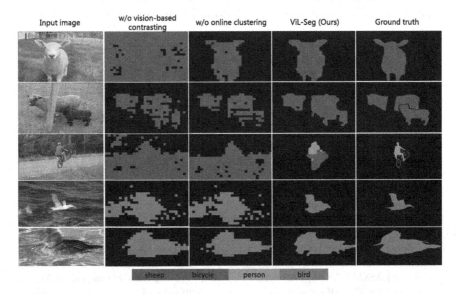

Fig. 4. Qualitative comparison among ViL-Seg, ViL-Seg without online clustering, and ViL-Seg without vision-based contrasting, with samples from PASCAL VOC dataset.

(3) Additionally, compared with the offline method, our online clustering design only increases 0.08M parameters to the model, which is less than 0.1% of the number of original network parameters.

Effect of Cluster Number in Online Clustering Head: The cluster number C is important in our method and affects the results of the online clustering head. Intuitively, fewer clusters might be incapable to cover the diverse semantics in the web-based image-caption data, while too many clusters might increase the learning difficulty given that the clustering head is only learned with an unsupervised objective of mutual information maximization. To validate the above intuitions and investigate the suitable choice of C, we repeated the experiment of ViL-Seg by varying $C \in \{5, 10, 15, 20, 25, 35\}$. As shown in Fig. 5, the model with middle-level of cluster number ($C = \{20, 25\}$) performs better than the model with smaller ($C = \{5, 10, 15\}$ or larger cluster number ($C = 30$). These results confirm our analysis above, and we finally adopt $C = 20$ in our method.

Performance on Different Unseen Classes: In Fig. 6, we show the mIoU of ViL-Seg on all 20 unseen classes of PASCAL VOC. It is observed that ViL-Seg can achieves more than 50% mIoU for classes like "bus", "cat", "horse" and "train", and attain mIoU larger than 20% on 14 out of 20 unseen classes. This owes to the diverse semantic information contained in the web-based data, which allows ViL-Seg to well segment these object categories even without using any of their training data with dense annotations. We also notice that the performance is relatively low in class like "person" or "car". This is probably caused by the imbalanced recognition capacity of vision-language models, which was also

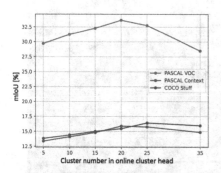
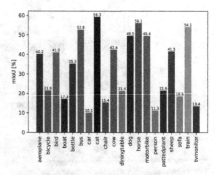

Fig. 5. Segmentation performance of ViL-Seg under different choices of cluster number C in the online clustering head, on PASCAL VOC, PASCAL Context and COCO stuff datasets.

Fig. 6. Segmentation performance of ViL-Seg on all 20 unseen classes of PASCAL VOC dataset. It is noticed that ViL-Seg can attain mIoU larger than 20% on 14 out of 20 unseen classe.

reported in previous studies [36]. For example, the image captions might usually use words like "man", "woman" to denote a person; and use the word of a brand name to denote a car, making the model less sensitive to these object categories. We may consider ensembling the results of different synonyms for an object category to alleviate this issue [27].

5 Conclusion

We have made the first attempt to learn to segment open-world object categories by purely leveraging the image-caption data from the Internet, without using any data with dense annotations. The proposed ViL-Seg attains the segmentation ability by employing two complementary contrastive learning strategies to promote the quality of visual embeddings, with an online clustering head to dynamically segment them into distinct semantic groups. Owing to the tremendous data resources on the Internet, our solution has outperformed zero-shot segmentation methods to segment the diverse semantic concepts in reality on three benchmark datasets, also opened a door for semantic segmentation task to reduce the human labeling to the greatest extent.

References

1. Bucher, M., Tuan-Hung, V., Cord, M., Pérez, P.: Zero-shot semantic segmentation. Adv. Neural. Inf. Process. Syst. **32**, 468–479 (2019)
2. Caesar, H., Uijlings, J., Ferrari, V.: COCO-Stuff: thing and stuff classes in context. In: Proceedings of the IEEE Conference on Computer Vision and Pattern Recognition, pp. 1209–1218 (2018)
3. Caron, M., Misra, I., Mairal, J., Goyal, P., Bojanowski, P., Joulin, A.: Unsupervised learning of visual features by contrasting cluster assignments. arXiv preprint arXiv:2006.09882 (2020)
4. Changpinyo, S., Sharma, P., Ding, N., Soricut, R.: Conceptual 12 m: pushing web-scale image-text pre-training to recognize long-tail visual concepts. In: Proceedings of the IEEE/CVF Conference on Computer Vision and Pattern Recognition, pp. 3558–3568 (2021)
5. Chen, L.-C., Papandreou, G., Kokkinos, I., Murphy, K., Yuille, A.L.: DeepLab: semantic image segmentation with deep convolutional nets, atrous convolution, and fully connected CRFs. IEEE Trans. Pattern Anal. Mach. Intell. **40**(4), 834–848 (2017)
6. Chen, T., Kornblith, S., Norouzi, M., Hinton, G.: A simple framework for contrastive learning of visual representations. In: International Conference on Machine Learning, pp. 1597–1607. PMLR (2020)
7. Chen, X., Shrivastava, A., Gupta, A.: Neil: extracting visual knowledge from web data. In: 2013 IEEE International Conference on Computer Vision, pp. 1409–1416 (2013)
8. Chen, Y.-C., et al.: UNITER: universal image-text representation learning. In: Vedaldi, A., Bischof, H., Brox, T., Frahm, J.-M. (eds.) ECCV 2020. LNCS, vol. 12375, pp. 104–120. Springer, Cham (2020). https://doi.org/10.1007/978-3-030-58577-8_7

9. Cheng, J., Nandi, S., Natarajan, P., Abd-Almageed, W.: Sign: spatial-information incorporated generative network for generalized zero-shot semantic segmentation (2021)

10. Everingham, M., Van Gool, L., Williams, C.K.I., Winn, J., Zisserman, A.: The pascal visual object classes (VOC) challenge. Int. J. Comput. Vis. **88**(2), 303–338 (2010)

11. Geng, C., Huang, S., Chen, S.: Recent advances in open set recognition: a survey. IEEE Trans. Pattern Anal. Mach. Intell. **43**(10), 3614–3631 (2020)

12. Gu, X., Lin, T.-Y., Kuo, W., Cui, Y.: Open-vocabulary object detection via vision and language knowledge distillation (2021)

13. Gu, Z., Zhou, S., Niu, L., Zhao, Z., Zhang, L.: Context-aware feature generation for zero-shot semantic segmentation. In: Proceedings of the 28th ACM International Conference on Multimedia, pp. 1921–1929 (2020)

14. Gupta, A., Dollar, P., Girshick, R.: LVIS: a dataset for large vocabulary instance segmentation. In: Proceedings of the IEEE/CVF Conference on Computer Vision and Pattern Recognition, pp. 5356–5364 (2019)

15. Hinton, G., Vinyals, O., Dean, J.: Distilling the knowledge in a neural network. arXiv preprint arXiv:1503.02531 (2015)

16. Huang, Z., Zeng, Z., Huang, Y., Liu, B., Fu, D., Fu, J.: Seeing out of the box: end-to-end pre-training for vision-language representation learning. In: Proceedings of the IEEE/CVF Conference on Computer Vision and Pattern Recognition, pp. 12976–12985 (2021)

17. Huo, Y., et al.: WenLan: bridging vision and language by large-scale multi-modal pre-training. arXiv preprint arXiv:2103.06561 (2021)

18. Hwang, J.-J., et al.: SegSort: segmentation by discriminative sorting of segments. In: Proceedings of the IEEE/CVF International Conference on Computer Vision, pp. 7334–7344 (2019)

19. Jia, C., et al.: Scaling up visual and vision-language representation learning with noisy text supervision. arXiv preprint arXiv:2102.05918 (2021)

20. Jocher, G.: ultralytics/yolov5: v3.1 - Bug Fixes and Performance Improvements, October 2020. https://github.com/ultralytics/yolov5

21. Kato, N., Yamasaki, T., Aizawa, K.: Zero-shot semantic segmentation via variational mapping. In: Proceedings of the IEEE/CVF International Conference on Computer Vision Workshops (2019)

22. Kodinariya, T.M., Makwana, P.R.: Review on determining number of cluster in k-means clustering. Int. J. **1**(6), 90–95 (2013)

23. Li, P., Wei, Y., Yang, Y.: Consistent structural relation learning for zero-shot segmentation. In: Advances in Neural Information Processing Systems, vol. 33 (2020)

24. Li, W., et al.: Unimo: towards unified-modal understanding and generation via cross-modal contrastive learning (2021)

25. Li, X., et al.: Oscar: object-semantics aligned pre-training for vision-language tasks. In: Vedaldi, A., Bischof, H., Brox, T., Frahm, J.-M. (eds.) ECCV 2020. LNCS, vol. 12375, pp. 121–137. Springer, Cham (2020). https://doi.org/10.1007/978-3-030-58577-8_8

26. Lin, G., Milan, A., Shen, C., Reid, I.: RefineNet: multi-path refinement networks for high-resolution semantic segmentation. In: Proceedings of the IEEE Conference on Computer Vision and Pattern Recognition, pp. 1925–1934 (2017)

27. Lin, J., et al.: M6: a Chinese multimodal pretrainer. arXiv preprint arXiv:2103.00823 (2021)

28. Long, J., Shelhamer, E., Darrell, T.: Fully convolutional networks for semantic segmentation. In: Proceedings of the IEEE Conference on Computer Vision and Pattern Recognition, pp. 3431–3440 (2015)
29. Loshchilov, I., Hutter, F.: Fixing weight decay regularization in Adam (2018)
30. Minaee, S., Boykov, Y.Y., Porikli, F., Plaza, A.J., Kehtarnavaz, N., Terzopoulos, D.: Image segmentation using deep learning: a survey. IEEE Trans. Pattern Anal. Mach. Intell. **44**(7), 3523–3542 (2021)
31. Mottaghi, R., et al.: The role of context for object detection and semantic segmentation in the wild. In: Proceedings of the IEEE Conference on Computer Vision and Pattern Recognition, pp. 891–898 (2014)
32. Oza, P., Patel, V.M.: C2AE: class conditioned auto-encoder for open-set recognition. In: Proceedings of the IEEE/CVF Conference on Computer Vision and Pattern Recognition, pp. 2307–2316 (2019)
33. Pakhomov, D., Hira, S., Wagle, N., Green, K.E., Navab, N.: Segmentation in style: unsupervised semantic image segmentation with styleGAN and clip. arXiv preprint arXiv:2107.12518 (2021)
34. Paninski, L.: Estimation of entropy and mutual information. Neural Comput. **15**(6), 1191–1253 (2003)
35. Perera, P., et al.: Generative-discriminative feature representations for open-set recognition. In: Proceedings of the IEEE/CVF Conference on Computer Vision and Pattern Recognition, pp. 11814–11823 (2020)
36. Radford, A., et al.: Learning transferable visual models from natural language supervision. arXiv preprint arXiv:2103.00020 (2021)
37. Ronneberger, O., Fischer, P., Brox, T.: U-Net: convolutional networks for biomedical image segmentation. In: Navab, N., Hornegger, J., Wells, W.M., Frangi, A.F. (eds.) MICCAI 2015. LNCS, vol. 9351, pp. 234–241. Springer, Cham (2015). https://doi.org/10.1007/978-3-319-24574-4_28
38. Scheirer, W.J., de Rezende Rocha, A., Sapkota, A., Boult, T.E.: Toward open set recognition. IEEE Trans. Pattern Anal. Mach. Intell. **35**(7), 1757–1772 (2012)
39. Su, W., et al.: VL-BERT: pre-training of generic visual-linguistic representations. arXiv preprint arXiv:1908.08530 (2019)
40. Thomee, B., et al.: YFCC100M: the new data in multimedia research. Commun. ACM **59**(2), 64–73 (2016)
41. Tschannen, M., Djolonga, J., Rubenstein, P.K., Gelly, S., Lucic, M.: On mutual information maximization for representation learning. arXiv preprint arXiv:1907.13625 (2019)
42. Van Gansbeke, W., Vandenhende, S., Georgoulis, S., Van Gool, L.: Unsupervised semantic segmentation by contrasting object mask proposals. arXiv preprint arXiv:2102.06191 (2021)
43. Wang, Z., Yu, J., Yu, A.W., Dai, Z., Tsvetkov, Y., Cao, Y.: SimVLM: simple visual language model pretraining with weak supervision. arXiv preprint arXiv:2108.10904 (2021)
44. Xian, Y., Choudhury, S., He, Y., Schiele, B., Akata, Z.: Semantic projection network for zero- and few-label semantic segmentation. In: 2019 IEEE/CVF Conference on Computer Vision and Pattern Recognition (CVPR), pp. 8248–8257 (2019)
45. Xian, Y., Choudhury, S., He, Y., Schiele, B., Akata, Z.: Semantic projection network for zero-and few-label semantic segmentation. In: Proceedings of the IEEE/CVF Conference on Computer Vision and Pattern Recognition, pp. 8256–8265 (2019)
46. Xie, J., Zheng, S.: ZSD-YOLO: zero-shot yolo detection using vision-language knowledge distillation (2021)

47. Xu, M., et al.: A simple baseline for zero-shot semantic segmentation with pre-trained vision-language model. arXiv preprint arXiv:2112.14757 (2021)
48. Ye, M., Zhang, X., Yuen, P.C., Chang, S.-F.: Unsupervised embedding learning via invariant and spreading instance feature. In: Proceedings of the IEEE/CVF Conference on Computer Vision and Pattern Recognition, pp. 6210–6219 (2019)
49. Zareian, A., Rosa, K.D., Hu, D.H., Chang, S.-F.: Open-vocabulary object detection using captions. In: Proceedings of the IEEE/CVF Conference on Computer Vision and Pattern Recognition, pp. 14393–14402 (2021)
50. Zhao, H., Shi, J., Qi, X., Wang, X., Jia, J.: Pyramid scene parsing network. In: Proceedings of the IEEE Conference on Computer Vision and Pattern Recognition, pp. 2881–2890 (2017)

Few-Shot Classification with Contrastive Learning

Zhanyuan Yang[1] , Jinghua Wang[2] , and Yingying Zhu[1(✉)]

[1] College of Computer Science and Software Engineering, Shenzhen University,
Shenzhen, China
yangzhanyuan2019@email.szu.edu.cn, zhuyy@szu.edu.cn
[2] School of Computer Science and Technology, Harbin Institute of Technology
(Shenzhen), Shenzhen, China
wangjinghua@hit.edu.cn

Abstract. A two-stage training paradigm consisting of sequential pre-training and meta-training stages has been widely used in current few-shot learning (FSL) research. Many of these methods use self-supervised learning and contrastive learning to achieve new state-of-the-art results. However, the potential of contrastive learning in both stages of FSL training paradigm is still not fully exploited. In this paper, we propose a novel contrastive learning-based framework that seamlessly integrates contrastive learning into both stages to improve the performance of few-shot classification. In the pre-training stage, we propose a self-supervised contrastive loss in the forms of feature vector *vs.* feature map and feature map *vs.* feature map, which uses global and local information to learn good initial representations. In the meta-training stage, we propose a cross-view episodic training mechanism to perform the nearest centroid classification on two different views of the same episode and adopt a distance-scaled contrastive loss based on them. These two strategies force the model to overcome the bias between views and promote the transferability of representations. Extensive experiments on three benchmark datasets demonstrate that our method achieves competitive results.

Keywords: Few-shot learning · Meta learning · Contrastive learning · Cross-view episodic training

1 Introduction

Thanks to the availability of a large amount of annotated data, deep convolutional neural networks (CNN) [15,21,39] yield impressive results on various visual recognition tasks. However, the time-consuming and costly collection process makes it a challenge for these deep learning-based methods to generalize

Supplementary Information The online version contains supplementary material available at https://doi.org/10.1007/978-3-031-20044-1_17.

in real-life scenarios with scarce annotated data. Inspired by the capability of human to learn new concepts from a few examples, few-shot learning (FSL) is considered as a promising alternative to meet the challenge, as it can adapt knowledge learned from a few samples of base classes to novel tasks.

Fig. 1. Distribution of feature embeddings of 64 base (*left*) and 20 novel (*right*) classes from miniImagenet in pre-train space by t-SNE [27].

Recently, popular FSL methods [10,28,31,36,40,42,47] mainly adopt the meta-learning strategy. These meta-learning based methods typically take episodic training mechanism to perform meta-training on base classes with abundant data. During meta-training, the episodes consist of a support set and a query set, which are used in few-shot classification to mimic the evaluation setting. The learned model is expected to be capable of generalizing across novel tasks of FSL. Besides, many other methods [5,25,45,48,49,54] achieve good classification accuracy by pre-training the feature extractor on base classes. These methods suggest that the transferable and discriminative representations learned through pre-training or meta-training is crucial for few-shot classification.

However, both the pre-training and meta-training procedures only minimize the standard cross-entropy (CE) loss with labels from base classes. The resulting models are optimized to solve the classification tasks of base classes. Due to this, these methods may discard the information that might benefit the classification tasks on the unseen classes. Figure 1 shows that the pre-trained model is able to identify samples from the base classes (left) well but performs poorly on samples from the novel classes (right). That is, the learned representations are somewhat overfitted on the base classes and not generalizable on the novel classes. Owing to the label-free nature of self-supervised learning methods, some recent works [7,11,41] have tried self-supervised pretext tasks to solve the FSL problem, while other works [9,25,26,32] focus on contrastive learning methods. Though promising, these approaches ignore the additional information from the self-supervised pretext tasks in meta-training or treat them as auxiliary losses simply in the FSL training paradigm.

In this work, we propose a contrastive learning-based framework that seamlessly integrates contrastive learning into the pre-training and meta-training stages to tackle the FSL problem. First, in the pre-training stage, we propose two types of contrastive losses based on self-supervised and supervised signals, respectively, to train the model. These losses consider the global and local information simultaneously. Our proposed self-supervised contrastive loss exploits local information in the forms of both feature vector *vs.* feature map (vector-map) and feature map *vs.* feature map (map-map), which differs from previous methods. Our supervised contrastive loss makes good use of the correlations among individual instances and the correlations among different instances of the same category. Second, in the meta-training stage, motivated by the idea of maximizing mutual information between features extracted from multiple views (*e.g.*, by applying different data augmentations on images) of the shared context (*e.g.*, original images) [1,16,30,50], we introduce a cross-view episodic training (CVET) mechanism to extract generalizable representations. Concretely, we randomly employ two different data augmentation strategies [3,14,54] to obtain the augmented episodes and treat them as different views of the original one. Note, the augmentation does not change the label of the data. We then conduct the nearest centroid classification between the augmented episodes to force the model to overcome the bias between views and generalize well to novel classes. As a complement to CVET, we take inter-instance distance scaling into consideration and perform query instance discrimination within the augmented episodes. These two methods effectively apply contrastive learning to the meta-training stage of FSL. Our proposed method learns meta-knowledge that can play a crucial role in recognizing novel classes. The key contributions of this work are as follows:

- We propose a contrastive learning-based FSL framework consisting of the pre-training and meta-training stages to improve the few-shot image classification. Our framework is easy to combine with other two-stage FSL methods.
- We adopt the self-supervised contrastive loss based on global and local information in the pre-training stage to enhance the generalizability of the resulting representations.
- We propose a CVET mechanism to force the model to find more transferable representations by executing classification between augmented episodes. Meanwhile, we introduce a distance-scaled contrastive loss based on the augmented episodes to ensure that the classification procedure is not affected by extreme bias between different views.
- Extensive experiments of few-shot classification on three benchmarks show that our proposed method achieves competitive results.

2   Related Work

2.1   Few-Shot Learning

FSL aims to learn patterns on a large number of labeled examples called base classes and adapt to novel classes with limited examples per class. Few-shot image classification has received great attention and many methods have been proposed. The existing methods can be broadly divided into two categories: *optimization-based* and *metric-based*. The *optimization-based* methods initialize the model on base classes and adapt to novel tasks efficiently within a few gradient update steps on a few labeled samples [10,24,28,33,36]. The *metric-based* methods aim to learn a generalizable representation space and use a well-defined metric to classify them [23,31,40,42,47,51,55]. The existing works have considered different metrics such as cosine similarity [47], Euclidean distance [40], a CNN-based relation module [42], a task-adaptive metric [31], a local descriptor based metric [23] and graph neural networks [37]. The Earth Mover's Distance [55] is employed as a metric to learn more discriminative structured representations. Many recent studies [5,38,45,48] have proposed a standard end-to-end pre-training framework to obtain feature extractors or classifiers on base classes. These pre-training based methods achieve competitive performance compared to episodic meta-training methods. Moreover, many papers [6,25,54,55] take advantage of a sequential combination of pre-training and meta-training stages to further enhance the performance. The methods [17,42,49,52,54] pay more attention to the transferability of representations through delicately designing task-specific modules in meta-training. Given the simplicity and effectiveness of these methods, we take FEAT [54] as our baseline, but drop its auxiliary loss.

2.2 Contrastive Learning

Recently, contrastive learning with the instance discrimination as a pretext task has become a dominant approach in self-supervised representation learning [1,3, 14,16,30,43,50]. These methods typically construct contrast pairs of instances with a variety of data augmentations and optimize a contrastive loss with the aim of keeping instances close to their augmented counterparts while staying away from other instances in the embedding space. The goal of contrastive learning using self-supervision from instances is to improve the generalizability of the representations and benefit various downstream tasks. Contrastive learning is also extended to group instances in a supervised manner [19] and achieves better performance than CE loss on standard classification tasks.

2.3 Few-Shot Learning with Contrastive Learning

In contrast to the works [7,11,41] that introduce self-supervised pretext tasks such as rotation [13] and jigsaw [29] into FSL as auxiliary losses, recent approaches [9,25,26,32] have explored contrastive learning of instance discrimination in different parts of the two-stage training pipeline of FSL. Methods [25,26,32] combine supervised contrastive loss [19] to the pre-training stage [26,32] and the meta-training stage [9,25], respectively. Unlike prior works, our proposed method boosts few-shot classification performance by seamlessly integrating instance-discriminative contrastive learning in both the pre-training and meta-training stages. In the pre-training stage, we conduct self-supervised contrastive loss in the forms of vector-map and map-map. In the meta-training stage, we combine contrastive learning with episodic training and define a distance-scaled contrastive loss to improve the transferability of the representations.

3   Method

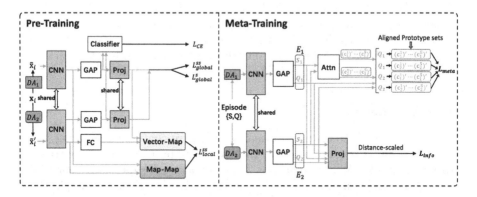

Fig. 2. Overview of our framework. Based on multiple views of an input through two random data augmentations DA_1 and DA_2, we compute contrastive losses at both global and local levels in the pre-training stage. In the meta-training stage, we enforce cross-view episodic training and compute a distance-scaled contrastive loss episodically. Here, GAP dentoes a global average pooling layer, Proj is a projection head, FC means a fully contected layer, Attn is the task-specific module from [54], $\mathcal{S}_r$, $\mathcal{Q}_r$ mean support and query set from different views of the episode $E = \{\mathcal{S}, \mathcal{Q}\}$ respectively, and $(c_r^k)'$ dentoes the aligned prototype in $\mathcal{S}_r$.

3.1   Preliminary

The few-shot classification task is slightly different from the standard supervised classification task. The meta-training set $\mathcal{D}_{train} = \{(x_i, y_i) \mid y_i \in \mathcal{C}_{base}\}$ consists of the samples from the base classes $\mathcal{C}_{base}$, and the meta-test set $\mathcal{D}_{test} = \{(x_i, y_i) \mid y_i \in \mathcal{C}_{novel}\}$ consists of the samples from the novel classes $\mathcal{C}_{novel}$. Here, y_i is the class label of sample x_i. In FSL, we aim to learn a model based on $\mathcal{D}_{train}$ and generalize it over $\mathcal{D}_{test}$, where $\mathcal{C}_{base} \cap \mathcal{C}_{novel} = \emptyset$. Following the prior meta-learning based methods [10,40,47], we adopt episodic mechanism to simulate the evaluation setting. Concretely, each $M-$way $K-$shot episode E consists of a support set and a query set. We first randomly sample M classes from $\mathcal{C}_{base}$ for meta-training (or from $\mathcal{C}_{novel}$ for meta-testing) and K instances per class to obtain the support set $\mathcal{S} = \{x_i, y_i\}_{i=1}^{M*K}$. Then, we sample Q instances in each of the selected classes to obtain the query set $\mathcal{Q} = \{x_i, y_i\}_{i=1}^{M*Q}$. Note that $y_i \in \{1, 2, \ldots, M\}$ and $\mathcal{S} \cap \mathcal{Q} = \emptyset$. The episodic training procedure classifies the samples in $\mathcal{Q}$ into the categories corresponding to the samples in $\mathcal{S}$.

3.2   Overview

In this work, we follow the two-stage training strategy and incorporate contrastive learning in both stages to learn more generalizable representations. Our proposed framework is illustrated in Fig. 2. In the pre-training stage, we adopt self-supervised and supervised contrastive losses to obtain a good initial representation. In the meta-training stage, we propose a novel cross-view episodic training (CVET) mechanism and a distance-scaled contrastive loss, which allows the model to overcome the bias between views of each episode and generalize well across novel tasks. Note that we take FEAT [54] without its auxiliary loss as the baseline due to its simple and effective task-specific module (a multi-head attention module [46]). We will detail our framework in the following subsections.

3.3   Pre-training

In this section, we introduce instance-discriminative contrastive learning [3,19] in the pre-training stage to alleviate the overfitting problem caused by training with CE loss only. As shown in Fig. 2, we propose self-supervised contrastive losses at the global and local levels, respectively. Using self-supervision in these losses helps produce more generalizable representations. Meanwhile, we also employ a global supervised contrastive loss [19] to capture the correlations among instances from the same category.

Global Self-supervised Contrastive Loss. This loss (*a.k.a* InfoNCE loss [3, 30]) aims to enhance the similarity between the views of the same image, while reducing the similarity between the views of different images. Formally, we randomly apply two data augmentation methods to a batch of samples $\{x_i, y_i\}_{i=1}^{N}$ from the meta-training set $\mathcal{D}_{train}$ and generate the augmented batch $\{\tilde{x}_i, \tilde{y}_i\}_{i=1}^{2N}$. Here, $\tilde{x}_i$ and $\tilde{x}'_i$ denote two different views of x_i, which are considered as a positive pair. We define f_ϕ as the feature extractor with learnable parameters ϕ to transform the sample $\tilde{x}_i$ into a feature map $\hat{x}_i = f_\phi(\tilde{x}_i) \in \mathbb{R}^{C \times H \times W}$ and further obtain the global feature $h_i \in \mathbb{R}^C$ after a global average pooling (GAP) layer. We use a MLP with one hidden layer to instantiate a projection head $proj(\cdot)$ [3] to generate the projected vector $z_i = proj(h_i) \in \mathbb{R}^D$. Then the global self-supervised contrastive loss can be computed as:

$$L_{global}^{ss} = -\sum_{i=1}^{2N} \log \frac{\exp\left(z_i \cdot z'_i / \tau_1\right)}{\sum_{j=1}^{2N} \mathbb{1}_{j \neq i} \exp\left(z_i \cdot z_j / \tau_1\right)}, \tag{1}$$

where the $\cdot$ operation denotes inner product after l_2 normalization, τ_1 is a scalar temperature parameter, and $\mathbb{1} \in \{0, 1\}$ is an indicator function. Here, the positive pair, z'_i and z_i, are extracted from the augmented versions of the same sample x_i.

Local Self-supervised Contrastive Loss. Though L_{global}^{ss} (Eq. (1)) favors transferable representations based on global feature vector h_i, it might ignore some local discriminative information in feature map $\hat{x}_i$ which could be beneficial in meta-testing. Inspired by [1,3,16,32], we compute self-supervised contrastive loss at the local level. Unlike previous approaches, we leverage map-map and

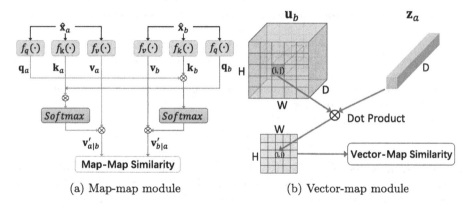

Fig. 3. Map-map and vector-map modules. (a) Operation $\otimes$ denotes dot product. Local feature maps $\hat{\mathbf{x}}_a$ and $\hat{\mathbf{x}}_b$ share three spatial projection heads $f_q(\cdot)$, $f_k(\cdot)$ and $f_v(\cdot)$. We first align $\hat{\mathbf{x}}_a$ with $\mathbf{q}_b$ and then align $\hat{\mathbf{x}}_b$ with $\mathbf{q}_a$. (b) We obtain $\mathbf{u}_b$ by adding a FC layer after $\hat{\mathbf{x}}_b$ and $\mathbf{z}_a$ denotes the projected vector from $proj(\hat{\mathbf{x}}_a)$.

vector-map modules to boost the robustness and generalizability of the representations. The **map-map module** is illustrated in Fig. 3a. Specifically, we use three f_q, f_k, f_v spatial projection heads to project local feature map $\hat{\mathbf{x}}_i$ into the query $\mathbf{q}_i = f_q(\hat{\mathbf{x}}_i)$, key $\mathbf{k}_i = f_k(\hat{\mathbf{x}}_i)$ and value $\mathbf{v}_i = f_v(\hat{\mathbf{x}}_i)$, respectively, where $\mathbf{q}_i, \mathbf{k}_i, \mathbf{v}_i \in \mathbb{R}^{HW \times D}$. For a pair of local feature map $\hat{\mathbf{x}}_a$ and $\hat{\mathbf{x}}_b$, we align the $\hat{\mathbf{x}}_a$ with $\hat{\mathbf{x}}_b$ to obtain $\mathbf{v}'_{a|b} = softmax\left(\frac{\mathbf{q}_b \mathbf{k}_a^\top}{\sqrt{d}}\right) \mathbf{v}_a$, and align $\hat{\mathbf{x}}_b$ with $\hat{\mathbf{x}}_a$ to obtain $\mathbf{v}'_{b|a} = softmax\left(\frac{\mathbf{q}_a \mathbf{k}_b^\top}{\sqrt{d}}\right) \mathbf{v}_b$. After l_2 normalization on each position (i, j) of the aligned results, we can compute the similarity between the two local feature maps $\hat{\mathbf{x}}_a$ and $\hat{\mathbf{x}}_b$ as follows:

$$sim_1\left(\hat{\mathbf{x}}_a, \hat{\mathbf{x}}_b\right) = \frac{1}{HW} \sum_{1 \leq i \leq H, 1 \leq j \leq W} \left(\mathbf{v}'_{a|b}\right)_{ij}^\top \left(\mathbf{v}'_{b|a}\right)_{ij}. \qquad (2)$$

Basically, Eq. (2) calculates the summation of the element-wise product of two feature maps $\mathbf{v}'_{a|b}$ and $\mathbf{v}'_{b|a} \in \mathbb{R}^{HW \times D}$. The self-supervised contrastive loss based on pairwise feature maps can be computed as follows:

$$L_{map-map}^{ss} = -\sum_{i=1}^{2N} \log \frac{\exp\left(sim_1\left(\hat{\mathbf{x}}_i, \hat{\mathbf{x}}_i'\right)/\tau_2\right)}{\sum_{j=1}^{2N} \mathbb{1}_{j \neq i} \exp\left(sim_1\left(\hat{\mathbf{x}}_i, \hat{\mathbf{x}}_j\right)/\tau_2\right)}, \qquad (3)$$

where $(\hat{\mathbf{x}}_i, \hat{\mathbf{x}}_i')$ is a positive pair and τ_2 denotes a temperature parameter, and $\mathbb{1}$ is an indicator function. Meanwhile, we adopt **vector-map module** to further exploit the local contrastive information between instances, which is shown in Fig. 3b. In specific, we use a fully connected (FC) layer to obtain $\mathbf{u}_i = g(\hat{\mathbf{x}}_i) = \sigma(\mathbf{W}\hat{\mathbf{x}}_i) \in \mathbb{R}^{D \times HW}$, where σ is a ReLU nonlinearity. We can compute the similarity between a contrast pair as $sim_2\left(\hat{\mathbf{x}}_a, \hat{\mathbf{x}}_b\right) = \frac{1}{HW} \sum_{1 \leq i \leq H, 1 \leq j \leq W} (\mathbf{u}_b)_{ij}^\top \mathbf{z}_a$, where $\mathbf{z}_a$ is the projected vector of $\hat{\mathbf{x}}_a$. The self-supervised contrastive loss based on pairs of feature vectors and feature maps can be computed as follows:

$$L_{vec-map}^{ss} = -\sum_{i=1}^{2N} \log \frac{\exp\left(sim_2\left(\hat{\mathbf{x}}_i, \hat{\mathbf{x}}_i'\right)/\tau_3\right)}{\sum_{j=1}^{2N} \mathbb{1}_{j\neq i} \exp\left(sim_2\left(\hat{\mathbf{x}}_i, \hat{\mathbf{x}}_j\right)/\tau_3\right)}, \tag{4}$$

where τ_3 and $\mathbb{1}$ act the same as in Eq. (3). Therefore, the local self-supervised contrastive loss can be defined as:

$$L_{local}^{ss} = L_{vec-map}^{ss} + L_{map-map}^{ss}. \tag{5}$$

Global Supervised Contrastive Loss. To exploit the correlations among individual instances and the correlations among different instances from the same category, we also adopt supervised contrastive loss [19] as follows:

$$L_{global}^{s} = \sum_{i=1}^{2N} \frac{1}{|P(i)|} \sum_{p \in P(i)} L_{ip}, \tag{6}$$

where $L_{ip} = -\log \frac{\exp(\mathbf{z}_i \cdot \mathbf{z}_p/\tau_4)}{\sum_{j=1}^{2N} \mathbb{1}_{j\neq i} \exp(\mathbf{z}_i \cdot \mathbf{z}_j/\tau_4)}$, and τ_4 is a temperature parameter. The set $P(i)$ contains the indexes of samples with the same label as x_i in the augmented batch except for index i.

To summarize, we minimize the following loss during pre-training:

$$L_{pre} = L_{CE} + \alpha_1 L_{global}^{ss} + \alpha_2 L_{local}^{ss} + \alpha_3 L_{global}^{s}, \tag{7}$$

where L_{CE} is the CE loss, and α_1, α_2 and α_3 are balance scalars. By optimizing L_{pre}, we promote the discriminability and generalizability of the representations, which is crucial for the following stage.

3.4   Meta-training

Cross-View Episodic Training. In order to capture information about the high-level concept of a shared context, a common practice in contrastive learning [3,14,30,50] is to maximize the mutual information between features extracted from multiple views of the shared context. Intuitively, given an episode $E = \{\mathcal{S}, \mathcal{Q}\}$ in episodic meta-learning, we can obtain two episodes $E_1 = \{\mathcal{S}_1, \mathcal{Q}_1\}$ and $E_2 = \{\mathcal{S}_2, \mathcal{Q}_2\}$ by applying two different data augmentation strategies on E respectively. Here, we consider E a shared context and treat the two augmented episodes as its two views. Inspired by the above idea, we propose a cross-view episodic training mechanism with the aim of forcing representations to learn meta-knowledge that can play a key role across various few-shot classification tasks. Specifically, we use the pre-trained feature extractor f_ϕ in Sect. 3.3 followed by a GAP to each data point x_i in both the episode E_1 and E_2, and derive the corresponding global vector $\mathbf{h}_i \in \mathbb{R}^C$. We then separately compute a prototype $\mathbf{c}_r^k$ for category k in support sets $\mathcal{S}_1$ and $\mathcal{S}_2$ as follows:

$$\mathbf{c}_r^k = \frac{1}{|\mathcal{S}_r^k|} \sum_{(\mathbf{h}_i, y_i) \in \mathcal{S}_r^k} \mathbf{h}_i, \tag{8}$$

where $r \in \{1, 2\}$ and $\mathcal{S}_r^k$ denotes the set of data points belonging to class $k \in \{1, 2, \ldots, M\}$ from $r-th$ support set. We denote the task-specific module (multi-head attention [46]) proposed in baseline [54] as $Attn(\cdot)$ and fix the number of heads to 1. With the task-specific module, we obtain an aligned prototype set $\mathcal{T}(r) = \{(\mathbf{c}_r^k)'\}_{k=1}^M = Attn(\mathbf{c}_r^1, \mathbf{c}_r^2, \ldots, \mathbf{c}_r^M)$ for each support set. Then, based on the aligned prototypes, a probability distribution of a data point x_i in the query set $\mathcal{Q}_r$ over M classes is defined as:

$$P(y = k \mid \mathbf{h}_i, \mathcal{T}(r)) = \frac{\exp\left(-d\left(\mathbf{h}_i, (\mathbf{c}_r^k)'\right)\right)}{\sum_{j=1}^M \exp\left(-d\left(\mathbf{h}_i, (\mathbf{c}_r^j)'\right)\right)}, \tag{9}$$

where $d(\cdot, \cdot)$ denotes Euclidean distance. Therefore the loss of the nearest centroid classifier on an episode can be computed as:

$$L_{mn} = \frac{1}{|\mathcal{Q}_m|} \sum_{(\mathbf{h}_i, y_i) \in \mathcal{Q}_m} -\log P(y = k \mid \mathbf{h}_i, \mathcal{T}(n)), \tag{10}$$

where $m, n \in \{1, 2\}$. Obviously different from the original episodic training, we classify $\mathcal{Q}_1$ on $\mathcal{S}_1$ and $\mathcal{S}_2$ respectively, and do the same for $\mathcal{Q}_2$. Equation(10) minimizes the differences between two views of instances of the same category. The whole process is illustrated in Fig. 2. Therefore, we computed the cross-view classification loss as follows:

$$L_{meta} = \frac{1}{4} \sum_{m,n} L_{mn}. \tag{11}$$

Distance-Scaled Contrastive Loss. Since contrastive learning approaches work solely at the instance level, it is superficial to simply add contrastive loss into meta-training without taking full advantage of the episodic training mechanism catered for FSL. To better apply contrastive learning to the meta-training, we perform query instance discrimination between two views of the shared episodes. Specifically, inspired by [4], we further inherit the pre-trained projection head $proj(\cdot)$ in Sect. 3.3 to map each sample x_i in the two episodes E_1 and E_2 into a projected vector $\mathbf{z}_i \in \mathbb{R}^D$. We similarly obtain the prototype $\mathbf{o}_r^k$ by averaging the projected vector of the same classes, that is $\mathbf{o}_r^k = \frac{1}{|\mathcal{S}_r^k|} \sum_{(\mathbf{z}_i, y_i) \in \mathcal{S}_r^k} \mathbf{z}_i$ where r is the same as in Eq. (8). For each query vector $\mathbf{z}_i$ in $\mathcal{Q}_1$ and $\mathcal{Q}_2$, we reconstruct its positive sample set by using corresponding augmented version $\mathbf{z}_i'$ and the samples of class y_i in both support sets. Therefore, we reformulate the supervised contrastive loss [19] in the form of episodic training as follows:

$$L(\mathbf{z}_i) = -\sum_{\mathbf{z}_H \in H(\mathbf{z}_i)} \log \frac{\lambda_{\mathbf{z}_i \mathbf{z}_H} \exp\left(\mathbf{z}_i \cdot \mathbf{z}_H / \tau_5\right)}{\sum_{\mathbf{z}_A \in A(\mathbf{z}_i)} \lambda_{\mathbf{z}_i \mathbf{z}_A} \exp\left(\mathbf{z}_i \cdot \mathbf{z}_A / \tau_5\right)}. \tag{12}$$

Here, operation $\cdot$ means the inner product between features after l_2 normalization. τ_5 is a temperature parameter. The $\lambda_{\mathbf{z}_i \mathbf{z}_j} = 2 - dist(\mathbf{z}_i, \mathbf{z}_j)$ is the coefficient that reflects the distance relationship between $\mathbf{z}_i$ and $\mathbf{z}_j$, where $dist(\cdot, \cdot)$ refers to cosine similarity. $H(\mathbf{z}_i) = \{\mathbf{z}_i'\} \cup \mathcal{S}_1^{y_i} \cup \mathcal{S}_2^{y_i}$ is the positive set of $\mathbf{z}_i$, and $A(\mathbf{z}_i) = \{\mathbf{z}_i'\} \cup \mathcal{S}_1 \cup \mathcal{S}_2 \cup \{\mathbf{o}_1^k\}_{k=1}^M \cup \{\mathbf{o}_2^k\}_{k=1}^M$. The distance coefficient $\lambda_{\mathbf{z}_i \mathbf{z}_j}$ and additional prototypes $\mathbf{o}_r^k$ on both views of the original episode are introduced to reduce the similarities of queries to their positives. Thus the model can learn more discriminative representations adapted to different tasks. Then we compute the distance-scaled contrastive loss as:

$$L_{info} = \sum_{\mathbf{z}_i \in \mathcal{Q}_1 \cup \mathcal{Q}_2} \frac{1}{|H(\mathbf{z}_i)|} L(\mathbf{z}_i). \tag{13}$$

By optimizing L_{info}, we force the representations to capture information about instance discrimination episodically and learn the interrelationships among samples of the same category in cross-view episodes.

Objective in Meta-training. In meta-training, we mainly build two losses as in Eq. (11) and Eq. (13) in order to enhance the transferability and discriminative ability of representations. Then the total objective function in meta-training is defined as:

$$L_{total} = L_{meta} + \beta L_{info}, \tag{14}$$

where β is a balance scalar and will be detailed in the following section.

4   Experiments

4.1   Datasets and Setup

Datasets. We evaluate our method on three popular benchmark datasets. The *miniImageNet* dataset [47] contains 100 classes with 600 images per class, and these classes are divided into 64, 16 and 20 for the training, validation and test sets, respectively. The *tieredImageNet* dataset [35] contains 608 classes grouped in 34 high-level categories with 779,165 images, where 351 classes are used for training, 97 for validation and 160 for testing. The *CIFAR-FS* dataset [2] contains 100 classes with 600 images per class. These classes are split into the training, validation and test sets in proportions of 64, 16 and 20.

Implementation Details. Following the baseline [54], we use ResNet-12 as backbone and a multihead attention as the task-specific module (number of heads is 1). All projection heads in our method has the same structure as in [3]. We adopt SGD optimizer with a weight decay of 5e-4 and a momentum of 0.9 for both the pre-training and meta-training stages. During pre-training, the learning rate is initialized to be 0.1 and adapted via cosine learning rate scheduler after warming up. Temperature parameters $\tau_{1,2,3,4}$ are set to 0.1 and the balance scalars $\alpha_{1,2,3}$ are set to 1.0. During meta-training, temperature τ_5 are set to 0.1. We use StepLR with a step size of 40 and gamma of 0.5 and set $\beta = 0.01$ for 1-shot. For 5-shot, StepLR is used with a step size of 50 and gamma of 0.5, and the β is set to 0.1.

Evaluation. We follow the 5-way 1-shot and 5-way 5-shot few-shot classification tasks. In meta-testing, our method (inductive) simply classify the query samples by computing the euclidean distance between the prototypes and query samples. We randomly sample 2000 episodes from test set in meta-testing with 15 query images per class and report the mean accuracy together with corresponding 95% confidence interval.

Data Augmentation. For all datasets, we empirically find that standard [54,55] and SimCLR-style [3,14] data augmentation strategies work best in pre-training and meta-training. During meta-testing, no data augmentation strategy is used. The image transformations used in standard strategy include randomresizedcrop, colorJitter and randomhorizontalflip, while SimCLR-style strategy contains randomresizedcrop, randomhorizontalflip, randomcolorjitter and randomgrayscale.

Table 1. The average 5-way few-shot classification accuracies(%) with 95% confidence interval on miniImageNet and tieredImageNet.

Method	Backbone	miniImageNet		tieredImageNet	
		1-shot	5-shot	1-shot	5-shot
MAML [10]	CONV-4	48.70 ± 1.75	63.11 ± 0.92	–	–
RelationNets [42]		50.44 ± 0.82	65.32 ± 0.70	54.48 ± 0.93	71.32 ± 0.78
MatchingNets [47]		48.14 ± 0.78	63.48 ± 0.66	–	–
ProtoNets [40]		44.42 ± 0.84	64.24 ± 0.72	53.31 ± 0.89	72.69 ± 0.74
LEO [36]	WRN-28-10	61.76 ± 0.08	77.59 ± 0.12	66.33 ± 0.05	82.06 ± 0.08
CC+rot [11]		62.93 ± 0.45	79.87 ± 0.33	62.93 ± 0.45	79.87 ± 0.33
wDAE [12]		61.07 ± 0.15	76.75 ± 0.11	68.18 ± 0.16	83.09 ± 0.12
PSST [7]		64.16 ± 0.44	80.64 ± 0.32	–	–
TADAM [31]	ResNet-12	58.5 ± 0.3	76.7 ± 0.3	–	–
MetaOptNet [22]		62.64 ± 0.61	78.63 ± 0.46	65.99 ± 0.72	81.56 ± 0.53
DeepEMD [55]		65.91 ± 0.82	82.41 ± 0.56	71.16 ± 0.87	86.03 ± 0.58
CAN [17]		63.85 ± 0.48	79.44 ± 0.34	69.89 ± 0.51	84.23 ± 0.37
FEAT [54]		66.78 ± 0.20	82.05 ± 0.14	70.80 ± 0.23	84.79 ± 0.16
RFS [45]		62.02 ± 0.63	79.64 ± 0.44	69.74 ± 0.72	84.41 ± 0.55
InfoPatch [25]		67.67 ± 0.45	82.44 ± 0.31	71.51 ± 0.52	85.44 ± 0.35
DMF [52]		67.76 ± 0.46	82.71 ± 0.31	71.89 ± 0.52	85.96 ± 0.35
RENet [18]		67.60 ± 0.44	82.58 ± 0.30	71.61 ± 0.51	85.28 ± 0.35
BML [56]		67.04 ± 0.63	83.63 ± 0.29	68.99 ± 0.50	85.49 ± 0.34
PAL [26]		69.37 ± 0.64	84.40 ± 0.44	72.25 ± 0.72	**86.95 ± 0.47**
TPMN [49]		67.64 ± 0.63	83.44 ± 0.43	72.24 ± 0.70	86.55 ± 0.63
Ours	ResNet-12	**70.19 ± 0.46**	**84.66 ± 0.29**	**72.62 ± 0.51**	86.62 ± 0.33

4.2   Main Results

In this subsection, we compare our method with competitors on three mainstream FSL datasets and the results are reported in Table 1 and Table 2. We can observe that our proposed method consistently achieves competitive results compared to the current state-of-the-art (SOTA) FSL methods on both the 5-way 1-shot and 5-way 5-shot tasks. For the miniImageNet dataset (Table 1), our proposed method outperforms the current best results by 0.82% in the 1-shot task and 0.26% in 5-shot task. For the tieredImageNet dataset (Table 1), our method improves over the current SOTA method by 0.37% for 1-shot and achieves the second best 5-shot result. Note that our method outperforms the original FEAT by 1.82% for 1-shot and 1.83% for 5-shot on tieredImageNet. For the CIFAR-FS dataset (Table 2), our method surpasses the current SOTA by 0.46% and 0.19% in the 1-shot and 5-shot tasks, respectively. Compared to those methods [7,11,25,26], our proposed method works better on most datasets. The consistent and competitive results on the three datasets indicate that our method can learn more transferable representations by incorporating contrastive learning in both the pre-training and meta-training stages.

Table 2. The average 5-way few-shot classification accuracies(%) with 95% confidence interval on CIFAR-FS. * results used our implementation.

Method	Backbone	1-shot	5-shot
Ravichandran *et al.* [34]	CONV-4	55.14 ± 0.48	71.66 ± 0.39
ConstellationNet [53]		69.3 ± 0.3	82.7 ± 0.2
CC+rot [11]	WRN-28-10	75.38 ± 0.31	87.25 ± 0.21
PSST [7]		77.02 ± 0.38	88.45 ± 0.35
Ravichandran *et al.* [34]	ResNet-12	69.15 ± -	84.7 ± -
MetaOptNet [22]		72.0 ± 0.7	84.2 ± 0.5
Kim *et al.* [20]		73.51 ± 0.92	85.65 ± 0.65
FEAT* [54]		75.41 ± 0.21	87.32 ± 0.15
RFS [45]		73.9 ± 0.8	86.9 ± 0.5
ConstellationNet [53]		75.4 ± 0.2	86.8 ± 0.2
RENet [18]		74.51 ± 0.46	86.60 ± 0.32
BML [56]		73.45 ± 0.47	88.04 ± 0.33
PAL [26]		77.1 ± 0.7	88.0 ± 0.5
TPMN [49]		75.5 ± 0.9	87.2 ± 0.6
Ours	ResNet-12	**77.56 ± 0.47**	**88.64 ± 0.31**

4.3   Ablation Study

In this subsection, we study the effectiveness of different components in our method on three datasets. The results in Table 3 show a significant improvement in the performance of our proposed method compared to the baseline pre-trained by CE loss (L_{CE}) only. Specifically, using L_{CE} and all contrastive losses L_{CL} in pre-training improves the accuracy by an average of 1.66% (1-shot) and 1.29% (5-shot) on the three datasets. This allows the representations to generalize better, rather than just focusing on the information only needed for the classification on base classes. We then apply the CVET and L_{info} respectively, and the consistent improvements suggest that CVET and L_{info} are effective. The results jointly based on CVET and L_{info} are further enhanced by 0.95% (1-shot) and 0.72% (5-shot) in average, indicating that our full method increases the transferability of representations on novel classes. Note that the gains obtained by CVET and L_{info} are relatively low on miniImageNet, as the number of training steps on this dataset is very small, leading to a sufficiently narrow gap between two different views of the same image at the end of pre-training. Additionally, more ablation experiments on the effectiveness of all parts in L_{CL} are included in the supplementary material.

Table 3. Ablation experiments on miniImageNet. L_{CE} means the baseline model is pre-trained with CE loss. L_{CL} denotes all contrastive losses used in pre-training. CVET and L_{info} are only available in meta-training.

L_{CE}	L_{CL}	CVET	L_{info}	miniImageNet		tieredImageNet		CIFAR-FS	
				1-shot	5-shot	1-shot	5-shot	1-shot	5-shot
√				66.58 ± 0.46	81.92 ± 0.31	70.41 ± 0.51	84.69 ± 0.36	75.54 ± 0.48	87.28 ± 0.32
√	√			69.53 ± 0.47	84.33 ± 0.29	71.83 ± 0.51	85.64 ± 0.35	76.15 ± 0.47	87.79 ± 0.33
√	√	√		69.89 ± 0.46	84.43 ± 0.29	72.39 ± 0.52	86.07 ± 0.35	77.02 ± 0.48	88.28 ± 0.32
√	√		√	69.78 ± 0.46	84.46 ± 0.29	72.51 ± 0.52	86.23 ± 0.34	77.37 ± 0.49	88.48 ± 0.32
√	√	√	√	70.19 ± 0.46	84.66 ± 0.29	72.62 ± 0.51	86.62 ± 0.33	77.56 ± 0.47	88.64 ± 0.31

Table 4. Comparison experiments of different data augmentation strategies on miniImageNet.

Augmentation	1-shot	5-shot
Standard [54,55]	67.84 ± 0.47	82.82 ± 0.31
SimCLR-style [3,14]	70.19 ± 0.46	84.66 ± 0.29
AutoAugment [8]	68.45 ± 0.46	83.94 ± 0.30
StackedRandAug [44]	68.31 ± 0.46	83.13 ± 0.31

4.4   Further Analysis

Data Augmentation. We investigate the impact of employing a different data augmentation strategy while maintaining the use of the standard one. Table 4 shows that the SimCLR-style [3,14] strategy works best while the AutoAugment [8] and StackedRandAug [44] perform slightly worse. However, all three strategies are better than the standard one. We believe that the standard strategy lacks effective random image transformations and thus always produces two similar views of the same image during the training process. Conversely, the AutoAugment and StackedRandAug strategies that intensely change the original image may lead to excessive bias between two views of the same image, thereby increasing the difficulty of performing contrastive learning.

Visualization. We give visualization to validate the transferability of representations produced by our framework on novel classes. Specifically, we randomly sample 5 classes from the meta-test set of miniImageNet with 100 images per class and obtain embeddings of all images using ProtoNet [40], ProtoNet+Ours, baseline and Ours (baseline + our framework), respectively. Then we use t-SNE [27] to project these embeddings into 2-dimensional space as shown in Fig. 4. The distributions of the embeddings obtained by ProtoNet + Ours and baseline + our framework are more separable and the class boundaries more precise and compact (see Fig. 4b, Fig. 4d *vs.* Fig. 4a, Fig. 4c). The visualization results indicate that our proposed framework generates more transferable and discriminative representations on novel classes. The complete visualization results are available in the supplementary material.

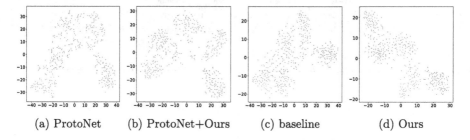

(a) ProtoNet   (b) ProtoNet+Ours   (c) baseline   (d) Ours

Fig. 4. Visualization of 100 randomly sampled images for each of the 5 meta-test classes from miniImageNet by t-SNE [27].

Table 5. Method combinations on miniImageNet.

Model	1-shot	5-shot
ProtoNet	63.77 ± 0.47	80.58 ± 0.32
ProtoNet+CL	66.17 ± 0.46	81.73 ± 0.30
ProtoNet+Ours	66.51 ± 0.47	81.97 ± 0.30
baseline	66.23 ± 0.47	80.81 ± 0.33
baseline+CL	69.53 ± 0.47	84.33 ± 0.29
Ours	70.19 ± 0.46	84.66 ± 0.29

Method Combination. We combine our proposed framework with the two-stage FSL methods ProtoNet [40] and baseline (FEAT [54]), respectively. The results are shown in Table 5. Integration with our framework improves the accuracy of ProtoNet by 2.74% (1-shot) and 1.39% (5-shot). Similarly, our framework improves the 1-shot accuracy of baseline by 3.96% and its 5-shot accuracy by 3.85%. The results in Fig. 4 and Table 5 show that our framework can be applied to other two-stage FSL methods and effectively enhance the performance.

5 Conclusions

In this paper, we apply contrastive learning to the two-stage training paradigm of FSL to alleviate the limitation on generalizability of representations. Concretely, we use vector-map and map-map modules to incorporate self-supervised contrastive losses in the pre-training stage. We further propose a CVET strategy and a distance-scaled contrastive loss to extend contrastive learning to the meta-training stage effectively. The comprehensive experimental results show that our proposed method achieves competitive performance on three well-known FSL datasets miniImageNet, tieredImageNet and CIFAR-FS.

Acknowledgement. This work are supported by: (i) National Natural Science Foundation of China (Grant No. 62072318 and No. 62172285); (ii) Natural Science Foundation of Guangdong Province of China (Grant No. 2021A1515012014); (iii) Science and Technology Planning Project of Shenzhen Municipality (Grant No. JCYJ20190808172007500).

References

1. Bachman, P., Hjelm, R.D., Buchwalter, W.: Learning representations by maximizing mutual information across views. In: NIPS, pp. 15509–15519 (2019)
2. Bertinetto, L., Henriques, J.F., Torr, P.H.S., Vedaldi, A.: Meta-learning with differentiable closed-form solvers. In: ICLR (2019)
3. Chen, T., Kornblith, S., Norouzi, M., Hinton, G.E.: A simple framework for contrastive learning of visual representations. In: ICML, pp. 1597–1607 (2020)
4. Chen, T., Kornblith, S., Swersky, K., Norouzi, M., Hinton, G.E.: Big self-supervised models are strong semi-supervised learners. In: NIPS, pp. 22243–22255 (2020)
5. Chen, W., Liu, Y., Kira, Z., Wang, Y.F., Huang, J.: A closer look at few-shot classification. In: ICLR (2019)
6. Chen, Y., Liu, Z., Xu, H., Darrell, T., Wang, X.: Meta-baseline: exploring simple meta-learning for few-shot learning. In: ICCV, pp. 9062–9071 (2021)
7. Chen, Z., Ge, J., Zhan, H., Huang, S., Wang, D.: Pareto self-supervised training for few-shot learning. In: CVPR, pp. 13663–13672 (2021)
8. Cubuk, E.D., Zoph, B., Mané, D., Vasudevan, V., Le, Q.V.: Autoaugment: learning augmentation strategies from data. In: CVPR, pp. 113–123 (2019)
9. Doersch, C., Gupta, A., Zisserman, A.: Crosstransformers: spatially-aware few-shot transfer. In: Larochelle, H., Ranzato, M., Hadsell, R., Balcan, M., Lin, H. (eds.) NIPS, pp. 21981–21993 (2020)
10. Finn, C., Abbeel, P., Levine, S.: Model-agnostic meta-learning for fast adaptation of deep networks. In: ICML, pp. 1126–1135 (2017)
11. Gidaris, S., Bursuc, A., Komodakis, N., Pérez, P., Cord, M.: Boosting few-shot visual learning with self-supervision. In: ICCV, pp. 8058–8067 (2019)
12. Gidaris, S., Komodakis, N.: Generating classification weights with GNN denoising autoencoders for few-shot learning. In: CVPR, pp. 21–30 (2019)
13. Gidaris, S., Singh, P., Komodakis, N.: Unsupervised representation learning by predicting image rotations. In: ICLR (2018)
14. He, K., Fan, H., Wu, Y., Xie, S., Girshick, R.B.: Momentum contrast for unsupervised visual representation learning. In: CVPR, pp. 9726–9735 (2020)

15. He, K., Zhang, X., Ren, S., Sun, J.: Deep residual learning for image recognition. In: CVPR, pp. 770–778 (2016)
16. Hjelm, R.D., et al.: Learning deep representations by mutual information estimation and maximization. In: ICLR (2019)
17. Hou, R., Chang, H., Ma, B., Shan, S., Chen, X.: Cross attention network for few-shot classification. In: NIPS, pp. 4005–4016 (2019)
18. Kang, D., Kwon, H., Min, J., Cho, M.: Relational embedding for few-shot classification. In: ICCV, pp. 8822–8833 (2021)
19. Khosla, P., et al.: Supervised contrastive learning. In: NIPS, pp. 18661–18673 (2020)
20. Kim, J., Kim, H., Kim, G.: Model-agnostic boundary-adversarial sampling for test-time generalization in few-shot learning. In: Vedaldi, A., Bischof, H., Brox, T., Frahm, J.-M. (eds.) ECCV 2020. LNCS, vol. 12346, pp. 599–617. Springer, Cham (2020). https://doi.org/10.1007/978-3-030-58452-8_35
21. Krizhevsky, A., Sutskever, I., Hinton, G.E.: Imagenet classification with deep convolutional neural networks. In: NIPS, pp. 1106–1114 (2012)
22. Lee, K., Maji, S., Ravichandran, A., Soatto, S.: Meta-learning with differentiable convex optimization. In: CVPR, pp. 10657–10665 (2019)
23. Li, W., Wang, L., Xu, J., Huo, J., Gao, Y., Luo, J.: Revisiting local descriptor based image-to-class measure for few-shot learning. In: CVPR, pp. 7260–7268 (2019)
24. Li, Z., Zhou, F., Chen, F., Li, H.: Meta-sgd: learning to learn quickly for few shot learning. arXiv preprint arXiv:1707.09835 (2017)
25. Liu, C., et al.: Learning a few-shot embedding model with contrastive learning. In: AAAI, pp. 8635–8643 (2021)
26. Ma, J., Xie, H., Han, G., Chang, S.F., Galstyan, A., Abd-Almageed, W.: Partner-assisted learning for few-shot image classification. In: ICCV, pp. 10573–10582 (2021)
27. Van der Maaten, L., Hinton, G.: Visualizing data using t-sne. JMLR **9**(11), 2579–2605 (2008)
28. Nichol, A., Achiam, J., Schulman, J.: On first-order meta-learning algorithms. arXiv preprint arXiv:1803.02999 (2018)
29. Noroozi, M., Favaro, P.: Unsupervised learning of visual representations by solving jigsaw puzzles. In: Leibe, B., Matas, J., Sebe, N., Welling, M. (eds.) ECCV 2016. LNCS, vol. 9910, pp. 69–84. Springer, Cham (2016). https://doi.org/10.1007/978-3-319-46466-4_5
30. Oord, A.V.D., Li, Y., Vinyals, O.: Representation learning with contrastive predictive coding. arXiv preprint arXiv:1807.03748 (2018)
31. Oreshkin, B.N., López, P.R., Lacoste, A.: TADAM: task dependent adaptive metric for improved few-shot learning. In: NIPS, pp. 719–729 (2018)
32. Ouali, Y., Hudelot, C., Tami, M.: Spatial contrastive learning for few-shot classification. In: ECML-PKDD, pp. 671–686 (2021)
33. Ravi, S., Larochelle, H.: Optimization as a model for few-shot learning. In: ICLR (2017)
34. Ravichandran, A., Bhotika, R., Soatto, S.: Few-shot learning with embedded class models and shot-free meta training. In: ICCV, pp. 331–339 (2019)
35. Ren, M., et al.: Meta-learning for semi-supervised few-shot classification. In: ICLR (2018)
36. Rusu, A.A., et al.: Meta-learning with latent embedding optimization. In: ICLR (2019)
37. Satorras, V.G., Estrach, J.B.: Few-shot learning with graph neural networks. In: ICLR (2018)

38. Shen, Z., Liu, Z., Qin, J., Savvides, M., Cheng, K.: Partial is better than all: revisiting fine-tuning strategy for few-shot learning. In: AAAI, pp. 9594–9602 (2021)
39. Simonyan, K., Zisserman, A.: Very deep convolutional networks for large-scale image recognition. In: ICLR (2015)
40. Snell, J., Swersky, K., Zemel, R.S.: Prototypical networks for few-shot learning. In: NIPS, pp. 4077–4087 (2017)
41. Su, J.-C., Maji, S., Hariharan, B.: When does self-supervision improve few-shot learning? In: Vedaldi, A., Bischof, H., Brox, T., Frahm, J.-M. (eds.) ECCV 2020. LNCS, vol. 12352, pp. 645–666. Springer, Cham (2020). https://doi.org/10.1007/978-3-030-58571-6_38
42. Sung, F., Yang, Y., Zhang, L., Xiang, T., Torr, P.H.S., Hospedales, T.M.: Learning to compare: relation network for few-shot learning. In: CVPR, pp. 1199–1208 (2018)
43. Tian, Y., Krishnan, D., Isola, P.: Contrastive multiview coding. In: Vedaldi, A., Bischof, H., Brox, T., Frahm, J.-M. (eds.) ECCV 2020. LNCS, vol. 12356, pp. 776–794. Springer, Cham (2020). https://doi.org/10.1007/978-3-030-58621-8_45
44. Tian, Y., Sun, C., Poole, B., Krishnan, D., Schmid, C., Isola, P.: What makes for good views for contrastive learning? In: NIPS, pp. 6827–6839 (2020)
45. Tian, Y., Wang, Y., Krishnan, D., Tenenbaum, J.B., Isola, P.: Rethinking few-shot image classification: a good embedding is all you need? In: Vedaldi, A., Bischof, H., Brox, T., Frahm, J.-M. (eds.) ECCV 2020. LNCS, vol. 12359, pp. 266–282. Springer, Cham (2020). https://doi.org/10.1007/978-3-030-58568-6_16
46. Vaswani, A., et al.: Attention is all you need. In: NIPS, pp. 5998–6008 (2017)
47. Vinyals, O., Blundell, C., Lillicrap, T., Kavukcuoglu, K., Wierstra, D.: Matching networks for one shot learning. In: NIPS, pp. 3630–3638 (2016)
48. Wang, Y., Chao, W.L., Weinberger, K.Q., van der Maaten, L.: Simpleshot: revisiting nearest-neighbor classification for few-shot learning. arXiv preprint arXiv:1911.04623 (2019)
49. Wu, J., Zhang, T., Zhang, Y., Wu, F.: Task-aware part mining network for few-shot learning. In: ICCV, pp. 8433–8442 (2021)
50. Wu, Z., Xiong, Y., Yu, S.X., Lin, D.: Unsupervised feature learning via non-parametric instance discrimination. In: CVPR, pp. 3733–3742 (2018)
51. Xing, C., Rostamzadeh, N., Oreshkin, B.N., Pinheiro, P.O.: Adaptive cross-modal few-shot learning. In: NIPS, pp. 4848–4858 (2019)
52. Xu, C., et al.: Learning dynamic alignment via meta-filter for few-shot learning. In: CVPR, pp. 5182–5191 (2021)
53. Xu, W., Xu, Y., Wang, H., Tu, Z.: Attentional constellation nets for few-shot learning. In: ICLR (2021)
54. Ye, H., Hu, H., Zhan, D., Sha, F.: Few-shot learning via embedding adaptation with set-to-set functions. In: CVPR, pp. 8805–8814 (2020)
55. Zhang, C., Cai, Y., Lin, G., Shen, C.: Deepemd: few-shot image classification with differentiable earth mover's distance and structured classifiers. In: CVPR, pp. 12200–12210 (2020)
56. Zhou, Z., Qiu, X., Xie, J., Wu, J., Zhang, C.: Binocular mutual learning for improving few-shot classification. In: ICCV, pp. 8402–8411 (2021)

Time-rEversed DiffusioN tEnsor Transformer: A New TENET of Few-Shot Object Detection

Shan Zhang[1], Naila Murray[2], Lei Wang[3], and Piotr Koniusz[4(✉)]

[1] Australian National University, Canberra, Australia
shan.zhang@anu.edu.au
[2] Meta AI, London, UK
murrayn@fb.com
[3] University of Wollongong, Wollongong, Australia
leiw@uow.edu.au
[4] Data61/CSIRO, Canberra, Australia
Piotr.Koniusz@anu.edu.au

Abstract. In this paper, we tackle the challenging problem of Few-shot Object Detection. Existing FSOD pipelines (i) use average-pooled representations that result in information loss; and/or (ii) discard position information that can help detect object instances. Consequently, such pipelines are sensitive to large intra-class appearance and geometric variations between support and query images. To address these drawbacks, we propose a Time-rEversed diffusioN tEnsor Transformer (TENET), which i) forms high-order tensor representations that capture multiway feature occurrences that are highly discriminative, and ii) uses a transformer that dynamically extracts correlations between the query image and the entire support set, instead of a single average-pooled support embedding. We also propose a Transformer Relation Head (TRH), equipped with higher-order representations, which encodes correlations between query regions and the entire support set, while being sensitive to the positional variability of object instances. Our model achieves state-of-the-art results on PASCAL VOC, FSOD, and COCO.

Keywords: Few-shot object detection · Transformer · Multiple order pooling · High order pooling · Heat diffusion process

---

SZ was mainly in charge of the pipeline/developing the transformer. PK (corresponding author) was mainly in charge of mathematical design of TENET & TSO.
Code: https://github.com/ZS123-lang/TENET.

Supplementary Information The online version contains supplementary material available at https://doi.org/10.1007/978-3-031-20044-1_18.

1   Introduction

Object detectors based on deep learning, usually addressed by supervised models, achieve impressive performance [8,13,27,31–33] but they rely on a large number of images with human-annotated class labels/object bounding boxes. Moreover, object detectors cannot be easily extended to new class concepts not seen during training. Such a restriction limits supervised object detectors to predefined scenarios. In contrast, humans excel at rapidly adapting to new scenarios by *"storing knowledge gained while solving one problem and applying it to a different but related problem"* [40], also called as *"transfer of practice"* [41].

Few-shot Object Detection (FSOD) [4,7,11,12,45,50,51] methods mimic this ability, and enable detection of test classes that are disjoint from training classes. They perform this adaptation using a few "support" images from test classes. Successful FSOD models must (i) find promising candidate regions-of-Interest (RoIs) in query images; and (ii) accurately regress bounding box locations and predict RoI classes, under large intra-class geometric and photometric variations.

To address the first requirement, approaches [7,45,50,51] use the region proposal network [33]. For example, FSOD-ARPN [7], PNSD [50] and KFSOD [51] cross-correlate query feature maps with a class prototype formed from average-pooled (*ie.*, first-order) features, second-order pooled representations and kernel-pooled representations, respectively. These methods use a single class prototype which limits their ability to leverage diverse information from different class samples. Inspired by Transformers [38], approach [23] uses average pooling over support feature maps to generate a vector descriptor per map. Attention mechanism is then used to modulate query image features using such descriptors.

The above methods rely on first- and second-order pooling, while so-called higher-order pooling is more discriminative [15,17,18]. Thus, we propose a non-trivial Time-rEversed diffusioN tEnsor Transformer (TENET). With TENET, higher-order tensors undergo a time-reversed heat diffusion to condense signal on super-diagonals of tensors, after which coefficients of these super-diagonals are passed to a Multi-Head Attention transformer block. TENET performs second-, third- and fourth-order pooling. However, higher-order pooling suffers from several issues, *ie.*, (i) high computational complexity of computing tensors with three/more modes, (ii) non-robust tensor estimates due to the limited number of vectors being aggregated, and (iii) tensor burstiness [17].

To this end, we propose a Tensor Shrinkage Operator (TSO) which generalizes spectral power normalization (SPN) operators [18], such as the Fast Spectral MaxExp operator (MaxExp(F)) [18], to higher-order tensors. As such, it can be used to reduce tensor burstiness. Moreover, by building on the linear algebra of the heat diffusion process (HDP) [34] and recent generalisation of HDP to SPN operators [18], we also argue that such operators can reverse the diffusion of signal in autocorrelation or covariance matrices, and high-order tensors, instead of just reducing the burstiness. Using a parametrization which lets us control the reversal of diffusion, TSO condenses signal captured by a tensor toward its super-diagonal, preserving information along it. This super-diagonal serves as our final representation, reducing the feature size from d^r to d, making our representation computationally tractable. Finally, shrinkage operators are known for their ability to estimate covariances well when only a small number of samples are available [22]. To the best of our knowledge, we are the first to show that MaxExp(F) is a shrinkage operator, and to propose TSO for orders $r \geq 2$.

To address the second requirement, FSOD-ARPN introduces a multi-relation head that captures global, local and patch relations between support and query objects, while PNSD passes second-order autocorrelation matrices to a similarity network. However, FSOD-ARPN and PNSD do not model spatial relations [9]. The QSAM [23] uses attention to highlight the query RoI vectors that are similar to the set of support vectors (obtained using only first-order spatial average pooling). Thus, we introduce a Transformer Relation Head (TRH) to improve modeling of spatial relations. TRH computes self-attention between spatially-aware features and global spatially invariant first-, second- and higher-order TENET representations of support and/or query RoI features. The second attention mechanism of TRH performs cross-attention between Z support embeddings (for Z-shot if $Z \geq 2$), and a set of global representations of query RoIs. This attention encodes similarities between query RoIs and support samples.

Our FSOD pipeline contains TENET Region Proposal Network (TENET RPN) and the TRH, both equipped with discriminative TENET representations, improving generation of RoI proposals and modeling of query-support relations.

Below are our contributions:

i. We propose a Time-rEversed diffusiON tEnsor Transformer, called TENET, which captures high-order patterns (multi-way feature cooccurrences) and decorrelates them/reduces tensor burstiness. To this end, we generalize the MaxExp(F) operator [18] for autocorrelation/covariance matrices to higher-order tensors by introducing the so-called Tensor Shrinkage Operator (TSO).
ii. We propose a Transformer Relation Head (TRH) that is sensitive both to the variability between the Z support samples provided in a Z-shot scenario, and to positional variability between support and query objects.
iii. In Subsect. 4.1, we demonstrate that TSO emerges from the MLE-style minimization over the Kullback-Leibler (KL) divergence between the input and output spectrum, with the latter being regularized by the Tsallis entropy [2]. Thus, we show that TSO meets the definition of shrinkage estimator whose target is the identity matrix (tensor).

Our proposed method outperforms the state of the art on novel classes by 4.0%, 4.7% and 6.1% mAP on PASCAL VOC 2007, FSOD, and COCO respectively.

2 Related Works

Below, we review FSOD models and vision transformers, followed by a discussion on feature grouping, tensor descriptors and spectral power normalization.

Few-shot Object Detection. A Low-Shot Transfer Detector (LSTD) [4] leverages rich source domain to construct a target domain detector with few training samples but needs to be fine-tuned to novel categories. Meta-learning-based approach [45] reweights RoI features in the detection head without fine-tuning. Similarly, MPSR [43] deals with scale invariance by ensuring the detector is trained over multiple scales of positive samples. NP-RepMet [46] introduces a negative- and positive-representative learning framework via triplet losses that bootstrap the classifier. FSOD-ARPN [7] is a general FSOD network equipped with a channel-wise attention mechanism and multi-relation detector that scores pairwise object similarity in both the RPN and the detection head, inspired by Faster R-CNN. PNSD [50], inspired by FSOD-ARPN [7], uses contraction of second-order autocorrelation matrix against query feature maps to produce attention maps. Single-prototype (per class) methods suffer information loss. Per-sample Prototype FSOD [23] uses the entire support set to form prototypes of a class but it ignores spatial information within regions. Thus, we employ TENET RPN and TRH to capture spatial and high-order patterns, and extract correlations between the query image and the Z-shot support samples for a class.

Transformers in Vision. Transformers, popular in natural language processing [38], have also become popular in computer vision. Pioneering works such as ViT [9] show that transformers can achieve the state of the art in image recognition. DETR [3] is an end-to-end object detection framework with a transformer encoder-decoder used on top of backbone. Its deformable variant [52] improves the performance/training efficiency. SOFT [29], the SoftMax-free transformer approximates the self-attention kernel by replacing the SoftMax function with Radial Basis Function (RBF), achieving linear complexity. In contrast, our TENET is concerned with reversing the diffusion of signal in high-order tensors via the shrinkage operation, with the goal of modeling spatially invariant high-order statistics of regions. Our attention unit, so-called Spatial-HOP in TRH, also uses RBF to capture correlations of between spatial and high-order descriptors.

Multi-path and Groups of Feature Maps. GoogleNet [36] has shown that multi-path representations (several network branches) lead to classification improvements. ResNeXt [44] adopts group convolution [20] in the ResNet bottleneck block. SK-Net [25], based on SE-Net [10], uses feature map attention across two network branches. However, these approaches do not model feature statistics. ReDRO [30] samples groups of features to apply the matrix square root over submatrices to improve the computational speed. In contrast, TENET forms fixed groups of features to form second-, third- and fourth-order tensors (simply using groups of features to form second-order matrices is not effective).

Second-Order Pooling (SOP). Region Covariance Descriptors for texture [37] and object recognition [17] use SOP. Approach [19] uses spectral pooling for fine-grained image classification, whereas SoSN [48] and its variants [47,49] leverage SOP and element-wise Power Normalization (PN) [17] for end-to-end few-shot learning. In contrast, we develop a multi-object few-shot detector. Similarly to SoSN, PNSD [50] uses SOP with PN as representations which are passed to the detection head. So-HoT [14] that uses high-order tensors for domain adaptation is also somewhat related to TENET but So-HoT uses multiple polynomial kernel matrices, whereas we apply TSO to achieve decorrelation and shrinkage. TENET without TSO reduces to polynomial feature maps and performs poorly.

Power Normalization (PN). Burstiness is "the property that a given visual element appears more times in an image than a statistically independent model would predict". PN [16] limits the burstiness of first- and second-order statistics due to the binomial PMF-based feature detection factoring out feature counts [16,17,19]. Element-wise MaxExp pooling [16] gives likelihood of "at least one particular visual word being present in an image", whereas SigmE pooling [19] is its practical approximation. Noteworthy are recent Fast Spectral MaxExp operator, MaxExp(F) [18], which reverses the heat diffusion on the underlying loopy graph of second-order matrix to some desired past state [34], and Tensor Power-Euclidean (TPE) metric [15]. TPE alas uses the Higher Order Singular Value Decomposition [21], which makes TPE intractable for millions of region proposals per dataset. Thus, we develop TENET and TSO, which reverses diffusion on high-order tensors by shrinking them towards the super-diagonal of tensor.

3 Background

Below, we detail notations and show how to calculate multiple higher-order statistics and Power Normalization, followed by revisiting the transformer block.

Notations. Let $\mathbf{x} \in \mathbb{R}^d$ be a d-dimensional feature vector. $\mathcal{I}_N$ stands for the index set $\{1, 2, ..., N\}$. We define a vector of all-ones as $\mathbf{1} = [1, ..., 1]^T$. Let $\mathcal{X}^{(r)} = \uparrow\otimes_r \mathbf{x}$ denote a tensor of order r generated by the r-th order outer-product of $\mathbf{x}$, and $\mathcal{X}^{(r)} \in \mathbb{R}^{d \times d ... \times d}$. Typically, capitalised boldface symbols such as $\boldsymbol{\Phi}$ denote matrices, lowercase boldface symbols such as $\boldsymbol{\phi}$ denote vectors and regular case such as $\Phi_{i,j}$, ϕ_i, n or Z denote scalars, *e.g.*, $\Phi_{i,j}$ is the (i, j)-th coefficient of $\boldsymbol{\phi}$.

High-Order Tensor Descriptors (HoTD). Below we formalize HoTD [14].

Proposition 1. *Let $\boldsymbol{\Phi} \equiv [\boldsymbol{\phi}_1, ..., \boldsymbol{\phi}_N] \in \mathbb{R}^{d \times N}$ and $\boldsymbol{\Phi}' \equiv [\boldsymbol{\phi}_1', ..., \boldsymbol{\phi}_M'] \in \mathbb{R}^{d \times M}$ be feature vectors extracted from some two image regions. Let $\boldsymbol{w} \in \mathbb{R}_+^N$, $\boldsymbol{w}' \in \mathbb{R}_+^M$ be some non-negative weights and $\boldsymbol{\mu}, \boldsymbol{\mu}' \in \mathbb{R}^d$ be the mean vectors of $\boldsymbol{\Phi}$ and $\boldsymbol{\Phi}'$, respectively. A linearization of the sum of polynomial kernels of degree r,*

$$\langle \mathcal{M}^{(r)}(\boldsymbol{\Phi}; \boldsymbol{w}, \boldsymbol{\mu}), \mathcal{M}^{(r)}(\boldsymbol{\Phi}'; \boldsymbol{w}', \boldsymbol{\mu}') \rangle = \frac{1}{NM} \sum_{n=1}^{N} \sum_{m=1}^{M} w_n^r w_m'^r \langle \boldsymbol{\phi}_n - \boldsymbol{\mu}, \boldsymbol{\phi}_m' - \boldsymbol{\mu}' \rangle^r, \quad (1)$$

yields the tensor feature map

$$\mathcal{M}^{(r)}(\boldsymbol{\Phi}; \boldsymbol{w}, \boldsymbol{\mu}) = \frac{1}{N} \sum_{n=1}^{N} w_n^r \uparrow\otimes_r (\boldsymbol{\phi}_n - \boldsymbol{\mu}) \in \mathbb{R}^{d \times d ... \times d}. \quad (2)$$

In our paper, we set $\boldsymbol{w} = \boldsymbol{w}' = 1$ and $\boldsymbol{\mu} = \boldsymbol{\mu}' = 0$, whereas orders $r = 2, 3, 4$.

(Eigenvalue) Power Normalization ((E)PN). For second-order matrices, MaxExp(F), a state-of-the-art EPN [18], is defined as .

$$g(\lambda; \eta) = 1 - (1 - \lambda)^\eta \tag{3}$$

on the ℓ_1-norm normalized spectrum from SVD $(\lambda_i := \lambda_i/(\varepsilon + \sum_{i'} \lambda_{i'}))$, and on symmetric positive semi-definite matrices as

$$\widehat{\mathcal{G}}_{\mathrm{MaxExp}}(\boldsymbol{M}; \eta) = \mathbb{I} - (\mathbb{I} - \boldsymbol{M})^\eta, \tag{4}$$

where $\boldsymbol{M}$ is a trace-normalized matrix, *ie.*, $\boldsymbol{M} := \boldsymbol{M}/(\varepsilon + \mathrm{Tr}(\boldsymbol{M}))$ with $\varepsilon \approx 1e{-}6$, and $\mathrm{Tr}(\cdot)$ denotes the trace. The time-reversed heat diffusion process is adjusted by integer $\eta \geq 1$. The larger the value of η is, the more prominent the time reversal is. $\mathcal{G}_{\mathrm{MaxExp}}$ is followed by the element-wise PN, called SigmE [18]:

$$\mathcal{G}_{\mathrm{SigmE}}(p; \eta') = 2/(1 + e^{-\eta' p}) - 1, \tag{5}$$

where p takes each output coefficient of Eq. (4), $\eta' \geq 1$ controls detecting feature occurrence *vs.* feature counting trade-off.

Transformers. An architecture based on blocks of attention and MLP layers forms a transformer network [38]. Each attention layer takes as input a set of query, key and value matrices, denoted $\boldsymbol{Q}$, $\boldsymbol{K}$ and $\boldsymbol{V}$, respectively. Let $\boldsymbol{Q} \equiv [\mathbf{q}_1, ..., \mathbf{q}_N] \in \mathbb{R}^{d \times N}$, $\boldsymbol{K} \equiv [\mathbf{k}_1, ..., \mathbf{k}_N] \in \mathbb{R}^{d \times N}$, and $\boldsymbol{V} \equiv [\mathbf{v}_1, ..., \mathbf{v}_N] \in \mathbb{R}^{d \times N}$, where N is the number of input feature vectors, also called tokens, and d is the channel dimension. A generic attention layer can then be formulated as:

$$\mathrm{Attention}(\boldsymbol{Q}, \boldsymbol{K}, \boldsymbol{V}) = \alpha\big(\gamma(\boldsymbol{Q}, \boldsymbol{K})\big) \boldsymbol{V}^T. \tag{6}$$

Self-attention is composed of $\alpha(\gamma(\cdot, \cdot))$, $\alpha(\cdot)$ is a non-linearity and $\gamma(\cdot, \cdot)$ computes similarity. A popular choice is SoftMax with the scaled dot product [38]:

$$\alpha(\cdot) = \mathrm{SoftMax}(\cdot) \quad \text{and} \quad \gamma(\boldsymbol{Q}, \boldsymbol{K}) = \frac{\boldsymbol{Q}^T \boldsymbol{K}}{\sqrt{d}}. \tag{7}$$

Note the LayerNorm and residual connections are added at the end of each block.

To facilitate the design of linear self-attention, approach [29] introduces a SoftMax-free transformer with the dot product replaced by the RBF kernel:

$$\alpha(\cdot) = \mathrm{Exp}(\cdot) \quad \text{and} \quad \gamma(\boldsymbol{Q}, \boldsymbol{K}) = -\frac{1}{2\sigma^2} \Big[\|\mathbf{q}_i - \mathbf{k}_j\|_2^2 \Big]_{i,j \in \mathcal{I}_N}, \tag{8}$$

where $[\cdot]$ stacks distances into a matrix, σ^2 is the kernel variance set by cross-validation, and $\mathbf{q}_i$ and $\mathbf{k}_j$ are ℓ_2-norm normalized.

The Multi-Head Attention (MHA) layer uses T attention units whose outputs are concatenated. Such an attention splits input matrices $\boldsymbol{Q}$, $\boldsymbol{K}$, and $\boldsymbol{V}$ along their channel dimension d into T groups and computes attention on each group:

$$\mathrm{MHA}(\boldsymbol{Q}, \boldsymbol{K}, \boldsymbol{V}) = [\mathbf{A}_1, ..., \mathbf{A}_T], \tag{9}$$

where $[\cdot]$ performs concatenation along the channel mode, the mth head is $\mathbf{A}_m = \mathrm{Attention}(\boldsymbol{Q}_m, \boldsymbol{K}_m, \boldsymbol{V}_m)$, and $\boldsymbol{Q}_m \in \mathbb{R}^{\frac{d}{T} \times N}$, $\boldsymbol{K}_m \in \mathbb{R}^{\frac{d}{T} \times N}$, and $\boldsymbol{V}_m \in \mathbb{R}^{\frac{d}{T} \times N}$.

4   Proposed Approach

Given a set of Z support crops $\{\boldsymbol{X}_z\}_{z\in\mathcal{I}_Z}$ and a query image $\boldsymbol{X}^*$ per episode, our approach learns a matching function between representations of query RoIs and and support crops. Figure 1 shows our pipeline comprised of three main modules:

i. **Encoding Network** (EN) extracts feature map $\boldsymbol{\Phi} \in \mathbb{R}^{d\times N}$ per image (of $N = W \times H$ spatial size) from query and support images via ResNet-50.

ii. **TENET RPN** extracts RoIs from the query image and computes embeddings for the query RoIs and the support crops. Improved attention maps are obtained by T-Heads Attention (THA) that operates on TENET descriptors.

iii. **Transformer Relation Head** captures relations between query and support features using self- and cross-attention mechanisms. This head produces representations for the classifier and bounding-box refinement regression loss.

Next, we describe TENET RPN, followed by the Transformer Relation Head.

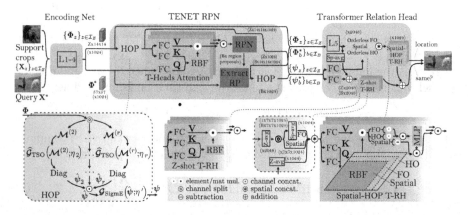

Fig. 1. Our pipeline. (*top*) We pass ground truth support crops $\{\boldsymbol{\mathcal{X}}_z\}_{z\in\mathcal{I}_Z}$ for Z-shot problem and a query image $\boldsymbol{\mathcal{X}}^*$ to the Encoding Network (EN). The resulting convolutional feature maps, $\{\boldsymbol{\Phi}_z\}_{z\in\mathcal{I}_Z}$ for support and $\boldsymbol{\Phi}^*$ for query, are passed to the TENET-RPN module to produce a set of B descriptors $\{\boldsymbol{\Phi}_b^*\}_{b\in\mathcal{I}_B}$ for B Region Proposals (RP a.k.a. RoIs) of query image, and high-order pooled (HOP) representations for both support crops $\{\psi_z\}_{z\in\mathcal{I}_Z}$ and query image RoIs $\{\psi_b^*\}_{b\in\mathcal{I}_B}$. TENET contains HOP units which compute high-order tensor descriptors and then apply a novel tensor shrinkage operator to them, yielding spatially orderless HOP representations. Sets of features $\{\psi_z\}_{z\in\mathcal{I}_Z}$ and $\{\psi_b^*\}_{b\in\mathcal{I}_B}$ are then passed to the Transformer Relation Head (TRH), along with the convolutional features $\{\boldsymbol{\Phi}_z\}_{z\in\mathcal{I}_Z}$ and $\{\boldsymbol{\Phi}_b^*\}_{b\in\mathcal{I}_B}$. The TRH consists of Z-shot and Spatial-HOP transformer heads for measuring similarities across support regions and query proposals, and refining the localization of target objects, respectively. (*bottom*) Details of HOP, "Orderless FO, Spatial, Orderless HO", Z-shot T-RH and Spatial-HOP T-RH blocks. See Subsect. 4.3 for details.

4.1 Extracting Representations for Support and Query RoIs

Figure 1 *(top)* shows our TENET RPN that produces embeddings and query RoIs. Central to this module is our HOP unit that produces Higher-order Tensor Descriptors (HoTDs). HOP splits features along the channel mode into multiple groups of feature maps, from which second-, third- and fourth- order tensors are aggregated over desired regions. HoTDs use a generalization of the MaxExp(F) to higher-order tensors, called the Tensor Shrinkage Operator (TSO).

Tensor Shrinkage Operator. Ledoit and Wolf [22] define autocorrelation/covariance matrix estimation as a trade-off between the sample matrix $\boldsymbol{M}$ and a highly structured operator $\boldsymbol{F}$, using the linear combination $(1-\delta)\boldsymbol{M}+\delta\boldsymbol{F}$. For symmetric positive semi-definite matrices and tensors, one can devise a convex shrinkage operator by minimizing some divergence $d(\boldsymbol{\lambda},\boldsymbol{\lambda}')$ between the source and target spectra, where $\boldsymbol{\lambda}'$ is regularized by $\Omega(\boldsymbol{\lambda}')$ with weight $\delta \geq 0$:

$$\boldsymbol{\lambda}^* = \arg\min_{\boldsymbol{\lambda}' \geq 0} d(\boldsymbol{\lambda}, \boldsymbol{\lambda}') + \delta\Omega(\boldsymbol{\lambda}'). \tag{10}$$

Below we derive the TSO as a generalization of MaxExp(F)[1].

With $\hat{\boldsymbol{\mathcal{G}}}_{\text{TSO}}(\cdot)$ we extract representations $\hat{\psi}_r$ from HoTDs $\boldsymbol{\mathcal{M}}^{(r)}$:

$$\hat{\boldsymbol{\mathcal{G}}}_{\text{TSO}}\left(\boldsymbol{\mathcal{M}}^{(r)}; \eta\right) = \boldsymbol{\mathcal{I}}_r - \left(\boldsymbol{\mathcal{I}}_r - \boldsymbol{\mathcal{M}}^{(r)}\right)^{\eta}, \tag{11}$$

$$\hat{\psi}_r = \text{Diag}\left(\hat{\boldsymbol{\mathcal{G}}}_{\text{TSO}}\left(\boldsymbol{\mathcal{M}}^{(r)}; \eta_r\right)\right), \tag{12}$$

$$\psi_r = \boldsymbol{\mathcal{G}}_{\text{SigmE}}\left(\hat{\psi}_r; \eta_r'\right), \tag{13}$$

where $\text{Diag}(\cdot)$ extracts the super-diagonal of tensor. The identity tensor $\boldsymbol{\mathcal{I}}_r$ of order r is defined such that $\mathcal{I}_{1,\dots,1} = \mathcal{I}_{2,\dots,2}\dots = \mathcal{I}_{d,\dots,d} = 1$ and $\mathcal{I}_{i_1,\dots,i_r} = 0$ if $i_j \neq i_k$ and $j \neq k$, $j,k \in \mathcal{I}_r$.

Theorem 1. *Let $\boldsymbol{\lambda}$ be the ℓ_1-norm normalized spectrum, as in Eq. (3). Then $g(\lambda; \eta) = 1 - (1 - \lambda)^{\eta}$ is a shrinkage operator as it is a solution to Eq. (10) if $\delta = \frac{\eta t^{1/\eta}}{s}(1 - \frac{1}{\eta})$, and the Kullback-Leibler divergence and the Tsallis entropy are substituted for the divergence term d and the regularization term Ω, respectively. In the above theorem, $\boldsymbol{\lambda} \in \mathbb{R}_+^d$, $s = d-1$, $t = d-t'$ and $t' > 0$ is the trace of $g(\boldsymbol{\lambda}; \eta)$.*

Proof. Let $d(\boldsymbol{\lambda}, \boldsymbol{\lambda}') = D_{\text{KL}}\left(\boldsymbol{\lambda}^\circ \| \boldsymbol{\lambda}^{\circ\prime}\right) = -\left(\sum_{i \in \mathcal{I}_d} \lambda_i^\circ \log \lambda_i^{\circ\prime}\right) + \left(\sum_{i \in \mathcal{I}_d} \lambda_i^\circ \log \lambda_i^\circ\right)$, where $\boldsymbol{\lambda}^\circ = \frac{1-\boldsymbol{\lambda}}{s}$ and $\boldsymbol{\lambda}^{\circ\prime} = \frac{1-\boldsymbol{\lambda}'}{t}$ are complements of $\boldsymbol{\lambda}$ and $\boldsymbol{\lambda}'$ normalized by $s = d-1$ and $t = d - t'$ to obtain normalized distributions to ensure valid use of the Kullback-Leibler divergence and the Tsallis entropy. Let $\Omega(\boldsymbol{\lambda}'; \alpha) = \frac{1}{\alpha-1}(1 - \sum_{i \in \mathcal{I}_d} \lambda_i^{\circ\prime\alpha})$.

---

[1] For $r = 2$, Eq. (12) yields $\text{Diag}(\mathbb{I} - (\mathbb{I} - \boldsymbol{M})^{\eta_2})$. $\text{Diag}(\text{Sqrtm}(\boldsymbol{M}))$ is its approximation.

Algorithm 1. Tensor Shrinkage Operator with Exponentiation by Squaring, left part for even orders and right part for odd orders r.

Input: $\mathcal{M}$, $\eta \geq 1$, $r = 2, 4, \ldots$

1: $\mathcal{M}_1^* = \mathcal{I}_r - \mathcal{M}$, $n = \text{int}(\eta)$, $t = 1$, $q = 1$
2: **while** $n \neq 0$:
3:    **if** $n \& 1$:
4:       **if** $t > 1$: $\mathcal{G}_{t+1} = \mathcal{G}_t \times_{1,\ldots,r/2} \mathcal{M}_q^*$,
5:       **else:** $\mathcal{G}_{t+1} = \mathcal{M}_q^*$
6:       $n \leftarrow n - 1$, $t \leftarrow t + 1$
7:    $n \leftarrow \text{int}(n/2)$
8:    **if** $n > 0$:
9:       $\mathcal{M}_{q+1}^* = \mathcal{M}_q^* \times_{1,\ldots,r/2} \mathcal{M}_q^*$
10:       $q \leftarrow q + 1$

Output: $\widehat{\mathcal{G}}_{\text{TSO}}(\mathcal{M}) = \mathcal{I}_r - \mathcal{G}_t$

Input: $\mathcal{M}$, $\eta = 3^0, 3^1, 3^2, \ldots$, $r = 3, 5, \ldots$

1: $\mathcal{M}_1^* = \mathcal{I}_r - \mathcal{M}$, $n = \text{int}(\eta)$, $q = 1$
2: **while** $n \neq 0$:
3:    $n \leftarrow \text{int}(n/3)$
4:    **if** $n > 0$:
5:       $\mathcal{M}_{q+1}^* = \mathcal{M}_q^* \times_{1,\ldots,\lfloor r/2 \rfloor} \mathcal{M}_q^*$
6:       $\times_{1,\ldots,\lceil r/2 \rceil} \mathcal{M}_q^*$
7:       $q \leftarrow q + 1$

Output: $\widehat{\mathcal{G}}_{\text{TSO}}(\mathcal{M}) = \mathcal{I}_r - \mathcal{M}_q^*$

Define $f(\boldsymbol{\lambda}, \boldsymbol{\lambda}') = -\left(\sum_{i \in \mathcal{I}_d} \lambda_i^\circ \log \lambda_i^{\circ\prime} \right) + \left(\sum_{i \in \mathcal{I}_d} \lambda_i^\circ \log \lambda_i^\circ \right) + \frac{1}{\alpha - 1} \left(1 - \sum_{i \in \mathcal{I}_d} \lambda_i^{\circ\prime\alpha} \right)$ (following Eq. (10)) which we minimize w.r.t. $\boldsymbol{\lambda}'$ by computing $\frac{\partial f}{\partial \lambda_i'} = 0$, that is,

$$\frac{\partial f}{\partial \lambda_i'} = 0 \Rightarrow \lambda_i' = 1 - t \left(\frac{\eta}{\delta s} \right)^\eta (1 - 1/\eta)^\eta (1 - \lambda_i)^\eta, \tag{14}$$

where $\frac{1}{\alpha} = \eta$. Solving $t \left(\frac{\eta}{\delta s} \right)^\eta (1 - \frac{1}{\eta})^\eta = 1$ for δ completes the proof.

Theorem 2. *The highly structured operator* $\boldsymbol{F}$ *(target of shrinkage) equals* $\mathbb{I}$.

Proof. Notice $\lim_{\eta \to \infty} 1 - (1 - \lambda_i)^\eta = 1$ if $\boldsymbol{\lambda} \neq 0$ is the ℓ_1-norm normalized spectrum from SVD, ie., $\boldsymbol{U} \boldsymbol{\lambda} \boldsymbol{U}^T = \boldsymbol{M} \succcurlyeq 0$ with $\lambda_i := \lambda_i / (\sum_{i'} \lambda_{i'} + \varepsilon)$, $\varepsilon > 0$. Thus, $\boldsymbol{U} \boldsymbol{U}^T = \mathbb{I}$.

Note $\mathcal{I}_r - (\mathcal{I}_r - \mathcal{M})^\eta \to \mathcal{I}_r$ if $\eta \to \infty$. Thus, for sufficiently large $1 \ll \eta \ll \infty$, the diffused heat reverses towards the super-diagonal, where the majority of the signal should concentrate. Thus, we limit the number of coefficients of feature representations by extracting the super-diagonals from the TSO-processed $\mathcal{M}^{(r)}$ as in Eq. 12, where r is the order of HoTD $\mathcal{M}$. In our experiments, $r = 2, 3, 4$.

As super-diagonals contain information obtained by multiplying η times the complement $1 - \lambda_i$ (spectral domain), TSO can be seen as $\eta - 1$ aggregation steps along the tensor product mode(s). THus, we pass $\hat{\psi}_r$ via the element-wise SigmE from Eq. (5) (as in Eq. (13)) to detect the presence of at least one feature being detected in $\hat{\psi}_r$ after such an aggregation. For brevity, we drop subscript r of ψ_r.

Complexity. For integers $\eta \geq 2$ and even orders $r \geq 2$, computing $\eta - 1$ tensor-tensor multiplications $(\mathcal{I}_r - \mathcal{M}^{(r)})^\eta$ has the complexity $\mathcal{O}(d^{\frac{3}{2}r} \eta)$. For odd orders $r \geq 3$, due to alternations between multiplications in $\lfloor \frac{r}{2} \rfloor$ and $\lceil \frac{r}{2} \rceil$ modes, the complexity is $\mathcal{O}(d^{\lfloor \frac{r}{2} \rfloor} d^{2 \lceil \frac{r}{2} \rceil} \eta) \approx \mathcal{O}(d^{\frac{3}{2}r} \eta)$. Thus, the complexity of Eq. (11) w.r.t.

integer $\eta \geq 2$ scales linearly. However, for even orders r, one can readily replace $(\mathcal{I}_r - \mathcal{M}^{(r)})^\eta$ with exponentiation by squaring [1], whose cost is $\log(\eta)$. This readily yields the sublinear complexity $\mathcal{O}(d^{\frac{3}{2}r} \log(\eta))$ w.r.t. η.

Implementation of TSO. Algorithm 1 shows fast TSO for even/odd orders r. We restrict the odd variant to $r = 3^0, 3^1, 3^2, \ldots$ for brevity. Finally, we note that matrix-matrix and tensor multiplications with cuBLAS are highly parallelizable so the $d^{\frac{3}{2}r}$ part of complexity can be reduced in theory even to $\log(d)$.

TENET-RPN. To extract query RoIs, we firstly generate a set $\mathcal{Z}$ of TSO representations $\boldsymbol{\Psi} \equiv \{\boldsymbol{\psi}_z\}_{z \in \mathcal{I}_Z}$ from the support images, with $Z = |\mathcal{Z}|$. We then perform cross-attention between $\boldsymbol{\Psi} \in \mathbb{R}^{d \times Z}$ and the N feature vectors $\boldsymbol{\Phi}^* \in \mathbb{R}^{d \times N}$ extracted from the query image. The attention input $\boldsymbol{Q}$ is then generated from $\boldsymbol{\Phi}^*$, while $\boldsymbol{K}$ and $\boldsymbol{V}$ are both generated from $\boldsymbol{\Psi}$. The output of the transformer block in Eq. 6 is fed into an RPN layer to output a set $\mathcal{B}$ of $B = |\mathcal{B}|$ query RoIs.

Query RoIs are represented by spatially ordered features $\{\boldsymbol{\Phi}_b^*\}_{b \in \mathcal{I}_B}$ and TSO orderless features $\{\boldsymbol{\psi}_b^*\}_{b \in \mathcal{I}_B}$, passed to the Transformer Relation Head.

4.2   Transformer Relation Head

TRH, in Fig. 1, models relations between support crops and query RoIs. TSO representations are derived from features of layer 4 of ResNet-50, leading to a channel dimension $d = 1024$. Spatially ordered representations of support features $\{\boldsymbol{\Phi}_z \in \mathbb{R}^{2d \times N}\}_{z \in \mathcal{I}_Z}$ and query RoIs $\{\boldsymbol{\Phi}_b^* \in \mathbb{R}^{2d \times N}\}_{b \in \mathcal{I}_B}$ are both extracted from layer 5 of ResNet-50, leading to a channel dimension $2d = 2048$.

They are then fed into 2 different transformers: (i) a **Z-shot transformer head**, which performs cross-attention between globally-pooled representations of the query images and support images; and (ii) a **Spatial-HOP transformer head**, which performs self-attention between spatially ordered representations and spatially orderless high-order representations for a given image.

Z-shot transformer head consists of a cross-attention layer formed with:

$$\begin{pmatrix} \boldsymbol{Q} \\ \boldsymbol{K} \\ \boldsymbol{V} \end{pmatrix} = \begin{pmatrix} [\mathbf{q}_1, ..., \mathbf{q}_B] \\ [\mathbf{k}_1, ..., \mathbf{k}_Z] \\ [\mathbf{v}_1, ..., \mathbf{v}_Z] \end{pmatrix} \quad \text{where} \quad \begin{pmatrix} \mathbf{q}_b \\ \mathbf{k}_z \\ \mathbf{v}_z \end{pmatrix} = \begin{pmatrix} \boldsymbol{W}^{(q)}(\bar{\boldsymbol{\phi}}_b^* + \boldsymbol{W}^{(p)}\boldsymbol{\psi}_b^*) \\ \boldsymbol{W}^{(k)}(\bar{\boldsymbol{\phi}}_z + \boldsymbol{W}^{(p)}\boldsymbol{\psi}_z) \\ \boldsymbol{W}^{(v)}(\bar{\boldsymbol{\phi}}_z + \boldsymbol{W}^{(p)}\boldsymbol{\psi}_z) \end{pmatrix}. \quad (15)$$

Moreover, $\bar{\boldsymbol{\phi}}_b^*$ and $\bar{\boldsymbol{\phi}}_z$ are average-pooled features $\frac{1}{N}\boldsymbol{\Phi}_b^*\mathbf{1}$ and $\frac{1}{N}\boldsymbol{\Phi}_z\mathbf{1}$, respectively. The matrices $\boldsymbol{W}^{(q)} \in \mathbb{R}^{2d \times 2d}$, $\boldsymbol{W}^{(k)} \in \mathbb{R}^{2d \times 2d}$, $\boldsymbol{W}^{(v)} \in \mathbb{R}^{2d \times 2d}$, and $\boldsymbol{W}^{(p)} \in \mathbb{R}^{2d \times d}$ is a linear projection mixing spatially orderless TSO representations with spatially ordered representations. Thus, each attention query vector $\mathbf{q}_b$ combines the extracted spatially orderless TSO representations with spatially ordered representations for a given query RoI. Similarly, each key vector $\mathbf{k}_z$ and value vector $\mathbf{v}_z$ combine such two types of representations for a support crop.

Spatial-HOP transformer head consists of a layer that performs self-attention on spatially orderless TSO representations and spatially ordered representations, extracted either from Z support crops, or B query RoIs. Below we take one support crop as an example. For the set $\mathcal{Z}$, we compute "Spatial", a spatially ordered Z-averaged representation $\boldsymbol{\Phi}^\dagger = \frac{1}{Z}\sum_{z\in\mathcal{I}_z}\boldsymbol{\Phi}_z \in \mathbb{R}^{2d\times N}$. We also compute HO, a spatially orderless High-Order Z-pooled representation $\boldsymbol{\psi}^\dagger = \frac{1}{Z}\sum_{z\in\mathcal{I}_z}\boldsymbol{\psi}_z \in \mathbb{R}^d$. We split $\boldsymbol{\Phi}^\dagger$ along the channel mode of dimension $2d$ to create two new matrices $\boldsymbol{\Phi}^{\dagger u} \in \mathbb{R}^{d\times N}$ and $\boldsymbol{\Phi}^{\dagger l} \in \mathbb{R}^{d\times N}$. We set $\boldsymbol{\Phi}^{\dagger l} = [\boldsymbol{\phi}_1^{\dagger l}, ..., \boldsymbol{\phi}_N^{\dagger l}] \in \mathbb{R}^d$. Also, we form FO (from $\boldsymbol{\Phi}^{\dagger u}$), a spatially orderless First-Order Z-pooled representation. Self-attention is then performed over the matrix of token vectors:

$$\boldsymbol{T} = [\boldsymbol{\phi}_1^{\dagger l}, ..., \boldsymbol{\phi}_N^{\dagger l}, \bar{\boldsymbol{\phi}}^{\dagger u}, \boldsymbol{W}^{(g)}\boldsymbol{\psi}^\dagger], \tag{16}$$

where $\boldsymbol{W}^{(g)} \in \mathbb{R}^{d\times d}$ a linear projection for HO. $\boldsymbol{T}$ is projected onto the query, key and value linear projections. Spatial-HOP attention captures relations among spatially-aware first-order and spatially orderless high-order representations. The outputs of Z-shot and Spatial-HOP transformer heads are combined and fed into the classifier/bounding-box regressor. See §B of **Suppl. Material** for details.

4.3   Pipeline Details (Fig. 1 (*Bottom*))

HOP unit uses ⊚ to split the channel mode into groups (*e.g.*, 2:1:1 split means two parts of the channel dimension are used to form $\mathcal{M}^{(2)}$, one part to form $\mathcal{M}^{(3)}$, and one part to form $\mathcal{M}^{(4)}$). TSOs with parameters $\eta_2 = ... = \eta_r = \eta$ are applied for orders $r = 2, ..., r$, and diagonal entries are extracted from each tensor and concatenated by ⊙. Element-wise SigmE with $\eta_2' = ... = \eta_r' = \eta'$ is applied.

"Orderless FO, Spatial, Orderless HO" block combines the First-Order (FO), spatial and High-Order (HO) representations. Operator "Z-avg" performs average pooling along Z-way mode, operator "Sp-avg" performs average pooling along the spatial modes of feature maps, operator ⊚ simply splits the channel mode into two equally sized groups (each is half of the channel dimension), and operator ⊛ performs concatenation of FO, spatial and HO representations along the spatial mode of feature maps, *e.g.*, we obtain $N+2$ fibers times 1024 channels.

Z-shot T-RH is a transformer which performs attention on individual Z-shots. The spatial representation $\boldsymbol{\Phi}$ is average pooled along spatial dimensions by "Sp-avg" and combined with high-order $\boldsymbol{\Psi}$ (passed by a FC layer) via addition ⊕. Another FC layer follows and subsequently the value, key and query matrices are computed, an RBF attention formed. Operator ● multiplies the value matrix with the RBF matrix. Head is repeated T times, outputs concatenated by ⊙.

Spatial-HOP T-RH takes inputs from the "Orderless FO, Spatial, Orderless HO" block, and computes the value, key and query matrices. The attention matrix (RBF kernel) has $(N+2)\times(N+2)$ size, being composed of spatial, FO-spatial and HO-spatial attention. After multiplying the attention matrix with the value matrix, we extract the spatial, FO and HO representations. Support and query first-order representations (FO) (and high-order representations (HO)) are element-wisely multiplied by ● (multiplicative relationship). Finally, support and query spatial representations use the subtraction operator $\ominus$. After the concatenation of FO and HO relational representations by $\odot$, passing via an MLP (FC+ReLU+ FC), and concatenation with the spatial relational representations, we get one attention block, repeated T times.

5   Experiments

Datasets and Settings. For PASCAL VOC 2007/12 [6], we adopt the 15/5 base/novel category split setting and use training/validation sets from PASCAL VOC 2007 and 2012 for training, and the testing set from PASCAL VOC 2007 for testing, following [7,11,23,50]. For MS COCO [28], we follow [45], and adopt the 20 categories that overlap with PASCAL VOC as the novel categories for testing, whereas the remaining 60 categories are used for training. For the FSOD dataset [7], we split its 1000 categories into 800/200 for training/testing.

Implementation Details. TENET uses ResNet-50 pre-trained on ImageNet [5] and MS COCO [28]. We fine-tune the network with a learning rate of 0.002 for the first 56000 iterations and 0.0002 for another 4000 iterations. Images are resized to 600 pixels (shorter edge) and the longer edge is capped at 1000 pixels. Each support image is cropped based on ground-truth boxes, bilinearly interpolated and padded to 320×320 pixels. We set, via cross-validation) SigmE parameter $\eta'=200$ and TSO parameters $\eta_2 = \eta_3 = \eta_4 = 7$. We report standard metrics for FSOD, namely mAP, AP, AP_{50} and AP_{75}.

Table 1. Evaluations (mAP %) on three splits of the VOC 2007 testing set.

Method/Shot		Split 1				Split 2				Split 3				Mean			
		1	3	5	10	1	3	5	10	1	3	5	10	1	3	5	10
FRCN	ICCV12	11.9	29.0	36.9	36.9	5.9	23.4	29.1	28.8	5.0	18.1	30.8	43.4	7.6	23.5	32.3	36.4
FR	ICCV19	14.8	26.7	33.9	47.2	15.7	22.7	30.1	39.2	19.2	25.7	40.6	41.3	16.6	25.0	34.9	42.6
Meta	ICCV19	19.9	35.0	45.7	51.5	10.4	29.6	34.8	45.4	14.3	27.5	41.2	48.1	14.9	30.7	40.6	48.3
FSOD	CVPR20	29.8	36.3	48.4	53.6	22.2	25.2	31.2	39.7	24.3	34.4	47.1	50.4	25.4	32.0	42.2	47.9
NP-RepMet	NeurIPS20	37.8	41.7	47.3	49.4	**41.6**	43.4	47.4	49.1	33.3	39.8	41.5	44.8	37.6	41.6	45.4	47.8
PNSD	ACCV20	32.4	39.6	50.2	55.1	30.2	30.3	36.4	42.3	30.8	38.6	46.9	52.4	31.3	36.2	44.5	49.9
MPSR	ECCV20	41.7	51.4	55.2	61.8	24.4	39.2	39.9	47.8	**35.6**	42.3	48.0	49.7	33.9	44.3	47.7	53.1
TFA	ICML20	39.8	44.7	55.7	56.0	23.5	34.1	35.1	39.1	30.8	42.8	49.5	49.8	31.4	40.5	46.8	48.3
FSCE	CVPR21	44.2	51.4	61.9	63.4	27.3	43.5	44.2	50.2	22.6	39.5	47.3	54.0	31.4	44.8	51.1	55.9
CGDP+FRCN	CVPR21	40.7	46.5	57.4	62.4	27.3	40.8	42.7	46.3	31.2	43.7	50.1	55.6	33.1	43.7	50.0	54.8
TIP	CVPR21	27.7	43.3	50.2	56.6	22.7	33.8	40.9	46.9	21.7	38.1	44.5	50.9	24.0	38.4	45.2	52.5
FSOD^{up}	ICCV21	43.8	50.3	55.4	61.7	31.2	41.2	44.2	48.3	35.5	43.9	50.6	53.5	36.8	45.1	50.1	54.5
QSAM	WACV22	31.1	39.2	50.7	59.4	22.9	32.1	35.4	42.7	24.3	35.0	50.0	53.6	26.1	35.4	45.4	51.9
TENET	(Ours)	**46.7**	**55.4**	**62.3**	**66.9**	40.3	**44.7**	**49.3**	**52.1**	35.5	**46.0**	**54.4**	**54.6**	**40.8**	**48.7**	**55.3**	**57.9**

Table 2. Evaluations on the MS COCO minival set (2a) and FSOD testset (2b).

(a)

Shot	Method		AP	AP_{50}	AP_{75}
	LSTD	AAAI18	3.2	8.1	2.1
	FR	ICCV12	5.6	12.3	4.6
	Meta	ICCV19	8.7	19.18	6.6
	MPSR	ECCV20	9.8	17.9	9.7
10	FSOD	CVPR20	11.1	20.4	10.6
	PNSD	ACCV20	12.3	21.7	11.7
	TFA	ICML20	9.6	10.0	9.3
	FSCE	CVPR21	10.7	11.9	10.5
	CGDP+FRCN	CVPR21	11.3	20.3	11.5
	FSOD[up]	ICCV21	11.6	23.9	9.8
	QSAM	WACV22	13.0	24.7	12.1
	TENET	**(Ours)**	**19.1**	**27.4**	**19.6**

(b)

Shot	Method		AP_{50}	AP_{75}
	LSTD (FRN)	AAAI18	23.0	12.9
	LSTD	AAAI18	24.2	13.5
5	FSOD	CVPR20	27.5	19.4
	PNSD	ACCV20	29.8	22.6
	QSAM	WACV22	30.7	25.9
	TENET	**(Ours)**	**35.4**	**31.6**

Table 3. Results on VOC2007 testset for applying TENET in RPN or TRH (3a, top panel of 3b). TRH ablation shown in bottom panel of 3b.

(a)

r (2 3 4)	dim. split	Shot(Novel) 5	10	Shot(Base) 5	10	Speed (img/ms)
✓		56.5	64.2	71.7	75.5	32
✓		55.7	63.2	67.0	72.1	69
✓		51.4	58.9	68.7	74.8	78
✓ ✓	3:1	58.3	63.2	69.3	75.1	42
✓ ✓	3:1	56.1	62.4	70.8	75.4	68
✓ ✓	2:2	51.8	61.7	68.1	73.6	71
	6:1:1	53.6	62.7	69.4	72.8	
	5:2:1	**62.3**	**66.9**	**73.8**	**77.9**	59
✓ ✓ ✓	5:1:2	53.9	63.1	69.7	73.3	
	4:2:2	61.4	65.0	70.4	74.9	
	4:3:1	59.1	63.6	71.8	75.2	
	4:1:3	61.0	64.1	68.9	72.5	

(b)

	RPN r	TRH r	Shot(Novel) 5	10	Shot(Base) 5	10
a	1	1	53.4	61.8	64.9	72.1
b	2,3,4	1	57.2	63.7	68.8	76.6
c	2,3,4	2,3,4	61.0	65.4	71.3	77.3
d	2,3,4	1,2,3,4	**62.3**	**66.9**	**73.8**	**78.2**

TRH Z-shot	Spatial-HOP	Shot(Novel) 5	10	Shot(Base) 5	10
✓		58.5	63.2	69.3	75.1
	✓	61.0	65.8	71.7	76.5
✓	✓	**62.3**	**66.9**	**73.8**	**78.2**

5.1   Comparisons with the State of the Art

PASCAL VOC 2007/12. We compare our method to QSAM [23], FSOD[up] [42], CGDP+FRCN [26], TIP [24], FSCE [35], TFA [39], Feature Reweighting (FR) [11], LSTD [4], FRCN [33], NP-RepMet [46], MPSR [43], PSND [50] and FSOD [7]. Table 1 shows that our TENET outperforms FSOD by a 7.1–15.4% margin. For the 1- and 10-shot regime, we outperform QSAM [23] by ∼14.7%.

MS COCO. Table 2a compares TENET with QSAM [23], FSOD^{up} [42], CGDP+ FRCN [26], TIP [24], FSCE [35], TFA [39], FR [11], Meta R-CNN [45], FSOD [7] and PNSD [50] on the MS COCO minival set (20 novel categories, 10-shot protocol). Although MS COCO is more challenging in terms of complexity and the dataset size, TENET boosts results to 19.1%, 27.4% and 19.6%, surpassing the SOTA method QSAM by 6.1%, 2.7% and 7.5% on AP, AP_{50} and AP_{75}.

FSOD. Table 2b compares TENET (5-shot) with PNSD [50], FSOD [7], LSTD [4] and LSTD (FRN [33]). We re-implement BD&TK, modules of LSTD, based on Faster-RCNN for fairness. TENET yields SOTA 35.4% AP_{50} and 31.6% AP_{75}.

5.2 Hyper-parameter and Ablation Analysis

TENET. Table 3a shows that among orders $r=2$, $r=3$ and $r=4$, variant $r=2$ is the best. We next consider pairs of orders, and the triplet $r=2,3,4$. As the number of tensor coefficients grows quickly w.r.t. r, we split the 1024 channels into groups, *e.g.*, $r=2,3$. A 3:1 split means that second- and third-order tensors are built from 768 and 256 channels ($768+256=1024$). We report only the best splits. For pairs of orders, variant $r=2,3$ was the best. Triplet $r=2,3,4$, the best performer, outperforms $r=2$ by 5.8% and 2.7% in novel classes (5- and 10-shot), and 2.1% and 2.4% in base classes. As all representations are 1024-dimensional, we conclude that multi-order variants are the most informative.

TSO. Based on the best channel-wise splits in Table 3a, we study the impact of η_r (shrinkage/decorrelation) of TSO to verify its effectiveness. Figure 2a shows mAP w.r.t. the individual η_2, η_3 and η_4 for $r=2$, $r=3$ and $r=4$. We then investigate the impact of η_r on pairwise representations, where we set the same η_r for pairwise variants, *e.g.*, $\eta_2 = \eta_3$. Again, the same value of η_r is used for triplet $r=2,3,4$. Note that for $\eta_r=1$, TSO is switched off and all representations reduce to the polynomial feature maps in So-HoT [14]. As shown in Fig. 2a, TSO is very beneficial ($\sim$ 5% gain for triplet $r=2,3,4$ over not using TSO).

TRH. Below we investigate the impact of Z-shot and Spatial-HOP T-RHs on results. Table 3b (bottom) shows that both heads are highly complementary.

Other Hyperparameters. We start by varying σ of the RBF kernel from 0.3 to 3. Fig 2b shows that $\sigma=0.5$ gives the best result. We now fix σ and investigate the impact of varying the number of heads used in T-Heads Attention (TA). Table 4a shows best performance with $TA=4$. Lastly, we vary the number of TENET blocks (TB). Table 4b shows that results are stable especially if $TB \geq 2$. Unless otherwise noted, $TA=2$ and $TB=4$, respectively, on VOC dataset. See §C of **Suppl. Material** for more results on FSOD and MS COCO.

Impact of TENET on RPN and TRH. Table 3b shows ablations w.r.t. TENET variants in: 1) either RPN or TRH, or 2) both RPN and TRH. Comparing results for settings a,b, and c confirms that using second-, third- and fourth-orders simultaneously benefits both RPN and TRH, achieving 3.8%/1.9%

as well as 3.8%/1.8% improvement on novel classes over the first-order-only variant. Results for settings c and d show that TRH encodes better the information carried within regions if leveraging both first- and higher-order representations.

Table 4. Effect of varying (a) group within MHA in Tab. 4a and (b) TENET block in Tab. 4b on PASCAL VOC 2007 (5/10-shot, novel classes). When varying TENET block, group number is fixed to 4 (best value).

TA	1	2	4	8	16	32	64	TB	1	2	3	4	5
Shot (Novel) 5	58.3	59.1	**61.8**	60.5	58.4	58.5	56.0	Shot (Novel) 5	61.8	**62.3**	62.1	61.8	61.9
10	61.2	62.8	**65.8**	64.2	61.2	61.7	60.4	10	65.8	**66.9**	66.4	66.1	66.4

(a)   (b)

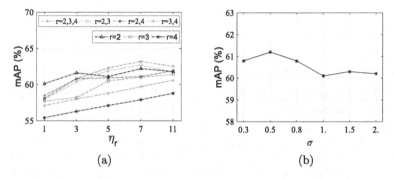

(a)   (b)

Fig. 2. mAP (VOC2007 dataset, novel classes, 10-shot) w.r.t. varying η_r in TSO (Fig. 2a) and the σ of RBF kernel in self-attention (Fig. 2b).

6   Conclusions

We have proposed TENET, which uses higher-order tensor descriptors, in combination with a novel Tensor Shrinkage Operator, to generate highly-discriminative representations with tractable dimensionality. We use these representations in our proposed Transformer Relation Head to dynamically extract correlations between query image regions and support crops. TENET has heightened robustness to large intra-class variations, leading to SOTA performance on all benchmarks.

References

1. Exponentiation by squaring. Wikipedia. https://en.wikipedia.org/wiki/Exponentiation_by_squaring. Accessed 12 Mar 2021
2. Tsallis entropy. Wikipedia. https://en.wikipedia.org/wiki/Tsallis_entropy. Accessed 12 Mar 2021
3. Carion, N., Massa, F., Synnaeve, G., Usunier, N., Kirillov, A., Zagoruyko, S.: End-to-end object detection with transformers. In: Vedaldi, A., Bischof, H., Brox, T., Frahm, J.-M. (eds.) ECCV 2020. LNCS, vol. 12346, pp. 213–229. Springer, Cham (2020). https://doi.org/10.1007/978-3-030-58452-8_13
4. Chen, H., Wang, Y., Wang, G., Qiao, Y.: LSTD: a low-shot transfer detector for object detection. In: McIlraith, S.A., Weinberger, K.Q. (eds.) Proceedings of the Thirty-Second AAAI Conference on Artificial Intelligence, (AAAI-18), the 30th innovative Applications of Artificial Intelligence (IAAI-18), and the 8th AAAI Symposium on Educational Advances in Artificial Intelligence (EAAI-18), New Orleans, Louisiana, USA, 2–7 February 2018, pp. 2836–2843. AAAI Press (2018)
5. Deng, J., Dong, W., Socher, R., Li, L., Li, K., Li, F.: Imagenet: a large-scale hierarchical image database. In: 2009 IEEE Computer Society Conference on Computer Vision and Pattern Recognition (CVPR 2009), 20–25 June 2009, Miami, Florida, USA, pp. 248–255. IEEE Computer Society (2009). https://doi.org/10.1109/CVPR.2009.5206848
6. Everingham, M., Gool, L.V., Williams, C.K.I., Winn, J.M., Zisserman, A.: The pascal visual object classes (VOC) challenge. Int. J. Comput. Vis. **88**(2), 303–338 (2010). https://doi.org/10.1007/s11263-009-0275-4
7. Fan, Q., Zhuo, W., Tai, Y.: Few-shot object detection with attention-rpn and multi-relation detector. CoRR abs/1908.01998 (2019)
8. Girshick, R.B.: Fast R-CNN. In: 2015 IEEE International Conference on Computer Vision, ICCV 2015, Santiago, Chile, 7–13 December 2015, pp. 1440–1448. IEEE Computer Society (2015). https://doi.org/10.1109/ICCV.2015.169
9. Hu, H., Zhang, Z., Xie, Z., Lin, S.: Local relation networks for image recognition. In: ICCV 2019, pp. 3463–3472. IEEE (2019). https://doi.org/10.1109/ICCV.2019.00356
10. Hu, J., Shen, L., Albanie, S., Sun, G., Wu, E.: Squeeze-and-excitation networks. IEEE Trans. Pattern Anal. Mach. Intell. **42**(8), 2011–2023 (2020). https://doi.org/10.1109/TPAMI.2019.2913372
11. Kang, B., Liu, Z., Wang, X., Yu, F., Feng, J., Darrell, T.: Few-shot object detection via feature reweighting. In: 2019 IEEE/CVF International Conference on Computer Vision, ICCV 2019, Seoul, Korea (South), October 27–November 2, 2019, pp. 8419–8428. IEEE (2019). https://doi.org/10.1109/ICCV.2019.00851
12. Karlinsky, L., et al.: Repmet: representative-based metric learning for classification and few-shot object detection. In: IEEE Conference on Computer Vision and Pattern Recognition, CVPR 2019, Long Beach, CA, USA, 16–20 June 2019, pp. 5197–5206. Computer Vision Foundation/IEEE (2019). https://doi.org/10.1109/CVPR.2019.00534
13. Kong, T., Yao, A., Chen, Y., Sun, F.: Hypernet: towards accurate region proposal generation and joint object detection. In: 2016 IEEE Conference on Computer Vision and Pattern Recognition, CVPR 2016, Las Vegas, NV, USA, 27–30 June 2016, pp. 845–853. IEEE Computer Society (2016). https://doi.org/10.1109/CVPR.2016.98

14. Koniusz, P., Tas, Y., Porikli, F.: Domain adaptation by mixture of alignments of second-or higher-order scatter tensors. In: 2017 IEEE Conference on Computer Vision and Pattern Recognition, CVPR 2017, Honolulu, HI, USA, 21–26 July 2017, pp. 7139–7148. IEEE Computer Society (2017). https://doi.org/10.1109/CVPR.2017.755

15. Koniusz, P., Wang, L., Cherian, A.: Tensor representations for action recognition. In: TPAMI (2020)

16. Koniusz, P., Yan, F., Gosselin, P.H., Mikolajczyk, K.: Higher-order occurrence pooling on mid-and low-level features: Visual concept detection. Tech, Report (2013)

17. Koniusz, P., Yan, F., Gosselin, P., Mikolajczyk, K.: Higher-order occurrence pooling for bags-of-words: visual concept detection. IEEE Trans. Pattern Anal. Mach. Intell. **39**(2), 313–326 (2017). https://doi.org/10.1109/TPAMI.2016.2545667

18. Koniusz, P., Zhang, H.: Power normalizations in fine-grained image, few-shot image and graph classification. In: TPAMI (2020)

19. Koniusz, P., Zhang, H., Porikli, F.: A deeper look at power normalizations. In: 2018 IEEE Conference on Computer Vision and Pattern Recognition, CVPR 2018, Salt Lake City, UT, USA, 18–22 June 2018, pp. 5774–5783. IEEE Computer Society (2018). https://doi.org/10.1109/CVPR.2018.00605

20. Krizhevsky, A., Sutskever, I., Hinton, G.E.: Imagenet classification with deep convolutional neural networks. Commun. ACM **60**(6), 84–90 (2017). https://doi.org/10.1145/3065386

21. Lathauwer, L.D., Moor, B.D., Vandewalle, J.: A multilinear singular value decomposition. SIAM J. Matrix Anal. Appl. **21**, 1253–1278 (2000)

22. Ledoit, O., Wolf, M.: Honey, i shrunk the sample covariance matrix. J. Portfolio Manage. **30**(4), 110–119 (2004). https://doi.org/10.3905/jpm.2004.110

23. Lee, H., Lee, M., Kwak, N.: Few-shot object detection by attending to per-sample-prototype. In: WACV, 2022, Waikoloa, HI, USA, 3–8 January 2022, pp. 1101–1110. IEEE (2022). https://doi.org/10.1109/WACV51458.2022.00117

24. Li, A., Li, Z.: Transformation invariant few-shot object detection. In: Proceedings of the IEEE/CVF Conference on Computer Vision and Pattern Recognition, pp. 3094–3102 (2021)

25. Li, X., Wang, W., Hu, X., Yang, J.: Selective kernel networks. In: IEEE Conference on Computer Vision and Pattern Recognition, CVPR 2019, Long Beach, CA, USA, 16–20 June 2019, pp. 510–519. Computer Vision Foundation/IEEE (2019). https://doi.org/10.1109/CVPR.2019.00060

26. Li, Y., et al.: Few-shot object detection via classification refinement and distractor retreatment. In: Proceedings of the IEEE/CVF Conference on Computer Vision and Pattern Recognition, pp. 15395–15403 (2021)

27. Lin, T., Dollár, P., Girshick, R.B., He, K., Hariharan, B., Belongie, S.J.: Feature pyramid networks for object detection. In: 2017 IEEE Conference on Computer Vision and Pattern Recognition, CVPR 2017, Honolulu, HI, USA, 21–26 July 2017, pp. 936–944. IEEE Computer Society (2017). https://doi.org/10.1109/CVPR.2017.106

28. Lin, T.-Y., et al.: Microsoft COCO: common objects in context. In: Fleet, D., Pajdla, T., Schiele, B., Tuytelaars, T. (eds.) ECCV 2014. LNCS, vol. 8693, pp. 740–755. Springer, Cham (2014). https://doi.org/10.1007/978-3-319-10602-1_48

29. Lu, J., et al.: SOFT: softmax-free transformer with linear complexity. CoRR abs/2110.11945 (2021)

30. Rahman, S., Wang, L., Sun, C., Zhou, L.: Redro: efficiently learning large-sized spd visual representation. In: European Conference on Computer Vision (2020)

31. Redmon, J., Farhadi, A.: YOLO9000: better, faster, stronger. In: 2017 IEEE Conference on Computer Vision and Pattern Recognition, CVPR 2017, Honolulu, HI, USA, 21–26 July 2017, pp. 6517–6525. IEEE Computer Society (2017). https://doi.org/10.1109/CVPR.2017.690

32. Redmon, J., Farhadi, A.: Yolov3: an incremental improvement. CoRR abs/1804.02767 (2018)

33. Ren, S., He, K., Girshick, R.B., Sun, J.: Faster R-CNN: towards real-time object detection with region proposal networks. In: Cortes, C., Lawrence, N.D., Lee, D.D., Sugiyama, M., Garnett, R. (eds.) Advances in Neural Information Processing Systems 28: Annual Conference on Neural Information Processing Systems 2015, 7–12 December 2015, Montreal, Quebec, Canada, pp. 91–99 (2015)

34. Smola, A.J., Kondor, R.: Kernels and regularization on graphs. In: Schölkopf, B., Warmuth, M.K. (eds.) COLT-Kernel 2003. LNCS (LNAI), vol. 2777, pp. 144–158. Springer, Heidelberg (2003). https://doi.org/10.1007/978-3-540-45167-9_12

35. Sun, B., Li, B., Cai, S., Yuan, Y., Zhang, C.: FSCE: few-shot object detection via contrastive proposal encoding. CoRR abs/2103.05950 (2021)

36. Szegedy, C., et al.: Going deeper with convolutions. In: IEEE Conference on Computer Vision and Pattern Recognition, CVPR 2015, Boston, MA, USA, 7–12 June 2015, pp. 1–9. IEEE Computer Society (2015). https://doi.org/10.1109/CVPR.2015.7298594

37. Tuzel, O., Porikli, F., Meer, P.: Region covariance: a fast descriptor for detection and classification. In: Leonardis, A., Bischof, H., Pinz, A. (eds.) ECCV 2006. LNCS, vol. 3952, pp. 589–600. Springer, Heidelberg (2006). https://doi.org/10.1007/11744047_45

38. Vaswani, A., et al.: Attention is all you need. In: Advances in Neural Information Processing Systems 30 2017, pp. 5998–6008 (2017)

39. Wang, X., Huang, T.E., Gonzalez, J., Darrell, T., Yu, F.: Frustratingly simple few-shot object detection. In: ICML 2020. Proceedings of Machine Learning Research, vol. 119, pp. 9919–9928. PMLR (2020)

40. West, J., Venture, D., Warnick, S.: Spring research presentation: a theoretical foundation for inductive transfer. Brigham Young Univ. College Phys. Math. Sci. (2007). https://web.archive.org/web/20070801120743/http://cpms.byu.edu/springresearch/abstract-entry?id=861

41. Woodworth, R.S., Thorndike, E.L.: The influence of improvement in one mental function upon the efficiency of other functions. Psychol. Rev. (I) 8(3), 247–261 (1901). https://doi.org/10.1037/h0074898

42. Wu, A., Han, Y., Zhu, L., Yang, Y.: Universal-prototype enhancing for few-shot object detection. In: Proceedings of the IEEE/CVF International Conference on Computer Vision (ICCV), pp. 9567–9576, October 2021

43. Wu, J., Liu, S., Huang, D., Wang, Y.: Multi-scale positive sample refinement for few-shot object detection. In: Vedaldi, A., Bischof, H., Brox, T., Frahm, J.-M. (eds.) ECCV 2020. LNCS, vol. 12361, pp. 456–472. Springer, Cham (2020). https://doi.org/10.1007/978-3-030-58517-4_27

44. Xie, S., Girshick, R.B., Dollár, P., Tu, Z., He, K.: Aggregated residual transformations for deep neural networks. In: 2017 IEEE Conference on Computer Vision and Pattern Recognition, CVPR 2017, Honolulu, HI, USA, 21–26 July 2017, pp. 5987–5995. IEEE Computer Society (2017). https://doi.org/10.1109/CVPR.2017.634

45. Yan, X., Chen, Z., Xu, A., Wang, X., Liang, X., Lin, L.: Meta R-CNN: towards general solver for instance-level low-shot learning. In: 2019 IEEE/CVF International Conference on Computer Vision, ICCV 2019, Seoul, Korea (South), October 27–November 2, 2019, pp. 9576–9585. IEEE (2019). https://doi.org/10.1109/ICCV.2019.00967

46. Yang, Y., Wei, F., Shi, M., Li, G.: Restoring negative information in few-shot object detection. In: Larochelle, H., Ranzato, M., Hadsell, R., Balcan, M., Lin, H. (eds.) Advances in Neural Information Processing Systems 33: Annual Conference on Neural Information Processing Systems 2020, NeurIPS 2020, 6–12 December 2020, virtual (2020)

47. Zhang, H., Zhang, L., Qi, X., Li, H., Torr, P.H.S., Koniusz, P.: Few-shot action recognition with permutation-invariant attention. In: Vedaldi, A., Bischof, H., Brox, T., Frahm, J.-M. (eds.) ECCV 2020. LNCS, vol. 12350, pp. 525–542. Springer, Cham (2020). https://doi.org/10.1007/978-3-030-58558-7_31

48. Zhang, H., Koniusz, P.: Power normalizing second-order similarity network for few-shot learning. In: IEEE Winter Conference on Applications of Computer Vision, WACV 2019, Waikoloa Village, HI, USA, 7–11 January 2019, pp. 1185–1193. IEEE (2019). https://doi.org/10.1109/WACV.2019.00131

49. Zhang, H., Koniusz, P., Jian, S., Li, H., Torr, P.H.S.: Rethinking class relations: absolute-relative supervised and unsupervised few-shot learning. In: Proceedings of the IEEE/CVF Conference on Computer Vision and Pattern Recognition (CVPR), pp. 9432–9441, June 2021

50. Zhang, S., Luo, D., Wang, L., Koniusz, P.: Few-shot object detection by second-order pooling. In: Proceedings of the Asian Conference on Computer Vision (2020)

51. Zhang, S., Wang, L., Murray, N., Koniusz, P.: Kernelized few-shot object detection with efficient integral aggregation. In: IEEE Conference on Computer Vision and Pattern Recognition (2022)

52. Zhu, X., Su, W., Lu, L., Li, B., Wang, X., Dai, J.: Deformable DETR: deformable transformers for end-to-end object detection. In: ICLR 2021. OpenReview.net (2021)

Self-Promoted Supervision for Few-Shot Transformer

Bowen Dong[1], Pan Zhou[2], Shuicheng Yan[2], and Wangmeng Zuo[1,3(✉)]

[1] Harbin Institute of Technology, Harbin, China
wmzuo@hit.edu.cn
[2] National University of Singapore, Singapore, Singapore
[3] Peng Cheng Laboratory, Shenzhen, China

Abstract. The few-shot learning ability of vision transformers (ViTs) is rarely investigated though heavily desired. In this work, we empirically find that with the same few-shot learning frameworks, replacing the widely used CNN feature extractor with a ViT model often severely impairs few-shot classification performance. Moreover, our empirical study shows that in the absence of inductive bias, ViTs often learn the low-qualified token dependencies under few-shot learning regime where only a few labeled training data are available, which largely contributes to the above performance degradation. To alleviate this issue, we propose a simple yet effective few-shot training framework for ViTs, namely Self-promoted sUpervisioN (SUN). Specifically, besides the conventional global supervision for global semantic learning, SUN further pretrains the ViT on the few-shot learning dataset and then uses it to generate individual location-specific supervision for guiding each patch token. This location-specific supervision tells the ViT which patch tokens are similar or dissimilar and thus accelerates token dependency learning. Moreover, it models the local semantics in each patch token to improve the object grounding and recognition capability which helps learn generalizable patterns. To improve the quality of location-specific supervision, we further propose: 1) background patch filtration to filtrate background patches out and assign them into an extra background class; and 2) spatial-consistent augmentation to introduce sufficient diversity for data augmentation while keeping the accuracy of the generated local supervisions. Experimental results show that SUN using ViTs significantly surpasses other few-shot learning frameworks with ViTs and is the first one that achieves higher performance than those CNN state-of-the-arts. Our code is publicly available at https://github.com/DongSky/few-shot-vit.

Keywords: Few-shot learning · Location-specific supervision

Supplementary Information The online version contains supplementary material available at https://doi.org/10.1007/978-3-031-20044-1_19.

1   Introduction

Vision transformers (ViTs) have achieved great success in the computer vision field, and even surpass corresponding state-of-the-art CNNs on many vision tasks, e.g. image classification [10,29,41,42,45,57], object detection [4,11] and segmentation [37,58]. One key factor contributing to ViTs' success is their powerful self-attention mechanism [43], which does not introduce any inductive bias and can better capture the long-range dependencies among local features in the data than the convolution mechanism in CNNs. This motivates us to investigate two important problems. First, we wonder whether ViTs can perform well under few-shot learning setting or not, which aims to recognize objects from novel categories with only a handful of labeled samples as reference. Second, if no, how to improve the few-shot learning ability of ViTs? The few-shot learning ability is heavily desired for machine intelligence, since in practice, many tasks actually have only a few labeled data due to data scarcity (e.g. disease data), high equipment cost to capture data and manual labeling cost (e.g. pixel-level labeling). In this work, we are particularly interested in few-shot classification [12,36,44] which is a benchmark task to evaluate few-shot learning capacity of a method.

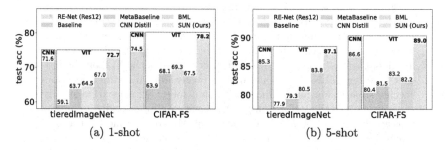

Fig. 1. Performance comparisons of different few-shot classification frameworks. Except the state-of-the-art RE-Net [19] which uses ResNet-12 [15] as the feature extractor, all methods use the same NesT transformer [57] as the feature extractor. With same ViT backbone, our SUN significantly surpasses other baselines.

For the first question, unfortunately, we empirically find that for representative few-shot learning frameworks, e.g. Meta-Baseline [6], replacing the CNN feature extractor by ViTs severely impairs few-shot classification performance. The most possible reason is the lack of inductive bias in ViTs—in absence of any prior inductive bias, ViTs needs a large amount of data to learn the dependencies among local features, i.e. patch tokens. Our analysis in Sect. 3 shows that introducing CNN-alike inductive bias can partly accelerate the token dependency learning in ViTs and thus partially mitigate this effect, which also accords with the observations in [22,56] under supervised learning. For example, in Fig. 1, the "Baseline" [5] and "Meta-Baseline" [6] use vanilla ViTs, while "CNN Distill" introduces CNN-alike inductive bias via distilling a pretrained CNN teacher network into the ViT and achieves higher performance. However, CNN-alike inductive bias cannot be inherently suitable for ViTs and cannot well enhance and accelerate token dependency learning in ViTs, because ViTs and CNNs have different network architectures, and also the inductive bias in CNNs often cannot well capture long-range dependencies. This naturally motivates us to consider how to improve the few-shot learning ability of ViTs.

We therefore propose an effective few-shot learning framework for ViTs, namely, *Self-promoted sUpervisioN* (SUN for short). SUN enables a ViT model to generalize well on novel unseen categories with a few labeled training data. The key idea of SUN is to enhance and accelerate dependency learning among different tokens, i.e., patch tokens, in ViTs. This idea is intuitive, since if ViTs can learn the token dependencies fast and accurately, and thus naturally can learn from less labeled training data, which is consistent with few-shot learning scenarios. In particular, at the meta-training phase, SUN provides global supervision to global feature embedding and further employs individual location-specific supervision to guide each patch token. Here SUN first trains the ViT on the training data in few-shot learning, and uses it to generate patch-level pseudo labels as location-specific supervision. This is why we call our method "Self-Promoted Supervision" as it uses the same ViT to generate local supervision. To improve the quality of patch-level supervision, we further propose two techniques, namely 1) background patch filtration and 2) spatial-consistent augmentation. The former aims to alleviate effects of bad cases where background patches are wrongly assigned to a semantic class and have incorrect local supervision; the latter is to introduce sufficient diversity to data augmentation while keeping the accuracy of generated location-specific supervisions. Our location-specific dense supervision benefits ViT on few-shot learning tasks from two aspects. Firstly, considering the location-specific supervisions on all tokens are consistent, i.e. similar pseudo labels for similar local tokens, this dense supervision tells ViT which patch tokens are similar or dissimilar and thus can accelerate ViT to learn high-quality token dependencies. Secondly, the local semantics in each patch token are also well modeled to improve the object grounding and recognition capabilities of ViTs. Actually, as shown in [18,60], modeling semantics in local tokens can avoid learning skewed and non-generalizable patterns and thus substantially improve the model generalization performance which is also heavily desired in few-shot learning. So both aspects can improve the few-shot learning capacity of ViTs. Next, at the meta-tuning phase, similar to existing methods [6,51], SUN adapts the knowledge of ViT trained at the meta-training phase to new tasks by fine-tuning ViT on the corresponding training data.

Extensive experimental results demonstrate that with the same ViT architecture, our SUN framework significantly outperforms all baseline frameworks on the few-shot classification tasks, as shown in Fig. 1. Moreover, to the best of our knowledge, SUN with ViT backbones is the first work that applies ViT backbones to few-shot classification, and is also the first ViT method that achieves comparable even higher performance than state-of-the-art baselines using similar-sized CNN backbones, as illustrated in Fig. 1. This well shows the superiority of SUN, since CNN has high inductive bias and is actually much more suitable than ViT on few-shot learning problems. Additionally, our SUN also provides a simple yet solid baseline for few-shot classification.

2   Related Work

Vision Transformer.   Unlike CNNs that use convolutions to capture local information in images, ViTs [10] model long-range interaction explicitly, and have shown great potential for vision tasks. However, to achieve comparable performance with CNN of a similar model size trained on ImageNet [8] , ViTs need huge-scale datasets to train [8,38], greatly increasing computational cost. To alleviate this issue and further improve performance, many works [18,29,41,49,53,57] [23,26,40,45,47,50,52,54] propose effective solutions. However, these ViTs still need to train on large-scale datasets, e.g. ImageNet [8] or JFT-300M [38], and often fail on small datasets [22,56,57]. Recently, a few works [3,22] also propose ViTs for small datasets. For instance, Li *et al.* [22] initialized ViT from a well-trained CNN to borrow inductive bias in CNN to ViT. Cao *et al.* [3] first used instance discriminative loss to pretrain ViT and then fine-tuned it on target datasets. In comparison, we propose self-promoted supervision to provide dense local supervisions, enhancing token dependency learning and alleviating the data-hungry issue. Moreover, we are interested in few-shot learning problems that focus on initializing ViTs with small training data from base classes and generalizing on novel unseen categories with a few labeled data.

Few-Shot Classification.   Few-shot classification methods can be coarsely divided into three groups: 1) *optimization-based methods* [12,20,35,61,62] for fast adaptation to new tasks; 2) *memory-based methods* [14,31,33] for storing important training samples; and 3) *metric-based methods* [6,36,44,51] which learn the similarity metrics among samples. This work follows the pipeline of metric-based methods for the simplicity and high accuracy. It is worth noting that some recent works [9,17,27,51,59] also integrate transformer layers with metric-based methods for few shot learning. Different from previous works [9,17,27,51,59] which leverage transformer layers as few-shot classifiers, our work focuses on improving the few-shot classification accuracy of ViT backbones.

3   Empirical Study of ViTs for Few-Shot Classification

Here we first introduce the few-shot classification task. Given a labeled base dataset $\mathbb{D}_{\text{base}}$ which contains base classes C_{base}, this task aims to learn a meta-learner f (feature extractor) such that f can be fast adapted to unseen classes C_{novel} (i.e. $C_{\text{base}} \cap C_{\text{novel}} = \varnothing$) in which each class has a few labeled training samples. This task is a benchmark to evaluate few-shot learning capacity of a method, since it needs to well learn the knowledge in the base data $\mathbb{D}_{\text{base}}$ and then fast adapt the learnt knowledge to the new data C_{novel} via only a few training data. Next, we investigate the ViTs' performance on this important task and also analyze the potential reasons for their performance.

Performance Investigation of ViTs. We investigate the few-shot learning performance of various ViTs, including LV-ViT [18] (single stage ViT), Le-ViT [13] (multi-stage ViT), Visformer [7] (CNN-enhanced ViT), Swin [29] and NesT [57]

Table 1. Classification accuracy (%) of the meta-baseline few-shot learning framework using ViTs and ResNet-12 as a feature extractor on *mini*ImageNet.

	Backbone	Params	Meta-training phase		Meta-tuning phase	
			5-way 1-shot	5-way 5-shot	5-way 1-shot	5-way 5-shot
CNN	ResNet-12	12.5M	**60.00 ± 0.44**	**80.55 ± 0.31**	**64.53 ± 0.45**	**81.41 ± 0.31**
ViT	LV-ViT	13.5M	43.08 ± 0.38	59.03 ± 0.39	43.07 ± 0.39	58.87 ± 0.39
	Le-ViT	12.6M	38.89 ± 0.39	53.51 ± 0.37	40.65 ± 0.40	54.23 ± 0.38
	Swin	12.3M	48.26 ± 0.42	65.72 ± 0.38	54.63 ± 0.45	70.60 ± 0.38
	Visformer	12.4M	43.55 ± 0.38	60.49 ± 0.39	47.61 ± 0.43	63.00 ± 0.39
	NesT	12.6M	49.23 ± 0.43	66.57 ± 0.39	54.57 ± 0.46	69.85 ± 0.38

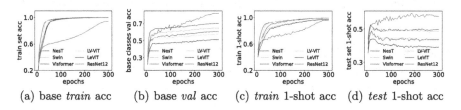

(a) base *train* acc   (b) base *val* acc   (c) *train* 1-shot acc   (d) *test* 1-shot acc

Fig. 2. Accuracy curve comparison of different feature extractors in meta-baseline.

(locality-enhanced ViT). For a fair comparison, we scale the depth and width of these ViTs such that their model sizes are similar to ResNet-12 [15] (∼12.5M parameters) which is the most commonly used architecture and achieves (nearly) state-of-the-art performance on few-shot classification tasks.

Here we choose a simple yet effective Meta-Baseline [6] as the few-shot learning framework to test the performance of the above five ViTs. Specifically, in the meta-training phase, Meta-Baseline performs conventional supervised classification; in meta-tuning phase, given a training task, it computes the class prototypes by averaging the sample features from the same class on the support set, and then minimizes the cosine distance between the sample feature and the corresponding class prototypes. For inference, similar to meta-tuning phase, given a test task τ, it first calculates the class prototypes on the support set of τ, and assigns a class label to a test sample according to cosine distance between the test sample feature and the class prototypes. See more details in Sect. 4.2.

Table 1 summarizes the experimental results on *mini*ImageNet [44]. Here, "Meta-Training Phase" means we test ViTs after meta-training phase in Meta-Baseline, while "Meta-Tuning Phase" denotes the test performance of ViTs after meta-tuning phase. Frustratingly, one can observe that even with the same Meta-Baseline framework, all ViTs perform much worse than ResNet-12 after meta-training phase. Moreover, after meta-tuning phase, all ViTs still suffer from much worse few-shot learning performance. These observations inspire us to investigate what happens during the meta-training period of ViTs and why ViTs perform much worse than CNNs.

Table 2. Empirical analysis of ViTs on two datasets. (Top) shows the effect of inductive bias (IB) to few-shot accuracy of ViTs; (Middle) studies whether CNN-alike inductive bias (CaIB) can enhance token dependency learning; and (Bottom) analyzes local attention (LA) in enhancing token dependency learning.

Method	ViT	Var	miniImageNet		tieredImageNet	
			5-way 1-shot	5-way 5-shot	5-way 1-shot	5-way 5-shot
(Top)		IB				
Meta-baseline [6]	NesT		54.57 ± 0.46	69.85 ± 0.38	$\mathbf{63.73 \pm 0.47}$	79.33 ± 0.38
[6]+CNN [46]	NesT	✓	$\mathbf{57.91 \pm 0.48}$	$\mathbf{73.31 \pm 0.38}$	63.46 ± 0.51	$\mathbf{80.13 \pm 0.38}$
(Middle)		CaIB				
Meta-baseline [6]	NesT		54.57 ± 0.46	69.85 ± 0.38	63.73 ± 0.47	79.33 ± 0.38
[6]+CNN Distill	NesT	✓	$\mathbf{55.79 \pm 0.45}$	$\mathbf{71.81 \pm 0.37}$	$\mathbf{64.48 \pm 0.50}$	$\mathbf{80.43 \pm 0.37}$
(Bottom)		LA				
Meta-baseline [6]	LeViT		38.89 ± 0.39	53.51 ± 0.37	55.75 ± 0.49	71.99 ± 0.39
Meta-baseline [6]	Swin	✓	$\mathbf{54.63 \pm 0.45}$	$\mathbf{70.60 \pm 0.38}$	62.68 ± 0.50	78.52 ± 0.38
Meta-baseline [6]	NesT	✓	54.57 ± 0.46	69.85 ± 0.38	$\mathbf{63.73 \pm 0.47}$	$\mathbf{79.33 \pm 0.38}$

Analysis. To analyze performance of ViTs in the meta-training phase, Fig. 2 plots accuracy curves on training and validation sets in base dataset $\mathbb{D}_{base}$, and reports the 5-way 1-shot accuracy on the training and test sets of the novel classes. According to Fig. 2(a) and 2(b), all ViTs converge well on training and validation sets in the base dataset $\mathbb{D}_{base}$. Figure 2(c) shows that ViTs also converge well on the training data of novel categories during the whole meta-training phase, while Fig. 2(d) indicates that ViTs generalize poorly to the test data of novel categories. For instance, though NesT [57] achieves ~52% accuracy in the first 30 meta-training epochs, its accuracy drops to 49.2% rapidly around the 30th meta-training epochs. In contrast, ResNet-12 maintains nearly 60% 1-shot accuracy in the last 100 meta-training epochs. These observations show that compared to CNN feature extractors, ViTs often suffer from generalization issues on novel categories though they enjoy good generalization ability on $\mathbb{D}_{base}$.

Next, we empirically analyze why ViTs perform worse than CNNs on few-shot classification. The most possible explanation is that the lack of inductive bias leads to the performance degeneration. To illustrate this point, we introduce inductive bias (in CNN) into ViT through three ways. **a)** For each stage, we use a ViT branch and a CNN branch to independently extract image features, and combine their features for fusion. See implementation details in the suppl.. In this way, this mechanism inherits the inductive bias from the CNN branch. Table 2 (Top) shows that by introducing this CNN inductive bias, the new NesT (i.e. "[6]+CNN") largely surpasses the vanilla NesT ("Meta-Baseline [6]"). **b)** We train a CNN model on $\mathbb{D}_{base}$, and use it to teach ViT via knowledge distillation [16]. This method, i.e. "[6]+CNN Distill" in Table 2 (Middle), can well introduce the CNN-alike inductive bias into ViT to enhance dependency learning, and improves the vanilla "Meta-Baseline [6]" by a significant margin. To analyze the quality of token dependency, we visual-

Fig. 3. Visualization of attention maps from vanilla ViTs with different training epochs.

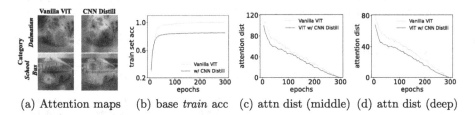

(a) Attention maps   (b) base *train* acc   (c) attn dist (middle) (d) attn dist (deep)

Fig. 4. (a) Attention maps from vanilla ViT and CNN-distilled ViT. (b) training accuracy curves between ViT and CNN-distilled ViT. (c&d) Attention score distance between ViT and CNN-distilled ViT at the middle layer (middle) and last layer (right).

ize the attention maps of the last block of vanilla ViT and CNN-distilled ViT in Fig. 4(a). One can find that CNN-distilled ViT can learn better token dependency than vanilla ViT, since the former captures almost all semantic tokens while the later one only captures a few semantic patches. **c)** Local attention enjoys better few-shot learning performance. Table 2 (Bottom) shows that with the same Meta-Baseline [6] framework, Swin and NesT achieve much higher accuracy than LeViT, since local attention in Swin and NesT introduces (CNN-alike) inductive bias and can enhance token dependency learning than the global attention in LeViT. All these results show that inductive bias benefits few-shot classification.

Moreover, we also observe that the convergence speed is not equivalent to the quality of token dependency learning. In comparison to vanilla ViT, CNN-distilled ViT can learn higher-qualified token dependency as shown in Fig. 4(a), but has slower convergence speed and lower training accuracy as illustrated in Fig. 4(b). So one can conclude that the convergence speed actually does not reflect the quality of token dependency learning. Besides, for vanilla ViT, its training accuracy increases rapidly in the first 30 epochs but becomes saturated in the following epochs in Fig. 4(b), while its attention maps at the last block still evolve in Fig. 3. Indeed, Fig. 4(c) (and 4(d)) indicates that attention score distance descends stably for all training epochs. So these results all show that convergence speed does not well reflect the quality of token dependency learning.

4   Self-Promoted Supervision for Few-Shot Classification

From the observations and analysis in Sect. 3, one knows that 1) directly replacing the CNN feature extractor with ViT in existing few-shot learning frameworks often leads to severe performance degradation; and 2) the few-shot learning frameworks that boost token dependency learning to remedy lack of inductive bias can improve performance of ViT on few-shot learning tasks. Accordingly, here we propose a Self-promoted sUpervisioN (SUN) framework which aims to improve the few-shot learning ability of ViTs via enhancing and accelerating token dependency learning in ViTs via a dense location-specific supervision. Similar to conventional few-shot learning frameworks, e.g. Meta-Baseline, our SUN framework also consists of two phases, i.e., meta-training and meta-tuning which are illustrated in Fig. 5. In the following, we will introduce these phases.

4.1 Meta-Training

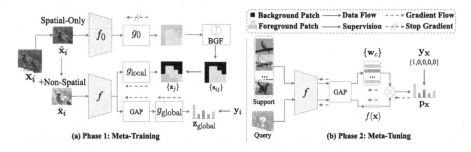

Fig. 5. Training pipeline of ViTs with self-promoted supervision (SUN). For meta-training phase (a), given an image $\mathbf{x}_i$, we use spatial-consistent augmentation to generate two crops $\bar{\mathbf{x}}_i$ and $\tilde{\mathbf{x}}_i$. Given $\tilde{\mathbf{x}}_i$, teacher (ViT f_0 & and classifier g_0) with background filtration (BGF) generates its location-specific supervision $\{\mathbf{s}_{ij}\}$. Given $\bar{\mathbf{x}}_i$, the meta-learner f extracts its token features, and then global classifier g_{global} and local classifier g_{local} respectively predict the global semantic label $\tilde{\mathbf{y}}_i$ of $\bar{\mathbf{x}}_i$ and the semantic labels $\{\tilde{\mathbf{s}}_{ij}\}$ of all patches. Finally, SUN uses ground-truth labels $\mathbf{y}_i$ and the location-specific label $\{\mathbf{s}_{ij}\}$ to supervise $\tilde{\mathbf{y}}_i$ and $\{\tilde{\mathbf{s}}_{ij}\}$ for optimizing f and $g_{\text{global/local}}$. During meta-tuning phase (b), SUN adopts an existing few-shot method (e.g., Meta-Baseline [6]) to fine-tune f and then predicts the label for each query $\mathbf{x}$ with a support set $\mathbf{S}$.

For meta-training phase in Fig. 5(a), similar to Meta-Baseline, our target is to learn a meta learner f mentioned in Sect. 3 such that f is able to fast adapt itself to new classes with a few training data. Here f is actually a feature extractor and is a ViT model in this work. The main idea of SUN is a dense location-specific supervision which aims to enhance and accelerate token dependency learning in ViT and thus boost the learning efficiency on the limited training data.

Specifically, to train the meta-learner f on the base dataset $\mathbb{D}_{\text{base}}$, SUN uses global supervision, i.e., the ground-truth label, to guide the global average token of all patch tokens in f for learning the global semantics of the whole image. More importantly, SUN further employs individual location-specific supervision to supervise each patch token in f. To generate patch-level pseudo labels as location-specific supervision, SUN first conducts the supervised classification pretraining to optimize a teacher model f_g consisting of a ViT f_0 with the same architecture as f and a classifier g_0 on the base dataset $\mathbb{D}_{\text{base}}$. This is reasonable, since teacher f_0 is trained on $\mathbb{D}_{\text{base}}$ and its predicted patch-level pseudo labels are semantically consistent with the ground-truth label of the whole image, i.e., similar pseudo labels for similar local and class tokens. This is why our method is so called *self-promoted* since f_0 has the same architecture as f. Formally, for each training image $\mathbf{x}_i$ in the base dataset $\mathbb{D}_{\text{base}}$, SUN first calculates the per-token classification confidence score $\hat{\mathbf{s}}_i$ as follows:

$$\hat{\mathbf{s}}_i = [\hat{\mathbf{s}}_{i1}, \hat{\mathbf{s}}_{i2}, \cdots, \hat{\mathbf{s}}_{iK}] = f_g(\mathbf{x}_i) = [g(\mathbf{z}_1), g(\mathbf{z}_2), \cdots, g(\mathbf{z}_K)] \in \mathbb{R}^{c \times K}, \quad (1)$$

where $\mathbf{z} = f(\mathbf{x}_i) = [\mathbf{z}_{\text{cls}}, \mathbf{z}_1, \mathbf{z}_2, \cdots, \mathbf{z}_K]$ denotes the class and patch tokens of image $\mathbf{x}_i$, and $\hat{\mathbf{s}}_{ij}$ is the pseudo label of the j-th local patch $\mathbf{x}_{ij}$ in sample $\mathbf{x}_i$. Accordingly, for each patch $\mathbf{x}_{ij}$, the position with relative high confidence in $\hat{\mathbf{s}}_{ij}$ indicates that this patch contains the semantics of corresponding categories.

To fully exploit the power of local supervision, we further propose background filtration (BGF) technique which targets at classifying background patches into a new unique class and improving quality of patch-level supervision. This technique is necessary, since being trained on $\mathbb{D}_{\text{base}}$ which has no background class, the teacher f_0 always wrongly assigns background patches into a base (semantic) class instead of background class and provides inaccurate location-specific supervision. To tackle this issue, we filtrate the local patches with very low confidence score as background patches and classify them into a new unique class. Specifically, given a batch of m patches with confidence score $\{\hat{\mathbf{s}}_{ij}\}$, we first select the maximum score for each patch, and then sort the m scores of patches by an ascending order. Next, we view the top $p\%$ patches with the lowest scores as background patches. Meanwhile, we add a new unique class, i.e., the background class, into the base classes C_{base}. Accordingly, we increase one dimension for the generated pseudo label $\hat{\mathbf{s}}_i \in \mathbb{R}^c$ to obtain $\mathbf{s}_{ij} \in \mathbb{R}^{c+1}$. For background patches, the last positions of their pseudo label $\mathbf{s}_{ij} \in \mathbb{R}^{c+1}$ are one and remaining positions are zero. For non-background patches, we only set the last positions of their pseudo label as zero and do not change other values. In this way, given one image $\mathbf{x}_i$ with ground-truth label $\mathbf{y}_i$, we define its overall training loss:

$$\mathcal{L}_{\text{SUN}} = H(g_{\text{global}}(\mathbf{z}_{\text{global}}), \mathbf{y}_i) + \lambda \sum_{j=1}^{K} H(g_{\text{local}}(\mathbf{z}_j), \mathbf{s}_{ij}), \quad (2)$$

where $\mathbf{z} = f(\mathbf{x}_i) = [\mathbf{z}_{\text{cls}}, \mathbf{z}_1, \mathbf{z}_2, \cdots, \mathbf{z}_K]$ denotes the class and patch tokens of image $\mathbf{x}_i$, and $\mathbf{z}_{\text{global}}$ is a global average pooling of all patch tokens $\{\mathbf{z}_j\}_{j=1}^{K}$. Here H denotes the cross entropy loss, and g_{global} and g_{local} are two trainable classifiers for global semantic classification and local patch classification respectively. For λ, we set it as 0.5 in all experiments for simplicity.

Now we discuss the two benefits of the dense supervision on few-shot learning tasks. Firstly, all location-specific supervisions are generated by the teacher f_0 and thus guarantees similar pseudo labels for similar local tokens. This actually tells ViT which patch tokens are similar or dissimilar and thus can accelerate token dependency learning. Secondly, in contrast to the global supervision on the whole image, the location-specific supervisions are at a fine-grained level, namely the patch level, and thus help ViTs to easily discover the target objects and improve the recognition accuracy. This point is also consistent with some previous literature. For instance, works [18,60] demonstrate that modeling semantics in local tokens can avoid learning skewed and non-generalizable patterns and thus substantially improve the model generalization performance. Therefore, both aspects are heavily desired in few-shot learning problems and can increase the few-shot learning capacity of ViTs.

To further improve the robustness of local supervision from teacher f_0 while keeping sufficient data diversity, we propose a "Spatial-Consistent Augmentation" (SCA) strategy to improve generalization. SCA consists of a *spatial-only* augmentation and a *non-spatial* augmentation. In spatial-only augmentation, we only introduce spatial transformation (e.g., random crop and resize, flip and rotation) to augment the input images x_i and obtain $\tilde{x}_i$. For non-spatial augmentation, it only leverages non-spatial augmentations, e.g. color jitter, on $\tilde{x}_i$ to obtain $\bar{x}_i$. During the meta-training phase, we feed $\tilde{x}_i$ into the teacher f_0 to generate s_{ij} used in Eqn. (2), and feed $\bar{x}_i$ into the target meta-leaner f. In this way, the samples used to train meta-learner f are of high diversity, while still enjoying very accurate location-specific supervisions s_{ij} since the teacher f_0 uses a weak augmentation $\tilde{x}_i$ to generate the location-specific supervisions. This also helps improve the generalization ability of ViTs.

4.2 Meta-Tuning

For meta-tuning phase in Fig. 5(b), our target is to finetune the meta-learner f via training it on multiple "N-way K-shot" tasks $\{\tau\}$ sampled from the base dataset $\mathbb{D}_{\text{base}}$, such that f can be adapted to a new task which contains unseen classes C_{novel} (i.e. $C_{\text{novel}} \notin \mathbb{D}_{\text{base}}$) with only a few labeled training samples. To this end, without loss of generality, we follow a simple yet effective meta-tuning method in Meta Baseline [6]. Besides, we also investigate different meta-tuning methods, such as FEAT [51] and DeepEMD [55], in the suppl.. With the same FEAT and DeepEMD, our SUN framework still shows superiority over other few-shot learning frameworks. See more details in the suppl.. For completeness, we introduce the meta-tuning in Meta-Baseline in the following. Specifically, given a task τ with support set S, SUN calculates the classification prototype w_k of class k via $w_k = \sum_{x \in S_k} GAP(f(x))/|S_k|$, where S_k denotes the support samples from class c and GAP means global average pooling operation. Then for each query image x, meta-learner f calculates the classification confidence score of the k-th class:

$$p_k = \frac{\exp(\gamma \cdot cos(GAP(f(x)), w_k))}{\sum_{k'} \exp(\gamma \cdot cos(GAP(f(x)), w_k))}, \tag{3}$$

where cos denotes cosine similarity and γ is a temperature parameter. Finally, it minimizes the cross-entropy loss $\mathcal{L}_{\text{few-shot}} = H(p_x, y_x)$ to fine-tune meta-leaner f on the various sampled tasks $\{\tau\}$, where $p_x = [p_1, \cdots, p_c]$ is the prediction and y_x is ground-truth label of x. After this meta-tuning, given a new test task τ' with support set $\mathbf{S}'$, we follow the above step to compute its classification prototypes, and then use Eqn. (3) to predict the labels of the test samples.

5 Experiments

Following [5,6,63], we evaluate SUN with various ViTs on three widely used few-shot benchmarks, i.e., *mini*ImageNet [44], *tiered*ImageNet [34] and CIFAR-FS [2] whose details are deferred to the supplement. For fairness, for each benchmark, we follow [5,6,63] and report the average accuracy with 95% confidence interval on 2,000 tasks under 5-way 1-shot and 5-way 5-shot classification settings. More training details can be found in the suppl..

Table 3. Comparison between SUN and meta baseline on *mini*ImageNet.

Backbone	Params	Meta-baseline [5]		SUN (Ours)	
		5-way 1-shot	5-way 5-shot	5-way 1-shot	5-way 5-shot
LV-ViT	12.6M	43.08 ± 0.38	59.03 ± 0.39	**59.00 ± 0.44**	**75.47 ± 0.34**
Swin	12.6M	54.63 ± 0.45	70.60 ± 0.38	**64.94 ± 0.46**	**80.40 ± 0.32**
Visformer	12.5M	47.61 ± 0.43	63.00 ± 0.39	**67.80 ± 0.45**	**83.25 ± 0.29**
NesT	12.8M	54.57 ± 0.46	69.85 ± 0.38	**66.54 ± 0.45**	**82.09 ± 0.30**

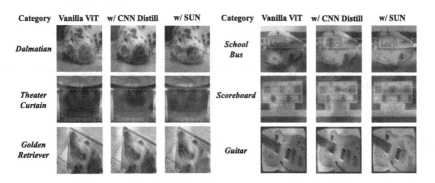

Fig. 6. Visualization of attention maps from vanilla ViT, CNN-distilled ViT and ViT with SUN. ViT with SUN performs better, thus obtains better token dependency.

5.1 Comparison on Different ViTs

We evaluate our SUN on four different ViTs, i.e., LV-ViT, Swin, Visformer and NesT, which cover most of existing ViT types. Table 3 shows that on *mini*ImageNet, SUN significantly surpasses Meta-Baseline [6] on the four ViTs. Specifically, it makes 15.9%, 10.3%, 20.2%, and 12.0% improvement over Meta-Baseline under 5-way 1-shot setting. These results demonstrate the superior generalization ability of our SUN on novel categories, and also its good compatibility with different kinds of ViTs. Besides, among all ViTs, Visformer achieves the best performance, since it uses many CNN modules and can better handle the few-shot learning task. In the following Sect. 5.2 and 5.4, we use the second best NesT as our ViT feature extractor, since NesT does not involve CNN modules and only introduce modules to focus on local features and thus better reveals few-shot learning ability of conventional ViT architecture. And in Sect. 5.3, we use both NesT and Visformer to make comparisons with the state-of-the-art CNN-based few-shot classification methods.

SUN v.s. ViT in Dependency Learning. Following the analysis in Sect. 3, we visualize the attention maps of vanilla ViT, CNN-distilled ViT and ViT with SUN to compare the quality of token dependency learning. Figure 6 demonstrate the comparison result. ViT with SUN can capture more semantic tokens than vanilla ViT and CNN-distilled ViT on various categories. The result indicates that ViT with SUN can learn better token dependency during the training.

Table 4. Comparison of different ViT based few-shot classification frameworks on *mini*ImageNet. "AC" means introducing CNN modules in ViT backbone to fuse features. "Distill" is knowledge distillation. "FT" is using meta-tuning to finetune meta learner (see Sect. 4.2). Except [22], all methods use NesT as ViT feature extractor.

Method	Backbone	AC	Distill	FT	*mini*ImageNet	
					5-way 1-shot	5-way 5-shot
Meta-Baseline [6]	*ResNet-12*			✓	64.53 ± 0.45	81.41 ± 0.31
Baseline [5]	*ViT*				49.23 ± 0.43	66.57 ± 0.39
Meta-Baseline [6]	*ViT*			✓	54.57 ± 0.46	69.85 ± 0.38
Re-Parameter [22]	*ViT*	✓		✓	46.59 ± 0.44	62.56 ± 0.41
[6]+CNN-Distill [46]	*ViT*	✓	✓	✓	55.79 ± 0.45	71.81 ± 0.37
Semiformer [46]	*CNN+ViT*	✓	✓		56.62 ± 0.46	72.91 ± 0.39
[6]+Semiformer [46]	*CNN+ViT*	✓	✓	✓	57.91 ± 0.48	73.31 ± 0.38
[6]+DrLoc [28]	*ViT*			✓	57.85 ± 0.46	74.03 ± 0.35
BML [63]	*ViT*		✓		59.35 ± 0.45	76.00 ± 0.35
SUN (Ours)	*ViT*		✓	✓	**66.54 ± 0.45**	**82.09 ± 0.30**

5.2   Comparison Among Different Few-shot Learning Frameworks

In Table 4, we compare SUN with other few-shot learning frameworks on the same ViT architecture. In Table 4, the method "[6]+CNN-Distill" pretrains a CNN to distill a ViT; "Semiformer" [46] introduces CNN modules into ViT; "Re-Parameter" [22] initializes a ViT by a trained CNN; "[6]+DrLoc" predicts the relative positions between local token pairs; "BML" [63] introduces mutual learning based few-shot learning framework. All these methods propose their own techniques to improve few-shot learning performance of ViT via introducing CNN-alike inductive bias or constructing tasks to better learn token dependency. Except [22] that uses its CNN-transformer-mixed architecture, all others adopt NesT [57] as a ViT feature extractor or use ResNet-12 as the CNN feature extractor. Table 4 shows that our SUN significantly surpasses all these methods in terms of accuracy on novel categories. Specifically, our SUN outperforms the runner-up by 7.19% and 6.09% under 5-way 1-shot and 5-shot settings, respectively. These results show that even on the same ViT, SUN achieves much better few-shot accuracy on novel categories than other few-shot learning frameworks.

5.3 Comparison with State-of-The-Arts

Here we compare SUN with state-of-the-arts (SoTAs), including CNN based methods and ViT based one, on *mini*ImageNet [44], *tiered*ImageNet [34] and CIFAR-FS [2]. Table 5 reports the evaluation results. Without introducing extremely complex few-shot learning methods like [19,48], our SUN achieves comparable performance with SoTA on *mini*ImageNet [44], and sets new SoTAs on *tiered*ImageNet [34] and CIFAR-FS [2]. Specifically, on *tiered*ImageNet under 5-way 1-shot and 5-shot settings, our SUN (Visformer) respectively obtains 72.99% and 86.74%, and respectively improves ~0.8% and ~0.2% over the SoTA TPMN [48]. On CIFAR-FS dataset, our SUN (NesT) obtains 78.17% and 88.98% in terms of 1-shot accuracy and 5-shot accuracy, which significantly outperforms all the state-of-the-art methods by at least 2.5% in terms of 1-shot accuracy. Meanwhile, our SUN (Visformer) also obtains 67.80% 1-shot accuracy on *mini*ImageNet

Table 5. Comparison with SoTA few-shot learning methods under 5-way few-shot classification setting. The results of the best 2 methods are in bold font.

Method	Classifier params	*mini*ImageNet		*tiered*ImageNet		CIFAR-FS	
		1-shot	5-shot	1-shot	5-shot	1-shot	5-shot
ResNet-12/18 as feature extractor							
MetaOptNet [20]	0	64.09 ± 0.62	80.00 ± 0.45	65.81 ± 0.74	81.75 ± 0.53	72.00 ± 0.70	84.20 ± 0.50
DeepEMD [55]	0	65.91 ± 0.82	82.41 ± 0.56	71.16 ± 0.80	86.03 ± 0.58	46.47 ± 0.70	63.22 ± 0.71
FEAT [51]	1.05M	66.78 ± 0.20	82.05 ± 0.14	70.80 ± 0.23	84.79 ± 0.16	–	–
TADAM [32]	1.23M	58.50 ± 0.30	76.70 ± 0.30	–	–	–	–
Rethink-Distill [39]	225K	64.82 ± 0.60	82.14 ± 0.43	71.52 ± 0.69	86.03 ± 0.49	73.90 ± 0.80	86.90 ± 0.50
DC [24]	224K	61.26 ± 0.20	79.01 ± 0.13	–	–	–	–
CloserLook++ [5]	131K	51.87 ± 0.77	75.68 ± 0.63	–	–	–	–
Meta-Baseline [6]	0	63.17 ± 0.23	79.26 ± 0.17	68.62 ± 0.27	83.29 ± 0.18	-	-
Neg-Cosine [25]	131K	63.85 ± 0.81	81.57 ± 0.56	–	–	–	–
AFHN [21]	359K	62.38 ± 0.72	78.16 ± 0.56	-	-	68.32 ± 0.93	81.45 ± 0.87
Centroid [1]	10K	59.88 ± 0.67	80.35 ± 0.73	69.29 ± 0.56	85.97 ± 0.49	–	–
RE-Net [19]	430K	67.60 ± 0.44	82.58 ± 0.30	71.61 ± 0.51	85.28 ± 0.35	74.51 ± 0.46	86.60 ± 0.32
TPMN [48]	16M	**67.64 ± 0.63**	**83.44 ± 0.43**	72.24 ± 0.70	86.55 ± 0.63	75.50 ± 0.90	87.20 ± 0.60
NesT ViT as feature extractor							
CloserLook++ [5]	180K	49.23 ± 0.43	66.57 ± 0.39	59.13 ± 0.46	77.88 ± 0.39	63.89 ± 0.49	80.43 ± 0.37
Meta-Baseline [6]	0	54.57 ± 0.46	69.85 ± 0.38	63.73 ± 0.47	79.33 ± 0.38	68.05 ± 0.48	81.53 ± 0.36
BML [63]	180K	59.35 ± 0.45	76.00 ± 0.35	66.98 ± 0.50	83.75 ± 0.34	67.51 ± 0.48	82.17 ± 0.36
SUN	0	66.54 ± 0.45	82.09 ± 0.30	**72.93 ± 0.50**	**86.70 ± 0.33**	**78.17 ± 0.46**	**88.98 ± 0.33**
Visformer ViT as feature extractor							
SUN	0	**67.80 ± 0.45**	**83.25 ± 0.30**	**72.99 ± 0.50**	**86.74 ± 0.33**	**78.37 ± 0.46**	**88.84 ± 0.32**

test set and surpasses the SoTA CNN-based few-shot classification methods by ~1.2%. These results well demonstrate the effectiveness of ViTs with our SUN in few-shot learning.

5.4   Ablation Study

Here we show the key ablation studies, more details can be found in the suppl..
Effect of Each Phase in SUN. We investigate the effect of each training phase in SUN. For fairness, we always use the same ViT feature extractor [57] as meta-learner f for few-shot classification. Table 6 shows that on *mini*ImageNet dataset, compared to the baseline (denoted by "Base" in Table 6), the teacher ViT f_g of meta-training phase significantly outperforms the baseline by 13.2% and 12.9% in terms of 1-shot and 5-shot accuracy. By introducing the location-specific supervision, type (b) obtains ~0.7% improvement for 1-shot accuracy. Then after introducing SCA and BGF, type (d) can further lead to a gain of ~1.8% 1-shot accuracy as well as ~1.0% 5-shot accuracy. A possible reason is that the SCA improves the quality of location-specific supervision from the pretrained transformer while keeping the strong augmentation ability for the target transformer in the meta-training phase. Meanwhile, BGF eliminates the negative effect from mislabeling background patches and improves the feature representation quality. These two techniques significantly improve location-specific supervision and improve the generalization ability of ViT feature extractors.

We also investigate meta-tuning phase. As shown in type (e) in Table 6, the meta-learner f after meta-training phase can be naturally adopted to the existing few-shot learning methods and obtain further improvement. Specifically, our SUN (denoted by (e)) also surpasses f after meta-training phase by 1.7% and 1.1% in terms of 1-shot and 5-shot accuracy.

Table 6. Ablation study of SUN on *mini*ImageNet. "Local" means location-specific supervision. SCA is spatial-consistent augmentation. BGF means background filtration. And f_g means the teacher ViT model in SUN meta-training phase.

Type	Meta-training				Meta-tuning	*mini*ImageNet	
	f_g	Local	SCA	BGF	Meta-Baseline [6]	5-way 1-shot	5-way 5-shot
Base						49.23 ± 0.43	66.57 ± 0.39
(a)	✓					62.40 ± 0.44	79.45 ± 0.32
(b)	✓	✓				63.06 ± 0.45	79.91 ± 0.32
(c)	✓	✓	✓			64.50 ± 0.45	80.59 ± 0.31
(d)	✓	✓	✓	✓		64.84 ± 0.45	80.96 ± 0.32
(e)	✓	✓	✓	✓	✓	**66.54 ± 0.45**	**82.09 ± 0.30**

Table 7. Ablation of meta-tuning, where "SUN-*" means ours and others are SoTAs.

Methods	MetaBaseline [6]	FEAT [51]	DeepEMD [55]	COSOC [30]	SUN-M	SUN-F	SUN-D
5-way 1-shot	64.53 ± 0.45	66.78 ± 0.20	68.77 ± 0.29	69.28 ± 0.49	67.80 ± 0.45	66.90 ± 0.44	$\mathbf{69.56 \pm 0.44}$
5-way 5-shot	81.41 ± 0.31	82.05 ± 0.14	84.13 ± 0.53	85.16 ± 0.42	83.25 ± 0.30	82.63 ± 0.30	$\mathbf{85.38 \pm 0.49}$

Meta-Tuning Methods. We also evaluate our SUN with different meta-tuning methods. To empirically analyze our SUN, we fix Visformer [7] as ViT backbone and choose Meta-Baseline [6], FEAT [51] and a dense prediction method Deep-EMD [55]. As shown in Table 7, all SUN-M/F/D improve the accuracy and perform better than corresponding CNN-based baselines. Moreover, SUN-D achieves 69.56% 5-way 1-shot accuracy, even slightly surpasses the SoTA COSOC [30].

6 Conclusion

In this work, we first empirically showed that replacing the CNN feature extractor as a ViT model leads to severe performance degradation on few-shot learning tasks, and the slow token dependency learning on limited training data largely contributes to this performance degradation. Then for the first time, we proposed a simple yet effective few-shot training framework for ViTs, i.e. Self-promoted sUpervisioN (SUN), to resolve this issue. By firstly pretraining the ViT on the few-shot learning dataset, SUN adopts it to generate individual location-specific supervision for 1) guiding the ViT on which patch tokens are similar or dissimilar and thus accelerating token dependency learning; 2) improving the object grounding and recognition ability. Experimental results on few-shot classification tasks demonstrate the superiority of SUN in accuracy over state-of-the-arts.

Acknowledgement. This work was supported in part by the National Key R&D Program of China under Grant No. 2021ZD0112100, and the Major Key Project of PCL under Grant No. PCL2021A12.

References

1. Afrasiyabi, A., Lalonde, J.-F., Gagné, C.: Associative alignment for few-shot image classification. In: Vedaldi, A., Bischof, H., Brox, T., Frahm, J.-M. (eds.) ECCV 2020. LNCS, vol. 12350, pp. 18–35. Springer, Cham (2020). https://doi.org/10.1007/978-3-030-58558-7_2
2. Bertinetto, L., Henriques, J.F., Torr, P., Vedaldi, A.: Meta-learning with differentiable closed-form solvers. In: International Conference on Learning Representations (2019). https://openreview.net/forum?id=HyxnZh0ct7
3. Cao, Y.H., Yu, H., Wu, J.: Training vision transformers with only 2040 images (2022)
4. Carion, N., Massa, F., Synnaeve, G., Usunier, N., Kirillov, A., Zagoruyko, S.: End-to-end object detection with transformers. In: Vedaldi, A., Bischof, H., Brox, T., Frahm, J. (eds.) Computer Vision - ECCV 2020–16th European Conference, Glasgow, UK, August 23–28, 2020, Proceedings, Part I. Lecture Notes in Computer Science, vol. 12346, pp. 213–229. Springer, Cham (2020). https://doi.org/10.1007/978-3-030-58452-8_13

5. Chen, W.Y., Liu, Y.C., Kira, Z., Wang, Y.C., Huang, J.B.: A closer look at few-shot classification. In: International Conference on Learning Representations (2019)
6. Chen, Y., Liu, Z., Xu, H., Darrell, T., Wang, X.: Meta-baseline: exploring simple meta-learning for few-shot learning. In: Proceedings of the IEEE/CVF International Conference on Computer Vision (ICCV), pp. 9062–9071, October 2021
7. Chen, Z., Xie, L., Niu, J., Liu, X., Wei, L., Tian, Q.: Visformer: the vision-friendly transformer. In: Proceedings of the IEEE/CVF International Conference on Computer Vision (ICCV), pp. 589–598, October 2021
8. Deng, J., Dong, W., Socher, R., Li, L.J., Li, K., Fei-Fei, L.: Imagenet: a large-scale hierarchical image database. In: 2009 IEEE Conference on Computer Vision and Pattern Recognition, pp. 248–255. IEEE (2009)
9. Doersch, C., Gupta, A., Zisserman, A.: Crosstransformers: spatially-aware few-shot transfer. In: Larochelle, H., Ranzato, M., Hadsell, R., Balcan, M.F., Lin, H. (eds.) Advances in Neural Information Processing Systems, vol. 33, pp. 21981–21993. Curran Associates, Inc. (2020). https://proceedings.neurips.cc/paper/2020/file/fa28c6cdf8dd6f41a657c3d7caa5c709-Paper.pdf
10. Dosovitskiy, A., et al.: An image is worth 16 x 16 words: transformers for image recognition at scale. In: International Conference on Learning Representations (2021)
11. Fang, Y., et al.: You only look at one sequence: rethinking transformer in vision through object detection. Adv. Neural Inf. Process. Syst. **34**, 26183–26197 (2021)
12. Finn, C., Abbeel, P., Levine, S.: Model-agnostic meta-learning for fast adaptation of deep networks. In: Proceedings of the 34th International Conference on Machine Learning (2017)
13. Graham, B., et al.: Levit: a vision transformer in convnet's clothing for faster inference. In: Proceedings of the IEEE/CVF International Conference on Computer Vision (ICCV), pp. 12259–12269, October 2021
14. Graves, A., Wayne, G., Danihelka, I.: Neural turing machines. arXiv preprint arXiv:1410.5401 (2014)
15. He, K., Zhang, X., Ren, S., Sun, J.: Deep residual learning for image recognition. In: 2016 IEEE Conference on Computer Vision and Pattern Recognition (CVPR), pp. 770–778 (2016)
16. Hinton, G., Vinyals, O., Dean, J.: Distilling the knowledge in a neural network. In: NIPS Deep Learning and Representation Learning Workshop (2015). http://arxiv.org/abs/1503.02531
17. Hou, R., Chang, H., MA, B., Shan, S., Chen, X.: Cross attention network for few-shot classification. In: Wallach, H., Larochelle, H., Beygelzimer, A., d'Alché-Buc, F., Fox, E., Garnett, R. (eds.) Advances in Neural Information Processing Systems, vol. 32. Curran Associates, Inc. (2019). https://proceedings.neurips.cc/paper/2019/file/01894d6f048493d2cacde3c579c315a3-Paper.pdf
18. Jiang, Z., et al.: All tokens matter: token labeling for training better vision transformers. In: Advances in Neural Information Processing Systems (2021)
19. Kang, D., Kwon, H., Min, J., Cho, M.: Relational embedding for few-shot classification. In: Proceedings of the IEEE/CVF International Conference on Computer Vision (ICCV), pp. 8822–8833, October 2021
20. Lee, K., Maji, S., Ravichandran, A., Soatto, S.: Meta-learning with differentiable convex optimization. 2019 IEEE/CVF Conference on Computer Vision and Pattern Recognition (CVPR), pp. 10649–10657 (2019)
21. Li, K., Zhang, Y., Li, K., Fu, Y.: Adversarial feature hallucination networks for few-shot learning. In: 2020 IEEE/CVF Conference on Computer Vision and Pattern Recognition (CVPR) (2020)

22. Li, S., Chen, X., He, D., Hsieh, C.J.: Can vision transformers perform convolution? (2021)
23. Li, Y., Zhang, K., Cao, J., Timofte, R., Van Gool, L.: Localvit: bringing locality to vision transformers. arXiv preprint arXiv:2104.05707 (2021)
24. Lifchitz, Y., Avrithis, Y., Picard, S., Bursuc, A.: Dense classification and implanting for few-shot learning. In: Proceedings of the IEEE/CVF Conference on Computer Vision and Pattern Recognition (CVPR), June 2019
25. Liu, B., et al.: Negative margin matters: Understanding margin in few-shot classification. In: European Conference on Computer Vision, pp. 438–455 (2020)
26. Liu, H., Dai, Z., So, D., Le, Q.: Pay attention to mlps. Adv. Neural Inf. Process. Syst. **34**, 9204–9215 (2021)
27. Liu, L., Hamilton, W.L., Long, G., Jiang, J., Larochelle, H.: A universal representation transformer layer for few-shot image classification. In: International Conference on Learning Representations (2021). https://openreview.net/forum?id=04cII6MumYV
28. Liu, Y., Sangineto, E., Bi, W., Sebe, N., Lepri, B., Nadai, M.D.: Efficient training of visual transformers with small datasets. In: Beygelzimer, A., Dauphin, Y., Liang, P., Vaughan, J.W. (eds.) Advances in Neural Information Processing Systems (2021). https://openreview.net/forum?id=SCN8UaetXx
29. Liu, Z., et al.: Swin transformer: Hierarchical vision transformer using shifted windows. In: Proceedings of the IEEE/CVF International Conference on Computer Vision (ICCV), pp. 10012–10022, October 2021
30. Luo, X., et al.: Rectifying the shortcut learning of background for few-shot learning. In: Beygelzimer, A., Dauphin, Y., Liang, P., Vaughan, J.W. (eds.) Advances in Neural Information Processing Systems (2021). https://openreview.net/forum?id=N1i6BJzouX4
31. Mishra, N., Rohaninejad, M., Chen, X., Abbeel, P.: A simple neural attentive metalearner. In: International Conference on Learning Representations (2018). https://openreview.net/forum?id=B1DmUzWAW
32. Oreshkin, B., Rodríguez López, P., Lacoste, A.: Tadam: task dependent adaptive metric for improved few-shot learning. In: Bengio, S., Wallach, H., Larochelle, H., Grauman, K., Cesa-Bianchi, N., Garnett, R. (eds.) Advances in Neural Information Processing Systems, vol. 31. Curran Associates, Inc. (2018). https://proceedings.neurips.cc/paper/2018/file/66808e327dc79d135ba18e051673d906-Paper.pdf
33. Ravi, S., Larochelle, H.: Optimization as a model for few-shot learning. In: ICLR (2017)
34. Ren, M., et al.: Meta-learning for semi-supervised few-shot classification. In: International Conference on Learning Representations (2018). https://openreview.net/forum?id=HJcSzz-CZ
35. Rusu, A.A., et al.: Meta-learning with latent embedding optimization. In: International Conference on Learning Representations (2019). https://openreview.net/forum?id=BJgklhAcK7
36. Snell, J., Swersky, K., Zemel, R.: Prototypical networks for few-shot learning. In: Proceedings of the 31st International Conference on Neural Information Processing Systems, pp. 4080–4090 (2017)
37. Strudel, R., Garcia, R., Laptev, I., Schmid, C.: Segmenter: transformer for semantic segmentation. In: Proceedings of the IEEE/CVF International Conference on Computer Vision (ICCV), pp. 7262–7272, October 2021
38. Sun, C., Shrivastava, A., Singh, S., Gupta, A.: Revisiting unreasonable effectiveness of data in deep learning era. In: Proceedings of the IEEE International Conference on Computer Vision, pp. 843–852 (2017)

39. Tian, Y., Wang, Y., Krishnan, D., Tenenbaum, J.B., Isola, P.: Rethinking few-shot image classification: a good embedding is all you need? In: ECCV (2020)
40. Tolstikhin, I.O., et al.: Mlp-mixer: an all-mlp architecture for vision. Adv. Neural Inf. Process. Syst. **34**, 24261–24272 (2021)
41. Touvron, H., Cord, M., Douze, M., Massa, F., Sablayrolles, A., Jegou, H.: Training data-efficient image transformers and distillation through attention. In: International Conference on Machine Learning, vol. 139, pp. 10347–10357, July 2021
42. Touvron, H., Cord, M., Sablayrolles, A., Synnaeve, G., Jégou, H.: Going deeper with image transformers. In: Proceedings of the IEEE/CVF International Conference on Computer Vision (ICCV), pp. 32–42 (October 2021)
43. Vaswani, A., et al.: Attention is all you need. In: Advances in Neural Information Processing Systems, pp. 5998–6008 (2017)
44. Vinyals, O., Blundell, C., Lillicrap, T., Wierstra, D., et al.: Matching networks for one shot learning. In: Advances in Neural Information Processing Systems, pp. 3630–3638 (2016)
45. Wang, W., et al.: Pyramid vision transformer: a versatile backbone for dense prediction without convolutions. In: Proceedings of the IEEE/CVF International Conference on Computer Vision (ICCV), pp. 568–578, October 2021
46. Weng, Z., Yang, X., Li, A., Wu, Z., Jiang, Y.G.: Semi-supervised vision transformers (2021)
47. Wu, H., et al.: Cvt: introducing convolutions to vision transformers. In: Proceedings of the IEEE/CVF International Conference on Computer Vision (ICCV), pp. 22–31, October 2021
48. Wu, J., Zhang, T., Zhang, Y., Wu, F.: Task-aware part mining network for few-shot learning. In: Proceedings of the IEEE/CVF International Conference on Computer Vision (ICCV), pp. 8433–8442, October 2021
49. Xiao, T., Dollar, P., Singh, M., Mintun, E., Darrell, T., Girshick, R.: Early convolutions help transformers see better. In: Beygelzimer, A., Dauphin, Y., Liang, P., Vaughan, J.W. (eds.) Advances in Neural Information Processing Systems (2021)
50. Xu, W., Xu, Y., Chang, T., Tu, Z.: Co-scale conv-attentional image transformers. In: Proceedings of the IEEE/CVF International Conference on Computer Vision (ICCV), pp. 9981–9990, October 2021
51. Ye, H.J., Hu, H., Zhan, D.C., Sha, F.: Few-shot learning via embedding adaptation with set-to-set functions. In: IEEE/CVF Conference on Computer Vision and Pattern Recognition (CVPR), pp. 8808–8817 (2020)
52. Yu, W., et al.: Metaformer is actually what you need for vision. arXiv preprint arXiv:2111.11418 (2021)
53. Yuan, L., et al.: Tokens-to-token vit: training vision transformers from scratch on imagenet. In: Proceedings of the IEEE/CVF International Conference on Computer Vision (ICCV), pp. 558–567, October 2021
54. Yuan, Y., Fu, R., Huang, L., Lin, W., Zhang, C., Chen, X., Wang, J.: Hrformer: high-resolution transformer for dense prediction. In: NeurIPS (2021)
55. Zhang, C., Cai, Y., Lin, G., Shen, C.: Deepemd: few-shot image classification with differentiable earth mover's distance and structured classifiers. In: IEEE/CVF Conference on Computer Vision and Pattern Recognition (CVPR), June 2020
56. Zhang, H., Duan, J., Xue, M., Song, J., Sun, L., Song, M.: Bootstrapping vits: towards liberating vision transformers from pre-training (2021)
57. Zhang, Z., Zhang, H., Zhao, L., Chen, T., Arik, S.O., Pfister, T.: Nested hierarchical transformer: towards accurate, data-efficient and interpretable visual understanding. In: AAAI Conference on Artificial Intelligence (AAAI) (2022)

58. Zheng, S., et al.: Rethinking semantic segmentation from a sequence-to-sequence perspective with transformers. In: Proceedings of the IEEE/CVF Conference on Computer Vision and Pattern Recognition (CVPR), pp. 6881–6890, June 2021

59. Zhmoginov, A., Sandler, M., Vladymyrov, M.: Hypertransformer: model generation for supervised and semi-supervised few-shot learning (2022)

60. Zhong, Z., Zheng, L., Kang, G., Li, S., Yang, Y.: Random erasing data augmentation. In: Proceedings of the AAAI Conference on Artificial Intelligence, vol. 34, pp. 13001–13008 (2020)

61. Zhou, P., Yuan, X., Xu, H., Yan, S., Feng, J.: Efficient meta learning via minibatch proximal update. In: Neural Information Processing Systems (2019)

62. Zhou, P., Zou, Y., Yuan, X.T., Feng, J., Xiong, C., Hoi, S.: Task similarity aware meta learning: Theory-inspired improvement on maml. In: Uncertainty in Artificial Intelligence, pp. 23–33. PMLR (2021)

63. Zhou, Z., Qiu, X., Xie, J., Wu, J., Zhang, C.: Binocular mutual learning for improving few-shot classification. In: Proceedings of the IEEE/CVF International Conference on Computer Vision (ICCV), pp. 8402–8411, October 2021

Few-Shot Object Counting and Detection

Thanh Nguyen[1], Chau Pham[1], Khoi Nguyen[1(✉)], and Minh Hoai[1,2]

[1] VinAI Research, Hanoi, Vietnam
ducminhkhoi@gmail.com
[2] Stony Brook University, Stony Brook, NY, USA

Abstract. We tackle a new task of few-shot object counting and detection. Given a few exemplar bounding boxes of a target object class, we seek to count and detect all objects of the target class. This task shares the same supervision as the few-shot object counting but additionally outputs the object bounding boxes along with the total object count. To address this challenging problem, we introduce a novel two-stage training strategy and a novel uncertainty-aware few-shot object detector: Counting-DETR. The former is aimed at generating pseudo ground-truth bounding boxes to train the latter. The latter leverages the pseudo ground-truth provided by the former but takes the necessary steps to account for the imperfection of pseudo ground-truth. To validate the performance of our method on the new task, we introduce two new datasets named FSCD-147 and FSCD-LVIS. Both datasets contain images with complex scenes, multiple object classes per image, and a huge variation in object shapes, sizes, and appearance. Our proposed approach outperforms very strong baselines adapted from few-shot object counting and few-shot object detection with a large margin in both counting and detection metrics. The code and models are available at https://github.com/VinAIResearch/Counting-DETR.

Keywords: Few-shot object counting · Few-shot object detection

1 Introduction

This paper addresses a new task of Few-Shot object Counting and Detection (FSCD) in crowded scenes. Given an image containing many objects of multiple classes, we seek to count and detect all objects of a target class of interest specified by a few exemplar bounding boxes in the image. To facilitate few-shot learning, in training, we are only given the supervision of few-shot object counting, i.e., dot annotations for the approximate centers of all objects and a few exemplar bounding boxes for object instances from the target class. It is worth noting that the test classes may or may not be present in training classes. The problem setting is depicted in Fig. 1.

FSCD is different from Few-Shot Object Counting (FSC) and Few-Shot Object Detection (FSOD). Compared to FSC, FSCD has several advantages: (1)

T. Nguyen and C. Pham—Equal contribution.

© The Author(s), under exclusive license to Springer Nature Switzerland AG 2022
S. Avidan et al. (Eds.): ECCV 2022, LNCS 13680, pp. 348–365, 2022.
https://doi.org/10.1007/978-3-031-20044-1_20

(a) Training   (b) Testing

Fig. 1. We address the task of few-shot counting and detection in a novel setting: (a) in training, each training image contains dot annotations for all objects and a few exemplar boxes. (b) In testing, given an image with a few exemplar boxes defining a target class, our goal is to count and detect all objects of that target class in the image.

obtaining object bounding boxes "for free", which is suitable for quickly annotating bounding boxes for a new object class with a few exemplar bounding boxes; (2) making the result of FSC more interpretable since bounding boxes are easier to verify than the density map. Compared to FSOD which requires bounding box annotation for all objects in the training phase of the base classes, FSCD uses significantly less supervision, i.e., only a few exemplar bounding boxes and dot annotations for all objects. This is helpful in crowded scenes where annotating accurate bounding boxes for all objects is ambiguously harder and significantly more expensive than the approximate dot annotation.

Consequently, FSCD is more challenging than both FSC and FSOD. FSCD needs to detect and count all the objects as for FSOD, but it is only trained with the supervision of FSC. This invalidates of most of available approaches used in these problems without significant changes in network architecture or loss function. Specifically, it is not trivial to extend the density map produced by FSC approaches to predict the object bounding boxes; and it is hard to train a few-shot object detector with few exemplar bounding boxes of the base classes.

A naive approach for FSCD is to extend FamNet [32], a density-map-based approach for FSC, whose counting number is obtained by summing over the predicted density map. To extend FamNet to detect objects, one can use a regression function on top of the features extracted from the peak locations (whose density values are highest in their respective local neighborhoods), the features extracted from the exemplars, and the exemplar boxes themselves. The process of this naive approach is illustrated in Fig. 2a. However, this approach has two limitations due to: 1) the imperfection of the predicted density map, and 2) the non-discriminative peak features. In the former, the density value is high in the environment locations whose color is similar to those of the exemplars, or the density map is peak-indistinguishable when the objects are packed in a dense region as depicted in Fig. 2b. In the latter, the extracted features are trained with counting objective (not object detection) so that they cannot represent for different shapes, sizes, and orientations, as illustrated in Fig. 2c.

To address the aforementioned limitations, we propose a new point-based approach, named Counting-DETR, treating objects as points. In particular, counting and detecting objects is equivalent to counting and detecting points, and the object bounding box is predicted directly from point features. Counting-

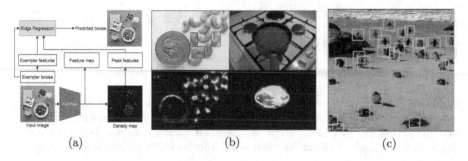

(a)   (b)   (c)

Fig. 2. Limitations of a naive approach for FSCD by extending FamNet [32] with a regression function for object detection. (a) Processing pipeline of this approach: a regressor takes as input exemplar boxes with their features, and features at peak density locations to predict bounding boxes for the peak locations. (b) Limitation 1: poor quality of the density map predicted by FamNet when the exemplars share similar appearance with background or densely packed region. The first row presents the input images with a few exemplars each, the second row presents the corresponding density map predicted by FamNet. (c) Limitation 2: Non-discriminative peak features cannot represent objects with significant differences in shape and size. The green boxes are predicted from the features extracted at the annotated dots.

DETR is based on an object detector, Anchor DETR [43], with improvements to better address FSCD. **First**, inspired by [5] we adopt a two-stage training strategy: (1) Counting-DETR is trained to generate pseudo ground-truth (GT) bounding boxes given the annotated points of training images; (2) Counting-DETR is further fine-tuned on the generated pseudo GT bounding boxes to detect objects on test images. **Second,** since the generated pseudo GT bounding boxes are imperfect, we propose to estimate the uncertainty for bounding box prediction in the second stage. The estimated uncertainty regularizes learning such that lower box regression loss is incurred on the predictions with high uncertainty. The overview of Counting-DETR is illustrated in Fig. 3.

In short, the contributions of our paper are: (1) we introduce a new problem of few-shot object counting and detection (FSCD); (2) we introduce two new datasets, FSCD-147 and FSCD-LVIS; (3) we propose a two-stage training strategy to first generate pseudo GT bounding boxes from the dot annotations, then use these boxes as supervision for training our proposed few-shot object detector; and (4) we propose a new uncertainty-aware point-based few-shot object detector, taking into account the imperfection of pseudo GT bounding boxes.

2   Related Work

In this section, we review some related work on object counting and detection.

Visual counting focuses on some predefined classes such as car [15,27], cell [2,46], and human [1,8,16,17,21,29,31,33,37,41,49,51]. The methods can be grouped into two types: density-map-based and detection-based. The former,

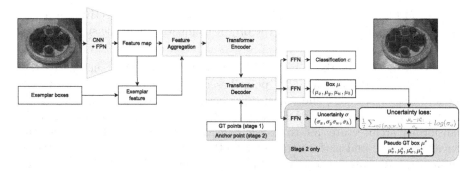

Fig. 3. The overview of our two-step training approach: (1) Counting-DETR is first trained on a few pairs of dot and bounding boxes and then used to predict pseudo GT boxes for the annotated dots; (2) Counting-DETR is trained to predict the object bounding boxes, with the prediction target being the pseudo GT boxes from the first stage. Specifically, the input image is first forwarded through a CNN+FPN backbone to extract its feature map. The exemplar features are extracted from their boxes to integrate with the feature map producing the exemplars-integrated feature map. This feature map is then taken as input to the encoder-decoder transformer along with either the annotated dots in the first stage or the anchor points in the second stage for foreground/background classification and bounding box regression. In the second stage, the estimated uncertainty is used to regularize the training with a new uncertainty loss to account for the imperfection of the pseudo GT bounding boxes.

e.g., [24,40], predicts and sums over density map from input image to get the final results. The latter (e.g., [11,15]) counts the number of objects based on the detected boxes. The latter is better at justifying the counting number, however, it requires the GT bounding boxes for training and its performance is exceeded by that of the former, especially for images of crowded scenes.

Few-shot counting (FSC) counts the number of objects in an image with some exemplar bounding boxes of a new object class. Since the number of exemplar boxes is so small that an object detector cannot be reliably learned, prior methods on FSC are all based on density-map regression. GMN [26] formulates object counting as object matching in video tracking such that a class agnostic counting module can be pretrained on a large-scale video object tracking dataset (ImageNet VID [36]). FamNet [32] correlates the features extracted from a few exemplars with the feature map to obtain the density map for object counting. VCN [30] improves upon [32] by augmenting the input image with different styles to make the counting more robust. LaoNet [22] combines self-attention and cross attention in the transformer to aggregate features from the exemplar to the image to facilitate density map prediction. ICFR [48] proposes an iterative framework to progressively refine the exemplar-related features, thus producing a better density map than a single correlation in [32]. However, these approaches do not output object bounding boxes. An extension for object detection from these approaches is depicted in Fig. 2a, but it has several limitations as illustrated

Fig. 2b and Fig. 2c. Whereas, our approach Counting-DETR effectively predicts object bounding boxes along with the object count with only the supervision of FSC.

Object detection methods include anchor-based approaches such as Faster-RCNN [34] and Retina Net [23], point-based approaches such as like FCOS [38] and Center-Net [52], and transformer-based approaches such as DETR [3], Point DETR [5] and Anchor DETR [43]. DETR is the first approach to apply transformer architecture [39] to object detection. Anchor DETR improves the convergence rate and performance of DETR by learnable anchor points representing the initial prediction of the objects in the image. However, these methods require thousands of bounding box annotations on some predefined classes for training and cannot generalize well on a new class in testing with a few box exemplars as in our few-shot setting. Point DETR [5] alleviates this requirement using two separate detectors: teacher (i.e., Point-DETR) and student (i.e., FCOS). The former learns from a small set of fully annotated boxes to generate pseudo-GT bounding boxes of a large amount of point-annotated images. Then the latter is trained with these pseudo-GT boxes to predict the bounding boxes of the test images. This approach is complicated and does not take into account the imperfect pseudo-GT bounding boxes. In contrast, our Counting-DETR is a unified single network with the uncertainty-aware bounding prediction.

Few-shot object detection (FSOD) approaches are mostly based on Faster-RCNN [34] and can be divided into two subgroups based on episodic training [6, 19,45,47,50] and fine-tuning [4,7,42,44]. The former leverages episodic training technique to mimic the evaluation setting in the training whereas the latter fine-tunes some layers while keeping the rest unchanged to preserve the knowledge learned from the training classes. However, all FSOD approaches require box annotations for all objects of the base classes in training. It is not the case in our setting where only a few exemplar bounding boxes are given in training. To address this problem, we propose a two-stage training strategy wherein the first stage, the pseudo GT bounding boxes for all objects are generated from the given exemplar boxes and the dot annotations.

Object detection with uncertainty accounts for the uncertainty in the input image due to blurring or indistinguishable boundaries between objects and the background. Prior work [14,20] assumes the object bounding boxes are characterized by a Gaussian distribution whose mean and standard deviation are predicted by a network trained with an uncertainty loss function derived from maximum the likelihood between the predicted distribution and the GT boxes. [9,10] apply the uncertainty loss for 3D object detection. Our uncertainty loss shares some similarities with prior work, however, we use the uncertainty loss to account for the imperfection of the pseudo GT bounding boxes (not the input image) such that the Laplace distribution works significantly better than the prior Gaussian distribution as shown in experiments.

3   Proposed Approach

Problem definition: In training, we are given a set of images containing multiple object categories. For each image I, a few exemplar bounding boxes B_k, $k = 1, \ldots, K$ where K is the number of exemplars, and the dot annotations for all object instances of a target class are annotated. This kind of supervision is the same as in few-shot object counting. In testing, given a query image with bounding boxes for a few exemplar objects in the target class, our goal is to detect and count all instances of the target class in the query image.

To address this problem, we propose a novel uncertainty-aware point-based few-shot object detector, named Counting-DETR trained with a novel two-stage training strategy to first generate pseudo GT bounding boxes from dot annotations and then train Counting-DETR on the generated pseudo GT boxes to predict bounding boxes of a new object class defined by a few bounding box exemplars in testing. The overview of Counting-DETR is illustrated in Fig. 3.

3.1   Feature Extraction and Feature Aggregation

Feature extraction: A CNN backbone is used to extract feature map $F^I \in \mathbb{R}^{H \times W \times D}$ from the input image I where H, W, D are height, width, and number of channels of the feature map, respectively. We then extract the exemplar feature vectors $f_k^B \in \mathbb{R}^{1 \times D}$, at the center of the exemplar bounding boxes B_k. Finally, the exemplar feature vector $f^B \in \mathbb{R}^{1 \times D}$ is obtained by averaging these feature vectors, or $f^B = \frac{1}{K} \sum_k f_k^B$.

Feature aggregation: We integrate the exemplar feature f^B to the feature map of the image F^I to produce the exemplar-integrated feature map F^A:

$$F^A = W_{proj} * [F^I; F^I \otimes f^B], \tag{1}$$

where $*, \otimes, [\cdot; \cdot]$ are the convolution, channel-wise multiplication, and concatenation operations, respectively. $W_{proj} \in \mathbb{R}^{2D \times D}$ is a linear projection weight. The first term in the concatenation preserves the original information of the feature map, while the second term aims at enhancing features at locations whose appearance are similar to those of the exemplars and suppressing the others.

3.2   The Encoder-Decoder Transformer

Inspired by DETR [3], we design our transformer of Counting-DETR to take as input the exemplar-integrated feature map F^A and M query points $\{p_m\}_{m=1}^M$ and predict the bounding box b_m for each query point p_m. The queries are the 2D points representing the initial guesses for the object locations rather than the learnable embeddings to achieve a faster training rate as shown in [43]. Thus, Counting-DETR is point-based approach that leverages the given dot annotations as the queries to predict the pseudo GT bounding boxes. Also, the transformer consists of two sub-networks: encoder and decoder. The former aims at enhancing features among the input set of features with the self-attention operation. The latter allows all the query points to interact with the enhanced features from the encoder with the cross-attention operation, thus capturing global information.

Next, the decoder is used to: (1) predict the classification score s representing the presence or absence of the object at a particular location, (2) regress the object's bounding box μ represented by the offset x, y from the GT object center to the query point along with its size w, h. Following [43], first, the Hungarian algorithm is used to match each of the GT bounding boxes with its corresponding predicted bounding boxes. Then for each pair of matched GT and predicted bounding boxes, the focal loss [23] and the combination of L_1 loss and GIoU loss [35] are used as training loss functions. In particular, at each query point, the following loss is computed:

$$L_{\text{DETR}} = \lambda_1 \text{Focal}(s, s^*) + \lambda_2 L_1(\mu, \mu^*) + \lambda_3 \text{GIoU}(\mu, \mu^*), \tag{2}$$

where s^*, μ^* are the GT class label and bounding box, respectively. $\lambda_1, \lambda_2, \lambda_3$ are the coefficients of focal, L_1, and GIoU loss functions, respectively.

Notably, Counting-DETR also estimates the uncertainty σ when training under the supervision of the imperfect pseudo GT bounding boxes $\tilde{\mu}$. This uncertainty is used to regularize the learning of bounding box μ such that a lower loss is incurred at the prediction with high uncertainty. We propose to use the following uncertainty loss:

$$L_{\text{uncertainty}} = \frac{1}{2} \sum_{o \in \{x, y, w, h\}} \frac{|\mu_o - \tilde{\mu}_o|}{\sigma_o} + \log \sigma_o, \tag{3}$$

where σ is the estimated uncertainty. This loss is derived from the maximum likelihood estimation (MLE) between the predicted bounding box distribution characterized by a Laplace distribution and the pseudo GT bounding box as evidence. Another option is the Gaussian distribution, however, in the experiments, we show that the Gaussian has the inferior performance to that of Laplace, and is even worse than the variant that does not employ uncertainty estimation.

3.3   The Two-Stage Training Strategy

The proposed few-shot object detector, Counting-DETR, can only be trained with the bounding box supervision for all objects. However, we only have bounding box annotation for a few exemplars and the point annotation for all objects as the setting of FSCD. Hence, we propose a two-stage training strategy as follows.

In Stage 1, we first pretrain Counting-DETR on a few exemplar bounding boxes with their centers as the query points (as described in Sect. 3.2). Subsequently, the pretrained network is used to predict the pseudo GT bounding boxes on the training images with the dot annotations as the query points. It is worth noting that, in this stage, we have the GT exemplar center as queries and their corresponding bounding boxes as supervision, so we do not use the

Hungarian matching, uncertainty estimation, and uncertainty loss, i.e., we only use L_{DETR} in Eq. (2) to train our Counting-DETR. The visualization of some generated pseudo-GT boxes is illustrated in Fig. 4.

In Stage 2, the generated pseudo GT bounding boxes on the training images are used to fine-tune the pretrained Counting-DETR. The fine-tuned model is then used to make predictions on the test images with the uniformly sampled anchor points as queries. Different from Stage 1, the supervision is the imperfect pseudo GT bounding boxes, hence, we additionally leverage the uncertainty estimation branch with uncertainty loss to train. Particularly, we use the following loss to train Counting-DETR in this stage:

$$L_{\text{combine}} = L_{\text{DETR}} + \lambda_4 L_{\text{uncertainty}}, \tag{4}$$

where λ_4 is the coefficient of $L_{\text{uncertainty}}$.

Fig. 4. Examples of pseudo GT bounding boxes generated by the 1-st stage our method.

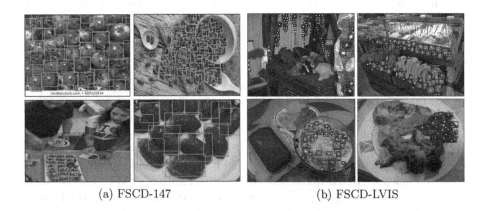

(a) FSCD-147   (b) FSCD-LVIS

Fig. 5. Sample images from our datasets and annotated bounding boxes.

4   New Datasets for Few-Shot Counting and Detection

A contribution of our paper is the introduction of two new datasets for few-shot counting and detection. In this section, we will describe these datasets.

Table 1. Comparison between the FSCD-147 and FSCD-LVIS datasets

Dataset	#Classes	Number of images			
		Total	Train	Val	Test
FSCD-147	147	6135	3659	1286	1190
FSCD-LVIS	372	6195	4000	1181	1014

Table 2. Number of images for each bin of the FSCD-LVIS dataset

Counting range	20–30	30–40	40–50	50–60	60–70	>70
# Images in range	2628	1212	684	402	286	959

4.1 The FSCD-147 Dataset

The FSC-147 dataset [32] was recently introduced for the few-shot object counting task with 6135 images across a diverse set of 147 object categories. In each image, the dot annotation for all objects and three exemplar boxes are provided. However, this dataset does not contain bounding box annotations for all objects. For evaluation purposes, we extend the FSC-147 dataset by providing bounding box annotations for all objects of the val and test sets. We name the new dataset FSCD-147. To be consistent with the counting, an object will be annotated with its bounding box only if it has a dot annotation. Figure 5a shows some samples of the FSCD-147 dataset. It is worth noting that annotating bounding boxes for many objects in crowded scenes of FSC-147 is a laborious process, and this is a significant contribution of our paper.

4.2 The FSCD-LVIS Dataset

Although the FSC-147 dataset contains images with a large number of objects in each image, the scene of each image is rather simple. Each image of FSC-147 shows the target object class so clearly that one can easily know which object class to count without having to specify any exemplars as shown in Fig. 1 and Fig. 5a. For real-world deployment of methods for few-shot counting and detection, we introduce a new dataset called FSCD-LVIS. Specifically, the scene is more complex with multiple object classes with multiple object instances each as illustrated in Fig. 5b. Without providing the exemplars for the target class, one cannot definitely guess which the target class is.

The FSCD-LVIS dataset contains 6196 images and 377 classes, extracted from the LVIS dataset [12]. For each image, we filter out all instances with an area smaller than 20 pixels, or a width or a height smaller than 4 pixels. The comparison between the FSCD-LVIS and FSC-147 datasets is shown in Table 1. The histogram of the number of labeled objects per image is illustrated in Table 2. The LVIS dataset has the box annotations for all objects, however, to be consistent with the setting of FSCD, we randomly choose three annotated bounding boxes of a selected object class as the exemplars for each image in the training set of FSCD-LVIS.

5   Experimental Results

Metrics. For object counting, we use Mean Average Error (MAE) and Root Mean Squared Error (RMSE), which are standard measures used in the counting literature. Besides, the Normalized Relative Error (NAE) and Squared Relative Error (SRE) are also adopted. In particular, $\text{MAE} = \frac{1}{J}\sum_{j=1}^{J}|c_j^* - c_j|; \text{RMSE} = \sqrt{\frac{1}{J}\sum_{j=1}^{J}(c_j^* - c_j)^2}; \text{NAE} = \frac{1}{J}\sum_{j=1}^{J}\frac{|c_j^*-c_j|}{c_j^*}; \text{SRE} = \sqrt{\frac{1}{J}\sum_{j=1}^{J}\frac{(c_j^*-c_j)^2}{c_j^*}}$ where J is the number of test images, c_j^* and c_j are GT and the predicted number of objects for image j, respectively. Unlike the absolute errors MAE and RMSE, the relative errors NAE and SRE reflect the practical usage of visual counting, i.e., with the same number of wrong objects counted (e.g., 10), it is more serious for images having a smaller number of objects (e.g., 20) than the ones having larger numbers of objects (e.g., 200).

For object detection, we use mAP and AP50. They are the average precision metrics with the IoU threshold between predicted and GT boxes for determining a correct prediction ranging from 0.5 to 0.95 for mAP and 0.5 for AP50.

Implementation Details. We implement our approach, baselines, and ablations in PyTorch [28]. Our backbone network is ResNet-50 [13] with the frozen Batch Norm layer [18]. We extract exemplar features f_k^B from exemplar boxes B_k from Layer 4 of the backbone. Our transformer network shares the same architecture as that of Anchor DETR [43] with the new uncertainty estimation and is trained with additional uncertainty loss as described in Sect. 3.2, while keeping the rest intact with six layers for both encoder and decoder. We use AdamW optimizer [25] with the learning rate of 10^{-5} for the backbone and 10^{-4} for the transformer to train Counting-DETR in 30 epochs with a batch size of one. We use the following training loss coefficients $\lambda_1 = 2, \lambda_2 = 5, \lambda_3 = 2, \lambda_4 = 2$, which were tuned based on the validation set. Also, the number of exemplar boxes is set to $K = 3$, as in FamNet [32] for a fair comparison.

5.1   Ablation Study

We conduct several experiments on the validation data of FSCD-147 to study the contribution of various components of our method.

Pseudo Box and Uncertainty Loss. From Table 3, we see that pseudo GT boxes (pseudo box) generated from our first stage are much better than the boxes generated by Ridge Regression (-6 in AP, +9 in MAE). Without using uncertainty loss (similar to Point DETR [5]), the performance drops substantially (-3 in AP, +3 in MAE). That justifies the effectiveness of our uncertainty loss. Without using both of them, the performance gets worst (-7 in AP, +11 in

Table 3. Ablation study on each component's contribution to the final results

Combination		Counting				Detection	
Pseudo box	Uncertainty	MAE ($\downarrow$)	RMSE($\downarrow$)	NAE($\downarrow$)	SRE ($\downarrow$)	AP($\uparrow$)	AP50($\uparrow$)
✓	✓	**20.38**	**82.45**	**0.19**	**3.38**	**17.27**	**41.90**
✗	✓	29.74	104.04	0.26	4.44	11.37	29.98
✓	✗	23.57	93.54	0.21	3.77	14.19	36.34
✗	✗	31.36	105.76	0.27	4.60	10.81	28.76

Table 4. Performance of Counting-DETR with different types of anchor points

Anchor type	Counting				Detection	
	MAE ($\downarrow$)	RMSE($\downarrow$)	NAE($\downarrow$)	SRE ($\downarrow$)	AP($\uparrow$)	AP50($\uparrow$)
Learnable	25.20	**81.94**	0.25	3.92	16.46	38.34
Fixed grid (proposed)	**20.38**	82.45	**0.19**	**3.38**	**17.27**	**41.90**

MAE). These results demonstrate the important contribution of our proposed pseudo GT box generation and uncertainty loss.

Types of Anchor Points. As described in Sect. 3.2, we follow the design of Anchor DETR whose anchor points can either be learnable or fixed-grid. The results of these two types are shown in Table 4, we can see that the fixed-grid anchor points are comparable to the learnable anchor points on counting metrics, but better on the detection metrics. Thus, the fixed-grid anchor points are chosen for the Counting-DETR.

Numbers of Anchor Points. Table 5 presents the results with different numbers of anchor points M. Both the detection and counting results increase as the number of anchor points increases, and they reach the highest points when the number of anchor points is $M = 600$. Hence, we choose 600 anchor points for Counting-DETR.

Types for the Uncertainty Loss. Instead of using the Laplace distribution as described in Sect. 3.2, we use Gaussian distribution to derive the uncertainty loss: $L_{\text{uncertainty}}^{\text{Gaussian}} = \frac{1}{2} \sum_{o \in \{x,y,w,h\}} \frac{(\mu_o - \mu_o^*)^2}{\sigma_o^2} + \log \sigma_o^2$. This loss is similar to [14]. The results are shown in Table 6. The uncertainty loss derived from the Gaussian distribution yields the worst results among the variants, even worse than the variant without using any uncertainty loss. On the contrary, our proposed uncertainty loss derived from the Laplace distribution gives the best results.

5.2   Comparison to Prior Work

Since there is no existing method for the new FSCD task, we compare Counting-DETR with several strong baselines adapted from few-shot object counting and few-shot object detection: FamNet [32]+RR, FamNet [32]+MLP, Attention-RPN [6]+RR, and FSDetView [45]+RR. Other few-shot object detectors are not chosen due to the unavailability of the source code or the requirement for fine-

Table 5. Performance of Counting-DETR with different numbers of anchor points

# Anchor points	Counting				Detection	
	MAE ($\downarrow$)	RMSE($\downarrow$)	NAE($\downarrow$)	SRE ($\downarrow$)	AP($\uparrow$)	AP50($\uparrow$)
100	30.22	113.24	0.24	4.55	11.26	28.62
200	26.76	103.11	0.22	4.15	14.06	34.33
300	23.57	93.54	0.21	3.77	14.19	36.34
400	22.62	88.45	0.21	3.69	14.91	37.30
500	21.72	85.20	0.20	3.52	16.03	39.66
600	**20.38**	**82.45**	**0.19**	**3.38**	**17.27**	**41.90**
700	21.19	83.70	0.22	3.47	15.10	37.85

Table 6. Ablation study for the uncertainty loss

Distribution type	Counting				Detection	
	MAE ($\downarrow$)	RMSE($\downarrow$)	NAE($\downarrow$)	SRE ($\downarrow$)	AP($\uparrow$)	AP50($\uparrow$)
W/o uncertainty loss	23.20	92.87	0.21	3.77	13.86	35.67
Gaussian loss	24.46	94.20	0.22	3.84	14.03	34.91
Laplacian loss (proposed)	**20.38**	**82.45**	**0.19**	**3.38**	**17.27**	**41.90**

tuning on the whole novel classes together (see Sect. 2). It is different from our setting, where each novel class is processed independently in a separate image. FamNet+RR is a method that uses Ridge Regression on top of the density map predicted by FamNet as depicted in Fig. 2a. FamNet+MLP is similar to FamNet+RR but replaces the ridge regression with a two-layer MLP with the Layer norm. Attention-RPN and FSDetView are detection-based methods, which require GT bounding boxes for all objects to train, thus, we generate the pseudo GT bounding boxes using either (1) the FamNet+RR with the features extracted from the dot annotations of training images instead of peak locations (called RR box) or (2) our first stage of training as described in Sect. 3.3 (called pseudo box). Table 7 and Table 8 show the comparison on the test sets of FSCD-147 and FSCD-LVIS, respectively.

Table 7. Comparison with strong baselines on the FSCD-147 test set

Method	Counting				Detection	
	MAE ($\downarrow$)	RMSE ($\downarrow$)	NAE($\downarrow$)	SRE ($\downarrow$)	AP($\uparrow$)	AP50($\uparrow$)
FamNet [32]+RR	22.09	**99.55**	0.44	6.45	9.44	29.73
FamNet [32]+MLP	22.09	**99.55**	0.44	6.45	1.21	6.12
Attention-RPN [6]+RR box	32.70	141.07	0.38	5.27	18.53	35.87
FSDetView [45]+RR box	37.83	146.56	0.48	5.47	13.41	32.99
Attention-RPN [6]+pseudo box	32.42	141.55	0.38	5.25	20.97	37.19
FSDetView [45]+pseudo box	37.54	147.07	0.44	5.40	17.21	33.70
Counting-DETR (proposed)	**16.79**	123.56	**0.19**	**5.23**	**22.66**	**50.57**

Table 8. Comparison with strong baselines on the FSCD-LVIS test set

Method	Counting				Detection	
	MAE (↓)	RMSE (↓)	NAE(↓)	SRE (↓)	AP(↑)	AP50(↑)
FamNet [32]+RR	60.53	84.00	1.82	14.58	0.84	2.04
Attention-RPN [6]+RR box	61.31	64.10	1.02	6.94	3.28	9.44
FSDetView [45]+RR box	26.81	33.18	0.56	4.51	1.96	6.70
Attention-RPN [6]+pseudo box	62.13	65.16	1.07	7.21	4.08	11.15
FSDetView [45]+pseudo box	24.89	31.34	0.54	4.46	2.72	7.57
Counting-DETR	**18.51**	**24.48**	**0.45**	**3.99**	**4.92**	**14.49**

Table 9. Comparison on FSCD-LVIS with unseen test classes

Method	Counting				Detection	
	MAE (↓)	RMSE (↓)	NAE(↓)	SRE (↓)	AP(↑)	AP50(↑)
FamNet [32]+RR	68.45	93.31	2.34	17.41	0.07	0.30
Attention-RPN [6]+RR box	35.55	42.82	1.21	7.47	2.52	7.86
FSDetView [45]+RR box	28.56	39.72	0.73	4.88	0.89	2.38
Attention-RPN [6]+pseudo box	39.16	46.09	1.34	8.18	3.15	7.87
FSDetView [45]+pseudo box	28.99	40.08	0.75	4.93	1.03	2.89
Counting-DETR	**23.50**	**35.89**	**0.57**	**4.17**	**3.85**	**11.28**

On FSCD-147, our method significantly outperforms others with a large margin for object detection. For counting, compared to a density-based approach like FamNet, Counting-DETR achieves worse results in RMSE metric but with much better results in other counting metrics MAE, NAE, and SRE. FamNet+MLP seems to overfit to the exemplar boxes so it performs the worst in detection.

On FSCD-LVIS, our method outperforms all others for both detection and counting tasks. This is because the image in FSCD-LVIS is much more complicated than those in FSCD-147, i.e., multiple object classes per image and significant differences in object size and shape. Also, the class of interest is usually packed and occluded by other classes, so the density map cannot be reliably predicted as shown in Fig. 6. More interestingly, we also evaluate the performance of Counting-DETR and other baselines on a special test set of unseen classes of the FSCD-LVIS dataset to show their generalizability to unseen classes during training in Table 9. It can be seen that our approach performs the best while the FamNet+RR performs the worst.

Figure 7 shows the qualitative comparison between our approach and the other methods, including FSDetView [45], Attention-RPN [6], and FamNet [32]+RR. Our method can successfully detect the objects of interest while other methods cannot, as shown in the first four rows of Fig. 7. The last row is a failure case for all methods, due to object truncation, perspective distortion, and scale variation.

Exemplar boxes   GT boxes   Predicted boxes   Predicted density map

Fig. 6. Results of FamNet on the FSCD-LVIS dataset. The objects of interest for Row 1 and 2 are bananas and chairs, respectively. FamNet fails to output good density maps for images containing objects with huge variations in shape and size.

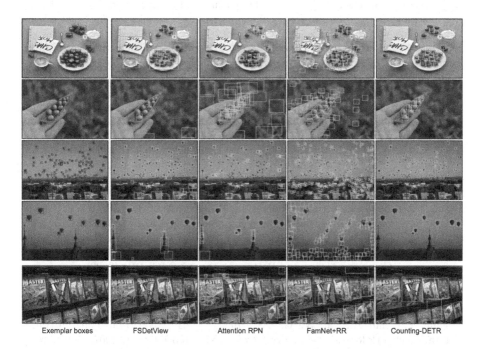

Exemplar boxes   FSDetView   Attention RPN   FamNet+RR   Counting-DETR

Fig. 7. Qualitative comparison. Each row shows an example with the exemplars (red) and GT bounding boxes (blue) on the first column. The first four rows show the superior performance of ours over the others, while the last row is a failure case. In the first row, our method can distinguish between the target class and other foreground classes, while other methods confuse between different foreground classes. In the next three rows, exemplar objects either share similar color with environment or contain many background pixels. These conditions lead to either under detect or over detect where other methods are unsure about if detected objects are foreground or background. The last row shows a failure case for all methods due to the huge variation in the scale, distortion, and truncation of the objects.

6 Conclusions

We have introduced a new task of few-shot object counting and detection that shares the same supervision with few-shot object counting but additionally predict object bounding boxes. To address this task, we have collected two new datasets, adopted a two-stage training strategy to generate pseudo bounding boxes for training, and developed a new uncertainty-aware few-shot object detector to adapt to the imperfection of pseudo label. Extensive experiments on the two datasets demonstrate that the proposed approach outperforms strong baselines adapted from few-shot object counting and few-shot object detection.

References

1. Abousamra, S., Hoai, M., Samaras, D., Chen, C.: Localization in the crowd with topological constraints. In: Proceedings of AAAI Conference on Artificial Intelligence (AAAI) (2021)
2. Arteta, C., Lempitsky, V., Noble, J.A., Zisserman, A.: Detecting overlapping instances in microscopy images using extremal region trees. Med. Image Anal. **27**, 3–16 (2016). discrete Graphical Models in Biomedical Image Analysis
3. Carion, N., Massa, F., Synnaeve, G., Usunier, N., Kirillov, A., Zagoruyko, S.: End-to-end object detection with transformers. In: Vedaldi, A., Bischof, H., Brox, T., Frahm, J.-M. (eds.) ECCV 2020. LNCS, vol. 12346, pp. 213–229. Springer, Cham (2020). https://doi.org/10.1007/978-3-030-58452-8_13
4. Chen, H., Wang, Y., Wang, G., Qiao, Y.: Lstd: a low-shot transfer detector for object detection. In: Proceedings of the AAAI Conference on Artificial Intelligence (2018)
5. Chen, L., Yang, T., Zhang, X., Zhang, W., Sun, J.: Points as queries: weakly semi-supervised object detection by points. In: Proceedings of the IEEE/CVF Conference on Computer Vision and Pattern Recognition, pp. 8823–8832 (2021)
6. Fan, Q., Zhuo, W., Tang, C.K., Tai, Y.W.: Few-shot object detection with attention-RPN and multi-relation detector. In: Proceedings of the IEEE/CVF Conference on Computer Vision and Pattern Recognition (CVPR) (2020)
7. Fan, Z., Ma, Y., Li, Z., Sun, J.: Generalized few-shot object detection without forgetting. In: Proceedings of the IEEE/CVF Conference on Computer Vision and Pattern Recognition (CVPR) (2021)
8. Fang, Y., Zhan, B., Cai, W., Gao, S., Hu, B.: Locality-constrained spatial transformer network for video crowd counting. arXiv preprint arXiv:1907.07911 (2019)
9. Feng, D., Rosenbaum, L., Timm, F., Dietmayer, K.: Labels are not perfect: improving probabilistic object detection via label uncertainty. arXiv preprint arXiv:2008.04168 (2020)
10. Feng, D., et al.: Labels are not perfect: inferring spatial uncertainty in object detection. IEEE Trans. Intell. Transp. Syst. **23**(8), 9981–9994 (2021)
11. Goldman, E., Herzig, R., Eisenschtat, A., Goldberger, J., Hassner, T.: Precise detection in densely packed scenes. In: Proceedings of the IEEE/CVF Conference on Computer Vision and Pattern Recognition (CVPR) (2019)
12. Gupta, A., Dollar, P., Girshick, R.: Lvis: a dataset for large vocabulary instance segmentation. In: Proceedings of the IEEE/CVF Conference on Computer Vision and Pattern Recognition (CVPR) (2019)

13. He, K., Zhang, X., Ren, S., Sun, J.: Deep residual learning for image recognition. In: Proceedings of the IEEE Conference on Computer Vision and Pattern Recognition (CVPR) (2016)
14. He, Y., Zhu, C., Wang, J., Savvides, M., Zhang, X.: Bounding box regression with uncertainty for accurate object detection. In: Proceedings of the IEEE/CVF Conference on Computer Vision and Pattern Recognition (CVPR) (2019)
15. Hsieh, M.R., Lin, Y.L., Hsu, W.H.: Drone-based object counting by spatially regularized regional proposal networks. In: The IEEE International Conference on Computer Vision (ICCV), IEEE (2017)
16. Hu, D., Mou, L., Wang, Q., Gao, J., Hua, Y., Dou, D., Zhu, X.X.: Ambient sound helps: audiovisual crowd counting in extreme conditions. arXiv preprint (2020)
17. Idrees, H., et al.: Composition loss for counting, density map estimation and localization in dense crowds. In: Proceedings of the European Conference on Computer Vision (ECCV) (2018)
18. Ioffe, S., Szegedy, C.: Batch normalization: accelerating deep network training by reducing internal covariate shift. In: International Conference on Machine Learning (ICML), PMLR (2015)
19. Kang, B., Liu, Z., Wang, X., Yu, F., Feng, J., Darrell, T.: Few-shot object detection via feature reweighting. In: Proceedings of the IEEE/CVF International Conference on Computer Vision (ICCV) (2019)
20. Lee, Y., Hwang, J.W., Kim, H.I., Yun, K., Kwon, Y.: Localization uncertainty estimation for anchor-free object detection. arXiv preprint arXiv:2006.15607 (2020)
21. Lian, D., Li, J., Zheng, J., Luo, W., Gao, S.: Density map regression guided detection network for rgb-d crowd counting and localization. In: The IEEE Conference on Computer Vision and Pattern Recognition (CVPR), June 2019
22. Lin, H., Hong, X., Wang, Y.: Object counting: you only need to look at one. arXiv preprint arXiv:2112.05993 (2021)
23. Lin, T.Y., Goyal, P., Girshick, R., He, K., Dollár, P.: Focal loss for dense object detection. In: Proceedings of the IEEE International Conference on Computer Vision (ICCV) (2017)
24. Liu, W., Salzmann, M., Fua, P.: Context-aware crowd counting. In: The IEEE Conference on Computer Vision and Pattern Recognition (CVPR), June 2019
25. Loshchilov, I., Hutter, F.: Decoupled weight decay regularization. arXiv preprint arXiv:1711.05101 (2017)
26. Lu, E., Xie, W., Zisserman, A.: Class-agnostic counting. In: Asian Conference on Computer Vision (ACCV) (2018)
27. Mundhenk, T.N., Konjevod, G., Sakla, W.A., Boakye, K.: A large contextual dataset for classification, detection and counting of cars with deep learning. In: Leibe, B., Matas, J., Sebe, N., Welling, M. (eds.) ECCV 2016. LNCS, vol. 9907, pp. 785–800. Springer, Cham (2016). https://doi.org/10.1007/978-3-319-46487-9_48
28. Paszke, A., et al.: Pytorch: an imperative style, high-performance deep learning library. In: Wallach, H., Larochelle, H., Beygelzimer, A., d'Alché-Buc, F., Fox, E., Garnett, R. (eds.) Advances in Neural Information Processing Systems (NeurIPS), vol. 32, pp. 8024–8035. Curran Associates, Inc. (2019)
29. Peng, D., Sun, Z., Chen, Z., Cai, Z., Xie, L., Jin, L.: Detecting heads using feature refine net and cascaded multi-scale architecture. In: 2018 24th International Conference on Pattern Recognition (ICPR), IEEE (2018)
30. Ranjan, V., Hoai, M.: Vicinal counting networks. In: Proceedings of the IEEE/CVF Conference on Computer Vision and Pattern Recognition, pp. 4221–4230 (2022)
31. Ranjan, V., Le, H., Hoai, M.: Iterative crowd counting. In: Proceedings of European Conference on Computer Vision (ECCV) (2018)

32. Ranjan, V., Sharma, U., Nguyen, T., Hoai, M.: Learning to count everything. In: Proceedings of the IEEE/CVF Conference on Computer Vision and Pattern Recognition (CVPR) (2021)
33. Ranjan, V., Wang, B., Shah, M., Hoai, M.: Uncertainty estimation and sample selection for crowd counting. In: Proceedings of the Asian Conference on Computer Vision (ACCV) (2020)
34. Ren, S., He, K., Girshick, R., Sun, J.: Faster r-cnn: Towards real-time object detection with region proposal networks. Adv. Neural Inf. Process. Syst. (NeurIPS) **28** (2015)
35. Rezatofighi, H., Tsoi, N., Gwak, J., Sadeghian, A., Reid, I., Savarese, S.: Generalized intersection over union: a metric and a loss for bounding box regression. In: Proceedings of the IEEE/CVF Conference on Computer Vision and Pattern Recognition, pp. 658–666 (2019)
36. Russakovsky, O., et al.: ImageNet large scale visual recognition challenge. Int. J. Comput. Vis. **115**(3), 211–252 (2015). https://doi.org/10.1007/s11263-015-0816-y
37. Sindagi, V.A., Yasarla, R., Patel, V.M.: Pushing the frontiers of unconstrained crowd counting: new dataset and benchmark method. In: Proceedings of the IEEE/CVF International Conference on Computer Vision (ICCV) (2019)
38. Tian, Z., Shen, C., Chen, H., He, T.: Fcos: fully convolutional one-stage object detection. In: Proceedings of the IEEE/CVF International Conference on Computer Vision (ICCV) (2019)
39. Vaswani, A., et al.: Attention is all you need. In: Advances in Neural Information Processing Systems (NeurIPS) (2017)
40. Wang, B., Liu, H., Samaras, D., Hoai, M.: Distribution matching for crowd counting. In: Advances in Neural Information Processing Systems (NeurIPS) (2020)
41. Wang, Q., Gao, J., Lin, W., Li, X.: Nwpu-crowd: a large-scale benchmark for crowd counting and localization. IEEE Trans. Pattern Anal. Mach. Intell. (TPAMI) **43**(6), 2141–2149 (2021)
42. Wang, X., Huang, T.E., Darrell, T., Gonzalez, J.E., Yu, F.: Frustratingly simple few-shot object detection. arXiv preprint arXiv:2003.06957 (2020)
43. Wang, Y., Zhang, X., Yang, T., Sun, J.: Anchor detr: query design for transformer-based detector. arXiv preprint arXiv:2109.07107 (2021)
44. Wu, J., Liu, S., Huang, D., Wang, Y.: Multi-scale positive sample refinement for few-shot object detection. In: Vedaldi, A., Bischof, H., Brox, T., Frahm, J.-M. (eds.) ECCV 2020. LNCS, vol. 12361, pp. 456–472. Springer, Cham (2020). https://doi.org/10.1007/978-3-030-58517-4_27
45. Xiao, Y., Marlet, R.: Few-shot object detection and viewpoint estimation for objects in the wild. In: Vedaldi, A., Bischof, H., Brox, T., Frahm, J.-M. (eds.) ECCV 2020. LNCS, vol. 12362, pp. 192–210. Springer, Cham (2020). https://doi.org/10.1007/978-3-030-58520-4_12
46. Xie, W., Noble, J.A., Zisserman, A.: Microscopy cell counting and detection with fully convolutional regression networks. Comput. Methods Biomech. Biomed. Eng. Imaging Vis. **6**(3), 283–292 (2018)
47. Yan, X., Chen, Z., Xu, A., Wang, X., Liang, X., Lin, L.: Meta r-cnn : towards general solver for instance-level low-shot learning. In: Proceedings of IEEE International Conference on Computer Vision (ICCV) (2019)
48. You, Z., Yang, K., Luo, W., Lu, X., Cui, L., Le, X.: Iterative correlation-based feature refinement for few-shot counting. arXiv preprint arXiv:2201.08959 (2022)
49. Zhang, C., Li, H., Wang, X., Yang, X.: Cross-scene crowd counting via deep convolutional neural networks. In: 2015 IEEE Conference on Computer Vision and Pattern Recognition (CVPR) (2015)

50. Zhang, G., Luo, Z., Cui, K., Lu, S.: Meta-detr: few-shot object detection via unified image-level meta-learning. arXiv preprint arXiv:2103.11731 (2021)
51. Zhang, Y., Zhou, D., Chen, S., Gao, S., Ma, Y.: Single-image crowd counting via multi-column convolutional neural network. In: 2016 IEEE Conference on Computer Vision and Pattern Recognition (CVPR) (2016)
52. Zhou, X., Wang, D., Krähenbühl, P.: Objects as points. In: arXiv preprint arXiv:1904.07850 (2019)

Rethinking Few-Shot Object Detection on a Multi-Domain Benchmark

Kibok Lee[1,2], Hao Yang[1(✉)], Satyaki Chakraborty[1], Zhaowei Cai[1],
Gurumurthy Swaminathan[1], Avinash Ravichandran[1], and Onkar Dabeer[1]

[1] AWS AI Labs, Palo Alto, USA
{kibok,haoyng,satyaki,zhaoweic,gurumurs,ravinash,onkardab}@amazon.com
[2] Yonsei University, Seoul, South Korea
kibok@yonsei.ac.kr

Abstract. Most existing works on few-shot object detection (FSOD) focus on a setting where both pre-training and few-shot learning datasets are from a similar domain. However, few-shot algorithms are important in multiple domains; hence evaluation needs to reflect the broad applications. We propose a Multi-dOmain Few-Shot Object Detection (MoFSOD) benchmark consisting of 10 datasets from a wide range of domains to evaluate FSOD algorithms. We comprehensively analyze the impacts of freezing layers, different architectures, and different pre-training datasets on FSOD performance. Our empirical results show several key factors that have not been explored in previous works: 1) contrary to previous belief, on a multi-domain benchmark, fine-tuning (FT) is a strong baseline for FSOD, performing on par or better than the state-of-the-art (SOTA) algorithms; 2) utilizing FT as the baseline allows us to explore multiple architectures, and we found them to have a significant impact on down-stream few-shot tasks, even with similar pre-training performances; 3) by decoupling pre-training and few-shot learning, MoF-SOD allows us to explore the impact of different pre-training datasets, and the right choice can boost the performance of the down-stream tasks significantly. Based on these findings, we list possible avenues of investigation for improving FSOD performance and propose two simple modifications to existing algorithms that lead to SOTA performance on the MoFSOD benchmark. The code is available here.

Keywords: Few-shot learning · Object detection

1 Introduction

Convolutional neural networks have led to significant progress in object detection by learning with a large number of training images with annotations [3,4,28,

K. Lee—Work done at AWS.

Supplementary Information The online version contains supplementary material available at https://doi.org/10.1007/978-3-031-20044-1_21.

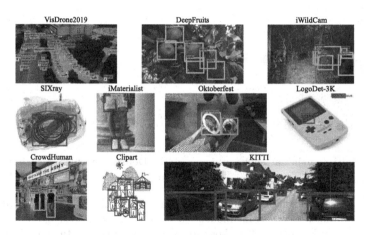

Fig. 1. Sample images in the proposed FSOD benchmark.

30]. However, humans can easily localize and recognize new objects with only a few examples. Few-shot object detection (FSOD) is a task to address this setting [8,19,26,35,41]. FSOD is desirable for many real-world applications in diverse domains due to lack of training data, difficulties in annotating them, or both, *e.g*let@tokeneonedot, identifying new logos, detecting anomalies in the manufacturing process or rare animals in the wild, *etc*let@tokeneonedot. These diverse tasks naturally have vast differences in class distribution and style of images. Moreover, large-scale pre-training datasets in the same domain are not available for many of these tasks. In such cases, we can only rely on existing natural image datasets, such as COCO [24] and OpenImages [21] for pre-training.

Despite the diverse nature of FSOD tasks, FSOD benchmarks used in prior works are limited to a homogeneous setting [8,19,26,35,41], such that the pre-training and few-shot test sets in these benchmarks are from the same domain, or even the same dataset, *e.g*let@tokeneonedot, VOC [7] 15 + 5 and COCO [24] 60+20 splits. The class distributions of such few-shot test sets are also fixed to be balanced. While they provide an artificially balanced environment for evaluating different algorithms, it might lead to skewed conclusions for applying them in more realistic scenarios. Note that few-shot classification suffered from the same problem in the past few years [29,39]; Meta-dataset [37] addressed the problem with 10 different domains and a sophisticated scheme to sample imbalanced few-shot episodes.

Inspired by Meta-dataset [37], we propose a Multi-dOmain FSOD (MoF-SOD) benchmark consisting of 10 datasets from 10 different domains, as shown in Fig. 1. The diversity of MoFSOD datasets can be seen in Fig. 2 where the domain distance of each dataset to COCO [24] is depicted. Our benchmark enables us to estimate the performance of FSOD algorithms across domains and settings and helps in better understanding of various factors, such as pre-training, architectures, *etc*let@tokeneonedot., that influence the algorithm per-

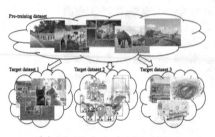 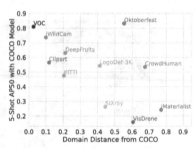

(a) Multi-domain datasets.

(b) Domain distance vs 5-shot performance.

Fig. 2. (a) Real-world applications of FSOD are not limited to the natural image domain; we propose to pre-train models on large-scale natural image datasets and transfer to target domains for few-shot learning. (b) We measure the domain distance between datasets (see Sect. 3.2 for details) in the benchmark and COCO, and plot against 5-shot AP50 of these datasets fine-tuned from a model pre-trained on COCO. VOC is added for reference. The figure shows the benchmark covers a wide range of domains.

formance. In addition, we propose a simple natural K-shot sampling algorithm that encourages more diversity in class distributions than balanced sampling.

Building on our benchmark, we extensively study the impact of freezing parameters, detection architectures, pre-training datasets, and the effectiveness of several state-of-the-art (SOTA) FSOD algorithms [26,35,41]. Our empirical study leads to rethinking conventions in the field and interesting findings that can guide future research on FSOD.

Conventionally, in FSOD or general few-shot learning, it is believed that freezing parameters to avoid overfitting is helpful or even crucial for good performance [34,35,41,43]. If we choose to tune more parameters, specific components or designs must be added, such as weight imprinting [44] or decoupled gradient [26], to prevent overfitting. Our experiments in the MoFSOD show that these design choices might be helpful when pre-training and few-shot learning are in similar domains, as in previous benchmarks. However, if we consider a broader spectrum of domains, unfreezing the whole network results in better overall performance, as the network has more freedom to adapt. We further demonstrate a correlation between the performance gain of tuning more parameters and domain distance (see Fig. 3a). Overall, *fine-tuning (FT) is a strong baseline* for FSOD on MoFSOD without any bells and whistles.

Using FT as a baseline allows us to explore the impact of different architectures on FSOD tasks. Previous FSOD methods [8,26,35,41] need to make architecture-specific design choices; hence focus on a single architecture – mostly Faster R-CNN [30], while we conduct extensive study on the impact of different architectures, *e.g*let@tokeneonedot, recent development of anchor-free [51,53] and transformer-based architectures [4,56], on few-shot performance. Surprisingly, we find that even with similar performance on COCO, different architectures have very different downstream few-shot performances. This finding suggests the potential benefits of specifically designed few-shot architectures for improved performance.

Moreover, unlike previous benchmarks, which split the pre-training and few-shot test sets from the same datasets (VOC or COCO), MoFSOD allows us to freely choose different pre-training datasets and explore the potential benefits of large-scale pre-training. To this end, we systematically study the effect of pre-training datasets with ImageNet [5], COCO [24], LVIS [13], FSOD-Dataset [8], Unified [52], and the integration of large-scale language-vision models. Similar to observations in recent works in image classification [20] and NLP [2,27], we find that large-scale pre-training can play a crucial role for downstream few-shot tasks.

Finally, motivated by the effectiveness of the unfreezing parameters and language-vision pre-training, we propose two extensions: FSCE+ and LVIS+. FSCE+ extends FSCE [35] to fine-tune more parameters with a simplified fully-connected (FC) detection head. LVIS+ follows the idea of using CLIP embedding of class names as the classifier, but instead of using it in zero-shot/open-vocabulary setting as in [11,50], we extend it to few-shot fine-tuning. Both methods achieve SOTA results with/without extra pre-training data.

We summarize our contributions as follows:

- We propose a Multi-dOmain Few-Shot Object Detection (MoFSOD) benchmark to simulate real-world applications in diverse domains.
- We conduct extensive studies on the effect of architectures, pre-training datasets, and hyperparameters with fine-tuning and several SOTA methods on the proposed benchmark. We summarize the observations below:
  - **Unfreezing more layers** do not lead to detrimental overfitting and improve the FSOD performance across different domains.
  - **Object detection model architectures** have a significant impact on the FSOD performance even when the architectures have a similar performance on the pre-training dataset.
  - **Pre-training datasets** play an important role in the downstream FSOD performance. Effective utilization of the pre-training dataset can significantly boost performance.
- Based on these findings, we propose two extensions that outperform SOTA methods by a significant margin on our benchmark.

2   Related Work

Meta-learning-based methods for FSOD are inspired by few-shot classification. Kang et al. [19] proposed a meta feature extractor with feature reweight module, which maps support images to mean features and reweight query features with the mean features, inspired by protoypical networks [34]. Meta R-CNN [46] extended the idea with an extra predictor-head remodeling network to extract class-attentive vectors. MetaDet [43] proposed meta-knowledge transfer with weight prediction module.

Two-stream methods take one query image and support images as inputs, and use the correlations between query and support features as the final features to the detection head and the Region Proposal Network (RPN). Several works in this direction [8, 14, 48] have shown competitive results. These methods require all classes to have at least one support image to be fed to the model, which makes the overall process slow.

Fine-tuning-based methods update only the linear classification and regression layers [41], the whole detection head and RPN with an additional contrastive loss [35], or decoupling the gradient of RPN and the detection head while updating the whole network [26]. These methods are simple yet have shown competitive results. We focus on benchmarking them due to their simplicity, efficiency, and higher performance than other types.

Multi-domain Few-Shot Classification Benchmarks. In few-shot classification, miniImageNet [39] and tieredImageNet [29] have been used as standard benchmarks. Similar to benchmarks in FSOD, they are divided into two splits and used for pre-training and few-shot learning, respectively, such that they are in the same natural image domain. Recent works have proposed new benchmarks to address this issue: Tseng *et al*let@tokeneonedot [38] proposed a cross-domain few-shot classification benchmark with five datasets from different domains. Triantafillou *et al*let@tokeneonedot [37] proposed Meta-Dataset, which is a large-scale few-shot classification benchmark with ten datasets and a sophisticatedly designed sampling algorithm to sample realistically imbalanced few-shot training datasets. Although not specifically catering to few-shot applications, Wang *et al*let@tokeneonedot [42] proposed universal object detection, which aims to cover multi-domain datasets with a single model for high-shot object detection.

3   MoFSOD: A Multi-domain FSOD Benchmark

In this section, we first describe existing FSOD benchmarks and their limitations. Then, we propose a Multi-dOmain FSOD (MoFSOD) benchmark.

3.1   Existing Benchmarks and Limitations

Recent FSOD works have evaluated their methods in PASCAL VOC 15 + 5 and MS COCO 60 + 20 benchmarks proposed by Kang *et al*let@tokeneonedot [19]. From the original VOC [7] with 20 classes and COCO [24] with 80 classes, they took 25% of classes as novel classes for few-shot learning, and the rest of them as base classes for pre-training. For VOC 15 + 5, three splits were made, where each of them consists of 15 base classes for pre-training, and the other 5 novel classes for few-shot learning. For each novel class, $K = \{1, 3, 5, 10\}$ object instances are sampled, which are referred to as shot numbers. For COCO, 20 classes overlapped with VOC are considered novel classes, and $K = \{10, 30\}$-shot settings are used. Different from classification, as an image usually contains multiple annotations in object detection, sampling exactly K annotations per class is difficult. Kang *et* *allet@tokeneonedot* [19] proposed pre-defined support sets for few-shot training, which would cause overfitting [17]. Wang *et allet@tokeneonedot* [41] proposed to sample few-shot training datasets with different random seeds to mitigate this issue, but the resulting sampled datasets often contain more than K instances. While these benchmarks contributed to the research progress in FSOD, they have several limitations.

First, these benchmarks do not capture the breadth of few-shot tasks and domains as they sample few-shot task instances from a single dataset, as we discussed in Sect. 1. Second, these benchmarks contain only a fixed number of classes, 5 or 20. However, real-world applications might have a varying number of classes, ranging from one class, *e.glet@tokeneonedot*, face/pedestrian detection [47,49], to thousands of classes, *e.glet@tokeneonedot*, logo detection [40]. Last but not least, these benchmarks are constructed with the balanced K-shot sampling. For example, in the 5-shot setting, a set of images containing exactly 5 objects [19] is pre-defined. Such a setting is unlikely in real-world few-shot tasks. We also demonstrate that such a sampling strategy can lead to high variances in the performance of multiple episodes (see Table 1b). Moreover, different from classification, object detection datasets tend to be imbalanced due to the multi-label nature of the datasets. For example, COCO [24] and OpenImages [21] have a more dominant number of person instances than any other objects. The benchmark datasets should also explore these imbalanced scenarios.

3.2   Multi-domain Benchmark Datasets

While FSOD applications span a wide range of domains, gathering enough pre-training data from these domains might be difficult. Hence, it becomes important to test the few-shot algorithm performance in settings where the pre-training and few-shot domains are different. Similar to Meta-dataset [37] in few-shot classification, we propose to extend the benchmark with datasets from a wide range of domains rather than a subset of natural image datasets. Our proposed benchmark consists of 10 datasets from 10 domains: VisDrone [54] in aerial images, DeepFruits [31] in agriculture, iWildCam [1] in animals in the wild, Clipart in cartoon, iMaterialist [12] in fashion, Oktoberfest [57] in food, LogoDet-3K [40] in logo, CrowdHuman [33] in person, SIXray [25] in security, and KITTI [10] in traffic/autonomous driving. We provide statistics of these datasets in Table 1a. The number of classes varies from 1 to 352, and that of boxes per image varies from 1.2 to 54.4, covering a wide range of scenarios.

In Fig. 2b, we illustrate the diversity of domains in our benchmark by computing the domain distances between these datasets and COCO [24] and plotting against the 5-shot performance of fine-tuning (FT) on each dataset from a model pre-trained on COCO. Specifically, we measure the domain similarity by calculating the recall of a pre-trained COCO model on each dataset in a class-agnostic fashion, similar to the measurement of unsupervised object proposals [16]. Intuitively, if a dataset is in a domain similar to COCO, then objects in the dataset are likely to be localized well by the model pre-trained on COCO. As a reference, VOC has a recall of 97%. For presentation purpose, we define $(1 - \text{recall})$

Table 1. Statistics of pre-training and MoFSOD datasets (Table 1a) and performance of different architectures on the pre-training datasets (Table 1b).

Domain	Dataset	# classes	# training images		Dataset	Architecture	AP
						Faster R-CNN	42.7
						Cascade R-CNN	45.1
	COCO	80	117k			CenterNet2	45.3
	FSODD	800	56k		COCO	RetinaNet	39.3
Image	LVIS	1203	100k			Deformable DETR	46.3
	Unified	723	2M			Cascade R-CNN-P67	45.9

Domain	Dataset	# classes	# bboxes per image		Dataset	Architecture	AP
						Faster R-CNN	24.2
					LVIS	CenterNet2	28.3
Aerial	VisDrone	10	54.4			Cascade R-CNN-P67	26.2
Agriculture	DeepFruits	7	5.6	
Animal	iWildCam	1	1.5		(b) Performance of benchmark architectures
Cartoon	Clipart	20	3.3		pre-trained on COCO and LVIS. FSODD and
Fashion	iMaterialist	46	7.3		Unified do not have a pre-defined validation/test
Food	Oktoberfest	15	2.4		set, so we do not measure their pre-training
Logo	LogoDet-3K	352	1.2		performances.
Person	CrowdHuman	2	47.1	
Security	SIXray	5	2.1	
Traffic	KITTI	4	7.0	

(a)**Top:** statistics of pre-training datasets.
Bottom: statistics of MoFSOD datasets.

as the domain distance. We can see diverse domain distances in the benchmark, ranging from 0.1 to 0.8. Interestingly, the domain distance also correlates with the FSOD performance. Although this is not the only deciding factor, as the intrinsic properties (such as the similarity between training and test datasets) of a dataset also play an important role, we can still see the linear correlation between the domain distance and 5-shot performance with the Pearson correlation coefficient -0.43. Oktoberfest and CrowdHuman are outliers in our analysis possibly as they are relatively easy.

Natural K-Shot Sampling. We use a natural K-shot sampling algorithm to maintain the original class distribution for this benchmark. Specifically, we sample $C \times K$ images from the original dataset without worrying about class labels, where C is the number of classes of the original dataset. Then, we check missing images to ensure we have at least one image for each class of all classes. We provide the details in the supplementary material. The comparison between the balanced K-shot and natural K-shot sampling shows that our conclusions do not change based on the sampling algorithm, but the performance of the natural K-shot sampling is more consistent on different episodes (see Table 2b) and covers imbalanced class distributions existing in the real-world applications.

Evaluation Protocol. To evaluate the scalability of methods, we experiment with four different average shot numbers, $K = \{1, 3, 5, 10\}$. We first sample a few-shot training dataset from the original training dataset with the natural K-shot sampling algorithm for each episode. Then, we initialize the object detection model with pre-trained model parameters and train the model with FSOD methods. For evaluation, we randomly sample 1k images from the original test set if the test set is larger than 1k. We repeat this episode 10 times with different random seeds for all multi-domain datasets and report the average of the mean and standard deviation of the performance.

Metrics. As evaluation metrics, we use AP50 and the average rank among compared methods [37]. AP50 stands for the average precision of predictions where the IoU threshold is 0.5, and the rank is an integer ranging from 1 to the number of compared methods, where the method with the highest AP50 gets rank 1. We first take the best AP50 among different hyperparameters at the end of training for each episode, compute the mean and standard deviation of AP50 and the rank over 10 episodes, and then average them over different datasets and/or shots, depending on the experiments.

4 Experiments

In this section, we conduct extensive experiments on MoFSOD and discuss the results. For better presentations, we highlight compared methods and architectures in *italics* and pre-training/few-shot datasets in **bold**.

4.1 Experimental Setup

Model Architecture. We conduct experiments on six different architectures. For simplicity, we use ResNet-50 [15] as the backbone of all architectures. We also employ deformable convolution v2 [55] in the last three stages of the backbone. Specifically, we benchmark two-stage object detection architectures: 1) *Faster R-CNN* [30] 2) *Cascade R-CNN* [3], and the newly proposed 3) *CenterNet2* [51], and one-stage architectures: 4) *RetinaNet* [23], as well as transformer-based 5) *Deformable-DETR* [56]. Note that all architectures utilize Feature Pyramid Networks (FPN) [22] or similar multi-scale feature aggregation techniques. In addition, we also experiment the combination of the FPN-P67 design from *RetinaNet* and *Cascade R-CNN*, dubbed 6) *Cascade R-CNN-P67* [52]. We conduct our architecture analysis pre-trained on **COCO** [24] and **LVIS** [13]. Table 1b summarizes the architectures and their pre-training performance.

Freezing Parameters. Based on the design of detectors, we can think of three different levels of fine-tuning the network: 1) only the last classification and regression fully-connected (FC) layer [41], 2) the detection head consisting of several FC and/or convolutional layers [8], and 3) the whole network, *i.e.* let@tokeneonedot, standard fine-tuning.[1] We study the effects of these three ways of tuning on different domains in MoFSOD with *Faster R-CNN* [30].

---

[1] When training an object detection model, the batch normalization layers [18] and the first two macro blocks of the backbone (**stem** and **res2**) are usually frozen, even for large-scale datasets. We follow this convention in our paper.

Pre-training Datasets. To explore the effect of pre-training dataset, we conduct experiments on five pre-training datasets: **ImageNet**[2] [5], **COCO** [24], **FSODD**[3] [8], **LVIS** [13], and **Unified**, which is a union of OpenImages v5 [21], Object365 v1 [32], Mapillary [6], and **COCO**, combined as in [52]. To reduce the combinations of different architectures and pre-training datasets, we conduct most of the studies on the best performing architecture *Cascade R-CNN-P67*. Also, inspired by [11,50], we experiment with the effect of CLIP [27] embeddings to initialize the final classification layer of detector when pre-training on **LVIS** dubbed **LVIS+**. In addition, **LVIS++** uses the backbone pre-trained on the ImageNet-21K classification task instead of ImageNet-1K before pre-training on **LVIS**. All experiments are done with Detectron2 [45].

Hyperparameters. For pre-training, we mostly follow standard hyperparameters of the corresponding method, with the addition of deformable convolution v2 [55]. On **COCO** and **LVIS**, for *Faster R-CNN*, *Cascade R-CNN*, and *Cascade R-CNN-P67*, we use the 3× scheduler with 270k iterations, the batch size of 16, the SGD optimizer with initial learning rate of 0.02 decaying by the factor of 0.1 at 210k and 250k. For *RetinaNet*, the initial learning rate is 0.01 [23]. For *CenterNet2*, following [51], we use the *CenterNet* [53] as the first stage and the Cascade R-CNN head as the second stage, where the other hyperparameters are the same as above. For *Deformable-DETR* [56], we follow the two-stage training of 50 epochs, the AdamW optimizer, and the initial learning rate of 0.0002 decaying by the factor of 0.1 at 40 epochs. On **FSODD**, we train for 60 epochs with the learning rate of 0.02 decaying by the factor of 0.1 at 40 and 54 epochs and the batch size of 32. On **Unified**, following [52], the label space of four datasets are unified, the dataset-aware sampling and equalization loss [36] are applied to handle long-tailed distributions, and training is done for 600k iterations with the learning rate of 0.02 decaying by the factor of 0.1 at 400k and 540k iterations, and the batch size of 32.

For few-shot training, we train models for 2k iterations with the batch size of 4 on a single V100 GPU.[4] For *Faster R-CNN*, *Cascade R-CNN*, *Cascade R-CNN-P67*, and *CenterNet2*, we train models with the SGD optimizer and different initial learning rates in {0.0025, 0.005, 0.01} and choose the best, where the learning rate is decayed by the factor of 0.1 after 80% of training. For *RetinaNet*, we halve the learning rates to {0.001, 0.0025, 0.005}, as we often observe training diverges with the learning rate of 0.01. For *Deformable-DETR* and *CenterNet2*

---

[2] This is ImageNet-1K for classification, which is commonly used for pre-training standard object detection methods, *i.e.*let@tokeneonedot, we omit pre-training on an object detection task.

[3] The name of the dataset is also FSOD, so we introduce an additional D to distinguish the dataset from the task.

[4] The batch size could be less than 4 if the sampled dataset size is less than 4, *e.g.*let@tokeneonedot, when the number of classes and shot number K is 1, and the batch size has to be 1.

with CLIP, we use the AdamW optimizer and initial learning rates in {0.0001, 0.0002, 0.0004}.[5]

Compared Methods. *TFA* [41] or *Two-stage Fine-tuning Approach* has shown to be a simple yet effective method for FSOD. TFA fine-tunes the box regressor and classifier on the few-shot dataset while freezing the other parameters of the model. For this method, we use the FC head, such that TFA is essentially the same as tuning the final FC layer.[6]

FSCE [35] or *Few-Shot object detection via Contrastive proposals Encoding* improves TFA by 1) additionally unfreezing detection head in the setting of TFA, 2) doubling the number of proposals kept after NMS and halving the number of sampled proposals in the RoI head, and 3) optimizing the contrastive proposal encoding loss. While the original work did not apply the contrastive loss for extremely few-shot settings (less than 3), we explicitly compare two versions in all shots: without (*FSCE-base*, the same as tuning the detection head) and with the contrastive loss (*FSCE-con*).

Table 2. The K-shot performance on MoFSOD with $K = \{1, 3, 5, 10\}$.

Method	Pre-training	1-shot	3-shot	5-shot	10-shot
TFA [41]		23.4 ± 4.6	29.2 ± 1.8	32.0 ± 1.3	35.2 ± 1.2
FSCE-base [35]		30.5 ± 5.3	39.4 ± 1.6	43.9 ± 1.4	50.2 ± 1.1
FSCE-con [35]		29.4 ± 2.5	38.8 ± 1.5	43.6 ± 1.2	50.4 ± 1.0
DeFRCN [26]	COCO	29.3 ± 4.2	37.8 ± 3.1	41.6 ± 1.9	48.2 ± 1.9
Ours-FT		**31.5** ± 2.0	41.1 ± 1.8	46.1 ± 1.4	52.6 ± 1.8
Ours-FSCE+		31.2 ± 2.4	**41.3** ± 1.5	**46.4** ± 1.1	**53.2** ± 1.5
Ours-FT+	COCO	35.4 ± 1.8	44.7 ± 1.6	49.9 ± 1.2	56.4 ± 1.1
Ours-FT++	LVIS++	**35.8** ± 3.4	**47.2** ± 2.1	**52.6** ± 1.4	**59.4** ± 1.1

(a) Performance with the natural K-shot.

Method	Pre-training	1-shot	3-shot	5-shot	10-shot
TFA [41]		22.9 ± 5.4	27.6 ± 2.1	28.3 ± 5.9	31.6 ± 2.7
FSCE-con [35]		26.8 ± 3.9	33.6 ± 4.9	36.3 ± 5.0	40.2 ± 5.4
DeFRCN [26]	COCO	27.5 ± 5.1	34.8 ± 2.2	37.4 ± 1.7	40.1 ± 6.5
Ours-FT		**28.8** ± 2.3	34.8 ± 3.1	37.4 ± 4.4	42.1 ± 5.4
Ours-FSCE+		28.7 ± 3.8	**36.5** ± 5.3	**38.2** ± 5.0	**42.7** ± 5.7

(b) Performance with the balanced K-shot.

Architecture	Pre-training	1-shot	3-shot	5-shot	10-shot
Faster R-CNN		31.5 ± 2.0	41.1 ± 1.8	46.1 ± 1.4	52.6 ± 1.8
Cascade R-CNN		31.5 ± 2.1	41.2 ± 1.5	45.6 ± 1.4	52.7 ± 1.2
CenterNet2	COCO	29.1 ± 5.2	40.2 ± 1.9	45.3 ± 1.8	52.5 ± 1.9
RetinaNet		25.4 ± 4.0	34.8 ± 2.4	40.9 ± 1.7	48.8 ± 1.2
Deformable-DETR		32.0 ± 2.9	42.3 ± 1.6	47.4 ± 1.4	54.7 ± 1.0
Cascade R-CNN-P67		**35.4** ± 1.8	**44.7** ± 1.6	**49.9** ± 1.2	**56.4** ± 1.1
Faster R-CNN		31.7 ± 1.7	41.6 ± 1.5	46.4 ± 1.2	53.6 ± 1.0
CenterNet2	LVIS	28.1 ± 3.3	39.0 ± 1.7	44.3 ± 1.4	51.6 ± 1.0
Cascade R-CNN-P67		34.4 ± 1.2	44.0 ± 1.6	48.7 ± 1.4	55.6 ± 1.0

(c) The effect of different architectures.

Architecture	Pre-training	1-shot	3-shot	5-shot	10-shot
	ImageNet	13.5 ± 1.6	23.2 ± 1.5	29.5 ± 1.2	37.7 ± 1.2
	COCO	**35.4** ± 1.8	44.7 ± 1.6	49.9 ± 1.2	56.4 ± 1.1
Cascade R-CNN-P67	FSODD	26.7 ± 2.9	36.9 ± 1.5	42.3 ± 1.2	49.1 ± 1.0
	LVIS	34.4 ± 2.0	44.0 ± 1.6	48.7 ± 1.4	55.6 ± 1.0
	Unified	33.3 ± 2.2	44.3 ± 1.4	49.6 ± 1.2	56.6 ± 1.1
	LVIS+	34.7 ± 4.2	**46.6** ± 1.6	**52.0** ± 1.2	**59.0** ± 1.0
	COCO	29.1 ± 5.2	40.2 ± 1.9	45.3 ± 1.8	52.5 ± 1.9
CenterNet2	LVIS	28.1 ± 3.3	39.0 ± 1.7	44.3 ± 1.2	51.6 ± 1.0
	LVIS+	34.9 ± 3.2	46.5 ± 1.9	51.8 ± 1.3	58.8 ± 1.0
	LVIS++	**35.8** ± 3.4	**47.2** ± 2.1	**52.6** ± 1.4	**59.4** ± 1.1

(d) The effect of different pre-training.

DeFRCN [26] or *Decoupled Faster R-CNN* can be distinguished with other methods by 1) freezing only the R-CNN head, 2) decoupling gradients to suppress gradients from RPN while scaling those from the R-CNN head, and 3) calibrating the classification score from an offline prototypical calibration block (PCB), which is a CNN-based prototype classifier pre-trained on ImageNet [5]. We note that PCB does not re-scale input images in their original implementation, unlike

---

[5] The learning rates are chosen from our initial experiments on three datasets. Note that training *CenterNet2* with AdamW results in worse performance than SGD.

[6] Replacing the FC head with the cosine-similarity results in a similar performance.

the object detector, so we manually scaled images to avoid GPU memory over-flow if they are too large.

FT or *Fine-tuning* does not freeze model parameters as done for other meth-ods. Though it is undervalued in prior works, we found that this simple baseline outperforms state-of-the-art methods in our proposed benchmark. All experiments are done with this method unless otherwise specified.

4.2   Experimental Results and Discussions

Effect of Tuning More Parameters. We first analyze the effect of tuning more or fewer parameters on MoFSOD. In Table 2a and Table 3a, we examine three methods freezing different number of parameters when fine-tuning: *TFA* as tuning the last FC layers only, *FSCE-base* as tuning the detection head, and *FT* as tuning the whole network. We observe that freezing fewer parameters improves the average performance: tuning the whole network (*FT*) shows better performance than others, while tuning the last FC layers only (*TFA*) shows lower performance than others. Also, the performance gap becomes larger as the size of few-shot training datasets increases. For example, *FT* outperforms *FSCE-base* and *TFA* by 1.0% and 8.1% in 1-shot, and 2.4%, 17.4% in 10-shot, respectively. This contrasts with the conventional belief that freezing most of the parameters generally improves the performance of few-shot learning, as it prevents overfitting [9,34,35,41]. However, this is not necessarily true for FSOD. For example, in the standard two-stage object detector training, RPN is class-agnostic, such that its initialization for training downstream few-shot tasks can be the one pre-trained on large-scale datasets, preserving the pre-trained knowledge on objectness. Also, the detection head utilizes thousands of examples even in few-shot scenarios, because RPN could generate 1–2k proposals per image. Hence, the risk of overfitting is relatively low.

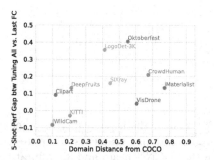

(a) Performance gain by tuning *the whole network FT*) rather than *the last FC layer only* (*TFA*) vs. domain distance.

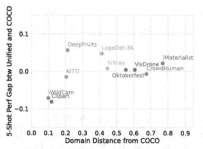

(b) Performance gap between **Unified** and **COCO** vs. domain distance.

Fig. 3. We demonstrate the correlation between tuning more parameters and domain distance in Fig. 3a and the correlation between pre-training datasets and domain distance in Fig. 3b.

However, fine-tuning more parameters does not always improve performance. Figure 3a illustrates the performance gain by tuning more parameters with respect to the domain distance. There is a linear correlation, *i.elet@tokeneonedot*, the performance gain by fine-tuning more parameters increases when the domain distance increases. This implies that fine-tuning fewer parameters to preserve the pre-trained knowledge is better when the few-shot dataset is close to the pre-training dataset. Hence, for datasets close to **COCO**, such as **KITTI** and **iWildCam**, *tuning Last FC layers (TFA)* is the best performing method. From these observations, an interesting research direction might be exploring a sophisticated tuning of layers based on the few-shot problem definition and the domain gap between pre-training and few-shot tasks.

One way to design such sophisticated tuning is to develop a better measure for domain distance. In fact, the proposed measure with class-agnostic object recall has limitations. If we decouple the object detection task into localization (background vs. foreground) and classification (among foreground classes), then the proposed domain distance is biased towards measuring localization gaps. Therefore, it ignores the potential classification gaps that would also require tuning more layers. For example, although we can get a good coverage with the proposals of **COCO** pre-training for **Clipart** (classic domain adaption dataset) and **DeepFruits** (infrared images), resulting in relatively small domain distances, there exist significant gaps of feature discrimination for fine-grained classification. In this case, we need to tune more parameters for better performance.

Effect of Model Architectures. A benefit of using *FT* as the baseline is that we can systematically study the effects of model architectures without the constraints of specifically designed components. Many different architectures have been proposed to solve object detection problems; each has its own merits and drawbacks. One-stage methods, such as YOLO [28] and RetinaNet [23], are known for their fast inference speed. However, different from two-stage methods, they do not have the benefit of inheriting the pre-trained class-agnostic RPN. Specifically, the classification layers for discriminating background/foreground and foreground classes need to be reinitialized, as they are often tied together in one-stage methods. We validate this hypothesis by comparing with two-stage methods in Table 2c. Compared to *Faster R-CNN*, *RetinaNet* has 4–6% low performance on MoFSOD. Per-dataset performance in Table 3b shows that *RetinaNet* is worse in all cases.

The same principle of preserving as much information as possible from pre-training also applies for two-stage methods, *i.elet@tokeneonedot*, reducing the number of randomly reinitialized parameters is better. Specifically, we look into the performance of *Cascade R-CNN* vs *Faster R-CNN*. For *Cascade R-CNN*, we need to reinitialize and learn three FC layers as there are three stages in the cascade detection head, while we only need to reinitialize the last FC layer for *Faster R-CNN*. However, *Cascade R-CNN* is known to have better performance, as demonstrated in Table 1b. In FSOD, these two factors cancelled out, such that their performance is on a par with each other as shown in Table 2c.

Table 3. Per-dataset 5-shot performance of the effects of tuning different parameters, different architectures and pre-training datasets.

5-shot	Aerial	Agriculture	Animal	Cartoon	Fashion	Food	Logo	Person	Security	Traffic	Mean	Rank
Unfrozen	VisDrone	DeepFruits	iWildCam	Clipart	iMaterialist	Oktoberfest	LogoDet-3K	CrowdHuman	SIXray	KITTI		
Last FC layers (TFA [41])	10.1 ± 1.8	47.5 ± 1.8	71.7 ± 2.4	40.2 ± 2.6	8.0 ± 0.9	41.1 ± 4.8	14.2 ± 3.7	30.6 ± 0.8	7.9 ± 2.0	48.3 ± 3.4	32.0 ± 1.3	2.7 ± 0.2
Detection head (FSCE-base [35])	13.0 ± 0.7	59.2 ± 3.2	70.6 ± 1.7	43.5 ± 2.5	20.5 ± 0.9	70.7 ± 3.5	47.2 ± 3.5	51.6 ± 2.0	15.9 ± 2.2	47.1 ± 4.3	43.9 ± 1.1	1.9 ± 0.2
Whole network (Ours-FT)	14.2 ± 0.7	60.7 ± 3.8	63.3 ± 5.7	49.3 ± 3.5	21.3 ± 0.8	81.6 ± 4.0	49.8 ± 3.5	51.5 ± 2.2	23.9 ± 3.9	45.3 ± 3.7	46.1 ± 1.4	1.5 ± 0.2

(a) Fine-tuning different number of parameters with Faster R-CNN pre-trained on COCO.

5-shot		Aerial	Agriculture	Animal	Cartoon	Fashion	Food	Logo	Person	Security	Traffic	Mean	Rank
Architecture	Pre-training	VisDrone	DeepFruits	iWildCam	Clipart	iMaterialist	Oktoberfest	LogoDet-3K	CrowdHuman	SIXray	KITTI		
Faster R-CNN		14.2 ± 0.7	60.7 ± 3.8	63.3 ± 5.7	49.3 ± 3.5	21.3 ± 0.8	81.6 ± 4.0	49.8 ± 3.5	51.5 ± 2.2	23.9 ± 3.9	45.3 ± 3.7	46.1 ± 1.4	3.4 ± 0.3
Cascade R-CNN		13.0 ± 0.8	58.9 ± 3.3	66.8 ± 3.7	50.8 ± 2.8	20.5 ± 0.8	80.5 ± 3.2	48.5 ± 4.7	51.4 ± 2.0	21.6 ± 4.4	44.4 ± 4.1	45.6 ± 1.4	4.1 ± 0.4
CenterNet2	COCO	13.5 ± 0.7	59.0 ± 4.7	61.6 ± 4.7	49.2 ± 7.6	22.2 ± 1.9	79.7 ± 3.5	51.0 ± 4.2	51.4 ± 3.7	22.2 ± 4.8	43.7 ± 4.9	45.3 ± 1.8	3.9 ± 0.3
RetinaNet		9.9 ± 0.6	55.8 ± 2.7	59.2 ± 6.8	26.8 ± 2.2	17.6 ± 0.4	79.5 ± 4.0	49.2 ± 3.5	47.6 ± 1.9	20.6 ± 3.3	39.7 ± 3.3	40.6 ± 1.7	5.4 ± 0.5
Deformable-DETR		15.0 ± 0.8	68.8 ± 4.3	66.0 ± 4.3	43.7 ± 1.6	22.4 ± 1.1	77.2 ± 4.6	50.3 ± 3.5	56.7 ± 2.4	26.3 ± 3.8	47.9 ± 3.2	47.4 ± 1.4	2.6 ± 0.5
Cascade R-CNN-P67		15.9 ± 0.8	63.1 ± 2.3	73.7 ± 2.8	56.6 ± 2.3	24.2 ± 0.8	83.2 ± 2.4	54.5 ± 4.4	53.6 ± 1.8	26.4 ± 3.9	47.6 ± 3.6	49.9 ± 1.2	1.5 ± 0.3
Faster R-CNN		14.2 ± 0.7	66.2 ± 3.4	69.3 ± 3.7	39.8 ± 1.8	28.0 ± 0.6	80.7 ± 2.7	51.3 ± 4.2	48.3 ± 2.2	24.7 ± 3.6	41.3 ± 3.5	46.4 ± 1.2	2.1 ± 0.3
CenterNet2	LVIS	13.3 ± 0.7	64.6 ± 3.5	50.2 ± 25.3	34.1 ± 2.5	20.5 ± 0.6	81.6 ± 2.6	53.7 ± 3.8	46.6 ± 2.0	22.8 ± 3.3	38.9 ± 3.9	43.1 ± 6.9	2.7 ± 0.3
Cascade R-CNN-P67		15.2 ± 0.7	65.4 ± 2.0	71.7 ± 2.0	46.0 ± 2.5	30.2 ± 0.5	81.9 ± 3.0	56.6 ± 5.0	49.6 ± 1.7	25.9 ± 4.3	44.7 ± 3.7	48.7 ± 1.4	1.2 ± 0.3

(b) Performance of different architectures pre-trained on COCO and LVIS.

5-shot		Aerial	Agriculture	Animal	Cartoon	Fashion	Food	Logo	Person	Security	Traffic	Mean	Rank
Architecture	Pre-training	VisDrone	DeepFruits	iWildCam	Clipart	iMaterialist	Oktoberfest	LogoDet-3K	CrowdHuman	SIXray	KITTI		
	ImageNet	9.8 ± 0.5	53.0 ± 3.5	7.4 ± 3.2	13.1 ± 2.5	19.2 ± 0.6	77.4 ± 3.4	46.4 ± 4.4	32.0 ± 3.1	13.1 ± 3.2	23.8 ± 3.4	29.5 ± 1.2	6.0 ± 0.0
	COCO	15.9 ± 0.8	63.1 ± 2.3	73.7 ± 2.8	56.6 ± 2.3	24.2 ± 0.8	83.2 ± 2.4	54.5 ± 4.4	53.6 ± 1.8	26.4 ± 3.9	47.6 ± 3.6	49.9 ± 1.2	2.8 ± 0.4
Cascade R-CNN-P67	FSODD	10.6 ± 0.6	67.5 ± 3.7	64.5 ± 3.4	26.5 ± 2.0	21.2 ± 0.5	82.9 ± 2.9	55.7 ± 3.8	38.4 ± 2.2	23.2 ± 3.7	32.3 ± 3.6	42.3 ± 1.2	4.3 ± 0.4
	LVIS	15.2 ± 0.7	65.4 ± 2.0	71.7 ± 2.0	46.0 ± 2.5	30.2 ± 0.5	81.9 ± 3.0	56.6 ± 5.0	49.6 ± 1.7	25.9 ± 4.3	44.7 ± 3.7	48.7 ± 1.4	3.4 ± 0.4
	Unified	16.3 ± 0.9	68.8 ± 3.6	66.0 ± 4.0	48.5 ± 2.9	26.4 ± 0.7	83.6 ± 2.9	59.3 ± 3.7	52.3 ± 1.6	27.2 ± 3.6	46.1 ± 3.9	49.6 ± 1.2	2.5 ± 0.4
	LVIS+	18.5 ± 1.0	76.8 ± 3.3	59.5 ± 4.1	52.6 ± 2.0	31.5 ± 0.7	82.5 ± 3.3	61.2 ± 2.7	52.3 ± 1.9	36.0 ± 2.5	49.6 ± 3.9	52.0 ± 1.1	1.9 ± 0.5
	COCO	13.5 ± 0.7	59.0 ± 4.7	61.6 ± 4.7	49.2 ± 7.6	22.2 ± 1.9	79.7 ± 3.5	51.0 ± 4.2	51.4 ± 3.7	22.2 ± 4.8	43.7 ± 4.9	45.3 ± 1.8	3.2 ± 0.3
CenterNet2	LVIS	13.3 ± 0.7	64.6 ± 3.5	61.5 ± 4.3	34.1 ± 2.5	25.6 ± 0.5	81.6 ± 2.6	53.7 ± 3.8	46.6 ± 2.0	22.8 ± 3.3	38.9 ± 3.9	44.3 ± 1.2	3.2 ± 0.2
	LVIS+	18.2 ± 0.9	74.0 ± 3.3	63.7 ± 3.8	50.5 ± 1.8	31.5 ± 0.8	80.4 ± 4.0	62.3 ± 3.2	54.0 ± 2.1	37.3 ± 4.1	46.5 ± 4.7	51.8 ± 1.3	2.0 ± 0.4
	LVIS++	18.1 ± 0.9	77.5 ± 2.4	64.4 ± 4.7	52.1 ± 1.8	33.1 ± 0.7	80.4 ± 3.5	61.0 ± 4.5	54.7 ± 2.0	37.8 ± 3.2	46.9 ± 4.1	52.6 ± 1.4	1.5 ± 0.3

(c) Performance of Cascade R-CNN-P67 and CenterNet2 pre-trained on different datasets.

Based on these insights, we extend *Cascade R-CNN* by applying the FPN-P67 architecture [23], similar to [51,52]. Specifically, assuming ResNet-like architecture [15], we use the last three stages of the backbone, namely [res3, res4, res5], instead of the last four in standard FPN. Then, we add P6 and P7 of the FPN features from P5 with two different FC layers, such that RPN takes in features from [P3, P4, P5, P6, P7] to improve the class-agnostic coverage, which can be inherited for down-stream tasks. While the detection head still uses [P3, P4, P5] only. As shown in Table 2c, the resulting architecture, *Cascade R-CNN-P67* improves *Cascade R-CNN* by 3–4% on the downstream few-shot tasks.

Moreover, recent works proposed new directions of improvement, such as utilizing point-based predictions [51,53] or transformer-based set predictions [4,56]. These methods are unknown quantities in FSOD, as no previous FSOD work has studied them. In our experiments, while *CenterNet2* [51] outperforms *Faster R-CNN* on **COCO** by 2.6% as shown in Table 1b, its FSOD performance on MoF-SOD is lower, *e.glet@tokeneonedot*, 2.4% in 1-shot. In the case of *Deformable-DETR* [56], it outperforms *Faster R-CNN* in both pre-training and few-shot learning, by 3.6% on **COCO** and 2.1% on MoFSOD in 10-shot. These results show that the upstream performance might not necessarily translate to the downstream FSOD performance. We note that we could not observe a significant correlation between the performance gap of different architectures and domain distances.

Effect of Pre-Training Datasets. MoFSOD consists of datasets from a wide range of domains, allowing us to freely explore different pre-training datasets while ensuring domain shifts between pre-training and few-shot learning. We examine the impact of the pre-training datasets with the best performing *Cascade R-CNN-P67* architecture.

In Table 2d, we first observe that pre-training on **ImageNet** for classification results in low performance, as it does not provide a good initialization for downstream FSOD, especially for RPN. On the other hand, compared to **COCO**, **FSODD**, **LVIS**, and **Unified** have more classes and/or more annotations, while they have a fewer, similar, and more number of images, respectively. Pre-training on these larger object detection datasets does not improve the FSOD performance significantly, as shown in Table 2d. For example, pre-training on **Unified** improves the performance over **COCO** when the domain distance from **COCO** is large, such as **Deepfruits** and **LogoDet-3K** as shown in Fig. 3b. However, pre-training on **Unified** results in lower performance for few-shot datasets close to **COCO**, such that the overall performance is similar. We hypothesize that this could be due to the non-optimal pre-training of **LVIS** and **Unified**, as these two datasets are highly imbalanced and difficult to train. It could also be the case that even **LVIS** and **Unified** do not have better coverage for these datasets.

On the other end of the spectrum, we can combine the idea of preserving knowledge and large-scale pre-training by utilizing a large-scale language-vision model. Following [11,50], we use CLIP to extract text features from each class name and build a classifier initialized with the text features. In this way, we initialize the classifier with the CLIP text embeddings for downstream few-shot tasks, such that it has strong built-in knowledge of text-image alignment, better than random initialization. As demonstrated in Table 2d, For **LVIS+**, we can see this improves performance significantly by 7.5% for *CenterNet2*, and 3.3% for *Cascade R-CNN-P67* in 5-shot. **LVIS++** pre-trains the backbone on ImageNet-21K instead of ImageNet-1K (before pre-training on **LVIS**), and it further improves over **LVIS+** by 0.8% in 5-shot. However, the benefit of CLIP initialization is valid only when class names are matched with texts presented in CLIP; an exceptional case is **Oktoberfest**, which has German class names, such that **LVIS+** does not help.

Comparison with SOTA Methods. Table 2a and Table 2b compare SOTA methods and our proposed methods. For balanced K-shot sampling, we follow Wang *et al*let@tokeneonedot [41] to sample K instances for each class whenever possible greedily. Here *FT* and *FSCE+* employ a similar backbone/architecture and pre-trained data as all SOTA methods for a fair comparison. We confirm that *FT* is indeed a strong baseline, such that it performs better than other SOTA methods in both natural K-shot and balanced K-shot settings. In addition to *FT*, based on the insights above, we propose several extensions: 1) *FSCE+* is an extension of *FSCE* by tuning the whole network parameters, similar to *FT*. We keep the contrastive proposal encoding loss, but we simplify the classification head from the cosine similarity head to the FC head. We can see the improvement by 2–3% compared to *FSCE* for both natural K-shot and balanced K-shot scenarios and performs slightly better than *FT*. 2) *FT+* replaces *Faster R-CNN* with *Cascade R-CNN-P67*, and it improves over *FT* by 3–4% without sacrificing inference speed or memory consumption much. 3) *FT++* replaces *Faster R-CNN* with *CenterNet2* and uses **LVIS++** for initialization, and it further improves the performance by around 3% in 3-, 5-, and 10-shot. We also observe that while the overall trend of performance is similar for both natural and balanced K-shot sampling, the standard deviation of the natural K-shot performance is less than that of the balanced K-shot.

5   Conclusion

We present the Multi-dOmain Few-Shot Object Detection (MoFSOD) benchmark consisting of 10 datasets from different domains to evaluate FSOD methods. Under the proposed benchmark, we conducted extensive experiments on the impact of freezing parameters, different architectures, and different pre-training datasets. Based on our findings, we proposed simple extensions of the existing methods and achieved state-of-the-art results on the benchmark. In the future, we would like to go beyond empirical studies and modifications, to designing architectures and smart-tuning methods for a wide range of FSOD tasks.

References

1. Beery, S., Agarwal, A., Cole, E., Birodkar, V.: The iwildcam 2021 competition dataset. arXiv preprint arXiv:2105.03494 (2021)
2. Brown, T.B., et al.: Language models are few-shot learners. In: NeurIPS (2020)
3. Cai, Z., Vasconcelos, N.: Cascade R-CNN: Delving into high quality object detection. In: CVPR (2018)
4. Carion, N., Massa, F., Synnaeve, G., Usunier, N., Kirillov, A., Zagoruyko, S.: End-to-end object detection with transformers. In: Vedaldi, A., Bischof, H., Brox, T., Frahm, J.-M. (eds.) ECCV 2020. LNCS, vol. 12346, pp. 213–229. Springer, Cham (2020). https://doi.org/10.1007/978-3-030-58452-8_13
5. Deng, J., Dong, W., Socher, R., Li, L.J., Li, K., Fei-Fei, L.: ImageNet: A large-scale hierarchical image database. In: CVPR. pp. 248–255. Ieee (2009)
6. Ertler, C., Mislej, J., Ollmann, T., Porzi, L., Kuang, Y.: Traffic sign detection and classification around the world. arXiv preprint arXiv:1909.04422 (2019)
7. Everingham, M., Van Gool, L., Williams, C.K., Winn, J., Zisserman, A.: The pascal visual object classes (VOC) challenge. IJCV **88**(2), 303–338 (2010)
8. Fan, Q., Zhuo, W., Tang, C.K., Tai, Y.W.: Few-shot object detection with attention-rpn and multi-relation detector. In: CVPR (2020)
9. Finn, C., Abbeel, P., Levine, S.: Model-agnostic meta-learning for fast adaptation of deep networks. In: ICML (2017)
10. Geiger, A., Lenz, P., Urtasun, R.: Are we ready for autonomous driving? the kitti vision benchmark suite. In: CVPR (2012)
11. Gu, X., Lin, T., Kuo, W., Cui, Y.: Open-vocabulary object detection via vision and language knowledge distillation. In: ICLR (2022)

12. Guo, S., et al.: The imaterialist fashion attribute dataset. arXiv preprint arXiv:1906.05750 (2019)
13. Gupta, A., Dollár, P., Girshick, R.B.: LVIS: a dataset for large vocabulary instance segmentation. In: CVPR (2019)
14. Han, G., He, Y., Huang, S., Ma, J., Chang, S.F.: Query adaptive few-shot object detection with heterogeneous graph convolutional networks. In: ICCV (2021)
15. He, K., Zhang, X., Ren, S., Sun, J.: Deep residual learning for image recognition. In: CVPR (2016)
16. Hosang, J.H., Benenson, R., Schiele, B.: How good are detection proposals, really? In: Valstar, M.F., French, A.P., Pridmore, T.P. (eds.) BMVC (2014)
17. Huang, G., Laradji, I., Vazquez, D., Lacoste-Julien, S., Rodriguez, P.: A survey of self-supervised and few-shot object detection. arXiv preprint arXiv:2110.14711 (2021)
18. Ioffe, S., Szegedy, C.: Batch normalization: accelerating deep network training by reducing internal covariate shift. In: ICML (2015)
19. Kang, B., Liu, Z., Wang, X., Yu, F., Feng, J., Darrell, T.: Few-shot object detection via feature reweighting. In: ICCV (2019)
20. Kolesnikov, A., et al.: Big Transfer (BiT): general visual representation learning. In: Vedaldi, A., Bischof, H., Brox, T., Frahm, J.-M. (eds.) ECCV 2020. LNCS, vol. 12350, pp. 491–507. Springer, Cham (2020). https://doi.org/10.1007/978-3-030-58558-7_29
21. Krasin, I., et al.: OpenImages: a public dataset for large-scale multi-label and multi-class image classification. Dataset available from https://storage.googleapis.com/openimages/web/index.html
22. Lin, T.Y., Dollár, P., Girshick, R., He, K., Hariharan, B., Belongie, S.: Feature pyramid networks for object detection. In: CVPR (2017)
23. Lin, T., Goyal, P., Girshick, R.B., He, K., Dollár, P.: Focal loss for dense object detection. In: ICCV (2017)
24. Lin, T.Y., et al.: Microsoft COCO: Common objects in context. arXiv:1405.0312 (2014)
25. Miao, C., et al.: Sixray: a large-scale security inspection x-ray benchmark for prohibited item discovery in overlapping images. In: CVPR (2019)
26. Qiao, L., Zhao, Y., Li, Z., Qiu, X., Wu, J., Zhang, C.: DeFRCN: decoupled faster r-cnn for few-shot object detection. In: ICCV (2021)
27. Radford, A., et al.: Learning transferable visual models from natural language supervision. In: ICML (2021)
28. Redmon, J., Farhadi, A.: Yolo9000: better, faster, stronger. In: CVPR, pp. 7263–7271 (2017)
29. Ren, M., et al.: Meta-learning for semi-supervised few-shot classification. In: ICLR (2018)
30. Ren, S., He, K., Girshick, R., Sun, J.: Faster R-CNN: towards real-time object detection with region proposal networks. NeurIPS (2015)
31. Sa, I., Ge, Z., Dayoub, F., Upcroft, B., Perez, T., McCool, C.: DeepFruits: a fruit detection system using deep neural networks. Sensors 16(8), 1222 (2016)
32. Shao, S., Li, Z., Zhang, T., Peng, C., Yu, G., Zhang, X., Li, J., Sun, J.: Objects365: a large-scale, high-quality dataset for object detection. In: ICCV (2019)
33. Shao, S., Zhao, Z., Li, B., Xiao, T., Yu, G., Zhang, X., Sun, J.: CrowdHuman: a benchmark for detecting human in a crowd. arXiv preprint arXiv:1805.00123 (2018)
34. Snell, J., Swersky, K., Zemel, R.S.: Prototypical networks for few-shot learning. In: NeurIPS (2017)

35. Sun, B., Li, B., Cai, S., Yuan, Y., Zhang, C.: Fsce: Few-shot object detection via contrastive proposal encoding. In: CVPR (2021)
36. Tan, J., Wang, C., Li, B., Li, Q., Ouyang, W., Yin, C., Yan, J.: Equalization loss for long-tailed object recognition. In: CVPR (2020)
37. Triantafillou, E., et al.: Meta-dataset: a dataset of datasets for learning to learn from few examples. In: ICLR (2020)
38. Tseng, H.Y., Lee, H.Y., Huang, J.B., Yang, M.H.: Cross-domain few-shot classification via learned feature-wise transformation. In: ICLR (2020)
39. Vinyals, O., Blundell, C., Lillicrap, T., Kavukcuoglu, K., Wierstra, D.: Matching networks for one shot learning. In: NeurIPS (2016)
40. Wang, J., Min, W., Hou, S., Ma, S., Zheng, Y., Jiang, S.: LogoDet-3K: a large-scale image dataset for logo detection. arXiv preprint arXiv:2008.05359 (2020)
41. Wang, X., Huang, T.E., Darrell, T., Gonzalez, J.E., Yu, F.: Frustratingly simple few-shot object detection. In: ICML (2020)
42. Wang, X., Cai, Z., Gao, D., Vasconcelos, N.: Towards universal object detection by domain attention. In: CVPR (2019)
43. Wang, Y.X., Ramanan, D., Hebert, M.: Meta-learning to detect rare objects. In: ICCV (2019)
44. Wu, X., Sahoo, D., Hoi, S.: Meta-rcnn: meta learning for few-shot object detection. In: Proceedings of the 28th ACM International Conference on Multimedia (2020)
45. Wu, Y., Kirillov, A., Massa, F., Lo, W.Y., Girshick, R.: Detectron2 (2019). https://github.com/facebookresearch/detectron2
46. Yan, X., Chen, Z., Xu, A., Wang, X., Liang, X., Lin, L.: Meta R-CNN: Towards general solver for instance-level low-shot learning. In: ICCV (2019)
47. Yang, S., Luo, P., Loy, C.C., Tang, X.: Wider face: a face detection benchmark. In: CVPR (2016)
48. Zhang, L., Zhou, S., Guan, J., Zhang, J.: Accurate few-shot object detection with support-query mutual guidance and hybrid loss. In: CVPR (2021)
49. Zhang, S., Xie, Y., Wan, J., Xia, H., Li, S.Z., Guo, G.: WiderPerson: a diverse dataset for dense pedestrian detection in the wild. IEEE Trans. Multimed. **22**(2), 380–393 (2019)
50. Zhou, X., Girdhar, R., Joulin, A., Krähenbühl, P., Misra, I.: Detecting twenty-thousand classes using image-level supervision. arXiv preprint arXiv:2201.02605 (2021)
51. Zhou, X., Koltun, V., Krähenbühl, P.: Probabilistic two-stage detection. arXiv preprint arXiv:2103.07461 (2021)
52. Zhou, X., Koltun, V., Krähenbühl, P.: Simple multi-dataset detection. In: CVPR (2022)
53. Zhou, X., Wang, D., Krähenbühl, P.: Objects as points. arXiv preprint arXiv:1904.07850 (2019)
54. Zhu, P., Wen, L., Du, D., Bian, X., Fan, H., Hu, Q., Ling, H.: Detection and tracking meet drones challenge. IEEE TPAMI (2021). https://doi.org/10.1109/TPAMI.2021.3119563
55. Zhu, X., Hu, H., Lin, S., Dai, J.: Deformable convnets V2: more deformable, better results. In: CVPR (2019)
56. Zhu, X., Su, W., Lu, L., Li, B., Wang, X., Dai, J.: Deformable DETR: Deformable transformers for end-to-end object detection. In: ICLR (2021)
57. Ziller, A., Hansjakob, J., Rusinov, V., Zügner, D., Vogel, P., Günnemann, S.: Oktoberfest food dataset. arXiv preprint arXiv:1912.05007 (2019)

Cross-Domain Cross-Set Few-Shot Learning via Learning Compact and Aligned Representations

Wentao Chen[1,2], Zhang Zhang[2,3(✉)], Wei Wang[2,3], Liang Wang[2,3], Zilei Wang[1], and Tieniu Tan[1,2,3]

[1] University of Science and Technology of China, Hefei, China
wentao.chen@cripac.ia.ac.cn, zlwang@ustc.edu.cn
[2] Center for Research on Intelligent Perception and Computing, NLPR, CASIA, Beijing, China
[3] University of Chinese Academy of Sciences, Beijing, China
{zzhang,wangwei,wangliang,tnt}@nlpr.ia.ac.cn

Abstract. Few-shot learning (FSL) aims to recognize novel queries with only a few support samples through leveraging prior knowledge from a base dataset. In this paper, we consider the domain shift problem in FSL and aim to address the domain gap between the support set and the query set. Different from previous cross-domain FSL work (CD-FSL) that considers the domain shift between base and novel classes, the new problem, termed cross-domain cross-set FSL (CDSC-FSL), requires few-shot learners not only to adapt to the new domain, but also to be consistent between different domains within each novel class. To this end, we propose a novel approach, namely *stab*PA, to learn prototypical compact and cross-domain aligned representations, so that the domain shift and few-shot learning can be addressed simultaneously. We evaluate our approach on two new CDCS-FSL benchmarks built from the DomainNet and Office-Home datasets respectively. Remarkably, our approach outperforms multiple elaborated baselines by a large margin, e.g., improving 5-shot accuracy by 6.0 points on average on DomainNet. Code is available at https://github.com/WentaoChen0813/CDCS-FSL.

Keywords: Cross-domain cross-set few-shot learning · Prototypical alignment

1 Introduction

Learning a new concept with a very limited number of examples is easy for human beings. However, it is quite difficult for current deep learning models, which usually require plenty of labeled data to learn generalizable and discriminative representations. To bridge the gap between humans and machines, few-shot learning (FSL) has been recently proposed [28, 42].

Supplementary Information The online version contains supplementary material available at https://doi.org/10.1007/978-3-031-20044-1_22.

Similar to human beings, most FSL algorithms leverage prior knowledge from known classes to assist recognizing novel concepts. Typically, a FSL algorithm is composed of two phases: (i) pre-train a model on a base set that contains a large number of seen classes (called meta-training phase); (ii) transfer the pre-trained model to novel classes with a small labeled support set and test it with a query set (meta-testing phase). Despite great progresses on FSL algorithms [9,29,37,42], most previous studies adopt a single domain assumption, where all images in both meta-training and meta-testing phases are from a single domain. Such assumption, however, may be easily broken in real-world applications. Considering a concrete example of online shopping, a clothing retailer commonly shows several high-quality pictures taken by photographers for each fashion product (support set), while customers may use their cellphone photos (query set) to match the displayed pictures of their expected products. In such case, there is a distinct domain gap between the support set and the query set. Similar example can be found in security surveillance: given the low-quality picture of a suspect captured at night (query set), the surveillance system is highly expected to recognize its identity based on a few high-quality registered photos (e.g., ID card). With such domain gap, FSL models will face more challenges besides limited support data.

In this paper, we consider the above problem in FSL and propose a new setting to address the domain gap between the support set and the query set. Following previous FSL work, a large base set from the source domain is available for meta-training. Differently, during meta-testing, only the support set or the query set is from the source domain, while the other is from a different target domain. Some recent studies also consider the cross-domain few-shot learning problem (CD-FSL) [15,26,39]. However, the domain shift in CD-FSL occurs between the meta-training and meta-testing phases. In other words, both the support and query sets in the meta-testing phase are still from the same domain (pictorial illustration is given in Fig. 1 (a)). To distinguish the considered setting from CD-FSL, we name this setting as cross-domain cross-set few-shot learning (CDCS-FSL), as the support set and the query set are across different domains. Compared to CD-FSL, the domain gap within each novel class imposes more requirements to learn a well-aligned feature space. Nevertheless, in terms of the above setting, it is nearly intractable to conquer the domain shift due to the very limited samples of the target domain, e.g., the target domain may contain only one support (or query) image. Thus, we follow the CD-FSL literature [26] to use unlabeled auxiliary data from the target domain to assist model training. Note that we do not suppose that the auxiliary data are from novel classes. Therefore, we can collect these data from some common-seen classes (e.g., base classes) without any annotation costs.

One may notice that re-collecting a few support samples from the same domain as the query set can 'simply' eliminate the domain gap. However, it may be intractable to re-collect support samples in some real few-shot applications, e.g., re-collecting ID photos for all persons is difficult. Besides, users sometimes not only want to get the class labels, but more importantly they'd like to retrieve the support images themselves (like the high-quality fashion pictures). Therefore, the CDCS-FSL setting can not be simply transferred into previous FSL and CD-FSL settings.

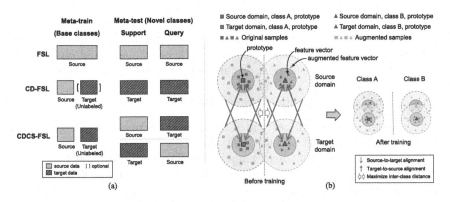

Fig. 1. Problem setup and motivation. (a) CD-FSL considers the domain shift between the meta-training and meta-testing phases, while CDCS-FSL considers the domain shift between the support set and the query set in the meta-testing phase. Following previous CD-FSL work [26], unlabeled target domain data are used in CDCS-FSL to assistant model training. (b) We propose a bi-directional prototypical alignment framework to address CDCS-FSL, which pushes feature vectors of one domain to be gathered around the corresponding prototype in the other domain bi-directionally, and separates feature vectors from different classes.

To address the CDCS-FSL problem, we propose a simple but effective bi-directional prototypical alignment framework to learn compact and cross-domain aligned representations, which is illustrated in Fig. 1 (b). The main idea of our approach is derived from two intuitive insights: (i) we need aligned representations to alleviate the domain shift between the source and target domains, and (ii) compact representations are desirable to learn a center-clustered class space, so that a small support set can better represent a new class. Specifically, given the labeled base set in the source domain and the unlabeled auxiliary set in the target domain, we first assign pseudo labels to the unlabeled data considering that pseudo labels can preserve the coarse semantic similarity with the visual concepts in source domain. Then, we minimize the point-to-set distance between the prototype (class center) in one domain and the corresponding feature vectors in the other domain bi-directionally. As results, the feature vectors of the source (or target) domain will be gathered around the prototype in the other domain, thus reducing the domain gap and intra-class variance simultaneously. Moreover, the inter-class distances are maximized to attain a more separable feature space. Furthermore, inspired by the fact that data augmentation even with strong transformations generally does not change sample semantics, we augment samples in each domain, and suppose that the augmented samples between different domains should also be aligned. Since these augmented samples enrich the data diversity, they can further encourage to learn the underlying invariance and strengthen the cross-domain alignment.

Totally, we summarize all the above steps into one approach termed "**St**rongly **A**ugmented **B**i-directional **P**rototypical **A**lignment", or *stab*PA. We evaluate its effectiveness on two new CDCS-FSL benchmarks built from the DomainNet [25] and Office-Home [40] datasets. Remarkably, the proposed *stab*PA achieves the best performance over both benchmarks and outperforms other baselines with a large margin, e.g., improving 5-shot accuracy by 6.0 points on average on the DomainNet dataset.

In summary, our contributions are three-fold:

- We consider a new FSL setting, CDCS-FSL, where a domain gap exists between the support set and the query set.
- We propose a novel approach, namely *stab*PA, to address the CDCS-FSL problem, of which the key is to learn prototypical compact and domain aligned representations.
- Extensive experiments demonstrate that *stab*PA can learn discriminative and generalizable representations and outperforms all baselines by a large margin on two CDCS-FSL benchmarks.

2   Related Work

FSL aims to learn new classes with very few labeled examples. Most studies follow a meta-learning paradigm [41], where a meta-learner is trained on a series of training tasks (episodes) so as to enable fast adaptation to new tasks. The meta-learner can take various forms, such as an LSTM network [28], initial network parameters [9], or closed-form solvers [29]. Recent advances in pre-training techniques spawn another FSL paradigm. In [4], the authors show that a simple pre-training and fine-tuning baseline can achieve competitive performance with respect to the SOTA FSL models. In [5,37], self-supervised pre-training techniques have proven to be useful for FSL. Our approach also follows the pre-training paradigm, and we further expect the learned representations to be compact and cross-domain aligned to address the CDCS-FSL problem.

CD-FSL [10,14,15,20,26,39,43] considers the domain shift problem between the base classes and the novel classes. Due to such domain gap, [4] show that meta-learning approaches fail to adapt to novel classes. To alleviate this problem, [39] propose a feature-wise transformation layer to learn rich representations that can generalize better to other domains. However, they need to access multiple labeled data sources with extra data collection costs. [26] solve this problem by exploiting additional unlabeled target data with self-supervised pre-training techniques. Alternatively, [14] propose to utilize the semantic information of class labels to minimize the distance between source and target domains. Without the need for extra data or language annotations, [43] augment training tasks in an adversarial way to improve the generalization capability.

Using target domain images to alleviate domain shift is related to the field of domain adaptation (DA). Early efforts align the marginal distribution of each domain by minimizing a pre-defined discrepancy, such as $\mathcal{H}\Delta\mathcal{H}$-divergence [1] or Maximum Mean Discrepancy (MMD) [13]. Recently, adversarial-based methods adopt a discriminator [12] to approximate the domain discrepancy, and learn domain-invariant distribution at image level [17], feature level [21] or output level [38]. Another line of studies assign pseudo labels to unlabeled target data [47–49], and directly align the feature distribution within each class. Although these DA methods are related to our work, they usually assume that the testing stage shares the same class space as the training stage, which is broken by the setting of FSL. Open-set DA [24,31] and Universal DA [30,45] consider the existence of unseen classes. However, they merely mark them as 'unknown'. In this work, we are more interested in addressing the domain shift for these unseen novel classes within a FSL assumption.

3   Problem Setup

Formally, a FSL task often adopts a setting of N-way-K-shot classification, which aims to discriminate between N novel classes with K exemplars per class. Given a support set $S = \{(x_i, y_i)\}_{i=1}^{N \times K}$ where $x_i \in \mathcal{X}_N$ denotes a data sample in novel classes and $y_i \in Y_N$ is the class label, the goal of FSL is to learn a mapping function $\phi : \phi(x_q) \to y_q$ which classifies a query sample x_q in the query set Q to the class label $y_q \in Y_N$. Besides S and Q, a large labeled dataset $B \subset \mathcal{X}_B \times \mathcal{Y}_B$ (termed base set) is often provided for meta-training, where $\mathcal{X}_B$ and $\mathcal{Y}_B$ do not overlap with $\mathcal{X}_N$ and $\mathcal{Y}_N$.

Conventional FSL studies assume the three sets S, Q and B are from the same domain. In this paper, we consider the domain gap between the support set and the query set (only one is from the same domain as the base set, namely the source domain $\mathcal{D}_s$, and the other is from a new target domain $\mathcal{D}_t$). Specifically, this setting has two situations:

(i) $\mathcal{D}_s - \mathcal{D}_t$: the support set is from the source domain and the query set is from the target domain, i.e., $S \subset \mathcal{D}_s$ and $Q \subset \mathcal{D}_t$.

(ii) $\mathcal{D}_t - \mathcal{D}_s$: the support set is from the target domain and the query set is from the source domain, i.e., $S \subset \mathcal{D}_t$ and $Q \subset \mathcal{D}_s$.

As the support set and the query set are across different domains, we name this setting as cross-domain cross-set few-shot learning (CDCS-FSL). Besides the above three sets, to facilitate crossing the domain gap, an unlabeled auxiliary set $\mathcal{U}$ from the target domain is available in the meta-training phase, where the data from novel classes are manually removed to promise they are not seen in meta-training.

4   Approach

Briefly, our approach contains two stages: 1) In the meta-training stage, we train a feature extractor $f : x_i \to f(x_i)$ with the base set B and the unlabeled auxiliary set $\mathcal{U}$; 2) In the meta-testing stage, we fix the feature extractor and train a linear classification head $g : f(x_i) \to y_i$ on the support set S, and the entire model $\phi = g \circ f$ is used to predict the labels for the query set Q. The framework of our approach is illustrated in Fig. 2.

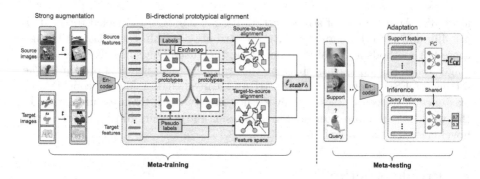

Fig. 2. Framework. In the meta-training stage, we train a feature extractor within the bi-directional prototypical alignment framework to learn compact and aligned representations. In the meta-testing stage, we fix the feature extractor and train a new classification head with the support set, and then evaluate the model on the query set.

4.1   Bi-directional Prototypical Alignment

A straightforward way to align feature distributions is through estimating class centers (prototypes) in both source and target domains. With the labeled base data, it is easy to estimate prototypes for the source domain. However, it is difficult to estimate prototypes in the target domain with only unlabeled data available. To address this issue, we propose to assign pseudo labels to the unlabeled data and then use the pseudo labels to approximate prototypes. The insight is that the pseudo labels can preserve the coarse semantic similarity even under domain or category shift (e.g., a painting tiger could be more likely to be pseudo-labeled as a cat rather than a tree). Aggregating samples with the same pseudo label can extract the shared semantics across different domains.

Specifically, given the source domain base set $\mathcal{B}$ and the target domain unlabeled set $\mathcal{U}$, we first assign pseudo labels to the unlabeled samples with an initial classifier ϕ_0 trained on the base set and obtain $\hat{\mathcal{U}} = \{(x_i, \hat{y}_i)|x_i \in \mathcal{U}\}$, where $\hat{y}_i = \phi_0(x_i)$ is the pseudo label. Then, we obtain the source prototypes $\{p_k^s\}_{k=1}^{|\mathcal{Y}_\mathcal{B}|}$ and the target prototypes $\{p_k^t\}_{k=1}^{|\mathcal{Y}_\mathcal{B}|}$ by averaging the feature vectors with the same label (or pseudo label). It should be noted that the prototypes are estimated on the entire datasets $\mathcal{B}$ and $\hat{\mathcal{U}}$, and adjusted together with the update of the feature extractor and pseudo labels (details can be found below).

With the obtained prototypes, directly minimizing the *point-to-point* distance between two prototypes p_k^s and p_k^t can easily reduce the domain gap for the class k. However, this may make the feature distribution of different classes mix together and the discrimination capability of the learned representations is still insufficient. To overcome these drawbacks, we propose to minimize the *point-to-set* distance across domains in a *bi-directional* way. That is, we minimize the distance between the *prototype in one domain* and the corresponding *feature vectors in the other domain*, and meanwhile maximize the feature distance between different classes. In this way, we can not only align features across domains, but also simultaneously obtain compact feature distributions for both domains to suit the requirement of few-shot learning.

Concretely, for a source sample $(x_i^s, y_i^s) \in \mathcal{B}$ of the q-th class (i.e., $y_i^s = q$), we minimize its feature distance to the prototype p_q^t in the target domain, and meanwhile maximize its distances to the prototypes of other classes. Here, a softmax loss function for the source-to-target alignment is formulated as:

$$\ell_{s-t}(x_i^s, y_i^s) = -\log \frac{\exp\left(-\|f(x_i^s) - p_q^t\|^2/\tau\right)}{\sum_{k=1}^{|\mathcal{Y}_B|} \exp\left(-\|f(x_i^s) - p_k^t\|^2/\tau\right)}, \tag{1}$$

where τ is a temperature factor. To get a better feature space for the target domain, a similar target-to-source alignment loss is applied for each target sample $(x_i^t, \hat{y}_i^t) \in \hat{\mathcal{U}}$ with $\hat{y}_i^t = q$:

$$\ell_{t-s}(x_i^t, \hat{y}_i^t) = -\log \frac{\exp\left(-\|f(x_i^t) - p_q^s\|^2/\tau\right)}{\sum_{k=1}^{|\mathcal{Y}_B|} \exp\left(-\|f(x_i^t) - p_k^s\|^2/\tau\right)}. \tag{2}$$

Since the initial pseudo labels are more likely to be incorrect, we gradually increase the weights of these two losses following the principle of curriculum learning [2]. For the source-to-target alignment, the loss weight starts from zero and converges to one, formulated as:

$$w(t) = \frac{2}{1 + \exp\left(-t/T_{max}\right)} - 1, \tag{3}$$

where t is the current training step and T_{max} is the maximum training step. For the target-to-source alignment, since the pseudo labels become more confident along with the training process, a natural curriculum is achieved by setting a confidence threshold to filter out the target samples with low confidence pseudo labels [33].

Together, the total loss for the bi-directional prototypical alignment is

$$\ell_{bPA} = \frac{1}{|\mathcal{B}|} \sum_{i=1}^{|\mathcal{B}|} w(t)\ell_{s-t}(x_i^s, y_i^s) + \frac{1}{|\hat{\mathcal{U}}|} \sum_{i=1}^{|\hat{\mathcal{U}}|} \mathbb{1}(p(\hat{y}_i^t) > \beta)\ell_{t-s}(x_i^t, \hat{y}_i^t), \tag{4}$$

where $p(\cdot)$ is the confidence of a pseudo label, and β is the confidence threshold below which the data samples will be dropped.

Updating Pseudo Label. The pseudo labels are initially predicted by a classifier ϕ_0 pre-trained on the base set $\mathcal{B}$. As the representations are updated, we update the pseudo labels by re-training a classifier $\phi_t = h \circ f$ based on the current feature extractor f, where h is a linear classification head for the base classes. The final pseudo labels are updated by linear interpolation between the predictions of the initial classifier ϕ_0 and the online updated classifier ϕ_t:

$$\hat{y}_i = \arg \max_{k \in \mathcal{Y}_B} \lambda\phi_0(k|x_i) + (1 - \lambda)\phi_t(k|x_i), \tag{5}$$

where λ is the interpolation coefficient. The combination of these two classifiers makes it possible to rectify the label noise of the initial classifier, and meanwhile inhibit the rapid change of pseudo labels of online classifier especially in the early training stage.

Generating Prototypes. Note that we are intended to estimate the prototypes on the entire dataset and update them with representation learning. For the source domain, instead of calculating the mean value of intra-class samples in the feature space, a cheaper way is to approximate prototypes with the normalized weights of the classification head h, as the classifier weights tend to align with class centers in order to reduce classification errors [27]. Specifically, we set the source prototypes as $p_k^s = W_k$, where W_k is the normalized classification weight for the k-th class. For the target domain, we adopt the momentum technique to update prototypes. The prototypes are initialized as zeros. At each training step, we first estimate the prototypes using target samples in current batch with their pseudo labels. Then, we update the target prototype p_k^t as:

$$p_k^t \longleftarrow m p_k^t + (1 - m) \frac{1}{n_k} \sum_{i=1}^{|\hat{\mathcal{U}}_b|} \mathbb{1}(\hat{y}_i^t = k) f(x_i^t), \qquad (6)$$

where n_k is the number of the target samples classified into the k-th class in a target batch $\hat{\mathcal{U}}_b$, and m is the momentum term controlling the update speed.

4.2   StabPA

Strong data augmentation has proved to be effective for learning generalizable representations, especially in self-supervised representation learning studies [3,16]. Given a sample x, strong data augmentation generates additional data points $\{\widetilde{x}_i\}_{i=1}^n$ by applying various intensive image transformations. The assumption behind strong data augmentation is that the image transformations do not change the semantics of original samples.

In this work, we further hypothesize that strongly augmented intra-class samples in different domains can also be aligned. It is expected that strong data augmentation can further strengthen the learning of cross-domain representations, since stronger augmentation provides more diverse data samples and makes the learned aligned representations more robust for various transformations in both the source and target domains.

Following this idea, we extend the bi-directional prototypical alignment with strong data augmentation and the entire framework is termed *stab*PA. Specifically, for a source sample (x_i^s, y_i^s) and a target sample $(x_i^t, \hat{y}_i^t)$, we generate their augmented versions $(\widetilde{x}_i^s, y_i^s)$ and $(\widetilde{x}_i^t, \hat{y}_i^t)$. Within the bi-directional prototypical alignment framework, we minimize the feature distance of a strongly augmented image to its corresponding prototype in the other domain, and maximize its distances to the prototypes of other classes. Totally, the *stab*PA loss is

$$\ell_{stabPA} = \frac{1}{|\widetilde{\mathcal{B}}|} \sum_{i=1}^{|\widetilde{\mathcal{B}}|} w(t) \ell_{s-t}(\widetilde{x}_i^s, y_i^s) + \frac{1}{|\widetilde{\mathcal{U}}|} \sum_{i=1}^{|\widetilde{\mathcal{U}}|} \mathbb{1}(p(\hat{y}_i^t) > \beta) \ell_{t-s}(\widetilde{x}_i^t, \hat{y}_i^t), \quad (7)$$

where $\widetilde{\mathcal{B}}$ and $\widetilde{\mathcal{U}}$ are the augmented base set and unlabeled auxiliary set, respectively.

To perform strong data augmentation, we apply random crop, Cutout [8], and RandAugment [7]. RandAugment comprises 14 different transformations and randomly selects a fraction of transformations for each sample. In our implementation, the magnitude for each transformation is also randomly selected, which is similar to [33].

5   Experiments

5.1   Datasets

DomainNet. DomainNet [25] is a large-scale multi-domain image dataset. It contains 345 classes in 6 different domains. In experiments, we choose the *real* domain as the source domain and choose one domain from *painting, clipart* and *sketch* as the target domain. We randomly split the classes into 3 parts: base set (228 classes), validation set (33 classes) and novel set (65 classes), and discard 19 classes with too few samples.

Office-Home. Office-Home [40] contains 65 object classes usually found in office and home settings. We randomly select 40 classes as the base set, 10 classes as the validation set, and 15 classes as the novel set. There are 4 domains for each class: *real, product, clipart* and *art*. We set the source domain as *real* and choose the target domain from the other three domains.

In both datasets, we construct the unlabeled auxiliary set by collecting data from the base and validation sets of the target domain and removing their labels. These unlabeled data combined with the labeled base set are used for meta-training. The validation sets in both domains are used to tune hyper-parameters. Reported results are averaged across 600 test episodes from the novel set.

5.2   Comparison Results

We compare our approach to a broad range of related methods. Methods in the first group [6,19,23,32,34,37,39,43,46] do not use the unlabeled auxiliary data during meta-training, while methods in the second group [11,18,26,33,35,36] utilize the unlabeled target images to facilitate crossing the domain gap. Note that methods in the second group are only different in representation learning, and adopt the same evaluation paradigm as ours, i.e., training a linear classifier on the support set. We also implement a baseline method, termed $stab$PA$^-$, where we do not apply domain alignment and only train the feature extractor on augmented source images, which is also equivalent to applying strong augmentation to Tian el al. [37]. We set $\beta = 0.5$, $\lambda = 0.2$ and $m = 0.1$ as default for our approach. All compared methods are implemented with the same backbone and optimizer. Implementation details including augmentation techniques can be found in the appendix.

The comparison results are shown in Tables 1 and 2. The 'r-r' setting denotes all images are from the source domain, and thus is not available for methods in the second group. In Table 1, we can see that the performance of conventional

Table 1. Comparison to baselines on the DomainNet dataset. We denote 'r' as *real*, 'p' as *painting*, 'c' as *clipart* and 's' as *sketch*. We report 5-way 1-shot and 5-way 5-shot accuracies with 95% confidence interval.

Method	r-r	r-p	p-r	r-c	c-r	r-s	s-r
5-way 1-shot							
ProtoNet [32]	$63.43_{\pm0.90}$	$45.36_{\pm0.81}$	$45.25_{\pm0.97}$	$44.65_{\pm0.81}$	$47.50_{\pm0.95}$	$39.28_{\pm0.77}$	$42.85_{\pm0.89}$
RelationNet [34]	$59.49_{\pm0.91}$	$42.69_{\pm0.77}$	$43.04_{\pm0.97}$	$44.12_{\pm0.81}$	$45.86_{\pm0.95}$	$36.52_{\pm0.73}$	$41.29_{\pm0.96}$
MetaOptNet [19]	$61.12_{\pm0.89}$	$44.02_{\pm0.77}$	$44.31_{\pm0.94}$	$42.46_{\pm0.80}$	$46.15_{\pm0.98}$	$36.37_{\pm0.72}$	$40.27_{\pm0.95}$
Tian et al. [37]	$67.18_{\pm0.87}$	$46.69_{\pm0.86}$	$46.57_{\pm0.99}$	$48.30_{\pm0.85}$	$49.66_{\pm0.98}$	$40.23_{\pm0.73}$	$41.90_{\pm0.86}$
DeepEMD [46]	$67.15_{\pm0.87}$	$47.60_{\pm0.87}$	$47.86_{\pm1.04}$	$49.02_{\pm0.83}$	$50.89_{\pm1.00}$	$42.75_{\pm0.79}$	$46.02_{\pm0.93}$
ProtoNet+FWT [39]	$62.38_{\pm0.89}$	$44.40_{\pm0.80}$	$45.32_{\pm0.97}$	$43.95_{\pm0.80}$	$46.32_{\pm0.92}$	$39.28_{\pm0.74}$	$42.18_{\pm0.95}$
ProtoNet+ATA [43]	$61.97_{\pm0.87}$	$45.59_{\pm0.84}$	$45.90_{\pm0.94}$	$44.28_{\pm0.83}$	$47.69_{\pm0.90}$	$39.87_{\pm0.81}$	$43.64_{\pm0.95}$
S2M2 [23]	$67.07_{\pm0.84}$	$46.84_{\pm0.82}$	$47.03_{\pm0.95}$	$47.75_{\pm0.83}$	$48.27_{\pm0.91}$	$39.78_{\pm0.76}$	$40.11_{\pm0.91}$
Meta-Baseline [6]	$\mathbf{69.46_{\pm0.91}}$	$\mathbf{48.76_{\pm0.85}}$	$48.90_{\pm1.12}$	$\mathbf{49.96_{\pm0.85}}$	$\mathbf{52.67_{\pm1.08}}$	$\mathbf{43.08_{\pm0.80}}$	$\mathbf{46.22_{\pm1.04}}$
*stab*PA$^-$ (*Ours*)	$68.48_{\pm0.87}$	$48.65_{\pm0.89}$	$\mathbf{49.14_{\pm0.88}}$	$45.86_{\pm0.85}$	$48.31_{\pm0.92}$	$41.74_{\pm0.78}$	$42.17_{\pm0.95}$
DANN [11]	–	$45.94_{\pm0.84}$	$46.85_{\pm0.97}$	$47.31_{\pm0.86}$	$50.02_{\pm0.94}$	$42.44_{\pm0.79}$	$43.66_{\pm0.92}$
PCT [35]	–	$47.14_{\pm0.89}$	$47.31_{\pm1.04}$	$\mathbf{50.04_{\pm0.85}}$	$49.83_{\pm0.98}$	$39.10_{\pm0.76}$	$39.92_{\pm0.95}$
Mean Teacher [36]	–	$46.92_{\pm0.83}$	$46.84_{\pm0.96}$	$48.48_{\pm0.81}$	$49.60_{\pm0.97}$	$43.39_{\pm0.81}$	$44.52_{\pm0.89}$
FixMatch [33]	–	$\mathbf{48.86_{\pm0.87}}$	$\mathbf{49.15_{\pm0.93}}$	$48.70_{\pm0.82}$	$49.18_{\pm0.93}$	$\mathbf{44.48_{\pm0.80}}$	$\mathbf{45.97_{\pm0.95}}$
STARTUP [26]	–	$47.53_{\pm0.88}$	$47.58_{\pm0.98}$	$49.24_{\pm0.87}$	$\mathbf{51.32_{\pm0.98}}$	$43.78_{\pm0.82}$	$45.23_{\pm0.96}$
DDN [18]	–	$48.83_{\pm0.84}$	$48.11_{\pm0.91}$	$48.25_{\pm0.83}$	$48.46_{\pm0.93}$	$43.60_{\pm0.79}$	$43.99_{\pm0.91}$
*stab*PA (*Ours*)	–	$\mathbf{53.86_{\pm0.89}}$	$\mathbf{54.44_{\pm1.00}}$	$\mathbf{56.12_{\pm0.83}}$	$\mathbf{56.57_{\pm1.02}}$	$\mathbf{50.85_{\pm0.86}}$	$\mathbf{51.71_{\pm1.01}}$
5-way 5-shot							
ProtoNet [32]	$82.79_{\pm0.58}$	$57.23_{\pm0.79}$	$65.60_{\pm0.95}$	$58.04_{\pm0.81}$	$65.91_{\pm0.78}$	$\mathbf{51.68_{\pm0.81}}$	$59.46_{\pm0.85}$
RelationNet [34]	$77.68_{\pm0.62}$	$52.63_{\pm0.74}$	$61.18_{\pm0.90}$	$57.24_{\pm0.80}$	$62.65_{\pm0.81}$	$47.32_{\pm0.75}$	$56.39_{\pm0.88}$
MetaOptNet [19]	$80.93_{\pm0.60}$	$56.34_{\pm0.76}$	$63.20_{\pm0.89}$	$57.92_{\pm0.79}$	$63.51_{\pm0.82}$	$48.20_{\pm0.79}$	$55.65_{\pm0.85}$
Tian et al. [37]	$84.50_{\pm0.55}$	$56.87_{\pm0.84}$	$63.90_{\pm0.95}$	$59.67_{\pm0.84}$	$65.33_{\pm0.80}$	$50.41_{\pm0.80}$	$56.95_{\pm0.84}$
DeepEMD [46]	$82.79_{\pm0.56}$	$56.62_{\pm0.78}$	$63.86_{\pm0.93}$	$60.43_{\pm0.82}$	$67.46_{\pm0.78}$	$51.66_{\pm0.80}$	$60.39_{\pm0.87}$
ProtoNet+FWT [39]	$82.42_{\pm0.55}$	$57.18_{\pm0.77}$	$65.64_{\pm0.93}$	$57.42_{\pm0.77}$	$65.11_{\pm0.83}$	$50.69_{\pm0.77}$	$59.58_{\pm0.84}$
ProtoNet+ATA [43]	$81.96_{\pm0.57}$	$57.69_{\pm0.83}$	$64.96_{\pm0.93}$	$56.90_{\pm0.84}$	$64.08_{\pm0.86}$	$51.67_{\pm0.80}$	$60.78_{\pm0.86}$
S2M2 [23]	$85.79_{\pm0.52}$	$58.79_{\pm0.81}$	$65.67_{\pm0.90}$	$\mathbf{60.63_{\pm0.83}}$	$63.57_{\pm0.88}$	$49.43_{\pm0.79}$	$54.45_{\pm0.89}$
Meta-Baseline [6]	$83.74_{\pm0.58}$	$56.07_{\pm0.79}$	$65.70_{\pm0.99}$	$58.84_{\pm0.80}$	$\mathbf{67.89_{\pm0.91}}$	$50.27_{\pm0.76}$	$\mathbf{61.88_{\pm0.94}}$
*stab*PA$^-$ (*Ours*)	$\mathbf{85.98_{\pm0.51}}$	$\mathbf{59.92_{\pm0.85}}$	$\mathbf{67.10_{\pm0.93}}$	$57.10_{\pm0.88}$	$62.90_{\pm0.83}$	$51.03_{\pm0.85}$	$57.11_{\pm0.93}$
DANN [11]	–	$56.83_{\pm0.86}$	$64.29_{\pm0.94}$	$59.42_{\pm0.84}$	$66.87_{\pm0.78}$	$53.47_{\pm0.75}$	$60.14_{\pm0.81}$
PCT [35]	–	$56.38_{\pm0.87}$	$64.03_{\pm0.99}$	$61.15_{\pm0.80}$	$66.19_{\pm0.82}$	$46.77_{\pm0.74}$	$53.91_{\pm0.90}$
Mean Teacher [36]	–	$57.74_{\pm0.84}$	$64.97_{\pm0.94}$	$61.54_{\pm0.84}$	$67.39_{\pm0.89}$	$54.57_{\pm0.79}$	$60.04_{\pm0.86}$
FixMatch [33]	–	$61.62_{\pm0.79}$	$67.46_{\pm0.89}$	$\mathbf{61.94_{\pm0.82}}$	$66.72_{\pm0.81}$	$\mathbf{55.26_{\pm0.83}}$	$\mathbf{62.46_{\pm0.87}}$
STARTUP [26]	–	$58.13_{\pm0.82}$	$65.27_{\pm0.92}$	$61.51_{\pm0.86}$	$\mathbf{67.95_{\pm0.78}}$	$54.89_{\pm0.81}$	$61.97_{\pm0.88}$
DDN [18]	–	$\mathbf{61.98_{\pm0.82}}$	$\mathbf{67.69_{\pm0.88}}$	$61.07_{\pm0.84}$	$65.58_{\pm0.79}$	$54.35_{\pm0.83}$	$60.37_{\pm0.88}$
*stab*PA (*Ours*)	–	$\mathbf{65.65_{\pm0.74}}$	$\mathbf{73.63_{\pm0.82}}$	$\mathbf{67.32_{\pm0.80}}$	$\mathbf{74.41_{\pm0.76}}$	$\mathbf{61.37_{\pm0.82}}$	$\mathbf{68.93_{\pm0.87}}$

FSL methods drops quickly when there is a domain shift between support and query sets. The proposed *stab*PA leveraging unlabeled target images for domain alignment can alleviate this problem, improving the previous best FSL baseline [6] by 7.05% across 6 CSCS-FSL situations. Similar results can be found on the Office-Home dataset in Table 2, where the *stab*PA outperforms the previous best FSL method, S2M2 [23], by 3.90% on average. When comparing our approach with methods in the second group, we find that the *stab*PA outperforms them in all situations, improving 5-shot accuracy by 5.98% over the previous best method FixMatch [33] on DomainNet. These improvements indicate that

Table 2. Comparison results on Office-Home. We denote 'r' as *real*, 'p' as *product*, 'c' as *clipart* and 'a' as *art*. Accuracies are reported with 95% confidence intervals.

Method	5-way 1-shot						
	r-r	r-p	p-r	r-c	c-r	r-a	a-r
ProtoNet [32]	35.24±0.63	30.72±0.62	30.27±0.62	28.52±0.58	28.44±0.63	26.80±0.47	27.31±0.58
RelationNet [34]	34.86±0.63	28.28±0.62	27.59±0.56	27.66±0.58	25.86±0.60	25.98±0.54	27.83±0.63
MetaOptNet [19]	36.77±0.65	33.34±0.69	33.28±0.65	28.78±0.53	28.70±0.64	29.45±0.69	28.36±0.64
Tian et al. [37]	39.53±0.67	33.88±0.69	33.98±0.67	30.44±0.60	30.86±0.66	30.26±0.57	30.30±0.62
DeepEMD [46]	41.19±0.71	34.27±0.72	35.19±0.71	30.92±0.62	31.82±0.70	31.05±0.59	31.07±0.63
ProtoNet+FWT [39]	35.43±0.64	32.18±0.67	30.92±0.61	28.75±0.62	27.93±0.63	27.58±0.52	28.37±0.65
ProtoNet+ATA [43]	35.67±0.66	31.56±0.68	30.40±0.62	27.20±0.56	26.61±0.62	27.88±0.55	28.48±0.65
S2M2 [23]	41.92±0.68	**35.46±0.74**	35.21±0.70	**31.84±0.66**	**31.96±0.66**	30.36±0.59	30.88±0.65
Meta-Baseline [6]	38.88±0.67	33.44±0.72	33.73±0.68	30.41±0.61	30.43±0.67	30.00±0.58	30.31±0.64
stabPA⁻ (Ours)	**43.43±0.69**	35.16±0.72	**35.74±0.68**	31.16±0.66	30.44±0.64	**32.09±0.62**	**31.71±0.67**
DANN [11]	–	33.41±0.71	33.60±0.66	30.98±0.64	30.81±0.70	31.67±0.60	32.07±0.64
PCT [35]	–	35.53±0.73	35.58±.71	28.83±0.58	28.44±0.67	31.56±0.58	31.59±0.65
Mean Teacher [36]	–	33.24±0.70	33.13±0.67	31.34±0.62	30.91±0.67	30.98±0.60	31.57±0.61
FixMatch [33]	–	**36.05±0.73**	35.83±0.76	**33.79±0.64**	**33.20±0.74**	31.81±0.60	32.32±0.66
STARTUP [26]	–	34.62±0.74	34.80±0.68	30.70±0.63	30.17±0.68	**32.06±0.59**	**32.40±0.66**
stabPA (Ours)	–	**38.02±0.76**	**38.09±0.82**	**35.44±0.76**	**34.74±0.76**	**34.81±0.69**	**35.18±0.72**
	5-way 5-shot						
ProtoNet [32]	49.21±0.59	39.74±0.64	38.98±0.64	34.81±0.59	35.85±0.59	34.56±0.58	36.27±0.66
RelationNet [34]	47.02±0.57	33.95±0.60	32.78±0.59	33.58±0.60	30.15±0.55	30.44±0.55	35.42±0.70
MetaOptNet [19]	52.00±0.59	43.21±0.69	42.97±0.63	36.48±0.57	36.56±0.65	36.75±0.63	38.48±0.68
Tian et al. [37]	56.89±0.61	45.79±0.69	44.27±0.63	38.27±0.64	38.99±0.63	38.80±0.61	41.56±0.72
DeepEMD [46]	58.76±0.61	47.47±0.71	45.39±0.65	38.87±0.63	40.06±0.66	39.20±0.58	41.62±0.72
ProtoNet+FWT [39]	51.40±0.61	41.50±0.68	40.32±0.60	36.07±0.62	35.80±0.60	34.60±0.56	37.36±0.67
ProtoNet+ATA [43]	51.19±0.63	41.19±0.68	38.06±0.61	32.74±0.56	33.98±0.67	35.36±0.56	36.87±0.68
S2M2 [23]	60.82±0.58	47.84±0.70	**46.32±0.67**	**40.09±0.66**	**41.63±0.64**	40.01±0.60	42.68±0.67
Meta-Baseline [6]	55.75±0.60	45.33±0.73	42.62±0.63	37.29±0.60	38.21±0.66	38.35±0.62	41.54±0.71
stabPA⁻ (Ours)	**61.87±0.57**	**48.02±0.73**	46.27±0.67	38.22±0.66	39.88±0.63	**41.75±0.59**	**44.09±0.69**
DANN [11]	–	45.09±0.48	42.71±0.65	39.11±0.61	39.49±0.69	41.40±0.59	43.68±0.73
PCT [35]	–	48.06±0.68	46.25±0.64	34.10±0.58	35.59±0.66	40.85±0.58	43.30±0.74
Mean Teacher [36]	–	44.80±0.69	43.16±0.61	39.30±0.61	39.37±0.66	39.98±0.60	42.50±0.68
FixMatch [33]	–	**48.45±0.70**	47.17±0.68	**43.13±0.67**	**43.20±0.69**	41.48±0.60	44.68±0.72
STARTUP [26]	–	47.18±0.71	45.00±0.64	38.10±0.62	38.84±0.70	**41.94±0.63**	**44.71±0.73**
stabPA (Ours)	–	**49.83±0.67**	**50.78±0.74**	**44.02±0.71**	**45.55±0.70**	**45.64±0.63**	**48.97±0.69**

the proposed bi-directional prototypical alignment is an effective approach to leveraging unlabeled images to reduce domain gap for CDCS-FSL.

5.3 Analysis

Has *stab*PA learned compact and aligned representations? To verify whether *stab*PA indeed learns compact and aligned representations, we visualize the feature distributions through the meta-training process using t-SNE [22]. From Fig. 3 (a)-(d), it can be seen that in the beginning, samples from different classes are heavily mixed. There are no distinct classification boundaries between classes. Besides, samples from two domains are far away from each other, indicating the existence of a considerable domain shift (such as the classes in green

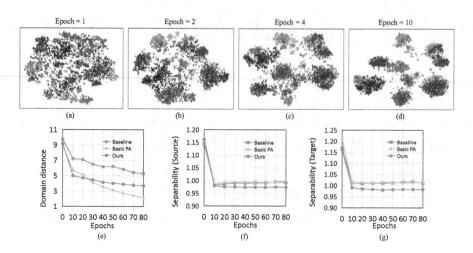

Fig. 3. (a)-(d) t-SNE visualization of feature distribution at different training epochs. Samples of the same class are painted in similar colors, where darker triangles represent source samples and lighter reverted triangles represent target samples (best viewed in color). Class centers are marked in black border. (e) Domain distance on novel classes. (f)-(g) Separability among novel classes in the source and target domains. Separability is represented by the average distance ratio, the lower the better.

and orange). However, as training continues, samples from the same class begin to aggregate together, and the margins between different classes are increasing. In other words, compact feature representation can be obtained by the *stab*PA. Moreover, we can see that samples from different domains are grouped into their ground-truth classes, even though no label information is given for the target domain data. These observations demonstrate that *stab*PA is indeed capable to learn compact and aligned representations.

Can *stab*PA learn generalizable representations for novel classes? To validate the generalization capability of the representations learned by *stab*PA, we propose two quantitative metrics, Prototype Distance (PD) and Average Distance Ratio (ADR), which indicate the domain distance and class separability among novel classes, respectively. A small PD value means the two domains are well aligned to each other, and a ADR less than 1 indicates most samples are classified into their ground-truth classes. Detailed definitions about these two metrics can be found in the appendix.

We compare *stab*PA with a FSL baseline [37] that does not leverage target images, and the BasicPA which aligns two domains by simply minimizing the point-to-point distance between prototypes in two domains [44]. The results are presented in Fig. 3 (e)-(g). It is noticed that all these methods can achieve lower domain distance as training processes, and BasicPA gets the lowest domain distance at the end. However, BasicPA does not improve the class separability as much as our approach, as shown in Fig. 3 (f)-(g). The inferior class separability can be understood that BasicPA merely aims to reduce the feature distance between two domains, without taking account of the intra-class variance

Table 3. The influence of the number of unlabeled samples and the number of base classes in the auxiliary set. We report average accuracy on DomainNet over 6 situations.

	number of samples				number of base classes				
FixMatch [33]	10%	40%	70%	100%	0%	10%	40%	70%	100%
1-shot 47.72	51.76	52.97	53.42	53.92	50.74	51.59	52.48	53.24	53.92
5-shot 62.58	65.96	67.56	67.96	68.55	65.04	65.68	67.07	67.87	68.55

Table 4. Pseudo label accuracy on DomainNet real-painting.

	fixed	epoch=0	10	20	30	40	50
Top-1	23.5	4.9	24.2	30.8	34.4	35.9	37.2
Top-5	40.0	14.4	41.9	48.2	51.4	52.8	53.9

and inter-class distances in each domain. Instead of aligning centers adopted by BasicPA, the proposed *stab*PA considers the feature-to-prototype distances across different domains and classes, so the domain alignment and class separability can be improved at the same time.

Number of Unlabeled Target Data. To test the robustness to the number of unlabeled samples, we gradually drop data from the auxiliary set in two ways: (i) randomly drop samples from the auxiliary set, (ii) select a subset of base classes and then manually remove samples that are from the selected classes. Table 3 shows the average accuracy of *stab*PA on DomainNet over 6 situations. Unsurprisingly, decreasing the number of samples will lead to performance drop (about 2.4 points from 100% to 10%). However, with only 10% samples remained, our approach still outperforms FixMatch which uses 100% auxiliary data. We can also see that removing whole classes leads to more performance drop than randomly removing samples, probably due to the class imbalance problem caused by the former. Nevertheless, the difference is very small (about 0.3 points), indicating that our approach is robust to the number of base classes.

Pseudo Label Accuracy. In Table 4, we show the pseudo label accuracies of the target domain images obtained by the fixed classifier and the online classifier during the training process. We can see that the fixed classifier is better than the online classifier at the early training epochs. However, as the training goes on, the online classier gets more accurate and outperforms the fixed classifier. This is because the online classifier is updated along the representation alignment process and gradually fits the data distribution of the target domain. After training with 50 epochs, the online classifier achieves 53.9% top-5 accuracy. To further improve the reliability of pseudo labels, we set a threshold to filter out pseudo labels with low confidence. Therefore, the actual pseudo label accuracy is higher than 53.9%.

Table 5. Ablation studies on DomainNet with 95% confidence interval.

ℓ_{s-t}	ℓ_{t-s}	aug	real-sketch		sketch-real	
			1-shot	5-shot	1-shot	5-shot
✗	✗	✗	$40.23_{\pm0.73}$	$50.41_{\pm0.80}$	$41.90_{\pm0.86}$	$56.95_{\pm0.84}$
✗	✗	✓	$41.74_{\pm0.78}$	$51.03_{\pm0.85}$	$42.17_{\pm0.95}$	$57.11_{\pm0.93}$
✓	✗	✗	$42.86_{\pm0.78}$	$52.16_{\pm0.78}$	$44.83_{\pm0.95}$	$60.87_{\pm0.91}$
✗	✓	✗	$44.20_{\pm0.77}$	$54.83_{\pm0.79}$	$44.45_{\pm0.92}$	$61.97_{\pm0.90}$
✓	✓	✗	$47.01_{\pm0.84}$	$56.68_{\pm0.81}$	$47.59_{\pm1.00}$	$64.32_{\pm0.86}$
✓	✓	✓	$\mathbf{50.85_{\pm0.86}}$	$\mathbf{61.37_{\pm0.82}}$	$\mathbf{51.71_{\pm1.01}}$	$\mathbf{68.93_{\pm0.87}}$

Ablation Studies. We conduct ablation studies on key components of the proposed *stab*PA. The results on DomainNet are shown in Table 5. As all key components are removed (the first row), our approach is similar to Tian et al. [37] that trains feature extractor with only the source data. When the unlabeled target data are available, applying either source-to-target alignment or target-to-source alignment can improve the performance evidently. Interestingly, we can see that the target-to-source alignment is more effective than the source-to-target alignment (about 1.2 points on average). This is probably because the source prototypes estimated by the ground truth labels are more accurate than the target prototypes estimated by the pseudo labels. Improving the quality of target prototypes may reduce this gap. Combing these two alignments together, we can get better results, indicating that the two alignments are complementary to each other. Finally, the best results are obtained by combining the strong data augmentation techniques, verifying that strong data augmentation can further strengthen the cross-domain alignment.

6   Conclusions

In this work, we have investigated a new problem in FSL, namely CDCS-FSL, where a domain gap exists between the support set and query set. To tackle this problem, we have proposed *stab*PA, a prototype-based domain alignment framework to learn compact and cross-domain aligned representations. On two widely-used multi-domain datasets, we have compared our approach to multiple elaborated baselines. Extensive experimental results have demonstrated the advantages of our approach. Through more in-depth analysis, we have validated the generalization capability of the representations learned by *stab*PA and the effectiveness of each component of the proposed model.

Acknowledgements. This work was supported in part by the National Natural Science Foundation of China under Grants 61721004, 61976214, 62076078, 62176246 and in part by the CAS-AIR.

References

1. Ben-David, S., Blitzer, J., Crammer, K., Kulesza, A., Pereira, F., Vaughan, J.W.: A theory of learning from different domains. Mach. Learn. **79**(1), 151–175 (2010)
2. Bengio, Y., Louradour, J., Collobert, R., Weston, J.: Curriculum learning. In: International Conference on Machine Learning (2009)
3. Chen, T., Kornblith, S., Norouzi, M., Hinton, G.: A simple framework for contrastive learning of visual representations. In: International Conference on Machine Learning, pp. 1597–1607. PMLR (2020)
4. Chen, W.Y., Liu, Y.C., Kira, Z., Wang, Y.C.F., Huang, J.B.: A closer look at few-shot classification. In: International Conference on Learning Representations (2019)
5. Chen, W., Si, C., Wang, W., Wang, L., Wang, Z., Tan, T.: Few-shot learning with part discovery and augmentation from unlabeled images. arXiv preprint arXiv:2105.11874 (2021)
6. Chen, Y., Liu, Z., Xu, H., Darrell, T., Wang, X.: Meta-baseline: Exploring simple meta-learning for few-shot learning. In: Proceedings of the IEEE/CVF International Conference on Computer Vision, pp. 9062–9071 (2021)
7. Cubuk, E.D., Zoph, B., Shlens, J., Le, Q.V.: RandAugment: practical automated data augmentation with a reduced search space. In: Proceedings of the IEEE/CVF Conference on Computer Vision and Pattern Recognition Workshops, pp. 702–703 (2020)
8. DeVries, T., Taylor, G.W.: Improved regularization of convolutional neural networks with cutout. arXiv preprint arXiv:1708.04552 (2017)
9. Finn, C., Abbeel, P., Levine, S.: Model-agnostic meta-learning for fast adaptation of deep networks. In: International Conference on Machine Learning, pp. 1126–1135. PMLR (2017)
10. Fu, Y., Fu, Y., Jiang, Y.G.: Meta-FDMixup: Cross-domain few-shot learning guided by labeled target data. In: ACMMM (2021)
11. Ganin, Y., et al.: Domain-adversarial training of neural networks. J. Mach. Learn. Res. **17**(1), 2096–2130 (2016)
12. Goodfellow, I., et al.: Generative adversarial nets. In: Advances in Neural Information Processing Systems, vol. 27 (2014)
13. Gretton, A., Borgwardt, K., Rasch, M., Schölkopf, B., Smola, A.: A kernel method for the two-sample-problem. NeurIPS (2006)
14. Guan, J., Zhang, M., Lu, Z.: Large-scale cross-domain few-shot learning. In: ACCV (2020)
15. Guo, Y., et al.: A broader study of cross-domain few-shot learning. In: Vedaldi, A., Bischof, H., Brox, T., Frahm, J.-M. (eds.) ECCV 2020. LNCS, vol. 12372, pp. 124–141. Springer, Cham (2020). https://doi.org/10.1007/978-3-030-58583-9_8
16. He, K., Fan, H., Wu, Y., Xie, S., Girshick, R.: Momentum contrast for unsupervised visual representation learning. In: Proceedings of the IEEE/CVF Conference on Computer Vision and Pattern Recognition, pp. 9729–9738 (2020)
17. Hoffman, J., et al.: CyCADA: Cycle-consistent adversarial domain adaptation. In: International Conference on Machine Learning, pp. 1989–1998. PMLR (2018)
18. Islam, A., Chen, C.F.R., Panda, R., Karlinsky, L., Feris, R., Radke, R.J.: Dynamic distillation network for cross-domain few-shot recognition with unlabeled data. Adv. Neural. Inf. Process. Syst. **34**, 3584–3595 (2021)
19. Lee, K., Maji, S., Ravichandran, A., Soatto, S.: Meta-learning with differentiable convex optimization. In: Proceedings of the IEEE/CVF Conference on Computer Vision and Pattern Recognition (2019)

20. Liang, H., Zhang, Q., Dai, P., Lu, J.: Boosting the generalization capability in cross-domain few-shot learning via noise-enhanced supervised autoencoder. In: ICCV (2021)
21. Long, M., CAO, Z., Wang, J., Jordan, M.I.: Conditional adversarial domain adaptation. In: Advances in Neural Information Processing Systems (2018)
22. Van der Maaten, L., Hinton, G.: Visualizing data using t-SNE. Journal of Machine Learning Research 9(11) (2008)
23. Mangla, P., Kumari, N., Sinha, A., Singh, M., Krishnamurthy, B., Balasubramanian, V.N.: Charting the right manifold: Manifold mixup for few-shot learning. In: Proceedings of the IEEE/CVF Winter Conference on Applications of Computer Vision, pp. 2218–2227 (2020)
24. Panareda Busto, P., Gall, J.: Open set domain adaptation. In: Proceedings of the IEEE International Conference on Computer Vision, pp. 754–763 (2017)
25. Peng, X., Bai, Q., Xia, X., Huang, Z., Saenko, K., Wang, B.: Moment matching for multi-source domain adaptation. In: Proceedings of the IEEE/CVF International Conference on Computer Vision, pp. 1406–1415 (2019)
26. Phoo, C.P., Hariharan, B.: Self-training for few-shot transfer across extreme task differences. In: International Conference on Learning Representations (2021)
27. Qiao, S., Liu, C., Shen, W., Yuille, A.L.: Few-shot image recognition by predicting parameters from activations. In: Proceedings of the IEEE Conference on Computer Vision and Pattern Recognition, pp. 7229–7238 (2018)
28. Ravi, S., Larochelle, H.: Optimization as a model for few-shot learning. In: International Conference on Learning Representations (2017)
29. Rusu, A.A., et al.: Meta-learning with latent embedding optimization. In: International Conference on Learning Representations (2019)
30. Saito, K., Kim, D., Sclaroff, S., Saenko, K.: Universal domain adaptation through self supervision. arXiv preprint arXiv:2002.07953 (2020)
31. Saito, K., Yamamoto, S., Ushiku, Y., Harada, T.: Open set domain adaptation by backpropagation. In: Proceedings of the European Conference on Computer Vision (ECCV), pp. 153–168 (2018)
32. Snell, J., Swersky, K., Zemel, R.: Prototypical networks for few-shot learning. In: Guyon, I. (ed.) Advances in Neural Information Processing Systems, vol. 30. Curran Associates Inc, New York (2017)
33. Sohn, K., et al.: FixMatch: Simplifying semi-supervised learning with consistency and confidence. arXiv preprint arXiv:2001.07685 (2020)
34. Sung, F., Yang, Y., Zhang, L., Xiang, T., Torr, P.H., Hospedales, T.M.: Learning to compare: relation network for few-shot learning. In: Proceedings of the IEEE Conference on Computer Vision and Pattern Recognition, pp. 1199–1208 (2018)
35. Tanwisuth, K., et al.: A prototype-oriented framework for unsupervised domain adaptation. In: NeurIPS (2021)
36. Tarvainen, A., Valpola, H.: Mean teachers are better role models: weight-averaged consistency targets improve semi-supervised deep learning results. In: Advances in Neural Information Processing Systems, vol. 30 (2017)
37. Tian, Y., Wang, Y., Krishnan, D., Tenenbaum, J.B., Isola, P.: Rethinking few-shot image classification: a good embedding is all you need? In: Proceedings of the European Conference on Computer Vision (ECCV) (2020)
38. Tsai, Y.H., Hung, W.C., Schulter, S., Sohn, K., Yang, M.H., Chandraker, M.: Learning to adapt structured output space for semantic segmentation. In: Proceedings of the IEEE Conference on Computer Vision and Pattern Recognition, pp. 7472–7481 (2018)

39. Tseng, H.Y., Lee, H.Y., Huang, J.B., Yang, M.H.: Cross-domain few-shot classification via learned feature-wise transformation. In: ICLR (2020)
40. Venkateswara, H., Eusebio, J., Chakraborty, S., Panchanathan, S.: Deep hashing network for unsupervised domain adaptation. In: Proceedings of the IEEE Conference on Computer Vision and Pattern Recognition, pp. 5018–5027 (2017)
41. Vilalta, R., Drissi, Y.: A perspective view and survey of meta-learning. Artif. Intell. Rev. **18**(2), 77–95 (2002)
42. Vinyals, O., Blundell, C., Lillicrap, T., Wierstra, D., et al.: Matching networks for one shot learning. Adv. Neural. Inf. Process. Syst. **29**, 3630–3638 (2016)
43. Wang, H., Deng, Z.H.: Cross-domain few-shot classification via adversarial task augmentation. In: IJCAI (2021)
44. Xie, S., Zheng, Z., Chen, L., Chen, C.: Learning semantic representations for unsupervised domain adaptation. In: International Conference on Machine Learning, pp. 5423–5432. PMLR (2018)
45. You, K., Long, M., Cao, Z., Wang, J., Jordan, M.I.: Universal domain adaptation. In: Proceedings of the IEEE/CVF conference on computer vision and pattern recognition, pp. 2720–2729 (2019)
46. Zhang, C., Cai, Y., Lin, G., Shen, C.: DeepEMD: Few-shot image classification with differentiable earth mover's distance and structured classifiers. In: IEEE/CVF Conference on Computer Vision and Pattern Recognition (2020)
47. Zhang, P., Zhang, B., Zhang, T., Chen, D., Wang, Y., Wen, F.: Prototypical pseudo label denoising and target structure learning for domain adaptive semantic segmentation. In: Proceedings of the IEEE/CVF Conference on Computer Vision and Pattern Recognition, pp. 12414–12424 (2021)
48. Zhang, Q., Zhang, J., Liu, W., Tao, D.: Category anchor-guided unsupervised domain adaptation for semantic segmentation. In: Advances in Neural Information Processing Systems, vol. 32 (2019)
49. Zheng, Z., Yang, Y.: Rectifying pseudo label learning via uncertainty estimation for domain adaptive semantic segmentation. Int. J. Comput. Vis. **129**(4), 1106–1120 (2021)

Mutually Reinforcing Structure with Proposal Contrastive Consistency for Few-Shot Object Detection

Tianxue Ma[1], Mingwei Bi[2], Jian Zhang[2], Wang Yuan[1], Zhizhong Zhang[1(✉)],
Yuan Xie[1], Shouhong Ding[2], and Lizhuang Ma[1(✉)]

[1] East China Normal University, Shanghai, China
{51205901016,51184501076,zzzhang}@stu.ecnu.edu.cn,
xieyuan8589@foxmail.com, ma-lz@cs.sjtu.edu.cn
[2] Tencent Youtu Lab, Shanghai, China
{mingweibi,timmmyzhang,ericshding}@tencent.com

Abstract. Few-shot object detection is based on the base set with abundant labeled samples to detect novel categories with scarce samples. The majority of former solutions are mainly based on meta-learning or transfer-learning, neglecting the fact that images from the base set might contain unlabeled novel-class objects, which easily leads to performance degradation and poor plasticity since those novel objects are served as the background. Based on the above phenomena, we propose a Mutually Reinforcing Structure Network (MRSN) to make rational use of unlabeled novel class instances in the base set. In particular, MRSN consists of a mining model which unearths unlabeled novel-class instances and an absorbed model which learns variable knowledge. Then, we design a Proposal Contrastive Consistency (PCC) module in the absorbed model to fully exploit class characteristics and avoid bias from unearthed labels. Furthermore,we propose a simple and effective data synthesis method undirectional-CutMix (UD-CutMix) to improve the robustness of model mining novel class instances, urge the model to pay attention to discriminative parts of objects and eliminate the interference of background information. Extensive experiments illustrate that our proposed approach achieves state-of-the-art results on PASCAL VOC and MS-COCO datasets. Our code will be released at https://github.com/MMatx/MRSN.

Keywords: Few-shot object detection · Contrastive learning · Data augmentation

S. Avidan et al. (Eds.): ECCV 2022, LNCS 13680, pp. 400–416, 2022.
https://doi.org/10.1007/978-3-031-20044-1_23

1   Introduction

Object detection [8,11,47,50], is a classical task in computer vision, which aims to identify and localize objects in an image. In recent years, the application of deep convolutional neural network [1] has accelerated the development of object detection [15–18,25,33,34,48,49]. However, the remarkable performance of object detection depends on abundant annotated data. Collecting annotated data is time-consuming and labor-intensive. As opposed to that, a few examples are sufficient for humans to learn a new concept. To bridge this gap, we focus on *few-shot object detection* (FSOD), which aims to adapt the model from the base class to the novel class based on a few labeled novel classes examples.

Based upon the base set, almost all methods for solving FSOD simply utilize the labeled base class objects in the base set as supervision information to train a detector. In the process of using base set, past methods overlook an important phenomenon: material neglect, in which unlabeled novel-class instances are explicitly learned as background. Material neglect has the following drawbacks. 1) An incorrect priori is introduced into the model, which limits the plasticity of the model on novel classes. The model explicitly takes all instances of novel-class that are unlabeled as the background when using the base set to train the detector. Due to incorrect supervision, subsequently, the detector is provided with labeled novel samples for adaptation, it is difficult for a small number of labeled novel class samples to correct the error knowledge. 2) Data augmentation has become a factor in training high-performance few-shot resolver. By providing a variety of samples for the model, the model can obtain a more robust representation space. But the current FSOD solution abandons the solution of data augmentation and even turns a blind eye to the existing unlabeled samples, which is undoubtedly a suboptimal solution to FSOD.

Thus, we advocate a new solution that uses data augmentation for resolving FSOD. Specifically, instead of using meta-learning or transfer-learning to obtain an adaptable model in the base set, we make rational use of unlabeled novel class instances and eliminate the negative effects. Inspired by [27], we use a semi-supervised framework to unearth unlabeled novel class instances and introduce a *Mutually Reinforcing Structure Network* (MRSN), which contains a mining model and an absorbed model.

To fully exploit class characteristics and avoid bias from noise labels, we design the *Proposal Contrastive Consistency* (PCC) module in the absorbed model. We use the mining model to unearth the unlabeled novel-class instances and use the absorbed model to learn the mined novel-class instances. The excavated novel instances may have noise. To prevent the consecutively detrimental effect of noisy pseudo-labels, we utilize PCC to keep the consistency of the mining model and the absorbed model by constructing positive and negative sample pairs between the two models. Meanwhile, we utilize PCC at the proposal level to compare the global and local information of the instance simultaneously, which compensates for the scarcity of training data. Intuitively, for classification head in Faster R-CNN, it ensures that the samples of the same category in the feature space are close to each other, and the regions related to different categories are

separable. This coarse goal leads to the loss of detailed features that are important for separating similar samples in the learning process. The PCC method we
proposed optimizes the features of "instance recognition", retains the information used to identify the subtle details between the classes and the instances.

Moreover, we propose an effective data synthesis method *undirectional-CutMix* (UD-CutMix) to improve the robustness of model mining novel class
instances, urge the model to pay attention to class discrimination features
and eliminate the interference of background. Extensive experiments show that
the proposed framework achieves significant improvements over state-of-the-art
methods on PASCAL VOC and MS-COCO.

2   Related Works

2.1   General Object Detection

At present, object detection methods based on deep learning can be divided into
two kinds according to different detection processes. One is two-stage approach
[2,7,21,31]. The detector first filters out proposals and then carries out regression and classification on them through the network. [31] uses RPN to generate
proposals and ROI head to detection. The other is the one-stage object detection method. The core of the one-stage object detection algorithm is regression,
which directly divides categories and regresses the border. [29,30] are based on
a single end-to-end network, which completes the input from the original image
to the output of object position and category. [26] combines the regression idea
and anchor mechanism, extracts the feature map of different scales, and then
regresses the multi-scale regional features of each position in the whole map.

Although the one-stage object detection algorithm has faster detection speed,
the two-stage algorithm are more reliable in terms of detection accuracy. In the
trade-off between accuracy and speed, we pick the Faster R-CNN with higher
accuracy as our benchmark framework.

2.2   Few-Shot Object Detection

Compared with general object detection, the scarcity of labeled novel-class data
brings huge difficulty to FSOD. The previous FSOD solving methods can be
divided into meta-learning [6,36] based methods and transfer-learning [28,35,39]
based methods. The idea of the former is that using the samples of support set
to learn the class representation, and then use the learned representation to
enhance the response of class related areas in the query sample features. [43]
aggregates features by reweighting the query features according to the class
encoder. [42] performs channel-wise multiplication of the extracted vector of
query features and extracted vector of class features. In the method based on
transfer-learning, [44] proposes a context transformer, which learns the ability to
integrate context knowledge in the source domain and migrates this ability to the
target domain to enhance the recognition ability of the detector, to reduce the

object confusion in the few-shot scene. [46] migrates shared changes in the base class to generate more diverse novel-class training examples. With the in-depth investigation of FSOD by researchers, some new methods have emerged. [19] proposes a class margin equilibrium approach to optimize the division of feature space. [32] puts forward a contrastive loss to alleviate the problem of classification error. [20] solves the task of FSOD from the perspective of classification structure enhancement and sample mining.

However, these methods almost ignore the "material neglect", where the image in the base set contains unlabeled novel-class instances.

2.3 Contrastive Learning

Recently, the success of the self-supervised model in various tasks mainly stems from the use of contrastive learning [3,4,9]. The goal of contrastive representation learning is to construct an embedding space in which similar pairs of samples remain close to each other while dissimilar ones remain far apart. Contrastive can be achieved by learning the similarity and differences of samples. [4] proposes a simclr framework to maximize the consistency between different views of the same image. [4] uses two neural networks to encode the data to construct positive and negative sample pairs, and encodes the image into query vectors and key vectors respectively. In the training process, try to improve the similarity between each query vector and its corresponding key vector, and reduce the similarity with the key vector of other images. The above methods are mainly used in classification tasks.

As far as we know, there is rare contrastive learning method specially designed for FSOD. Under the proposed MRSN, we propose a contrastive-consistency learning method PCC for FSOD.

3   Method

3.1   Problem Definition

Followed by the widely adopted FSOD setting in [37,38,43], our framework aims to perform adaptation on novel categories with the aid of a large and labeled base set D_{base} and a small-scale novel support set D_{novel}. Here, the base set $D_{base} = \{(I_i^{base}, C_i^{base}, B_i^{base})\}$ contains abundant annotated instances, where I_i^{base} denotes the i-th image, $C_i^{base} = \{(c_i^{base})_j\}$ is the entire class set in this image, and $B_i^{base} = \{(x_i, y_i, w_i, h_i)_j\}$ indicates the position of each instance j in this image. While the novel set $D_{novel} = \{(I_i^{novel}, C_i^{novel}, B_i^{novel})\}$ shares the same structure with the base set, the class set C^{base} and C^{novel} of those two collections are non-overlapping. More specifically, the ultimate goal of FSOD task is to classify and locate all instances belonging to the novel-class and maintain the performance on base classes.

3.2   Training Phase

The initialization of our MRSN is very important since the MRSN is an iterative process to unearth unlabeled novel-class instances. Therefore, we first introduce the acquisition of the initialization model of the MRSN, and then introduce the learning process of MRSN in detail.

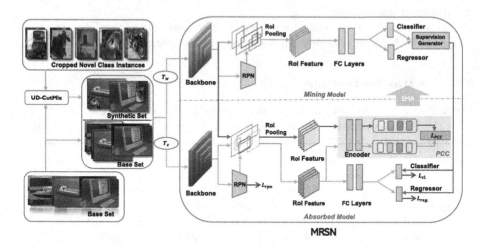

Fig. 1. Framework of our method. First, cropped novel class instances and base set construct a synthetic set through UD-CutMix. Then the images are processed by two different data augmentations (where T_w and T_s denotes different data augmentations) and sent to two models. In MRSN, the outputs generated by the mining model are provided to the absorbed model and the absorbed model transfers the learned knowledge to the mining model through EMA. Meanwhile, to fully exploit class characteristics and avoid bias from noise labels, the PCC module is designed in the absorbed model

Training plain detector. We call the model used to initialize the MRSN as M_{plain}. First, we use D_{base} to train the whole detector. The whole detector $F(\theta_{emb}, \theta_{rpn}, \theta_{roi})$ include the network parameters of feature extraction, RPN module and ROI module, which are expressed as $\theta_{emb}, \theta_{rpn}$ and θ_{roi}, respectively. During the experiment, we found that due to the scarcity of data in novel set, the overfitting problem is easily incurred if we only use the limited data of novel classes to fine-tune the whole detector. To tackle with this problem, we sample a balanced training dataset $D_{balance}$ from D_{base} and D_{novel} as [37]. Specifically for each category in C^{base}, we only take k samples to form $D_{balance}$ together with the whole D_{novel}. The combined training dataset $D_{balance}$ is eventually utilized to fine-tune the classifier and regressor of the detector. In this way, we get a new detector $M_{plain} = F(\theta_{emb}, \theta_{rpn}, \theta_{roi}^*)$, which can detect C^{novel} and C^{base} at the same time.

Training mutually reinforcing structure network. To solve material neglect, we propose a method to excavate unlabeled instances of novel classes C^{novel} in base set D_{base}. The specific method is shown in Fig. 1.

Our MRSN is a dual model construction, including M_{mine} and M_{absorb}, where M_{mine} is used to mine unlabeled novel-class instances in D_{base} and give them pseudo labels including corresponding categories and locations, and M_{absorb} is used to learn the mined instances. We utilize the mined instances as explicit supervision for both backbone, RPN and ROIhead. The initialization values of network parameters of M_{mine} and M_{absorb} are both inherited from $M_{plain} = F(\theta_{emb}, \theta_{rpn}, \theta^*_{roi})$. Based on the above definition, the evolutionary mechanism of MRSN will be introduced in Sect. 3.3. To further impose model discrimination between different categories and avoid the negative impact of noise labels, we design PCC module for M_{absorb} as shown in Sect. 3.4.

3.3   Mutually Reinforcing Structure Network

As illustrated in Fig. 1, our MRSN consists of a two-stream detection architecture. We first execute UD-CutMix by recombining labeled novel instances with base images to construct a new synthetic set D_{syn}. Taking images in the D_{syn} and D_{base} with weak data transformation as input, the upper mining stream explores the augmented novel information and further reserves the most credible prediction. The lower absorbed stream gradually adopts the novel knowledge delivered from the upper stream, without forgetting the base knowledge by utilizing base supervision at the same time. Our two-stream framework thus evolves continuously in the process of mutually reinforcing learning.

UD-CutMix. A huge challenge of FSOD is the limited annotations and diversity scarce in the novel set. To address this issue, we propose UD-CutMix combine the cropped novel classes instances and the selected images in D_{base}. UD-CutMix adopts the detector M_{plain} to select the base image, which prediction doesn't contain any novel objects, to be mixed. Specifically, we first crop a novel-class instance from a novel set image, then scale and paste it to a selected base image. By repeating this operation, we can construct a new synthetic set $D_{syn} = \{(I_i^{syn}, C_i^{syn}, B_i^{syn})\}$ as an enlarged and labeled set. The visual comparisons of the original and synthetic images are illustrated in Fig. 2.

The proposed UD-CutMix is distinct from the CutMix [45] in the following two aspects. First, CutMix is agnostic to the category, as it randomly samples data among all categories. UD-CutMix is category-specific, whice samples images in the base dataset and only pastes the novel class instances. We find this strategy can well cope with the lack of novel class data and the imbalance between the base class and novel class labelled data. Second, UD-CutMix crops the complete bounding box of the novel class instance, while CutMix randomly selects a patch to crop, which destroys the instance's global information.

It is notable that some base images involve novel instances. However, those unlabeled novel objects serve as background or even distractors during the base classes training process, hindering the transfer capability to novel classes. We can make full use of these base images by generating pseudo labels. Thus in our framework, the above synthetic set D_{syn} and the original base set D_{base} are both included as supervision in subsequent novel-classes adaptation pipeline.

Novel knowledge mining and absorbing. The novel knowledge mining stream excavates the possible novel instances and assigns them pseudo labels for further supervision.

Given an image $I_i \in D_{base} \cup D_{syn}$ with true label $GT_i = (C_i, B_i)$. We use two different data enhancements for I_i, one is denoted as strong data enhancement T_s, and the other is weak data enhancement T_w. The images applied with T_s are sent to the absorb stream M_{absorb}, and the images applied with T_w are sent to the mining stream M_{mine}. The corresponding predicted value P_{absorb} and P_{mine} are formulated as the following Eq. 1.

$$
\begin{aligned}
P_{absorb} &= M_{absorb}(T_s(I_i)) \\
&= \{(c, s, box)_j\}, box = (x, y, w, h), \\
P_{mine} &= M_{mine}(T_w(I_i)) \\
&= \{(c, s, box)_j\}, box = (x, y, w, h),
\end{aligned}
\tag{1}
$$

where c, s and box in $(c, s, box)_j$ respectively represent the category corresponding to the maximum classification probability in the j-th prediction result of image I_i, the corresponding score and the position of the bounding box.

To boost the credibility of pseudo labels and reduce the negative impact caused by noise prediction, we propose a supervision generator to exclude those distractors. We first refer to a confidence score threshold ϕ to get a trusted prediction subset P_{tr}, as expressed in Eq. 2.

$$
P_{tr} = \{\mathbb{I}(s \geq \phi)(c, s, box)_j | (c, s, box)_j \in P_{mine}\},
\tag{2}
$$

where the filtered prediction P_{tr} is the high-confidence predicted bounding box.

To provide deterministic novel supervision and reduce the learning bias of the base classes, we then set an overlap threshold δ to discard those predictions whose location is at the neighborhood of ground truth, as expressed in Eq. 3.

$$
\begin{aligned}
S_j &= MAX(IOU(box_j, box_k)), j \in P_{tr}, k \in GT_i, \\
P_{final} &= \{\mathbb{I}(S_j \leq \delta)(c, s, box)_j | (c, s, box)_j \in P_{tr}\},
\end{aligned}
\tag{3}
$$

where the final prediction P_{final} is the subset of those proposals whose overlapping with the GT bounding boxes of base classes is below a certain threshold.

Finally, we choose the images that novel class instances in the final prediction P_{final} as the supervision images. To stabilize the training process and accelerate the convergence speed, we combine the ground truth labels of selected images with P_{final}. In general, our supervision generator contains bounding box-level confidence filtering as well as overlap filtering, image-level instance filtering, label combining.

Interaction between two stream. By applying supervision generator, the most convincing images subset is obtained for subsequent novel knowledge learning. For M_{absorb}, we use those most convincing images as supervisory information to update the whole model by back-propagation. As expressed in Eq. 4, instead of using gradient back propagation to update M_{mine}, we transfer the parameters of M_{absorb} to iterate M_{mine}. In this way, we can alleviate the accumulation effect on the unreliability of pseudo labels.

$$\theta^t_{mine} = \alpha\theta^{t-1}_{mine} + (1 - \alpha)\theta^t_{absorb}, \tag{4}$$

where α is the hyperparameter to balance the update ratio of M_{mine}. Both θ_{mine} and θ_{absorb} include the network parameters of backbone, RPN module, and ROI module. When M_{absorb} is updated after a certain number of iterations, we use the absorb model parameters θ_{absorb} to update the freeze mining model parameters θ_{mine}. Here, t and $(t-1)$ represent the network parameters at the current time and before updating, respectively.

3.4   Proposal Contrastive Consistency

The self-supervised method [3,4,9] performs well in classification, but not in intensive prediction tasks, such as object detection. The main reason is that the current self-supervised method is global-level feature extraction, and the prediction ability of local is not considered. After in-depth research, we propose a contrastive learning method Proposal Contrastive Consistency (PCC) suitable for our MRSN in FSOD. We introduce a PCC branch to the primary RoI head, parallel to the classification and regression branches. In M_{absorb}, RPN takes feature maps as inputs and generates region proposals, then a mini-batch RoIs [31] is sampled for training. We express RoIs sampled by M_{absorb} as $Roi^{absorb} = \{(x, y, w, h)_j\}$. f_{mine} and f_{absorb} are the feature maps obtained by the feature extractors of M_{mine} and M_{absorb} respectively. According to Roi^{absorb}, RoIPooling is performed on f_{mine} and f_{absorb} to obtain the features of the proposals, and then through the identical encoder E, features in the two feature spaces are mapped to the same area.

$$\begin{aligned} z^{mine}_j &= E(RoIPooling(f_{mine}, Roi^{absorb}_j)), \\ z^{absorb}_j &= E(RoIPooling(f_{absorb}, Roi^{absorb}_j)). \end{aligned} \tag{5}$$

We take the contrastive learning features from the same proposal and different models as positive sample pairs and others as negative sample pairs. And we measure the cosine similarity between two proposal features in the projected hypersphere.

Our Proposal Contrastive Consistency loss L_{PCC} is formulated as Eq. 6:

$$\begin{aligned} L_{PCC} &= \sum_j -log\frac{Pos_j/\tau}{Neg^1_j/\tau + Neg^2_j/\tau + Pos_j/\tau}, \\ Neg^1_j &= \sum_{k=1}^N \mathbb{I}(k \neq j)exp(sim(z^{absorb}_j, z^{absorb}_k)/\tau, \\ Neg^2_j &= \sum_{k=1}^N \mathbb{I}(k \neq j)exp(sim(z^{mine}_j, z^{absorb}_k)/\tau, \\ Pos_j &= exp(sim(z^{mine}_j, z^{absorb}_j)). \end{aligned} \tag{6}$$

Table 1. Performance on the PASCAL VOC dataset. We evaluate the performance on three different sets of novel classes

Method/Shot	Novel Set 1					Novel Set 2					Novel Set 3				
	1-shot	2-shot	3-shot	5-shot	10-shot	1-shot	2-shot	3-shot	5-shot	10-shot	1-shot	2-shot	3-shot	5-shot	10-shot
MetaDet [38]	17.1	19.1	28.9	35.0	48.8	18.2	20.6	25.9	30.6	41.5	20.1	22.3	27.9	41.9	42.9
RepMet [14]	26.1	32.9	34.4	38.6	41.3	17.2	22.1	23.4	28.3	35.8	27.5	31.1	31.5	34.4	37.2
CRDR [20]	40.7	45.1	46.5	57.4	62.4	27.3	31.4	40.8	42.7	46.3	31.2	36.4	43.7	50.1	55.6
FRCN-ft [38]	13.8	19.6	32.8	41.5	45.6	7.9	15.3	26.2	31.6	39.1	9.8	11.3	19.1	35.0	45.1
Meta R-CNN [43]	19.9	25.5	35.0	45.7	51.5	10.4	19.4	29.6	34.8	45.4	14.3	18.2	27.5	41.2	48.1
TFA w/ fc [37]	36.8	29.1	43.6	55.7	57.0	18.2	29.0	33.4	35.5	39.0	27.7	33.6	42.5	48.7	50.2
MPSR [41]	41.7	-	51.4	55.2	61.8	24.4	-	39.2	39.9	47.8	**35.6**	-	42.3	48.0	49.7
FSCE [32]	32.9	44.0	46.8	52.9	59.7	23.7	30.6	38.4	43.0	48.5	22.6	33.4	39.5	47.3	54.0
CME [19]	41.5	47.5	50.4	58.2	60.9	27.2	30.2	41.4	42.5	46.8	34.3	39.6	45.1	48.3	51.5
DCNet [12]	33.9	37.4	43.7	51.1	59.6	23.2	24.8	30.6	36.7	46.6	32.3	34.9	39.7	42.6	50.7
UP [40]	43.8	47.8	50.3	55.4	61.7	31.2	30.5	41.2	42.2	48.3	35.5	**39.7**	43.9	50.6	53.5
Ours	**47.6**	**48.6**	**57.8**	**61.9**	**62.6**	**31.2**	**38.3**	**46.7**	**47.1**	**50.6**	35.5	30.9	**45.6**	**54.4**	**57.4**

Finally, for our absorb stream M_{absorb}, the total loss is as follows:

$$L = L_{rpn} + \lambda_1(L_{cl.} + L_{reg.}) + \lambda_2 L_{PCC}, \tag{7}$$

where the L_{rpn} contains the cross entropy and regression loss of RPN, the middle two terms $L_{cl.}$, $L_{reg.}$ are the focal loss [23] of classification, smoothed-L1 loss of regression in roi head respectively, L_{PCC} is the proposal contrastive consistency loss. λ_1 and λ_2 are used to balance the loss functions.

4   Experiments

4.1   Datasets

For a fair comparison with previous work, we use MS-COCO [24] and PASCAL VOC [5] benchmarks to verify the effectiveness of our method.

PASCAL VOC. PASCAL VOC contains 20 categories. According to the previous FSOD experimental settings [13], we divide the 20 categories into 15 base classes and 5 novel classes under three different divisions. We take the trainval sets of the 2007 and 2012 as the training set and PASCAL VOC 2007 test set as the evaluation set. For each category in the novel set, k-shot novel instances are sampled. In FSOD scenarios, we set $k = (1, 2, 3, 5, 10)$.

MS-COCO. MS-COCO contains 80 categories, of which 20 categories are identical to PASCAL VOC. We choose 20 categories in the PASCAL VOC dataset as the novel set, and the rest 60 classes as the base set. For FSOD scenarios in MS-COCO, we set $k = (10, 30)$.

4.2   Implementation Details

We use Faster R-CNN [31] as the basic detector, in which ResNet-101 [10] is served as the backbone and a 4-layer Feature Pyramid Network [22] is utilized for boosting the multi-scale learning process. Models are trained with standard SGD optimizer and batch size of 2 in 1 GPU. We set the learning rate, the momentum, and the weight decay to 0.001, 0.9, and 0.0001, respectively.

Table 2. Few-shot object detection performance on MS-COCO. We report the AP and AP75 on the 20 novel categories

Method/Shot	10-shot		30-shot	
	AP	AP75	AP	AP75
MetaDet 2019 [38]	7.1	6.1	11.3	8.1
Meta R-CNN 2019 [43]	8.7	6.6	12.4	10.8
TFA w/fc 2020 [37]	9.1	8.8	12.1	12.0
MPSR 2020 [41]	9.8	9.7	14.1	14.2
Viewpoint [42]	12.5	9.8	14.7	12.2
FSCE 2021 [32]	11.1	9.8	15.3	14.2
CRDR 2021 [20]	11.3	-	15.1	-
UP 2021 [40]	11.0	10.7	15.6	15.7
Ours	**15.7**	**14.8**	**17.5**	**17.9**

4.3 Comparison Experiments

Results on PASCAL VOC. For all three random novel splits from PASCAL VOC, the evaluation results AP50 are presented in Table 1. The experimental results include $k = (1, 2, 3, 5, 10)$ shot under three different base/novel divisions. Our method is greatly improved over the previous method, which shows the effectiveness of our method. Meanwhile, this indicates that focusing on material neglect plays a key role in solving FSOD. Basically, the best results have been achieved in any shots and any splits. Our method makes the performance of split 2 reach a new level.

Results on MS-COCO. Compared with PASCAL VOC, the MS-COCO contains more categories and more instances in each image, which indicates that there is more serious material neglects in the MS-COCO. Therefore, the improvement of our method on MS-COCO is significant. FSOD results of MS-COCO are shown in Table 2. Compared with baseline methods, TFA [37], our method consistently outperforms its performance. In particular, under the settings of 10 shot and 30 shot, our method is 6.6% and 5.4% higher than TFA on AP, respectively. Meanwhile, in 10-shot setting, our proposed methods gain +3.2% AP and +4.1% AP75 above the current SOTA on the 20 novel classes.

4.4 Ablation

We analyze the effectiveness of proposed modules in our method. Unless otherwise specified, the experiments are carried out on the split1 of PASCAL VOC.

Module effectiveness. We analyze the effects of different modules and show the results in Table 3. From Table 3, we can see that the different designed modules are important to improve the effect. MRSN improves the effect by 3% - 7% under different shot. This shows that MRSN can effectively unearth unlabeled

Table 3. Ablation study of different modules for FSOD on PASCAL VOC novel classes (split-1). "MRSN" denotes Mutually Reinforcing Structure Network, "PCC" Proposal Contrastive Consistency, and"avg.$\triangle$" average performance improvements

MRSN	PCC	UD-CutMix	1-shot	2-shot	3-shot	5-shot	10-shot	avg.$\triangle$
			39.4	41.5	48.4	52.9	53.7	
✓			43.6	44.5	54.3	59.9	59.7	+5.2
✓	✓		44.9	47.7	57.2	61.1	62.1	+7.4
✓	✓	✓	**47.6**	**48.6**	**57.8**	**61.9**	**62.6**	**+8.5**

novel-class instances in the base set and generate trusted pseudo labels, to provide richer knowledge for the network to learn. We show unlabeled novel-class instances discovered by MRSN in Fig. 5. With the help of PCC, novel class AP50 can be improved by up to 3.2%. The improvement of effect strongly illustrates the importance of imposing model discrimination between different categories and avoiding the negative impact of noise labels. Although UD-CutMix only improves the effectiveness of the model by 0.5% in 10-shot scenario, it can improve the effect by 2.7% in 1-shot scenario, indicating that in the scenario with extremely scarce data, the MRSN has a weak ability to discover novel instance, and UD-CutMix can provide a supplement to the MRSN, allowing the model to locate and identify novel classes of different scales in different backgrounds. We show images obtained by UD-CutMix in Fig. 2.

Fig. 2. Data obtained by UD-CutMix. The first line is labeled novel instance. The second line is the selected base image. And the third line is synthetic data

Proposal contrastive consistency. The Table 4 shows that the PCC has brought a huge performance improvement. We have tried another form of contrastive learning, which we call Same Class Contrastive Consistency (SCCC).

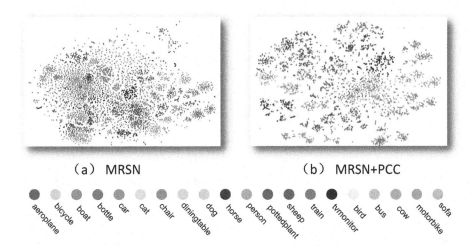

(a) MRSN   (b) MRSN+PCC

aeroplane bicycle boat bottle car cat chair diningtable dog horse person pottedplant sheep train tvmonitor bird bus cow motorbike sofa

Fig. 3. t-SNE visualization of (a) Object proposal features learned without PCC and (b) Object proposal features learned with PCC

SCCC takes the proposals of the same prediction category as positive samples and the proposals of different categories as negative samples. From the result in Table 4, we can see that the combination of SCCC and our MRSN framework is suboptimal. The reason we analyze is that under the MRSN framework, the generated pseudo labels have noise. When the category prediction of the proposal deviates, the effect of SCCC is to forcibly close the samples of different categories in the feature space, which will conflict with the optimization goal of the classifier. The PCC regards proposals at the same location as a positive sample pair, and proposals at different locations as a negative sample pair. The learning process is independent of the specific prediction category, which can effectively resist the negative impact caused by noise pseudo labels.

Table 4. Ablation for contrastive learning method, results of novel class AP50

MRSN	UD-CutMix	SCCC	PCC	3-shot	5-shot	10-shot
✓	✓			55.0	60.0	60.2
✓	✓	✓		57.0	60.7	60.9
✓	✓		✓	**57.8**	**61.9**	**62.6**

Figure 3 shows the object proposal features learned with and without PCC. From t-SNE visualization, we can see that with the help of PCC, the features of different classes are pulled away in the embedded space, while the features of the same class are closer in the embedded space.

Mutually reinforcing structure network. We use M_{mine} in the MRSN to mine unlabeled novel-class instances and generate pseudo label. M_{absorb} learns

Table 5. Few-shot object detection results on PASCAL VOC base classes (bAP50) under 1,2,3,5,10-shot settings

Method\shot	1-shot	2-shot	3-shot	5-shot	10-shot
MPSR [41]	59.4	67.8	68.4	41.7	51.4
TFA [37]	78.9	78.6	77.3	77.6	75.2
FSCE [32]	72.2	71.8	70.1	74.3	73.6
Ours	**80.2**	**79.7**	**79.3**	**79.5**	**79.2**

new knowledge according to the pseudo label provided by M_{mine}, and then M_{mine} accepts the new knowledge learned by M_{absorb} in a certain ratio.

We analyze the influence of the update interval and ratio of M_{mine} on the results. We carry out the experiment under the 3-shot setting of split 1. From the Fig. 4, we can see that the impact of too small update interval and too high update ratio on the M_{mine} is similar, which will reduce the performance. When the interval is too small or the ratio is too high, M_{mine} will shift rapidly to M_{absorb}, which will lead to the drift of the pseudo labels generated by M_{mine}, and reduce the ability of the mutually reinforcing structure to resist noise labels. From the Fig. 4, we can see that when the update interval of M_{mine} is too large, M_{mine} does not receive new knowledge for a long time, which will degrade the performance. In our experiment, we set the update ratio to 0.001 and the update interval to 1000 iterations.

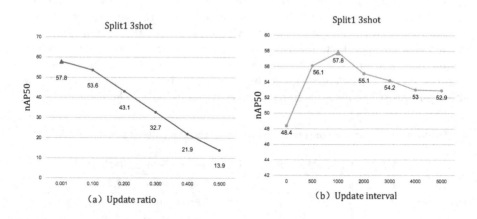

Fig. 4. Ablation study on the mining model update ratio $(1-\alpha)$ and update iterations in the novel classes of the first split of PASCAL VOC

To more intuitively see the role of MRSN, we show the pseudo labels generated by M_{mine} in Fig. 5. From columns (a) and (b), we can see that even in complex scenarios, M_{mine} can mine unlabeled novel-class instances. Columns

(c) and (d) show that M_{mine} can give full play to its mining ability in difficult-to-detect scenes (dark) and multi-object scenes. In column (e), the unlabeled novel-class instance is a cow, but M_{mine} is not confused by a similar object horse, and it accurately classifies and locates the cow.

Not forgetting. The previous works almost focus on the performance of the novel classes, while ignoring the non-forgetting effect on the base class. In the real scene, we prefer the model to adapt to the novel classes on the premise that the detection ability of the base class is basically unchanged. The method in this paper achieves the above ideal goal. We show the bAP50 of the model on the base class in Table 5.

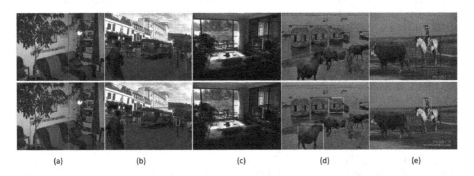

(a)   (b)   (c)   (d)   (e)

Fig. 5. Visualization of unearthed novel-class instances. The first line is the images from the base set where the novel instances are treated as background, such as "sofa" in columns (a) and (c), "bus" in column (b), "cow" in columns (d) and (e). The second line is the images processed by our mining model

5   Conclusions

In this paper, we discover that previous FSOD solving methods overlook material neglect . Towards solving this problem, a Mutually Reinforcing Structure Network (MRSN) is introduced to mine and absorb unlabeled novel instances. We design a Proposal Contrastive Consistency (PCC) module in MRSN to help learn more detailed features and resist the influence of noise labels. Meanwhile, we design a data synthesis method undirectional-CutMix (UD-CutMix) that combines the novel class instances and the images in the base set. Our method can effectively solve the problem of material neglect in FSOD. Experiments show that our method outperforms the previous methods by a large margin.

Acknowledgment. This work is supported by National Key Research and Development Program of China (2019YFC1521104, 2021ZD0111000), National Natural Science Foundation of China (72192821, 61972157, 62176092, 62106075), Shanghai Municipal Science and Technology Major Project (2021SHZDZX0102), Art major project of National Social Science Fund (I8ZD22), Shanghai Science and Technology Commission (21511101200, 22YF1420300, 21511100700), Natural Science Foundation of Shanghai (20ZR1417700), CAAI-Huawei MindSpore Open Fund.

References

1. Albawi, S., Mohammed, T.A., Al-Zawi, S.: Understanding of a convolutional neural network. In: 2017 international Conference on Engineering and Technology (ICET), pp. 1–6. IEEE (2017)

2. Cai, Z., Vasconcelos, N.: Cascade R-CNN: delving into high quality object detection. In: Proceedings of the IEEE Conference on Computer Vision and Pattern Recognition, pp. 6154–6162 (2018)

3. Chen, T., Kornblith, S., Norouzi, M., Hinton, G.: A simple framework for contrastive learning of visual representations. In: International Conference on Machine Learning, pp. 1597–1607. PMLR (2020)

4. Chen, X., Fan, H., Girshick, R., He, K.: Improved baselines with momentum contrastive learning. arXiv preprint arXiv:2003.04297 (2020)

5. Everingham, M., Van Gool, L., Williams, C.K., Winn, J., Zisserman, A.: The pascal visual object classes (VOC) challenge. Int. J. Comput. Vis. **88**(2), 303–338 (2010)

6. Finn, C., Abbeel, P., Levine, S.: Model-agnostic meta-learning for fast adaptation of deep networks. In: International Conference on Machine Learning, pp. 1126–1135. PMLR (2017)

7. Girshick, R.: Fast R-CNN. In: Proceedings of the IEEE International Conference on Computer Vision, pp. 1440–1448 (2015)

8. Gu, Q., et al.: PIT: Position-invariant transform for cross-FoV domain adaptation. In: Proceedings of the IEEE/CVF International Conference on Computer Vision, pp. 8761–8770 (2021)

9. He, K., Fan, H., Wu, Y., Xie, S., Girshick, R.: Momentum contrast for unsupervised visual representation learning. In: Proceedings of the IEEE/CVF Conference on Computer Vision and Pattern Recognition, pp. 9729–9738 (2020)

10. He, K., Zhang, X., Ren, S., Sun, J.: Deep residual learning for image recognition. In: Proceedings of the IEEE Conference on Computer Vision and Pattern Recognition, pp. 770–778 (2016)

11. He, L., et al.: End-to-end video object detection with spatial-temporal transformers. In: Proceedings of the 29th ACM International Conference on Multimedia, pp. 1507–1516 (2021)

12. Hu, H., Bai, S., Li, A., Cui, J., Wang, L.: Dense relation distillation with context-aware aggregation for few-shot object detection. In: Proceedings of the IEEE/CVF Conference on Computer Vision and Pattern Recognition, pp. 10185–10194 (2021)

13. Kang, B., Liu, Z., Wang, X., Yu, F., Feng, J., Darrell, T.: Few-shot object detection via feature reweighting. In: Proceedings of the IEEE/CVF International Conference on Computer Vision, pp. 8420–8429 (2019)

14. Karlinsky, L., et al.: RepMet: Representative-based metric learning for classification and few-shot object detection. In: Proceedings of the IEEE/CVF Conference on Computer Vision and Pattern Recognition, pp. 5197–5206 (2019)

15. Li, B., Sun, Z., Guo, Y.: SuperVAE: Superpixelwise variational autoencoder for salient object detection. In: The Thirty-Third AAAI Conference on Artificial Intelligence, AAAI 2019, Honolulu, Hawaii, USA, January 27 - February 1, 2019, pp. 8569–8576. AAAI Press (2019). https://doi.org/10.1609/aaai.v33i01.33018569

16. Li, B., Sun, Z., Li, Q., Wu, Y., Hu, A.: Group-wise deep object co-segmentation with co-attention recurrent neural network. In: 2019 IEEE/CVF International Conference on Computer Vision, ICCV 2019, Seoul, Korea (South), October 27 - November 2, 2019, pp. 8518–8527. IEEE (2019). https://doi.org/10.1109/ICCV.2019.00861

17. Li, B., Sun, Z., Tang, L., Sun, Y., Shi, J.: Detecting robust co-saliency with recurrent co-attention neural network. In: Kraus, S. (ed.) Proceedings of the Twenty-Eighth International Joint Conference on Artificial Intelligence, IJCAI 2019, Macao, China, August 10–16, 2019, pp. 818–825. ijcai.org (2019). https://doi.org/10.24963/ijcai.2019/115

18. Li, B., Sun, Z., Wang, Q., Li, Q.: Co-saliency detection based on hierarchical consistency. In: Amsaleg, L., et al. (eds.) Proceedings of the 27th ACM International Conference on Multimedia, MM 2019, Nice, France, October 21–25, 2019, pp. 1392–1400. ACM (2019). https://doi.org/10.1145/3343031.3351016

19. Li, B., Yang, B., Liu, C., Liu, F., Ji, R., Ye, Q.: Beyond max-margin: Class margin equilibrium for few-shot object detection. In: Proceedings of the IEEE/CVF Conference on Computer Vision and Pattern Recognition, pp. 7363–7372 (2021)

20. Li, Y., et al.: Few-shot object detection via classification refinement and distractor retreatment. In: Proceedings of the IEEE/CVF Conference on Computer Vision and Pattern Recognition, pp. 15395–15403 (2021)

21. Lin, T.Y., Dollár, P., Girshick, R., He, K., Hariharan, B., Belongie, S.: Feature pyramid networks for object detection. In: Proceedings of the IEEE Conference on Computer Vision and Pattern Recognition, pp. 2117–2125 (2017)

22. Lin, T.Y., Dollár, P., Girshick, R., He, K., Hariharan, B., Belongie, S.: Feature pyramid networks for object detection. In: Proceedings of the IEEE Conference on Computer Vision and Pattern Recognition, pp. 2117–2125 (2017)

23. Lin, T.Y., Goyal, P., Girshick, R., He, K., Dollár, P.: Focal loss for dense object detection. In: Proceedings of the IEEE International Conference on Computer Vision, pp. 2980–2988 (2017)

24. Lin, T.Y., et al.: Microsoft COCO: common objects in context. In: Fleet, D., Pajdla, T., Schiele, B., Tuytelaars, T. (eds.) ECCV 2014. LNCS, vol. 8693, pp. 740–755. Springer, Cham (2014). https://doi.org/10.1007/978-3-319-10602-1_48

25. Liu, L., et al.: Deep learning for generic object detection: a survey. Int. J. Comput. Vis. **128**(2), 261–318 (2020)

26. Liu, W., et al.: SSD: single shot multibox detector. In: Leibe, B., Matas, J., Sebe, N., Welling, M. (eds.) ECCV 2016. LNCS, vol. 9905, pp. 21–37. Springer, Cham (2016). https://doi.org/10.1007/978-3-319-46448-0_2

27. Liu, Y.C., et al.: Unbiased teacher for semi-supervised object detection. In: International Conference on Learning Representations (2021)

28. Pan, S.J., Yang, Q.: A survey on transfer learning. IEEE Trans. Knowl. Data Eng. **22**(10), 1345–1359 (2009)

29. Redmon, J., Divvala, S., Girshick, R., Farhadi, A.: You only look once: unified, real-time object detection. In: Proceedings of the IEEE Conference on Computer Vision and Pattern Recognition, pp. 779–788 (2016)

30. Redmon, J., Farhadi, A.: Yolo9000: better, faster, stronger. In: Proceedings of the IEEE conference on computer vision and pattern recognition, pp. 7263–7271 (2017)

31. Ren, S., He, K., Girshick, R., Sun, J.: Faster R-CNN: towards real-time object detection with region proposal networks. Adv. Neural. Inf. Process. Syst. **28**, 91–99 (2015)

32. Sun, B., Li, B., Cai, S., Yuan, Y., Zhang, C.: FSCE: Few-shot object detection via contrastive proposal encoding. In: Proceedings of the IEEE/CVF Conference on Computer Vision and Pattern Recognition, pp. 7352–7362 (2021)

33. Tang, L., Li, B.: CLASS: cross-level attention and supervision for salient objects detection. In: Ishikawa, H., Liu, C.-L., Pajdla, T., Shi, J. (eds.) ACCV 2020. LNCS, vol. 12624, pp. 420–436. Springer, Cham (2021). https://doi.org/10.1007/978-3-030-69535-4_26

34. Tang, L., Li, B., Zhong, Y., Ding, S., Song, M.: Disentangled high quality salient object detection. In: 2021 IEEE/CVF International Conference on Computer Vision, ICCV 2021, Montreal, QC, Canada, October 10–17, 2021, pp. 3560–3570. IEEE (2021). https://doi.org/10.1109/ICCV48922.2021.00356

35. Torrey, L., Shavlik, J.: Transfer learning. In: Handbook of Research on Machine Learning Applications and Trends: Algorithms, Methods, and Techniques, pp. 242–264. IGI global (2010)

36. Vilalta, R., Drissi, Y.: A perspective view and survey of meta-learning. Artificial Intell. Rev. **18**(2), 77–95 (2002)

37. Wang, X., Huang, T.E., Darrell, T., Gonzalez, J.E., Yu, F.: Frustratingly simple few-shot object detection. arXiv preprint arXiv:2003.06957 (2020)

38. Wang, Y.X., Ramanan, D., Hebert, M.: Meta-learning to detect rare objects. In: Proceedings of the IEEE/CVF International Conference on Computer Vision, pp. 9925–9934 (2019)

39. Weiss, K., Khoshgoftaar, T.M., Wang, D.D.: A survey of transfer learning. J. Big Data **3**(1), 1–40 (2016). https://doi.org/10.1186/s40537-016-0043-6

40. Wu, A., Han, Y., Zhu, L., Yang, Y., Deng, C.: Universal-prototype augmentation for few-shot object detection. arXiv preprint arXiv:2103.01077 (2021)

41. Wu, J., Liu, S., Huang, D., Wang, Y.: Multi-scale positive sample refinement for few-shot object detection. In: Vedaldi, A., Bischof, H., Brox, T., Frahm, J.-M. (eds.) ECCV 2020. LNCS, vol. 12361, pp. 456–472. Springer, Cham (2020). https://doi. org/10.1007/978-3-030-58517-4_27

42. Xiao, Y., Marlet, R.: Few-shot object detection and viewpoint estimation for objects in the wild. In: Vedaldi, A., Bischof, H., Brox, T., Frahm, J.-M. (eds.) ECCV 2020. LNCS, vol. 12362, pp. 192–210. Springer, Cham (2020). https://doi. org/10.1007/978-3-030-58520-4_12

43. Yan, X., Chen, Z., Xu, A., Wang, X., Liang, X., Lin, L.: Meta R-CNN: Towards general solver for instance-level low-shot learning. In: Proceedings of the IEEE/CVF International Conference on Computer Vision, pp. 9577–9586 (2019)

44. Yang, Z., Wang, Y., Chen, X., Liu, J., Qiao, Y.: Context-transformer: tackling object confusion for few-shot detection. In: Proceedings of the AAAI Conference on Artificial Intelligence, vol. 34, pp. 12653–12660 (2020)

45. Yun, S., Han, D., Chun, S., Oh, S.J., Yoo, Y., Choe, J.: CutMix: Regularization strategy to train strong classifiers with localizable features. In: 2019 IEEE/CVF International Conference on Computer Vision, ICCV 2019, Seoul, Korea (South), October 27 - November 2, 2019, pp. 6022–6031. IEEE (2019)

46. Zhang, W., Wang, Y.X.: Hallucination improves few-shot object detection. In: Proceedings of the IEEE/CVF Conference on Computer Vision and Pattern Recognition, pp. 13008–13017 (2021)

47. Zhao, Z.Q., Zheng, P., Xu, S.T., Wu, X.: Object detection with deep learning: a review. IEEE Trans. Neural Netw. Learn. Syst. **30**(11), 3212–3232 (2019)

48. Zhong, Y., Li, B., Tang, L., Kuang, S., Wu, S., Ding, S.: Detecting camouflaged object in frequency domain. In: Proceedings of the IEEE/CVF Conference on Computer Vision and Pattern Recognition (CVPR), pp. 4504–4513 (2022)

49. Zhong, Y., Li, B., Tang, L., Tang, H., Ding, S.: Highly efficient natural image matting. CoRR abs/2110.12748 (2021), https://arxiv.org/abs/2110.12748

50. Zhou, Q., et al.: TransVOD: end-to-end video object detection with spatial-temporal transformers. arXiv preprint arXiv:2201.05047 (2022)

Dual Contrastive Learning with Anatomical Auxiliary Supervision for Few-Shot Medical Image Segmentation

Huisi Wu[(✉)], Fangyan Xiao, and Chongxin Liang

College of Computer Science and Software Engineering,
Shenzhen University, Shenzhen, China
hswu@szu.edu.cn, {2070276124,2060271074}@email.szu.edu.cn

Abstract. Few-shot semantic segmentation is a promising solution for scarce data scenarios, especially for medical imaging challenges with limited training data. However, most of the existing few-shot segmentation methods tend to over rely on the images containing target classes, which may hinder its utilization of medical imaging data. In this paper, we present a few-shot segmentation model that employs anatomical auxiliary information from medical images without target classes for dual contrastive learning. The dual contrastive learning module performs comparison among vectors from the perspectives of prototypes and contexts, to enhance the discriminability of learned features and the data utilization. Besides, to distinguish foreground features from background features more friendly, a constrained iterative prediction module is designed to optimize the segmentation of the query image. Experiments on two medical image datasets show that the proposed method achieves performance comparable to state-of-the-art methods. Code is available at: https://github.com/cvszusparkle/AAS-DCL_FSS.

Keywords: Few-shot segmentation · Medical image segmentation · Contrastive learning · Anatomical information

1 Introduction

Automatic segmentation of medical images is widely used in clinical applications such as lesion localization, disease diagnosis, and prognosis. Recently, although segmentation networks based on fully supervised deep learning have achieved excellent performance [7,34,48,49], they rely on large amounts of pixel-level annotations and are difficult to directly be applied to segment unseen classes. For medical images, especially CT and MRI, labeling images is often expensive

Supplementary Information The online version contains supplementary material available at https://doi.org/10.1007/978-3-031-20044-1_24.

and even requires years of clinical experience by experts. And it is generally impractical to retrain a model for each tissue or organ class. These problems lead to the challenges of medical image segmentation with few manual annotations and poor model generalization ability.

Few-shot learning is proposed as a potential developable scheme to alleviate those challenges, and the segmentation method based on few-shot learning is called few-shot semantic segmentation (FSS) [32]. The idea is to adopt the prior knowledge distilled on labeled samples (denoted as *support*) to segment unlabeled samples (denoted as *query*). Therefore, FSS often learns few-shot *tasks* composed of *base* classes in an episodic training manner, and segments *unseen* classes in the form of *tasks* in the inference stage.

Few-shot segmentation methods have made considerable progress in recent years, but most of them are applied to natural images [17,21,29,39,40,44,46,50], while FSS approaches for medical imaging are still developing [9,24,28,31,35, 36,51]. There are the following possible factors: due to the low contrast, the fundamental difference in modalities and the less amount of data of medical images (such as CT or MRI), methods based on natural images cannot be directly applied to medical images; due to most of the current FSS methods for medical imaging rely on images with target classes, while medical image datasets contain limited target classes images, and some images without target classes are discarded, resulting in low data availability. However, those images without target classes (denoted as *non-target* images in this paper) may be rich in some anatomical knowledge. Therefore, how to efficiently employ the existing medical image data (i.e. try not to discard any data) and design a segmentation network suitable for the characteristics of medical images are quite significant. Nevertheless, so far, it seems that only SSL-ALPNet [28] considered applying those medical images without target classes to FSS, other methods still only depend on images containing target classes [9,31,35,36].

In order to enhance data availability and improve segmentation performance under the bottleneck of scarce training data, we consider the acquisition of meaningful knowledge from *non-target* images for contrastive learning. Intuitively, the information of irrelevant tissues contained in non-target slices are the anatomical background cues of the target classes. Taking those non-target slices into consideration to construct contrastive learning can greatly increase the capability of the network for learning discriminative features. At present, contrastive learning has been widely used in visual tasks recently, such as [4,5,11,12]; it has also been gradually introduced into few-shot classification, such as [19,27]; but it is developing slowly in few-shot segmentation, especially for medical imaging. The possible reason is that it is not as simple to construct positive and negative samples for medical images as compared to natural images, and there are challenges in both the number of data samples and the characteristics of the image modalities. How to construct effective contrastive learning to help few-shot medical image segmentation may be a developable thought.

In this paper, we propose a scheme for designing *dual contrastive learning with anatomical auxiliary supervision*, called *AAS-DCL*. The purpose is to adopt the tissue knowledge in the images without target classes as negative samples, and the support guidance information as positive samples, to construct prototypical contrastive learning and contextual contrastive learning, to strengthen the discriminability of the features learned by the network and the data utilization. Furthermore, in order to more effectively discriminate foreground and background features, we also build a *constrained iterative prediction* (CIP) module to optimize the query segmentation. That is, by exploring the distribution consistency between the feature similarity map and mask similarity map to constrain the update of the prediction results of the classifier.

Our contributions are summarized as follows:

- We propose to make full use of the anatomical information in the medical images without target classes to construct *dual contrastive learning* (DCL), from both prototypical and contextual perspectives, to encourage the network to learn more discriminative features and improve the data utilization.
- We present a *constrained iterative prediction* module to optimize the prediction mask of the classifier, which can effectively discriminate the query features learned by the network equipped with the DCL module.
- Our proposed method set the start-of-the-art performance of few-shot segmentation in two famous medical image datasets.

2   Related Work

Few-Shot Semantic Segmentation. The FSS methods for **natural images** are emerging in endlessly [6,17,21,32,37,39,40,44,46,50]. OSLSM [32] proposed the pioneering two branches and generated weights from support images for few-shot segmentation; PL [6] proposed a prototypical framework tailored for few-shot natural image segmentation; PANet [40] designed an alignment loss to help the network learn the consistency between the support and query feature spaces. PMMs [44] applied the *Expectation-Maximization* algorithm to extract multiple partial prototypes to help the query image's segmentation; PPNet [21] proposed to build local prototypes based on superpixels; PFENet [37] designed to construct prior masks and effective feature enrichment module to reconcile the spatial inconsistency of query and support features. More recently, both RePRI [3] and CWT [22] considered some unique training schemes, such as transductive inference or multi-stage training, opening up some thought-provoking few-shot segmentation insights. And some of those innovative units for natural images may be conductive to few-shot medical image segmentation.

As for **medical imaging**, the FSS methods have also progressed in recent years, and most of them are implemented based on 2D frameworks [28,31,35,36,43,45]. As a pioneering work, sSENet [31] designed a strong interactive segmentation network tailored for the traits of medical images; GCN-DE [35] improved part of the structure in sSENet by applying a global correlation module with discriminative embedding; PoissonSeg [33] applied *Poisson* learning to propagate supervised signals and spatial consistency calibration for representative learning; SSL-ALPNet [28] proposed to adopt the idea of self-supervision without annotations, designed adaptive local prototypical network and obtained state-of-the-art effect. RP-Net [36] designed a clever model combined class-level prototypical network with iterative refinement to segment the query image and got the state-of-the-art performance. Besides, AKFNet [41] proposed to subtly use the anatomical knowledge from support images to guide the query segmentation. There are also FSS methods based on 3D frameworks, such as BiGRU [14], which proposed to capture the correlation between adjacent slices of medical volumes. It can be seen that most of the current few-shot segmentation methods rely on images with target classes. Differently, we propose to apply slices without target classes for auxiliary supervision to construct contrastive learning, which will help the network to learn more discriminative features.

Contrastive Learning. Contrastive learning aims to learn the similarity of various samples in embedding space, that is, to pull closer the more similar samples and push farther the dissimilar ones. Recent works have been increasingly used in computer vision tasks, including self-supervised ideas [5,11,12] and supervised methods [13,42]. SimCLR [4] applied different data augmentations on same images for self-supervised contrastive learning; MoCo [12] proposed a momentum encoder to increase memory banks to store considerable views of different data transformations. Recently, contrastive learning has also been adopted to few-shot learning tasks [10,19,23,27]. Liu et al. [19] proposed to use distinct data augmentations into a few-shot embedding network for contrastive learning; Ouali et al. [27] employed spatial features to build contrastive learning for few-shot classification. For few-shot segmentation, there is a slow development of contrastive learning. Liu et al. [20] presented to apply global and local contrastive losses to pre-train feature extractor for query prior features to help few-shot segmentation of natural images. DPCL [15] was designed as a dual prototypical contrastive learning network suitable for few-shot natural image segmentation. By exploring the traits and challenges of medical image datasets, we design contrastive learning for few-shot medical image segmentation.

Superpixel Segmentation. Superpixels are pixel groups generated by clustering pixels using statistical methods according to image features. Based on superpixels, one image can be segmented to different degrees, which can generate pseudo-labels to provide effective supervision for unlabeled scenarios. There are some popular methods for superpixel segmentation, such as SLIC [1], SEEDS [2], Graph-cuts [25] and Felzenszwalb's [8]. SSL-ALPNet [28] applied the way [8] to produce pseudo-labels of unlabeled images by superpixel segmentation for self-supervision; Li et al. [18] employed the same method [8] to generate superpixels as pseudo- classes and construct self-supervised tasks. In this work, different superpixel segmentation algorithms will be studied to generate pseudo-labels for *non-target* medical slices to construct contrastive learning.

3   Method

3.1   Problem Definition

The general formulation of few-shot semantic segmentation (FSS) is mainly followed by [32]. In the FSS setting, the idea is to train a model on an annotated

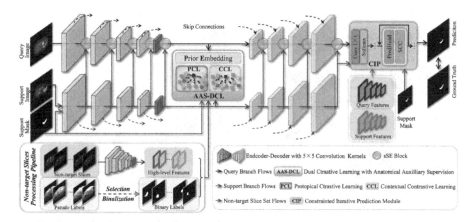

Fig. 1. Workflow of the proposed framework for few-shot medical image segmentation, which is based on the baseline sSENet [31] with skip connections, and instrumental schemes are designed including AAS-DCL and CIP modules

training dataset $\mathcal{D}_{train}$ that can perform segmentation well on the testing data $\mathcal{D}_{test}$ with a few labeled samples, without re-training. The training classes $\mathcal{C}_{train}$ have no overlap with the testing classes $\mathcal{C}_{test}$, *i.e.*, $\mathcal{C}_{train} \cap \mathcal{C}_{test} = \emptyset$. The episodic training is adopted in FSS, where the model is trained with many epochs and one epoch contains many episodes. To be specific, in each episode, the few-shot learning consists of a support and query data pair, *i.e.*, $\mathcal{D}_{train}$ is composed of the support set $\mathcal{S}_{tr}$ and query set $\mathcal{Q}_{tr}$. Here, $\mathcal{S}_{tr} = \{(\mathbf{x}^s(c_i), \mathbf{y}^s(c_i))\}$ and $\mathcal{Q}_{tr} = \{(\mathbf{x}^q(c_i))\}$, where $x \in \mathcal{X}$ refers to the image and $\mathbf{y} \in \mathcal{Y}$ refers to corresponding binary mask, $i = 1, 2, ..., |c|$ is the number index of class $c \in \mathcal{C}_{train}$. In *inference*, $\mathcal{D}_{test}$ is defined in the same mode but for testing images and masks with the unseen class set $\mathcal{C}_{test}$. The background class (denoted as c_0) is not counted in $\mathcal{C}_{train}$ and $\mathcal{C}_{test}$.

3.2 Network Overview

Inspired by the strong interactive structure designed by sSENet [31] for medical images, we adopt it as a basic network and improve it subtly. As shown in Fig. 1, the framework includes the following parts: (1) Encoder-decoder structures: feature extraction and reconstruction referring to the sSENet with added skip connections; (2) *Non-target slices processing pipeline*: generating pseudo-labels for non-target slices and extracting features for randomly selected partial slices of *non-target slice set*; (3) *Dual contrastive learning with anatomical auxiliary supervision* (AAS-DCL): prototypes generation and feature processing for prototypical and contextual contrastive learning, respectively; (4) *Constrained iterative prediction* (CIP): a designed classifier for iterative query prediction.

Non-target Slices Processing Pipeline. For CT and MRI datasets, most raw scans have several slices without target organs, and their masks are all black. We define the slices as *non-target slice set* $\mathcal{D}_{nt}$. Since those non-target slices are rich

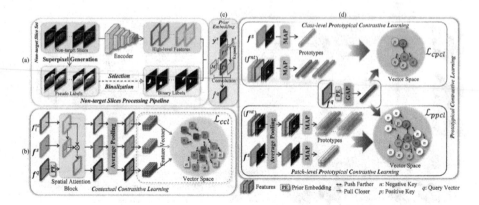

Fig. 2. The diagram of dual contrastive learning with anatomical auxiliary supervision (AAS-DCL). (a) Non-target slices processing pipeline: applying superpixel segmentation to generate pseudo-labels and then selecting randomly some slices to extract features by the encoder. (b) The workflow of contextual contrastive learning (CCL). (c) The structure of the prior embedding (PE) module. (d) The workflow of the prototypical contrastive learning, including two sub-modules (CPCL and PPCL)

in anatomical knowledge about irrelevant organs or tissues, they are inherently discriminative from target classes. To effectively utilize these information, as shown in Fig. 2(a), we first use a superpixel segmentation algorithm ([1,2] or [8]) to produce pseudo-labels offline for $\mathcal{D}_{nt}$. Besides, K non-target slices are randomly selected in each training episode, defined as $\mathcal{X}_{nt} = \{(\mathbf{x}^{nt})\}^K$. For their pseudo-labels, one superpixel (denoted as a pseudo-class) in each pseudo-label is randomly selected to binarize the label, and the K binary pseudo-labels are defined as $\mathcal{Y}_{nt} = \{(\mathbf{y}^{nt})\}^K$. Then the non-target slices $\mathcal{X}_{nt}$ are sent to the network encoder to extract non-target features $\mathcal{F}_{nt} = \{(\boldsymbol{f}^{nt})\}^K$; finally, the pseudo-labels $\mathcal{Y}_{nt}$ and the feature maps $\mathcal{F}_{nt}$ are input into the AAS-DCL scheme.

3.3 Dual Contrastive Learning with Anatomical Auxiliary Supervision

Due to the low contrast of medical images, the demarcation between the target class and the background tissues is not obvious, which makes it difficult to accurately segment target organs. Therefore, we explore the utilization of non-target slices with rich anatomical knowledge to provide more background guidance, which can constitute contrastive learning with query and support features. For one thing, that will help to enhance the discriminability of learned features of the model and improve the segmentation performance; for another, it will also take better advantage of medical image datasets in few-shot segmentation.

As shown in Fig. 2, in the AAS-DCL scheme, we design *prototypical contrastive learning* (PCL) and *contextual contrastive learning* (CCL) to form the DCL module in each training episode, from the perspectives of semantic class and spatial information, respectively, to make the features similar to the representation information of the target class closer, and the dissimilar ones are farther away. The contrastive losses applied in the DCL module all refer to infoNCE [26]:

$$\mathcal{L}_{cl}(q, k^+, k^-) = -\log \frac{\exp(q \cdot k^+ / \tau)}{\sum_{i=0}^{K} \exp(q \cdot k_i^- / \tau)} \tag{1}$$

where K denotes the number of negative keys and τ is a temperature factor.

Prior Embedding. To further guide the activation of foreground class-specific information in query features, inspired by PFENet [37], we design the *Prior Embedding* (PE) (Fig. 2(c)). The module needs to obtain the confidence map $\mathcal{M}$ by the normalized similarity map between the support and query feature [37], and the class-level prototype p^s generated from the support feature and its mask. Then the similarity map $\mathcal{M}$, the expanded prototype $\mathcal{E}(p^s)$ and the query feature f^q are concatenated and convolved to get an enhanced query feature $\hat{f}^q$.

Prototypical Contrastive Learning. Prototypes are vectors rich in semantic information; and most few-shot methods obtain prototypes by global average pooling of features and corresponding masks, which are often denoted as class-level prototypes to guide the segmentation of the query image. However, since class-level prototypes are prone to lose intra-class information, many few-shot methods [21,28,44] proposed to produce partial prototypes to retain sufficient intra-class cues. Our PCL includes two sub-modules: *class-level prototypical contrastive learning* (CPCL) and *patch-level prototypical contrastive learning* (PPCL), which comprehensively discriminate semantic information from the global and local prototypes, respectively.

(i) **Class-level Prototypical Contrastive Learning.** Class-level prototype-based learning can distinguish foreground and background features from the perspective of the overall semantic class. To be specific, the support feature and support mask, non-target features and the corresponding binary pseudo-labels are used to generate their class-level prototypes by masked averaged pooling (MAP) operation [50], respectively. And the query feature $\hat{f}^q$ is used to obtain a mean vector v^q by global average pooling (GAP). In addition, the vector is regarded as the *query vector*; the support prototype is regarded as a positive key; and those non-target prototypes are regarded as negative keys. As shown in the CPCL of Fig. 2(d), the contrastive loss is formulated as:

$$\mathcal{L}_{cpcl} = \mathcal{L}_{cl}(v^q, p^s, p^{nt}) = -\log \frac{\exp(v^q \cdot p^s / \tau)}{\sum_{i=0}^{K} \exp(v^q \cdot p_i^{nt} / \tau)} \tag{2}$$

(ii) **Patch-level Prototypical Contrastive Learning.** We propose the PPCL for two reasons: for one thing, since the class-level prototypes obtained by MAP will filter out other background information around the target class, which may be utilized as negative samples to enhance the discriminativeness of the semantics around the target class, so the PPCL is used to remedy this problem; for another, because contrastive learning largely requires amounts of effective negative samples, it is not sufficient to only consider class-level prototypes. The generation of patch-level prototypes is inspired by the local prototypes in [28]. First, the entire feature map is evenly divided based on patches, and then MAP is applied to each feature patch and the mask patch of the corresponding position to obtain the local prototypes, we called them patch-level prototypes. Given the patch size (D_H, D_H) and a feature map $f \in \mathbb{R}^{C \times H \times W}$, the process is formulated as follows:

$$p_{d_j} = \frac{\sum_{(h,w) \in d_j} f_{d_j}(h,w) \cdot y_{d_j}(h,w)}{\sum_{(h,w) \in d_j} y_{d_j}(h,w)} \tag{3}$$

where the (h, w) are spatial coordinates in each patch, p_{d_j} refers to the prototype in a patch d_j and $j = 1, 2, ..., \frac{H}{D_H}$ denotes the patch index.

The support feature and its mask can obtain both some patch-level background prototypes $\{p_d^{s-}\}$ and the patch-level foreground prototype p_d^{s+} by setting a threshold [28]. Similarly, each non-target feature and corresponding pseudo-label are also used to produce mickle patch-level prototypes $\{p_d^{nt}\}$. The query feature $\hat{f}^q$ is used to generate a mean vector v^q by GAP operation. When building the contrastive loss, the support foreground prototype p_d^{s+} is regarded as a positive key, and other support patch-level prototypes $\{p_d^{s-}\}$ and all non-target patch-level prototypes $\{p_d^{nt}\}$ are regarded as negative keys. This scheme increases the number of prototype samples, which is beneficial to learning the correlation between local features of similar classes and the discriminativeness among local features of dissimilar semantic classes. As shown in the PPCL of Fig. 2(d), the contrastive loss is written as:

$$\mathcal{L}_{ppcl} = \mathcal{L}_{cl}\left(v^q, p_d^{s+}, (p_d^{s-}, p_d^{nt})\right)$$

$$= -\log \frac{\exp\left(v^q \cdot p_d^{s+}/\tau\right)}{\sum_{i=0}^{K} \exp\left(v^q \cdot p_d^{nt}(i)/\tau\right) + \sum_{i=0}^{(\frac{D_H}{H}-1)} \exp\left(v^q \cdot p_d^{s-}(i)/\tau\right)} \tag{4}$$

Contextual Contrastive Learning. In order to increase the discriminativeness of context features, we propose the contextual contrastive learning (CCL), as shown in Fig. 2(b). Specifically, the support feature, enhanced query feature $\hat{f}^q$ and non-target features are first processed by a spatial attention block [30], to make all feature maps focus on richer contextual information. Then, these processed features are averagely pooled to obtain different feature vector sets, which will be used to construct contrastive learning. When calculating once a CCL loss, a query feature vector $\overline{f}_j^q$ is regarded as the *query vector*, the support feature vector $\overline{f}_j^s$ at the same position is regarded as a positive key, and all non-target feature vectors are regarded as negative keys. The number of loss calculations N is equal to the number of query feature vectors, and finally the mean value of the cumulative sum of all CCL losses obtained is adopted as the output contextual contrastive loss. The formula is as follows:

$$\mathcal{L}_{ccl} = \frac{1}{N} \sum_{j=0}^{N} \mathcal{L}_{cl}^j(\overline{f}_j^q, \overline{f}_j^s, \overline{f}^{nt}) = \frac{1}{N} \sum_{j=0}^{N} \left[-\log \frac{\exp\left(\overline{f}_j^q \cdot \overline{f}_j^s/\tau\right)}{\sum_{i=0}^{K} \exp\left(\overline{f}_j^q \cdot \overline{f}_i^{nt}/\tau\right)} \right] \tag{5}$$

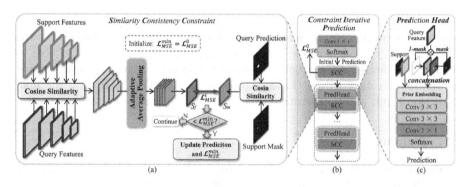

Fig. 3. The diagram of constrained iterative prediction (CIP). (a) The details of similarity consistency constraint (SCC). (b) The main iteration optimization process of CIP. (c) The structure of the designed classifier to generate the query prediction

3.4   Constrained Iterative Prediction

The design of the dual contrastive learning module is beneficial to enhance the features' discriminability of the encoder-decoder. To encourage the foreground and background of query feature to be segmented more effectively on the final classifier, so the CIP module is proposed to optimize the part, which can also as an assistant to the DCL module. As shown in Fig. 3(b), the basic components include *similarity consistency constraint* (SCC) and *prediction head*.

Prediction Head. As shown in Fig. 3(c), unlike only a generic classifier (1×1 convolution and softmax), in order to predict more meaningfully, we consider integrating the query prediction into the query feature by convolution, which is derived from the iterative optimization idea of RP-Net [36] and CANet [47]. RP-Net fuses the prediction and the high-level query feature by a correlation matrix, where a prototype-based classifier is further employed to obtain a new prediction. And CANet fuses the prediction and the middle-level query feature by residual concatenation, where an ASPP is further used to produce a new prediction. Differently, we first utilize the result of the generic classifier as the initial prediction mask, fuse it with the query feature of the decoder's tail and send into the *prior embedding*(PE) module, where the new segmentation result is obtained by a series of convolution operations. The application of PE is to make use of the guidance information of the support feature to promote the fusion of the predicted query mask. The new query prediction will be determined to be updated through the following constraints in SCC.

Similarity Consistency Constraint. As shown in Fig. 3(a), we consider the information distribution constraints and first compute two similarity maps, S_m and S_f. The S_m is calculated between the predicted query mask and the support mask; S_f is calculated from the support and query features obtained by the encoder. Intuitively, the probability value distributions on $\mathcal{S}_m$ and $\mathcal{S}_f$ should be as consistent as possible, because a more consistent distribution means that the predicted query mask is closer to ground truth. We minimize the MSE loss to quantify the better distribution consistency between $\mathcal{S}_m$ and $\mathcal{S}_f$, and employ an iterative strategy to use this loss to constrain whether to update the final prediction result and the iteration times I is experimentally set to 5.

3.5   Training Strategy

For each episode, one support and query pair is selected according to the strategy in [31] to form a 1-way 1-shot few-shot setting, and K non-target slices and corresponding pseudo-labels are randomly selected. All pairs are input to the network for end-to-end learning. For loss function, we adopt cross-entropy loss to supervise the segmentation of query image. And three contrastive losses are employed to learn the feature discriminability of the model, so the total loss is:

$$\mathcal{L} = \mathcal{L}_{ce} + \lambda_{cpcl} \cdot \mathcal{L}_{cpcl} + \lambda_{ppcl} \cdot \mathcal{L}_{ppcl} + \lambda_{ccl} \cdot \mathcal{L}_{ccl} \qquad (6)$$

where the $\lambda_{cpcl}, \lambda_{ppcl}$ and λ_{ccl} are experimentaly set as 0.08, 0.12 and 0.06, respectively.

4   Experiments

4.1   Dataset and Evaluation Metric

In order to evaluate the effectiveness of the proposed method, we conducted experiments on two public medical image datasets: (1) **CHAOS-T2** is from the challenge (Task 5) of ISBI 2019 [38], and it contains 20 MRI scans of *T2SPIR* with 623 axial 2D slices in total, including 492 slices with target classes and 131 non-target slices. (2) **Synapse** is from the public challenge [16], and it contains 30 CT scans with 3779 axial abdominal slices, including 1755 slices with target classes involved in our experiments, the remaining are non-target slices.

Due to the 2D framework of our model, experiments are performed with images sliced from 3D scans along the 2D axis. Then five-fold cross-validation is applied in our experiments. To simulate the scarcity of annotated data in realistic situations, we performed 1-way 1-shot setting experiments for four organ classes, including the *liver, left kidney*(LK), *right kidney*(RK) and *spleen*. We employ one of the four classes as the unseen class and the other three classes as the base classes, so that four few-shot tasks are constructed. To evaluate the segmentation performance of 2D slices on 3D scans, we follow the protocol in the work [31] and apply the Dice coefficient as the major evaluation metric. In addition, we also conduct the statistical significance analysis, which is reported in the supplementary material.

Table 1. Statistical comparison (in Dice score %) of different methods on the validation sets for CHAOS-T2 and Synapse datasets

Method	CHAOS-T2					Synapse				
	Liver	RK	LK	Spleen	Mean	Liver	RK	LK	Spleen	Mean
PANet [40]	41.87	44.55	50.64	49.94	46.75	36.18	25.09	32.41	22.13	28.95
sSENet [31]	27.25	59.22	54.89	48.45	47.45	41.87	34.55	40.60	39.94	39.24
PoissonSeg [33]	61.03	53.57	50.58	52.85	54.51	58.74	47.02	50.11	52.33	52.05
GCN-DE [35]	49.47	83.03	76.07	60.63	67.30	46.77	**75.50**	**68.13**	56.53	61.73
SSL-ALPNet [28]	68.71	79.53	71.14	61.11	70.12	73.26	57.73	64.34	64.89	65.05
RP-Net [36]	69.73	82.06	**77.55**	73.90	75.81	**74.11**	68.32	63.75	65.24	67.85
Ours	**69.94**	**83.75**	76.90	**74.86**	**76.36**	71.61	69.95	64.71	**66.36**	**68.16**

4.2 Implementation Details

The network is implemented by Pytorch on an Nvidia RTX 2080Ti GPU. All 2D slices are resized to the resolution of 256 × 256. In PPCL, the number of patches is experimentally set to 4. By using an initial learning rate of 0.0001, we use the loss function (Equ. 6) for episodic training with a batch size of 1. And the learning rate is based on a polynomial decay learning rate policy. Besides, the Adam algorithm is employed to optimize the learning process with power = 0.95 and weight decay = 1e-7. The model is trained by 70 epochs, each of which contains the number of episodes equal to the scale of the training set. For data augmentation, we focus more on random modifying the image sharpness, with the visibility factor range of [0.2, 0.5] and the lightness factor range of [0.5, 1.0].

4.3 Comparisons with State-of-the-art Methods

To demonstrate the superiority of the proposed method, we compare our method with the classical and current state-of-the-art FSS methods, as shown in Table 1 and Fig. 4. The PANet [40] proposed an alignment regularization loss, which is utilized widely by other few-shot methods; sSENet [31] is the pioneering work of FSS for medical imaging; GCN-DE [35] employs global correlation module to optimize the structure of sSENet; PoissonSeg [33] considers *Poisson* learning and spatial consistency calibration for few-shot medical image segmentation; SSL-ALPNet [28] utilizes self-supervision and adaptive local prototypes for FSS; RP-Net [36] applies global prototypes with recurrent refinement for prediction. Since GCN-DE and PoissonSeg are not open source, we directly compare the statistical results from their original works; besides, the results of other FSS methods are obtained by re-experiments in a unified experimental environment.

Table 1 shows that compared with other few-shot methods of medical imaging, for the **CHAOS-T2** dataset, our method achieves relatively best performance in most target classes, including right kidney, liver and spleen, as well as the mean Dice score; for the **Synapse** dataset, our method also get the best Dice scores on left kidney, spleen and mean Dice. It is precisely because our method makes full use of anatomical auxiliary knowledge of non-target slices to help the

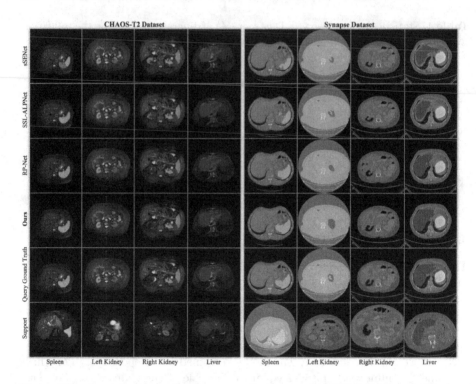

Fig. 4. Visual comparison of different methods on the CHAOS-T2 and Synapse datasets

network learn more discriminative features through contrastive learning, and applies the CIP module for more effective prediction to make the segmentation performance better.

4.4   Ablation Studies

Ablation experiments are performed on the CHAOS-T2 dataset. Based on the baseline sSENet [31], we focus on the roles of the proposed dual contrastive learning (DCL) and constrained iterative prediction (CIP), the impact of the sub-modules (CPCL, PPCL and CCL) in DCL, and other factors of the network.

Effect of the DCL Module. To demonstrate the effect of the DCL module, we performed ablation experiments on it, the data results are shown in Table 2 and Table 3. And Fig. 5 shows the visualization results of the DCL module, which applies the visualized feature map after 1×1 convolution and before the *softmax* function. Table 2 shows that the DCL can effectively enhance the Dice performance due to its promotion to the network's ability of learning discriminative features. For the sub-modules CPCL and PPCL in PCL, Table 3 and Fig. 5 both show that when the two kinds of prototypical contrastive losses are

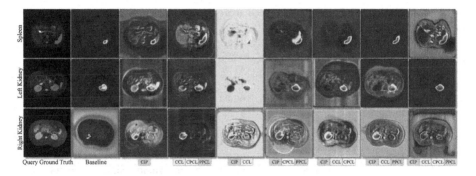

Fig. 5. Visual feature maps of different components based on the baseline‡(on the CHAOS-T2 dataset). The warmer colors represent the better discriminative features

Table 2. Ablation study results (in Dice score %) on the CHAOS-T2 dataset for the DCL and CIP modules. Baseline†: sSENet [31]; Baseline‡: sSENet with skip connections

Methods	Liver	RK	LK	Spleen	Mean
Baseline†[31]	27.25	59.22	54.89	48.45	47.45
Baseline‡	30.21	62.47	55.96	47.68	49.08
Baseline‡+CIP	51.02	69.84	63.77	63.65	62.07
Baseline‡+DCL	56.91	75.23	67.19	69.52	67.21
Baseline‡+DCL+CIP	**69.94**	**83.75**	**76.90**	**74.86**	**76.36**

Table 3. Ablation study results (in Dice score %) on the CHAOS-T2 dataset for sub-modules (CPCL, PPCL and CCL) of the DCL module

CCL	CPCL	PPCL	Liver	RK	LK	Spleen	Mean
✓			59.25	71.64	69.04	65.66	66.40
	✓	✓	67.63	79.14	71.09	70.79	72.16
✓	✓		68.00	80.26	72.22	71.85	73.08
✓		✓	68.53	80.80	73.17	72.54	73.76
✓	✓	✓	**69.94**	**83.75**	**76.90**	**74.86**	**76.36**

combined concurrently, the performance gets better. Because the two can assist each other: CPCL is conducive to learning the gap among semantic classes, and PPCL effectively utilizes local prototypes to improve the identifiability among the local features within classes. Then the CCL can increase the discriminability of contextual features and also help improve the segmentation performance.

Importance of the CIP Module. Table 2 and Fig. 5 also show the ablation experimental results of CIP. It can be seen that the design of CIP can help the network to further improve the segmentation performance. The PerdHead and SCC cooperate to perform iterative optimization, which is quite significant for discriminative features learned by DCL in the network. More Dice performances under different iteration numbers are shown in Fig. 6.

Influence of Other Factors. (1) For the **superpixel segmentation algorithms** generating pseudo-labels for non-target slices, we adopt the SLIC [1], SEEDS [2] and Felzenszwalb's [8] to produce pseudo-labels respectively, the results are shown in Table 4. It is shown that Felzenszwalb's method gets the relatively best Dice score; because it could generate more irregularly shaped superpixel pseudo-labels that better fit anatomical contours, enabling our PCL

Table 4. Ablation study results (in Dice score) on the CHAOS-T2 dataset for other factors

Alation Factors	Settings	Liver(%)	RK(%)	LK(%)	Spleen(%)	Mean(%)
Superpixel Segmentation	SLIC [1]	45.23	70.45	68.72	61.00	61.35
	SEEDS [2]	49.05	80.83	71.19	67.22	67.07
	Felzenszwalb's [8]	**69.94**	**83.75**	**76.90**	**74.86**	**76.36**
Iteration Numbers	I = 3	67.53	78.47	72.39	71.93	72.58
	I = 5	69.94	**83.75**	**76.90**	**74.86**	**76.36**
	I = 7	**70.11**	83.58	76.17	74.62	76.12
Number of Slices	K = 1	64.60	76.34	71.08	69.33	70.34
	K = 3	**69.94**	**83.75**	**76.90**	**74.86**	**76.36**
	K = 5	69.35	82.83	75.81	71.22	74.80

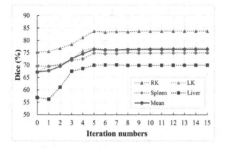

Fig. 6. Dice performance under different iteration numbers in the CIP module (on the CHAOS-T2 dataset)

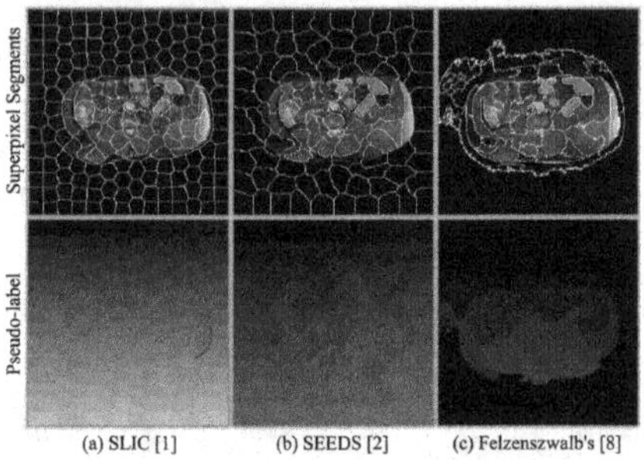

Fig. 7. Examples of superpixels and pseudo-labels obtained by different methods on a same non-target slice

to efficiently utilize them. The superpixels generated by SLIC are too regular and not friendly to the learning of anatomical information. The SEEDs-based method has a moderate performance. Moreover, the superpixels produced by the latter two would contain more all-black backgrounds, which may also affect the contrastive learning. So we employ the Felzenszwalb's as main algorithm for our experiments. Examples of related superpixels and pseudo-labels are shown in Fig. 7. (2) For the **iteration numbers** of CIP module, we employ different numbers for ablation experiments, the results are shown in Table 4. To get a trade-off between the performance and memory usage, we utilize five iterations in our experiments. (3) For **the number of randomly selected non-target slices**, we consider the memory usage and performance and apply $K = 3$ for main experiments, the results are shown in Table 4.

5 Conclusion

We presented a network of dual contrastive learning with anatomical auxiliary information from medical images without target classes, including prototypical and contextual contrastive learning, and enhanced the discriminability of learned features and the data utilization. Besides, the designed constrained iterative prediction module optimized the query segmentation result. Experiments show the superiority of the proposed method for CT and MRI datasets. Moreover, The contrastive learning with the anatomical supervision from non-target images may provide a developable insight for few-shot medical image segmentation.

Acknowledgments. This work was supported in part by the National Natural Science Foundation of China under Grant 61973221 and Grant 62273241, the Natural Science Foundation of Guangdong Province of China under Grant 2018A030313381 and Grant 2019A1515011165.

References

1. Achanta, R., Shaji, A., Smith, K., Lucchi, A., Fua, P., Süsstrunk, S.: Slic superpixels. Technical report (2010)
2. Van den Bergh, M., Boix, X., Roig, G., de Capitani, B., Van Gool, L.: SEEDS: superpixels extracted via energy-driven sampling. In: Fitzgibbon, A., Lazebnik, S., Perona, P., Sato, Y., Schmid, C. (eds.) ECCV 2012. LNCS, vol. 7578, pp. 13–26. Springer, Heidelberg (2012). https://doi.org/10.1007/978-3-642-33786-4_2
3. Boudiaf, M., Kervadec, H., Masud, Z.I., Piantanida, P., Ben Ayed, I., Dolz, J.: Few-shot segmentation without meta-learning: a good transductive inference is all you need? In: Proceedings of the IEEE/CVF Conference on Computer Vision and Pattern Recognition, pp. 13979–13988 (2021)
4. Chen, T., Kornblith, S., Norouzi, M., Hinton, G.: A simple framework for contrastive learning of visual representations. In: International Conference on Machine Learning, pp. 1597–1607. PMLR (2020)
5. Chen, X., Fan, H., Girshick, R., He, K.: Improved baselines with momentum contrastive learning. arXiv preprint arXiv:2003.04297 (2020)
6. Dong, N., Xing, E.P.: Few-shot semantic segmentation with prototype learning. In: BMVC, vol. 3 (2018)
7. Fang, X., Yan, P.: Multi-organ segmentation over partially labeled datasets with multi-scale feature abstraction. IEEE Trans. Med. Imaging **39**(11), 3619–3629 (2020)
8. Felzenszwalb, P.F., Huttenlocher, D.P.: Efficient graph-based image segmentation. Int. J. Comput. Vis. **59**(2), 167–181 (2004)
9. Feng, R., et al.: Interactive few-shot learning: limited supervision, better medical image segmentation. IEEE Trans. Med. Imaging (2021)
10. Gao, Y., Fei, N., Liu, G., Lu, Z., Xiang, T.: Contrastive prototype learning with augmented embeddings for few-shot learning. In: Uncertainty in Artificial Intelligence, pp. 140–150. PMLR (2021)
11. Grill, J.B., et al.: Bootstrap your own latent-a new approach to self-supervised learning. Adv. Neural. Inf. Process. Syst. **33**, 21271–21284 (2020)
12. He, K., Fan, H., Wu, Y., Xie, S., Girshick, R.: Momentum contrast for unsupervised visual representation learning. In: Proceedings of the IEEE/CVF Conference on Computer Vision and Pattern Recognition, pp. 9729–9738 (2020)
13. Khosla, P., et al.: Supervised contrastive learning. Adv. Neural. Inf. Process. Syst. **33**, 18661–18673 (2020)

14. Kim, S., An, S., Chikontwe, P., Park, S.H.: Bidirectional RNN-based few shot learning for 3D medical image segmentation. arXiv preprint arXiv:2011.09608 (2020)
15. Kwon, H., Jeong, S., Kim, S., Sohn, K.: Dual prototypical contrastive learning for few-shot semantic segmentation. arXiv preprint arXiv:2111.04982 (2021)
16. Landman, B., Xu, Z., Igelsias, J.E., Styner, M., Langerak, T., Klein, A.: Miccai multi-atlas labeling beyond the cranial vault-workshop and challenge. In: Proceedings of the MICCAI: Multi-Atlas Labeling Beyond Cranial Vault-Workshop Challenge (2015)
17. Li, G., Jampani, V., Sevilla-Lara, L., Sun, D., Kim, J., Kim, J.: Adaptive prototype learning and allocation for few-shot segmentation. In: Proceedings of the IEEE/CVF Conference on Computer Vision and Pattern Recognition, pp. 8334–8343 (2021)
18. Li, Y., Data, G.W.P., Fu, Y., Hu, Y., Prisacariu, V.A.: Few-shot semantic segmentation with self-supervision from pseudo-classes. arXiv preprint arXiv:2110.11742 (2021)
19. Liu, C., et al.: Learning a few-shot embedding model with contrastive learning. In: Proceedings of the AAAI Conference on Artificial Intelligence, vol. 35, pp. 8635–8643 (2021)
20. Liu, W., Wu, Z., Ding, H., Liu, F., Lin, J., Lin, G.: Few-shot segmentation with global and local contrastive learning. arXiv preprint arXiv:2108.05293 (2021)
21. Liu, Y., Zhang, X., Zhang, S., He, X.: Part-aware prototype network for few-shot semantic segmentation. In: Vedaldi, A., Bischof, H., Brox, T., Frahm, J.-M. (eds.) ECCV 2020. LNCS, vol. 12354, pp. 142–158. Springer, Cham (2020). https://doi.org/10.1007/978-3-030-58545-7_9
22. Lu, Z., He, S., Zhu, X., Zhang, L., Song, Y.Z., Xiang, T.: Simpler is better: few-shot semantic segmentation with classifier weight transformer. In: Proceedings of the IEEE/CVF International Conference on Computer Vision, pp. 8741–8750 (2021)
23. Majumder, O., Ravichandran, A., Maji, S., Achille, A., Polito, M., Soatto, S.: Supervised momentum contrastive learning for few-shot classification. arXiv preprint arXiv:2101.11058 (2021)
24. Mondal, A.K., Dolz, J., Desrosiers, C.: Few-shot 3D multi-modal medical image segmentation using generative adversarial learning. arXiv preprint arXiv:1810.12241 (2018)
25. Moore, A.P., Prince, S.J., Warrell, J., Mohammed, U., Jones, G.: SuperPixel lattices. In: 2008 IEEE Conference on Computer Vision and Pattern Recognition, pp. 1–8. IEEE (2008)
26. Van den Oord, A., Li, Y., Vinyals, O.: Representation learning with contrastive predictive coding. arXiv e-prints pp. arXiv-1807 (2018)
27. Ouali, Y., Hudelot, C., Tami, M.: Spatial contrastive learning for few-shot classification. In: Oliver, N., Pérez-Cruz, F., Kramer, S., Read, J., Lozano, J.A. (eds.) ECML PKDD 2021. LNCS (LNAI), vol. 12975, pp. 671–686. Springer, Cham (2021). https://doi.org/10.1007/978-3-030-86486-6_41
28. Ouyang, C., Biffi, C., Chen, C., Kart, T., Qiu, H., Rueckert, D.: Self-supervision with superpixels: training few-shot medical image segmentation without annotation. In: Vedaldi, A., Bischof, H., Brox, T., Frahm, J.-M. (eds.) ECCV 2020. LNCS, vol. 12374, pp. 762–780. Springer, Cham (2020). https://doi.org/10.1007/978-3-030-58526-6_45
29. Rakelly, K., Shelhamer, E., Darrell, T., Efros, A.A., Levine, S.: Few-shot segmentation propagation with guided networks. arXiv preprint arXiv:1806.07373 (2018)

30. Roy, A.G., Navab, N., Wachinger, C.: Recalibrating fully convolutional networks with spatial and channel "squeeze and excitation" blocks. IEEE Trans. Med. Imaging **38**(2), 540–549 (2018)

31. Roy, A.G., Siddiqui, S., Pölsterl, S., Navab, N., Wachinger, C.: 'squeeze & excite'guided few-shot segmentation of volumetric images. Med. Image Anal. **59**, 101587 (2020)

32. Shaban, A., Bansal, S., Liu, Z., Essa, I., Boots, B.: One-shot learning for semantic segmentation. arXiv preprint arXiv:1709.03410 (2017)

33. Shen, X., Zhang, G., Lai, H., Luo, J., Lu, J., Luo, Y.: PoissonSeg: semi-supervised few-shot medical image segmentation via poisson learning. In: 2021 IEEE International Conference on Bioinformatics and Biomedicine (BIBM), pp. 1513–1518. IEEE (2021)

34. Shi, G., Xiao, L., Chen, Y., Zhou, S.K.: Marginal loss and exclusion loss for partially supervised multi-organ segmentation. Med. Image Anal. **70**, 101979 (2021)

35. Sun, L., et al.: Few-shot medical image segmentation using a global correlation network with discriminative embedding. Comput. Biol. Med. **140**, 105067 (2022)

36. Tang, H., Liu, X., Sun, S., Yan, X., Xie, X.: Recurrent mask refinement for few-shot medical image segmentation. arXiv preprint arXiv:2108.00622 (2021)

37. Tian, Z., Zhao, H., Shu, M., Yang, Z., Li, R., Jia, J.: Prior guided feature enrichment network for few-shot segmentation. IEEE Trans. Pattern Anal. Mach. Intell. (2020)

38. Valindria, V.V., et al.: Multi-modal learning from unpaired images: application to multi-organ segmentation in CT and MRI. In: 2018 IEEE Winter Conference on Applications of Computer Vision (WACV), pp. 547–556. IEEE (2018)

39. Wang, H., Zhang, X., Hu, Y., Yang, Y., Cao, X., Zhen, X.: Few-shot semantic segmentation with democratic attention networks. In: Vedaldi, A., Bischof, H., Brox, T., Frahm, J.-M. (eds.) ECCV 2020. LNCS, vol. 12358, pp. 730–746. Springer, Cham (2020). https://doi.org/10.1007/978-3-030-58601-0_43

40. Wang, K., Liew, J.H., Zou, Y., Zhou, D., Feng, J.: PANet: Few-shot image semantic segmentation with prototype alignment. In: Proceedings of the IEEE/CVF International Conference on Computer Vision, pp. 9197–9206 (2019)

41. Wei, Y., Tian, J., Zhong, C., Shi, Z.: AKFNET: an anatomical knowledge embedded few-shot network for medical image segmentation. In: 2021 IEEE International Conference on Image Processing (ICIP), pp. 11–15. IEEE (2021)

42. Wu, Z., Efros, A.A., Yu, S.X.: Improving generalization via scalable neighborhood component analysis. In: European Conference on Computer Vision, pp. 685–701 (2018)

43. Xiao, J., Xu, H., Zhao, W., Cheng, C., Gao, H.: A prior-mask-guided few-shot learning for skin lesion segmentation. Computing, pp. 1–23 (2021)

44. Yang, B., Liu, C., Li, B., Jiao, J., Ye, Q.: Prototype mixture models for few-shot semantic segmentation. In: Vedaldi, A., Bischof, H., Brox, T., Frahm, J.-M. (eds.) ECCV 2020. LNCS, vol. 12353, pp. 763–778. Springer, Cham (2020). https://doi.org/10.1007/978-3-030-58598-3_45

45. Yu, Q., Dang, K., Tajbakhsh, N., Terzopoulos, D., Ding, X.: A location-sensitive local prototype network for few-shot medical image segmentation. In: 2021 IEEE 18th International Symposium on Biomedical Imaging (ISBI), pp. 262–266. IEEE (2021)

46. Zhang, B., Xiao, J., Qin, T.: Self-guided and cross-guided learning for few-shot segmentation. In: Proceedings of the IEEE/CVF Conference on Computer Vision and Pattern Recognition, pp. 8312–8321 (2021)

47. Zhang, C., Lin, G., Liu, F., Yao, R., Shen, C.: CANet: Class-agnostic segmentation networks with iterative refinement and attentive few-shot learning. In: Proceedings of the IEEE/CVF Conference on Computer Vision and Pattern Recognitionm, pp. 5217–5226 (2019)
48. Zhang, D., Huang, G., Zhang, Q., Han, J., Han, J., Yu, Y.: Cross-modality deep feature learning for brain tumor segmentation. Pattern Recogn. **110**, 107562 (2021)
49. Zhang, L., et al.: Generalizing deep learning for medical image segmentation to unseen domains via deep stacked transformation. IEEE Trans. Med. Imaging **39**(7), 2531–2540 (2020)
50. Zhang, X., Wei, Y., Yang, Y., Huang, T.S.: SG-One: similarity guidance network for one-shot semantic segmentation. IEEE Trans. Cybern. **50**(9), 3855–3865 (2020)
51. Zhao, A., Balakrishnan, G., Durand, F., Guttag, J.V., Dalca, A.V.: Data augmentation using learned transformations for one-shot medical image segmentation. In: Proceedings of the IEEE/CVF Conference on Computer Vision and Pattern Recognition, pp. 8543–8553 (2019)

Improving Few-Shot Learning Through Multi-task Representation Learning Theory

Quentin Bouniot[1,2(✉)] ⓘ, Ievgen Redko[2] ⓘ, Romaric Audigier[1] ⓘ,
Angélique Loesch[1] ⓘ, and Amaury Habrard[2,3] ⓘ

[1] Université Paris-Saclay, CEA, List, 91120 Palaiseau, France
{quentin.bouniot,romaric.audigier,angelique.loesch}@cea.fr
[2] Université de Lyon, UJM-Saint-Etienne, CNRS, IOGS, Laboratoire Hubert Curien
UMR 5516, 42023 Saint-Etienne, France
{quentin.bouniot,ievgen.redko,amaury.habrard}@univ-st-etienne.fr
[3] Institut Universitaire de France (IUF), Paris, France

Abstract. In this paper, we consider the framework of multi-task representation (MTR) learning where the goal is to use source tasks to learn a representation that reduces the sample complexity of solving a target task. We start by reviewing recent advances in MTR theory and show that they can provide novel insights for popular meta-learning algorithms when analyzed within this framework. In particular, we highlight a fundamental difference between gradient-based and metric-based algorithms in practice and put forward a theoretical analysis to explain it. Finally, we use the derived insights to improve the performance of meta-learning methods via a new spectral-based regularization term and confirm its efficiency through experimental studies on few-shot classification benchmarks. To the best of our knowledge, this is the first contribution that puts the most recent learning bounds of MTR theory into practice for the task of few-shot classification.

Keywords: Few-shot learning · Meta-learning · Multi-task learning

1 Introduction

Even though many machine learning methods now enjoy a solid theoretical justification, some more recent advances in the field are still in their preliminary state which requires the hypotheses put forward by the theoretical studies to be implemented and verified in practice. One such notable example is the success of *meta-learning*, also called *learning to learn* (LTL), methods where the goal is to produce a model on data coming from a set of (meta-train) source tasks to use it

Supplementary Information The online version contains supplementary material available at https://doi.org/10.1007/978-3-031-20044-1_25.

as a starting point for learning successfully a new previously unseen (meta-test) target task. The success of many meta-learning approaches is directly related to their capacity of learning a good representation [37] from a set of tasks making it closely related to multi-task representation learning (MTR). For this latter, several theoretical studies [2,5,29,34,53] provided probabilistic learning bounds that require the amount of data in the meta-train source task *and* the number of meta-train tasks to tend to infinity for it to be efficient. While capturing the underlying general intuition, these bounds do not suggest that all the source data is useful in such learning setup due to the additive relationship between the two terms mentioned above and thus, for instance, cannot explain the empirical success of MTR in few-shot classification (FSC) task. To tackle this drawback, two very recent studies [15,45] aimed at finding deterministic assumptions that lead to faster learning rates allowing MTR algorithms to benefit from all the source data. Contrary to probabilistic bounds that have been used to derive novel learning strategies for meta-learning algorithms [2,53], there has been no attempt to verify the validity of the assumptions leading to the fastest known learning rates in practice or to enforce them through an appropriate optimization procedure.

In this paper, we aim to use the recent advances in MTR theory [15,45] to explore the inner workings of these popular meta-learning methods. Our rationale for such an approach stems from a recent work [47] proving that the optimization problem behind the majority of meta-learning algorithms can be written as an MTR problem. Thus, we believe that looking at meta-learning algorithms through the recent MTR theory lens, could provide us a better understanding for the capacity to work well in the few-shot regime. In particular, we take a closer look at two families of meta-learning algorithms, notably: gradient-based algorithms [6,10,26,28,31–33,37,38] including MAML [16] and metric-based algorithms [1,22,27,41–43,46] with its most prominent example given by PROTONET [42].

Our main contributions are then two-fold:

1. We empirically show that tracking the validity of assumptions on optimal predictors used in [15,45] reveals a striking difference between the behavior of gradient-based and metric-based methods in how they learn their optimal feature representations. We provide elements of theoretical analysis that explain this behavior and explain the implications of it in practice. Our work is thus complementary to Wang et al. [47] and connects MTR, FSC and Meta-Learning from both theoretical and empirical points of view.
2. Following the most recent advances in the MTR field that leads to faster learning rates, we show that theoretical assumptions mentioned above can be forced through simple yet effective learning constraints which improve performance of the considered algorithms for FSC baselines: gradient- and metric-based methods using episodic training, as well as non-episodic algorithm such as Multi-Task Learning (MTL [47]).

The rest of the paper is organized as follows. We introduce the MTR problem and the considered meta-learning algorithms in Sect. 2. In Sect. 3, we investigate and explain how they behave in practice. We further show that one can force meta-learning algorithms to satisfy such assumptions through adding an appropriate spectral regularization term to their objective function. In Sect. 4, we provide an experimental evaluation of several state-of-the-art meta-learning and MTR methods. We conclude and outline the future research perspectives in Sect. 5.

2   Preliminary Knowledge

2.1   Multi-task Representation Learning Setup

Given a set of T source tasks observed through finite size samples of size n_1 grouped into matrices $\mathbf{X}_t = (\mathbf{x}_{t,1}, \ldots, \mathbf{x}_{t,n_1}) \in \mathbb{R}^{n_1 \times d}$ and vectors of outputs $y_t = (y_{t,1}, \ldots, y_{t,n_1}) \in \mathbb{R}^{n_1}$, $\forall t \in [[T]] := \{1, \ldots, T\}$ generated by their respective distributions μ_t, the goal of MTR is to learn a shared representation ϕ belonging to a certain class of functions $\Phi := \{\phi \mid \phi : \mathbb{X} \to \mathbb{V}, \ \mathbb{X} \subseteq \mathbb{R}^d, \ \mathbb{V} \subseteq \mathbb{R}^k\}$, generally represented as (deep) neural networks, and linear predictors $\mathbf{w}_t \in \mathbb{R}^k$, $\forall t \in [[T]]$ grouped in a matrix $\mathbf{W} \in \mathbb{R}^{T \times k}$. More formally, this is done by solving the following optimization problem:

$$\hat{\phi}, \widehat{\mathbf{W}} = \arg\min_{\phi \in \Phi, \mathbf{W} \in \mathbb{R}^{T \times k}} \frac{1}{Tn_1} \sum_{t=1}^{T} \sum_{i=1}^{n_1} \ell(y_{t,i}, \langle \mathbf{w}_t, \phi(\mathbf{x}_{t,i}) \rangle), \tag{1}$$

where $\ell : \mathbb{Y} \times \mathbb{Y} \to \mathbb{R}_+$, with $\mathbb{Y} \subseteq \mathbb{R}$, is a loss function. Once such a representation is learned, we want to apply it to a new previously unseen target task observed through a pair $(\mathbf{X}_{T+1} \in \mathbb{R}^{n_2 \times d}, y_{T+1} \in \mathbb{R}^{n_2})$ containing n_2 samples generated by the distribution μ_{T+1}. We expect that a linear classifier $\mathbf{w}$ learned on top of the obtained representation leads to a low true risk over the whole distribution μ_{T+1}. For this, we first use $\hat{\phi}$ to solve the following problem:

$$\hat{\mathbf{w}}_{T+1} = \arg\min_{\mathbf{w} \in \mathbb{R}^k} \frac{1}{n_2} \sum_{i=1}^{n_2} \ell(y_{T+1,i}, \langle \mathbf{w}, \hat{\phi}(\mathbf{x}_{T+1,i}) \rangle). \tag{2}$$

Then, we define the true target risk of the learned linear classifier $\hat{\mathbf{w}}_{T+1}$ as: $\mathcal{L}(\hat{\phi}, \hat{\mathbf{w}}_{T+1}) = \mathbb{E}_{(\mathbf{x},y) \sim \mu_{T+1}}[\ell(y, \langle \hat{\mathbf{w}}_{T+1}, \hat{\phi}(\mathbf{x}) \rangle)]$ and want it to be as close as possible to the ideal true risk $\mathcal{L}(\phi^*, \mathbf{w}_{T+1}^*)$ where $\mathbf{w}_{T+1}^*$ and ϕ^* satisfy:

$$\forall t \in [[T+1]] \text{ and } (\mathbf{x}, y) \sim \mu_t, \quad y = \langle \mathbf{w}_t^*, \phi^*(\mathbf{x}) \rangle + \varepsilon, \quad \varepsilon \sim \mathcal{N}(0, \sigma^2). \tag{3}$$

Equivalently, most of the works found in the literature seek to upper-bound the *excess risk* defined as $\mathrm{ER}(\hat{\phi}, \hat{\mathbf{w}}_{T+1}) := \mathcal{L}(\hat{\phi}, \hat{\mathbf{w}}_{T+1}) - \mathcal{L}(\phi^*, \mathbf{w}_{T+1}^*)$.

2.2   Learning Bounds and Assumptions

First studies in the context of MTR relied on the probabilistic assumption [2,5,29,34,53] stating that meta-train and meta-test tasks distributions are all sampled i.i.d. from the same random distribution. A natural improvement to this bound was then proposed by [15] and [45] that obtained the bounds on the excess risk behaving as

$$\mathrm{ER}(\hat{\phi}, \hat{\mathbf{w}}_{T+1}) \leq O \left(\frac{C(\Phi)}{n_1 T} + \frac{k}{n_2} \right), \tag{4}$$

where $C(\Phi)$ is a measure of the complexity of Φ. Both these results show that all the source and target samples are useful in minimizing the excess risk. Thus, in the FSC regime where target data is scarce, all source data helps to learn well. From a set of assumptions made by the authors in both of these works, we note the following two:

Assumption 1: Diversity of the Source Tasks. The matrix of optimal predictors $\mathbf{W}^*$ should cover all the directions in $\mathbb{R}^k$ evenly. More formally, this can be stated as $\kappa(\mathbf{W}^*) := \frac{\sigma_1(\mathbf{W}^*)}{\sigma_k(\mathbf{W}^*)} = O(1)$,

where $\sigma_i(\cdot)$ denotes the ith singular value of $\mathbf{W}^*$. As pointed out by the authors, such an assumption can be seen as a measure of diversity between the source tasks that are expected to be complementary to each other to provide a useful representation for a previously unseen target task. In the following, we will refer to $\kappa(\mathbf{W})$ as the *condition number* for matrix $\mathbf{W}$.

Assumption 2: Consistency of the Classification Margin. The norm of the optimal predictors $\mathbf{w}^*$ should not increase with the number of tasks seen during meta-training[1]. This assumption says that the classification margin of linear predictors should remain constant thus avoiding over- or under-specialization to the seen tasks.

An intuition behind these assumptions and a detailed review can be found in the Appendix. While being highly insightful, the authors did not provide any experimental evidence suggesting that verifying these assumptions in practice helps to learn more efficiently in the considered learning setting.

2.3   Meta-learning Algorithms

Meta-learning algorithms considered below learn an optimal representation sequentially via the so-called episodic training strategy introduced by [46], instead of jointly minimizing the training error on a set of source tasks as done in MTR. Episodic training mimics the training process at the task scale with each task data being decomposed into a training set *(support set S)* and a testing set *(query set Q)*. Recently, [8] showed that the episodic training setup used in meta-learning leads to a generalization bounds of $O(\frac{1}{\sqrt{T}})$. This bound is independent of the task sample size n_1, which could explain the success of this training strategy for FSC in the asymptotic limit. However, unlike the results obtained by [15]

---

[1] While not stated separately, this assumption is used in [15] to derive the final result on p.5 after the discussion of Assumption 4.3.

studied in this paper, the lack of dependence on n_1 makes such a result uninsightful in practice as we are in a finite-sample size setting. This bound does not give information on other parameters to leverage when the task number cannot increase. We now present two major families of meta-learning approaches below.

Metric-Based Methods. These methods learn an embedding space in which feature vectors can be compared using a similarity function (usually a L_2 distance or cosine similarity) [1,22,27,41–43,46]. They typically use a form of contrastive loss as their objective function, similarly to Neighborhood Component Analysis (NCA) [18] or Triplet Loss [20]. In this paper, we focus our analysis on the popular Prototypical Networks [42] (PROTONET) that computes prototypes as the mean vector of support points belonging to the same class: $c_i = \frac{1}{|S_i|} \sum_{s \in S_i} \phi(s)$, with S_i the subset of support points belonging to class i. PROTONET minimizes the negative log-probability of the true class i computed as the softmax over distances to prototypes c_i:

$$\mathcal{L}_{proto}(S, Q, \phi) := \mathbb{E}_{q \sim Q} \left[- \log \frac{\exp(-d(\phi(q), c_i))}{\sum_j \exp(-d(\phi(q), c_j))} \right] \tag{5}$$

with d being a distance function used to measure similarity between points in the embedding space.

Gradient-Based Methods. These methods learn through end-to-end or two-step optimization [6,10,26,28,31–33,37,38] where given a new task, the goal is to learn a model from the task's training data specifically adapted for this task. MAML [16] updates its parameters θ using an end-to-end optimization process to find the best initialization such that a new task can be learned quickly, *i.e.* with few examples. More formally, given the loss ℓ_t for each task $t \in [[T]]$, MAML minimizes the expected task loss after an *inner loop* or *adaptation* phase, computed by a few steps of gradient descent initialized at the model's current parameters:

$$\mathcal{L}_{\text{MAML}}(\theta) := \mathbb{E}_{t \sim \eta}[\ell_t(\theta - \alpha \nabla \ell_t(\theta))], \tag{6}$$

with η the distribution of the meta-training tasks and α the learning rate for the adaptation phase. For simplicity, we take a single step of gradient update in this equation.

In what follows, we establish our theoretical analysis for the popular methods PROTONET and MAML. We add their improved variations respectively called Infinite Mixture Prototypes [1] (IMP) and Meta-Curvature [33] (MC) in the experiments to validate our findings.

3   Understanding Meta-learning Algorithms Through MTR Theory

3.1   Link Between MTR and Meta-learning

Recently, [47] has shown that meta-learning algorithms that only optimize the last layer in the inner-loop, solve the same underlying optimization procedure as

Fig. 1. Evolution of $\kappa(\mathbf{W}_N)$, $\|\mathbf{W}_N\|_F$ and $\kappa(\mathbf{W})$ (*in log scale*) during the training of PROTONET (*red, left axes*) and MAML (*blue, right axes*) on miniImageNet (*mini, solid lines*) and tieredImageNet (*tiered, dashed lines*) with 5-way 1-shot episodes. (Color figure online)

multi-task learning. In particular, their contributions have the following implications:

1. For *Metric-based algorithms*, the majority of methods can be seen as MTR problems. This is true, in particular, for PROTONET and IMP algorithms considered in this work.
2. In the case of *Gradient-based algorithms*, such methods as ANIL [37] and MetaOptNet [26] that do not update the embeddings during the inner-loop, can be also seen as multi-task learning. However, MAML and MC in principal do update the embeddings even though there exists strong evidence suggesting that the changes in the weights during their inner-loop are mainly affecting the last layer [37]. Consequently, we follow [47] and use this assumption to analyze MAML and MC in MTR framework as well.
3. In practice, [47] showed that the mismatch between the multi-task and the actual episodic training setup leads to a negligible difference.

In the following section, we start by empirically verifying that the behavior of meta-learning methods reveals very distinct features when looked at through the prism of the considered MTR theoretical assumptions.

3.2   What Happens in Practice?

To verify whether theoretical results from MTR setting are also insightful for episodic training used by popular meta-learning algorithms, we first investigate the natural behavior of MAML and PROTONET when solving FSC tasks on the popular *miniImageNet* [38] and *tieredImageNet* [39] datasets. The full experimental setup is detailed in Sect. 4.1 and in the Appendix. Additional experiments for *Omniglot* [25] benchmark dataset portraying the same behavior are also postponed to the Appendix.

To verify Assumption 1 from MTR theory, we want to compute singular values of $\mathbf{W}$ during the meta-training stage and to follow their evolution. In practice, as T is typically quite large, we propose a more computationally efficient solution that is to calculate the condition number only for the last batch of N

predictors (with $N \ll T$) grouped in the matrix $\mathbf{W}_N \in \mathbb{R}^{N \times k}$ that capture the latest dynamics in the learning process. We further note that $\sigma_i(\mathbf{W}_N \mathbf{W}_N^\top) = \sigma_i^2(\mathbf{W}_N)$, $\forall i \in [[N]]$ implying that we can calculate the SVD of $\mathbf{W}_N \mathbf{W}_N^\top$ (or $\mathbf{W}_N^\top \mathbf{W}_N$ for $k \leq N$) and retrieve the singular values from it afterwards. We now want to verify whether $\mathbf{w}_t$ cover all directions in the embedding space and track the evolution of the ratio of singular values $\kappa(\mathbf{W}_N)$ during training. For the first assumption to be satisfied, we expect $\kappa(\mathbf{W}_N)$ to decrease gradually during the training thus improving the generalization capacity of the learned predictors and preparing them for the target task. To verify the second assumption, the norm of the linear predictors should not increase with the number of tasks seen during training, *i.e.*, $\|\mathbf{w}\|_2 = O(1)$ or, equivalently, $\|\mathbf{W}\|_F^2 = O(T)$ and $\|\mathbf{W}_N\|_F = O(1)$.

For gradient-based methods, linear predictors are directly the weights of the last layer of the model. Indeed, for each task, the model learns a batch of linear predictors and we can directly take the weights for $\mathbf{W}_N$. Meanwhile, metric-based methods do not use linear predictors but compute a similarity between features. In the case of PROTONET, the similarity is computed with respect to class prototypes that are mean features of the instances of each class. For the Euclidean distance, *this is equivalent to a linear model* with the prototype of a class acting as the linear predictor of this class [42]. This means that we can apply our analysis directly to the *prototypes* computed by PROTONET. In this case, the matrix $\mathbf{W}^*$ will be the matrix of the optimal prototypes and we can then take the *prototypes* computed for each task as our matrix $\mathbf{W}_N$.

From Fig. 1, we can see that for MAML (*blue*), both $\|\mathbf{W}_N\|_F$ (*left*) and $\kappa(\mathbf{W}_N)$ (*middle*) increase with the number of tasks seen during training, whereas PROTONET (*red*) naturally learns prototypes with a good coverage of the embedding space, and minimizes their norm. Since we compute the singular values of the last N predictors in $\kappa(\mathbf{W}_N)$, we can only compare the *overall behavior* throughout training between methods. For the sake of completeness, we also compute $\kappa(\mathbf{W})$ (*right*) at different stages in the training. To do so, we fix the encoder ϕ_T learned after T episodes and recalculate the linear predictors of the T past training episodes with this fixed encoder. We can see that $\kappa(\mathbf{W})$ of PROTONET also decreases during training and reach a lower final value than $\kappa(\mathbf{W})$ of MAML. This confirms that the dynamics of $\kappa(\mathbf{W}_N)$ and $\kappa(\mathbf{W})$ are similar whereas the values $\kappa(\mathbf{W}_N)$ between methods should not be directly compared. The behavior of $\kappa(\mathbf{W})$ also validate our finding that PROTONET learns to cover the embedding space with prototypes. This behavior is rather peculiar as neither of the two methods explicitly controls the theoretical quantities of interest, and still, PROTONET manages to do it implicitly.

3.3   The Case of Meta-learning Algorithms

The differences observed above for the two methods call for a deeper analysis of their behavior.

PROTONET. We start by first explaining why PROTONET learns prototypes that cover the embedding space efficiently. This result is given by the following theorem (cf. Appendix for the proof):

Theorem 1. (Normalized PROTONET)
If $\forall i \; \|\mathbf{c}_i\| = 1$, then $\forall \hat{\phi} \in \arg\min\limits_{\phi} \mathcal{L}_{proto}(S, Q, \phi)$, the matrix of the optimal prototypes $\mathbf{W}^*$ is well-conditioned, i.e. $\kappa(\mathbf{W}^*) = O(1)$.

This theorem explains the empirical behavior of PROTONET in FSC task: the minimization of its objective function naturally minimizes the condition number when the norm of the prototypes is low.

In particular, it implies that norm minimization seems to initiate the minimization of the condition number seen afterwards due to the contrastive nature of the loss function minimized by PROTONET. We confirm this latter implication through experiments in Sect. 4 showing that norm minimization is enough for considered metric-based methods to obtain the well-behaved condition number and that minimizing both seems redundant.

MAML. Unfortunately, the analysis of MAML in the most general case is notoriously harder, as even expressing its loss function and gradients in the case of an overparametrized linear regression model with only 2 parameters requires using a symbolic toolbox for derivations [3]. To this end, we resort to the linear regression model considered in this latter paper and defined as follows. We assume for all $t \in [[T]]$ that the task parameters $\boldsymbol{\theta}_t$ are normally distributed with $\boldsymbol{\theta}_t \sim \mathcal{N}(\mathbf{0}_d, \boldsymbol{I}_d)$, the inputs $\mathbf{x}_t \sim \mathcal{N}(\mathbf{0}_d, \boldsymbol{I}_d)$ and the output $y_t \sim \mathcal{N}(\langle \boldsymbol{\theta}_t, \mathbf{x}_t \rangle, 1)$. For each t, we consider the following learning model and its associated square loss:

$$\hat{y}_t = \langle \mathbf{w}_t, \mathbf{x}_t \rangle, \quad \ell_t = \mathbb{E}_{p(\mathbf{x}_t, y_t | \boldsymbol{\theta}_t)}(y_t - \langle \mathbf{w}_t, \mathbf{x}_t \rangle)^2. \tag{7}$$

We can now state the following result.

Proposition 1. *Let $\forall t \in [[T]]$, $\boldsymbol{\theta}_t \sim \mathcal{N}(\mathbf{0}_d, \boldsymbol{I}_d)$, $\mathbf{x}_t \sim \mathcal{N}(\mathbf{0}_d, \boldsymbol{I}_d)$ and $y_t \sim \mathcal{N}(\langle \boldsymbol{\theta}_t, \mathbf{x}_t \rangle, 1)$. Consider the learning model from Eq. (7), let $\boldsymbol{\Theta}_i := [\boldsymbol{\theta}_i, \boldsymbol{\theta}_{i+1}]^T$, and denote by $\widehat{\mathbf{W}}_2^i$ the matrix of last two predictors learned by MAML at iteration i starting from $\widehat{\mathbf{w}}_0 = \mathbf{0}_d$. Then, we have that:*

$$\forall i, \quad \kappa(\widehat{\mathbf{W}}_2^{i+1}) \geq \kappa(\widehat{\mathbf{W}}_2^i), \quad if \; \sigma_{min}(\boldsymbol{\Theta}_i) = 0. \tag{8}$$

This proposition provides an explanation of why MAML may tend to increase the ratio of singular values during the iterations. Indeed, the condition when this happens indicates that the optimal predictors forming matrix $\boldsymbol{\Theta}_i$ are linearly dependent implying that its smallest singular values becomes equal to 0. While this is not expected to be the case for all iterations, we note, however, that in FSC task the draws from the dataset are in general not i.i.d. and thus may correspond to co-linear optimal predictors. In every such case, the condition number is expected to remain non-decreasing, as illustrated in Fig. 1 (*left*) where for MAML, contrary to PROTONET, $\kappa(\mathbf{W}_N)$ exhibits plateaus but also intervals where it is increasing.

This highlights a major difference between the two approaches: MAML does not specifically seek to diversify the learned predictors, while PROTONET does. Proposition 1 gives a hint to the essential difference between the methods studied. On the one hand, PROTONET constructs its classifiers directly from the data of the current task and they are independent of the other tasks. On the other hand, for MAML, the weights of the classifiers are reused between tasks and only slightly adapted to be specific to each task. This limits the generalization capabilities of the linear predictors learned by MAML since they are based on predictors from previous tasks.

3.4   Enforcing the Assumptions

Why Should We Force the Assumptions? From the results obtain by [15], and with the same assumptions, we can easily make appear $\kappa(\mathbf{W}^*)$ to obtain a more explicit bound:

Proposition 2. *If $\forall t \in [[T]], \|\mathbf{w}_t^*\| = O(1)$ and $\kappa(\mathbf{W}^*) = O(1)$, and $\mathbf{w}_{T+1}$ follows a distribution ν such that $\|\mathbb{E}_{\mathbf{w} \sim \nu}[\mathbf{w}\mathbf{w}^\top]\| \leq O\left(\frac{1}{k}\right)$, then*

$$ER(\hat{\phi}, \hat{\mathbf{w}}_{T+1}) \leq O\left(\frac{C(\Phi)}{n_1 T} \cdot \kappa(\mathbf{W}^*) + \frac{k}{n_2}\right). \tag{9}$$

Proposition 2 suggests that the terms $\|\mathbf{w}_t^*\|$ and $\kappa(\mathbf{W}^*)$ underlying the assumptions directly impact the tightness of the established bound on the excess risk. The full proof can be found in the Appendix.

Can We Force the Assumptions? According to the previous result, satisfying the assumptions from MTR theory is expected to come in hand with better performance. However all the terms involved refer to optimal predictors, that we cannot act upon. Thus, we aim to answer the following question:

Given $\mathbf{W}^*$ such that $\kappa(\mathbf{W}^*) \gg 1$, can we learn $\widehat{\mathbf{W}}$ with $\kappa(\widehat{\mathbf{W}}) \approx 1$ while solving the underlying classification problems equally well?

While obtaining such a result for any distribution seems to be very hard in the considered learning setup, we provide a constructive proof for the existence of a distribution for which the answer to the above-mentioned question is positive in the case of two tasks. The latter restriction comes out of the necessity to analytically calculate the singular values of $\mathbf{W}$ but we expect our example to generalize to more general setups and a larger number of tasks as well.

Proposition 3. *Let $T = 2$, $\mathbb{X} \subseteq \mathbb{R}^d$ be the input space and $\mathbb{Y} = \{-1, 1\}$ be the output space. Then, there exist distributions μ_1 and μ_2 over $\mathbb{X} \times \mathbb{Y}$, representations $\hat{\phi} \neq \phi^*$ and matrices of predictors $\widehat{\mathbf{W}} \neq \mathbf{W}^*$ that satisfy the data generating model (Eq. (3)) with $\kappa(\widehat{\mathbf{W}}) \approx 1$ and $\kappa(\mathbf{W}^*) \gg 1$.*

See Appendix for full proof and illustration. The established results show that even when $\mathbf{W}^*$ does not satisfy Assumptions 1–2 in the ϕ^* space, it may still be possible to learn $\hat{\phi}$ such that the optimal predictors do satisfy them.

How to Force the Assumptions? This can be done either by considering the constrained problem (Eq. (10)) or by using a more common strategy that consists in adding $\kappa(\mathbf{W})$ and $\|\mathbf{W}\|_F^2$ as regularization terms (Eq. (11)):

$$\widehat{\phi}, \widehat{\mathbf{W}} = \arg\min_{\phi, \mathbf{W}} \frac{1}{Tn_1} \sum_{t=1}^{T} \sum_{i=1}^{n_1} \ell(y_{t,i}, \langle \mathbf{w}_t, \phi(\mathbf{x}_{t,i}) \rangle) \text{ s.t. } \kappa(\mathbf{W}) = O(1), \|\mathbf{w}_t\| = 1,$$

$$(10)$$

$$\widehat{\phi}, \widehat{\mathbf{W}} = \arg\min_{\phi, \mathbf{W}} \frac{1}{Tn_1} \sum_{t=1}^{T} \sum_{i=1}^{n_1} \ell(y_{t,i}, \langle \mathbf{w}_t, \phi(\mathbf{x}_{t,i}) \rangle) + \lambda_1 \kappa(\mathbf{W}) + \lambda_2 \|\mathbf{W}\|_F^2.$$

$$(11)$$

To the best of our knowledge, such regularization terms based on insights from the advances in MTR theory have never been used in the literature before. We refer the reader to the Appendix for more details about them.

3.5   Positioning with Respect to Previous Work

Understanding Meta-learning. While a complete theory for meta-learning is still lacking, several recent works aim to shed light on phenomena commonly observed in meta-learning by evaluating different intuitive heuristics. For instance, [37] investigate whether MAML works well due to rapid learning with significant changes in the representations when deployed on target task, or due to feature reuse where the learned representation remains almost intact. In [19], the authors explain the success of meta-learning approaches by their capability to either cluster classes more tightly in feature space (task-specific adaptation approach), or to search for meta-parameters that lie close in weight space to many task-specific minima (full fine-tuning approach). Finally, the effect of the number of shots on the FSC accuracy was studied in [7] for PROTONET . More recently, [51] studied the impact of the permutation of labels when training MAML. Our paper investigates a new aspect of meta-learning that has never been studied before and, unlike [7,19,37] and [51], provides a more complete experimental evaluation with the two different approaches of meta-learning.

Normalization. Multiple methods in the literature introduce a normalization of their features either to measure cosine similarity instead of Euclidean distance [9,17,35] or because of the noticed improvement in their performance [44,47,49]. In this work, we proved in Sect. 3.3 above that for PROTONET prototypes normalization is enough to achieve a good coverage of the embedding space, and we empirically show in Sect. 4.2 below that it leads to better performance. Since we only normalize the prototypes and not all the features, we do not measure cosine similarity. Moreover, with our Theorem 1, we give explanations through MTR theory regarding the link between the coverage of the representation space and performance.

Common Regularization Strategies. In general, we note that regularization in meta-learning (i) is applied to either the weights of the whole neural network [4,53], or (ii) the predictions [19,21] or (iii) is introduced via a prior hypothesis biased regularized empirical risk minimization [12–14,24,34]. Contrary to the first group of methods and weight decay approach [23], we do not regularize the whole weight matrix learned by the neural network but the linear predictors of its last layer. While weight decay is used to avoid overfitting by penalizing large magnitudes of weights, our goal is to keep the classification margin unchanged during the training to avoid over-/under-specialization to source tasks. Similarly, spectral normalization proposed by [30] does not affect the condition number κ. Second, we regularize the singular values of the matrix of linear predictors obtained in the last batch of tasks instead of the predictions used by the methods of the second group (*e.g.*, using the theoretic-information quantities in [21]). Finally, the works of the last group are related to the online setting with convex loss functions only and do not specifically target the spectral properties of the learned predictors.

4 Impact of Enforcing Theoretical Assumptions

4.1 Experimental Setup

We consider on three benchmark datasets for FSC, namely: 1) **Omniglot** [25] consisting of 1,623 classes with 20 images/class of size 28×28; 2) **miniImageNet** [38] consisting of 100 classes with 600 images of size 84×84/class; 3) **tieredImageNet** [39] consisting of 779,165 images divided into 608 classes.

For each dataset, we follow a common experimental protocol used in [9,16] and use a four-layer convolution backbone with 64 filters (C64) as done by [9]. We also provide experiments with the ResNet-12 architecture (R12) [26] and we follow the recent practice to initialize the models with the weights pretrained on the entire meta-training set [36,40,52]. For PROTONET, we use the released code of [52] with a temperature of 1 instead of tuning it for all settings to be fair with other settings and methods. For MAML, we follow the exact implementation details from [51] since the code was not available. We measure the performance using the top-1 accuracy with 95% confidence intervals, reproduce the experiments with 4 different random seeds and average the results over 2400 test tasks. For all FSC experiments, unless explicitly stated, we use the regularization parameters $\lambda_1 = \lambda_2 = 1$ in the regularized problem (Eq. (11)). We refer the reader to the Appendix for all the hyperparameters used.[2]

---

[2] Code for the experiments is available at https://github.com/CEA-LIST/MetaMTReg.

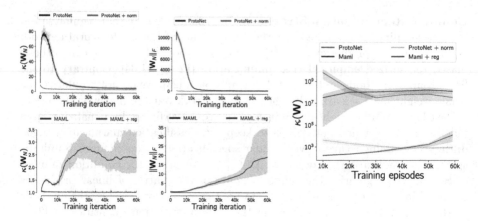

Fig. 2. Evolution of $\kappa(\mathbf{W}_N)$ (*left*), $\|\mathbf{W}_N\|_F$ (*middle*) and $\kappa(\mathbf{W})$ (*right*) on 5-way 1-shot episodes from *miniImageNet*, for PROTONET (*top*), MAML (*bottom*) and their regularized or normalized counterparts.

4.2   Metric-based Methods

Theorem 1 tells us that with normalized class prototypes that act as linear predictors, PROTONET naturally decreases the condition number of their matrix. Furthermore, since the prototypes are directly the image features, adding a regularization term on the norm of the prototypes makes the model collapse to the trivial solution which maps all images to 0. To this end, we choose to ensure the theoretical assumptions for metric-based methods (PROTONET and IMP) only with prototype normalization, by using the normalized prototypes $\tilde{\mathbf{w}} = \frac{\mathbf{w}}{\|\mathbf{w}\|}$. According to Theorem 1, the normalization of the prototypes makes the problem similar to the constrained problem given in Eq. (10).

As can be seen in Fig. 2, the normalization of the prototypes has the intended effect on the condition number of the matrix of predictors. Indeed, $\kappa(\mathbf{W}_N)$ (*left*) stay constant and low during training, and we achieve a much lower $\kappa(\mathbf{W})$ (*right*) than without normalization. From Fig. 3, we note that normalizing the prototypes from the very beginning of the training process has an overall positive effect on the obtained performance, and this gain is statistically significant in most of the cases according to the *Wilcoxon signed-rank* test (p < 0.05) [11,50].

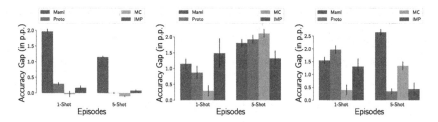

Fig. 3. Accuracy gap (in p.p.) when adding the normalization of prototypes for PRO-TONET (*red*) and IMP (*green*), and both spectral and norm regularization for MAML (*blue*) and MC (*purple*) enforcing the theoretical assumptions on Omniglot (*left*), mini-Imagenet (*middle*) and tieredImagenet (*right*) datasets. (Color figure online)

In Table 1a, we compare the performance obtained against state-of-the-art algorithms behaving similarly to Instance Embedding algorithms [52] such as PROTONET, depending on the architecture used. Even with a ResNet-12 architecture, the proposed normalization still improves the performance to reach competitive results with the state-of-the-art. On the miniImageNet 5-way 5-shot benchmark, our normalized PROTONET achieves 80.95%, better than DSN (78.83%), CTM (78.63%) and SimpleShot (80.02%). We refer the reader to the Appendix for more detailed training curves.

4.3   Gradient-based Methods

Gradient-based methods learn a batch of linear predictors for each task, and we can directly take them as $\mathbf{W}_N$ to compute its SVD. In the following experiments, we consider the regularized problem of Eq. (11) for MAML as well as Meta-Curvature (MC). As expected, the dynamics of $\|\mathbf{W}_N\|_F$ and $\kappa(\mathbf{W}_N)$ during the training of the regularized methods remain bounded and the effect of the regularization is confirmed with the lower value of $\kappa(\mathbf{W})$ achieved (cf. Fig. 2).

The impact of our regularization on the results is quantified in Fig. 3 where a statistically significant accuracy gain is achieved in most cases, according to the *Wilcoxon signed-rank* test (p < 0.05) [11,50]. In Table 1a, we compare the performance obtained to state-of-the-art gradient-based algorithms. We can see that our proposed regularization is globally improving the results, even with a bigger architecture such as ResNet-12 and with an additional pretraining. On the miniImageNet 5-way 5-shot benchmark, with our regularization MAML achieves 82.45%, better than TADAM (76.70%), MetaOptNet (78.63%) and MATE with MetaOptNet (78.64%). We include in the Appendix ablative studies on the effect of each term in our regularization scheme for gradient-based methods, and more detailed training curves.

Table 1. Performance comparison on FSC benchmarks. (a) FSC models. (b) MTL models [47]. *For a given architecture,* **bold** values are the highest accuracy and underlined values are near-highest accuracies (less than 1-point lower).

Model	Arch.	miniImageNet 5-way		tieredImageNet 5-way	
		1-shot	5-shot	1-shot	5-shot
Gradient-based algorithms					
Maml[16]	C64	48.70 ± 1.84	63.11 ± 0.92	-	-
Anil [37]	C64	48.0 ± 0.7	62.2 ± 0.5	-	-
Meta-SGD [28]	C64	**50.47 ± 1.87**	64.03 ± 0.94	-	-
TADAM [32]	R12	58.50 ± 0.30	76.70 ± 0.30	-	-
MC [33]	R12	61.22 ± 0.10	75.92 ± 0.17	66.20 ± 0.10	82.21 ± 0.08
MetaOptNet [26]	R12	62.64 ± 0.61	78.63 ± 0.46	65.99 ± 0.72	81.56 ± 0.53
MATE[10]	R12	62.08 ± 0.64	78.64 ± 0.46	-	-
Maml (Ours)	C64	47.93 ± 0.83	64.47 ± 0.69	50.08 ± 0.91	67.5 ± 0.79
Maml + reg. (Ours)	C64	49.15 ± 0.85	**66.43 ± 0.69**	51.5 ± 0.9	70.16 ± 0.76
MC (Ours)	C64	49.28 ± 0.83	63.74 ± 0.69	55.16 ± 0.94	71.95 ± 0.77
MC + reg. (Ours)	C64	49.64 ± 0.83	65.67 ± 0.70	**55.85 ± 0.94**	**73.34 ± 0.76**
Maml (Ours)	R12	63.52 ± 0.20	81.24 ± 0.14	63.96 ± 0.23	81.79 ± 0.16
Maml + reg. (Ours)	R12	**64.04 ± 0.22**	**82.45 ± 0.14**	64.32 ± 0.23	81.28 ± 0.11
Metric-based algorithms					
ProtoNet [42]	C64	46.61 ± 0.78	65.77 ± 0.70	–	–
IMP [1]	C64	49.6 ± 0.8	**68.1 ± 0.8**	–	–
SimpleShot [49]	C64	49.69 ± 0.19	66.92 ± 0.17	51.02 ± 0.20	68.98 ± 0.18
Relation Nets [43]	C64	50.44 ± 0.82	65.32 ± 0.70	–	–
SimpleShot [49]	R18	62.85 ± 0.20	80.02 ± 0.14	69.09 ± 0.22	84.58 ± 0.16
CTM [27]	R18	62.05 ± 0.55	78.63 ± 0.06	64.78 ± 0.11	81.05 ± 0.52
DSN [41]	R12	62.64 ± 0.66	78.83 ± 0.45	66.22 ± 0.75	82.79 ± 0.48
ProtoNet (Ours)	C64	49.53 ± 0.41	65.1 ± 0.35	51.95 ± 0.45	71.61 ± 0.38
ProtoNet+ norm. (Ours)	C64	50.29 ± 0.41	67.13 ± 0.34	**54.05 ± 0.45**	71.84 ± 0.38
IMP (Ours)	C64	48.85 ± 0.81	66.43 ± 0.71	52.16 ± 0.89	71.79 ± 0.75
IMP + norm. (Ours)	C64	**50.69 ± 0.8**	67.29 ± 0.68	53.46 ± 0.89	**72.38 ± 0.75**
ProtoNet (Ours)	R12	59.25 ± 0.20	77.92 ± 0.14	41.39 ± 0.21	83.06 ± 0.16
ProtoNet+ norm. (Ours)	R12	**62.69 ± 0.20**	**80.95 ± 0.14**	**68.44 ± 0.23**	**84.20 ± 0.16**

(a)

Model	Arch.	miniImageNet 5-way		tieredImageNet 5-way	
		1-shot	5-shot	1-shot	5-shot
MTL	R12	55.73 ± 0.18	76.27 ± 0.13	62.49 ± 0.21	81.31 ± 0.15
MTL + norm.	R12	59.49 ± 0.18	**77.3 ± 0.13**	66.66 ± 0.21	83.59 ± 0.14
MTL + reg. (Ours)	R12	**61.12 ± 0.19**	76.62 ± 0.13	66.28 ± 0.22	81.68 ± 0.15

(b)

4.4   Multi-task Learning Methods

We implement our regularization on a recent Multi-Task Learning (MTL) method[47], following the same experimental protocol. The objective is to empirically validate our analysis on a method using the MTR framework. As mentioned in Sect. 3.5, the authors introduce feature normalization in their method, speculating that it improves coverage of the representation space [48]. Using their code, we reproduce their experiments on three different settings compared in Table 1b: the vanilla MTL, the MTL with feature normalization, and MTL with our proposed regularization on the condition number and the norm of the linear predictors. We use $\lambda_1 = 1$ in all the settings, and $\lambda_2 = 1$ in the 1-shot setting and $\lambda_2 = 0.01$ in the 5-shot settings. We include in the Appendix an ablative study on the effect of each term of the regularization. Our regularization, as well as the normalization, globally improve the performance over the non-normalized models. Notably, our regularization is the most effective when there is the less data which is well-aligned with the MTR theory in few-shot setting. We can also note that in most of the cases, the normalized models and the regularized ones achieve similar results, hinting that they may have a similar effect. All of these results show that our analysis and our proposed regularization are also valid in the MTL framework.

5 Conclusion

In this paper, we studied the validity of the theoretical assumptions made in recent papers of Multi-Task Representation Learning theory when applied to popular metric- and gradient-based meta-learning algorithms. We found a striking difference in their behavior and provided both theoretical and experimental arguments explaining that metric-based methods satisfy the considered assumptions, while gradient-based don't. We further used this as a starting point to implement a regularization strategy ensuring these assumptions and observed that it leads to faster learning and better generalization.

While this paper proposes an initial approach to bridging the gap between theory and practice for Meta-Learning, some questions remain open on the inner workings of these algorithms. In particular, being able to take better advantage of the particularities of the training tasks during meta-training could help improve the effectiveness of these approaches. The similarity between the source and test tasks was not taken into account in this work, which is an additional assumption in the theory of [15]. We provide a preliminary study using different datasets between the meta-training and meta-testing in the Appendix to foster future work on this topic. Self-supervised meta-learning and multiple target tasks prediction are also important future perspectives for the application of meta-learning.

Acknowledgements. This work was made possible by the use of the Factory-AI supercomputer, financially supported by the Ile-de-France Regional Council.

References

1. Allen, K., Shelhamer, E., Shin, H., Tenenbaum, J.: Infinite mixture prototypes for few-shot learning. In: Proceedings of the 36th International Conference on Machine Learning (2019)

2. Amit, R., Meir, R.: Meta-learning by adjusting priors based on extended PAC-Bayes theory. In: International Conference on Machine Learning (2018)
3. Arnold, S., Iqbal, S., Sha, F.: When MAML can adapt fast and how to assist when it cannot. In: AISTATS (2021)
4. Balaji, Y., Sankaranarayanan, S., Chellappa, R.: MetaReg: towards domain generalization using meta-regularization. In: Advances in Neural Information Processing Systems, pp. 998–1008 (2018)
5. Baxter, J.: A model of inductive bias learning. J. Artif. Intell. Res. **12**, 149–198 (2000)
6. Bertinetto, L., Henriques, J.F., Torr, P., Vedaldi, A.: Meta-learning with differentiable closed-form solvers. In: International Conference on Learning Representations (2018)
7. Cao, T., Law, M.T., Fidler, S.: A theoretical analysis of the number of shots in few-shot learning. In: ICLR (2020)
8. Chen, J., Wu, X.M., Li, Y., Li, Q., Zhan, L.M., Chung, F.l.: A closer look at the training strategy for modern meta-learning. In: Advances in Neural Information Processing Systems (2020)
9. Chen, W.Y., Wang, Y.C.F., Liu, Y.C., Kira, Z., Huang, J.B.: A closer look at few-shot classification. In: ICLR (2019)
10. Chen, X., Wang, Z., Tang, S., Muandet, K.: MATE: plugging in model awareness to task embedding for meta learning. In: Advances in Neural Information Processing Systems (2020)
11. David, F.N., Siegel, S.: Nonparametric Statistics for the Behavioral Sciences. Biometrika (1956)
12. Denevi, G., Ciliberto, C., Grazzi, R., Pontil, M.: Learning-to-learn stochastic gradient descent with biased regularization. In: International Conference on Machine Learning (2019)
13. Denevi, G., Ciliberto, C., Stamos, D., Pontil, M.: Incremental learning-to-learn with statistical guarantees. In: Conference on Uncertainty in Artificial Intelligence, pp. 457–466 (2018)
14. Denevi, G., Ciliberto, C., Stamos, D., Pontil, M.: Learning to learn around a common mean. In: Advances in Neural Information Processing Systems, pp. 10169–10179 (2018)
15. Du, S.S., Hu, W., Kakade, S.M., Lee, J.D., Lei, Q.: Few-shot learning via learning the representation, provably. In: International Conference on Learning Representations (2020)
16. Finn, C., Abbeel, P., Levine, S.: Model-agnostic meta-learning for fast adaptation of deep networks. In: International Conference on Machine Learning (2017)
17. Gidaris, S., Komodakis, N.: Dynamic few-shot visual learning without forgetting. In: CVPR, pp. 4367–4375 (2018)
18. Goldberger, J., Hinton, G.E., Roweis, S., Salakhutdinov, R.R.: Neighbourhood components analysis. In: Saul, L., Weiss, Y., Bottou, L. (eds.) NeurIPS (2005)
19. Goldblum, M., Reich, S., Fowl, L., Ni, R., Cherepanova, V., Goldstein, T.: Unraveling meta-learning: Understanding feature representations for few-shot tasks. In: International Conference on Machine Learning (2020)
20. Hoffer, E., Ailon, N.: Deep metric learning using triplet network. In: Feragen, A., Pelillo, M., Loog, M. (eds.) SIMBAD 2015. LNCS, vol. 9370, pp. 84–92. Springer, Cham (2015). https://doi.org/10.1007/978-3-319-24261-3_7
21. Jamal, M.A., Qi, G.J.: Task agnostic meta-learning for few-shot learning. In: CVPR (2019)

22. Koch, G., Zemel, R., Salakhutdinov, R.: Siamese neural networks for one-shot image recognition. In: International Conference on Machine Learning Deep Learning Workshop (2015)
23. Krogh, A., Hertz, J.A.: A simple weight decay can improve generalization. In: Advances in Neural Information Processing Systems (1992)
24. Kuzborskij, I., Orabona, F.: Fast rates by transferring from auxiliary hypotheses. Mach. Learn. **106**(2), 171–195 (2016). https://doi.org/10.1007/s10994-016-5594-4
25. Lake, B.M., Salakhutdinov, R., Tenenbaum, J.B.: Human-level concept learning through probabilistic program induction. Science **350**, 1332–1338 (2015)
26. Lee, K., Maji, S., Ravichandran, A., Soatto, S.: Meta-learning with differentiable convex optimization. In: CVPR (2019)
27. Li, H., Eigen, D., Dodge, S., Zeiler, M., Wang, X.: Finding task-relevant features for few-shot learning by category traversal. In: 2019 IEEE/CVF Conference on Computer Vision and Pattern Recognition (CVPR) (2019)
28. Li, Z., Zhou, F., Chen, F., Li, H.: Meta-SGD: learning to learn quickly for few-shot learning. arXiv:1707.09835 (2017)
29. Maurer, A., Pontil, M., Romera-Paredes, B.: The benefit of multitask representation learning. J. Mach. Learn. Res. **17**, 1–32 (2016)
30. Miyato, T., Kataoka, T., Koyama, M., Yoshida, Y.: Spectral normalization for generative adversarial networks. In: ICLR (2018)
31. Nichol, A., Achiam, J., Schulman, J.: On first-order meta-learning algorithms. arXiv:1803.02999 [cs] (2018)
32. Oreshkin, B., Rodríguez López, P., Lacoste, A.: TADAM: task dependent adaptive metric for improved few-shot learning. In: Advances in Neural Information Processing Systems (2018)
33. Park, E., Oliva, J.B.: Meta-curvature. In: Wallach, H.M., Larochelle, H., Beygelzimer, A., d'Alché-Buc, F., Fox, E.B., Garnett, R. (eds.) NeurIPS, pp. 3309–3319 (2019)
34. Pentina, A., Lampert, C.H.: A PAC-Bayesian bound for lifelong learning. In: International Conference on Machine Learning (2014)
35. Qi, H., Brown, M., Lowe, D.G.: Low-Shot Learning with Imprinted Weights. In: 2018 IEEE/CVF Conference on Computer Vision and Pattern Recognition, Salt Lake City, UT (2018)
36. Qiao, S., Liu, C., Shen, W., Yuille, A.L.: Few-shot image recognition by predicting parameters from activations. In: IEEE Conference on Computer Vision and Pattern Recognition, pp. 7229–7238 (2018)
37. Raghu, A., Raghu, M., Bengio, S., Vinyals, O.: Rapid learning or feature reuse? Towards understanding the effectiveness of MAML. In: ICLR (2020)
38. Ravi, S., Larochelle, H.: Optimization as a model for few-shot learning. In: ICLR (2017)
39. Ren, M., et al.: Meta-learning for semi-supervised few-shot classification. In: ICLR (2018)
40. Rusu, A.A., et al.: Meta-learning with latent embedding optimization. In: International Conference on Learning Representations (2018)
41. Simon, C., Koniusz, P., Nock, R., Harandi, M.: Adaptive subspaces for few-shot learning. In: 2020 IEEE/CVF Conference on Computer Vision and Pattern Recognition (CVPR) (2020)
42. Snell, J., Swersky, K., Zemel, R.S.: Prototypical networks for few-shot learning. In: Advances in Neural Information Processing Systems (2017)
43. Sung, F., Yang, Y., Zhang, L., Xiang, T., Torr, P.H., Hospedales, T.M.: Learning to compare: relation network for few-shot learning. In: CVPR, pp. 1199–1208 (2018)

44. Tian, Y., Wang, Y., Krishnan, D., Tenenbaum, J.B., Isola, P.: Rethinking few-shot image classification: a good embedding is all you need? In: Vedaldi, A., Bischof, H., Brox, T., Frahm, J.-M. (eds.) ECCV 2020. LNCS, vol. 12359, pp. 266–282. Springer, Cham (2020). https://doi.org/10.1007/978-3-030-58568-6_16
45. Tripuraneni, N., Jin, C., Jordan, M.: Provable meta-learning of linear representations. In: International Conference on Machine Learning. PMLR (2021)
46. Vinyals, O., Blundell, C., Lillicrap, T., Kavukcuoglu, K., Wierstra, D.: Matching networks for one shot learning. In: Advances in Neural Information Processing Systems, pp. 3630–3638 (2016)
47. Wang, H., Zhao, H., Li, B.: Bridging multi-task learning and meta-learning: towards efficient training and effective adaptation. In: Proceedings of the 38th International Conference on Machine Learning. Proceedings of Machine Learning Research. PMLR (2021)
48. Wang, T., Isola, P.: Understanding contrastive representation learning through alignment and uniformity on the hypersphere. In: ICML, pp. 9929–9939. Proceedings of Machine Learning Research (2020)
49. Wang, Y., Chao, W.L., Weinberger, K.Q., van der Maaten, L.: SimpleShot: revisiting nearest-neighbor classification for few-shot learning. arXiv:1911.04623 (2019)
50. Wilcoxon, F.: Individual comparisons by ranking methods. Biometrics Bull. 1, 80–83 (1945)
51. Ye, H.J., Chao, W.L.: How to train your MAML to excel in few-shot classification. In: International Conference on Learning Representations (2022)
52. Ye, H.J., Hu, H., Zhan, D.C., Sha, F.: Few-shot learning via embedding adaptation with set-to-set functions. In: Computer vision and pattern recognition (CVPR) (2020)
53. Yin, M., Tucker, G., Zhou, M., Levine, S., Finn, C.: Meta-learning without memorization. In: ICLR (2020)

Tree Structure-Aware Few-Shot Image Classification via Hierarchical Aggregation

Min Zhang[1,2,4], Siteng Huang[2,4], Wenbin Li[3], and Donglin Wang[2,4(✉)]

[1] Zhejiang University, Hangzhou, China
[2] Westlake University, Hangzhou, China
{zhangmin,huangsiteng,wangdonglin}@westlake.edu.cn
[3] State Key Laboratory for Novel Software Technology, Nanjing University, Nanjing, China
liwenbin@nju.edu.cn
[4] Institute of Advanced Technology, Westlake Institute for Advanced Study, Hangzhou, China

Abstract. In this paper, we mainly focus on the problem of how to learn additional feature representations for few-shot image classification through pretext tasks (*e.g.*, rotation or color permutation and so on). This additional knowledge generated by pretext tasks can further improve the performance of few-shot learning (FSL) as it differs from human-annotated supervision (*i.e.*, class labels of FSL tasks). To solve this problem, we present a plug-in *Hierarchical Tree Structure-aware (HTS)* method, which not only learns the relationship of FSL and pretext tasks, but more importantly, can adaptively select and aggregate feature representations generated by pretext tasks to maximize the performance of FSL tasks. A hierarchical tree constructing component and a gated selection aggregating component is introduced to construct the tree structure and find richer transferable knowledge that can rapidly adapt to novel classes with a few labeled images. Extensive experiments show that our HTS can significantly enhance multiple few-shot methods to achieve new state-of-the-art performance on four benchmark datasets. The code is available at: https://github.com/remiMZ/HTS-ECCV22.

Keywords: Hierarchical tree structure · Few-shot learning · Pretext tasks

1 Introduction

Few-shot learning (FSL), especially few-shot image classification [2,17,22,37,44], has attracted a lot of machine learning community. FSL aims to learn transferable feature representations by training the model with a collection of FSL tasks on base (seen) classes and generalizing the representations to novel (unseen)

Supplementary Information The online version contains supplementary material available at https://doi.org/10.1007/978-3-031-20044-1_26.

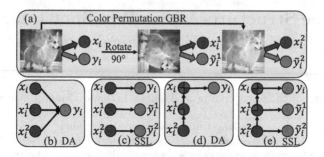

Fig. 1. Differences in the learning process for few-shot image classification using pretext tasks between previous and our works. (a) shows the process of generating augmented images using FSL images. (b) and (c) show the learning process for previous works under the DA or SSL setting, which uses all images indiscriminately. (d) and (e) show the learning process for our work under the DA or SSL setting, which can exploit the hierarchical tree structure to adaptively select useful feature representations.

classes by accessing an extremely few labeled images [1,4,7,10,30,38]. However, due to the data scarcity, the learned supervised representations mainly focus on the differences between the base class while ignoring the valuable semantic features within images for novel classes, weakening the model's generalization ability. Therefore, more feature representations should be extracted from the limited available images to improve the generalization ability of the FSL model.

One effective way of extracting more useful feature representations is to use pretext tasks, such as rotation with multiple angles or color permutation among different channels [5,11,16,24,34]. Because these pretext tasks can generate additional augmented images and the semantic representations of these augmented images are a good supplementary to the normal human-annotated supervisions (*i.e.*, class labels of FSL images), which is beneficial to the few-shot learning model generalization on novel classes. The standard training process of using pretext tasks to assist in FSL can be roughly classified into two settings, *i.e.*, data augmentation (DA) and self-supervised learning (SSL), following existing works [11,24,26,34,37]. As shown in Fig. 1, (a) shows that pretext tasks are used to generate multiple augmented images (x_i^1, x_i^2) and x_i is FSL images. (b) and (c) are the learning process of using pretext tasks to improve the FSL performance under the DA or SSL setting in previous works. However, in the DA setting, all augmented and raw images are placed in the same label space (*e.g.*, $y_i = \tilde{y}_i^1 = \tilde{y}_i^2 = $ dog) and the empirical risk minimization (ERM) (see Eq. (3)) is used to optimize the model, making the model use all images indiscriminately.

We find that when augmented images are generated by using inappropriate pretext tasks, this optimization method (*i.e.*, considering the information of all images on average) may destroy the performance of the FSL task (see Fig. 3). This is because the augmented images bring ambiguous semantic information (*e.g.*, rotation for symmetrical object) [6,9,25]. Although it can be solved using expert experience to select appropriate pretext tasks for different datasets, which is very labor consuming [26]. To this end, we believe that it is very important that the model can adaptively select augmented image features to improve the performance of the FSL task. In the SSL setting (Fig. 1 (c)), it retains FSL as the main task and uses pretext tasks as additional auxiliary tasks (*i.e.*, SSL tasks). From Fig. 3, we find that SSL using independent label spaces (*e.g.*, y_i = dog, $\tilde{y}_i^1 = 90°$, $\tilde{y}_i^2 = $ GBR) to learn these tasks separately (see Eq. (4)) can alleviate the problem caused by DA training based on one single label space, but it is not enough to fully learn the knowledge hidden in these augmented images only through the sharing network. This is because there are similarities among augmented images generated under the same raw image and different pretext tasks, and the relationships among these augmented images should be modeled.

To effectively learn knowledge from augmented images, we propose a plug-in *Hierarchical Tree Structure-aware (HTS)* method for few-shot classification. The core of the proposed HTS method: (1) using a tree structure to model the relationships of raw and augmented images; (2) using a gated aggregator to adaptively select the feature representations to improve the performance of the FSL task. Next, we outline two key components of the proposed HTS method.

Modeling the Relationships via a Hierarchical Tree Constructing Component. This component aims to construct a tree structure for each raw image, and thus we can use the edge of the tree to connect the feature information among different augmented images and use the level of the tree to learn the feature representations from different pretext tasks. In addition, when pretext tasks or augmented images change (*e.g.*, adding, deleting or modifying), it is very flexible for our HTS method to change the number of levels or nodes.

Adaptively Learning Features via a Gated Selection Aggregating Component. In this paper, we use *Tree-based Long Short Term Memory (TreeLSTM)* [36] as the gated aggregator following the reasons as below: (1) On the above trees, we find that the augmented images (*i.e.*, nodes) from different pretext tasks can be further formulated as a sequence with variable lengths from the bottom to the top level. (2) TreeLSTM generates a forgetting gate for each child node, which is used to filter the information of the corresponding child nodes (different colors are shown in Fig. 1 (d) and (e)). This indicates that the representations of the lower-level nodes can be sequentially aggregated and enhance the upper-level nodes' outputs. Finally, these aggregated representations will be used in training and testing phases. The main contributions of HTS are:

1. We point out the limitations of using pretext tasks to help the few-shot model learn richer and transferable feature representations. To solve these limitations, we propose a hierarchical tree structure-aware method.
2. We propose a hierarchical tree constructing component to model the relationships of augmented and raw images and a gated selection aggregating component to adaptively learn and improve the performance of the FSL task.
3. Extensive experiments on four benchmark datasets demonstrate that the proposed HTS is significantly superior to the state-of-the-art FSL methods.

2   Related Work

2.1   Few-Shot Learning

The recent few-shot learning works are dominated by meta-learning based methods. They can be roughly categorized into two groups: (1) *Optimization-based methods* advocate learning a suitable initialization of model parameters from base classes and transferring these parameters to novel classes in a few gradient steps [3,10,27,31,45]. (2) *Metric-based methods* learn to exploit the feature similarities by embedding all images into a common metric space and using well-designed nearest neighbor classifiers [14,19,20,42]. In this paper, our HTS can equip with an arbitrary meta-learning based method and improve performance.

2.2   Pretext Tasks

Pretext tasks have succeeded to learn useful representations by focusing on richer semantic information of images to significantly improve the performance of image classification. In this paper, we are mainly concerned with the works of using pretext tasks to improve the performance of few-shot classification [5,11,12,16,43]. However, these works are often shallow, *e.g.*, the original FSL training pipeline is intact and an additional loss (self-supervised loss) on each image is introduced, leading to the learning process not being able to fully exploit the augmented image representations. Different from these works, we introduce a hierarchical tree structure (HTS) to learn the pretext tasks. Specifically, the relationships of each image are modeled to learn more semantic knowledge. Moreover, HTS can adaptively select augmented features to avoid the interference of ambiguous information. Our experimental results show that thanks to the reasonable learning of pretext tasks, our HTS method clearly outperforms these works (see Table 2).

3   Preliminaries

3.1   Problem Setting in Few-Shot Learning

We consider that meta-learning based methods are used to solve the few-shot classification problem, and thus follow the episodic (or task) training paradigm. In the meta-training phase, we randomly sample episodes from a base class set $\mathcal{D}_b$ to imitate the meta-testing phase sampled episodes from a novel class set $\mathcal{D}_n$. Note that $\mathcal{D}_b$ contains a large number of labeled images and classes but has a *disjoint* label space with $\mathcal{D}_n$ (*i.e.* $\mathcal{D}_b \cap \mathcal{D}_n = \varnothing$). Each n-way k-shot episode $\mathcal{T}_e$ contains a support set $\mathcal{S}_e$ and a query set $\mathcal{Q}_e$. Concretely, we first randomly sample a set of n classes $\mathcal{C}_e$ from $\mathcal{D}_b$, and then generate $\mathcal{S}_e = \{(x_i, y_i)|y_i \in \mathcal{C}_e, i = 1, \cdots, n \times k\}$ and $\mathcal{Q}_e = \{(x_i, y_i)|y_i \in \mathcal{C}_e, i = 1, \cdots, n \times q\}$ by sampling k support and q query images from each class in $\mathcal{C}_e$, and $\mathcal{S}_e \cap \mathcal{Q}_e = \varnothing$. For simplicity, we denote $l_k = n \times k$ and $l_q = n \times q$. In the meta-testing phase, the trained few-shot learning model is fine-tuned using the support set $\mathcal{S}_e$ and is tested using the query set $\mathcal{Q}_e$, where the two sets are sampled from the novel class set $\mathcal{D}_n$.

3.2   Few-Shot Learning Classifier

We employ ProtoNet [33] as the few-shot learning (FSL) model for the main instantiation of our HTS framework due to its simplicity and popularity. However, we also show that any meta-learning based FSL method can be combined with our proposed HTS method (see results in Table 3). ProtoNet contains a feature encoder E_ϕ with learnable parameters ϕ (e.g., CNN) and a simple non-parametric classifier. In each episode $\mathcal{T}_e = \{\mathcal{S}_e, \mathcal{Q}_e\}$, ProtoNet computes the mean feature embedding of the support set for each class $c \in \mathcal{C}_e$ as the prototype $\tilde{p}_c$:

$$\tilde{p}_c = \frac{1}{k} \sum_{(x_i, y_i \in \mathcal{S}_e)} E_\phi(x_i) \cdot \mathbb{I}(y_i = c), \tag{1}$$

where $\mathbb{I}$ is the indicator function with its output being 1 if the input is true or 0 otherwise. Once the class prototypes are obtained from the support set, ProtoNet computes the distances between the feature embedding of each query set image and that of the corresponding prototypes. The final loss function over each episode using the empirical risk minimization (ERM) is defined as follows:

$$
\begin{aligned}
L_{FSL}(\mathcal{S}_e, \mathcal{Q}_e) &= \frac{1}{|\mathcal{Q}_e|} \sum_{(x_i, y_i \in \mathcal{Q}_e)} -log\ p_{y_i}, \\
p_{y_i} &= \frac{exp(-d(E_\phi(x_i), \tilde{p}_{y_i}))}{\sum_{c \in \mathcal{C}_e} exp(-d(E_\phi(x_i), \tilde{p}_c))},
\end{aligned}
\tag{2}
$$

where $d(\cdot, \cdot)$ denotes a distance function (e.g., the squared euclidean distance for ProtoNet method following the original paper [33]).

4   Methodology

4.1   Pretext Tasks in FSL

Pretext tasks assisting in few-shot learning have two settings: data augmentation (DA) and self-supervised learning (SSL) (see the schematic in Fig. 1). We first define a set of pretext-task operators $\mathcal{G} = \{g_j | j = 1, \cdots, J\}$, where g_j means the operator of using the j-th pretext task and J is the total number of pretext tasks. Moreover, we also use M_j to represent the number of augmented images generated by using the j-th pretext task for each raw image and the pseudo label set of this task is defined as $\tilde{Y}^j = \{0, \cdots, M_j - 1\}$. For example, for a 2D-rotation operator, each raw image will be rotated with multiples of 90° angles (e.g., 90°, 180°, 270°), where the augmented images are $M_{rotation} = 3$ and the pseudo label set is $\tilde{Y}^{rotation} = \{0, 1, 2\}$. Given a raw episode $\mathcal{T}_e = \{\mathcal{S}_e, \mathcal{Q}_e\}$ as described in Sect. 3.1, we utilize these pretext-task operators from $\mathcal{G}$ in turn to augment each image in $\mathcal{T}_e$. This results in a set of J augmented episodes are $\mathcal{T}_{aug} = \{(x_i, y_i, \tilde{y}_i, j) | y_i \in \mathcal{C}_e, \tilde{y}_i \in \tilde{Y}^j, i = 1, \cdots, M_j \times l_k, M_j \times (l_k + 1), \cdots, M_j \times (l_k + l_q), j = 1, \cdots, J\}$, where the first images $M_j \times l_k$ are from the augmented support set $\mathcal{S}_e^j$ and the rest of images $M_j \times l_q$ from the augmented query set $\mathcal{Q}_e^j$.

Data Augmentation. For DA setting, we use the combined episodes $\mathcal{T} = \{\{\mathcal{S}_e^r, \mathcal{Q}_e^r\}|r = 0, \cdots, J\}$, where $\{\mathcal{S}_e^0, \mathcal{Q}_e^0\}$ is the raw episode and $\{\{\mathcal{S}_e^r, \mathcal{Q}_e^r\}|r = 1, \cdots, J\}$ is the augmented episodes. In this paper, when $r \geq 1$, the value of r is equal to j unless otherwise stated. Each image (x_i, y_i) in $\mathcal{T}$ takes the same class label y_i (from the human annotation) for supervised learning to improve the performance of the FSL. The objective is to minimize a cross-entropy loss:

$$L_{DA} = \frac{1}{J+1} \sum_{r=0}^{J} L_{FSL}(\mathcal{S}_e^r, \mathcal{Q}_e^r). \tag{3}$$

L_{DA} uses the empirical risk minimization (ERM) algorithm based on the same label space (*e.g.*, y_i) to learn raw and augmented feature representations. However, if the augmented images have ambiguous representations, this optimization method may interfere with the semantic learning of the FSL model.

Self-Supervised Learning. For SSL setting, each raw image (x_i, y_i) in $\mathcal{T}_e$ uses a class label y_i for supervised learning, while each augmented image $(x_i, \tilde{y}_i)$ in $\mathcal{T}_{aug}$ carries a pseudo label $\tilde{y}_i$ for self-supervised learning. A multi-task learning loss (FSL main task and SSL auxiliary task) is normally adopted as below:

$$L_{SSL} = L_{FSL}(\mathcal{S}_e, \mathcal{Q}_e) + \sum_{j=1}^{J} \beta_j L_j,$$

$$L_j = \frac{1}{E} \sum_{(x_i, \tilde{y}_i \in \mathcal{T}_e^j)} -\log \frac{exp([\theta^j(E_\phi(x_i))]_{\tilde{y}_i})}{\sum_{\tilde{y}'}^{\tilde{Y}^j} exp([\theta^j(E_\phi(x_i))]_{\tilde{y}'})}, \tag{4}$$

where $E = M_j \times (l_k + l_q)$, $[\theta^j(E_\phi(x_i))]$ denotes the j-th pretext task scoring vector and $[.]_{\tilde{y}}$ means taking the $\tilde{y}$-th element. L_{SSL} learns the knowledge of the few-shot learning and multiple sets of pretext tasks in different label spaces, but only uses a shared feature encoder E_ϕ to exchange these semantic information.

4.2   Pretext Tasks in HTS

In this paper, we propose a hierarchical tree structure-aware (HTS) method that uses a tree to model relationships and adaptively selects knowledge among different image features. There are two key components in HTS: hierarchical tree constructing component and gated selection aggregating component[1].

4.2.1   Hierarchical Tree Constructing Component
Given augmented episodes $\mathcal{T}_{aug} = \{\mathcal{T}_e^j = \{\mathcal{S}_e^j, \mathcal{Q}_e^j\}|j = 1, \cdots, J\}$ as described in Sect. 4.1, each augmented episode in $\mathcal{T}_{aug}$ corresponds to one specific pretext task of the same set of images from the raw episode $\mathcal{T}_e$. Therefore, we believe that these augmented images with different pretext tasks should be modeled to capture the correlations and further learn more semantic feature representations.

---

[1] Note that our method mainly focuses on how to adaptively learn the knowledge of pretext tasks and improve the performance of few-shot image classification.

To this end, we construct a tree structure for each FSL image and its corresponding multiple sets of augmented images generated by using different pretext tasks in each episode. Specifically, (1) we extract the feature vectors of the raw and augmented episodes by using the shared feature encoder E_ϕ and denote the feature set as $\mathcal{T}_{emd}$, where $\mathcal{T}_{emd} = \{E_\phi(x_i)|(x_i, y_i, \tilde{y}_i, r) \in \mathcal{T}_e^r, r = 0, \cdots, J, i = 1, \cdots, M_j \times (l_k + l_q)\}$. (2) The feature vectors of the raw episode $\mathcal{T}_{emd}^0$ are taken as the root nodes and each raw image has its own tree structure. (3) The augmented feature vectors of these augmented episodes $\mathcal{T}_{emd}^j$ with the j-th pretext task are put in the $(j + 1)$-th level of the tree. (4) We take a raw image x_i (i.e., a tree structure) and its multiple sets of augmented images $\{x_i^j\}$ as an example to indicate how to construct this tree, and repeat the process for other raw images. The form of the tree is $\{E_\phi(x_i) \xrightarrow{g_1} E_\phi(x_i^1) \xrightarrow{g_2} \cdots \xrightarrow{g_j} E_\phi(x_i^j) \cdots \xrightarrow{g_J} E_\phi(x_i^J)\}$, where the raw feature set $E_\phi(x_i)$ is in the 1-st level (root nodes) and the augmented feature set is in the $(j + 1)$-th level sharing the same pretext task g_j. For each episode, we construct $(l_k + l_q)$ hierarchical tree structures, and every level has M_j child nodes with the j-th pretext task. In these tree structures, the edge information is used to model the relationships of different augmented or raw images. The level knowledge is used to learn the representations from different pretext tasks. In Sect. 4.2.2, we introduce how to better aggregate image features.

4.2.2 Gated Selection Aggregating Component

As mentioned above, we have constructed a tree structure for each raw image x_i in each randomly sampled episode. The following question is how to effectively use this tree structure for learning and inference. Firstly, our intuition is to preserve the tree structure information, because it models the relationships among images (e.g., FSL and augmentations). Secondly, we should selectively aggregate the features of all child nodes from the bottom to the top level, because the information aggregated process aims to maximize the performance of the parent node after aggregation. Thirdly, since the feature information in the hierarchical propagation can be regarded as sequential inputs divided by levels

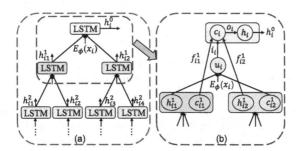

Fig. 2. The learning process of gated selection aggregating component. (a) shows that the aggregator to sequentially and hierarchically aggregate the information from bottom to top level. (b) details the internal aggregation of TreeLSTM (e.g., two levels). The subscript marked in this figure represents the number of child nodes, the superscript represents the number of levels and different colors are different LSTM cells.

but meta-learning can not directly process the sequential data. Finally, we adopt the *tree-based Long Short Term Memory (TreeLSTM)* as our gated aggregator to encode the lower-level information into an upper-level output. In this way, we can mine much richer features from the tree structure (see Fig. 2). Next, we will detail the aggregation and propagation process of TreeLSTM on these trees.

For simplicity, we take a tree structure as an example to introduce how to selectively aggregate information by using TreeLSTM aggregator and repeat the process for other tree structures. The $(J+1)$-level tree structure are constructed based on raw image features $E_\phi(x_i)$ and its multiple sets of augmented features $\{E_\phi(x_i^j)|i = 1, \cdots, M_j, j = 1, \cdots, J\}$. The TreeLSTM aggregates information step by step from bottom level (*i.e.*, $(J+1)$-th level) to top level (*i.e.*, 1-st or root node level). We use $\{h_i^r|r = 0, \cdots, J\}$ to represent the aggregated node representations of each level except the bottom-level nodes in this tree. Because the bottom node has no child nodes, its aggregate information is itself $\{E_\phi(x_i^J)\}$. The aggregation process can be formalized as: $\{h_i^0 \xleftarrow{agg} h_i^1 \xleftarrow{agg} \cdots \xleftarrow{agg} h_i^r \cdots \xleftarrow{agg} E_\phi(x_i^J)\}$, where agg denotes the aggregation operation by using the TreeLSTM. The aggregated output h_i^r of each level is represented as:

$$h_i^r = TreeLSTM(s_i, \{h_m\}), m \in \mathcal{M}_i, \tag{5}$$

where $s_i \in \{h_i^r\}$ is any node in the tree and $\mathcal{M}_i$ is the set of child nodes of the i-th node in $(r+1)$ level of the tree.

Since this form of child-sum in TreeLSTM conditions its components on the sum of child hidden states h_m, it is a permutation invariant function and is well-suited for trees whose children are unordered. But, we find that using the child-mean replaces the child-sum in our implementation for better normalization, compared with the original paper [36]. The formulation of the TreeLSTM is:

$$f_m = \sigma(W_f s_i + U_f h_m + b_f),$$

$$h_{me} = \frac{\sum_{m \in \mathcal{M}_i} h_m}{|\mathcal{M}_i|}, \quad u_i = tanh(W_u s_i + U_u h_{me} + b_u),$$

$$o_i = \sigma(W_o s_i + U_o h_{me} + b_o), \quad i_i = \sigma(W_i s_i + U_i h_{me} + b_i),$$

$$c_i = i_i \odot u_i + \frac{\sum_{m \in \mathcal{M}_i} f_m \odot c_m}{|\mathcal{M}_i|}, \quad h_i^r = o_i \odot tanh(c_i),$$

$$\tag{6}$$

where $\odot$ denotes the element-wise multiplication, σ is sigmoid function, W_a, U_a, b_a, $a \in \{i, o, f, u\}$ are trainable parameters, c_i, c_k are memory cells, h_i, h_m are hidden states and h_{me} denotes the child nodes mean in $(r+1)$-th level. From Eq. (6) and Fig. 2, the aggregator learns a forgetting gate for each child and uses it to adaptively select the child features that are beneficial to its parent node based on the trained parameters. Finally, the aggregated episodes are denoted as $\mathcal{T}_{agg} = \{\{\mathcal{S}_{agg}^r, \mathcal{Q}_{agg}^r\}|r = 0, \cdots, J\}$, where $\mathcal{S}_{agg}^r = \{h_i^r|i = 1, \cdots, M_r * l_k)\}$ and $\mathcal{Q}_{agg}^r = \{h_i^r|i = 1, \cdots, M_r * l_q)\}$ with $M_0 = 1, M_r = M_j$ when $r \geq 1$.

4.3 Meta-training Phase with HTS

The meta-training models with HTS use previous settings (DA or SSL) to train the FSL model. Given the aggregated episodes $\mathcal{T}_{agg} = \{\{\mathcal{S}_{agg}^r, \mathcal{Q}_{agg}^r\} | r = 0, \cdots, J\}$, where $\{\mathcal{S}_{agg}^0, \mathcal{Q}_{agg}^0\}$ is the raw representations aggregated by using its all augmented images (*i.e.*, from $(J + 1)$-th to 2-nd of the tree) and $\{\mathcal{S}_{agg}^r, \mathcal{Q}_{agg}^r\}_{r=1}^J$ is the augmented representations with each pretext task aggregated by using other augmented images (*e.g.*, from $(J + 1)$-th to r-th of the tree).

Data Augmentation. For data augmentation (DA) setting, we only use the aggregated root node (raw images) $\{\mathcal{S}_{agg}^0, \mathcal{Q}_{agg}^0\}$ to train the FSL model. The cross-entropy loss can be defined as:

$$L_{DA}^{HTS} = L_{FSL}(\mathcal{S}_{agg}^0, \mathcal{Q}_{agg}^0). \tag{7}$$

L_{HTS}^{DA} uses the aggregated representations with more knowledge to train the few-shot image classification model, different from Eq. (3).

Self-Supervised Learning. For SSL setting, the aggregated root nodes train the FSL main task and each aggregated augmented node $(h_i^r, \tilde{y}_i)$ in $\{\mathcal{T}_{agg}^r = \{\mathcal{S}_{agg}^r, \mathcal{Q}_{agg}^r\} | r = 1, \cdots, J\}$ trains the SSL auxiliary task. With the pseudo label $\tilde{y}_i$ and every level aggregated features in the tree, the multi-task learning loss is:

$$L_{SSL}^{HTS} = L_{FSL}(\mathcal{S}_{agg}^0, \mathcal{Q}_{agg}^0) + \sum_{r=1}^J \beta_r L_r,$$
$$L_r = \frac{1}{E} \sum_{(h_i^r, \tilde{y}_i \in \mathcal{T}_{agg}^r)} -log \frac{exp([\theta^r(h_i^r)]_{\tilde{y}_i})}{\sum_{\tilde{y}'}^{\tilde{Y}^j} exp([\theta^r(h_i^r)]_{\tilde{y}'})}, \tag{8}$$

where E and $[\theta^r(\cdot)]_{\tilde{y}}$ have the same meanings as Eq. (4). And when $r \geq 1$, $\theta_r = \theta_j$. For easy reproduction, we summarize the overall algorithm for FSL with HTS in Algorithm 1 in Appendix A.1. Notably, our HTS is a lightweight method, consisting of a parameter-free tree constructing component and a simple gated aggregator component. Therefore, as a plug-in method, HTS will not introduce too much additional computing overhead[2]. Once trained, with the learned network parameters, we can perform the testing over the test episodes.

4.4 Meta-testing Phase with HTS

During the meta-testing stage, we find that using pretext tasks for both the query set and support set, and then aggregating features based on the trained aggregator can further bring gratifying performance with only a few raw images (see Fig. 4). Therefore, the predicted label for $x_i \in \mathcal{Q}_e$ can be computed with Eq. (1) and aggregated raw features h_i^0 as:

$$y_i^{pred} = \underset{y \in \mathcal{C}_e}{argmax} \frac{exp(-d(h_i^0, \tilde{p}_{y_i}))}{\sum_{c \in \mathcal{C}_e} exp(-d(h_i^0, \tilde{p}_c))}. \tag{9}$$

---

[2] During training, for a 5-way 1-/5-shot setting, one episode time is 0.45/0.54 s (0.39/0.50 s for baseline) with 75 query images over 500 randomly sampled episodes.

4.5   Connections to Prior Works

Hierarchical Few-Shot Learning. Recently, some works have been proposed to solve the problem of few-shot learning using the hierarchical structure. One representative work aims to improve the effectiveness of meta-learning by hierarchically clustering different tasks based on the level of model parameter [40], while our method is based on the feature level with less running time and computation. Another work [23] learns the relationship between fine-grained and coarse-grained images through hierarchical modeling, but it requires the data itself to be hierarchical, while our method can adapt to any dataset.

Long Short Term Memory Few-Shot Learning. Some works [21,38,41] use chain-based long short-term memory (ChainLSTM) to learn the relationship of images. However, our work uses tree-based long short-term memory (TreeLSTM) to learn structured features and meanwhile preserve tree-structured information.

5   Experimental Results

5.1   Experimental Setup

Research Questions. Research questions guiding the remainder of the paper are as follows: **RQ1.** What is the real performance of the heuristic combination of FSL main task and pretext auxiliary tasks? **RQ2.** How effective is the proposed HTS framework for the FSL image classification task in both single-domain and cross-domain settings? **RQ3.** Could the proposed HTS framework adaptively select augmented features for better improving the performance of the FSL main task? **RQ4.** How does the proposed HTS method work (ablation study)?

Benchmark Datasets. All experiments are conducted on four FSL benchmark datasets, *i.e.*, *mini*ImageNet [38], *tiered*ImageNet [29], CUB-200-2011 [39] and CIFAR-FS [3]. The *mini*ImageNet and *tiered*ImageNet are the subsets of the ILSVRC-12 dataset [8]. CUB-200-2011 is initially designed for fine-grained classification. The resolution of all images in the three datasets is resized to 84×84. CIFAR-FS is a subset of CIFAR-100 and each image is resized to 32×32. See Appendix A.3 for more details on the four few-shot benchmark datasets.

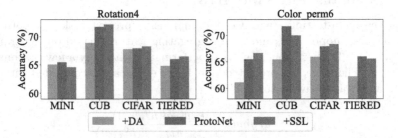

Fig. 3. The results (5-shot) for motivation on *mini*ImageNet (MINI), CUB-200-2011 (CUB), CIFAR-FS (CIFAR) and *tiered*ImageNet (TIERED) with two pretext tasks.

Implementation Details. We adopt the episodic training procedure under 5-way 1-shot and 5-shot settings [38]. In each episode, 15 unlabeled query images per class are used for the training and testing phases. We apply Conv4 (filters:64) and ResNet12 (filters: [64, 160, 320, 640]) as the encoder. Our model is trained from scratch and uses the Adam optimizer with an initial learning rate 10^{-3}. The hyperparameter $\beta_j = 0.1$ for all experiments. Each mini-batch contains four episodes and we use a validation set to select the best-trained model. For all methods, we train 60,000 episodes for 1-shot and 40,000 episodes for 5-shot [4]. We use PyTorch with one NVIDIA Tesla V100 GPU to implement all experiments and report the ***average accuracy (%)*** with 95% *confidence interval* over the 10,000 randomly sampled testing episodes. For our proposed HTS framework, we consider ProtoNet as our FSL model unless otherwise stated, but note that our framework is broadly applicable to other meta-learning based FSL models. More implementation details are given in Appendix A.3.

Pretext Tasks. Following [16], as the entire input images during training is important for image classification, we choose two same pretext tasks: *rotation* and *color permutation*. We also give some subsets of the two tasks to meet the needs of experiments. The rotation task has **rotation1** (90°), **rotation2** (90°, 180°), **rotation3** (90°, 180°, 270°) and **rotation4** (0°, 90°, 180°, 270°). The color permutation task has **color_perm1** (GBR), **color_perm2** (GBR, BRG), **color_perm3** (RGB, GBR, BRG) and **color_perm6** (RGB, GBR, BRG, GBR, GRB, BRG, BGR). Note that our method can use arbitrary pretext tasks as it adaptively learns efficient features and improves the performance of FSL task.

5.2 RQ1. Performance of Pretext Tasks in FSL

The experimental results in Fig. 3 illustrate the key motivation of this paper and answer RQ1. Intuitively, when we use the pretext tasks to generate multiple sets of augmented images, it can learn more knowledge and improve the performance of the FSL model, like [26]. However, we find that not arbitrary pretext task has good performance on downstream datasets, and we need to find suitable tasks

Table 1. Classification accuracy (%) results comparison with 95% confidence intervals for cross-domain evaluation with **rotation3**. More results see Appendix A.4.

Training →Testing	Method		1-shot	5-shot
*mini*Imagenet→ CUB-200-2011	ProtoNet		32.18±0.25	55.95±0.21
		+HTS	39.57±0.17	60.56±0.18
*tiered*ImageNet→ CUB-200-2011	ProtoNet		39.47±0.22	56.58±0.25
		+HTS	42.24±0.20	60.71±0.18
*tiered*ImageNet→ *mini*ImageNet	ProtoNet		47.01±0.26	66.82±0.25
		+HTS	55.29±0.20	72.67±0.16

Table 2. Accuracy (%) with **rotation3**. Best results are displayed in boldface. Train/test-SSL/-DA mean using data augmentation and self-supervised learning during training/testing. KD means knowledge distillation and & means "and" operation.

Method	Backbone	Tricks	miniImageNet		CUB-200-2011		CIFAR-FS		tieredImageNet	
			1-shot	5-shot	1-shot	5-shot	1-shot	5-shot	1-shot	5-shot
MAML [10]	Conv4-32	Fine-tuning	48.70 ± 1.84	55.31 ± 0.73	55.92 ± 0.95	72.09 ± 0.76	58.90 ± 1.90	71.50 ± 1.00	51.67 ± 1.81	70.30 ± 1.75
PN [11]	Conv4-64	Train-SSL	53.63 ± 0.43	71.70 ± 0.36	–	–	64.69 ± 0.32	80.82 ± 0.24	–	–
CC [11]	Conv4-64	Train-SSL	54.83 ± 0.43	71.86 ± 0.33	–	–	63.45 ± 0.31	79.79 ± 0.24	–	–
closer [4]	Conv4-64	Train-DA	48.24 ± 0.75	66.43 ± 0.63	60.53 ± 0.83	79.34 ± 0.61	–	–	–	–
CSS [1]	Conv4-64	Train-SSL	50.85 ± 0.84	68.08 ± 0.73	66.01 ± 0.90	81.84 ± 0.59	56.49 ± 0.93	74.59 ± 0.72	–	–
SLA [16]	Conv4-64	Train-SSL	44.95 ± 0.79	63.32 ± 0.68	48.43 ± 0.82	71.30 ± 0.72	45.94 ± 0.87	68.62 ± 0.75	–	–
PSST [5]	Conv4-64	Train-SSL	57.04 ± 0.51	73.17 ± 0.48	–	–	64.37 ± 0.33	80.42 ± 0.32	–	–
CAN [13]	Conv4-64	Train-DA	52.04 ± 0.00	65.54 ± 0.00	–	–	–	–	52.45 ± 0.00	66.81 ± 0.00
HTS (ours)	Conv4-64	Train&test-SSL	58.96 ± 0.18	75.17 ± 0.14	67.32 ± 0.24	78.09 ± 0.15	64.71 ± 0.21	76.45 ± 0.17	53.20 ± 0.22	72.38 ± 0.19
shot-Free [28]	ResNet12	Train-DA	59.04 ± 0.43	77.64 ± 0.39	–	–	69.20 ± 0.40	84.70 ± 0.40	66.87 ± 0.43	82.64 ± 0.39
MetaOpt [18]	ResNet12	Train-DA	62.64 ± 0.61	78.63 ± 0.46	–	–	72.00 ± 0.70	84.20 ± 0.50	65.81 ± 0.74	82.64 ± 0.39
Distill [37]	ResNet12	Train&test-DA&KD	64.82 ± 0.60	82.14 ± 0.43	–	–	–	–	71.52 ± 0.69	86.03 ± 0.49
HTS (ours)	ResNet12	Train&test-SSL	64.95 ± 0.18	83.89 ± 0.15	72.88 ± 0.22	85.32 ± 0.13	73.95 ± 0.22	85.36 ± 0.14	68.38 ± 0.23	86.34 ± 0.18

for different datasets. The results of ProtoNet+DA are significantly lower than the baseline (ProtoNet), which proves our concern that it is unreasonable for empirical risk minimization (ERM) to treat all samples equally (see Eq. (3)). For ProtoNet+SSL, the results are slightly higher than the baseline in most cases, which indicates that only using a shared encoder is not enough (see Eq. (4)).

5.3   RQ2. Performance of Pretext Tasks in HTS

To answer RQ2, we conduct experiments on both single domain and cross domain with comparison to few-shot learning methods using the benchmark datasets.

Performance on Single Domain. Table 2 reports the average classification accuracy. For fairness, we use the same backbone network to compare to state-of-the-art meta-learning based FSL methods. From Table 2, we have the following findings: (1) HTS improves the performance of the ProtoNet in most settings to achieve a new state of the arts. This is because that our framework has the advantage of modeling relationships among these images and adaptively learning augmented features via a gated selection aggregating component. This observation demonstrates our motivation and the effectiveness of our framework. (2)

Table 3. Accuracy (%) by incorporating HTS (two-level trees) into each method with rotation3 and the Conv4-64. Best results are displayed in **boldface** and performance improvement in red text. Appendix A.2 shows the formulation of these methods.

Methods under 5-way		miniImageNet		CUB-200-2011		CIFAR-FS		tieredImageNet	
		1-shot	5-shot	1-shot	5-shot	1-shot	5-shot	1-shot	5-shot
ProtoNet [33]		44.42	64.24	51.31	70.77	51.90	69.95	49.35	67.28
	+HTS_DA	57.39	74.25	66.88	77.37	65.01 (+13.11)	75.23	52.92	70.18
	+HTS_SSL	58.96 (+14.54)	75.17 (+10.93)	67.32 (+16.01)	78.09 (+7.32)	64.71	76.45 (+6.50)	53.20 (+3.85)	72.38 (+5.10)
RelationNet [35]		49.31	66.30	62.45	76.11	55.00	69.30	54.48	71.32
	+HTS_DA	52.78	73.22	65.63	80.67 (+4.56)	57.68	73.09	55.44	72.88 (+1.56)
	+HTS_SSL	53.24 (+3.93)	73.57 (+7.27)	67.38 (+4.93)	79.00	58.60 (+3.60)	73.15 (+3.85)	56.09 (+0.61)	71.78
MatchingNet [38]		48.14	63.48	61.16	72.86	53.00	60.23	54.02	70.11
	+HTS_DA	53.64 (+5.50)	63.67	63.29	73.17	55.57 (+2.57)	62.58	56.39	72.01
	+HTS_SSL	52.29	64.36 (+0.88)	63.54 (+4.93)	74.76 (+1.90)	55.18	63.87 (+3.64)	57.16 (+3.14)	72.60 (+2.49)
GNN [32]		49.02	63.50	51.83	63.69	45.59	65.62	43.56	55.31
	+HTS_DA	59.06	72.80	61.69	73.46	52.35	71.66	55.32	69.48
	+HTS_SSL	60.52 (+11.50)	74.63 (+11.13)	62.85 (+11.02)	77.58 (+13.89)	58.31 (+12.72)	73.24 (+7.62)	55.73 (+12.17)	70.42 (+15.11)

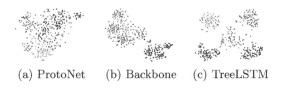

(a) ProtoNet   (b) Backbone   (c) TreeLSTM

Fig. 4. (a)–(c) represent the t-SNE of the five class features. (b) and (c) are our method.

One observation worth highlighting is that HTS not only outperforms traditional meta-learning based methods, but also is superior to methods using pretext tasks under DA or SSL settings. (3) Compared to [37], the results further show that augmenting the query set can bring more benefits during the testing phase.

Performance on Cross Domain. In Table 1, we show the testing results on testing domains using the model trained on training domains. The setting is challenging due to the domain gap between training and testing datasets. The results clearly show that (1) HTS has significant performance in all cross-domain settings and obtains consistent improvements. (2) When CUB-200-2011 is selected as the testing domain, the transferred performance of different training domains has a large difference. It is caused by the degrees of domain gap. From the last two rows, we find that all methods from *tiered*ImageNet to *mini*ImageNet have a large improvement compared with CUB-200-2011, because the two datasets come from the same database (ILSVRC-12), leading to a small domain gap.

Performance with Other Meta-learning Based Methods. To further verify the effectiveness of HTS, we embed it into four meta-learning based methods: ProtoNet [33], RelationNet [35], MatchingNet [38] and GNN [32]. Table 3 reports the accuracy and we have the following findings: (1) our HTS is flexible and can be combined with any meta-learning method, making the performance of these methods significantly improved in all datasets. (2) In terms of our method, the HTS_SSL performs better than HTS_DA in most cases, indicating that a single label space is not best for learning the knowledge carried by these pretext tasks.

T-SNE Visualization. For the qualitative analysis, we also apply t-SNE [15] to visualize the embedding distribution obtained before and after being equipped with HTS in ProtoNet. As shown in Fig. 4, (b) and (c) represent the features obtained without and with using the TreeLSTM aggregator. The results show that (1) our HTS can learn more compact and separate clusters indicating that the learned representations are more discriminative. (2) Considering our method, (c) is better than (b), which once again proves the effectiveness of the aggregator.

5.4   RQ3. Adaptive Selection Aggregation

One of the most important properties of our framework is that the learned augmented features are different for different pretext tasks or child nodes. Thus, in this subsection, we investigate if the proposed framework can adaptively learn

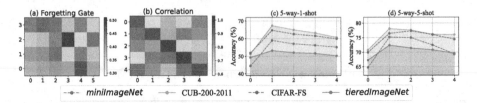

Fig. 5. (a) is the forgetting gate for different child nodes (vertical axis) with the same raw image (horizontal axis). (b) is the correlations of different augmented images (vertical axis is rotation1, horizontal axis is color_perm1), where the diagonal (or off-diagonal) line is based on the same raw (or different raw) images. (c)-(d) are classification accuracy (%) for different numbers of levels under ProtoNet with HTS.

different forgetting gates for different child nodes, which aims to answer RQ3. We visualize the average forgetting value (*i.e.*, f_m in Eq. (6)) for each child node with the same raw image randomly sampled from CUB-200-2011. In Fig. 5 (a), we show the results of six raw images as we have similar observations on other images and datasets. The darkness of a step represents the probability that the step is selected for aggregation. We can find that (1) different types of child nodes exhibit different forgetting values based on the same raw image. (2) Different raw images focus on different child node features. These results further explain that the contributions of different pretext tasks to the raw images are not equal. In Fig. 5 (b), we also provide the correlation, *i.e.*, the cosine similarity of learned features of different types of augmentation based on the same raw image. We clearly observe the correlations between augmentation with the same raw image are large while the correlation between different raw images is small, which meets our expectation that multiple sets of augmentation generated based on the same raw image generally have similar relationships.

5.5   RQ4. Ablation Study

To answer RQ4, we conduct experiments to show the performance under different pretext tasks, child nodes and levels. More experiments see Appendix A.5.

The Effect of Different Pretext Tasks. To show the performance of different pretext tasks, we construct experiments based on the two-level trees (*i.e.* only using a pretext task) for each specific pretext task with the same number of child nodes on the *mini*ImageNet and CUB-200-2011 under the 5-way 1-shot setting. Figure 6 (a) and (b) show the results. Along the horizontal axis, the number 0 indicates the baseline (*i.e.*, ProtoNet) only using raw (FSL) images. The rotation1/2/3 and color_perm1/2/3 are used for the numbers 1/2/3, respectively. We follow these findings: (1) all pretext tasks significantly improve the classification performance and outperform the baseline. These results indicate that our HTS does not need to manually define a suitable pretext task, and it can adaptively learn useful information. (2) The color permutation task brings a higher boost

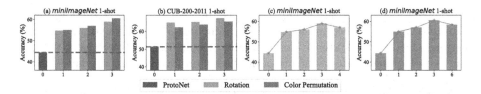

Fig. 6. (a) and (b) indicate different pretext tasks with the same number of child nodes. (c) and (d) indicate different numbers of child nodes with same pretext task.

on *mini*ImageNet and the rotation task has a good performance on CUB-200-2011. It is because for fine-grained image classification, rotation angles can carry affluent representations, but for a large of images, the color gain is higher.

The Effect of Different Numbers of Child Nodes. We verify the performance of the same pretext task with different numbers of child nodes on the *mini*ImageNet under the 5-way 1-shot setting. Figure 6 (c) and (d) show the results. Along the horizontal axis, these integers are introduced in Sect. 5.1. From Fig. 6 (c)–(d), we have the following findings: (1) our proposed HTS method with the two-level trees outperforms the ProtoNet in terms of all settings, which demonstrates the effectiveness of our method to learn pretext tasks. (2) The performance with different numbers of child nodes is different. The reason is that the increase of child nodes is accompanied by large model complexity, leading to overfitting and affecting the generalization ability of the FSL model.

The Effect of Different Numbers of Tree Levels. We further verify the performance of HTS with different numbers of levels on the four datasets under the 5-way 1-shot and 5-shot settings. Figure 5 (c) and (d) show the results. Along the horizontal axis, the integer 1 indicates only using FSL images, 2 using [rotation3], 3 using [rotation3; color_perm2], 4 using [rotation3; color_perm2; color_perm3] and 5 using [rotation3; color_perm2; color_perm3; rotation2]. We find that our HTS method with different tree levels outperforms the original baseline (ProtoNet) using only raw (FSL) images. The different number of tree levels bring model complexity and over-fitting problems, resulting in performance differences among different levels. To save computational time and memory, two-level trees are used, which achieves the new state-of-the-art performance.

6   Conclusion

In this paper, we study the problem of learning richer and transferable representations for few-shot image classification via pretext tasks. We propose a plug-in *hierarchical tree structure-aware (HTS)* method that constructs tree structures to address this problem. Specifically, we propose the hierarchical tree constructing component, which uses the edge information to model the relationships of different augmented and raw images, and uses the level to explore the knowledge among different pretext tasks. On the constructed tree structures, we also introduce the gated selection aggregating component, which uses the forgetting gate to adaptively select and aggregate the augmented features and improve the performance of FSL. Extensive experimental results demonstrate that our HTS can achieve a new state-of-the-art performance on four benchmark datasets.

Acknowledgments. This work was supported by the National Science and Technology Innovation 2030 - Major Project (Grant No. 2022ZD0208800), and NSFC General Program (Grant No. 62176215). We thank Dr. Zhitao Wang for helpful feedback and discussions.

References

1. An, Y., Xue, H., Zhao, X., Zhang, L.: Conditional self-supervised learning for few-shot classification. In: International Joint Conference on Artificial Intelligence, IJCAI (2021)
2. Bateni, P., Goyal, R., Masrani, V., Wood, F., Sigal, L.: Improved few-shot visual classification. In: IEEE Conference on Computer Vision and Pattern Recognition, CVPR, pp. 14481–14490 (2020)
3. Bertinetto, L., Henriques, J.F., Torr, P.H.S., Vedaldi, A.: Meta-learning with differentiable closed-form solvers. In: International Conference on Learning Representations, ICLR (2019)
4. Chen, W., Liu, Y., Kira, Z., Wang, Y.F., Huang, J.: A closer look at few-shot classification. In: International Conference on Learning Representations, ICLR (2019)
5. Chen, Z., Ge, J., Zhan, H., Huang, S., Wang, D.: Pareto self-supervised training for few-shot learning. In: IEEE Conference on Computer Vision and Pattern Recognition, CVPR (2021)
6. Cubuk, E.D., Zoph, B., Mané, D., Vasudevan, V., Le, Q.V.: AutoAugment: learning augmentation strategies from data. In: IEEE Conference on Computer Vision and Pattern Recognition, CVPR (2019)
7. Cui, W., Guo, Y.: Parameterless transductive feature re-representation for few-shot learning. In: International Conference on Machine Learning, ICML (2021)
8. Deng, J., Dong, W., Socher, R., Li, L., Li, K., Li, F.: ImageNet: a large-scale hierarchical image database. In: IEEE Conference on Computer Vision and Pattern Recognition, CVPR (2009)
9. Feng, Z., Xu, C., Tao, D.: Self-supervised representation learning by rotation feature decoupling. In: IEEE Conference on Computer Vision and Pattern Recognition, CVPR (2019)
10. Finn, C., Abbeel, P., Levine, S.: Model-agnostic meta-learning for fast adaptation of deep networks. In: International Conference on Machine Learning, ICML (2017)
11. Gidaris, S., Bursuc, A., Komodakis, N., Pérez, P., Cord, M.: Boosting few-shot visual learning with self-supervision. In: International Conference on Computer Vision, ICCV (2019)
12. Gidaris, S., Singh, P., Komodakis, N.: Unsupervised representation learning by predicting image rotations. In: International Conference on Learning Representations, ICLR (2018)
13. Hou, R., Chang, H., Ma, B., Shan, S., Chen, X.: Cross attention network for few-shot classification. In: Advances in Neural Information Processing Systems, NeurIPS (2019)

14. Kang, D., Kwon, H., Min, J., Cho, M.: Relational embedding for few-shot classification. In: International Conference on Computer Vision, ICCV (2021)
15. Laurens, V.D.M., Hinton, G.: Visualizing data using t-SNE. J. Mach. Learn. Res. **9**, 2579–2605 (2008)
16. Lee, H., Hwang, S.J., Shin, J.: Self-supervised label augmentation via input transformations. In: International Conference on Machine Learning, ICML (2020)
17. Li, F., Fergus, R., Perona, P.: A bayesian approach to unsupervised one-shot learning of object categories. In: International Conference on Computer Vision, ICCV (2003)
18. Li, H., Eigen, D., Dodge, S., Zeiler, M., Wang, X.: Finding task-relevant features for few-shot learning by category traversal. In: IEEE Conference on Computer Vision and Pattern Recognition, CVPR (2019)
19. Li, W., Wang, L., Xu, J., Huo, J., Gao, Y., Luo, J.: Revisiting local descriptor based image-to-class measure for few-shot learning. In: IEEE Conference on Computer Vision and Pattern Recognition, CVPR (2019)
20. Li, W., Xu, J., Huo, J., Wang, L., Gao, Y., Luo, J.: Distribution consistency based covariance metric networks for few-shot learning. In: Association for the Advancement of Artificial Intelligence, AAAI (2019)
21. Li, Z., Zhou, F., Chen, F., Li, H.: Meta-SGD: learning to learn quickly for few shot learning. CoRR abs/1707.09835 (2017)
22. Liu, L., Hamilton, W.L., Long, G., Jiang, J., Larochelle, H.: A universal representation transformer layer for few-shot image classification. In: International Conference on Learning Representations, ICLR. OpenReview.net (2021)
23. Liu, L., Zhou, T., Long, G., Jiang, J., Zhang, C.: Learning to propagate for graph meta-learning. In: Advances in Neural Information Processing Systems, NeurIPS, pp. 1037–1048 (2019)
24. Liu, S., Davison, A.J., Johns, E.: Self-supervised generalisation with meta auxiliary learning. In: Advances in Neural Information Processing Systems, NeurIPS (2020)
25. Misra, I., van der Maaten, L.: Self-supervised learning of pretext-invariant representations. In: IEEE Conference on Computer Vision and Pattern Recognition, CVPR (2020)
26. Ni, R., Goldblum, M., Sharaf, A., Kong, K., Goldstein, T.: Data augmentation for meta-learning. In: International Conference on Machine Learning, ICML (2021)
27. Qiao, S., Liu, C., Shen, W., Yuille, A.L.: Few-shot image recognition by predicting parameters from activations. In: IEEE Conference on Computer Vision and Pattern Recognition, CVPR (2018)
28. Ravichandran, A., Bhotika, R., Soatto, S.: Few-shot learning with embedded class models and shot-free meta training. In: International Conference on Computer Vision, ICCV (2019)
29. Ren, M., et al.: Meta-learning for semi-supervised few-shot classification. In: International Conference on Learning Representations, ICLR (2018)
30. Requeima, J., Gordon, J., Bronskill, J., Nowozin, S., Turner, R.E.: Fast and flexible multi-task classification using conditional neural adaptive processes. In: Advances in Neural Information Processing Systems, NeurIPS, pp. 7957–7968 (2019)
31. Rusu, A.A., et al.: Meta-learning with latent embedding optimization. In: International Conference on Learning Representations, ICLR (2019)
32. Satorras, V.G., Estrach, J.B.: Few-shot learning with graph neural networks. In: International Conference on Learning Representations, ICLR (2018)
33. Snell, J., Swersky, K., Zemel, R.S.: Prototypical networks for few-shot learning. In: Advances in Neural Information Processing Systems, NeurIPS (2017)

34. Su, J.-C., Maji, S., Hariharan, B.: When does self-supervision improve few-shot learning? In: Vedaldi, A., Bischof, H., Brox, T., Frahm, J.-M. (eds.) ECCV 2020. LNCS, vol. 12352, pp. 645–666. Springer, Cham (2020). https://doi.org/10.1007/978-3-030-58571-6_38

35. Sung, F., Yang, Y., Zhang, L., Xiang, T., Torr, P.H.S., Hospedales, T.M.: Learning to compare: relation network for few-shot learning. In: IEEE Conference on Computer Vision and Pattern Recognition, CVPR (2018)

36. Tai, K.S., Socher, R., Manning, C.D.: Improved semantic representations from tree-structured long short-term memory networks. In: Association for Computational Linguistics, ACL (2015)

37. Tian, Y., Wang, Y., Krishnan, D., Tenenbaum, J.B., Isola, P.: Rethinking few-shot image classification: a good embedding is all you need? In: Vedaldi, A., Bischof, H., Brox, T., Frahm, J.-M. (eds.) ECCV 2020. LNCS, vol. 12359, pp. 266–282. Springer, Cham (2020). https://doi.org/10.1007/978-3-030-58568-6_16

38. Vinyals, O., Blundell, C., Lillicrap, T., Kavukcuoglu, K., Wierstra, D.: Matching networks for one shot learning. In: Advances in Neural Information Processing Systems, NeurIPS (2016)

39. Wah, C., Branson, S., Welinder, P., Perona, P., Belongie, S.: The Caltech-UCSD Birds-200-2011 Dataset. California Institute of Technology (CNS-TR-2011-001) (2011)

40. Yao, H., Wei, Y., Huang, J., Li, Z.: Hierarchically structured meta-learning. In: International Conference on Machine Learning, ICML. Proceedings of Machine Learning Research, vol. 97, pp. 7045–7054. PMLR (2019)

41. Ye, H., Hu, H., Zhan, D., Sha, F.: Few-shot learning via embedding adaptation with set-to-set functions. In: IEEE Conference on Computer Vision and Pattern Recognition, CVPR (2020)

42. Zhang, C., Cai, Y., Lin, G., Shen, C.: DeepEMD: few-shot image classification with differentiable earth mover's distance and structured classifiers. In: IEEE Conference on Computer Vision and Pattern Recognition, CVPR, pp. 12200–12210 (2020)

43. Zhang, M., Zhang, J., Lu, Z., Xiang, T., Ding, M., Huang, S.: IEPT: instance-level and episode-level pretext tasks for few-shot learning. In: International Conference on Learning Representations, ICLR (2020)

44. Zhang, M., Huang, S., Wang, D.: Domain generalized few-shot image classification via meta regularization network. In: ICASSP, pp. 3748–3752 (2022)

45. Zhang, M., Wang, D., Gai, S.: Knowledge distillation for model-agnostic meta-learning. In: European Conference on Artificial Intelligence, ECAI (2020)

Inductive and Transductive Few-Shot Video Classification via Appearance and Temporal Alignments

Khoi D. Nguyen[1], Quoc-Huy Tran[2], Khoi Nguyen[1], Binh-Son Hua[1], and Rang Nguyen[1]

[1] VinAI Research, Hanoi, Vietnam
[2] Retrocausal, Inc., Redmond, USA

Abstract. We present a novel method for few-shot video classification, which performs appearance and temporal alignments. In particular, given a pair of query and support videos, we conduct appearance alignment via frame-level feature matching to achieve the appearance similarity score between the videos, while utilizing temporal order-preserving priors for obtaining the temporal similarity score between the videos. Moreover, we introduce a few-shot video classification framework that leverages the above appearance and temporal similarity scores across multiple steps, namely prototype-based training and testing as well as inductive and transductive prototype refinement. To the best of our knowledge, our work is the first to explore transductive few-shot video classification. Extensive experiments on both Kinetics and Something-Something V2 datasets show that both appearance and temporal alignments are crucial for datasets with temporal order sensitivity such as Something-Something V2. Our approach achieves similar or better results than previous methods on both datasets. Our code is available at https://github.com/VinAIResearch/fsvc-ata.

Keywords: Few-shot learning · Video classification · Appearance alignment · Temporal alignment · Inductive inference · Transductive inference

1 Introduction

Recognizing video contents plays an important role in many real-world applications such as video surveillance [34,51], anomaly detection [17,41], video retrieval [13,39], and action segmentation [21,22]. In the modern era of deep learning, there exist a large number of studies focusing on learning to classify videos by fully supervising a neural network with a significant amount of labeled data [6,43,44,47]. While these fully-supervised approaches provide

Supplementary Information The online version contains supplementary material available at https://doi.org/10.1007/978-3-031-20044-1_27.

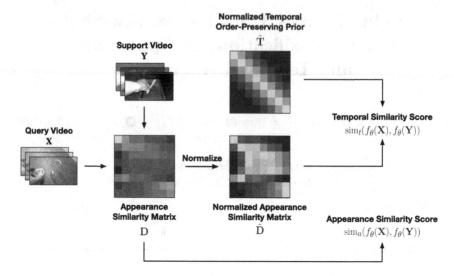

Fig. 1. Our Few-Shot Video Classification Approach. Given a pair of query and support videos, we first perform appearance alignment (i.e., via frame-level feature matching) to compute the appearance similarity score, and then temporal alignment (i.e., by leveraging temporal order-preserving priors) to calculate the temporal similarity score. The final alignment score between the two videos is the weighted sum of appearance and temporal similarity scores.

satisfactory results, the high costs of data collection and annotation make it unrealistic to transfer an existing network to new tasks. To reduce such high costs, few-shot learning [11,25,31,32,36–38,42,45] is an emerging trend that aims to adapt an existing network to recognize new classes with limited training data. Considerable research efforts have been invested in few-shot learning on images [9,10,14,24,29,46,50]. Extending few-shot learning to videos has been rather limited.

The main difference between a video and an image is the addition of temporal information in video frames. To adapt few-shot learning for videos, recent emphasis has been put on temporal modeling that allows the estimate of the similarity between two videos via frame-to-frame alignment. In the few-shot setting, this similarity function is crucial because it helps classify a query video by aligning it with the given support videos. Early methods [1,2,4,12,53] achieve low performance as they neglect temporal modeling and simply collapse the temporal information in their video representation learning. Recent methods [5,28] jointly consider appearance and temporal information by using Dynamic Time Warping [8] or Optimal Transport [7] to align the videos.

We propose to separately consider appearance and temporal alignments, yielding robust similarity functions for use in training and testing in both inductive and transductive few-shot learning. Following previous works [5,57], we sparsely sample a fixed number of frames from each video and extract their corresponding features using a neural network-based feature extractor. We then compute the pairwise cosine similarity between frame features of the two videos, yielding the appearance similarity matrix. To compute the appearance similarity score between the videos, we first match each frame from one video to the most similar frame in the other, ignoring their temporal order. We then define our appearance similarity score between the two videos as the total similarity of all matched frames between them. Next, motivated by the use of temporal order-preserving priors in different video understanding tasks [5,16,23,40] (i.e., initial frames from a source video should be mapped to initial frames in a target video, and similarly the subsequent frames from the source and target video should match accordingly), we encourage the appearance similarity matrix to be as similar as possible to the temporal order-preserving matrix. Our temporal similarity score between the two videos is then computed as the negative Kullback-Leibler divergence between the above matrices. Furthermore, we show how to apply the above appearance and temporal similarity scores in different stages of few-shot video classification, from the prototype-based training and testing procedures to the finetuning of prototypes in inductive and transductive settings. Our method achieves the state-of-the-art performance on both Kinetics [20] and Something-Something V2 [15] datasets. Figure 1 illustrates our ideas.

In summary, our contributions include:

– We introduce a novel approach for few-shot video classification leveraging appearance and temporal alignments. Our main contribution includes an appearance similarity score based on frame-level feature matching and a temporal similarity score utilizing temporal order-preserving priors.
– We incorporate the above appearance and temporal similarity scores into various steps of few-shot video classification, from the prototype-based training and testing procedures to the refinement of prototypes in inductive and transductive setups. To our best knowledge, our work is the first to explore transductive few-show video classification.
– Extensive evaluations demonstrate that our approach performs on par with or better than previous methods in few-shot video classification on both Kinetics and Something-Something V2 datasets.

2   Related Work

Few-Shot Learning. Few-shot learning aims to extract task-level knowledge from seen data while effectively generalizing learned meta-knowledge to unknown tasks. Based on the type of meta-knowledge they capture, few-shot learning techniques can be divided into three groups. Memory-based methods [30,37] attempt to solve few-shot learning by utilizing external memory. Metric-based methods [18,19,32,38,42,45,49,52] learn an embedding space such that samples of the same class are mapped to nearby points whereas samples of different classes are mapped to far away points in the embedding space. Zhang et al. [52] introduce the earth mover's distance to few-shot learning, which performs spatial alignment between images. Optimization-based approaches [11,25,31,36,55] train a model to quickly adapt to a new task with a few gradient descent iterations. One notable work is MAML [11], which learns an initial representation that can effectively be finetuned for a new task with limited labelled data. A new group of works [9,24,50] utilize information from meta-training dataset explicitly during meta-testing to boost the performance. While Le et al. [24] and Das et al. [9] use base samples as distractors for refining classifiers, Yang et al. [50] finds similar features from base data to augment the support set.

Few-Shot Video Classification. Zhu and Yang [56] extend the notion of memory-based meta-learning for video categorization by learning many key-value pairs to represent a video. However, the majority of existing few-shot video classification methods are based on metric-based meta-learning, in which a generalizable video metric is learned from seen tasks and applied to novel unseen tasks. The main difference between a video and an image is the extra temporal dimension of the video frames. Extending few-shot learning to videos thus requires the temporal modeling of this extra dimension. This field of research can be divided in two main groups: aggregation-based and matching-based. The former prioritizes semantic contents, whereas latter is concerned with temporal orderings. Both of these groups start by sampling frames or segments from a video and obtaining their representations using a pretrained encoder. However, aggregation-based methods generate a video-level representation for distance calculation via pooling [1,4], adaptive fusion [2,12], or attention [53], whereas matching-based methods explicitly align two sequences and define distance between the videos as their matching cost via different matching techniques such as Dynamic Time Warping (DTW) [5] and Optimal Transport (OT) [28]. While [5] performs temporal alignment between videos via DTW, which strictly enforces a monotonic frame sequence ordering, [28] handles permutations using OT with the iterative Sinkhorn-Knopp algorithm, which can be practically slow and poorly scaled [48]. Our proposed method balances between appearance and temporal alignments which allows permutations to some extent. Moreover, it is non-iterative, hence more computationally efficient. Lastly, [5] and [28] explore the inductive setting only.

Transductive Inference. Transductive learning approaches exploit both the labeled training data and the unlabelled testing data to boost the performance. Many previous works [3,18,24,33,58] explicitly utilize unsupervised information from the query set to augment the supervised information from the support set. With access to more data, transductive inference methods typically outperform their inductive counterparts. However, existing transductive methods are designed mostly for few-shot image classification. In this work, we develop a cluster-based transductive learning approach with a novel assignment function leveraging both temporal and appearance information to address few-shot video classification.

3   Our Method

We present, in this section, the details of our approach. We first describe the problem of few-shot video classification in Sect. 3.1. Next, we introduce our appearance and temporal similarity scores in Sect. 3.2, which are then used in our prototype-based training and testing presented in Sect. 3.3. Lastly, we discuss how the prototypes are refined in both inductive and transduction settings in Sect. 3.4.

3.1   Problem Formulation

In few-shot video classification, we are given a base set $\mathcal{D}_b = \{(\mathbf{X}_i, y_i)\}_{i=1}^{T_b}$, where $\mathbf{X}_i$ and y_i denote a video sample and its corresponding class label for N_b classes, respectively, while T_b is the number of samples. This base set is used for training a neural network which is subsequently adapted to categorize unseen videos with novel classes. At test time, we are given a set of support videos $\mathcal{D}_n^s = \{(\mathbf{X}_i, y_i)\}_{i=1}^{N \times K}$, where N and K denote the number of novel classes and the number of video samples per novel class, respectively. Note that the novel classes do not overlap with the base classes. The support set provides a limited amount of data to guide the knowledge transfer from the base classes to the novel classes. The goal at test time is to classify the query videos $\mathcal{D}_n^q$ into one of these novel classes. Such configuration is called an N-way K-shot video classification task. In this paper, we explore two configurations: 5-way 1-shot and 5-way 5-shot video classification.

There are two settings for few-shot learning: inductive and transductive learning. In the former, each of the query videos are classified independently, whereas, in the latter, the query videos are classified collectively, allowing unlabeled visual cues to be shared and leveraged among the query videos, which potentially improves the overall classification results. In this work, we consider both inductive and transductive settings. To our best knowledge, our work is the first to explore transductive learning for few-shot video classification.

Following prior works [5,57], we represent a video by a fixed number of M frames randomly sampled from equally separated M video segments, or $\mathbf{X} = [\mathbf{x}^1, \ldots, \mathbf{x}^M]$ with $\mathbf{x}^i \in \mathbb{R}^{3 \times H \times W}$ and $i \in \{1, \ldots, M\}$, while H and W are the width and height of the video frame, respectively. Next, a neural network f with parameters θ is used to extract a feature vector $f_\theta(\mathbf{x}^i) \in \mathbb{R}^C$ (C is the number of channels) for each sampled frame $\mathbf{x}^i$. Lastly, the features of a video $\mathbf{X}$ is represented by $f_\theta(\mathbf{X}) = [f_\theta(\mathbf{x}^1), \ldots, f_\theta(\mathbf{x}^M)] \in \mathbb{R}^{C \times M}$.

3.2   Appearance and Temporal Similarity Scores

Existing distance/similarity functions are either computationally inefficient [28], imposing too strong constraint [5], or oversimplified that they neglect temporal information [57]. To capture both appearance and temporal information with low computational costs, we propose to explore appearance and temporal cues via two simple yet novel similarity functions. The final prediction is then a linear

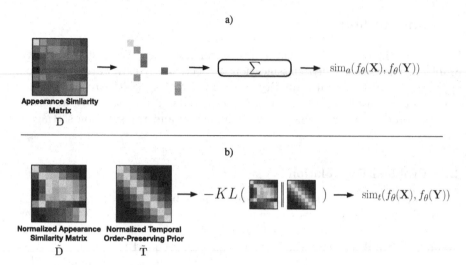

Fig. 2. Appearance and Temporal Similarity Scores. a) The proposed appearance similarity score is computed as the sum of the frame-level maximum appearance similarity scores between frames in $\mathbf{X}$ and frames in $\mathbf{Y}$. b) The proposed temporal similarity score is based on the Kullback-Leibler divergence between the row-wise normalized appearance similarity matrix $\tilde{\mathbf{D}}$ and the row-wise normalized temporal order-preserving prior $\tilde{\mathbf{T}}$.

combination of the predictions from the two functions. We detailed our similarity functions, below.

Appearance Similarity Score. Given a pair of videos $(\mathbf{X}, \mathbf{Y})$, we first compute the appearance similarity matrix $\mathbf{D} \in \mathbb{R}^{M \times M}$ between them. The element $\mathbf{D}(i,j)$ of $\mathbf{D}$ is the pairwise cosine similarity between frame $\mathbf{x}^i$ in $\mathbf{X}$ with frame $\mathbf{y}^j$ in $\mathbf{Y}$ as follows:

$$\mathbf{D}(i,j) = \frac{f_\theta(\mathbf{x}^i)^T f_\theta(\mathbf{y}^j)}{\|f_\theta(\mathbf{x}^i)\| \, \|f_\theta(\mathbf{y}^j)\|}. \tag{1}$$

For every frame $\mathbf{x}^i$ in $\mathbf{X}$, we can align it with the frame $\mathbf{y}^k$ in $\mathbf{Y}$ which has the highest appearance similarity score with $\mathbf{x}^i$ (i.e., $k = \operatorname{argmax}_j \mathbf{D}(i,j)$), ignoring their relative ordering. We define the appearance similarity score between $\mathbf{X}$ and $\mathbf{Y}$ as the sum of the optimal appearance similarity scores for all frames in $\mathbf{X}$ as:

$$\operatorname{sim}_a(f_\theta(\mathbf{X}), f_\theta(\mathbf{Y})) = \sum_{i=1}^{M} \max_j \mathbf{D}(i,j) \approx \sum_{i=1}^{M} \lambda \log \sum_{j=1}^{M} \exp^{\frac{\mathbf{D}(i,j)}{\lambda}}, \tag{2}$$

where we use *log-sum-exp* (with the smoothing temperature $\lambda = 0.1$) to continuously approximate the *max* operator. Intuitively, the $\operatorname{sim}_a(f_\theta(\mathbf{X}), f_\theta(\mathbf{Y}))$ shows how similar in appearance frames in $\mathbf{X}$ to frames in $\mathbf{Y}$. Note that sim_a is not a symmetric function. Figure 2(a) shows the steps to compute our appearance similarity score.

Temporal Similarity Score. Temporal order-preserving priors have been employed in various video understanding tasks such as sequence matching [40], few-shot video classification [5], video alignment [16], and activity segmentation [23]. In particular, given two videos $\mathbf{X}$ and $\mathbf{Y}$ of the same class, it encourages initial frames in $\mathbf{X}$ to be aligned with initial frames in $\mathbf{Y}$, while subsequent frames in $\mathbf{X}$ are encouraged to be aligned with subsequent frames in $\mathbf{Y}$. Mathematically, it can be modeled by a 2D distribution $\mathbf{T} \in \mathbb{R}^{M \times M}$, whose marginal distribution along any line perpendicular to the diagonal is a Gaussian distribution centered at the intersection on the diagonal, as:

$$\mathbf{T}(i,j) = \frac{1}{\sigma\sqrt{2\pi}} \exp^{-\frac{l^2(i,j)}{2\sigma^2}}, \quad l(i,j) = \frac{|i-j|}{\sqrt{2}}, \tag{3}$$

where $l(i,j)$ is the distance from the entry (i,j) to the diagonal line and the standard deviation parameter $\sigma = 1$. The values of $\mathbf{T}$ peak on the diagonal and gradually decrease along the direction perpendicular to the diagonal.

In this work, we adopt the above temporal order-preserving prior for few-shot video classification. Let $\tilde{\mathbf{D}}$ and $\tilde{\mathbf{T}}$ denote the row-wise normalized version of $\mathbf{D}$ and $\mathbf{T}$ respectively, as:

$$\tilde{\mathbf{D}}(i,j) = \frac{\exp^{\mathbf{D}(i,j)}}{\sum_{k=1}^{M} \exp^{\mathbf{D}(i,k)}}, \quad \tilde{\mathbf{T}}(i,j) = \frac{\mathbf{T}(i,j)}{\sum_{k=1}^{M} \mathbf{T}(i,k)}. \tag{4}$$

We define the temporal similarity score between $\mathbf{X}$ and $\mathbf{Y}$ as the negative Kullback-Leibler (KL) divergence between $\tilde{\mathbf{D}}$ and $\tilde{\mathbf{T}}$:

$$\text{sim}_t(f_\theta(\mathbf{X}), f_\theta(\mathbf{Y})) = -KL(\tilde{\mathbf{D}}||\tilde{\mathbf{T}}) = -\frac{1}{M}\sum_{i=1}^{M}\sum_{j=1}^{M}\tilde{\mathbf{D}}(i,j)\log\frac{\tilde{\mathbf{D}}(i,j)}{\tilde{\mathbf{T}}(i,j)}. \tag{5}$$

Intuitively, $\text{sim}_t(f_\theta(\mathbf{X}), f_\theta(\mathbf{Y}))$ encourages the temporal alignment between $\mathbf{X}$ and $\mathbf{Y}$ to be as similar as possible to the temporal order-preserving alignment $\mathbf{T}$. Figure 2(b) summarizes the steps to compute our temporal similarity score.

3.3 Training and Testing

Training. We employ the global training with prototype-based classifiers [38] in our approach. A set of prototypes $\mathbf{W} = [\mathbf{W}_1, \ldots, \mathbf{W}_{N_b}]$ are initialized randomly, where each prototype $\mathbf{W}_p = [\mathbf{w}_p^1, \ldots, \mathbf{w}_p^M] \in \mathbb{R}^{C \times M}$ represents a class in the base set. Note that in contrast to few-shot image classification [38], where each prototype is a single feature vector, in our approach for few-shot video classification, each prototype is a sequence of M feature vectors. To learn $\mathbf{W}$, we adopt the below supervised loss:

$$\mathcal{L}_{Sup} = -\mathbb{E}_{(\mathbf{X},y)\sim\mathcal{D}_b}\log\frac{\exp^{\text{sim}_a(f_\theta(\mathbf{X}),\mathbf{W}_y))}}{\sum_{p=1}^{N_b}\exp^{\text{sim}_a(f_\theta(\mathbf{X}),\mathbf{W}_p))}} - \alpha\mathbb{E}_{(\mathbf{X},y)\sim\mathcal{D}_b}\text{sim}_t(f_\theta(\mathbf{X}),\mathbf{W}_y). \tag{6}$$

Here, α is the balancing weight between the two terms, and the appearance and temporal similarity scores are computed between the features of a video and the prototype of a class. We empirically observe that the temporal order-preserving prior is effective for Something-Something V2 but not much for Kinetics. This is likely because the actions in Something-Something V2 are order-sensitive, whereas those in Kinetics are not. Therefore, we use both terms in Eq. 6 for training on Something-Something V2 (i.e., $\alpha = 0.05$) but only the first term for training on Kinetics (i.e., $\alpha = 0$).

Solely training with the above supervised loss can lead to a trivial solution that the model only learns discriminative features for each class (i.e., the M feature vectors within each prototype are similar). To avoid such cases, we add another loss which minimizes the entropy of $\tilde{\mathbf{D}}$:

$$\mathcal{L}_{Info} = -\frac{1}{M}\sum_{i=1}^{M}\sum_{j=1}^{M}\tilde{\mathbf{D}}(i,j)\log\tilde{\mathbf{D}}(i,j). \tag{7}$$

The final training loss is a combination of the above losses and is written as:

$$\mathcal{L} = \mathcal{L}_{Sup} + \nu\mathcal{L}_{Info}. \tag{8}$$

Here, $\nu = 0.1$ is the balancing weight.

Testing. At test time, we discard the global prototype-based classifiers, and keep the feature extractor. Given an N-way K-shot episode, we first extract the features of all support and query samples. We then initialize N prototypes by the average features of support samples from the corresponding classes, i.e., $\mathbf{W}_c = \frac{1}{K}\sum_{\mathbf{X}\in\mathcal{S}_c} f_\theta(\mathbf{X})$, where $\mathcal{S}_c$ is the support set of the c-th class. Given the prototypes, we define the predictive distribution over classes of each video sample $p(c|\mathbf{X}, \mathbf{W}), c \in \{1, \dots, N\}$, as:

$$p(c|\mathbf{X},\mathbf{W}) = (1-\beta)\frac{\exp^{sim_a(f_\theta(\mathbf{X}),\mathbf{W}_c)}}{\sum_{j=1}^{N}\exp^{sim_a(f_\theta(\mathbf{X}),\mathbf{W}_j)}} + \beta\frac{\exp^{sim_t(f_\theta(\mathbf{X}),\mathbf{W}_c)}}{\sum_{j=1}^{N}\exp^{sim_t(f_\theta(\mathbf{X}),\mathbf{W}_j)}}, \tag{9}$$

where β is the balancing weight between the two terms. The above predictive distribution is a combination of the *softmax* predictions based on appearance and temporal similarity scores. As we discussed previously, since the actions on Something-Something V2 are order-sensitive but those on Kinetics are not, the temporal order-preserving is effective for Something-Something V2 but not for Kinetics. Therefore, we set $\beta = 0.5$ for Something-Something V2 and $\beta = 0$ for Kinetics. As we empirically show in Sect. 4.1, these settings yield good results.

3.4   Prototype Refinement

The prototypes can be further refined with support samples (in the inductive setting) or with both support and query samples (in the transductive setting) before being used for classification.

Inductive Setting. For the inductive inference, we finetune the prototypes on the support set with the following cross-entropy loss:

$$\mathcal{L}_{inductive} = -\mathbb{E}_{(\mathbf{X},y)\sim\mathcal{D}^s} \log \frac{\exp^{\text{sim}_a(f_\theta(\mathbf{X}),\mathbf{W}_y)}}{\sum_{i=1}^{N} \exp^{\text{sim}_a(f_\theta(\mathbf{X}),\mathbf{W}_i)}}. \tag{10}$$

Transductive Setting. We introduce a transductive inference step that utilizes unsupervised information from the query set to finetune the prototypes. The transductive inference typically has a form of Soft K-means [33] with a novel assignment function. For each update iteration, we refine the prototypes with the weighted sums of the support and query samples:

$$\mathbf{W}_c = \frac{\sum_{\mathbf{X}\sim\mathcal{S}} f_\theta(\mathbf{X})z(\mathbf{X},c) + \sum_{\mathbf{X}\sim\mathcal{Q}} f_\theta(\mathbf{X})z(\mathbf{X},c)}{\sum_{\mathbf{X}\sim\mathcal{S}} z(\mathbf{X},c) + \sum_{\mathbf{X}\sim\mathcal{Q}} z(\mathbf{X},c)}, \tag{11}$$

where $\mathcal{S}$ and $\mathcal{Q}$ are the support and query sets, respectively. The assignment function $z(\mathbf{X},c)$ simply returns the c-th element of the one-hot label vector of $\mathbf{X}$ if the sample is from the support set. For query samples, we directly use the predictive distribution in Eq. 9, i.e., $z(\mathbf{X},c) = p(c|\mathbf{X},\mathbf{W})$ for $\mathbf{X} \in \mathcal{Q}$. In our experiments, the prototypes are updated for 10 iterations.

4   Experiments

There are two parts of experiments in this section. In the first part (Sect. 4.1), we conduct ablation studies to show the effectiveness of appearance and temporal similarity scores. In the second part (Sect. 4.2), we compare our proposed method to state-of-the art methods on two widely used datasets: Kinetics [20], and Something-Something V2 [15]. The evaluations are conducted on both inductive and transductive settings.

Datasets. We perform experiments on two standard few-shot video classification datasets: the few-shot versions of Kinetics [20] and Something-Something V2 [15]. Kinetics [20] contains 10,000 videos, while Something-Something V2 [15] has 71,796 videos. These two datasets are split into 64 training classes, 16 validation classes, and 20 testing classes, following the splits from [56] and [5].

Table 1. Comparison to different aggregation methods in the **inductive** setting on the Kinetics and Something-Something V2 datasets.

Method	Kinetics		Something V2	
	1-shot	5-shot	1-shot	5-shot
OT	61.71 ± 0.41	77.16 ± 0.37	35.60 ± 0.41	47.82 ± 0.44
Max	73.46 ± 0.38	$\mathbf{87.67 \pm 0.29}$	40.97 ± 0.41	58.77 ± 0.43
Ours	$\mathbf{74.26 \pm 0.38}$	87.40 ± 0.30	$\mathbf{43.82 \pm 0.42}$	$\mathbf{61.07 \pm 0.42}$

Implementation Details. For fair comparisons, we follow the preprocessing steps from prior works [5,54,57]. In particular, we first resize the frames of a particular video to 256×256 and perform random cropping (for training) or central cropping (for testing) a 224×224 region from the frames. The number of sampled segments or frames for each video is $M = 8$. We use the same network architecture as previous works, which is a ResNet-50 pretrained on ImageNet [35]. Stochastic gradient descent (SGD) with a momentum of 0.9 is adopted as our optimizer. The model is trained for 25 epochs with an initial learning rate of 0.001 and a weight decay of 0.1 at epoch 20. In the inference stage, we report mean accuracy with 95% confidence interval on 10,000 random episodes.

4.1 Ablation Study

The Effectiveness of Appearance and Temporal Similarity Scores. In this experiment, we compare our appearance and temporal similarity scores with several methods for aggregating the appearance similarity matrix $\mathbf{D}$ in Eq. 2, namely the optimal transport (OT) and the max operator. For OT, we use the negative appearance similarity matrix $-\mathbf{D}$ as the cost matrix and employ the equal partition constraint. It optimally matches frames of two videos without considering temporal information. For the max operator, we replace the log-sum-exp operator in Eq. 2 with the max operator. Table 1 shows their results on Kinetics and Something-Something V2 datasets. Our method outperforms the other two similarity functions by large margins on both datasets with the exception that the max variant performs slightly better than our method in the Kinetics 5-way 5-shot setting (87.67% as compared to 87.40% of our method).

Table 2. Comparison of different training losses in the **inductive** and **transductive** settings on the Kinetics and Something-Something V2 datasets.

$\mathcal{L}_{Sup}$	$\mathcal{L}_{Info}$	Transd	Kinetics		Something V2	
			1-shot	5-shot	1-shot	5-shot
✓			74.25 ± 0.38	87.15 ± 0.30	41.84 ± 0.42	57.26 ± 0.43
✓	✓		74.26 ± 0.38	87.40 ± 0.30	43.82 ± 0.42	61.07 ± 0.42
✓		✓	82.71 ± 0.44	92.91 ± 0.31	47.02 ± 0.54	67.56 ± 0.56
✓	✓	✓	$\mathbf{83.08 \pm 0.44}$	$\mathbf{93.33 \pm 0.30}$	$\mathbf{48.67 \pm 0.54}$	$\mathbf{69.42 \pm 0.55}$

The Effectiveness of Training Losses. We perform analysis of our training losses in Sect. 3.3 on both Kinetics and Something-Something V2 datasets. Results are shown in Table 2. As mentioned earlier, adding $\mathcal{L}_{Info}$ in the training phase helps avoiding trivial solutions and hence consistently improves the performance on both Kinetics and Something-Something V2 datasets. Using the complete training loss with the transductive setting yields the best performance on both Kinetics and Something-Something V2 datasets.

The Relative Importance of Appearance and Temporal Cues. In this experiment, we investigate the effectiveness of the hyperparameter β in the inductive and transductive settings. More specifically, we train our model with the loss in Eq. 6 and evaluate the result on the validation set with different values of β. The results of the inductive and transductive settings are shown in Tab. 3.

Table 3. Ablation study on the relative importance of the appearance and temporal terms in computing the predictive distribution and the assignment function of the **inductive** and **transductive** inference on the Kinetics and Something-Something V2 datasets.

β	Transd.	Kinetics 1-shot	Kinetics 5-shot	Something V2 1-shot	Something V2 5-shot
1.00	✗	21.09 ± 0.34	27.00 ± 0.39	25.28 ± 0.37	33.57 ± 0.41
0.90		24.02 ± 0.36	41.75 ± 0.44	28.09 ± 0.38	43.21 ± 0.43
0.80		28.94 ± 0.38	62.66 ± 0.43	32.24 ± 0.40	54.45 ± 0.43
0.70		39.75 ± 0.42	79.99 ± 0.35	39.13 ± 0.42	62.73 ± 0.42
0.60		58.87 ± 0.43	86.37 ± 0.30	45.40 ± 0.43	65.99 ± 0.41
0.50		74.05 ± 0.37	87.33 ± 0.29	48.05 ± 0.43	$\mathbf{66.91 \pm 0.40}$
0.40		75.16 ± 0.37	$\mathbf{87.38 \pm 0.29}$	$\mathbf{48.55 \pm 0.42}$	66.83 ± 0.40
0.30		75.34 ± 0.36	87.38 ± 0.29	48.54 ± 0.42	66.64 ± 0.41
0.20		75.37 ± 0.36	87.37 ± 0.29	48.36 ± 0.43	66.32 ± 0.41
0.10		$\mathbf{75.38 \pm 0.36}$	87.36 ± 0.29	48.21 ± 0.43	66.06 ± 0.41
0.00		75.37 ± 0.36	87.35 ± 0.29	48.13 ± 0.43	65.77 ± 0.41
1.00	✓	54.83 ± 0.49	66.61 ± 0.50	39.53 ± 0.48	51.46 ± 0.51
0.90		63.88 ± 0.49	80.48 ± 0.43	43.00 ± 0.50	58.45 ± 0.53
0.80		72.38 ± 0.48	88.71 ± 0.36	46.56 ± 0.53	65.33 ± 0.54
0.70		78.09 ± 0.45	91.49 ± 0.32	50.03 ± 0.54	70.46 ± 0.53
0.60		80.73 ± 0.44	92.66 ± 0.31	52.36 ± 0.55	73.42 ± 0.52
0.50		82.34 ± 0.43	93.24 ± 0.30	53.65 ± 0.55	74.77 ± 0.52
0.40		83.29 ± 0.43	93.51 ± 0.29	54.46 ± 0.55	75.42 ± 0.52
0.30		83.81 ± 0.43	93.59 ± 0.29	$\mathbf{54.67 \pm 0.55}$	$\mathbf{75.47 \pm 0.52}$
0.20		84.06 ± 0.43	$\mathbf{93.60 \pm 0.29}$	54.75 ± 0.55	75.25 ± 0.52
0.10		84.33 ± 0.42	93.55 ± 0.29	54.66 ± 0.55	74.77 ± 0.52
0.00		$\mathbf{84.36 \pm 0.41}$	93.42 ± 0.29	54.4 ± 0.54	73.97 ± 0.51

The two table shows that for order-sensitive actions in Something-Something V2, balancing between the importance of appearance and temporal scores gives

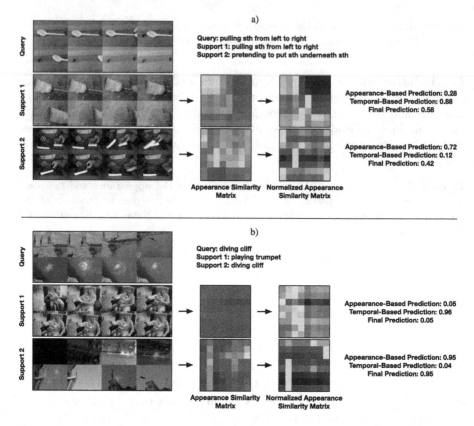

Fig. 3. Qualitative Results of 2-way 1-shot Tasks on Something-Something V2 and Kinetics. For each task, we present the appearance similarity matrix **D** between the query video and each support video in the second column. In the third column, we show the row-wise normalized version **D̃**. Finally, we show the predictions of the two similarity scores and the final prediction. Ground truth class labels are shown at the top. a) Results from Something-Something V2. b) Results from Kinetics.

the best performance. In particular, for the inductive setting, our approach optimally achieves 48.55% for 1-shot with $\beta = 0.4$, whereas the best result for 5-shot is 66.91% with $\beta = 0.5$. Next, $\beta = 0.3$ gives the best results for both 1-shot and 5-shot in the transductive setting, achieving 54.76% and 75.47% respectively. Jointly considering appearance and temporal cues consistently improves the model performance as compared to the appearance-only version ($\beta = 0.0$). In contrast, temporal information does not show much benefits for an order-insensitive dataset like Kinetics. For small values of β, there are minor changes in the performance of our approach for both inductive and transductive settings on Kinetics. In the worst case, the temporal-only version ($\beta = 1.0$) produces nearly random guesses of 21.09% in the 1-shot inductive setting.

We show some qualitative results in Fig. 3. We respectively perform inductive inferences on 2-way 1-shot tasks of Something-Something V2 and Kinetics datasets with $\beta = 0.5$ and $\beta = 0.0$ respectively. On Something-Something V2 (Fig. 3(a)), we observe that appearance or temporal cues can misclassify query samples sometimes, but utilizing both appearance or temporal cues gives correct classifications. On the other hand, the actions on Kinetics (Fig. 3(b)) are not order-sensitive, and hence the temporal similarity score is not meaningful, which agrees with the results in Table 3.

Table 4. Comparison to the state-of-the-art methods in the **inductive** setting on the Kinetics and Something-Something V2 datasets. Results of Meta-Baseline [38] and CMN [56] are reported from [57]. † denotes results from our re-implementation.

Method	Kinetics		Something V2	
	1-shot	5-shot	1-shot	5-shot
Meta-Baseline [38]	64.03 ± 0.41	80.43 ± 0.35	37.31 ± 0.41	48.28 ± 0.44
CMN [56]	65.90 ± 0.42	82.72 ± 0.34	40.62 ± 0.42	51.90 ± 0.44
OTAM [5]	$73.00 \pm n/a$	$85.80 \pm n/a$	$42.80 \pm n/a$	$52.30 \pm n/a$
Baseline Plus [57]†	70.48 ± 0.40	82.67 ± 0.33	43.05 ± 0.41	57.50 ± 0.43
ITANet [54]	73.60 ± 0.20	84.30 ± 0.30	$\mathbf{49.20 \pm 0.20}$	$\mathbf{62.30 \pm 0.30}$
Ours	$\mathbf{74.26 \pm 0.38}$	$\mathbf{87.40 \pm 0.30}$	43.82 ± 0.42	61.07 ± 0.42

4.2 Comparison with Previous Methods

We compare our approach against previous methods [5,26,33,38,54,56,57] in both inductive and transductive settings. More specifically, in the inductive setting, we first consider Prototypical Network [38] from few-shot image classification, which is re-implemented by [57]. In addition, CMN [56], OTAM [5], Baseline Plus [57], and ITANet [54], which are previous methods designed to tackle few-shot video classification, are also considered. The results of Prototypical Network (namely, Meta-Baseline) and CMN are taken from [57]. We re-implement Baseline Plus [57]. The results of OTAM and ITANet are taken from the original papers. Competing methods in the transductive setting include clustering-based methods from few-shot image classification, namely Soft K-means [33], Bayes K-means [26], and Mean-shift [26].

Inductive. We first consider the inductive setting. The results are presented in Tab. 4. As can be seen from the table, our method achieves the best results on the Kinetics dataset. It outperforms all the competing methods by around 1% for 1-shot setting and over 3% for 5-shot setting, establishing the new state of the art. In addition, our method performs comparably with previous works on Something-Something V2, outperforming all the competing methods except for ITANet, which further adopts additional layers on top of the ResNet-50 for self-attention modules.

Transductive. Next, we consider the transductive setting (shown in Tab. 5). As can be seen in Table 5, our method outperforms all the competing methods by large margins, which are around 9% for both 1-shot and 5-shot settings on Kinetics and 2% and 5% for 1-shot and 5-shot settings respectively on Something-Something V2.

Table 5. Comparison to the state-of-the-art methods in **transductive** setting on the Kinetics and Something-Something V2 datasets. Results of other methods are from our re-implementation on the trained feature extractor of [57].

Method	Kinetics		Something V2	
	1-shot	5-shot	1-shot	5-shot
Soft K-means [33]	74.21 ± 0.40	84.13 ± 0.33	46.46 ± 0.46	64.93 ± 0.45
Bayes K-means [26]	70.66 ± 0.40	81.21 ± 0.34	43.15 ± 0.41	59.48 ± 0.42
Mean-shift [26]	70.52 ± 0.40	82.31 ± 0.34	43.15 ± 0.41	60.03 ± 0.43
Ours	$\mathbf{83.08 \pm 0.44}$	$\mathbf{93.33 \pm 0.30}$	$\mathbf{48.67 \pm 0.54}$	$\mathbf{69.42 \pm 0.55}$

5 Limitation Discussion

We propose two similarity functions for aligning appearance and temporal cues of videos. While our results are promising in both inductive and transductive experiments, there remain some limitations. Firstly, our temporal order-preserving prior does not work for all datasets. Utilizing a permutation-aware temporal prior [27] would be an interesting next step. Secondly, we have not leveraged spatial information in our approach yet. Such spatial information could be important for scenarios like modeling left-right concepts. We leave this investigation as our future work.

6 Conclusion

We propose, in this paper, a novel approach for few-shot video classification via appearance and temporal alignments. Specifically, our approach performs frame-level feature alignment to compute the appearance similarity score between the query and support videos, while utilizing temporal order-preserving priors to calculate the temporal similarity score between the videos. The proposed similarity scores are then used across different stages of our few-shot video classification framework, namely prototype-based training and testing, and inductive and transductive prototype enhancement. We show that our similarity scores are most effective on temporal order-sensitive datasets such as Something-Something V2, while our approach produces comparable or better results than previous few-shot video classification methods on both Kinetics and Something-Something V2 datasets. To the best of our knowledge, our work is the first to explore transductive few-shot video classification, which could facilitate more future works in this direction.

References

1. Ben-Ari, R., Nacson, M.S., Azulai, O., Barzelay, U., Rotman, D.: TAEN: temporal aware embedding network for few-shot action recognition. In: CVPR (2021)
2. Bo, Y., Lu, Y., He, W.: Few-shot learning of video action recognition only based on video contents. In: WACV (2020)
3. Boudiaf, M., Ziko, I., Rony, J., Dolz, J., Piantanida, P., Ben Ayed, I.: Information maximization for few-shot learning. NeurIPS (2020)
4. Cao, C., Li, Y., Lv, Q., Wang, P., Zhang, Y.: Few-shot action recognition with implicit temporal alignment and pair similarity optimization. In: CVIU (2021)
5. Cao, K., Ji, J., Cao, Z., Chang, C.Y., Niebles, J.C.: Few-shot video classification via temporal alignment. In: CVPR (2020)
6. Carreira, J., Zisserman, A.: Quo Vadis, action recognition? A new model and the kinetics dataset. In: CVPR (2017)
7. Cuturi, M.: Sinkhorn distances: lightspeed computation of optimal transport. In: NeurIPS (2013)
8. Cuturi, M., Blondel, M.: Soft-DTW: a differentiable loss function for time-series. In: ICML (2017)
9. Das, R., Wang, Y.X., Moura, J.M.: On the importance of distractors for few-shot classification. In: ICCV (2021)
10. Fei, N., Gao, Y., Lu, Z., Xiang, T.: Z-score normalization, hubness, and few-shot learning. In: ICCV (2021)
11. Finn, C., Abbeel, P., Levine, S.: Model-agnostic meta-learning for fast adaptation of deep networks. arXiv preprint arXiv:1703.03400 (2017)
12. Fu, Y., Zhang, L., Wang, J., Fu, Y., Jiang, Y.G.: Depth guided adaptive meta-fusion network for few-shot video recognition. In: ACM MM (2020)
13. Geetha, P., Narayanan, V.: A survey of content-based video retrieval (2008)
14. Ghaffari, S., Saleh, E., Forsyth, D., Wang, Y.X.: On the importance of firth bias reduction in few-shot classification. In: ICLR (2022)
15. Goyal, R., et al.: The "something something" video database for learning and evaluating visual common sense. In: ICCV (2017)
16. Haresh, S., et al.: Learning by aligning videos in time. In: CVPR (2021)
17. Haresh, S., Kumar, S., Zia, M.Z., Tran, Q.H.: Towards anomaly detection in dash-cam videos. In: 2020 IEEE Intelligent Vehicles Symposium (IV), pp. 1407–1414. IEEE (2020)
18. Hou, R., Chang, H., Ma, B., Shan, S., Chen, X.: Cross attention network for few-shot classification. arXiv preprint arXiv:1910.07677 (2019)
19. Kang, D., Kwon, H., Min, J., Cho, M.: Relational embedding for few-shot classification. In: ICCV (2021)
20. Kay, W., et al.: The kinetics human action video dataset. arXiv preprint arXiv:1705.06950 (2017)
21. Khan, H., et al.: Timestamp-supervised action segmentation with graph convolutional networks. In: IEEE/RSJ International Conference on Intelligent Robots and Systems (IROS) (2022)
22. Konin, A., Syed, S.N., Siddiqui, S., Kumar, S., Tran, Q.H., Zia, M.Z.: Retroactivity: rapidly deployable live task guidance experiences. In: IEEE International Symposium on Mixed and Augmented Reality Demonstration (2020)
23. Kumar, S., Haresh, S., Ahmed, A., Konin, A., Zia, M.Z., Tran, Q.H.: Unsupervised activity segmentation by joint representation learning and online clustering. In: CVPR (2022)

24. Le, D., Nguyen, K.D., Nguyen, K., Tran, Q.H., Nguyen, R., Hua, B.S.: POODLE: improving few-shot learning via penalizing out-of-distribution samples. In: NeurIPS (2021)
25. Li, Z., Zhou, F., Chen, F., Li, H.: Meta-SGD: learning to learn quickly for few-shot learning. arXiv preprint arXiv:1707.09835 (2017)
26. Lichtenstein, M., Sattigeri, P., Feris, R., Giryes, R., Karlinsky, L.: TAFSSL: task-adaptive feature sub-space learning for few-shot classification. In: Vedaldi, A., Bischof, H., Brox, T., Frahm, J.-M. (eds.) ECCV 2020. LNCS, vol. 12352, pp. 522–539. Springer, Cham (2020). https://doi.org/10.1007/978-3-030-58571-6_31
27. Liu, W., Tekin, B., Coskun, H., Vineet, V., Fua, P., Pollefeys, M.: Learning to align sequential actions in the wild. arXiv preprint arXiv:2111.09301 (2021)
28. Lu, S., Ye, H.J., Zhan, D.C.: Few-shot action recognition with compromised metric via optimal transport. arXiv preprint arXiv:2104.03737 (2021)
29. Ma, C., Huang, Z., Gao, M., Xu, J.: Few-shot learning via Dirichlet tessellation ensemble. In: ICLR (2022)
30. Munkhdalai, T., Sordoni, A., Wang, T., Trischler, A.: Metalearned neural memory. In: NeurIPS (2019)
31. Oreshkin, B., Rodríguez López, P., Lacoste, A.: TADAM: task dependent adaptive metric for improved few-shot learning. In: NeurIPS (2018)
32. Qiao, S., Liu, C., Shen, W., Yuille, A.L.: Few-shot image recognition by predicting parameters from activations. In: CVPR (2018)
33. Ren, M., et al.: Meta-learning for semi-supervised few-shot classification. arXiv preprint arXiv:1803.00676 (2018)
34. Rodriguez, M., Sivic, J., Laptev, I., Audibert, J.Y.: Data-driven crowd analysis in videos. In: 2011 International Conference on Computer Vision, pp. 1235–1242. IEEE (2011)
35. Russakovsky, O., et al.: ImageNet large scale visual recognition challenge. Int. J. Comput. Vis. 115(3), 211–252 (2015). https://doi.org/10.1007/s11263-015-0816-y
36. Rusu, A.A., et al.: Meta-learning with latent embedding optimization. arXiv preprint arXiv:1807.05960 (2018)
37. Santoro, A., Bartunov, S., Botvinick, M., Wierstra, D., Lillicrap, T.: Meta-learning with memory-augmented neural networks. In: ICML (2016)
38. Snell, J., Swersky, K., Zemel, R.: Prototypical networks for few-shot learning. In: NeurIPS (2017)
39. Snoek, C.G., Worring, M.: Concept-Based Video Retrieval. Now Publishers Inc. (2009)
40. Su, B., Hua, G.: Order-preserving wasserstein distance for sequence matching. In: CVPR (2017)
41. Sultani, W., Chen, C., Shah, M.: Real-world anomaly detection in surveillance videos. In: Proceedings of the IEEE Conference on Computer Vision and Pattern Recognition, pp. 6479–6488 (2018)
42. Sung, F., Yang, Y., Zhang, L., Xiang, T., Torr, P.H., Hospedales, T.M.: Learning to compare: relation network for few-shot learning. In: CVPR (2018)
43. Tran, D., Bourdev, L., Fergus, R., Torresani, L., Paluri, M.: Learning spatiotemporal features with 3D convolutional networks. In: ICCV (2015)
44. Tran, D., Wang, H., Torresani, L., Ray, J., LeCun, Y., Paluri, M.: A closer look at spatiotemporal convolutions for action recognition. In: CVPR (2018)
45. Vinyals, O., Blundell, C., Lillicrap, T., Wierstra, D., et al.: Matching networks for one shot learning. In: NeurIPS (2016)
46. Wang, R., Pontil, M., Ciliberto, C.: The role of global labels in few-shot classification and how to infer them. In: NeurIPS (2021)

47. Wang, X., Girshick, R., Gupta, A., He, K.: Non-local neural networks. In: CVPR (2018)
48. Wertheimer, D., Tang, L., Hariharan, B.: Few-shot classification with feature map reconstruction networks. In: Proceedings of the IEEE/CVF Conference on Computer Vision and Pattern Recognition, pp. 8012–8021 (2021)
49. Wu, J., Zhang, T., Zhang, Y., Wu, F.: Task-aware part mining network for few-shot learning. In: ICCV (2021)
50. Yang, S., Liu, L., Xu, M.: Free lunch for few-shot learning: Distribution calibration. arXiv preprint arXiv:2101.06395 (2021)
51. Zhan, B., Monekosso, D.N., Remagnino, P., Velastin, S.A., Xu, L.Q.: Crowd analysis: a survey. Mach. Vis. Appl. **19**(5), 345–357 (2008)
52. Zhang, C., Cai, Y., Lin, G., Shen, C.: DeepEMD: differentiable earth mover's distance for few-shot learning. arXiv preprint arXiv:2003.06777 (2020)
53. Zhang, H., Zhang, L., Qi, X., Li, H., Torr, P.H.S., Koniusz, P.: Few-shot action recognition with permutation-invariant attention. In: Vedaldi, A., Bischof, H., Brox, T., Frahm, J.-M. (eds.) ECCV 2020. LNCS, vol. 12350, pp. 525–542. Springer, Cham (2020). https://doi.org/10.1007/978-3-030-58558-7_31
54. Zhang, S., Zhou, J., He, X.: Learning implicit temporal alignment for few-shot video classification. arXiv preprint arXiv:2105.04823 (2021)
55. Zhang, X., Meng, D., Gouk, H., Hospedales, T.M.: Shallow bayesian meta learning for real-world few-shot recognition. In: ICCV (2021)
56. Zhu, L., Yang, Y.: Compound memory networks for few-shot video classification. In: Ferrari, V., Hebert, M., Sminchisescu, C., Weiss, Y. (eds.) ECCV 2018. LNCS, vol. 11211, pp. 782–797. Springer, Cham (2018). https://doi.org/10.1007/978-3-030-01234-2_46
57. Zhu, Z., Wang, L., Guo, S., Wu, G.: A closer look at few-shot video classification: a new baseline and benchmark. arXiv preprint arXiv:2110.12358 (2021)
58. Ziko, I., Dolz, J., Granger, E., Ayed, I.B.: Laplacian regularized few-shot learning. In: ICML (2020)

Temporal and Cross-modal Attention for Audio-Visual Zero-Shot Learning

Otniel-Bogdan Mercea[1]([⊠]) [iD], Thomas Hummel[1] [iD], A. Sophia Koepke[1] [iD],
and Zeynep Akata[1,2,3] [iD]

[1] University of Tübingen, Tübingen, Germany
{otniel-bogdan.mercea,thomas.hummel,a-sophia.koepke,
zeynep.akata}@uni-tuebingen.de
[2] MPI for Informatics, Tübingen, Germany
[3] MPI for Intelligent Systems, Stuttgart, Germany

Abstract. Audio-visual generalised zero-shot learning for video classification requires understanding the relations between the audio and visual information in order to be able to recognise samples from novel, previously unseen classes at test time. The natural semantic and temporal alignment between audio and visual data in video data can be exploited to learn powerful representations that generalise to unseen classes at test time. We propose a multi-modal and Temporal Cross-attention Framework (TCaF) for audio-visual generalised zero-shot learning. Its inputs are temporally aligned audio and visual features that are obtained from pre-trained networks. Encouraging the framework to focus on cross-modal correspondence across time instead of self-attention within the modalities boosts the performance significantly. We show that our proposed framework that ingests temporal features yields state-of-the-art performance on the UCF-GZSL^{cls}, VGGSound-GZSL^{cls}, and ActivityNet-GZSL^{cls} benchmarks for (generalised) zero-shot learning. Code for reproducing all results is available at https://github.com/ExplainableML/TCAF-GZSL.

Keywords: Zero-shot learning · Audio-visual learning

1 Introduction

Learning task-specific audio-visual representations commonly requires a great number of annotated data samples. However, annotated datasets are limited in size and in the labelled classes that they contain. If a model which was trained with supervision on such a dataset is applied in the real world, it encounters classes that it has never seen. To recognise those novel classes, it would not be feasible to train a new model from scratch. Therefore, it is essential to analyse the behaviour of a trained model in new settings. Ideally, a model should be able

---

O. -B. Mercea and T. Hummel—Equal contribution.

Supplementary Information The online version contains supplementary material available at https://doi.org/10.1007/978-3-031-20044-1_28.

S. Avidan et al. (Eds.): ECCV 2022, LNCS 13680, pp. 488–505, 2022.
https://doi.org/10.1007/978-3-031-20044-1_28

to transfer knowledge obtained from classes seen during training to previously unseen categories. This ability is probed in the zero-shot learning (ZSL) task. In addition to zero-shot capabilities, a model should retain the class-specific information from seen training classes. This is challenging and is investigated in the so-called generalised ZSL (GZSL) setting which considers the performance on both, seen and unseen classes.

Prior works [46,47,55] have proposed frameworks that address the (G)ZSL task for video classification using audio-visual inputs. Those methods learn a mapping from the audio-visual input data to textual label embeddings, enabling the classification of samples from unseen classes. At test time, the class whose word embedding is closest to the predicted audio-visual output embedding is selected. Similar to this, we use the textual label embedding space to allow for information transfer from training classes to previously unseen classes.

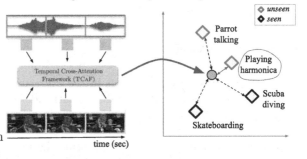

However, [46,47,55] used temporally averaged features as inputs that were extracted from networks pre-trained on video data. The averaging disregarded the temporal dynamics in videos. We propose a Temporal Cross-attention Framework (TCAF) which builds on [47] and additionally exploits temporal information by using temporal audio and visual data as inputs. This gives a significant boost in performance for the audio-visual (G)ZSL task com-

Fig. 1. Our temporal cross-attention framework for audio-visual (G)ZSL learns a multi-modal embedding (green circle) by exploiting the temporal alignment between audio and visual data in videos. Textual label embeddings (grey squares) are used to transfer information from seen training classes (black) to unseen test classes (pink). The correct class is playing harmonica (red). (Color figure online)

pared to using temporally averaged input features. Different from computationally expensive methods that operate directly on raw visual inputs [13,33,40], our TCAF uses features extracted from networks pre-trained for audio and video classification as inputs. This leads to an efficient setup that uses temporal information instead of averaging across time.

The natural alignment between audio and visual information in videos, e.g. a frog being visible in a frame while the sound of a frog croaking is audible, provides a rich training signal for learning video representations. This can be attributed to the semantic and temporal correlation between the audio and visual information when comparing the two modalities. We encourage our TCAF to put special emphasis on the correlation across the two modalities by employing repeated cross-attention. This attention mechanism only allows attention to tokens from the other modality. This effectively acts as a bottleneck which results in cheaper computations and gives a boost in performance over using full self-attention across all tokens from both modalities.

We perform a detailed model ablation study to show the benefits of using temporal inputs and our proposed cross-attention. Furthermore, we confirm that our training objective is well-suited to the task at hand. We also analyse the learnt audio-visual embeddings with t-SNE visualisations which confirm that training our TCAF improves the class separation for both seen and unseen classes.

To summarise, our contributions are as follows: (1) We propose a temporal cross-attention framework TCAF for audio-visual (G)ZSL. (2) Our proposed model achieves state-of-the-art results on the UCF-GZSL^{cls}, VGGSound-GZSL^{cls}, and ActivityNet-GZSL^{cls} datasets, demonstrating that using temporal information is extremely beneficial for improving the (generalised) zero-shot classification accuracy compared to using temporally averaged features as model inputs. (3) We perform a detailed analysis of the use of enhanced cross-attention across modalities and time, demonstrating the benefits of our proposed model architecture and training setup.

2   Related Work

Our work relates to several themes in the literature: audio-visual learning, ZSL with side information, audio-visual ZSL with side information, and multi-modal transformer architectures. We discuss those in more detail in the following.

Audio-Visual Learning. The temporal alignment between audio and visual data in videos is a strong learning signal which can be exploited for learning audio-visual representations. [7,10,37,53,54,56]. In addition to audio and video classification, numerous other tasks benefit from audio-visual inputs, such as the separation and localisation of sounds in video data [1,4,8,15,24,52,65], audio-driven synthesis of images [31,70], audio synthesis driven by visual information [23,25,35,36,50,59,77], and lip reading [2,3]. Some approaches use class-label supervision between modalities [16,20] which does not require the temporal alignment between the input modalities. In contrast to full class-label supervision, we train our model only on the subset of seen training classes.

ZSL with Side Information. Visual ZSL methods commonly map the visual inputs to class side information [5,6,21], e.g. word2vec [48] class label embeddings. This allows to determine the class with the side information that is closest at test time as the class prediction. Furthermore, attribute annotations have been used as side information [19,68,71,74]. Recent non-generative methods identify key visual attributes [76], use attention to find discriminative regions [75], or disambiguate class embeddings [43]. In contrast, feature generation methods train a classifier on generated and real features [51,72,73,78]. Unlike methods for ZSL with side information with unimodal (visual) inputs, our proposed framework uses multi-modal audio-visual inputs.

Audio-Visual ZSL with Side Information. The task of GZSL from audio-visual data was introduced by [46,55] on the AudioSetZSL dataset [55] using

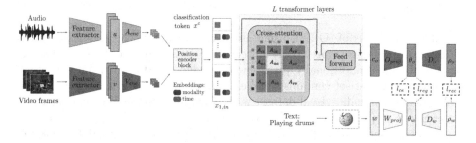

Fig. 2. TCAF takes audio and visual features extracted from video data as inputs. Those are embedded and equipped with modality and time embeddings before passing through a sequence of L transformer layers with cross-attention. The output classification token c_o is then projected to embedding spaces that are shared with the textual information. The loss functions operate on the joint embedding spaces. At test time, the class prediction c is obtained by determining the word label embedding θ_w^j that is closest to θ_o.

class label word embeddings as side information. Recently, [47] proposed the AVCA framework which uses cross-attention to fuse information from the averaged audio and visual input features for audio-visual GZSL. Our proposed framework builds on [47], but instead of using temporally averaged features as inputs [46,47,55], we explore the benefits of using temporal cross-attention. Unlike [47]'s two-stream architecture, we propose the fusion into a single output branch with a classification token that aggregates multi-modal information. Furthermore, we simplify the training objective, and show that the combination of using temporal inputs, our architecture, and training setup leads to superior zero-shot classification performance.

Multi-Modal Transformers. The success of transformer models in the language domain [17,57,67] has been translated to visual recognition tasks with the Vision Transformer [18]. Multi-modal vision-language representations have been obtained with a masked language modelling objective, and achieved state-of-the-art performance on several text-vision tasks [38,39,44,60–63]. In this work, we consider audio-visual multi-modality. Transformer-based models that operate on audio and visual inputs have recently been proposed for text-based video retrieval [22,42,69], dense video captioning [28], audio-visual event localization [41], and audio classification [12]. Different to vanilla transformer-based attention, our TCAF puts special emphasis on cross-attention between the audio and visual modalities in order to learn powerful representations for the (G)ZSL task.

3 TCAF Model

In this section, we describe the problem setting (Sect. 3.1), our proposed model architecture (Sect. 3.2), and the loss functions used to train TCAF (Sect. 3.3).

3.1 Problem Setting

We address the task of (G)ZSL using audio-visual inputs. The aim of ZSL is to be able to generalise to previously unseen test classes at test time. For GZSL, the model should additionally preserve knowledge about seen training classes, since the GZSL test set contains samples from both, seen and unseen classes.

We denote an audio-visual dataset with N samples and K (seen and unseen) classes by $\mathcal{V} = \{\mathcal{X}_{a[i]}, \mathcal{X}_{v[i]}, y_{[i]}\}_{i=1}^{N}$, consisting of audio data $\mathcal{X}_{a[i]}$, visual data $\mathcal{X}_{v[i]}$, and ground-truth class labels $y_{[i]} \in \mathbb{R}^{K}$. Naturally, video data contains temporal information. In the following, we use T_a and T_v to denote the number of audio and visual segments in a video clip.

A pre-trained audio classification CNN is used to extract a sequence of audio features $\boldsymbol{a}_{[i]} = \{a_1, \ldots, a_t, \ldots, a_{T_a}\}_i$ to encode the audio information $\mathcal{X}_{a[i]}$. The visual data $\mathcal{X}_{v[i]}$ is encoded into a temporal sequence of features $\boldsymbol{v}_{[i]} = \{v_1, \ldots, v_t, \ldots, v_{T_v}\}_i$ by representing visual segments with features extracted from a pre-trained video classification network.

3.2 Model Architecture

In the following, we describe the architecture of our proposed TCaF (see Fig. 2).

Embedding the Inputs and Position Encoder Block. TCaF takes pre-extracted audio and visual features $\boldsymbol{a}_{[i]}$ and $\boldsymbol{v}_{[i]}$ as inputs. For readability, we will drop the subscript i in the following which denotes the $i-$th sample. In order to project audio and visual features to the same feature dimension, $\boldsymbol{a}$ and $\boldsymbol{v}$ are passed through two modality-specific embedding blocks, giving embeddings:

$$\phi_a = A_{enc}(\boldsymbol{a}) \text{ and } \phi_v = V_{enc}(\boldsymbol{v}), \tag{1}$$

with $\phi_a \in \mathbb{R}^{T_a * d_{dim}}$ and $\phi_v \in \mathbb{R}^{T_v * d_{dim}}$. The embedding blocks are composed of two linear layers f_1^m, f_2^m for $m \in \{\boldsymbol{a}, \boldsymbol{v}\}$, where $f_1^m : \mathbb{R}^{T_m * d_{in_m}} \rightarrow \mathbb{R}^{T_m * d_{fhidd}}$ and $f_2^m : \mathbb{R}^{T_m * d_{fhidd}} \rightarrow \mathbb{R}^{T_m * d_{dim}}$. f_1^m, f_2^m are each followed by batch normalisation [29], a ReLU [49], and dropout [58] with dropout rate $drop_{enc}$.

The position encoder block adds learnt modality and temporal positional embeddings to the outputs of the modality-specific embedding blocks. We explain this in detail below. To handle different frame rates in the audio and visual modalities, we use Fourier features [64] $pos_t \in \mathbb{R}^{d_{pos}}$ for the temporal embeddings that encode the actual point in time in the video which corresponds to an audio or visual representation. This allows to capture the relative temporal position of the audio and visual features across the modalities.

For an audio embedding ϕ_{a_t} at time t, a linear map $g_a : \mathbb{R}^{d_{pos}+d_{dim}} \rightarrow \mathbb{R}^{d_{dim}}$, and a dropout layer g^D with dropout probability $drop_{prob,pos}$, we obtain position-aware audio feature tokens

$$a_t^p = g^D(g_a(concat(\phi_{a_t}, pos_{at}))) \quad \text{with} \quad pos_{at} = pos_a + pos_t, \tag{2}$$

with modality and temporal embeddings $pos_a, pos_t \in \mathbb{R}^{d_{pos}}$ respectively. Position-aware visual tokens v_t^p are obtained analogously.

Furthermore, we prepend a learnt classification token $x^c \in \mathbb{R}^{d_{dim}}$ to the sequence of feature tokens. The corresponding output classification token c_o is used by our output projection O_{proj} to obtain the final prediction.

Audio-Visual Transformer Layers. TCAF contains L stacked audio-visual transformer layers that allow for enhanced cross-attention. Each of our transformer layers consists of an attention function $f_{l,Att}$, followed by a feed forward function $g_{l,FF}$. The output of the l-th transformer layer is given as

$$x_{l,out} = x_{l,ff} + x_{l,att} = g_{l,FF}(x_{l,att}) + x_{l,att}, \tag{3}$$

with

$$x_{l,att} = f_{l,Att}(x_{l,in}) + x_{l,in}, \tag{4}$$

where

$$x_{l,in} = \begin{cases} [x^c, a_1^p, \cdots, a_{T_a}^p, v_1^p, \cdots, v_{T_v}^p] & \text{if } l = 1, \\ x_{l-1,out} & \text{if } 2 \geq l \leq L. \end{cases}$$

We explain the cross-attention used in our transformer layers in the following.

Transformer Cross-Attention. TCAF primarily exploits cross-modal audio-visual attention to combine the information across the audio and visual modalities. All attention mechanisms in TCAF consist of multi-head attention [67] with H heads and a dimension of d_{head} per head.

We describe the first transformer layer $\mathcal{M}_1$, the transformer layer $\mathcal{M}_l$ operates analogously. We project the position-aware input features x^c, $\{a_t^p\}_{t \in [1,T_a]}$, $\{v_t^p\}_{t \in [1,T_v]}$ to queries, keys, and values with linear maps $g_s : \mathbb{R}^{d_{dim}} \rightarrow \mathbb{R}^{d_{head}H}$ for $s \in \{q, k, v\}$. We can then write the outputs of the projection as zero-padded query, key, and value features. We write those out for the queries below, the keys and values are padded in the same way:

$$\mathbf{q}_c = [g_q(x^c), 0, \cdots, 0], \tag{5}$$
$$\mathbf{q}_a = [0, \cdots, 0, g_q(a_1^p), \cdots, g_q(a_{T_a}^p), 0, \cdots, 0], \tag{6}$$
$$\mathbf{q}_v = [0, \cdots, 0, g_q(v_1^p), \cdots, g_q(v_{T_v}^p)]. \tag{7}$$

The full query, key, and value representations, $\mathbf{q}$, $\mathbf{k}$, and $\mathbf{v}$, are the sums of their modality-specific components

$$\mathbf{q} = \mathbf{q}_c + \mathbf{q}_a + \mathbf{q}_v, \quad \mathbf{k} = \mathbf{k}_c + \mathbf{k}_a + \mathbf{k}_v, \quad \text{and } \mathbf{v} = \mathbf{v}_c + \mathbf{v}_a + \mathbf{v}_v. \tag{8}$$

The output of the first attention block $x_{1,att}$ is the aggregation of the per-head attention with a linear mapping $g_h : \mathbb{R}^{d_{head}H} \rightarrow \mathbb{R}^{d_{dim}}$, g^{DL} dropout with dropout probability $drop_{prob}$ and layer normalisation g^{LN} [11], such that

$$x_{1,att} = f_{1,Att}(x_{l,in}) = g^{DL}(g_h(f_{1,att}^1(g^{LN}(x_{1,in})), \cdots, f_{1,att}^H(g^{LN}(x_{1,in})))), \tag{9}$$

with the attention f_{att}^h for the attention head h. We can write the attention for the head h as

$$f_{att}^h(x_{1,in}) = softmax\left(\frac{\mathbf{A}}{\sqrt{d_{head}}}\right)\mathbf{v}, \tag{10}$$

where $\mathbf{A}$ can be split into its cross-attention and self-attention components:

$$\mathbf{A}_c = \mathbf{q}_c\,\mathbf{k}^T + \mathbf{k}\,\mathbf{q}_c^T, \qquad \mathbf{A}_x = \mathbf{q}_a\,\mathbf{k}_v^T + \mathbf{q}_v\,\mathbf{k}_a^T, \tag{11}$$

$$\mathbf{A}_{self} = \mathbf{q}_a\,\mathbf{k}_a^T + \mathbf{q}_v\,\mathbf{k}_v^T.$$

We then get

$$\mathbf{A} = \mathbf{A}_c + \mathbf{A}_x + \mathbf{A}_{self} = \begin{pmatrix} A_{cc}\ A_{ca}\ A_{cv} \\ A_{ac} \quad \ddots \quad \vdots \\ A_{vc} \ \cdots \ 0 \end{pmatrix} + \begin{pmatrix} 0 \ \cdots \ 0 \\ \vdots \ \ddots \ A_{av} \\ 0 \ A_{va} \ 0 \end{pmatrix} + \begin{pmatrix} 0 \ \cdots \ 0 \\ \vdots \ A_{aa} \ \vdots \\ 0 \ \cdots \ A_{vv} \end{pmatrix}, \tag{12}$$

where the A_{mn} with $m, n \in \{c, a, v\}$ describe the attention contributions from the classification token, the audio and the visual modalities respectively.

Our TCAF uses the cross-attention $\mathbf{A}_c + \mathbf{A}_x$ to put special emphasis on the attention across modalities. Results for different model variants that use only the within-modality self-attention ($\mathbf{A}_c + \mathbf{A}_{self}$) or the full attention which combines self-attention and cross-attention are presented in Sect. 4.3.

Feed Forward Function. The feed forward function $g_{l,FF} : \mathbb{R}^{d_{dim}} \rightarrow \mathbb{R}^{d_{dim}}$ is applied to the output of the attention function

$$x_{l,ff} = g_{l,FF}(x_{l,att}) = g^{DL}(g_{l,F2}(g^{DL}(g^{GD}(g_{l,F1}(g^{LN}(x_{l,att})))))) \tag{13}$$

where $g_{l,F1} : \mathbb{R}^{d_{dim}} \rightarrow \mathbb{R}^{d_{ff}}$ and $g_{l,F2} : \mathbb{R}^{d_{ff}} \rightarrow \mathbb{R}^{d_{dim}}$ are linear mappings, g^{GD} is a GELU layer [26] and a dropout layer with dropout probability $drop_{prob}$, g^{DL} is dropout with $drop_{prob}$ and g^{LN} is layer normalisation.

Output Prediction. To determine the final class prediction, the audio-visual embedding is projected to the same embedding space as the textual class label representations. We project the output classification token c_o of the temporal cross-attention to $\theta_o = O_{proj}(c_o)$ where $\theta_o \in \mathbb{R}^{d_{out}}$. The projection block is composed of a sequence of two linear layers f_3 and f_4, where $f_3 : \mathbb{R}^{d_{dim}} \rightarrow \mathbb{R}^{d_{fhidd}}$ and $f_4 : \mathbb{R}^{d_{fhidd}} \rightarrow \mathbb{R}^{d_{out}}$. f_3, f_4 are each followed by batch normalisation, a ReLU, and dropout with rate $drop_{proj_o}$. We project the word2vec class label embedding w^j for class j using the projection block $W_{proj}(w^j) = \theta_w^j$, where $\theta_w^j \in \mathbb{R}^{d_{out}}$. W_{proj} consists of a linear projection followed by batch normalisation, ReLU, and dropout with dropout rate $drop_{proj_w}$. The class prediction c is obtained by determining the projected word2vec embedding which is closest to the output embedding:

$$c = \underset{j}{argmin}(\|\theta_w^j - \theta_o\|_2). \tag{14}$$

3.3   Loss Functions

Our training objective l combines a cross-entropy loss l_{ce}, a reconstruction loss l_{rec}, and a regression loss l_{reg}:

$$l = l_{ce} + l_{rec} + l_{reg}. \tag{15}$$

Cross-Entropy Loss. For the ground-truth label y_i with corresponding class index $k_{gt} \in \mathbb{R}^{K_{seen}}$, the output of our temporal cross-attention θ_{o_i}, and a matrix containing the textual label embeddings for the K_{seen} seen classes $\theta_{w_{seen}}$, we define the cross-entropy loss for n training samples as

$$l_{ce} = -\frac{1}{n} \sum_i^n y_i \log \left(\frac{\exp\left(\theta_{w_{seen},k_{gt}}\theta_{o_i}\right)}{\sum_{k_j}^{K_{seen}} \exp\left(\theta_{w_{seen},k_j}\theta_{o_i}\right)} \right). \tag{16}$$

Regression Loss. While the cross-entropy loss updates the probabilities for both the correct and incorrect classes, our regression loss directly focuses on reducing the distance between the output embedding for a sample and the corresponding projected word2vec embedding. The regression loss is based on the mean squared error metric with the following formulation:

$$l_{reg} = \frac{1}{n} \sum_{i=1}^n (\theta_{o_i} - \theta_{w_i})^2, \tag{17}$$

where θ_{o_i} is the audio-visual embedding, and θ_{w_i} is the projection of the word2vec embedding corresponding to the i-th sample.

Reconstruction Loss. The goal of the reconstruction loss is to ensure that the embeddings θ_o and θ_w contain semantic information from the word2vec embedding w. We use $D_u : \mathbb{R}^{d_{out}} \mapsto \mathbb{R}^{d_{dim}}$ with $\rho_u = D_u(\theta_u)$ for $u \in \{o, w\}$. D_w is a sequence of one linear layer, batch normalisation, a ReLU, and dropout with rate $drop_{proj_w}$. D_o is composed of a sequence of two linear layers each followed by batch normalisation, a ReLU, and dropout with dropout rate $drop_{proj_o}$. Our reconstruction loss encourages the reconstruction of the output embedding, ρ_{o_i}, and the reconstruction of the word2vec projection, ρ_{w_i}, to be close to the original word2vec embedding w_i:

$$l_{rec} = \frac{1}{n} \sum_{i=1}^n (\rho_{o_i} - w_i)^2 + \frac{1}{n} \sum_{i=1}^n (\rho_{w_i} - w_i)^2. \tag{18}$$

4   Experiments

In this section, we detail our experimental setup (Sect. 4.1), and compare to state-of-the-art methods for audio-visual GZSL (Sect. 4.2). Furthermore, we present an ablation study in Sect. 4.3 which shows the benefits of using our proposed attention scheme and training objective. Finally, we present t-SNE visualisations of our learnt audio-visual embeddings in Sect. 4.4.

4.1   Experimental Setup

Here, we describe the datasets used, the evaluation metrics, and the implementation details for all models.

Datasets. We use the UCF-GZSL^{cls}, VGGSound-GZSL^{cls}, and ActivityNet-GZSL^{cls} datasets [47] for audio-visual (G)ZSL for training and testing all models. [47] introduced benchmarks for two sets of features, the first uses a model pre-trained using self-supervision on the VGGSound dataset from [9], the second takes features extracted from pre-trained VGGish [27] and C3D [66] audio and video classification networks. Since the VGGSound dataset is also used for the zero-shot learning task (VGGSound-GZSL), we selected the second option (using VGGish and C3D) and use the corresponding dataset splits proposed in [47].

In particular, the audio features are extracted using VGGish [27] to obtain one 128-dimensional feature vector for each 0.96 s snippet. The visual features are obtained using C3D [66] pre-trained on Sports-1M [32]. For this, all videos are resampled to 25 fps. A 4096-dimensional feature vector is then extracted for 16 consecutive video frames.

Evaluation Metrics. We follow [47,71] and use the mean class accuracy to evaluate all models. The ZSL performance is obtained by considering only the subset of test samples from the unseen test classes. For the GZSL performance, the models are evaluated on the full test set which includes seen and unseen classes. We then report the performance on the subsets of seen (S) and unseen (U) classes, and also report their harmonic mean (HM).

Implementation Details. For TCAF, we use $d_{in_a} = 128$, $d_{in_v} = 4096$, $d_{fhidd} = 512$, $d_{dim} = 300$ and $d_{out} = 64$. Furthermore, TCAF has $L = 6$ transformer layers layers for UCF-GZSL^{cls} and ActivityNet-GZSL^{cls}, and $L = 8$ for VGGSound-GZSL^{cls}. We set $d_{pos} = 64$, $d_{ff} = 128$. For ActivityNet-GZSL^{cls}/ UCF-GZSL^{cls}/ VGGSound-GZSL^{cls} we use dropout rates $drop_{enc} = 0.1/0.3/0.2$, $drop_{prob,pos} = 0.2/0.2/0.1$, $drop_{prob} = 0.4/0.3/0.5$, $drop_{proj_w} = 0.1/0.1/0.1$, and $drop_{proj_o} = 0.1/0.1/0.2$. All attention blocks use $H = 8$ heads with a dimension of $d_{head} = 64$ per head. We train all models using the Adam optimizer [34] with running average coefficients $\beta_1 = 0.9$, $\beta_2 = 0.999$, and weight decay 0.00001. We use a batch size of 64 for all datasets. In order to efficiently train on ActivityNet-GZSL^{cls}, we randomly trim the features to a maximum sequence length of 60 during training, and we evaluate on features that have a maximum sequence length of 300 and which are centered in the middle of the video. We note, that TCAF can be efficiently trained on a single Nvidia 2080-Ti GPU. All models are trained for 50 epochs. We use a base learning rate of 0.00007 for UCF-GZSL^{cls} and ActivityNet-GZSL^{cls}, and 0.00006 for VGGSound-GZSL^{cls}. For UCF-GZSL^{cls} and ActivityNet-GZSL^{cls} we use a scheduler that reduces the learning rate by a factor of 0.1 when the HM on the validation set has not improved for 3 epochs. To eliminate the bias that the ZSL methods have towards seen classes, we used calibrated stacking [14] on the search space composed of the interval $[0, 3]$ with a step size of 0.2.

We train all models with a two-stage training protocol [47]. In the first stage, we determine the calibrated stacking [14] and the epoch with the best HM performance on the validation set. In the second stage, using the hyperparameters from the first stage, we re-train the models on the union of the training and validation sets. We evaluate the final models on the test set.

4.2 Quantitative Results

We compare our proposed TCAF to state-of-the-art audio-visual ZSL frameworks and to audio-visual frameworks that we adapted to the ZSL task.

Audio-Visual ZSL Baselines. We compare our TCAF to three audio-visual ZSL frameworks. **CJME** [55] consists of a relatively simple architecture which maps both input modalities to a shared embedding space. The modality-specific embeddings in the shared embedding space are input to an attention predictor module that determines the dominant modality which is used for the output prediction. **AVGZSLNet** [46] builds on CJME by adding a shared decoder and introducing additional loss functions to improve the performance. AVGZSLNet removes the attention predictor network and replaces it with a simple average between the output from the head of each modality. **AVCA** [47] is a recent state-of-the-art method for audio-visual G(ZSL). It uses a simple cross-attention mechanism on the temporally averaged audio and visual input features to combine the information from the two modalities. Our proposed TCAF improves upon the closely related AVCA framework by additionally ingesting temporal information in the audio and visual inputs with an enhanced cross-attention mechanism that gathers information across time and modalities.

Audio-Visual Baselines Adapted to ZSL. We adapt two attention-based audio-visual frameworks to the ZSL setting. **Attention Fusion** [20] is a method for audio-visual classification which is trained to classify unimodal information. It then fuses the unimodal predictions with learnt attention weights. The **Perceiver** [30] is a scalable multi-modal transformer framework for flexible learning with arbitrary modality information. It uses a latent bottleneck to encode input information by repeatedly attending to the input with transformer-style attention. The Perceiver allows for a comparison to another transformer-based architecture with focus on multi-modality. We adapt the Perceiver to use the same positional encodings and model capacity as TCAF. We use 64 latent tokens and the same number of layers and dimensions as TCAF. Both Attention Fusion and Perceiver use the same input features, input embedding functions A_{enc} and V_{enc}, learning rate and loss functions as TCAF. For Attention Fusion, we temporally average the input features after A_{enc} and V_{enc} to deal with non-synchronous modality sequences due to different feature extraction rates.

All baselines, except for the Perceiver, operate on temporally averaged audio and visual features. This decreases the amount of information contained in the inputs, in particular regarding the dynamics in a video. In contrast to methods that use temporally averaged inputs, TCAF exploits the temporal dimension which boosts the (G)ZSL performance.

Table 1. Performance of our TCAF and of state-of-the-art methods for audio-visual (G)ZSL on the VGGSound-GZSL^{cls}, UCF-GZSL^{cls}, and ActivityNet-GZSL^{cls} datasets. The mean class accuracy for GZSL is reported on the seen (S) and unseen (U) test classes, and their harmonic mean (HM). For the ZSL performance, only the test subset of unseen classes is considered.

Model	VGGSound-GZSL^{cls}				UCF-GZSL^{cls}				ActivityNet-GZSL^{cls}			
	S	U	HM	ZSL	S	U	HM	ZSL	S	U	HM	ZSL
Attention Fusion	14.13	3.00	4.95	3.37	39.34	18.29	24.97	20.21	11.15	3.37	5.18	4.88
Perceiver	13.25	3.03	4.93	3.44	46.85	26.82	34.11	28.12	18.25	4.27	6.92	4.47
CJME	10.86	2.22	3.68	3.72	33.89	24.82	28.65	29.01	10.75	5.55	7.32	6.29
AVGZSLNet	**15.02**	3.19	5.26	4.81	**74.79**	24.15	36.51	31.51	13.70	5.96	8.30	6.39
AVCA	12.63	6.19	8.31	6.91	63.15	30.72	41.34	37.72	16.77	7.04	9.92	7.58
TCAF	12.63	**6.72**	**8.77**	**7.41**	67.14	**40.83**	**50.78**	**44.64**	30.12	**7.65**	**12.20**	**7.96**

Results. We compare the results obtained with our TCAF to state-of-the-art baselines for audio-visual (G)ZSL and for audio-visual learning in Table 1. TCAF outperforms all previous methods on the VGGSound-GZSL^{cls}, UCF-GZSL^{cls}, and ActivityNet-GZSL^{cls} datasets for both, GZSL performance (HM) and ZSL performance. For ActivityNet-GZSL^{cls}, our proposed model is significantly better than its strongest competitor AVCA, with a HM of 12.20% compared to 9.92% and a ZSL performance of 7.96% compared to 7.58%. The CJME and AVGZSLNet frameworks are weaker than the AVCA model. Similar patterns are exhibited for the VGGSound-GZSL^{cls} and UCF-GZSL^{cls} datasets. Interestingly, the GZSL performance for TCAF is improved by a more significant margin than the ZSL performance compared to AVCA across all three datasets. This shows that using temporal information and allowing our model to attend across time and modalities is especially beneficial for the GZSL task.

Furthermore, we observe that the audio-visual Attention Fusion framework and the Perceiver give worse results than AVGZSLNet and AVCA on all three datasets. In particular, our TCAF yields stronger ZSL and GZSL performances than the Perceiver which also takes temporal audio and visual features as inputs, with a HM of 8.77% on VGGSound-GZSL^{cls} for TCAF compared to 4.93% for the Perceiver. Attention Fusion and the Perceiver architecture were not designed for the (G)ZSL setting that uses text as side information. Our proposed training objective, used to also train the Perceiver, aims to regress textual embeddings which might be challenging for the Perceiver given its tight latent bottlenecks.

4.3 Ablation Study on the Training Loss and Attention Variants

Here, we analyse different components of our proposed TCAF. We first compare the performance of our model when trained using different loss functions. We then investigate the influence of the attention mechanisms used in the model architecture on the (G)ZSL performance. Finally, we show that using multi-modal inputs is beneficial and results in outperforming unimodal baselines.

Comparing Different Training Losses. We show the contributions of the different components in our training loss function to the (G)ZSL performance

Table 2. Influence of using different components of our proposed training objective for training TCAF on the (G)ZSL performance on the VGGSound-GZSL^{cls}, UCF-GZSL^{cls}, and ActivityNet-GZSL^{cls} datasets.

Loss	VGGSound-GZSL^{cls}				UCF-GZSL^{cls}				ActivityNet-GZSL^{cls}			
	S	U	HM	ZSL	S	U	HM	ZSL	S	U	HM	ZSL
l_{reg}	0.10	2.41	0.19	2.50	14.30	18.82	16.25	30.17	1.09	0.27	0.43	2.11
$l_{reg} + l_{ce}$	**13.67**	4.06	6.26	4.31	**75.31**	37.15	49.76	41.75	11.36	5.28	7.21	5.31
$l = l_{reg} + l_{ce} + l_{rec}$	12.63	**6.72**	**8.77**	**7.41**	67.14	**40.83**	**50.78**	**44.64**	**30.12**	**7.65**	**12.20**	**7.96**

in Table 2. Using only the regression loss l_{reg} to train our model results in the weakest performance across all datasets, with HM/ZSL performances of 16.25%/30.17% on UCF-GZSL^{cls} compared to 50.78%/44.64% for our full TCAF. Interestingly, the seen performance (S) when using only l_{reg} is relatively weak, likely caused by the calibrated stacking. Similarly, on ActivityNet-GZSL^{cls}, using only l_{reg} yields a low test performance of 0.43% HM. Jointly training with the regression and cross-entropy loss functions ($l_{reg} + l_{ce}$) improves the GZSL and ZSL performance significantly, giving a ZSL of 4.31% compared to 2.50% for l_{reg} on VGGSound-GZSL^{cls}. The best results are obtained when training with our full training objective l which includes a reconstruction loss term, giving the best performance on all three datasets.

Comparing Different Attention Variants. We study the use of different attention patterns in Table 3. In particular, we analyse the effect of using within-modality ($\mathbf{A}_{self}$) and cross-modal ($\mathbf{A}_x$) attention (cf. Eq. 11), on the GZSL and ZSL performance. Additionally, we investigate models that use a classification token x^c with corresponding output token c_o (*with class. token*) and models for which we simply average the output of the transformer layers which is then used as input to O_{proj} (*w/o class. token*).

Interestingly, we observe that with no global token, using the full attention $\mathbf{A}_{self} + \mathbf{A}_x$ gives better results than using only cross-attention on UCF-GZSL^{cls} and ActivityNet-GZSL^{cls} for ZSL and GZSL, but is slightly worse on VGGSound-GZSL^{cls}. This suggests that the bottleneck introduced by limiting the information flow in the attention when using only cross-attention is beneficial for (G)ZSL on VGGSound-GZSL^{cls}. When not using the classification token and only self-attention $\mathbf{A}_{self}$, representations inside the transformer are created solely within their respective modalities.

Using a classification token (*with class. token*) and the cross-attention variant ($\mathbf{A}_c + \mathbf{A}_x$) yields the strongest ZSL and GZSL results across all three datasets. The most drastic improvements over full attention can be observed on the UCF-GZSL^{cls} dataset, with a HM of 50.78% for the cross-attention with classification token ($\mathbf{A}_c + \mathbf{A}_x$) compared to 39.18% for the full attention ($\mathbf{A}_c + \mathbf{A}_{self} + \mathbf{A}_x$). Furthermore, when using x_c, cross-attention $\mathbf{A}_x$ instead of self-attention $\mathbf{A}_{self}$ leads to a better performance on all three datasets. For $\mathbf{A}_x$ and x_c, we obtain HM scores of 8.77% and 50.78 % on VGGSound-GZSL^{cls} and UCF-GZSL^{cls}

compared to 6.71% and 37.37% with $\mathbf{A}_{self}$ and x_c. This shows that using information from both modalities is important for creating strong and transferable video representations for (G)ZSL. Using the global token relaxes the pure cross-attention setting to a certain extent, since $\mathbf{A}_c$ allows for attention between all tokens from both modalities and the global token. The results in Table 3 have demonstrated the clear benefits of our cross-attention variant used in TCAF.

The Influence of Multi-Modality. We compare using only a single input modality for training TCAF to using multiple input modalities in Table 4. For the unimodal baselines TCAF- audio and TCAF- visual, we train TCAF only with the corresponding input modality. Using only audio inputs gives stronger GZSL and ZSL results than using only visual inputs on VGGSound-GZSL^{cls} and ActivityNet-GZSL^{cls}. We obtain a HM of 5.84% for audio compared to 2.92% for visual inputs on ActivityNet-GZSL^{cls}. Interestingly this pattern is reversed for the UCF-GZSL^{cls} dataset where using visual inputs only results in a slightly higher performance than using the audio inputs with HM scores of 31.49% compared to 25.38%, and ZSL scores of 27.25% and 24.24%. However, using both modalities (TCAF) increases the HM to 50.78% and ZSL to 44.64% on UCF-GZSL^{cls}. Similar trends can be observed for VGGSound-GZSL^{cls} and ActivityNet-GZSL^{cls} which highlights the importance of the tight multi-modal coupling in our TCAF.

Table 3. Ablation of different attention variants with and without a classification token on the VGGSound-GZSL^{cls}, UCF-GZSL^{cls}, and ActivityNet-GZSL^{cls} datasets.

Model	VGGSound-GZSL^{cls}				UCF-GZSL^{cls}				ActivityNet-GZSL^{cls}			
	S	U	HM	ZSL	S	U	HM	ZSL	S	U	HM	ZSL
w/o class. token												
$\mathbf{A}_{self} + \mathbf{A}_x$	18.40	3.78	6.27	4.25	31.70	32.57	32.13	33.26	11.87	3.80	5.75	3.90
$\mathbf{A}_{self}$	16.08	3.56	5.83	4.00	42.59	24.04	30.73	27.49	9.51	4.33	5.95	4.39
$\mathbf{A}_x$	14.62	4.22	6.55	4.59	19.52	29.80	23.62	31.35	1.85	3.50	2.42	3.50
with class. token												
$\mathbf{A}_c + \mathbf{A}_{self} + \mathbf{A}_x$	11.36	5.50	7.41	5.97	36.73	**41.99**	39.18	42.56	17.75	6.79	9.83	6.89
$\mathbf{A}_c + \mathbf{A}_{self}$	12.23	4.63	6.71	5.25	40.14	34.95	37.37	35.74	4.24	3.23	3.67	3.25
$\mathbf{A}_c + \mathbf{A}_x$ (TCAF)	12.63	**6.72**	**8.77**	**7.41**	**67.14**	40.83	**50.78**	**44.64**	**30.12**	**7.65**	**12.20**	**7.96**

Table 4. Influence of using multiple modalities for training and evaluating our proposed model on the (G)ZSL performance on the VGGSound-GZSL^{cls}, UCF-GZSL^{cls}, and ActivityNet-GZSL^{cls} datasets.

Model	VGGSound-GZSL^{cls}				UCF-GZSL^{cls}				ActivityNet-GZSL^{cls}			
	S	U	HM	ZSL	S	U	HM	ZSL	S	U	HM	ZSL
TCAF-audio	5.11	4.06	4.53	4.28	35.51	19.75	25.38	24.24	9.28	4.26	5.84	4.65
TCAF-visual	3.97	3.12	3.50	3.19	38.10	26.84	31.49	27.25	2.75	3.11	2.92	3.11
TCAF	**12.63**	**6.72**	**8.77**	**7.41**	**67.14**	**40.83**	**50.78**	**44.64**	**30.12**	**7.65**	**12.20**	**7.96**

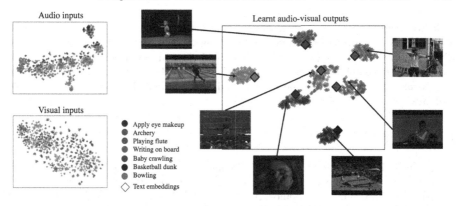

Fig. 3. t-SNE visualisation for five seen (*apply eye makeup, archery, baby crawling, basketball dunk, bowling*) and two unseen (*playing flute, writing on board*) test classes from the UCF-GZSL^{cls} dataset, showing audio and visual input embeddings extracted with C3D and VGGish, and audio-visual output embeddings learned with TCAF. Textual class label embeddings are visualised with a square.

4.4   Qualitative Results

We present a qualitative analysis of the learnt audio-visual embeddings in Fig. 3. For this, we show t-SNE [45] visualisations for the audio and visual input features and for the learnt multi-modal embeddings from 7 classes in the UCF-GZSL^{cls} test set. We averaged the input features for both modalities across time. We observe that the audio and visual input features are poorly clustered. In contrast, the audio-visual embeddings (θ_o) are clearly clustered for both, seen and unseen classes. This suggests that our network is actually learning useful representations for unseen classes, too. Furthermore, the word2vec class label embeddings (θ_w^j) lie inside the corresponding audio-visual clusters. This confirms that the learnt audio-visual embeddings are mapped to locations that are close to the corresponding word2vec embeddings, showing that our embeddings capture semantic information from the word2vec representations.

5   Conclusion

We presented a cross-attention transformer framework that addresses (G)ZSL for video classification using audio-visual input data with temporal information. Our proposed model achieves state-of-the-art performance on the three audio-visual (G)ZSL datasets UCF-GZSL^{cls}, VGGSound-GZSL^{cls}, and ActivityNet-GZSL^{cls}. The use of pre-extracted audio and visual features as inputs results in a computationally efficient framework compared to using raw data. We demonstrated that using cross-modal attention on temporal audio and visual input features and suppressing the contributions from the within-modality self-attention is beneficial for obtaining strong audio-visual embeddings that can transfer information from classes seen during training to novel, unseen classes at test time.

Acknowledgements. This work was supported by BMBF FKZ: 01IS18039A, DFG: SFB 1233 TP 17 - project number 276693517, by the ERC (853489 - DEXIM), and by EXC number 2064/1 - project number 390727645. The authors thank the International Max Planck Research School for Intelligent Systems (IMPRS-IS) for supporting O.-B. Mercea and T. Hummel. The authors would like to thank M. Mancini for valuable feedback.

References

1. Afouras, T., Asano, Y.M., Fagan, F., Vedaldi, A., Metze, F.: Self-supervised object detection from audio-visual correspondence. In: CVPR (2022)
2. Afouras, T., Chung, J.S., Senior, A., Vinyals, O., Zisserman, A.: Deep audio-visual speech recognition. IEEE TPAMI (2018)
3. Afouras, T., Chung, J.S., Zisserman, A.: ASR is all you need: cross-modal distillation for lip reading. In: ICASSP (2020)
4. Afouras, T., Owens, A., Chung, J.S., Zisserman, A.: Self-supervised learning of audio-visual objects from video. In: Vedaldi, A., Bischof, H., Brox, T., Frahm, J.-M. (eds.) ECCV 2020. LNCS, vol. 12363, pp. 208–224. Springer, Cham (2020). https://doi.org/10.1007/978-3-030-58523-5_13
5. Akata, Z., Perronnin, F., Harchaoui, Z., Schmid, C.: Label-embedding for image classification. IEEE TPAMI (2015)
6. Akata, Z., Reed, S., Walter, D., Lee, H., Schiele, B.: Evaluation of output embeddings for fine-grained image classification. In: CVPR (2015)
7. Alwassel, H., Mahajan, D., Torresani, L., Ghanem, B., Tran, D.: Self-supervised learning by cross-modal audio-video clustering. In: NeurIPS (2020)
8. Arandjelović, R., Zisserman, A.: Objects that sound. In: Ferrari, V., Hebert, M., Sminchisescu, C., Weiss, Y. (eds.) ECCV 2018. LNCS, vol. 11205, pp. 451–466. Springer, Cham (2018). https://doi.org/10.1007/978-3-030-01246-5_27
9. Asano, Y., Patrick, M., Rupprecht, C., Vedaldi, A.: Labelling unlabelled videos from scratch with multi-modal self-supervision. In: NeurIPS (2020)
10. Aytar, Y., Vondrick, C., Torralba, A.: Soundnet: Learning sound representations from unlabeled video. In: NeurIPS (2016)
11. Ba, J.L., Kiros, J.R., Hinton, G.E.: Layer normalization. arXiv preprint arXiv:1607.06450 (2016)
12. Boes, W., Van hamme, H.: Audiovisual transformer architectures for large-scale classification and synchronization of weakly labeled audio events. In: ACM MM (2019)
13. Brattoli, B., Tighe, J., Zhdanov, F., Perona, P., Chalupka, K.: Rethinking zero-shot video classification: End-to-end training for realistic applications. In: CVPR (2020)
14. Chao, W.-L., Changpinyo, S., Gong, B., Sha, F.: An empirical study and analysis of generalized zero-shot learning for object recognition in the wild. In: Leibe, B., Matas, J., Sebe, N., Welling, M. (eds.) ECCV 2016. LNCS, vol. 9906, pp. 52–68. Springer, Cham (2016). https://doi.org/10.1007/978-3-319-46475-6_4
15. Chen, H., Xie, W., Afouras, T., Nagrani, A., Vedaldi, A., Zisserman, A.: Localizing visual sounds the hard way. In: CVPR (2021)
16. Chen, Y., Xian, Y., Koepke, A.S., Shan, Y., Akata, Z.: Distilling audio-visual knowledge by compositional contrastive learning. In: CVPR (2021)
17. Devlin, J., Chang, M.W., Lee, K., Toutanova, K.: BERT: pre-training of deep bidirectional transformers for language understanding. In: ACL (2019)

18. Dosovitskiy, A., et al.: An image is worth 16x16 words: transformers for image recognition at scale. In: ICLR (2021)
19. Farhadi, A., Endres, I., Hoiem, D., Forsyth, D.: Describing objects by their attributes. In: CVPR (2009)
20. Fayek, H.M., Kumar, A.: Large scale audiovisual learning of sounds with weakly labeled data. In: IJCAI (2020)
21. Frome, A., et al.: Devise: a deep visual-semantic embedding model. In: NeurIPS (2013)
22. Gabeur, V., Sun, C., Alahari, K., Schmid, C.: Multi-modal transformer for video retrieval. In: Vedaldi, A., Bischof, H., Brox, T., Frahm, J.-M. (eds.) ECCV 2020. LNCS, vol. 12349, pp. 214–229. Springer, Cham (2020). https://doi.org/10.1007/978-3-030-58548-8_13
23. Gan, C., Huang, D., Chen, P., Tenenbaum, J.B., Torralba, A.: Foley music: learning to generate music from videos. In: Vedaldi, A., Bischof, H., Brox, T., Frahm, J.-M. (eds.) ECCV 2020. LNCS, vol. 12356, pp. 758–775. Springer, Cham (2020). https://doi.org/10.1007/978-3-030-58621-8_44
24. Gao, R., Grauman, K.: Co-separating sounds of visual objects. In: ICCV (2019)
25. Goldstein, S., Moses, Y.: Guitar music transcription from silent video. In: BMVC (2018)
26. Hendrycks, D., Gimpel, K.: Gaussian error linear units (GELUs). arXiv preprint arXiv:1606.08415 (2016)
27. Hershey, S., et al.: CNN architectures for large-scale audio classification. In: ICASSP (2017)
28. Iashin, V., Rahtu, E.: A better use of audio-visual cues: dense video captioning with bi-modal transformer. In: BMVC (2020)
29. Ioffe, S., Szegedy, C.: Batch normalization: accelerating deep network training by reducing internal covariate shift. In: ICML (2015)
30. Jaegle, A., Gimeno, F., Brock, A., Vinyals, O., Zisserman, A., Carreira, J.: Perceiver: general perception with iterative attention. In: ICML (2021)
31. Jamaludin, A., Chung, J.S., Zisserman, A.: You said that?: synthesising talking faces from audio. In: IJCV (2019)
32. Karpathy, A., Toderici, G., Shetty, S., Leung, T., Sukthankar, R., Fei-Fei, L.: Large-scale video classification with convolutional neural networks. In: CVPR (2014)
33. Kerrigan, A., Duarte, K., Rawat, Y., Shah, M.: Reformulating zero-shot action recognition for multi-label actions. In: NeurIPS (2021)
34. Kingma, D.P., Ba, J.: Adam: a method for stochastic optimization. arXiv preprint arXiv:1412.6980 (2014)
35. Koepke, A.S., Wiles, O., Moses, Y., Zisserman, A.: Sight to sound: an end-to-end approach for visual piano transcription. In: ICASSP (2020)
36. Koepke, A.S., Wiles, O., Zisserman, A.: Visual pitch estimation. In: SMC (2019)
37. Korbar, B., Tran, D., Torresani, L.: Cooperative learning of audio and video models from self-supervised synchronization. In: NeurIPS (2018)
38. Li, G., Duan, N., Fang, Y., Gong, M., Jiang, D.: Unicoder-VL: a universal encoder for vision and language by cross-modal pre-training. In: AAAI (2020)
39. Li, L.H., Yatskar, M., Yin, D., Hsieh, C.J., Chang, K.W.: VisualBERT: a simple and performant baseline for vision and language. arXiv preprint arXiv:1908.03557 (2019)
40. Lin, C.C., Lin, K., Wang, L., Liu, Z., Li, L.: Cross-modal representation learning for zero-shot action recognition. In: CVPR (2022)
41. Lin, Y.B., Wang, Y.C.F.: Audiovisual transformer with instance attention for audio-visual event localization. In: ACCV (2020)

42. Liu, S., Fan, H., Qian, S., Chen, Y., Ding, W., Wang, Z.: Hit: Hierarchical transformer with momentum contrast for video-text retrieval. In: ICCV (2021)
43. Liu, Y., Guo, J., Cai, D., He, X.: Attribute attention for semantic disambiguation in zero-shot learning. In: CVPR (2019)
44. Lu, J., Batra, D., Parikh, D., Lee, S.: Vilbert: pretraining task-agnostic visiolinguistic representations for vision-and-language tasks. In: NeurIPS (2019)
45. Van der Maaten, L., Hinton, G.: Visualizing data using t-SNE. In: JMLR (2008)
46. Mazumder, P., Singh, P., Parida, K.K., Namboodiri, V.P.: AVGZSLNet: audio-visual generalized zero-shot learning by reconstructing label features from multimodal embeddings. In: WACV (2021)
47. Mercea, O.B., Riesch, L., Koepke, A.S., Akata, Z.: Audio-visual generalised zero-shot learning with cross-modal attention and language. In: CVPR (2022)
48. Mikolov, T., Chen, K., Corrado, G., Dean, J.: Efficient estimation of word representations in vector space. In: ICLR (2013)
49. Nair, V., Hinton, G.E.: Rectified linear units improve restricted Boltzmann machines. In: ICML (2010)
50. Narasimhan, M., Ginosar, S., Owens, A., Efros, A.A., Darrell, T.: Strumming to the beat: audio-conditioned contrastive video textures. arXiv preprint arXiv:2104.02687 (2021)
51. Narayan, S., Gupta, A., Khan, F.S., Snoek, C.G.M., Shao, L.: Latent embedding feedback and discriminative features for zero-shot classification. In: Vedaldi, A., Bischof, H., Brox, T., Frahm, J.-M. (eds.) ECCV 2020. LNCS, vol. 12367, pp. 479–495. Springer, Cham (2020). https://doi.org/10.1007/978-3-030-58542-6_29
52. Owens, A., Efros, A.A.: Audio-visual scene analysis with self-supervised multisensory features. In: Ferrari, V., Hebert, M., Sminchisescu, C., Weiss, Y. (eds.) ECCV 2018. LNCS, vol. 11210, pp. 639–658. Springer, Cham (2018). https://doi.org/10.1007/978-3-030-01231-1_39
53. Owens, A., Wu, J., McDermott, J.H., Freeman, W.T., Torralba, A.: Learning sight from sound: ambient sound provides supervision for visual learning. Int. J. Comput. Vis. **126**(10), 1120–1137 (2018). https://doi.org/10.1007/s11263-018-1083-5
54. Owens, A., Wu, J., McDermott, J.H., Freeman, W.T., Torralba, A.: Learning sight from sound: Ambient sound provides supervision for visual learning. In: IJCV (2018)
55. Parida, K., Matiyali, N., Guha, T., Sharma, G.: Coordinated joint multimodal embeddings for generalized audio-visual zero-shot classification and retrieval of videos. In: WACV (2020)
56. Patrick, M., Asano, Y.M., Fong, R., Henriques, J.F., Zweig, G., Vedaldi, A.: Multimodal self-supervision from generalized data transformations. In: NeurIPS (2020)
57. Radford, A., Wu, J., Child, R., Luan, D., Amodei, D., Sutskever, I.: Language models are unsupervised multitask learners. In: OpenAI blog (2019)
58. Srivastava, N., Hinton, G., Krizhevsky, A., Sutskever, I., Salakhutdinov, R.: Dropout: a simple way to prevent neural networks from overfitting. In: JMLR (2014)
59. Su, K., Liu, X., Shlizerman, E.: Multi-instrumentalist net: unsupervised generation of music from body movements. arXiv preprint arXiv:2012.03478 (2020)
60. Su, W., Zhu, X., Cao, Y., Li, B., Lu, L., Wei, F., Dai, J.: VL-BERT: pre-training of generic visual-linguistic representations. arXiv preprint arXiv:1908.08530 (2019)
61. Sun, C., Baradel, F., Murphy, K., Schmid, C.: Learning video representations using contrastive bidirectional transformer. arXiv preprint arXiv:1906.05743 (2019)
62. Sun, C., Myers, A., Vondrick, C., Murphy, K., Schmid, C.: VideoBERT: a joint model for video and language representation learning. In: ICCV (2019)

63. Tan, H., Bansal, M.: LXMERT: learning cross-modality encoder representations from transformers. In: EMNLP (2019)
64. Tancik, M., et al.: Fourier features let networks learn high frequency functions in low dimensional domains. In: NeurIPS (2020)
65. Tian, Y., Shi, J., Li, B., Duan, Z., Xu, C.: Audio-visual event localization in unconstrained videos. In: Ferrari, V., Hebert, M., Sminchisescu, C., Weiss, Y. (eds.) ECCV 2018. LNCS, vol. 11206, pp. 252–268. Springer, Cham (2018). https://doi.org/10.1007/978-3-030-01216-8_16
66. Tran, D., Bourdev, L., Fergus, R., Torresani, L., Paluri, M.: Learning spatiotemporal features with 3d convolutional networks. In: ICCV (2015)
67. Vaswani, A., et al.: Attention is all you need. In: NeurIPS (2017)
68. Wah, C., Branson, S., Welinder, P., Perona, P., Belongie, S.: The caltech-UCSD birds-200-2011 dataset (2011)
69. Wang, X., Zhu, L., Yang, Y.: T2VLAD: global-local sequence alignment for text-video retrieval. In: CVPR (2021)
70. Wiles, O., Koepke, A.S., Zisserman, A.: X2Face: a network for controlling face generation using images, audio, and pose codes. In: Ferrari, V., Hebert, M., Sminchisescu, C., Weiss, Y. (eds.) ECCV 2018. LNCS, vol. 11217, pp. 690–706. Springer, Cham (2018). https://doi.org/10.1007/978-3-030-01261-8_41
71. Xian, Y., Lampert, C.H., Schiele, B., Akata, Z.: Zero-shot learning-a comprehensive evaluation of the good, the bad and the ugly. IEEE TPAMI (2018)
72. Xian, Y., Lorenz, T., Schiele, B., Akata, Z.: Feature generating networks for zero-shot learning. In: CVPR (2018)
73. Xian, Y., Sharma, S., Schiele, B., Akata, Z.: f-vaegan-d2: A feature generating framework for any-shot learning. In: CVPR (2019)
74. Xiao, J., Hays, J., Ehinger, K.A., Oliva, A., Torralba, A.: Sun database: large-scale scene recognition from abbey to zoo. In: CVPR (2010)
75. Xie, G.S., et al.: Attentive region embedding network for zero-shot learning. In: CVPR (2019)
76. Xu, W., Xian, Y., Wang, J., Schiele, B., Akata, Z.: Attribute prototype network for zero-shot learning. In: NeurIPS (2020)
77. Zhou, H., Liu, Z., Xu, X., Luo, P., Wang, X.: Vision-infused deep audio inpainting. In: ICCV (2019)
78. Zhu, Y., Xie, J., Liu, B., Elgammal, A.: Learning feature-to-feature translator by alternating back-propagation for generative zero-shot learning. In: ICCV (2019)

HM: Hybrid Masking for Few-Shot Segmentation

Seonghyeon Moon[1]([✉]), Samuel S. Sohn[1], Honglu Zhou[1], Sejong Yoon[2],
Vladimir Pavlovic[1], Muhammad Haris Khan[3], and Mubbasir Kapadia[1]

[1] Rutgers University, Newark, NJ, USA
{sm2062,samuel.sohn,honglu.zhou,vladimir,mubbasir.kapadia}@rutgers.edu
[2] The College of New Jersey, Trenton, NJ, USA
[3] Mohamed Bin Zayed University of Artificial Intelligence, Abu Dhabi, UAE
muhammad.haris@mbzuai.ac.ae

Abstract. We study few-shot semantic segmentation that aims to segment a target object from a query image when provided with a few annotated support images of the target class. Several recent methods resort to a feature masking (FM) technique to discard irrelevant feature activations which eventually facilitates the reliable prediction of segmentation mask. A fundamental limitation of FM is the inability to preserve the fine-grained spatial details that affect the accuracy of segmentation mask, especially for small target objects. In this paper, we develop a simple, effective, and efficient approach to enhance feature masking (FM). We dub the enhanced FM as hybrid masking (HM). Specifically, we compensate for the loss of fine-grained spatial details in FM technique by investigating and leveraging a complementary basic input masking method. Experiments have been conducted on three publicly available benchmarks with strong few-shot segmentation (FSS) baselines. We empirically show improved performance against the current state-of-the-art methods by visible margins across different benchmarks. Our code and trained models are available at: https://github.com/moonsh/HM-Hybrid-Masking

Keywords: Few-shot segmentation · Semantic segmentation · Few-shot learning

1 Introduction

Deep convolutional neural networks (DCNNs) have enabled remarkable progress in various important computer vision (CV) tasks, such as image recognition [8, 10,13,31], object detection [17,27,28], and semantic segmentation [3,20,43]. Despite proving effective for various CV tasks, DCNNs require a large amount of labeled training data, which is quite cumbersome and costly to acquire for

Supplementary Information The online version contains supplementary material available at https://doi.org/10.1007/978-3-031-20044-1_29.

dense prediction tasks, such as semantic segmentation. Furthermore, these models often fail to segment novel (unseen) objects when provided with very few annotated training images. To counter the aforementioned problems, few shot segmentation (FSS) methods, that rely on a few annotated support images, have been actively studied [1,14,18,22,24–26,30,33,35–38,40,42].

After the pioneering work of OSLSM [29], many few-shot segmentation methods have been proposed in recent years [1,14,18,22,24–26,30,33,35–38,40,42]. Among others, an important challenge in few-shot segmentation is how to use support images towards capturing more meaningful information. Many recent state-of-the-art methods [9,11,24,33,36,39,40,42] rely on feature masking (FM) [42] to discard irrelevant feature activations for reliable segmentation mask prediction. However, when masking a feature map, some crucial information in support images, such as the target object boundary, is partially lost. In particular, when the size of the target object is relatively small, this lost fine-grained spatial information renders it rather difficult to obtain accurate segmentation (see Fig. 1).

In this paper, we propose a simple, effective, and efficient technique to enhance feature masking (FM) [42]. We dub the enhanced FM as hybrid masking (HM). In particular, we compensate for the loss of target object details in FM technique through leveraging a simple input masking (IM) technique [29]. We note that IM is capable of preserving the fine details, especially around object boundaries, however, it lacks discriminative information, as such, after the removal of background information. To this end, we investigate the possibility of transferring object details in the IM to enrich FM technique. We instantiate the proposed hybrid masking (HM) into two recent strong baselines: HSNet [24] and VAT [9]. Results reveal more accurate segmentation masks by recovering the fine-grained details, such as target boundaries and target textures (see Fig. 1). Following are the main contributions of this paper:

- We propose a simple, effective, and efficient way to enhance a de-facto feature masking technique (FM) in several recent few-shot segmentation methods.
- We perform extensive experiments to validate the effectiveness of proposed hybrid masking across two strong FSS baselines, namely HSNet [24] and VAT [9] on three publicly available datasets: Pascal-5^i [29], COCO-20^i [16], and FSS-1000 [15]. Results show notable improvements against the state-of-the-art methods in all datasets.
- We note that our HM facilitates improving the training efficiency. When integrated into HSNet [24] with ResNet101 [8], it speeds up its training convergence by around 11x times on average on COCO-20^i [16].

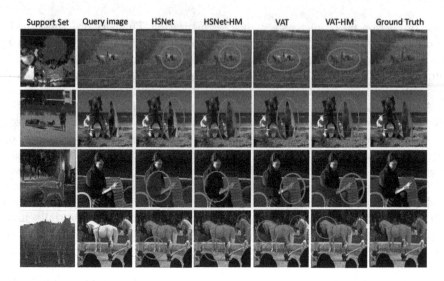

Fig. 1. Our HM allows generation of more accurate segmentation masks by recovering the fine-grained details (marked in cyan ellipse) when integrated into the current state-of-the-art methods, HSNet [24] and VAT [9] on COCO-20^i [16].

2   Related Work

Few-Shot Segmentation. The work of Shaban *et al.* [29] is believed to introduce the few shot segmentation task to the community. It generated segmentation parameters by using the conditioning branch on the support set. Later, we observe steady progress in this task, and so several methods were proposed [1,14,18,22,24–26,30,33,35–38,40,42]. CANet [40] modified the cosine similarity with the additive alignment module and enhanced the performance by performing various iterations. To improve segmentation quality, PFENet [33] designed a pyramid module and used a prior map. Inspired by prototypical networks [32], PANet [36] leveraged novel prototype alignment network. Along similar lines, PPNet [19] utilized part-aware prototypes to get the detailed object features and PMM [38] used the expectation-maximization (EM) algorithm to generate multiple prototypes. ASGNet [14] proposed two modules, superpixel-guided clustering (SGC) and guided prototype allocation (GPC) to extract and allocate multiple prototypes. In pursuit of improving correspondence between support and query images, DAN [35] democratized graph attention. Yang *et al.* [39] introduced a method to mine latent classes in the background, and CWT [22] designed a simple classifier with transformer. CyCTR [41] mined information from the whole support image using transformer. ASNet [11] proposed the integrative few-shot learning framework (iFSL) overcoming limitations of few-shot classification and few-shot segmentation. HSNet [24] utilized efficient 4D convolution to analyze deeply accumulated features and achieved remarkable performance. Recently, VAT [9] proposed a cost aggregation network, based on transformers, to model dense semantic correspondence between images and capture intra-class variations. We validate the effectiveness of our hybrid masking approach by instantiating it in two strong FSS baselines: HSNet [24] and VAT [9].

Feature Masking. Zhang *et al.* [42] proposed Masked Average Pooling (MAP) to eliminate irrelevant feature activations which facilitates reliable mask prediction. In MAP, feature masking (FM) was introduced and utilized before average pooling. Afterward, FM was widely adopted as the de-facto technique to achieve feature masking [9,11,24,33,36,39,40,42]. We note that the FM method loses information about the target object in the process of feature masking. Specifically, it is prone to losing the fine-grained spatial details, which can be crucial for generating a precise segmentation mask. In this work, we compensate for the loss of target object details in FM technique via leveraging a simple input masking (IM) technique [29].

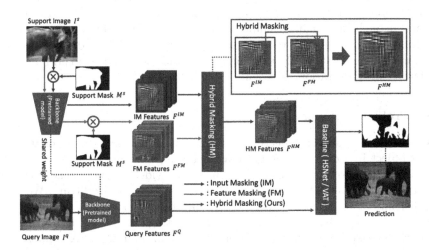

Fig. 2. The overall architecture when our proposed hybrid masking (HM) approach is integrated into FSS baselines. At its core, it contains a feature backbone, a feature masking (FM) technique. After extracting support and query features, the feature masking suppresses irrelevant activations in the support features. We introduce a simple, effective, and efficient way to enhance feature masking (FM), termed as *hybrid masking* (HM). It compensates for the loss of target object details in the FM technique by leveraging a simple input masking (IM) technique [29].

Input Masking. Input masking (IM) [29] is a technique to eliminate background pixels by multiplying the support image with its corresponding support mask. There were two key motivations behind erasing the background pixels. First, the largest object in the image has a tendency to dominate the network. Second, the variance of the output parameters increased when the background information was included in the input. We observe that IM can preserve the fine details, however, it lacks target discriminative information, important for distinguishing between the foreground and the background. In this work, we investigate the possibility of transferring object details present in the IM to enrich the FM technique, thereby exploiting the complementary strengths of both.

3   Methodology

Figure 2 displays the overall architecture when our proposed *hybrid masking* (HM) approach is introduced into the FSS baselines, such as HSNet [24] and VAT [9]. Fundamentally, it comprises of a feature backbone for extracting support and query features, a feature masking (FM) technique for suppressing irrelevant support activations, and FSS model (i.e. HSNet/VAT) for predicting the segmentation mask from the relevant activations. In this work, we propose a simple, effective, and efficient way to enhance feature masking (FM), termed as hybrid masking (HM). It compensates for the loss of target object details in the FM technique by leveraging a simple input masking (IM) technique [29]. In what follows, we first lay out the problem setting (sect. 3.1), next we describe feature masking technique (sect. 3.2), and finally we detail the proposed hybrid masking for few-shot segmentation (sect. 3.3).

3.1   Problem Setting

Few-shot segmentation's objective is to train a model that can recognize the target object in a query image given a small sample of annotated images from the target class. We tackle this problem using the widely adopted episodic training scheme [9,24,34,36], which has been shown to reduce overfitting. We have the disjoint sets of training classes C_{train} and testing classes C_{test}. The training data D_{train} belongs to C_{train} and the testing data D_{test} is from C_{test}. Multiple episodes are constructed using the D_{train} and D_{test}. A support set, $S = (I^s, M^s)$, and a query set, $Q = (I^q, M^q)$, are the two components that make up each episode. I and M represent an image and its mask. We have N_{train} episodes for training $D_{train} = \{(S_i, Q_i)\}_{i=1}^{N_{train}}$ and N_{test} episodes for testing $D_{test} = \{(S_i, Q_i)\}_{i=1}^{N_{test}}$. Sampled episodes from D_{train} are used to train a model to predict query mask M^q. Afterward, the learned model is evaluated by randomly sampling episodes from the testing data D_{test} in the same manner and comparing the predicted query masks to the ground truth.

3.2   Feature Masking

Zhang *et al.* [42] argued that IM greatly increases the variance of the input data for a unified network, and according to Long *et al.* [21], the relative positions of input pixels can be preserved by fully convolutional networks. These two ideas motivated Masked Average Pooling (MAP), which ideally extracts the features of a target object while eliminating the background content. Although MAP falls short of this in practice, it remains helpful for learning better object features [3] while keeping the input structure of the network unchanged. In particular, the feature masking part is still widely used.

Given a support RGB image $I^s \in \mathbb{R}^{3 \times w \times h}$ and a support mask $\mathbf{M}^s \in$

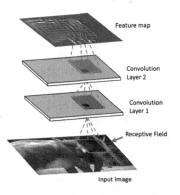

Fig. 3. Impact of growing receptive field on feature masking. The input image's elephant is a target object and the other pixels are background. One pixel at the feature is generated from lots of pixels' information from previous layer. The background and target object information both can be present in one feature map pixel.

$\{0,1\}^{w \times h}$, where w and h are the width and height of the image, the support feature maps of I^s are $F^s \in \mathbb{R}^{c \times w' \times h'}$, where c is the number of channels, w' and h' are the width and height of the feature maps. Feature masking is performed after matching the mask to the feature size using the bilinear interpolation. We denote a function resizing the mask as $\tau(\cdot) : \mathbb{R}^{w \times h} \rightarrow \mathbb{R}^{c \times w' \times h'}$. Then, the feature masking features $F^{FM} \in \mathbb{R}^{c \times w' \times h'}$ are computed according to Eq. 1,

$$F^{FM} = F^s \odot \tau(M^s), \tag{1}$$

where $\odot$ denotes the Hadamard product. Zhang *et al.* [42] fit the feature size to the mask size, but we conversely fit the mask to the feature size.

Feature masking (FM), which forms the core part of Masked Average Pooling (MAP) [42], is utilized to eliminate background information from support features and has become the de facto technique for masking feature maps, appearing in several recent few-shot segmentation methods [33,36,39,40,42] even in current state-of-the-art [24]. However, FM inadvertently eliminates both the background and the target object information because one pixel from the last layer's feature map corresponds to many pixels in the input image.

Figure 3 shows that one pixel of the feature map could contain background and target object information together [23]. This is further analyzed in Fig. 5, which clearly shows that FM loses useful information through its masking and progressively worsens with deeper layers. If a target object in the support set appears very small, the segmentation of the query image becomes even more challenging because the features are fed into network with relatively large proportion of undesired background features. Figure 4 shows the limitation of FM.

3.3   Hybrid Masking for Few-shot Segmentation (HM)

We aim to maximize target information from the support set so that the network can efficiently learn to provide more accurate segmentation of a target object.

The overall architecture when our proposed HM is integrated into a FSS baseline is shown in Fig. 2. We obtain two feature maps, F^{IM} and F^{FM}, using IM and FM respectively. These two feature maps are merged by hybrid masking (Algorithm 1) to generate HM feature map F^{HM}. This HM feature map will be used as input for HSNet and VAT, which takes full advantage of the given features to predict the target object mask.

Input Masking (IM). IM [29] eliminates background pixels by multiplying the support image with its corresponding support mask because of two empirical reasons. (1) The network has a tendency to favor the largest object in the image, which is not the object we want to segment. (2) The background information will result in an increase in the variance of the output parameters.

Suppose we have the RGB support image $I^s \in \mathbb{R}^{3 \times w \times h}$ and a support mask $\mathbf{M}^s \in \{0,1\}^{w \times h}$ in the image space, where w and h are the width and height of the image. IM, computed as

$$I^{s'} = I^s \odot \tau(M^s), \tag{2}$$

contains the target object alone. We use the function $\tau(\cdot)$ for resizing the mask M^s to fit the image I^s.

Hybrid Masking (HM). We propose an alternative masking approach, which takes advantage of the features generated by both FM and IM. First, FM and IM features are computed according to the existing methods. The unactivated values in the FM features are then replaced with IM features. Other activated values remain without replacing to maintain FM features. We name this process as hybrid masking (Algorithm 1). HM prioritizes the information from FM features and supplements the lacking information, such as the precise target boundaries and fine-grained texture information, from IM features, which is superior for delineating the boundaries of target objects and the missing texture information. The method is as follows even if feature maps are stacked to have a sequence of intermediate feature maps. Only one more loop is needed for the deep feature maps.

Algorithm 1: Hybrid Masking

Input : IM feature maps F^{IM} and FM features maps F^{FM}
Each channel i, $f_i^{IM} \in F^{IM}$ and $f_i^{FM} \in F^{FM}$
for $i = 1, \ldots, c$ **do**
    Set $f_i^{HM} = f_i^{FM}$
    for *Entire pixels* $\in f_i^{HM}$ **do**
        Find an inactive pixel, $p \in f_i^{HM}$
        if $p < 0$ **then**
           | Replace the pixel, p, with corresponding pixel $\in f_i^{IM}$
        end
    end
end
Output: HM feature maps F^{HM}

The generated HM feature maps are used as inputs to two strong FSS models (HSNet and VAT). These two models are best-suited for fully utilizing the features generated by hybrid masking, because they build multi-level correlation maps taking advantage of the rich semantics that are provided at different feature levels.

4   Experiment

4.1   Setup

Datasets. We evaluate the efficacy of our hybrid masking technique on three publicly available segmentation benchmarks: PASCAL-5^i [29] , COCO-20^i [16], and FSS-1000 [15]. PASCAL-5^i was produced from PASCAL VOC 2012 [6] with additional mask annotations [7]. PASCAL-5^i contains 20 types of object classes, COCO-20^i contains 80 classes, and FSS-1000 contains 1000 classes. The PASCAL and COCO data sets were divided into four folds following the training and evaluation methods of other works [9,11,19,24,25,33,35,38], where each fold of PASCAL-5^i consisted of 5 classes, and each fold of COCO-20^i had 20. We conduct cross-validation using these four folds. When evaluating a model on fold^i, all other classes not belonging to fold^i are used for training. 1000 episodes are sampled from the other fold^i to evaluate the trained model. For FSS-1000, the training, validation, and test datasets are divided into 520, 240, and 240 classes.

Implementation Details. We integrate our hybrid masking technique into two FSS baselines: HSNet [24] and VAT [9], and the resulting methods are denoted by HSNet-HM and VAT-HM. We use ResNet50 [8] and ResNet101 [8] backbone networks pre-trained on ImageNet [5] with their weights frozen to extract features, following HSNet [24] and VAT [9]. From conv3_x to conv5_x of ResNet (i.e., the three layers before global average pooling), the features in the bottleneck part before the ReLU activation of each layer were stacked up to create deep features. We follow the HSNet [24] and the VAT [9] default settings for optimizer [12] and learning rate. A batch size of 20 is used for HSNet-HM training for all benchmarks. For VAT-HM training, 8, 4, and 4 batch sizes are utilized for COCO-20^i, PASCAL-5^i and FSS-1000 respectively. We used data augmentation for HSNet-HM training on PASCAL-5^i following [2,4,9]. For COCO-20^i and FSS-1000 benchmarks, no data augmentation was employed when training HSNet-HM.

Evaluation Metrics. Following [9,24,33,35], we adopt two evaluation metrics, mean intersection over union (mIoU) and foreground-background IoU (FB-IoU) for model evaluation. The mIoU averages the IoU values for all classes in each fold. FB-IoU calculates the foreground and background IoU values ignoring object classes and averages them. Note that, mIoU is a better indicator of model generalization than FB-IoU [24].

4.2 Comparison with the State-of-the-Art (SOTA)

PASCAL-5^i. Table 1 compares our methods, HSNet-HM and VAT-HM, with other methods on PASCAL-5^i datasets. In the 1-shot test, HSNet-HM provides a gain of 0.7% mIoU compared to HSNet [24] with ResNet50 backbone and performs on par with HSNet [24] using ReNet101 backbone. In the 5-shot test, HSNet-HM shows slightly inferior performance in mIoU and FB-IoU. VAT-HM shows a similar pattern to HSNet-HM. In the 1-shot test, VAT-HM shows a gain of 0.5% mIoU with ResNet50 and a gain of 0.3% mIoU with ResNet101. In the 5-shot test, VAT-HM provides an improvement of 0.8% mIoU with ResNet50.

COCO-20^i. Table 2 reports results on the COCO-20^i dataset. In 1-shot test, HSNet-HM, VAT-HM, and ASNet-HM show a significant improvement over HSNet [24], VAT [9], and ASNet [11]. HSNet-HM delivers a gain of 5.1% and 5.3% in mIoU with ResNet50 and ResNet101 backbones, respectively. VAT-HM provides a gain of 2.9% mIoU with ResNet50. ASNet-HM provides a gain of 2.5% and 2.8% mIoU with ResNet50 and ResNet101. Similarly, in 5-shot test, HSNet-HM outperforms HSNet [24] by 3.5% and 1.1% with ResNet50 and ResNet101 backbones, respectively. VAT-HM provides 0.4% mIoU improvement with ResNet50 in 5-shot test. ASNet-HM shows slightly worse performance with ResNet50 but ASNet-HM delivers a gain of 1.1% with ResNet101. Figure 1 draws visual comparison with HSNet [24] and VAT [9] under several challenging segmentation instances. Note that, compared to HSNet and VAT, HSNet-HM and VAT-HM produce more accurate segmentation masks that recover fine-grained details under appearance variations and complex backgrounds.

Table 1. Performance comparison with the existing methods on Pascal-5^i [6]. Superscript asterisk denotes that data augmentation was applied during training. Best results are bold-faced and the second best are underlined.

Backbone feature	Methods	1-shot						5-shot					
		5^0	5^1	5^2	5^3	mIoU	FB-IoU	5^0	5^1	5^2	5^3	mIoU	FB-IoU
ResNet50 [8]	PANet [36]	44.0	57.5	50.8	44.0	49.1	–	55.3	67.2	61.3	53.2	59.3	–
	PFENet [33]	61.7	69.5	55.4	56.3	60.8	73.3	63.1	70.7	55.8	57.9	61.9	73.9
	ASGNet [14]	58.8	67.9	56.8	53.7	59.3	69.2	63.4	70.6	64.2	57.4	63.9	74.2
	CWT [22]	56.3	62.0	59.9	47.2	56.4	–	61.3	68.5	68.5	56.6	63.7	–
	RePRI [1]	59.8	68.3	62.1	48.5	59.7	–	64.6	71.4	**71.1**	59.3	66.6	–
	CyCTR [41]	67.8	**72.8**	58.0	58.0	64.2	–	71.1	<u>73.2</u>	60.5	57.5	65.6	–
	HSNet [24]	64.3	70.7	60.3	60.5	64.0	76.7	70.3	<u>73.2</u>	67.4	**67.1**	69.5	<u>80.6</u>
	HSNet*	63.5	70.9	<u>61.2</u>	60.6	64.3	**78.2**	70.9	73.1	68.4	65.9	<u>69.6</u>	<u>80.6</u>
	VAT [9]	67.6	<u>71.2</u>	**62.3**	60.1	<u>65.3</u>	<u>77.4</u>	<u>72.4</u>	**73.6**	<u>68.6</u>	65.7	**70.0**	80.9
	HSNet*-HM	**69.0**	70.9	59.3	<u>61.0</u>	65.0	76.5	69.9	72.0	63.4	63.3	67.1	77.7
	VAT-HM	<u>68.9</u>	70.7	61.0	**62.5**	**65.8**	77.1	**71.1**	<u>72.5</u>	62.6	<u>66.5</u>	68.2	78.5
ResNet101 [8]	FWB [25]	51.3	64.5	56.7	52.2	56.2	–	54.8	67.4	62.2	55.3	59.9	–
	DAN [35]	54.7	68.6	57.8	51.6	58.2	71.9	57.9	69.0	60.1	54.9	60.5	72.3
	PFENet [33]	60.5	69.4	54.4	55.9	60.1	72.9	62.8	70.4	54.9	57.6	61.4	73.5
	ASGNet [14]	59.8	67.4	55.6	54.4	59.3	71.7	64.6	71.3	64.2	57.3	64.4	75.2
	CWT [22]	56.9	65.2	61.2	48.8	58.0	–	62.6	70.2	**68.8**	57.2	64.7	–
	RePRI [1]	59.6	68.6	62.2	47.2	59.4	–	66.2	71.4	67.0	57.7	65.6	–
	CyCTR [41]	69.3	**72.7**	56.5	58.6	64.3	72.9	<u>73.5</u>	74.0	58.6	60.2	66.6	75.0
	HSNet [24]	67.3	72.3	62.0	63.1	66.2	77.6	71.8	74.4	67.0	68.3	70.4	80.6
	HSNet*	67.5	**72.7**	<u>63.5</u>	63.2	66.7	77.7	71.7	74.8	68.2	<u>68.7</u>	70.8	80.9
	VAT [9]	68.4	<u>72.5</u>	**64.8**	64.2	<u>67.5</u>	**78.8**	73.3	<u>75.2</u>	<u>68.4</u>	**69.5**	**71.6**	**82.0**
	HSNet*-HM	<u>69.8</u>	72.1	60.4	<u>64.3</u>	66.7	77.8	72.2	73.3	64.0	67.9	69.3	79.7
	VAT-HM	**71.2**	**72.7**	62.7	**64.5**	**67.8**	<u>79.4</u>	**74.0**	**75.5**	65.4	68.6	<u>70.9</u>	<u>81.5</u>

FSS-1000. Table 3 compares HSNet-HM, VAT-HM, and competing methods on the FSS-1000 dataset [15]. In the 1-shot test, HSNet-HM yields a gain of 1.6% and 1.3% in mIoU over [24] with ResNet50 and ResNet101 backbones, respectively. In the 5-shot test, we observe an improvement of 0.2% in mIoU over [24] with the ResNet50 backbone. In the 1-shot test, VAT-HM shows slightly inferior mIoU compared to VAT [9] with ResNet50 but it performs a little better than VAT [9] with ResNet101.

Generalization Test. Following previous works [1,24], we perform a domain shift test to evaluate the generalization capability of the proposed method. We trained HSNet-HM and VAT-HM on the COCO-20^i dataset and tested this model on the PASCAL-5^i dataset. The training/testing folds were constructed following [1,24]. The objects in training classes do not overlap with the object in the testing classes. As shown in Table 4, HSNet-HM outperforms the current state-of-the-art approaches under both 1-shot and 5-shot tests. In 1-shot test, it delivers a 2% mIoU gain over RePRI [1] and a 2.4% mIoU gain over HSNet [24] with ResNet50 and ResNet101 backbones, respectively. In the 5-shot test,

Table 2. Performance comparison on COCO-20^i [16] in mIoU and FB-IoU. Best results are bold-faced and the second best are underlined.

Backbone feature	Methods	1-shot						5-shot					
		20^0	20^1	20^2	20^3	mIoU	FB-IoU	20^0	20^1	20^2	20^3	mIoU	FB-IoU
ResNet50 [8]	PMM [38]	29.3	34.8	27.1	27.3	29.6	–	33.0	40.6	30.3	33.3	34.3	–
	RPMM [38]	29.5	36.8	28.9	27.0	30.6	–	33.8	42.0	33.0	33.3	35.5	–
	PFENet [33]	36.5	38.6	34.5	33.8	35.8	–	36.5	43.3	37.8	38.4	39.0	–
	ASGNet [14]	–	–	–	–	34.6	60.4	–	–	–	–	42.5	67.0
	RePRI [1]	32.0	38.7	32.7	33.1	34.1	–	39.3	45.4	39.7	41.8	41.6	–
	HSNet [24]	36.3	43.1	38.7	38.7	39.2	68.2	43.3	51.3	48.2	45.0	46.9	70.7
	CyCTR [41]	38.9	43.0	39.6	39.8	40.3	–	41.1	48.9	45.2	47.0	45.6	–
	VAT [9]	39.0	43.8	42.6	39.7	41.3	68.8	44.1	51.1	50.2	46.1	47.9	_72.4_
	ASNet [11]	41.5	44.1	42.8	40.6	42.2	69.4	**48.0**	_52.1_	49.7	_48.2_	**49.5**	72.7
	HSNet-HM	41.0	_45.7_	**46.9**	_43.7_	_44.3_	70.8	45.3	**53.1**	52.1	47.0	_49.4_	72.2
	VAT-HM	_42.2_	44.3	_45.0_	42.2	43.2	70.0	45.2	51.0	_50.7_	46.4	48.3	71.8
	ASNet-HM	**42.8**	**46.0**	44.8	**45.0**	**44.7**	_70.4_	_46.3_	50.2	48.4	**48.6**	48.4	72.2
ResNet101 [8]	FWB [25]	17.0	18.0	21.0	28.9	21.2	–	19.1	21.5	23.9	30.1	23.7	–
	DAN [35]	–	–	–	–	24.4	62.3	–	–	–	–	29.6	63.9
	PFENet [33]	36.8	41.8	38.7	36.7	38.5	63.0	40.4	46.8	43.2	40.5	42.7	65.8
	HSNet [24]	37.2	44.1	42.4	41.3	41.2	69.1	45.9	_53.0_	_51.8_	47.1	_49.5_	72.4
	ASNet [11]	_41.8_	45.4	43.2	41.9	43.1	69.4	**48.0**	52.1	49.7	48.2	_49.5_	72.7
	HSNet-HM	41.2	**50.0**	**48.8**	_45.9_	**46.5**	71.5	46.5	**55.2**	_51.8_	_48.9_	50.6	_72.9_
	ASNet-HM	**43.5**	_46.4_	_47.2_	**46.4**	_45.9_	_71.1_	_47.7_	51.6	**52.1**	**50.8**	50.6	**73.3**

Table 3. Performance comparison with other methods on FSS-1000 [15] dataset. Best results are bold-faced and the second best are underlined.

Backbone feature	Methods	mIoU		Backbone feature	Methods	mIoU	
		1-shot	5-shot			1-shot	5-shot
ResNet50 [8]	FSOT [18]	82.5	83.8	ResNet101 [8]	DAN [35]	85.2	88.1
	HSNet [24]	85.5	87.8		HSNet [24]	86.5	88.5
	VAT [9]	**89.5**	90.3		VAT [9]	_90.0_	**90.6**
	HSNet-HM	87.1	88.0		HSNet-HM	87.8	88.5
	VAT-HM	_89.4_	_89.9_		VAT-HM	**90.2**	_90.5_

Table 4. Comparison of generalization performance with domain shift test. A model was trained on COCO-20^i [16] and then evaluated on PASCAL-5^i [6].

Backbone feature	Methods	1-shot					5-shot				
		5^0	5^1	5^2	5^3	mIoU	5^0	5^1	5^2	5^3	mIoU
ResNet50 [8]	RPMM [38]	36.3	55.0	52.5	54.6	49.6	40.2	58.0	55.2	61.8	53.8
	PFENet [33]	43.2	_65.1_	66.5	69.7	61.1	45.1	66.8	68.5	73.1	63.4
	RePRI [1]	**52.2**	64.3	64.8	71.6	63.2	56.5	_68.2_	70.0	76.2	67.7
	HSNet [24]	45.4	61.2	63.4	75.9	61.6	_56.9_	65.9	71.3	80.8	_68.7_
	VAT	_52.1_	64.1	67.4	74.2	64.5	**58.5**	68.0	_72.5_	79.9	**69.7**
	HSNet-HM	43.4	**68.2**	69.4	79.9	65.2	50.7	**71.4**	73.4	83.1	**69.7**
	VAT-HM	48.3	64.9	_67.5_	_79.8_	65.1	55.6	68.1	72.4	_82.8_	**69.7**
ResNet101 [8]	HSNet [24]	**47.0**	_65.2_	_67.1_	_77.1_	_64.1_	**57.2**	_69.5_	_72.0_	_82.4_	_70.3_
	HSNet-HM	_46.7_	**68.6**	**71.1**	**79.7**	**66.5**	_53.7_	**70.7**	**75.2**	**83.9**	**70.9**

HSNet-HM outperforms HSNet [24] by 1.0% and 0.6% in mIoU with ResNet50 and ResNet101 backbones, respectively.

4.3 Ablation Study and Analysis

Comparison of the Three Different Masking Approaches. We compare all three masking approaches, IM [29], FM [42], and the proposed HM after incorporating them into HSNet [24] and evaluate them on the COCO-20^i dataset (Table 5). We can see that in both 1-shot and 5-shot tests, the proposed HM approach provides noticeable gains over either individual FM and IM techniques.

Table 5. Ablation study of the three different masking methods on COCO-20^i [16].

Backbone feature	Masking methods	1-shot						5-shot					
		20^0	20^1	20^2	20^3	mIoU	FB-IoU	20^0	20^1	20^2	20^3	mIoU	FB-IoU
ResNet50 [8]	HSNet-FM [24]	36.3	43.1	38.7	38.7	39.2	68.2	43.3	51.3	48.2	45.0	46.9	70.7
	HSNet-IM	39.8	45.0	46.0	43.2	43.5	70.0	43.4	50.9	49.5	48.0	47.6	71.7
	HSNet-HM	41.0	45.7	46.9	43.7	44.3	70.8	45.3	53.1	52.1	47.0	49.4	72.2
ResNet101 [8]	HSNet-FM [24]	37.2	44.1	42.4	41.3	41.2	69.1	45.9	53.0	51.8	47.1	49.5	72.4
	HSNet-IM	41.0	48.3	47.3	44.5	45.2	70.9	46.6	54.5	50.4	47.7	49.8	72.7
	HSNet-HM	41.2	50.0	48.8	45.9	46.5	71.5	46.5	55.2	51.8	48.9	50.6	72.9

Table 6. Ablation study of the three different merging methods on COCO-20^i

Feature backbone	Methods	1-shot				
		20^0	20^1	20^2	20^3	mIoU
ResNet50	HSNet-HM (Simple Add.)	40.0	43.5	43.4	43.2	42.5
	HSNet-HM (Reverse)	39.4	45.2	42.3	41.6	42.1
	HSNet-HM	41.0	45.7	46.9	43.7	44.3

Table 7. Run-time comparison at inference stage on COCO-20^i [16].

Inference Time	Secs/ Image	Additional Overhead in %
HSNet-FM	0.27	–
HSNet-HM	0.34	25.9

Figure 4 shows the qualitative results from the three masking methods. The blue objects in the support set are the target objects for segmentation. The red pixels are the segmentation results. FM can coarsely segment the objects from the background but fails to precisely recover target details, such as target boundaries. IM is capable of recovering precise object boundaries, but struggles in distinguishing objects from the background. The proposed approach, HM, clearly distinguishes between the target objects and the background and also recovers precise details such as, target boundaries.

| Support Set | Query image | FM Prediction | IM prediction | HM prediction | Ground Truth |

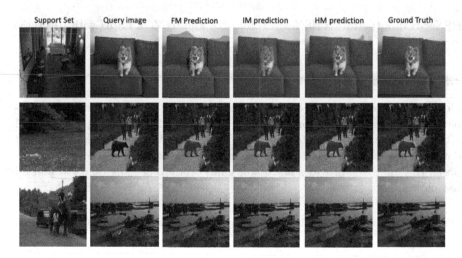

Fig. 4. Qualitative comparison of three different masking approaches on COCO-20^i [16] with HSNet. The blue objects in the support set are the target objects for segmentation. The red pixels are the segmentation results. HSNet-FM can coarsely segment the objects from the background but fails to precisely recover target details, such as target boundaries. HSNet-IM is capable of recovering precise object boundaries, but struggles in distinguishing objects from the background. The proposed approach, HSNet-HM, clearly distinguishes between the target objects and the background and also recovers precise details such as, target boundaries.

Figure 5 shows the visual comparison between the feature maps of IM and FM features. The feature maps inside the red rectangles reveal that the two features produced from the two masking approaches are different. Looking at the area where activations occur in the IM feature map at layer 50, we can see it is more indicative of the target object boundaries than the FM feature. Additionally, looking at the IM feature map at layer 34, we observe that there is a strong signal around the edge and even in side of the target object. This happens because FM performs masking after extracting features, and so this results in less precise target boundaries and loss of texture information.

Other Combination Proposals for Obtaining HM. We apply various masking methods to create HM features. Various mask sizes are tested by applying dilation to the mask of the support set, but the most plausible result is obtained with the IM masking method. Also, the method in [42] obtains a mask using the bounding box and applies the average pooling method, but fails to achieve better performance than FM. We provide two ablation studies to understand the effectiveness of replacement operation (see Table 6). First, we simply add the corresponding feature maps of IM and FM, denoted as HM(Simple Add). Second, we perform the reverse procedure of the proposed HM. We initialize the HM by IM, and supplement the inactivated features with FM features, denoted

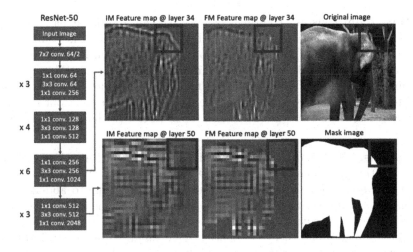

Fig. 5. Visual comparison between the feature maps of IM and FM. These are from ResNet50 at layer34 and layer50. We visualize the first channel of the feature map in grayscale. The feature maps inside the red rectangles reveal that the features from the two feature masking methods are different. Observe activations in the IM feature map at layer 50, it is more indicative of the target object boundaries than the FM feature. Additionally, the IM feature map at layer 34 displays a strong signal around the edge.

as HM(Reverse). Note that, our proposed HM is more effective for FSS compared to HM(Simple Add) and HM(Reverse).

Training Efficiency. Figure 6 shows the training profiles of HSNet-HM on COCO-20^i. We see that HM results in faster training convergence compared to HSNet, reducing the training time by a factor of 11x on average. To reach the best model with ResNet101 on COCO-20^3, 296.5 epochs are required for HSNet [24] but HSNet-HM only needs 26.8 epochs on average. A similar trend was observed in the PASCAL-5^i and FSS-1000 datasets, for which the results are reported in the supplementary material.

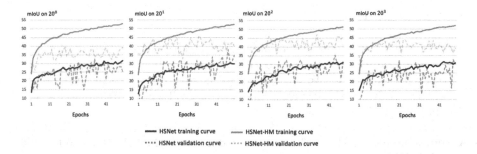

Fig. 6. Training profiles of HSNet [24] and HSNet-HM on COCO-20^i.

Runtime Comparison. Hybrid masking takes an additional pass over the pixel values to choose between FM and IM. We measure the computation time of the HM method and other methods for comparison. IM/FM take 0.05 secs/image, and their throughput is 20 images/sec. Whereas HM takes 0.07 secs/image. Therefore, HM induces 40% more computational time when compared to IM/FM. However, in terms of the model's inference time, our HM adds a relatively less extra overhead (25.9%) on top of HSNet (with FM) (see Table 7).

Limitations. We found that HSNet-HM performance in PASCAL-5^i [6] was inferior to the performance of the COCO-20^i [16] dataset. A potential reason is that HSNet-HM quickly enters the over-fitting phase due to abundance of information about the target object. The following data augmentation method [2, 4, 9] was able to alleviate this problem to some extent, but it did not solve the problem completely. Further, we identify some failure cases for HM (Fig. 7). HM struggles when the target is occluded due to small objects. Also, when the appearance/shape of the target image of the support set and the target image of the query image are radically different.

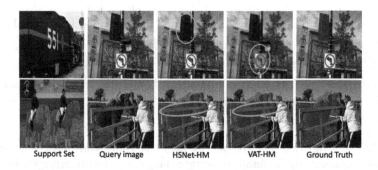

Support Set   Query image   HSNet-HM   VAT-HM   Ground Truth

Fig. 7. Although HSNet-HM/VAT-HM improves mIoU compared to baselines on COCO-20^i [16], its performance can be further improved. We identify cases where it struggles to produce accurate segmentation masks (shown as cyan ellipse).

5   Conclusion

We proposed a new effective masking approach, termed as hybrid masking. It aims to enhance the feature masking (FM) technique, that is commonly used in existing SOTA methods. We instantiate HM in strong baselines and the results reveal that utilizing HM surpasses the existing SOTA by visible margins and also improves training efficiency.

Acknowledgement. This work was supported in part by NSF IIS Grants #1955404 and #1955365.

References

1. Boudiaf, M., Kervadec, H., Masud, Z.I., Piantanida, P., Ben Ayed, I., Dolz, J.: Few-shot segmentation without meta-learning: a good transductive inference is all you need? In: Proceedings of the IEEE/CVF Conference on Computer Vision and Pattern Recognition (CVPR), pp. 13979–13988 (2021)
2. Buslaev, A., Iglovikov, V., Khvedchenya, E., Parinov, A., Druzhinin, M., Kalinin, A.: Albumentations: fast and flexible image augmentations. Information 11(2), 125 (2020). https://doi.org/10.3390/info11020125
3. Chen, L.C., Papandreou, G., Kokkinos, I., Murphy, K., Yuille, A.L.: DeepLab: semantic image segmentation with deep convolutional nets, atrous convolution, and fully connected CRFs. IEEE Trans. Pattern Anal. Mach. Intell. 40(4), 834–848 (2018). https://doi.org/10.1109/TPAMI.2017.2699184
4. Cho, S., Hong, S., Jeon, S., Lee, Y., Sohn, K., Kim, S.: Cats: cost aggregation transformers for visual correspondence. In: Thirty-Fifth Conference on Neural Information Processing Systems (2021)
5. Deng, J., Dong, W., Socher, R., Li, L.J., Li, K., Fei-Fei, L.: Imagenet: a large-scale hierarchical image database. In: 2009 IEEE Conference on Computer Vision and Pattern Recognition (CVPR), pp. 248–255 (2009). https://doi.org/10.1109/CVPR.2009.5206848
6. Everingham, M., Van Gool, L., Williams, C., Winn, J., Zisserman, A.: The pascal visual object classes (voc) challenge. Int. J. Comput. Vis. 88, 303–338 (2010). https://doi.org/10.1007/s11263-009-0275-4
7. Hariharan, B., Arbeláez, P., Girshick, R., Malik, J.: Simultaneous detection and segmentation. In: Fleet, D., Pajdla, T., Schiele, B., Tuytelaars, T. (eds.) ECCV 2014. LNCS, vol. 8695, pp. 297–312. Springer, Cham (2014). https://doi.org/10.1007/978-3-319-10584-0_20
8. He, K., Zhang, X., Ren, S., Sun, J.: Deep residual learning for image recognition. In: 2016 IEEE Conference on Computer Vision and Pattern Recognition (CVPR), pp. 770–778. IEEE Computer Society (2016). https://doi.org/10.1109/CVPR.2016.90
9. Hong, S., Cho, S., Nam, J., Kim, S.: Cost aggregation is all you need for few-shot segmentation. arXiv preprint arXiv:2112.11685 (2021)
10. Huang, G., Liu, Z., Van Der Maaten, L., Weinberger, K.Q.: Densely connected convolutional networks. In: Proceedings of the IEEE Conference on Computer Vision and Pattern Recognition (CVPR), pp. 4700–4708 (2017)
11. Kang, D., Cho, M.: Integrative few-shot learning for classification and segmentation. In: Proceedings of the IEEE/CVF Conference on Computer Vision and Pattern Recognition (CVPR) (2022)
12. Kingma, D.P., Ba, J.: Adam: a method for stochastic optimization. In: Bengio, Y., LeCun, Y. (eds.) 3rd International Conference on Learning Representations, ICLR 2015, San Diego, CA, USA, 7–9 May 2015, Conference Track Proceedings (2015). https://arxiv.org/abs/1412.6980
13. Krizhevsky, A., Sutskever, I., Hinton, G.E.: Imagenet classification with deep convolutional neural networks. Advances in Neural Information Processing Systems 25 (2012)
14. Li, G., Jampani, V., Sevilla-Lara, L., Sun, D., Kim, J., Kim, J.: Adaptive prototype learning and allocation for few-shot segmentation. In: CVPR (2021)
15. Li, X., Wei, T., Chen, Y.P., Tai, Y.W., Tang, C.K.: Fss-1000: a 1000-class dataset for few-shot segmentation. In: CVPR (2020)
16. Lin, T.Y., et al.: Microsoft COCO: Common objects in context (2015)

17. Liu, W., et al.: SSD: Single shot multibox detector. In: Leibe, B., Matas, J., Sebe, N., Welling, M. (eds.) ECCV 2016. LNCS, vol. 9905, pp. 21–37. Springer, Cham (2016). https://doi.org/10.1007/978-3-319-46448-0_2

18. Liu, W., Zhang, C., Ding, H., Hung, T.Y., Lin, G.: Few-shot segmentation with optimal transport matching and message flow. arXiv preprint arXiv:2108.08518 (2021)

19. Liu, Y., Zhang, X., Zhang, S., He, X.: Part-aware prototype network for few-shot semantic segmentation (2020)

20. Long, J., Shelhamer, E., Darrell, T.: Fully convolutional networks for semantic segmentation. In: 2015 IEEE Conference on Computer Vision and Pattern Recognition (CVPR), pp. 3431–3440. IEEE Computer Society, Los Alamitos, CA, USA (2015). https://doi.org/10.1109/CVPR.2015.7298965

21. Long, J., Shelhamer, E., Darrell, T.: Fully convolutional networks for semantic segmentation. In: 2015 IEEE Conference on Computer Vision and Pattern Recognition (CVPR), pp. 3431–3440. IEEE Computer Society, Los Alamitos, CA, USA (2015). https://doi.org/10.1109/CVPR.2015.7298965

22. Lu, Z., He, S., Zhu, X., Zhang, L., Song, Y.Z., Xiang, T.: Simpler is better: few-shot semantic segmentation with classifier weight transformer. In: ICCV (2021)

23. Luo, W., Li, Y., Urtasun, R., Zemel, R.: Understanding the effective receptive field in deep convolutional neural networks. In: Proceedings of the 30th International Conference on Neural Information Processing Systems, pp. 4905–4913. NIPS2016, Curran Associates Inc., Red Hook, NY, USA (2016)

24. Min, J., Kang, D., Cho, M.: Hypercorrelation squeeze for few-shot segmentation. In: Proceedings of the IEEE/CVF International Conference on Computer Vision (ICCV), pp. 6941–6952 (2021)

25. Nguyen, K., Todorovic, S.: Feature weighting and boosting for few-shot segmentation. In: Proceedings of the IEEE/CVF International Conference on Computer Vision (ICCV) (2019)

26. Rakelly, K., Shelhamer, E., Darrell, T., Efros, A.A., Levine, S.: Conditional networks for few-shot semantic segmentation. In: 6th International Conference on Learning Representations, ICLR 2018, Vancouver, BC, Canada, 30 April - 3 May 2018, Workshop Track Proceedings. OpenReview.net (2018). https://openreview.net/forum?id=SkMjFKJwG

27. Redmon, J., Divvala, S., Girshick, R., Farhadi, A.: You only look once: Unified, real-time object detection. In: Proceedings of the IEEE conference on computer vision and pattern recognition (CVPR), pp. 779–788 (2016)

28. Ren, S., He, K., Girshick, R., Sun, J.: Faster R-CNN: towards real-time object detection with region proposal networks. Advances in neural information processing systems 28 (2015)

29. Shaban, A., Bansal, S., Liu, Z., Essa, I., Boots, B.: One-shot learning for semantic segmentation. In: Kim, T.-K., Zafeiriou, G.B.S., Mikolajczyk, K. (eds.) Proceedings of the British Machine Vision Conference (BMVC), pp. 1–13. BMVA Press (2017). https://doi.org/10.5244/C.31.167

30. Siam, M., Oreshkin, B.N., Jägersand, M.: AMP: adaptive masked proxies for few-shot segmentation. In: 2019 IEEE/CVF International Conference on Computer Vision, ICCV 2019, Seoul, Korea (South), 27 Oct - 2 Nov 2019, pp. 5248–5257. IEEE (2019). https://doi.org/10.1109/ICCV.2019.00535

31. Simonyan, K., Zisserman, A.: Very deep convolutional networks for large-scale image recognition. In: Bengio, Y., LeCun, Y. (eds.) 3rd International Conference on Learning Representations, ICLR 2015, San Diego, CA, USA, 7–9 May 2015, Conference Track Proceedings (2015). https://arxiv.org/abs/1409.1556

32. Snell, J., Swersky, K., Zemel, R.: Prototypical networks for few-shot learning. Advances in neural information processing systems 30 (2017)
33. Tian, Z., Zhao, H., Shu, M., Yang, Z., Li, R., Jia, J.: Prior guided feature enrichment network for few-shot segmentation. In: TPAMI (2020)
34. Vinyals, O., Blundell, C., Lillicrap, T., kavukcuoglu, k., Wierstra, D.: Matching networks for one shot learning. In: Lee, D., Sugiyama, M., Luxburg, U., Guyon, I., Garnett, R. (eds.) Advances in Neural Information Processing Systems, vol. 29. Curran Associates, Inc. (2016). https://proceedings.neurips.cc//paper/2016/file/90e1357833654983612fb05e3ec9148c-Paper.pdf
35. Wang, H., Zhang, X., Hu, Y., Yang, Y., Cao, X., Zhen, X.: Few-Shot Semantic Segmentation with Democratic Attention Networks. In: Vedaldi, A., Bischof, H., Brox, T., Frahm, J.-M. (eds.) ECCV 2020. LNCS, vol. 12358, pp. 730–746. Springer, Cham (2020). https://doi.org/10.1007/978-3-030-58601-0_43
36. Wang, K., Liew, J.H., Zou, Y., Zhou, D., Feng, J.: PANet: few-shot image semantic segmentation with prototype alignment. In: The IEEE International Conference on Computer Vision (ICCV) (2019)
37. Xie, G.S., Liu, J., Xiong, H., Shao, L.: Scale-aware graph neural network for few-shot semantic segmentation. In: Proceedings of the IEEE/CVF Conference on Computer Vision and Pattern Recognition (CVPR), pp. 5475–5484 (2021)
38. Yang, B., Liu, C., Li, B., Jiao, J., Ye, Q.: Prototype mixture models for few-shot semantic segmentation. In: Vedaldi, A., Bischof, H., Brox, T., Frahm, J.-M. (eds.) ECCV 2020. LNCS, vol. 12353, pp. 763–778. Springer, Cham (2020). https://doi.org/10.1007/978-3-030-58598-3_45
39. Yang, L., Zhuo, W., Qi, L., Shi, Y., Gao, Y.: Mining latent classes for few-shot segmentation. In: ICCV (2021)
40. Zhang, C., Lin, G., Liu, F., Yao, R., Shen, C.: CANet: class-agnostic segmentation networks with iterative refinement and attentive few-shot learning. In: Proceedings of the IEEE/CVF Conference on Computer Vision and Pattern Recognition (CVPR) (2019)
41. Zhang, G., Kang, G., Yang, Y., Wei, Y.: Few-shot segmentation via cycle-consistent transformer (2021)
42. Zhang, X., Wei, Y., Yang, Y., Huang, T.: SG-One: similarity guidance network for one-shot semantic segmentation. IEEE Trans. Cybern. 50, 3855–3865 (2020)
43. Zhao, H., Shi, J., Qi, X., Wang, X., Jia, J.: Pyramid scene parsing network. In: 2017 IEEE Conference on Computer Vision and Pattern Recognition (CVPR), pp. 6230–6239 (2017). https://doi.org/10.1109/CVPR.2017.660

TransVLAD: Focusing on Locally Aggregated Descriptors for Few-Shot Learning

Haoquan Li[1], Laoming Zhang[2], Daoan Zhang[1], Lang Fu[2],
Peng Yang[2,3], and Jianguo Zhang[1,4]

[1] Research Institute of Trustworthy Autonomous Systems, Department of Computer Science and Engineering, Southern University of Science and Technology, Shenzhen, China
{12032492,12032503}@mail.sustech.edu.cn, zhangjg@sustech.edu.cn
[2] Guangdong Provincial Key Laboratory of Brain-inspired Intelligent Computation, Department of Computer Science and Engineering, Southern University of Science and Technology, Shenzhen, China
{12032505,12032485}@mail.sustech.edu.cn, yangp@sustech.edu.cn
[3] Department of Statistics and Data Science, Southern University of Science and Technology, Shenzhen, China
[4] Peng Cheng Lab, Shenzhen, China

Abstract. This paper presents a transformer framework for few-shot learning, termed TransVLAD, with one focus showing the power of locally aggregated descriptors for few-shot learning. Our TransVLAD model is simple: a standard transformer encoder following a NeXtVLAD aggregation module to output the locally aggregated descriptors. In contrast to the prevailing use of CNN as part of the feature extractor, we are the first to prove self-supervised learning like masked autoencoders (MAE) can deal with the overfitting of transformers in few-shot image classification. Besides, few-shot learning can benefit from this general-purpose pre-training. Then, we propose two methods to mitigate few-shot biases, supervision bias and simple-characteristic bias. The first method is introducing masking operation into fine-tuning, by which we accelerate fine-tuning (by more than 3x) and improve accuracy. The second one is adapting focal loss into soft focal loss to focus on hard characteristics learning. Our TransVLAD finally tops 10 benchmarks on five popular few-shot datasets by an average of more than 2%.

Keywords: Few-shot learning · Transformers · Self-supervised learning · NeXtVLAD · Focal loss

---

H. Li and L. Zhang—made equal contributions to this work

Supplementary Information The online version contains supplementary material available at https://doi.org/10.1007/978-3-031-20044-1_30.

1   Introduction

The success of deep learning largely depends on the expansion of data scaling and model capacity [25]. However, the extreme hungry for data hampered its wide application. This limitation has promoted a wide range of research fields such as transfer learning [38], domain adaptation [50], semi-supervised [4] and unsupervised learning [19]. Few-shot learning is also one of them working on the low data regime [12,18,42,48]. It aims to get adaptive models capable of learning new objects or concepts with only a few labeled samples, just like humans do.

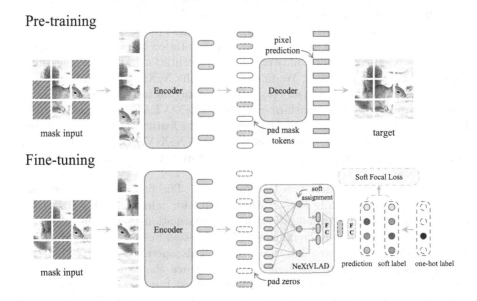

Fig. 1. Pre-training and fine-tuning use the same training set (base classes). In fine-tuning, the feature extractor (TransVLAD) consists of a transformer-based encoder and a NeXtVLAD aggregation module. Soft focal loss is proposed for reducing bias on simple characteristics

While transformer-based models [47] have many nice properties and have topped a bunch of benchmarks in many fields of computer vision, e.g., instance discrimination [34], object detection [7], and semantic segmentation [34], convolutional neural networks [25] still seem to be an inevitable choice for few-shot image classification. This is partly because transformers is more prone to overfitting on the scale of popular few-shot classification datasets. A recent study shows a new paradigm of self-supervised learning based on the idea of masking image modeling (MIM) [2,21,52] is robust to the type and size of the pre-training data [17], and even pre-training on target task data can still obtain comparable performance. This inspires us to use such an approach called *masked autoencoders* (MAE) [21] to pre-train transformer-based models on the target dataset. In addition, the MAE pre-training also benefits few-shot learning from two aspects.

First, self-supervised learning can provide unbiased features for classes. Second, the prediction of patch pixels enables models to be with *spatial awareness*. The features output by the pre-trained MAE encoder is aware of spatial information, so we call them *patch features*. In fine-tuning, without the supervision of patch pixels, the output features will become more abstract and we call them *local features*. The understanding of new concepts usually derives from the reorganization of existing concepts (centaur can be seen as recombination of a human part and a horse part). Therefore, dividing complete image information into discrete local features is beneficial to generalize to new concepts or objects.

To get the global features of an image, the most common way of aggregating all of the local features is by "mean pooling" [16]. This method, a little bit reckless, will lose abundant local characteristics. Another opposite choice is by keeping all of the local features as the complete representation of an image. This operation will contrarily lose the ability of global representation and result in learning incompact data distributions. We finally choose the NeXtVLAD module [29], a compromise choice between mean pooling and doing nothing, to aggregate local features provided by the transformer encoder. NeXtVLAD softly assigns local features to each cluster center, and concatenates the features of all clusters as the image-level feature, Fig. 1. The output features of NeXtVLAD are called *locally aggregated descriptors*. So far, our proposed feature extractor, TransVLAD, has been established.

Few-shot learning exists two learning biases. The first bias is *supervision bias* [15]. Models will only keep features useful for minimizing the losses for base classes (visible for training) but discard information that might help with novel classes. To eliminate the influence of supervision bias, we transfer masking operation from pre-training into fine-tuning. This operation is somewhat like adding "dropout" [43] to the input. Local features that are crucial for base classes may be masked, then the classification must rely on the minor features to infer the class. According to the experimental results, we finally choose the masking ratio of 70% and it accelerates our fine-tuning process by more than 3 times.

The second bias is *simple-characteristic bias*. To efficiently fit data to the real distribution, the model will tend to learn simpler features or fit simpler classes first. Then many hard features or those features for hard samples might be underestimated. This is not good for the generalization of novel classes. To avoid simple-characteristic bias, we replace the cross-entropy loss with soft focal loss, a soft version of focal loss [30] we proposed, which gives more weight to hard samples. The soft focal loss works at the sample level, but together with the masking operation, it will be able to work at the feature level. This is because the masking operation transforms a training sample into some uncertain parts which only contain part features of an image.

Our contributions can be summarized as follows:

1. In contrast to the prevailing use of CNN as part of the feature extractor, we are the first to prove self-supervised learning like MAE can deal with the overfitting of transformers in few-shot image classification.

2. We introduce NeXtVLAD to aggregate local features for few-shot learning. Masking operation and soft focal loss are yielded to solve the supervision bias and simple-characteristic bias, respectively.
3. Our TransVLAD tops 10 benchmarks on five popular few-shot datasets by an average of more than 2%. It also shows a great effect on the cross-domain few-shot scenario.

2 Related Work

2.1 Transformers in Few-Shot Learning

Since ViT introduced transformers to computer vision, transformer-based algorithms have conquered lots of visual tasks [7,34]. Recently, some researchers tried to transfer it in few-shot learning and achieved significant improvements [9,15,20,31]. CrossTransformers [15] designed a transformer block to capture the spatial-alignment relationship between images. SSFormer [9] proposed a sparse spatial transformer layer to select key patches for few-shot classification. These studies all regard the transformer block as an auxiliary module following CNN to enhance accuracies. As far as we know, we are the first to use transformer blocks without CNN as the feature extractor for few-shot learning.

2.2 Self-supervised Learning

Self-supervised learning [2,8,10,11,21,22] often designs surrogate supervision signal with image intrinsic properties. Recently, inspired by natural language processing's great success in masked pre-training methods [13,14,33,39], like Bert [14], a similar implementation for image patches have been studied [2,17, 21,52]. BEiT [2] predicts the corresponding discrete tokens for masked patches. Those tokens are generated by another autoencoder trained in advance. MAE [21] simply masks random patches and reconstructs the missing pixels. SplitMask [17] deeply studies these kinds of approaches and concludes that a large-scale dataset is not necessary for masked pre-training. In addition, MAE has lower computing resource requirements than contrastive learning which often needs a large batch size for best performance.

2.3 Traditional Few-Shot Learning

Traditional few-shot learning can be roughly divided into the following classes. (1) Optimized-based methods [18,26]. MAML [18] tries to find a set of initialization parameters where the model can converge fast and effectively to novel tasks. (2) Metric-based methods [3,27,42,45,54]. Since prototypical network [42] was proposed, Metric-based methods have become the most common methods in few-shot learning. The key to this method is how to get a feature extractor with better generalization. In evaluation, the average of features for each class in the support set (labeled few samples) is viewed as a class prototype (center).

Query sample (test sample) will be finally classified into a class whose prototype is nearest to this feature. (3) Methods based on data augmentation [1,51,53]. [53] expands data scale at the feature level by treating each value as a Gaussian sampling. In our paper, we use a metric-based evaluation method as most recent papers do.

Recently, some methods based on local features are proposed in the few-shot image classification field. They usually generate local features by a fully convolutional network without the final global average pooling. DeepEMD [55] measures the Earth Mover's Distance of local features as image-to-image distance; DN4 [28] revisits NBNN(Naive-Bayes Nearest-Neighbor) [6] approach and suggests a method based on k-nearest neighbors for producing image-to-class distance at local-level features.

3   Methodology

Our method is a two-stage method that contains the MAE pre-training and a designed fine-tuning. After the pre-training, our TransVLAD model is constructed by the MAE encoder followed by a NeXtVLAD aggregation module. Subsequently, we optimize the fine-tuning process by two simple improvements, masking input patches and applying soft focal loss, under the assumptions of supervision bias and simple-characteristic bias in few-shot learning.

3.1   Problem Definition

Few-shot classification can be defined by two data sets with different classes. Training data set contains base classes with abundant labeled data, $D_{train} = \{(x_i, y_i)|y_i \in C_{base}\}$. Testing data set contains novel classes with scarce labeled data, $D_{eval} = \{(x_i, y_i)|y_i \in C_{novel}\}$. The two data sets do not intersect, $C_{base} \cap C_{novel} = \varnothing$. The standard N-way K-shot testing condition is that given N novel classes with K labeled samples per class (termed as support set), to classify the same N classes with Q unlabeled samples per class (termed as query set). A classic idea to this problem is to train a feature extractor on D_{train} and then test by assigning classes to query samples with the feature difference between query and support set. Then the problem has changed to be how to train a more generalized feature extractor.

3.2   Masked Autoencoders Pre-training

Masked Autoencoders (MAE) pre-training is a simple approach to learn general features by reconstructing missing pixels from masked patches. Its main contributions include yielding a very big masking ratio to eliminate redundancy and an ingenious framework design. These creativities make MAE training efficiently and effectively.

Encoder. MAE encoder is a standard ViT. It projects patches into linear embeddings by a linear projection before adding positional embeddings. Subsequently, transformer layers will process them into more representative patch-wise embeddings. The only difference is that MAE encoder just processes 25% patches to reduce image redundancy while increasing computing speed.

Decoder. MAE decoder is designed to be lightweight to process full set of patch tokens (padded with mask tokens) as Fig. 1 shows. Decoder is also composed of several transformer layers following a linear projection to project patch embeddings to original pixels of corresponding masked patches. Before inputting decoder, tokens need to be added positional embeddings again for noticing positional information of masked tokens. The small fraction of input to encoder and the lightweight design of decoder together speed up the training of MAE model by three times or more.

3.3 NeXtVLAD Module

In fine-tuning, we yield NeXtVLAD [29] module to aggregate features output by the transformer-based encoder, Fig. 1. NeXtVLAD is a neural network version of VLAD [23] which focuses on the distribution differences between local features and cluster centers. It was first proposed to aggregate frame-level features into a compact video feature. This method models learnable cluster centers and computes their differences to local features, in our trials, patch-level features. With this module, we can get aggregated features of local features. The whole principle of NeXtVLAD is shown in Fig. 2.

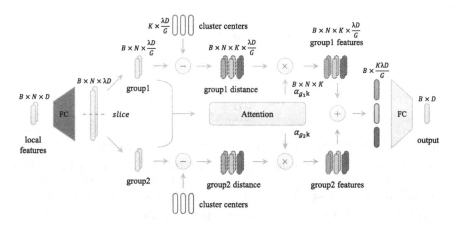

Fig. 2. The principle of NeXtVLAD module. In this case, $B = 1$, $N = 2$, $G = 2$, $K = 3$. It works by computing weighted sum of differences between cluster centers and grouped features. Finally, aggregated features will be flattened into a vector and projected to dimension 768

Considering an image with N patches, after transformer-based encoder processing, the D-dimensional local features can be denoted as $\{x_i\}_{i=1}^{N}$. First, local features will be expanded to higher dimensions λD as $\{\widetilde{x}_i\}_{i=1}^{N}$ by a linear projection L_e, λ is a multiplier:

$$\widetilde{x}_i = L_e(x_i|\theta_e),$$ (1)

in which θ_e denotes the parameters of L_e. Then it will be equally sliced into G groups $\{\widetilde{x}_i^{(g)}\}_{i=1}^{N}, g \in \{1, 2, ..., G\}$. Grouping here is meant to decompose features into more independent parts before aggregating together. The underlying assumption is that each local feature can incorporate multiple concepts or objects. In order to calculate the differences between grouped features and cluster centers $\{c_k\}_{k=1}^{K}$, cluster centers are initialized to $\lambda D/G$ dimensions. For each cluster, it will finally provide a compact feature F_k by weighted summing all corresponding differences:

$$F_k = \mathbb{N}(X|C, \theta_e, \theta_g, \theta_k) = \sum_{i,g} \alpha_{gk}(\widetilde{x}_i|\theta_g, \theta_k)\left(\widetilde{x}_i^{(g)} - c_k\right)$$ (2)

$$k \in \{1, ..., K\}, i \in \{1, ..., N\}, g \in \{1, ..., G\},$$

where $\mathbb{N}$ is the NeXtVLAD module, θ means corresponding parameters and C means all learnable cluster centers. α_{gk} decides the soft assignment of the corresponding difference in the final representation and is calculated by two parts:

$$\alpha_{gk}(\widetilde{x}_i|\theta_g, \theta_k) = \alpha_g(\widetilde{x}_i|\theta_g)\alpha_k^{(g)}(\widetilde{x}_i|\theta_k)$$ (3)

where

$$\alpha_g(\widetilde{x}_i|\theta_g) = \sigma(L_g(\widetilde{x}_i|\theta_g)),$$ (4)

$$\alpha_k^{(g)}(\widetilde{x}_i|\theta_k) = \frac{e^{L_k^{(g)}\left(\widetilde{x}_i|\theta_k^{(g)}\right)}}{\sum_{k'=1}^{K} e^{L_{k'}^{(g)}\left(\widetilde{x}_i|\theta_{k'}^{(g)}\right)}}.$$ (5)

α_g represents the attention for each group while $\alpha_k^{(g)}$ measures the soft assignment of $\widetilde{x}_i^{(g)}$ to the cluster k in probabilistic form. $\sigma(.)$ is a sigmoid function to scale group attention into $(0, 1)$. L_g and L_k are different linear layers with parameters θ_g and θ_k respectively. To show formula concisely, we split L_k into $L_k^{(g)}, g \in \{1, 2, ..., G\}$. In practice, a single layer L_k is used.

Finally, we will get K features of $\lambda D/G$ dimensions for an image. We flatten and project it into a lower dimension with another linear layer, followed by a batch normalization and a ReLU activation function. Until now, NeXtVLAD have compacted patch-level features into an image-level feature.

3.4   Masking Operation in Fine-tuning

In MAE pre-training, encoder only observes part of patches (25%). This brings great benefits on both efficiency and effectiveness. The idea of prompt [32], a new idea about narrowing the gap between pre-training and fine-tuning, inspired us to design similar operation. Therefore, we decided to transfer masking operation in fine-tuning as well. However, in evaluation, masking patches will hinder full access to image information which is obviously unreasonable. To deal with this problem, a variable-length processing for our model is necessary. Fortunately, transformers and NeXtVLAD were created for text or video field and they naturally support for variable-length processing.

The difference between MAE and our masking operation is that we need to restore the original order for visible patches and pad zeros to masked ones before input to NeXtVLAD module. In evaluation, we just let all patches get through our whole feature extractor, including transformer blocks and NeXtVLAD module. Similar to MAE, we surprisingly find 70% masking ratio is a more suitable choice according to our ablation experiments, Fig. 5.

Masking operation in our TransVLAD model has four benefits:

1. It eliminates information redundancy in images [21].
2. It can be regarded as a kind of data augmentation.
3. It speeds up the training process by about three times.
4. It shortens the difference of tasks between pre-training and fine-tuning.

3.5   Soft Focal Loss

To avoid simple-characteristic bias and enhance attention on difficult characteristics or hard samples in training, we use focal loss [30] to replace the cross-entropy loss. Focal loss is a changed version of cross-entropy loss and proposed to address extreme imbalance between foreground and background classes in one-stage object detection scenario. It is defined as:

$$\mathcal{FL}(p) = -\alpha_i (1 - p)^\gamma \log (p) \qquad (6)$$

in which p is the prediction of the class that the sample really belongs to. $\gamma \geq 0$ is the tunable focusing parameter. The term $(1 - p)$ measures the learning degree of each sample. The bigger γ is, the more focusing on hard classified samples. α_i is appied to balance the classes with different sample sizes.

This formula uses one-hot encoded labels by default. However, in our settings, mixup [56] and label smoothing [36] have been used to reduce overfitting. They encode labels into a soft version which may have target values for all classes. To suitable for soft targets, we expand focal loss to a soft version, called soft focal loss:

$$\mathcal{SFL}(p, t) = -\sum_{i=1}^{C} \alpha_i \left[(t_i - p_i)^2 \right]^{\frac{\gamma}{2}} t_i \log (p_i), \qquad (7)$$

in which p_i, t_i are the prediction and target to the i_{th} class respectively. The term $(t_i - p_i)$ measures the difference between them to influence the weight of relative loss term. C is the number of classes to predict. In our setting, α_i is always set to be one because our training classes are balanced.

Soft focal loss is suggested to be used with masking operation. Because with masking, the difficulty varies by the remaining part of images, so the soft focal loss can work at the local feature level.

3.6   Whole Classification

Our paper uses whole classification task, the most common setting for classification with cross-entropy loss (soft focal loss in our method), instead of meta-training method which is popular in few-shot learning. This is because some recent studies have found that whole classification can obtain comparable results to meta-training and has better generalization in novel classes while meta-training concerns more on testing condition [12].

4   Experiments

4.1   Datasets

During the evaluation, we compared our methods on five standard datasets for few-shot classification, *mini*ImageNet [48], *tiered*ImageNet [40], CIFAR-FS [5], FC100 [37], and CUB [49].

*mini*ImageNet and *tiered*ImageNet are subsets of ImageNet [41]. *mini*ImageNet consists of 100 classes with 600 samples per class and is randomly divided into three disjoint sets of the training set (64 classes), validation set (16 classes), and testing set (20 classes). *tiered*ImageNet, a bigger version of *mini*ImageNet, contains 608 classes with 1200 samples per class and is randomly split into 351/97/160 for train/val/test. CIFAR-FS and FC100 both are variants of CIFAR-100. CIFAR-FS is randomly split into 64/16/20 classes for train/val/test, while FC100 is divided into 60/20/20 classes according to 20 superclasses. Each superclass contains five similar classes with a more generalized concept. So, FC100 is a harder dataset created for reducing *semantical overlaps* between training and evaluation. CUB-200-2011 (CUB) is a fine-grained dataset of 200 bird species with total 11,788 images. It is randomly split into 100/50/50 for train/val/test. The image size of *mini*ImageNet, *tiered*ImageNet, CUB is 84×84 while for CIFAR-FS and FC100 is 32×32.

4.2   Implementation Details

Training Details. For each dataset, we use self-supervised pre-training and supervised fine-tuning to train the model with base classes and test the performance using novel classes. In the pre-training phase, we use the MAE strategy and parameters. The encoder is a 12-layer transformer model with dimension

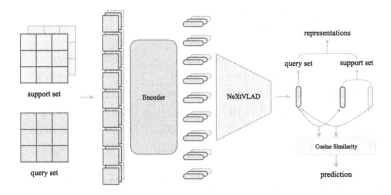

Fig. 3. Given a few-shot task, we compute an average feature for samples of each class in *support set*, then we classify the sample in *query set* by nearest neighbor method with cosine similarity

768 and the decoder is a 4-layer transformer model with dimension 384. The masking ratio is 75% and training for 1600 epochs. In the fine-tuning stage, we just keep the encoder and add NeXtVLAD module ($\lambda = 4, K = 64, G = 8$) after it. We follow the default training settings for BEiT [2] except for the initial learning rate is 7e-4. Our masking ratio here is 70%, focusing parameter γ is 2 and fine-tuning for 100 epochs. More specific settings can be found in supplemental material.

Evaluation. The overall test flow is shown in Fig. 3. It is worth noting that we do not perform masking operation during validation and testing. We evaluate our method based on ProtoNet [42] setting with randomly sample 600 N-way K-shot tasks from the novel set, and take averaged top-1 classification accuracy. To get a fair comparison with the previous works, we perform model selection based on the validation set.

4.3   Compare to the State-of-the-art Methods

Table 1 and Table 2 compare our method with the current state-of-the-art methods on five datasets. As shown in the tables our method achieves the best results on all five datasets. Since we only tune parameters on *mini*ImageNet and copy them for the other four datasets, these results demonstrate the excellent generalization of our model.

We find that the increase of performance on *tiered*ImageNet (1-shot +3.5%, 5-shot +2.2%) is apparently larger than on *mini*ImageNet (1-shot +0.6%, 5-shot +1.2%), while the only difference is the data scale. We hypothesize that this is mainly because pre-training on a bigger dataset can get better generalization which is beneficial to distinguish new objects. Secondly, our model presents great transferability across superclasses on FC100 (1-shot +3.9%, 5-shot +5.3%). It inspired us to do cross-domain few-shot studies.

Table 1. Comparison with the prior and current state-of-the-art methods on *mini*ImageNet, *tiered*ImageNet datasets

Year	Methods	*mini*ImageNet		*tiered*ImageNet	
		1-shot	5-shot	1-shot	5-shot
2017	ProtoNet [42]	54.16± 0.82	73.68±0.65	53.31±0.89	72.69±0.74
2018	RelationNet [44]	52.19±0.83	70.20±0.66	54.48±0.93	71.32±0.78
2019	DCO [26]	62.64±0.61	78.63±0.46	65.99±0.72	81.56±0.53
2020	Meta-baseline [12]	63.17±0.23	79.26±0.17	68.62±0.27	83.29±0.18
2020	DeepEMD [55]	65.91±0.82	82.41±0.56	71.16±0.87	83.95±0.58
2020	S2M2 [35]	64.93±0.18	**83.18±0.11**	**73.71±0.22**	**88.59±0.14**
2021	RENet [24]	**67.60±0.44**	82.58±0.30	71.61±0.51	85.28±0.35
2021	SSFormers [9]	67.25±0.24	82.75±0.20	72.52±0.25	86.61±0.18
2022	TransVLAD	**68.24±0.59**	**84.42±0.23**	**77.20±0.60**	**90.74±0.32**

Table 2. Comparison with the prior and current state-of-the-art methods on CIFAR-FS, FC100, CUB datasets

Year	Methods	CIFAR-FS		FC100		CUB	
		1-shot	5-shot	1-shot	5-shot	1-shot	5-shot
2017	ProtoNet [42]	55.50±0.70	72.00±0.60	37.50±0.60	52.50±0.60	71.88±0.91	87.42±0.48
2018	RelationNet [44]	55.00±1.00	69.30±0.80	–	–	68.65±0.91	81.12±0.63
2019	DCO [26]	72.00±0.70	84.20±0.50	41.10±0.60	55.50±0.60	–	–
2020	DeepEMD [55]	–	–	–	–	75.65±0.83	88.69±0.50
2020	S2M2 [35]	74.81±0.19	87.47±0.13	–	–	**80.68±0.81**	90.85±0.44
2021	RENet [24]	74.51±0.46	86.60±0.32	–	–	79.49±0.44	**91.11±0.24**
2021	SSFormers [9]	74.50±0.21	86.61±0.23	43.72±0.21	58.92±0.18	–	–
2022	TransVLAD	**77.48±0.41**	**89.82±0.42**	**47.66±0.12**	**64.25±0.18**	**82.66±1.22**	**92.45±0.80**

4.4 Cross-Domain Few-Shot Learning

Cross-domain few-shot image classification, where unseen classes and examples come from diverse data sources, has seen growing interest [46]. To further validate the transferability of our method, we also conducted cross-dataset evaluation experiments in a simple way, training with one dataset and testing with another different dataset. From Table 3, We can see that the transferring effect of *mini*ImageNet on CUB is better than the previous method, and the mutual evaluation effect of *mini*ImageNet and CIFAR-FS is close to the result of the intra-domain training, partly because they both are randomly divided. The poor transfer of the CUB is expected, because there are only pictures of birds in the CUB, resulting in difficulty to learn contents from other fields. These results demonstrate that our model has great cross-domain transferability.

Table 3. Few-shot cross-domain evaluation results between *mini*ImageNet, CIFAR-FS and CUB where grey numbers denote intra-domain results

Methods	Training Dataset	*mini*ImageNet		CIFAR-FS		CUB	
		1-shot	5-shot	1-shot	5-shot	1-shot	5-shot
DCO [26]	*mini*ImageNet	62.64±0.61	78.63±0.46	–	–	44.79±0.75	64.98±0.68
S2M2 [35]	*mini*ImageNet	64.93±0.18	83.18±0.11	–	–	48.42±0.84	70.44±0.75
TransVLAD	*mini*ImageNet	68.24±0.59	84.42±0.23	66.00±0.85	83.26±0.50	50.45±0.12	72.20±0.34
TransVLAD	CIFAR-FS	54.56±1.35	72.55±1.00	77.48±0.41	89.82±0.42	46.29±1.23	65.81±1.01
TransVLAD	CUB	39.36±0.86	52.78±1.34	34.12±0.59	45.91±1.56	82.66±1.22	92.45±0.80

5   Discussion and Ablation Studies

5.1   Overall Analysis

Our method consists of four main components: MAE pre-training, NeXtVLAD feature aggregation module, masked fine-tuning, and soft focal loss. In this section, we conduct ablation studies to analyze how each component affects the few-shot recognition performance. We show the individual and combined effects of these components in Table 4. Specifically, if the NeXtVLAD is not selected, mean pooling will replace it for feature aggregation. Similarly, the cross-entropy loss is selected as the replacement for soft focal loss.

Table 4. The individual and combined effects of MAE pre-training, NeXtVLAD module, masked fine-tuning, and soft focal loss are studied. The experiments are conducted on *mini*ImageNet. We can find that each part of our model has an important contribution

Pre-train	NeXtVLAD	Mask	Soft focal loss	1-shot	5-shot
				34.55	45.73
	✓	✓	✓	44.56	61.34
✓				64.10	81.40
✓		✓		65.46	82.05
✓		✓	✓	66.79	83.40
✓	✓			65.62	81.70
✓	✓	✓		66.91	83.32
✓	✓	✓	✓	**68.24**	**84.42**

We can see that the transformer-based model is seriously overfitting compared to the CNN baseline [12] when directly trained. But with the join of MAE pre-training, the performance of the model will be comparable to the CNN baseline. Interestingly, our method can still improve the performance with no pre-training. In all of our experiments, NeXtVLAD module provides a steady boost. With the masking operation, it further improves the *mini*ImageNet by 2.8% (1-shot), while the consuming time of fine-tuning is shortened by one-third. The soft focal loss also plays an important role in our TransVLAD. It surely improves the generalization of novel classes and shortens the prediction bias for different classes which we will discuss it later.

5.2 NeXtVLAD Feature Aggregation

We perform a t-SNE visualization of embeddings generated by Meta-Baseline [12], MAE++ (MAE adds masked fine-tuning and soft focal loss), and our TransVLAD (MAE ++ adds NeXtVLAD module) on the test set of *mini*ImageNet (see Fig. 4). We observe that our methods, both MAE++ and TransVLAD, produce more compact features for the same class. Besides, TransVLAD only adds the NeXtVLAD module compared to MAE++, and apparently gets a more discriminative feature distribution with larger boundaries between clusters. We hypothesize that this behavior is due to the learned cluster centers of NeXtVLAD. Features are guided to be close to those related centers.

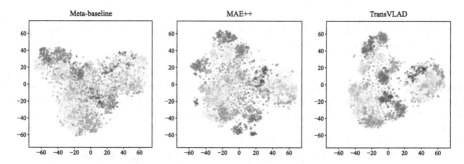

Fig. 4. t-SNE visualization of features on *mini*ImageNet test set produced by Meta-Baseline [12], MAE++ (added masking operation and soft focal loss) and TransVLAD (added NeXtVLAD over MAE++)

5.3 Masking Ratio

Our feature extractor, including transformer blocks and NeXtVLAD module, is very flexible with the input size while the output size is fixed. So, we can mask input patches at any ratio, called masking ratio. Figure 5 shows the influence of the masking ratio. Similar to the MAE's result, the optimal masking ratio is surprisingly high, up to 70%. This behavior makes sense. Masking operation is somewhat like *dropout* at the input. It prevents our model from over-relying on some key features but ignoring the learning of other features, namely, overfitting at base classes.

By skipping the masked patches, we accelerate the training process by more than 3 times. In addition, the memory usage is greatly reduced (about 70%), which allows us to train larger models with the same batch size. In fact, fine-tuning our TransVLAD for 100 epochs with 128 batch size on *mini*ImageNet only needs less than 4 h on one RTX3090, and the memory usage is less than 8G. This is extremely fast for a ViT-Base [16] encoder.

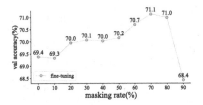

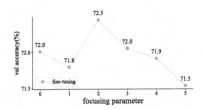

Fig. 5. Validation accuracy for changing masking ratio

Fig. 6. Validation accuracy for changing focusing parameter γ

5.4 Focusing Parameter

Similar to the masking ratio, we conduct a separate experiment to evaluate the influence of the soft focal loss with different values of focusing parameter γ. Figure 6 shows that the best result is obtained when γ is equal to 2. When the focusing parameter is small, the effect of focusing on hard examples is not obvious. When it is too big, the model will continue to overreact to unpredictable samples and ignore what it could have learned.

Soft focal loss is suggested to be used with masking operation. Because if there is no mask, soft focal loss only works at the sample level. With masking operation, the difficulty varies by the remaining part of images, so the soft focal loss can work at the local feature level.

6   Conclusion

The transformer-based network has strong potential for few-shot learning, and the core problem that restricts it to dominate few-shot learning is its serious overfitting. Our paper gives an efficient solution. The new self-supervised per-training paradigm, such as masked autoencoders, is able to reduce its overfitting and simultaneously improve the model generalization.

The model we designed, TransVLAD, takes full advantage of the nature of transformers and few-shot tasks, and it not only significantly improves the performance but also improves the training speed.

Acknowledgement. This work is supported in part by National Key Research and Development Program of China (Grant No. 2021YFF1200800), and the Stable Support Plan Program of Shenzhen Natural Science Fund (Grant No. 20200925154942002).

References

1. Antoniou, A., Storkey, A., Edwards, H.: Data augmentation generative adversarial networks. arXiv preprint. arXiv:1711.04340 (2017)
2. Bao, H., Dong, L., Wei, F.: Beit: bert pre-training of image transformers. arXiv preprint. arXiv:2106.08254 (2021)
3. Bateni, P., Goyal, R., Masrani, V., Wood, F., Sigal, L.: Improved few-shot visual classification. In: Proceedings of the IEEE/CVF Conference on Computer Vision and Pattern Recognition, pp. 14493–14502 (2020)
4. Berthelot, D., Carlini, N., Goodfellow, I., Papernot, N., Oliver, A., Raffel, C.A.: Mixmatch: a holistic approach to semi-supervised learning. In: Advances in Neural Information Processing Systems, vol. 32 (2019)
5. Bertinetto, L., Henriques, J.F., Torr, P.H., Vedaldi, A.: Meta-learning with differentiable closed-form solvers. arXiv preprint. arXiv:1805.08136 (2018)
6. Boiman, O., Shechtman, E., Irani, M.: In defense of nearest-neighbor based image classification. In: 2008 IEEE Conference on Computer Vision and Pattern Recognition, pp. 1–8. IEEE (2008)
7. Carion, N., Massa, F., Synnaeve, G., Usunier, N., Kirillov, A., Zagoruyko, S.: End-to-end object detection with transformers. In: Vedaldi, A., Bischof, H., Brox, T., Frahm, J.-M. (eds.) ECCV 2020. LNCS, vol. 12346, pp. 213–229. Springer, Cham (2020). https://doi.org/10.1007/978-3-030-58452-8_13
8. Caron, M., et al.: Emerging properties in self-supervised vision transformers. In: Proceedings of the IEEE/CVF International Conference on Computer Vision, pp. 9650–9660 (2021)
9. Chen, H., Li, H., Li, Y., Chen, C.: Shaping visual representations with attributes for few-shot learning. arXiv preprint. arXiv:2112.06398 (2021)
10. Chen, T., Kornblith, S., Norouzi, M., Hinton, G.: A simple framework for contrastive learning of visual representations. In: International Conference on Machine Learning, pp. 1597–1607. PMLR (2020)
11. Chen, X., Xie, S., He, K.: An empirical study of training self-supervised vision transformers. In: Proceedings of the IEEE/CVF International Conference on Computer Vision, pp. 9640–9649 (2021)
12. Chen, Y., Liu, Z., Xu, H., Darrell, T., Wang, X.: Meta-baseline: exploring simple meta-learning for few-shot learning. In: Proceedings of the IEEE/CVF International Conference on Computer Vision, pp. 9062–9071 (2021)
13. Clark, K., Luong, M.T., Le, Q.V., Manning, C.D.: Electra: pre-training text encoders as discriminators rather than generators. arXiv preprint. arXiv:2003.10555 (2020)
14. Devlin, J., Chang, M.W., Lee, K., Toutanova, K.: Bert: pre-training of deep bidirectional transformers for language understanding. arXiv preprin. arXiv:1810.04805 (2018)
15. Doersch, C., Gupta, A., Zisserman, A.: Crosstransformers: spatially-aware few-shot transfer. Adv. Neural. Inf. Process. Syst. **33**, 21981–21993 (2020)
16. Dosovitskiy, A., et al.: An image is worth 16x16 words: Transformers for image recognition at scale. arXiv preprint. arXiv:2010.11929 (2020)
17. El-Nouby, A., Izacard, G., Touvron, H., Laptev, I., Jegou, H., Grave, E.: Are large-scale datasets necessary for self-supervised pre-training? arXiv preprint. arXiv:2112.10740 (2021)
18. Finn, C., Abbeel, P., Levine, S.: Model-agnostic meta-learning for fast adaptation of deep networks. In: International conference on machine learning, pp. 1126–1135. PMLR (2017)

19. Ganin, Y., Lempitsky, V.: Unsupervised domain adaptation by backpropagation. In: International conference on machine learning, pp. 1180–1189. PMLR (2015)

20. Gidaris, S., Bursuc, A., Komodakis, N., Pérez, P., Cord, M.: Boosting few-shot visual learning with self-supervision. In: Proceedings of the IEEE/CVF International Conference on Computer Vision, pp. 8059–8068 (2019)

21. He, K., Chen, X., Xie, S., Li, Y., Dollár, P., Girshick, R.: Masked autoencoders are scalable vision learners. arXiv preprint. arXiv:2111.06377 (2021)

22. He, K., Fan, H., Wu, Y., Xie, S., Girshick, R.: Momentum contrast for unsupervised visual representation learning. In: Proceedings of the IEEE/CVF conference on computer vision and pattern recognition, pp. 9729–9738 (2020)

23. Jégou, H., Douze, M., Schmid, C., Pérez, P.: Aggregating local descriptors into a compact image representation. In: 2010 IEEE Computer Society Conference on Computer Vision and Pattern Recognition, pp. 3304–3311. IEEE (2010)

24. Kang, D., Kwon, H., Min, J., Cho, M.: Relational embedding for few-shot classification. In: Proceedings of the IEEE/CVF International Conference on Computer Vision, pp. 8822–8833 (2021)

25. Krizhevsky, A., Sutskever, I., Hinton, G.E.: Imagenet classification with deep convolutional neural networks. In: Advances in Neural Information Processing Systems, vol. 25 (2012)

26. Lee, K., Maji, S., Ravichandran, A., Soatto, S.: Meta-learning with differentiable convex optimization. In: Proceedings of the IEEE/CVF Conference on Computer Vision and Pattern Recognition, pp. 10657–10665 (2019)

27. Li, W.H., Liu, X., Bilen, H.: Improving task adaptation for cross-domain few-shot learning. arXiv preprint. arXiv:2107.00358 (2021)

28. Li, W., Wang, L., Xu, J., Huo, J., Gao, Y., Luo, J.: Revisiting local descriptor based image-to-class measure for few-shot learning. In: Proceedings of the IEEE/CVF Conference on Computer Vision and Pattern Recognition, pp. 7260–7268 (2019)

29. Lin, R., Xiao, J., Fan, J.: Nextvlad: an efficient neural network to aggregate frame-level features for large-scale video classification. In: Proceedings of the European Conference on Computer Vision (ECCV) Workshops (2018)

30. Lin, T.Y., Goyal, P., Girshick, R., He, K., Dollár, P.: Focal loss for dense object detection. In: Proceedings of the IEEE International Conference on Computer Vision, pp. 2980–2988 (2017)

31. Liu, L., Hamilton, W., Long, G., Jiang, J., Larochelle, H.: A universal representation transformer layer for few-shot image classification. arXiv preprint. arXiv:2006.11702 (2020)

32. Liu, P., Yuan, W., Fu, J., Jiang, Z., Hayashi, H., Neubig, G.: Pre-train, prompt, and predict: A systematic survey of prompting methods in natural language processing. arXiv preprint. arXiv:2107.13586 (2021)

33. Liu, Y., et al.: Roberta: a robustly optimized bert pretraining approach. arXiv preprint. arXiv:1907.11692 (2019)

34. Liu, Z., et al.: Swin transformer v2: Scaling up capacity and resolution. arXiv preprint. arXiv:2111.09883 (2021)

35. Mangla, P., Kumari, N., Sinha, A., Singh, M., Krishnamurthy, B., Balasubramanian, V.N.: Charting the right manifold: Manifold mixup for few-shot learning. In: Proceedings of the IEEE/CVF Winter Conference on Applications of Computer Vision, pp. 2218–2227 (2020)

36. Müller, R., Kornblith, S., Hinton, G.E.: When does label smoothing help? In: Advances in Neural Information Processing Systems, vol. 32 (2019)

37. Oreshkin, B., Rodríguez López, P., Lacoste, A.: Tadam: task dependent adaptive metric for improved few-shot learning. In: Advances in Neural Information Processing Systems, vol. 31 (2018)
38. Pan, S.J., Yang, Q.: A survey on transfer learning. IEEE Trans. Knowl. Data Eng. **22**(10), 1345–1359 (2009)
39. Radford, A., Narasimhan, K., Salimans, T., Sutskever, I.: Improving language understanding with unsupervised learning (2018)
40. Ren, M., et al.: Meta-learning for semi-supervised few-shot classification. arXiv preprint. arXiv:1803.00676 (2018)
41. Russakovsky, O., et al.: ImageNet large scale visual recognition challenge. Int. J. Comput. Vis. **115**(3), 211–252 (2015). https://doi.org/10.1007/s11263-015-0816-y
42. Snell, J., Swersky, K., Zemel, R.: Prototypical networks for few-shot learning. In: Advances in Neural Information Processing Systems, vol. 30 (2017)
43. Srivastava, N., Hinton, G., Krizhevsky, A., Sutskever, I., Salakhutdinov, R.: Dropout: a simple way to prevent neural networks from overfitting. J. Mach. Learn. Res. **15**(1), 1929–1958 (2014)
44. Sung, F., Yang, Y., Zhang, L., Xiang, T., Torr, P.H., Hospedales, T.M.: Learning to compare: relation network for few-shot learning. In: Proceedings of the IEEE Conference on Computer Vision and Pattern Recognition, pp. 1199–1208 (2018)
45. Tian, Y., Wang, Y., Krishnan, D., Tenenbaum, J.B., Isola, P.: Rethinking few-shot image classification: a good embedding is all you need? In: Vedaldi, A., Bischof, H., Brox, T., Frahm, J.-M. (eds.) ECCV 2020. LNCS, vol. 12359, pp. 266–282. Springer, Cham (2020). https://doi.org/10.1007/978-3-030-58568-6_16
46. Triantafillou, E., et al.: Meta-dataset: a dataset of datasets for learning to learn from few examples. arXiv preprint. arXiv:1903.03096 (2019)
47. Vaswani, A., et al.: Attention is all you need. In: Advances in Neural Information Processing systems, vol. 30 (2017)
48. Vinyals, O., Blundell, C., Lillicrap, T., Wierstra, D., et al.: Matching networks for one shot learning. In: Advances in Neural Information Processing Systems, vol. 29 (2016)
49. Wah, C., Branson, S., Welinder, P., Perona, P., Belongie, S.: The caltech-ucsd birds-200-2011 dataset (2011)
50. Wang, M., Deng, W.: Deep visual domain adaptation: a survey. Neurocomputing **312**, 135–153 (2018)
51. Wang, Y.X., Girshick, R., Hebert, M., Hariharan, B.: Low-shot learning from imaginary data. In: Proceedings of the IEEE Conference on Computer Vision and Pattern Recognition, pp. 7278–7286 (2018)
52. Wei, C., Fan, H., Xie, S., Wu, C.Y., Yuille, A., Feichtenhofer, C.: Masked feature prediction for self-supervised visual pre-training. arXiv preprint. arXiv:2112.09133 (2021)
53. Yang, S., Liu, L., Xu, M.: Free lunch for few-shot learning: Distribution calibration. arXiv preprint. arXiv:2101.06395 (2021)
54. Zhang, B., Li, X., Ye, Y., Huang, Z., Zhang, L.: Prototype completion with primitive knowledge for few-shot learning. In: Proceedings of the IEEE/CVF Conference on Computer Vision and Pattern Recognition, pp. 3754–3762 (2021)
55. Zhang, C., Cai, Y., Lin, G., Shen, C.: Deepemd: few-shot image classification with differentiable earth mover's distance and structured classifiers. In: Proceedings of the IEEE/CVF Conference on Computer Vision and Pattern Recognition, pp. 12203–12213 (2020)
56. Zhang, H., Cisse, M., Dauphin, Y.N., Lopez-Paz, D.: Mixup: beyond empirical risk minimization. arXiv preprint. arXiv:1710.09412 (2017)

Kernel Relative-prototype Spectral Filtering for Few-Shot Learning

Tao Zhang[1]([✉])[iD] and Wu Huang[2][iD]

[1] Chengdu Techman Software Co., Ltd., Chengdu, Sichuan, China
zhangtao@tme.com.cn
[2] Sichuan University, Chengdu, Sichuan, China
huangwu@scu.edu.cn

Abstract. Few-shot learning performs classification tasks and regression tasks on scarce samples. As one of the most representative few-shot learning models, Prototypical Network represents each class as sample average, or a prototype, and measures the similarity of samples and prototypes by Euclidean distance. In this paper, we propose a framework of spectral filtering (shrinkage) for measuring the difference between query samples and prototypes, or namely the relative prototypes, in a reproducing kernel Hilbert space (RKHS). In this framework, we further propose a method utilizing Tikhonov regularization as the filter function for few-shot classification. We conduct several experiments to verify our method utilizing different kernels based on the *mini*ImageNet dataset, *tiered*-ImageNet dataset and CIFAR-FS dataset. The experimental results show that the proposed model can perform the state-of-the-art. In addition, the experimental results show that the proposed shrinkage method can boost the performance. Source code is available at https://github.com/zhangtao2022/DSFN.

Keywords: Few-shot learning · Relative-prototype · Spectral filtering · Shrinkage · Kernel

1 Introduction

Humans have an innate ability to quickly learn from one or several labeled pictures and infer the category of new pictures. In contrast, deep learning, despite its breakthrough success in computer vision, still needs huge data to drive it. This shortcoming seriously hinders its applications in some practical situations where data is scanty. Therefore, it is an important and challenging problem for machines to acquire the human-like ability to make inferences about unknown samples based on too few samples. Inspired by this ability of humans, few-shot learning is proposed and has become a hot spot [9,19,43]. Vinyals et al. [44]

Supplementary Information The online version contains supplementary material available at https://doi.org/10.1007/978-3-031-20044-1_31.

proposed a training paradigm that few-shot learning models should learn new categories of unseen examples from query set using very few examples from support set. Meta-learning methods can be well applied to the few-shot settings that need to complete the task that contains both support set and query set.

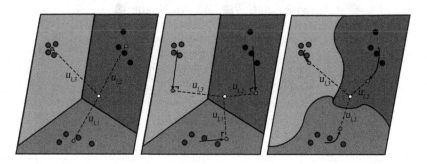

Fig. 1. Comparison of the relative-prototypes (dotted line) in the Prototypical Networks (ProtoNet) [41], Deep Subspace Networks (DSN) [40] and the proposed DSFN. **Left**: ProtoNet in Euclidean space; **Middle**: DSN in Euclidean space; **Right**: DSFN in a reproducing kernel Hilbert space. For them, $u_{l,1}, u_{l,2}$ and $u_{l,3}$ are the relative-prototypes with class 1(brown), class 2(green) and class 3(blue), respectively. (Color figure online)

Recently, a series of meta-learning models for few-shot setting have been proposed, which can be divided into the metric-based models and the optimization-based models [2,10,13,17,20,23,24,29,35,37,41,42,47,48]. Prototypical Networks (ProtoNet), as one of the metric-based representatives, proposes to use prototypes to represent each category, and to measure the similarity between a sample and a prototype by using Euclidean distance [41]. Based on this idea, some prototype-related models have been proposed in recent years [5,11,32,33].

The prototype in ProtoNet is estimated by support sample mean in each class, which may deviate from the true prototype [22]. In [22], Two bias elimination mechanisms are proposed to eliminate the difference between the prototype estimated value and the true value. In addition, a method of using a combination of semantic information and visual information to better estimate the prototype was proposed, and a significant improvement was obtained [45].

In measurement of sample similarity, Mahalanobis distance would be better than Euclidean distance to capture the information from the distribution within the class [3]. In addition, similarity measurement in hyperbolic space that learns the features of a hierarchical structure could be better than those in Euclidean space for few-shot image classification [8,16].

In this paper, we propose a method called Deep Spectral Filtering Networks (DSFN) aiming to better estimate relative prototypes, or the difference between prototypes and query samples (Fig. 1). Euclidean distance can not fully capture the information of intra-class difference as it is applied to measure the difference between sample and prototype. Thus, the main components of the inner class distance are taken into account in the similarity assessment between the sample and the prototype, which to some extent interferes with this assessment. In our approach, the influence of these components is weakened via spectral filtering. It is similar to the kernel mean shrinkage estimation that aims to reduce the expected risk of mean estimation [25–28], but the estimated relative prototypes don't have to be close to the true mean.

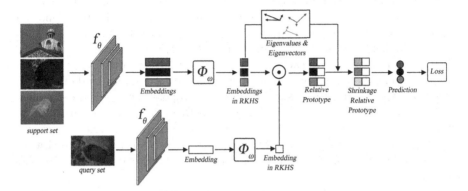

Fig. 2. The overview of our proposed approach. The features of support set and query set extracted from f_θ are mapped into RKHS by the function ϕ_ω. The relative prototypes are shrunk based on the eigenvalues and eigenvectors from the support set.

Recent work has shown that kernel embedding can significantly improve the performance of few-shot learning [8]. Based on kernel mean embedding theory, our approach can measure the relative prototypes in a reproducing kernel Hilbert space (RKHS). The advantage of this is that we can deal with the projection problem of feature space by means of the kernel embedding. The overview of our proposed approach is shown in Fig. 2.

The contributions of our work are summarized as four folds:

1) To our knowledge, this work is the first to estimate relative-prototypes using a kernel shrinkage method for few-shot learning.
2) We propose to estimate relative prototypes instead of prototypes, aiming to better capture the probability information that the query samples belongs to a class.
3) We propose a framework of spectral filtering to estimate the relative prototypes. This approach can filter the interference within cluster variation while measuring the relative prototypes and boost the performance.
4) We introduce kernel embedding for measuring the relative prototypes via spectral filtering in RKHS, which allows us to apply the kernels that achieve the state-of-the-art performance.

2   Related Work

2.1   Metric-based Few-shot Learning

Recently, some metric-based meta-learning algorithms for few-shot setting, one-shot setting and zero-shot setting were proposed, and the representatives of metric-based meta learning are Matching Networks (MatchingNet) [44] and ProtoNet [41]. MatchingNet proposed an attention mechanism for rapid learning. In addition, it pointed out that in the training procedure, the test condition should match with the train condition. ProtoNet proposed a simple strategy to improve the performance of MatchingNet. The strategy is using prototype, the mean of support samples in each class, to represent its support set [41]. From the perspective of classifier design, some works proposed a different kind of classifier with MatchingNet and ProtoNet. Simon et al. proposed the dynamic classifiers by using subspaces for few-shot learning called Deep Subspace Networks (DSN) [40]. These classifiers are defined as the projection of the difference between a prototype and a query sample onto a subspace. Lee et al. [20] proposed a linear support vector machine instead of nearest neighbor classifiers for few-shot learning.

Our work is related to the rectification of the prototype. For ProtoNet, the prototype is simply calculated by averaging the support sample values. Some kinds of work show that, for improving the classification performance, the prototype need to be better estimated. For example, compared with the simple visual information, the combination of cross-modal information and visual information can better represent the prototype [33,45]. In addition, a method of transductive setting has been proposed to rectify prototype, which diminishes the intra-class bias via label propagation and diminishes the cross-class bias via feature shifting [22]. Gao et al. used a combination of the instance-level attention and the feature-level attention for the noisy few-shot relation classification [12]. In this method, prototype is a weighted mean of support samples after conducting by the instance-level attention.

Our work is also related to the similarity measurement of two samples. Similarity measurement between samples in ProtoNet is carried out by using Euclidean distance. Euclidean distance implies two assumptions, that is, the characteristic dimensions are not correlated and have consistent covariance. However, in the real world these two assumptions are not necessarily true. Boris N. Oreshkin et al. found that, by using the metric scaling, the performance of few-shot algorithms can be improved by optimizing the metric scaling, and showed that the performance using scaled cosine distance nears that using Euclidean distance [32]. The metric scaling can also be learned from a Bayesian method perspective. [5]. Sung et al. proposed the Relation Networks (Relation Nets) that learns to learn a transferrable deep metric [42]. In addition, Mahalanobis distance is proposed to overcome the defect of Euclidean distance on measuring the similarity between two samples [3,11]. Recently, Khrulkov et al. studied how to measure sample similarity in a hyperbolic space instead of Euclidean space. They found that, compared with Euclidean space, hyperbolic embeddings can benefit the embedding of images and provide a better performance [16]. However, it is more difficult to operate in hyperbolic space, such as sample averaging, which further hinders the use of hyperbolic geometry. Fang et al. [8] provides several positive definite kernel functions of hyperbolic spaces, which enable one to operate in hyperbolic spaces.

2.2 Kernel Mean Shrinkage Estimation

The method of kernel mean shrinkage estimation is relevant to our work, which is employed to estimate the relative prototype. In recent years, some work has been done on the kernel mean shrinkage estimation. Muandet et al. [25] pointed out that the estimation of an empirical average in RKHS can be improved by employing Stein effect. Like James-Stein estimator, they propose several kernel mean shrinkage estimators, e.g., empirical-bound kernel mean shrinkage estimator (B-KMSE) and spectral kernel mean shrinkage estimator (S-KMSE) [27]. B-KMSE is easily optimized but the degree of shrinkage is the same for all coordinates. S-KMSE can shrink differently on each coordinate on the basis of eigenspectrum of covariance matrix, but it is difficult to optimize. Furthermore, "Shrinkage" can be achieved by a generalized filter function that can be embodied by various forms, such as Truncated SVD [28]. For these estimators, the shrinkage parameters are commonly obtained by a cross-validation score.

3 Methodology

3.1 Preliminary

In metric-based few-shot learning, before measuring the difference between query and classes, one often firstly represent each class by using support samples belonging to them, respectively. Here we consider two different geometric types of the class representation.

Point Type. In this type, a class is often represented as a point, e.g., a prototype, which is calculated by averaging of support samples in each class [41]. In addition, the point can also be calculated by summing over the support samples [42]. In these cases, The probability of measuring a sample belonging to a class is usually based on the distance between two points. For example, for ProtoNets, the distance square is expressed as

$$d_{l,c}^2 = \|f_\theta\left(\boldsymbol{q}_l\right) - \boldsymbol{\mu}_c\|^2, \text{ with } \boldsymbol{\mu}_c = \frac{1}{n}\sum_{i=1}^n f_\theta\left(\mathbf{s}_{i,c}\right), \tag{1}$$

where $\boldsymbol{q}_l$ is the lth sample in query set, and $\boldsymbol{s}_{i,c}$ is the ith sample in the cth class of the support set. In addition, $f_\theta(\cdot)$ with the parameter θ is a network.

Subspace Type. In addition to the point type, a class can also be represented as a subspace, e.g., the subspace spanned by the feature vectors created from the support samples in a class [40]. In this case, a sample should be close to its own class subspace and stay away from the subspaces of other classes. For example, for DSN, the distance square is expressed as

$$d_{l,c}^2 = \|(\boldsymbol{I} - \boldsymbol{P}_c\boldsymbol{P}_c^T)(f_\theta\,(\boldsymbol{q}_l) - \boldsymbol{\mu}_c)\|^2, \text{ with } \boldsymbol{\mu}_c = \frac{1}{n}\sum_{i=1}^{n} f_\theta\,(\mathbf{s}_{i,c}), \tag{2}$$

where P_c the truncated matrix of W_c the eigenvector matrix of empirical covariance matrix of support set. For a SVM classifier in few-shot learning, the representation of each class is one or more subspaces whose boundaries are so-called hyperplanes, which are determined by the support vectors [20]. In this case, a sample should belong to its own class subspace and stay away from the hyperplanes.

In the following, we show that the class representations of the two types are not entirely different. For example, we demonstrate that our framework can be embodied as either of ProtoNet and DSN with different filter functions (see the Sect. 3.4).

3.2 Kernel Shrinkage Relative-Prototype

Here we propose the definition of kernel shrinkage relative-prototype. Given the C-way n-shot support set $\boldsymbol{S} = \{\boldsymbol{S}_1, \boldsymbol{S}_2, ..., \boldsymbol{S}_C\}$ with $\boldsymbol{S}_c = \{\boldsymbol{s}_{1,c}, \boldsymbol{s}_{2,c}, ..., \boldsymbol{s}_{n,c}\}$, and the query set $\boldsymbol{Q} = \{\boldsymbol{q}_1, \boldsymbol{q}_2, ..., \boldsymbol{q}_m\}$. We firstly use function ϕ_ω to map the observations of support samples of class c in feature space, and calculate the prototype as:

$$\boldsymbol{\mu}_c = \frac{1}{n}\sum_{i=1}^{n} \phi_\omega\,(f_\theta\,(\mathbf{s}_{i,c})), \tag{3}$$

where $\phi_\omega(\cdot)$ with the parameter ω is a mapping function, and $f_\theta(\cdot)$ with the parameter θ is a network. Similar to the concepts in [15,46], we propose the relative-prototype, the difference between a sample in the query set and a prototype of class as a mean form:

$$\boldsymbol{\mu}_{l,c} = \phi_\omega\,(f_\theta\,(\boldsymbol{q}_l)) - \boldsymbol{\mu}_c = \frac{1}{n}\sum_{i=1}^{n} \boldsymbol{m}_{l,i,c}, \tag{4}$$

where

$$\boldsymbol{m}_{l,i,c} = \phi_\omega\,(f_\theta\,(\boldsymbol{q}_l)) - \phi_\omega\,(f_\theta\,(\boldsymbol{s}_{i,c})). \tag{5}$$

Simon et al. proposed a method that using adaptive subspaces instead of prototypes for few-shot learning [40]. In their work, the sample similarity is measured via a distance between a query sample and a subspace created from a support set, which can be seen as a "shrinkage" of the distance between a query sample and a prototype. Enlightened by it, we apply the shrinkage estimation theory to extend their idea and measure the sample similarity in a RKHS, or express the kernel shrinkage estimation of $\mu_{l,c}$ as

$$\mu_{l,c}(\lambda_c) = \mu_{l,c} - \sum_{i=1}^{n} h(\gamma_{i,c}, \lambda_{l,c}) \gamma_{i,c} \langle \mu_{l,c}, w_{i,c} \rangle w_{i,c}, \tag{6}$$

where $\lambda_{l,c}$ is the shrinkage coefficient with the class c, $(\gamma_{i,c}, w_{i,c})$ are respectively the eigenvalues and eigenvectors of the empirical covariance matrix $C = \sum_{i=1}^{n} r_{i,c} \otimes r_{i,c}$ where

$$r_{i,c} = \phi_\omega \left(f_\theta \left(s_{i,c} \right) \right) - \mu_c. \tag{7}$$

In Eq. 6, $h(\gamma_{i,c}, \lambda_{l,c})$ is a shrinkage function that approaches $1/\gamma_{i,c}$ as $\lambda_{l,c}$ decreases to 0, implying that the relative prototype is fully shrunk; As $\lambda_{l,c}$ increases, the shrinkage of relative prototype decreases or remain the same. There exist different ways of kernel mean shrinkage estimation that can be realized by constructing $h(\gamma_{i,c}, \lambda_{l,c})$ differently, such as Tikhonov regularization and Truncated SVD [28]. In this work we apply the Tikhonov regularization as the filter function that

$$h(\gamma_{i,c}, \lambda_{l,c}) = \frac{1}{\gamma_{i,c} + \lambda_{l,c}}. \tag{8}$$

In other ways, however, the kernel relative-prototype shrinkage estimation $\mu_{l,c}(\lambda_{l,c})$ in Eq. 6 is not a computable form as the mapping function ϕ_ω is not or in some cases can not be known, e.g., $\phi_\omega : \mathbb{R}^d \to \mathbb{R}^\infty$. Thus, we propose a computable form of $\mu_{l,c}(\lambda_{l,c})$ based on the work by Muandet et al. [28].

Theorem 1. *Denote the $n \times n$ metrix K_{ss}^c whose entry at the row i and the column j ($\forall i, j$) can be expressed the kernel form $k(f_\theta(s_{i,c}), f_\theta(s_{j,c})) = \phi_\omega(f_\theta(s_{i,c}))^T \phi_\omega(f_\theta(s_{i,c}))$, and the $n \times n$ metrix $K_{qs}^{l,c}$ whose entry at the row i and the column j ($\forall i, j$) can be expressed the kernel form $k(f_\theta(s_{i,c}), f_\theta(q_{l,c})) = \phi_\omega(f_\theta(s_{i,c}))^T \phi_\omega(f_\theta(q_{l,c}))$. The kernel kean shrinkage estimation of $\mu_{l,c}$ in Eq. 6 can be expressed as:*

$$\mu_{l,c}(\lambda_{l,c}) = \mu_{l,c} - \sum_{i=1}^{n} \alpha_{l,i,c}(\lambda_{l,c}) r_{i,c}, \tag{9}$$

with

$$\alpha_{l,c}(\lambda_{l,c}) = g^h(\tilde{K}_{ss}^c, \lambda_{l,c}) \tilde{K}_{qs}^{l,c} I_n, \tag{10}$$

$$\tilde{K}_{ss}^c = K_{ss}^c - \hat{I}_n K_{ss}^c - K_{ss}^c \hat{I}_n + \hat{I}_n K_{ss}^c \hat{I}_n, \tag{11}$$

$$\tilde{K}_{qs}^{l,c} = K_{qs}^{l,c} - \hat{I}_n K_{qs}^{l,c} - K_{ss}^c + \hat{I}_n K_{ss}^c, \tag{12}$$

where $\alpha_{l,c}(\lambda_{l,c}) = [\alpha_{l,1,c}(\lambda_{l,c}), ..., \alpha_{l,n,c}(\lambda_{l,c})]^T$ and $I_n = [1/n, 1/n, ..., 1/n]^T$, and $\{\hat{I}_n\}_{i,j} = 1/n$. Suppose the eigen-decomposition that $\tilde{K}_{ss}^c = V \Psi V^T$, then

$$g(\tilde{K}_{ss}^c, \lambda_{l,c}) = V g^h(\Psi, \lambda_{l,c}) V^T, \tag{13}$$

where $g(\boldsymbol{\Psi}, \lambda_{l,c}) = \mathrm{diag}(h(\gamma_{1,c}, \lambda_{l,c}), ..., h(\gamma_{n,c}, \lambda_{l,c}))$ with the $\tilde{\boldsymbol{K}}_{ss}^{c}$'s eigenvalues $\gamma_1, \gamma_2, ..., \gamma_n$.

Proof. Suppose that $v_{i,j}$ is the entry at the row i and column j of $\boldsymbol{V}$. According to [39], we have $\mathbf{w}_{i,c} = (1/\sqrt{\gamma_i}) \sum_j^n v_{i,j} \boldsymbol{r}_{j,c}$,

$$
\begin{aligned}
\boldsymbol{\mu}_{l,c}(\lambda_{l,c}) &= \boldsymbol{\mu}_{l,c} - \sum_{i=1}^{n} h(\gamma_{i,c}, \lambda_{l,c}) \gamma_{i,c} \langle \boldsymbol{\mu}_{l,c}, \mathbf{w}_{i,c} \rangle \mathbf{w}_{i,c} \\
&= \boldsymbol{\mu}_{l,c} - \sum_{j=1}^{n} \sum_{i=1}^{n} v_{i,j} h(\gamma_{i,c}, \lambda_{l,c}) \langle \boldsymbol{\mu}_{l,c}, \sum_{k}^{n} v_{i,k} \boldsymbol{r}_{k,c} \rangle \boldsymbol{r}_{j,c} \quad (14) \\
&= \boldsymbol{\mu}_{l,c} - \sum_{j=1}^{n} \alpha_{l,j,c}(\lambda_{l,c}) \boldsymbol{r}_{j,c}.
\end{aligned}
$$

where

$$
\begin{aligned}
\alpha_{l,j,c}(\lambda_{l,c}) &= \sum_{i=1}^{n} v_{i,j} h(\gamma_{i,c}, \lambda_{l,c}) \langle \boldsymbol{\mu}_{l,c}, \sum_{k=1}^{n} v_{i,k} \boldsymbol{r}_{k,c} \rangle \\
&= \sum_{i=1}^{n} v_{i,j} h(\gamma_{i,c}, \lambda_{l,c}) \sum_{k}^{n} v_{i,k} \langle \boldsymbol{\mu}_{l,c}, \boldsymbol{r}_{k,c} \rangle
\end{aligned}
\quad (15)
$$

Suppose that $\boldsymbol{v}_i = [v_{i,1}, v_{i,2}, ..., v_{i,n}]^T$,

$$
\begin{aligned}
\boldsymbol{\alpha}_{l,c}(\lambda_{l,c}) &= \sum_{i=1}^{n} \boldsymbol{v}_i h(\gamma_{i,c}, \lambda_{l,c}) \sum_{k}^{n} v_{i,k} \langle \boldsymbol{\mu}_{l,c}, \boldsymbol{r}_{k,c} \rangle \\
&= \sum_{i=1}^{n} \boldsymbol{v}_i h(\gamma_{i,c}, \lambda_{l,c}) \boldsymbol{v}_i^T \tilde{\boldsymbol{K}}_{qs}^{l,c} \boldsymbol{I}_n \\
&= g(\tilde{\boldsymbol{K}}_{ss}^{c}) \tilde{\boldsymbol{K}}_{qs}^{l,c} \boldsymbol{I}_n.
\end{aligned}
\quad (16)
$$

Based on Theorem 1, we can calculate the similarity between each sample in query set and prototype of each class using $\boldsymbol{\mu}_{l,c}(\lambda_{l,c})$.

3.3 Shrinkage Base Classifiers

For the convenience of calculation, we suppose that all the shrinkage parameters are the same, or $\lambda_{l,c} = \lambda$. The similarity of sample-pairs can be measured using the square of distance between the relative-prototype and original point in RKHS

$$
\begin{aligned}
d_{l,c}^2(\lambda, \boldsymbol{S}_c, \boldsymbol{q}_l) &= \|\boldsymbol{\mu}_{l,c}(\lambda)\|^2 \\
&= (\boldsymbol{\alpha}_{l,c}(\lambda))^T \tilde{\boldsymbol{K}}_{ss}^{c} \boldsymbol{\alpha}_{l,c}(\lambda) + \boldsymbol{I}_n^T \tilde{\boldsymbol{K}}_{qq}^{l,c} \boldsymbol{I}_n - 2(\boldsymbol{\alpha}_{l,c}(\lambda))^T \tilde{\boldsymbol{K}}_{qs}^{l,c} \boldsymbol{I}_n
\end{aligned}
\quad (17)
$$

where $\tilde{\boldsymbol{K}}_{qq}^{l,c}$ can be written as

$$
\tilde{\boldsymbol{K}}_{qq}^{l,c} = \boldsymbol{K}_{qq}^{l,c} + \boldsymbol{K}_{ss}^{c} - \boldsymbol{K}_{qs}^{l,c} - (\boldsymbol{K}_{qs}^{l,c})^T, \quad (18)
$$

with the $n \times n$ metrix $\boldsymbol{K}_{qq}^{l,c}$ whose entry at the row i and the column j ($\forall i, j$) can be expressed the kernel form $k(f_\theta(\boldsymbol{q}_l), f_\theta(\boldsymbol{q}_l)) = \phi_\omega(f_\theta(\boldsymbol{q}_l))^T \phi_\omega(f_\theta(\boldsymbol{q}_l))$. The probability of the sample $\boldsymbol{q}_l$ in query set belonging to class c is

$$P_{\omega,\theta}\left(Y = c | \boldsymbol{q}_l\right) = \frac{\exp\left(-\zeta d_{l,c}^2\left(\lambda, \boldsymbol{S}_c, \boldsymbol{q}_l\right)\right)}{\sum_{c=1}^{C} \exp\left(-\zeta d_{l,c}^2\left(\lambda, \boldsymbol{S}_c, \boldsymbol{q}_l\right)\right)}, \tag{19}$$

where ζ is the metric scaling parameter. The loss function of DSFN is

$$\mathcal{L}(\omega, \theta) = -\frac{1}{m} \sum_{l=1}^{m} \log P_{\omega,\theta}\left(y_l | \boldsymbol{q}_l\right) \quad, \tag{20}$$

where y_l is the label of $\boldsymbol{q}_l$. The few-shot learning process with the proposed DSFN is shown as Algorithm 1.

Algorithm 1. Few-shot learning with the proposed DSFN

Input: Support set $\mathbf{S}$ and query set $\mathbf{Q}$, learning rate α.
Output: θ
1:   Initialize θ randomly;
2:   **for** $t = 1$ to T **do**
3:     Generate episode by randomly sampling $\mathbf{S}^{(t)}$ from $\mathbf{S}$ and $\mathbf{Q}^{(t)}$ from $\mathbf{Q}$;
4:     **for** $c = 1$ to C **do**
5:       **for** $l = 1$ to m **do**
6:         Compute $\tilde{\boldsymbol{K}}_{ss}^c$, $\tilde{\boldsymbol{K}}_{qs}^{l,c}$ and $\tilde{\boldsymbol{K}}_{qq}^{l,c}$ using Eq. 11, Eq. 12 and Eq. 18, respectively, where $\boldsymbol{K}_{ss}^c$, $\boldsymbol{K}_{qs}^{l,c}$ and $\boldsymbol{K}_{qq}^{l,c}$ are calculated using the samples in $\mathbf{S}^{(t)}$ and $\mathbf{Q}^{(t)}$;
7:         Compute $\boldsymbol{\alpha}_{l,c}(\lambda_{l,c})$ with Eq. 10, using $\tilde{\boldsymbol{K}}_{qs}^{l,c}$ and eigenvalue decomposition of $\tilde{\boldsymbol{K}}_{ss}^c$;
8:         Compute $d_{l,c}\left(\lambda^{(t)}, \boldsymbol{S}_c^{(t)}, \boldsymbol{q}_l^{(t)}\right)$ using Eq. 17;
9:       **end for**
10:     **end for**
11:     Compute the loss function using Eq. 20;
12:     Update θ with $\theta - \alpha \nabla_\theta \mathcal{L}(\omega, \theta)$.
13:   **end for**

3.4 Relationship to Other Methods

Here we discuss the connection of our class representation to the point type (e.g., ProtoNet) and subspace type (e.g., DSN) representations. In fact, they are different mainly because they use different filter functions.

Relationship to ProtoNet. As the filter function $h(\gamma_{i,c}, \lambda_{l,c}) = 0$ instead of Tikhonov regularization that causes the disappearance of shrinkage effect, and the map function ϕ_ω is identical, the proposed framework (Eq. 6) is embodied as ProtoNet.

Relationship to DSN. While using the Truncated SVD as the filter function that $h(\gamma_{i,c}, \lambda_{l,c}) = \mathbb{I}_{(\gamma_{i,c} \geq \lambda_{l,c})} \gamma_{i,c}^{-1}$ instead of Tikhonov regularization, where $\mathbb{I}_{(\gamma_{i,c} \geq \lambda_{l,c})} \gamma_{i,c}^{-1}$ is the indicative function that is 1 if $\gamma_{i,c} \geq \lambda_{l,c}$ else 0, and the map function ϕ_ω is identical, the proposed framework (Eq. 6) is embodied as DSN. In this case, by setting different values of $\lambda_{l,c}$, different dimension of subspace can be selected. Formally, the relationship of DSFN and DSN is shown in Theorem 2 (see detailed proof in supplementary material).

Theorem 2. *Suppose that: 1)* $h(\gamma_{i,c}, \lambda_{l,c}) = \mathbb{I}_{(\gamma_{i,c} \geq \lambda_{l,c})} \gamma_{i,c}^{-1}$; *2)* $\zeta = 1$; *3)* $\lambda_{l,c} = constant$ *for all l and c; 4) the map function* ϕ_ω *is identical. Equation 17 is reduced to* $d_{l,c}^2(\lambda, \mathbf{S}_c, \mathbf{q}_l) = \|(I - P_c P_c^T)(f_\theta(\mathbf{q}_l) - \boldsymbol{\mu}_c)\|^2$ *with* P_c *the truncated matrix of* W_c, *where* W_c *is the eigenvector matrix of empirical covariance matrix* C.

Theorem 2 implies that, while using Truncated SVD as the filter function, the loss function of our proposed framework (Eq. 19) can be reduced to the loss of DSN with no regularization (See Eq. 5 in the work by Simon et al. [40]).

4   Experiments Setup

4.1   Datasets

*mini***ImageNet.** *mini*ImageNet dataset [44] was often used for few-shot learning, which contains a total of 60,000 color images in 100 classes randomly selected from ILSVRC-2012, with 600 samples in each class. The size of each image is 84 × 84. In the data set, the training set, validation set and test set contains the number of classes with 64:16:20.

*tiered***-ImageNet.** The *tiered*-ImageNet dataset [36] is a benchmark image dataset that is also selected from ILSVRC-2012 but contains 608 classes that is more than that in *mini*ImageNet dataset. These classes are divided into 34 high-level categories, can each category contains 10 to 30 classes. The size of each image is 84 × 84. Further, the categories are divided into the training set, validation set and test set with 20:6:8.

CIFAR-FS. The CIFAR-FS dataset [4] is a few-shot learning benchmark containing all 100 classes from CIFAR-100 [18], and each class contains 600 samples. The size of each image is 32 × 32. The classes are divided into the training set, validation set and test set with 64:16:20.

4.2   Implementation

In training stage, 15-shot 10-query samples are chosen on *mini*ImageNet dataset; 10-shot 15-query samples are chosen on *tiered*ImageNet dataset; 2-shot 20-query samples are chosen for 1-shot task and 15-shot 10-query samples are chosen for 5-shot task on CIFAR-FS dataset. λ is set to the best by choosing 0.01,0.1,1,10 or 100. For these datasets, the setting of 8 episodes per batch is utilized in the experiments. The total number of training epochs is 80, and in each epoch

1000 batches are sampled. In testing stage, 1000 episodes are used to assess our model. For the 1-shot K-way learning, another support sample was created by flipping the original support sample, and two support samples are used for spectral filtering in the validation and testing stages. Our model is trained and tested in the PyTorch machine learning package [34].

Two backbones, the Conv-4 and Resnet-12, are utilized as the backbones in our model. For Conv-4, the Adam optimizer with default momentum values $([\beta_1, \beta_2] = [0.9, 0.999])$ is applied for the training. The learning rate is initially set as 0.005 then decayed to 0.0025, 0.00125, 0.0005 and 0.00025 at 8, 30, 45 and 50 epochs, respectively. For ResNet-12, the SGD optimizer is applied for the training, and the learning rate is initially set as 0.1 then decayed to 0.0025, 0.00032, 0.00014 and 0.000052 at 12, 30, 45 and 57 epochs, respectively.

The identity kernel $k^{ide}(z_i, z_j) = \langle z_i, z_j \rangle$ and the RBF kernel $k^{rbf}(z_i, z_j) = \exp\left(-\|z_i - z_j\|^2/(2\sigma^2)\right)$ are chosen as the kernel functions, where σ^2 is assigned as the dimension of embeddings. In addition, the scaling parameter ζ is learned as a variable. For the filter function, we set the shrinkage coefficient λ as the fixed multiple of the maximum eigenvalue with each class.

5   Experiments and Discussions

5.1   Comparison with State-of-the-art Methods

Results on *mini*ImageNet Dataset. Firstly, the proposed DSFN and the state-of-the art methods for 5-way classification tasks on *mini*ImageNet dataset are compared in Table 1. Table 1 shows that the proposed DSFN with identity kernel can achieve the bests on 5-way 5-shot classification tasks using both Conv-4 and ResNet-12 backbones, which are higher than DSN with 3.3% and 1.3%, respectively. The RBF kernel can achieve the second best on 5-way 5-shot classification tasks. These results illustrate that the proposed DSFN can achieve the state-of-the-art performance for 5-way 5-shot classification tasks on the dataset.

Results on *tiered*-ImageNet dataset. We compare the proposed DSFN with the state-of-the-art methods for 5-way classification tasks on *tiered*ImageNet dataset, as shown in Table 2. It can be seen that, the performance of DSFN with identity kernel and RBF kernel is slightly lower than DSN on 5-way 5-shot classification task. However, they are better than the others on 5-way 5-shot classification task.

Results on CIFAR-FS Dataset. A further comparison is made on CIFAR-FS dataset, as shown in Table 3. Table 3 shows that the proposed DSFN with RBF kernel performs the best on 5-way 5-shot classification task, whose test accuracy is about 1.2% and 2.8% higher than those of DSN and ProtoNet, respectively. Thus, the proposed DSFN performs the state-of-the-art for 5-way 5-shot classification task on the dataset.

Table 1. Test accuracies (%) from the proposed DSFN and the state-of-the art methods for 5-way tasks on *mini*ImageNet dataset with 95% confidence intervals. ‡ means that training set and validation set are used for training the corresponding model.

Model	Backbone	5-way	
		1-shot	5-shot
MatchingNet [44]	Conv-4	43.56 ± 0.84	55.31 ± 0.73
MAML [10]	Conv-4	48.70 ± 1.84	63.11 ± 0.92
Reptile [31]	Conv-4	49.97 ± 0.32	65.99 ± 0.58
ProtoNet [41]	Conv-4	44.53 ± 0.76	65.77 ± 0.66
Relation Nets [42]	Conv-4	50.44 ± 0.82	65.32 ± 0.70
DSN [40]	Conv-4	51.78 ± 0.96	68.99 ± 0.69
DSFN(identity kernel)	Conv-4	**50.21 ± 0.64**	**72.20 ± 0.51**
DSFN(RBF kernel)	Conv-4	**49.97 ± 0.63**	**72.04 ± 0.51**
SNAIL [24]	ResNet-12	55.71 ± 0.99	68.88 ± 0.92
TADAM [32]	ResNet-12	58.50 ± 0.30	76.70 ± 0.30
AdaResNet [30]	ResNet-12	56.88 ± 0.62	71.94 ± 0.57
LEO‡ [38]	WRN-28-10	61.76 ± 0.08	77.59 ± 0.12
LwoF [13]	WRN-28-10	60.06 ± 0.14	76.39 ± 0.11
wDAE-GNN‡ [14]	WRN-28-10	62.96 ± 0.15	78.85 ± 0.10
MetaOptNet-SVM [20]	ResNet-12	62.64 ± 0.61	78.63 ± 0.46
DSN [40]	ResNet-12	62.64 ± 0.66	78.83 ± 0.45
CTM [21]	ResNet-18	62.05 ± 0.55	78.63 ± 0.06
Baseline [6]	ResNet-18	51.75 ± 0.80	74.27 ± 0.63
Baseline++ [6]	ResNet-18	51.87 ± 0.77	75.68 ± 0.63
Hyper ProtoNet [16]	ResNet-18	59.47 ± 0.20	76.84 ± 0.14
Hyperbolic RBF kernel [8]	ResNet-18	60.91 ± 0.21	77.12 ± 0.15
DSFN(identity kernel)	ResNet-12	**61.27 ± 0.71**	**80.13 ± 0.17**
DSFN(RBF kernel)	ResNet-12	**59.43 ± 0.66**	**79.60 ± 0.46**

Table 2. Test accuracies (%) from the proposed DSFN and the state-of-the art methods for 5-way tasks on *tiered*-ImageNet dataset with 95% confidence intervals. ‡ means that training set and validation set are used for training the corresponding model.

Model	Backbone	5-way	
		1-shot	5-shot
ProtoNet [41]	ResNet-12	61.74 ± 0.77	80.00 ± 0.55
CTM [21]	ResNet-18	64.78 ± 0.11	81.05 ± 0.52
LEO‡ [38]	WRN-28-10	66.33 ± 0.05	81.44 ± 0.09
MetaOptNet-RR [20]	ResNet-12	65.36 ± 0.71	81.34 ± 0.52
MetaOptNet-SVM [20]	ResNet-12	65.99 ± 0.72	81.56 ± 0.53
DSN [40]	ResNet-12	66.22 ± 0.75	82.79 ± 0.48
DSFN(identity kernel)	ResNet-12	**65.46 ± 0.70**	**82.41 ± 0.53**
DSFN(RBF kernel)	ResNet-12	**64.27 ± 0.70**	**82.26 ± 0.52**

Table 3. Test accuracies (%) from the proposed DSFN and some state-of-the art methods for 5-way classification tasks on CIFAR-FS dataset with 95% confidence intervals.

Model	Backbone	5-way	
		1-shot	5-shot
ProtoNet [41]	ResNet-12	72.2 ± 0.7	83.5 ± 0.5
MetaOptNet-RR [20]	ResNet-12	72.6 ± 0.7	84.3 ± 0.5
MetaOptNet-SVM [20]	ResNet-12	72.0 ± 0.7	84.2 ± 0.5
DSN [40]	ResNet-12	72.3 ± 0.7	85.1 ± 0.5
DSFN(identity kernel)	ResNet-12	$\mathbf{70.62 \pm 0.79}$	$\mathbf{86.11 \pm 0.58}$
DSFN(RBF kernel)	ResNet-12	$\mathbf{71.28 \pm 0.70}$	$\mathbf{86.30 \pm 0.58}$

5.2 Ablation Study

The impact of shrinkage parameter. The influence of different values of shrinkage parameter λ on the performances of the proposed DSFN is shown in Fig. 3. Figure 3 shows that the general trends drop for these datasets as the shrinkage parameter increases from 0.01 to 100, and the descending trends with 5 shot is more obvious that those with 1 shot. These results indicate that smaller shrinkage parameters (e.g., 1, 0.1, 0.01) or stronger shrinkage effect can better improve the performance of the proposed model.

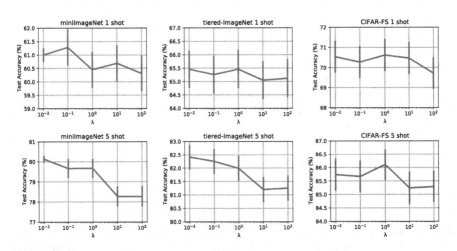

Fig. 3. Test accuracies on few-shot classification tasks from the proposed DSFN against different values of shrinkage parameter, where ResNet-12 is used.

The Effectiveness of Shrinkage. We show an ablation study to illustrate the effectiveness of shrinkage in our work, as shown in Table 4. In this experiment, the performances of identity kernel and RBF kernel with and without shrinkage

Table 4. Accuracies (%) from models with and without shrinkge for 5-way 1-shot and 5-way 5-shot classification tasks, w S: with shrinkge, w/o S: without shrinkge.

Dataset	Kernel	w S	w/o S	1-shot	5-shot
*mini*ImageNet	identity		✓	60.14 ± 0.67	77.65 ± 0.52
	identity	✓		61.00 ± 0.25	80.13 ± 0.17
	RBF		✓	58.74 ± 0.65	79.18 ± 0.46
	RBF	✓		59.43 ± 0.66	79.60 ± 0.46
tiered-ImageNet	identity		✓	65.05 ± 0.72	81.14 ± 0.55
	identity	✓		65.46 ± 0.70	82.41 ± 0.53
	RBF		✓	64.23 ± 0.70	82.07 ± 0.53
	RBF	✓		64.27 ± 0.70	82.26 ± 0.52
CIFAR-FS	identity		✓	70.54 ± 0.82	85.30 ± 0.59
	identity	✓		70.62 ± 0.79	86.11 ± 0.58
	RBF		✓	71.18 ± 0.73	86.09 ± 0.47
	RBF	✓		71.28 ± 0.70	86.30 ± 0.46

in few-shot classification tasks are compared. Table 4 shows that, the proposed model with shrinkage performs better than those without shrinkage for both kernels. In addition, the improvement of shrinkage on 5-way 5-shot classification tasks is more obvious than those on 5-way 1-shot classification tasks, probably because the eigenvalues and eigenvectors with one shot is hard to learn.

5.3 Time Complexity

Our proposed DSFN approach has the time complexity of $\mathcal{O}(\max(CN^3, CN^2D))$, where C, N, D are the number of way, shot and feature dimensionality, respectively. The proposed DSFN approach is slower than DSN ($\mathcal{O}(\min(CND^2, CN^2D))$) and ProtoNet ($\mathcal{O}(CND)$) due to the kernel matrix calculation, eigen-decomposition and multiple matrix multiplication. A way to reduce the time complexity is using some efficient algorithms, such as the fast adaptive eigenvalue decomposition [7] and faster matrix multiplication [1].

6 Conclusion

In this work, we propose a framework called DSFN for few-shot learning. In this framework, one can represent the similarity between a query and a prototype as the distance after spectral filtering of support set in each class in RKHS. DSFN is an extension of some mainstream methods, e.g., ProtoNet and DSN, and with appropriate filter function, the framework of DSFN can be embodies as those methods. In addition, we also showed that in this framework, one can explore new methods by applying diverse forms, e.g., Tikhonov regularization, as the filter function, and diverse forms of kernels. Several experiments verified the effectiveness of various specific forms of the proposed DSFN. Future works should take a closer look at the selection of filter function and the role of shrinkage parameter in the proposed framework.

References

1. Alman, J., Williams, V.V.: A refined laser method and faster matrix multiplication. In: Proceedings of the 2021 ACM-SIAM Symposium on Discrete Algorithms (SODA), pp. 522–539. SIAM (2021)
2. Andrychowicz, M., et al.: Learning to learn by gradient descent by gradient descent. In: Advances in Neural Information Processing Systems, pp. 3981–3989 (2016)
3. Bateni, P., Goyal, R., Masrani, V., Wood, F., Sigal, L.: Improved few-shot visual classification. In: Proceedings of the IEEE/CVF Conference on Computer Vision and Pattern Recognition, pp. 14493–14502 (2020)
4. Bertinetto, L., Henriques, J.F., Torr, P.H., Vedaldi, A.: Meta-learning with differentiable closed-form solvers. arXiv preprint arXiv:1805.08136 (2018)
5. Chen, J., Zhan, L., Wu, X., Chung, F.: Variational metric scaling for metric-based meta-learning. In: The Thirty-Fourth AAAI Conference on Artificial Intelligence, pp. 3478–3485 (2020)
6. Chen, W.Y., Liu, Y.C., Kira, Z., Wang, Y.C.F., Huang, J.B.: A closer look at few-shot classification. arXiv preprint arXiv:1904.04232 (2019)
7. Chonavel, T., Champagne, B., Riou, C.: Fast adaptive eigenvalue decomposition: a maximum likelihood approach. Signal Process. **83**(2), 307–324 (2003)
8. Fang, P., Harandi, M., Petersson, L.: Kernel methods in hyperbolic spaces. In: Proceedings of the IEEE/CVF International Conference on Computer Vision, pp. 10665–10674 (2021)
9. Feifei, L., Fergus, R., Perona, P.: One-shot learning of object categories. IEEE Trans. Pattern Anal. Mach. Intell. **28**(4), 594–611 (2006)
10. Finn, C., Abbeel, P., Levine, S.: Model-agnostic meta-learning for fast adaptation of deep networks. In: Proceedings of the 34th International Conference on Machine Learning, p. 1126–1135. JMLR. org. (2017)
11. Fort, S.: Gaussian prototypical networks for few-shot learning on omniglot. arXiv preprint arXiv:1708.02735 (2017)
12. Gao, T., Han, X., Liu, Z., Sun, M.: Hybrid attention-based prototypical networks for noisy few-shot relation classification. In: Proceedings of the AAAI Conference on Artificial Intelligence, vol. 33, pp. 6407–6414 (2019)
13. Gidaris, S., Komodakis, N.: Dynamic few-shot visual learning without forgetting. In: Proceedings of the IEEE Conference on Computer Vision and Pattern Recognition, pp. 4367–4375 (2018)
14. Gidaris, S., Komodakis, N.: Generating classification weights with GNN denoising autoencoders for few-shot learning. In: Proceedings of the IEEE/CVF Conference on Computer Vision and Pattern Recognition, pp. 21–30 (2019)
15. Kang, D., Kwon, H., Min, J., Cho, M.: Relational embedding for few-shot classification. In: Proceedings of the IEEE/CVF International Conference on Computer Vision, pp. 8822–8833 (2021)
16. Khrulkov, V., Mirvakhabova, L., Ustinova, E., Oseledets, I., Lempitsky, V.: Hyperbolic image embeddings. In: Proceedings of the IEEE/CVF Conference on Computer Vision and Pattern Recognition, pp. 6418–6428 (2020)

17. Koch, G., Zemel, R., Salakhutdinov, R., et al.: Siamese neural networks for one-shot image recognition. In: ICML Deep Learning Workshop, vol. 2 (2015)
18. Krizhevsky, A., Hinton, G., et al.: Learning multiple layers of features from tiny images (2009)
19. Lake, B.M., Salakhutdinov, R., Tenenbaum, J.B.: Human-level concept learning through probabilistic program induction. Science 350(6266), 1332–1338 (2015)
20. Lee, K., Maji, S., Ravichandran, A., Soatto, S.: Meta-learning with differentiable convex optimization. 10657–10665. arXiv:1904.03758 (2019)
21. Li, H., Eigen, D., Dodge, S., Zeiler, M., Wang, X.: Finding task-relevant features for few-shot learning by category traversal. In: Proceedings of the IEEE/CVF Conference on Computer Vision and Pattern Recognition, pp. 1–10 (2019)
22. Liu, J., Song, L., Qin, Y.: Prototype rectification for few-shot learning. In: Vedaldi, A., Bischof, H., Brox, T., Frahm, J.-M. (eds.) ECCV 2020. LNCS, vol. 12346, pp. 741–756. Springer, Cham (2020). https://doi.org/10.1007/978-3-030-58452-8_43
23. Mangla, P., Kumari, N., Sinha, A., Singh, M., Krishnamurthy, B., Balasubramanian, V.N.: Charting the right manifold: manifold mixup for few-shot learning. In: Proceedings of the IEEE/CVF Winter Conference on Applications of Computer Vision, pp. 2218–2227 (2020)
24. Mishra, N., Rohaninejad, M., Chen, X., Abbeel, P.: A simple neural attentive meta-learner. In: 6th International Conference on Learning Representations (2018)
25. Muandet, K., Fukumizu, K., Sriperumbudur, B., Gretton, A., Scholkopf, B.: Kernel mean estimation and stein effect. In: International Conference on Machine Learning, pp. 10–18 (2014)
26. Muandet, K., Fukumizu, K., Sriperumbudur, B., Scholkopf, B.: Kernel mean embedding of distributions: a review and beyond. arXiv preprint arXiv:1605.09522 (2016)
27. Muandet, K., Sriperumbudur, B., Fukumizu, K., Gretton, A., Scholkopf, B.: Kernel mean shrinkage estimators. J. Mach. Learn. Res. 17, 1–41 (2016)
28. Muandet, K., Sriperumbudur, B., Scholkopf, B.: Kernel mean estimation via spectral filtering. arXiv preprint arXiv:1411.0900 (2014)
29. Munkhdalai, T., Yu, H.: Meta networks. In: Proceedings of the 34th International Conference on Machine Learning, pp. 2554–2563 (2017)
30. Munkhdalai, T., Yuan, X., Mehri, S., Trischler, A.: Rapid adaptation with conditionally shifted neurons. In: International Conference on Machine Learning, pp. 3664–3673 (2018)
31. Nichol, A., Achiam, J., Schulman, J.: On first-order meta-learning algorithms. arXiv preprint arXiv:1803.02999 (2018)
32. Oreshkin, B.N., Lopez, P.R., Lacoste, A.: Tadam: task dependent adaptive metric for improved few-shot learning. In: Advances in Neural Information Processing Systems, pp. 721–731 (2018)
33. Pahde, F., Puscas, M., Klein, T., Nabi, M.: Multimodal prototypical networks for few-shot learning. In: Proceedings of the IEEE/CVF Winter Conference on Applications of Computer Vision, pp. 2644–2653 (2021)
34. Paszke, A., et al.: Automatic differentiation in pytorch (2017)
35. Ravi, S., Larochelle, H.: Optimization as a model for few-shot learning (2016)
36. Ren, M., et al.: Meta-learning for semi-supervised few-shot classification. arXiv preprint arXiv:1803.00676 (2018)
37. Rodríguez, P., Laradji, I., Drouin, A., Lacoste, A.: Embedding propagation: smoother manifold for few-shot classification. In: Vedaldi, A., Bischof, H., Brox, T., Frahm, J.-M. (eds.) ECCV 2020. LNCS, vol. 12371, pp. 121–138. Springer, Cham (2020). https://doi.org/10.1007/978-3-030-58574-7_8

38. Rusu, A.A., et al.: Meta-learning with latent embedding optimization. arXiv preprint arXiv:1807.05960 (2018)
39. Scholkopf, B., Smola, A., Muller, K.R.: Nonlinear component analysis as a kernel eigenvalue problem. Neural Comput. **10**(5), 1299–1319 (1998)
40. Simon, C., Koniusz, P., Nock, R., Harandi, M.: Adaptive subspaces for few-shot learning. In: Proceedings of the IEEE/CVF Conference on Computer Vision and Pattern Recognition, pp. 4136–4145 (2020)
41. Snell, J., Swersky, K., Zemel, R.S.: Prototypical networks for few-shot learning. In: Advances in Neural Information Processing Systems, pp. 4077–4087 (2017)
42. Sung, F., Yang, Y., Zhang, L., Xiang, T., Torr, P.H.S., Hospedales, T.M.: Learning to compare: relation network for few-shot learning. In: 2018 IEEE Conference on Computer Vision and Pattern Recognition, pp. 1199–1208 (2018)
43. Vanschoren, J.: Meta-learning: a survey. arXiv:1810.03548 (2018)
44. Vinyals, O., Blundell, C., Lillicrap, T., Kavukcuoglu, K., Wierstra, D.: Matching networks for one shot learning. In: Advances in Neural Information Processing Systems, pp. 3637–3645 (2016)
45. Xing, C., Rostamzadeh, N., Oreshkin, B., Pinheiro, P.O.: Adaptive cross-modal few-shot learning. In: Advanced Neural Information Processing System, vol. 32, pp. 4847–4857 (2019)
46. Ye, H.J., Hu, H., Zhan, D.C., Sha, F.: Few-shot learning via embedding adaptation with set-to-set functions. In: Proceedings of the IEEE/CVF Conference on Computer Vision and Pattern Recognition, pp. 8808–8817 (2020)
47. Zhang, H., Koniusz, P.: Zero-shot kernel learning. In: Proceedings of the IEEE Conference on Computer Vision and Pattern Recognition, pp. 7670–7679 (2018)
48. Ziko, I., Dolz, J., Granger, E., Ayed, I.B.: Laplacian regularized few-shot learning. In: International Conference on Machine Learning, pp. 11660–11670 (2020)

"This Is My Unicorn, Fluffy":
Personalizing Frozen Vision-Language Representations

Niv Cohen[1,2]([⊠])([iD]), Rinon Gal[1,3]([iD]), Eli A. Meirom[1], Gal Chechik[1,4]([iD]),
and Yuval Atzmon[1]([iD])

[1] NVIDIA Research, Tel Aviv, Israel
[2] The Hebrew University of Jerusalem, Jerusalem, Israel
`cohen.niv@cse.huji.ac.il`
[3] Tel Aviv University, Tel Aviv, Israel
[4] Bar Ilan University, Ramat Gan, Israel

Abstract. Large Vision & Language models pretrained on web-scale data provide representations that are invaluable for numerous V&L problems. However, it is unclear how they can be extended to reason about *user-specific* visual concepts in unstructured language. This problem arises in multiple domains, from personalized image retrieval to personalized interaction with smart devices. We introduce a new learning setup called *Personalized Vision & Language* (PerVL) with two new benchmark datasets for retrieving and segmenting user-specific ("personalized") concepts "in the wild". In PerVL, one should learn personalized concepts (1) *independently* of the downstream task (2) allowing a pretrained model to reason about them with free language, and (3) without providing personalized negative examples. We propose an architecture for solving PerVL that operates by *expanding* the input vocabulary of a pretrained model with new word embeddings for the personalized concepts. The model can then simply employ them as part of a sentence. We demonstrate that our approach learns personalized visual concepts from a few examples and effectively applies them in image retrieval and semantic segmentation using rich textual queries. For example the model improves MRR by 51.1% (28.4% vs 18.8%) compared to the strongest baseline.

The code and benchmark are available on github under NVlabs/ PALAVRA (https://github.com/NVlabs/PALAVRA) and NVlabs/ PerVLBenchmark (https://github.com/NVlabs/PerVLBenchmark).

Keywords: Personalization · Few-shot learning · CLIP · Vision and language · Zero-shot

Supplementary Information The online version contains supplementary material available at https://doi.org/10.1007/978-3-031-20044-1_32.

1   Introduction

Large Vision & Language (V&L) models pre-trained on web-scale data made a breakthrough in computer vision [6,50,70]. These models provide a multimodal vision-language representation, and are used in a multitude of downstream tasks, from image captioning [47] and video retrieval [19], through image generation [22,48] and segmentation [36,71], to robotic manipulation [54]. All these tasks benefit from the "open-world" capabilities of large V&L models, enabling the use of rich, free-form text with a long "tail" vocabulary of visual categories.

However, even with these powerful representations, an important challenge remains: How can these models be leveraged to reason about *user-specific*, "*personalized*" object instances in open-world vision problems? For example, we may wish to find an image that portrays us wearing a *specific* sweater, ask a robot assistant to make us coffee in *our* "best-mom mug", or synthesize an image of our child's treasured toy Fluffy in an entirely new context.

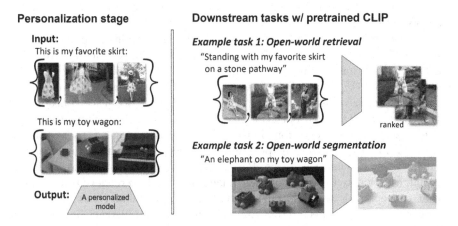

Fig. 1. The Personalized Vision & Language (PerVL) learning setup. Left: A user provides a few image examples of their personalized visual concepts: a favorite skirt (top), or a toddler's toy wagon (bottom). Examples are provided independently of the downstream tasks. **(Right)** the personalized model can be used in various downstream tasks. **Top-right:** An image retrieval task: given a textual query and a collection of images, rank and retrieve the queried image. **(Bottom-right)** Open-world semantic segmentation task. Segment a personalized object referred by a textual query. This example illustrates multiple ambiguities. First, there are two wagons that carry an elephant. Second, there are two wagons that correspond to the personalized concept. Resolving the ambiguity requires reasoning in both the visual and text modalities.

Clearly, pretrained V&L models cannot be used directly to reason about new personal items. Luckily, it is typically easy for a user to collect a few image examples for a personalized concept. It then remains to develop methods that extend the pretrained models to new concepts using these examples. One challenge is that typically it is easy for people to provide positive image examples for a concept, but harder to provide consistent negative distractor examples [29,53].

Learning from a few examples is considered a hallmark of intelligence. When people learn novel concepts from a few examples [7,34,44–46], they can seamlessly employ them in their semantic mental state and *reason jointly* both over the personalized concepts and over a large body of prior knowledge. Could a computational approach learn in a similar way using a pretrained V&L model?

Previous efforts [23,56,73,76] focused on learning a transformation module on top of CLIP's output space. However, as we explain below, these approaches risk forgetting prior knowledge, or face difficulties in accessing it concurrently with newly learned concepts. In addition, these previous approaches take a multiclass approach, discriminating between several new concepts. They are not designed for learning a single new personalized concept, which is natural in the context of personalization. Therefore, it is unknown how to learn a single personalized concept from few image examples in a way that (1) allows the pretrained model to reason about the concept with free language, and (2) uses only "positive" image examples of the target concept.

Here, we address the question of personalizeing a pretrained model using few samples while maintaining its performance on the original vocabulary. We study a new representation learning setup, which we call *"Personalized Vision & Language"* (PerVL) (Fig. 1). In PerVL, we are given a pretrained V&L model, one or more personalized visual concepts, a few training images of each concept, and a string describing the concept type, like "a mug" or "a short sleeve top". The goal is to learn a representation that can later be used to solve a set of downstream V&L tasks involving the personalized concept. No further supervision is given for these downstream tasks. PerVL arises in various scenarios. In image retrieval, a user may tag a few of their images and wish to retrieve other photos of that concept in a visual specific context [2,10]; in human-robot interaction, a worker may show a specific tool to a robotic arm, and instruct how to use it [42,54,64]; in video applications, an operator may search for a specific known item in the context of other items or people doing activities that are described with language.

Unlike previous efforts, instead of modifying a V&L model *output*, we propose a framework for expanding its *input* vocabulary. Specifically, we learn new word embeddings for the new personalized concepts by intervening with the model input space. The concept of "my best-mom mug" would be associated with a new symbol [MY BEST-MOM MUG] that has its own *dense* word embedding. The model could later represent sentences that use it, like *"Sipping tea from my best-mom mug on a porch"* by detecting "my best mom mug" and mapping its symbol [MY BEST-MOM MUG] to its new embedding vector. Such tokens trivially preserve the structure of the original model, since the encoder model itself remains unmodified. Moreover, as we show below, the new concepts can also be easily integrated into existing downstream V&L tasks. In summary, we address the question of using a small number of samples to personalize a pretrained V&L model, while maintaining its performance in the original vocabulary.

This paper makes the following novel contributions: (1) A new representation learning setup, *PerVL*, for personalizing V&L representations, while keeping their "zero-shot" reasoning capabilities. (2) Two new benchmark datasets for *PerVL*. (3) A novel approach, PALAVRA[1], to *expand* and personalize the vocabulary of the V&L representation *inputs*. PALAVRA uses a cycle-consistent loss, learned with positive image examples only. (4) A technique for using a *textual* encoder to improve the generalization of a network to new *visual* concepts.

2 Related Work

The success of CLIP led to diverse work that leverage its powerful representation for few-shot learning. Most works [23,43,56,73] are based on learning a residual "adapter" layer [28] over the output of CLIP encoders. Taking a different approach, [76] proposes learning a soft prefix to improve accuracy in a classification task. Our work differs from these approaches in two key aspects: (1) They focus solely on classifying images using a narrow vocabulary. In contrast, our setup learns a representation which is then used in any downstream tasks. Moreover, our method *expands* CLIP's vocabulary rather than narrowing it. (2) Adapter-based methods override the output representation of the encoders, leading to a change in their input $\rightarrow$ output mappings. Our method does not change the pretrained mapping but enriches its input vocabulary with new concepts.

Recently, [26,32,65] have shown that fine-tuning can actively harm out-of-distribution generalization, even when tested on the same downstream task for which the model was tuned. Our method does not fine-tune the pretrained model and does not leverage in-distribution labeled examples for the downstream tasks.

Other approaches [27,61] study "fast" concept learning combined with "slow-learned" concepts, showing that the new concepts can be applied to "slowly-learned" downstream tasks. However, the "fast" learned concepts are stored implicitly in the network activations, rather than grounded in the vocabulary.

A related set of studies can be found in tasks of image captioning and generator inversion. These studies extract meaningful semantic information from images and map them to tokens that represent the concepts, such as words or latent codes. Of these, [13,15,17,20,30,40,55] focus on personalizing image captions to a user writing style. Alternatively, [16,25,41,62,67,75] extend image captions with novel concepts using "slot filling".

Our model differs from zero and few-shot learning (FSL) based on meta-learning [1,4,5,11,21,35,49,57,59,63,68] or incremental learning [12,18,31,51, 60,66] in three aspects. First, we impose stronger generalization requirements. Our model can reason about new concepts in diverse downstream tasks, which may be unknown at training time. Second, in common FSL, the concept distribution used during meta-learning ("support set") is also used during the

---

[1] Palavra means "word" in Portuguese, *as we learn new word-embeddings*. For acronym lovers, PALAVRA also stands for "Personalizing LAnguage Vision RepresentAtions".

FSL stage. For example, meta-learn on birds, then do FSL with new types of birds. While our technique for training with text allows to generalize beyond the domain of concepts in the training images. Third, our approach improves upon CLIP's zero-shot perceptual capabilities, and is compatible with many CLIP-based downstream tasks.

Finally, in existing FSL benchmarks [52,63,74] there is *no* instance level annotations, and there is only a single task. As a result, existing FSL benchmarks do not directly fit our setting. Our work addresses *rich text* query *of a specific instance*, that can be used in a flexible way with many downstream tasks.

3   A New Setup, *Personalized Vision & Language*

We propose *"Personalized Vision & Language"* (PerVL), a new representation learning setup, to personalize a pretrained model with few positive image examples, without supervision for the downstream task.

In PerVL, we are given a pretrained model $h(S_V, I)$ that accepts a sentence S and an image I. The sentences that the model accepts are defined in a vocabulary V. We wish to update h so that it can accept sentences from an expanded vocabulary $V' = V \cup C$ where C is a new set of concepts $C = \{c_1, ...c_k\}$, which results in an extended model $h'(S_{V'}, I)$. In general, we expect that adapting the model would not strongly affect the original vocabulary, namely $h'(S_V, I) \approx h(S_V, I)$.

At training (personalization) time, we adapt the model given a small set of images $\{I_i\}_{i=1}^{N_c}$ for every concept c, without assuming access to negative training *images*. We are also provided with a string describing the type of the new concept, such as a "mug" or a "short sleeve top". Stating the type is a natural way for non-expert users to provide prior knowledge about the personalized concept. The type can be used to guide learning to distinguish the personalized concept from the general concept type. Concepts describing coarser classes from a hierarchy of concepts (e.g. "dog" for "poodle") have been used for this purpose [14]. We denote the concept type by S_c.

During inference, we are given a downstream V&L task T that can be inferred using the pretrained model h for the vocabulary V, and we wish to solve it for an instance x that contains the new concept c. The instance may contain images and sentences pertaining to c.

Encoder PerVL: Here we focus on the special case of CLIP [50]. The model h applies a cosine similarity between a sentence S and an image I: $h(S, I) = \cos(h^T(S), h^I(I))$, where h^I and h^T are CLIP image and text encoders.

4   Methods

Before describing our approach, we first explain the reasons for expanding the *input* vocabulary of the V&L model and how it differs from previous approaches.

Several studies extend CLIP by learning an "Adapter" module on top of the CLIP representation [23,28,43,56,73]. That module is applied to the *output* of a CLIP encoder network that is kept frozen. It is trained for a classification task with labeled data and a templated text query ("a photo of a [concept-type]").

We show below (Sect. 6 and Appendix B) that this approach tends to be brittle and fails when its input sentences deviate from the template used for training. This is probably because the adapter *overrides* the output representation of the encoder, so training it with very few examples hurts its generalization power.

Conversely, our approach does not overrides the encoder outputs. Our working hypothesis is that the text input space of a web-scale V&L model is rich enough for reasoning about new personalized concepts. We just need to find the right word embedding representation for any new personalized concept. We illustrate this architectural distinction in Fig. 2.

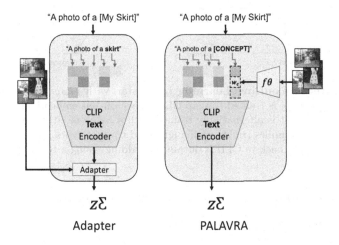

Fig. 2. Visualization of an adapter-based approach (left) and PALAVRA (right). Adapters change CLIP's output space by appending additional layers following the encoder. Our method defines new tokens in CLIP's existing input space, leaving the output space unchanged.

Finally, one could fully retrain a CLIP model with the expanded vocabulary set. However, retraining CLIP requires $\sim 400M$ images. Our approach is trained with a tiny fraction of that, $< 1M$ samples, and once it is trained, different users can use it, each with their own vocabulary.

Notation: For brevity, we describe adding a single concept c. Adding multiple concepts can be done iteratively. We use the notation [CONCEPT] to refer to a learned concept (c) within a textual query. $\mathcal{I}$ denotes the CLIP embedded image space, $\mathcal{T}$ the CLIP embedded textual space, $z_k = h^{\mathcal{I}}(I_k)$ is the embedding of an image I_k into $\mathcal{I}$, and similarly $h^{\mathcal{T}}(S)$ is the embedding of a sentence S into $\mathcal{T}$. Finally, $\mathcal{W}$ denotes the space used to embed input word tokens into CLIP.

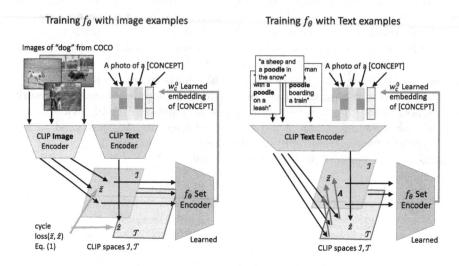

Fig. 3. Architecture outline: Learning f_θ. We start with a *large-scale-data* training step. A set encoder f_θ is trained to map CLIP-space output embeddings to a code in CLIP's input space. It is alternatively trained with a batch of either image examples (left), or sentence examples (right) with augmented concept types. We use a cycle loss by mapping the code back to CLIP's output embedding, using a template sentence.

Architecture and Workflow: At a high level, our workflow has three steps.

(1) **Learn an inversion mapping f_θ** from a *set* of points in CLIP image space $\mathcal{I}$ to a point in its word embedding *input* space $\mathcal{W}$ (Fig. 3). Formally, $f_\theta : \{z_k \in \mathcal{I}\}_{k=1}^K \rightarrow \mathcal{W}$. It is trained with *non-personalized, large-scale* data.
(2) **Initial personalization** (Fig. 4). Learn a word embedding $\mathbf{w}_c$ of a new personalized concept c. Thus, given a set of image examples $I_1, ..., I_K$ we map them to CLIP image space, then map them using f_θ to obtain an initial word embedding $\mathbf{w}_c^0 = f_\theta(\{h^{\mathcal{I}}(I_k)\})$. Formally, $\{I_k\}_{k=1}^K \rightarrow \{h^{\mathcal{I}}(I_k)\}_{k=1}^K \rightarrow \mathbf{w}_c^0 \in \mathcal{W}$.
(3) **Fine-tuning.** The initial embedding $\mathbf{w}_c^0$ is then updated using gradient steps to maximize the similarity of the template text embeddings to the image examples, while contrasting it with an embedding of a "super-concept".

Next, we describe the learning of each component in more detail.

4.1   Learning the Inversion Mapping f_θ

We now describe how we learn an "inversion" map f_θ from a set of points in CLIP space $z_1, ..., z_k \in \mathcal{I}$, to a word embedding $\mathbf{w}_c^0 \in \mathcal{W}$, where W is the input space of the language encoder. We base f_θ's architecture on "Deep Sets" [72]. We now discuss the loss and how to train f_θ with two types of *large-scale, non-personalized* data: images and text. See Fig. 3.

A Contrastive Cycle Loss. f_θ maps from CLIP space to $\mathbf{w}_c^0 \in \mathcal{W}$. Then, by pairing $\mathbf{w}_c^0$ to the word embedding for [CONCEPT] we can feed $\mathbf{w}_c^0$ into h^T with a template sentence T_c like "A photo of a [CONCEPT]". We can then define a cycle consistency loss to match the input of f_θ with the output of h^T (see Fig. 3 left). Specifically, let $\bar{z}_c$ be the average over samples in $\mathcal{I}$ from the concept c, $\bar{z}_c = \sum_{k=1}^{K} z_k / K$ and let $\hat{z}_c$ be the CLIP embedding of a template sentence, $\hat{z}_c = h^T(T_c)$. We wish to tune f_θ so that $\hat{z}_c$ is close to $\bar{z}_c$ for the concept c and far from other concepts. We therefore define a symmetric contrastive loss for a concept c, with a formulation similar to SimCLR [9]:

$$\ell_{Cycle}\left(c; \{\bar{z}_{c'}, \hat{z}_{c'}\}_{c'=1}^{C}\right) = -\log \frac{e^{\cos(\bar{z}_c, \hat{z}_c)}}{\sum_{c'=1}^{C} e^{\cos(\bar{z}_c, \hat{z}_{c'})} + \sum_{c' \neq c} e^{\cos(\hat{z}_c, \bar{z}_{c'})}}$$
$$-\log \frac{e^{\cos(\bar{z}_c, \hat{z}_c)}}{\sum_{c'=1}^{C} e^{\cos(\hat{z}_c, \bar{z}_{c'})} + \sum_{c' \neq c} e^{\cos(\bar{z}_c, \hat{z}_{c'})}}, \quad (1)$$

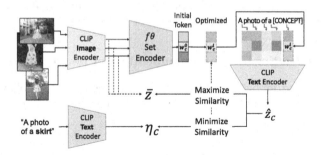

Fig. 4. Architecture outline: Personalization. Given a set of examples of a personalized concept and its type (skirt), we embed them with CLIP and predict an initial code ($\mathbf{w}_0$) for the concept using a frozen f_θ. We then further tune the code with a contrastive loss.

where $\cos(\hat{z}, \bar{z})$ denotes cosine similarity, C is the number of concepts in a batch.

We also use a regularization term ℓ_{GT} that maximizes the similarity of the predicted $\mathbf{w}_c^0$ with its ground truth. See details in the Appendix. Finally, the cycle loss and ground-truth regularization terms are combined with a hyperparameter $\lambda_{gt} \geq 0$, and the total loss is $\ell_{total} = \ell_{Cycle} + \lambda_{gt} \cdot \ell_{GT}$.

Training f_θ with Images. We use a variant of COCO [38] that extracted the subject and object from each caption as in [3], and take the 1000 most frequent concepts. In every training batch, we draw at random C concepts, then draw

K images for each concept. We then map them to CLIP image space, yielding $\{z_k\}_1^K = \{h^{\mathcal{I}}(I_k)\}_1^K$ in CLIP image space for each concept.

Training f_θ with Text. When training with COCO data, f_θ learns concepts that are frequent in COCO captions. However, our goal is to have f_θ generalize to widely diverse concepts. Yet, naively training with the COCO images does not generalize well to out-of-COCO-vocabulary concepts (see Appendix C).

To generalize to out-of-vocabulary concepts, we propose synthesizing *textual* descriptions with an expanded vocabulary and embed them into the shared embedding space. Specifically, we use COCO *captions* of a concept to generate additional training examples for new concepts by replacing the concept type with the most similar concept type from a large predefined vocabulary of 20K types [33], where (cosine) similarity is measured in CLIP text space. Finally, we embed the augmented captions by taking their CLIP-text feature representation (Fig. 3, right). Overall, we found that training with augmented text representation significantly improved the performance of the model (see Table A.2).

As in [37], we observed that the encoding distribution of text and images does not overlap in CLIP space. As a result, training f_θ with CLIP embeddings of captions does not generalize well to image inputs. We address this problem by learning an alignment matrix **A** that maps CLIP representations of texts to their presumed image counterpart (Fig. 3, right). **A** is learned jointly with f_θ, and is only used when learning the personalization tokens. It is not used at inference time. Formally, a set of captions is first encoded by $h^{\mathcal{T}}$, then mapped to the image area of the CLIP space using **A** and then fed to f_θ.

4.2 *Personalization*: Learn an embedding of personalized concepts

To learn the word embedding of a personalized concept, we follow a similar process to training f_θ, but instead of tuning the parameters of f_θ, we optimize the actual embedding vector $\mathbf{w}_c$.

Fig. 5. Examples of textual queries and evaluation images of the new benchmarks, [CONCEPT] denotes the personalized concepts. **(a)** YTVOS frames. The queried concept is highlighted in a yellow box. In YTVOS segmentation, one should segment the correct concept in the frame. In YTVOS retrieval, each evaluation image is a box cropped around the concept. **(b)** Deepfashion2 examples.

Specifically, let $\{I_k\}_{k=1}^{N_c}$ be a set of input images for the new concept c, we (1) map them to $\mathcal{I}$ using CLIP encoder $h^{\mathcal{I}}$, (2) map to $\mathbf{w}_c^0$ using f_θ, (3) plug the embedding $\mathbf{w}_c^0$ in a template sentence and (4) map to CLIP text space using $h^{\mathcal{T}}$. Once again, we define a contrastive cycle-consistent loss, matching the estimated text embedding of the template sentence $\hat{z}_c$, and average image embedding $\bar{z}_c$. However, here we contrast it with the embedding of the concept *type*, say "a short sleeve top", denoted by $\eta_c = h^{\mathcal{T}}(S_c)$. Since no negative image examples are provided in the personalization stage, the concept type can be viewed as a "super-concept" in the hierarchy. It allows the learning process to focus on the specific features that make the object unique from the general population of similar concept types. Similar to the SimCLR [9] loss, our loss is:

$$\ell\big(\hat{z}_c, \bar{z}_c, \eta_c\big) = -\log\frac{\exp\big(\cos(\bar{z}_c, \hat{z}_c)\big)}{\exp\big(\cos(\bar{z}_c, \hat{z}_c)\big) + 2 \cdot \exp\big(\cos(\eta_c, \hat{z}_c)\big)} \tag{2}$$

The factor 2 results from contrasting η_c with both visual and a text embedding.

4.3   Inference

Our approach expands the vocabulary of word-embedding tokens with personalized tokens, without modifying the underlying V&L model h. Therefore, for a given downstream task T and a sentence S, we use the pretrained V&L model h as it would have been used with T. But when we encounter an input sentence S that includes a [CONCEPT] token, we apply its learned embedding $\mathbf{w}_c$. Also, we found that having [CONCEPT] followed by the concept type S_c is beneficial.

5   Evaluation Datasets for PerVL

We created two new personalization benchmark datasets for the evaluation of PerVL. (1) We collected captions for images from DeepFashion2 [24], which serve as search queries in an image retrieval task. (2) We collected captions for frames from Youtube-VOS [69], and also collected their corresponding segmentation maps for a referring-expression segmentation task.

5.1   Personalized Fashion-Item Retrieval with DeepFashion2

We used the DeepFashion2 dataset [24] to create an image retrieval benchmark of personalized fashion items given a textual query. It contains photos of people wearing unique fashion items from 13 popular clothing categories, like *skirt* or *long-sleeve dress*, which we use as concept types. See the examples in Fig. 1, 5.

We created a dataset of 1700 images from 100 unique fashion items (concepts). Each item was assigned a unique [CONCEPT] tag. We assigned 450 images (out of the above 1700) to an *evaluation set*, and used raters to collect a textual description referring to each item appearing in the images. For instance, *The [CONCEPT] is facing a glass store display.*". In Appendix F.1 we describe the steps we took to select context-rich items.

Short Versus Detailed Captions: We collected two types of captions for each image. First, *detailed* captions like *"White cabinets, some with open drawers, are alongside and behind the [CONCEPT]."*. These describe extensive context about the image and can facilitate retrieval. Second are *short* captions like *"White cabinets are behind the [CONCEPT]."*. These pose a greater challenge, because they describe less detail, and therefore are more ambiguous.

Finally, we randomly split the data to 50 val. concepts and 50 test concepts.

5.2   Youtube-VOS for Personalized Retrieval and Segmentation

We created an image segmentation benchmark of personalized visual concepts given a textual query using Youtube-VOS (YTVOS) [69]. YTVOS is a dataset for instance segmentation in video, which includes 4000+ videos, 90+ categories, and 7800+ unique object instances. To transform it to an image personalization benchmark, we take the last frame of each video (scene) for evaluation and the object instances that appear in the frame as the target concepts. Earlier frames are used as candidate frames for training. See the examples in Fig. 5 (left).

This benchmark is challenging as it contains ambiguities about both the textual queries and the appearance of the personalized concept. Hence, only a model that is successful in both personalization and image-text reasoning can succeed in this task. For that, we only select videos such that their object concept appears at least twice in an evaluation frame.

Finally, we annotated the instances in the evaluation frame with captions using AMT. We instructed the AMT workers to concisely describe what makes a specific entity distinct, compared to similar entities in the image. We provide more details and examples in Appendix F.3.

In total, this benchmark includes ~500 unique personalized concepts, with ~6300 training samples. For evaluation, we split according to unique scenes (videos), resulting in 246 validation concepts and 251 test concepts.

Personalized Image Retrieval: We also created an image retrieval variant of YTVOS. We extracted a set of images that correspond to the AMT captions collected for segmentation. Every image in the retrieval set was extracted from a wide box cropped around every instance in each evaluation frame. The goal is to retrieve the image of the correct instance given its textual query. Compared to the segmentation task, there are fewer distractors from the same scene for every instance, since not all instances were labeled in the data, but there are many more distractors coming from different scenes.

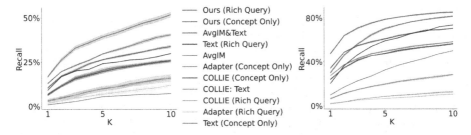

Fig. 6. Recall at K for our approach and baselines. DeepFashion2 (left), Youtube-VOS (right), PALAVRA achieves the highest rates for all retrieval metrics. On both benchmarks and metrics it achieves a significant improvement compared with "AvgIm&Text", which is the strongest baseline. The experiments with "Concept-only query" demonstrate that the information in the textual query is essential for telling the target image from distractors, since the performance of both PALAVRA and "Text Only" substantially degrades with "Concept-only" queries.

6   Experiments

We tested PALAVRA with two PerVL benchmarks and compared with several leading baselines (Sect. 6.1, 6.2). We then study in greater depth the properties of PALAVRA, by an ablation study (Appendix C). All design decisions and hyperparameter tuning were performed on the validation set to avoid overfitting the test set. The experiments were carried out on NVIDIA V100 and A100 GPU. We provide additional implementation details and results in Appendix A,B.

6.1   Personalized Image Retrieval with a Textual Query

The objective of this task is to retrieve the correct image given a text query that includes the new concept (Fig. 1, top-right). We use AMT captions as textual queries describing a single image from the dataset. The challenge in this setup is to overcome two types of distractors. (a) visually similar distractors: images of the same personalized concept but in a different context than the context described by the textual query (e.g. the two instances of "my favorite skirt" in Fig. 1), (b) semantically similar distractors: images which include an item of a similar concept type (e.g. "a skirt"), in a similar context as described in the textual query.

To rank images according to a textual query, we rank images according to their cosine similarity with the embedded text query.

Compared Methods. We compare our approach **PALAVRA**, with 5 baselines and their variants: **Text Only:** score an image-query pair using CLIP embedding of the text query $h^T(S)$. Using the concept type for [CONCEPT] instead of its learned word embedding. **AvgIM:** Ignoring the text query and replace it by the average over the embedding of its concept training images. This is equivalent to the FSL baseline in [11]. **IM&Text:** Represent the query as the average between *AvgIM* and *Text*. **Random:** Test images are ranked in random order.

COLLIE [56]: Learn an adapter module over the output of CLIP text encoder, with an additional scaler function $Scaler(h^T(S)) \in [0,1]$ that softly applies the adapter layer. COLLIE is closest to our method, because it may preserve some capabilities of the underlying pretrained model, when $Scaler(\cdot) = 0$. **Adapter:** As in COLLIE, but replace the scaler with a constant value of 1, making the "Adapter" layer always active. **COLLIE:Text:** COLLIE, when the text query uses the concept type for [CONCEPT], rather than the trained concept.

Table 1. MRR retrieval metrics.

	DeepFashion2 MRR	YTVOS MRR
Random	2.9 ± 0.4	2.8 ± 0.2
Concept-only query		
Text Only	4.2 ± 0.0	21.5 ± 0.0
Adapter	13.4 ± 0.5	35.5 ± 0.3
COLLIE	13.8 ± 0.5	35.6 ± 0.3
AvgIm	13.8 ± 0.5	38.2 ± 0.3
PALAVRA (Ours)	19.4 ± 0.6	53.4 ± 0.8
Rich query		
Adapter	5.9 ± 0.7	5.3 ± 0.3
COLLIE	7.9 ± 0.7	6.2 ± 0.3
COLLIE: Text	8.0 ± 1.0	7.2 ± 0.3
Text Only	17.6 ± 0.0	37.6 ± 0.0
AvgIm+Text	18.8 ± 0.4	47.2 ± 0.3
PALAVRA w.o. tuning	22.1 ± 0.2	47.1 ± 0.8
PALAVRA (Ours)	**28.4 ± 0.7**	**61.2 ± 0.4**

Evaluation Metrics and Queries.
For image retrieval, we report two metrics (1) **Recall at K**: The rate of successful retrievals among the top-K scoring images. (2) **MRR** (Mean Reciprocal Rank): Average of 1 divided by the rank of the correct image. Errors denote the standard error of the mean (SEM) across 5 model initialization seeds. We use two types of queries: **(1) Rich query** uses the free-formed text annotated by AMT workers: "[CONCEPT] is leaning on a rock". **(2) Concept-only queries** overrides the "Rich query" by a template that focuses only on the concept: "A photo of a [CONCEPT]". Note that the baseline "AvgIM" is more related to the "Concept-only query" because the rich query embedding is overridden by the average embedding of the training examples.

Retrieval Results. Table 1 and Fig. 6 describe the retrieval rates of PALAVRA and the compared methods when using challenging *short* captions as our **Rich Queries**. We report the retrieval rates with *detailed* captions in the appendix. We note that both *short* queries and *detailed* queries are rich queries, containing known concepts in addition to the personalized ones. The *detailed* version possibly contains more of them. PALAVRA achieves the highest rate in all the retrieval metrics. On both benchmarks and metrics, it achieves significant improvement (between ∼30% to ∼50%) compared with "AvgIm+Text", which is the strongest baseline.

Comparing the results of "Concept-only query" with the "Rich query" results demonstrate that: (1) Information in the "Rich query" is essential for retrieval. (2) Adapter baselines (Adapter & COLLIE) improve over vanilla CLIP when only the concept is queried.

Their performance degrades when using the "Rich query". This happens as the adaptation layer trained for the personalized [CONCEPT] does not perform well with free-form text it has not seen during training. In fact, we find that Adapter and COLLIE are even sensitive to the prompt *prefix* of the query.

When changing their prefix to a prefix not used in training, their performance substantially degrades. We report this finding quantitatively in Appendix B.

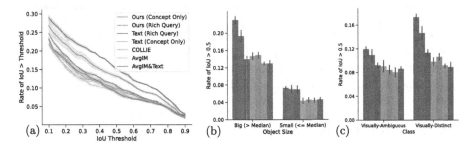

Fig. 7. Segmentation results on YTVOS. **(a)** Percent of images where IoU values exceeded a threshold, as a function of the threshold. Our approach dominates across the full range **(b, c)** Investigating model robustness under 2 levels of task complexity. We consider two scenarios that influence difficulty: Object size (b), and intra-class class ambiguity (c). When a clear visual signal is available (large objects, high intra-class visual variance), our model significantly outperforms the alternatives. Even in more challenging scenarios, our model still leverages textual descriptions, mitigating the loss of quality seen in models that ignore or corrupt CLIP's embedding space.

6.2   Semantic Segmentation with a Textual Query

The second downstream task, aims to segment an instance of a personalized concept in an image, based on a textual query that refers to the concept (Fig. 1, right bottom). The challenge here is to overcome two types of distractors: First, *visual distractors* that look similar to the concept and can be disambiguated with the text query. Second, *semantic distractors* that include a concept of a similar type (e.g. another type of a "toy wagon"), but CLIP has difficulty to resolve using just the concept type, like "an elephant on a toy wagon".

Here we investigate the performance of PALAVRA and baseline models on YTVOS dataset using a recent CLIP-based semantic segmentation [71]. In brief, [71] creates a set of query-driven relevance maps for the image, coupled with transformer interpretability methods [8]. The maps then serve as pseudolabels for single-image semantic segmentation [58].

Compared Methods. We compare PALAVRA with a set of baselines in two setups: "Rich query" and "Concept-only query", as described in Sect. 6.1. **(1) Text (CLIP)**, using both the rich and concept only queries, **(2) AvgIM, (3) IM&Text** and **(4) COLLIE**. All baselines are described in Sect. 6.1.

Evaluation Metrics. We calculate the intersection-over-union (IOU) between the predicted segment and the ground-truth segment. We report the **Rate of IOU > threshold**, which is the fraction of segments with IOU > threshold. Error bars denote the standard error of the mean (SEM) across 5 model seeds.

Semantic Segmentation Results

Figure 7a shows the percent of test-set images for which IoU exceeds a given threshold. Our model consistently outperforms the baselines with wide margins (e.g. a 44.69% improvement over the best competitor at an IoU threshold of 0.5). These results demonstrate that a personalized prompt can extend even to localized image tasks. Moreover, as our method only expands CLIP's input space - it can be readily integrated with existing models for downstream tasks.

Surprisingly, our method performs better when using the concept-only query. We hypothesize that this is a result of CLIP's difficulty in reasoning over complex referring expressions: By mentioning the context within the text, CLIP's attention is diverted to the context itself which leads to false negatives. In contrast, in retrieval, context objects rarely appear in "negative" (non-target) images. Since they appear in the target image, they actually help to detect the correct image. Appendix E.2 provides quantitative evidence supporting this hypothesis.

We further examine cases where we expect existing methods to falter. In Fig. 7b, we examine a scenario where objects are small, so their crops may not provide sufficient signal to the CLIP image encoder. Indeed, when for objects below the median size, segmentation fares worse in general, and image based-methods suffer even more. Our method, however, can rely on the signal derived from the text and degrades less. Figure 7c examines a scenario in which objects in a scene are less visually distinct. We divide the evaluation set to object which are usually visually distinct and images which often contain visually ambiguous objects, where a few instances of the same object appear in the same image. In practice, we split to animal and non-animal categories, as animals are mostly visually ambiguous (e.g. Fig. 5 left). Our model substantially outperforms the baselines when the concepts are visually distinct and also improves when the concepts are mostly ambiguous.

In the Appendix B we provide and discuss qualitative segmentation results.

7   Discussion

We described an approach to leverage large pre-trained V&L models like CLIP, for learning a representation of new "personal" classes from a handful of samples. Our key idea is to expand the input space of V&L models by finding a representation of the new concept. The extended model can then be used for V&L tasks with a rich language that "understands" both novel and known concepts. A limitation of the approach is that it suffers from the limitations of the underlying V&L model. For instance, CLIP struggles with understanding spatial relations within a photo [39], and extended representations based on CLIP suffer from the same problem. We expect that our approach can be extended to other V&L models. See an example in Appendix D.

To conclude, we hope that the method presented in this paper will pave the way to using pretrained models in problems that involve user-specific concepts, like home robotics and organizing personal data.

References

1. Akata, Z., Perronnin, F., Harchaoui, Z., Schmid, C.: Label-embedding for image classification. IEEE Trans. Pattern Anal. Mach. Intell. **38**, 1425–1438 (2016)
2. Anwaar, M.U., Labintcev, E., Kleinsteuber, M.: Compositional learning of image-text query for image retrieval. In: Proceedings of the IEEE/CVF Winter Conference on Applications of Computer Vision, pp. 1140–1149 (2021)
3. Atzmon, Y., Berant, J., Kezami, V., Globerson, A., Chechik, G.: Learning to generalize to new compositions in image understanding. arXiv preprint arXiv:1608.07639 (2016)
4. Atzmon, Y., Chechik, G.: Probabilistic and-or attribute grouping for zero-shot learning. In: Proceedings of the Thirty-Forth Conference on Uncertainty in Artificial Intelligence (2018)
5. Atzmon, Y., Chechik, G.: Adaptive confidence smoothing for generalized zero-shot learning. In: Proceedings of the IEEE/CVF Conference on Computer Vision and Pattern Recognition, pp. 11671–11680 (2019)
6. Bommasani, R., et al.: On the opportunities and risks of foundation models. arXiv preprint arXiv:2108.07258 (2021)
7. Carey, S., Bartlett, E.: Acquiring a single new word (1978)
8. Chefer, H., Gur, S., Wolf, L.: Transformer interpretability beyond attention visualization. In: Proceedings of the IEEE/CVF Conference on Computer Vision and Pattern Recognition (CVPR), pp. 782–791, June 2021
9. Chen, T., Kornblith, S., Norouzi, M., Hinton, G.: A simple framework for contrastive learning of visual representations. In: International Conference on Machine Learning, pp. 1597–1607. PMLR (2020)
10. Chen, Y., Gong, S., Bazzani, L.: Image search with text feedback by visiolinguistic attention learning. In: Proceedings of the IEEE/CVF Conference on Computer Vision and Pattern Recognition, pp. 3001–3011 (2020)
11. Chen, Y., Liu, Z., Xu, H., Darrell, T., Wang, X.: Meta-baseline: exploring simple meta-learning for few-shot learning. In: Proceedings of the IEEE/CVF International Conference on Computer Vision, pp. 9062–9071 (2021)
12. Cheraghian, A., Rahman, S., Fang, P., Roy, S.K., Petersson, L., Harandi, M.: Semantic-aware knowledge distillation for few-shot class-incremental learning. In: Proceedings of the IEEE/CVF Conference on Computer Vision and Pattern Recognition, pp. 2534–2543 (2021)
13. Chunseong Park, C., Kim, B., Kim, G.: Attend to you: personalized image captioning with context sequence memory networks. In: Proceedings of the IEEE Conference on Computer Vision and Pattern Recognition, pp. 895–903 (2017)
14. Dekel, O., Keshet, J., Singer, Y.: Large margin hierarchical classification. In: Proceedings of the Twenty-First International Conference on Machine Learning, p. 27 (2004)
15. Del Chiaro, R., Twardowski, B., Bagdanov, A., Van de Weijer, J.: RATT: recurrent attention to transient tasks for continual image captioning. In: Advances in Neural Information Processing Systems 33, pp. 16736–16748 (2020)
16. Demirel, B., Cinbis, R.G., Ikizler-Cinbis, N.: Image captioning with unseen objects. arXiv preprint arXiv:1908.00047 (2019)
17. Denton, E., Weston, J., Paluri, M., Bourdev, L., Fergus, R.: User conditional hashtag prediction for images. In: Proceedings of the 21th ACM SIGKDD International Conference on Knowledge Discovery and Data Mining, pp. 1731–1740 (2015)

18. Fan, L., Xiong, P., Wei, W., Wu, Y.: FLAR: a unified prototype framework for few-sample lifelong active recognition. In: Proceedings of the IEEE/CVF International Conference on Computer Vision, pp. 15394–15403 (2021)

19. Fang, H., Xiong, P., Xu, L., Chen, Y.: CLIP2Video: mastering video-text retrieval via image clip. arXiv preprint arXiv:2106.11097 (2021)

20. Feng, F., Liu, R., Wang, X., Li, X., Bi, S.: Personalized image annotation using deep architecture. IEEE Access **5**, 23078–23085 (2017)

21. Finn, C., Abbeel, P., Levine, S.: Model-agnostic meta-learning for fast adaptation of deep networks. In: International Conference on Machine Learning, pp. 1126–1135. PMLR (2017)

22. Gal, R., Patashnik, O., Maron, H., Chechik, G., Cohen-Or, D.: StyleGAN-NADA: clip-guided domain adaptation of image generators. arXiv preprint arXiv:2108.00946 (2021)

23. Gao, P., et al.: Clip-adapter: better vision-language models with feature adapters. arXiv preprint arXiv:2110.04544 (2021)

24. Ge, Y., Zhang, R., Wu, L., Wang, X., Tang, X., Luo, P.: A versatile benchmark for detection, pose estimation, segmentation and re-identification of clothing images. In: Proceedings of the IEEE/CVF Conference on Computer Vision and Pattern Recognition (CVPR) (2019)

25. Hendricks, L.A., Venugopalan, S., Rohrbach, M., Mooney, R., Saenko, K., Darrell, T.: Deep compositional captioning: Describing novel object categories without paired training data. In: Proceedings of the IEEE Conference on Computer Vision and Pattern Recognition, pp. 1–10 (2016)

26. Hewitt, J., Li, X.L., Xie, S.M., Newman, B., Liang, P.: Ensembles and cocktails: robust finetuning for natural language generation. In: NeurIPS 2021 Workshop on Distribution Shifts: Connecting Methods and Applications (2021)

27. Hill, F., Tieleman, O., von Glehn, T., Wong, N., Merzic, H., Clark, S.: Grounded language learning fast and slow. arXiv preprint arXiv:2009.01719 (2020)

28. Houlsby, N., et al.: Parameter-efficient transfer learning for NLP. In: International Conference on Machine Learning, pp. 2790–2799. PMLR (2019)

29. Hsieh, Y.G., Niu, G., Sugiyama, M.: Classification from positive, unlabeled and biased negative data. In: International Conference on Machine Learning, pp. 2820–2829. PMLR (2019)

30. Jia, X., Zhao, H., Lin, Z., Kale, A., Kumar, V.: Personalized image retrieval with sparse graph representation learning. In: Proceedings of the 26th ACM SIGKDD International Conference on Knowledge Discovery and Data Mining, pp. 2735–2743 (2020)

31. Khan, M., Srivatsa, P., Rane, A., Chenniappa, S., Hazariwala, A., Maes, P.: Personalizing pre-trained models. arXiv preprint arXiv:2106.01499 (2021)

32. Kumar, A., Raghunathan, A., Jones, R., Ma, T., Liang, P.: Fine-tuning can distort pretrained features and underperform out-of-distribution. arXiv preprint arXiv:2202.10054 (2022)

33. Kuznetsova, A., et al.: The open images dataset V4. Int. J. Comput. Vis. **128**(7), 1956–1981 (2020). https://doi.org/10.1007/s11263-020-01316-z

34. Lake, B.M., Piantadosi, S.T.: People infer recursive visual concepts from just a few examples. Comput. Brain Behav. **3**(1), 54–65 (2020)

35. Lampert, C., Nickisch, H., Harmeling, S.: Learning to detect unseen object classes by between-class attribute transfer. In: Proceedings of the IEEE/CVF Conference on Computer Vision and Pattern Recognition (CVPR) (2009)

36. Li, B., Weinberger, K.Q., Belongie, S., Koltun, V., Ranftl, R.: Language-driven semantic segmentation. In: International Conference on Learning Representations (2022). https://openreview.net/forum?id=RriDjddCLN
37. Liang, W., Zhang, Y., Kwon, Y., Yeung, S., Zou, J.: Mind the gap: understanding the modality gap in multi-modal contrastive representation learning. arXiv preprint arXiv:2203.02053 (2022)
38. Lin, T.-Y., et al.: Microsoft COCO: common objects in context. In: Fleet, D., Pajdla, T., Schiele, B., Tuytelaars, T. (eds.) ECCV 2014. LNCS, vol. 8693, pp. 740–755. Springer, Cham (2014). https://doi.org/10.1007/978-3-319-10602-1_48
39. Liu, N., Li, S., Du, Y., Tenenbaum, J., Torralba, A.: Learning to compose visual relations. In: Advances in Neural Information Processing Systems 34 (2021)
40. Long, C., Yang, X., Xu, C.: Cross-domain personalized image captioning. Multimedia Tools Appl. **79**(45), 33333–33348 (2020). https://doi.org/10.1007/s11042-019-7441-7
41. Lu, J., Yang, J., Batra, D., Parikh, D.: Neural baby talk. In: Proceedings of the IEEE Conference on Computer Vision and Pattern Recognition, pp. 7219–7228 (2018)
42. Lynch, C., Sermanet, P.: Language conditioned imitation learning over unstructured data. arXiv preprint arXiv:2005.07648 (2020)
43. Ma, T., et al.: A simple long-tailed recognition baseline via vision-language model. arXiv preprint arXiv:2111.14745 (2021)
44. Malaviya, M., Sucholutsky, I., Oktar, K., Griffiths, T.L.: Can humans do less-than-one-shot learning? arXiv preprint arXiv:2202.04670 (2022)
45. Markman, E.M.: Constraints children place on word meanings. Cogn. Sci. **14**(1), 57–77 (1990)
46. Markman, E.M., Wasow, J.L., Hansen, M.B.: Use of the mutual exclusivity assumption by young word learners. Cogn. Psychol. **47**(3), 241–275 (2003)
47. Mokady, R., Hertz, A., Bermano, A.H.: ClipCap: clip prefix for image captioning. arXiv preprint arXiv:2111.09734 (2021)
48. Patashnik, O., Wu, Z., Shechtman, E., Cohen-Or, D., Lischinski, D.: StyleCLIP: text-driven manipulation of StyleGAN imagery. In: Proceedings of the IEEE/CVF International Conference on Computer Vision, pp. 2085–2094 (2021)
49. Paz-Argaman, T., Atzmon, Y., Chechik, G., Tsarfaty, R.: ZEST: zero-shot learning from text descriptions using textual similarity and visual summarization (2020)
50. Radford, A., et al.: Learning transferable visual models from natural language supervision. In: International Conference on Machine Learning, pp. 8748–8763. PMLR (2021)
51. Ren, M., Iuzzolino, M.L., Mozer, M.C., Zemel, R.S.: Wandering within a world: online contextualized few-shot learning. arXiv preprint arXiv:2007.04546 (2020)
52. Ren, M., et al.: Meta-learning for semi-supervised few-shot classification. In: International Conference on Learning Representations (2018). https://openreview.net/forum?id=HJcSzz-CZ
53. Shinoda, K., Kaji, H., Sugiyama, M.: Binary classification from positive data with skewed confidence. arXiv preprint arXiv:2001.10642 (2020)
54. Shridhar, M., Manuelli, L., Fox, D.: CLIPort: What and where pathways for robotic manipulation. In: Proceedings of the 5th Conference on Robot Learning (CoRL) (2021)
55. Shuster, K., Humeau, S., Hu, H., Bordes, A., Weston, J.: Engaging image captioning via personality. In: Proceedings of the IEEE/CVF Conference on Computer Vision and Pattern Recognition, pp. 12516–12526 (2019)

56. Skantze, G., Willemsen, B.: Collie: Continual learning of language grounding from language-image embeddings. arXiv preprint arXiv:2111.07993 (2021)
57. Snell, J., Swersky, K., Zemel, R.: Prototypical networks for few-shot learning. In: Advances in Neural Information Processing Systems 30 (2017)
58. Sofiiuk, K., Petrov, I., Konushin, A.: Reviving iterative training with mask guidance for interactive segmentation. arXiv preprint arXiv:2102.06583 (2021)
59. Sung, F., Yang, Y., Zhang, L., Xiang, T., Torr, P.H., Hospedales, T.M.: Learning to compare: relation network for few-shot learning. In: Proceedings of the IEEE Conference on Computer Vision and Pattern Recognition, pp. 1199–1208 (2018)
60. Tao, X., Hong, X., Chang, X., Dong, S., Wei, X., Gong, Y.: Few-shot class-incremental learning. In: Proceedings of the IEEE/CVF Conference on Computer Vision and Pattern Recognition, pp. 12183–12192 (2020)
61. Tsimpoukelli, M., Menick, J., Cabi, S., Eslami, S., Vinyals, O., Hill, F.: Multimodal few-shot learning with frozen language models. In: Advances in Neural Information Processing Systems 34 (2021)
62. Venugopalan, S., Anne Hendricks, L., Rohrbach, M., Mooney, R., Darrell, T., Saenko, K.: Captioning images with diverse objects. In: Proceedings of the IEEE Conference on Computer Vision and Pattern Recognition, pp. 5753–5761 (2017)
63. Vinyals, O., Blundell, C., Lillicrap, T., Wierstra, D., et al.: Matching networks for one shot learning. In: Advances in Neural Information Processing Systems 29 (2016)
64. Wang, L., Meng, X., Xiang, Y., Fox, D.: Hierarchical policies for cluttered-scene grasping with latent plans. IEEE Robot. Autom. Lett. **7**, 2883–2890 (2022)
65. Wortsman, M., et al.: Robust fine-tuning of zero-shot models. arXiv preprint arXiv:2109.01903 (2021)
66. Wu, G., Gong, S., Li, P.: Striking a balance between stability and plasticity for class-incremental learning. In: Proceedings of the IEEE/CVF International Conference on Computer Vision, pp. 1124–1133 (2021)
67. Wu, Y., Zhu, L., Jiang, L., Yang, Y.: Decoupled novel object captioner. In: Proceedings of the 26th ACM International Conference on Multimedia, pp. 1029–1037 (2018)
68. Xian, Y., Schiele, B., Akata, Z.: Zero-shot learning - the good, the bad and the ugly. In: Proceedings of the IEEE/CVF Conference on Computer Vision and Pattern Recognition (CVPR) (2017)
69. Xu, N., et al.: YouTube-VOS: sequence-to-sequence video object segmentation. In: Ferrari, V., Hebert, M., Sminchisescu, C., Weiss, Y. (eds.) ECCV 2018. LNCS, vol. 11209, pp. 603–619. Springer, Cham (2018). https://doi.org/10.1007/978-3-030-01228-1_36
70. Yuan, L., et al.: Florence: a new foundation model for computer vision. arXiv preprint arXiv:2111.11432 (2021)
71. Zabari, N., Hoshen, Y.: Semantic segmentation in-the-wild without seeing any segmentation examples (2021)
72. Zaheer, M., Kottur, S., Ravanbakhsh, S., Poczos, B., Salakhutdinov, R.R., Smola, A.J.: Deep sets. In: Advances in Neural Information Processing Systems 30 (2017)
73. Zhang, R., et al.: Tip-adapter: training-free clip-adapter for better vision-language modeling. arXiv preprint arXiv:2111.03930 (2021)
74. Zhang, Y., Zhang, C.B., Jiang, P.T., Cheng, M.M., Mao, F.: Personalized image semantic segmentation. In: Proceedings of the IEEE/CVF International Conference on Computer Vision, pp. 10549–10559 (2021)

75. Zheng, Y., Li, Y., Wang, S.: Intention oriented image captions with guiding objects. In: Proceedings of the IEEE/CVF Conference on Computer Vision and Pattern Recognition, pp. 8395–8404 (2019)
76. Zhou, K., Yang, J., Loy, C.C., Liu, Z.: Learning to prompt for vision-language models. arXiv preprint arXiv:2109.01134 (2021)

CLOSE: Curriculum Learning on the Sharing Extent Towards Better One-Shot NAS

Zixuan Zhou[1,3], Xuefei Ning[1,2], Yi Cai[1], Jiashu Han[1], Yiping Deng[2], Yuhan Dong[3], Huazhong Yang[1], and Yu Wang[1(✉)]

[1] Department of Electronic Engineering, Tsinghua University, Beijing, China
zhouzx21@mails.tsinghua.edu.cn, yu-wang@tsinghua.edu.cn
[2] Huawei TCS Lab, Shanghai, China
[3] Tsinghua Shenzhen International Graduate School, Shenzhen, China

Abstract. One-shot Neural Architecture Search (NAS) has been widely used to discover architectures due to its efficiency. However, previous studies reveal that one-shot performance estimations of architectures might not be well correlated with their performances in stand-alone training because of the excessive sharing of operation parameters (i.e., large sharing extent) between architectures. Thus, recent methods construct even more over-parameterized supernets to reduce the sharing extent. But these improved methods introduce a large number of extra parameters and thus cause an undesirable trade-off between the training costs and the ranking quality. To alleviate the above issues, we propose to apply Curriculum Learning On Sharing Extent (CLOSE) to train the supernet both efficiently and effectively. Specifically, we train the supernet with a large sharing extent (an easier curriculum) at the beginning and gradually decrease the sharing extent of the supernet (a harder curriculum). To support this training strategy, we design a novel supernet (CLOSENet) that decouples the parameters from operations to realize a flexible sharing scheme and adjustable sharing extent. Extensive experiments demonstrate that CLOSE can obtain a better ranking quality across different computational budget constraints than other one-shot supernets, and is able to discover superior architectures when combined with various search strategies. Code is available at https://github.com/walkerning/aw_nas.

Keywords: Neural Architecture Search (NAS) · One-shot estimation · Parameter sharing · Curriculum learning · Graph-based encoding

---

Z. Zhou and X. Ning—Equal contribution.

Supplementary Information The online version contains supplementary material available at https://doi.org/10.1007/978-3-031-20044-1_33.

S. Avidan et al. (Eds.): ECCV 2022, LNCS 13680, pp. 578–594, 2022.
https://doi.org/10.1007/978-3-031-20044-1_33

1   Introduction

Neural Architecture Search (NAS) [39] has achieved great success in automatically designing deep neural networks (DNN) in the past few years. However, traditional NAS methods are extremely time-consuming for discovering the optimal architectures, since each architecture sampled in the search process needs to be trained from scratch separately. To alleviate the severe problem of search inefficiency, one-shot NAS proposes to share operation parameters among candidate architectures in a "supernet" and train this supernet to evaluate all sampled candidate architectures [1,4,18,22], which reduces the overall search cost from thousands of GPU days to only a few GPU hours.

Despite its efficiency, previous studies reveal that one-shot NAS suffers from the poor ranking correlation between one-shot estimations and stand-alone estimations, which leads to unfair comparisons between the candidate architectures [19,20,34,35]. Ning et al. [20] give some insights on the failure of one-shot estimations. They conclude that one of the main causes of the poor ranking quality is the **large sharing extent of the supernet**. Several recent studies try to improve one-shot NAS by addressing the large sharing extent issue. Zhao et al. [37] confirm the negative impact of co-adaption of parameters in one supernet. Thus, they split the whole search space into several smaller ones, and train a supernet for each subspace. Su et al. [29] reveal that only one copy of parameters is hard to be maintained for massive architectures. Therefore, they duplicate each parameter of the supernet into several copies, train and then estimate with all the duplicates. However, these improved methods introduce a large number of parameters, which barricades the supernet training. As a result, they have to make a trade-off between the training costs and the ranking quality.

In this paper, we propose to adopt the Curriculum Learning On Sharing Extent (CLOSE) to train the one-shot supernet efficiently and effectively. The underlying intuition behind our method is that training with a large sharing extent can **efficiently bootstrap** the supernet, since the number of parameters to be optimized is much smaller. While in the later training stage, using a supernet with a smaller sharing extent (i.e., a more over-parameterized supernet) can **improve the saturating ranking quality**. Thus, CLOSE uses a relatively large sharing extent in the early training stage of the supernet, then gradually decreases the supernet sharing extent. To support this training strategy, we design a new supernet with an adjustable sharing extent, namely CLOSENet, of which the sharing extent can be flexibly adjusted in the training process. The difference between CLOSENet and the vanilla supernet is that, the construction of vanilla supernets presets the sharing scheme between any architecture pairs, i.e., designates which parameter is shared by which operations in different architectures. In contrast, CLOSENet could flexibly adjust the sharing scheme and extent between architecture pairs, during the training process.

In summary, the contributions of our work are as follows:

1. We propose to apply Curriculum Learning On Sharing Extent (CLOSE) to efficiently and effectively train the one-shot supernet. Specifically, we use a

larger sharing extent in the early stages to accelerate the training process, and gradually switch to smaller ones to boost the saturating performances.

2. To fit the CLOSE strategy, we design a novel supernet (CLOSENet) with an adjustable sharing extent. Different from the vanilla supernet with an unadjustable sharing scheme and sharing extent, CLOSENet can flexibly adapt its sharing scheme and sharing extent during the training process.

3. Extensive experiments on four NAS benchmarks show that CLOSE can achieve a better ranking quality under any computational budgets. When searching for the optimal architectures, CLOSE enables one-shot NAS to find superior architectures compared to existing one-shot NAS methods.

2   Related Work

2.1   One-Shot Neural Architecture Search (NAS)

Neural Architecture Search (NAS) is proposed to find optimal architectures automatically. However, the vanilla sample-based NAS methods [39] are extremely time-consuming. To make it more efficient, Pham et al. [22] propose the parameter sharing technique by constructing an over-parameterized network, namely supernet, to share parameters among the candidate architectures. Based on the parameter sharing technique, various one-shot NAS methods are proposed to efficiently search for optimal architectures by only training "one" supernet. Bender et al. [1] propose to directly train the whole supernet with a path-dropout strategy. Liu et al. [18] develop a differentiable search strategy and use it in conjunction with the parameter sharing technique. Guo et al. [10] propose to separate the stages of supernet training and architecture search.

2.2   Weakness and Improvement of One-Shot NAS

Despite its high efficiency, one-shot NAS suffers from the poor ranking correlation between the architecture performances using one-shot training and stand-alone training. Sciuto et al. [34] discover that the parameter-sharing rankings do not correlate with stand-alone rankings by conducting a series of experiments in a toy search space. Zela et al. [35] confirm the poor ranking quality in a much larger NAS-Bench-1shot1 search space. Luo et al. [19] make a further investigation of the one-shot NAS, and attribute the poor ranking quality to the insufficient and imbalanced training, and the coupling of training and search phases. Ning et al. [20] provide comprehensive evaluations on multiple NAS benchmarks, and conclude three perspectives to improve the ranking quality of the one-shot NAS, i.e., reducing the temporal variance, sampling bias or parameter sharing extent.

Recent studies adopt the direction of sharing extent reduction to improve the one-shot NAS. Ning et al. [20] prune the search space to reduce the number of candidate architectures, and reveal the improvement of the ranking quality in the pruned search space. But the ranking quality in the overall search space is not improved. Zhao et al. [37] propose *Few-shot NAS* to split the whole search space into several subspaces, and train a single supernet for each subspace. Su et al. [29] propose *K-shot NAS* to duplicate each parameter of the supernet into several copies, and estimate architectures' performances with all of them. However, these two methods reduce sharing extent with even more over-parameterized supernets, which brings extra computational costs.

2.3 Curriculum Learning

Bengio et al. [2] first propose curriculum learning (CL) strategy based on the learning process of humans and animals in the real world. The basic idea of the CL strategy is to guide models to learn from easier data (tasks) to harder data (tasks). In the past few years, many studies have successfully applied CL strategy in various applications [7–9, 14, 23, 26, 30], and demonstrated that CL can improve the models' generalization capacity and convergence speed. Besides common CL methods that adjust the data, there exist CL methods that conduct curriculum learning on the model capacity. Karras et al. [15] propose to progressively increase the model capacity of the GAN to speed up and stabilize the training. Soviany et al. [28] propose a general CL framework at the model level that adjusts the curriculum by gradually increasing the model capacity.

2.4 NAS Benchmarks

NAS benchmarks enable researchers to reproduce the NAS experiments easily and compare different NAS methods fairly. NAS-Bench-201 [6] constructs a cell-based NAS search space containing 15625 architectures and provides their complete training information. NAS-Bench-301 [27] uses a surrogate model to predict the performances of approximately 10^{18} architectures in a more generic search space, with the stand-alone performance of 60k landmark architectures. Different from NAS-Bench-201 and NAS-Bench-301 that focus on topological search spaces, NDS [24] provides benchmarks on two non-topological search spaces (e.g., ResNet [11] and ResNeXt [31]).

3 Method

3.1 Motivation and Preliminary Experiments

In one-shot NAS, many operations in different architectures share the same parameter, while their desired parameters are not necessarily the same. The excessive sharing of parameters, i.e., the large sharing extent, has been widely regarded as the most important factor causing the unsatisfying performance estimation [3, 20, 29, 36, 37]. The most recent studies [29, 37] improve the ranking quality by reducing the sharing extent. But their methods cause an inevitable trade-off between the training cost and ranking quality at the same time.

Supernets with larger sharing extents (i.e., more parameters) are easier to train in the early training stage. We verify this statement with an experiment

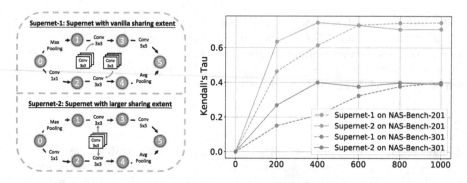

Fig. 1. Comparison of supernets with small and large sharing extent. The left part shows a cell-architecture that uses Supernet-1 (top) and Supernet-2 (bottom). The right part shows the Kendall's Tau of Supernet-1 and Supernet-2 throughout the training process on NAS-Bench-201 and NAS-Bench-301.

on two popular cell-based NAS benchmarks. We construct two supernets with different sharing extents: Supernet-1 (a cell shown in top-left of Fig. 1) is a vanilla supernet adopted by many one-shot NAS methods (e.g., DARTS [18]), in which the compound edges in one cell use different copies of parameters. While Supernet-2 (a cell shown in the bottom-left of Fig. 1) shares only one copy of parameters for all the compound edges in each cell, which leads to a much larger sharing extent than Supernet-1. For example, the parameters of $conv_{3\times3}$ in edge (1,3) and (2,4) are different when using Supernet-1, but the same when using Supernet-2. We train them to convergence and use Kendall's Tau (see Sect. 4.1 for definition), to evaluate the ranking correlation between the estimated performances by the supernets and the groud-truth performances.

Figure 1 (right) shows that, on one hand, using a smaller sharing extent (more parameters, larger supernet capacity) can alleviate the undesired coadaptation between architectures, and has the potential to achieve higher saturating performances. The Kendall's Tau of Supernet-2 is slightly worse than that of Supernet-1 when the supernets are relatively well-trained (epoch 800 to 1000). On the other hand, training the supernet with higher sharing extent than the vanilla one (fewer parameters, smaller supernet capacity) greatly accelerates the training process of parameters, and help the supernet obtain a good ranking quality faster. In the early stage of the training process (epoch 0 to 600), Supernet-2 has a much higher Kendall's Tau than Supernet-1.

Based on the above results and analysis, **a natural idea to achieve a win-win scenario of supernet training efficiency and high ranking quality is to adapt the sharing extent during the supernet training process.** We can draw parallels between this idea and the CL methods on model capacity [15,28] (see Sect. 2.3), as they all progressively increase the capacity of the model or supernet to achieve training speedup and better performances in the mean time. Based on the above idea, we propose to employ Curriculum Learning On the Sharing Extent (CLOSE) of the supernet. And to enable the adaption of the sharing extent during the training process, we design a novel supernet, CLOSENet, whose sharing extent can be easily adjusted.

In the following, we first demonstrate the construction of CLOSENet in Sect. 3.2. Then, in Sect. 3.3, we describe CLOSE with some necessary training techniques to achieve the best ranking quality.

3.2 CLOSENet: A Supernet with an Adjustable Sharing Extent

The design of CLOSENet is illustrated in Fig. 2. The key idea behind CLOSENet is to decouple the parameters from operations to enable flexible sharing scheme and adjustable sharing extent. Specifically, we design the GLobal Operation Weight (GLOW) Block to store the parameters, and design a GATE module for assigning the proper GLOW block to each operation.

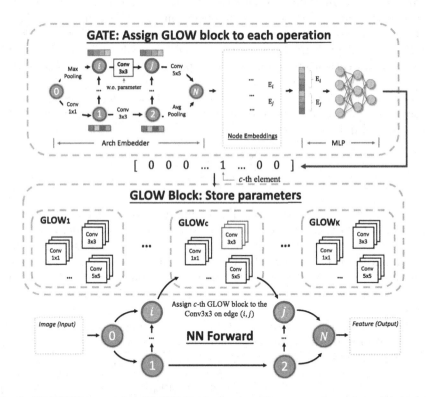

Fig. 2. CLOSENet contains GLOW blocks to store the parameters of candidate operations, and the GATE module to assign GLOW blocks to operations.

For a better understanding, we take the generic cell-based topological search space as an example. We will show in the appendix that CLOSENet can also adapt to other types of search spaces, such as ResNet-like search spaces.

Generic Cell-Based Topological Search Space. In cell-based search space, a complete architecture is stacked by a cell-architecture for multiple times (e.g., 15 times on NAS-Bench-201). A cell-architecture can be represented as a *directed acyclic graph (DAG)*. Each node x^i represents a feature map, while each edge $o^{(i,j)}$ represents an operation that transforms x^i to x^j with the corresponding parameters. For each node j, the feature x_j is defined as:

$$x^j = \sum_{i<j} o^{(i,j)}(x^i, W^{(i,j)}), \tag{1}$$

where $o(x, W)$ denotes that the operation o transforms the feature x with the parameters from W.

GLobal Operation Weight (GLOW) Block. The function of GLOW blocks is to store the parameters of candidate operations, as shown in Fig. 2 (middle). In the forward pass (as shown in the bottom of Fig. 2), the GLOW blocks are assigned to each operation via the GATE module (will be introduced in the following), and each operation can use the parameters from its assigned block to process the input feature map.

Specifically, we denote G_i as the i-th GLOW block, and $c^{(i,j)}$ as the index of the assigned block for the operation in edge (i, j). Then, the computation of the feature x^j in Eq. 1 can be rewritten as:

$$x^j = \sum_{i<j} o^{(i,j)}(x^i, G_{c^{(i,j)}}) \tag{2}$$

GATE Module. We design a GATE module for assigning GLOW blocks to operations. The GATE module consists of an architecture embedder and a MLP module, as shown in Fig. 2 (top). We construct a GCN-based architecture embedder [21], and use it to compute the node embeddings in the architecture. Then, we concatenate the embeddings of the input and output nodes of each operation and feed it into the MLP to get the assignment of the GLOW block.

Specifically, we denote E_i as the embedding of node i, and K as the number of GLOW blocks. For a cell-architecture a with N nodes, we first obtain the node embeddings by the architecture embedder as Eq. 3, and then calculate the probability distribution in edge (i, j) as Eq. 4 and Eq. 5.

$$[E_1, E_2, E_3, ..., E_N] = \text{ArchEmb}(a) \tag{3}$$

$$[\lambda_1^{(i,j)}, \lambda_2^{(i,j)}, \lambda_3^{(i,j)}, ..., \lambda_K^{(i,j)}] = \text{MLP}(concat(E_i, E_j)) \tag{4}$$

$$\text{Pr}(c^{(i,j)} = k) = \frac{exp(\lambda_k^{(i,j)})}{\sum_{k'=1}^{K} exp(\lambda_{k'}^{(i,j)})} \tag{5}$$

To allow the back-propagation of gradients, we apply the reparameterization trick on Eq. 2 and Eq. 5, and rewrite the computation of x^j as:

$$x^j = \sum_{i<j} \sum_{k=1}^{K} h_k^{(i,j)} o^{(i,j)}(x^i, G_k), \tag{6}$$

$$h^{(i,j)} = \arg\max_k (\lambda_k^{(i,j)} + g_k), \tag{7}$$

where $h^{(i,j)}$ is a one-hot vector of dimension K, and g_k are i.i.d samples from Gumbel(0, 1). To make Eq. 7 differentiable, we relax the $\arg\max$ function to a softmax function as:

$$\hat{h}_k^{(i,j)} = \frac{exp((\lambda_k^{(i,j)} + g_k)/\tau)}{\sum_{k'=1}^{K} exp((\lambda_{k'}^{(i,j)} + g_{k'})/\tau)}, \tag{8}$$

where τ is the Gumbel-Softmax temperature. We use Eq. 7 in the forward pass, and use Eq. 8 in the backward pass to allow gradient propagation.

Adjustment of Sharing Extent Denote $E^{(i,j)}$ as the set of cell-architectures that contain the edge from node i to node j. For edge (i,j), we define its sharing extent $s^{(i,j)}$ as the average number of architectures that share one GLOW block in CLOSENet. The sharing extent of the supernet s equals the sum of sharing extent of all the edges:

$$s = \sum_{i,j} s^{(i,j)} = \sum_{i,j} \frac{|E^{(i,j)}|}{K} = \frac{\sum_{i,j} |E^{(i,j)}|}{K}. \tag{9}$$

Therefore, we can naturally adjust the sharing extent of CLOSENet by adding or reducing the GLOW blocks. CLOSENet with more GLOW blocks (larger K) has a smaller sharing extent and vice versa.

Strengths Compared to Vanilla Supernets. The vanilla supernet and its variants (e.g., *K-shot* and *Few-shot* supernets [29,37]) preset the sharing scheme and extent by attaching a fixed set of parameters to each operation. On the contrary, CLOSENet decouples the parameters from the operations and enables **the dynamic decision of sharing scheme** based on a graph-based encoding of architecture operations. Specifically, the vanilla supernet shares parameters according to the position specified by the node indexes, i.e., the operations in the same "position" share the same parameters across different architectures. This sharing scheme is not flexible and can be suboptimal in some cases. For example, as shown in Fig. 3(upper), the 1×1 convolutions on the 0–2 edge share the same parameters between the two architectures, while they should have vastly different optimal parameters. Intuitively, if two operations in two architectures have similar data processing functionality, it might be more reasonable to share their parameters. The design of CLOSENet matches this intuition: The GATE

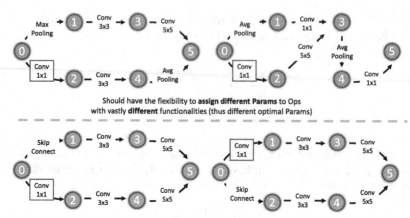

Fig. 3. Two examples that show the strengths of CLOSENet. The sharing scheme of the vanilla supernet, i.e., sharing parameters between operations with the same position indexes, is improper in these two cases. In contrast, CLOSENet designates a more proper sharing pattern between operations according to their graph-based encoding given by GATE.

module learns to pick the right GLOW block for each operation based on the graph-based encoding of all operations and topology in the cell architecture. Instead of presetting the sharing scheme according to the position information, CLOSENet takes a more flexible and reasonable way to dynamically determine which block each operation should use, and thereby designates which operations in different architectures should share their parameters. For example, as shown in Fig. 3(bottom), since the two 1×1 convolutions are equivalent in two isomorphic architectures despite having different position indexes, it is reasonable for them to share parameters. The vanilla supernet uses different parameters for these two convolutions, while CLOSENet assigns the same GLOW block for them.

Moreover, this decoupling enables us to **flexibly adjust the sharing extent** by changing K in Eq. 9. Thus, CLOSENet enables us to apply our curriculum learning-like training strategy. In summary, both the dynamic sharing scheme and the adjustable sharing extent make CLOSENet a more powerful supernet.

3.3 CLOSE: Curriculum Learning on Sharing Extent

We borrow the idea of curriculum learning to design a novel supernet training strategy CLOSE. Specifically, we initialize the CLOSENet with only one GLOW block at the beginning. This large sharing extent helps us to train the supernet much faster. Then, we gradually add GLOW blocks at preset epochs to reduce the sharing extent. In this way, CLOSE not only accelerates the supernet training, but also improves the saturating ranking quality of the supernet.

When switching the curriculum (i.e., increasing the sharing extent), we add a new GLOW block into CLOSENet and a corresponding MLP output unit to the GATE module. How to initialize the newly added parameters is critical to the performance of CLOSENet. Additionally, the regular schedule for the learning rate does not fit for CLOSE (see below). Correspondingly, we propose two techniques, the Weight Inherited Technique and the Schedule Restart Technique.

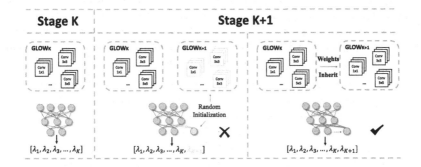

Fig. 4. WIT for GLOW blocks (top) and the MLP output units (bottom).

Weight Inherit Technique (WIT). Instead of randomly initializing the new GLOW block and MLP output unit, we make their weights inherit from those of previous GLOW blocks and MLP output units, as shown in Fig. 4. This helps with the more efficient training of the new GLOW block and MLP unit.

Schedule Restart Technique (SRT). In the training process, the learning rate is reduced gradually to approach the optimal solution. That is to say, it will become quite small after many epochs. However, following this schedule, CLOSE might fail to jump out of the local optimal solution of the preceding curriculum. To overcome this problem, we propose to restart the learning rate and schedule at preset epochs. With SRT, CLOSE can quickly reach the new optimal solution after switching to a new curriculum.

4   Experiments

4.1   Evaluation of Ranking Quality

We evaluate our method on four NAS search spaces, including NAS-Bench-201 [6], NAS-Bench-301 [27], NDS ResNet [24] and NDS ResNeXt-A [24]. The training configurations are shown in the appendix. Following previous studies [20,21], we use two evaluation criteria as follows:

– Kendall's Tau (KD): The relative difference of the number of concordant pairs and discordant pairs, which reflects the overall ranking correlation.

– Precision@topK (P@topK): The proportion of true top-K architectures in the
top-K architectures according to the one-shot estimations, which reflects the
ability of identifying the top-performing architectures.

Comparison with Vanilla One-Shot Baselines. We compare CLOSENet
with vanilla supernets on four NAS benchmarks. As shown in Fig. 5, CLOSENet
achieves a higher KD and P@top5% on all the NAS benchmarks. Moreover, we
can see that throughout the training process, CLOSENet consistently achieves
higher ranking quality, which implies CLOSENet's superiority to the vanilla
supernet under any budget for supernet training.

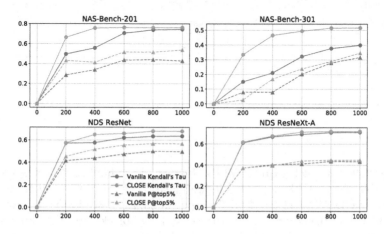

Fig. 5. Comparison of different criteria with the vanilla one-shot supernet on four NAS
benchmarks. X-axis: Training epochs. Y-axis: Evaluation criteria.

Comparison with Improved One-Shot Methods. Figure 6 compares
CLOSE with previous work on improving NAS evaluation strategy, including
EPEE [20], *AngleNet* [13], *K-shot NAS* [29] and *Few-shot NAS* [37]. Results
show that CLOSE reaches SOTA KDs on all the three datasets of NAS-Bench-
201.

4.2   Evaluation of Search Performance

We combine CLOSE with various search strategies, including DARTS [18],
SNAS [32] and CARS [33], to evaluate whether it improves the search performance.

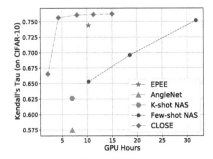

Method	Kendall's Tau	
	C-100	IN-16
EPEE	0.5600	0.5400
AngleNet	0.6040	0.5445
K-shot NAS	0.6122	0.5633
CLOSE	**0.6693**	**0.6632**

Fig. 6. Comparison with previous improved methods on NAS-Bench-201 for three datasets, i.e. CIFAR-10, CIFAR-100 (C-10) and ImageNet-16 (IN-16).

Results. We run DARTS and SNAS search with CLOSE on NAS-Bench-301 and show the derived architecture accuracy in Fig. 7. We can see that CLOSE benefits the search process significantly. In particular, it can alleviate the collapse issue of DARTS caused by the improper preference of parameter-free operations (i.e., *skip_connect*) in early training stages [12,16], as it provides a less biased estimation (see Appendix 1.1).

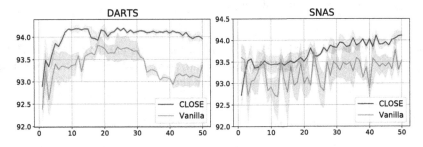

Fig. 7. Evaluation of CLOSE with two search strategies in the NAS-Bench-301 search space. X-axis: Training epochs. Y-axis: Test accuracy.

We run CARS with CLOSE in the DARTS search space and Table 1 shows the performances of the discovered architecture. As can be seen, CLOSE achieves a competitive test error of 2.72% in CIFAR-10. And when transferred to the ImageNet, the found architecture achieves a low test error of 24.7%.

4.3 Ablation Studies

Effect of Number of Curriculums. We conduct an ablation study on the number of curriculums in two different types of search spaces, NAS-Bench-301 (a topological search space) and NDS ResNet (a non-topological search space). Results in Fig. 8 show that, in most cases, using more curriculums can improve the ranking quality of CLOSE.

Table 1. Comparison of architecture performances on CIFAR-10 and ImageNet.

Method	CIFAR-10			ImageNet	
	Top-1 Error (%)	Param (M)	Search Cost (GPU days)	Top-1 Error (%)	Param (M)
NASNet-A [40]	2.65	3.3	2000	26.0	5.3
AmoebaNet-B [25]	2.55	2.8	3150	26.0	5.3
PNAS [17]	3.41	5.1	225	25.8	5.1
ENAS [22]	2.89	4.6	0.5	–	–
DARTS [18]	2.76	3.3	1.5	26.9	4.9
SNAS [32]	2.85	2.8	1.5	27.3	4.3
BayesNAS [38]	2.81	3.4	0.2	26.5	3.9
GDAS [5]	2.82	2.5	0.17	27.5	4.4
CLOSE (Ours)	2.72 ± 0.04	4.1	0.6	24.7	4.8

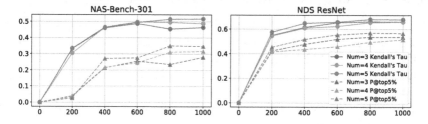

Fig. 8. The ranking quality of CLOSE with different numbers of curriculums. X-axis: Training epochs. Y-axis: Values of criteria.

Effect of WIT and SRT. Table 2 demonstrates the effect of WIT and SRT adopted by CLOSE on NAS-Bench-301 and NDS ResNet. Results show that these two techniques are both necessary for the ranking quality of CLOSE.

Table 2. The ranking quality of CLOSE w./w.o. the proposed techniques.

WIT	SRT	NAS-Bench-301		NDS ResNet	
		KD	P@top5%	KD	P@top5%
		0.1104	0.1145	0.6339	0.5387
✓		0.1047	0.1122	0.6550	0.5520
	✓	0.2004	0.1610	0.6448	0.5280
✓	✓	**0.5168**	**0.3470**	**0.6786**	**0.5667**

Effect of GATE Module. We compare the ranking quality of using the GATE module with randomly assigning GLOW blocks to operations. Results in Table 3 reveal that our learnable GATE module plays an essential role in CLOSENet.

Table 3. The ranking quality of CLOSE w./w.o. the GATE module

GATE	NAS-Bench-201		NAS-Bench-301	
	KD	P@top5%	KD	P@top5%
w/o	0.3627	0.2014	0.2236	0.1924
w	**0.7622**	**0.5387**	**0.5168**	**0.3470**

Effect of Gradually Adding GLOW Blocks. To show the benefit of gradually adding GLOW blocks, we conduct two contrast experiments. In the first experiment, we keep a fixed number of GLOW blocks in the training process. Results in Table 4 demonstrate CLOSE performs better than fixing the sharing extent. In the second experiment, we gradually add blocks and stop after adding a certain number of blocks. Results in Table 5 show that the final ranking quality at 1000 epoch will degrade if GLOW blocks are not sufficiently added.

Table 4. The ranking quality of supernets that use a *fixed* number of blocks

Benchmark	Fixed number of blocks				CLOSE
	2	3	4	5	
NB201	0.7320	0.7247	0.7073	–	**0.7622**
NB301	0.4533	0.3427	0.3301	0.3106	**0.5168**

Table 5. The ranking quality of supernets that add *fewer* number of blocks.

Benchmark	Number of added blocks in total				
	1	2	3	4	5
NB201	0.7050	0.7072	0.7502	**0.7622**	–
NB301	0.3990	0.4500	0.4641	0.4879	**0.5168**

5 Conclusions

This work borrows the idea of curriculum learning and proposes a novel training strategy CLOSE to train the NAS supernet both efficiently and effectively. Specifically, CLOSE adopts a curriculum learning-like schedule on the parameter sharing extent of supernets. To support this strategy, we design a novel one-shot supernet, namely CLOSENet, of which the sharing extent can be flexibly adjusted and the sharing scheme is decided based on a graph-based encoding. Extensive experiments demonstrate that equipped with CLOSENet, our proposed method CLOSE reaches a SOTA ranking quality on four NAS benchmarks. When searching in large search spaces, CLOSE can help to discover architectures with superior performances.

Acknowledgments. This work was supported by National Natural Science Foundation of China (No. U19B2019, 61832007), National Key Research and Development Program of China (No. 2019YFF0301500), Tsinghua EE Xilinx AI Research Fund, Beijing National Research Center for Information Science and Technology (BNRist), and Beijing Innovation Center for Future Chips.

References

1. Bender, G., Kindermans, P.J., Zoph, B., Vasudevan, V., Le, Q.: Understanding and simplifying one-shot architecture search. In: International Conference on Machine Learning (ICML), pp. 550–559. PMLR (2018)
2. Bengio, Y., Louradour, J., Collobert, R., Weston, J.: Curriculum learning. In: International Conference on Machine Learning (ICML), pp. 41–48 (2009)
3. Benyahia, Y., et al.: Overcoming multi-model forgetting. In: International Conference on Machine Learning (ICML), pp. 594–603. PMLR (2019)
4. Brock, A., Lim, T., Ritchie, J.M., Weston, N.: Smash: one-shot model architecture search through hypernetworks. In: International Conference on Learning Representations (ICLR) (2018)
5. Dong, X., Yang, Y.: Searching for a robust neural architecture in four GPU hours. In: IEEE Conference on Computer Vision and Pattern Recognition (CVPR), pp. 1761–1770 (2019)
6. Dong, X., Yang, Y.: NAS-bench-201: Extending the scope of reproducible neural architecture search. In: International Conference on Learning Representations (ICLR) (2020)
7. Gong, C., Yang, J., Tao, D.: Multi-modal curriculum learning over graphs. ACM Trans. Intell. Syst. Technol. (TIST) **10**(4), 1–25 (2019)
8. Guo, S., et al.: CurriculumNet: weakly supervised learning from large-scale web images. In: Ferrari, V., Hebert, M., Sminchisescu, C., Weiss, Y. (eds.) ECCV 2018. LNCS, vol. 11214, pp. 139–154. Springer, Cham (2018). https://doi.org/10.1007/978-3-030-01249-6_9
9. Guo, Y., et al.: Breaking the curse of space explosion: towards efficient NAS with curriculum search. In: International Conference on Machine Learning (ICML), pp. 3822–3831. PMLR (2020)
10. Guo, Z., et al.: Single path one-shot neural architecture search with uniform sampling. In: Vedaldi, A., Bischof, H., Brox, T., Frahm, J.-M. (eds.) ECCV 2020. LNCS, vol. 12361, pp. 544–560. Springer, Cham (2020). https://doi.org/10.1007/978-3-030-58517-4_32
11. He, K., Zhang, X., Ren, S., Sun, J.: Deep residual learning for image recognition. In: IEEE Conference on Computer Vision and Pattern Recognition (CVPR), pp. 770–778 (2016)
12. Hong, W., et al.: Dropnas: grouped operation dropout for differentiable architecture search. In: International Joint Conference on Artificial Intelligence (IJCAI), pp. 2326–2332 (2020)

13. Hu, Y., et al.: Angle-based search space shrinking for neural architecture search. In: Vedaldi, A., Bischof, H., Brox, T., Frahm, J.-M. (eds.) ECCV 2020. LNCS, vol. 12364, pp. 119–134. Springer, Cham (2020). https://doi.org/10.1007/978-3-030-58529-7_8

14. Jiang, L., Meng, D., Mitamura, T., Hauptmann, A.G.: Easy samples first: self-paced reranking for zero-example multimedia search. In: ACM International Multimedia Conference (MM), pp. 547–556 (2014)

15. Karras, T., Aila, T., Laine, S., Lehtinen, J.: Progressive growing of GANs for improved quality, stability, and variation. In: International Conference on Learning Representations (ICLR). OpenReview.net (2018)

16. Liang, H., et al.: Darts+: improved differentiable architecture search with early stopping. arXiv preprint arXiv:1909.06035 (2019)

17. Liu, C., et al.: Progressive neural architecture search. In: Ferrari, V., Hebert, M., Sminchisescu, C., Weiss, Y. (eds.) ECCV 2018. LNCS, vol. 11205, pp. 19–35. Springer, Cham (2018). https://doi.org/10.1007/978-3-030-01246-5_2

18. Liu, H., Simonyan, K., Yang, Y.: Darts: differentiable architecture search. In: International Conference on Learning Representations (ICLR) (2019)

19. Luo, R., Qin, T., Chen, E.: Understanding and improving one-shot neural architecture optimization. CoRR abs/1909.10815 (2019)

20. Ning, X., et al.: Evaluating efficient performance estimators of neural architectures. In: Annual Conference on Neural Information Processing Systems (NIPS) (2021)

21. Ning, X., Zheng, Y., Zhao, T., Wang, Yu., Yang, H.: A generic graph-based neural architecture encoding scheme for predictor-based NAS. In: Vedaldi, A., Bischof, H., Brox, T., Frahm, J.-M. (eds.) ECCV 2020. LNCS, vol. 12358, pp. 189–204. Springer, Cham (2020). https://doi.org/10.1007/978-3-030-58601-0_12

22. Pham, H., Guan, M., Zoph, B., Le, Q., Dean, J.: Efficient neural architecture search via parameters sharing. In: International Conference on Machine Learning (ICML), pp. 4095–4104. PMLR (2018)

23. Platanios, E.A., Stretcu, O., Neubig, G., Póczos, B., Mitchell, T.: Competence-based curriculum learning for neural machine translation. In: Proceedings of the 2019 Conference of the North American Chapter of the Association for Computational Linguistics: Human Language Technologies, Volume 1 (Long and Short Papers), pp. 1162–1172 (2019)

24. Radosavovic, I., Kosaraju, R.P., Girshick, R., He, K., Dollár, P.: Designing network design spaces. In: IEEE Conference on Computer Vision and Pattern Recognition (CVPR), pp. 10428–10436 (2020)

25. Real, E., Aggarwal, A., Huang, Y., Le, Q.V.: Regularized evolution for image classifier architecture search. In: AAAI Conference on Artificial Intelligence, vol. 33, pp. 4780–4789 (2019)

26. Ren, Z., Dong, D., Li, H., Chen, C.: Self-paced prioritized curriculum learning with coverage penalty in deep reinforcement learning. IEEE Trans. Neural Networks Learn. Syst. 29(6), 2216–2226 (2018)

27. Siems, J., Zimmer, L., Zela, A., Lukasik, J., Keuper, M., Hutter, F.: NAS-bench-301 and the case for surrogate benchmarks for neural architecture search. arXiv preprint arXiv:2008.09777 (2020)

28. Soviany, P., Ionescu, R.T., Rota, P., Sebe, N.: Curriculum learning: a survey. Int. J. Comput. Vis. (IJCV) 130, 1526–1565 (2022)

29. Su, X., et al.: K-shot NAS: learnable weight-sharing for NAS with k-shot supernets. In: International Conference on Machine Learning (ICML), pp. 9880–9890. PMLR (2021)

30. Tay, Y., et al.: Simple and effective curriculum pointer-generator networks for reading comprehension over long narratives. In: Proceedings of the 57th Annual Meeting of the Association for Computational Linguistics, pp. 4922–4931 (2019)
31. Xie, S., Girshick, R., Dollár, P., Tu, Z., He, K.: Aggregated residual transformations for deep neural networks. In: IEEE Conference on Computer Vision and Pattern Recognition (CVPR), pp. 1492–1500 (2017)
32. Xie, S., Zheng, H., Liu, C., Lin, L.: SNAS: stochastic neural architecture search. In: International Conference on Learning Representations (ICLR) (2019)
33. Yang, Z., et al.: Cars: continuous evolution for efficient neural architecture search. In: IEEE Conference on Computer Vision and Pattern Recognition (CVPR), pp. 1829–1838 (2020)
34. Yu, K., Sciuto, C., Jaggi, M., Musat, C., Salzmann, M.: Evaluating the search phase of neural architecture search. In: International Conference on Learning Representations (ICLR) (2020)
35. Zela, A., Siems, J., Hutter, F.: NAS-bench-1shot1: benchmarking and dissecting one-shot neural architecture search. In: International Conference on Learning Representations (ICLR) (2019)
36. Zhang, M., Li, H., Pan, S., Chang, X., Su, S.: Overcoming multi-model forgetting in one-shot NAS with diversity maximization. In: IEEE Conference on Computer Vision and Pattern Recognition (CVPR), pp. 7809–7818 (2020)
37. Zhao, Y., Wang, L., Tian, Y., Fonseca, R., Guo, T.: Few-shot neural architecture search. In: International Conference on Machine Learning (ICML), pp. 12707–12718. PMLR (2021)
38. Zhou, H., Yang, M., Wang, J., Pan, W.: BayesNAS: a Bayesian approach for neural architecture search. In: International Conference on Machine Learning (ICML), pp. 7603–7613. PMLR (2019)
39. Zoph, B., Le, Q.V.: Neural architecture search with reinforcement learning. In: International Conference on Learning Representations (ICLR) (2017)
40. Zoph, B., Vasudevan, V., Shlens, J., Le, Q.V.: Learning transferable architectures for scalable image recognition. In: IEEE Conference on Computer Vision and Pattern Recognition (CVPR), pp. 8697–8710 (2018)

Streamable Neural Fields

Junwoo Cho[1], Seungtae Nam[1], Daniel Rho[1], Jong Hwan Ko[1,2],
and Eunbyung Park[1,2(✉)]

[1] Department of Artificial Intelligence, Sungkyunkwan University,
Suwon, South Korea
{jwcho000,stnamjef,daniel231,jhko,epark}@skku.edu
[2] Department of Electrical and Computer Engineering, Sungkyunkwan University,
Suwon, South Korea

Abstract. Neural fields have emerged as a new data representation
paradigm and have shown remarkable success in various signal representations. Since they preserve signals in their network parameters, the
data transfer by sending and receiving the entire model parameters prevents this emerging technology from being used in many practical scenarios. We propose streamable neural fields, a single model that consists
of executable sub-networks of various widths. The proposed architectural and training techniques enable a single network to be streamable
over time and reconstruct different qualities and parts of signals. For
example, a smaller sub-network produces smooth and low-frequency signals, while a larger sub-network can represent fine details. Experimental
results have shown the effectiveness of our method in various domains,
such as 2D images, videos, and 3D signed distance functions. Finally, we
demonstrate that our proposed method improves training stability, by
exploiting parameter sharing. Our code is available at https://github.
com/jwcho5576/streamable_nf.

1 Introduction

Neural fields [47] have emerged as a powerful representation of real-world signals. It uses a multilayer perceptron (MLP) that takes inputs as the spatial
or temporal coordinates and produces signal values in arbitrary resolutions.
Thanks to recent advances such as input feature encoding [26,42,53] and periodic activation functions [37], it can faithfully reconstruct complex and high-
frequency signals. It has achieved great success in various signal representations
such as images [4,20,27], 3D shapes [2,5,7,11,15,25,29,36], and novel view synthesis [3,18,21,26,28,34,51].

There are still many challenges that prevent this emerging technique from
being used in practical scenarios. In neural fields, the network itself is a data

---

J. Cho and S. Nam—Equal contribution, alphabetically ordered.

Supplementary Information The online version contains supplementary material
available at https://doi.org/10.1007/978-3-031-20044-1_34.

S. Avidan et al. (Eds.): ECCV 2022, LNCS 13680, pp. 595–612, 2022.
https://doi.org/10.1007/978-3-031-20044-1_34

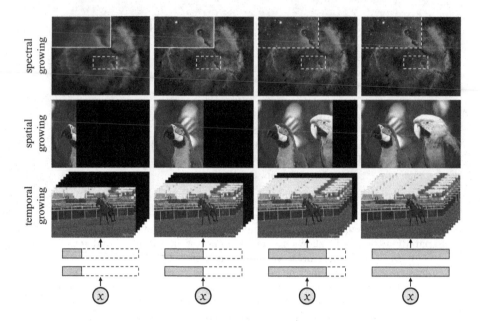

Fig. 1. Representing a varying signal with streamable neural fields. It is a *single* neural network executable at different widths that can reconstruct varying signal domains. *Spectral growing*: larger network reconstructs more high frequency details. *Spatial growing*: larger network reconstructs more pixel locations. *Temporal growing*: larger network reconstructs more video frames.

representation [6] (the signals are stored as the parameters of neural networks), and the signal transmissions are done by sending and receiving the entire model parameters. Thus, finding the optimal model size is crucial for lower latency and higher throughput. A naive approach would train different size networks multiple times by increasing depths and widths. However, it would not be an affordable solution since training even a single network takes a long time to converge for deep neural networks. It is tempting to predetermine various network configurations (e.g., widths and depths) for different sizes and types of signals, yet it is not a feasible solution either since the required size of the parameters is determined by the complexity of the signals, not the size or type of the data.

In addition, the raw signals often need to be transmitted in different resolutions or qualities. For example, in media streaming services, users want to receive various quality signals according to their circumstances, e.g., high-resolution videos at home and lower quality on mobile devices. Real-time encoding on-demand is not viable since it requires a long latency gradient descent on deep neural networks. As an alternative approach, we could locally store multiple sizes of the networks representing different qualities of the signals beforehand. However, it is a waste of storage space and not an acceptable solution given the exponential growth of media data.

Unlike most of the standard compression algorithms, such as JPEG [30] and MPEG [10] which are designed to be easily broken down into smaller pieces for potential use cases, including streaming service or partial reconstruction in poor network connections, a neural field cannot be decoupled into meaningful chunks. All the weight parameters are highly intertwined, and missing a small part of them would result in catastrophic failures for signal reconstruction.

We propose *streamable neural fields* to overcome the issues mentioned above. We suggest training techniques and architectural designs that enable a *single* trained network to be separated into executable sub-networks of various widths. With *single training* procedure, the proposed algorithm can generate parameters of a single network that are *streamable over time* and capable of reconstructing *various qualities of signals* (Fig. 1). Each sub-network is in charge of representing some portion of the signals. For example, a small sub-network can only generate the signals in specific quality or a specific temporal (or spatial) range of the signals. A wider network that subsumes narrower sub-networks can represent the additional signals that are not encoded in the narrower sub-networks. By streaming the network parameters (from narrower to wider sub-networks), the signals will be progressively reconstructed in visual quality and temporal (or spatial) orders, which is desirable in many useful scenarios.

In sum, we present a single neural network that can represent multiple visual qualities and spatial (or temporal) ranges and decode signals in a streamline. The proposed network architecture and training strategy maximize the use of learned partial signals preserved in the small sub-networks. The larger networks explicitly make use of them, resulting in a more stable training procedure, improving reconstruction performance, and increasing parameter efficiency. We show the proposed method's effectiveness in various signals, including images, videos, and 3D shapes.

2 Related Work

Neural Fields and Spectral Bias. Neural fields, also known as coordinate-based neural representations or implicit neural representations have shown great success in representing natural signals such as images [4,20,27], videos [17,37], audios [37], 3D shapes [2,5,7,11,15,25,29,36] and view synthesis [3,18,21, 26,28,34,51]. They struggled to represent high-frequency details due to low-dimensional inputs and spectral bias in training procedure [31,42]. Fourier feature encodings [26,42] and periodic non-linear activation function [37] enabled networks to represent fine details and have been successful. Although spectral bias is an unpleasant training behavior in many practical tasks, our work exploits this phenomenon to implement a neural field that can decode signals in various qualities with a single neural network.

Learning Decomposed Signals. Several studies on neural fields represent spatially partitioned signals using voxel grids [9,38,48], latent codes [4,23], and a group of neural networks [32]. Voxel-based approaches [9,38,48] directly bake a radiance field into the feature grids. The feature grids can be transmitted in a streamline, but their size is very large, which is unfavorable in streaming and compression. Another line of works [4,23] divides an image into tiles and encodes them as latent vectors. While the latent vectors are much smaller than the feature grids, a decoder is required on the client-side. Our method allows the network parameters to be streamed instead of the latent vectors and does not require an additional decoder to reconstruct the signal. KiloNeRF [32] subdivides a single network into numerous tiny ones. Similar to our method, the parameters are streamable. However, thousands of networks should be trained independently. In addition, we also found that learning a spatially partitioned scene with independent models yields line artifacts (Fig. 10), while our method seamlessly reconstructed the scenes.

On the other branch of work, a signal is partitioned in a frequency domain and learned hierarchically. Takikawa et al. [39] proposed to learn multiresolution codebooks similar to [40], allowing variable bitrate streaming. In a similar spirit to ours, recent works [16,19,35] have suggested a single network representing a signal with various bandwidths. Input layers are laterally connected to each intermediate layer, and the intermediate [16] (or additional output [19,35]) layers reconstruct band-limited signals. To constrain the bandwidth, [19,35] initialize and fix the parameters of each input layer such that they are uniformly distributed over a certain frequency range. Progressive implicit networks (PINs) [16] sort sampled frequencies in ascending order, divide them into subsets, and use each subset as the Fourier encoding [26,42] frequencies. Although these works [16,19,35] share some similarities with our method, there are significant differences. First, we exploit the spectral bias to learn the optimal frequency bandwidths given the limited network capacity while [19,35] manually constrains the bandwidths, which may result in inefficient use of the network capacity. Second, our method is agnostic to input encoding methods while [16] designed customized algorithms for particular input encoding methods. Finally, we also proposed to grow widths of networks instead of depths [16,19,35].

Dynamic Neural Networks. Unlike traditional neural networks with static architecture and size, dynamic neural networks can expand or shrink in size on the fly during training and inference. They can adapt to various computing environments and achieve a trade-off between efficiency and accuracy. One branch of dynamic neural networks related to our work is slimmable neural networks (SNN) [49,50], which dynamically expands the channel width of convolution filters during training. In SNN, without re-training each different network architecture, the model is executable for multiple predefined widths. Each trained sub-network has similar or better performance than individually trained models with the help of knowledge distillation [14] and parameter sharing.

In the field of lifelong learning (or incremental learning) [43], neural networks learn from a sequence of multiple tasks. Progressive neural network [33] dynamically expands while transferring knowledge from prior tasks to new ones to handle more tasks and overcome catastrophic forgetting behavior [12,22]. In this work, we focus on representing signals and propose a neural field that can dynamically grow network size to represent a higher-quality or broader range of signals while preserving the representations from small sub-networks.

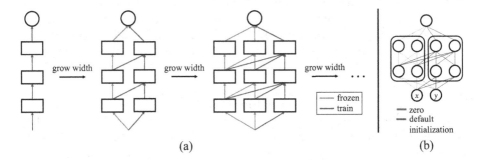

(a)   (b)

Fig. 2. (a) Network architecture and training procedure of streamable neural fields. When training for each sub-network is done, it can grow the width into arbitrary size. To keep the output of each trained sub-network, previous weights are kept frozen. (b) Parameter initialization for newly added weights. Leaving the lateral connections (red) to default initialization, the remaining weights (green) are initialized to zero for fast convergence. We used SIREN initialization [37] as default. (Color figure online)

3 Streamable Neural Fields

This section explains training techniques and architectural designs of *streamable neural fields* that consist of executable sub-networks of various widths. Once training is completed, a single network can represent signals in various qualities without re-training (spectral growing). Narrower sub-networks preserve low-frequency signals and wider sub-networks contain high-frequency details. This functionality can support many practical applications by selecting a width to offer requested quality signals on demand. In general, there is no straightforward solution to manually divide weights to achieve this goal since the network parameters are highly entangled.

In addition, streamable neural fields also support spatial and temporal growing. Each runnable sub-network can only represent a specific part of the entire signal. As depicted in Fig. 1, one would expect to send or receive a video sequentially over time, and one would prefer to receive a small part of a large image. It would be beneficial for transmitting large-size signals, such as high-resolution images or videos.

3.1 Network Architecture and Progressive Training

Our network architecture and training procedure are illustrated in Fig. 2 (a). The model starts training with a small and narrow MLP to predict a target signal. Once converged, it grows its width by arbitrary size. Similar to progressive neural networks architecture [33], we remove the weights that connect from newly added hidden units to the previous units, which prevents the added units from affecting the small network's output. We also freeze weights in the small network and only update the newly added network parameters. This progressive training strategy encourages the large network to use the knowledge learned by the previous small network and only learn the residual signals that the small network cannot capture. We keep iterating this process until the desired signal quality or spatial/temporal size is fulfilled.

Progressive Training vs Slimmable Training. Our general purpose is to create a single network that is executable at various widths. We found that the training technique in slimmable networks [50] can also achieve the goal for image and video fitting tasks. Unlike the proposed progressive training, it iterates over predefined widths, takes a sub-network of the corresponding width, and computes loss using the target signal prediction. The gradients of the sub-networks are accumulated until it visits every width, and weights are updated all at once. Experimental results have shown that progressive training outperforms slimmable training in reconstruction quality and convergence speed. One possible explanation is that the target residual signals for wider networks change over the training course, which results in slowing down convergence speed. The details of the progressive and slimmable training algorithms are described in the supplementary materials.

Training Loss. In every task, the training objective is to minimize the mean squared error (MSE) between prediction and ground truth target value:

$$\min_{\theta} \frac{1}{N} \sum_{i=1}^{N} \| f_\theta(x_i) - y_i \|_2^2, \tag{1}$$

where $\{(x_i, y_i)\}_{i=1}^{N}$ is coordinate and corresponding signal value pair, e.g., $x_i \in \mathbb{R}^2$ and $y_i \in \mathbb{R}^3$ for image signals, e.g., RGB colors. f_θ is a neural network parameterized by θ. For spectral growing, the target signal is fixed to y_i. On the other hand, for spatial and temporal growing, we divided the input coordinate and the ground truth signal into the desired size for each network to represent. For example, to only represent a specific part of a certain sub-network, our objective becomes:

$$\min_{\theta} \frac{1}{N} \left(\sum_{i \notin S} \| f_\theta(x_i) \|_2^2 + \sum_{i \in S} \| f_\theta(x_i) - y_i \|_2^2 \right), \tag{2}$$

where S is the set of indices that belongs to the part. We train the network to predict zero for coordinate locations that do not belong to the part of interest. We empirically found that the first term in Eq. 2 is critical to seamlessly reconstructing the entire signals. The network usually predicts garbage outputs for the locations that are not part of training coordinates. It causes severe artifacts in the boundary areas when we stitch different parts of signals.

Initialization. A careful initialization is required when using periodic activation, in order to achieve high performance and convergence speed. Figure 2 (b) shows the suggested initialization scheme. We used the initialization method in SIREN [37] for lateral connections: $w \sim \mathcal{U}(-\sqrt{6/n}, \sqrt{6/n})$, where n is the number of input neurons in a particular layer. It helps the network maintain the activation distribution throughout layers. The remaining weights that connect between newly added neurons are set to zero. In a consequence, only information from the small sub-networks flows at the beginning of the training process. It would encourage the large network to utilize the small networks' knowledge first and keep newly added parameters from learning redundant signals already learned by the small networks. We empirically found that this technique increases network parameter efficiency and improves the convergence speed.

3.2 Spectral Decomposition of Streamable Neural Fields

In this section, we analyze a trained streamable neural field through the lens of spectral bias [31,42]. In neural fields using a simple MLP, the final output is a weighted sum of the previous layer's outputs (assuming no activation functions on the output layer). Any signals can be represented as a sum of different frequency components, and it is theoretically possible to assign values to the weight matrix in the final layer in a way that the final output can progressively represent higher frequency components as increasing the width.

The proposed progressive training scheme does not modify previously learned sub-network outputs, implying that the newly added hidden units only represent the residual signals (Fig. 3) and spectral bias [31,42] claims that the network prioritizes learning low-frequency parts of signals. Therefore, a narrow sub-network trained by the progressive training would represent low-frequency signal, and wide network will preserve high-frequency details. The output value of a fully-connected neural network is a linear combination of the final hidden activation. Our model progressively accumulates the final hidden activation learned by each sub-network. More formally, we can formulate the output layer of an MLP as follows:

$$y = f_\theta(x) = \sum_{j=1}^{d} w_j \phi_j(x) = \underbrace{\sum_{j=1}^{s} w_j \phi_j(x)}_{\text{low frequency reconstruction}} + \underbrace{\sum_{k=s+1}^{d} w_k \phi_k(x)}_{\text{high frequency residual}}, \qquad (3)$$

where $x \in \mathbb{R}^n$ is an n-dimensional coordinate, $y \in \mathbb{R}$ is a single-channel signal, $w \in \mathbb{R}^d$ is a weight vector in the final layer of the network, $\phi_j(x)$ is the j-th hidden unit in the final layer. The right-hand side in Eq. 3 is a decomposition into two partial sums, splitting the summation by index s (width of a sub-network). We assume there is no bias term in the final layer.

We can interpret ϕ_j as basis functions and w_j as coefficients. Unlike well-known basis functions, such as Fourier basis or Chebyshev polynomials, we learn basis functions through progressive training and do not have any constraints e.g., orthogonality and periodicity on a Fourier basis. We solely rely on the inductive bias of MLP architecture and spectral bias of training neural networks to obtain spectral growing neural fields.

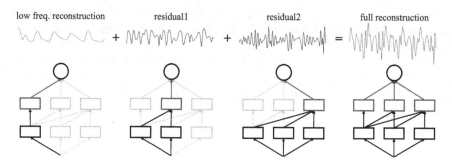

Fig. 3. Spectral growing and residual representation of streamable neural fields. Due to spectral bias, a neural network trained with a standard MSE loss would naturally build up signals in increasing frequency orders. Narrower sub-networks will first learn low-frequency components and wider sub-networks will increasingly represent higher frequency signals as long as the capacity of the network permit. Pruning the partial weights of the final hidden layer outputs residual signals and linearly combining them (red boxes) gives the full reconstruction. (Color figure online)

Quality Control. The central premise of modern image compression algorithms is that the human eye cannot detect high-frequency components. JPEG [30] uses discrete cosine transform (DCT) and quantization matrix to remove the high-frequency parts in an image that are not recognizable to human eyes. It uses a quality factor that determines the compression rate and the reconstruction quality depending on the user's demand. Our streamable neural fields gradually eliminate the high-frequency component as the width decreases. Thus, choosing the width of the network to reconstruct a signal of desired quality is analogous to choosing a quality factor in image compression algorithms.

4   Experiments

We tested our model on various signal reconstruction tasks: 1D sinusoidal functions, 2D images, videos, and 3D signed distance functions (SDF). Our experimental setting is described in Fig. 4. For spectral growing, since there is no ground-truth residual signal, every individually trained model is trained to reconstruct the original target. On the other hand, for spatial/temporal growing, we

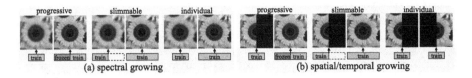

(a) spectral growing   (b) spatial/temporal growing

Fig. 4. Description of our experimental setting. Every shown image above is the desired output signal and the number of parameters for each model is matched for a fair comparison. (a) Spectral growing experiment. The ground truth signal is fixed. (b) Spatial growing experiment. Individually trained models have constant network capacity.

divided the ground truth signal into multiple patches/frames and make individual models that have identical network sizes. Each model is then trained to represent specific image patches or video frames. Note that streaming spatial/temporal growing individual models can also achieve sequential signal transmission, while spectral growing individual models don't.

4.1 1D Sinusoidal Function Reconstruction

In this section, we show the spectral growing of a signal by simple 1D scalar function fitting followed by [31]. This experiment intuitively shows the residual representation of our model. The target function is a mapping $f : [0, 1] \rightarrow \mathbb{R}$, constructed by summation of sinusoids with different frequencies and phase angles. Starting with a width of 10 (the channel size of hidden layers), we gradually increased the size up to 40. Each sub-network was trained for 150 epochs. Figure 5 shows the output signals and residuals learned by each sub-network. As expected, the summation of low-frequency signals from small sub-networks and residual outputs give the same signal learned by larger sub-network.

4.2 Spectral Growing in Images and 3D Shapes

Images. We trained on 24 images in the Kodak dataset for spectral growing. The network size grows three times, using the same ground-truth images for four different executable sub-networks. We compared to baseline method, denoted as *individual*, which trains MLPs with different sizes. The number of hidden units in *individual* models are adjusted to match the total number of parameters in sub-networks of the *streamable* network. *streamable (progressive)* denotes a streamable model trained with the proposed progressive training, and *streamable (slimmable)* for the slimmable training.

Figure 6 (a) shows the averaged PSNR, SSIM [44] and LPIPS [52]. As expected, enlarging the network capacity gives a higher quality image. We found that the progressive training strategy outperforms the slimmable training. We believe the fact that during the slimmable training, large sub-networks affect the outputs of the small sub-networks, and vice versa would hurt the final reconstruction performance and convergence speed. In terms of the final reconstruction quality compared to *individual*, our method performs comparably given the

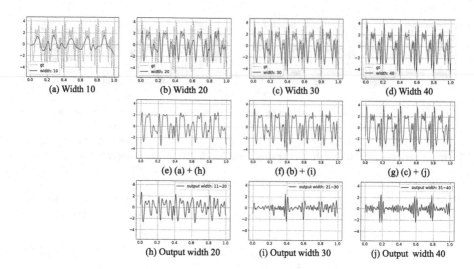

Fig. 5. The results of the 1D sinusoidal function experiment. From (a) to (d): Output of each sub-network. As width grows, the model represents high-frequency details. Yellow dashed lines are the target ground truth signal. From (e) to (g): Summation of each sub-network and residual output. It is identical to the network outputs from (b) to (d). From (h) to (j): Residual output of each sub-network. The newly added weights reconstruct high-frequency residuals. (Color figure online)

same number of parameters. Especially, LPIPS gives a good score to *stream-able (progressive)* model, even though when PSNR and SSIM do not. Since raw Kodak images contain undetectable high frequency components, our model gets rid of these due to spectral bias [31]. Although our model does not *exactly* reconstruct the target signal in terms of PSNR, low LPIPS implies that it sufficiently represents important features in terms of human's visual perception.

Moreover, the *individual* model in spectral growing does not have a stream-able functionality, which means that the *individual* requires much more parameters to represent various qualities, e.g., four *individual* models for four different qualities. Table 1 shows the memory comparison between the *streamable (progressive)* model and the *individual* model. We trained on Kodak image 23 and divided it into 15 sub-networks that reconstruct 15 spectral growing images. The memory requirement of our *streamable (progressive)* model increases much more slowly compared to the *individual*. The difference between the two models becomes larger as more sub-networks are created.

In Fig. 7, we show a qualitative result on a large size image. The original image presents many stars of various sizes and brightness. After training each sub-network to the same target, we obtained residual signals by the method described in Fig. 3. The figure shows that the residual output images only contain small stars not captured by smaller sub-network.

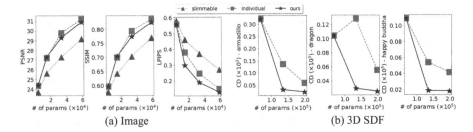

(a) Image   (b) 3D SDF

Fig. 6. Quantitative result of the spectral growing experiment. (a): Averaged PSNR↑, SSIM↑, and LPIPS↓ of 24 Kodak images. (b): Chamfer distance (multiplied by 10^3) of three 3D shapes.

Table 1. The total number of parameters required to cover the indicated range (PSNRs in the first row) of image quality. The second row shows the number of parameters of each *sub-network* in our model to reconstruct the PSNR range above, and the rightmost value (195K) subsumes all the numbers left. The values in the third row are obtained by accumulating the number of parameters of *individual* models.

PSNR	25.2	$25.2 \sim 27.8$	$25.2 \sim 29.8$	...	$25.2 \sim 38.5$	$25.2 \sim 39.2$	$25.2 \sim 39.8$
Ours	1.8K	5.2K	10.2K	...	148K	171K	195K
Individual	1.8K	6.5K$_{(\times 1.25)}$	15.0K$_{(\times 1.47)}$	...	418K$_{(\times 2.82)}$	499K$_{(\times 2.92)}$	590K$_{(\times 3.03)}$

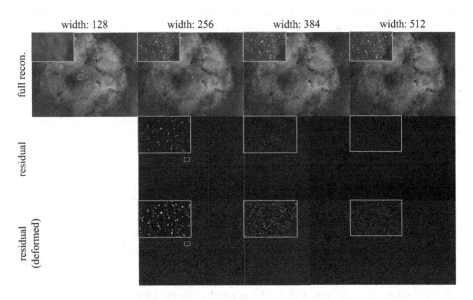

Fig. 7. Visualizing residual representation of elephant's trunk nebula image (2560 × 1984). As the network enlarges, darker stars gradually appear. The third row shows the deformed residual images for better visualization. Note that all images shown are the output of a single network.

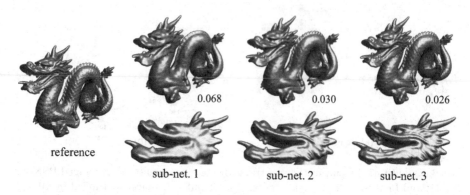

Fig. 8. Spectral growing of *Dragon* in Stanford 3D scanning repository. Indicated numbers are chamfer distances against the reference shape.

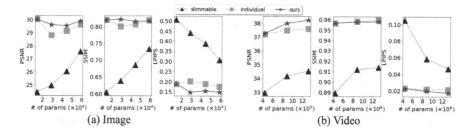

Fig. 9. (a) Averaged performance of spatial growing of 8 Kodak images. (b) Averaged performance of temporal growing of 7 UVG videos.

3D Shapes. We also tested our method on 3D shapes represented by SDF. We trained on *Armadillo*, *Dragon*, and *Happy Buddha* from the Stanford 3D Scanning Repository. Following [29], we used bidirectional chamfer distance (CD) against the true shape to quantitatively compare different approaches. The results are shown in Fig. 6 (b) and the qualitative result of the *Dragon* shape is illustrated in Fig. 8.

Again, our single network outputs the same shape in various qualities. Despite many hyperparameter searches, the slimmable training approach did not converge on 3D shapes, so we only compared against the *individual*. Even with the streamable functionality in our method, *streamable (progressive)* shows better reconstruction quality given the same number of parameters. Again, *individual* needs multiple independent models to support various quality outputs, which requires enormous storage space for 3D shapes.

In 3D shapes, training vanilla SIREN [37] was very unstable, so some models fail on reconstruction (see Fig. 6 (b) second column blue line). However, our streamable model shows a steady performance improvement when the model capacity increases. We provide an analysis of this phenomenon in terms of stable training behavior of the proposed progressive training (Sect. 4.4).

4.3 Spatial and Temporal Growing

Images. We trained on 8 images in the Kodak dataset for spatial growing. The network size grows three times and a total of four sub-networks represent spatially growing (horizontal direction) images. We compared to baselines described in Fig. 4 (b). For a fair comparison, model prediction is compared against the signal domain that contains the desired signal in the evaluation stage. The quantitative result in Fig. 9 (a) shows that *streamable (progressive)* models have higher representation power compared to the *individual* models. The performance drop along with the width increase implies that high-frequency details are concentrated in the center of the images. Though the *streamable (slimmable)* model falls behind, due to the sandwich rule [49], the representation power keeps increasing as the width grows.

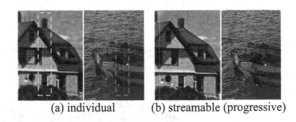

(a) individual   (b) streamable (progressive)

Fig. 10. Individually trained model shows line artifacts while our model doesn't in spatial growing of images.

We also found that training independent models to spatially separate scenes yields line artifacts between each patch (Fig. 10 (a)) because the entire scene is not taken into account during the training of each individual network. Since these line artifacts are easily recognized, we believe our method has advantages over using individual models when learning spatially decomposed signals (Fig. 10 (b)). Seamlessly combining individually trained scenes is an active research area, and the recent Block-NeRF [41] had to use additional alignment techniques, e.g., inverse distance weighting interpolation.

Videos. We trained on 7 videos in UVG [24] dataset, resized to 480×270 for temporal growing. The network size grows two times and a total of three sub-networks represent temporally growing videos. Each sub-network reconstructs 8 frames, giving the largest network a total of 24 frames to reconstruct. We compared to baselines described in Fig. 4 (b). As shown in Fig. 9, for the same network capacity, *streamable (progressive)* models outperform the *individual* models.

4.4 Our Model Stabilizes the Training

As shown in Fig. 11, the PSNR curves of *individual* models drop significantly and then rebound during training, whereas those of *streamable (progressive)* models show no abrupt changes. We reconstructed the predicted RGB values of the largest *individual* model at the drop points and found severe artifacts on the images (Fig. 11 right). This phenomenon is not limited to the example provided. When we trained the *individual* models on 24 Kodak images, we found 19 similar cases. This suggests that the *streamable (progressive)* model can maintain the training process stable.

Since high-frequency components are sensitive to the perturbations in network parameters [31], fine-tuning the weights is required for a neural network to represent the high-frequency parts. Updating the entire parameters will significantly change the output signal and the loss value. This might be one reason why the individual model's PSNR curve fluctuates a lot. On the other hand, our progressive training method freezes pre-trained sub-network parameters and only updates the newly added weights. Updating partial weights results in small changes in the network output and encourages stable fine-tuning.

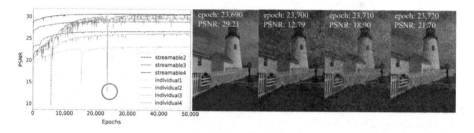

Fig. 11. PSNR curve and intermediate reconstructions of *individual4* model during training. The *individual* model's PSNR curve (blue) fluctuates during the entire training process while *streamable (progressive)* (red) doesn't. (Color figure online)

5   Limitation and Discussion

Visiting each width during the training stage is unavoidable if a neural network is to be runnable at varied architectures, and it requires a demanding training time. There are several suggested parameter initialization methods on dynamically growing networks for fast convergence [8,45,46]. Further study on optimization and initialization techniques to improve the training dynamics can make our model more applicable for real-world tasks.

Conventional MLPs used in neural fields are black box models so that we cannot analyze how exactly the model reconstructs the desired signal. The predicted signal of our model can be decoupled into partial reconstructions and the additive sense of predicting the output can give interpretability to the general machine learning and deep learning models [1,13]. We believe our work can also bring insight into the interpretability of neural fields and encourage further research on neural field architectures in the community.

6   Conclusion

We showed the possibility of decoding neural signal representation in a streamline. Our model is applicable to various signals such as images, videos, and 3D shapes which can vary in quality or grow spatially/temporally. Compared with individually trained models, the streamable neural field has similar or higher representation power, along with efficient memory cost by utilizing parameter sharing. Without re-training multiple individual models, our model can dynamically grow during the training procedure to search for optimal network capacity. We have also shown that partially training the model weights stabilizes training.

Acknowledgements. This research was supported by the Ministry of Science and ICT (MSIT) of Korea, under the National Research Foundation (NRF) grant (2021R1F1A1061259), Institute of Information and Communication Technology Planning Evaluation (IITP) grants for the AI Graduate School program (IITP-2019-0-00421) and Artificial Intelligence Innovation Hub program (2021-0-02068).

References

1. Agarwal, R., et al.: Neural additive models: Interpretable machine learning with neural nets. In: Advances in Neural Information Processing Systems, vol. 34, pp. 4699–4711. Curran Associates, Inc. (2021)
2. Chabra, R., et al.: Deep local shapes: learning local SDF priors for detailed 3D reconstruction. In: Vedaldi, A., Bischof, H., Brox, T., Frahm, J.-M. (eds.) ECCV 2020. LNCS, vol. 12374, pp. 608–625. Springer, Cham (2020). https://doi.org/10.1007/978-3-030-58526-6_36
3. Chan, E.R., Monteiro, M., Kellnhofer, P., Wu, J., Wetzstein, G.: Pi-GAN: periodic implicit generative adversarial networks for 3D-aware image synthesis. In: Proceedings of the IEEE/CVF Conference on Computer Vision and Pattern Recognition (CVPR), pp. 5799–5809, June 2021
4. Chen, Y., Liu, S., Wang, X.: Learning continuous image representation with local implicit image function. In: Proceedings of the IEEE/CVF Conference on Computer Vision and Pattern Recognition (CVPR), pp. 8628–8638, June 2021
5. Chen, Z., Zhang, H.: Learning implicit fields for generative shape modeling. In: Proceedings of the IEEE/CVF Conference on Computer Vision and Pattern Recognition (CVPR), pp. 5939–5948, June 2019
6. Dupont, E., Kim, H., Eslami, S.M.A., Rezende, D.J., Rosenbaum, D.: From data to functa: Your data point is a function and you should treat it like one. arXiv preprint arXiv:2201.12204 (2022)
7. Erler, P., Guerrero, P., Ohrhallinger, S., Mitra, N.J., Wimmer, M.: POINTS2SURF learning implicit surfaces from point clouds. In: Vedaldi, A., Bischof, H., Brox, T., Frahm, J.-M. (eds.) ECCV 2020. LNCS, vol. 12350, pp. 108–124. Springer, Cham (2020). https://doi.org/10.1007/978-3-030-58558-7_7
8. Evci, U., van Merrienboer, B., Unterthiner, T., Pedregosa, F., Vladymyrov, M.: Gradmax: growing neural networks using gradient information. In: International Conference on Learning Representations (2022)
9. Fridovich-Keil, S., Yu, A., Tancik, M., Chen, Q., Recht, B., Kanazawa, A.: Plenoxels: radiance fields without neural networks. In: Proceedings of the IEEE/CVF

Conference on Computer Vision and Pattern Recognition (CVPR), pp. 5501–5510, June 2022

10. Gall, D.L.: MPEG: a video compression standard for multimedia applications. Commun. ACM **34**(4), 46–58 (1991)

11. Genova, K., Cole, F., Vlasic, D., Sarna, A., Freeman, W.T., Funkhouser, T.: Learning shape templates with structured implicit functions. In: Proceedings of the IEEE/CVF International Conference on Computer Vision (ICCV), pp. 7154–7164 (October 2019)

12. Goodfellow, I.J., Mirza, M., Xiao, D., Courville, A., Bengio, Y.: An empirical investigation of catastrophic forgetting in gradient-based neural networks. arXiv preprint arXiv:1312.6211 (2013)

13. Hastie, T.J.: Generalized additive models. In: Statistical models in S, pp. 249–307. Routledge (2017)

14. Hinton, G., Vinyals, O., Dean, J., et al.: Distilling the knowledge in a neural network. arXiv preprint arXiv:1503.02531 (2015)

15. Jiang, C.M., Sud, A., Makadia, A., Huang, J., Niessner, M., Funkhouser, T.: Local implicit grid representations for 3D scenes. In: Proceedings of the IEEE/CVF Conference on Computer Vision and Pattern Recognition (CVPR), pp. 6001–6010, June 2020

16. Landgraf, Z., Hornung, A.S., Cabral, R.S.: Pins: progressive implicit networks for multi-scale neural representations. In: Proceedings of the International Conference on Machine Learning (ICML), pp. 11969–11984. PMLR, July 2022

17. Li, Z., Niklaus, S., Snavely, N., Wang, O.: Neural scene flow fields for space-time view synthesis of dynamic scenes. In: Proceedings of the IEEE/CVF Conference on Computer Vision and Pattern Recognition (CVPR), pp. 6498–6508, June 2021

18. Lin, C.H., Ma, W.C., Torralba, A., Lucey, S.: Barf: Bundle-adjusting neural radiance fields. In: Proceedings of the IEEE/CVF International Conference on Computer Vision (ICCV), pp. 5741–5751, October 2021

19. Lindell, D.B., Van Veen, D., Park, J.J., Wetzstein, G.: Bacon: band-limited coordinate networks for multiscale scene representation. In: Proceedings of the IEEE/CVF Conference on Computer Vision and Pattern Recognition (CVPR), pp. 16252–16262, June 2022

20. Martel, J.N.P., Lindell, D.B., Lin, C.Z., Chan, E.R., Monteiro, M., Wetzstein, G.: Acorn: adaptive coordinate networks for neural scene representation. ACM Trans. Graph. **40**(4), 58:1-58:13 (2021)

21. Martin-Brualla, R., Radwan, N., Sajjadi, M.S.M., Barron, J.T., Dosovitskiy, A., Duckworth, D.: Nerf in the wild: neural radiance fields for unconstrained photo collections. In: Proceedings of the IEEE/CVF Conference on Computer Vision and Pattern Recognition (CVPR), pp. 7210–7219, June 2021

22. McCloskey, M., Cohen, N.J.: Catastrophic interference in connectionist networks: the sequential learning problem. Psychol. Learn. Motivat. **24**, 109–165 (1989)

23. Mehta, I., Gharbi, M., Barnes, C., Shechtman, E., Ramamoorthi, R., Chandraker, M.: Modulated periodic activations for generalizable local functional representations. In: Proceedings of the IEEE/CVF International Conference on Computer Vision (ICCV), pp. 14214–14223, October 2021

24. Mercat, A., Viitanen, M., Vanne, J.: UVG dataset: 50/120fps 4k sequences for video codec analysis and development. In: Proceedings of the 11th ACM Multimedia Systems Conference, MMSys, pp. 297–302. ACM, June 2020

25. Mescheder, L., Oechsle, M., Niemeyer, M., Nowozin, S., Geiger, A.: Occupancy networks: Learning 3d reconstruction in function space. In: Proceedings of the

IEEE/CVF Conference on Computer Vision and Pattern Recognition (CVPR), pp. 4460–4470, June 2019

26. Mildenhall, B., Srinivasan, P.P., Tancik, M., Barron, J.T., Ramamoorthi, R., Ng, R.: NeRF: representing scenes as neural radiance fields for view synthesis. In: Vedaldi, A., Bischof, H., Brox, T., Frahm, J.-M. (eds.) ECCV 2020. LNCS, vol. 12346, pp. 405–421. Springer, Cham (2020). https://doi.org/10.1007/978-3-030-58452-8_24

27. Müller, T., Evans, A., Schied, C., Keller, A.: Instant neural graphics primitives with a multiresolution hash encoding. ACM Trans. Graph. **41**(4), 102:1-102:15 (2022)

28. Niemeyer, M., Geiger, A.: Giraffe: representing scenes as compositional generative neural feature fields. In: Proceedings of the IEEE/CVF Conference on Computer Vision and Pattern Recognition (CVPR), pp. 11453–11464, June 2021

29. Park, J.J., Florence, P., Straub, J., Newcombe, R., Lovegrove, S.: DeepSDF: learning continuous signed distance functions for shape representation. In: Proceedings of the IEEE/CVF Conference on Computer Vision and Pattern Recognition (CVPR), pp. 165–174, June 2019

30. Pennebaker, W.B., Mitchell, J.L.: JPEG: Still Image Data Compression Standard. Springer, New York (1992)

31. Rahaman, N., et al.: On the spectral bias of neural networks. In: Proceedings of the International Conference on Machine Learning (ICML), pp. 5301–5310. PMLR, June 2019

32. Reiser, C., Peng, S., Liao, Y., Geiger, A.: Kilonerf: speeding up neural radiance fields with thousands of tiny MLPS. In: Proceedings of the IEEE/CVF International Conference on Computer Vision (ICCV), pp. 14335–14345, October 2021

33. Rusu, A.A., et al.: Progressive neural networks. arXiv preprint arXiv:1606.04671 (2016)

34. Schwarz, K., Liao, Y., Niemeyer, M., Geiger, A.: GRAF: Ggnerative radiance fields for 3D-aware image synthesis. In: Advances in Neural Information Processing Systems, vol. 33, pp. 20154–20166. Curran Associates, Inc. (2020)

35. Shekarforoush, S., Lindell, D.B., Fleet, D.J., Brubaker, M.A.: Residual multiplicative filter networks for multiscale reconstruction. arXiv preprint arXiv:2206.00746 (2022)

36. Sitzmann, V., Chan, E., Tucker, R., Snavely, N., Wetzstein, G.: MetaSDF: meta-learning signed distance functions. In: Advances in Neural Information Processing Systems, vol. 33, pp. 10136–10147. Curran Associates, Inc. (2020)

37. Sitzmann, V., Martel, J., Bergman, A., Lindell, D., Wetzstein, G.: Implicit neural representations with periodic activation functions. In: Advances in Neural Information Processing Systems, vol. 33, pp. 7462–7473. Curran Associates, Inc. (2020)

38. Sun, C., Sun, M., Chen, H.T.: Direct voxel grid optimization: super-fast convergence for radiance fields reconstruction. In: Proceedings of the IEEE/CVF Conference on Computer Vision and Pattern Recognition (CVPR), pp. 5459–5469, June 2022

39. Takikawa, T., et al.: Variable bitrate neural fields. arXiv preprint arXiv:2206.07707 (2022)

40. Takikawa, T., et al.: Neural geometric level of detail: real-time rendering with implicit 3D shapes. In: Proceedings of the IEEE/CVF Conference on Computer Vision and Pattern Recognition (CVPR), pp. 11358–11367, June 2021

41. Tancik, M., et al.: Block-nerf: Scalable large scene neural view synthesis. In: Proceedings of the IEEE/CVF Conference on Computer Vision and Pattern Recognition (CVPR), pp. 8248–8258, June 2022

42. Tancik, M., et al.: Fourier features let networks learn high frequency functions in low dimensional domains. In: Advances in Neural Information Processing Systems, vol. 33, pp. 7537–7547. Curran Associates, Inc. (2020)

43. Thrun, S.: A lifelong learning perspective for mobile robot control. In: Proceedings of IEEE/RSJ International Conference on Intelligent Robots and Systems, pp. 23–30. IEEE, October 1994

44. Wang, Z., Bovik, A.C., Sheikh, H.R., Simoncelli, E.P.: Image quality assessment: from error visibility to structural similarity. IEEE Trans. Image Process. **13**(4), 600–612 (2004)

45. Wu, L., Liu, B., Stone, P., Liu, Q.: Firefly neural architecture descent: a general approach for growing neural networks. In: Advances in Neural Information Processing Systems, vol. 33, pp. 22373–22383. Curran Associates, Inc. (2020)

46. Wu, L., Wang, D., Liu, Q.: Splitting steepest descent for growing neural architectures. In: Advances in Neural Information Processing Systems, vol. 32, pp. 10655–10665. Curran Associates, Inc. (2019)

47. Xie, Y., et al.: Neural fields in visual computing and beyond. Comput. Graph. Forum **41**(2), 641–676 (2022). https://neuralfields.cs.brown.edu/

48. Yu, A., Li, R., Tancik, M., Li, H., Ng, R., Kanazawa, A.: Plenoctrees for real-time rendering of neural radiance fields. In: Proceedings of the IEEE/CVF International Conference on Computer Vision (ICCV), pp. 5752–5761, October 2021

49. Yu, J., Huang, T.S.: Universally slimmable networks and improved training techniques. In: Proceedings of the IEEE/CVF International Conference on Computer Vision (ICCV), pp. 1803–1811 (October 2019)

50. Yu, J., Yang, L., Xu, N., Yang, J., Huang, T.: Slimmable neural networks. In: International Conference on Learning Representations (2019)

51. Zhang, K., Riegler, G., Snavely, N., Koltun, V.: Nerf++: analyzing and improving neural radiance fields. arXiv preprint arXiv:2010.07492 (2020)

52. Zhang, R., Isola, P., Efros, A.A., Shechtman, E., Wang, O.: The unreasonable effectiveness of deep features as a perceptual metric. In: Proceedings of the IEEE/CVF Conference on Computer Vision and Pattern Recognition (CVPR), pp. 586–595, June 2018

53. Zhong, E.D., Bepler, T., Davis, J.H., Berger, B.: Reconstructing continuous distributions of 3d protein structure from cryo-em images. In: International Conference on Learning Representations (2020)

Gradient-Based Uncertainty
for Monocular Depth Estimation

Julia Hornauer[1(✉)] and Vasileios Belagiannis[2]

[1] Institute of Measurement, Control and Microtechnology, Ulm University, Ulm,
Germany
`julia.hornauer@uni-ulm.de`
[2] Department of Simulation and Graphics, Otto von Guericke University,
Magdeburg, Germany

Abstract. In monocular depth estimation, disturbances in the image context, like moving objects or reflecting materials, can easily lead to erroneous predictions. For that reason, uncertainty estimates for each pixel are necessary, in particular for safety-critical applications such as automated driving. We propose a post hoc uncertainty estimation approach for an already trained and thus fixed depth estimation model, represented by a deep neural network. The uncertainty is estimated with the gradients which are extracted with an auxiliary loss function. To avoid relying on ground-truth information for the loss definition, we present an auxiliary loss function based on the correspondence of the depth prediction for an image and its horizontally flipped counterpart. Our approach achieves state-of-the-art uncertainty estimation results on the KITTI and NYU Depth V2 benchmarks without the need to retrain the neural network. Models and code are publicly available at https://github.com/jhornauer/GrUMoDepth.

Keywords: Depth estimation · Uncertainty estimation · Training-free

1 Introduction

Deep neural networks have shown astonishing performance in 3D perception tasks such as depth prediction [9,15]. Depth estimation from a single image has particularly attracted attention because RGB cameras are cheaper compared to LiDAR sensors while offering higher resolution and frame rates. Nevertheless, disturbances in the image context, like occlusions, moving objects, or reflecting materials, can easily affect the neural network and therefore lead to erroneous

---

Most of this work was done while Vasileios Belagiannis was with Ulm University.

Supplementary Information The online version contains supplementary material available at https://doi.org/10.1007/978-3-031-20044-1_35.

predictions [20]. It is thus crucial to estimate the uncertainty of the depth estimates, especially for safety-critical applications such as automated driving.

Bootstrapped ensembles [16] and Monte Carlo Dropout [6], which are the well-known uncertainty estimation approaches, have been explored for computationally expensive tasks, such as depth estimation [22]. Both approaches rely on sampling from the model distribution, which is not suitable for real-time applications due to the high computational cost. Predictive methods, such as maximum likelihood maximization [13,14], on the other hand, require the adaptation of the training procedure to learn an estimate of the uncertainty contained in the data. Also, it is not always wished or even possible to retrain the neural network with the objective of uncertainty estimation. For instance, the parameters of models provided by external sources are usually not accessible and therefore not modifiable. The parameter modification is not feasible for neural network models that are either specialized to the target system (e.g., by pruning) or have to meet certain system requirements. For these reasons, we target the problem of uncertainty estimation of an already trained model in a *post hoc* manner.

One approach to training-free uncertainty estimation is the usage of dropout only during inference [18]. Although this approach does not require any adaptation of the training protocol, the uncertainty estimation is still based on sampling from the model distribution and is therefore not suitable for real-time applications. In contrast, we propose to estimate uncertainty based on gradients extracted from the neural network, which is much less computationally intensive and requires only one additional backward pass.

Inspired by gradient-based approaches for out-of-distribution detection in classification [11,17], we use gradients to estimate the uncertainty associated with pixel-wise depth, obtained from monocular depth estimation models. Given a trained model, we aim to extract meaningful gradients without relying on the ground-truth depth values. Therefore, we define an auxiliary loss function based on the correspondence of the depth prediction for an image and its horizontally flipped counterpart. With the loss defined as the error between both depth predictions, we calculate the derivative *w.r.t.* feature maps, by back-propagating through the neural network. We then rely on the gradients extracted from the feature maps of a decoder layer of the depth estimation model to obtain the final uncertainty score. Importantly, our approach achieves state-of-the-art uncertainty estimation results without the need to retrain the neural network.

Overall, we summarize the contributions of our paper as follows: Firstly, we propose a *post hoc* uncertainty estimation approach for depth prediction based on the gradients extracted from already trained models. Essentially, our approach is independent of how the model was trained. Secondly, for the gradient generation, we define an auxiliary loss function without relying on ground-truth depth values. In this context, we empirically show the meaningfulness of the gradient-based uncertainty estimation. Lastly, in an extensive comparison to existing approaches on the two common depth estimation benchmarks KITTI [7] and NYU Depth V2 [19], we demonstrate *state-of-the-art* uncertainty estimation results.

2 Related Work

Depth Estimation Uncertainty. Compared to classification tasks, it is more difficult to determine the uncertainty of high-dimensional predictions, such as in depth estimation. In general, a distinction can be made between epistemic and aleatoric uncertainty. While epistemic uncertainty arises from model weights and is reducible with more training data, aleatoric uncertainty is due to noise in the input data [13]. Bootstrapped ensembles [16] and Monte Carlo (MC) Dropout [6] estimate the epistemic uncertainty by modeling the distribution over parameters. For bootstrapped ensembles, this is achieved by training multiple models, sampling initial weights from a specified distribution, while for MC Dropout dropout layers are applied during training and inference. The depth and uncertainty estimates are obtained by sampling from the model distribution and computing the mean and variance, respectively. A predictive approach that accounts for aleatory uncertainty is to learn a distribution with mean and variance that represents the data-dependent error instead of a single output value by maximizing the negative log-likelihood [14]. Kendall and Gal [13] demonstrate how to combine both types of uncertainty. Recent works explore the integration of uncertainty for computationally intensive tasks such as depth estimation [4,28], semantic segmentation [10], optical flow [26], or multi-task learning [31]. Moreover, Poggi et al. [22] extensively compare different empirical and predictive uncertainty estimation approaches for self-supervised depth estimation. Among others, image flipping post-processing proposed by Godard et al. [8] serves as a simple baseline to obtain the variance over two outputs as an uncertainty measure. In addition, Poggi et al. [22] present their approach of self-training, where variance learned with log-likelihood maximization is improved by using knowledge obtained from a teacher model using its predictions as labels. One downside of those approaches is the adaptation of the model design [14] or the specific training pipeline [6,16]. Furthermore, empirical approaches have additional computational overhead [6,16] and increased memory footprint [16], making them unsuitable for real-time applications such as autonomous driving or robotics. By contrast, in this work, we explore the usage of gradients as a *post hoc* uncertainty estimation approach that is independent of the conducted training procedure. Since model re-training is not always feasible, Mi et al. [18] explore different training-free strategies to generate a distribution over the model output by data augmentation, inference-time dropout, and additive noise in intermediate network layers. We also explore training-free uncertainty estimation but forgo computationally intensive sampling by relying on gradients extracted with an auxiliary loss function.

Model Robustness by Gradient Analysis. Deep neural networks are mostly trained using gradient-based optimization. The informativeness of gradients for the task of out-of-distribution detection is explored in [11,17,21]. To obtain the gradients, a loss function must be defined before backpropagating through the model. Oberdiek et al. [21] use the negative log-likelihood at the predicted class since no labels are present during testing. Lee and AlRegib [17], on the other hand, define confounding labels with only positive entries for the cross-entropy loss to show whether the model can associate features with any of the learned classes. Huang et al. [11] determine the KL divergence between the softmax output and the Uniform distribution, which indicates whether the predicted probabilities are distributed across all classes or concentrated in one class. Unlike, we use gradients of a regression neural network to determine the pixel-wise uncertainty for depth estimation models. More precisely, we need an uncertainty score for each input pixel and not just one score for the entire input image. For this purpose, we define the pixel-wise distance between the depth prediction and a reference depth obtained by image transformation as an auxiliary loss function.

Monocular Depth Estimation. The recent monocular depth estimation approaches train deep neural networks with supervision [1,2,5,15,23]. While Eigen et al. [5] use information from local and global features, Laina et al. [15] propose a fully convolutional deep neural network and Bauer et al. [1] integrate novel view synthesis. However, a large amount of data, which is time-consuming and expensive to obtain, is necessary for neural network training. Especially the generation of ground truth depth requires different well calibrated sensors. Therefore, self-supervised approaches leveraging monocular sequences [9,32], or stereo image pairs [8,29,30] are proposed. In the works of Godard et al. [8] as well as Yang et al. [29], learning depth from stereo image pairs by leveraging the scene geometry is introduced. In contrast, the use of monocular sequences for supervision by image reprojection [32], which is beneficial in practical applications, is more challenging due to the unknown scale and camera position that must be learned simultaneously with depth. Godard et al. [9] adapt the reprojection loss to handle occlusions and moving objects, while semantic or scale consistency is proposed by Toasi et al. [24] and Bian et al. [3], respectively. Xu et al. [27], on the other hand, introduce Region Deformer Networks to take moving objects into account. We propose an uncertainty estimation approach for monocular depth estimation models that are trained in a supervised or self-supervised manner. For the self-supervised case, the supervision can be provided by both monocular as well as stereo image pairs.

3   Method

Consider a deep neural network $\mathbf{d} = f(\mathbf{x}; \theta)$, parameterized by θ, that takes an image $\mathbf{x} \in \mathbb{R}^{w \times h \times 3}$ with width w and height h as input and predicts the pixel-wise depth $\mathbf{d} \in \mathbb{R}^{w \times h \times 1}$. Based on the trained depth prediction model f_θ, we aim to predict the uncertainty $\mathbf{u} \in \mathbb{R}^{w \times h \times 1}$ for each predicted depth value.

Post Hoc Uncertainty Estimation. In this work, we target *post hoc* uncertainty estimation. We do not adjust the model parameters, but only assume access to the internal model representation given by the feature maps $\mathbf{a}_i$, where i denotes the feature map obtained from the i-th layer of the depth decoder. Essentially, we estimate the uncertainty of the already trained depth estimation model and therefore do not need to adjust the training protocol or the network

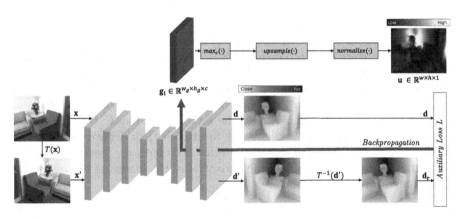

Fig. 1. First, the image $\mathbf{x}$ and its flipped counterpart $\mathbf{x}'$ are forwarded through the neural network. Then, the depth prediction $\mathbf{d}'$ of the flipped counterpart is flipped back to obtain the reference depth $\mathbf{d}_r$. Afterwards an auxiliary loss based on the two depth predictions is calculated to back-propagate through the neural network. Finally, the gradients $\mathbf{g}_i$ of the i-th intermediate layer are extracted. To determine the pixel-wise uncertainty score $\mathbf{u}$, the channel number is reduced from c to 1 by taking the maximum value of each pixel in $\mathbf{g}_i$ in the channel dimension. Then, the gradient map is upsampled with nearest neighbor interpolation to match the depth map resolution.

design. Moreover, our approach is independent of the underlying model training strategy. For instance, the depth estimation model may have been trained in a supervised or self-supervised manner with monocular or stereo supervision. Figure 1 illustrates an overview of our method.

3.1 Gradient-Based Uncertainty

Gradient Generation. We propose to use gradients extracted from the feature maps $\mathbf{a}_i$ to estimate the pixel-wise uncertainty of depth predictions. For the gradient generation, the pixel-level loss function must be defined first, followed by the derivative of the loss function *w.r.t* the respective feature map using backpropagation. Our goal is to convert the model loss to pixel-wise uncertainty estimates. We assume that the pixel-wise uncertainty is in accordance with the depth estimation error. We later empirically show our claims in our evaluations.

In general, the most meaningful loss function is the error between the depth prediction $\mathbf{d}$ and the ground-truth depth $\mathbf{y} \in \mathbb{R}^{w \times h \times 1}$. Nevertheless, we have access to ground-truth information during training, but not during inference. Therefore, we define an auxiliary loss function $\mathcal{L}$ for the gradient generation. For the definition of this loss function, we consider the reference image $\mathbf{x}'$ whose structure equals the structure of the original image $\mathbf{x}$. We make the assumption that the depth maps $\mathbf{d}$ and $\mathbf{d}'$ predicted for $\mathbf{x}$ and $\mathbf{x}'$ match. To generate the reference image $\mathbf{x}'$, we define the horizontal flip operation $T(\cdot)$, which performs a left-right flip. The inverse function $T^{-1}(\cdot)$ reverts the flip operation. The depth prediction on an image $\mathbf{x}$ and the horizontally flipped version $\mathbf{x}' = T(\mathbf{x})$ should be consistent, since the linear transformation preserves the pixel information of the original image and thus the structure of the scene.

Overall, the reference depth $\mathbf{d}_r$ is the depth $\mathbf{d}'$ predicted for the flipped image $\mathbf{x}'$, which is flipped back to match the depth $\mathbf{d}$ obtained for the original input image $\mathbf{x}$. More precisely, we conduct two forward passes, one with the original image $\mathbf{x}$ to obtain $\mathbf{d} = f(\mathbf{x}; \theta)$ and one with the flipped version $\mathbf{x}' = T(\mathbf{x})$ to obtain $\mathbf{d}' = f(\mathbf{x}'; \theta)$. Then, we apply the inverse function $T^{-1}(\mathbf{d}')$ to obtain the reference depth $\mathbf{d}_r$. The auxiliary loss function is then defined based on the difference between the depth $\mathbf{d}$ and the reference depth $\mathbf{d}_r$ (see Sec. 3.2). To obtain the gradients, we calculate the derivative of the loss $w.r.t$ the respective feature map $\mathbf{a}_i$:

$$\mathbf{g}_i = \frac{\partial \mathcal{L}}{\partial \mathbf{a}_i}, \tag{1}$$

where $\mathbf{g}_i \in \mathbb{R}^{w_g \times h_g \times c}$ with w_g, h_g and c being the feature map width, feature map height and the channel number, respectively.

Uncertainty Score. Next, the pixel-wise uncertainty estimate $\mathbf{u}$ is determined from the gradients $\mathbf{g}_i$ extracted from the neural network. Since an uncertainty value is required for each pixel, the gradient map must match the resolution of the depth map. First, we define the function $max_c(\cdot) : \mathbb{R}^{w_g \times h_g \times c} \rightarrow \mathbb{R}^{w_g \times h_g \times 1}$ to perform max-pooling over the channel dimensions:

$$\mathbf{g}_i^{(\max)} = max_c(\mathbf{g}_i), \tag{2}$$

where $\mathbf{g}_i^{(\max)} \in \mathbb{R}^{w_g \times h_g \times 1}$. We are interested in gathering the gradients with the largest magnitude for the uncertainty estimation using the max-pooling operation. Afterward, we define the function $upsample(\cdot) : \mathbb{R}^{w_g \times h_g \times 1} \rightarrow \mathbb{R}^{w \times h \times 1}$ to upsample $\mathbf{g}_i^{(\max)}$ by nearest-neighbor interpolation. Finally, we propose the uncertainty estimate as the self-normalized gradient map:

$$\mathbf{u} = \frac{upsample(\mathbf{g}_i^{(\max)}) - \min \mathbf{g}_i^{(\max)}}{\max \mathbf{g}_i^{(\max)} - \min \mathbf{g}_i^{(\max)}}, \tag{3}$$

where $\max \mathbf{g}_i^{(\max)}$ and $\min \mathbf{g}_i^{(\max)}$ are the maximum and the minimum value of the gradient map, respectively. The final uncertainty is then normalized to a range of $[0, 1]$.

3.2   Training Strategy

We instantiate the auxiliary loss function for gradient generation based on the predicted depth $\mathbf{d}$ and the reference depth $\mathbf{d}_r$. First, we define our loss function for depth estimation models that only predict the depth $\mathbf{d} = f_r(\mathbf{x}; \theta_r)$. Then, we consider a Bayesian depth estimation model $\mathbf{d}, \sigma = f_b(\mathbf{x}; \theta_b)$ with $\mathbf{d}$ being the depth and σ the variance as a measure of uncertainty.

Depth Estimation Model. For standard depth estimation models f_{θ_r}, that do not predict the uncertainty but only the depth, the auxiliary loss function is defined as follows:

$$\mathcal{L}_r = (\mathbf{d} - \mathbf{d}_r)^2. \tag{4}$$

An uncertainty estimate can be generated by a post-processing step in which image flipping is applied to estimate the uncertainty as the variance over two outputs (*Post*) [9]. Here, the variance is also the pixel-wise difference of the two depth predictions. We claim that the information given by the gradients of the feature maps is higher compared to the uncertainty solely given by the loss. This is supported by the results in Sect. 4.2, where we empirically demonstrate that the information given by the gradients of the feature maps is more effective for uncertainty estimation compared to simply relying on image flipping as a post-processing step.

Bayesian Depth Estimation Model. Since our approach is independent of the underlying training strategy, we do not only consider conventional depth estimation models but also the Bayesian version f_{θ_b} that already predicts the variance as an uncertainty measure. Different strategies are proposed to train Bayesian models. One is training with log-likelihood maximization objective (*Log*), while another is the self-teaching paradigm (*Self*) as proposed by Poggi et al. [22]. For Bayesian depth prediction networks, we aim to improve the uncertainty estimate given by the variance. Since these models do not only output the pixel-wise depth $\mathbf{d}$ but also the pixel-wise variance σ, we also make use of the variance for the gradient generation. In this context, we add the squared variance to the loss term. The auxiliary loss for the Bayesian model is given in the following:

$$\mathcal{L}_b = \mathcal{L}_r + \lambda\sigma^2, \tag{5}$$

where λ controls the influence of the variance loss term. Since σ represents the error inherent in the data, high variance leads to larger gradients, while small variance leads to small gradients.

4 Experiments

We evaluate our approach on two standard monocular depth estimation benchmarks. First, we describe the experimental setup including datasets, models, metrics, and implementation. Then, we demonstrate our uncertainty estimation results and provide visual illustrations. Ultimately, we demonstrate the informativeness of the gradient space when ground truth is given for the loss computation and validate horizontal flipping as the chosen image transformation.

4.1 Experimental Setup

Datasets. Since there is no other work that performs a detailed analysis of different approaches to uncertainty estimation using a well-defined protocol, we follow the same evaluation as proposed by Poggi et al. [22]. Therefore, we evaluate our approach on the KITTI [7] dataset. KITTI is an autonomous driving dataset taken at 61 scenes with an average image resolution of 375×1242. Similar to Poggi et al. [22], we use the Eigen split [5] with maximum depth set to 80 m and use the improved ground truth depth provided in [25] for evaluation. In addition, we propose to evaluate the uncertainty estimation on the NYU Depth V2 dataset. This is an indoor depth estimation dataset taken at 494 scenes. The original image resolution is 480×640. Here, we adapt the evaluation protocol proposed by Poggi et al. [22] but set the maximum depth to 10 m.

Models. We choose Monodepth2 [9] as the depth estimation model to make a fair comparison to the numerous uncertainty estimation approaches in [22]. Similar to Poggi et al. [22], we use both Monodepth2 trained with monocular and stereo supervision. Importantly, networks trained with monocular sequences are not only subject to depth uncertainty but also the uncertainty of the estimated camera pose. For NYU Depth V2 [19] we train Monodepth2 [9] in a supervised manner with the provided ground truth depth maps.

Evaluation Metrics. Similar to prior work [12,22], our goal is to estimate the uncertainty with respect to the prediction error. Therefore, we evaluate the estimated uncertainty on in-distribution data by *sparsification* plots that measure if the uncertainty agrees with the true error. Given a data sample and an error metric, the uncertainty is ranked in descending order. Then, the most uncertain pixels are removed to calculated the error metric for the remaining pixels to obtain the sparsification over the fraction of removed pixels. As in [12,22], we compute the Area Under the Sparsification Error (AUSE) that is the difference between the sparsification and the oracle sparsification. The oracle sparsification is given if the uncertainty ranking corresponds to the ranking of the true error. We evaluate the AUSE in terms of the depth estimation metrics Absolute Relative Error (Abs Rel), Root Mean Squared Error (RMSE), and Accuracy ($\delta \geq 1.25$). In addition, as proposed by Poggi et al. [22], we compute the Area Under the Random Gain (AURG) that considers the difference between the sparsification and the random sparsification that quantifies whether the estimated uncertainty is better than not modeling the uncertainty. Again, we calculate the AURG in terms of Abs Rel, RMSE, and $\delta \geq 1.25$.

Comparison to Related Work. Since our approach is a *post hoc* uncertainty estimation method, we apply our approach to already trained depth estimation models. As described in Sect. 3.2, the base models (Base) are a standard depth estimation model with post-processing applied (*Post*) [9], and the two Bayesian depth estimation models *Log* [22] and *Self* [22], that also predict the variance. We compare our method to the uncertainties obtained from the base models themselves as well as inference only dropout (*In-Drop*) [18], MC Dropout (*Drop*) [6] and bootstrapped ensembles (*Boot*) [16]. Note that *In-Drop* is also applied to the base models as a training-free approach, while *Drop* and *Boot* are separate models. Furthermore, we compare our approach to the uncertainty obtained with the variance over different test-time augmentations (*Var*) as additional *post hoc* baseline.

Implementation Details. All base models are implemented based on Monodepth2 [9] as depth estimation network. Since Poggi et al. [22] provide trained models for the uncertainly estimation baselines *Post*, *Log*, *Self*, *Boot*, and *Drop* trained with monocular as well as stereo supervision, we evaluate our model with the provided weights for a fair comparison. For *In-Drop* we apply dropout with probability 0.2 at the same locations as used for *Drop* but only during inference. As in [22], we perform 8 forward passes for *Drop* and *In-Drop*. For NYU Depth V2 [19], we train Monodepth2 [9] as the base model with random rotation, random scaling, horizontal flipping, color jittering and cropped to a final resolution of 224 × 288. For this setup, we additionally implement the Bayesian model trained with log-likelihood maximization (*Log*) as well as MC Dropout (*Drop*) next to the standard depth estimation network. Moreover, *In-Drop* is implemented as for KITTI [7]. For *Var*, we apply grayscale, flipping, noise and rotation as augmentations to calculate the variance over the resulting outputs. For all setups, we choose the 6th layer of the decoder to extract the gradients for our gradient-based uncertainty estimation approach. Results with gradients extracted from different layers are provided in the supplementary material. In the case of the Bayesian depth estimation networks, we choose λ to be 2.0.

4.2   Uncertainty Estimation Results

KITTI Monocular Supervision. The uncertainty estimation results for Monodepth2 trained with monocular sequences from KITTI are reported in Tab. 1. For the monocular setup, where the uncertainty stems not only from the depth estimation model but also from the pose prediction, our method outperforms the base models as well as *In-Drop*. Especially, AUSE and AURG in terms of RMSE are clearly improved compared to the base models *Post* [9], *Log* [22], and *Self* [22]. Moreover, all methods obtain better uncertainty estimation results compared to the two empirical methods *Drop* [6] and *Boot* [16]. In terms of inference time, the base methods are slightly faster, but our gradient-based uncertainty estimation achieves better uncertainty estimation results in comparison. *In-Drop*, on the other hand, has significantly higher inference times due to the sampling procedure and does not always outperform the baseline model when considering AUSE and AURG in terms of Abs Rel and $\delta \geq 1.25$.

KITTI Stereo Supervision. In Table 2, the uncertainty estimation results for Monodepth2 trained with stereo pair supervision, where the uncertainty results only from the depth estimation model, are listed. In this setup, the worst results are obtained from the computationally expensive MC Dropout [6]. For *Post* [9] and *Log* [22], the gradient-based uncertainty estimation especially improves over the base model in terms of AUSE and AURG for Abs Rel and RMSE. Moreover, our method is on par with the *Self* [22] model while being a *post hoc* approach. One downside of self-teaching is the elaborate training procedure that is required to obtain the final depth estimation model. Furthermore, the training-free *In-Drop* [18] shows no improvement over the base models while requiring a significantly higher inference time. Furthermore, *In-Drop* [18] is barely better than random chance when observing AURG.

Table 1. Uncertainty evaluation results for Monodepth2 [9] trained with monocular supervision on KITTI [7].

Model	Method	Abs Rel		RMSE		$\delta \geq 1.25$		Inf [ms]
		AUSE ↓	AURG ↑	AUSE ↓	AURG ↑	AUSE ↓	AURG ↑	
Drop	Base [22]	0.056	0.000	2.568	0.944	0.097	0.002	47.12
Boot	Base [22]	0.058	0.001	3.982	−0.743	0.084	−0.001	47.12
Post	Base [9]	0.044	0.012	2.864	0.412	0.056	0.022	11.78
	Var	0.055	0.003	3.575	−0.210	0.074	0.006	29.45
	In-Drop [18]	0.031	0.027	0.871	2.495	0.029	0.051	53.01
	Ours	**0.029**	**0.029**	**0.540**	**2.825**	**0.025**	**0.054**	18.84
Log	Base [22]	0.039	0.020	2.562	0.916	0.044	0.038	5.89
	Var	0.055	0.004	3.686	−0.209	0.075	0.008	29.45
	In-Drop [18]	0.045	0.014	1.916	1.561	0.056	0.027	53.01
	Ours	**0.026**	**0.033**	**0.819**	**2.658**	**0.024**	**0.059**	18.84
Self	Base [22]	0.030	0.026	2.009	1.266	0.030	0.045	5.89
	Var	0.055	0.001	3.632	−0.357	0.073	0.002	29.45
	In-Drop [18]	0.033	0.024	1.091	2.184	0.031	0.044	53.01
	Ours	**0.024**	**0.032**	**0.494**	**2.780**	**0.017**	**0.057**	18.84

Table 2. Uncertainty evaluation results for Monodepth2 [9] trained with stereo pair supervision on KITTI [7].

Model	Method	Abs Rel		RMSE		$\delta \geq 1.25$		
		AUSE ↓	AURG ↑	AUSE ↓	AURG ↑	AUSE ↓	AURG ↑	Inf [ms]
Drop	Base [22]	0.103	−0.029	6.163	−2.169	0.231	−0.080	47.12
Boot	Base [22]	0.028	0.029	2.291	0.964	0.031	0.048	47.12
Post	Base [9]	0.036	0.020	2.523	0.736	0.044	0.034	11.78
	Var	0.054	0.003	3.672	−0.327	0.074	0.006	29.45
	In-Drop [18]	0.061	−0.004	3.366	−0.021	0.089	−0.009	53.01
	Ours	**0.022**	**0.035**	**0.510**	**2.835**	**0.023**	**0.057**	18.84
Log	Base [22]	0.022	0.036	0.938	2.402	**0.018**	**0.061**	5.89
	Var	0.054	0.003	3.639	−0.299	0.073	0.006	29.45
	In-Drop [18]	0.073	−0.016	4.063	−0.724	0.107	−0.028	53.01
	Ours	**0.019**	**0.038**	**0.490**	**2.849**	**0.018**	**0.061**	18.84
Self	Base [22]	**0.022**	**0.035**	1.679	1.642	**0.022**	**0.056**	5.89
	Var	0.056	0.001	3.728	−0.408	0.076	0.002	29.45
	In-Drop [18]	0.076	−0.019	4.173	−0.852	0.109	−0.031	53.01
	Ours	**0.022**	**0.035**	**0.515**	**2.806**	0.023	0.055	18.84

Table 3. Uncertainty evaluation results for Monodepth2 [9] trained on NYU Depth V2 [19].

Model	Method	Abs Rel		RMSE		$\delta \geq 1.25$		
		AUSE ↓	AURG ↑	AUSE ↓	AURG ↑	AUSE ↓	AURG ↑	Inf [ms]
Drop	Base [6]	0.086	0.002	0.214	0.160	0.169	−0.001	26.32
Post	Base [9]	0.066	0.018	0.267	0.096	0.116	0.037	6.58
	Var	0.063	0.024	0.221	0.151	0.110	0.045	16.45
	In-Drop [18]	0.086	0.000	**0.223**	**0.149**	0.163	0.000	29.61
	Ours	**0.061**	**0.025**	0.252	0.120	**0.106**	**0.048**	12.05
Log	Base [22]	0.055	0.030	**0.159**	**0.210**	0.089	0.066	3.29
	Var	0.061	0.024	0.220	0.149	0.106	0.049	16.45
	In-Drop [18]	0.087	−0.002	0.215	0.154	0.166	−0.012	29.61
	Ours	**0.053**	**0.032**	0.176	0.193	**0.086**	**0.069**	12.05

NYU Depth V2. The uncertainty estimation performance of Monodepth2 trained on NYU Depth V2 is reported in Table 3. Again, the uncertainty only stems from the depth estimation itself due to the supervised training. We implement our approach on the *Post* [9] and *Log* [22] models and state the MC Dropout [6] model as baseline. As in the previous setups, uncertainty estimation with dropout sampling results in the worst uncertainty estimation performance. Here, both versions, with dropout during training as well as only during inference, do obtain worse uncertainty estimates compared to the base models in almost all metrics. By contrast, our gradient-based method does improve the performance for AUSE and ARUG in terms of Abs Rel as well as $\delta \geq 1.25$.

Table 4. Comparison of the different uncertainty estimation methods regarding number of trained models (#Train), whether a specific training strategy is required (Specialized Training), number of models for uncertainty estimation (#Models), number of forward passed required to obtain depth and uncertainty (#Forward) and number of backward passes (#Backward). Note that N is set to 8.

Method	#Train	Specialized Training	#Models	#Forward	#Backward
Drop [6]	1	Yes	1	N	–
Boot [16]	N	Yes	N	N	–
Post [9]	1	No	1	2	–
Log [22]	1	Yes	1	1	–
Self [22]	2	Yes	1	1	–
Var	–	No	–	4 + 1	–
In-Drop [18]	–	No	–	N + 1	–
Ours	–	No	–	2	1

Sparsification Error Plots. In Fig. 2, we illustrate the sparsification error curves in terms of RMSE for KITTI with monocular supervision (Fig. 2a) and with stereo pair supervision (Fig. 2b) averaged over the KITTI test set. The sparsification error represents the deviation of the sparsification curve from the respective oracle sparsification. Therefore, a smaller area under the curve means better performance in uncertainty estimation. For our *post hoc* approach, the uncertainty best matches the true error, while *In-Drop* [18] does not improve on the baseline model for the stereo setup. In the monocular setup, *Boot* [16] performs the worst, while the uncertainty modeled with *Drop* [6] in the stereo case is the least consistent with the true error.

Summary. Overall, our method improves the uncertainty estimates for most of the cases over the base model while only adding a slight overhead in inference time. The empirical approaches *Boot* [16], *Drop* [6], and *In-Drop* [18], on the other hand, add a huge computational overhead and therefore require inference times that are not feasible in real-time applications such as autonomous driving or robotics. In addition, these methods do not outperform the uncertainty

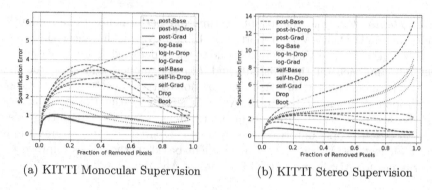

(a) KITTI Monocular Supervision   (b) KITTI Stereo Supervision

Fig. 2. The sparsification error in terms of RMSE over the fraction of removed pixels is shown for Monodepth2 [9] trained on KITTI [7] with monocular (a) and stereo (b) supervision.

of the *Post* [9], *Log* [22], and *Self* [22] models. The comparison of our method with the *Post* baseline highlights the higher significance of gradients extracted from the model compared to using image flipping alone as the difference in pixel space. In Tab. 4, the main characteristics of the different methods are compared regarding the number of trained models (#Train), the adaptation of the training strategy (Specialized Training), the number of models that must be saved (#Models), the number of forward passes necessary to obtain the depth as well as the uncertainty estimate (#Forward) and the number of required backward passes (# Backward). In general, only *In-Drop*, *Var* and our method are *post hoc* approaches applicable to all kinds of depth estimation networks. Moreover, all approaches besides post-processing require an adaptation to the training strategy or the network architecture. While the empirical methods result in a large computational overhead, our approach only needs one additional backward pass to obtain the gradients. Moreover, empirical methods are known to decrease the depth estimation performance [22]. When comparing our approach with the other training-free methods *In-Drop* [18] and *Var*, these approaches need one forward pass to obtain the depth prediction and N or 4 (depending on the number of augmentations) forward passes to get the uncertainty estimate, while our approach only needs two forward and one backward pass to obtain depth and uncertainty estimates.

Visual Results. In Fig. 3, an example from Monodepth2 trained with the standard depth estimation protocol on NYU Depth V2 is illustrated. Overall, the estimated uncertainty for high error regions should be high, while the uncertainty for low error regions should be low. When comparing the depth prediction (Fig. 3b) and the RMSE (Fig. 3c), it becomes apparent that the region around the two lamps has the greatest error. Both, the post-processing [9] (Fig. 3d) and our gradient-based method (Fig. 3f) highlight the high uncertainty at the lamp. Moreover, our approach improves over the uncertainty estimation by post-processing in regions with low error. *In-Drop* [18] (Fig. 3e), on the other hand, overestimates uncertainty in most regions, especially in the lower right corner. More visual results can be found in the supplementary material.

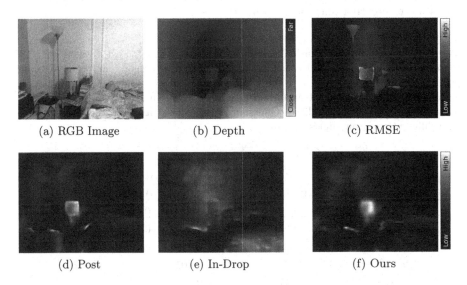

(a) RGB Image   (b) Depth   (c) RMSE

(d) Post   (e) In-Drop   (f) Ours

Fig. 3. Uncertainty estimation example from Monodepth2 [9] trained on NYU Depth V2 [19]. In (a), the input image is shown. The depth prediction (b) and the root mean squared error (c) highlight that the lamp and the region around it are not estimated correctly. The second row demonstrates the uncertainty estimated by post-processing (d), inference dropout (e) and gradients (f).

4.3 Ablation Study

In Table 5, the uncertainty estimation results of our gradient-based approach on Monodepth2 trained with NYU Depth V2 are reported for the *Post* model. We demonstrate different configurations to define the loss for the gradient generation. We consider the squared difference of the prediction to the ground truth depth (GT) and to transformed images by image flipping (Flip), gray-scale conversion (Gray), additive Gaussian noise (Noise) and rotation (Rot) with different rotation angles. Note that the selected augmentations are not necessarily within the training time augmentation. The uncertainty estimation results for all three depth estimation metrics clearly show that the loss between the prediction and the ground truth depth results in the best uncertainty estimation performance. Nevertheless, the ground truth is not available during inference in real-world application. However, results in Sect. 4.2 demonstrate promising results for uncertainty estimation based on the information given by the extracted gradients. Comparing the different transformation operations, the flipping operation performs best. Compared to the other augmentations considered, when flipping horizontally, the model observes the scene from a different context, while preserving the geometry of the scene. In case of the rotation operation, the uncertainty cannot be determined for each pixel. This confirms the use of the flipping operation to create a reference depth.

Table 5. Uncertainty estimation results for Monodepth2 [9] Post trained on NYU Depth V2 [19] when using different loss functions for the gradient generation. We compare the error of the prediction to the ground truth depth (GT) and depth predictions obtained by different image transformations. We consider image flipping (Flip), grayscale conversion (Gray), additive Gaussian noise (Noise) and rotation (Rot-[angle]) with different rotation angles.

	Abs Rel		RMSE		$\delta \geq 1.25$	
Loss	AUSE ↓	AURG ↑	AUSE ↓	AURG ↑	AUSE ↓	AURG ↑
GT	**0.020**	**0.066**	**0.106**	**0.266**	**0.031**	**0.123**
Flip	<u>0.061</u>	<u>0.025</u>	<u>0.252</u>	<u>0.120</u>	<u>0.106</u>	<u>0.048</u>
Gray	0.067	0.020	0.253	0.119	0.111	0.044
Noise	0.066	0.020	0.261	0.111	0.108	0.046
Rot-5°	0.066	0.020	0.267	0.105	0.114	0.041
Rot-10°	0.069	0.017	0.269	0.103	0.124	0.031
Rot-20°	0.076	0.011	0.270	0.102	0.143	0.012

5   Conclusions

We present the usage of gradients for uncertainty estimation of depth predictions. Given an already trained model, we introduce an auxiliary loss function, that is independent of ground truth depth, based on the correspondence of the depth prediction for an image and its horizontally flipped counterpart. Note that our method does not adjust the neural network weights but estimates the uncertainty in a *post hoc* manner. With the loss defined as the error between the depth predictions, we calculate the derivative *w.r.t.* feature maps by backpropagation. Finally, we determine the pixel-wise uncertainty based on the gradients of an inter-mediate feature map extracted from the depth decoder. We performed a comprehensive comparison with related work using two depth estimation benchmarks, namely the KITTI and NYU Depth V2 datasets. Our evaluation shows the high potential to estimate uncertainty based on gradients extracted from an already trained model.

Acknowledgement. The research leading to these results is funded by the German Federal Ministry for Economic Affairs and "Climate Action" within the project "KI Delta Learning" (Förderkennzeichen 19A19013A). The authors would like to thank the consortium for the successful cooperation.

References

1. Bauer, Z., Li, Z., Orts-Escolano, S., Cazorla, M., Pollefeys, M., Oswald, M.R.: NVS-monodepth: improving monocular depth prediction with novel view synthesis. In: 2021 International Conference on 3D Vision (3DV), pp. 848–858 (2021)
2. Bhat, S., Alhashim, I., Wonka, P.: Adabins: depth estimation using adaptive bins. In: 2021 IEEE/CVF Conference on Computer Vision and Pattern Recognition (CVPR), pp. 4008–4017 (2021)
3. Bian, J., et al.: Unsupervised scale-consistent depth and ego-motion learning from monocular video. In: NeurIPS (2019)
4. Chanduri, S.S., Suri, Z.K., Vozniak, I., Müller, C.: Camlessmonodepth: monocular depth estimation with unknown camera parameters. arXiv:abs/2110.14347 (2021)
5. Eigen, D., Puhrsch, C., Fergus, R.: Depth map prediction from a single image using a multi-scale deep network. In: NIPS (2014)
6. Gal, Y., Ghahramani, Z.: Dropout as a bayesian approximation: representing model uncertainty in deep learning. In: Proceedings of The 33rd International Conference on Machine Learning, June 2015
7. Geiger, A., Lenz, P., Stiller, C., Urtasun, R.: Vision meets robotics: the kitti dataset. In: International Journal of Robotics Research (IJRR) (2013)
8. Godard, C., Aodha, O.M., Brostow, G.J.: Unsupervised monocular depth estimation with left-right consistency. In: 2017 IEEE Conference on Computer Vision and Pattern Recognition (CVPR), pp. 6602–6611 (2017)
9. Godard, C., Aodha, O.M., Brostow, G.J.: Digging into self-supervised monocular depth estimation. In: 2019 IEEE/CVF International Conference on Computer Vision (ICCV), pp. 3827–3837 (2019)
10. Gustafsson, F.K., Danelljan, M., Schön, T.B.: Evaluating scalable Bayesian deep learning methods for robust computer vision. In: 2020 IEEE/CVF Conference on Computer Vision and Pattern Recognition Workshops (CVPRW), pp. 1289–1298 (2020)
11. Huang, R., Geng, A., Li, Y.: On the importance of gradients for detecting distributional shifts in the wild. arXiv:abs/2110.00218 (2021)
12. Ilg, E., et al.: Uncertainty estimates and multi-hypotheses networks for optical flow. In: Ferrari, V., Hebert, M., Sminchisescu, C., Weiss, Y. (eds.) ECCV 2018. LNCS, vol. 11211, pp. 677–693. Springer, Cham (2018). https://doi.org/10.1007/978-3-030-01234-2_40
13. Kendall, A., Gal, Y.: What uncertainties do we need in Bayesian deep learning for computer vision? In: NIPS (2017)
14. Klodt, M., Vedaldi, A.: Supervising the new with the old: learning SFM from SFM. In: Ferrari, V., Hebert, M., Sminchisescu, C., Weiss, Y. (eds.) ECCV 2018. LNCS, vol. 11214, pp. 713–728. Springer, Cham (2018). https://doi.org/10.1007/978-3-030-01249-6_43
15. Laina, I., Rupprecht, C., Belagiannis, V., Tombari, F., Navab, N.: Deeper depth prediction with fully convolutional residual networks. In: 2016 Fourth International Conference on 3D Vision (3DV), pp. 239–248 (2016)
16. Lakshminarayanan, B., Pritzel, A., Blundell, C.: Simple and scalable predictive uncertainty estimation using deep ensembles. In: NIPS (2017)
17. Lee, J., AlRegib, G.: Gradients as a measure of uncertainty in neural networks. In: 2020 IEEE International Conference on Image Processing (ICIP), pp. 2416–2420 (2020)

18. Mi, L., Wang, H., Tian, Y., He, H., Shavit, N.: Training-free uncertainty estimation for dense regression: sensitivity as a surrogate (2019)
19. Silberman, N., Hoiem, D., Kohli, P., Fergus, R.: Indoor segmentation and support inference from RGBD images. In: Fitzgibbon, A., Lazebnik, S., Perona, P., Sato, Y., Schmid, C. (eds.) ECCV 2012. LNCS, vol. 7576, pp. 746–760. Springer, Heidelberg (2012). https://doi.org/10.1007/978-3-642-33715-4_54
20. Nguyen, A., Yosinski, J., Clune, J.: Deep neural networks are easily fooled: High confidence predictions for unrecognizable images. In: Proceedings of the IEEE Conference on Computer Vision and Pattern Recognition (CVPR), June 2015
21. Oberdiek, P., Rottmann, M., Gottschalk, H.: Classification uncertainty of deep neural networks based on gradient information. In: ANNPR (2018)
22. Poggi, M., Aleotti, F., Tosi, F., Mattoccia, S.: On the uncertainty of self-supervised monocular depth estimation. In: 2020 IEEE/CVF Conference on Computer Vision and Pattern Recognition (CVPR), pp. 3224–3234 (2020)
23. Song, M., Lim, S., Kim, W.: Monocular depth estimation using laplacian pyramid-based depth residuals. IEEE Trans. Circuits Syst. Video Technol. **31**, 4381–4393 (2021)
24. Tosi, F., Aleotti, F., Poggi, M., Mattoccia, S.: Learning monocular depth estimation infusing traditional stereo knowledge. In: 2019 IEEE/CVF Conference on Computer Vision and Pattern Recognition (CVPR), pp. 9791–9801 (2019)
25. Uhrig, J., Schneider, N., Schneider, L., Franke, U., Brox, T., Geiger, A.: Sparsity invariant CNNs. In: 2017 International Conference on 3D Vision (3DV), pp. 11–20 (2017)
26. Wannenwetsch, A.S., Keuper, M., Roth, S.: Probflow: joint optical flow and uncertainty estimation. In: 2017 IEEE International Conference on Computer Vision (ICCV), pp. 1182–1191 (2017)
27. Xu, H., Zheng, J., Cai, J., Zhang, J.: Region deformer networks for unsupervised depth estimation from unconstrained monocular videos. In: IJCAI (2019)
28. Yang, N., von Stumberg, L., Wang, R., Cremers, D.: D3vo: deep depth, deep pose and deep uncertainty for monocular visual odometry. In: 2020 IEEE/CVF Conference on Computer Vision and Pattern Recognition (CVPR), pp. 1278–1289 (2020)
29. Yang, N., Wang, R., Stückler, J., Cremers, D.: Deep virtual stereo odometry: leveraging deep depth prediction for monocular direct sparse odometry. In: Ferrari, V., Hebert, M., Sminchisescu, C., Weiss, Y. (eds.) ECCV 2018. LNCS, vol. 11212, pp. 835–852. Springer, Cham (2018). https://doi.org/10.1007/978-3-030-01237-3_50
30. Zhou, H., Greenwood, D., Taylor, S.: Self-supervised monocular depth estimation with internal feature fusion. In: British Machine Vision Conference (BMVC) (2021)
31. Zhou, H., Taylor, S., Greenwood, D.: Sub-depth: self-distillation and uncertainty boosting self-supervised monocular depth estimation. arXiv:abs/2111.09692 (2021)
32. Zhou, T., Brown, M.A., Snavely, N., Lowe, D.G.: Unsupervised learning of depth and ego-motion from video. In: 2017 IEEE Conference on Computer Vision and Pattern Recognition (CVPR), pp. 6612–6619 (2017)

Online Continual Learning
with Contrastive Vision Transformer

Zhen Wang[(✉)], Liu Liu, Yajing Kong, Jiaxian Guo, and Dacheng Tao

The University of Sydney, Camperdown, Australia
{zwan4121,liu.liu1,ykon9947,jguo5934,dacheng.tao}@sydney.edu.au

Abstract. Online continual learning (online CL) studies the problem of learning sequential tasks from an online data stream without task boundaries, aiming to adapt to new data while alleviating catastrophic forgetting on the past tasks. This paper proposes a framework Contrastive Vision Transformer (CVT), which designs a focal contrastive learning strategy based on a transformer architecture, to achieve a better stability-plasticity trade-off for online CL. Specifically, we design a new external attention mechanism for online CL that implicitly captures previous tasks' information. Besides, CVT contains learnable focuses for each class, which could accumulate the knowledge of previous classes to alleviate forgetting. Based on the learnable focuses, we design a focal contrastive loss to rebalance contrastive learning between new and past classes and consolidate previously learned representations. Moreover, CVT contains a dual-classifier structure for decoupling learning current classes and balancing all observed classes. The extensive experimental results show that our approach achieves state-of-the-art performance with even fewer parameters on online CL benchmarks and effectively alleviates the catastrophic forgetting.

Keywords: Online continual learning · Vision transformer · Supervised contrastive learning

1 Introduction

One of the major challenges in research on artificial neural networks is developing the ability to accumulate knowledge over time from a non-stationary data stream [7,29,38]. Although most successful deep learning techniques can achieve excellent results on pre-collected and fixed datasets, they are incapable of adapting their behavior to the non-stationary environment over time [18,75]. When streaming data comes in continuously, training on the new data can severely interfere with the model's previously learned knowledge of past data, resulting in a drastic drop in performance on the previous tasks. This phenomenon is known as *catastrophic forgetting* or *catastrophic interference* [54,61]. Continual learning (also known as lifelong learning or incremental learning) [18,62,63,75,79,80,95] aims to solve this problem by maintaining and accumulating the acquired knowledge over time from a stream of non-stationary data.

© The Author(s), under exclusive license to Springer Nature Switzerland AG 2022
S. Avidan et al. (Eds.): ECCV 2022, LNCS 13680, pp. 631–650, 2022.
https://doi.org/10.1007/978-3-031-20044-1_36

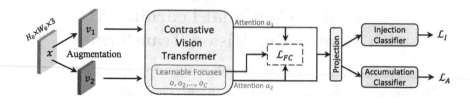

Fig. 1. The overall framework of Contrastive Vision Transformer (CVT).

Continual learning requires neural networks to be both stable to prevent forgetting as well as plastic to learn new concepts, which is referred to as the *stability-plasticity* dilemma [27,55]. The early works have focused on the Task-aware protocol, which selects the corresponding classifier using oracle knowledge of the task identity at inference time [2,18,46,65,69]. For example, regularization-based methods penalize the changes of important parameters when learning new tasks and typically assign a separate classifier for each task [12,42,64,85,93]. Recent studies have focused on a more practical Task-free protocol, which evaluates the network on all classes observed during training without requiring the identity of the task [3,8,13,37,81,90,95]. Among them, rehearsal-based methods that store a small set of seen examples in a limited memory for replaying have demonstrated promising results [4,9,62,76]. This paper focuses on a more realistic and challenging setting: online continual learning (online CL) [34,50,67], where the model learns a sequence of classification tasks with a single pass over the data and without task boundaries. There are also Task-free and Task-aware protocols in online CL.

Inspired by recent breakthroughs in self-supervised learning [11,16,30,72], we find that the knowledge learned by supervised contrastive learning [28,31,41] reveals greater robustness and transferability. The general and transferable knowledge is the essence of what online CL seeks to explore, which could effectively help mitigate forgetting. Unfortunately, it is challenging to employ contrastive learning in the continual setting. There are two main issues as follows: 1) Contrastive learning requires informative negative samples to learn and distinguish different clusters. However, in online CL, previous task data is unavailable or very limited, which causes severe imbalanced contrast between past and new classes. 2) The contrastively learned knowledge may suffer from forgetting since the distribution of the data stream is continually changing.

Considering the superior modeling capability of Vision Transformers [21] recently demonstrated on computer vision tasks [17,26,57,84,92], we take the utilization of the potential of attention mechanism [74] to develop online CL. Overall, we strategically integrate contrastive learning and transformer to model the online data stream. We propose a novel framework, Contrastive Vision Transformer (CVT), to alleviate the forgetting problem and tackle the above imbalance issue of contrastive learning in online CL. An overview of the framework is illustrated in Fig. 1. Specifically, we newly design an effective and efficient transformer architecture with external attention to implicitly capture previous tasks' information and reduce the number of parameters. CVT contains learnable focuses for each class, which could accumulate the knowledge of previous classes to alleviate forgetting. Based on the learnable focuses, we design a focal contrastive loss at the attention level to rebalance contrastive learning between new and past classes and improve the inter-class distinction and intra-class aggregation. Moreover, CVT contains a dual-classifier structure: an *injection classifier* is used to inject representation of stream data into the model, mitigating interference with previous knowledge; and an *accumulation classifier* focuses on integrating the previous and new knowledge in a balanced manner.

We systematically compare state-of-the-art and well-established methods for the online CL problem in both the Task-free and Task-aware protocols. Experimental results show that the CVT framework significantly outperforms other approaches in terms of accuracy and forgetting, even with fewer parameters. Ablation study validates each component of the proposed framework.

The main contributions of this paper are three-fold:

- We propose a novel framework Contrastive Vision Transformer (CVT) to achieve a better stability-plasticity trade-off for online CL. CVT contains class-wise learnable focuses, which can accumulate the knowledge of previous classes to alleviate forgetting.
- We design a focal contrastive loss to rebalance contrastive learning between new and past classes and learn more robust representations.
- The extensive experimental results show that CVT achieves state-of-the-art performance with even fewer parameters on online continual learning benchmarks.

2   Related Works

2.1   Continual Learning Methods

Continual learning (CL) methods have been developed to alleviate catastrophic forgetting in neural networks. These methods can be divided into three main categories: expansion-based, regularization-based, and rehearsal-based methods.

As new tasks arrive, expansion-based methods dynamically expand networks and keep sub-networks related to previous tasks fixed [1,45,53,59,69,82,86,89]. However, most expansion-based methods require task identity during inference in order to allocate distinct sets of parameters to distinct tasks. Regularization-based methods limit the changes in important parameters during the learning of new tasks by estimating the importance of each network parameter for prior tasks [2,12,42,43,64,66,93]. These works differ in how they compute the importance of network parameters.

Rehearsal-based methods [8,15,22,37,49,52,60,68,79,81] alleviate catastrophic forgetting by replaying a subset of data of past tasks stored in limited buffer. iCaRL [62] trains a nearest-class-mean classifier while limiting the change of the representation in later tasks through a self-distillation loss. In addition to replaying past experiences, HAL [13] keeps predictions on some *anchor* points by an additional objective. IExpressNet [95] introduces representative expression memory while employing a novel Center-expression-distilled loss and having shown satisfactory performance for the task of facial expression recognition. DER++ [9] combines rehearsal with distillation loss [36,47,83] to retrain past experience and obtain state-of-the-art performance. RM [6] proposes an uncertainty-based sampling approach by using uncertainty and data augmentation to improve rehearsal. The proposed method in this paper belongs to the rehearsal-based method.

2.2 Online Continual Learning

Online continual learning (online CL) is a more realistic [49,50,67,78,87,96] and difficult setup [34], where the model learns from a non-i.i.d data stream online, without the help of task identifiers or task boundaries both at training and inference stages. Online CL methods are mainly based on rehearsal.

Experience Replay (ER) [63] employs reservoir sampling for memory management and jointly optimizes a network by mixing memory data with online stream data. ERT [10] improves ER by balanced sampling and bias control. AGEM [14] and GEM [49] use episodic memory to compute past task gradients to constrain the online update step. GSS [4] presents a gradient-based sampling to store diversified data for learning more information. ASER [67] is based on the Shapley Value theory to improve the memory buffer update and sampling. CLS [5] uses two extra models to maintain long-term and short-term semantic memories for knowledge consolidation. SCR [52] uses supervised contrastive loss for representation learning and employs the nearest class mean to classify.

2.3 Vision Transformers

Transformer model is firstly applied into machine translation tasks [74], and then, Transformers have become the state-of-the-art models for most natural language processing tasks [20,24,25,51,58]. Attention modules are the core components of transformers, which aggregate information from the entire input sequence. Recently, Vision Transformer (ViT) [21] is proposed to makes Transformer architecture scalable for image classification when the data is large enough. Since then, a lot of effort has been dedicated to improving Vision Transformers' data efficiency and model efficiency [32,40,91,92], where an effective direction is to strategically integrate properties of convolution into the Transformer architecture [17,23,26,48,73,84,91,94]. CoAT [88] proposes a conv-attention module that realizes relative position embeddings with convolutions. LeViT [26] builds pyramid attention structure with pooling to learn convolutional-like features. By leveraging sequence pooling and convolutions, CCT [33] eliminates the need for class tokens and positional embeddings.

Nevertheless, current vision transformers may not be applicable to modeling the online data stream, and existing continual learning algorithms developed for CNNs may not be ideal for vision transformers as well. To this end, we propose a lightweight Contrastive Vision Transformer (CVT) with a focal contrastive loss for online continual learning and achieve better performance than other transformers and CNN baselines.

3   Preliminary

Problem Setup. Formally, an online continual learning problem is split in a sequence of T supervised learning tasks $\mathcal{T}_t$, $t \in \{1, ..., T\}$, where T is the total number of tasks. Let $\mathcal{D} = \{\mathcal{D}_1, ..., \mathcal{D}_T\}$ be the corresponding online data stream, where $\mathcal{D}_t$ is the dataset of task $\mathcal{T}_t$. For task $\mathcal{T}_t$, input samples $x \in \mathcal{X}_t$ and the corresponding ground truth labels $y \in \mathcal{Y}_t$ are drawn from the i.i.d. distribution $\mathcal{D}_t$. A mini-batch of training data $\mathcal{B}$ from $\mathcal{D}$ comes gradually in an online stream (each sample is seen only once). Besides, a limited memory buffer $\mathcal{M}$ saves a small set of training data of seen tasks. The model is trained on $\mathcal{B} \cup \mathcal{M}$ at each iteration. At task $\mathcal{T}_t$, The label space of the model is all observed classes $\cup_{i=1}^{t} \mathcal{Y}_i$, and the model is expected to predict well on all classes at the inference stage.

Supervised Contrastive Learning (SCL). SCL [16,28,41] aims to push the representation of samples with different classes farther apart while tightly clustering representation of samples with the same class. Suppose that the classification model can be decomposed into two components: an encoder f and a classifier w. Encoder f maps an image sample x to a vectorial embedding (representation) $z = f(x)$. Classifier w maps the representation z to a classification vector $\hat{y} = w(z)$. Without training w, SCL focuses on training f as follows: given a batch of b samples $\mathcal{B} = \{(x_k, y_k)\}_{k=1}^{b}$, SCL first generates an augmented batch $\widetilde{\mathcal{B}} = \{(\tilde{x}_k, \tilde{y}_k)\}_{k=1}^{2b}$ by making two random augmentations of $\mathcal{B}$, with $y_k = \tilde{y}_{2k-1} = \tilde{y}_{2k}$. The SCL loss takes the following form:

$$\mathcal{L}_{\text{SCL}} = \sum_{i \in \mathcal{I}} \frac{-1}{|\mathcal{P}(i)|} \sum_{z_p \in \mathcal{P}(i)} \log \frac{\exp(z_i \cdot z_p / \tau)}{\sum_{z_j \in \mathcal{A}(i)} \exp(z_i \cdot z_j / \tau)}, \tag{1}$$

where $\mathcal{I}$ represents the set of indices of $\widetilde{\mathcal{B}}$; $\mathcal{A}(i) \equiv \{z_i : i \in \mathcal{I} \backslash \{i\}\}$ is the set of representations of samples in $\widetilde{\mathcal{B}}$ except for that of x_i; $\mathcal{P}(i) := \{z_p \in \mathcal{A}(i) : \tilde{y}_p = \tilde{y}_i\}$ is the set of representations of positive samples (i.e., samples with the same labels) with respect to the anchor $\tilde{x}_i$; $\tau \in \mathbb{R}^+$ is a temperature hyperparameter; $|\mathcal{P}(i)|$ is its cardinality.

Although SCL could learn the transferable representation to help prevent forgetting in online CL, SCL will face new challenges: 1) previous task data is unavailable or very limited due to the streaming fashion, which causes severe imbalanced contrast between past and new classes; 2) the contrastively learned knowledge may suffer from forgetting since the distribution of the data stream is continually changing. For visualization clarity we use a 2D feature space. As illustrated in Fig. 2(a1) and Fig. 2(a2), the imbalanced data stream in online CL makes the representation of previous tasks drift and difficult to be accurately clustered by SCL.

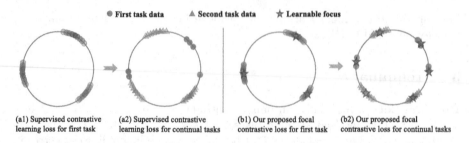

Fig. 2. Continually arriving samples are clustered on the hypersphere by contrastive learning (for visualization clarity, we use a 2D representation space). Different colors represent different classes. Supervised contrastive learning loss on online stream data fails to obtain good inter-class distinction and intra-class aggregation caused by the class imbalance. Focal contrastive loss effectively mitigates class imbalance in online CL and accumulates class-wise knowledge by the learnable *focuses*.

4   Methodology

To alleviate the forgetting problem in online continue learning, we propose a framework Contrastive Vision Transformer (CVT), which designs a new focal contrastive learning strategy based on the transformer architecture. An overview of the framework is depicted in Fig. 1. CVT plays the strengths of the attention mechanism in online CL, which design an effective transformer architecture with external attention. We tackle the imbalance issue of SCL in online CL by proposing a focal contrastive loss at the attention level. The learnable focuses in CVT can accumulate class-specific knowledge to alleviate the forgetting of previous tasks. Besides, a dual-classifier is used to decouple learning current classes and balancing all seen classes, improving the stability-plasticity trade-off.

In the following, we will go through the description of CVT in terms of both model architecture (Sect. 4.1) and focal contrastive continual learning strategy (Sect. 4.2), respectively.

4.1   Model Architecture

Figure 3 illustrates the CVT architecture. The major contributing components in the architecture include 1) *external attention*, which implicitly captures previous tasks' information and reduces the number of parameters, and 2) *learnable focuses*, which could maintain and accumulate the knowledge of previous classes.

External Attention. CVT plays the strengths of the attention mechanism in online CL. Unlike the vanilla self-attention in vision transformers [21,26] that derives the attention map by computing similarity between self-queries and self-keys [74], we introduce an external attention mechanism [79] to obtain the attention map by computing the affinities between self-queries and a learnable external

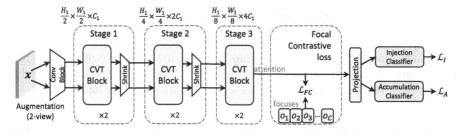

Fig. 3. The architecture of Contrastive Vision Transformer (CVT). CVT architecture is composed of stacked transformer blocks after a simple convolutional block. Shrink module performs downsampling to reduce the resolution of the activation maps and increase their number of channels between CVT stages. Focuses are a set of learnable attention vectors. After a projection layer, two classifiers serve for knowledge injection and accumulation, respectively.

key K_W with an attention bias B, which implicitly injects previous task information in the attention mechanism. Moreover, the proposed architecture can save the number of parameters compared to self-attention.

Let the input tensor be X, we apply linear transformation with weights W_q and W_v to get the vanilla self-query $Q_X=W_qX$ and self-value $V_X=W_vX$, respectively. We employ a linear layer K_W to replace the input-depended self-key, and explicitly add a learnable attention bias B to attention maps. Consider H attention heads, which are uniformly split into H segments Q_X^h, K_W^h, V_X^h, and B^h. The external attention mechanism computes the head-specific attention map A^h and concatenates the multi-head attention as follows:

$$A^h = \text{Softmax}\left(\frac{\text{Norm}(Q_X^h(K_W^h)^\top)+B^h}{\sqrt{d/H}}\right), \quad X_{out}^h = A^hV_X^h, \quad h = 1, ..., H,$$

(2)

where Norm() denotes batch normalization; d is the dimension of the key.

Learnable Focuses. CVT contains learnable *focuses* for each class, which could maintain and accumulate the knowledge of previous classes to alleviate forgetting for online CL. The class-wise learnable focuses $\mathcal{O} = \{o_1, o_2, ..., o_C\}$ is a set of learnable attention vectors, as shown in Fig. 3, where a focus o_c corresponds to class c, and C is the number of the seen classes. The size of the learnable focuses is negligible in relation to the overall model.

When a new class c appears in the data stream, the corresponding focus o_c to class c starts to participate in the training (refer to Eq. (3)). Even if class c no longer appears in the data stream afterward, the focus o_c always participates in the online CL training and serves as a negative sample for the other classes, as illustrated in Fig. 2(b1) and Fig. 2(b2). Thus, focuses $\mathcal{O}$ preserve and accumulate the previously learned class-specific knowledge and acts as a forgetting mitigation in online continual learning.

4.2 Focal Contrastive Continual Learning

We propose a rehearsal-based focal contrastive learning scheme to 1) tackle the imbalance issue of SCL in online CL and 2) accumulate class-specific knowledge, to alleviate the interference with previous tasks. The learning scheme includes two losses: a focal contrastive loss and a dual-classifier loss, as following.

Focal Contrastive Loss. For learning representation continually, we propose a focal contrastive loss $\mathcal{L}_{FC}$ in online CL. As mentioned in Sect. 3, during the training phase, the model observes a mini-batch $\mathcal{B}$ at a time sampled from task $\mathcal{T}_t$ in the data stream $\mathcal{D}$. An input batch for the model is composed of $\mathcal{B}$ and $\mathcal{B}_{\mathcal{M}}$ sampled from the memory buffer $\mathcal{M}$. The input batch and its augmented view are encoded by CVT blocks to generate attention z, as shown in Fig. 3. As mentioned previously, a set of class-wise learnable *focuses* $\mathcal{O} = \{o_1, o_2, ..., o_C\}$ is utilized by focal contrastive loss $\mathcal{L}_{FC}$, where a focus o_c is a learnable attention vector for class c and C is the number of the seen classes. The FC loss function is defined as:

$$\mathcal{L}_{FC} = \sum_{i \in \mathcal{I}} \frac{-1}{|\mathcal{P}(i) \cup o_{\tilde{y}_i}|} \sum_{z_p \in \mathcal{P}(i) \cup o_{\tilde{y}_i}} \delta_{z_p} \log \frac{\exp(z_i \cdot z_p / \tau)}{\sum_{j \in \mathcal{A}(i) \cup \mathcal{O}} \exp(z_i \cdot z_j / \tau)}, \quad (3)$$

where

$$\delta_{z_p} = \begin{cases} \mu, & z_p \in \mathcal{O} \\ 1.0, & z_p \in \mathcal{P}(i) \end{cases}, \quad (4)$$

and μ is the weight of focuses. We set $\mu > 1$ to make focuses play a more important role in contrastive learning; $\mathcal{P}(i)$ and $\mathcal{A}(i)$ are the same with supervised contrastive learning in Eq. (1); $\tau \in \mathbb{R}^+$ is a temperature hyperparameter.

The benefits of using focal contrastive loss are two-fold. First, it alleviates the imbalance issue in online CL by employing the class-wise focuses. Second, it accumulates the previous knowledge by the learnable focuses which will continually serve as the prototypes of classes to maintain class-specific information. With a proposed focal contrastive loss $\mathcal{L}_{FC}$ in training, CVT rebalances contrastive learning between new and past classes and improve the inter-class distinction and intra-class aggregation, as illustrated in Fig. 2. In Sect. 5.3, we empirically observe that $\mathcal{L}_{FC}$ outperforms the original $\mathcal{L}_{SCL}$ and boost the online CL.

Dual-classifier Loss. We propose a dual-classifier structure to decouple learning current classes and balancing all seen classes, which contains an *injection classifier* to inject new task representation into the model, alleviating interference to previously learned knowledge, and an *accumulation classifier* to integrate past and new knowledge in a balanced manner.

Let $g(\boldsymbol{x})$ be the representation of a sample $\boldsymbol{x}$ outputted from the Projection of CVT before the classifier. When new data stream batch arrives, we utilize the output from an independent injection classifier to compute a classification loss:

$$\mathcal{L}_I = \sum_{(\boldsymbol{x},y)\in\mathcal{B}} \ell(y, f_I(g(\boldsymbol{x}))), \tag{5}$$

where f_I denotes the injection classifier and ℓ adopts a cross-entropy loss. f_I is only trained on stream data and does not participate in the inference stage.

Besides, we employ an accumulation classifier to focus on improving the stability-plasticity trade-off by integrating previous and new knowledge in a balanced manner. The accumulation classifier is used at the inference stage for outputting the prediction. Rehearsing the limited memory data during learning new tasks is a crucial way to maintain previous knowledge. We can replay the exemplars stored in the memory buffer with their ground truth labels. In addition, the accumulation classifier also needs a supervised signal from current task data. Therefore, we give the accumulation classifier loss:

$$\mathcal{L}_A = \alpha\mathbb{E}_{(\boldsymbol{x}',y')\sim\mathcal{M}}\left[\ell(y', f_A(g(\boldsymbol{x}')))\right] + \beta \sum_{(\boldsymbol{x},y)\in\mathcal{B}} \ell(y, f_A(g(\boldsymbol{x}))), \tag{6}$$

where f_A denotes the accumulation classifier; α and β are the coefficients balancing knowledge consolidation. We approximate the expectation by computing gradients on batches sampled from the memory buffer.

Overall, the total loss used in CVT is the sum of Eq. (3), Eq. (5), and Eq. (6):

$$\mathcal{L} = \mathcal{L}_A + \mathcal{L}_I + \gamma\mathcal{L}_{\text{FC}}, \tag{7}$$

where γ is the coefficient balancing $\mathcal{L}_{\text{FC}}$. After updating the whole CVT model on a mini-batch $\mathcal{B}$, the memory buffer $\mathcal{M}$ will be updated with $\mathcal{B}$ by Reservoir sampling [9,77].

5   Experiment

5.1   Experimental Setup and Implementation

We consider a strict evaluation setting [38,75] for online continual learning, including task-aware protocol [56] and task-free protocol [39]. For task-aware protocol [56], the task identities are required for each time evaluation. For task-free protocol [39], the task identities are unavailable at inference time.

Datasets. Online continual learning benchmarks evaluate the capacity of an algorithm to learn on not independent and identically distributed (non-iid) data. CIFAR-100 [44] contains 100 classes and each class has 100 testing and 500 training color images. TinyImageNet [70] consists 200 classes that include 100,000 images for training and 10,000 images for testing. ImageNet100 [62] contains 100 classes randomly chosen from ILSVRC [19], including about 120,000 images for training and 5,000 images for validation.

Table 1. Results (overall accuracy %) on CIFAR100 benchmark which is averaged over five runs. #Paras means the number of parameters in the model, which is counted by million.

Memory Buffer	Method	#Paras	10 splits		20 splits	
			Task-free	Task-aware	Task-free	Task-aware
–	SGD	11.2	5.77±0.35	37.44±1.83	3.53±0.12	36.81±2.63
500	ER [63]	11.2	15.59±1.20	59.72±0.83	12.51±0.77	62.72±1.70
	GEM [49]	11.2	14.34±1.63	50.39±0.28	5.98±0.33	57.15±1.77
	AGEM [14]	11.2	6.35±0.13	39.18±0.27	3.62±0.08	39.55±0.17
	iCaRL [62]	11.2	15.18±0.31	48.95±0.33	12.79±0.26	60.53±0.38
	FDR [8]	11.2	5.97±0.20	32.18±1.60	3.60±0.07	39.98±1.32
	GSS [4]	11.2	10.91±0.36	59.10±0.26	6.33±0.30	62.80±2.31
	DER++ [9]	11.2	15.72±1.29	54.45±1.59	11.29±0.25	63.62±1.05
	HAL [13]	22.4	10.51±0.63	33.70±1.37	7.09±0.39	52.89±2.36
	ERT [10]	11.2	16.28±0.73	60.11±2.18	17.92±0.42	68.08±0.37
	ASER$_\mu$ [67]	11.2	12.42±0.68	53.77±1.30	9.63±0.51	58.91±1.62
	RM [6]	11.2	14.32±0.49	58.76±1.82	13.73±0.60	64.73±2.01
	CLS [5]	33.8	15.06±0.57	59.82±1.24	14.84±0.83	65.74±1.94
	CVT (ours)	**8.9**	**24.45±0.62**	**62.52±1.43**	**21.81±0.52**	**71.23±1.40**
1000	ER [63]	11.2	20.41±1.46	63.39±1.37	17.02±1.63	68.52±1.09
	GEM [49]	11.2	16.49±1.08	52.30±0.51	8.40±1.05	62.59±1.89
	AGEM [14]	11.2	6.57±0.12	40.38±0.30	3.74±0.05	42.39±0.42
	iCaRL [62]	11.2	16.31±0.26	50.49±0.18	13.03±0.26	61.13±0.20
	FDR [8]	11.2	6.58±0.31	36.99±0.45	3.72±0.04	42.45±0.39
	GSS [4]	11.2	12.38±0.59	60.75±0.28	7.40±0.23	66.06±1.25
	DER++ [9]	11.2	21.27±1.69	61.80±1.24	13.42±0.50	71.26±0.61
	HAL [13]	22.4	11.81±0.79	39.67±2.63	13.14±0.72	60.03±1.20
	ERT [10]	11.2	23.43±0.58	62.25±1.33	24.58±0.36	72.61±0.67
	ASER$_\mu$ [67]	11.2	14.38±0.43	58.91±1.76	12.79±0.60	62.47±1.82
	RM [6]	11.2	22.41±1.28	61.82±2.13	18.91±1.15	67.30±1.34
	CLS [5]	33.8	19.73±1.17	62.54±1.52	17.06±1.47	70.08±0.85
	CVT (ours)	**8.9**	**28.83±0.86**	**65.86±1.24**	**28.15±0.51**	**75.76±0.93**

Baselines. We compare CVT with state-of-the-art and well-established Online CL baselines, including 11 rehearsal-based methods (ER [63], GEM [49], AGEM [14], GSS [4], FDR [8], HAL [13], ASER$_\mu$ [67], ERT [10], RM [6], SCR [52], and CLS [5]), 2 methods leveraging Knowledge Distillation (iCaRL [62] and DER++ [9]). Besides, we also compare vision transformers (ViT [21], LeViT [26], CoAT [88], and CCT [33]) with rehearsal strategy for continual learning. We also provide the results of simply performing SGD without any countermeasure to alleviate forgetting.

Metrics. We evaluate online CL methods in terms of accuracy and forgetting following [9,12,14]. The accuracy is defined by $\mathbf{A}_T = \frac{1}{T} \sum_{t=1}^{T} a_{T,t}$, and the forgetting is defined by $\mathbf{F}_T = \frac{1}{T-1} \sum_{t=1}^{T-1} \max_{i \in \{1,...,T-1\}} (a_{i,t} - a_{T,t})$, where $a_{T,t}$ is the inference accuracy on task $\mathcal{T}_t$ when the model finished learning task $\mathcal{T}_T$.

Table 2. Results (overall accuracy %) on TinyImageNet and ImageNet100, which are averaged over three runs. #Paras means the number of parameters in the model, which is counted by million.

Memory Buffer	Method	#Paras	TinyImageNet		#Paras	ImageNet100	
			Task-free	Task-aware		Task-free	Task-aware
–	SGD	11.2	4.54±0.03	26.25±0.16	11.2	3.98±0.02	19.05±0.17
500	ER [63]	11.2	9.71±0.18	42.76±0.35	11.2	9.88±0.46	32.38±0.63
	AGEM [14]	11.2	4.63±0.08	27.86±0.13	11.2	3.38±0.08	21.80±0.15
	iCaRL [62]	11.2	6.17±1.03	27.22±1.52	11.2	7.70±0.32	20.45±0.80
	FDR [8]	11.2	5.19±0.18	28.23±0.46	11.2	3.34±0.16	19.24±0.07
	DER++ [9]	11.2	9.56±0.69	40.52±0.47	11.2	10.30±0.24	29.20±0.38
	ERT [10]	11.2	9.95±0.72	40.42±1.57	11.2	10.28±0.25	28.53±0.48
	ASER$_\mu$ [67]	11.2	9.22±0.25	41.09±0.68	11.2	9.75±0.57	31.71±0.79
	RM [6]	11.2	8.39±1.37	41.63±0.74	11.2	8.53±0.68	28.30±0.52
	SCR [52]	11.2	9.08±0.74	39.85±0.93	11.2	8.81±0.79	29.62±1.24
	CVT (ours)	**9.0**	**14.71±1.04**	**43.93±1.42**	**9.4**	**14.82±0.31**	**36.74±0.46**
1000	ER [63]	11.2	12.46±0.45	45.50±0.61	11.2	10.42±0.51	34.26±0.43
	AGEM [14]	11.2	4.92±0.13	28.38±0.15	11.2	3.66±0.05	23.56±0.19
	iCaRL [62]	11.2	6.91±0.52	28.56±0.37	11.2	8.93±0.48	22.37±1.04
	FDR [8]	11.2	5.27±0.12	28.94±0.36	11.2	3.58±0.09	21.28±0.11
	DER++ [9]	11.2	12.97±0.42	47.21±0.33	11.2	13.94±0.71	40.02±0.39
	ERT [10]	11.2	13.84±0.77	44.65±0.79	11.2	12.26±0.23	33.88±0.60
	ASER$_\mu$ [67]	11.2	12.26±0.38	46.02±0.82	11.2	11.38±0.54	35.76±1.28
	RM [6]	11.2	11.73±0.89	45.89±0.64	11.2	11.85±1.17	32.72±0.85
	SCR [52]	11.2	10.19±1.14	43.58±1.20	11.2	10.74±0.85	31.84±2.27
	CVT (ours)	**9.0**	**16.54±1.22**	**48.50±0.88**	**9.4**	**18.02±0.25**	**42.61±0.72**

Implementation Details. In order to compare each method fairly, we train all networks using stochastic gradient descent (SGD) optimizer. The images used for training are randomly cropped and flipped for each method following [9,10,68]. We adopt 1 epoch with mini-batch size of 10 for all datasets, following [9,10,62,93]. Online continual learning baselines use ResNet18 [35] as backbone and cross-entropy as the classification loss, following [6,9,13,14,68,71]. The implementation of transformer block is based on LeViT [26] and ViT [21]. CVT framework employs GELU activation and dropout in transformer blocks and uses a global average pooling to the last activation map.

5.2 Comparison to State-of-the-Art Methods

Evaluation on CIFAR100. Following the setting proposed in [62,89], we trains all 100 classes in several splits, including 10, 20 incremental tasks. Table 1 summarizes the overall accuracy on CIFAR100 with 500 and 1000 memory sizes. It is demonstrated that CVT outperforms other baselines by a considerable margin in different incremental splits, e.g., CVT can improve the accuracy of continual learning by more than 8% in 10-split with 500 memory capacity. Especially in the case of small memory, the advantage of CVT is more obvious, which indicates CVT can effectively alleviate the imbalance issue in online CL. It is worth noting that although CVT uses fewer parameters (8.9M) than other methods (11.2M∼33.8M), it can still achieve superior performance. One reason is that

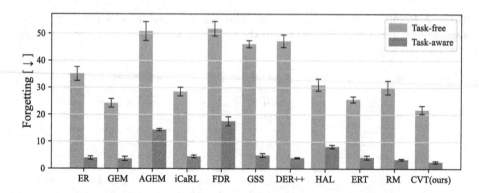

Fig. 4. Forgetting results (%) on CIFAR100 (lower is better).

CVT inherits the merits of transformers for modeling the stream of tasks without stacking a lot of parameters, and besides, the number of parameters for the proposed learnable focuses is extremely small.

Evaluation on ImageNet Datasets. Table 2 summarizes the evaluation results for the TinyImageNet and ImageNet100 datasets with 10 splits. It is demonstrated that CVT consistently surpasses other methods with a considerable margin for Task-free and Task-aware on TinyImageNet and ImageNet100 datasets. Specifically, our method outperforms the state-of-the-art with about **4.3%** for the Task-aware accuracy on the ImageNet100 benchmark. For TinyImageNet benchmark, the Task-free accuracy is improved from 9.95% to 14.71%(**+4.76%**). Moreover, CVT takes fewer parameters compared to other CNN-based methods.

Forgetting. To compare the alleviating forgetting capability, we assess the *average forgetting* [9,12] that measures the performance degradation in subsequent tasks. As shown in Fig. 4, CVT suffers from less forgetting than all the other baselines in both of Task-free and Task-aware settings with memory buffer 1000 on CIFAR100. This is because CVT utilizes the focal contrastive loss and dual-classifier loss, which improve the stability of the vision transformer network.

Incremental Performance. We also evaluate the *average incremental performance* [9,62] under the Task-free protocol with 500 memory buffer, which is the result of evaluating on all the tasks observed so far after completing each task. As illustrated in Fig. 5, the results are curves of accuracy and forgetting after each task. It is observed that during the learning process, most methods degrade rapidly as new tasks arrive, while our method consistently outperforms the state-of-the-art methods in both accuracy and forgetting.

5.3   Ablation Study and Analysis

Comparison to Transformer and CNN Backbone. We compare CVT to Vision Transformer networks (ViT [21], LeViT [26], CoAT [88], and CCT [33])

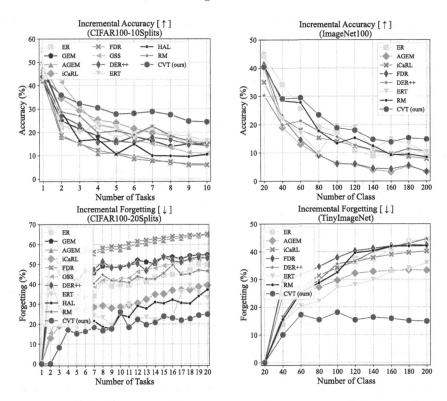

Fig. 5. Incremental performance evaluated on tasks observed so far. [↑] higher is better, [↓] lower is better *(best seen in color)*.

and the CNN benchmark ResNet18 [35] under the proposed rehearsal strategy in online continual learning. Table 3 demonstrates the results of accuracy and forgetting on CIFAR100 and TinyImageNet with 500 memory. It is observed that ViT is not up to the task of online continual learning, since it is "data-hungry" and only fits i.i.d. large datasets. Besides, LeViT, CoAT, and CCT contain CNN structures to obtain inductive biases, which still suffer from catastrophic forgetting in online CL. We can find that using Vision Transformer directly for online CL can not consistently outperform CNN-based networks. Our proposed CVT architecture essentially inherits the merits of CNN and transformers and thus works well in online streaming data and modeling long-dependencies in the input data. Moreover, CVT even takes fewer parameters to achieve better performance for online CL, which also benefits from the focal contrastive loss and the dual-classifier structure.

Effect of Each Component. To assess the effects of the components in CVT, we perform ablation study in terms of accuracy and forgetting. From Table 3 we can observe that the proposed focal contrastive loss $\mathcal{L}_{FC}$ plays an important role in alleviating catastrophic forgetting and accumulating knowledge. However,

Table 3. Ablation study on backbone and each component of CVT. "-" indicates the removal operation. "+" represents an add component operation.

Method	#Paras	CIFAR100		#Paras	TinyImageNet	
		Accuracy[↑]	Forgetting[↓]		Accuracy[↑]	Forgetting[↓]
ViT [21]	16.2	8.48	35.56	16.3	7.91	42.26
LeViT [26]	10.9	14.55	44.53	12.1	9.02	41.41
CoAT [88]	10.3	13.17	47.64	11.3	8.78	39.80
CCT [33]	**4.3**	13.86	51.06	**4.4**	9.97	40.21
ResNet18 [35]	11.2	15.72	43.82	11.2	9.56	42.13
ResNet18 + $\mathcal{L}_{FC}$	11.3	17.49	38.73	11.3	10.17	36.49
ResNet18 + dual-classifier	11.2	18.84	35.38	11.2	10.91	33.55
CVT - $\mathcal{L}_{FC}$	8.8	19.92	26.16	8.9	12.47	19.43
CVT - $\mathcal{L}_{FC}$ + $\mathcal{L}_{SCL}$	8.9	20.73	25.52	9.0	12.59	20.47
CVT - dual-classifier	8.9	22.08	29.81	9.0	13.63	18.95
CVT (ours)	8.9	**24.45**	**21.86**	9.0	**14.71**	**16.32**

if we simply use supervised contrastive learning loss $\mathcal{L}_{SCL}$ to replace $\mathcal{L}_{FC}$, we find that the forgetting problem is not mitigated compared to not using any contrastive loss. This is because using $\mathcal{L}_{SCL}$ directly could cause a severe imbalance between new and past classes in online CL, which limits the learning of transferable representation. While $\mathcal{L}_{FC}$ can overcome the issue by utilizing the learnable focuses to boost the performance of online CL. This supports that $\mathcal{L}_{FC}$ can **rebalance contrastive learning between new and past classes and improve the inter-class distinction and intra-class aggregation**. Besides, we can see the dual-classifier loss obtains 2.37% and 7.97% gain in terms of accuracy and forgetting on CIFAR100, respectively. The results of Table 3 demonstrate the effectiveness of each component of CVT.

6   Conclusion

In this paper, we propose a novel attention-based framework, Contrastive Vision Transformer (CVT), to effectively mitigate the catastrophic forgetting for online CL. To the best of our knowledge, this paper is the first in the literature to design a Transformer for online CL. CVT contains external attention and learnable focuses to accumulate previous knowledge and maintain class-specific information. With a proposed focal contrastive loss in training, CVT rebalances contrastive continual learning between new and past classes and improves the inter-class distinction and intra-class aggregation. Moreover, CVT designs a dual-classifier structure to decouple learning current classes and balancing all seen classes. Extensive experimental results show that our approach significantly outperforms current state-of-the-art methods with fewer parameters. Ablation study validates the effectiveness of each proposed component.

Acknowledgment. This work is supported by ARC FL-170100117, DP-180103424, IC-190100031, and LE-200100049.

References

1. Abati, D., Tomczak, J., Blankevoort, T., Calderara, S., Cucchiara, R., Bejnordi, B.E.: Conditional channel gated networks for task-aware continual learning. In: Proceedings of the IEEE/CVF Conference on Computer Vision and Pattern Recognition, pp. 3931–3940 (2020)
2. Aljundi, R., Babiloni, F., Elhoseiny, M., Rohrbach, M., Tuytelaars, T.: Memory aware synapses: learning what (not) to forget. In: Ferrari, V., Hebert, M., Sminchisescu, C., Weiss, Y. (eds.) ECCV 2018. LNCS, vol. 11207, pp. 144–161. Springer, Cham (2018). https://doi.org/10.1007/978-3-030-01219-9_9
3. Aljundi, R., Chakravarty, P., Tuytelaars, T.: Expert gate: lifelong learning with a network of experts. In: CVPR, pp. 3366–3375 (2017)
4. Aljundi, R., Lin, M., Goujaud, B., Bengio, Y.: Gradient based sample selection for online continual learning. In: Advances in Neural Information Processing Systems (2019)
5. Arani, E., Sarfraz, F., Zonooz, B.: Learning fast, learning slow: A general continual learning method based on complementary learning system. In: International Conference on Learning Representations (ICLR) (2022)
6. Bang, J., Kim, H., Yoo, Y., Ha, J.W., Choi, J.: Rainbow memory: continual learning with a memory of diverse samples. In: Proceedings of the IEEE/CVF Conference on Computer Vision and Pattern Recognition (CVPR), pp. 8218–8227 (2021)
7. Bengio, Y., Goodfellow, I., Courville, A.: Deep Learning, vol. 1. MIT press Massachusetts, USA (2017)
8. Benjamin, A.S., Rolnick, D., Kording, K.P.: Measuring and regularizing networks in function space. In: International Conference on Learning Representations (2019)
9. Buzzega, P., Boschini, M., Porrello, A., Abati, D., Calderara, S.: Dark experience for general continual learning: a strong. In: Advances in Neural Information Processing Systems (NeurIPS), Simple Baseline (2020)
10. Buzzega, P., Boschini, M., Porrello, A., Calderara, S.: Rethinking experience replay: a bag of tricks for continual learning. In: International Conference on Pattern Recognition (ICPR), pp. 2180–2187. IEEE (2021)
11. Caron, M., Misra, I., Mairal, J., Goyal, P., Bojanowski, P., Joulin, A.: Unsupervised learning of visual features by contrasting cluster assignments. Adv. Neural. Inf. Process. Syst. **33**, 9912–9924 (2020)
12. Chaudhry, A., Dokania, P.K., Ajanthan, T., Torr, P.H.S.: Riemannian walk for incremental learning: understanding forgetting and intransigence. In: Ferrari, V., Hebert, M., Sminchisescu, C., Weiss, Y. (eds.) ECCV 2018. LNCS, vol. 11215, pp. 556–572. Springer, Cham (2018). https://doi.org/10.1007/978-3-030-01252-6_33
13. Chaudhry, A., Gordo, A., Dokania, P.K., Torr, P.H., Lopez-Paz, D.: Using hindsight to anchor past knowledge in continual learning. In: AAAI (2021)
14. Chaudhry, A., Ranzato, M., Rohrbach, M., Elhoseiny, M.: Efficient lifelong learning with a-GEM. In: International Conference on Learning Representations (ICLR) (2019)
15. Chen, S., Wang, L., Wang, Z., Yan, Y., Wang, D.H., Zhu, S.: Learning meta-adversarial features via multi-stage adaptation network for robust visual object tracking. Neurocomputing **491**, 365–381 (2022)
16. Chen, T., Kornblith, S., Norouzi, M., Hinton, G.: A simple framework for contrastive learning of visual representations. In: International Conference on Machine Learning, pp. 1597–1607. PMLR (2020)

17. Dai, Z., Liu, H., Le, Q.V., Tan, M.: Coatnet: Marrying convolution and attention for all data sizes. In: Advances in Neural Information Processing Systems (2021)
18. Delange, M., et al.: A continual learning survey: Defying forgetting in classification tasks. IEEE Tran. Pattern Anal. Mach. Intell., 1 (2021)
19. Deng, J., Dong, W., Socher, R., Li, L.J., Li, K., Fei-Fei, L.: Imagenet: a large-scale hierarchical image database. In: Proceedings of the IEEE Conference on Computer Vision and Pattern Recognition (CVPR), pp. 248–255 (2009)
20. Devlin, J., Chang, M.W., Lee, K., Toutanova, K.: Bert: Pre-training of deep bidirectional transformers for language understanding. arXiv preprint arXiv:1810.04805 (2018)
21. Dosovitskiy, A., et al.: An image is worth 16x16 words: Transformers for image recognition at scale. In: International Conference on Learning Representations (2021)
22. Douillard, A., Cord, M., Ollion, C., Robert, T., Valle, E.: PODNet: pooled outputs distillation for small-tasks incremental learning. In: Vedaldi, A., Bischof, H., Brox, T., Frahm, J.-M. (eds.) ECCV 2020. LNCS, vol. 12365, pp. 86–102. Springer, Cham (2020). https://doi.org/10.1007/978-3-030-58565-5_6
23. Douillard, A., Ramé, A., Couairon, G., Cord, M.: Dytox: Transformers for continual learning with dynamic token expansion. Conference on Computer Vision and Pattern Recognition (CVPR) (2022)
24. Duan, Y., Wang, Z., Wang, J., Wang, Y.K., Lin, C.T.: Position-aware image captioning with spatial relation. Neurocomputing **497**, 28–38 (2022)
25. Girdhar, R., Carreira, J., Doersch, C., Zisserman, A.: Video action transformer network. In: Proceedings of the IEEE/CVF Conference on Computer Vision and Pattern Recognition (CVPR). pp. 244–253 (2019)
26. Graham, B., et al.: Levit: a vision transformer in convnet's clothing for faster inference. In: Proceedings of the IEEE/CVF International Conference on Computer Vision (ICCV), pp. 12259–12269, October 2021
27. Grossberg, S.: Adaptive resonance theory: how a brain learns to consciously attend, learn, and recognize a changing world. Neural Netw. **37**, 1–47 (2013)
28. Gunel, B., Du, J., Conneau, A., Stoyanov, V.: Supervised contrastive learning for pre-trained language model fine-tuning. arXiv preprint arXiv:2011.01403 (2020)
29. Guo, J., Gong, M., Liu, T., Zhang, K., Tao, D.: LTF: A label transformation framework for correcting label shift. In: ICML, vol. 119, pp. 3843–3853 (2020)
30. Guo, J., Li, J., Fu, H., Gong, M., Zhang, K., Tao, D.: Alleviating semantics distortion in unsupervised low-level image-to-image translation via structure consistency constraint. In: Proceedings of the IEEE/CVF Conference on Computer Vision and Pattern Recognition (CVPR), pp. 18249–18259 (2022)
31. Hadsell, R., Chopra, S., LeCun, Y.: Dimensionality reduction by learning an invariant mapping. In: 2006 IEEE Computer Society Conference on Computer Vision and Pattern Recognition (CVPR 2006), vol. 2, pp. 1735–1742. IEEE (2006)
32. Han, K., et al.: A survey on visual transformer. ArXiv abs/2012.12556 (2020)
33. Hassani, A., Walton, S., Shah, N., Abuduweili, A., Li, J., Shi, H.: Escaping the big data paradigm with compact transformers. arXiv preprint arXiv:2104.05704 (2021)
34. He, J., Mao, R., Shao, Z., Zhu, F.: Incremental learning in online scenario. In: Proceedings of the IEEE/CVF Conference on Computer Vision and Pattern Recognition (CVPR), pp. 13926–13935 (2020)
35. He, K., Zhang, X., Ren, S., Sun, J.: Deep residual learning for image recognition. In: Proceedings of the IEEE Conference on Computer Vision and Pattern Recognition, pp. 770–778 (2016)

36. Hinton, G., Vinyals, O., Dean, J.: Distilling the Knowledge in a Neural Network. In: NeurIPS Workshop (2014)
37. Hou, S., Pan, X., Loy, C.C., Wang, Z., Lin, D.: Learning a unified classifier incrementally via rebalancing. In: CVPR, pp. 831–839 (2019)
38. Hsu, Y.C., Liu, Y.C., Ramasamy, A., Kira, Z.: Re-evaluating continual learning scenarios: a categorization and case for strong baselines. In: NeurIPS Continual Learning Workshop (2018)
39. Jin, X., Sadhu, A., Du, J., Ren, X.: Gradient-based editing of memory examples for online task-free continual learning. In: Advances in Neural Information Processing Systems 34 (2021)
40. Khan, S., et al.: Transformers in vision: a survey. arXiv preprint arXiv:2101.01169 (2021)
41. Khosla, P., et al.: Supervised contrastive learning. In: Advances in Neural Information Processing Systems 33 (2020)
42. Kirkpatrick, J., et al.: Overcoming catastrophic forgetting in neural networks. Proc. Natl. Acad. Sci. **114**(13), 3521–3526 (2017)
43. Kong, Y., Liu, L., Wang, J., Tao, D.: Adaptive curriculum learning. In: Proceedings of the IEEE/CVF International Conference on Computer Vision, pp. 5067–5076 (2021)
44. Krizhevsky, A., Hinton, G., et al.: Learning multiple layers of features from tiny images (2009)
45. Li, X., Zhou, Y., Wu, T., Socher, R., Xiong, C.: Learn to grow: a continual structure learning framework for overcoming catastrophic forgetting. In: International Conference on Machine Learning (ICML) (2019)
46. Li, Z., Hoiem, D.: Learning without forgetting. IEEE Trans. Pattern Anal. Mach. Intell. (TPAMI) **40**(12) (2017)
47. Liu, X., Wang, X., Matwin, S.: Improving the interpretability of deep neural networks with knowledge distillation. In: 2018 IEEE International Conference on Data Mining Workshops (ICDMW). IEEE (2018)
48. Liu, Z., et al.: Swin transformer: hierarchical vision transformer using shifted windows. In: Proceedings of the IEEE/CVF International Conference on Computer Vision (ICCV), pp. 10012–10022 (2021)
49. Lopez-Paz, D., Ranzato, M.: Gradient episodic memory for continual learning. In: Advances in Neural Information Processing Systems (NeurIPS) (2017)
50. Losing, V., Hammer, B., Wersing, H.: Incremental on-line learning: a review and comparison of state of the art algorithms. Neurocomputing **275**, 1261–1274 (2018)
51. Luong, T., Pham, H., Manning, C.D.: Effective approaches to attention-based neural machine translation. In: Proceedings of the 2015 Conference on Empirical Methods in Natural Language Processing, pp. 1412–1421. Association for Computational Linguistics (September 2015)
52. Mai, Z., Li, R., Kim, H., Sanner, S.: Supervised contrastive replay: Revisiting the nearest class mean classifier in online class-incremental continual learning. In: 2021 IEEE/CVF Conference on Computer Vision and Pattern Recognition Workshops (CVPRW), pp. 3584–3594. IEEE (2021)
53. Mallya, A., Davis, D., Lazebnik, S.: Piggyback: adapting a single network to multiple tasks by learning to mask weights. In: Ferrari, V., Hebert, M., Sminchisescu, C., Weiss, Y. (eds.) ECCV 2018. LNCS, vol. 11208, pp. 72–88. Springer, Cham (2018). https://doi.org/10.1007/978-3-030-01225-0_5
54. McCloskey, M., Cohen, N.J.: Catastrophic interference in connectionist networks: the sequential learning problem. Psychology of Learning and Motivation, vol. 24, pp. 109–165. Academic Press (1989)

55. Mermillod, M., Bugaiska, A., Bonin, P.: The stability-plasticity dilemma: investigating the continuum from catastrophic forgetting to age-limited learning effects. Front. Psychol. **4**, 504 (2013)
56. Pham, Q., Liu, C., Hoi, S.: Dualnet: continual learning, fast and slow. In: Advances in Neural Information Processing Systems 34 (2021)
57. Radford, A., et al.: Learning transferable visual models from natural language supervision. arXiv preprint arXiv:2103.00020 (2021)
58. Radford, A., Wu, J., Child, R., Luan, D., Amodei, D., Sutskever, I., et al.: Language models are unsupervised multitask learners. OpenAI blog **1**(8), 9 (2019)
59. Rajasegaran, J., Hayat, M., Khan, S.H., Khan, F.S., Shao, L.: Random path selection for continual learning. In: Advances in Neural Information Processing Systems (NeurIPS) (2019)
60. Rannen, A., Aljundi, R., Blaschko, M.B., Tuytelaars, T.: Encoder based lifelong learning. In: Proceedings of the IEEE International Conference on Computer Vision, pp. 1320–1328 (2017)
61. Ratcliff, R.: Connectionist models of recognition memory: constraints imposed by learning and forgetting functions. Psychol. Rev. **97**(2), 285 (1990)
62. Rebuffi, S.A., Kolesnikov, A., Sperl, G., Lampert, C.H.: icarl: incremental classifier and representation learning. In: Proceedings of the IEEE Conference on Computer Vision and Pattern Recognition (CVPR) (2017)
63. Riemer, M., et al.: Learning to learn without forgetting by maximizing transfer and minimizing interference. In: International Conference on Learning Representations (ICLR) (2019)
64. Schwarz, J., et al.: Progress & Compress: a scalable framework for continual learning. In: International Conference on Machine Learning (2018)
65. Serra, J., Suris, D., Miron, M., Karatzoglou, A.: Overcoming catastrophic forgetting with hard attention to the task. In: International Conference on Machine Learning, vol. 80, pp. 4548–4557 (2018)
66. Shi, Y., Yuan, L., Chen, Y., Feng, J.: Continual learning via bit-level information preserving. In: Proceedings of the IEEE/CVF Conference on Computer Vision and Pattern Recognition (CVPR), pp. 16674–16683 (2021)
67. Shim, D., Mai, Z., Jeong, J., Sanner, S., Kim, H., Jang, J.: Online class-incremental continual learning with adversarial shapley value. In: Proceedings of the AAAI Conference on Artificial Intelligence, vol. 35, pp. 9630–9638 (2021)
68. Simon, C., Koniusz, P., Harandi, M.: On learning the geodesic path for incremental learning. In: Proceedings of the IEEE/CVF Conference on Computer Vision and Pattern Recognition (CVPR), pp. 1591–1600 (2021)
69. Singh, P., Mazumder, P., Rai, P., Namboodiri, V.P.: Rectification-based knowledge retention for continual learning. In: Proceedings of the IEEE/CVF Conference on Computer Vision and Pattern Recognition (CVPR), pp. 15282–15291 (2021)
70. Stanford: Tiny ImageNet Challenge (CS231n) (2015). http://tiny-imagenet.herokuapp.com/
71. Tang, S., Chen, D., Zhu, J., Yu, S., Ouyang, W.: Layerwise optimization by gradient decomposition for continual learning. In: Proceedings of the IEEE/CVF Conference on Computer Vision and Pattern Recognition (CVPR), pp. 9634–9643 (2021)
72. Tian, Y., Krishnan, D., Isola, P.: Contrastive multiview coding. arXiv preprint arXiv:1906.05849 (2019)
73. Vaswani, A., Ramachandran, P., Srinivas, A., Parmar, N., Hechtman, B., Shlens, J.: Scaling local self-attention for parameter efficient visual backbones. In: Proceedings of the IEEE/CVF Conference on Computer Vision and Pattern Recognition, pp. 12894–12904 (2021)

74. Vaswani, A., et al.: Attention is all you need. In: Advances in neural information processing systems, pp. 5998–6008 (2017)

75. van de Ven, G.M., Tolias, A.S.: Three continual learning scenarios. NeurIPS Continual Learning Workshop (2018)

76. Verwimp, E., De Lange, M., Tuytelaars, T.: Rehearsal revealed: the limits and merits of revisiting samples in continual learning. In: ICCV (2021)

77. Vitter, J.S.: Random sampling with a reservoir. ACM Trans. Math. Softw. (TOMS) **11**(1), 37–57 (1985)

78. Wang, Z., Duan, Y., Liu, L., Tao, D.: Multi-label few-shot learning with semantic inference. In: Proceedings of the AAAI Conference on Artificial Intelligence, vol. 35, pp. 15917–15918 (2021)

79. Wang, Z., Liu, L., Duan, Y., Kong, Y., Tao, D.: Continual learning with lifelong vision transformer. In: Proceedings of the IEEE/CVF Conference on Computer Vision and Pattern Recognition (CVPR), pp. 171–181, June 2022

80. Wang, Z., Liu, L., Duan, Y., Tao, D.: Continual learning with embeddings: Algorithm and analysis. In: ICML 2021 Workshop on Theory and Foundation of Continual Learning (2021)

81. Wang, Z., Liu, L., Duan, Y., Tao, D.: Continual learning through retrieval and imagination. In: Proceedings of the AAAI Conference on Artificial Intelligence, vol. 36(8), pp. 8594–8602 (2022)

82. Wang, Z., Liu, L., Duan, Y., Tao, D.: Sin: semantic inference network for few-shot streaming label learning. IEEE Trans. Neural Networks Learn. Syst., 1–14 (2022)

83. Wang, Z., Liu, L., Tao, D.: Deep streaming label learning. In: International Conference on Machine Learning (ICML), vol. 119, pp. 9963–9972 (2020)

84. Wu, H., Xiao, B., Codella, N., Liu, M., Dai, X., Yuan, L., Zhang, L.: Cvt: introducing convolutions to vision transformers. In: Proceedings of the IEEE/CVF International Conference on Computer Vision (ICCV), pp. 22–31, October 2021

85. Xi, H., Aussel, D., Liu, W., Waller, S.T., Rey, D.: Single-leader multi-follower games for the regulation of two-sided mobility-as-a-service markets. Europ. J. Oper. Res. (2022)

86. Xi, H., He, L., Zhang, Y., Wang, Z.: Bounding the efficiency gain of differentiable road pricing for EVS and GVS to manage congestion and emissions. PLoS ONE **15**(7), e0234204 (2020)

87. Xi, H., Liu, W., Rey, D., Waller, S.T., Kilby, P.: Incentive-compatible mechanisms for online resource allocation in mobility-as-a-service systems. arXiv preprint arXiv:2009.06806 (2020)

88. Xu, W., Xu, Y., Chang, T., Tu, Z.: Co-scale conv-attentional image transformers. In: Proceedings of the IEEE/CVF International Conference on Computer Vision (ICCV), pp. 9981–9990, October 2021

89. Yan, S., Xie, J., He, X.: Der: dynamically expandable representation for class incremental learning. In: Proceedings of the IEEE Conference on Computer Vision and Pattern Recognition (CVPR) (2021)

90. Yu, L., et al.: Semantic drift compensation for class-incremental learning. In: Proceedings of the IEEE/CVF Conference on Computer Vision and Pattern Recognition, pp. 6982–6991 (2020)

91. Yu, P., Chen, Y., Jin, Y., Liu, Z.: Improving vision transformers for incremental learning. arXiv preprint arXiv:2112.06103 (2021)

92. Yuan, L., et al.: Tokens-to-token vit: training vision transformers from scratch on imagenet. In: Proceedings of the IEEE/CVF International Conference on Computer Vision (ICCV), pp. 558–567, October 2021

93. Zenke, F., Poole, B., Ganguli, S.: Continual learning through synaptic intelligence. In: International Conference on Machine Learning (ICML) (2017)
94. Zhao, X., Zhu, J., Luo, B., Gao, Y.: Survey on facial expression recognition: History, applications, and challenges. IEEE Multimedia **28**(4), 38–44 (2021)
95. Zhu, J., Luo, B., Zhao, S., Ying, S., Zhao, X., Gao, Y.: Iexpressnet: facial expression recognition with incremental classes. In: Proceedings of the 28th ACM International Conference on Multimedia, pp. 2899–2908 (2020)
96. Zhu, J., Wei, Y., Feng, Y., Zhao, X., Gao, Y.: Physiological signals-based emotion recognition via high-order correlation learning. ACM Trans. Multimed. Comput. Commun. Appl. (TOMM) **15**(3s), 1–18 (2019)

CPrune: Compiler-Informed Model Pruning for Efficient Target-Aware DNN Execution

Taeho Kim[1], Yongin Kwon[2]([✉]), Jemin Lee[2], Taeho Kim[2],
and Sangtae Ha[1]

[1] University of Colorado Boulder, Boulder, USA
{taeho.kim,sangtae.ha}@colorado.edu
[2] Electronics and Telecommunications Research Institute, Daejeon, South Korea
{yongin.kwon,leejaymin,taehokim}@etri.re.kr

Abstract. Mobile devices run deep learning models for various purposes, such as image classification and speech recognition. Due to the resource constraints of mobile devices, researchers have focused on either making a lightweight deep neural network (DNN) model using model pruning or generating an efficient code using compiler optimization. Surprisingly, we found that the straightforward integration between model compression and compiler auto-tuning often does not produce the most efficient model for a target device. We propose CPrune, a compiler-informed model pruning for efficient target-aware DNN execution to support an application with a required target accuracy. CPrune makes a lightweight DNN model through informed pruning based on the structural information of subgraphs built during the compiler tuning process. Our experimental results show that CPrune increases the DNN execution speed up to 2.73× compared to the state-of-the-art TVM auto-tune while satisfying the accuracy requirement.

1 Introduction

Deep neural networks (DNNs) have become increasingly popular for various applications like image processing and speech recognition. DNN requires heavy computation due to large and complex neural networks, so recent approaches proposed intelligent offloading using the cloud for resource-constrained mobile devices. Unfortunately, they often suffer from network overhead and disruptions, necessitating running DNNs directly on devices. Furthermore, making DNN inferences on devices is often desirable for user experience and privacy.

Several efforts have focused on improving DNN inference on mobile devices by considering energy efficiency, accuracy, and execution time. These efforts are primarily divided into two approaches: (1) making a lightweight DNN model

---

Supplementary Information The online version contains supplementary material available at https://doi.org/10.1007/978-3-031-20044-1_37.

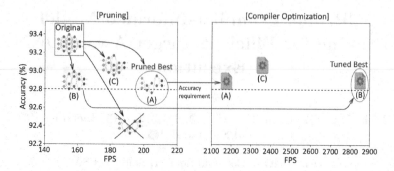

Fig. 1. An experiment to find the fastest model whose accuracy is higher than 92.80% accuracy for the target device among various pruned models of VGG-16 for CIFAR-10. We use the number of processed figures per second (FPS), given that indirect metrics like FLOPs cannot replace actual execution speed. The best-pruned model does not guarantee the best after compiler optimization. For example, the best model with pruning achieves only 2174 FPS while the suboptimal one obtains a higher 2857 FPS after compiler optimization.

using model compression and (2) generating optimized code for the target device using a DNN compiler. The model compression, for example, optimizes the neural networks by pruning, quantization, or neural architecture search (NAS). In particular, model pruning reduces the size of a model by removing non-critical or redundant neurons from a DNN model [26], while quantization does it by reducing the precision of the datatype [23]. In addition, NAS optimizes the structure of a DNN model for maximizing performance [19]. The DNN compiler, on the other hand, takes a DNN model and generates an optimized execution code for each target device.

Efficient execution through a lightweight model is critical for many applications, such as AR/VR applications, autonomous driving [24], drone or robot control [8], medical health monitoring [16], malware detection [40], and user authentication such as Face ID [27,50]. While one might think that the straightforward integration between model compression and compiler optimization could generate the most efficient code for a specific target device, this is not the case. Figure 1 shows experimental results of direct integration between model compression and compiler optimization. As shown, the fastest model meeting the required accuracy found by pruning is often not the best model after compiler optimization, demanding joint optimization between them.

Based on this observation, we propose CPrune, a new pruning technique for an efficient DNN execution for target devices that jointly considers model compression and compiler optimization. Instead of optimizing DNN compression and compiler optimization independently, leading to suboptimal optimization, CPrune exploits the information extracted during the compiler optimization process. In particular, CPrune uses the subgraph structures of a neural network and their execution times on the target device during compiler optimization to create an efficient target-aware pruned model fulfilling accuracy requirements.

To the best of our knowledge, this is the first work using the information collected during the compiler optimization to create the most efficient target-aware compressed model meeting accuracy requirements.

Our contributions are:

- We report from our experiments that the fastest model that meets the required accuracy found by pruning is often not the best model after compiler optimization, necessitating joint optimization.
- We propose CPrune, which incorporates the information extracted during the compiler optimization process into creating a target-oriented compressed model fulfilling accuracy requirements. This information also significantly reduces the search space for parameter tuning such that CPrune can make a compressed model substantially faster than NetAdapt [44], the state-of-the-art hardware-aware model compression framework.
- Our experimental results show that CPrune achieves target-oriented performance improvement and increases the speed up to 2.73× compared to the de facto DNN compiler framework TVM auto-tune while satisfying the accuracy requirement.

While CPrune's approach is generally applicable, we can utilize CPrune for computer vision tasks requiring fast processing but a slight reduction in accuracy is acceptable, such as object detection and image classification for autonomous driving and biosignal image processing for seizure detection on a mobile device. We implement CPrune on top of an open deep learning compiler stack Apache TVM [1] and Microsoft NNI [2]. Our source code can be found at https://github.com/taehokim20/CPrune.

2 Related Works

2.1 Model Compression

Model compression reduces the model size and improves speed while meeting the performance requirements: pruning compresses a model while NAS creates a much lighter model. There are also approaches using the performance measurements on real hardware as feedback for model compression.

Pruning. Pruning removes non-critical, redundant neurons from a DNN model [26]. Figure 2 shows an example of a convolution operation in DNN. c_i is the number of input channels, each of which comprises height (h_i) and width (w_i) of the input. After convolving c_i input channels with filters of layer l_i, the input becomes abstracted to a feature map with output channels c_{i+1}, where c_{i+1} is determined by the number of filters in layer l_i. Structured pruning [14,15,21,39,48] removes filters or channels according to various algorithms to judge the redundancy of weights. For example, removing a few input channels c_i reduces the overall sizes of the input and filter shape, and pruning a filter reduces the sizes because it eliminates a channel in the next layer (both cases are highlighted as shaded regions in Fig. 2). On the other hand, non-structured

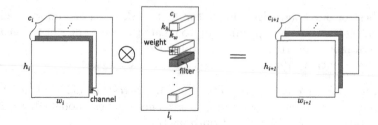

Fig. 2. Structured pruning removes channels or filters to reduce the model's overall size. In contrast, non-structured pruning selectively prunes less important weights to minimize the sizes without impacting accuracy loss.

pruning [11,47] selects individual weights to prune. As a result, they achieve very high compression without much accuracy loss even though it is hard to gain as much as speedups of structured ones [9,10,28].

NAS. NAS optimizes the structure of a DNN model for maximizing performance [7]. It uses various search strategies to explore the search space of neural architectures, such as random search, Bayesian optimization [35,41], evolutionary method [6,30], reinforcement learning (RL) [51], and gradient-based methods [7,25].

Hardware-Aware Model Compression. Recently, several approaches have utilized performance measurements on real hardware to make a compressed model [22,37,42,44,45]. Especially, NetAdapt [44] points out that indirect metrics (e.g., the number of MACs or weights) may not necessarily reduce the direct metrics (e.g., execution time and energy consumption) [43,46], so they incorporate performance metrics of target hardware into its pruning adaptation algorithm.

Note that our approach CPrune is different from NetAdapt in that it jointly considers model compression and compiler optimization.

2.2 DNN Compilers with Auto-Tuning

Various DNN compilers generate code for DNN models on different hardware architectures [4,20,33]. For example, TVM [4] translates the input DNN models into the intermediate representation (IR) during the compilation process [32]. After performing hardware-independent optimizations with the IR, it translates high-level IR into low-level IR, which provides interfaces to tune the computation and memory access. In addition, auto-tuning approaches such as AutoTVM [5] and AutoScheduler [49] optimize various low-level hardware-dependent parameters since the optimal parameters vary depending on the DNN operations and the target hardware.

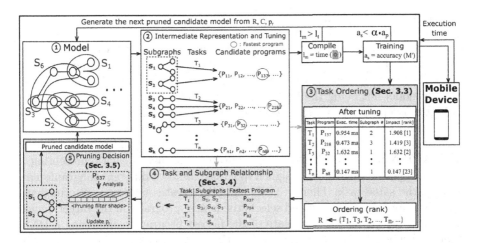

Fig. 3. CPrune overview. CPrune generates the fastest DNN execution model on the target device by using the structural information extracted during compiler optimization and the on-device performance of a DNN model.

3   CPrune: Compiler-Informed Model Pruning

Our motivation starts from a question: *Can an optimally pruned model guarantee the best performance after compiler optimization for a specific device?* To check this question, we prune VGG-16 [36] randomly and create 20 different models. We use the CIFAR-10 [17] dataset, and the original model's accuracy is 93.29%. We check if the model showing the highest number of processed figures per second (FPS) with an accuracy higher than 92.80% would show the highest FPS after compiler optimization. We use FPS, given that indirect metrics like FLOPs cannot replace actual execution speed. For this experiment, we run and measure the FPS on a PC with an NVIDIA RTX-3080 GPU.

As shown in Fig. 1, when we measure the FPS of each pruned model, the fastest model that meets the accuracy requirement is model A (92.85%, 205 FPS). However, after compiler optimization on the pruned models, the fastest model is model B (92.86%, 2857 FPS). It is different from what we expect: the most efficient model after pruning should be the most efficient one after compiler optimization. We also find no strong correlation between the performance of pruned models before and after compiler optimization, emphasizing the need for joint optimization between model compression and compiler optimization.

Inspired by this finding, we design a new pruning scheme, CPrune, that generates an efficient DNN execution model by pruning a model based on the extracted information during compiler optimization for the target device. Below we present the details of our approach.

3.1   CPrune Overview

The goal of CPrune is to find and prune neurons that primarily impact the execution of the DNN model on the target device while meeting the accuracy requirements. This informed pruning effectively prevents the issue of the best pruning model not being the best on the target device after compilation.

Figure 3 depicts how CPrune leverages the information collected during compiler optimization to prune neurons from the DNN model effectively. The shaded boxes indicate what CPrune adds to the existing compiler optimization framework. A DNN model consists of many convolutional layers, each representing a subgraph (①). A subgraph is assigned to a task, and two different subgraphs can point to the same task. Then, a DNN compiler creates numerous intermediate representations for each task and selects the fastest program on the target device (②). After compiling the aggregate of the fastest programs comprising a DNN model, CPrune checks if the model meets the minimum execution and accuracy requirements. After that, CPrune sorts tasks based on their execution times in descending order to find the most efficient model with further pruning. Since pruning a task with a longer execution time could significantly reduce a model's execution time, CPrune selects the most time-consuming task as a candidate for further pruning (③). CPrune now needs to know which subgraph(s) is associated with this task as pruning candidates. CPrune also store the fastest program for each task. For this purpose, CPrune builds a table keeping the relationship among tasks, subgraphs, and programs (④). Finally, CPrune prunes subgraphs of the selected task while ensuring their code structures follow the structure of the fastest program of that task (⑤). Since the computation structure impacts the execution time, preserving the same computation structure after pruning is critical for efficient pruning. This process continues to make the most efficient pruned DNN model satisfying the accuracy requirement.

In the following subsections, we describe the details of an algorithm (Sect. 3.2), task ordering (Sect. 3.3), task and subgraph relationship table (Sect. 3.4) and pruning decision (Sect. 3.5).

3.2   CPrune Algorithm

CPrune prunes a DNN model gradually and iteratively. In each iteration, CPrune analyzes subgraphs of the model and creates relationships between subgraphs and each task. Then, it takes a pre-trained model M and the minimum accuracy requirement a_g and returns the efficient target-aware DNN executable program, as shown in Algorithm 1. The computational complexity of Algorithm 1 and a table that summarizes variables used are in the Supplementary Materials.

Initialization Step (Line 1). It initializes tuning-related parameters, including the pruning rate of each subgraph p_r, the target execution time of the following iteration l_t, and the short-term accuracy of the previous best model a_p in Algorithm 1. It also initializes a task/subgraph table C, storing the relationship among tasks, subgraphs, and fastest programs and the list of tasks prioritized by tuning R.

Algorithm 1 CPrune Algorithm

Input: Pre-trained model M and a_g
Output: An efficient DNN executable file of M
 1: Tune M and initialize p_r, l_t, a_p, C and R
 2: **while** $a_p > a_g$ **and** $R \neq \{\}$ **do**
 3:     **for** r in R **do**
 4:         For r, obtain associated subgraphs (S) and the fastest program (P) from C
 5:         Update pruning rate (p_r) by analyzing the structure of P
 6:         Create a pruned candidate model (M') by pruning S with pruning rate p_r
 7:         Extract tasks from M' and create a candidate task/subgraph table (C')
 8:         Tune these tasks and create a candidate list of prioritized tasks (R')
 9:         Compile the aggregate of tuned tasks and measure l_m
10:         Go line 3 if $l_m \geq l_t$
11:         Short-term train M' and measure a_s
12:         Remove r from R and go line 3 if $a_s < \alpha \cdot a_p$
13:         $M \leftarrow M'$, $R \leftarrow R'$, $C \leftarrow C'$, update p_r, $l_t = \beta \cdot l_m$, $a_p = a_s$
14:         **break**
15:     **end for**
16: **end while**
17: Final long-term training, tuning and compilation for M

Main Step: Pruning Based on the Compiler Optimization (Line 2–16). In each pruning iteration (the while loop in Algorithm 1) of this step, CPrune generates a pruned candidate model (① in Fig. 3) and moves to the intermediate representation (IR) and tuning stage (②). When a task in order of R is picked up (③), CPrune extracts associated subgraphs and the fastest program of the task from C (④). It decides the number of filters to prune by analyzing the arrangement of filters of the program and prunes filters of the subgraphs to create a pruned candidate model M' (⑤, details in Sect. 3.5). CPrune uses the relationship between subgraphs of the model M' and each task and creates a task/subgraph table C'. It also puts the fastest program of each task in C' (details in Sect. 3.4). Furthermore, the IR and tuning process maintains a candidate list of prioritized tasks R' for the next iteration (details in Sect. 3.3). If the measured execution time of M' (l_m) is less than l_t, it trains M' shortly and measures the short-term accuracy a_s. When a_s meets the requirement of the current iteration, it updates parameters and goes to the next iteration. α is the ratio to represent the minimum allowable accuracy after pruning, and β is the ratio to define the target execution time of the next pruning iteration.

If l_m is greater than or equal to l_t, the system selects the next task from R to create the next pruned candidate model. Once a_s is less than the target accuracy of the current iteration $(\alpha \cdot a_p)$, CPrune removes the current task from R and does not consider this task as a candidate for pruning in the rest iterations.

Final Step: Final Training and Tuning (Line 17). If there are no more tasks in R that can be pruned while meeting the accuracy requirement, CPrune progresses to the final step. In this step, CPrune trains and tunes the final model to achieve optimal accuracy and execution time.

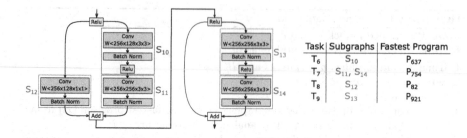

Fig. 4. The graph analysis of CPrune for the ResNet-18 model

3.3   Task Ordering

CPrune needs to select the most promising task of the current model M for pruning. Therefore, CPrune sorts tasks according to the pruning impact (task's execution time × number of subgraphs associated with the task). The higher the pruning impact, the more likely the task satisfies the current iteration's target execution time l_t.

The rightmost shaded box in Fig. 3 shows the result of task ordering as an example. The pruning impact of T_1 is $0.954 \times 2 = 1.908$, T_2 is $0.473 \times 3 = 1.419$, T_3 is $1.632 \times 1 = 1.632$, and so on. After CPrune performs task ordering including the results of other tasks, the ordered list R' becomes $\{T_1, T_3, T_2, ..., T_n, ...\}$. In this example, the subgraphs associated to T_1 will be pruned first. If this choice does not satisfy the execution time and accuracy requirements, T_3 and then T_2 will be selected for the next pruning candidate.

3.4   Task and Subgraph Relationship

CPrune creates a table storing the relationship among tasks, subgraphs, and fast programs to determine which subgraphs are pruned and which program is analyzed when selecting a task in the next pruning iteration. CPrune analyzes a pruned candidate model by checking the layer connectivity, weight shape, Rectified Linear Unit (ReLU), and Batch Normalization (BN). Figure 4 shows CPrune's graph analysis for part of the ResNet-18 model [3]. As ResNet-18 consists of multiple convolutions of the same shape, the compiler partitions the large computational graph of a DNN into multiple subgraphs, each of which can be associated with other subgraphs [32]. For example, subgraphs S_{11} and S_{14} are connected to the same task T_7 due to the same properties of BN and input shapes. On the other hand, subgraphs S_{10}, S_{12}, and S_{13} are connected to different tasks, respectively. In addition, each task is connected to its fastest program during the tuning.

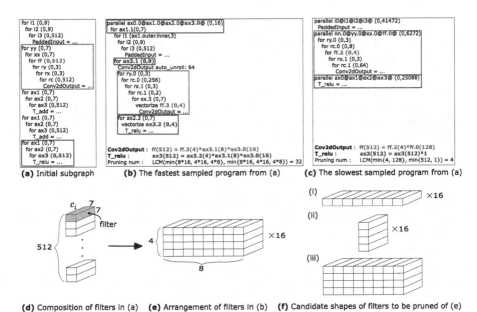

(a) Initial subgraph   **(b)** The fastest sampled program from (a)   **(c)** The slowest sampled program from (a)

(d) Composition of filters in (a)   **(e)** Arrangement of filters in (b)   **(f)** Candidate shapes of filters to be pruned of (e)

Fig. 5. Sampled program comparison extracted from one subgraph among several subgraphs of ResNet-18. c_i is the number of input channels.

3.5 Pruning Decision

CPrune prunes filters while maintaining the structure of the fastest program created during the IR tuning process of compiler optimization. Figures 5(b) and (c) compare the structures of the fastest and slowest programs generated from the same subgraph in Fig. 5(a). As we can see, the fastest program Fig. 5(b) allows a subgraph to perform parallel processing of computations most efficiently. In contrast, the slowest program Fig. 5(c) shows the most inefficient parallel processing of computations. If CPrune prunes four filters suitable for maintaining the structure of the program Fig. 5(c), a DNN compiler may generate an inefficient program for the corresponding subgraph of a DNN model. Instead, pruning the subgraph to follow the program structure of the fastest program would lead a DNN compiler to generate an efficient program. In addition, since the execution time of the convolution layer increases in a step pattern rather than linear with the number of filters [38], pruning an insufficient number of filters would only increase the tuning time without improving FPS.

In particular, CPrune splits a given model into multiple subgraphs and determines the number of filters to prune for each subgraph. We elaborate on calculating the number of filters to be pruned by taking one subgraph as an example, as shown in Fig. 5(a). We can see that the total number of filters is 512, and the kernel shape is 7 × 7 like Fig. 5(d) by checking the 2D convolution layer (Conv2d) and the ReLU activation function layer (T_relu) in Fig. 5(a). Iterators related to the number of filters are ff and ax3. With this information, CPrune checks how these iterators are further split or merged. For example, Fig. 5(b) shows that 512 of ff in Fig. 5(a) is converted to 4×8×16 by the argument ff.3. 512 of ax3 is also converted to 4×8×16. Therefore, both iterators arrange filters like Fig. 5(e).

Finally, CPrune calculates the minimum number of filters to prune based on the arrangement shapes of filters obtained from the related iterators. If CPrune reduces the number in width, height, or depth in the arrangement of filters of Fig. 5(e), the subgraph can maintain the structure of Fig. 5(b) after pruning the filters. However, if CPrune prunes many filters at once, it cannot achieve optimal pruning due to the high difference from the performance goal in the current pruning iteration. Therefore, we prune filters by the step size to improve performance by pruning in the convolution layer [38]. Let $L_1 = \{a_1, a_2, ..., a_m\}$ be the set of product combinations of the first iterator, and $L_2 = \{b_1, b_2, ..., b_n\}$ be the set of product combinations of the second iterator. CPrune determines the least common multiple (LCM) for the two pruning numbers as the minimum number of pruning filters using L_1 and L_2.

$$LCM(\min_{l \in L_1} \frac{\prod_{k=1}^{m} a_k}{l}, \min_{l \in L_2} \frac{\prod_{k=1}^{n} b_k}{l}) = LCM(\frac{\prod_{k=1}^{m} a_k}{\max_{l \in L_1} l}, \frac{\prod_{k=1}^{n} b_k}{\max_{l \in L_2} l})$$

For example, we calculate the minimum number of filters that can be pruned while maintaining the program structure in Fig. 5(b) and Fig. 5(c), respectively. In Fig. 5(b), the product combination of iterators ff and ax3 is the same as 4×8×16. CPrune can prune 8×16, 4×16, or 4×8 filters like Fig. 5(f) while maintaining the product combination. Thus, the minimum number of pruning filters of Fig. 5(b) is $LCM(\min\{8 \times 16, 4 \times 16, 4 \times 8\}, \min\{8 \times 16, 4 \times 16, 4 \times 8\}) = 32$. On the other hand, in Fig. 5(c), ff and ax3 are converted to 4×128 and 512×1, respectively. Therefore, the minimum number of pruning filters of Fig. 5(c) is $LCM(\min\{4, 128\}, \min\{512, 1\}) = 4$.

After determining the number of filters to be pruned, CPrune decides which filters to prune. It calculates the sum of each filter's absolute weights (i.e., l_1-norm) and prunes filters starting with the smallest sum [2,21].[1] CPrune also prunes the input channels of the next layer by the determined number as described in Sect. 2.1 to maintain the connection consistency. In the case of the shortcut part in a ResNet model, we prune the same output channels using the dependency-aware mode supported by Microsoft NNI [2].

4 Experiments

4.1 Experimental Setup

In this section, we carry out various experiments to evaluate the performance of CPrune. We conduct experiments over different pre-trained DNN models, including ResNet-18 [12], MobileNetV2 [34], and MnasNet1.0 [37], with ImageNet [18]

---

[1] Using other metrics can improve the performance as well.

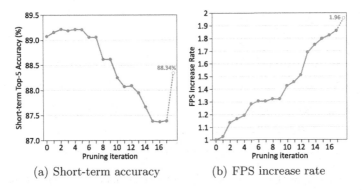

(a) Short-term accuracy   (b) FPS increase rate

Fig. 6. The FPS increase rate and short-term accuracy during the iterative CPrune process compared to using only TVM auto-tune. After completing the compiler-informed pruning process, CPrune performs final long-term training and tuning (Line 17 of Algorithm 1). Therefore, we marked the final results with dotted lines to distinguish them from other short-term measurements.

or CIFAR-10 [17] datasets on various resource-constrained mobile devices (Samsung Galaxy S8 (Kryo 280 CPU), S9 (Kryo 385 CPU and Mali-G72 GPU), S20+ (Kryo 585 CPU), or Google Pixel 3 XL (Kryo 385 CPU)). We also use multiple host PCs with NVIDIA GeForce RTX 1080 Ti or 2080 Ti. All pruned models are optimized by stochastic gradient descent (SGD) [31]. The number of training epochs considered varies according to the choice of the dataset and the specific phase of training (i.e., either short-term or final training). For the CIFAR-10 dataset, the short-term and final training epochs are 5 and 100, respectively. For the ImageNet dataset, the training epochs are one-fifth of CIFAR-10 due to the enormous data size. Due to space limitations, the details of finding reasonable α and β values and additional experiments about the impact of tuning and selective search are in the Supplementary Materials.

4.2 Overall Performance

This experiment shows CPrune's FPS and short-term accuracy during the iterative pruning process compared to the case of using only a DNN compiler and its auto-tuning. For this experiment, we select TVM auto-tune [49] for the DNN compiler and use the ResNet-18 model on Kyro 385 CPU using the ImageNet dataset. Figure 6 shows the results. In each pruning iteration, CPrune modifies the current model by pruning selected subgraphs. For any given pruning iteration, if a pruned candidate model satisfies the condition of the given iteration (l_t and $\alpha \cdot a_p$), the candidate model is selected, and the iteration progresses. The final accuracy is 88.34%, and the FPS increase rate is 1.96 times faster than using TVM auto-tune alone. We emphasize that our CPrune's FPS increase rate is around two times that obtained by only using the TVM auto-tune, while the final accuracy is within tolerable limits. Note that CPrune can stop its pruning

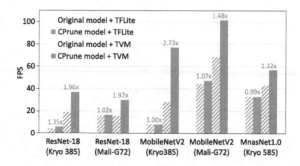

Fig. 7. Comparison of FPS when we execute CPrune model with the target compiler framework TVM and a target-agnostic deep learning framework TFLite.

around the 6th pruning iteration if the accuracy requirement is more than 89%. Then the FPS increase rate is 1.3 times faster than TVM auto-tune.

In a practical scenario, user applications can provide the accuracy and execution time requirements to CPrune.

4.3 Performance on Different Target Devices

This section evaluates CPrune's performance on different target devices. For this experiment, we integrate CPrune with the TVM compiler to compare its performance on different target devices. We also convert the final pruned model to a widely used TFLite executable for performance comparison. We compare the FPS of different DNN models using the ImageNet dataset on mobile CPU (Kryo 385 or Kryo 585) and GPU (Mali-G72), as shown in Fig. 7. CPrune, along with the target compiler framework, shows a significantly higher FPS than the cases of running only a DNN compiler (e.g., TVM) and a target agnostic DNN library (e.g., TFLite). Furthermore, regardless of the type of processor and model, the FPS increase rate when executing the CPrune model on a target processor is significantly higher than when we run it on other processors, as shown in Fig. 8.

4.4 Comparison with Other Pruning Schemes

This section compares CPrune with other pruning schemes on different mobile platforms. We also integrate CPrune into TVM for a fair comparison. Table 1 shows the results. CPrune shows a higher FPS than the model-based pruning models (e.g., PQF [29], FPGM [13], AMC [14]). CPrune also shows similar or better performance than the hardware-aware pruning model (e.g., NetAdapt [44]) with TVM. It also shows that indirect metrics such as FLOPS and parameters do not fully reflect actual performance. While FLOPS is a suitable indirect measure of the extent of compression obtained during the pruning process, FPS suitably reflects the pruning gains in terms of execution times or speeds.

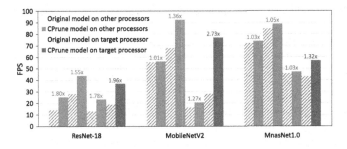

Fig. 8. Comparison of FPS when we execute CPrune model on a target processor and other processors.

Table 1. Mobile CPU (Kryo 385 and 585) and GPU (Mali-G72) performance test (ResNet-18, MobileNetV2, MnasNet1.0 with ImageNet dataset)

Model	Method	FPS (Increase rate)	FLOPS	Params	Top-1 Acc	Top-5 Acc
ResNet-18 (Kryo 385)	Original (TVM)	18.86	1.81B	11.7M	69.76%	89.08%
	PQF [29]+TVM	18.73 (0.99×)	166M	8.26M	66.74%	87.16%
	FPGM [13]+TVM	22.93 (1.22×)	1.10B	7.30M	68.37%	88.43%
	NetAdapt [44]+TVM	35.17 (1.86×)	1.24B	9.59M	68.45%	88.37%
	CPrune	36.92 (1.96×)	1.17B	10.3M	68.30%	88.34%
ResNet-18 (Mali-G72)	Original	15.65	1.81B	11.7M	69.76%	89.08%
	PQF+TVM	24.14 (1.54×)	166M	8.26M	66.74%	87.16%
	FPGM+TVM	26.62 (1.70×)	1.10B	7.30M	68.37%	88.43%
	CPrune	30.02 (1.92×)	1.55B	10.4M	69.83%	89.24%
MobileNetV2 (Kryo 385)	Original	28.20	301M	3.47M	71.88%	90.29%
	AMC [14]+TVM	67.62 (2.40×)	211M	2.31M	70.85%	89.91%
	CPrune	76.92 (2.73×)	255M	3.29M	70.33%	89.57%
MobileNetV2 (Mali-G72)	Original	68.68	301M	3.47M	71.88%	90.29%
	AMC+TVM	90.58 (1.32×)	211M	2.31M	70.85%	89.91%
	CPrune	101.56 (1.48×)	281M	3.31M	71.39%	90.16%
MnasNet1.0 (Kryo 585)	Original	42.92	314M	4.35M	73.46%	91.51%
	CPrune	56.85 (1.32×)	284M	3.82M	72.90%	91.16%

4.5 Effect of Pruning on Associated Subgraphs

This experiment checks the effectiveness of pruning filters of all subgraphs related to a task as in CPrune than pruning filters of only one subgraph in each iteration. For a task associated with greater than one subgraph, we have a design choice to prune filters of a single subgraph at a time (NetAdapt [44]) or prune filters of all the subgraphs associated with the task (like CPrune). Associated subgraphs pruning can shorten the CPrune process in proportional to the number of related subgraphs in the model. We observe that the associated subgraphs pruning consumes relatively less time in the Main step of CPrune than the single subgraph pruning, as shown in Fig. 9(a). In addition, the associated subgraphs pruning improves the FPS by more than 13 FPS compared to the single subgraph pruning strategy without significantly reducing the accuracy, as shown in Fig. 9(b) and Table 2.

Table 2. Mobile CPU performance test (ResNet-18 with CIFAR-10 dataset)

Model	Method	FPS (Increase rate)	FLOPS	Params	Top-1 Acc
ResNet-18 (Kryo 280)	Original (TVM)	33.82	555M	11.2M	94.37%
	CPrune	109.45 (3.24×)	161M	2.62M	93.74%
ResNet-18 (Kryo 585)	Original	40.50	555M	11.2M	94.37%
	CPrune	93.63 (2.31×)	297M	3.54M	94.14%
	CPrune (w/o tuning)	57.77 (1.43×)	390M	5.08M	94.51%
	CPrune (single subgraph pruning)	79.62 (1.97×)	294M	4.55M	94.27%

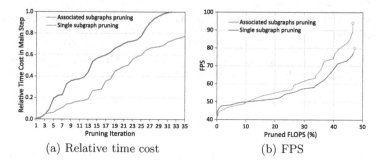

(a) Relative time cost   (b) FPS

Fig. 9. The effect of pruning all the subgraphs associated with the task (ResNet-18, Kryo 585, CIFAR-10). (a) is the relative time cost comparison with pruning a single subgraph at a time

5   Conclusion

Existing methods generate a compressed model by focusing on a model itself for fast DNN model execution on resource-constrained target devices. However, we have confirmed that knowing which model is best from pruning is impossible without considering compiler optimization. Therefore, we propose CPrune, a compiler-informed model pruning for efficient target-aware DNN execution. CPrune ensures the actual performance with pruning on the target device by using the relationship between the task and subgraphs during the compiler optimization. CPrune generates a target-aware DNN execution model by pruning a model based on on-device performance and compiler optimization. We verified that the pruned model generated by CPrune improves a DNN model's execution speed significantly while meeting the accuracy requirement.

Acknowledgments. This work was supported by the Institute of Information & communications Technology Planning & Evaluation (IITP) grant funded by the Korea government (MSIT) (No. 2018-0-00769, Neuromorphic Computing Software Platform for Artificial Intelligence Systems).

References

1. Apache tvm. https://github.com/apache/tvm. Accessed 03 July 2022
2. Microsoft nni. https://github.com/microsoft/nni. Accessed 03 July 2022
3. Torchvision models. https://pytorch.org/vision/stable/models.html. Accessed 03 July 2022
4. Chen, T., et al.: TVM: an automated end-to-end optimizing compiler for deep learning. In: 13th USENIX Symposium on Operating Systems Design and Implementation (OSDI 18), pp. 578–594 (2018)
5. Chen, T., et al.: Learning to optimize tensor programs. arXiv preprint arXiv:1805.08166 (2018)
6. Chen, Y., et al.: Renas: reinforced evolutionary neural architecture search. In: Proceedings of the IEEE/CVF Conference on Computer Vision and Pattern Recognition, pp. 4787–4796 (2019)
7. Elsken, T., Metzen, J.H., Hutter, F.: Neural architecture search: a survey. J. Mach. Learn. Res. **20**(1), 1997–2017 (2019)
8. Fang, B., Zeng, X., Zhang, M.: Nestdnn: Resource-aware multi-tenant on-device deep learning for continuous mobile vision. In: Proceedings of the 24th Annual International Conference on Mobile Computing and Networking, pp. 115–127 (2018)
9. Gale, T., Zaharia, M., Young, C., Elsen, E.: Sparse gpu kernels for deep learning. IEEE Press (2020)
10. Gong, Z., Ji, H., Fletcher, C.W., Hughes, C.J., Baghsorkhi, S., Torrellas, J.: Save: sparsity-aware vector engine for accelerating DNN training and inference on CPUs. In: 2020 53rd Annual IEEE/ACM International Symposium on Microarchitecture (MICRO), pp. 796–810 (2020)
11. Guo, Y., Yao, A., Chen, Y.: Dynamic network surgery for efficient DNNs. In: Proceedings of the 30th International Conference on Neural Information Processing Systems, pp. 1387–1395 (2016)
12. He, K., Zhang, X., Ren, S., Sun, J.: Deep residual learning for image recognition. In: Proceedings of the IEEE Conference on Computer Vision and Pattern Recognition, pp. 770–778 (2016)
13. He, Y., Liu, P., Wang, Z., Hu, Z., Yang, Y.: Filter pruning via geometric median for deep convolutional neural networks acceleration. In: Proceedings of the IEEE/CVF Conference on Computer Vision and Pattern Recognition, pp. 4340–4349 (2019)
14. He, Y., Lin, J., Liu, Z., Wang, H., Li, L.-J., Han, S.: AMC: AutoML for model compression and acceleration on mobile devices. In: Ferrari, V., Hebert, M., Sminchisescu, C., Weiss, Y. (eds.) ECCV 2018. LNCS, vol. 11211, pp. 815–832. Springer, Cham (2018). https://doi.org/10.1007/978-3-030-01234-2_48
15. He, Y., Zhang, X., Sun, J.: Channel pruning for accelerating very deep neural networks. In: Proceedings of the IEEE International Conference on Computer Vision, pp. 1389–1397 (2017)
16. Kim, T., et al.: Epileptic seizure detection and experimental treatment: a review. Front. Neurol. **11**, 701 (2020)
17. Krizhevsky, A., Hinton, G., et al.: Learning multiple layers of features from tiny images (2009)
18. Krizhevsky, A., Sutskever, I., Hinton, G.E.: Imagenet classification with deep convolutional neural networks. Adv. Neural. Inf. Process. Syst. **25**, 1097–1105 (2012)
19. Kyriakides, G., Margaritis, K.: An introduction to neural architecture search for convolutional networks. arXiv preprint arXiv:2005.11074 (2020)

20. Lattner, C., et al.: Mlir: scaling compiler infrastructure for domain specific computation. In: 2021 IEEE/ACM International Symposium on Code Generation and Optimization (CGO), pp. 2–14. IEEE (2021)
21. Li, H., Kadav, A., Durdanovic, I., Samet, H., Graf, H.P.: Pruning filters for efficient convnets. arXiv preprint arXiv:1608.08710 (2016)
22. Li, Z., et al.: Npas: a compiler-aware framework of unified network pruning and architecture search for beyond real-time mobile acceleration. In: Proceedings of the IEEE/CVF Conference on Computer Vision and Pattern Recognition, pp. 14255–14266 (2021)
23. Liang, T., Glossner, J., Wang, L., Shi, S., Zhang, X.: Pruning and quantization for deep neural network acceleration: a survey. Neurocomputing **461**, 370–403 (2021)
24. Liu, H., He, Y., Yu, F.R., James, J.: Flexi-compression: a flexible model compression method for autonomous driving. In: Proceedings of the 11th ACM Symposium on Design and Analysis of Intelligent Vehicular Networks and Applications, pp. 19–26 (2021)
25. Liu, H., Simonyan, K., Yang, Y.: Darts: Differentiable architecture search. arXiv preprint arXiv:1806.09055 (2018)
26. Liu, J., Tripathi, S., Kurup, U., Shah, M.: Pruning algorithms to accelerate convolutional neural networks for edge applications: a survey. arXiv preprint arXiv:2005.04275 (2020)
27. Lu, L., Yu, J., Chen, Y., Liu, H., Zhu, Y., Kong, L., Li, M.: Lip reading-based user authentication through acoustic sensing on smartphones. IEEE/ACM Trans. Networking **27**(1), 447–460 (2019)
28. Ma, X., et al.: Pconv: the missing but desirable sparsity in DNN weight pruning for real-time execution on mobile devices. In: Proceedings of the AAAI Conference on Artificial Intelligence, vol. 34, pp. 5117–5124 (2020)
29. Martinez, J., Shewakramani, J., Liu, T.W., Bârsan, I.A., Zeng, W., Urtasun, R.: Permute, quantize, and fine-tune: Efficient compression of neural networks. In: Proceedings of the IEEE/CVF Conference on Computer Vision and Pattern Recognition, pp. 15699–15708 (2021)
30. Periaux, J., Gonzalez, F., Lee, D.S.C.: Evolutionary methods. In: Evolutionary Optimization and Game Strategies for Advanced Multi-Disciplinary Design, pp. 9–20. Springer (2015)
31. Robbins, H., Monro, S.: A stochastic approximation method. In: The Annals of Mathematical Statistics, pp. 400–407 (1951)
32. Roesch, J., et al.: Relay: a new IR for machine learning frameworks. In: Proceedings of the 2nd ACM SIGPLAN International Workshop on Machine Learning and Programming Languages, pp. 58–68 (2018)
33. Rotem, N., et al.: Glow: graph lowering compiler techniques for neural networks. arXiv preprint arXiv:1805.00907 (2018)
34. Sandler, M., Howard, A., Zhu, M., Zhmoginov, A., Chen, L.C.: Mobilenetv 2: Inverted residuals and linear bottlenecks. In: Proceedings of the IEEE Conference on Computer Vision and Pattern Recognition, pp. 4510–4520 (2018)
35. Shahriari, B., Swersky, K., Wang, Z., Adams, R.P., De Freitas, N.: Taking the human out of the loop: a review of bayesian optimization. Proc. IEEE **104**(1), 148–175 (2015)
36. Simonyan, K., Zisserman, A.: Very deep convolutional networks for large-scale image recognition. arXiv preprint arXiv:1409.1556 (2014)
37. Tan, M., et al.: Mnasnet: platform-aware neural architecture search for mobile. In: Proceedings of the IEEE/CVF Conference on Computer Vision and Pattern Recognition, pp. 2820–2828 (2019)

38. Tang, X., Han, S., Zhang, L.L., Cao, T., Liu, Y.: To bridge neural network design and real-world performance: a behaviour study for neural networks. Proc. Mach. Learn. Syst. **3**, 21–37 (2021)
39. Wang, Z., Li, C., Wang, X.: Convolutional neural network pruning with structural redundancy reduction. In: Proceedings of the IEEE/CVF Conference on Computer Vision and Pattern Recognition, pp. 14913–14922 (2021)
40. Wei, L., Luo, W., Weng, J., Zhong, Y., Zhang, X., Yan, Z.: Machine learning-based malicious application detection of android. IEEE Access **5**, 25591–25601 (2017)
41. White, C., Neiswanger, W., Savani, Y.: Bananas: Bayesian optimization with neural architectures for neural architecture search. arXiv preprint arXiv:1910.11858 1(2) (2019)
42. Wu, B., et al.: Fbnet: Hardware-aware efficient convnet design via differentiable neural architecture search. In: Proceedings of the IEEE/CVF Conference on Computer Vision and Pattern Recognition, pp. 10734–10742 (2019)
43. Yang, T.J., Chen, Y.H., Sze, V.: Designing energy-efficient convolutional neural networks using energy-aware pruning. In: Proceedings of the IEEE Conference on Computer Vision and Pattern Recognition, pp. 5687–5695 (2017)
44. Yang, T.-J., et al.: NetAdapt: platform-aware neural network adaptation for mobile applications. In: Ferrari, V., Hebert, M., Sminchisescu, C., Weiss, Y. (eds.) ECCV 2018. LNCS, vol. 11214, pp. 289–304. Springer, Cham (2018). https://doi.org/10.1007/978-3-030-01249-6_18
45. Yang, T.J., Liao, Y.L., Sze, V.: Netadaptv2: efficient neural architecture search with fast super-network training and architecture optimization. In: Proceedings of the IEEE/CVF Conference on Computer Vision and Pattern Recognition, pp. 2402–2411 (2021)
46. Yu, J., Lukefahr, A., Palframan, D., Dasika, G., Das, R., Mahlke, S.: Scalpel: customizing DNN pruning to the underlying hardware parallelism. ACM SIGARCH Comput. Architecture News **45**(2), 548–560 (2017)
47. Zhang, T., Ye, S., Zhang, K., Tang, J., Wen, W., Fardad, M., Wang, Y.: A systematic DNN weight pruning framework using alternating direction method of multipliers. In: Ferrari, V., Hebert, M., Sminchisescu, C., Weiss, Y. (eds.) ECCV 2018. LNCS, vol. 11212, pp. 191–207. Springer, Cham (2018). https://doi.org/10.1007/978-3-030-01237-3_12
48. Zhao, C., Ni, B., Zhang, J., Zhao, Q., Zhang, W., Tian, Q.: Variational convolutional neural network pruning. In: Proceedings of the IEEE/CVF Conference on Computer Vision and Pattern Recognition, pp. 2780–2789 (2019)
49. Zheng, L., et al.: Ansor: generating high-performance tensor programs for deep learning. In: 14th USENIX Symposium on Operating Systems Design and Implementation (OSDI 2020), pp. 863–879 (2020)
50. Zhou, B., Lohokare, J., Gao, R., Ye, F.: Echoprint: two-factor authentication using acoustics and vision on smartphones. In: Proceedings of the 24th Annual International Conference on Mobile Computing and Networking, pp. 321–336 (2018)
51. Zoph, B., Le, Q.V.: Neural architecture search with reinforcement learning. arXiv preprint arXiv:1611.01578 (2016)

EAutoDet: Efficient Architecture Search for Object Detection

Xiaoxing Wang[1], Jiale Lin[1], Juanping Zhao[2], Xiaokang Yang[1],
and Junchi Yan[1(✉)]

[1] Department of Computer Science and Engineering & MoE Key Lab
of Artificial Intelligence, Shanghai Jiao Tong University, Shanghai, China
{figure1_wxx,linjiale,xkyang,yanjunchi}@sjtu.edu.cn
[2] Guangdong OPPO Mobile Telecommunications Co., Ltd., Shanghai, China
zhaojuanping1325@oppo.com

Abstract. Training CNN for detection is time-consuming due to the
large dataset and complex network modules, making it hard to search
architectures on detection datasets directly, which usually requires vast
search costs (usually tens and even hundreds of GPU-days). In contrast,
this paper introduces an efficient framework, named EAutoDet, that can
discover practical backbone and FPN architectures for object detection
in 1.4 GPU-days. Specifically, we construct a supernet for both backbone and FPN modules and adopt the differentiable method. To reduce
the GPU memory requirement and computational cost, we propose a
kernel reusing technique by sharing the weights of candidate operations
on one edge and consolidating them into one convolution. A dynamic
channel refinement strategy is also introduced to search channel numbers. Extensive experiments show significant efficacy and efficiency of
our method. In particular, the discovered architectures surpass state-of-
the-art object detection NAS methods and achieve 40.1 mAP with 120
FPS and 49.2 mAP with 41.3 FPS on COCO test-dev set. We also transfer the discovered architectures to rotation detection task, which achieve
77.05 mAP_{50} on DOTA-v1.0 test set with 21.1M parameters. The code
is publicly available at https://github.com/vicFigure/EAutoDet.

1 Introduction and Related Work

Handcrafted neural architectures that designed by experts with large amounts
of trials and errors have achieved promising performance across computer vision
tasks [13,19,44]. Automated architecture search methods have been recently
explored, including reinforcement learning [45], evolutionary algorithm [25],
Bayesian optimization [35], maximum flow in graph theory [38], as well as the
more cost-effective one-shot NAS [1] that builds a supernet as the surrogate

Supplementary Information The online version contains supplementary material
available at https://doi.org/10.1007/978-3-031-20044-1_38.

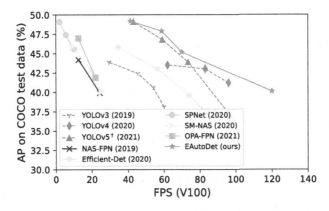

Fig. 1. Results of various detection models. Solid or dashed lines indicate NAS or hand-crafted architectures. †: obtained by our experiments, otherwise from the references.

model to predict the performance of candidate architectures. DARTS [21] further introduces a differentiable method that reduces the search cost to a few GPU-days. However, DARTS requires vast GPU memory since it has to train an over-parameterized supernet, making it impossible to search on large datasets or complex tasks. Some works [5,34,37] are dedicated to reducing the memory requirement.

Though NAS has achieved great success on classification tasks, it is still an open question on how to directly search detection architectures for detection tasks with two major difficulties: **1) It is time-consuming to train detection models from scratch** due to its complex architecture, which consists of multiple modules, including backbone and feature pyramid network (FPN). So that many works [2,8,24] pre-train the backbone on ImageNet; **2) Training a detection model requires vast GPU memory cost**, especially for those NAS works [3,10] that need to build an over-parameterized supernet. They even have to pre-train it on ImageNet, further increasing the search cost. To simplify the supernet and lower the search difficulty, they usually restrict the search space by either searching backbone [3,6,14] or FPN [7,33,36], which, however, actually ignores the relationship between the two modules. In contrast, this paper introduces kernel reusing technique and dynamic channel refinement to speed up the convergence to train a supernet and reduce GPU memory requirement. We thus propose an efficient search method, named EAutoDet, which can jointly search architectures of backbone and FPN on MS-COCO [20] detection dataset in a few GPU-days, with no need to pre-train a supernet on ImageNet (Fig. 1).

Additionally, the prior NAS detection methods are based on RetinaNet (one-stage) or Faster-RCNN (two-stage) framework that adopts ResNet-like architectures. Few have explored to search for YOLO (one-stage) framework that could leverage its known fast speed and outstanding performance. The hand-crafted YOLO models even outperform many NAS methods in similar inference speed (shown in Table 1), which implies the potential of a combination of NAS and YOLO. Nevertheless, a vanilla combination is undesirable. On the one

Table 1. Our method enables authentic fine-grained search w.r.t. operations, number of channels, and connections between layers, and is much faster than SM-NAS and Hit-Detector thanks to our kernel reusing and dynamic channel refinement techniques.

Search space	Method
Backbone alone	DetNAS [3], SpineNet [6], SPNet [14]
FPN alone	NAS-FPN [7], NAS-FCOS [33], OPA-FPN [17], Auto-FPN [36]
Joint search	SM-NAS [42], Hit-Detector [10], EAutoDet (ours)

hand, the handcrafted architecture of YOLO is subtle and elaborate. Such a nearly impeccable baseline puts forward higher requests for the ability of search method discovering the optimal architecture. On the other hand, it is better to absorb the knowledge of those well-designed architectures to design a sophisticated and large search space, which further asks for a flexible and efficient search method. However, our method offers a practical solution thanks to its low memory requirement and rapid convergence rate to train a supernet. Besides, our technique ameliorates the computation of convolutions in the supernet rather than restricting sub-architectures of supernet, making our method flexible to suit various search spaces. Experiments show the outstanding performance of our method on MS-COCO and DOTA-v1.0. Our contributions are summarized as follows:

1) Efficient Architecture Search Method for Object Detection. We propose kernel reusing and dynamic channel refinement techniques for the fine-grained search of backbone and FPN modules in 1.4 GPU-days on a single V100 GPU, significantly outperforming prior NAS methods, e.g., 28 GPU-days [33] and 44 GPU-days [3]. Unlike other NAS methods [3,10,42] that have to pre-train an over-parameterized supernet on the ImageNet, our supernet is trained from scratch on MS-COCO thanks to the property of our method: low memory requirement and rapid convergence rate to train a supernet.

2) Sophisticated and Large Search Space for Object Detection. By absorbing the knowledge of well-designed YOLO models, we design a complex search space for detection, including convolution types, channel numbers, and connection of layers for backbone and FPN. It puts forward higher requests for the flexibility and efficiency of search methods. This paper offers a solution, and to the best of our knowledge, it is the first NAS method that outperforms YOLO models with competitive high inference speed. Notice that our method can be easily applied to other detectors, e.g., Faster-RCNN [28] and RetinaNet [19].

3) Strong Performance and Fast Speed. Our design allows for direct search and evaluation without pre-training. The discovered architectures achieve outstanding performance on classic horizontal, as well as rotation detection tasks: 40.1 mAP with 120 FPS on MS-COCO where the bounding box is always assumed horizontal, and 77.05 mAP_{50} on DOTA which is a dominant rotation detection benchmark but has not been used in NAS literature.

Below we briefly discuss the related works.

Object Detection. Existing detection frameworks usually consist of four modules: backbone, feature fusion neck, region proposal network (in two-stage detectors), and detection head. For real-time detection, [2,19,24,27,32] design efficient architectures for the four modules. There are also emerging manually-designed detectors for rotation detection whereby different loss functions are carefully devised ranging from regression [11,41] to classification [39,40] models. Unlike the above works requiring massive trials and expert experience to design CNNs, we aim to search architectures for detection automatically.

Neural Architecture Search. Researchers have been dedicated to efficient search algorithms for neural architectures in recent years. NASNet [45] utilizes reinforcement learning (RL) and proposes to generate candidate architectures by an RNN controller. [25] adopt evolutionary algorithms (EA) to derive new architectures by crossover and mutation. Besides the above time-consuming methods, one-shot NAS [1] is introduced and can reduce the search cost to a few GPU-days. DARTS [21] regards NAS as a bi-level optimization problem and proposes to solve it by a differentiable method. This paper adopts the differentiable method in DARTS due to its efficacy and high efficiency.

NAS for Object Detection. Recent NAS methods for object detection can be briefly categorized into three streams: 1) Search backbone architecture and fix FPN, e.g. DetNAS [3] and SP-NAS [14]. 2) Search FPN architecture and fix backbone, e.g. NAS-FPN [7], Auto-FPN [36], and NAS-FCOS [33]. 3) Jointly search backbone and FPN, e.g., SM-NAS [42], Hit-Detector [10] and our method. Unlike SM-NAS and Hit-Detector that require vast GPUs to search, this work introduces kernel reusing and dynamic channel refinement techniques that can significantly reduce the GPU memory requirement during the search process. Specifically, our EAutoDet can search under a more extensive search space on MS-COCO dataset directly on a single V100 GPU in 1.4 days. Moreover, unlike many NAS detection methods [3,10,30,42] that need to pre-train supernets on the ImageNet, our EAutoDet trains the supernet from scratch on the MS-COCO during the search process, demonstrating its outstanding convergence ability.

2   The Proposed EAutoDet

Referring to DARTS [21] that regards NAS as a bi-level optimization task, we also build a supernet and define architecture parameters α to denote the importance of candidate operations. However, it is intractable to search on detection dataset directly by DARTS since the supernet is an over-parameterized model, making it much more challenging to train it from scratch on detection datasets, e.g., MS-COCO. We illustrate one super-edge in the supernet of DARTS in Fig. 2 (left), which contains all independent candidate operations and thus requires massive GPU memory during the search stage. To address the above memory explosion issue, we introduce two techniques to reduce memory requirement and computational cost: *Kernel Reusing Technique* to search operation types, and *Dynamic Channel Refinement* to search channel numbers.

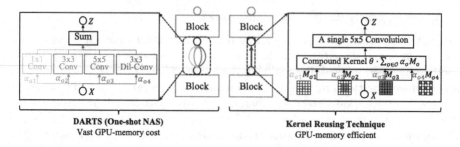

Fig. 2. Compared to DARTS, the kernel reusing technique compounds multiple convolutions into a single 5×5 convolution, which can reduce the memory cost and enables efficient search for backbone and FPN.

2.1 Kernel Reusing for Operation Type Search

Each edge in the supernet contains multiple convolutions with various kernel sizes. Suppose X and Z are the input and output of convolutions on one edge, then $Z = \sum_{o \in \mathcal{O}} \alpha_o X \otimes \theta_o$, where '$\otimes$' denotes convolution operation, $\mathcal{O}$ is the candidate set of convolutions, and θ is the kernel weights.

To reduce the parameters, we reuse the weights of different convolutions as shown in Fig. 2 (right), that is, kernels of all convolutions can be extracted from unified weights by a binary mask M. Moreover, since convolutions are linear operations, the weighted sum of multiple convolutions on the same edge can be compounded into one convolution. Therefore, the output of each edge can be simplified as Eq. 1, where θ is the unified weights, and kernels of candidate convolution o can be obtained by $M_o \cdot \theta$.

$$
Z = \underbrace{\sum_{o \in \mathcal{O}} \alpha_o X \otimes [M_o \cdot \theta]}_{|\mathcal{O}| \text{ convolutions}} = X \otimes \underbrace{\left[\theta \cdot \sum_{o \in \mathcal{O}} \alpha_o M_o \right]}_{\text{One convolution}} \tag{1}
$$

Advantage of Our Kernel Reusing Technique. Apart from our kernel reusing technique, sampling-based methods [5,37] are also popular to reduce the memory and computational cost for classification tasks. However, they involve a dynamic network structure by sampling a sub-network of the supernet at each iteration, which will affect the supernet training. Though such an issue can be tolerated on classification tasks, it worsens on detection tasks. In contrast, our kernel reusing technique holds a stable network structure and reduces GPU memory and computational cost by compounding multiple kernels into one without interfering with the supernet training.

We illustrate the mAP of supernets on MS-COCO validation set during the search stage in Fig. 3. EAutoDet-s/m/l/x denotes four supernets under various search spaces (details are introduced in Sect. 2.3). They are built based on our kernel reusing technique. EAutoDet-s-sample denotes that an s-level supernet is built without kernel reusing and trained by sampling operation based on Gumbel reparameterization technique at each iteration [5]. We observe that: 1) Our four

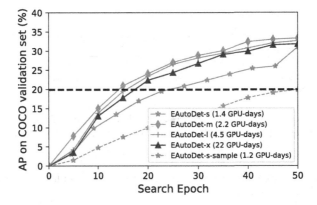

Fig. 3. MAP of supernets trained for 50 epochs. Search cost on a V100 GPU is given in the legend. EAutoDet-s-sample denotes to search by sampling-based method [5]. The horizontal dash line is the final AP of EAutoDet-s-sample.

supernets can converge to 30% mAP; 2) Our kernel reusing technique converges better and faster than sampling-based method, which confirms the above analysis on better convergence property of our method. Notice that Fig. 3 shows mAP of supernets during 50 epochs' training in the search stage. The ultimate performance of the discovered models is evaluated by training them from scratch for 300 epochs and is reported in Table 1. The search cost is also illustrated in the legend. Specifically, an effective s-level model can be discovered in 1.4 GPU-days on a single V100 GPU, showing the remarkable efficiency of our method.

Difference from the Prior Works. Unlike the prior NAS works [29,34] that focus on classification tasks, our method aims to search detection models that are more complex and requires more computation resources. Our kernel reusing technique can significantly reduce the memory and computational cost, making it possible to discover an effective architecture in a few GPU-days. Apart from NAS works, RepVGG [4] is also related to our approach, which introduces a re-param strategy to merge skip-connection, 3×3 and 1×1 convolutions for a plain inference-time model. However, the motivation of RepVGG is to stabilize the training of VGG, and the merge process is applied after the training stage. In contrast, our kernel reusing technique is applied during the search process and aims to reduce the memory and computational cost.

2.2 Dynamic Channel Refinement for Channel Number Search

Layer channels are essential hyperparameters for neural network architectures affecting the model size and FLOPs. Unfortunately, few differentiable NAS works have explored to search channel numbers, especially for detection tasks. In this work, we introduce the dynamic channel refinement technique based on Gumbel reparameterization technique [9] to search for optimal expansion rates for layers. Specifically, we sample an expansion rate for each layer at every iteration and refine the operation weights θ dynamically to fit the changeable channel numbers.

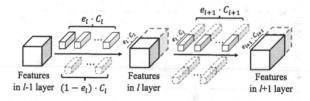

Fig. 4. Refinement for convolution weights θ on adjacent layers in our dynamic channel refinement technique. e indicates the sampled expansion rate. Dotted blocks in light gray indicate the inactivated channels.

Sampling for Expansion Rate. Expansion rate for one layer can be sampled by Gumbel-argmax technique:

$$E = \text{one\_hot} \left[\arg\max_i (\log \tilde{\alpha}_e^i + g^i) \right], \tag{2}$$

where $\tilde{\alpha}_e = \text{softmax}(\alpha_e)$ is the normalized weights for candidate expansion rates, and g^i are random variables sampled from Gumbel$(0, 1)$ distribution. To make E differentiable w.r.t. α_e, we adopt Gumbel-softmax to relax the sampled vector as follows, where τ is a gradually decayed temperature.

$$\tilde{E} = \frac{\exp\left[(\log \tilde{\alpha}_e^i + g^i)/\tau\right]}{\sum_j \exp\left[(\log \tilde{\alpha}_e^j + g^j)/\tau\right]}, \tag{3}$$

We adopt the one-hot vector E (Eq. 2) to activate one candidate expansion rate during the forward pass and utilize the relaxed vector $\tilde{E}$ (Eq. 3) to obtain gradients for α_e during the back-propagation, which is a popular reparameterization technique that has been widely-use in many recent works [5,31].

Dynamic Channel Refinement for Operation Weights. After sampling an expansion rate e_l for layer l with base channel number C_l, the output channel becomes $e_l C_l$. The operation weights on layer l can be refined by preserving the first $e_l C_l$ filters. Besides, it affects the input channel of convolution on layer $l+1$. Generally, sampling channels on one layer will affect the operation weights on the current and next layer. We, therefore, dynamically refine channels for weights of adjacent layers, as shown in Fig. 4. FBNet-V2 [31] is related to our method, which also utilizes the Gumbel technique to search for classification model architectures. However, FBNet-V2 has to pad zero on channels to obtain a unified dimension due to short-cut connections, resulting in useless computation. In contrast, we discard the short-cut connection and dynamically refine the channel numbers for each layer at every iteration. There is no useless computation resulting from padding zeros on channels.

Transformation for Concatenation Layers. The channel number for each layer alters dynamically during the search process, which brings difficulty to refine weights for convolutions after a concatenation layer. Specifically, suppose two features with expansion rates and base channels (e_1, C_1) and (e_2, C_2) are

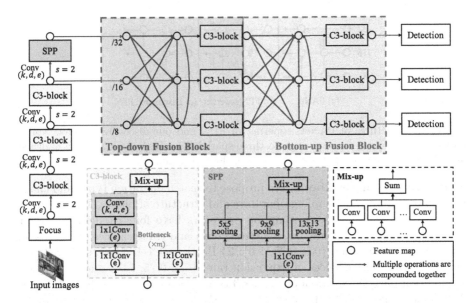

Fig. 5. The architecture of supernet, containing all candidate operations and connections in the search space. A red edge indicates candidate operations compounded by the kernel reusing technique, which is illustrated in Fig. 2. In backbone, C3-block, and SPP module, parentheses under 'Conv' indicate hyper-parameters to search: kernel size k, dilation ratio d, expansion rate of output channels e. (Color figure online)

concatenated. A convolution is applied after the concatenation layer, then the activated input channels of weight are separated ($0 \sim e_1C_1$ and $C_1 \sim C_1+e_2C_2$), making it hard to extract. To this end, we first give the following proposition, which is proved in the supplementary material.

Proposition 1. *The output of a concatenation layer followed by a convolution layer is equivalent to the sum of separate convolutions on the inputs.*

2.3   Detection-Oriented Search Space Design

Unlike ResNet-like and Mobile-like backbones in RetinaNet and Faster-RCNN that are transferred from classification tasks, YOLO models are specifically designed for detection tasks by experts considering both speed and performance. We would like to absorb the knowledge of the elaborate architectures and design a sophisticated and large detection-oriented search space. In particular, our method ameliorates the computation of convolutions in a supernet rather than restricting architectures of sub-models, making it flexible to suit such complex and large search spaces. Specifically, we resort to YOLOv5 [15] and PANet [22] and construct four types of supernets with various widths and depths, denoted as s (small), m (medium), l (large), and x (extra large), whose details are in the supplementary. In the following, we separately introduce the search spaces for backbone and FPN for their fundamentally different roles in the detection pipeline. The size of search space and details are given in the supplementary.

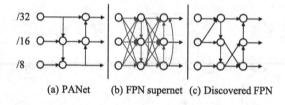

/32

/16

/8

(a) PANet   (b) FPN supernet (c) Discovered FPN

Fig. 6. Architectures of PANet, supernet of FPN and our discovered model. Nodes on three rows denote feature maps on three spatial sizes. Red edges indicate multiple operations are compounded on that edge. (Color figure online)

Search Space for Backbone. We propose to search operation types and channel numbers for down-sampling operators and structure of blocks. The supernet is shown in Fig. 5. **1)** Down-sampling operators have four candidates: $\{1 \times 1$ conv, 3×3 conv, 5×5 conv, 3×3 dilated conv$\}$ and three choices of expansion rate for output channel: $\{0.5, 0.75, 1.0\}$; **2)** Bottleneck cell consists of two convolutions with three choices of expansion rate: $\{0.5, 0.75, 1.0\}$, and the second convolution have three candidates: $\{3 \times 3$ conv, 5×5 conv, 3×3 dilated conv$\}$; **3)** C3-block has two 1×1 convolutions with two choices of expansion rates: $\{0.75, 1.0\}$. Architectures of different layers are independently searched.

Search Space for Feature Pyramid Network. The supernet for top-down and bottom-up fusion blocks is shown in Fig. 5, which enables to search connections of features in three spatial scales, operation types and channel numbers of each connection, and the structure of C3-blocks. **1)** Nodes indicate feature maps and connect with all their predecessors in the supernet. After the search stage, only two predecessors will be selected for each node; **2)** Each red edge contains four candidate operations: $\{1 \times 1$ conv, 3×3 conv, 5×5 conv, 3×3 dilated conv$\}$ with three possible expansion rates for the output channel: $\{0.5, 0.75, 1.0\}$; **3)** C3-blocks, whose architectures are also searched, are concatenated at the end of each fusion block to independently extract multi-scale features, as shown in Fig. 5. For each fusion block, we introduce architecture parameters α_e and α_o to denote the importance of edges and operations. Suppose $\tilde{\alpha} = \mathrm{softmax}(\alpha)$ is the normalized weight. The fused feature $z_j = \sum_{i<j} \left[\tilde{\alpha}_e^{(i,j)} \cdot \sum_{o \in \mathcal{O}} \tilde{\alpha}_o^{(i,j)} \cdot o(x_i) \right]$, where $\mathcal{O}$ is the candidate operation set, x_i is the features of predecessors.

Deriving the Final Architecture. We utilize the magnitude of architecture parameters as the importance estimation for operations. The final backbone architecture is derived by preserving the best operation. While for the feature pyramid network, nodes in each fusion block preserve top-2 connections, and each connection will preserve the best operation. Figure 6 compares the architecture of PANet, our supernet, and the discovered FPN module.

3 Experiments

We conduct experiments on two popular detection tasks: classic detection and rotation detection. The former is typical in many detection contests to locate

Table 2. Comparison with prior works on the COCO test-dev. FPS for YOLOv5 and our method are calculated on a single V100 GPU, and results for other methods are directly obtained from their papers. Different blocks indicate models with various inference speeds and prediction performance. '†': The results are obtained by our experiments. '–': The value is not provided by the original paper. '*': The unit of search cost is TPU-days, while the unit of other methods is GPU-days. '‡': SPNet [14] shows the search cost on VOC is 26 GPU-days, and is six times lower than that on COCO.

Method	FPS	#Params (M)	mAP (%)	AP$_{50}$ (%)	AP$_{75}$ (%)	AP$_S$ (%)	AP$_M$ (%)	AP$_L$ (%)	Search Cost
YOLOv4 [2]	96	-	41.2	62.8	44.3	20.4	44.4	56.0	–
YOLOv5s† [15]	113	7.3	36.9	56.0	40.0	19.9	41.1	46.0	–
EfficientDet-D0 [30]	98	3.9	33.8	52.2	35.8	12.0	38.3	51.2	–
NAS-FPN [7]	24	60.3	39.9	–	–	–	–	–	333*
NAS-FCOS@128 [33]	–	27.8	37.9	–	–	–	–	–	28
SpineNet-49S [6]	–	11.9	39.5	59.3	43.1	20.9	42.2	54.3	–
SM-NAS:E2 [42]	25	–	40.0	58.2	43.4	21.1	42.4	51.7	187
EAutoDet-s (ours)	**120**	**9.1**	**40.1**	**58.7**	**43.5**	**21.7**	**43.8**	**50.5**	**1.4**
YOLOv3 + ASFF [23]	54	–	40.6	60.6	45.1	20.3	44.2	54.1	–
YOLOv4 [2]	83	–	43.0	64.9	46.5	24.3	46.1	55.2	–
YOLOv4-csp [32]	80†	43	46.2	64.8	50.2	24.6	50.4	61.9	–
YOLOv5m† [15]	88	21.4	43.9	62.5	47.6	25.1	48.1	54.9	–
EfficientDet-D1 [30]	74	6.6	39.6	58.6	42.3	17.9	44.3	56.0	–
DetNAS [3]	–	–	42.0	63.9	45.8	24.9	45.1	56.8	44
NAS-FPN [7]	13	60.3	44.2	–	–	–	–	–	333*
Auto-FPN [36]	–	32.6	40.5	61.5	43.8	25.6	44.9	51.0	16
NAS-FCOS@256 [33]	–	57.3	43.0	–	–	–	–	–	28
SpineNet-49 [6]	–	28.5	42.8	62.3	46.1	23.7	45.2	57.3	–
SM-NAS:E3 [42]	20	–	42.8	61.2	46.5	23.5	45.5	55.6	187
Hit-Detector [10]	–	27.1	41.4	62.4	45.9	25.2	45.0	54.1	–
OPA-FPN@64 [17]	22	29.5	41.9	–	–	–	–	–	4
EAutoDet-m (ours)	**70**	**28.1**	**45.2**	**63.5**	**49.1**	**25.7**	**49.1**	**57.3**	**2.2**
YOLOv3 + ASFF [23]	46	–	42.4	63.0	47.4	25.5	45.7	52.3	–
YOLOv4 [2]	62	–	43.5	65.7	47.3	26.7	46.7	53.3	–
YOLOv4-csp [32]	65†	53	47.5	66.2	51.7	28.2	51.2	59.8	–
EAutoDet-csp (ours)	**55**	**49.8**	**47.8**	**66.1**	**51.9**	**28.6**	**51.5**	**60.1**	**4.2**
YOLOv5l† [15]	59	47.1	46.8	65.4	50.9	27.7	51.0	58.5	–
EfficientDet-D2 [30]	57	8.1	43.0	62.3	46.2	22.5	47.0	58.4	–
SPNet(BNB) [14]	10	–	45.6	64.3	49.6	28.4	48.4	60.1	156‡
SM-NAS:E5 [42]	9	–	45.9	64.6	48.6	27.1	49.0	58.0	187
OPA-FPN@160 [17]	13	60.6	47.0	–	–	–	–	–	4
EAutoDet-l (ours)	**59**	**34.4**	**47.9**	**66.3**	**52.0**	**28.3**	**52.0**	**59.9**	**4.5**
YOLOv3 + ASFF [23]	29	–	43.9	64.1	49.2	27.0	46.6	53.4	–
YOLOv5x† [15]	43	87.8	49.1	67.5	53.6	30.2	53.4	61.4	–
EfficientDet-D3 [30]	35	12	45.8	65.0	49.3	26.6	49.4	59.8	–
SPNet(XB) [14]	6	–	47.4	65.7	51.9	29.6	51.0	60.4	156‡
EAutoDet-x (ours)	**41**	**86.0**	**49.2**	**67.5**	**53.6**	**30.4**	**53.4**	**61.5**	**22**

Table 3. Joint search VS. independent search for backbone and FPN on MS-COCO validation set. Results show the superiority of jointly search against independent search.

Architecture		s-level (small)		m-level (medium)	
Backbone	FPN	mAP$_{(\%)}$	$\Delta_{(\%)}$	mAP$_{(\%)}$	$\Delta_{(\%)}$
Default	Default	36.9	+0.0	44.0	+0.0
Searched	Default	37.4	+0.5	44.6	+0.6
Default	*Searched*	38.9	+2.0	45.0	+1.0
Searched	*Searched*	**40.1**	**+3.2**	**45.5**	**+1.5**

common objects; The latter has been widely used in aerial images aiming to locate the ground object instances with an oriented bounding box (OBB). For the classic detection task, we adopt MS-COCO 2017 benchmark with 80 common object categories. For the rotation detection task, we adopt DOTA-v1.0 benchmark, one of the largest aerial detection benchmarks.

Search Settings. The training set of MS-COCO is divided into two parts to train architecture parameters and network weights. The final architecture is derived after alternately optimizing architecture parameters and network weights for 50 epochs by an SGD optimizer.

Evaluation Settings. Firstly, the discovered architectures are trained on MS-COCO training set from scratch for 300 epochs by an SGD optimizer and evaluated on its validation and test sets. We directly utilize the hyper-parameters in YOLOv5. To fairly compare the speed (FPS) with YOLO methods, we convert the trained models to the style of YOLOv4 [2] and evaluate the FPS on the Darknet platform [26], which is written in C and CUDA. Secondly, we train the architectures on rotation detection task on DOTA-v1.0 training set from scratch for 300 epochs and evaluate them on the validation and test sets. Notice that we train and test on a single input scale unlike previous works [11,41] that adopt multi-scale training technique and random rotation augmentation.

3.1   Results on Classic Detection Benchmark: MS-COCO

Table 2 reports the performance of our methods and compares with other state-of-the-art works on the COCO test-dev dataset. Different blocks indicate models with various inference speeds and prediction performance. We observe that our discovered models (EAutoDet) achieve the best performance. Specifically, EAutoDet-s achieves 40.1 mAP with 120 FPS, outperforming EfficientDet-D0 by 6.3% mAP with similar inference speed. Compared to manually-designed detectors (YOLO series), our method has competitive and even better performance. EAutoDet-m achieves 45.2% mAP with 69.9 FPS, surpassing YOLOv4 by 2.2% mAP. Moreover, we compare to YOLOv4-csp [32] by inheriting its backbone CD-53 and searching FPN architecture ('EAutoDet-csp' in Table 2). The discovered architecture outperforms YOLOv4-csp by 0.3% mAP with fewer parameters.

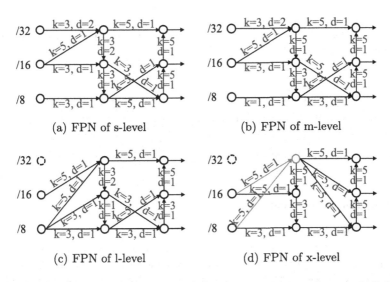

(a) FPN of s-level   (b) FPN of m-level

(c) FPN of l-level   (d) FPN of x-level

Fig. 7. FPN architecture of our searched models. 'k' denotes kernel size and 'd' denotes dilation ratio. Note that s-level and m-level have the same topology, while l-level and x-level discard the 32× down-sampled features (dashed node), which are the output of spatial pyramid pooling layer (SPP). In (d), we use blue lines and nodes to highlight the discarded SPP layer. We further analyze the effect of SPP by replacing one of the blue lines with the edge between dashed and blue nodes. (Color figure online)

Table 4. Transferred models VS. directly searched models on the MS-COCO validation set. 'Tf-x' denotes that we transfer the discovered x-level (extra large) model to s/m/l-level. 'Tf-m' denotes that we transfer the discovered m-level (medium) model to s-level. 'Search' denotes the directly searched models.

Model	s-level (small)				m-level (medium)			l-level (large)		
	YOLOv5	Tf-x	Tf-m	Search	YOLOv5	Tf-x	Search	YOLOv5	Tf-x	Search
#Params(M)	7.3	8.0	8.0	9.1	21.4	21.8	28.1	47.1	48.6	42.2
mAP(%)	36.9	37.4	39.5	40.1	44.0	44.6	45.5	47.0	47.3	47.9

3.2 Ablation Study of Backbone and FPN Search

We compare the performance of joint and independent search for backbone and FPN in Table 3, where 'default' indicates that we directly adopt the architecture of YOLOv5, and 'searched' indicates that we search for the architectures. We observe that 1) Joint search achieves the best performance, showing the effectiveness of our algorithm and the necessity of joint search for detection models. 2) The search performance is much more sensitive to the architecture of FPN compared to the backbone module.

Table 5. Study of SPP on MS-COCO validation set. 's-16-8' is the original transferred model from x-level, while 's-32-8' and 's-32-16' are the modified models by adding connections from the SPP layer.

Model	w/ SPP	mAP	Δ
YOLOv5-s	✓	36.9	+0.0
s-16-8 (transferred from x)	×	37.4	+0.5
s-32-16 (transferred from x	✓	39.0	+2.1
s-32-8 (transferred from x)	✓	38.8	+1.9

3.3   Transferablity Evaluation

We transfer the discovered x-level model to s, m, and l levels to evaluate the transferability of the discovered architecture. Table 4 compares the performance of the discovered models and the transferred models on the validation set of MS-COCO. We observe that: 1) Though transferred s-level and m-level can outperform baselines (YOLOv5), significant gaps exist between them and the directly searched models; 2) The transferred l-level model and the searched one achieve competitive performance. We attribute it to the different architecture preferences for small and large neural networks. By comparing the discovered architectures, we find that they have similar backbone structures but rather different FPN structures, as shown in Fig. 7. Specifically, s-level and m-level have the same FPN topology but differ in operation types (kernel size and dilation ratio of convolutions). Besides, both l-level and x-level discard the 32× downsampled features, which is the output of SPP.

To verify our analysis, we transfer the searched m-level model to s-level since they have similar FPN structures. Table 4 shows that the s-level model transferred from m achieves 39.5% mAP, significantly surpassing the one transferred from x by more than 2%.

3.4   Discussion on Spatial Pyramid Pooling

Spatial pyramid pooling (SPP) [12] is designed to integrate various receptive fields and extract multi-scale features with the same spatial size. Most recent manually-designed detectors adopt it by default, including YOLOv4 and YOLOv5. However, our experiments show that the value of SPP degrades when the network gets deeper. As shown in Fig. 7, models on l-level and x-level discard the SPP layer and choose to enlarge the receptive field by 5×5 convolution. While models on s-level and m-level still prefer the SPP layer.

In our analysis, SPP is vital for shallow networks as it can increase the receptive field to extract global information. However, the receptive field is enough for deep networks, making the SPP layer dispensable with the increment of network depth. Results in Table 4 supports our analysis: When transferring x-level models to s-level, the performance degrades significantly. To verify the effectiveness of SPP for shallow networks, we manually add SPP for the transferred

Table 6. Comparison to YOLOv5 on the test set of oriented bounding box (OBB) task in DOTA-v1.0.

Model	s-level (small)		m-level (medium)	
	YOLOv5	EAutoDet	YOLOv5	EAutoDet
#Params(M)	7.6	21.7	8.7	22.7
FLOPS(G)	17.5	50.6	21.5	50.8
mAP_{50}(%)	74.00	76.33	75.72	77.05

s-level model. In Fig. 7(d), the blue node connects with 16× and 8× down-sampled features. We construct two models by removing one of the connections (blue lines) and connecting the blue node with the dashed node (output of SPP layer), whose performance on the COCO validation set is reported in Table 5. We observe that after recovering the SPP layer, the performance of transferred model can be improved significantly.

3.5 Results on Rotation Detection Benchmark: DOTA

We utilize Circular Smooth Label (CSL) technique [40] to obtain robust angular prediction through classification without suffering boundary conditions. Our models are trained with rotation classification loss used in CSL [40] from scratch for 300 epochs. We compare to YOLOv5 based on the open-sourced codes [16]. Results are shown in Table 6. We also compare with other rotation detection baselines that adopts RetinaNet [19] detection framework with ResNet152 [13] backbone and FPN [18] module are given in the supplementary.

The above results show the generalization of our discovered architectures, which also verify the effectiveness of our search method. Notice that the discovered architectures are not limited to the specific CSL as tested in our experiment, as the search paradigm is agnostic to the choice of rotation detection loss, e.g., GWD [41] and BBAVectors [43], which we leave for future work.

4 Conclusion

This paper introduces kernel and dynamic channel refinement techniques and proposes a fast and memory-efficient search method for detection. We also design a sophisticated and large search space for detection by absorbing the knowledge of well-designed architectures of YOLO models. Our method can discover light-weighted models in 1.4 GPU-days, achieving 40.1 mAP on COCO test-dev with 120 FPS surpassing state-of-the-art NAS methods. Besides, our ablation studies suggest that the SPP plays a more vital role in shallow models than in deep models in the hope of facilitating future network design for detection. Moreover, the discovered architecture archives 77.05% mAP_{50} on DOTA-v1.0 benchmark, outperforming most of the manually-designed models e.g. CSL [40] (76.24%), further verifying the effectiveness of our search method.

Acknowledgements. This work was supported in part by National Key Research and Development Program of China (2020AAA0107600), National Science of Foundation China (U19B2035, 61972250, 72061127003), and Shanghai Municipal Science and Technology Major Project (2021SHZDZX0102).

References

1. Bender, G., Kindermans, P., Zoph, B., Vasudevan, V., Le, Q.V.: Understanding and simplifying one-shot architecture search. In: ICML (2018)
2. Bochkovskiy, A., Wang, C.Y., Liao, H.Y.M.: Yolov4: Optimal speed and accuracy of object detection. arXiv preprint arXiv:2004.10934 (2020)
3. Chen, Y., Yang, T., Zhang, X., Meng, G., Xiao, X., Sun, J.: Detnas: Backbone search for object detection. NeurIPS (2019)
4. Ding, X., Zhang, X., Ma, N., Han, J., Ding, G., Sun, J.: Repvgg: making vgg-style convnets great again. In: CVPR (2021)
5. Dong, X., Yang, Y.: Searching for a robust neural architecture in four gpu hours. In: CVPR (2019)
6. Du, X., et al.: Spinenet: learning scale-permuted backbone for recognition and localization. In: CVPR (2020)
7. Ghiasi, G., Lin, T.Y., Le, Q.V.: Nas-fpn: Learning scalable feature pyramid architecture for object detection. In: CVPR (2019)
8. Girshick, R.: Fast r-cnn. In: ICCV (2015)
9. Gumbel, E.J.: Statistical theory of extreme values and some practical applications: a series of lectures, vol. 33. US Government Printing Office (1954)
10. Guo, J., et al.: Hit-detector: hierarchical trinity architecture search for object detection. In: CVPR (2020)
11. Han, J., Ding, J., Xue, N., Xia, G.: Redet: a rotation-equivariant detector for aerial object detection. In: CVPR (2021)
12. He, K., Zhang, X., Ren, S., Sun, J.: Spatial pyramid pooling in deep convolutional networks for visual recognition. TPAMI (2015)
13. He, K., Zhang, X., Ren, S., Sun, J.: Deep residual learning for image recognition. In: CVPR (2016)
14. Jiang, C., Xu, H., Zhang, W., Liang, X., Li, Z.: Sp-nas: Serial-to-parallel backbone search for object detection. In: CVPR (2020)
15. Jocher, G.: Yolov5 documentation, May 2020. https://docs.ultralytics.com/
16. Kaixuan, H.: Yolov5 for oriented object detection (2020). https://github.com/hukaixuan19970627/yolov5_obb
17. Liang, T., Wang, Y., Tang, Z., Hu, G., Ling, H.: Opanas: one-shot path aggregation network architecture search for object detection. In: CVPR (2021)
18. Lin, T.Y., Dollár, P., Girshick, R., He, K., Hariharan, B., Belongie, S.: Feature pyramid networks for object detection. In: CVPR (2017)
19. Lin, T., Goyal, P., Girshick, R.B., He, K., Dollár, P.: Focal loss for dense object detection. TPAMI **42**(2), 318–327 (2020)
20. Lin, T.-Y., Maire, M., Belongie, S., Hays, J., Perona, P., Ramanan, D., Dollár, P., Zitnick, C.L.: Microsoft COCO: common objects in context. In: Fleet, D., Pajdla, T., Schiele, B., Tuytelaars, T. (eds.) ECCV 2014. LNCS, vol. 8693, pp. 740–755. Springer, Cham (2014). https://doi.org/10.1007/978-3-319-10602-1_48
21. Liu, H., Simonyan, K., Yang, Y.: DARTS: differentiable architecture search. In: ICLR (2019)

22. Liu, S., Qi, L., Qin, H., Shi, J., Jia, J.: Path aggregation network for instance segmentation. In: CVPR (2018)
23. Liu, S., Huang, D., Wang, Y.: Learning spatial fusion for single-shot object detection. arXiv preprint arXiv:1911.09516 (2019)
24. Liu, W., Anguelov, D., Erhan, D., Szegedy, C., Reed, S., Fu, C.-Y., Berg, A.C.: SSD: single shot MultiBox detector. In: Leibe, B., Matas, J., Sebe, N., Welling, M. (eds.) ECCV 2016. LNCS, vol. 9905, pp. 21–37. Springer, Cham (2016). https://doi.org/10.1007/978-3-319-46448-0_2
25. Real, E., Aggarwal, A., Huang, Y., Le, Q.V.: Regularized evolution for image classifier architecture search. In: AAAI (2019)
26. Redmon, J.: Darknet: Open source neural networks in c (2013–2016). https://pjreddie.com/darknet/
27. Redmon, J., Divvala, S.K., Girshick, R.B., Farhadi, A.: You only look once: Unified, real-time object detection. In: CVPR (2016)
28. Ren, S., He, K., Girshick, R.B., Sun, J.: Faster R-CNN: towards real-time object detection with region proposal networks. In: Cortes, C., Lawrence, N.D., Lee, D.D., Sugiyama, M., Garnett, R. (eds.) NeurIPS (2015)
29. Stamoulis, D., et al.: Single-path NAS: designing hardware-efficient convnets in less than 4 hours. In: ECML (2019)
30. Tan, M., Pang, R., Le, Q.V.: Efficientdet: scalable and efficient object detection. In: CVPR (2020)
31. Wan, A., et al.: Fbnetv2: differentiable neural architecture search for spatial and channel dimensions. In: CVPR (2020)
32. Wang, C.Y., Bochkovskiy, A., Liao, H.Y.M.: Scaled-yolov4: scaling cross stage partial network. In: CVPR (2021)
33. Wang, N., et al.: Nas-fcos: fast neural architecture search for object detection. In: CVPR (2020)
34. Wang, X., Xue, C., Yan, J., Yang, X., Hu, Y., Sun, K.: Mergenas: merge operations into one for differentiable architecture search. In: IJCAI (2020)
35. White, C., Neiswanger, W., Savani, Y.: BANANAS: bayesian optimization with neural architectures for neural architecture search. In: AAAI (2021)
36. Xu, H., Yao, L., Li, Z., Liang, X., Zhang, W.: Auto-fpn: automatic network architecture adaptation for object detection beyond classification. In: ICCV (2019)
37. Xu, Y., et al.: Pc-darts: partial channel connections for memory-efficient architecture search. In: ICLR (2020)
38. Xue, C., Wang, X., Yan, J., Li, C.G.: A flow-based approach for neural architecture search. In: ECCV (2022)
39. Yang, X., Hou, L., Zhou, Y., Wang, W., Yan, J.: Dense label encoding for boundary discontinuity free rotation detection. In: Proceedings of the IEEE Conference on Computer Vision and Pattern Recognition, pp. 15819–15829 (2021)
40. Yang, X., Yan, J.: Arbitrary-oriented object detection with circular smooth label. In: Vedaldi, A., Bischof, H., Brox, T., Frahm, J.-M. (eds.) ECCV 2020. LNCS, vol. 12353, pp. 677–694. Springer, Cham (2020). https://doi.org/10.1007/978-3-030-58598-3_40
41. Yang, X., Yan, J., Qi, M., Wang, W., Xiaopeng, Z., Qi, T.: Rethinking rotated object detection with gaussian wasserstein distance loss. In: International Conference on Machine Learning (2021)
42. Yao, L., Xu, H., Zhang, W., Liang, X., Li, Z.: Sm-nas: structural-to-modular neural architecture search for object detection. In: AAAI (2020)

43. Yi, J., Wu, P., Liu, B., Huang, Q., Qu, H., Metaxas, D.: Oriented object detection in aerial images with box boundary-aware vectors. In: Proceedings of the IEEE Winter Conference on Applications of Computer Vision, pp. 2150–2159 (2021)
44. Zhao, Z., Wu, Z., Zhuang, Y., Li, B., Jia, J.: Tracking objects as pixel-wise distributions (2022)
45. Zoph, B., Vasudevan, V., Shlens, J., Le, Q.V.: Learning transferable architectures for scalable image recognition. In: CVPR (2018)

A Max-Flow Based Approach for Neural Architecture Search

Chao Xue[1,2(✉)], Xiaoxing Wang[3], Junchi Yan[3], and Chun-Guang Li[1]

[1] School of Artificial Intelligence, Beijing University of Posts and
Telecommunications, Beijing, People's Republic of China
{xch,lichunguang}@bupt.edu.cn
[2] JD Explore Academy, Beijing, People's Republic of China
[3] Department of Computer Science and Engineering and MoE Key Lab of Artificial
Intelligence, Shanghai Jiao Tong University, Shanghai, People's Republic of China
{figure1_wxx,yanjunchi}@sjtu.edu.cn

Abstract. Neural Architecture Search (NAS) aims to automatically
produce network architectures suitable to specific tasks on given datasets.
Unlike previous NAS strategies based on reinforcement learning, genetic
algorithm, Bayesian optimization, and differential programming, we formulate the NAS task as a Max-Flow problem on search space consisting
of Directed Acyclic Graph (DAG) and thus propose a novel NAS approach, called MF-NAS, which defines the search space and designs the
search strategy in a fully graphic manner. In MF-NAS, parallel edges
with capacities are induced by combining different operations, including
skip connection, convolutions and pooling, and the weights and capacities
of the parallel edges are updated iteratively during the search process.
Moreover, we interpret MF-NAS from the perspective of non-parametric
density estimation and show the relationship between the flow of a graph
and the corresponding classification accuracy of a neural network architecture. We evaluate the competitive efficacy of our proposed MF-NAS
across different datasets with different search spaces that are used in
DARTS/ENAS and NAS-Bench-201.

1 Introduction

Recent advances in deep neural networks result in growing interests in automated
machine learning (AutoML), whose goal is to optimize hyper-parameters and to
identify network architectures suitable to specific datasets without much human
intervention. The target of AutoML can be generally formalized as follows:

$$x^* \in \arg\min_{x \in \mathcal{X}} f(x), \tag{1}$$

where $f : \mathcal{X} \to \mathbb{R}$ is a function defined over a search space $\mathcal{X}$.

In the past few years, white-box-based and black-box-based approaches have been dedicated to developing algorithms for AutoML. In a white-box formulation [27,47], the form of function f is explicitly known, so that $x \in \mathcal{X}$ can be optimized in a differential programming manner. In comparison, in a black-box assumption [8,37], the function f can only be evaluated at $x \in \mathcal{X}$, yielding noisy observations.

One important factor for AutoML is the definition and construction of search space $\mathcal{X}$. Different search spaces $\mathcal{X}$ give rise to different settings for AutoML, and thus promote a variety of search strategies. Given a convex set $\mathcal{X} \subset \mathbb{R}^d$, then Eq. (1) can be viewed as a hyper-parameter optimization (HPO) problem. And Bayesian Optimization (BO) [6,37,39] can provide an elegant compromise in terms of capturing a surrogate model to indicate the likelihood of function f and maximizing an acquisition function to trade off exploration and exploitation. Given a tree-based search space $\mathcal{X} \subset \mathcal{T}$, which is often used for either building the hyper-parameters' dependency or searching for a macro neural architecture where the layers are stacked sequentially, some works [38,40] manage to capture the dependency among layers and to decide a proper exploration-exploitation balance by using Monte Carlo Tree Search (MCTS) strategy. Moreover, if the searching space is extended to a Directed Acyclic Graph (DAG), i.e. $\mathcal{X} \subset \mathcal{G}$, where $\mathcal{G}$ is a DAG search space, then the multi-branch and the skip relationship of hyper-parameters can be established. This elastic expression leads to the research direction called Neural Architecture Search (NAS). In [12,15,27,43,49] the architecture is represented as a super-net, and a white-box approach (i.e., differential programming based approach) is used to update architecture's importance and network's weight. However, due to the approximation in bi-level optimization, the differential based methods (both one-shot and single-path approach) suffer instability, i.e.during the search process, the architecture collapsing occurs and thus the skip-layer tends to dominate [10,41]. On the contrary, some black-box BO methods are also introduced to design NAS strategy. Nevertheless, because of the inconsistency between the vector space $\mathbb{R}^d$ and the DAG space $\mathcal{G}$, BO solutions need extra efforts for encoding the neural architecture [19,30,35,45].

Another important factor for AutoML is search strategy, which has been explored by most of the existing methods for NAS based on reinforcement learning, genetic algorithm, Bayesian optimization, and differential programming. These methods decouple the search space and search strategy, leading to lower efficiency due to ignorance on graph property. To bridge the gap, we focus on designing an efficient and stable search strategy on a DAG search space in a fully graphic manner, which is akin to using MCTS in a tree-based search space or adopting Gaussian process (GP) in an Euclidean space. A few prior works [21,46] have pioneered to perform NAS using graph theory. However, [21] is limited to linear search space and lacks evaluation results on real datasets; whereas [46] lacks feedback reward, making it an open-loop strategy—rather than a NAS method it is virtually a method to define search space. Recently, GFlowNet [4,5] views a Markov Decision Processes (MDP) as a flow network, and connects the flow-matching (conservation) conditions to the generated policy with the target reward function. Though GFlowNet targets at sampling a diverse set of candidates, which may be not the case of NAS, GFlowNet builds a solid foundation for achieving the black-box optimization with the flow network.

In this paper, by introducing the maximum-flow calculation in a flow network, we formulate the NAS task as a multi-graph maximum-flow problem directly defined on a DAG search space with edge capacity indicating the contribution of the corresponding operation to the model performance. To be specific, our contributions are three-fold:

1. We propose a Max-Flow based NAS approach, called MF-NAS, which defines the search space and the search strategy both in a fully graphic manner.
2. To our best knowledge, we make the first attempt to address the NAS task from a multi-graph maximum-flow perspective.
3. We conduct extensive experiments to evaluate the proposed search strategy across multiple datasets on multiple search spaces and show competitive performance.

2   Preliminaries

2.1   Multi-graph Flow in Graph Theory

Consider a directed graph $G := (V, E)$, where V is the vertex set and E is the edge set. A multi-graph is defined as a graph with multiple edges (i.e., parallel edges) between two vertices. A flow graph is a directed graph where each edge has a capacity and receives a flow that is limited by the edge capacity. The formal definition of the problem for finding a feasible and maximal flow on a multi-graph is given as follows.

Definition 1 (Maximum-Flow on Multi-Graph). *Given a directed graph $G = (V, E)$ with a source node $s \in V$, a sink node $t \in V$, edge set $E = \{e_{u,v}^k | u, v \in V, k \in \mathcal{K}\}$, edge capacity function $c : V \times V \times \mathcal{K} \to \mathbb{R}$, where $\mathcal{K} := \{0, 1, \ldots, K-1\}$ and K is number of parallel edges, then the max-flow on the multi-graph G is defined as a feasible s-t-flow function $f : V \times V \times \mathcal{K} \to \mathbb{R}$ that maximizes the flow value $|f| := \sum_{u \in V, k \in \mathcal{K}} f(e_{u,t}^k)$ on G, where $f(e_{u,t}^k)$ denotes the flow on the k-th edge from node u to node t.*

Flow graph [2,44] has been a useful tool for modeling network traffic, circulation, and etc. In this paper, we demonstrate that NAS can be modeled as a maximum-flow selection problem on a multi-graph.

2.2   Hyperband and ENAS

Both Hyperband [24] and ENAS [32] are black-box methods. Hyperband [24] speeds up random search by using an early-stopping strategy to allocate resources adaptively. NAS and ENAS [32,52] both involve training a RNN controller in a loop, where the controller samples a child model (i.e., candidate architecture) for training and its achieved performance is fed back to the controller as a reward to train the LSTM controller. Specifically, in NAS, the child models are trained from scratch to convergence, whereas in ENAS, the child models share their weights to reduces the cost of architecture search. Our proposed MF-NAS applies to the pipeline of NAS: the controller is responsible for generating candidate architectures (i.e., child models), then the parameters of the controller and the weights of child models are updated alternately. Unlike NAS and ENAS, our MF-NAS adopts a max-flow-based controller.

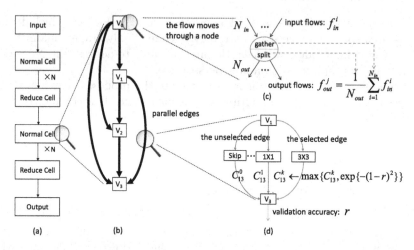

Fig. 1. Illustration of NAS as a max-flow problem. a) A cell-based search space used in DARTS and NAS-Bench-201. b) Connections within a cell in NAS-Bench-201, where bold lines indicate a group of parallel edges (i.e., candidate operations). c) The flow moves through a node following the conservation law, where ihe input flows are first collected by a gather function (e.g., summation function) and then the converged flow is split into output flows. d) Parallel edges with limited capacities between the node pair. In each search step, only one path is selected to construct a child model, and the capacity of the selected edge may be updated with respect to validation accuracy.

3   Methodology

In this section, we formulate the NAS task as a max-flow problem on a multigraph at first, and then show that MF-NAS can be interpreted from a non-parametric density estimation perspective. Finally, we integrate MF-NAS into the classical AutoML pipelines, demonstrating that MF-NAS enjoys wide applicability.

We regard the architecture as a directed acyclic graph (DAG) referring to NAS [52], ENAS [32], DARTS [27] and its subsequent works [7,16,41]. For clarity, we give a sketch of the search space in DARTS and NAS-Bench-201 in a multigraph perspective in Fig. 1. Specifically, operations are regarded as parallel edges between a pair nodes with learned capacities indicating the importance of the architecture, and the moving flow across nodes follows the conservation law [4].

3.1   Neural Architecture Search with Maximum-Flow

Given a training dataset $\mathcal{D}$ for a certain task, NAS aims to find an architecture $m \in \mathcal{A}$ that maximizes a posterior probability, i.e.,

$$m^* = \arg\min_{m \in \mathcal{A}} p(m|\mathcal{D}). \tag{2}$$

To formulate the task of NAS, we use a directed multi-graph to represent the cell-based search space, where parallel edges are the operations, including convolutions, skip connection, pooling, and no connection.

Instead of finding the optimal solution with maximum posterior probability as in Eq. (2), in this paper, we consider the flow value on the operation-induced multi-graph to be the candidate architecture's fitness, which will be further interpreted in Sect. 3.2. Consequently, we convert the task of searching for the best architecture to a task of finding a feasible s-t-flow function f that achieves the maximum-flow value, that is:

$$f^* := \arg\min_{f} \sum_{u \in V, k \in \mathcal{K}} f(e_{u,t}^k). \tag{3}$$

Then, the optimal architecture m^* is selected as a set of edges whose flows are nonzero according to the optimal s-t-flow function f:

$$m^* := \{e_{u,v}^k | u, v \in V \text{ and } k \in \mathcal{K}, \text{ where } f^*(e_{u,v}^k) > 0\}. \tag{4}$$

In this way, the optimization problem in (2) can be solved by addressing the optimal s-t-flow problem in (3) and (4).

Next, we give a proposition to show that the maximum-flow of the flow network in Fig. 1 (a) can be obtained by calculating the maximum-flow of the *normal cell* and the *reduce cell*, separately.

Proposition 1. *Suppose that the architecture is formed by connecting the normal cells m_{normal} and the reduce cells m_{reduce} in a chain manner. Then, the set of edges m^* that maximize the flow value $|f|$ as defined in Definition 1 can be achieved by calculating the union set of that maximizes the flow value of the normal cell m_{normal}^* and that maximizes the flow value of the reduce cell m_{reduce}^*, i.e., $m^* = m_{normal}^* \cup m_{reduce}^*$* [1].

Proof. This can be proved by mathematical induction. **Base case:** if the layer of architecture is only one (formed by one cell), obviously, the statement holds. **Inductive step:** assume the statement holds for n-layers architecture (i.e., $m_n^* = m_{normal}^* \cup m_{reduce}^*$)—this is the induction hypothesis (IH), then the statement will be proved to hold for $n + 1$-layers architecture as shown in Fig. 2(c). Without loss of generality, we consider the $n + 1$-th cell as a normal cell. Denote $|f_{normal}|$, $|f_n|$ and $|f_{n+1}|$ as the flow value of the normal cell in Fig. 2(a), the

---

[1] If there are only the normal cells in the architecture, as NAS-Bench-201, then $m^* = m_{normal}^*$.

flow of the n-layers and $n + 1$-layer network in Fig. 2(c), respectively. According to Max-Flow Min-Cut Theorem [44], the flow of the $n + 1$-layer network in the chain can be represented as max $|f_{n+1}| = min\{$max $|f_n|$, max $|f_{normal}|\}$. If max $|f_n| >= $ max $|f_{normal}|$, then maximizing the flow of $n + 1$-layer network is equivalent to maximize that of the normal cell, and thus $m^*_{n+1} = m^*_n \cup m^*_{normal} \overset{IH}{=} m^*_{normal} \cup m^*_{reduce}$. If max $|f_n| < $ max $|f_{normal}|$, then the maximum flow problem of $n + 1$-layers network is equivalent to maximize the flow of n-layers network. Any m_{normal} that satisfies max $|f_{normal}| > $ max $|f_n|$ will not affect the result, hence m^*_{normal} can be chosen as the optimal solution $m^*_{n+1} = m^*_n \cup m^*_{normal} \overset{IH}{=} m^*_{normal} \cup m^*_{reduce}$. This completes the proof.

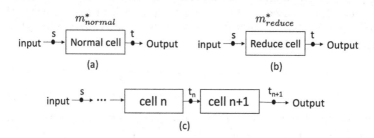

Fig. 2. Illustration of mathematical induction for flow network analysis (see Proposition 1): a) base case for normal cell; b) base case for reduce cell; c) inductive step.

Hence, Eq. (3) can be solved in a greedy manner by separately optimizing inside each cell:

$$\underset{f}{\text{argmax}} \sum_{u \in V_{cell}, k \in \mathcal{K}} f(e^k_{u,t})$$

$$\text{s.t.} \ \ f(e^k_{u,v}) \leq c(e^k_{u,v}), \ \ \forall u, v \in V_{cell}, \ \forall k \in \mathcal{K}$$

$$\sum_{u \in V_{cell}} \sum_{k \neq K-1} \mathbb{I}(e^k_{u,v}) = M, \ \ \forall v \in V_{cell}$$

$$\mathbb{I}(e^{K-1}_{u,v}) = \prod_{k \neq K-1} (1 - \mathbb{I}(e^k_{u,v})), \ \ \forall u, v \in V_{cell}, \tag{5}$$

where V_{cell} is the node set of a normal cell or a reduce cell, t is the sink node of the cell, the index of the sink node is related to the number of nodes N inside the cell, M is the input degree of node, K and $f(e^k_{u,v})$ are defined in Definition 1, and the operation whose index is $K - 1$ denotes *no connection*[2], and $\mathbb{I}(e^k_{u,v})$ is an indicator which is 0 if $f(e^k_{u,v}) = 0$ and otherwise 1. The capacities $c(\cdot)$ of

[2] For the search space in ENAS and DARTS, we set $N = 4$, $M = 2$, $K = 8$; for the search space in NAS-Bench-201, we set $N = 3$ and $K = 5$ without constraining M.

the parallel edges in the multi-graph flow network are updated by the candidate model's reward (i.e., validation accuracy) as follows:

$$c(e_{u,v}^k) = \begin{cases} e^{-(1-r)^2} & \text{if } e_{u,v}^k \in m^* \& e^{-(1-r)^2} > c(e_{u,v}^k) \\ c(e_{u,v}^k) & \text{otherwise,} \end{cases} \quad (6)$$

where m^* is the edge set defined in Eq. (4), r is the accuracy on validation set for image classification, thus the capacity $c(\cdot)$ will be scaled in $[0, 1]$.

MF-NAS benefits from the update rule in two aspects. On one hand, removing less important operations has little influence on final prediction accuracy, while deleting some important operations can lead to a significant drop [3]. We use Eq. (6) to update the capacities, making important operations less prone to be removed. On the other hand, the update formula of Eq. (6) can be viewed as a contraction mapping of the accuracy, which promotes the exploration capability.

We can use dynamic programming (DP) to solve Eq. (5). Take the DARTS search space as an example. In the DARTS search space, we assume that there are two previous nodes of each cell whose initial flow value can be set to one (i.e., the upper-bound of the accuracy), and the feature maps of different nodes in one cell are concatenated as the output. As shown in Fig. 3, there are two states in each node within the cell of the search space: 1) there exists a link between the $(n-1)$-th node and the n-th node 2) there does not exist a link between the $(n-1)$-th node and the n-th node. Denote V_{cell}^n as the sub-cell which is terminated by the n-th node (i.e., the nodes in the sub-cell are $1, 2, ..., n$), whose max-flow value can be defined as:

$$F_n^* := \max_f \sum_{u \in V_{cell}^{n-1}, k \in \mathcal{K}} f(e_{u,n}^k). \quad (7)$$

We initialize $F_{-1}^* = 1$ and $F_0^* = 1$, and then recursively compute the flow value as follows:

$$F_n^* = \max\{F_n^{(1)}, F_n^{(2)}\}, \quad (8)$$

where

$$F_n^{(1)} = 2F_{n-1}^* - F_{n-2}^* + \max_{u \in V_{cell}^{n-2}, k \in \mathcal{K}} c(e_{u,n}^k), \quad (9)$$

$$F_n^{(2)} = F_{n-1}^* + \max_{u \in V_{cell}^{n-2}, k \in \mathcal{K}} c(e_{u,n}^k) + \max_{u \in V_{cell}^{n-2} \setminus \{u^*\}, k \in \mathcal{K} \setminus \{k^*\}} c(e_{u,n}^k), \quad (10)$$

in which $(u^*, k^*) := \arg\max_{u \in V_{cell}^{n-2}, k \in \mathcal{K}} c(e_{u,n}^k)$. We can see that updating the two states corresponds to computing $F_n^{(1)}$ and $F_n^{(2)}$, respectively. By using the DP algorithm as in Eq. (8) and the selection method in Eq. (4), we get the candidates architecture m^* for the next running. There is an equivalent representation of Eq. (8), where the flow value is normalized before moving to the next nodes, making the factors slightly different. If the hypothesis space in Eq. (5) is small enough, random search can be a practical alternative to solve Eq. (5).

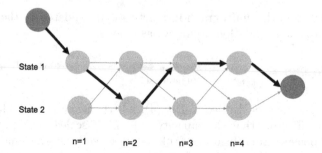

Fig. 3. Illustration for the dynamic programming (DP) to solve Eq. (5).

3.2   MF-NAS: A Probabilistic Perspective

We now describe the motivation for the maximum-flow formulation as well as the relationship between the flow of a graph and the performance (i.e., classification accuracy) of a network architecture.

Let Ω_1 be the set of available operations, Ω_2 be the set of feasible edges in a search space, $\Omega = \Omega_1 \times \Omega_2$, and $\mathcal{A}$ be a class (i.e., set of sets). Then each element $A \in \mathcal{A}$ represents a feasible architecture in the search space, and it is a subset of Ω, i.e., $A \subset \omega$. Let $\phi : \mathcal{A} \to \mathbb{R}$ be a set function, indicating the reward of an architecture. We define a probability measure of an architecture[3] that can achieve the best reward, i.e.,

$$p(A) := p(\phi(A) = r^*), \tag{11}$$

where $r^* = \max_{A' \in \mathcal{A}} \phi(A')$. Furthermore, the conditional probability over an architecture given a specific edge in it can be derived by $p(A|a) := p(A|a \in A) = p(\phi(A) = r^*|a \in A)$. Here, a is the same denotation as $e_{u,v}^k$ in Definition 1.

To make less assumptions about the conditional distribution, we use a non-parametric approach to estimate the density. Choosing a Gaussian kernel function gives rise to the following kernel density estimation model (KDE):

$$p(\phi(A) = r^*|a \in A) = \frac{1}{N} \sum_{\substack{i=1 \\ a \in A_i}}^{N} \frac{\exp\left\{-\frac{\|r^* - \phi(A_i)\|^2}{2\sigma^2}\right\}}{\sqrt{2\pi}\sigma}, \tag{12}$$

where N architectures containing edge a are sampled. For a small enough σ, in the exponential term the one for which $\|r^* - \phi(A_i)\|^2$ is the smallest term will approach zero most slowly, and hence the sample architectures with the best reward—$A_s^* = \arg\max_{A_i} \phi(A_i)$ will dominate the density. That is, Eq. (12) can be approximated by:

$$p(\phi(A) = r^*|a \in A) \propto \exp\left\{-(r^* - \phi(A_s^*))^2\right\}. \tag{13}$$

---

[3] Many architectures can get the same best reward.

For image classification tasks, the reward is measured by the accuracy on validation set, and the upper bound of the reward can be set as $r^* = 1$. Obviously, Eq. (13) holds the same form as the equation of capacity update in Eq. (6). As a consequence, the information flow on edge can be interpreted by an approximated conditional probability as shown in Eq. (13), which reveals the preference for different architectures that can achieve the best performance given the contained edge. We further show that the propagation of the conditional probability can be viewed as a flow moving from the input edges to the output edges. In Fig. 4, three basic types of the network topology are illustrated: a) one-input/one-output, where an architecture A containing the output edge has only one input edge, and thus the conditional probabilities conditioned on the output and input are identical; b) many-input/one-output, where an architecture containing any input edge will flow through the output edge, and thus the probability conditioned on the output edge is the summation of the probabilities conditioned on the input edges; c) one-input/many-output, where an architecture containing the input edge spreads to multiple outputs which are chosen uniformly as a priori, and thus the probability is divided equally by the number of output edges. Note that all the network topology including residual or multi-branches graphs can be derived from the three basic types of topology mentioned above. In a nutshell, the probability defined in Eq. (11) can be propagated as a flow, therefore finding an architecture with the highest validation accuracy is equivalent to finding an architecture with the maximum conditional probability on the output edge, which is also equivalent to finding a maximum flow on the DAG graph with the capacity defined in Eq. (6).

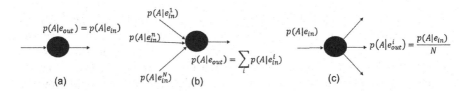

Fig. 4. Three types of network topology. a) one-to-one; b) many-to-one; c) one-to-many.

3.3   MF-NAS Pipeline

The proposed MF-NAS provides a candidate architecture generation method in a single-path NAS solution. To demonstrate the efficiency and the stability, we apply MF-NAS to a multi-fidelity pipeline. For example, Hyperband [24] selects top-k from its candidate pool; whereas MF-NAS dynamically generates new promising candidates as the search progresses by maximizing the network flow. The reason is that the maximum-flow controller involves the edge capacity constraint; and the flow value on one edge can be different for the same net flow value of a cell. In other words, multiple architectures may result in the

---

Algorithm 1 Max-Flow based Neural Architecture Search (MF-NAS)

Input:
1: Search space S, Capacity matrix with zeros initialization, and the fixed hyper-parameters H, e.g., the learning rate and its decay policy, scale of the chain cell structure, input degree of node in the cell, N search rounds, n_i candidates, each candidate trains r_i steps in round i;
2: Initialization: ϵ-greedy factor ϵ;

Output:
3: The optimal architecture m^*;
4: **for** $i \in \{0, 1, ..., N-1\}$ **do**
5:     **if** i = 0 **then**
6:         Generate n_0 architectures $m_0^0, ..., m_0^{n_0-1}$ randomly;
7:     **end if**
8:     Train $m_i^0, ..., m_i^{n_i-1}$ with r_i steps;
9:     Evaluate validation accuracy, update network capacity c by Eq. 6;
10:    **for** $j \in \{0, 1, ..., n_{i+1}-1\}$ **do**
11:        **if** $random() < \epsilon$ **then**
12:            $m_{i+1}^j \leftarrow$ an architecture generated randomly;
13:        **else**
14:            $m_{i+1}^j \leftarrow$ an architecture with the maximum-flow calculated by Eq. 5 and Eq. 4;
15:        **end if**
16:    **end for**
17: **end for**
18: Model selection: return the best architecture m^* with the maximum validation accuracy seen so far.

---

same net flow value. To enhance the exploration capability of MF-NAS, we choose the candidate architecture using an ϵ-greedy strategy [28]. With this strategy, a random architecture is taken with probability ϵ, and the max-flow architecture is chosen with probability $1-\epsilon$. We anneal ϵ from 1 to 0 such that the controller begins from an exploration phase and slowly starts to moving towards the exploitation phase. We also keep a small portion of top architectures during different rounds to stabilize the searching process. We refer to Algorithm 1 for giving an implementation for MF-NAS, which aims at finding an architecture whose classification accuracy on validation set is maximized on the operation-induced DAG search space.

4   Experiments

To evaluate the effectiveness of our proposed MF-NAS approach, we conduct experiments with different DAG search space on several benchmark datasets, including CIFAR-10 [20], CIFAR-100 [20], STL-10 [13], FRUITS [29], FLOWER-102 [31], Caltech-256 [17] and ImageNet [34].

Settings in Experiments. Specifically, we evaluate the efficiency and stability of MF-NAS in three settings: a) the micro search space used in ENAS [32] and

DARTS [27], b) search space of NAS-Bench-201 [16], and c) performance on ImageNet classification. Furthermore, we set the initial learning rate as 0.025 with a cosine scheduler, use SGD with a momentum 0.9. The maximal search round is $N = 4$, and there are n = [30, 20, 10, 5] candidates that are trained r = [30, 60, 70, 80] epochs in different rounds. ϵ = [1.0, 0.6, 0.5, 0.25] indicates it decays for every search round. Experiments are conducted on RTX 3090 and NVIDIA V100 GPUs.

Table 1. Comparison with classification architectures on CIFAR-10. Similar to other NAS algorithms, the search cost of MF-NAS does not include the final evaluation cost.

Methods	Test error %	Params (M)	FLOPs (M)	Search cost GPU-days	Search Method
AmoebaNet-B [33]	2.55 ± 0.05	2.8	506	3150	Evolution
Hierarchical Evo [26]	3.75 ± 0.12	15.7	–	300	Evolution
PNAS [25]	3.41 ± 0.09	3.2	730	225	SMBO
DARTS (1st order) [27]	3.00 ± 0.14	3.3	519	0.4	Gradient
DARTS (2nd order) [27]	2.76 ± 0.09	3.4	547	1.0	Gradient
SNAS (moderate) [47]	2.85 ± 0.02	2.8	441	1.5	Gradient
P-DARTS [11]	2.50	3.4	551	0.3	Gradient
PC-DARTS (1st order) [48]	2.57 ± 0.07	3.6	576	0.1	Gradient
BayseNAS [51]	2.81 ± 0.04	3.4	–	0.2	Gradient
SGAS [22]	2.66 ± 0.24	3.7	–	0.25	Gradient
DARTS+PT [41]	2.61 ± 0.08	3.0	–	0.8	Gradient
NASNet-A [53]	2.65	3.3	624	1800	RL
ENAS [32]	2.89	4.6	626	0.5	RL
ENAS+ϵ-greedy	2.82	3.7	578	0.5	RL
Hyperband [24]	2.96 ± 0.19	2.9	476	2.2	Random
BANANAS [45]	2.64	–	–	11.8	BO
MF-NAS	2.63 ± 0.16	3.3	529	1.0	Max-Flow
MF-NAS (best)	2.40	4.0	653	1.0	Max-Flow

4.1 Results on Micro Cells Based Search Space in ENAS/DARTS

We evaluate MF-NAS on micro cells-based DAG search space used in ENAS [32] and DARTS [27][4]. The main difference between MF-NAS and other algorithms lies in the selection method for candidate networks. To be specific, MF-NAS applies optimization algorithms with graph frameworks to choose candidates; whereas others generate candidates by means of Bayesian optimization, reinforcement learning, evolution or gradient-based methods. For demonstrating the efficiency of MF-NAS, we implement MF-NAS following the controller-child pipeline [32], but do not use the weight-sharing. As shown in the Table 1, max-flow based method gets a 2.63% error rate with 3.3M parameters by taking about one search days, which is more efficient than RL/Evolution/Random/BO based

---

[4] Precisely, MF-NAS uses the search space of DARTS.

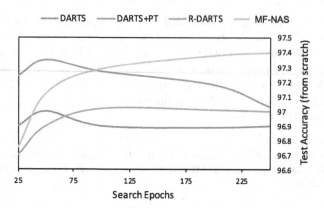

Fig. 5. Stability of DARTS (w/ its amendment versions) and MF-NAS on CIFAR-10.

methods on DAG search space. This comes from the fact that the proposed MF-NAS are coupled with DAG search space, and thus they can take advantage of the graph structure, i.e. the relation of the edges and nodes. Also, unlike the gradient-based methods, MF-NAS does not suffer the skip domination issue or the discretization problem [41]. To validate the stability of MF-NAS, we evaluate the performance at different search steps in Fig. 5.

Figure 5 compares the stability of DARTS [27] and its amendment versions [41,50] with MF-NAS. Note that the search epoch is not related to real search time. For DARTS+PT [41], we get our results on the super-nets of DARTS pre-trained from 25 to 250 training epochs. Both vanilla DARTS and DARTS+PT suffer from the skip layer dominated problem [7,50], i.e. there are about 6 skip connections in the normal cell at 250 epochs. The ultimate performance of differential methods usually decreases over search epochs because they use some approximations to solve the bi-level optimization. In contrast, thanks to the maximum-flow solution, MF-NAS can steadily achieve improved performance.

Table 2. Top-1 test set accuracy with one GPU day budget.

NAS method	STL-10 (96 × 96)	FRUITS (100 × 100)	FLOWER-102 (256 × 256)	Caltech-256 (256 × 256)
Hyperband	0.79 ± 0.02	0.99 ± 0.0006	0.95 ± 0.002	0.50 ± 0.06
ENAS	0.79 ± 0.02	0.97 ± 0.0012	0.93 ± 0.007	0.49 ± 0.05
DARTS	0.80 ± 0.01	0.99 ± 0.0018	0.94 ± 0.006	0.60 ± 0.03
MF-NAS	0.80 ± 0.02	0.99 ± 0.0009	0.95 ± 0.002	0.65 ± 0.02

Based on this search space, we evaluate MF-NAS's generalization ability. We use other four datasets with different image resolutions, including STL-10 [13], FRUITS [29], FLOWER-102 [31], and Caltech-256 [17]. ENAS [32], DARTS [27], and Hyperband [24] are evaluated as the baselines whose search protocols cover reinforcement learning, differential method and closed-loop random search.

Similar to ENAS, we extend Hyperband to support neural network search in the same search space of DARTS. The experiments are under the budget constraints of one GPU day. As shown in Table 2, MF-NAS outperforms other methods on the last dataset. We analyze that Caltech-256 is a challenging dataset, on which the early performance of an architecture does not align with its ultimate performance precisely, and thus Hyperband fails. Additionally, the limited GPU memory restricts the batch size for DARTS on large-scale images, which can cause a performance drop [36].

4.2   Results on Search Space of NAS-Bench-201

In NAS-Bench-201, the architectures are stacked by 17 cells with five operations, and three datasets are evaluated, including CIFAR-10 [20], CIFAR-100 [20], and ImageNet-16-120 [16]. We follow the training settings and split protocol as the paper [16]. We search for the architecture based on MF-NAS three times with different random seeds, and the results are reported in Table 3, showing that our method achieves competitive results with the state-of-art methods. We observe that ENAS can also benefit from the ϵ-greedy strategy, indicating that ENAS may suffer from the weight-sharing problem: it is hard to figure out whether the improvement under the weight-sharing strategy is attributed to a better-chosen architecture or comes from well-trained weights. Thus, there will still be a big room to improve the weight-sharing strategy in the NAS area.

Table 3. Top-1 accuracy (%) on NAS-Bench-201.

Methods	CIFAR-10		CIFAR-100		ImageNet-16-120	
	Valid	Test	Valid	Test	Valid	Test
DARTS-V1 [27]	39.77 ± 0.00	54.30 ± 0.00	15.03 ± 0.00	15.61 ± 0.00	16.43 ± 0.00	16.32 ± 0.00
DARTS-V2 [27]	39.77 ± 0.00	54.30 ± 0.00	15.03 ± 0.00	15.61 ± 0.00	16.43 ± 0.00	16.32 ± 0.00
GDAS [15]	89.89 ± 0.08	93.61 ± 0.09	71.34 ± 0.04	70.70 ± 0.30	41.59 ± 1.33	41.71 ± 0.98
SETN [14]	84.04 ± 0.28	87.64 ± 0.00	58.86 ± 0.06	59.05 ± 0.24	33.06 ± 0.02	32.52 ± 0.21
RSPS [23]	82.25 ± 4.90	86.09 ± 4.90	56.56 ± 8.91	56.39 ± 8.70	31.17 ± 6.28	30.27 ± 5.80
DARTS-PT [41]	–	88.11	–	–	–	–
DARTS-PT (fix α) [41]	–	93.80	–	–	–	–
ENAS [32]	37.55 ± 3.14	53.80 ± 0.71	13.47 ± 2.21	14.00 ± 2.27	14.88 ± 2.19	14.87 ± 2.05
ENAS+ϵ-greedy	77.37 ± 3.08	84.10 ± 3.16	56.85 ± 4.47	57.31 ± 4.95	32.16 ± 2.95	32.93 ± 3.14
MF-NAS	27437	93.72 ± 0.55	71.23 ± 1.07	71.86 ± 0.64	44.16 ± 0.10	44.33 ± 0.18
ResNet [18]	90.83	93.97	70.42	70.86	44.53	43.63
Optimal	91.61	94.37	73.49	73.51	46.77	47.31

4.3   Results on ImageNet Classification

For ImageNet [34], the one-shot NAS approaches cost too much memory to build a super-net. Works [9,27,53] directly transfer the architecture searched on CIFAR-10 to ImageNet with little modification. Specifically, the search phase runs on CIFAR-10, while the evaluation phase uses the architecture transferred from the searched cells with deeper and wider scales as suggested in DARTS [27]. Learning rate decay policy is different between the last two blocks of Table 4. Table 4 evaluates the performance on ImageNet. In general, MF-NAS can get competitive results in transfer cases.

Table 4. Study on full ImageNet. The first block represents the direct search case while the last two blocks show the transfer cases. † and ‡ represent lr decay strategy as DARTS and P-DARTS/PC-DARTS, respectively.

Methods	Params	FLOPs	GPU (days)	Top-1 Acc
NASNet-A	5.3M	620M	1800	0.740
AmoebaNet-A	5.1M	541M	3150	0.745
DARTS†	4.9M	545M	1	0.731
MF-NAS†	4.9M	549M	0.4	0.744
P-DARTS‡	4.9M	557M	0.3	0.756
PC-DARTS‡	5.3M	586M	0.1	0.749
MF-NAS‡	4.9M	549M	0.4	0.753

4.4   Ablation Study on ϵ-greedy

The proposed MF-NAS follows the reinforcement learning based NAS method [1] to involve the ϵ-greedy strategy into the algorithm pipeline. To eliminate the influence of ϵ-greedy, we show the ablation study in Table. 5.

Table 5. Test accuracy (%) with different ϵ-greedy strategies on CIFAR-10. $\epsilon \equiv 0$ means the ϵ-greedy strategy has no effect on the corresponding NAS method; Note that $\epsilon \equiv 1$ degrades to Random Search.

Methods	$\epsilon \equiv 0$	$\epsilon \equiv 1$	ϵ-greedy used in our paper
Hyperband	97.04	96.75	97.06
ENAS	97.11	96.75	97.18
MF-NAS	97.32	96.75	97.40

The results show that MF-NAS and ENAS [32] can benefit from the ϵ-greedy strategy. In fact, ϵ-greedy strategy can enhance the exploration capacity, which may be more critical in a maximum-flow/RL strategy than a random search solution. As a consequence, max-flow-based and reinforcement learning-based NAS methods turn out to be more efficient than random search schemes considering the pure random search solution only gets 96.75 ± 0.2 % accuracy.

5   Conclusion and Future Work

We have proposed a multi-graph maximum-flow approach for NAS, called MF-NAS, which casts the problem of optimal architecture search as finding a path on DAG. In the search phase, the weights and the capacities of the parallel edges in the multi-graph are updated alternately. Extensive experiments on the image classification task demonstrated the effectiveness of our proposal. As the future work, we will explore for using larger search space, with non-linear gather functions in a general maximum-flow network, and test for other vision tasks, e.g., segmentation and detection [42].

Acknowledgments. J. Yan is supported by the National Key Research and Development Program of China under grant 2020AAA0107600. C.-G. Li is supported by the National Natural Science Foundation of China under grant 61876022.

References

1. Baker, B., Gupta, O., Naik, N., Raskar, R.: Designing neural network architectures using reinforcement learning. In: ICLR (2017)
2. Bein, W.W., Brucker, P., Tamir, A.: Minimum cost flow algorithms for series-parallel networks. Discrete Appl. Math. **10**, 117–124 (1985)
3. Bender, G., Kindermans, P., Zoph, B., Vasudevan, V., Le, Q.V.: Understanding and simplifying one-shot architecture search. In: ICML (2018)
4. Bengio, E., Jain, M., Korablyov, M., Precup, D., Bengio, Y.: Flow network based generative models for non-iterative diverse candidate generation. In: NeurIPS (2021)
5. Bengio, Y., Deleu, T., Hu, E.J., Lahlou, S., Tiwari, M., Bengio, E.: GFlowNet foundations. arXiv preprint arXiv:2111.09266 (2021)
6. Bergstra, J., Bardenet, R., Bengio, Y., Kégl, B.: Algorithms for hyper-parameter optimization. In: NeurIPS (2011)
7. Bi, K., Hu, C., Xie, L., Chen, X., Wei, L., Tian, Q.: Stabilizing DARTS with amended gradient estimation on architectural parameters. arXiv:1910.11831 (2019)
8. Bonilla, E.V., Chai, K.M.A., Williams, C.K.I.: Multi-task gaussian process prediction. In: NeurIPS (2007)
9. Chao, X., Mengting, H., Xueqi, H., Chun-Guang, L.: Automated search space and search strategy selection for AutoML. Pattern Recognit. **124**, 108474 (2022)
10. Chen, X., Hsieh, C.J.: Stabilizing differentiable architecture search via perturbation-based regularization. In: ICLR (2020)
11. Chen, X., Xie, L., Wu, J., Tian, Q.: Progressive differentiable architecture search: bridging the depth gap between search and evaluation. In: ICCV (2019)
12. Chu, X., Wang, X., Zhang, B., Lu, S., Wei, X., Yan, J.: Darts-: robustly stepping out of performance collapse without indicators. In: ICLR (2021)
13. Coates, A., Ng, A.Y., Lee, H.: An analysis of single-layer networks in unsupervised feature learning. In: AISTATS (2011)
14. Dong, X., Yang, Y.: One-shot neural architecture search via self-evaluated template network. In: ICCV (2019)
15. Dong, X., Yang, Y.: Searching for a robust neural architecture in four GPU hours. In: CVPR (2019)

16. Dong, X., Yang, Y.: An algorithm-agnostic NAS benchmark. In: ICLR (2020)
17. Griffin, G., Holub, A., Perona, P.: Caltech-256 object category dataset (2007)
18. He, K., Zhang, X., Ren, S., Sun, J.: Deep residual learning for image recognition. In: CVPR (2016)
19. Kandasamy, K., Neiswanger, W., Schneider, J., Póczos, B., Xing, E.P.: Neural architecture search with Bayesian optimisation and optimal transport. In: NeurIPS (2018)
20. Krizhevsky, A., Hinton, G., et al.: Learning multiple layers of features from tiny images. Technical report, Citeseer (2009)
21. de Laroussilhe, Q., Jastrzkebski, S., Houlsby, N., Gesmundo, A.: Neural architecture search over a graph search space. CoRR (2018)
22. Li, G., Qian, G., Delgadillo, I.C., Muller, M., Thabet, A., Ghanem, B.: SGAS: sequential greedy architecture search. In: CVPR (2020)
23. Li, L., Talwalkar, A.: Random search and reproducibility for neural architecture search. In: UAI (2019)
24. Li, L., Jamieson, K.G., DeSalvo, G., Rostamizadeh, A., Talwalkar, A.: Hyperband: a novel bandit-based approach to hyperparameter optimization. J. Mach. Learn. Res., 185:1–185:52 (2017)
25. Liu, C., et al.: Progressive neural architecture search. In: Ferrari, V., Hebert, M., Sminchisescu, C., Weiss, Y. (eds.) ECCV 2018. LNCS, vol. 11205, pp. 19–35. Springer, Cham (2018). https://doi.org/10.1007/978-3-030-01246-5_2
26. Liu, H., Simonyan, K., Vinyals, O., Fernando, C., Kavukcuoglu, K.: Hierarchical representations for efficient architecture search. In: ICLR (2018)
27. Liu, H., Simonyan, K., Yang, Y.: DARTS: differentiable architecture search. In: ICLR (2019)
28. Mnih, V., et al.: Nature (2015)
29. Muresan, H., Oltean, M.: Fruit recognition from images using deep learning. Acta Universitatis Sapientiae Informatica (2018)
30. Nguyen, V., Le, T., Yamada, M., Osborne, M.A.: Optimal transport kernels for sequential and parallel neural architecture search. In: ICML (2021)
31. Nilsback, M., Zisserman, A.: Automated flower classification over a large number of classes. In: ICVGIP (2008)
32. Pham, H., Guan, M.Y., Zoph, B., Le, Q.V., Dean, J.: Efficient neural architecture search via parameter sharing. In: ICML (2018)
33. Real, E., Aggarwal, A., Huang, Y., Le, Q.V.: Regularized evolution for image classifier architecture search. In: AAAI (2019)
34. Russakovsky, O., et al.: ImageNet large scale visual recognition challenge. In: IJCV (2015)
35. Shi, H., Pi, R., Xu, H., Li, Z., Kwok, J., Zhang, T.: Bridging the gap between sample-based and one-shot neural architecture search with BONAS. In: NeurIPS (2020)
36. Smith, S.L., Kindermans, P., Ying, C., Le, Q.V.: Don't decay the learning rate, increase the batch size. In: ICLR (2018)
37. Snoek, J., Larochelle, H., Adams, R.P.: Practical Bayesian optimization of machine learning algorithms. In: NeurIPS (2012)
38. Su, X., et al.: Prioritized architecture sampling with Monto-Carlo tree search. In: CVPR (2021)
39. Swersky, K., Snoek, J., Adams, R.P.: Multi-task Bayesian optimization. In: NeurIPS (2013)
40. Wang, L., Fonseca, R., Tian, Y.: Learning search space partition for black-box optimization using Monte Carlo tree search. In: NeurIPS (2020)

41. Wang, R., Cheng, M., Chen, X., Tang, X., Hsieh, C.J.: Rethinking architecture selection in differentiable NAS. In: ICLR (2021)
42. Wang, X., Lin, J., Zhao, J., Yang, X., Yan, J.: EAutoDet: efficient architecture search for object detection. In: Farinella, T. (ed.) ECCV 2022. LNCS, vol. 13680, pp. 668–684 (2022)
43. Wang, X., Xue, C., Yan, J., Yang, X., Hu, Y., Sun, K.: MergeNAS: merge operations into one for differentiable architecture search. In: IJCAI (2020)
44. West, D.B., et al.: Introduction to Graph Theory, vol. 2. Prentice Hall, Upper Saddle River (1996)
45. White, C., Neiswanger, W., Savani, Y.: Bananas: Bayesian optimization with neural architectures for neural architecture search. In: AAAI (2021)
46. Xie, S., Kirillov, A., Girshick, R.B., He, K.: Exploring randomly wired neural networks for image recognition. In: ICCV (2019)
47. Xie, S., Zheng, H., Liu, C., Lin, L.: SNAS: stochastic neural architecture search. In: ICLR (2019)
48. Xu, Y., et al.: PC-DARTS: partial channel connections for memory-efficient architecture search. In: ICLR (2019)
49. Xue, C., Wang, X., Yan, J., Hu, Y., Yang, X., Sun, K.: Rethinking Bi-level optimization in neural architecture search: a gibbs sampling perspective. In: AAAI (2021)
50. Zela, A., Elsken, T., Saikia, T., Marrakchi, Y., Brox, T., Hutter, F.: Understanding and robustifying differentiable architecture search. In: ICLR (2020)
51. Zhou, H., Yang, M., Wang, J., Pan, W.: BayesNAS: a Bayesian approach for neural architecture search. In: ICML (2019)
52. Zoph, B., Le, Q.V.: Neural architecture search with reinforcement learning. In: ICLR (2017)
53. Zoph, B., Vasudevan, V., Shlens, J., Le, Q.V.: Learning transferable architectures for scalable image recognition. In: CVPR (2018)

OccamNets: Mitigating Dataset Bias by Favoring Simpler Hypotheses

Robik Shrestha[1]([✉])[ID], Kushal Kafle[2][ID], and Christopher Kanan[1,3][ID]

[1] Rochester Institute of Technology, Rochester, USA
rss9369@rit.edu, ckanan@cs.rochester.edu
[2] Adobe Research, San Jose, USA
kkafle@adobe.com
[3] University of Rochester, Rochester, USA

Abstract. Dataset bias and spurious correlations can significantly impair generalization in deep neural networks. Many prior efforts have addressed this problem using either alternative loss functions or sampling strategies that focus on rare patterns. We propose a new direction: modifying the network architecture to impose inductive biases that make the network robust to dataset bias. Specifically, we propose OccamNets, which are biased to favor simpler solutions by design. OccamNets have two inductive biases. First, they are biased to use as little network depth as needed for an individual example. Second, they are biased toward using fewer image locations for prediction. While OccamNets are biased toward simpler hypotheses, they can learn more complex hypotheses if necessary. In experiments, OccamNets outperform or rival state-of-the-art methods run on architectures that do not incorporate these inductive biases. Furthermore, we demonstrate that when the state-of-the-art debiasing methods are combined with OccamNets (https://github.com/erobic/occam-nets-v1) results further improve.

1 Introduction

Frustra fit per plura quod potest fieri per pauciora

William of Occam, *Summa Totius Logicae (1323 CE)*

Spurious correlations and dataset bias greatly impair generalization in deep neural networks [2,6,23,62]. This problem has been heavily studied. The most common approaches are re-sampling strategies [8,15,22,57], altering optimization to mitigate bias [55], adversarial unlearning [1,20,53,76], learning invariant representations [5,11,67], and ensembling with bias-amplified models [7,12,47]. Here, we propose a new approach: incorporating architectural inductive biases that combat dataset bias.

Supplementary Information The online version contains supplementary material available at https://doi.org/10.1007/978-3-031-20044-1_40.

© The Author(s), under exclusive license to Springer Nature Switzerland AG 2022
S. Avidan et al. (Eds.): ECCV 2022, LNCS 13680, pp. 702–721, 2022.
https://doi.org/10.1007/978-3-031-20044-1_40

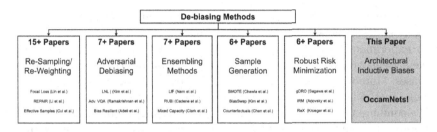

Fig. 1. OccamNets focus on architectural inductive biases, which is an orthogonal direction to tackling dataset biases compared to the existing works.

In a typical feedforward network, each layer can be considered as computing a function of the previous layer, with each additional layer making the hypothesis more complex. Given a system trained to predict multiple categories, with some being highly biased, this means the network uses the same level of complexity across all of the examples, even when some examples should be classified with simpler hypotheses (e.g., less depth). Likewise, pooling in networks is typically uniform in nature, so every location is used for prediction, rather than only the minimum amount of information. In other words, typical networks violate Occam's razor. Consider the Biased MNIST dataset [62], where the task is to recognize a digit while remaining invariant to multiple spuriously correlated factors, which include colors, textures, and contextual biases. The most complex hypothesis would exploit every factor during classification, including the digit's color, texture, or background context. A simple hypothesis would instead be to focus on the digit's shape and to ignore these spuriously correlated factors that work very well during training but do not generalize. We argue that a network should be capable of adapting its hypothesis space for each example, rather than always resorting to the most complex hypothesis, which would help it to ignore extraneous variables that hinder generalization.

Here, we propose convolutional OccamNets which have architectural inductive bias that favor using the minimal amount of network depth and the minimal number of image locations during inference for a given example. The first inductive bias is implemented using early exiting, which has been previously studied for speeding up inference. The network is trained such that later layers focus on examples earlier layers find hard, with a bias toward exiting early. The second inductive bias replaces global average pooling before a classification layer with a function that is regularized to favor pooling with fewer image locations from class activation maps (CAMs). We hypothesize this would be especially useful for combating background and contextual biases [3,63]. OccamNets are complementary to existing approaches and can be combined with them.

In this paper, we demonstrate that architectural inductive biases are effective at mitigating dataset bias. Our specific contributions are:

- We introduce the OccamNet architecture, which has architectural inductive biases for favoring simpler solutions to help overcome dataset biases. Occam-

Nets do not require the biases to be explicitly specified during training, unlike many state-of-the-art debiasing algorithms.

- In experiments using biased vision datasets, we demonstrate that Occam-Nets greatly outperform architectures that do not use the proposed inductive biases. Moreover, we show that OccamNets outperform or rival existing debiasing methods that use conventional network architectures.
- We combine OccamNets with four recent debiasing methods, which all show improved results compared to using them with conventional architectures.

2   Related Work

Dataset Bias and Bias Mitigation. Deep networks trained with empirical risk minimization (ERM) tend to exploit training set biases resulting in poor test generalization [23,45,62,70]. Existing works for mitigating this problem have focused on these approaches: 1) focusing on rare data patterns through re-sampling [8,40], 2) loss re-weighting [15,57], 3) adversarial debiasing [20,34], 4) model ensembling [7,12], 5) minority/counterfactual sample generation [8,9,35] and 6) invariant/robust risk minimization [5,36,56]. Most of these methods require bias variables, e.g., sub-groups within a category, to be annotated [20,34,40,57,62]. Some recent methods have also attempted to detect and mitigate biases without these variables by training separate bias-amplified models for de-biasing the main model [13,47,58,71]. This paper is the first to explore architectural inductive biases for combating dataset bias.

Early Exit Networks. OccamNet is a multi-exit architecture designed to encourage later layers to focus on samples that earlier layers find difficult. Multi-exit networks have been studied in past work to speed up average inference time by minimizing the amount of compute needed for individual examples [10,31,66,73], but their impact on bias-resilience has not been studied. In [59], a unified framework for studying early exit mechanisms was proposed, which included commonly used training paradigms [26,38,64,72] and biological plausibility [46,49,50]. During inference, multi-exit networks choose the earliest exit based on either a learned criterion [10] or through a heuristic, e.g., exit if the confidence score is sufficiently high [19], exit if there is low entropy [66], or exit if there is agreement among multiple exits [78]. Recently, [19] proposed early exit networks for long-tailed datasets; however, they used a class-balanced loss and did not study robustness to hidden covariates, whereas, OccamNets generalize to these hidden variables without oracle bias labels during training.

Exit Modules and Spatial Maps. OccamNets are biased toward using fewer spatial locations for prediction, which we enable by using spatial activation maps [24,44,54]. While most recent convolutional neural networks (CNNs) use global average pooling followed by a linear classification layer [25,29,32], alternative pooling methods have been proposed, including spatial attention [4,21,30,74] and dynamic pooling [30,33,37]. However, these methods have not been explored for their ability to combat bias mitigation, with existing bias mitigation methods adopting conventional architectures that use global average pooling instead. For OccamNets, each exit produces a class activation map, which is biased toward using fewer visual locations.

3   OccamNets

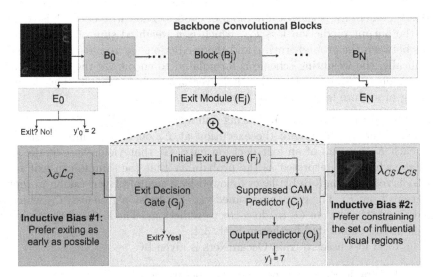

Fig. 2. OccamNets are multi-exit architectures capable of exiting early through the exit decision gates. The exits yield class activation maps that are trained to use a constrained set of visual regions.

3.1   OccamNet Architecture for Image Classification

OccamNets have two inductive biases: a) they prefer exiting as early as possible, and b) they prefer using fewer visual regions for predictions. Following Occam's principles, we implement these inductive biases using simple, intuitive ideas: early exit is based on whether or not a sample is correctly predicted during training and visual constraint is based on suppressing regions that have low confidence towards the ground truth class. We implement these ideas in a CNN. Recent CNN architectures, such as ResNets [25] and DenseNets [32], consist of multiple blocks of convolutional layers. As shown in Fig. 2, these inductive biases are enabled by attaching an exit module E_j to block B_j of the CNN, as the blocks serve as natural endpoints for attaching them. Below, we describe how we implement these two inductive biases in OccamNets.

In an OccamNet, each exit module E_j takes in feature maps produced by the backbone network and processes them with F_j, which consists of two convolutional layers, producing feature maps used by the following components:

Suppressed CAM Predictors (C_j). Each C_j consists of a single convolutional layer, taking in the feature maps from F_j to yield class activation maps, c. The maps provide location wise predictions of classes. Following Occam's principles, the usage of visual regions in these CAMs is suppressed through the CAM suppression loss $\mathcal{L}_{CS}$ described in Sect. 3.2.

Output Predictors (O_j). The output predictor applies global average pooling on the suppressed CAMs predicted by C_j to obtain the output prediction vector, $\hat{y}_j \in \mathbb{R}^{n_Y}$, where n_Y is the total number of classes. The entire network is trained with the output prediction loss $\mathcal{L}_O$, which is a weighted sum of cross entropy losses between the ground truth y and the predictions $\hat{y}_j$ from each of the exits. Specifically, the weighting scheme is formulated to encourage the deeper layers to focus on the samples that the shallower layers find difficult. The detailed training procedure is described in Sect. 3.3.

Exit Decision Gates (G_j). During inference, OccamNet needs to decide whether or not to terminate the execution at E_j on a per-sample basis. For this, each E_j consists of an exit decision gate, G_j that yields an exit decision score g_j, which is interpreted as the probability that the sample can exit from E_j. G_j is realized via a ReLU layer followed by a sigmoid layer, taking in representations from F_j. The gates are trained via exit decision gate loss, $\mathcal{L}_G$ which is based on whether or not O_j made correct predictions. The loss and the training procedure are elaborated further in Sect. 3.4.

The total loss used to train OccamNets is given by:

Frustra fit per plura quod potest fieri per pauciora

William of Occam, *Summa Totius Logicae (1323 CE)*

3.2   Training the Suppressed CAMs

To constrain the usage of visual regions, OccamNets regularize the CAMs so that only some of the cells exhibit confidence towards the ground truth class, whereas rest of the cells exhibit inconfidence i.e., have uniform prediction scores for all the classes. Specifically, let $c_y \in \mathbb{R}^{h \times w}$ be the CAM where each cell encodes the score for the ground truth class. Then, we apply regularization on the locations that obtain softmax scores lower than the average softmax score for the ground truth class. That is, let $\bar{c}_y$ be the softmax score averaged over all the cells in c_y, then the cells at location l, $c^l \in \mathbb{R}^{n_Y}$ are regularized if the softmax score for the ground truth class, c_y^l is less than $\bar{c}_y$. The CAM suppression loss is:

$$\mathcal{L}_{CS} = \sum_{l=1}^{hw} \mathbb{1}(c_y^l < \bar{c}_y)\, KLD(c^l, \frac{1}{n_Y}\mathbf{1}), \tag{1}$$

where, $KLD(c^l, \frac{1}{n_Y}\mathbf{1})$ is the KL-divergence loss with respect to a uniform class distribution and $\mathbb{1}(c_y^l < \bar{c}_y)$ ensures that the loss is applied only if the ground truth class scores lower than $\bar{c}_y$. The loss weight for $\mathcal{L}_{CS}$ is λ_{CS}, which is set to 0.1 for all the experiments.

3.3   Training the Output Predictors

The prediction vectors $\hat{y}$ obtained by performing global average pooling on the suppressed CAMs c are used to compute the output prediction losses. Specifically, we train a bias-amplified first exit E_0, using a loss weight of: $W_0 = p_0^{\gamma_0}$, where, p_0 is the softmax score for the ground truth class. Here, $\gamma_0 > 0$ encourages E_0 to amplify biases i.e., it provides higher loss weights for the samples that already have high scores for the ground truth class. This encourages E_0 to focus on the samples that it already finds easy to classify correctly. For all the experiments, we set $\gamma_0 = 3$ to sufficiently amplify the biases. The subsequent exits are then encouraged to focus on samples that the preceding exits find difficult. For this, the loss weights are defined as:

$$W_j = (1 - g_{j-1} + \epsilon), \text{ if j } > 0, \tag{2}$$

where, g_{j-1} is the exit decision score predicted by $(j-1)^{th}$ exit decision gate and $\epsilon = 0.1$ is a small offset to ensure that all the samples receive a minimal, non-zero loss weight. For the samples where g_{j-1} is low, the weight loss for j^{th} exit, W_j becomes high. The total output loss is then:

$$\mathcal{L}_O = \sum_{j=0}^{n_E-1} W_j \, CE(\hat{y}_j, y), \tag{3}$$

where, $CE(\hat{y}_j, y)$ is the cross-entropy loss and n_E is the total number of exits. Note that E_j's are 0-indexed and the first bias-amplified exit E_0 is not used during inference. Furthermore, during training, we prevent the gradients of E_0 from passing through $B_0(.)$ to avoid degrading the representations available for the deeper blocks and exits.

3.4   Training the Exit Decision Gates

Each exit decision gate $G_j(.)$ yields an exit probability score $\hat{g}_j = G_j(.)$. During inference, samples with $\hat{g}_j \geq 0.5$ exit from E_j and samples with $\hat{g}_j < 0.5$ continue to the next block B_{j+1}, if available. During training, all the samples use the entire network depth and g_j is used to weigh losses as described in Sect. 3.3. Now, we specify the exit decision gate loss used to train G_j:

$$\mathcal{L}_G = \sum_{k \in \{0,1\}} \frac{\mathbb{1}(g_j = k) BCE(g_j, \hat{g}_j)}{\sqrt{\sum \mathbb{1}(g_j = k)}}, \tag{4}$$

where g_j is the ground truth value for the jth gate, which is set to 1 if the predicted class y' is the same as the ground truth class y and 0 otherwise. That is, G_j is trained to exit if the sample is correctly predicted at depth j, else it is trained to continue onto the next block. Furthermore, the denominator: $\sqrt{\sum \mathbb{1}(g_j = k)}$ balances out the contributions from the samples with $g = 1$ and $g = 0$ to avoid biasing one decision over the other. With this setup, sufficiently parameterized models that obtain 100% training accuracy will result in a trivial solution where g_j is always set to 1 i.e., the exit will learn that all the samples can exit. To avoid this issue, we stop computing g_j once E_j's mean-per-class training accuracy reaches a predefined threshold $\tau_{acc,j}$. During training, we stop the gradients from G from passing through $F_j(.)$ and $B(.)$, since this improved the training stability and overall accuracy in the preliminary experiments. The loss weight λ_G is set to 1 in all the experiments.

4 Experimental Setup

4.1 Datasets

(a) Biased MNIST   (b) COCO-on-Places   (c) BAR

Fig. 3. For each dataset, the first two columns show bias-aligned (majority) samples, and the last column shows bias-conflicting (minority) samples. For BAR, the train set does not contain any bias-conflicting samples.

Biased MNIST [62]. As shown in Fig. 3a, Biased MNIST requires classifying MNIST digits while remaining robust to multiple sources of biases, including color, texture, scale, and contextual biases. This is more challenging than the widely used Colored MNIST dataset [5,34,40], where the only source of bias is the spuriously correlated color. In our work, we build on the version created in [62]. We use 160×160 images with 5×5 grids of cells, where the target digit is placed in one of the grid cells and is spuriously correlated with: a) digit size/scale (number of cells a digit occupies), b) digit color, c) type of background texture, d) background texture color, e) co-occurring letters, and f) colors of the co-occurring letters. Following [62], we denote the probability with which each digit co-occurs with its biased property in the training set by p_{bias}. For instance, if $p_{bias} = 0.95$, then 95% of the digit 1s are red, 95% of digit 1s co-occur with letter 'a' (not necessarily colored red) and so on. We set p_{bias} to 0.95 for all the experiments. The validation and test sets are unbiased. Biased MNIST has 10 classes and 50K train, 10K validation, and 10K test samples.

COCO-on-Places [3]. As shown in Fig. 3b, COCO-on-Places puts COCO objects [41] on spuriously correlated Places backgrounds [77]. For instance, buses mostly appear in front of balloons and birds in front of trees. The dataset provides three different test sets: a) biased backgrounds (in-distribution), which reflects the object-background correlations present in the train set, b) unseen backgrounds (non-systematic shift), where the objects are placed on backgrounds that are absent from the train set and c) seen, but unbiased backgrounds (systematic shift) where the objects are placed on backgrounds that were not spuriously correlated with the objects in the train set. Results in [3] show it is difficult to maintain high accuracy on both the in-distribution and shifted-distribution test sets. Apart from that, COCO-on-Places also includes an anomaly detection task, where anomalous samples from unseen object class need to be distinguished from the in-distribution samples. COCO-on-Places has 9 classes with 7200 train, 900 validation, and 900 test images.

Biased Action Recognition (BAR) [47]. BAR reflects real world challenges where bias attributes are not explicitly labeled for debiasing algorithms, with the test set containing additional correlations not seen during training. The dataset consists of correlated action-background pairs, where the train set consists of selected action-background pairs, e.g., climbing on a rock, whereas the evaluation set consists of differently correlated action-background pairs, e.g., climbing on snowy slopes (see Fig. 3c). The background is not labeled for the debiasing algorithms, making it a challenging benchmark. BAR has 6 classes with 1941 train and 654 test samples.

4.2   Comparison Methods, Architectures and Other Details

We compare OccamNets with four state-of-the-art bias mitigation methods, apart from the vanilla empirical risk minimization procedure:

- **Empirical Risk Minimization (ERM)** is the default method used by most deep learning models and it often leads to dataset bias exploitation since it minimizes the train loss without any debiasing procedure.
- **Spectral Decoupling (SD)** [51] applies regularization to model outputs to help decouple features. This can help the model focus more on the signal.
- **Group Upweighting (Up Wt)** balances the loss contributions from the majority and the minority groups by multiplying the loss by $\frac{1}{n_g^\gamma}$, where n_g is the number of samples in group g and γ is a hyper-parameter.
- **Group DRO (gDRO)** [55] is an instance of a broader family of distributionally robust optimization techniques [18,48,52], that optimizes for the difficult groups in the dataset.
- **Predictive Group Invariance (PGI)** [3] is another grouping method, that encourages matched predictive distributions across easy and hard groups within each class. It penalizes the KL-divergence between predictive distributions from within-class groups.

Dataset Sub-groups. For debiasing, Up Wt, gDRO, and PGI require additional labels for covariates (sub-group labels). Past work has focused on these labels being supplied by an oracle; however, having access to all relevant sub-group labels is often impractical for large datasets. Some recent efforts have attempted to infer these sub-groups. Just train twice (JTT) [42] uses a bias-prone ERM model by training for a few epochs to identify the difficult groups. Environment inference for invariant learning (EIIL) [14] learns sub-group assignments that maximize the invariant risk minimization objective [5]. Unfortunately, inferred sub-groups perform worse in general than when they are supplied by an oracle [3,42]. For the methods that require them, which *excludes* Occam-Nets, we use oracle group labels (i.e., for Biased MNIST and COCO-on-Places). Inferred group labels are used for BAR, as oracle labels are not available.

For Biased MNIST, all the samples having the same class and the same value for all of the spurious factors are placed in a single group. For COCO-on-Places, objects placed on spuriously correlated backgrounds form the majority group, while the rest form the minority group. BAR does not specify oracle group labels, so we adopt the JTT method. Specifically, we train an ERM model for single epoch, reserving 20% of the samples with the highest losses as the difficult group and the rest as the easy group. We chose JTT over EIL for its simplicity. OccamNets, of course do not require such group labels to be specified.

Architectures. ResNet-18 is used as the standard baseline architecture for our studies. We compare it with an OccamNet version of ResNet-18, i.e., OccamResNet-18. To create this architecture, we add early exit modules to each of ResNet-18's convolutional blocks. To keep the number of parameters in OccamResNet-18 comparable to ResNet-18, we reduce the feature map width from 64 to 48. Assuming 1000 output classes, ResNet-18 has 12M parameters compared to 8M in OccamResNet-18. Further details are in the appendix.

Metrics and Model Selection. We report the means and standard deviations of test set accuracies computed across five different runs for all the datasets. For Biased MNIST, we report the unbiased test set accuracy (i.e., $p_{bias} = 0.1$) alongside the majority and minority group accuracies for each bias variable. For COCO-on-Places, unless otherwise specified, we report accuracy on the most challenging test split: with seen, but unbiased backgrounds. We also report the average precision score to measure the ability to distinguish 100 anomalous samples from the in-distribution samples for the anomaly detection task of COCO-on-Places. For BAR, we report the overall test accuracies. We use unbiased validation set of Biased MNIST and validation set with unbiased backgrounds for COCO-on-Places for hyperparameter tuning. The hyperparameter search grid and selected values are specified in the appendix.

5   Results and Analysis

5.1   Overall Results

OccamNets vs. ERM and Recent Bias Mitigation Methods. To examine how OccamNets fare against ERM and state-of-the-art bias mitigation methods, we run the comparison methods on ResNet and compare the results with

Table 1. Unbiased test set accuracies comparing OccamResNet to the more conventional ResNet architectures without early exits and constrained class activation maps. We format the <u>**first**</u>, **second** and <u>*third*</u> best results.

Architecture+Method	Biased MNIST	COCO-on-Places	BAR
Results on Standard ResNet-18			
ResNet+ERM	36.8 ± 0.7	<u>35.6</u> ± 1.0	<u>51.3</u> ± 1.9
ResNet+SD [51]	37.1 ± 1.0	35.4 ± 0.5	<u>51.3</u> ± 2.3
ResNet+Up Wt	<u>37.7</u> ± 1.6	35.2 ± 0.4	51.1 ± 1.9
ResNet+gDRO [56]	19.2 ± 0.9	35.3 ±0.1	38.7 ± 2.2
ResNet+PGI [3]	**48.6** ± 0.7	**42.7** ± 0.6	<u>**53.6**</u> ± 0.9
Results on OccamResNet-18			
OccamResNet	<u>**65.0**</u> ±1.0	<u>**43.4**</u> ± 1.0	**52.6** ± 1.9

OccamResNet. Results are given in Table 1. OccamResNet outperforms state-of-the-art methods on Biased MNIST and COCO-on-Places and rivals PGI on BAR, demonstrating that architectural inductive biases alone can help mitigate dataset bias. The gap between OccamResNet and other methods is large on Biased MNIST (16.4–46.0% absolute difference). For COCO-on-Places, PGI rivals OccamResNet, and clearly outperforms all other methods, in terms of accuracy on the test split with seen, but unbiased backgrounds. OccamResNet's results are impressive considering that Up Wt, gDRO, and PGI all had access to the bias group variables, unlike OccamNet, ERM, and SD.

Combining OccamNets with Recent Bias Mitigation Methods. Because OccamNets are a new network architecture, we used OccamResNet-18 with each of the baseline methods instead of ResNet-18. These results are shown in Table 2, where we provide unbiased accuracy along with any improvement or impairment of performance when OccamResNet-18 is used instead of ResNet-18. All methods benefit from using the OccamResNet architecture compared to ResNet-18, with gains of 10.6%–28.2% for Biased MNIST, 0.9%–7.8% for COCO-on-Places, and 1.0%–14.2% for BAR.

Table 2. Unbiased accuracies alongside improvement/impairment when the comparison methods are run on OccamResNet instead of ResNet.

Architecture+Method	Biased MNIST	COCO-on-Places	BAR
OccamResNet	<u>65.0</u> (+28.2)	**43.4** (+7.8)	<u>52.6</u> (+1.3)
OccamResNet+SD [51]	55.2 (+18.1)	39.4 (+4.0)	52.3 (+1.0)
OccamResNet+Up Wt	**65.7** (+28.0)	<u>42.9</u> (+7.7)	52.2 (+1.1)
OccamResNet+gDRO [56]	29.8 (+10.6)	40.7 (+5.4)	**52.9** (+14.2)
OccamResNet+PGI [3]	<u>**69.6**</u> (+21.0)	<u>**43.6**</u> (+0.9)	<u>**55.9**</u> (+2.3)

5.2   Analysis of the Proposed Inductive Biases

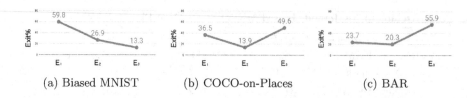

(a) Biased MNIST   (b) COCO-on-Places   (c) BAR

Fig. 4. Percentage of samples exited (Exit%) from each exit (barring E_0).

In this section, we analyze the impacts of each of the proposed modifications in OccamNets and their success in achieving the desired behavior.

Analysis of Early Exits. OccamResNet has four exits, the first exit is used for bias amplification and the rest are used to potentially exit early during inference. To analyze the usage of the earlier exits, we plot the percentage of samples that exited from each exit in Fig. 4. For Biased MNIST dataset, a large portion of the samples, i.e., 59.8% exit from the shallowest exit of E_1 and only 13.3% exit from the final exit E_3. For COCO-on-Places and BAR, 50.4% and 44.1% samples exit before E_3, with 49.6% and 55.9% samples using the full depth respectively. These results show that the OccamNets favor exiting early, but that they do use the full network depth if necessary.

Fig. 5. Original image, and Grad-CAM visualizations for ERM and PGI on ResNet, and CAM visualizations on OccamResNet. The visualizations are for the ground truth.

CAM Visualizations. To compare the localization capabilities of OccamResNets to ResNets, we present CAM visualizations in Fig. 5. For ResNets that were run with ERM and PGI, we show Grad-CAM visualizations [60], whereas for OccamResNets, we directly visualize the CAM heatmaps obtained from the earliest exit used for each sample. As shown in the figure, OccamResNet generally prefers smaller regions that include the target object. On the other hand, comparison methods tend to focus on larger visual regions that include irrelevant object/background cues leading to lower accuracies.

Ablations. To study the importance of the proposed inductive biases, we perform ablations on Biased MNIST and COCO-on-Places. First, to examine if the

Table 3. Ablation studies on OccamResNet

Ablation description	Biased MNIST	COCO-on-Places
Using only one exit at the end	35.9	35.0
Weighing all the samples equally i.e., $\mathcal{W}_j = 1$	**66.3**	**40.8**
Without the CAM suppression loss i.e., $\lambda_{CS} = 0$	<u>48.2</u>	<u>37.1</u>
Full OccamResNet	**65.0**	<u>**43.4**</u>

multi-exit setup is helpful, we train networks with single exit attached to the end of the network. This caused accuracy drops of 29.1% on Biased MNIST and 8.4% on COCO-on-Places, indicating that the multi-exit setup is critical. To examine if the weighted output prediction losses are helpful or harmful, we set all the loss weights ($\mathcal{W}_j$) to 1. This resulted in an accuracy drop of 2.6% on COCO-on-Places. However, it improved accuracy on Biased MNIST by 1.3%. We hypothesize that since the earlier exits suffice for a large number of samples in Biased MNIST (as indicated in Fig. 4a), the later exits may not receive sufficient training signal with the weighted output prediction losses. Finally, we ran experiments without the CAM suppression loss by setting λ_{CS} to 0. This caused large accuracy drops of 16.8% on Biased MNIST and 6.3% on COCO-on-Places. These results show that both inductive biases are vital for OccamNets.

5.3 Robustness to Different Types of Shifts

A robust system must handle different types of bias shifts. To test this ability, we examine the robustness to each bias variable in Biased MNIST and we also compare the methods on the differently shifted test splits of COCO-on-Places.

Table 4. Accuracies on majority (maj)/minority (min) groups for each bias variable in Biased MNIST ($p_{bias} = 0.95$). We **embolden** the results with the lowest differences between the groups.

Architecture+Method	Test Acc.	Digit scale maj/min	Digit color maj/min	Texture maj/min	Texture color maj/min	Letter maj/min	Letter color maj/min
ResNet+ERM	36.8	87.2/31.3	78.5/32.1	76.1/32.4	41.9/36.3	46.7/35.7	45.7/35.9
ResNet+PGI	48.6	91.9/43.8	84.8/44.6	79.5/45.1	51.3/48.3	67.2/46.5	55.8/47.9
OccamResNet	65.0	94.6/61.7	96.3/61.5	81.6/63.1	**66.8/64.8**	**64.7/65.1**	64.7/65.1
OccamResNet+PGI	69.6	95.4/66.7	97.0/66.5	88.6/67.4	**71.4/69.4**	**69.6/69.6**	70.5/69.5

In Table 4, we compare how ResNet and OccamResNet are affected by the different bias variables in Biased MNIST. For this, we present majority and

Table 5. Accuracies on all three test splits of COCO-on-Places, alongside mean average precision for the anomaly detection task.

Architecture+Method	Biased backgrounds	Unseen backgrounds	Seen, but non-spurious backgrounds	Anomaly detection
Results on Standard ResNet-18				
ResNet+ERM	**84.9** ± 0.5	53.2 ± 0.7	35.6 ± 1.0	20.1 ± 1.5
ResNet+PGI [3]	77.5 ± 0.6	52.8 ± 0.7	42.7 ± 0.6	20.6 ± 2.1
Results on OccamResNet-18				
OccamResNet	84.0 ± 1.0	**55.8** ± 1.2	43.4 ± 1.0	**22.3** ± 2.8
OccamResNet+PGI [3]	82.8 ± 0.6	55.3 ± 1.3	43.6 ± 0.6	21.6 ± 1.6

minority group accuracies for each variable. Bias variables with large differences between the majority and the minority groups, i.e., large majority/minority group discrepancy (MMD) are the most challenging spurious factors. OccamResNets, with and without PGI, improve on both majority and minority group accuracies across all the bias variables. OccamResNets are especially good at ignoring the distracting letters and their colors, obtaining MMD values between 0–1%. ResNets, on the other hand are susceptible to those spurious factors, obtaining MMD values between 7.9–20.7%. Among all the variables, digit scale and digit color are the most challenging ones and OccamResNets mitigate their exploitation to some extent. Next, we show accuracies for base method and PGI on both ResNet-18 and OccamResNet-18 on all of the test splits of COCO-on-Places in Table 5. The different test splits have different kinds of object-background combinations, and ideally the method should work well on all three test splits. PGI run on ResNet-18 improves on the split with seen, but non-spurious backgrounds but incurs a large accuracy drop of 7.4% on the in-distribution test set, with biased backgrounds. On the other hand, OccamResNet-18 shows only 0.9% drop on the biased backgrounds, while showing 2.6% accuracy gains on the split with unseen backgrounds and 7.8% accuracy gains on the split with seen, but non-spurious backgrounds. It further obtains 2.2% gains on the average precision metric for the anomaly detection task. PGI run on OccamResNet exhibits a lower drop of 2.1% on the in-distribution split as compared to the PGI run on ResNet, while obtaining larger gains on rest of the splits. These results exhibit that OccamNets obtain high in-distribution and shifted-distribution accuracies.

5.4   Evaluation on Other Architectures

To examine if the proposed inductive biases improve bias-resilience in other architectures too, we created OccamEfficientNet-B2 and OccamMobileNet-v3 by modifying EfficientNet-B2 [65] and MobileNet-v3 [28,29]. OccamNet variants outperform standard architectures on both BiasedMNIST (OccamEfficientNet-B2: 59.2 vs. EfficientNet-B2: 34.4 and OccamMobileNet-v3: 49.9 vs. MobileNet-v3: 40.4) and COCO-on-Places (OccamEfficientNet-B2: 39.2 vs. EfficientNet-B2: 34.2 and OccamMobileNet-v3: 40.1 vs. MobileNet-v3: 34.9). The gains show that the proposed modifications help other architectures too.

5.5   Do OccamNets Work on Less Biased Datasets?

To examine if OccamNets also work well on datasets with less bias, we train ResNet-18 and OccamResNet-18 on 100 classes of the ImageNet dataset [16]. OccamResNet-18 obtains competitive numbers compared to the standard ResNet-18 (OccamResNet-18: 92.1, vs. ResNet-18: 92.6, top-5 accuracies). However, as described in rest of this paper, OccamResNet-18 achieves this with improved resistance to bias (e.g., if the test distributions were to change in future). Additionally, it also reduces the computations with 47.6% of the samples exiting from E_1 and 13.3% of samples exiting from E_2. As such, OccamNets have the potential to be the *de facto* network choice for visual recognition tasks regardless of the degree of bias.

6   Discussion

Relation to Mixed Capacity Models. Recent studies show that sufficiently simple models, e.g., with fewer parameters [13,27] or models trained for a few epochs [42,71] amplify biases. Specifically, [27] shows that model compression disproportionately hampers minority samples. Seemingly, this is an argument against smaller models (simpler hypotheses), i.e., against Occam's principles. However, [27] does not study network depth, unlike our work. Our paper suggests that using only the necessary capacity for each example yields greater robustness.

Relation to Multi-Hypothesis Models. Some recent works generate multiple plausible hypotheses [39,68] and use extra information at test time to choose the best hypothesis. The techniques include training a set of models with dissimilar input gradients [68] and training multiple prediction heads that disagree on a target distribution [39]. An interesting extension of OccamNets could be making diverse predictions through the multiple exits and through CAMs that focus on different visual regions. This could help avoid discarding complex features in favor of simpler ones [61]. This may also help with under-specified tasks, where there are equally viable ways of making the predictions [39].

Other Architectures and Tasks. We tested OccamNets implemented with CNNs; however, they may be beneficial to other architectures as well. The ability to exit dynamically could be used with transformers, graph neural networks, and feed-forward networks more generally. There is some evidence already for this on natural language inference tasks, where early exits improved robustness in a transformer architecture [78]. While the spatial bias is more vision specific, it could be readily integrated into recent non-CNN approaches for image classification [17,43,69,75].

Conclusion. In summary, the proposed OccamNets have architectural inductive biases favoring simpler solutions. The experiments show improvements over state-of-the-art bias mitigation techniques. Furthermore, existing methods tend to do better with OccamNets as compared to the standard architectures.

Acknowledgments. This work was supported in part by NSF awards #1909696 and #2047556. The views and conclusions contained herein are those of the authors and should not be interpreted as representing the official policies or endorsements of any sponsor.

References

1. Adeli, E., Zhao, Q., Pfefferbaum, A., Sullivan, E., Fei-Fei, L., Niebles, J.C., Pohl, K.: Bias-resilient neural network. arXiv abs/1910.03676 (2019)
2. Agrawal, A., Batra, D., Parikh, D., Kembhavi, A.: Don't just assume; look and answer: overcoming priors for visual question answering. In: 2018 IEEE Conference on Computer Vision and Pattern Recognition, CVPR 2018, Salt Lake City, UT, USA, 18–22 June 2018, pp. 4971–4980. IEEE Computer Society (2018). https://doi.org/10.1109/CVPR.2018.00522
3. Ahmed, F., Bengio, Y., van Seijen, H., Courville, A.: Systematic generalisation with group invariant predictions. In: International Conference on Learning Representations (2020)
4. Anderson, P., et al.: Bottom-up and top-down attention for image captioning and visual question answering. In: 2018 IEEE Conference on Computer Vision and Pattern Recognition, CVPR 2018, Salt Lake City, UT, USA, 18–22 June 2018, pp. 6077–6086. IEEE Computer Society (2018). https://doi.org/10.1109/CVPR.2018.00636
5. Arjovsky, M., Bottou, L., Gulrajani, I., Lopez-Paz, D.: Invariant risk minimization. arXiv preprint arXiv:1907.02893 (2019)
6. Bolukbasi, T., Chang, K., Zou, J.Y., Saligrama, V., Kalai, A.T.: Man is to computer programmer as woman is to homemaker? Debiasing word embeddings. In: Lee, D.D., Sugiyama, M., von Luxburg, U., Guyon, I., Garnett, R. (eds.) Advances in Neural Information Processing Systems 29: Annual Conference on Neural Information Processing Systems 2016, Barcelona, Spain, 5–10 December 2016, pp. 4349–4357 (2016)
7. Cadène, R., Dancette, C., Ben-younes, H., Cord, M., Parikh, D.: RUBi: Reducing unimodal biases for visual question answering. In: Wallach, H.M., Larochelle, H., Beygelzimer, A., d'Alché-Buc, F., Fox, E.B., Garnett, R. (eds.) Advances in Neural Information Processing Systems 32: Annual Conference on Neural Information Processing Systems 2019, NeurIPS 2019, Vancouver, BC, Canada, 8–14 December 2019, pp. 839–850 (2019)
8. Chawla, N.V., Bowyer, K.W., Hall, L.O., Kegelmeyer, W.P.: Smote: synthetic minority over-sampling technique. J. Artif. Intell. Res. **16**, 321–357 (2002)
9. Chen, L., Yan, X., Xiao, J., Zhang, H., Pu, S., Zhuang, Y.: Counterfactual samples synthesizing for robust visual question answering. In: Proceedings of the IEEE/CVF Conference on Computer Vision and Pattern Recognition, pp. 10800–10809 (2020)
10. Chen, X., Dai, H., Li, Y., Gao, X., Song, L.: Learning to stop while learning to predict. In: International Conference on Machine Learning, pp. 1520–1530. PMLR (2020)
11. Choe, Y.J., Ham, J., Park, K.: An empirical study of invariant risk minimization. In: ICML 2020 Workshop on Uncertainty and Robustness in Deep Learning (2020)

12. Clark, C., Yatskar, M., Zettlemoyer, L.: Don't take the easy way out: ensemble based methods for avoiding known dataset biases. In: Proceedings of the 2019 Conference on Empirical Methods in Natural Language Processing and the 9th International Joint Conference on Natural Language Processing (EMNLP-IJCNLP), Hong Kong, China, pp. 4069–4082. Association for Computational Linguistics (2019). https://doi.org/10.18653/v1/D19-1418

13. Clark, C., Yatskar, M., Zettlemoyer, L.: Learning to model and ignore dataset bias with mixed capacity ensembles. In: Findings of the Association for Computational Linguistics: EMNLP 2020, pp. 3031–3045. Association for Computational Linguistics (2020). https://doi.org/10.18653/v1/2020.findings-emnlp.272

14. Creager, E., Jacobsen, J.H., Zemel, R.: Environment inference for invariant learning. In: International Conference on Machine Learning, pp. 2189–2200. PMLR (2021)

15. Cui, Y., Jia, M., Lin, T., Song, Y., Belongie, S.J.: Class-balanced loss based on effective number of samples. In: IEEE Conference on Computer Vision and Pattern Recognition, CVPR 2019, Long Beach, CA, USA, 16–20 June 2019, pp. 9268–9277. Computer Vision Foundation/IEEE (2019). https://doi.org/10.1109/CVPR.2019.00949

16. Deng, J., Dong, W., Socher, R., Li, L.J., Li, K., Fei-Fei, L.: ImageNet: a large-scale hierarchical image database. In: 2009 IEEE conference on Computer Vision and Pattern Recognition, pp. 248–255. IEEE (2009)

17. Dosovitskiy, A., et al.: An image is worth 16x16 words: transformers for image recognition at scale. arXiv preprint arXiv:2010.11929 (2020)

18. Duchi, J.C., Hashimoto, T., Namkoong, H.: Distributionally robust losses against mixture covariate shifts. Under review (2019)

19. Duggal, R., Freitas, S., Dhamnani, S., Horng, D., Sun, J., et al.: ELF: an early-exiting framework for long-tailed classification. arXiv preprint arXiv:2006.11979 (2020)

20. Grand, G., Belinkov, Y.: Adversarial regularization for visual question answering: strengths, shortcomings, and side effects. In: Proceedings of the Second Workshop on Shortcomings in Vision and Language, Minneapolis, Minnesota , pp. 1–13. Association for Computational Linguistics (2019). https://doi.org/10.18653/v1/W19-1801

21. Guo, M.H., et al.: Attention mechanisms in computer vision: a survey. arXiv preprint arXiv:2111.07624 (2021)

22. He, H., Bai, Y., Garcia, E.A., Li, S.: ADASYN: adaptive synthetic sampling approach for imbalanced learning. In: 2008 IEEE International Joint Conference on Neural Networks (IEEE world congress on Computational Intelligence), pp. 1322–1328. IEEE (2008)

23. He, H., Garcia, E.A.: Learning from imbalanced data. IEEE Trans. Knowl. Data Eng. 21(9), 1263–1284 (2009)

24. He, K., Gkioxari, G., Dollár, P., Girshick, R.: Mask R-CNN. In: Proceedings of the IEEE International Conference on Computer Vision, pp. 2961–2969 (2017)

25. He, K., Zhang, X., Ren, S., Sun, J.: Deep residual learning for image recognition. In: Proceedings of the IEEE Conference on Computer Vision and Pattern Recognition, pp. 770–778 (2016)

26. Hettinger, C., Christensen, T., Ehlert, B., Humpherys, J., Jarvis, T., Wade, S.: Forward thinking: building and training neural networks one layer at a time. arXiv preprint arXiv:1706.02480 (2017)

27. Hooker, S., Moorosi, N., Clark, G., Bengio, S., Denton, E.: Characterising bias in compressed models. arXiv preprint arXiv:2010.03058 (2020)

28. Howard, A., et al.: Searching for mobileNetV3. In: Proceedings of the IEEE/CVF International Conference on Computer Vision, pp. 1314–1324 (2019)

29. Howard, A.G., et al.: MobileNets: efficient convolutional neural networks for mobile vision applications. arXiv preprint arXiv:1704.04861 (2017)

30. Hu, J., Shen, L., Albanie, S., Sun, G., Vedaldi, A.: Gather-excite: exploiting feature context in convolutional neural networks. In: Advances in Neural Information Processing Systems 31 (2018)

31. Hu, T.K., Chen, T., Wang, H., Wang, Z.: Triple wins: boosting accuracy, robustness and efficiency together by enabling input-adaptive inference. arXiv preprint arXiv:2002.10025 (2020)

32. Huang, G., Liu, Z., Van Der Maaten, L., Weinberger, K.Q.: Densely connected convolutional networks. In: Proceedings of the IEEE Conference on Computer Vision and Pattern Recognition, pp. 4700–4708 (2017)

33. Jaderberg, M., Simonyan, K., Zisserman, A., et al.: Spatial transformer networks. In: Advances in Neural Information Processing Systems, vol. 28 (2015)

34. Kim, B., Kim, H., Kim, K., Kim, S., Kim, J.: Learning not to learn: training deep neural networks with biased data. In: IEEE Conference on Computer Vision and Pattern Recognition, CVPR 2019, Long Beach, CA, USA, 16–20 June 2019, pp. 9012–9020. Computer Vision Foundation/IEEE (2019). https://doi.org/10.1109/CVPR.2019.00922

35. Kim, E., Lee, J., Choo, J.: BiaSwap: removing dataset bias with bias-tailored swapping augmentation. In: Proceedings of the IEEE/CVF International Conference on Computer Vision, pp. 14992–15001 (2021)

36. Krueger, D., et al.: Out-of-distribution generalization via risk extrapolation (REx). In: International Conference on Machine Learning. pp, 5815–5826. PMLR (2021)

37. Lee, C.Y., Gallagher, P.W., Tu, Z.: Generalizing pooling functions in convolutional neural networks: mixed, gated, and tree. In: Artificial Intelligence and Statistics, pp. 464–472. PMLR (2016)

38. Lee, C.Y., Xie, S., Gallagher, P., Zhang, Z., Tu, Z.: Deeply-supervised nets. In: Artificial Intelligence and Statistics, pp. 562–570. PMLR (2015)

39. Lee, Y., Yao, H., Finn, C.: Diversify and disambiguate: learning from underspecified data. arXiv preprint arXiv:2202.03418 (2022)

40. Li, Y., Vasconcelos, N.: REPAIR: removing representation bias by dataset resampling. In: IEEE Conference on Computer Vision and Pattern Recognition, CVPR 2019, Long Beach, CA, USA, 16–20 June 2019, pp. 9572–9581. Computer Vision Foundation/IEEE (2019). https://doi.org/10.1109/CVPR.2019.00980

41. Lin, T.-Y., Maire, M., Belongie, S., Hays, J., Perona, P., Ramanan, D., Dollár, P., Zitnick, C.L.: Microsoft COCO: common objects in context. In: Fleet, D., Pajdla, T., Schiele, B., Tuytelaars, T. (eds.) ECCV 2014. LNCS, vol. 8693, pp. 740–755. Springer, Cham (2014). https://doi.org/10.1007/978-3-319-10602-1_48

42. Liu, E.Z., et al.: Just train twice: Improving group robustness without training group information. In: International Conference on Machine Learning, pp. 6781–6792. PMLR (2021)

43. Liu, Z., Lin, Y., Cao, Y., Hu, H., Wei, Y., Zhang, Z., Lin, S., Guo, B.: Swin transformer: hierarchical vision transformer using shifted windows. In: Proceedings of the IEEE/CVF International Conference on Computer Vision, pp. 10012–10022 (2021)

44. Long, J., Shelhamer, E., Darrell, T.: Fully convolutional networks for semantic segmentation. In: Proceedings of the IEEE Conference on Computer Vision and Pattern Recognition, pp. 3431–3440 (2015)

45. Mehrabi, N., Morstatter, F., Saxena, N., Lerman, K., Galstyan, A.: A survey on bias and fairness in machine learning. ACM Comput. Surv. (CSUR) **54**(6), 1–35 (2021)
46. Mostafa, H., Ramesh, V., Cauwenberghs, G.: Deep supervised learning using local errors. Front. Neurosci. **12**, 608 (2018)
47. Nam, J., Cha, H., Ahn, S., Lee, J., Shin, J.: Learning from failure: training debiased classifier from biased classifier. In: Advances in Neural Information Processing Systems (2020)
48. Namkoong, H., Duchi, J.C.: Stochastic gradient methods for distributionally robust optimization with F-divergences. In: Lee, D.D., Sugiyama, M., von Luxburg, U., Guyon, I., Garnett, R. (eds.) Advances in Neural Information Processing Systems 29: Annual Conference on Neural Information Processing Systems 2016, Barcelona, Spain, 5–10 December 2016, pp. 2208–2216 (2016)
49. Nøkland, A.: Direct feedback alignment provides learning in deep neural networks. In: Advances in Neural Information Processing Systems, vol. 29 (2016)
50. Nøkland, A., Eidnes, L.H.: Training neural networks with local error signals. In: International Conference on Machine Learning, pp. 4839–4850. PMLR (2019)
51. Pezeshki, M., Kaba, S.O., Bengio, Y., Courville, A., Precup, D., Lajoie, G.: Gradient starvation: a learning proclivity in neural networks. arXiv preprint arXiv:2011.09468 (2020)
52. Rahimian, H., Mehrotra, S.: Distributionally robust optimization: a review. arXiv preprint arXiv:1908.05659 (2019)
53. Ramakrishnan, S., Agrawal, A., Lee, S.: Overcoming language priors in visual question answering with adversarial regularization. In: Bengio, S., Wallach, H.M., Larochelle, H., Grauman, K., Cesa-Bianchi, N., Garnett, R. (eds.) Advances in Neural Information Processing Systems 31: Annual Conference on Neural Information Processing Systems 2018, NeurIPS 2018, Montréal, Canada, 3–8 December 2018, pp. 1548–1558 (2018)
54. Ronneberger, O., Fischer, P., Brox, T.: U-Net: convolutional networks for biomedical image segmentation. In: Navab, N., Hornegger, J., Wells, W.M., Frangi, A.F. (eds.) MICCAI 2015. LNCS, vol. 9351, pp. 234–241. Springer, Cham (2015). https://doi.org/10.1007/978-3-319-24574-4_28
55. Sagawa, S., Koh, P.W., Hashimoto, T.B., Liang, P.: Distributionally robust neural networks for group shifts: on the importance of regularization for worst-case generalization. CoRR abs/1911.08731 (2019), https://arxiv.org/abs/1911.08731
56. Sagawa, S., Koh, P.W., Hashimoto, T.B., Liang, P.: Distributionally robust neural networks for group shifts: on the importance of regularization for worst-case generalization. arXiv preprint arXiv:1911.08731 (2019)
57. Sagawa, S., Raghunathan, A., Koh, P.W., Liang, P.: An investigation of why overparameterization exacerbates spurious correlations. In: Proceedings of the 37th International Conference on Machine Learning, ICML 2020, 13–18 July 2020, Virtual Event. Proceedings of Machine Learning Research, vol. 119, pp. 8346–8356. PMLR (2020)
58. Sanh, V., Wolf, T., Belinkov, Y., Rush, A.M.: Learning from others' mistakes: avoiding dataset biases without modeling them. arXiv preprint arXiv:2012.01300 (2020)
59. Scardapane, S., Scarpiniti, M., Baccarelli, E., Uncini, A.: Why should we add early exits to neural networks? Cognit. Comput. **12**(5), 954–966 (2020)
60. Selvaraju, R.R., Cogswell, M., Das, A., Vedantam, R., Parikh, D., Batra, D.: Grad-CAM: visual explanations from deep networks via gradient-based localization. In:

Proceedings of the IEEE International Conference on Computer Vision, pp. 618–626 (2017)

61. Shah, H., Tamuly, K., Raghunathan, A., Jain, P., Netrapalli, P.: The pitfalls of simplicity bias in neural networks. In: Advances in Neural Information Processing Systems, vol. 33 (2020)

62. Shrestha, R., Kafle, K., Kanan, C.: An investigation of critical issues in bias mitigation techniques. In: Proceedings of the IEEE/CVF Winter Conference on Applications of Computer Vision, pp. 1943–1954 (2022)

63. Singh, K.K., Mahajan, D., Grauman, K., Lee, Y.J., Feiszli, M., Ghadiyaram, D.: Don't judge an object by its context: learning to overcome contextual bias. In: 2020 IEEE/CVF Conference on Computer Vision and Pattern Recognition, CVPR 2020, Seattle, WA, USA, 13–19 June 2020, pp. 11067–11075. IEEE (2020). https://doi.org/10.1109/CVPR42600.2020.01108

64. Szegedy, C., Liu, W., Jia, Y., Sermanet, P., Reed, S., Anguelov, D., Erhan, D., Vanhoucke, V., Rabinovich, A.: Going deeper with convolutions. In: Proceedings of the IEEE Conference on Computer Vision and Pattern Recognition, pp. 1–9 (2015)

65. Tan, M., Le, Q.: EfficientNet: rethinking model scaling for convolutional neural networks. In: International Conference on Machine Learning, pp. 6105–6114. PMLR (2019)

66. Teerapittayanon, S., McDanel, B., Kung, H.T.: BranchyNet: fast inference via early exiting from deep neural networks. In: 2016 23rd International Conference on Pattern Recognition (ICPR), pp. 2464–2469. IEEE (2016)

67. Teney, D., Abbasnejad, E., van den Hengel, A.: Unshuffling data for improved generalization. arXiv preprint arXiv:2002.11894 (2020)

68. Teney, D., Abbasnejad, E., Lucey, S., van den Hengel, A.: Evading the simplicity bias: training a diverse set of models discovers solutions with superior ood generalization. arXiv preprint arXiv:2105.05612 (2021)

69. Tolstikhin, I.O., Houlsby, N., Kolesnikov, A., Beyer, L., Zhai, X., Unterthiner, T., Yung, J., Steiner, A., Keysers, D., Uszkoreit, J., et al.: Mlp-mixer: An all-mlp architecture for vision. Advances in Neural Information Processing Systems 34 (2021)

70. Torralba, A., Efros, A.A.: Unbiased look at dataset bias. In: The 24th IEEE Conference on Computer Vision and Pattern Recognition, CVPR 2011, Colorado Springs, CO, USA, 20–25 June 2011. pp. 1521–1528. IEEE Computer Society (2011). https://doi.org/10.1109/CVPR.2011.5995347

71. Utama, P.A., Moosavi, N.S., Gurevych, I.: Towards debiasing NLU models from unknown biases. In: Proceedings of the 2020 Conference on Empirical Methods in Natural Language Processing (EMNLP), pp. 7597–7610. Association for Computational Linguistics (2020). https://doi.org/10.18653/v1/2020.emnlp-main.613. https://www.aclweb.org/anthology/2020.emnlp-main.613

72. Venkataramani, S., Raghunathan, A., Liu, J., Shoaib, M.: Scalable-effort classifiers for energy-efficient machine learning. In: Proceedings of the 52nd Annual Design Automation Conference. pp. 1–6 (2015)

73. Wołczyk, M., et al.:: Zero time waste: recycling predictions in early exit neural networks. In: Advances in Neural Information Processing Systems 34 (2021)

74. Xu, K., Ba, J., et al.: Show, attend and tell: neural image caption generation with visual attention. In: International conference on Machine Learning, pp. 2048–2057. PMLR (2015)

75. Yu, W., et al.: Metaformer is actually what you need for vision. arXiv preprint arXiv:2111.11418 (2021)

76. Zhang, B.H., Lemoine, B., Mitchell, M.: Mitigating unwanted biases with adversarial learning. In: Proceedings of the 2018 AAAI/ACM Conference on AI, Ethics, and Society, pp. 335–340 (2018)
77. Zhou, B., Lapedriza, A., Khosla, A., Oliva, A., Torralba, A.: Places: A 10 million image database for scene recognition. IEEE T. Pattern Anal. Mach. Intell. **40**(6), 1452–1464 (2017)
78. Zhou, W., et al.: BERT loses patience: fast and robust inference with early exit. In: Advances in Neural Information Processing Systems, vol. 33, pp. 18330–18341 (2020)

ERA: Enhanced Rational Activations

Martin Trimmel[1]([✉])[iD], Mihai Zanfir[2][iD] , Richard Hartley[3,4][iD],
and Cristian Sminchisescu[1,4][iD]

[1] Lund University, Lund, Sweden
martin.trimmel@math.lth.se
[2] Google Research, Bucharest, Romania
[3] Australian National University, Canberra, Australia
[4] Google Research, Zurich, Switzerland

Abstract. Activation functions play a central role in deep learning since they form an essential building stone of neural networks. In the last few years, the focus has been shifting towards investigating new types of activations that outperform the classical Rectified Linear Unit (ReLU) in modern neural architectures. Most recently, rational activation functions (RAFs) have awakened interest because they were shown to perform on par with state-of-the-art activations on image classification. Despite their apparent potential, prior formulations are either not safe, not smooth, or not "true" rational functions, and they only work with careful initialisation. Aiming to mitigate these issues, we propose a novel, enhanced rational function, *ERA*, and investigate how to better accommodate the specific needs of these activations, to both network components and training regime. In addition to being more stable, the proposed function outperforms other standard ones across a range of lightweight network architectures on two different tasks: image classification and 3d human pose and shape reconstruction.

Keywords: Rational activation · Activation function · Deep learning · Neural networks

1 Introduction

Neural networks keep ever-increasing in size and modern architectures can have billions of parameters. While larger models often perform better than smaller ones, their training entails heavy requirements on the computing infrastructure, resulting in increased costs and environmental concerns due to the electricity

M. Trimmel—Denotes equal contribution. | Code available at: github.com/martrim/rational.

Supplementary Information The online version contains supplementary material available at https://doi.org/10.1007/978-3-031-20044-1_41.

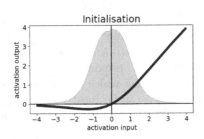

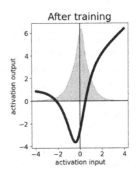

Fig. 1. Plots of the first *ERA* of a CNN before and after training on CIFAR10. The *ERA* is initialised as a Swish activation and has degree 5/4. The plots show that the learned activations differ drastically from the initialization. The shaded curve in the background shows the probability density distribution of the input to the activation (not true scale on the y-axis). The training of the network decreases the variance of the input. Plots of the derivatives are in the appendix.

they consume [21,36]. The trend towards larger architectures has been reinforced in recent times by the advent of the Transformer [39], which has since its introduction moved into the field of computer vision [9], reducing the usage of more parameter-efficient CNN architectures like VGG [35] and ResNet [11]. The trend away from convolutions to dense linear layers continues with [37] who recently introduced a full-MLP architecture that performs on par with the state-of-the-art on large vision datasets.

Since the mathematical form of the learned network function depends heavily on the employed activation functions, choosing the right activation may be a key to increasing the network capacity while limiting the number of training parameters. Research has for a long time focused on non-parametric activation functions like ReLU [10,31], Leaky ReLU [27], GELU [12], Swish [34] and Mish [28]. On the other hand, parameterized activation functions may be auspicious because they have the potential to better adapt to the task at hand.

Recently, [29] showed that rational activation functions (previously employed by [5] in the context of graph convolutional networks) can perform on par with more traditional activation functions on image classification and proved that rational neural networks are universal function approximators. Moreover, rational functions can almost perfectly approximate all standard activation functions. Using rational functions as activations adds only around 10 parameters per network layer, which is negligible compared to the total number of network weights. [3] employ rational activations to train a GAN on the MNIST dataset and mathematically show that rational networks need exponentially smaller depth compared to ReLU networks to approximate smooth functions. [33] propose an interpretable network architecture based on continued fractions, which is a rational function of the input, despite not employing rational activations.

On the other hand, the applicability of rational activations is impeded by the fact that initializing them like the rest of the network via a normal or a uniform distribution typically results in the divergence of training. To mitigate this fact, [29] and [3] fit the activation's parameters to standard activation functions like Leaky ReLU and ReLU. Since each of [3,29] and [7] use different rational functions, it is unclear whether there is a specific version which should be preferred in practice. Furthermore, rational activations have so far mostly been employed in large architectures to solve simple tasks. Their larger expressivity and adaptability thus begs the question of whether they have additional benefits when used in smaller networks or on more complex tasks.

Our Contributions

- We propose an enhanced rational activation (i.e. *ERA*), incorporating a factorised denominator, which negates the requirement of using absolute values, as in prior work. This results in a safe, smooth rational function that exhibits better performance than previous versions.
- *ERA* allows us to derive a partial fractions representation, reducing memory requirements and further benefiting network performance.
- We investigate different ways of normalizing the input of the activation functions. We show that the right normalization has a great effect on the stability of training, allowing for the first time to investigate rational activations with random initialisation. In line with the recent paper [26], we find that layer normalization benefits all types of activations in CNNs, when compared to batch normalization. In the case of lightweight networks that we employ, we find that instance normalization achieves even better results.
- We study how *ERA* changes during training and conclude that neural networks are able to learn non-trivial activations that drastically differ from their initialisation.
- We ablate *ERA* across different types of neural network architectures, ranging from FCNs, CNNs, MLP-Mixers and Transformers on two different tasks: image classification and 3d human pose and shape reconstruction. For the latter task, we show that a very lightweight, real-time network, of just 3.9M parameters, can achieve results on par with state-of-the-art methods that employ networks in the order of tens or hundreds of millions of parameters.

2   Background

We start with a formal definition of rational functions:

Definition 1. *We say that $f : \mathbb{R} \to \mathbb{R}$ is a rational function if there are numbers $a_0, \ldots a_m, b_0, \ldots, b_n \in \mathbb{R}$ such that*

$$f(x) = \frac{P(x)}{Q(x)} = \frac{a_0 + a_1 x + \cdots + a_m x^m}{b_0 + b_1 x + \cdots + b_n x^n} \tag{1}$$

and $b_j \neq 0$ for at least one j. The numbers a_i, b_j are called the coefficients of f. If m and n are the largest indices such that $a_m \neq 0, b_n \neq 0$, we say that $\deg(f) := m/n$ is the degree of f.

There are a number of points to consider when applying rational functions as activations. Firstly, the function f is undefined on poles (numbers x for which $Q(x) = 0$) and numerically unstable close to poles. In order to make the rational function *safe*, [7] and [29] propose the following modified denominators, respectively:

$$Q_A(x) = 1 + \sum_{i=1}^{n} |b_i x^i| \tag{2}$$

$$Q_B(x) = 1 + \left| \sum_{i=1}^{n} b_i x^i \right| \tag{3}$$

Since neither Q_A nor Q_B are polynomials, using them in the denominator does not produce a rational function. Apart from lacking the interpretability of a rational function, this makes gradient computation more costly. As an alternative, [3] do not consider the possibility of numerical instabilities and report that using a *non-safe* quadratic denominator is stable in their experiments. However, there persists the risk of numerical instabilities in more complex settings and deeper networks.

Another important point is the degree of the rational function. [29] uses a rational function of degree 5/4, whereas [3] uses a degree of 3/2, which may have helped them to avoid instabilities, due to the lower order of the denominator. They also note that superdiagonal types behave like a non-constant linear function at $\pm\infty$, which is intuitively desirable to avoid exploding, diminishing or vanishing values.

3   Enhanced Rational Function

We now investigate mathematically how to define a rational function without poles in $\mathbb{R}$, thus avoiding the problems mentioned in the previous section.

By a corollary to the Fundamental Theorem of Algebra, every univariate polynomial Q of degree n with coefficients in $\mathbb{C}$ has n complex roots. If the polynomial coefficients are real, the roots can moreover be shown to appear in complex conjugates, implying a factorization of the form:

$$Q(x) = \prod_i (x - x_i) \prod_j (x - z_j)(x - \bar{z}_j). \tag{4}$$

with $x_i \in \mathbb{R}; z_j \in \mathbb{C} \setminus \mathbb{R}$. Since we want Q to have no real roots (i.e. no poles in $\mathbb{R}$), we discard the first factor and obtain a factorisation of Q of the form:

$$Q_C(x) = \prod_{i=1}^{n/2} \left((x - c_i)^2 + d_i^2 \right) \tag{5}$$

for some $c_i \in \mathbb{R}, d_i \in \mathbb{R}_{>0}$. This gives us a rational function of the form:

$$\sigma(x) = \frac{P(x)}{Q_C(x)}, \tag{6}$$

which we call *ERA*. Since the numerator P has its degree one higher than the denominator, we may perform polynomial division. Moreover, by Theorem 2.4 in [4] we can make use of partial fractions to get an expression of the form

$$\sigma(x) = ax + b + \sum_{i=1}^{n/2} \frac{P_i(x)}{Q_i(x)} \tag{7}$$

with $\deg(P_i) = 1, \deg(Q_i) = 2$ for all i. The numbers $a, b \in \mathbb{R}$ denote learnable parameters. Our main motivation for using partial fractions is to make rational activations more computationally efficient: e.g. a rational function of degree 5/4 requires 9 additions/subtractions in Eq. (6) and Eq. (7), but the former requires 13 multiplications/divisions and the latter only 6 multiplications/divisions. Table 4 shows that the use of partial fractions also increases the accuracy of the network.

Table 1. Number of arithmetic operations to compute different types of rational functions. Using partial fractions decreases the number of costly multiplication/division operations from 14 to 6.

Denominator	Add	Subtract	Multiply	Divide
Non-factorised [29]	9	0	13	1
ERA no partial frac	7	2	12	1
ERA	7	2	4	2

Using a polynomial Q as in Eq. (5) as the denominator may result in numerical instabilities when the d_i are small. Hence, we propose a numerically safe version by adding a small number $\epsilon > 0$:

$$Q_C(x) = \epsilon + \prod_{i=1}^{n} \left((x - c_i)^2 + d_i^2 \right) \tag{8}$$

In practice, we choose $\epsilon = 10^{-6}$.

The proposed activation functions *ERA* in Eqs. (6) and (7) are not polynomials. Hence, it follows from the following theorem that networks using *ERA* as activation function are universal function approximators. (We note that [29] use the same argument to show that the use of their proposed rational activation function yields networks that are universal function approximators.)

Theorem 1 (Theorem 3.1, [32]). *For any i, let $C(\mathbb{R}^i)$ denote the set of continuous functions $\mathbb{R}^i \to \mathbb{R}$. Let $\sigma \in C(\mathbb{R})$. Then*

$$M(\sigma) = span\{\sigma(\mathbf{w} \cdot \mathbf{x} - \theta) : \theta \in \mathbb{R}, \mathbf{w} \in \mathbb{R}^n\} \tag{9}$$

is dense in $C(\mathbb{R}^n)$ in the topology of uniform converge on compacta, if and only if σ is not a polynomial.

4   Experiments

4.1   Image Classification

Setup. We train neural networks to perform image classification on the MNIST [22], CIFAR10 [19] and ImageNet [8] datasets, using the Tensorflow [1] framework. Since we are aiming to improve lightweight networks, we focus on two small architectures. We consider fully-connected (FCN) and convolutional neural (CNN) networks, defined recursively by

$$\mathbf{x}_{i+1} = (\sigma_i \circ \ell_i)(\mathbf{x}_i), \quad 1 \leq i \leq 4; \tag{10}$$

where $\mathbf{x}_1$ is the input, ℓ_i is the ith linear layer of the network, followed by an activation function σ_i. In the fully-connected case, ℓ_i is a dense layer and in the convolutional case, ℓ_i is a convolutional layer. We train the networks for 100 epochs using SGD, a batch size of 64 and an initial learning rate of 0.01 which decays by a factor of 10 after having completed 25, 50 and 75 training epochs. All network layers use l2 weight regularization with parameter 0.001. We apply data augmentation with horizontal and vertical shifts of up to 6 pixels and enabling horizontal flipping of images.

Input Normalization. There is a range of existing normalization methods that are commonly being employed in neural networks [2,14,38,40]. In general, the input $\mathbf{x}$ to a normalisation layer undergoes two steps: A *standardization step*

$$\mathbf{y} = \frac{\mathbf{x} - \boldsymbol{\mu}}{\sigma} \tag{11}$$

that makes $\mathbf{x}$ have zero mean and unit standard deviation along one or multiple *standardization axes*, and a *learnable step*

$$\mathbf{z} = \boldsymbol{\beta} + \boldsymbol{\gamma} \cdot \mathbf{y} \tag{12}$$

where the distribution of the output $\mathbf{z}$ the learned mean $\boldsymbol{\beta}$ and the learned standard deviation $\boldsymbol{\gamma}$ along some *learnable axes*. By exploring multiple options, we find that instance normalization [38] works best for CNNs and layer normalization [2] works best for fully-connected networks. Our ablations on the standardization and learnable axes are available in the supplementary material.

Initialization and Stability. Whereas previous works [3,29] have each focused on a single activation (ReLU and Leaky ReLU) for initializing the rational function, we perform a broader ablation study, encompassing also the newer GELU and Swish activations. Initialising rational activation functions as any standard activation always leads to convergence in our image classification experiments (except when the activation from [29] was initialised as Swish). This leads us to ask whether such an initialisation is necessary or if standard random Glorot initialization can also be used. Unfortunately, random initialization always leads to divergence of the network training in the standard setting. We notice that this is due to 2 reasons:

1. The distribution of the input to the activation shifts during training, which hampers training because the activation parameters need to be constantly adjusted to the input.
2. Random initialisation can lead to unstable derivatives, which can lead to very large gradient values.

   Problem 1 is solved via the normalization techniques that we investigated previously. In order to mitigate problem 2, we apply weight clipping. Both techniques together greatly stabilise the training of randomly initialised activations and lead to network convergence in all of our image classification experiments.

Table 2. Test accuracy on CIFAR10 in percent, conditioned on the network type and the activation function. Our version outperforms all the baselines for both of the network types.

Network	Activation	Test accuracy
FCN	Leaky ReLU	51.8 ± 0.1
FCN	ReLU	52.6 ± 0.1
FCN	Swish	52.3 ± 0.2
FCN	Rational, [29]	55.0 ± 0.2
FCN	**ERA**	$\mathbf{58.8 \pm 0.2}$
CNN	Leaky ReLU	76.0 ± 0.2
CNN	ReLU	77.3 ± 0.1
CNN	Swish	76.8 ± 0.2
CNN	Rational, [29]	82.1 ± 0.2
CNN	**ERA**	$\mathbf{84.4 \pm 0.2}$

Results

Outperforming the Baselines. As shown in Table 2, our best results outperform non-rational activation functions considerably by a margin of more than 6% in both fully-connected and convolutional neural networks on CIFAR10. In addition, the test accuracy of the modified rational function we proposed is significantly higher than the one achieved by the function proposed by [29], achieving a margin of more than 2% for the CNN and almost 4% for the fully-connected neural network. The fact that both our rational function and the baseline one clearly outperform standard activations, shows that rational activation functions are indeed qualified for increasing the network capacity.

Ablation Studies. We perform multiple ablations, in particular on the initialisation of the network, the normalisation of the input of the activation functions and the use of partial fractions. Table 3 shows consistent benefits of initialising *ERA*s as Swish activations. We note that although randomly initialised *ERA*s perform somewhat worse than the other initialisations, they still outperform all the standard activations that we investigated (cf. Table 2). As shown in Table 4 also note that normalising the input of the activations gives consistent benefits across different settings and multiple network types. Table 4 also indicates a small benefit in performance when using partial fractions instead of standard fractions.

We refer the reader to the supplementary material for a more comprehensive overview of our ablation studies.

Table 3. Test accuracy on CIFAR10 in percent, conditioned on the network type and the initialisation of the rational activation. Swish consistently outperforms the other initialisations, whereas random initialisation performs slightly worse. Layer Normalisation was applied to the input of the activation.

Initialization	FCN	CNN
GELU	58.7 ± 0.3	82.8 ± 0.5
Leaky ReLU	57.8 ± 0.2	82.7 ± 0.2
Random	57.5 ± 0.4	82.5 ± 0.5
ReLU	57.9 ± 0.1	82.9 ± 0.2
Swish	$\mathbf{58.8 \pm 0.2}$	$\mathbf{84.4 \pm 0.2}$

Table 4. Test accuracy on CIFAR10 in percent, conditioned on the network type and if partial fractions are used. Both layer normalization and partial fractions give consistent benefits. All networks were initialised with the Swish activation.

Configuration	FCN	CNN
Ours (baseline)	57.0 ± 0.4	81.8 ± 0.3
+ layer normalization	57.9 ± 0.2	84.1 ± 0.2
+ partial fraction	$\mathbf{58.8 \pm 0.2}$	$\mathbf{84.4 \pm 0.2}$

Analysis

Plotting of *ERA*s. In Fig. 1 we compare one of our the best *ERA*s before and after training. The activation function was initialised as a Swish activation and after training vaguely resembles a quadratic function where most of the probability mass of the data lies. We observe similar results for many other *ERA*s that were initialised as Swish activations. On the other hand, the shapes of randomly initialised *ERAs* vary a lot, depending on initial parameter configuration and the convergence of training. Figure 2 displays a successfully trained (final test accuracy higher than 80%) randomly initialised *ERA* of degree 5/4 and its first derivative before and after training. In this particular example, we observe an activation function that is close to linear on the data distribution which is something that we observe regularly in the case of random initialisation. We also note that the relatively small values of the first derivative do not seem to hurt network performance.

Due to space limitations, we present a more in-depth analysis of plots of rational activations and their derivatives in the appendix.

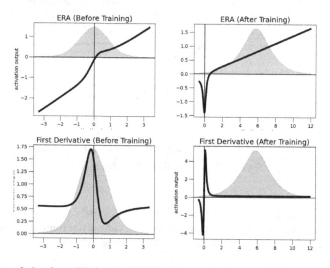

Fig. 2. Plots of the first *ERA* of a CNN before and after training on CIFAR10. The *ERA* is randomly initialised and has degree 5/4. Both the activation and its derivatives differ drastically from the initialization. The shaded curve in the background shows the probability density distribution of the input to the activation (not true to scale on the y-axis). In contrast to Swish initialised *ERAs*, random initialisation can result in very small or large derivative values which can make the training more difficult.

Loss Landscapes. Inspired by the analysis in [28], we use the methodology of [23], based on filter normalisation, to create 3D visualisations of the loss functions. Figure 3 shows the plots of two Convolutional Neural Networks trained on CIFAR10. Here the xy-plane corresponds to configurations of network parameters and the z-axis corresponds to the corresponding value of the loss function evaluated on the training data. The *ERA* used in the top row is randomly initialised, whereas the one used in the bottom row was initialised with the Swish activation. In both cases, the rational activation had degree 5/4. Both networks

Random Initialization, degree 5/4

Swish Initialization, degree 5/4

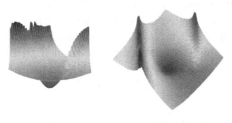

Fig. 3. Plots of the loss surfaces of a CNN trained on CIFAR10, using *ERA*. **Top**: randomly initialized *ERA* of degree 5/4. **Bottom**: Swish-initialized *ERA* of degree 5/4. In each row, the two images show the same loss surface from different viewing angles. *The plots are best viewed digitally in 3d. The 3d versions are included in the supplementary material.*

achieve a high test accuracy of more than 80%. We see that despite the different initalisations, the loss landscapes are similar and reasonably well-behaved (i.e. there is a single local minimum in the "valleys"). We perform a more detailed comparison in the supplementary material, where we also show the loss landscapes of randomly initialised networks that did not converge or achieved a mediocre performance.

4.2   3D Human Pose and Shape Reconstruction

Table 5. Number of parameters for our lightweight versions of the architecture introduced in THUNDR [41], for the task of 3d pose and shape reconstruction.

Backbone	Head	Hidden dim	#Parameters
MobileNet	MLP-Mixer	64	6.6M
MobileNet	MLP-Mixer	256	10.0M
MobileNet	Transformer	64	3.9M
MobileNet	Transformer	256	6.7M
ResNet50	Transformer	256	*25.0M* [41]

In this section, we experiment with and ablate our activation function *ERA* for the task of 3d pose and shape reconstruction of humans from a monocular image. We choose an implementation of *ERA* initialised from a Swish activation and using a partial fraction formulation as suggested in our previous experiments on image classification. We adopt and adapt the architecture previously introduced in THUNDR [41]. At its core, THUNDR is a hybrid convolutional-transformer architecture [9]. In its original implementation, the backbone consists of a ResNet50 [11] and the regression head is a Transformer [39] architecture. We replace the backbone with a lightweight MobileNet [13] architecture and further ablate the choice of the head architecture, by also considering an MLP-Mixer [37]. Our focus here is to assess the impact of *ERA* on lightweight networks and, more specifically, in more recent architectures, i.e. Transformers and MLP-Mixers. We do not replace the standard activations used in the backbone convolutional architecture, which is pre-trained on ImageNet. This could have an impact on the performance even further, but this is not in the scope of this paper and we leave it aside for future work.

Setup. As in the original THUNDR paper, we experiment with two different training regimes, both weakly supervised and a mixed regime, weakly and fully supervised. For weak supervision, we use images from both the COCO2017 [25] and OpenImages [20] datasets. We use a mix of ground-truth 2d keypoint annotations, where available, and 2d detections from pre-trained models. For fully supervised training we consider the standard Human3.6M [15] dataset, where ground-truth 3d is available. For all experiments we report results on protocol 2 (P2) of the Human3.6M test set.

Ablations. MLP-Mixers and Transformers both have a hidden dimension hyperparameter that has a big impact on the number of network parameters. We ablate two settings, one with low capacity (64) and one with higher capacity, as proposed in [41] (256). Different from [41], we do not share the parameters across the encoder layers, to better assess the stability of training when multiple instances of *ERA* are learned. We ablate *ERA* with two denominator settings, one with lower expressivity ($\deg(Q) = 2$) and one with higher ($\deg(Q) = 4$). For a comparison with other activation functions, we use ReLU (as in [41]), GELU and Swish.

We follow the training schedule and hyper-parameters described in [41]. We do not investigate further the impact of these choices on the performance of the different activation functions, and leave this for future work. Note that the parameters are tailored in the original paper for the specific choice of ReLU as the activation function. For each different choice of head-model architecture, hidden size and activation function we run 3 different trials and report averaged metrics over the trials. The metrics we use are the standard ones for 3d reconstruction. For 3d joint errors we report mean per joint position error (MPJPE) and MJPE-PA, which is MPJPE after rigid alignment via Procrustes Analysis. For evaluating 3d shape we use the mean per vertex position error (MPVPE) between the vertices of the predicted and ground-truth meshes, respectively.

Results

Number of Parameters and Running Time. In Table 5 we show the number of parameters for our lightweight networks as a function of the backbone and head architectures, and the hidden dimension. Our networks have significantly fewer parameters than the one used in [41], with the smallest one having around 3.9M parameters. Using *ERA* as activation only adds an insignificant number of parameters (depending on the degree of the denominator), between 10 and 20 more for each *ERA* instance. On a desktop computer with an NVIDIA GeForce RTX 2080 Ti graphics card, all of our networks run in real-time at inference, with a frame-per-second rate of around 100–110.

Numerical Results. We present results in Tables 6, 7 for different training regimes, weakly supervised and mixed, respectively. In the former, our proposed activation *ERA* achieves the best results across the board, with the most noticeable performance boost in the case of lowest capacity (i.e. hidden dimension of 64) for both head-networks type. In the mixed training regime, this observation still holds. But, as the network capacity increases, the gap between *ERA* and other standard activations shrinks, and even reverses, as in the case of MLP-Mixers with higher capacity. Nevertheless, we see in Table 7 that the best performing setting is *ERA* with a high capacity Transformer. Overall, the biggest improvement in performance is achieved in the setting with the lowest capacity, a Transformer with a hidden dimension of 64, where *ERA* has a significantly lower error than all other activations. This holds true for both training regimes, either weakly or with mixed supervision. We compare this setting to the state of the art, and show that this lightweight network is on par or better than other methods with considerably bigger architecture sizes. For this comparison we selected the network that achieved the best training error out of 3 different trials.

The effect of the degree of the denominator Q for *ERA*, seems to be quite minimal, with a slight edge for the setting in which the denominator is set to 2. In general, Transformers seem to perform better than MLP-Mixers for this task, but we did not perform any hyperparameter search for either of the architectures. Additional details, experiments and ablations are included in the supplementary.

5   Conclusions

We have presented *ERA*, a learnable rational activation function that can replace standard activations in a wide range of neural network architectures. Under a theoretical framework, we analysed and showed that *ERA* is a smooth, safe, rational function and can be refactored using a partial fraction representation, which empirically exhibits improved performance. We also showed that the right normalization of inputs has a great effect on the stability and the performance of *ERA*, and, as a byproduct, we can mitigate some of the effects of initialization, even from a random distribution, something not possible before in prior work on rational layers. Across different types of lightweight neural network architectures and for two different tasks, image classification, and 3d human pose and shape reconstruction, *ERA* increases network performance and closes the gap between different sized architectures by increasing the expressivity of the network. Specifically, on the task of 3d human pose and shape reconstruction, we show that a real-time, lightweight network, with only 3.9M parameters, can achieve results that are competitive with state-of-the-art methods, where network parameters are in the order of tens or hundreds of millions.

Table 6. Ablation for networks with **MLP-Mixer (Top)** and **Transformer (Bottom)** architectures, trained in a **weakly supervised** regime. Results are in mm, averaged over 3 trials, reported on the Human3.6M test dataset, protocol P2. We mark the best and second best results.

Activation	deg Q	Hidden dim	MPJPE-PA	MPJPE	MPVPE
GELU	-	64	72.2 ±1.3	109.4 ±2.1	117.1 ±2.0
ReLU	-	64	69.2 ±1.8	102.8 ±2.9	109.8 ±3.2
Swish	-	64	72.8 ±0.8	109.5 ±0.5	117.2 ±0.3
ERA	2	64	**67.7** ±0.5	**98.0** ±1.0	**104.3** ±1.5
ERA	4	64	69.0 ±1.2	100.3 ±1.6	107.0 ±2.1
GELU	-	256	66.4 ±0.5	95.0 ±1.0	101.1 ±1.1
ReLU	-	256	66.8 ±1.2	97.2 ±1.7	103.5 ±2.0
Swish	-	256	67.0 ±0.2	97.0 ±0.6	103.6 ±0.6
ERA	2	256	**65.2** ±0.2	93.2 ±0.9	99.4 ±1.4
ERA	4	256	**65.2** ±0.4	**92.7** ±0.5	**98.7** ±0.8

MLP-Mixer

Activation	deg Q	Hidden dim	MPJPE-PA	MPJPE	MPVPE
GELU	-	64	80.6 ±0.3	125.5 ±0.6	132.8 ±0.9
ReLU	-	64	78.8 ±0.9	122.9 ±2.7	130.4 ±2.8
Swish	-	64	81.5 ±0.7	127.4 ±2.8	134.9 ±2.3
ERA	2	64	73.8 ±2.0	112.3 ±3.5	119.5 ±2.7
ERA	4	64	**73.6** ±1.5	**112.1** ±3.5	119.5 ±3.9
GELU	-	256	63.0 ±0.2	89.1 ±1.1	94.5 ±0.7
ReLU	-	256	69.8 ±0.7	101.9 ±1.9	109.5 ±1.8
Swish	-	256	64.0 ±0.5	91.1 ±1.6	96.7 ±1.3
ERA	2	256	**62.5** ±0.8	**88.5** ±1.8	**93.6** ±2.0
ERA	4	256	63.5 ±0.5	89.6 ±1.9	95.0 ±2.1

Transformer

Table 7. Ablation for networks with **MLP-Mixer (Top)** and **Transformer (Middle)** architectures, pre-trained in a weakly supervised regime and fine-tuned in a **fully-supervised** regime. Results are in mm, averaged over 3 trials, reported on the Human3.6M test dataset, protocol P2. We mark the best and second best results. **(Bottom)** We report our *best performing lightest network with *ERA* out of the 3 trials (selected based on training error), relative to the state-of-the-art methods. Our network is the lightest of the architectures, by a large margin, with around 3.9M parameters (e.g. THUNDR has 25M parameters).

Activation	deg Q	Hidden dim	MPJPE-PA	MPJPE	MPVPE
GELU	-	64	53.5 ±1.2	72.1 ±1.5	76.1 ±1.3
ReLU	-	64	51.6 ±0.7	71.8 ±1.5	75.8 ±1.1
Swish	-	64	54.2 ±0.8	73.9 ±2.5	78.2 ±3.0
ERA	2	64	51.4 ±0.2	70.9 ±1.6	74.4 ±1.7
ERA	4	64	**51.2** ±1.1	69.6 ±1.5	**73.4** ±1.5
GELU	-	256	**49.6** ±0.8	68.9 ±1.2	72.9 ±1.6
ReLU	-	256	49.9 ±0.9	68.3 ±1.5	72.2 ±1.6
Swish	-	256	**49.6** ±0.4	**68.2** ±0.8	**72.1** ±1.6
ERA	2	256	50.1 ±0.8	70.6 ±1.8	74.3 ±2.4
ERA	4	256	50.8 ±2.1	70.7 ±4.8	74.8 ±4.6

MLP-Mixer

Activation	deg Q	Hidden dim	MPJPE-PA	MPJPE	MPVPE
GELU	-	64	50.9 ±1.6	71.6 ±2.7	76.2 ±3.0
ReLU	-	64	51.5 ±0.5	72.8 ±1.2	77.2 ±0.6
Swish	-	64	51.8 ±0.7	71.9 ±0.6	76.4 ±0.5
ERA	2	64	**47.2** ±1.5	**66.7** ±1.9	**71.3** ±2.6
ERA	4	64	47.5 ±0.6	67.4 ±1.4	72.1 ±1.4
GELU	-	256	44.8 ±0.6	**63.1** ±1.5	**67.3** ±2.0
ReLU	-	256	45.4 ±0.3	64.2 ±0.5	67.8 ±0.1
Swish	-	256	44.9 ±0.2	63.7 ±1.3	67.7 ±1.4
ERA	2	256	**43.7** ±0.4	63.4 ±0.3	67.4 ±0.5
ERA	4	256	44.8 ±0.9	63.5 ±1.4	67.6 ±1.6

Transformer

Method	MPJPE-PA	MPJPE
HMR [16]	56.8	88.0
GraphCMR [18]	50.1	-
Pose2Mesh [6]	47.0	64.9
I2L-MeshNet [30]	41.7	55.7
SPIN [17]	41.1	-
METRO [24]	36.7	54.0
THUNDR [41]	34.9	48.0
[41] + **ERA*** (lightest network)	45.9	64.0

State of the art

Acknowledgements. This work was supported in part by the European Research Council Consolidator grant SEED, CNCSUEFISCDI PN-III-PCCF-2016-0180, Swedish Foundation for Strategic Research (SSF) Smart Systems Program, as well as the Wallenberg AI, Autonomous Systems and Software Program (WASP) funded by the Knut and Alice Wallenberg Foundation.

References

1. Abadi, M., et al.: TensorFlow: large-scale machine learning on heterogeneous systems (2015). www.tensorflow.org/. software available from tensorflow.org
2. Ba, J., Kiros, J.R., Hinton, G.E.: Layer normalization. arXiv abs/1607.06450 (2016)
3. Boulle, N., Nakatsukasa, Y., Townsend, A.: Rational neural networks (2020). https://proceedings.neurips.cc/paper/2020/file/a3f390d88e4c41f2747bfa2f1b5f87db-Paper.pdf
4. Bradley, W.T., Cook, W.J.: Two proofs of the existence and uniqueness of the partial fraction decomposition. In: International Mathematical Forum, pp. 1517–1535 (2012)
5. Chen, Z., Chen, F., Lai, R., Zhang, X., Lu, C.: Rational neural networks for approximating jump discontinuities of graph convolution operator. CoRR abs/1808.10073 (2018). arxiv.org/abs/1808.10073
6. Choi, H., Moon, G., Lee, K.M.: Pose2Mesh: graph convolutional network for 3D human pose and mesh recovery from a 2D human pose. In: Vedaldi, A., Bischof, H., Brox, T., Frahm, J.-M. (eds.) ECCV 2020. LNCS, vol. 12352, pp. 769–787. Springer, Cham (2020). https://doi.org/10.1007/978-3-030-58571-6_45
7. Delfosse, Q., et al.: Rational activation functions (2020)
8. Deng, J., Dong, W., Socher, R., Li, L.J., Li, K., Fei-Fei, L.: ImageNet: a large-scale hierarchical image database (2009). https://doi.org/10.1109/CVPR.2009.5206848
9. Dosovitskiy, A., et al.: An image is worth 16x16 words: transformers for image recognition at scale (2021). https://openreview.net/forum?id=YicbFdNTTy
10. Fukushima, K.: Neocognitron: a self-organizing neural network model for a mechanism of pattern recognition unaffected by shift in position. Biol. Cybern. **36**, 193–202 (1980)
11. He, K., Zhang, X., Ren, S., Sun, J.: Deep residual learning for image recognition (2016). https://doi.org/10.1109/CVPR.2016.90
12. Hendrycks, D., Gimpel, K.: Gaussian error linear units (GELUs). arXiv preprint arXiv:1606.08415 (2016)
13. Howard, A.G., et al.: MobileNets: efficient convolutional neural networks for mobile vision applications. CoRR abs/1704.04861 (2017). arxiv.org/abs/1704.04861
14. Ioffe, S., Szegedy, C.: Batch normalization: Accelerating deep network training by reducing internal covariate shift. CoRR abs/1502.03167 (2015). arxiv.org/abs/1502.03167
15. Ionescu, C., Papava, D., Olaru, V., Sminchisescu, C.: Human3.6M: large scale datasets and predictive methods for 3D human sensing in natural environments. IEEE Trans. Pattern Anal. Mach. Intell. **36**(7), 1325–1339 (2014)
16. Kanazawa, A., Black, M.J., Jacobs, D.W., Malik, J.: End-to-end recovery of human shape and pose. In: CVPR (2018)
17. Kolotouros, N., Pavlakos, G., Black, M.J., Daniilidis, K.: Learning to reconstruct 3D human pose and shape via model-fitting in the loop. In: Proceedings of the IEEE International Conference on Computer Vision, pp. 2252–2261 (2019)

18. Kolotouros, N., Pavlakos, G., Daniilidis, K.: Convolutional mesh regression for single-image human shape reconstruction. In: Proceedings of the IEEE Conference on Computer Vision and Pattern Recognition, pp. 4501–4510 (2019)

19. Krizhevsky, A.: Learning multiple layers of features from tiny images. Master's thesis, University of Toronto (2009)

20. Kuznetsova, A., et al.: The open images dataset V4. Int. J. Comput. Vis. **128**(7), 1956–1981 (2020)

21. Lacoste, A., Luccioni, A., Schmidt, V., Dandres, T.: Quantifying the carbon emissions of machine learning (2019). http://arxiv.org/1910.09700

22. LeCun, Y., Cortes, C., Burges, C.: Mnist handwritten digit database. ATT Labs (2010). http://yann.lecun.com/exdb/mnist

23. Li, H., Xu, Z., Taylor, G., Studer, C., Goldstein, T.: Visualizing the loss landscape of neural nets. In: Bengio, S., Wallach, H., Larochelle, H., Grauman, K., Cesa-Bianchi, N., Garnett, R. (eds.) Advances in Neural Information Processing Systems. vol. 31. Curran Associates, Inc. (2018). https://proceedings.neurips.cc/paper/2018/file/a41b3bb3e6b050b6c9067c67f663b915-Paper.pdf

24. Lin, K., Wang, L., Liu, Z.: End-to-end human pose and mesh reconstruction with transformers. In: CVPR, pp. 1954–1963 (2021)

25. Lin, T.-Y., et al.: Microsoft COCO: common objects in context. In: Fleet, D., Pajdla, T., Schiele, B., Tuytelaars, T. (eds.) ECCV 2014. LNCS, vol. 8693, pp. 740–755. Springer, Cham (2014). https://doi.org/10.1007/978-3-319-10602-1_48

26. Liu, Z., Mao, H., Wu, C.Y., Feichtenhofer, C., Darrell, T., Xie, S.: A convnet for the 2020s (2022)

27. Maas, A.L., Hannun, A.Y., Ng, A.Y.: Rectifier nonlinearities improve neural network acoustic models (2013)

28. Misra, D.: Mish: A self regularized non-monotonic neural activation function. arXiv preprint arXiv:1908.08681 (2019)

29. Molina, A., Schramowski, P., Kersting, K.: Padé activation units: end-to-end learning of flexible activation functions in deep networks (2019)

30. Moon, G., Lee, K.M.: I2L-MeshNet: image-to-lixel prediction network for accurate 3D human pose and mesh estimation from a single RGB image. In: Vedaldi, A., Bischof, H., Brox, T., Frahm, J.-M. (eds.) ECCV 2020. LNCS, vol. 12352, pp. 752–768. Springer, Cham (2020). https://doi.org/10.1007/978-3-030-58571-6_44

31. Nair, V., Hinton, G.E.: Rectified linear units improve restricted Boltzmann machines (2010)

32. Pinkus, A.: Approximation theory of the MLP model in neural networks. Acta Numerica **8**, 143–195 (1999). https://doi.org/10.1017/S0962492900002919

33. Puri, I., Dhurandhar, A., Pedapati, T., Shanmugam, K., Wei, D., Varshney, K.R.: CoFrNets: interpretable neural architecture inspired by continued fractions (2021). https://openreview.net/forum?id=kGXlIEQgvC

34. Ramachandran, P., Zoph, B., Le, Q.V.: Searching for activation functions (2018). https://openreview.net/forum?id=SkBYYyZRZ

35. Simonyan, K., Zisserman, A.: Very deep convolutional networks for large-scale image recognition. arXiv preprint arXiv:1409.1556 (2014)

36. Strubell, E., Ganesh, A., McCallum, A.: Energy and policy considerations for modern deep learning research, April 2020. https://doi.org/10.1609/aaai.v34i09.7123, https://ojs.aaai.org/index.php/AAAI/article/view/7123

37. Tolstikhin, I.O., et al.: MLP-mixer: an all-MLP architecture for vision. CoRR abs/2105.01601 (2021). https://arxiv.org/abs/2105.01601

38. Ulyanov, D., Vedaldi, A., Lempitsky, V.S.: Instance normalization: the missing ingredient for fast stylization. CoRR abs/1607.08022 (2016). http://arxiv.org/1607.08022

39. Vaswani, A., et al.: Attention is all you need (2017). https://proceedings.neurips.cc/paper/2017/file/3f5ee243547dee91fbd053c1c4a845aa-Paper.pdf

40. Wu, Y., He, K.: Group normalization. CoRR abs/1803.08494 (2018). http://arxiv.org/1803.08494

41. Zanfir, M., Zanfir, A., Bazavan, E.G., Freeman, W.T., Sukthankar, R., Sminchisescu, C.: THUNDR: transformer-based 3D human reconstruction with markers. In: Proceedings of the IEEE/CVF International Conference on Computer Vision (ICCV), pp. 12971–12980, October 2021

Convolutional Embedding Makes Hierarchical Vision Transformer Stronger

Cong Wang[1,2] , Hongmin Xu[1(✉)], Xiong Zhang[4], Li Wang[2], Zhitong Zheng[1],
and Haifeng Liu[1,3]

[1] Data & AI Engineering System, OPPO, Beijing, China
xhmjimmy@gmail.com, {blade,liam}@oppo.com
[2] Beijing Key Lab of Urban Intelligent Traffic Control Technology, North China
University of Technology, Beijing, China
li.wang@ncut.edu.cn
[3] University of Science and Technology of China, Hefei, China
[4] Neolix Autonomous Vehicle, Beijing, China
zhangxiong@neolix.cn

Abstract. Vision Transformers (ViTs) have recently dominated a range of computer vision tasks, yet it suffers from low training data efficiency and inferior local semantic representation capability without appropriate inductive bias. Convolutional neural networks (CNNs) inherently capture regional-aware semantics, inspiring researchers to introduce CNNs back into the architecture of the ViTs to provide desirable inductive bias for ViTs. However, *is the locality achieved by the micro-level CNNs embedded in ViTs good enough?* In this paper, we investigate the problem by profoundly exploring how the macro architecture of the hybrid CNNs/ViTs enhances the performances of hierarchical ViTs. Particularly, we study the role of token embedding layers, alias *convolutional embedding* (CE), and systemically reveal how CE injects desirable inductive bias in ViTs. Besides, we apply the optimal CE configuration to 4 recently released state-of-the-art ViTs, effectively boosting the corresponding performances. Finally, a family of efficient hybrid CNNs/ViTs, dubbed **CETNets**, are released, which may serve as generic vision backbones. Specifically, CETNets achieve 84.9% Top-1 accuracy on ImageNet-1K (training from scratch), 48.6% box mAP on the COCO benchmark, and 51.6% mIoU on the ADE20K, substantially improving the performances of the corresponding state-of-the-art baselines.

Keywords: Vision Transformers · Convolutional neural networks · Convolutional embedding · Micro and macro design

1 Introduction

Over the last decades, convolutional neural networks (CNNs) significantly succeeded in the computer vision community due to their inherent properties,

Supplementary Information The online version contains supplementary material available at https://doi.org/10.1007/978-3-031-20044-1_42.

S. Avidan et al. (Eds.): ECCV 2022, LNCS 13680, pp. 739–756, 2022.
https://doi.org/10.1007/978-3-031-20044-1_42

including the translation invariance, the locality attention, and the sharing weight design. Those characteristics prove critical for many tasks, such as image recognition [16,25], semantic image segmentation [7,70], and object detection [11,42]. At the same time, researchers take a very different way in the natural language processing (NLP) field. Since the seminal work [54] demonstrated the extraordinary capability of the transformer by employing a unified yet simple architecture to tackle the machine translation task, transformers have become the de-facto architectures to resolve NLP tasks [17,40] (Fig. 1).

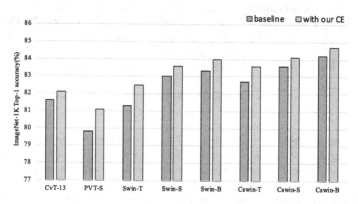

Fig. 1. Performance Improvements of the Convolutional Embedding (CE). The figure presents the performance gains of the CE among 4 SOTA ViTs, *i.e., CvT* [59], *PVT* [56], *SWin* [35], and *CSWin* [18] on the ImageNet-1K dataset, indicating that the CE indeed significantly improves the baseline methods

In the computer vision domain, certain works [3,29,46,57] successfully brought the key idea of transformer, *i.e.,* attention paradigm, into CNNs and achieved remarkable improvements. Naively transferring the transformer to image recognition, Dosovitskiy *et al.* [19] demonstrated that the vanilla Vision Transformers (ViTs) structure could achieve comparable performance compared with the state-of-the-art (SOTA) approaches on the ImageNet-1K dataset [19]. Further, pre-trained on the huge dataset JFT-300M [48], the vanilla ViTs outperformed all the SOTA CNNs by a large margin and substantially advanced the SOTA, suggesting that ViTs may have a higher performance ceiling.

ViTs rely on highly flexible multi-head self-attention layers to favor dynamic attention, capture global semantics, and achieve good generalization ability. Yet, recent works find that lacking proper inductive bias, locality, ViTs are substandard optimizability [60], low training sample-efficient [15], and poor to model the complex visual features and the local relation in an image [39,59]. Most existing works attempt to introduce local mechanisms to ViTs in two paths. One line of work relives the inductive bias problem via non-convolutional ways. Liu *et al.* [18,35,53] limits the attention computation in a local window to enable local receptive field for attention layers, making the overall network still maintain nearly pure attention-based architecture. Concurrently, as the CNNs are inherently efficient for their sliding-window manner, local receptive field, and inductive bias [2], another line of work directly integrates CNNs into ViTs design to bring convolutional inductive bias into ViTs in a hard [33,59,65] or soft [15,52] way. However, most of those works focus on modifying the micro design of ViTs to enable locality, which raises the question:

Is the inductive bias obtained via the micro design of ViTs good enough to empower the locality to ViTs? or Can the macro architecture design of the network further introduce the desirable inductive bias to ViTs?

We ask the questions mentioned above based on the following findings. The previous work, EarlyConv [60], notices that simply replacing the original patchify stem with a 5-layer convolutional stem can yield 1–2% top-1 accuracy on ImageNet-1K and improve the training stability of ViTs. Subsequently, CoAtNet [14] further explore the hybrid design of CNNs and ViTs based on the observation: the depth-wise convolution can be naturally integrated into the attention block. Meanwhile, those works suggest that the convolutional layers can efficiently introduce inductive bias in a shallow layer of the whole network. However, when we retrospect the roadmap of modern CNNs, at the late of 2012, after AlexNet [32] exhibited the potential of CNNs, subsequent studies, for instance, VGG [45], ResNet [25], DenseNet [30], EfficientNet [50,51] ConvNet2020 [36], and *etc.* reveal that: convolutions can represent the complex visual features even in a deep layer of network effectively and efficiently. Our research explores the macro network design with hybrid CNNs/ViTs. We want to bridge the gap between the pure CNNs network and the pure ViT network and extend the limitation of the hybrid CNNs/ViTs network.

To test this hypothesis, we start from CNNs' effective receptive field (ERF). As previous work of Luo *et al.* [38] points out, the output of CNNs considerably depends on their ERF. With larger ERF, CNNs can leave out no vital information, which leads to better prediction results or visual features. Under this perspective, our exploration is imposing strong and effective inductive bias for attention layers via the macro design of network architecture. We specifically focus on *patch embedding*, alias *convolutional embedding* (CE), of the hierarchical ViTs architecture. CE locate at the beginning of each stage, as shown in Fig. 2. The CE aims to adjust the dimension and the number of tokens. Most following works also apply one or two convolutions embedding layers [18,33,56,59,68]. However, these embedding layers cannot offer enough ERF to capture complex visual representations with desirable inductive bias. Since stacking more convolutional layers can increase the ERF [38], we construct a simple baseline with only 1-layer CE and gradually increase the number of convolutions layers in CE to get more variants. Meanwhile, keep changing FLOPs and parameter numbers as small as possible. We observe that the small change of CE in each stage results in a significant performance increase in the final model.

Based on extensive experiments, we further understand how CE affects the hybrid network design of CNNs/ViTs by injecting desirable inductive bias. We make several observations. 1) CNNs bring strong inductive bias even in deep layers of a network, making the whole network easier to train and capturing much complex visual features. At the same time, ViTs allow the entire network has a higher generalization ceiling. 2) CE can impose effective inductive bias, yet different convolution layers show variable effectiveness. Besides, the large ERF is essential to designing CE or injecting the desirable inductive bias to ViTs, even though it is a traditional design in a pure CNNs network [38,51]. 3) CNNs can help ViTs see better even in deep networks, providing valuable insights to guide how to design the hybrid CNNs/ViTs network. 4) It is beneficial to combine the macro and micro introduce inductive bias to obtain a higher generalization ceiling of ViTs-based network.

Our results advocate the importance of CE and deep hybrid CNNs/ViTs design for vision tasks. ViT is a general version of CNN [10], and tons of works have proven the high generalization of ViTs-based networks, which spurs researchers to line up to chase the performance ceiling with pure attention networks. After inductive bias is found to be crucial to significantly improve the training speed and sample-efficiency of ViTs, people's efforts are mainly devoted to creating the micro design of ViTs to enhance it [15,35]. Concurrently, EarlyConv [60] and CoAtNet [14] verify the efficiency of convolutions in shallow layers of the ViT-based network. Our study further pushes the boundary of the macro design of the hybrid CNNs/ViTs network. Our results also suggest that even in deep layers of the ViTs network, correctly choosing the combination of CNNs/ViTs design, one can further improve the upper-performance limitation of the whole network. Finally, we propose a family of models of hybrid CNNs/ViTs as a generic vision backbone.

To sum up, we hope our findings and discussions presented in this paper will deliver possible insights to the community and encourage people to rethink the value of CE in the hybrid CNNs/ViTs network design.

2   Related Work

Convolutional Neural Networks. Since the breakthrough performance of AlexNet [32], the computer vision field has been dominated by CNNs for many years. In the past decade, we have witnessed a steady stream of new ideas being proposed to make CNNs more effective and efficient [22,25,27,29,30,41,43,45, 49–51,69]. One line of work focuses on improving the individual convolutional layer except for the architectural advances. For example, the depthwise convolution [62] is widely used due to its lower computational cost and smaller parameter numbers, and the deformable convolution [12] can adapt to shape, size, and other geometric deformations of different objects by adding displacement variables. The dilated convolution introduces a new parameter called "dilation rate" into the convolution layer, which can arbitrarily expand the receptive field without additional parameters cost. These CNNs architectures are still the primary backbones for computer vision tasks such as object detection [42], instance segmentation [24], semantic segmentation [37], and so on.

Vision Transformers. Transformer [54] has become a prevalent model architecture in natural language processing (NLP) [4,17] for years. Inspired by the success of NLP, increasing effort on adapting Transformer to computer vision tasks. Dosovitskiy *et al.* [19] is the pioneering work that proves pure Transformer-based architectures can attain very competitive results on image classification, shows strong potential of the Transformer architecture for handling computer vision tasks. The success of [19] further inspired the applications of Transformer to various vision tasks, such as image classification [23,52,59,63,66], object detection [5,13,71,73] and semantic segmentation [47,55]. Furthermore, some recent works [35,56,67] focusing on technique a general vision Transformer backbone for general-purpose vision tasks. They all follow a hierarchical architecture and develop different self-attention mechanisms. The hierarchical design can produce multi-scale features that are beneficial for dense prediction tasks. An efficient self-attention mechanism reduced the computation complexity and enhanced modeling ability.

Integrated CNNs and Transformer. CNNs are good at capturing local features and have the advantages of shift, scale, and distortion invariance, while Transformer have the properties of dynamic attention, global receptive field, and large model capacity. Combining convolutional and Transformer layers can achieve better model generalization and efficiency. Many researchers are trying to integrate CNNs and Transformer. Some methods [3,28,44,46,57,58] attempt to augment CNNs backbones with self-attention modules or replace part of convolution blocks with Transformer layers. In comparison, inspired by the success of ViT [19], recent trials attempt to leverage some appropriate convolution properties to enhance Transformer backbones. ConViT [15] introduce a parallel convolution branch to impose convolutional inductive biases into ViT [19]. Localvit [33] adding a depth-wise convolution in Feed-Forward Networks (FFN) component to extract the locality, and CvT [59] employs convolutional projection to calculate self-attention matrices to achieve additional modeling of local spatial context. Besides the "internal" fusion, some method [14,60] focus on structural combinations of Transformer and CNNs.

3   Hybrid CNNs/ViTs Network Design

A general hierarchical ViTs model architecture is illustrated in Fig. 2. The convolutional stem is applied for an input image to extract the low-level feature, followed typically four stages to extract diverse-scale deep representations gradually. Each stage consists of a *convolutional embedding* (CE) block and a set of ViTs blocks. More specifically, the CE block is located at the beginning of each stage and aims to adjust the dimension, change the number, and reduce the resolution of input tokens. For two adjacent stages, the reduction factor is set to 2. After that, those tokens are fed to the following ViTs blocks to generate a global context. To sum up, the proposed network contains five stages $(S0, S1, S2, S3, S4)$, where $S0$ is the convolutional stem.

The original hierarchical ViTs architecture [35] is following the traditional design of the CNNs network, VGG [45], targeting giving the tokens the ability to represent increasingly complex visual patterns over increasingly larger spatial

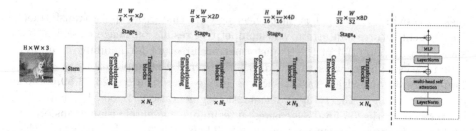

Fig. 2. Left: the overall architecture of a general hierarchical ViTs network, which consists of two main parts: *Stem* and *Stages*. Each stage contains a *convolutional embedding* and a set of *ViTs blocks*. Right: an example of a standard *ViTs block* [19]

footprints. Our goal is to impose the desirable inductive bias for ViTs blocks of the hierarchical architecture. Thus, our explorations are directed to two paths with pure CNNs structure: *Macro Design* and *Micro Design*.

3.1   Macro Design

Convolutional Stem. In ViTs-based network design, the *Stem* is concerned to extract the right inductive bias to following global attention modules. It is necessary that maintain enough effective receptive field (ERF) of CE to extract rich visual features at the early stage [60]. In consideration of parameter numbers and FLOPs budget, as well as we notice that $S0$ can be merged with the CE of $S1$, our final *Stem* consists of 4 Fused-MBConvs [21] layers and 1 regular convolutional layer. Specifically, $S0$ contains 2 Fused-MBConv with stride 2 and stride 1, respectively. The CE of $S1$ is consists of the same as $S0$ and followed by one 1×1 convolution at the end to match the channels and normalized by a layer normalization [1]. Another reason that we choose Fused-MBConv to compose *Stem* is that EfficientNetV2 [51] shows that Fused-MBConv is surprisingly effective in achieving better generalization and capacity as well as makes training convergence rate faster [60]. Besides, like EfficientNetV2, the hidden state's expand ratios of Fused-MBConv and MBConv [43] are arranged to 1 or 2. Such a setting allows the convolutional *Stem* to have fewer parameters and FLOPs without the loss of accuracy. Please refer to the supplementary materials for more details about Fused-MBConv and MBConv.

Convolutional Embedding. In the following stages, $S2$, $S3$, and $S4$, each stage contains a CE block and a set of ViTs blocks. The CE block captures diverse deep representation with a convolutional inductive bias for subsequent attention modules. It is worth noting that EarlyConv [60] and CoAtNet [14] point out stacking convolutional structure may enhance the ViTs in early stage. However, as we show in Table 8, we argue that CNNs is also able to represent the same, if not better, deep feature as ViTs in the deep layer of a network. Meanwhile, maintaining the CNNs design of the embedding layer naturally introduces proper inductive bias to following ViTs and retains the sample-efficient learning property. For the same computation resource constraint consideration, the CE adopts only effective and efficient convolutions. The CE of $S2$, $S3$, and $S4$ adopts MBConv as the basic unit.

3.2 Micro Design

Locally Enhanced Window Self-attention. Previous work [35] restricts the attention calculation with a local window to reduce computational complexity from quadratic to linear. Meanwhile, to some extent, the local window injects some inductive bias into ViTs. Concurrently, a set of works [56,59] attempt to directly integrate CNNs into ViTs to bring in convolutional inductive bias, yet their computational complexity is still quadratic. We present a simple Locally Enhanced Window Self-Attention (LEWin) to take advantage of fast attention calculating and locality inductive bias. Inspired by CvT, we performed a *convolutional projection* on input tokens in a local shift window. The *convolutional projection* is implemented by a depth-wise separable convolution with kernel size 3×3, stride 1, and padding 1. The LEWin can be formulated as:

$$x^{i-1} = \text{Flatten}\left(\text{Conv2D}\left(\text{Reshape}\left(z^{i-1}\right)\right)\right), \tag{1}$$

$$\hat{z}^i = \text{WHMSA}\left(\text{LN}\left(x^{i-1}\right)\right) + x^{i-1}, \tag{2}$$

$$z^i = \text{MLP}\left(\text{LN}\left(\hat{z}^i\right)\right) + \hat{z}^i, \tag{3}$$

$$x^i = \text{Flatten}\left(\text{Conv2D}\left(\text{Reshape}\left(z^i\right)\right)\right), \tag{4}$$

$$\hat{z}^{i+1} = \text{SWMSA}\left(\text{LN}\left(x^i\right)\right) + x^i, \tag{5}$$

$$z^{i+1} = \text{MLP}\left(\text{LN}\left(\hat{z}^{i+1}\right)\right) + \hat{z}^{i+1} \tag{6}$$

where x^i is the unperturbed token in local window prior to the *convolutional projection*, $\hat{z}^i$ and z^i denote the output features of the WMSA or S-WMSA module and the MLP module for the $i-th$ block, respectively. W-MSA and SW-MSA define window-based multi-head self-attention based on the regular and shifted windows from SWin [35], respectively.

3.3 CETNet Variants

We consider three different network configurations for CETNet to compare with other ViTs backbones under similar model size and computation complexity conditions. By changing the base channel dimension and the number of ViTs blocks of each stage, we build three variants, tiny, small, and base models, namely CETNet-T, CETNet-S, and CETNet-B. For more detailed configurations, please refer to our supplementary materials.

4 Experiments

To verify the ability of our CETNet as a general vision backbone. We conduct experiments on ImageNet-1K classification [16], COCO object detection [34], and ADE20K semantic segmentation [72]. In addition, comprehensive ablation studies are performed to validate the design of the proposed architecture.

Table 1. Comparison of image classification on ImageNet-1K for different models. The models are grouped based on the similar model size and computation complexity

ImageNet-1K 224² trained model											
Model	Params	FLOPs	Top-1 (%)	Model	Params	FLOPs	Top-1 (%)	Model	Params	FLOPs	Top-1 (%)
ResNet-50 [25]	25M	4.1G	76.2	ResNet-101 [25]	45M	7.9G	77.4	ResNet-152 [25]	60M	11.0G	78.3
RegNetY-4G [41]	21M	4.0G	80.0	RegNetY-8G [41]	39M	8.0G	81.7	RegNetY-16G [41]	84M	16.0G	82.9
DeiT-S [52]	22M	4.6G	79.8	PVT-M [56]	44M	6.7G	81.2	DeiT-B [52]	87M	17.5G	81.8
PVT-S [56]	25M	3.8G	79.8	T2T-19 [66]	39M	8.9G	81.5	PiT-B [26]	74M	12.5G	82.0
T2T-14 [66]	22M	5.2G	81.5	T2T$_t$ -19 [66]	39M	9.8G	82.2	T2T-24 [66]	64M	14.1G	82.3
ViL-S [67]	25M	4.9G	82.0	ViL-M [67]	40M	8.7G	83.3	T2T$_t$ -24 [66]	64M	15.0G	82.6
TNT-S [23]	24M	5.2G	81.3	MViT-B [20]	37M	7.8G	81.0	CPVT-B [9]	88M	17.6G	82.3
CViT-15 [6]	27M	5.6G	81.0	CViT-18 [6]	43M	9.0G	82.5	TNT-B [23]	66M	14.1G	82.8
LViT-S [33]	22M	4.6G	80.8	CViT$_c$-18 [6]	44M	9.5G	82.8	ViL-B [67]	56M	13.4G	83.2
CPVT-S [9]	23M	4.6G	81.9	Twins-B [8]	56M	8.3G	83.2	Twins-L [8]	99M	14.8G	83.7
Swin-T [35]	29M	4.5G	81.3	Swin-S [35]	50M	8.7G	83.0	Swin-B [35]	88M	15.4G	83.5
CvT-13 [59]	20M	4.5G	81.6	CvT-21 [59]	32M	7.1G	82.5	CETNet-B	75M	15.1G	**83.8**
CETNet-T	23M	4.3G	**82.7**	CETNet-S	34M	6.8G	**83.4**				
ImageNet-1K 384² finetuned model											
CvT-13 [59]	25M	16.3G	83.0	CvT-21 [59]	32M	24.9G	83.3	ViT-B/16 [19]	86M	49.3G	77.9
T2T-14 [66]	22M	17.1G	83.3	CViT$_c$-18 [6]	45M	32.4G	83.9	DeiT-B [52]	86M	55.4.3G	83.1
CViT$_c$-15 [6]	28M	21.4G	83.5	CETNet-S	34M	19.9G	**84.6**	Swin-B [35]	88M	47.0G	84.5
CETNet-T	24M	12.5G	**84.2**					CETNet-B	75M	44.50G	**84.9**

4.1 Image Classification

Settings. For Image classification, we compare different methods on ImageNet-1K [16], with about 1.3M images and 1K classes, the training set, and the validation set containing 1.28M images and 50K images respectively. The top 1 accuracy on the validation set is reported to show the capacity of our CETNet. For a fair comparison, the experiment setting follows the training strategy in [35]. All model variants are trained for 300 epochs with a batch size of 1024 or 2048 and using a cosine decay learning rate scheduler with 20 epochs of linear warm-up. We adopt an AdamW [31] optimizer with an initial learning rate of 0.001, and a weight decay of 0.05 is used. We use the same data augmentation methods, and regularization strategies used in [35] for training. All models are trained with 224 × 224 input size, while a center crop is used during evaluation on the validation set. When fine-tuning on 384 × 384 input, we train the models for 30 epochs with a learning rate of 2e−5, batch size of 512, and no center crop for evaluation.

Results. Table 1 presents comparisons to other models on the image classification task, and the models are split into three groups based on the similar model size (Params) and computation complexity (FLOPs). It can be seen from the table, our models consistently exceed other methods by large margins. More specifically, CETNet-T achieves 82.7% Top-1 accuracy with only 23M parameters and 4.3G FLOPs, surpassing CvT-13, Swin-T, and DeiT-S by 1.1%, 1.4% and 2.9% separately. For small and base models, our CETNet-S and CETNet-B still achieve better performance with comparable Params and FLOPs. Compared with the state-of-the-art CNNs RegNet [41] which also trained with input size 224 × 224 with out extra data, our CETNet achieves better accuracy. On the 384

× 384 input, Our CETNet's performance is still better than other Backbone, achieve 84.2%, 84.6% and 84.9% respectively, well demonstrates the powerful learning capacity of our CETNet. As a result, we can conclude that an effective combination of CNNs and ViTs can optimize well in small (or middle) scale datasets.

Table 2. Object detection and instance segmentation performance on COCO 2017 validation set with the Mask R-CNN framework. The FLOPs are measured at a resolution of 800 × 1280, and the backbones are pre-trained on the ImageNet-1K

Mask R-CNN 1× schedule									Mask R-CNN 3× schedule					
Backbone	Params	FLOPs	AP^b	AP^b_{50}	AP^b_{75}	AP^m	AP^m_{50}	AP^m_{75}	AP^b	AP^b_{50}	AP^b_{75}	AP^m	AP^m_{50}	AP^m_{70}
Res50 [25]	44M	260G	38	58.6	41.4	34.4	55.1	36.7	41	61.7	44.9	37.1	58.4	40.1
PVT-S [56]	44M	245G	40.4	62.9	43.8	37.8	60.1	40.3	43	65.3	46.9	39.9	62.5	42.8
ViL-S [67]	45M	218G	44.9	67.1	49.3	**41.0**	64.2	**44.1**	**47.1**	68.7	51.5	**42.7**	**65.9**	**46.2**
TwinsP-S [8]	44M	245G	42.9	65.8	47.1	40.0	62.7	42.9	46.8	**69.3**	**51.8**	42.6	66.3	46
Twins-S [8]	44M	228G	43.4	66.0	47.3	40.3	63.2	43.4	46.8	69.2	51.2	42.6	66.3	45.8
Swin-T [35]	48M	264G	43.7	64.6	46.2	39.1	61.6	42.0	46	68.2	50.2	41.6	65.1	44.8
CETNet-T	43M	261G	**45.5**	**67.7**	**50.0**	40.7	**64.4**	43.7	46.9	67.9	51.5	41.6	65	44.7
Res101 [25]	63M	336G	40.4	61.1	44.2	36.4	57.7	38.8	42.8	63.2	47.1	38.5	60.1	41.3
X101-32 [62]	63M	340G	41.9	62.5	45.9	37.5	59.4	40.2	42.8	63.2	47.1	38.5	60.1	41.3
PVT-M [56]	64M	302G	42.0	64.4	45.6	39.0	61.6	42.1	42.8	63.2	47.1	38.5	60.1	41.3
ViL-M [67]	60M	261G	43.4	–	–	39.7	–	–	44.6	66.3	48.5	40.7	63.8	43.7
TwinsP-B [8]	64M	302G	44.6	66.7	48.9	40.9	63.8	44.2	47.9	70.1	52.5	43.2	67.2	46.3
Twins-B [8]	76M	340G	45.2	67.6	49.3	41.5	64.5	44.8	48	69.5	52.7	43	66.8	46.6
Swin-S [35]	69M	354G	44.8	66.6	48.9	40.9	63.4	44.2	48.5	**70.2**	53.5	**43.3**	**67.3**	**46.6**
CETNet-S	53M	315G	**46.6**	**68.7**	**51.4**	**41.6**	**65.4**	**44.8**	48.6	69.8	**53.5**	43	66.9	46
X101-64 [62]	101M	493G	42.8	63.8	47.3	38.4	60.6	41.3	44.4	64.9	48.8	39.7	61.9	42.6
PVT-L [56]	81M	364G	42.9	65.0	46.6	39.5	61.9	42.5	44.5	66.0	48.3	40.7	63.4	43.7
ViL-B [67]	76M	365G	45.1	–	–	41.0	–	–	45.7	67.2	49.9	41.3	64.4	44.5
TwinsP-L [8]	81M	364G	45.4	–	–	41.6	–	–	–	–	–	–	–	–
Twins-L [8]	111M	474G	45.9	–	–	41.6	–	–	–	–	–	–	–	–
Swin-B [35]	107M	496G	46.9	–	–	42.3	–	–	48.5	**69.8**	53.2	**43.4**	66.8	**46.9**
CETNet-B	94M	495G	**47.9**	**70.3**	**53.0**	42.5	**67.2**	**45.6**	48.6	69.5	**53.7**	43.1	**66.9**	46.4

4.2   Object Detection and Instance Segmentation

Settings. For object detection and instance segmentation experiments, we evaluate our CETNet with the Mask R-CNN [24] framework on COCO 2017, which contains over 200K images with 80 classes. The models pre-trained on the ImageNet-1K dataset are used as visual backbones. We follow the standard to use two training schedules, 1× schedule(12 epochs with the learning rate decayed by 10× at epochs 8 and 11) and 3× schedule(36 epochs with the learning rate decayed by 10× at epochs 27 and 33). We utilize the multi-scale training strategy [5], resizing the input that the shorter side is between 480 and 800 while the longer side is no more than 1333. AdamW [31] optimizer with initial learning rate of 0.0001, weight decay of 0.05. All models are trained on the 118K training images with a batch size of 16, and the results are reported on 5K images in COCO 2017 validation set.

Results. As shown in Table 2, the results of the Mask R-CNN framework show that our CETNet variants clearly outperform all counterparts with 1× schedule. In detail, our CETNet-T outperforms Swin-T by +1.8 box AP, +1.6 mask AP. On the small and base configurations, the performance gain can also be achieved with +1.8 box AP, +0.7 mask AP and +1.0 box AP, +0.2 mask AP respectively. For 3× schedule, our model, CETNet-S and CETNet-B, can achieve

Table 3. Comparison of semantic segmentation on ADE20K with the Upernet framework, both single and multi-scale evaluations are reported in the last two columns. FLOPs are calculated with a resolution of 512 2048, and the backbones are pre-trained on the ImageNet-1K

Upernet 160k trained models				
Backbone	Params	FLOPs	mIoU(%)	MS mIoU(%)
TwinsP-S [8]	54.6M	919G	46.2	47.5
Twins-S [8]	54.4M	901G	46.2	47.1
Swin-T [35]	59.9M	945G	44.5	45.8
CETNet-T	53.2M	935G	**46.5**	**47.9**
Res101 [25]	86.0M	1029G	–	44.9
TwinsP-B [8]	74.3M	977G	47.1	48.4
Twins-B [8]	88.5M	1020G	47.7	48.9
Swin-S [35]	81.3M	1038G	47.6	49.5
CETNet-S	63.4M	990G	**48.9**	**50.6**
TwinsP-L [8]	91.5M	1041G	48.6	49.8
Twins-L [8]	133.0M	1164G	48.8	50.2
Swin-B [35]	121.0M	1188G	48.1	49.7
CETNet-B	106.3M	1176G	**50.2**	**51.6**

competitive results with lower Params and FLOPs than the current state-of-the-art ViTs methods in small and base scenarios. However, like SWin, CETNet-T can't beat the SOTA performance. Also, in the small and base configurations, we notice the variants of our CETNet and SWin do not improve the performance. We conjecture such inferior performance may be because of insufficient data.

4.3 Semantic Segmentation

Settings. We further use the pre-trained models as the backbone to investigate the capability of our models for Semantic Segmentation on the ADE20K [72] dataset. ADE20K is a widely-used semantic segmentation dataset that contains 150 fine-grained semantic categories, with 20,210, 2,000, and 3,352 images for training, validation, and testing, respectively. We follow previous works [35] to employ UperNet [61] as the basic framework and follow the same setting for a fair comparison. In the training stage, we employ the AdamW [31] optimizer and set the initial learning rate to 6e-5 and use a polynomial learning rate decay, and the weight decay is set to 0.01 and train Upernet 160k iterations with batch size of 16. The data augmentations adopt the default setting in mmsegmentation of random horizontal flipping, random re-scaling within ratio range [0.5, 2.0], and random photometric distortion. Both single and multi-scale inference are used for evaluation.

Results. In Table 3, we list the results of Upernet 160k trained model on ADE20K. It can be seen that our CETNet models significantly outperform previous state-of-the-arts under different configurations. In details, CETNet-T achieves 46.5 mIoU and 47.9 multi-scale tested mIoU, +2.0 and +2.1 higher than Swin-T with similar computation cost. On the small and base configurations, CETNet-S and CETNet-B still achieve +1.3 and +2.1 higher mIOU and +1.1 and +1.9 multi-scale tested mIoU than the Swin counterparts. The performance gain is promising and demonstrates the effectiveness of our CETNet design.

4.4 Ablation Study

In this section, we ablate the critical elements of the proposed CETNet backbone using ImageNet-1K image classification. The experimental settings are the same as the settings in Sect. 4.1. For the attention mechanism, when replacing the micro design LEWin self-attention with the origin shifted window self-attention, the performance dropped 0.2%, demonstrating the effectiveness of our micro design. Furthermore, we find the "deep-narrow" architecture is better than the "shallow-wide" counterpart. Specifically, the deep-narrow model with [2, 2, 18, 2] Transformer blocks for four stages with the base channel dimensions $D = 64$ and the shallow-wide model with [2, 2, 6, 2] blocks for four stages with $D = 96$. As we can see from the 1st and 3rd rows, even with larger Params and FLOPs, the shallow-wide model performs worse than the deep-narrow design. The CE module is the key element in our models. To verify the effectiveness CE module, we compare it with the existing method used in hierarchical ViTs backbones, including the patch embedding and patch merging modules in SWin [35] and convolutional token embedding modules described in CvT [59]. For a fair comparison, we use the shallow-wide design mentioned above and apply these three methods in all these models with all other factors kept the same. As shown in the last three rows of Table 4, our CE module performs better than other existing methods.

Table 4. Ablation study of CETNet's key elements. 'Without LEAtten' denotes replacing the micro design LEWin self-attention with the shifted window self-attention in SWin [35]. The 'D' and 'S' represents the deep-narrow and the shallow-wide model, respectively. The 1st and 3rd rows are the baseline 'D' and 'S' models.

Models	Param	FLOPs	Top-1 (%)
CETNet-T (D with CE)	23.4M	4.3G	**82.7**
Without LEAtten (D with CE)	23.3M	4.2G	82.5
Shallow-wide (S with CE)	29M	4.6G	82.5
Patch embedding+patch merging [35] (S)	27M	4.6G	81.5
Convolutional token embedding [59] (S)	27M	4.6G	82.0

5 Investigating the Role of Convolutional Embedding

This section systematically understands how the hybrid CNNs/ViTs network benefits from the CE. As we need carefully make the trade-off among the computation budget, the model generalization, and the model capacity, the CE, two design choices are mainly explored in this work: the number of CNNs layers of CE and the basic unit of CE. Besides, we show the superiority of CE via integrate CE into 4 popular hierarchical ViTs models. Finally, we further investigate the role of CNNs in hybrid CNNs/ViTs network design. Except for specific declaration, all configuration is follow the Image Classification setting as Sect. 4.

5.1   Effect of the Stacking Number

We first explore how large effective receptive field (ERF) affect the computation budget and the performance of the network. As we mentioned in Sect. 3.1, the $S0$ and the CE of $S1$ are combined as one whole CNNs stem. We refer it as the first CE (CE^{1st}) in rest of this section, CE^{2nd}, CE^{3rd}, and CE^{4th} represent the CE of $S2$, $S3$, $S4$, respectively. To explore the effect of

Table 5. Performance of model with different layer numbers of CE. The baseline is slightly modified from SWin-Tiny [35]

Models	Param	FLOPs	Top-1 (%)
Swin-T [35]	29M	4.5G	81.3
1-layer CE	28.1M	4.5G	81.8
3-layer CE	28.1M	4.6G	82.2
5-layer CE	29.1M	4.9G	82.4
7-layer CE	30.1M	5.4G	82.6

the stacking number of the basic unit, we slightly modify the SWin-Tiny [35] model by replacing its patch embedding and patch merging modules with pure CNNs. We choose MBConv as the basic unit of CE and gradually increase the layer number of CE from 1 to 3, 5, and 7. As shown in Table 5, as the CE contains more MBConvs, the Param and FLOPs grow, and the performance increases from 81.8% to 82.6%. The 3-layer CE model has slightly higher FLOPs than the 1-layer CE (4.48G vs. 4.55G) with near the same Param (28.13M vs. 28.11M). Besides, it is worth noticing that the performance grows negligibly from 5-layer CE to 7-layer CE. Thus, we employ 5-layer setting, offering large ERF [51,64], as CETNet's final configuration.

5.2   Effect of Different CNNs Blocks

We next explore how well convolutional inductive bias inject by different CNNs architectures. The CE layers aim to offer rich features for later attention modules. CNNs' effective receptive field (ERF) determines the information it can cover and process. As Luo *et al.* mentioned, stacking more layers, subsampling, and CNNs with large ERF, such as dilated convolution [64] can enlarge the ERF of CNNs networks. For find an efficient basic unit for CE, we use the CETNet-Tiny (CETNet-T) model as a baseline and replace the CE with some

Table 6. Transfer some popular CNNs blocks to CETNet-T, including the DenseNet block, ShuffleNet block, ResNet block, SeresNet block, and GhostNet block. PureMBConv represents to use MBConv block to replace the Fused-MBConv block in the early stage of CETNet-T

CNNs type	Params	FLOPs	Top-1(%)
CETNet-T	23.4M	4.3G	**82.7**
PureMBConvs [43]	23.4M	4.2G	82.6
GhostNetConvs [22]	24.6M	4.3G	82.6
DenseNetConvs [30]	22.3M	4.4G	82.5
ShuffleNetConvs [69]	23.4M	4.3G	82.4
ResNetConvs [25]	23.8M	4.3G	82.0
SeresNetConvs [29]	24.0M	4.3G	82.0

recent CNNs building blocks, such as MBConv [43], DenseNet [30], ShuffleNet [69], ResNetConvs [25], SeresNetConvs [29], and GhostNetConvs [22]. All candidate convolutions layers stack to 5 layers based on previous finding. For a fair comparison, all model variants are constructed to have similar parameter numbers and FLOPs. As we can see from Table 6, when replaced the Fused-MBConvs in the early stage of CETNet-T by MBConvs, the top-1 accuracy increased 0.1%, and we observed a 12% decrease in training speed. Also, CETNet-T and PureMB model achieve higher performance than other candidate convolutions. We argue that may be the internal relation between depthwise convolution and ViTs as pointed out by CoAtNet [14], which is further verified by the GhostNet and shuffleNet model, which archive 82.6% and 82.4% top-1 accuracy. Besides, we noticed that under that CNNs help ViTs see better perspective, in CE case, dense connections in convolutions may not necessarily hurt performance. Since the result shows that our DenseNet model can also archive 82.5% top-1 accuracy, which is comparable with the performance of CETNet-T, PureMB, GhostNet, and shuffleNet model. However, ResNet and SeresNet show inferior performance. We conjecture that the basic units have different ERF with the same stacking number.

5.3   Generalization of CE

Then, we attempt to generalize the CE design to more ViTs backbones of CV. Here, we apply our CE design, 5-layer Fused-MBConv of CE^{1st}, and 5-layer MBConv of CE^{2nd}, CE^{3rd}, and CE^{4th} respectively, to 4 prevalent backbones, CvT [59], PVT [56], SWin [35], and CSWin [18]. For a fair comparison, we slightly change the structure, removing some ViTs blocks, of 4 models to keeps their parameter numbers and FLOPs maintaining the similar level as their original version. Also, we modify the small-scale model variant CvT-13 and PVT-S of CvT and PVT. As shown in Table 7, those modified models outperform the original model 0.5% and 1.3% separately.

Table 7. Generalize the CE module to 4 ViT backbones. All models are trained on ImageNet-1K dataset and compared with the original model under the same training scheme. Depths indicate the number of Transformer layers of each stage. FLOPs are calculated with a resolution of 224 × 224

Framework	Models	Channels	Depths	Param	FLOPs	Top-1
CvT [59]	CvT-13	64	[1, 2, 10]	20.0M	4.5G	81.6
	CE-CvT-13	64	[1, 2, 10]	20.0M	4.4G	82.1(0.5↑)
PVT [56]	PVT-S	64	[3, 4, 6, 3]	24.5M	3.8G	79.8
	CE-PVT-S	64	[3, 4, 4, 3]	22.6M	3.7G	81.1(1.3↑)
SWin [35]	Swin-T	96	[2, 2, 6, 2]	28.3M	4.5G	81.3
	CE-Swin-T	96	[2, 2, 4, 2]	23.4M	4.3G	82.5(1.2↑)
	Swin-S	96	[2, 2, 18, 2]	50.0M	8.7G	83.0
	CE-Swin-S	96	[2, 2, 16, 2]	48.2M	8.8G	83.6(0.6↑)
	Swin-B	128	[2, 2, 18, 2]	88.0M	15.4G	83.3
	CE-Swin-B	128	[2, 2, 16, 2]	85.2M	15.5G	84.0(0.7↑)
CSWin [18]	CSWin-T	64	[1, 2, 21, 1]	23.0M	4.3G	82.7
	CE-CSWin-T	64	[1, 2, 20, 1]	21.6M	4.2G	83.6(0.9↑)
	CSWin-S	64	[2, 4, 32, 2]	35.0M	6.9G	83.6
	CE-CSWin-S	64	[2, 4, 31, 2]	33.9M	6.6G	84.1(0.5↑)
	CSWin-B	96	[2, 4, 32, 2]	78.0M	15.0G	84.2
	CE-CSWin-B	96	[2, 4, 31, 2]	75.8M	14.7G	84.7(0.5↑)

Furthermore, when introducing our design into SWin and CSWin, the top-1 accuracy of all counterparts is improved even under lower parameter numbers and FLOPs scenarios. For details, the modified models of Swin counterparts gain 1.2%, 0.6% and 0.7%, and the CSwin counterparts gain 0.9%, 0.5% and 0.5% respectively. Those results demonstrated that CE could be easily integrated with other ViT models and significantly improve the performance of those ViT models.

5.4   Understanding the Role of CNNs in Hybrid CNNs/ViTs Design

Finally, we explore how well CNNs in the deep layer of the hybrid CNNs/ViTs network improves ViTs. Previous works [14,39,60] show the *shallow* CNNs structure is enough to bring the convolutional inductive bias to all following ViTs blocks. However, one may notice that the CE^{2nd}, CE^{3rd}, and CE^{4th} are **not** locate the *shallow* layer of network. To fully understand: 1) whether CNNs in the deep layer enhances the inductive bias for subsequent ViTs blocks; 2) how hybrid CNNs/ViTs design affects the final performance of the network. We conduct the following experiments. From macro view, CETNet can be view as 'C-T-C-T-C-T-C-T', where C and T denote CE and ViTs blocks respectively, where CE^{1st} is Fused-MBConv, CE^{2nd}, CE^{3rd}, and CE^{4th} are MBConv. We conduct three main experiments: **CNNs to ViTs, ViTs to CNNs,** and **Others**. In **CNNs to ViTs** group, we gradually replace the convolutions with transformers. In **ViTs to CNNs** group, we do the reverse. As we can see, only adopting CNNs in *early* stage is not optimal. In addition, all hybrid models outperform the pure ViTs model in **CNNs to ViTs**. Besides, in comparison with **ViTs to CNNs**, one may notice that in deep layer architecture with more ViTs is superior to more CNNs. In addition, we have: hybrid CNNs/ViTs $\geq$ pure ViT $\geq$ pure CNNs, in deep layer of network. In **Others** group, we further list some variants' experiment results to the audience and hope that any possible insights may raise a rethinking of the hybrid CNNs/ViTs network design.

Table 8. Comparison of different hybrid CNNs/ViTs designs. 'Arch' represents architecture for short. C represents MBConvs(Fused-MBConvs in the early stage), and T represents the ViTs block mentioned in Sect. 3.2

CNNs to ViTs				ViTs to CNNs				Others			
Arch	Param	FLOPs	Top-1	Arch	Param	FLOPs	Top-1	Arch	Param	FLOPs	Top-1
C-T-C-T-C-T-C-T	23.4M	4.3G	82.7	C-T-C-T-C-T-C-T	23.4M	4.3G	82.7	C-T-C-T-C-T-C-T	23.4M	4.3G	82.7
C-T-C-T-C-T-T-T	24.0M	4.2G	82.8	C-T-C-T-C-T-C-C	23.7M	4.2G	82.0	C-C-C-T-C-T-C-T	23.5M	4.3G	82.7
C-T-C-T-T-T-T-T	24.1M	4.2G	82.5	C-T-C-T-C-C-C-C	24.4M	4.2G	79.6	C-C-C-C-T-T-T-T	25.5M	4.4G	81.8
C-T-T-T-T-T-T-T	24.1M	4.4G	82.3	C-T-C-C-C-C-C-C	24.3M	4.2G	79.2	T-T-T-C-C-C-C	23.4M	4.8G	76.3
T-T-T-T-T-T-T-T	24.3M	4.2G	80.1	C-C-C-C-C-C-C-C	24.6M	5.1G	79.0	T-C-T-C-T-C-T-C	24.5M	4.2G	79.8

6   Conclusions

This paper proposes a principled way to produce a hybrid CNNs/ViTs architecture. With the idea of injecting desirable inductive bias in ViTs, we present 1) a conceptual understanding of combining CNNs/ViTs into a single architecture, based on using a *convolutional embedding* and its effect on the inductive bias of the architecture. 2) a conceptual framework of micro and macro detail of an hybrid architecture, where different design decisions are made at the small and large levels of detail to impose an inductive bias into the architecture. Besides, we deliver a family of models, dubbed CETNets, which serve as a generic vision backbone and achieve the SOTA performance on various vision tasks under constrained data size. We hope that what we found could raise a rethinking of the network design and extend the limitation of the hybrid CNNs/ViTs network.

References

1. Ba, J.L., Kiros, J.R., Hinton, G.E.: Layer normalization. arXiv preprint arXiv:1607.06450 (2016)
2. Battaglia, P.W., et al.: Relational inductive biases, deep learning, and graph networks. arXiv preprint arXiv:1806.01261 (2018)
3. Bello, I., Zoph, B., Vaswani, A., Shlens, J., Le, Q.V.: Attention augmented convolutional networks. In: Proceedings of the IEEE/CVF International Conference on Computer Vision, pp. 3286–3295 (2019)
4. Brown, T.B., et al.: Language models are few-shot learners. arXiv preprint arXiv:2005.14165 (2020)
5. Carion, N., Massa, F., Synnaeve, G., Usunier, N., Kirillov, A., Zagoruyko, S.: End-to-end object detection with transformers. In: Vedaldi, A., Bischof, H., Brox, T., Frahm, J.-M. (eds.) ECCV 2020. LNCS, vol. 12346, pp. 213–229. Springer, Cham (2020). https://doi.org/10.1007/978-3-030-58452-8_13
6. Chen, C.F., Fan, Q., Panda, R.: CrossViT: cross-attention multi-scale vision transformer for image classification. arXiv preprint arXiv:2103.14899 (2021)
7. Chen, L.C., Papandreou, G., Kokkinos, I., Murphy, K., Yuille, A.L.: DeepLab: semantic image segmentation with deep convolutional nets, atrous convolution, and fully connected CRFs. IEEE Trans. Pattern Anal. Mach. Intell. **40**(4), 834–848 (2017)
8. Chu, X., et al.: Twins: revisiting spatial attention design in vision transformers. arXiv preprint arXiv:2104.13840 (2021)
9. Chu, X., et al.: Conditional positional encodings for vision transformers. arXiv preprint arXiv:2102.10882 (2021)
10. Cordonnier, J.B., Loukas, A., Jaggi, M.: On the relationship between self-attention and convolutional layers. arXiv preprint arXiv:1911.03584 (2019)
11. Dai, J., Li, Y., He, K., Sun, J.: R-FCN: object detection via region-based fully convolutional networks. In: Advances in Neural Information Processing Systems vol. 29 (2016)
12. Dai, J., et al.: Deformable convolutional networks. In: Proceedings of the IEEE International Conference on Computer Vision, pp. 764–773 (2017)
13. Dai, Z., Cai, B., Lin, Y., Chen, J.: UP-DETR: unsupervised pre-training for object detection with transformers. In: Proceedings of the IEEE/CVF Conference on Computer Vision and Pattern Recognition, pp. 1601–1610 (2021)
14. Dai, Z., Liu, H., Le, Q.V., Tan, M.: CoAtNet: marrying convolution and attention for all data sizes. arXiv preprint arXiv:2106.04803 (2021)
15. d'Ascoli, S., Touvron, H., Leavitt, M., Morcos, A., Biroli, G., Sagun, L.: Convit: Improving vision transformers with soft convolutional inductive biases. arXiv preprint arXiv:2103.10697 (2021)
16. Deng, J., Dong, W., Socher, R., Li, L.J., Li, K., Fei-Fei, L.: ImageNet: a large-scale hierarchical image database. In: 2009 IEEE Conference on Computer Vision and Pattern Recognition (CVPR), pp. 248–255. IEEE (2009)
17. Devlin, J., Chang, M.W., Lee, K., Toutanova, K.: BERT: pre-training of deep bidirectional transformers for language understanding. arXiv preprint arXiv:1810.04805 (2018)
18. Dong, X., et al.: CSWin transformer: a general vision transformer backbone with cross-shaped windows. arXiv preprint arXiv:2107.00652 (2021)
19. Dosovitskiy, A., et al.: An image is worth 16x16 words: transformers for image recognition at scale. arXiv preprint arXiv:2010.11929 (2020)

20. Fan, H., Xiong, B., Mangalam, K., Li, Y., Yan, Z., Malik, J., Feichtenhofer, C.: Multiscale vision transformers. arXiv preprint arXiv:2104.11227 (2021)
21. Gupta, S., Tan, M.: EfficientNet-EdgeTPU: creating accelerator-optimized neural networks with AutoML. Google AI Blog **2**, 1 (2019)
22. Han, K., Wang, Y., Tian, Q., Guo, J., Xu, C., Xu, C.: GhostNet: more features from cheap operations. In: Proceedings of the IEEE/CVF Conference on Computer Vision and Pattern Recognition, pp. 1580–1589 (2020)
23. Han, K., Xiao, A., Wu, E., Guo, J., Xu, C., Wang, Y.: Transformer in transformer. arXiv preprint arXiv:2103.00112 (2021)
24. He, K., Gkioxari, G., Dollár, P., Girshick, R.: Mask R-CNN. In: Proceedings of the IEEE International Conference on Computer Vision, pp. 2961–2969 (2017)
25. He, K., Zhang, X., Ren, S., Sun, J.: Deep residual learning for image recognition. In: Proceedings of the IEEE Conference on Computer Vision and Pattern Recognition, pp. 770–778 (2016)
26. Heo, B., Yun, S., Han, D., Chun, S., Choe, J., Oh, S.J.: Rethinking spatial dimensions of vision transformers. arXiv preprint arXiv:2103.16302 (2021)
27. Howard, A.G., et al.: MobileNets: efficient convolutional neural networks for mobile vision applications. arXiv preprint arXiv:1704.04861 (2017)
28. Hu, H., Zhang, Z., Xie, Z., Lin, S.: Local relation networks for image recognition. In: Proceedings of the IEEE/CVF International Conference on Computer Vision, pp. 3464–3473 (2019)
29. Hu, J., Shen, L., Sun, G.: Squeeze-and-excitation networks. In: Proceedings of the IEEE Conference on Computer Vision and Pattern Recognition, pp. 7132–7141 (2018)
30. Huang, G., Liu, Z., Van Der Maaten, L., Weinberger, K.Q.: Densely connected convolutional networks. In: Proceedings of the IEEE Conference on Computer Vision and Pattern Recognition, pp. 4700–4708 (2017)
31. Kingma, D.P., Ba, J.: Adam: a method for stochastic optimization. arXiv preprint arXiv:1412.6980 (2014)
32. Krizhevsky, A., Sutskever, I., Hinton, G.E.: ImageNet classification with deep convolutional neural networks. Adv. Neural Inf. Process. Syst. **25**, 1097–1105 (2012)
33. Li, Y., Zhang, K., Cao, J., Timofte, R., Van Gool, L.: LocalViT: bringing locality to vision transformers. arXiv preprint arXiv:2104.05707 (2021)
34. Lin, T.-Y., et al.: Microsoft COCO: common objects in context. In: Fleet, D., Pajdla, T., Schiele, B., Tuytelaars, T. (eds.) ECCV 2014. LNCS, vol. 8693, pp. 740–755. Springer, Cham (2014). https://doi.org/10.1007/978-3-319-10602-1_48
35. Liu, Z., et al.: Swin transformer: hierarchical vision transformer using shifted windows. In: Proceedings of the IEEE/CVF International Conference on Computer Vision (ICCV) (2021)
36. Liu, Z., Mao, H., Wu, C.Y., Feichtenhofer, C., Darrell, T., Xie, S.: A convnet for the 2020s. arXiv preprint arXiv:2201.03545 (2022)
37. Long, J., Shelhamer, E., Darrell, T.: Fully convolutional networks for semantic segmentation. In: Proceedings of the IEEE Conference on Computer Vision and Pattern Recognition, pp. 3431–3440 (2015)
38. Luo, W., Li, Y., Urtasun, R., Zemel, R.: Understanding the effective receptive field in deep convolutional neural networks. In: Advances in Neural Information Processing Systems vol. 29 (2016)
39. Marquardt, T.P., Jacks, A., Davis, B.L.: Token-to-token variability in developmental apraxia of speech: three longitudinal case studies. Clin. Linguist. Phon. **18**(2), 127–144 (2004)

40. Radford, A., Narasimhan, K., Salimans, T., Sutskever, I.: Improving language understanding by generative pre-training (2018)
41. Radosavovic, I., Kosaraju, R.P., Girshick, R., He, K., Dollár, P.: Designing network design spaces. In: Proceedings of the IEEE/CVF Conference on Computer Vision and Pattern Recognition, pp. 10428–10436 (2020)
42. Ren, S., He, K., Girshick, R., Sun, J.: Faster R-CNN: towards real-time object detection with region proposal networks. Adv. Neural Inf. Process. Syst. **28**, 91–99 (2015)
43. Sandler, M., Howard, A., Zhu, M., Zhmoginov, A., Chen, L.C.: MobileNetV 2: inverted residuals and linear bottlenecks. In: Proceedings of the IEEE Conference on Computer Vision and Pattern Recognition, pp. 4510–4520 (2018)
44. Shen, Z., Zhang, M., Zhao, H., Yi, S., Li, H.: Efficient attention: attention with linear complexities. In: Proceedings of the IEEE/CVF Winter Conference on Applications of Computer Vision, pp. 3531–3539 (2021)
45. Simonyan, K., Zisserman, A.: Very deep convolutional networks for large-scale image recognition. arXiv preprint arXiv:1409.1556 (2014)
46. Srinivas, A., Lin, T.Y., Parmar, N., Shlens, J., Abbeel, P., Vaswani, A.: Bottleneck transformers for visual recognition. In: Proceedings of the IEEE/CVF Conference on Computer Vision and Pattern Recognition, pp. 16519–16529 (2021)
47. Strudel, R., Garcia, R., Laptev, I., Schmid, C.: Segmenter: transformer for semantic segmentation. arXiv preprint arXiv:2105.05633 (2021)
48. Sun, C., Shrivastava, A., Singh, S., Gupta, A.: Revisiting unreasonable effectiveness of data in deep learning era. In: Proceedings of the IEEE International Conference on Computer Vision, pp. 843–852 (2017)
49. Szegedy, C., et al.: Going deeper with convolutions. In: Proceedings of the IEEE Conference on Computer Vision and Pattern Recognition, pp. 1–9 (2015)
50. Tan, M., Le, Q.: EfficientNet: rethinking model scaling for convolutional neural networks. In: International Conference on Machine Learning, pp. 6105–6114. PMLR (2019)
51. Tan, M., Le, Q.V.: EfficientNetV2: smaller models and faster training. arXiv preprint arXiv:2104.00298 (2021)
52. Touvron, H., Cord, M., Douze, M., Massa, F., Sablayrolles, A., Jégou, H.: Training data-efficient image transformers & distillation through attention. In: International Conference on Machine Learning, pp. 10347–10357. PMLR (2021)
53. Vaswani, A., Ramachandran, P., Srinivas, A., Parmar, N., Hechtman, B., Shlens, J.: Scaling local self-attention for parameter efficient visual backbones. In: Proceedings of the IEEE/CVF Conference on Computer Vision and Pattern Recognition, pp. 12894–12904 (2021)
54. Vaswani, A., et al.: Attention is all you need. In: Advances in Neural Information Processing Systems, pp. 5998–6008 (2017)
55. Wang, H., Zhu, Y., Adam, H., Yuille, A., Chen, L.C.: Max-DeepLab: end-to-end panoptic segmentation with mask transformers. In: Proceedings of the IEEE/CVF Conference on Computer Vision and Pattern Recognition, pp. 5463–5474 (2021)
56. Wang, W., et al.: Pyramid vision transformer: a versatile backbone for dense prediction without convolutions. arXiv preprint arXiv:2102.12122 (2021)
57. Wang, X., Girshick, R., Gupta, A., He, K.: Non-local neural networks. In: Proceedings of the IEEE Conference on Computer Vision and Pattern Recognition, pp. 7794–7803 (2018)
58. Wu, B., et al.: Visual transformers: token-based image representation and processing for computer vision. arXiv preprint arXiv:2006.03677 (2020)

59. Wu, H., et al.: CVT: introducing convolutions to vision transformers. arXiv preprint arXiv:2103.15808 (2021)
60. Xiao, T., Dollar, P., Singh, M., Mintun, E., Darrell, T., Girshick, R.: Early convolutions help transformers see better. In: Advances in Neural Information Processing Systems vol. 34 (2021)
61. Xiao, T., Liu, Y., Zhou, B., Jiang, Y., Sun, J.: Unified perceptual parsing for scene understanding. In: Ferrari, V., Hebert, M., Sminchisescu, C., Weiss, Y. (eds.) ECCV 2018. LNCS, vol. 11209, pp. 432–448. Springer, Cham (2018). https://doi.org/10.1007/978-3-030-01228-1_26
62. Xie, S., Girshick, R., Dollár, P., Tu, Z., He, K.: Aggregated residual transformations for deep neural networks. In: Proceedings of the IEEE Conference on Computer Vision And Pattern Recognition, pp. 1492–1500 (2017)
63. Xu, W., Xu, Y., Chang, T., Tu, Z.: Co-scale conv-attentional image transformers. arXiv preprint arXiv:2104.06399 (2021)
64. Yu, F., Koltun, V.: Multi-scale context aggregation by dilated convolutions. arXiv preprint arXiv:1511.07122 (2015)
65. Yuan, K., Guo, S., Liu, Z., Zhou, A., Yu, F., Wu, W.: Incorporating convolution designs into visual transformers. In: Proceedings of the IEEE/CVF International Conference on Computer Vision, pp. 579–588 (2021)
66. Yuan, L., et al.: Tokens-to-token ViT: training vision transformers from scratch on imagenet. arXiv preprint arXiv:2101.11986 (2021)
67. Zhang, P., et al.: Multi-scale vision longformer: a new vision transformer for high-resolution image encoding. arXiv preprint arXiv:2103.15358 (2021)
68. Zhang, Q., Yang, Y.: Rest: an efficient transformer for visual recognition (2021)
69. Zhang, X., Zhou, X., Lin, M., Sun, J.: ShuffleNet: an extremely efficient convolutional neural network for mobile devices. In: Proceedings of the IEEE Conference on Computer Vision and Pattern Recognition, pp. 6848–6856 (2018)
70. Zhang, X., et al.: DCNAS: densely connected neural architecture search for semantic image segmentation. In: Proceedings of the IEEE/CVF Conference on Computer Vision and Pattern Recognition, pp. 13956–13967 (2021)
71. Zheng, M., et al.: End-to-end object detection with adaptive clustering transformer. arXiv preprint arXiv:2011.09315 (2020)
72. Zhou, B., Zhao, H., Puig, X., Xiao, T., Fidler, S., Barriuso, A., Torralba, A.: Semantic understanding of scenes through the ade20k dataset. Int. J. Comput. Vis. **127**(3), 302–321 (2019)
73. Zhu, X., Su, W., Lu, L., Li, B., Wang, X., Dai, J.: Deformable DETR: deformable transformers for end-to-end object detection. arXiv preprint arXiv:2010.04159 (2020)

Author Index

Printed in the United States
by Baker & Taylor Publisher Services